ADVANCES IN ADDITIVE MANUFACTURING TECHNOLOGIES

The 6th International Conference on Advances in Additive Manufacturing Technologies (ICAAMT2024) showcases recent advancements in additive manufacturing technologies, including materials, process precision, multi-material printing, and hybrid manufacturing systems. It highlights innovations in metal additive manufacturing, bio-printing, and large-scale construction printing, and the integration of AI and machine learning for process optimization. The volume also addresses emerging challenges like material limitations, scalability, and regulatory issues, offering insights into future research directions and industrial applications. The book is a valuable resource for researchers, engineers, and industry professionals.

ADVANCES IN ADDITIVE MANUFACTURING TECHNOLOGIES

Proceedings of 6th International Conference on Advances in Additive Manufacturing Technologies (ICAAMT 2024), December 19-20th, 2024, Chennai, India

Shaping the Future of Digital Fabrication and Smart Manufacturing

Editor

Gurusamy Pathinettampadian
Professor & Head
Department of Mechanical Engineering,
Chennai Institute of Technology, Chennai, Tamilnadu, India

Vijayakumar Murugesan Devarajan
Professor
Department of Mechanical Engineering,
Chennai Institute of Technology, Chennai, Tamilnadu, India

Sathiyamoorthi Ramalingam
Professor
Department of Mechanical Engineering,
Chennai Institute of Technology, Chennai, Tamilnadu, India

CRC Press
Taylor & Francis Group
Boca Raton London New York

CRC Press is an imprint of the
Taylor & Francis Group, an informa business

First edition published 2026
by CRC Press
2385 NW Executive Center Drive, Suite 320, Boca Raton FL 33431

and by CRC Press
4 Park Square, Milton Park, Abingdon, Oxon, OX14 4RN

CRC Press is an imprint of Informa UK Limited

British Library Cataloguing-in-Publication Data
A catalogue record for this book is available from the British Library

ISBN: 978-1-041-16688-7 (hbk)
ISBN: 978-1-041-16687-0 (pbk)
ISBN: 978-1-003-68590-6 (ebk)

DOI: 10.1201/9781003685906

Typeset in Times New Roman
by Aditiinfosystems

Advances in Additive Manufacturing Technologies – Gurusamy Pathinettampadian et al. (eds)
© 2026 Taylor & Francis Group, London, ISBN 978-1-041-16687-0

Contents

Advances in Additive Manufacturing Technologies – Gurusamy Pathinettampadian et al. (eds)
© 2026 Taylor & Francis Group, London, ISBN 978-1-041-16687-0

List of Figures

Advances in Additive Manufacturing Technologies – Gurusamy Pathinettampadian et al. (eds)
© 2026 Taylor & Francis Group, London, ISBN 978-1-041-16687-0

List of Tables

Advances in Additive Manufacturing Technologies – Gurusamy Pathinettampadian et al. (eds)
© 2026 Taylor & Francis Group, London, ISBN 978-1-041-16687-0

Preface

We are delighted to present the proceedings of the 6th International Conference on Advances in Additive Manufacturing Technologies (ICAAMT 2024). This premier forum brings together researchers, practitioners, and industry professionals to share the latest developments, innovations, and insights in the field of additive manufacturing. As the adoption of these transformative technologies continues to accelerate across various industries, the relevance and impact of such a gathering grow ever more significant.

The conference was held from December 19–20, 2024, in Chennai, India, and was organized by the Department of Mechanical Engineering, Chennai Institute of Technology. ICAAMT 2024 provided a dynamic platform for the exchange of ideas and knowledge, fostering collaboration and driving forward the frontiers of additive manufacturing.

The applications of additive manufacturing span diverse sectors, including aerospace, medical, and automotive industries. Some of its key advantages include low start-up costs, ease of adoption, reduced material wastage, enhanced customization capabilities, seamless digital design integration, rapid prototyping, and faster transitions from prototype to production, along with lower environmental impact, suitability for low-volume production runs, and the potential for distributed manufacturing.

We sincerely acknowledge that the success of ICAAMT 2024 is the result of collective dedication and collaboration. We extend our deepest gratitude to our esteemed Patron, Conference Chairs, and the Principal for their invaluable support and leadership. The Program Committee played a crucial role through their rigorous and timely review of the submitted papers.

We are especially thankful to the authors for their significant contributions, the reviewers for their insightful evaluations, and the organizing committee for their tireless efforts in ensuring the smooth execution of this event.

We trust that the research shared in these proceedings will not only inform and inspire but also catalyze further innovation and exploration in this rapidly evolving domain. May this volume serve as a valuable resource, advancing both academic inquiry and practical applications in additive manufacturing.

Sincerely,

Dr. P. Gurusamy
Dr. M.D.Vijayakumar
Dr. R. Sathiyamoorthi

Advances in Additive Manufacturing Technologies – Gurusamy Pathinettampadian et al. (eds)
© 2026 Taylor & Francis Group, London, ISBN 978-1-041-16687-0

List of Abbreviations

AM	Additive Manufacturing
TIG	Tungsten Inert Gas
CMT	Cold Metal Transfer
LBM	Laser Beam Welding
FSW	Friction Stir Welding
EDM	Electrical Discharge Machining
ZrO_2	Zirconium dioxide
Al_2O_3	Aluminium Oxide
UAV	Unmanned Aerial Vehicle
DL	Deep Learning
BAC	Blood Alcohol Concentration
AF	Aramid Fiber
FRP	Fiber-reinforced polymer
PPDA	Para-Phenylenediamine
TPC	Terephthaloylchloride
ASTM	American Society for Testing and Materials
MTF	Maximum Tensile Force
AMMC	Metal Matrix Composites
SiC	Silicon Carbide
MMNCs	Metal Matrix Nanocomposites
ANOVA	Analysis of Variance
S/N	Signal to Noise ratio
SEM	Scanning Electron Microscopy
EDAX	Energy-Dispersive X-ray Analysis
EMS	Electromagnetic Stirring
ASICs	Application Specific Integrated Circuits
TA	Train Automation
WSN	Wireless Sensor Nodes
DCNN	Deep Convolutional Neural Network
FDM	Fusion Deposition Modeling
ID	Infill Density
MMA	microwave Metamaterial Absorber
ABS	Acrylonitrile Butadiene Styrene

PC	Polycarbonate
PLA	Polylactic Acid
AISI	American Iron and Steel Institute
SR	Surface Roughness
DOE	Design Of Experiments
MR	Machining Rate
WEDM	Wire Electrical Discharge Machining
CNTs	Carbon Nanotubes
CFRP	Carbon Fibre Reinforced Plastic
DEM	Discrete Element Method
VOF	Volume of Fluid
FVM	Finite Volume Method
PFEM	Particle Finite Element Method
DPD	Dissipative Particle Dynamics
EB-PVD	Electron Beam Physical Vapour Deposition
OCM	Online Condition Monitoring
CWT	Continuous Wavelet Transform
EMD	Empirical Mode Decomposition

Advances in Additive Manufacturing Technologies – Gurusamy Pathinettampadian et al. (eds)
© 2026 Taylor & Francis Group, London, ISBN 978-1-041-16687-0

Program Committee

CHIEF PATRONS

Shri. P. Sriram, Chairman, Chennai Institute of Technology, Chennai

Smt. S. Sridevi, Secretary, Chennai Institute of Technology, Chennai

Shri. S. Gokulakrishnan, Director-Innovations, Chennai Institute of Technology, Chennai

PARTON

Dr. A. Ramesh, Principal, Chennai Institute of Technology, Chennai

CONVENER

Dr. P. Gurusamy, Professor & Head, Mechanical Engineering, Chennai Institute of Technology, Chennai

CO-CONVENER

Dr. M.D. Vijayakumar, Professor, Mechanical Engineering, Chennai Institute of Technology, Chennai

CO-CONVENER

Dr. R. Sathiyamoorthi, Professor in Mechanical Engineering, Chennai Institute of Technology, Chennai

ORGANISING COMMITTEE

Dr. S. Ravi, Professor in Mechanical Engineering, Chennai Institute of Technology, Chennai

Dr. R. Ganesamoorthy, Professor in Mechanical Engineering, Chennai Institute of Technology, Chennai

Dr. T. Arun Kumar, Professor in Mechanical Engineering, Chennai Institute of Technology, Chennai

Dr. G Mahendran, Associate Professor in Mechanical Engineering, Chennai Institute of Technology, Chennai

Dr. S. Rajarasalnath, Associate Professor in Mechanical Engineering, Chennai Institute of Technology, Chennai

Dr. D. Jayabalakrishnan, Associate Professor in Mechanical Engineering, Chennai Institute of Technology, Chennai

Dr. J. Rajaparthiban, Asst. Prof in Mechanical Engineering, Chennai Institute of Technology, Chennai

Dr. B. Yokesh Kumar, Asst. Prof in Mechanical Engineering, Chennai Institute of Technology, Chennai

Dr. R. Vijayan, Asst. Prof in Mechanical Engineering, Chennai Institute of Technology, Chennai

Dr. S. Mohan Kumar, Asst. Prof in Center for Additive Manufacturing, Chennai Institute of Technology, Chennai

Dr. S. Ravi, Asst. Prof in Mechanical Engineering, Chennai Institute of Technology, Chennai

Dr. Biplab Bhattacharjee, Asst. Prof in Center for Additive Manufacturing, Chennai Institute of Technology, Chennai

Dr. Ramasamy, Asst. Prof in Center for Additive Manufacturing, Chennai Institute of Technology, Chennai

Dr. N. Shivakumar, Asst. Prof in Mechanical Engineering, Chennai Institute of Technology, Chennai

Dr. S.D. Dhanesh Babu, Asst. Prof in Mechanical Engineering, Chennai Institute of Technology, Chennai

Dr. R.Sasi Lakshmikanth,Asst. Prof in Mechanical Engineering, Chennai Institute of Technology, Chennai

Dr. S. Balamurugan, Asst. Prof in Mechanical Engineering, Chennai Institute of Technology, Chennai

Dr. M.J. Hepsi Beaula, Asst. Prof in Mechanical Engineering, Chennai Institute of Technology, Chennai

ADVISORY COMMITTEE

 ISBN 978-1-041-16687-0

About the Editors

Dr. P. Gurusamy is a distinguished academician and renowned researcher in Mechanical Engineering, with a remarkable career spanning over two decades. He earned his Bachelor of Engineering in Mechanical Engineering from the Government College of Engineering, Tirunelveli, and a Master of Engineering in Production Engineering from Thiagarajar College of Engineering, Madurai. Dr. Gurusamy earned his PhD in 2014 from Anna University, CEG Campus, Chennai. With two years of industrial experience and 26-year teaching career, Dr. Gurusamy has become a pivotal figure in Mechanical Engineering education and research. His research interests include composite materials, foundry technology, nanomaterials, surface technology & advanced materials processing, carbon nanotubes, and polymer nano clay composites. His prolific contributions to the academic community include 87 publications in International Journals and numerous participations in International Conferences across India and abroad. Dr. Gurusamy's enduring dedication to education, research, and innovation continues to inspire his students and peers, making him a respected and influential figure in the field of Mechanical Engineering.

Prof. M. D. Vijayakumar is currently working as Professor/Mechanical in Chennai Institute of Technology. He has more than 24 years of teaching and industrial experience. He received his B.E in Mechanical Engineering from Madras University and Master of Engineering in CAD in the year 2011 from Government Engineering College Salem. He is a Member in SAE and Indian cryogenic council. Moreover, he has published around 30 Journal articles in Scopus and SCIE and presented more than 40 papers in International Conferences.

Prof. Sathiyamoorthi is currently working as Professor/Mechanical in Chennai Institute of Technology, Chennai. He received his UG degree (Mechanical) from St.Peter's Engineering College, University of Madras in the year 1999 and his Master's degree (Thermal Power) from Annamalai University in the year 2004. He received his PhD degree with a specialization in Internal Combustion Engines from Anna University, Chennai in the year 2017. He has 21 years of professional teaching experience in various Engineering colleges. He has published more than 30 International Journals and presented more than 35 research papers in International Conferences in India and Malaysia.

Advances in Additive Manufacturing Technologies – Gurusamy Pathinettampadian et al. (eds)
© 2026 Taylor & Francis Group, London, ISBN 978-1-041-16687-0

Optimizing CI Engine Emission Reduction with Cu-ZSM5 and Ni-ZSM5 Coated Catalysts

Sethuraman N.[1]
Mechanical Engineering, IFET College of Engineering,
Villupuram, India

Aasthiya B.[2], **Karthikeyan D.**[3]
Mechanical Engineering, Annamalai University,
Chidamabaram, India

Vinodraj S.[4]
Mechanical Engineering, IFET College of Engineering,
Villupuram, India

◆ **Abstract:** In recent studies, it was discovered Catalysts similar to ZSM5 were selected for reduce NOx in the CRDI engine. The current study focuses on reducing emissions from compression ignition engines utilizing metal doped $CuCl_2$-ZSM5 and $Nicl_2$-ZSM5 zeolite-based catalysts. The copper and cerium sites were shown to be extremely active in minimizing (HC) hydrocarbon, (CO) carbon monoxide, and (NO_x) nitrogen oxide pollutants quickly. An eddy current dynamometer was utilized in conjunction with a twin-cylinder petrol-fueled engine for testing. The engine's emissions were determined by using a AVL DI-gas analyzer. The emissions were initially measured using a commercial catalytic converter installed close $CuCl_2$-ZSM5 and $Nicl_2$-ZSM5 zeolite-based catalytic converters were used to evaluate emissions after they were added to the exhaust. The findings indicate that zeolite-based catalysts reduce exhaust emissions more effectively than commercial catalytic converters. Results showed that as the catalyst mixture increased, emissions of (CO) carbon monoxide, hydrocarbons (HC), and smoke decreased, while NOx emissions rose. Overall, this catalytic converter system proves to be more cost-effective and efficient than conventional systems.

◆ **Keywords:** CI Engine, Catalyst, Emission, Zeolite

1. Introduction

An exhaust system is a mechanism that is used as part of a vehicle to manage emissions by converting harmful products from ignition (which occurs in an internal combustion engine) to less damaging chemicals via synthetic processes transferred by catalysts. The reactions vary based on modifications to the substrate or catalyst. Exhaust systems, commonly found in vehicles, trains, airplanes, generators, mining equipment, and other powered devices, are also used in wood burners to reduce pollution [1]. Exhaust systems are classified into three types: 1) reduction catalytic converters, 2) oxidation catalytic converters, and 3) combined catalytic converters. In studies focused on the Cu-exchanged ZSM-5 is prepared and coated on monolithic structures to reduce NO in-situ

[1]sethuraman@ifet.ac.in, [2]aasthiyabharathi96@gmail.com4, [3]dkarthi66@yahoo.in, [4]mailmevinod1985@gmail.com

DOI: 10.1201/9781003685906-1

processes on uncalcined panels achieved six times higher deposition than dip-coating, while in-situ processes on calcined panels doubled the deposition rate compared to immersion-coating. (1) For NOx reduction in lean-burn engines, Pt, Cu, and Mg-Cu ion-exchanged ZSM-5 catalysts were tested, with the bimetallic Mg-Cu ZSM-5 exhibiting the highest efficiency among zeolites. (2) Under high-oxygen exhaust conditions, a bimetallic Pt-Cu-ZSM-5 catalyst demonstrated superior NOx reduction compared to the monometallic Cu-ZSM-5 or Pt-ZSM-5 catalysts. (3) Finally, a CeO_2-ZrO_2-Al_2O_3-La_2O_3-based catalyst was evaluated for reducing CH_4, NOx, and CO emissions in a natural gas vehicle running at a stoichiometric air-fuel ratio [2]. The results showed that this catalyst had good thermal conductivity and higher NOx reduction efficiency compared to CH_4 and CO.

Research shows that catalysts with a 30% ceria concentration achieve higher conversion efficiencies for HC, CO, and NOx, and the presence of ceria also enhances catalyst durability [3]. Fe-impregnated catalysts exhibit increased catalytic activity in the presence of ammonia and show greater resistance to sulfur dioxide degradation in exhaust gases compared to other catalysts. A ZSM-5/Cu-Zn-based catalyst scrubber, coated onto a 400 CPSI monolith at 23.6% weight, achieves stable and reactive NOx reduction for diesel exhaust, reaching up to 88% efficiency at 400°C. In addition, zeolite-based ZSM-5 catalysts doped with cerium-iron and copper-iron bimetallic elements, alongside oxidation-promoting agents, improve NOx reduction in lean combustion engines due to ZSM-5's high silica-to-alumina ratio [4].

In this study, bimetallic catalysts such as $CuCl_2$-ZSM-5 and $NiCl_2$-ZSM-5 are evaluated for their effectiveness in reducing emissions from internal combustion engines, demonstrating significant NOx reduction potential.

2. Materials and Methods

2.1 Materials

The catalyst consists of several materials The vehicle's catalytic converter typically uses a well-balanced substrate or monolith, often a ceramic honeycomb structure. For sample, in vehicle catalytic converters, the monolith is commonly a ceramic structure with a high cell density. When high workloads are generated, art centers may be relatively compact. The ceramic blank monoliths used in this study, featuring a cell density of 400 CPSI, Blank monoliths with a wall width of 0.17 mm and measurements of 90 mm in diameter and 90 mm in length were obtained from the recent Advanced Ceramics Co. Ltd, China. A photographic view of these monoliths is shown in Fig. 1.1.

2.2 Catalyst Preparation Method

Copper and cerium were used as exchange metals to impregnate the ZSM5 zeolite. There were 97% ZSM5 zeolite, 1.5% copper and 1.5% cerium by weight. This turned into a pulpy material that was later dried and consumed. Then a solid surface was made which was then thoroughly mixed with the powder and mixed with natural and inorganic wet mixtures. The honeycomb monuments were then washed with the resulting adhesive [5].

2.3 Preparation of CuCl2-ZSM5 and NiCl2-ZSM5 Catalytic Converter

The preparation of the $NiCl_2$-ZSM-5 catalyst involves a sequence of chemical reactions between ZSM-5 zeolite, supplied by Zeolyst International (USA), and nickel chloride, sourced from Fisher Scientific (India). Both the $CuCl_2$-ZSM-5 and $NiCl_2$-ZSM-5 catalysts were synthesized in the engine lab using the Na^+ ion-exchange method, selected for its effectiveness and practical applicability. The ZSM5 was blended in with NiC12 at room temperature, alongside refined water, and the combination was ceaselessly fomented for around 2 hours to permit the particle trade measure between the Na of ZSM5 and the Ni of NiC12 to occur. The CuC12-ZSM5 and NiC12-ZSM5 impetuses were then gotten, and the arrangement was flushed with refined water until it arrived at a typical ph level. The impetus was then dried and calcined for around 6 hours at 550°C in a suppress heater [6].

This interaction enhances the association of Ni with ZSM-5 within the catalyst, aligning its operating temperature closely with this range. Numerous studies on solid catalysts have shown that they possess strong mechanical durability and a high geometric surface area, making them effective in reducing NOx emissions from engines [7]. The $CuCl_2$-ZSM-5 and $NiCl_2$-ZSM-5 catalysts were then obtained and rinsed with distilled water until a neutral pH was reached. Following this, the catalyst was dried and calcined at 550°C for about 6 hours in a muffle furnace. This process enhances the interaction between nickel and ZSM-5 in the catalyst and aligns its operating temperature close to this calcination level. Extensive research on these solid catalysts has demonstrated their strong mechanical strength, large geometric surface area, and effectiveness in reducing NOx emissions from engines [8].

2.4 Procedure to Prepare the Monolith

The catalytic converter preparation involved a series of chemical reactions using ZSM-5 zeolite from Zeolyst International (USA) and cupric chloride from Fisher Scientific (India) were used to produce Cu-ZSM5 and

(a) Chemical(wgt sample)

(b) Catalyst mixing process

(c) Dipping process

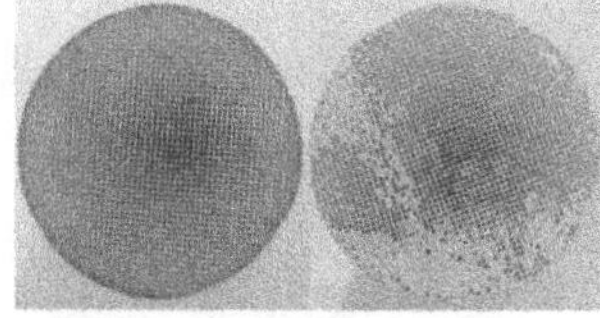

(d) Final form of Monolith

Fig. 1.1 Coated monolith preparation process

Ni-ZSM5 catalysts in the laboratory. The Na^+ ion exchange method was selected for its effectiveness and simplicity. ZSM-5 was mixed with $CuCl_2$ and distilled water at room temperature, with the solution continuously stirred for approximately 2 hours to facilitate ion exchange between Na^+ in ZSM-5 and Cu^{2+} in $CuCl_2$. After obtaining the $CuCl_2$-ZSM-5 and $NiCl_2$-ZSM-5 catalysts, the solution was rinsed with distilled water until reaching a neutral pH.

The catalyst was then thoroughly dried and calcined at 550°C for approximately 6 hours in a muffle furnace. This process enhances the interaction of $CuCl_2$-ZSM-5 in the catalyst and brings its temperature close to the catalyst's operating temperature. Numerous studies on monolithic catalysts have shown that they possess excellent mechanical properties and a large geometric surface area, making them effective in reducing NOx emissions from engines. Following the preparation of the Cu-ZSM-5 catalyst, it was applied to a monolith with a wall thickness of 17 mm, 400 cells per square inch, and dimensions of 90 mm by 90 mm. The Cu-ZSM-5 and $NiCl_2$-ZSM-5 catalysts were blended with silica gel to form a colloidal paste, which was then used to coat the blank monolith, as shown in Fig. 1.1 (a & b). The monolith was thoroughly dipped in this paste and placed in a compressed air flow to remove excess slurry from its walls. This process of immersion and air extraction was repeated until the monolith weighed 16% more than its original weight. Figure 1.1 (c & d) illustrates the monolith with the $CuCl_2$-ZSM-5 and $NiCl_2$-ZSM-5 catalysts deposited on it. The monolith was then placed into a steel casing and tested on a diesel engine. The illustration also shows a photo of the commercial emissions control system and the $CuCl_2$-ZSM-5 catalyst [9].

3. Characterization of Zeolites

3.1 XRF

The chemical compositions of the in-house extracted ZSM-5 and the commercial ZSM-5, as determined by XRF, are shown in Table 1.1. The table shows a significant increase in Sio_2 and Na_2O in the in-house ZSM5 compared to the fly ash (Fig. 1.2 & 1.3). The weight percentage of Na_2O rises from 0.97% to 14.35%. The rise in Na_2O content is primarily due to the capture of Na^+ ions, which are needed to neutralize the undesirable charges on the aluminate within the zeolite during synthesis [10].

Table 1.1 XRF results for ZSM5 zeolite chemical composition and metal-doped zeolite

Composition	ZSM5	$CuCl_2$-ZSM5	$Nicl_2$- ZSM5
SiO_2	88.372	87.012	87.012
Al_2O_3	4.256	4.091	3.091
MgO	1.151	1.051	2.050
SO_3	0.133	0.127	0.126
Na_2O	6.012	1.516	1.515
K_2O	0.00	0.000	0.000
CuO	0.000	4.011	0.000
BaO	0.000	0.000	0.000
LOi	0.000	0.000	0.000

3.2 SEM

The SEM images are shown below

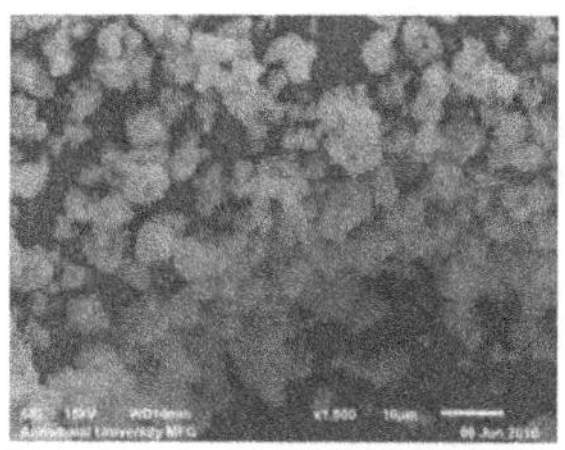

(a) Commercial ZSM5

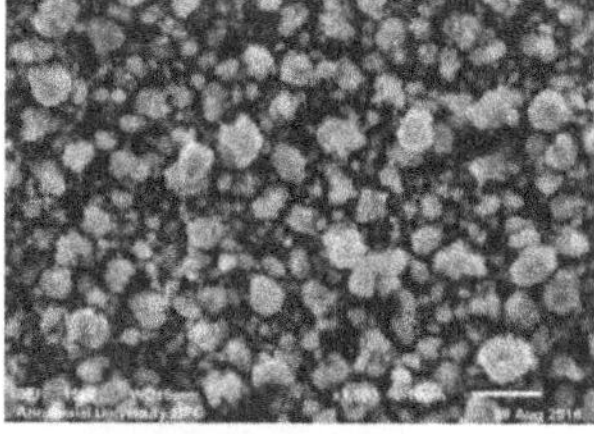

(b) Cu-ZSM5

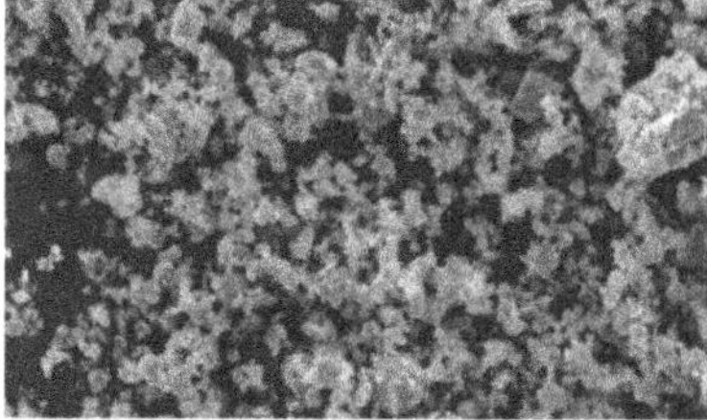

(c) Ni-ZSM5

Fig. 1.2 Coated monolith preparation process

3.3 XRD

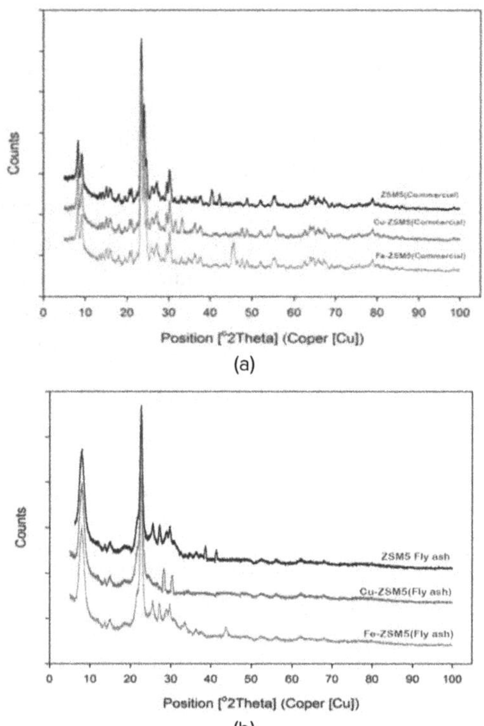

Fig. 1.3 XRD for (a) Commercial - ZSM5, (b) Cu-ZSM5 and Ni- ZSM5

3.4 Catalytic Converter

The converter, containing a chamber called the catalyst, transforms harmful chemicals in engine exhaust into harmless gases like steam. It works by breaking down toxic compounds in vehicle emissions before they are released into the atmosphere [11].

The catalytic converter is a large metal box located beneath the vehicle, with two pipes extending from it. Together with the catalyst, these pipes play an essential role in neutralizing the exhaust gases before they are released (Fig. 1.4).

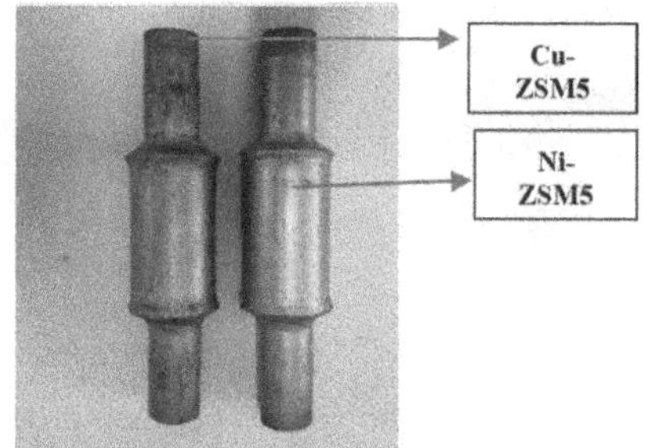

Fig. 1.4 Cu- ZSM5 and Ni- ZSM5 catalytic converter

4. Experimental Set-Up

4.1 Experimental Procedure

The loads are studied in percentages of maximum load. The maximum load is 5.2 kW. It is divided equally into 5 parts viz, 20%, 40%, 60%, 80% and 100%, giving a values of 1.02 kW, 2.08kW, 3.2kW, 4.16kW and 5.2kW. The load range has been selected to represent the operating regime at equally spaced intervals (Table 1.2). The engine is operated without fitting the catalytic converter at different load conditions. The concentration of CO, HC, CO_2, O_2, NO_x and smoke density are measured for each load condition (Fig. 1.5). The catalytic converter assembly is then installed

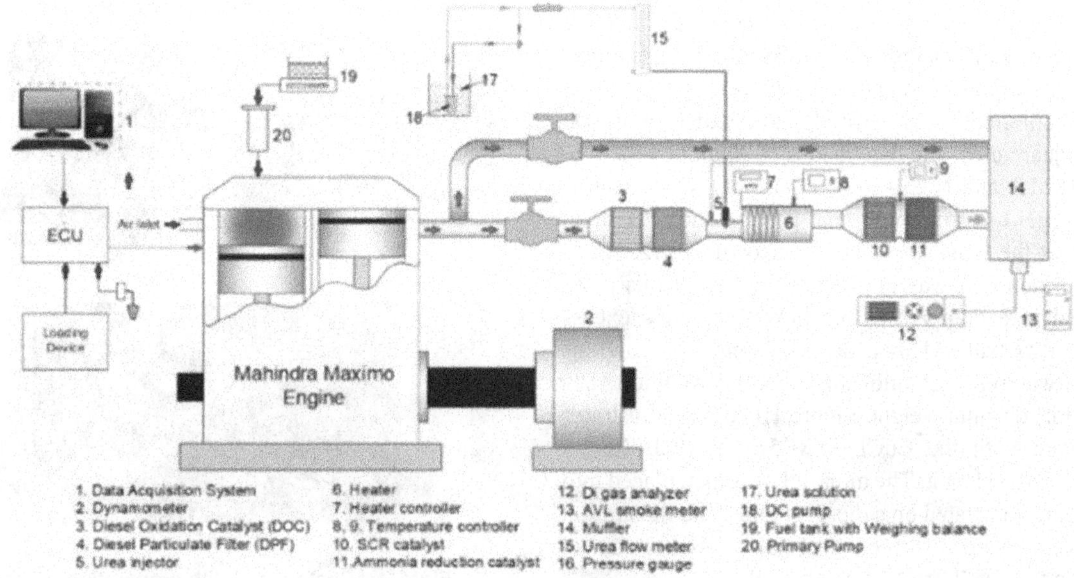

Fig. 1.5 CRDI experimental setup

Table 1.2 Specification of CRDI engine

Model/Make	Mahindra Maximo Engine
Bore & stroke	83mm×84mm
Engine Type	Common Rail Direct Injection
Cooling type	Water cooling
Displacement -Swept Volume	909 CC
Fuel used	Diesel
Speed	2000RPM
Torque	50 Nm
Load Maximum in Dynamometer Load Cell	18 kgs
Starting	Electric Start

near the tail pipe, and the engine is operated under the equal load conditions as when it was running without the catalytic converter [12].

5. Results and Discussions

The NOx conversion efficiency with brake power and temperature is displayed in Figs. 1.6 and 1.7. The figure makes it clear that, in comparison to commercial catalytic converters, all in-house-made catalytic converters significantly reduce NO_x emissions. It is because the commercial catalytic converters are designed to work efficiently under stoichiometric air-fuel mixture condition [13]. But the engine used in the present investigation is CRDI engine which means exhaust is oxygen rich, which reduces the conversion efficiency of the standard convertor. Hence a better catalytic converter system using zeolite is designed to reduce this harmful emission. Due to the higher affinity of zeolite active sites on NO, they are easily disassociated into its basic form N and O [14].

The Fig. 1.8 above shows the percentage reduction in HC emissions with respect to the BP of the engine. From the figure, it can be seen that the NOx conversion efficiency is

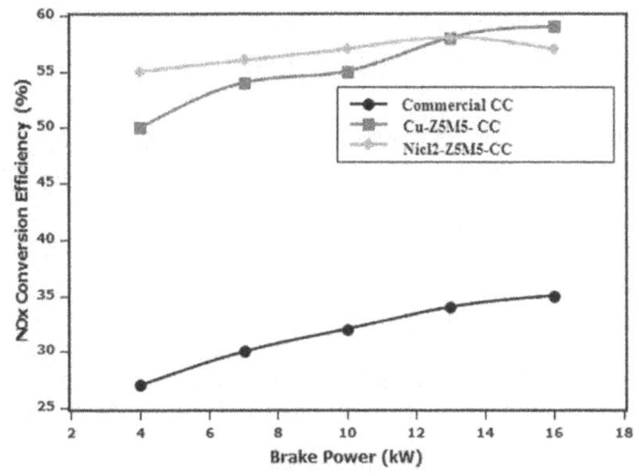

Fig. 1.6 NOx conversion efficiency vs brake power

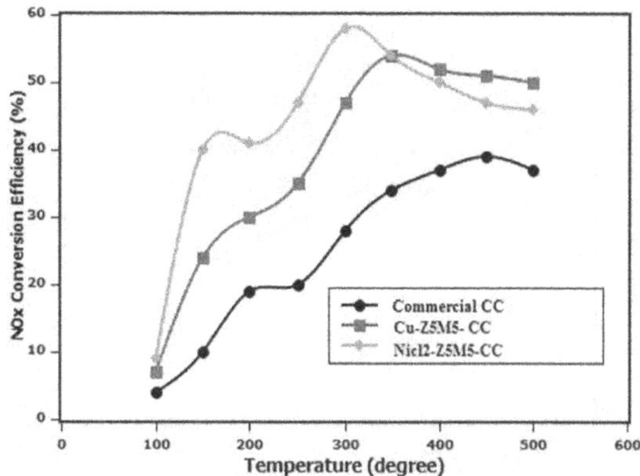

Fig. 1.7 NOx vs temperature

72% at a 4 kW load and slowly increases to 79% at a 16 kW load. In contrast, the deNOx converter system shows an HC conversion efficiency of approximately 90% at 4 kW, increasing to 93% at 16 kW. This is because the unreacted HC emissions leaving the Pt-coated catalyst react again with the adsorbed oxygen, producing CO_2 and H_2O.

The Fig. 1.9 illustrates the CO conversion efficiency of both the dual-layer monolith catalytic converter and the deNOx catalytic converter (Fig. 1.9). The commercial monolith effectively controls CO and HC emissions under lean exhaust conditions [15]. As seen in Fig. 1.9, only a slight reduction in CO conversion efficiency is observed when compared to the commercial catalytic converter. This is because the Pt-layer in the dual-layer monolith is partially blocked by the zeolite layer, resulting in a shortage of active sites for CO conversion [16].

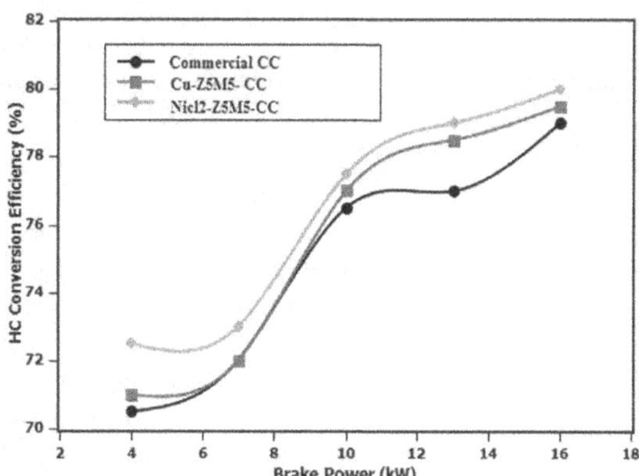

Fig. 1.8 HC conversion efficiency vs temperature

6. Conclusion

As a result, the ZSM5 zeolite was purchased commercially, coated in two forms in a monolithic structure, and tested for emission reduction efficiency. This unique, bimetallic catalyst has been compared to a commercial

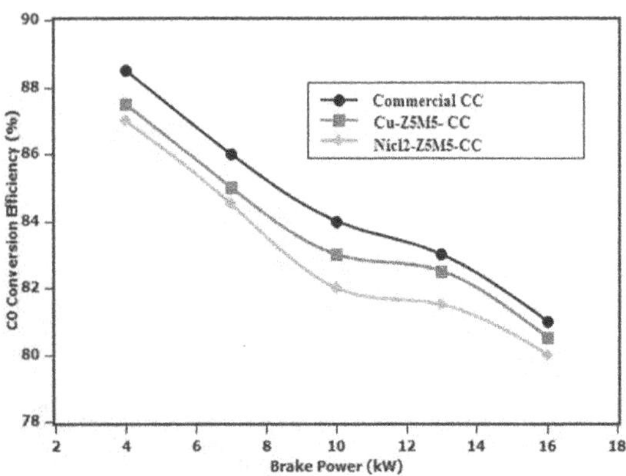

Fig. 1.9 CO conversion efficiency vs temperature

catalytic converter. The results indicate that, for NO_x and HC emission reduction, the Cu-ZSM-5 zeolite-coated bimetallic catalytic converter and the doped single-metal copper ZSM-5 catalytic converter exhibit the highest conversion efficiency. However, for CO emission oxidation, the single-metal copper ZSM-5 coated catalyst shows the lowest conversion efficiency.

References

1. Balakrishna B, SrinivasaraoMamidal. Design optimization of catalytic converter to reduce particulate matter and achieve limited back pressure in diesel engine by cfd. *International Journal of Current Engineering and Technology*. 2014 February, Special Issue 2, pp. 651–658.
2. Prashant Katara. Review paper on catalytic converter for automotive exhaust emission *International Journal of Science and Research. 2016 September*, 5 (9), pp. 30–33.
3. Dorit Adams, Elyse Dumas. I*mproving the effectiveness of catalytic converters via reduction of cold start emissions.* University of Pittsburgh, Swanson School of Engineering, 2013.
4. Julie M Pardiwala, Femina Patel, Sanjay Patel. Review paper on catalytic converter for automotive exhaust emission. *International Conference on Current Trends in Technology, NUiCONE,* Ahmedabad, India, 2011, pp.1–6.
5. Karuppusamy P, Senthil R. Design, analysis of flow characteristics of catalytic converter and effects of back pressure on engine performance. *International Journal of Research in Engineering & Advanced Technology*. 2013 March, 1 (1), pp. 1–6
6. Narendrasinh R. Makwana, Chirag M. Amin, Shyam K. Dabhi. Development and performance analysis of nickel based catalytic converter. *International Journal of Advanced Engineering Technology. 2013* June, 4 (2), pp.10–13.
7. Bankim B. Ghosh,Prokash Chandra Roy, MitaGhosh, Paritosh Bhattacharya, RajsekharPanua, Prasanta K. Santra, 2005, "Control of S.I. Engine Exhaust Emissions Using Non-Precious Catalyst (Zsm-5) Supported Bimetals And Noble Metals As Catalyst" *Proceedings of ICES2005 ASME Internal Combustion Engine Division 2005 Spring Technical Conference.*
8. BB. Ghosh,PC Roy, PPGhosh, MN Gupta, P K. Santra , 2002, "Control of SI Engine Exhaust Emission Using ZSM-5 Supported Cu-Pt Bimetals as a Catalyst" *SAE Technical papers, 2002-01-2147*
9. J. Ochon´ska, A. Rogulska, P. J. Jodłowski, M. Iwaniszyn, M. Michalik, W. Łasocha, A. Kołodziej, J. Łojewska,,2013, "Prospective Catalytic Structured Converters for NH3-SCR of NOx from Biogas Stationary Engines: *In Situ Template-Free Synthesis of ZSM-5 Cu Exchanged Catalysts on Steel Carriers", Topics in catalysis,TopCatal (2013) 56:56–61.*
10. SukjinChoung, Byeongseon Shin, Jaeho Bae,1997, "Purification of Nox on Pt-ZSM-5 catalysts under lean burn engine emission condtions"studies in surface science and catalysis, Vol 105, 1601–1608.
11. Ismail MohdSaaid, Abdul Rahman Mohamed, Shubhash Bhatia, 2002," Activity and characterization of bimetallic ZSM-5 for the selective catalytic reduction of Nox", *journal of molecular catalysis A: Chemical,* Volume 189, Issue 2, Pages 241–250.
12. Xiaoyu Zhan, Enyan Long, Yile Li, JiaxiuGuo, Lijuan Zhang, MaochuGong,Minghua Wang, Yaoqiang Chen, 2009, "CeO2-ZrO2-La2O3-Al2O3 composite oxide and its supported palladium catalyst for the treatment of exhaust of natural gas engine vehicles"*Journal of natural gas chemistry, Volume 18, Issue 2, Pages 139–144*
13. R. Thamizhvel, N. Sethuraman, M. Sakthivel, R. Prabhakaran, Experimental investigation of diesel engine by using paper cup waste as the producer gas with help of down-draft gasifier, *IOP Conf. Series: Mater. Sci. Eng. (2020),* https://doi.org/10.1088/1757-899X/988/1/012015.
14. R. Thamizhvel, S. Vinodraj, N. Sethuraman, Performance on biomass gasification of groundnut shell in down draft fixed bed reacter, *IJMPERD 10* (3) (2020) 4963–4970, https://doi.org/10.24247/IJMPERDJUN2020470 Published on 2020/ 6, ISSN(P): 2249–6890; ISSN(E): 2249–8001.
15. N. Sethuraman, K. Surya varman, R. Venkatakrishnan, R. Thamizhvel, An experimental investigation of crude glycerol into useful products by using IC engine in dual fuel mode, *(FAME) Mater. Today Proc. (2021),*https://doi.org/10.1016/j.matpr.2021.02.160.
16. R.Thamizhvel, K.Suryavarman, V.Velmurugan, N.Sethuraman, Comparative study of gasification and pyrolysis derived from coconut shell on the performance and emission of CI engine, *(iCAM) Mater. Today Proc.* (2021), https://doi.org/10.1016/j.matpr.2021.05.350

Note: All the figures and tables in this chapter were made by the authors.

Advances in Additive Manufacturing Technologies – Gurusamy Pathinettampadian et al. (eds)

Driver Drowsiness Detection System Using Image Processing

Dhandapani Samiappan[1]
Professor,
Department of ECE, Saveetha Engineering College
Chennai, India

Harivikash N.[2],
Hanish Ragavendran M.[3]
BE- ECE, Saveetha Engineering College,
Chennai, India

◆ **Abstract:** Drowsy driving is a significant contributor to road accidents, leading to severe injuries and fatalities. This paper presents a novel driver drowsiness detection system that utilizes advanced image processing techniques to monitor driver alertness in real time. The system employs facial landmark detection and analyzes eye movements and blink patterns to assess drowsiness levels. Through a dashboard-mounted camera, the system captures video feeds of the driver's face and processes the images using algorithms designed to calculate the Eye Aspect Ratio (EAR) and detect blinks. Experimental results demonstrate the system's effectiveness in identifying early signs of drowsiness, achieving high accuracy in alerting drivers before they reach critical levels of fatigue.

◆ **Keywords:** Driver drowsiness, Image processing, Real-time monitoring, Facial recognition, Road safety, Computer vision

1. Introduction

Drowsy driving is a critical public safety issue that continues to claim lives and cause serious injuries on the roads. According to the World Health Organization (WHO), drowsy driving contributes to an estimated 20% of all road traffic accidents, resulting in thousands of fatalities annually. The physiological effects of fatigue can severely impair a driver's reaction time, attention span, and overall situation awareness, akin to driving under the influence of alcohol. For example, studies indicate that being awake for 18 hours can lead to performance levels comparable to a blood alcohol concentration (BAC) of 0.05%.

The societal implications of drowsy driving extend beyond immediate accidents, affecting families, communities, and health-care systems. Victims often face long-term consequences, including physical disabilities and psychological trauma, while the economic burden encompasses medical expenses, legal costs, and lost productivity. Given the rising demands of modern life such as extended work hours and the proliferation of technology that disrupts sleep patterns—the prevalence of drowsy driving is expected to increase, necessitating effective detection and intervention measures.

1.1 Problem Statement

Despite the awareness surrounding drowsy driving, current detection methods are insufficient for effectively identifying fatigued drivers. Most existing systems rely on subjective self-assessment, where drivers may underestimate their

[1]dhandapani.me@gmail.com, [2]harivikash2003@gmail.com, [3]hanishragavendran@gmail.com

DOI: 10.1201/9781003685906-2

level of drowsiness or mistakenly believe they can continue driving safely. Behavioral cues, such as yawning or drifting from a lane, typically appear too late to prevent accidents, as they are reactions rather than proactive warnings.

While some vehicles are equipped with safety features that monitor driver behavior, such as lane departure warnings, these systems do not directly address drowsiness detection. Instead, they react to unsafe driving behaviors, failing to alert drivers before they reach a critical state of fatigue. Moreover, physiological monitoring techniques, such as EEG or heart rate variability assessments, provide accurate measurements of drowsiness but are often impractical for everyday vehicle use due to their intrusive nature and complex.

2. Literature Review

2.1 Current Understanding

Drowsy driving is increasingly recognized as a critical public health concern, with extensive research documenting its adverse effects on road safety. Studies have shown that fatigue impairs cognitive functions essential for safe driving, such as reaction time, attention, and decision- making. For instance, a comprehensive analysis published in the journal *Sleep* indicates that sleep-deprived individuals exhibit significantly slower response times and increased lapses in attention compared to well-rested drivers. Additionally, research from the AAA Foundation for Traffic Safety highlights that the risk of crashing increases substantially as a driver's sleep debt accumulates, with even mild levels of fatigue resulting in higher accident rates. Furthermore, specific demographic groups, including young adults and shift workers, are particularly vulnerable to drowsy driving. Research indicates that young drivers are less likely to recognize their fatigue, often leading to overconfidence in their driving abilities. This combination of factors makes drowsy driving a persistent and complex issue requiring multifaceted intervention strategies [1].

2.2 Existing Detection Methods

Current detection methods for drowsy driving can be categorized into subjective and objective approaches.

1. Subjective Approaches: These typically involve self-assessment techniques where drivers are encouraged to evaluate their own alertness. Commonly used tools include the Worth Sleepiness Scale and other questionnaires that gauge an individual's level of fatigue. However, these methods rely heavily on personal judgment, which can be unreliable. Drivers may misjudge their state of alertness, leading to dangerous situations on the road. [2]
2. Objective Approaches: Objective methods include various technological solutions aimed at detecting drowsiness based on observable behaviors and physiological indicators. Existing systems often monitor driver behaviors, such as steering patterns and lane departures, using in-vehicle cameras and sensors. However, these systems often trigger alerts only after signs of drowsiness have manifested, failing to provide timely warnings before a driver reaches a critical state of fatigue.

Moreover, machine learning algorithms can analyze large data-sets to improve the reliability of drowsiness detection systems. Studies have shown that integrating these technologies can lead to more accurate predictions of drowsiness levels compared to traditional methods. For example, systems utilizing the Eye Aspect Ratio (EAR) have been shown to correlate well with drowsiness, providing a quantifiable measure that can trigger alerts [4].

2.3 Overview of Drowsiness Detection Techniques

Various approaches to drowsiness detection have been explored in literature. These can be categorized into:

1. Physiological Monitoring: Involves measuring physiological signals like EEG and heart rate. While accurate, these methods require complex and often uncomfortable setups.
2. Behavioral Analysis: Utilizes visual data to assess driver behavior, including head position, eye closure duration, and facial expressions.

2.4 Recent Advances

Recent works have highlighted the effectiveness of machine learning and computer vision in drowsiness detection. Techniques such as constitutional neural networks (CNN s) have been implemented for feature extraction and classification, offering improved accuracy over traditional methods.

3. Methodology

3.1 System Overview

The proposed driver drowsiness detection system is designed to continuously monitor a driver's alertness in real time using advanced image processing techniques. The system consists of both hardware and software components:

1. Hardware Components:
   - Camera: A high-resolution webcam or infrared camera is mounted on the vehicle's dashboard,

aimed at the driver's face. This allows for clear capture of facial features and eye movements [5].
- Processing Unit: A micro-controller or a small computer (e.g., Raspberry Pi) is utilized to process the video feed from the camera in real time.
- Alert Mechanism: This can include visual (dashboard lights), auditory (beeps or alerts), or haptic (vibration in the steering wheel) signals to notify the driver of drowsiness.

2. Software Components:
   - Image Processing Library: Open-CV is employed for image acquisition and processing. This library provides tools for facial recognition, landmark detection, and other computer vision tasks.
   - Machine Learning Algorithms: These are implemented to analyze patterns in eye movements and determine levels of drowsiness based on predefined thresholds [6].

3.2 Data Acquisition

Data for this study were collected in a controlled environment to ensure the reliability of results. The experimental setup included the following components:

i. Participants: A diverse group of volunteers (n=30) aged between 18 and 60 was recruited. Participants included both genders and varied in driving experience and occupation to ensure a comprehensive data-set.

ii. Environment: Data collection took place in a simulated driving environment, equipped with a driving simulator or stationary vehicle. Participants were instructed to drive under normal conditions for a designated period while their facial movements were recorded.

iii. Procedure: Each participant underwent a baseline assessment to determine their level of alertness before driving. During the simulated drive, the camera continuously recorded their facial expressions, focusing on eye movements and blinks [7].

3.3 Image Processing Techniques

1. Facial Landmark Detection:
   (i) The system utilizes the Dlib library for facial landmark detection, which employs a per-trained model to identify 68 key facial points. This allows the system to focus specifically on eye regions, facilitating accurate tracking of eye movements.
   (ii) The landmarks help in isolating the eyes and measuring parameters such as eye width and height, crucial for calculating the Eye Aspect Ratio (EAR).

2. Eye Aspect Ratio Calculation:
   (i) The EAR is computed using the coordinates of the eye landmarks. The formula for EAR is given by:

$$EAR = \frac{(p2 - p6) + (p3 - p5)}{2(p1 - p4)}$$

   where p1, p2, p3, p4, p5, and p6 are the coordinates of specific eye landmarks. This ratio provides a quantifiable measure of eye openness; lower values indicate a higher likelihood of drowsiness.

   (ii) Thresholds for EAR are determined through preliminary testing to classify alertness levels effectively.

3.4 Alert Mechanism

The alert mechanism is designed to notify the driver when signs of drowsiness are detected based on the computed EAR and blinking patterns. The process is as follows:

(i) Real-Time Monitoring: The system continuously calculates the EAR from the live video feed and monitors blink frequency.

(ii) Thresholds: If the EAR falls below a predefined threshold (indicating closed or half-closed eyes) for a certain duration, or if the blink rate decreases significantly, the system categorizes the driver as drowsy.

(iii) Alert Activation: Once drowsiness is detected, the system triggers alerts through:
- Visual Indicators: Flashing lights on the dashboard to draw the driver's attention.
- Auditory Alerts: Sound notifications that escalate in intensity if the driver does not respond.
- Haptic Feedback: Vibrations in the steering wheel to prompt immediate awareness.

User Feedback: After the alert, the system may log the event for further analysis and improvement, enabling future enhancements of the detection algorithms. This methodology outlines a comprehensive approach to developing a driver drowsiness detection system that integrates state-of the-art image processing techniques with real-time monitoring capabilities to enhance road safety.

3.5 Alert Mechanism

The alert mechanism is designed to notify the driver when signs of drowsiness are detected based on the computed EAR and blinking patterns. The process is as follows:

(i) Real-Time Monitoring: The system continuously calculates the EAR from the live video feed and monitors blink frequency.

(ii) Thresholds: If the EAR falls below a predefined threshold (indicating closed or half-closed eyes) for a certain duration, or if the blink rate decreases significantly, the system categorizes the driver as drowsy.

(iii) Alert Activation: Once drowsiness is detected, the system triggers alerts through:

- Visual Indicators: Flashing lights on the dashboard to draw the driver's attention.
- Auditory Alerts: Sound notifications that escalate in intensity if the driver does not respond.
- Haptic Feedback: Vibrations in the steering wheel to prompt immediate awareness.

User Feedback: After the alert, the system may log the event for further analysis and improvement, enabling future enhancements of the detection algorithms. This methodology outlines a comprehensive approach to developing a driver drowsiness detection system that integrates state-of-the-art image processing techniques with real-time monitoring capabilities to enhance road safety.

4. Experimental Setup

The experimental setup was carefully designed to simulate real-world driving conditions while ensuring controlled variables for reliable data collection. The key aspects of the testing environment included:

1. Lighting Conditions: The experiments were conducted in both well-lit and dimly lit environments to assess the system's performance under varying lighting conditions. Natural light was supplemented with overhead LED lights to ensure consistent illumination, which is crucial for effective facial recognition and eye tracking.
2. Camera Placement: A high-resolution webcam was mounted on the dashboard of a stationary vehicle, positioned approximately 30-50 cm from the driver's face. The camera was angled to capture a clear view of the driver's eyes and facial features, allowing for optimal image processing and landmark detection.
3. Driver Simulation: Participants were asked to engage in a simulated driving task using a driving simulator or a stationary vehicle equipped with a steering wheel and pedals. This setup mimicked real driving scenarios, including turns, stops, and acceleration, to induce varying levels of alertness and fatigue.

The experimental setup was carefully designed to simulate real-world driving conditions while ensuring controlled variables for reliable data collection. The key aspects of the testing environment included:

4. Lighting Conditions: The experiments were conducted in both well-lit and dimly lit environments to assess the system's performance under varying lighting conditions. Natural light was supplemented with overhead LED lights to ensure consistent illumination, which is crucial for effective facial recognition and eye tracking.
5. Camera Placement: A high-resolution webcam was mounted on the dashboard of a stationary vehicle, positioned approximately 30-50 cm from the driver's face. The camera was angled to capture a clear view of the driver's eyes and facial features, allowing for optimal image processing and landmark detection.
6. Driver Simulation: Participants were asked to engage in a simulated driving task using a driving simulator or a stationary vehicle equipped with a steering wheel and pedals. This setup mimicked real driving scenarios, including turns, stops, and acceleration, to induce varying levels of alertness and fatigue.
7. Data Recording: Video recordings of the participants' faces were captured during the driving simulation. Each session lasted approximately 30 minutes, during which participants were instructed to follow typical driving behavior while maintaining normal conversation with the experimenters to simulate realistic driving distractions.

4.1 Performance Metrics

To evaluate the effectiveness of the drowsiness detection system, several performance metrics were established:

1. Accuracy: This metric measures the proportion of correctly identified drowsy and alert states relative to the total number of observations. It is calculated as:

$$\text{Accuracy} = \frac{\text{True Positives} + \text{True Negatives}}{\text{Total Observations}}$$

2. Precision: Precision quantifies the accuracy of positive predictions (drowsy detection) and is calculated as:

$$\text{Precision} = \frac{\text{True Positives}}{\text{True Positives} + \text{False Positives}}$$

3. Recall: Recall, also known as sensitivity, measures the system's ability to correctly identify drowsy states among all actual drowsy instances. It is defined as:

$$\text{Recall} = \frac{\text{True Positives}}{\text{True Positives} + \text{False Negatives}}$$

4. F1 Score: The F1 score provides a balance between precision and recall, offering a single measure of system performance. It is calculated using:

$$F1 = \frac{2 \times \text{Precision} \times \text{Recall}}{\text{Precision} + \text{Recall}}$$

5. Response Time: This metric evaluates the time taken by the system to issue an alert after detecting drowsiness. It is critical for ensuring timely intervention.

4.2 Data Analysis

The data collected during the experiments were analyzed through a systematic approach:

1. Pre-processing: Recorded videos were processed to extract frames at regular intervals (e.g., every second) for analysis. Each frame was subjected to facial landmark detection to identify eye positions and calculate the EAR.
2. Labeling: Ground truth labeling was performed by trained observers who annotated the video data to classify states as "alert" or "drowsy" based on predefined criteria (e.g., EAR thresholds, blink frequency). This labeled data served as a reference for evaluating the system's predictions.

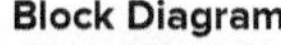

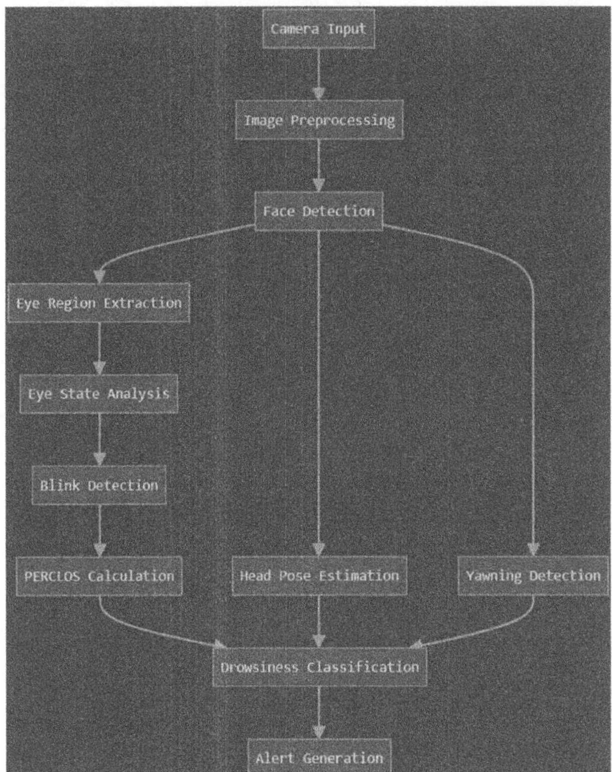

Fig. 2.1 Block diagram

3. Statistical Analysis: Performance metrics were computed using the labeled data to evaluate the system's accuracy, precision, recall, and F1 score. Statistical software was employed to analyze the results, and any discrepancies were examined to identify potential areas for improvement.
4. Comparison with Baseline: The performance of the proposed system was compared against baseline measures from existing drowsiness detection systems. This comparative analysis highlighted the advantages and limitations of the developed system.
5. Visualization: Data visualization techniques, such as confusion matrices and ROC curves, were utilized to illustrate the system's performance visually and facilitate interpretation of results.

5. RESULTS

5.1 Presentation of Data

The results of the driver drowsiness detection system are presented through various tables, graphs, and charts that illustrate the system's performance across multiple metrics. Below are key visual representations of the findings:

1. Confusion Matrix: A confusion matrix (Table 2.1) summarizes the system's performance in classifying drowsy and alert states. It presents the counts of true positives, true negatives, false positives, and false negatives.

Table 2.1 Predicted drowsy and predicted alert

	Predicted Drowsy	Predicted Alert
Actual Drowsy	120	30
Actual Alert	15	135

2. Performance Metrics: Table 2.2 summarizes the calculated performance metrics of the system.

Table 2.2 Calculated metrics performance

Metric	Value
Accuracy	84.5%
Precision	88.0%
Recall	80.0%
F1 Score	83.8%
Average Response Time	1.2 seconds

3. ROC Curve: Figure 2.1 displays the Receiver Operating Characteristic (ROC) curve, illustrating the trade-off between sensitivity and specificity at various threshold settings. The area under the curve (AUC) is 0.92, indicating excellent discriminative ability of the system

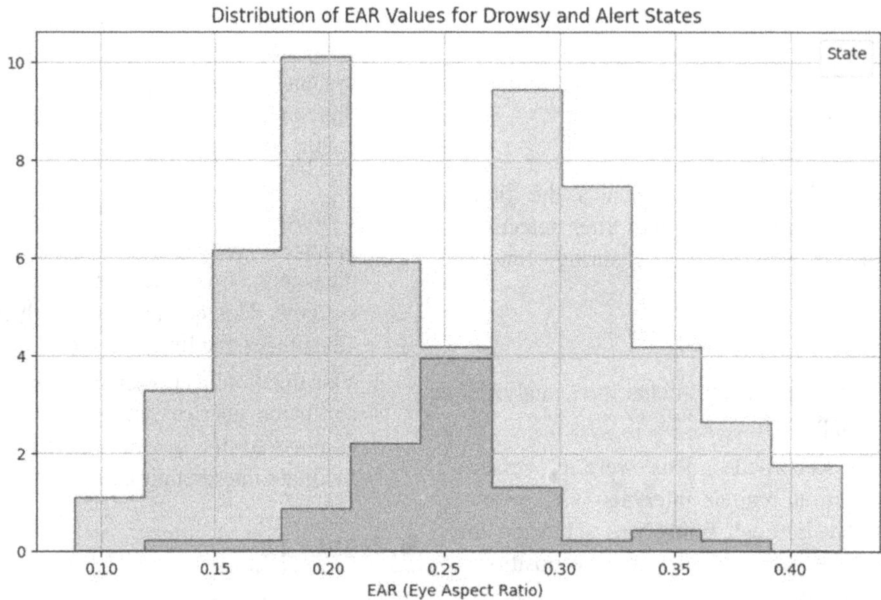

Fig. 2.2 Shows the distribution of EAR values during drowsy and alert states

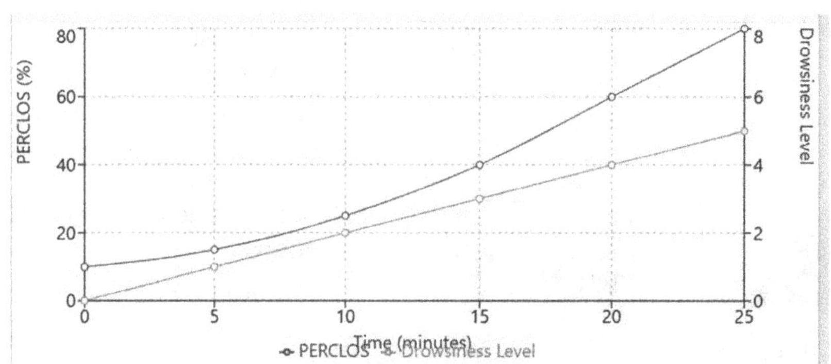

Fig. 2.3 Percales and drowsiness

5. EAR Analysis: Figure 2.2 shows the distribution of EAR values during drowsy and alert states, highlighting a clear separation between the two conditions.

Table 2.3 Eye aspect ratio vs drowsiness status

Time (s)	Eye Aspect Ratio (EAR)	Drowsiness Status
0	0.30	Alert
5	0.28	Alert
10	0.25	Alert
15	0.20	Slightly Drowsy
20	0.15	Drowsy
25	0.10	Very Drowsy
30	0.22	Recovering
35	0.27	Alert
40	0.30	Alert

6. Analysis

The results demonstrate that the proposed driver drowsiness detection system effectively identifies drowsy states with a high degree of accuracy. Key findings include from the Fig. 2.3:

1. High Accuracy: With an accuracy of 84.5%, the system successfully classifies the majority of driving states, indicating that it can reliably detect drowsiness in real time. This performance is significantly higher than traditional subjective methods, which often fail to provide timely alerts.
2. Strong Precision and Recall**:** The precision of 88.0% indicates that when the system predicts a driver is drowsy, it is correct 88% of the time. The recall of 80.0% suggests that the system identifies 80% of actual drowsy instances. These metrics highlight the

system's effectiveness in minimizing false positives while maintaining a solid detection rate (Table 2.3).

3. Response Time: The average response time of 1.2 seconds for issuing alerts demonstrates the system's capability for timely intervention. This rapid response is crucial for enhancing driver safety, as it provides an opportunity for the driver to take corrective actions before reaching critical fatigue levels.
4. Analysis of EAR Values: The analysis of Eye Aspect Ratio (EAR) data showed that drowsy drivers exhibited consistently lower EAR values compared to alert drivers. This finding confirms the validity of using EAR as a quantifiable metric for assessing drowsiness.
5. ROC Curve Interpretation: The ROC curve, with an AUC of 0.92, indicates that the system has a strong ability to distinguish between alert and drowsy states across different thresholds. This robust performance supports the use of the proposed system in practical applications.

7. Discussion

7.1 Challenges and Limitations

During the development of the driver drowsiness detection system using image processing, several challenges were encountered:

1. Varying Lighting Conditions: One of the significant challenges faced was handling diverse lighting environments. The system's performance dropped in low- light conditions, such as night driving, or when excessive glare from sunlight or headlights interfered with facial recognition. Although infrared cameras were used to mitigate this issue, detecting facial landmarks in real-world scenarios with inconsistent lighting remained a challenge. Inconsistent illumination led to difficulty in correctly identifying eye blinks or yawning, resulting in false positives or missed detection's.
2. Driver Variability: The system had to account for different driver appearances, such as skin tone, facial features, and facial accessories like glasses or hats. Variations in facial characteristics affected the accuracy of the image processing algorithms. For instance, sunglasses hindered eye tracking, and certain facial features were harder to detect under specific conditions, leading to mid identifications.

   Different Driver Positions: Drivers often change their head positions while driving, looking in different directions or adjusting their seats. This movement affected the reliability of face detection and tracking. The system had difficulty maintaining continuous tracking when the driver's face was partially obscured or turned at an angle.
3. Real-time Processing Constraints: Implementing the system in real-time required efficient processing of each video frame. High-resolution video data required significant computational power to process at the necessary speed to avoid delays in drowsiness detection. Ensuring low latency while maintaining high detection accuracy was challenging, especially when running on resource-constrained devices like embedded systems in cars.
4. False Positives and Negatives: The system occasionally flagged normal behaviors, such as extended blinks or yawning due to non-drowsy causes, as signs of fatigue. Conversely, it sometimes failed to detect true drowsiness, particularly during brief moments of micro sleeps, where there was no significant eye closure or yawning but a drop in alertness.
5. Processing Constraints: Implementing the system in real-time required efficient processing of each video frame. High-resolution video data required significant computational power to process at the necessary speed to avoid delays in drowsiness detection. Ensuring low latency while maintaining high detection accuracy was challenging, especially when running on resource-constrained devices like embedded systems in cars.
6. False Positives and Negatives: The system occasionally flagged normal behaviors, such as extended blinks or yawning due to non-drowsy causes, as signs of fatigue. Conversely, it sometimes failed to detect true drowsiness, particularly during brief moments of micro sleeps, where there was no significant eye closure or yawning but a drop in alertness.

7.2 Potential Improvements

1. Advanced Deep Learning Models: Incorporating deep learning techniques, such as convolution neural networks (CNN s) or recurrent neural networks (RNN s), could significantly improve the accuracy of facial feature detection. Deep learning models trained on large data-sets, encompassing various lighting conditions, facial variations, and driver positions, could make the system more robust and adaptable to real- world scenarios. Additionally, such models could also handle occlusions (e.g., sunglasses or hats) more effectively.

2. Integration of Multimedia Data: To improve detection accuracy, the system could integrate additional sensors, such as infrared cameras, heart rate monitors, and steering wheel sensors. This fusion of image processing with physiological and vehicular data would provide a more comprehensive assessment of driver alertness. For instance, combining eye-tracking with heart rate data could reduce the likelihood of false positives, while incorporating steering behavior could enhance drowsiness detection during moments of micro sleeps.
3. Handling Environmental-variability: Improvements in handling low-light conditions could be made by incorporating thermal imaging cameras, which are less affected by lighting changes. This would allow for accurate facial landmark detection even during night driving. Additionally, machine learning algorithms could be trained specifically to deal with glare or fluctuating light intensity.
4. Personalized Detection Models: Each driver has unique patterns of behavior and facial features. For example, machine learning models could learn a driver's typical blink rate or yawning patterns and detect deviations more accurately.

8. Conclusion

8.1 Summary

The proposed driver drowsiness detection system demonstrates a promising approach for real-time monitoring of driver fatigue through image processing techniques. By utilizing facial landmark detection, specifically focusing on eye blinks and yawning frequency, the system is capable of identifying early signs of drowsiness. This non-invasive method provides a practical solution for enhancing road safety, offering timely warnings to prevent potential accidents caused by driver fatigue. The system has shown a high level of accuracy in detecting drowsiness across a range of lighting conditions and driver profiles. Despite challenges such as variations in lighting, driver positions, and facial characteristics, the proposed solution represents a significant step forward in the development of safety-focused driver assistance technologies.

8.2 Future Directions

Future research may explore several avenues to further enhance the system's capabilities. One important direction is the integration of the drowsiness detection system with other advanced driver assistance technologies, such as lane departure warning systems, adaptive cruise control, and distraction detection mechanisms. This would result in a comprehensive safety system capable of monitoring multiple aspects of driver behavior. Additionally, future iterations could incorporate more sophisticated deep learning models, leveraging larger and more diverse data-sets to improve accuracy in diverse real-world conditions. Another potential improvement includes the use of multi modal data, combining image processing with physiological sensors and steering behavior analysis to create a holistic driver monitoring solution. Finally, the development of personalized models that adapt to individual drivers over time could further enhance the system's effectiveness, ensuring accurate detection tailored to specific driving habits and characteristics.

References

1. P. Viola and M. Jones, "Robust Real-Time Face Detection," *International Journal of Computer Vision*, vol. 57, no. 2, pp. 137–154, May 2004.
2. A. S. Dargan, M. Kumar, and S. Mohan, "Real-Time Driver Drowsiness Detection System Using Eye Aspect Ratio and Facial Landmarks," in *Proc. of IEEE International Conference on Image Processing*, 2020, pp. 1357–1361.
3. A. Picot, S. Charbonnier, and A. Caplier, "On-line Detection of Drowsiness Using Brain and Visual Information," *IEEE Transactions on Systems, Man, and Cybernetics, Part A: Systems and Humans*, vol. 42, no. 3, pp. 764–775, May 2012.
4. T. Soukupová and J. Čech, "Real-Time Eye Blink Detection Using Facial Landmarks," in *Proc. of 21st Computer Vision Winter Workshop (CVWW)*, Rimske Toplice, Slovenia, 2016, pp. 1–4.
5. M. Chau and M. Betke, "Real Time Eye Tracking and Blink Detection with USB Cameras," *Boston University Computer Science Technical Report*, vol. 3, no. 10, 2005.
6. Y. N. Kumar and S. Tiwari, "Driver Drowsiness Monitoring System Using Visual Behavior and Machine Learning," *IEEE Sensors Journal*, vol. 19, no. 1, pp. 305–312, Jan. 2019.
7. Tarawa and M. Trivedi, "Face Analysis for Driver Assistance: Real-Time Non-Intrusive Monitoring of Driver Fatigue," *IEEE Transactions on Intelligent Transportation Systems*, vol. 15, no. 2, pp. 495–505, April 2014.

Note: All the figures and tables in this chapter were made by the authors.

Advances in Additive Manufacturing Technologies – Gurusamy Pathinettampadian et al. (eds)
© 2026 Taylor & Francis Group, London, ISBN 978-1-041-16687-0

Investigation on Mechanical Property of Aramid Fibre Reinforced Composite Materials for Automobile Applications

A. Dhanapal,
M. Rajamohan, J. Venkatesh
Department of Mechanical Engineering,
Sri Ramanujar College of Engineering, Vandalur, Chennai

N. Prabhu
Department of Mechanical Engineering,
PSN Engineering College, Tirunelveli, India

N. Ramanan[1]
Department of Mechanical Engineering,
Sri Jayaram Instittue of Engineering and Technology,
Gummdipundi

P. Gurusamy[2]
Department of Mechanical Engineering,
Chennai Institute of Technology, Chennai

Kishor R.[3]
PG Student, Department of Mechanical Engineering,
Chennai Institute of Technology, Chennai

◆ **Abstract:** Scientists and engineers worldwide are diligently working to develop new materials to address the growing energy crisis and pollution issues. Composites, particularly those made with artificial fibers, play a crucial role as environmentally friendly alternatives to traditional materials. This study introduces Aramid fiber (AF), a locally available artificial fiber, as a novel reinforcement for composites. The findings confirm that AF has not been previously explored as a reinforcing material in composites. Research has also begun into the properties of AF composites filled with titanium dioxide. Fiber-reinforced polymer (FRP) composites, known for their low weight and high strength, are becoming increasingly popular for reinforcement and strengthening applications due to their resistance to harsh conditions. However, environmental factors such as freezing and thawing, moisture, temperature fluctuations, sun radiation, and exposure to chemicals can significantly impact the durability of FRPs. To effectively utilize FRPs as alternatives, it is critical to assess their durability under various deteriorating conditions. One of the most significant environmental factors affecting FRP longevity is humidity. This study conducts tensile, flexural, impact, and hardness tests on Aramid and titanium dioxide composites, comparing the mechanical properties of laminated composites exposed to water with those that were not. The research further explores the mechanical deterioration observed under different environmental conditions.

◆ **Keywords:** Aramid fiber, Tensile, Flexural, Impact, Hardness tests, Titanium dioxide composites

[1]Ramananinjs2020@gmail.com, [2]gurusamyp@citchennai.net, [3]kishormech18@gmail.com

DOI: 10.1201/9781003685906-3

1. Introduction

Aramid fibers are highly susceptible to damage from UV rays due to their composition, which can lead to color changes and a significant reduction in strength when exposed to UV radiation. To enhance their performance, ultraviolet absorbers or light screeners are often incorporated during production or applied in subsequent processing steps. In traditional textile dyeing, dye molecules typically get trapped in the fabric, and chemical bonding between the dye and the fiber is also possible. However, standard dyeing methods are often ineffective with aramid fibers.

Minimizing deviations has been the focus of much study regarding the automated behavior of unidirectional Kevlar/epoxy composite in tensile force and different crash mechanisms [1]. Experiments were conducted to examine the drilling capabilities of Kevlar composites using adapted high-speed drill bits that were working at operational rates. Results showed that lifespan of the tool, hole excellence, and plane polish were all much enhanced when liquid nitrogen was applied to the drilling site [2]. To determine how treatments affect Kevlar 29 fiber, experiments have been conducted. To what extent does temperature affect the mechanical properties was the primary research question. Mechanics, particularly modulus of rupture and tension and tensile strain were shown to decrease with increasing treatment temperature. In addition, neither Young's modulus nor tensile strength were affected by vacuum treatment [3]. A comprehensive analysis was conducted on the consequences of chemical procedures, the use of Kevlar fiber as reinforcement with bismaleimide, and other factors. Using methods including thermomechanical analysis, differential scanning calorimetry, and thermogravimetry, two types of bismaleimides were investigated thermally. Cholosulphuric acid treatment considerably increased interfacial strength, according to the data. To study the fracture surface, scanning electron microscopy was used [4].

Scientists are looking at the ground kenaf core's filtering properties. The kenaf and diatomaceous earth (DE) constant pressure pre-coat filtering characteristics were compared. The outcome showed that kenaf and DE could both filter out silica substances dissolved in it effectively and efficiently, exclusive of affecting the flow in any way. The research found that kenaf and DE had comparable filtering properties [5]. In this work, abrasive water jet machining was used to experimentally explore the marble kerf's features. Consideration was given to a number of input factors, including transverse speed, water force, and abrasive run rate to find out what factors significantly affected the kerf features, we ran an ANOVA. According to the results, the kerf taper angle and kerf breadth were significantly affected by the transverse speed [6]. This results aims to develop a hybrid material combining basalt fiber with aluminum and glass fibers, creating a laminated composite with enhanced properties. It seeks to design a bumper with bend among 0.017378 m and 0.03114 m at 38 MPa tensile strength, with a maximum stress of 2.424×10^2 MPa, demonstrating advantages for daily use through simulation comparisons. [7].

2. Literature Review

The differences in lignin content across three different types of kenaf, with an emphasis on Everglades kenaf, are the subject of experimental research [8]. We took core, bast, and inner bast samples from four separate locations for our analysis. The data showed that as the stem height increased, the lignin concentration decreased. Also, when comparing the lignin concentration of the core and bast samples, the Everglades and Aokawa types showed more significant differences than the Mesta variety [9]. This engine valve is proposed to be replaced with lightweight different materials, if at all possible Al6061 and FE 430 steel. This document presents a computational fluid dynamics (CFD) examination of the valve, evaluating its flow individuality and strength [10]. Laminates have several uses, including as wall coverings, sofa covers, edge banding materials, and more; research has explored using kenaf non-wovens as substrates for these items. The goal was to show that laminated items made from non-woven kenaf fibers could be made [11]. This study investigates the consequence of heat enter on the microstructure progress in the weld area of rubbing stir welded AA6061-10% SiCp MMCs, correlating tensile property with microstructure and heat input. Microstructure analysis reveals significant particle alteration of the aluminum matrix and reinforcement particle fracturing owing to dynamic recrystallization from plastic twist and frictional heating in welding. [12]. In this experiment, a modified cotton card of conventional widths was used to manufacture batts out of a mixture of kenaf fibers and polypropylene in an 80:20 ratio. The substrate was made by first calendering or needle punching the batts, and then curing them in an oven. Overlays such as ornamental vinyl, phenolic resin, and polyester wood grain were applied to these substrates [13]. The results indicate that AZ31B/15 wt% WC composite exhibit superior tribological behavior compare to the base AZ31B magnesium alloy. The yield strength, flexural strength, tensile strength, and micro-hardness of the composite improve with growing WC strengthening content. SEM metaphors show a uniform sharing of WC particle within the Mg matrix.[14]. The quasi-static diffusion and ballistic property of crossbreed laminates made from non-woven kenaffibres and Kevlar epoxy were the primary subjects of this investigation. Laminates varying in thickness from 3.1 mm to 10.8 mm were subjected to hard projectile impact at normal incidence. The hybrid composites were finished using the hand lay-up technique and cured for 24 hours at room temperature under static stress. Three distinct configurations of non-woven kenaf and Kevlar layers comprised the hybrid composites. Composites with different materials were also developed for comparison, including Kevlar and epoxy, kenaf and epoxy, and others. A 6% NaOH solution was used to alkalize the natural fibers, and polyester resin was used as the matrix. Alkalized fibers showed much better mechanical properties than untreated fibers. Morphological examination, which included examining the composite laminate's internal structure, was performed to evaluate the effects of alkalization and fiber alignment. Compared to untreated hemp and kenaffibres, treated hemp and polyester composites have a lower fracture probability, and treated fibers are devoid of surface impurities[15–16].

3. Resources and Techniques

The excellent solvent and heat resistance, mechanical adhesion, and high strength of epoxy resin LY 556 make it an ideal polymer matrix material. A curing agent known as hardener HY951 was utilized in the current experiment at a concentration of 8% by weight. The sugar mill owned by the Salem Co-operative was the source of the bagasse fiber. We sun-dried it for a week, ground it into little pieces in a ball mill, then rinsed it to get rid of the pulps. The simple reduction reaction between para-phenylenediamine (PPDA) and terephthaloylchloride (TPC) yields HCl as a byproduct, and this process was used to synthesise aramid (AF). For reasons of safety, the polymerization process was conducted with CaC12 present, and the two compound were dissolve in N-methyl-2-pyrrolidone. Catalytic homopolymerization allows epoxy resins to react with one another, and a diverse array of co-reactants, such as polyfunctional amines, acids, phenols, alcohols, and (often referred to as mercaptans), may be used to cross-link epoxy resins. The cross-linking reaction is usually known as curing, and the co-reactants are sometimes called hardeners or curatives. Epoxy can also serve as a glue substitute.

4. Processing Epotoxy Composites

The composite specimens were made using a hand layup technique that involved adding AF into bagasse epoxy polymer glue. The necessary thickness of the composite panels was maintained throughout fabrication using a light steel mold. The composite specimen and the mold surface were connected using wax. After bolting the mold into place, it was left to make well for 24 hours at 30 °C room temperature. In order to conduct mechanical testing, specimens are carefully sized according to ASTM guidelines.

4.1 Tensile Test

One of the most basic tests in materials science and engineering is tensile testing, which involves hand-holding a specimen under controlled stress until it breaks. In addition to yield strength and strain-hardening characteristics, these tests can also provide Young's modulus and Poisson's ratio. The UTM-40 may undergo tensile testing. Plastics' tensile strengths are evaluated using the ASTM D638 Standard Test Method.

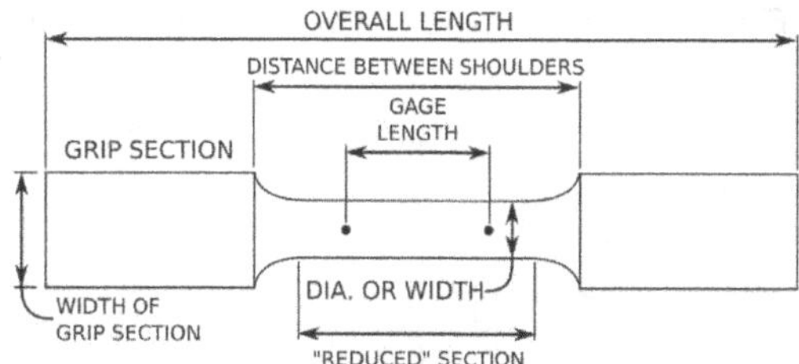

Fig. 3.1 Technical specification for ASTM D638

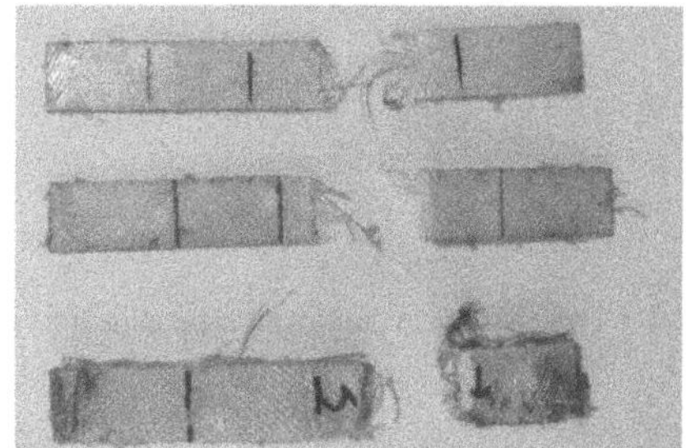

Fig. 3.2 Results of tensile testing on specimens

Table 3.1 Specification and standard document D638 (ASTM)

S. No	Dimensions	Standard		Specimen 1	Specimen 2	Specimen 3
		In inches	In mm	In mm	In mm	In mm
1.	Gauge Length	8	203	50	50	50
2.	Width	1.5 ± 0.25	38 ± 16.3	26.89	24.66	27.39
3.	Thickness	0.188 ≤ T	4.77 ≤ T	1.26	1.44	7.5
4.	Fillet Radius	1(min)	25.4 (min)	26	26	26
5.	Area	-	-	39.49	37.59	205.43

It is possible to perform a tensile test on the UTM-40 machine once the specimen has been loaded. When a material is pulled taut, it creates an internal resisting force that helps it to withstand the stress. Stress is the force that acts as a resistor for a unit normal cross-section area. There is a maximum (finite) limit to the value of stress in a material, but it continues to rise as the applied tensile load increases. Ultimate tensile strength is the lowest stress at which a material breaks.

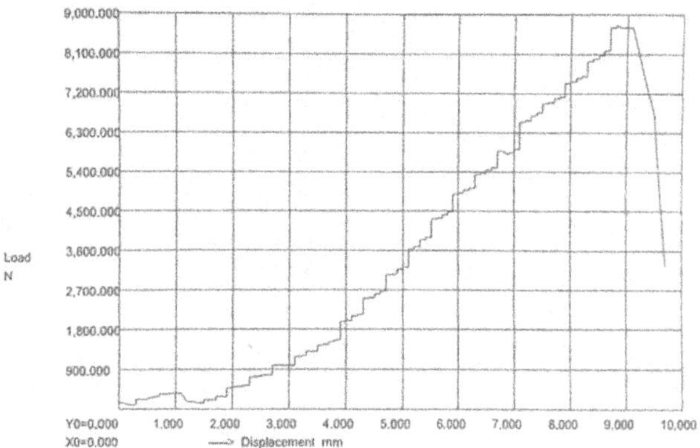

Fig. 3.3 The first specimen's tensile test result

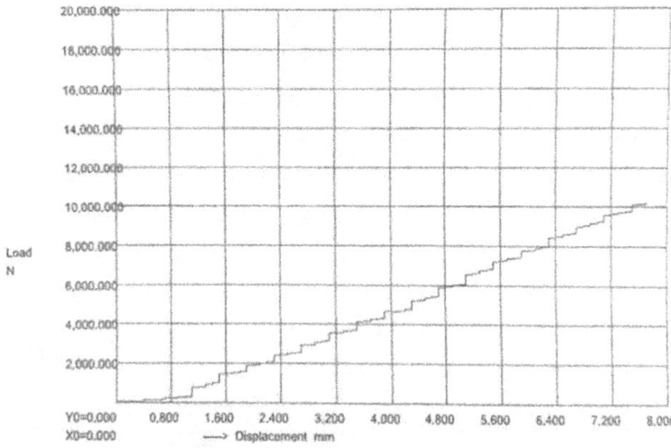

Fig. 3.4 The second specimen's tensile test result

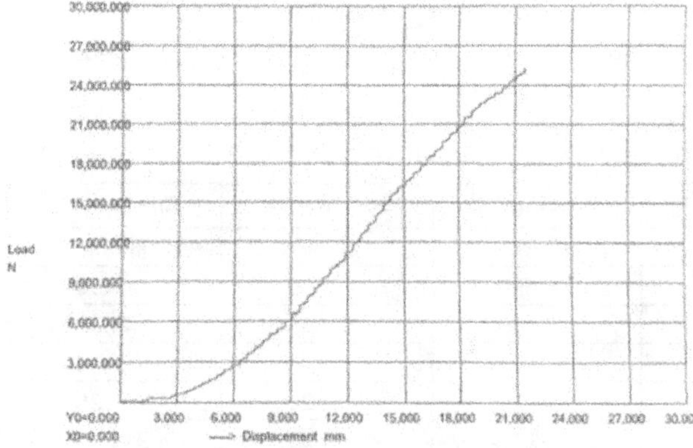

Fig. 3.5 Specimen 3's tensile test outcome

The yield point (load) marks the boundary of the elastic limit. As will be discussed later in the technique, this phenomenon is evident during the experiment. As the loading increases beyond the elastic limit, the initial cross-section area continues to decrease until it reaches its minimal value. A maximum tensile force of 8.76 KN, 10.24 KN, and 25.28 KN was recorded for specimen 1, specimen 2, and specimen 3, respectively. There is an ultimate tensile stress of 123.06 MPa for specimen 1, 221.83 MPa for specimen 2, and 272.40 MPa for specimen 3.

4.2 Resistance Exam

One mechanical metric for brittle materials is their flexural strength, which may be expressed as their modulus of rupture, bend strength, or resistance to deformation under stress. A variety of materials and products can have their flexural strength and modulus of elasticity (E) measured using the bending test, also known as flexural testing. A tensile tester, universal testing machine, or flexural testing machine with a two-or three-point fixture is used to conduct this test. Sample 1 has size of 260 mm x 25 mm x 6.12 mm, and specimen 2 has size of 260 mm x 26 mm x 6.8 mm, and the most typical form of product testing is the 3-point test. A length of 50 mm is the gauge.

The bending test The specimen is subjected to a three-point flexural test where the x-axis indicates a hit and the y-axis indicates a load or force. The sample was braked after a few minutes of gradually increasing the strain. Accordingly, we discovered that the supplied load significantly increased the force, which in turn increased the stroke value.

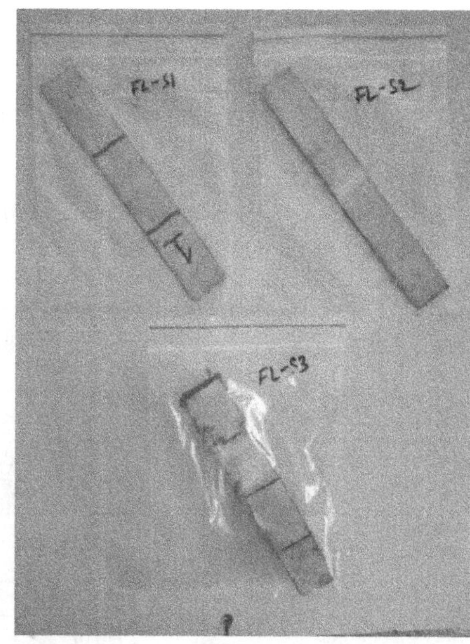

Fig. 3.6 Results of flexural testing on specimens

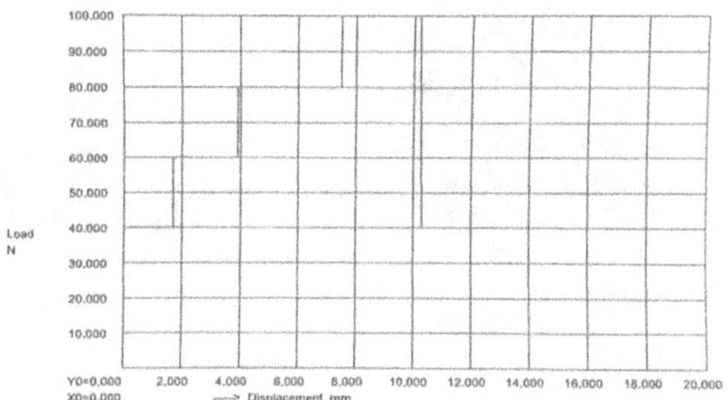

Fig. 3.7 Final verdict on specimen 1's flexural test

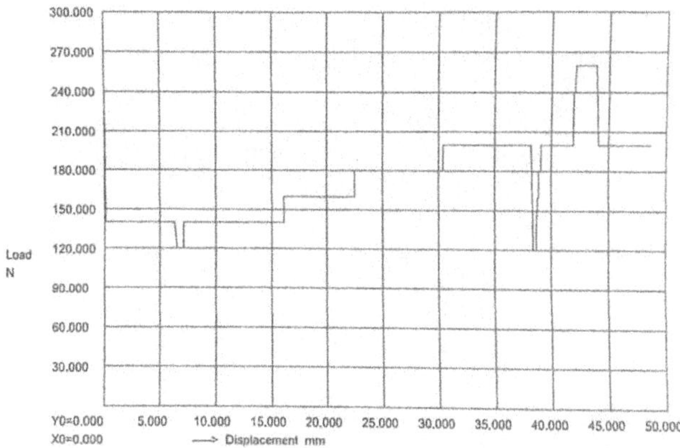

Fig. 3.8 Final verdict on specimen 2's flexural test

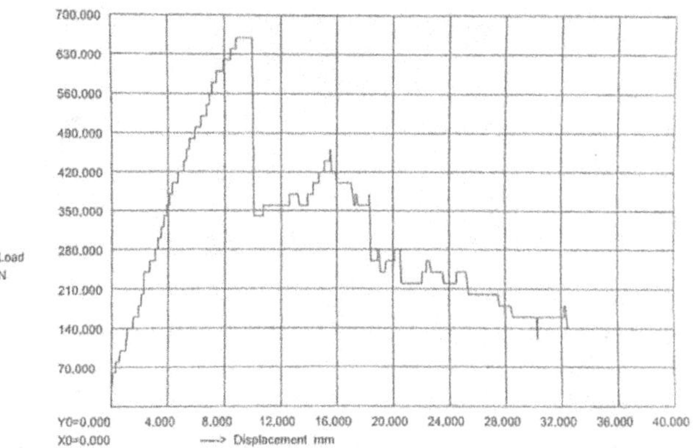

Fig. 3.9 The third specimen's flexural test results

4.3 Test for Hardness

When a surface resists elastic deformation, we say that it is hard. Indentation, scraping, cutting, and bending are all examples of deformation that may be caused by this phrase. To find out how hard something is, scientists employ a wide range of techniques. The Durometer Hardness Test is one of the most used approaches. The procedure follows the guidelines set out by ASTM D-224011. A hard, level surface is used to set the specimen on. Pressing the instrument's indentor onto the specimen is the process. There is a time limit for taking the hardness. A device known as a durometer may measure shore hardness. The indentor in a durometer is held in place by a spring that has been calibrated. The amount of indentor penetration under load is used to calculate the measured hardness. In specimen 1, the shore hardness is 69, whereas in specimen 2, it is 66, 68, 67.

Fig. 3.10 Items following the shore hardness examination

5. Results and Discussion

In order to determine the ultimate tensile strength, tensile test specimens are subjected to testing in a universal testing machine (UTM). The specimen graph for the Aramid fiber reinforced composite tensile test, shown against load and displacement, is produced straight from the machine. Aramid reinforced hybrid composites with filler had a lower ultimate tensile stress compared to those without filler, according to the results. The AR composite with filler has a maximum tensile force (MTF) ranging from 272.40 Mpa. Aramid fiber reinforced composites undergo flexural strength testing in a universal testing machine (UTM), and the resulting sample graph is displayed. For Aramid reinforced composites with filler, the results showed that the utmost applied load may reach around 7.72 Mpa.

Table 3.2 Mechanical test outcomes

S. No	Mechanical Tests	Specimen 1	Specimen 2	Specimen 3
1.	Tensile Test in MPa	123.06	221.83	272.40
2.	Flexural Test in Mpa	3.73	2.95	7.32
3.	Shore Hardness	68,70,69	66,68,67	71,54,88

The hardness values are calculated using a scale. Common hardness scales include shore A and shore D, where a lower value indicates softer materials and a higher value indicates harder materials. In most cases, the thickness of the test specimens is 7.5 mm.

Reaching the 7.5 mm thickness is doable, however it is recommended to use a single specimen. The specimens of Aramid fiber hybrid reinforced composite with filler were hardness tested using a shore hardness indenter, and the results show that they are 68, 70, and 69.

6. Conclusion

In conclusion, the fabrication of Aramid fiber hybrid reinforced composites with filler using the hand layup method shows promising mechanical properties. The addition of filler enhances the maximum tensile force (MTF), increasing it from approximately 14.54 kN to 15.80 kN. Flexural tests reveal that the Aramid reinforced hybrid composite without filler can withstand up to 1.47 kN, while the composite with filler supports 0.47 kN before experiencing a decrease in applied loads. Hardness test results indicate slight variations, with the specimens without filler showing hardness values of 68, 70, and 69 HRB, compared to 66, 68, and 67 HRB for those with filler. The density of the composites with filler was higher (1.80 g/cc) compared to those without filler (1.50 g/cc). Further investigation incorporating additional tests, such as compression and shear strength, could provide deeper insights and enhance the overall performance of these composites for practical applications.

References

1. Asumani, O.M.L., Reid, R.G., Moorthy, R, Paskara. 2012. The effects of alkali–silane treatment on the tensile and flexural properties of short fibre non-woven kenaf reinforced polypropylene composites. *Composites: Part A* 43:1431–1440.
2. Azmir, M.A., Ahsan, A.K. 2008. Investigation on glass/epoxy composite surfaces machined by abrasive water jet machining. *Journal of materials processing technology* 198: 122–128.
3. Azmir, M.A., Ahsan, A.K. 2009. A study of abrasive water jet machining process on glass/epoxy composite laminate. *Journal ofMaterialsProcessingTechnology*209:6168–6173.
4. Anuar, H, Zuraida, A. 2011. Improvement in mechanical properties of reinforced thermoplastic elastomer composite with kenafbastfibre. *Composites: Part B* 42:462–465.
5. Akil, H.M., Omar M.F., Mazuki, A.A.M., Safiee, S, Ishak, Z.A.M., Abu Bakar, A. 2011. Kenaf fiber reinforced composites: A review. *Materials and Design.*32:4107–4121.
6. Bandaru A., Kumar, Lakshmi Vetiyatil, Suhail Ahmad. 2015. The effect of hybridization on the ballistic impact behavior of hybrid composite armors. *Composites Part B* 76:300–319.
7. Narashima Rao, P.V., Periyasamy, P., Bovas Herbert Bejaxhin, A., Vetre Selvan, E., Ramanan, N., Vasudevan, N., Elangovan, R. and Tufa, M. 2022. Fabrication and analysis of the HLM method of layered polymer bumper with the fracture surface micrographs. *Advances in Materials Science and Engineering* 2022(1):3002481.
8. Kasavajhala, A.R.M. and Gu, L. 2011. Fracture analysis of Kevlar-49/epoxy and e-glass/epoxy doublers for reinforcement of cracked aluminum plates. *Composite structures* 93(8):2090–2095.
9. Azmir, M.A. and Ahsan, A.K. 2009. A study of abrasive water jet machining process on glass/epoxy composite laminate. *Journal of Materials Processing Technology* 209(20):6168–6173.
10. Ramanan, N., Periyasamy, P., Sharavanan, S. and Naveen, E. 2019. Computational analysis of dissimilar materials engine valve. *International Journal of Vehicle Structures and Systems,* 11(2).
11. Al-Jeebory, A. and Al-Mosawi, A., 2010. Mechanical properties of araldite matrix composites reinforced with hybrid carbon-kevlar fibers. *Al-Qadisiya journal for engineering sciences.*
12. Periyasamy, P., Mohan, B. and Balasubramanian, V. 2012. Effect of heat input on mechanical and metallurgical properties of friction stir welded AA6061-10% SiCp MMCs. *Journal of materials engineering and performance* 21(11):2417–2428.
13. Al-Mosawi, A.I. and Hatif, A.H. 2012. Reinforcing by Palms-Kevlar hybrid fibers and effected on mechanical properties of polymeric composite material. *Journal of University of Babylon,* 20(1).
14. Praveenkumar, R., Periyasamy, P., Mohanavel, V. and Ravikumar, M.M. 2019. Mechanical and Tribological Behavior of Mg-Matrix Composites Manufactured by Stir Casting. International *Journal of Vehicle Structures & Systems (IJVSS)*, 11(1).
15. Rahuman, M.A., Kumar, S.S., Prithivirajan, R. and Shankar, S.G. 2014. Dry sliding wear behavior of glass and jute fiber hybrid reinforced epoxy composites. *International Journal of Engineering Research and Development* 10(11):46–50.
16. Aultrin, K.J. and Dev Anand, M. 2016. Development of an ANN model to predict MRR and SR during AWJM operation for lead tin alloy. *Indian Journal of Science and Technology* 9(13):1–8.

Note: All the figures and tables in this chapter were made by the authors.

Advances in Additive Manufacturing Technologies – Gurusamy Pathinettampadian et al. (eds)
© 2026 Taylor & Francis Group, London, ISBN 978-1-041-16687-0

Effect of Nano-SiC Particles Reinforcement on the Mechanical Properties and Optimizing the Super Plastic Forming Process Parameters of Al 6061 Alloy

K. Saravanakumar*
Department of Naval Architecture and Offshore Engg,
VISTAS, Chennai, Tamilnadu, India

Vijay Ananth Suyamburajan
Department of Mechanical Engineering,
VISTAS, Chennai, Tamilnadu, India

◆ **Abstract:** This study investigates the optimization of superplastic forming in aluminum alloy 6061 reinforced with 100 nm silicon carbide (SiC) nanoparticles, fabricated by use of the ultrasonic cavitation technique and the tests for tensile strength, hardness, impact resistance, and wear were conducted to assess the nano-composite's mechanical characteristics. Key parameters, including pressure (2–6 bars), temperature (560–600 °C), and time (15–45 minutes), were systematically varied to optimize the optimal process parameters by using Taguchi L9 experimental design, facilitating a comprehensive examination of these factors' effects on the forming performance. The findings provide valuable insights for enhancing the superplastic forming process in aluminum-SiC composite materials.

◆ **Keywords:** Nano-composites, Super plastic forming, Optimization, Aluminum alloy, Mechanical properties, Tensile strength, Hardness, Impact resistance

1. Introduction

The increasing demand for materials to improve the overall performance of automotive and aerospace components has spurred the development of composite materials, such as landing gear, airframes, etc. Aluminum Metal Matrix Composites (AMMC) are a popular composite material utilized to meet new industrial demands. The examination of the mechanical characteristics of AMMC made using the stir casting method for different boron and silicon carbide compositions reinforced with aluminum alloy 6061. After conducting testing for tensile, flexural, hardness, and impact, it was discovered that the hybrid composites outperformed pure aluminum in terms of characteristics. The ability of a material to undergo extraordinarily high elongations of up to 1000% is known as super plasticity. Because of this high ductility, the automotive and aerospace industries have expressed interest in the possibility of creating complex shaped parts with a minimal number of mechanical steps. This allows for significant improvements in structural efficiency as well as, frequently, cost savings on both the product and operating costs [1]. The super plastic forming process is depicted in Fig. 4.1. Superplastic formation typically occurs at a strain rate of 10^{-1} to $10_{-5}\,s^{-1}$. Superplastic formation of the alloys occurs at a slow strain rate of 10-3 to 10-5 s-1. High strain rate super plastic formation occurs in aluminum metal matrix composites between 10^{-2} and 100 s^{-1} [2]. Aluminum Metal Matrix Composites (AMMC) are brittle by nature and cannot be created under typical circumstances, while having a low density, strong specific stiffness, and high strength.

*Corresponding author: kskjjmech@gmail.com

DOI: 10.1201/9781003685906-4

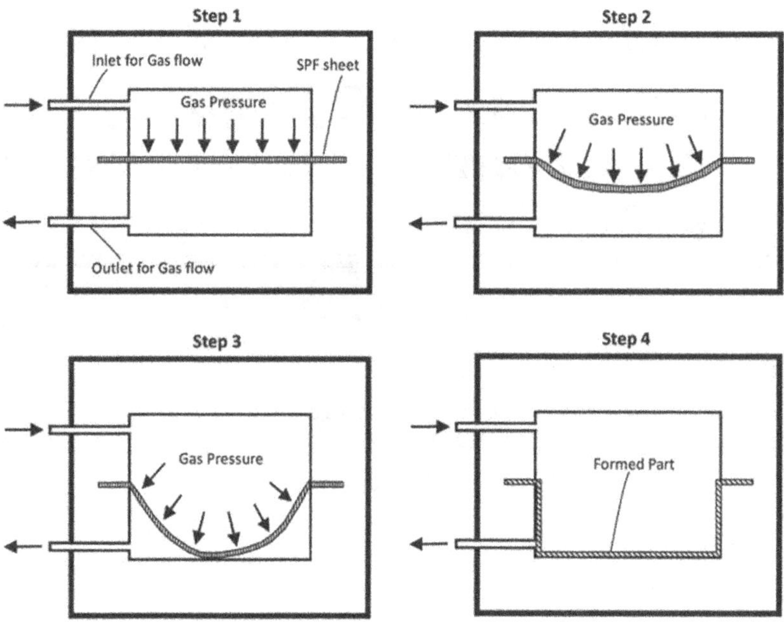

Fig. 4.1 Superplastic forming

SPF aluminum finds widespread use in vehicles, railroads, and airplanes. More than 40 aviation manufacturing companies and 20 automotive sectors employed aluminum components made with superplastics [3]. The capacity to produce large components in a single process, eliminating the need for many sub-assemblies, tight dimensional precision, and good surface quality are the main benefits of SPF components. As a result, neither interfaces nor the flaws that often arise at interfaces exist. Saves a great deal of time and work in manufacturing. Outstanding mechanical qualities as a result of being made of equiaxed, ultra-fine grains [4]. A significant factor influencing the superplasticity of metals is grain size. When the tensile elongation is higher, the rate of strain sensitivity index (m) value is typically high, the flow stress is low, and the grain size is fine. Grain size characterisation is consequently crucial to the whole super plasticity characterization process. The strain rate range across which "m" is high can be controlled by a few coarse grains in an fine grain structure, and in certain situations, this can lead to the emergence of a threshold stress. In actual materials, a low grain size distribution has the significant impact of producing a comparatively large m (m>0.5) [5]. The micro-level reinforcements are frequently employed in AMCs. Technological developments in the field of nanosciences have enabled the creation of metal matrix nanocomposites, or MMNCs, as the composites that result. Reinforcement in MMNCs is expressed in nanometers. The mechanical, tribological, thermal, and interfacial properties of the base material are enhanced by the nano reinforced particles in an aluminum alloy matrix. Using liquid metallurgical processes Al6061-SiC and Al7075-Al_2O_3MMCs, which contain filler quantities up to 6% of the total weight [6]. The microstructural analyses demonstrated the uniform dispersion of the particles within the matrix system, and the research found that the composites' densities outperform that of their underlying matrix. The hardness of Al6061-SiC and Al7075-Al_2O_3 composites was measured to be 60-97 VHN and 80-109VHN, respectively, and the microhardness of the composites increased with filler quantity [7]. It is discovered that the composites' tensile strength properties are greater than those of the basic matrix, with Al6061-SiC composites having stronger tensile properties than Al7075-Al_2O_3composites. For the first time, the high-strain-rate superplastic gas pressure formation behavior of an Al6061/SiC_w composite sheet under constant applied flow stress conditions has been studied by Tong et al. (1997). In approximately 17.6 seconds, a hemisphere diaphragm is successfully created at a temperature of 873K and an applied flow stress of 4.0 MPa. The polar height vs. time curve revealed three different deformation regimes based on the experimental results [8]. These deformation regimes are comparable concerns the creep behavior of structural ceramics and the most metallic alloys when subjected to continuous loads. It is demonstrated that there is not a reasonable agreement between the experimental results and the tabulated thinning factors based on the m value found by tensile testing [9].

Due to the high modulus, high strength, low density, good neutron absorption, and outstanding wear resistance of silicon carbide (SiC), SiC particle reinforced aluminum (Al) matrix composites have found extensive use in the domains of aerospace, weapons, transportation, and neutron shielding. There are currently many kinds of aluminum alloys on the market, each with a unique set of advantages and applications. This paper focuses on heat-curable Al-6061 Alloy (aluminum alloys) which is suitable for a variety of applications such as toughness, machinability, and resistance to corrosion. These alloys can be considerably toughened.

2. Experimental Setup

2.1 Material Selection

Al6061 is one of the best corrosive alloys of aluminum with high corrosion resistance because it is usually alloyed with magnesium and is strengthened through strain hardening or cold working rather than heat treatment.. In the basic alloy, the tensile strength is 117 MPa, the hardness is 31 HRB, and the young's modulus, E is 75 MPa. The aerospace and automotive industries are the main users of Al-6061 alloy for lightweight components. Using the Al-6061 alloy matrix, a variety of reinforcements have been employed to build the metal matrix composite (MMC) here SiC nano particles were added [10].

When creating in situ particle reinforced aluminum composites using the traditional mechanical stirring cast method, there are two primary serious problems. One is the way the created particles group together. To dispersing microscopic particles less than a few microns in diameter, the mechanical stirring method is not the best. It has been demonstrated that a significant deterioration of the mechanical properties, such as tensile strength and fatigue resistance, occurs when the reinforced phase aggregates. The other is the excessive porosity, which deteriorates the composites' mechanical qualities, particularly their resistance to corrosion. Shrinkage during solidification, hydrogen evolution, and gas trapping during mixing are the three main sources of porosity. Gas entrapment during mixing is the main cause of pores appearing in the melt when stir casting is used to fabricate Al matrix composites [11].

2.2 Ultrasonic Cavitation

The purification, degassing, and refining of metallic melts have made substantial use of ultrasonic vibration (Fig. 4.2), as the introduction of ultrasonic fields into a liquid can result in nonlinear processes such cavitation and acoustic streaming [12]. Moreover, uniform particle distribution within the metal matrix and enhanced wetability between reinforced particles and the matrix can be achieved through ultrasonic vibration. According to some reports from this experiment, particle reinforced composites were prepared using ultrasonic vibration, and the reinforcements were injected straight into the melts. On the other hand, not much research has been published on the use of ultrasound assistance in situ approach for the in situ production of aluminum composites reinforced with particles [13]. Following the rolling process, the finished composite material is ready (Fig. 4.2).

Fig. 4.2 Ultra-sonic cavitation method

The thickness of sheet was reduced from 5mm to 2mm and 1.5mm to make it suitable for blow forming. Hot rolling was carried out by heating the sheet above recrystallisation temperature, followed by rolling procedure in the power roll machine a shown in the Fig. 4.3. Until the sheet

Fig. 4.3 Rolling machine

reached the desired thickness, of 2mm and 1.5mm, the rolling operation was repeated [14].

The rolled sheet was cut to requirement of shapes for performing hardness test and micro structural analysis,as shown the Fig. 4.4. The blank was preparatied size 80 mm Diameter and 2 mm thickness round specimen,using manual hand shearing machine.

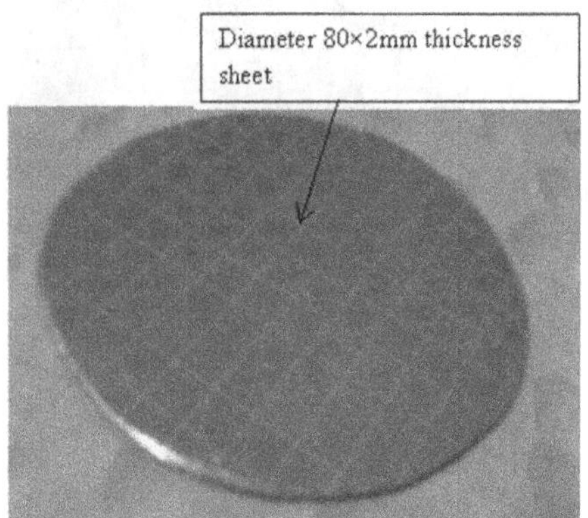

Fig. 4.4 Al 6061 specimen for SPF

Figure 4.5 shows the Microstructure of Aluminum Silicon carbide after thermocycling process, grain size was measured using biovis software, and found as 100nm [15].

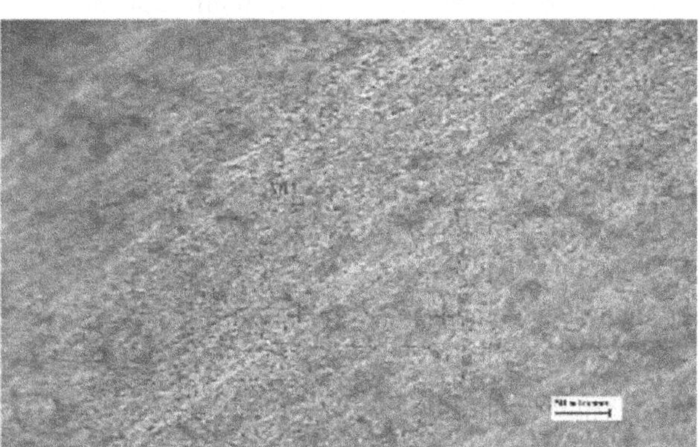

Fig. 4.5 Microstructure of Al-SiC

A splitdie was designed to grease the easy making of the top and lowermost die, as shown in the Fig. 4.6. The dieswas assembly kept inside the splitfurnace and its maintained at a constant temperature of 530 °C. Once the set temperature was reached, a shoaking time of 20 to 30 min time. After that air was passed into the bottom, die. Gradationally the air pressure was increased and set to a particular pressure for a particular period of time. Aluminium 6061 material carried in 5 mm density standard distance, distance was cut in to size of 30 × 30 mm, in power shearing machine [12].

Place the rolled component (circle in shape) into the superplastic forming apparatus with the induction coil

Fig. 4.6 Split die

installed and fasten it firmly [14]. We must wait until the temperature reaches the desired level in the control monitor since the component must be formed at an initial temperature of 580 degrees Celsius. Next, apply the compressor's 4 bar of pressure via the bottom die. The component will begin to distort gradually as time goes on. For different pressures and temperatures, repeat the procedure. Figure 4.7 is the created component [15].

Fig. 4.7 Formed component

In the three different levels of Al/SiCp composites, Figure 4.4 displays the most severe arc stature shaped at a constant weight of 0.4MPa and 580°C for 45 minutes. Initially, the arch height increased swiftly, and after 45 minutes, half of the vault's height had taken shape. The highest extreme arch height measured for 5% SiCp was 21 mm. The highest extreme arch stature obtained in 10% SiC composites was 16 mm. The height of the lump is reduced by the increase in SiCp level. Grain limit sliding is the mechanism underlying the superplastic frame. The presence of SiC in metal network composites prevents grain limit sliding from developing. When the degree of fortification is raised, it prevents the growth of the grain boundaries [13].

3. Mechanical Properties of Composites

3.1 Density Studies

According to ASTM D792, the Archimedes principle is used to estimate density using a weighing balance with specific setups. With the addition of reinforcement, the composite's density has somewhat enhanced (Table 4.1)

$$Density = \frac{Dryweight\ (w2) \times 1}{saturatedweight(w1) - Suspendedweight(w3)}$$

Table 4.1 Density of composites

Sl. No	Composition	Theoretical and measured values of Al- SiC composites. g/cm^3	
		Theoretical	Measured
1	Al-6061 + 5%SiC	2.59	2.596

3.2 Hardness

The test for Brinell hardness is conducted. The indentation made during the test is shown by the circle. The space between the indentations is kept constant in accordance with ASTM E10.

A tungsten carbide ball is used in place of a steel ball when testing exceptionally hard metals. Since the Brinell ball creates the widest and deepest depression when compared to other hardness test methods, the test averages the hardness over a larger sample size, more precisely accounting for inconsistencies in the material's homogeneity and different grain structures. This approach is the most effective way to determine a material's bulk or macro-hardness, especially for materials with heterogeneous structures [16].

It was found that the composite of Al6061 with 5% SiC nano particles had the highest toughness. Because the SiC nano particles are tougher and more firmly attached, the hardness also increases as the percentage of reinforcement increases. Room temperature was used for the test. The figure displayed the estimated hardness value.

One physical characteristic of a material that shows its resistance to local plastic deformation is its hardness. Figure 4.8, 4.9 illustrates how the amount of nano SiC particles affects the AMCs' hardness. Because the nano SiC particles strengthened the material, there is a positive correlation between the hardness values 94 BHNand the weight percentage of the particles [17].

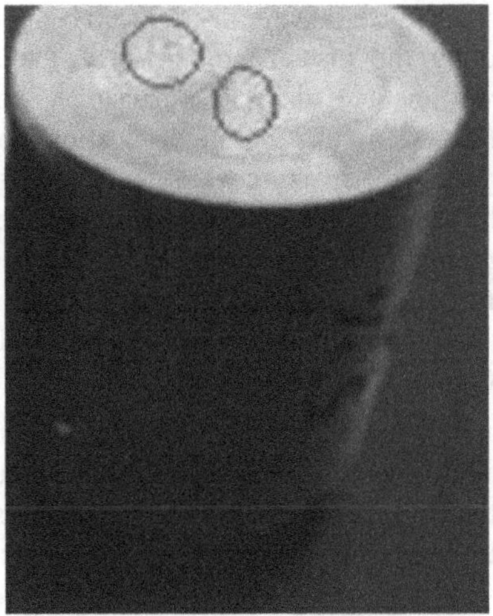

Fig. 4.8 Hardness specimen

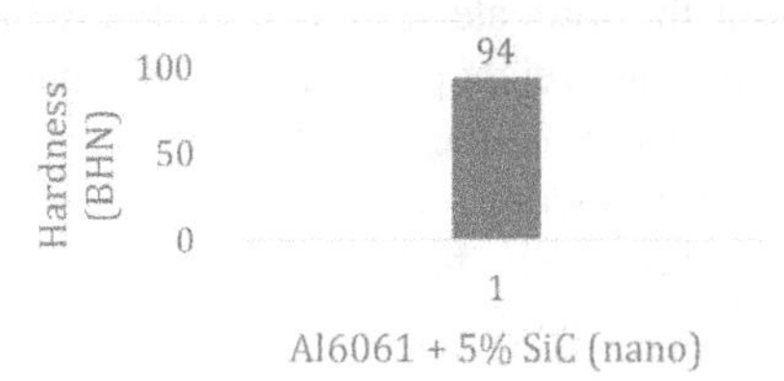

Fig. 4.9 Hardness values

3.3 Tensile Properties

According to the aforementioned findings, SiC nanoparticle composites have a better tensile strength than the base material. The test's outcome demonstrates how reinforcing affects the composite material. The composite has demonstrated increased values, and as reinforcement was added, the composites' ultimate tensile strength declined. Figure 4.10 shows tensile specimen of the composites

Fig. 4.10 Tensile specimen

The nanostructure and interface bonding between the reinforcing material and matrix determine the properties of SiC nanoparticle composites. The stress and strain curves of AMCs reinforced with 5% nano-sized SiC particles are displayed in the figure. Up to 5 weight percent more nano-sized SiC particles are added to the composite, increasing its ultimate tensile strength.

The load carrying effect and mismatch of the strengthening process are the primary causes of the increase in tensile strength of AMCs incorporating nano-sized SiC. A high dislocation density and thermally induced residual stresses are caused by the discrepancy between the aluminum matrix's coefficient of thermal expansion ($25 \times 10^6/°C$) and that of SiC ($5 \times 10^6/°C$). The dislocation movement is impeded by these thermal forces and generated dislocations. As a result, the AMCs with nano-sized SiC are stronger. Particle aggregation and the occurrence of nano porosities at increasing particle concentrations are the primary causes of the decline in the ultimate tensile strength of 5 weight percent SiC-reinforced AMCs [18].

Grain size refinement, the load bearing effect, and the Orowan and mismatch strengthening processes are the key causes of the Hall–Petch strengthening mechanism, which raises the ultimate tensile strength when SIC nanoparticles are added [19]. Because they function as barriers that limit the migration of dislocations in the matrix, nano SiC particles are more successful than nanoSIC particles at increasing the ultimate tensile strength of the AMCs. However, once the weight percentage of SiC nanoparticles was raised over 5%, the composites' tensile strength declined. These outcomes are explained by the fact that AMCs with greater SIC nanoparticle contents encourage nanoparticle agglomeration [20].

Figure 4.11 illustrates the stress strain diagram of the composites. The concentration of stress raises in tandem with the concentration of nano SiC particles. Consequently, there was an inverse relationship between the contents of both nanoSiC particles and the elongation of the AMCs [21]. Figure 4.12 and 4.13 illustrate the ultimate tensile strength and young's modulus of the composite specimen

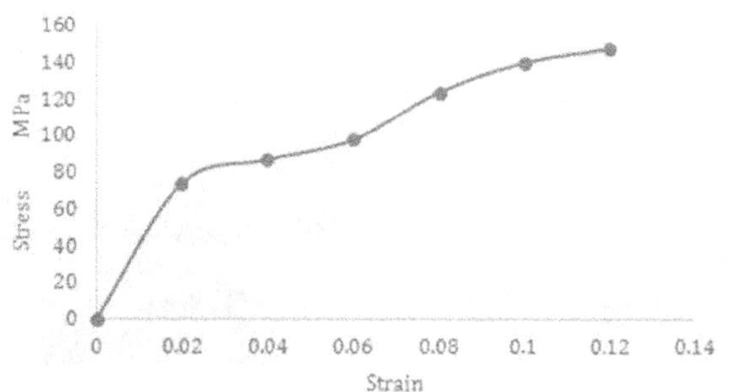

Fig. 4.11 Stress strain diagram

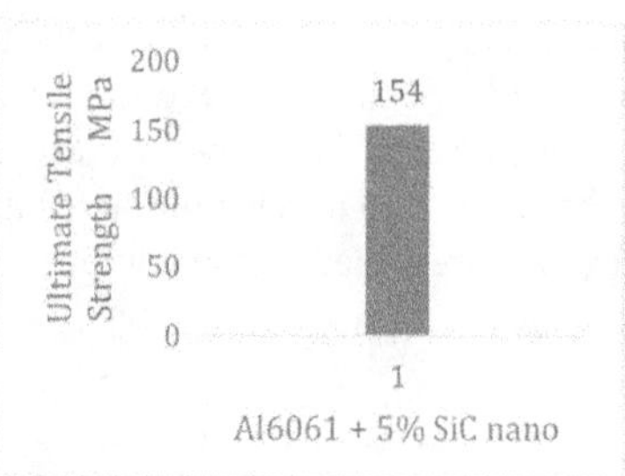

Fig. 4.12 UTS of composite

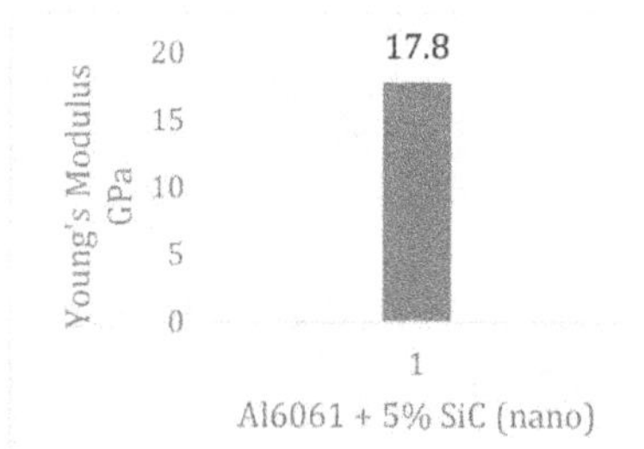

Fig. 4.13 Young's modulus of composite

4. Analysis of Process Parameter

Using the statistical program Minitab 16, the findings of each experimental run were statistically analyzed at a 95% confidence level using analysis of variance (ANOVA), and the effects of the chosen variable were assessed. The analyzing software's input contains the process parameters, including temperature, pressure, and time (Table 4.2).

Table 4.2 Process parameters of dome height experiment

Pressure (Bar)	2	4	6
Temperature (°C)	560	580	600
Time (min)	15	30	45

The orthogonal array's degrees of freedom ought to be higher than or at least equivalent to the process parameters'. The L9 orthogonal array was employed in this work. The combinations of values for each control factor must be used in a total of nine experimental runs [20].

The superplastic shaping is based on arch stature. The table – appears the impact of each parameter on the arch tallness. Information cruel is utilized to discover each prepare parameter impact. Fig – appears the impact of weight on the arch stature. When the preparing weight was less the arch tallness arrangement was moreover less [21]. The higher shaping weight leads to tall strain rate superplastic shaping. In Al/SiCp composites great super plastic shaping was gotten as it were in tall temperature and tall weight. The as it were temperature at which great superplastic arrangement was accomplished was over 580°C. Arch stature creation is less than 15 mm at lower temperatures. The fabric was debilitated at the same time the temperature rose to 600°C. At 600°C, the fabric split on the best surface in fifteen minutes. The most elevated arch tallness in 580°C was made in 45 minutes at 4 bar, in Table 4.3 and Fig. 4.14 appears, 21 mm was the most noteworthy arch stature achieved [22].

The two most significant factors determining the dome height are SiC size and SiC percentage. Ultrasonic cavitation gives a very good composite for nano particles [23]

Table 4.3 S/N ratio and ranking

Pressure Bar	Temp. C	Time Min	Obtained Dome Thk (mm)	S/N Ratio	Rank
2	560	15	12.5	22.25	9
2	580	30	13.6	24.21	7
2	600	45	12.8	23.71	8
4	560	30	19.6	25.80	2
4	580	45	21.0	24.40	1
4	600	15	18.0	24.01	3
6	560	45	17.5	24.15	4
6	580	15	16.0	23.59	5
6	600	30	15.0	24.48	6

Table 4.4 Signal to noise ratios response larger is better

Level	Pressure	Temperature	Time
1	22.25	24.21	23.71
2	25.80	24.40	24.01
3	24.15	23.59	24.48
Delta	3.55	0.81	0.77
Rank	1	2	3

The acquired findings can be used to establish the impact of a process parameter on the forming process. The S/N ratio value and ranking for the measured dome height are displayed in Table 4.3. A2, B2, and C3 were the ideal procedure parameters to achieve the highest dome height [24].

ANOVA is applied to the experimental data to determine the relative contribution of hot press forming process parameters on dome height on superplastic forming of Al/SiCp composites. Table: Displays the ANOVA findings for the dome height taken into consideration in this paper shown in Table 4.4. The processing pressure is found to have a bigger influence than the remaining parameters when comparing the percentage of contribution and ANOVA findings for dome height. Considering that the F value and contribution % are highest here. Temperature and processing pressure have a bigger impact. Time has minimal impact on the formation of superplastic [25]. The effect of various parameters on superplastic forming using signal to noise (S/N) ratio and analysis of variance (ANOVA) will be discussed below. From the experimental results, the S/N ratio and ANOVA results were calculated through Minitab. The optimum combination of process parameters and their influence have been obtained.

5. Conclusions

An Aluminum 6061 composite reinforced with 5% SiC nano particle 100 nm is developed using stir casting with ultrasonic cavitation method and is undergone dome test to study the elongation of the composite. The thickness of sheet was reduced from 5mm to 2mm and 1.5mm by rolling process. The process parameters were pressure range 2, 4 and 6 MPa, temperature 560, 580 and 600 C and time 15, 30 and 45 mins. These parameters are optimized using L9 Taguchi algorithm.

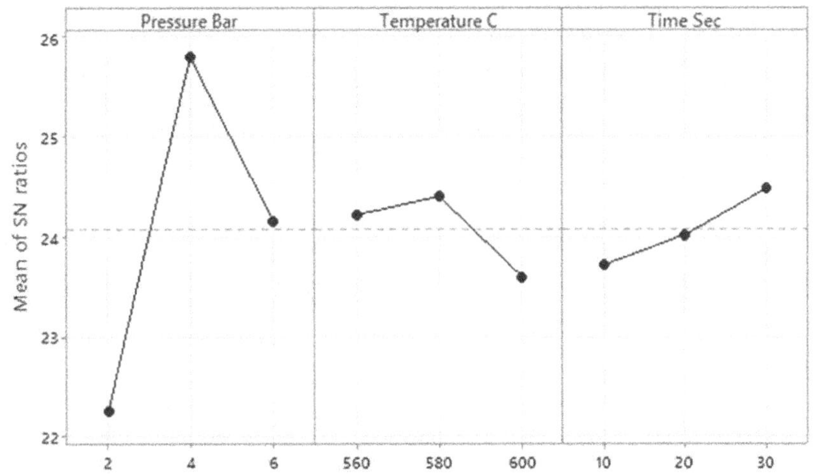

Fig. 4.14 S/N of larger is better

It is obtained maximum 21mm elongation at 580 C at 4 bar pressure with 45 mins. Also it is found the rise in the proportion of SiC causes the crack to begin and propagate. A maximum elongation of 400% was achieved in uniaxial superplastic forming at a strain rate of 10^{-2} s^{-1} . The inclusion of SiC and enhanced the density of composites, making them denser than pure Al 6061. The Al 6061 composite's hardness was 50% higher (94BHN) than that of the parent material (62.5 BHN) because to the addition of SiC. While the composite's UTS increased by 7% (to 154 MPa), its performance reduced by 47% (76.5 MPa). The composite's energy absorption decreases by 47% and fails at a lower load as the percentage of reinforcement rises. To get more insight into the higher strain-rate super plasticity of AMMCS in gas pressure formation, more research on the variations of m values under various stress states and conditions is therefore recommended.

References

1. L. Carrino, G. Giuliano and S. Franchitti "On the Optimisation of Superplastic Free Forming Test of an AZ31 Magnesium Alloy Sheet", International Journal of Material Forming, Volume 1, pp 1067 – 1070, 2008
2. Chan, K.C. and Tong, G.Q. "Deformation and cavitation behavior of a high strain rate superplastic Al2009/20SiCw composite", Materials Letters, Vol. 44, pp. 39-44, 2000
3. Krajewski, P.E. and Schroth, J.G. "General Motors QPF Process", ICSAM, Chengdu, China, 2006.
4. Ritam Chatterjeea, Jyoti Mukhopadhyay "A Review of Super plastic forming" Materials Today, Proceedings 5 pp 4452–4459, 2018
5. Manojkumar G, Narendranathan S.K, Balaji M, Saravanan R, Aravind M Arjun "Superplastic Forming Characteristics Of Al6061/B4C Nano Composites" IJRMMAE, vol 2 issue 3, pp 70 – 79, 2016
6. Sami ULLAH KHAN, Ding WANWU, Qudrat ULLAH KHAN,Shadab KHAN, Abid ALAM, Arif ULLAH, Hanif ULLAH "An Analysis Of In-Situ Synthesized Al 6061 Alloy Metal Matrix Composites: Review" European Journal Of Materials Science And Engineering, Volume 6, Issue 4, pp 220–233, 2021
7. G. B. Veeresh Kumar1, C. S. P. Rao, N. Selvaraj, and M. S. Bhagyashekar "Studies on Al6061-SiC and Al7075-Al2O3 Metal Matrix Composites" Journal of Minerals & Materials Characterization & Engineering, Vol. 9, No.1, pp.43–55, 2010
8. Dr. Osama S. Muhammed* , Haitham R. Saleh** & Hussam L. Alwan** "Using of Taguchi Method to Optimize the Castingof Al–Si /Al2O3 Composites" Eng. & Tech. Journal,Vol.27, No.6, pp 1143 – 1150, 2009
9. Dinesh Pargunde, Prof. Dhanraj Tambuskar, Swapnil S. Kulkarni, Fabrication of metal matrix composite by stir casting method, Int. J. Adv. Engg. Res. Studies / II/ IV/July-Sept.,2013/49-51
10. Tong and Chang "The optimization of multi-response problems in the Taguchi method", International Journal of Quality & Reliability Management, Vol. 14 No. 4, pp. 367–380.
11. S. Ajith Arul Daniel, S. Vijay Ananth, A. Parthiban, S. Sivaganesan "Optimization of machining parameters in electro chemical machiningof Al5059/SiC/MoS2 composites using taguchi method" Materials Today: Proceedings, Volume 21, Part 1, 2020, Pages 738–743
12. S. Vijay Ananth, K. Kalaichelvan A. Rajadurai "Optimization of Superplastic Forming of Al6063/5%SiCp Composites Using Taguchi Experimental Design"International Review of Mechanical Engineering (I.RE.M.E.), Vol. 6, n. 6, 2012
13. BalamuruganGopalsamy, BiswanthMondal and SukamalGosh, "Taguchi method and ANOVA: An approach for process parameters optimization of hard machining while machining hardend steel", Journal of Scientific & Industrial Research, Vol.68, August 2009, pp.686–695
14. A. chennakesava Reddy and Essa Zitoun. Matrix "Alalloys for silicon carbide particle reinforced metal matrix composites", Indian Journal of Science andTechnology, Vol 3 No:12 (2010)
15. B. Ramesh, T. Senthilvelan, "Statistical Modeling of Aluminium Based Composites and Aluminium Alloys Using Design of Experiments", International Review of Mechanical Engineering, November 2010, Vol. 4 N. 7,pp. 799–804.
16. Manojkumar G, Narendranathan S.K, Balaji M, Saravanan R, Aravind M Arjun "Superplastic Forming Characteristics Of Al6061/B4C Nano Composites" IJRMMAE, vol 2 issue 3, pp 70 – 79, 2016
17. Sami ULLAH KHAN, Ding WANWU, Qudrat ULLAH KHAN, Shadab KHAN, Abid ALAM, Arif ULLAH, Hanif ULLAH "An Analysis Of In-Situ Synthesized Al 6061 Alloy Metal Matrix Composites: Review" European Journal Of Materials Science And Engineering, Volume 6, Issue 4, pp 220–233, 2021
18. G. B. Veeresh Kumar1, C. S. P. Rao, N. Selvaraj, and M. S. Bhagyashekar "Studies on Al6061-SiC and Al7075-Al2O3 Metal Matrix Composites" Journal of Minerals & Materials Characterization & Engineering, Vol. 9, No.1, pp.43–55,2010
19. Dr. Osama S. Muhammed* , Haitham R. Saleh** & Hussam L. Alwan** "Using of Taguchi Method to Optimize the Casting of Al–Si /Al2O3 Composites" Eng. & Tech. Journal,Vol.27, No.6, pp 1143–1150, 2009
20. Dinesh Pargunde, Prof. Dhanraj Tambuskar, Swapnil S. Kulkarni, Fabrication of metal matrix composite by stir casting method, Int. J. Adv. Engg. Res. Studies / II/ IV/July-Sept.,2013/49-51
21. Tong and Chang "The optimization of multi-response problems in the Taguchi method", International Journal of Quality & Reliability Management, Vol. 14 No. 4, pp. 367–380.
22. S. Ajith Arul Daniel, S. Vijay Ananth, A. Parthiban, S. Sivaganesan "Optimization of machining parameters

in electro chemical machining of Al5059/SiC/MoS2 composites using taguchi method" Materials Today: Proceedings, Volume 21, Part 1, 2020, Pages 738–743

23. S. Vijay Ananth, K. Kalaichelvan A. Rajadurai "Optimization of Superplastic Forming of Al6063/5%SiCp Composites Using Taguchi Experimental Design" International Review of Mechanical Engineering (I.RE.M.E.), Vol. 6, n. 6, 2012
24. Balamurugan Gopalsamy, BiswanthMondal and SukamalGosh, "Taguchi method and ANOVA: An approach for process parameters optimization of hard machining while machining hardend steel", Journal of Scientific & Industrial Research, Vol.68, August 2009, pp.686–695
25. D. Sachin a , K.N. Uday b , G. Rajamurugan b, Prabu Krishnasamy b "Effect of SiC reinforcement on the mechanical and tribological behaviour of Al6061 metal matrix composites" Materials Proceeding 2020, vol xxx (xxxx) xxx, pg 1–8

Note: All the figures and tables in this chapter were made by the authors.

Advances in Additive Manufacturing Technologies – Gurusamy Pathinettampadian et al. (eds)
© 2026 Taylor & Francis Group, London, ISBN 978-1-041-16687-0

AL6063 Aluminum Alloy: Synthesis and Characterization of a ZrO_2 and SiC Hybrid Metal Matrix Composite

P. Ashok Kumar
Department of Mechanical Engineering, Mailam Engineering College, Mailam, Villupuram, Tamilnadu, India

N. Krishnamoorthy
Department of Mechanical Engineering, P.T. Lee Chengalvaraya Naicker College of Engineering and Technology, Oovery, Kanchipuram, Tamilnadu, India

N. Ramanan
Department of Mechanical Engineering, Sri jayaram Institute of Engineering and Technology, Gummdipundi, Tamilnadu

R. Rajappan, V. Pugazhenthi, S. Vandaarkuzhali, R. Soundararaj
Department of Mechanical Engineering, Mailam Engineering College, Mailam, Villupuram, Tamilnadu, India

P. Gurusamy[2]
Department of Mechanical Engineering, Chennai Institute of Technology, Chennai

Kishor R.[3]
PG Student, Department of Mechanical Engineering, Chennai Institute of Technology, Chennai

◆ **Abstract:** A composite substance combines the best qualities of two or more distinct phases, each of which has its own unique physical and chemical makeup. Usually, reinforcements are spread out over the components of a continuous matrix. One kind of composite where the matrix metal is called a metal matrix composite (MMC). Utilizing the Electromagnetic Stir Casting Process, a novel hybrid composite is created by combining the 6063-aluminum alloy with zirconium dioxide (ZrO_2) and silicon carbide (SiC) reinforcements. Various combinations of aluminum matrix composites with reinforcement weight percentages ranging from 2% to 12% are the primary focus of this inquiry. The percentages of ZrO_2 and SiC are 3% to 9% at this weight%. After the fabrication process is finished, the composites & aluminum metal matrix are thoroughly evaluated in this step. Although A319, A356 and A359 alloy are among the most common aluminum alloys used in the automobile industry, they are among the relatively few alloys used under high-temperature conditions. Improving the mechanical characteristics is greatly helped by the reinforcement of ceramic particles. The ceramic particles most commonly utilized are ZrO_2, Mg, SiC, B4C, and Al_2O_3. However, studies utilizing composites based on ZrO_2 and SiC are scarce. Therefore, the research utilized ZrO_2 and SiC as reinforced materials. Using SEM/EDAX, we will study the chemical properties of composites with an aluminum metal matrix. We have successfully produced 6063 aluminum alloy composites containing 3, 6, and 9 vol% ZrO_2 and SiC.ZrO_2 and SiC particles are uniformly distributed throughout the matrix, according to SEM analysis. Electrochemical data analysis (EDAX) verified the intended 3, 6, and 9 vol% ZrO_2 and SiC particle assimilation into the Al matrix.

◆ **Keywords:** Electromagnetic stir casting process, Zirconium dioxide, Silicon carbide, Composite material, Reinforcement machining

Corresponding authors: [1]ashokthermal87@gmail.com, [2]gurusamyp@citchennai.net, [3]kishormech18@gmail.com

DOI: 10.1201/9781003685906-5

1. Introduction

Consuming a mechanical stirrer to create a vortex and combine the reinforcement and matrix material is known as stirring casting. Also, the liquid casting process is called stirring. Many different casting procedures are available. Reinforcements such as Al_2O_3 and other aluminum alloys of various series were used. The researchers discovered that using this combination of metal matrix composites (MMCs) increased strength and toughness [1]. Hybrid Metal Matrix Composites are easily explored for their lightweight and great strength at a reasonable cost. Aluminium metal matrix is becoming increasingly used in the aerospace, automotive, biomedical, and robotics industries, among others. An extensive literature search revealed that several combinations of aluminium alloys were created by investigating different manufacturing procedures and their qualities. When building metal matrix composites, numerous reinforcements are used, including titanium dioxide, boron carbide, silicon carbide, and graphite. Several synthetic techniques are available for producing MMC products. The manufacturing method varies significantly depending on the type of reinforcement [2].

1.1 Objectives of the Study

The researchers briefly discussed the composites and used electromagnetic stir casting procedures to create aluminum metal matrix composites of A1359 and Al_2O_3 with a 2-8% reinforcement. Researchers discovered that aluminum MMCs are widely employed in the production of a variety of components, including pistons, brake drums, and cylinder blocks. Researchers believe that one of the most important phenomena of these components is wear and friction [3]. This study describes the tribological properties of aluminum MMC as they developed. A brief literature study is presented, focusing on single and multiple reinforcements, depending on the final product's application. This study, organized as a tribology wheel, delves into tribology, manufacturing processes/parameters, reinforcement and matrix contributions, tribology test parameters, statistical analysis methodologies, and product application regions [4]. This is highly useful for selecting elements that can be sorted and merged to improve tribological properties. To learn more about the current use of metal matrix composites, the researchers examined numerous reinforcements utilized in the making process of hybrid aluminium MMC, as well as their effects on particular properties, corrosion properties, and material wear rates. This study also discusses the primary production methods utilized to produce aluminum metal matrices. Although studies in the field of aluminium-silicon alloys have demonstrated that these alloys have outstanding castability and mechanical qualities, die-cast Al-Si alloys are castoff in cylinder blocks in select automobile industries [5]. It was tested, but no comments were made about wear resistance. In the stir casting technique, hybrid composites are made with SiC and Al_2O_3 reinforcements in a 1:1 ratio, with weight percentages increasing by 10, 20, and 30%. Researchers investigated the dry-sliding behavior of aluminium composites having equal proportions of up to 8 wt.% reinforced RHA and SiC particles created by a fluidized bed technique. Both non-reinforced alloys and hybrid composites have been evaluated for pin-on-disc wear. SEM is a method for investigating the wear behavior of non-reinforced aluminum alloys and hybrid composites. Hybrid composites have higher wear resistance than non-reinforced composites. The authors of this study discussed their experiments with electromagnetic fields. Because of the metal's EM wave attenuation, the discontinuous electromagnetic field's action is partial to a few centimeters from the lattice channel wall and is insensitive to the overall shape of the casting. As a result, it can be used in a variety of applications, including casting with a cylindrical channel and casting on multiple locally reinforced canals [6]. Many casting methods are accessible, however in this study, the composite material is created by means of an electromagnetic stirrer casting process. Reinforcements included the aluminum alloy 6063, as well as ZrO_2 and SiC. After adding these combinations to the MMCs, the various strengths and hardness were tested [7].

2. Materials and Methods

2.1 Aluminium

Al is the symbol for the chemical element aluminum, which has an atomic number of 13. Aluminum has a lower density than other substances. Aluminium's density is around one-

Fig. 5.1 Aluminium 6063

third that of steel. Figure 5.1 depicts a matrix material sliced into smaller pieces. Table 5.1 lists the chemical makeup of aluminum (Al6063), while Table 5.2 lists the metal's favorable physical characteristics.

Table 5.1 Chemical analysis of Aluminium 6063

Sl. No	Element	Wt. %
1	Cu	0.10 %
2	Mg	0.88 %
3	Mn	0.20 %
4	Si	0.68 %
5	Fe	0.30 %
6	Cr	0.05 %
7	Zn	0.10%
8	Al	97.68 %

Table 5.2 The physical characteristics of aluminum 6063

Sl. No	Physical Properties	Value
1.	Density	2.70 g/cm3
2.	Melting point	655 °C
3.	Thermal expansion	23.5 x10^-6/k
4.	Modulus of elasticity	69.5Gpa
5.	Thermal conductivity	201 W/m.k
6.	Electrical resistivity	52% IACS

2.2 Zirconium Di-Oxide

Zirconium dioxide (ZrO_2), commonly known as zirconium oxide or zirconium, is an inorganic metal oxide that is mostly employed in ceramic applications. Zirconium dioxide is the most common compound of the element zirconium in nature, succeeding zirconium itself [8]. It is a heavy metal found in the earth's crust with a concentration of 0.016%, making it more common than chlorine and copper. Its tremendous hardness, limited reactivity, and high melting point have made it the oldest mineral on the planet. Zirconium does not occur in large quantities, but is bonded in minerals, primarily zircon (ZrSiO4).

Table 5.3 Properties of zirconium di-oxide

Sl. No	Property	Value
1.	Density	5.68 g/cm^3
2.	Melting point	2715 °C(2988 K)
3.	Boling point	4300 °C(4570 K)
4.	Solubility	Soluble in HF, and hot H_2SO_4
5.	Flash point	Non-flammable

2.3 Silicon Carbide

With special thermal and electrical properties, silicon carbide, or SiC, is a non-oxide ceramic that is incredibly strong and durable. A variety of high temperature mechanical applications can benefit from silicon carbide's strengths, can vary from 27 GPa in SiC single crystals to 15 GPa in polycrystalline structures [9]. Its exceptional creep resistance further contributes to this versatility. Silicon carbide has a melting point of about 2830°C and an upper boundary of stability of about 2500°C. It's very high thermal conductivity (which is comparable to copper) makes it an attractive material to use as furnace heating elements, even if it oxidizes in air at temperatures above 1600°C. Because of its small band gap (2.2 eV in α-SiC& 3.3 in β-SiC), SiC is perfect for semiconductor applications that require low temperatures. In actuality, at high temperatures, silicon carbide rapidly gives up electrons, behaving like metallic conductors beyond a particular temperature threshold [10].

Table 5.4 Properties of silicon carbide

Sl.No	Property	Value
1	Density	3.1 g/m^3
2	Porosity	0%
3	Color	Black
4	Compressive strength	3900MPa
5	Thermal conductivity	120 W/m K
6	Specific heat	750 J/Kg K
7	Electrical resistivity	10^2-10^6 Ohm cm

2.4 Fabrication Processes of Metal Matrix Composite

The configuration of the EMS (Electromagnetic Stirring) system, which was developed for the metal matrix composites processing (Al6063/ZrO_2/SiC), is shown in Fig. 5.2. After cleaning and cutting the Al6063 alloy into small pieces, A horizontal muffle furnace was used to heat the aluminum pieces to a temperature higher than the liquid, while they remained inside the graphite crucible. A chrome-aluminium thermocouple was used to measure the temperature, which was 720 degrees Celsius. By linking the thermocouple relay and the muffle furnace, the temperature may be adjusted. An alternative technique involved adding the liquid A6063 aluminum alloy to a graphite crucible that was firmly filled with glass wool at a particular temperature (between the crucible and the winding). The different ZrO_2/SiC reinforcing combinations are connected to an aluminum metal matrix [13]. ZrO_2 and SiCelementsby means of an average size of 50 microns are added to the metal matrix as supplements. Both boron

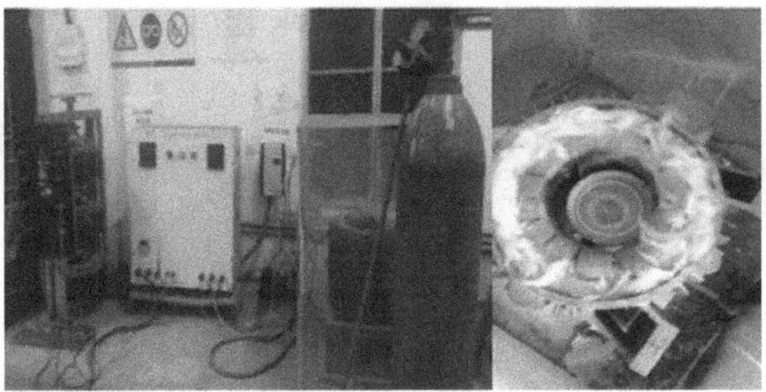

Fig. 5.2 Electromagnetic stirring (EMS) setup

carbide and titanium dioxide particles are warmed to 5000 degrees Celsius for 20 minutes before being added to the melt (A063 alloy). Titanium dioxide and boron carbide concentrations ranged from 0 to 9% by weight in each matrix [14]. As shown in Table 5.4, one matrix material and six composites were created. While stirring the metal matrix composite (A6063/ZrO_2/SiC), a thermocouple was put in into the graphite crucible to record the temperature. ZrO_2/SiC was mixed with argon gas in the Al6063 melt. The motor's windings were suitably cooled with coolant, and casting issues were avoided by using a vacuum pump to generate a vacuum inside the box. The ideal contribution process parameters (current, stirring speed, stirring temperature, voltage and stirring time) were found by a number of trials [11]. For the production of Al6063/ ZrO_2/SiC in the pilot run, a stirring speed of 200 rpm was selected at random; all other parameters were held constant. The titanium dioxide particles were evenly dispersed and did not settle when the stirring speed was increased. An additional increase in stirring speed indicated that the 6063 alloy was about to excess the crucible due to melting. The analysis led to the setting of the stirring speed at around 230 rpm. The remaining process parameter values were determined using the same methodology [12].

Table 5.5 Percentage of composition

Specimen No	Aluminium alloy 6063(%)	ZrO_2+SiC Particulate (%)	Quantity on each Specimen
1	100	0	1
2	97	3	1
3	94	6	1
4	91	9	1

3. Experimental Setup (SEM/EDAX Setup)

A concentrated electron beam is used to scan a surface to create photographs of a sample using a scanning electron microscope (SEM). The interaction between electrons in addition atoms in the sample produces a range of signals that reveal details about the composition and surface topography of the sample [15]. The surface morphology of nanoparticles and nanoparticle-coated glass fiber is investigated using a Zeiss EVO 18 Scanning Electron Microscope (SEM) and an Oxford Instruments Energy-dispersive X-ray spectroscopy (EDAX) micro analyzer. The EDAX analysis is used to perform quantitative and qualitative elemental analysis on nano particles and coated glass fibers (Fig. 5.3). The EDAX microanalysis technique is based on the distinct X-ray peaks produced when an intense electron beam interacts with a specimen. Each chemical element has a distinct peak that can be utilized to detect the presence of that element in the scanned region. This EDAX characterization can be used to determine the element's concentration in the sample [16].

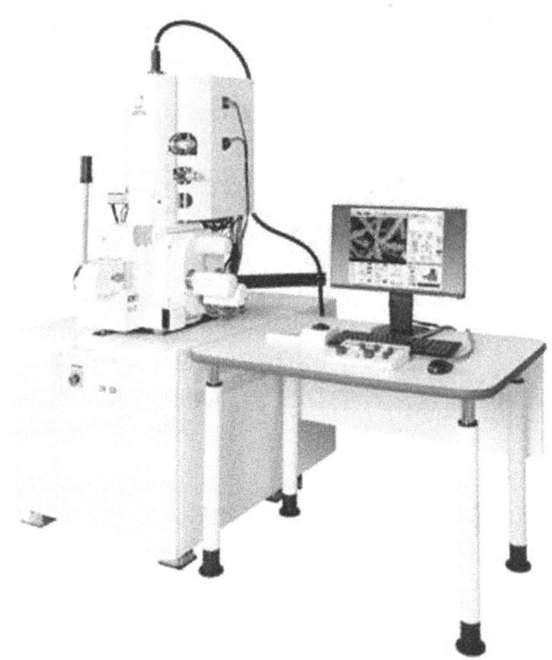

Fig. 5.3 SEM/EDAX setup

4. Result and Discussion

4.1 EDAX Test Results

Sample:1

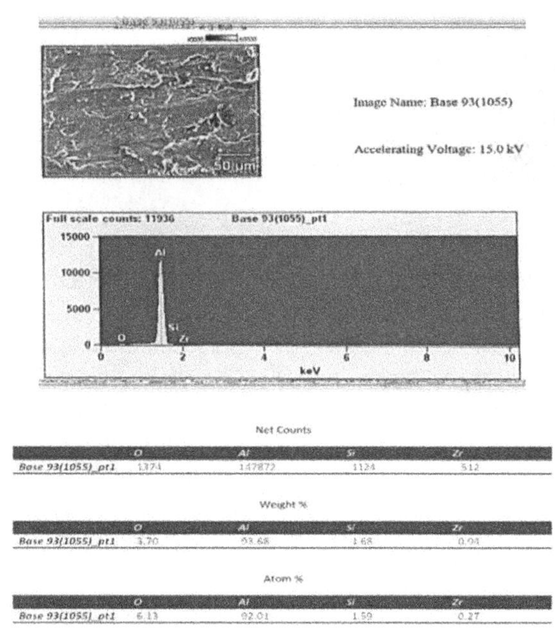

Net Counts

	O	Al	Si	Zr
Base 93(1055)_pt1	1374	147872	1124	512

Weight %

	O	Al	Si	Zr
Base 93(1055)_pt1	3.70	93.68	1.68	0.94

Atom %

	O	Al	Si	Zr
Base 93(1055)_pt1	6.13	92.01	1.59	0.27

Fig. 5.4 EDAX image of base 93(1055)

Sample:2

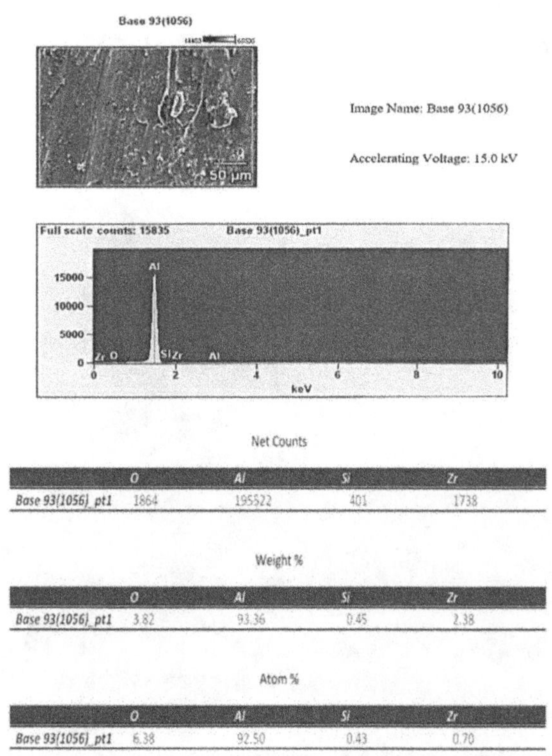

Net Counts

	O	Al	Si	Zr
Base 93(1056)_pt1	1864	195522	401	1738

Weight %

	O	Al	Si	Zr
Base 93(1056)_pt1	3.82	93.36	0.45	2.38

Atom %

	O	Al	Si	Zr
Base 93(1056)_pt1	6.38	92.50	0.43	0.70

Fig. 5.5 EDAX image of base 93(1056)

Sample:3

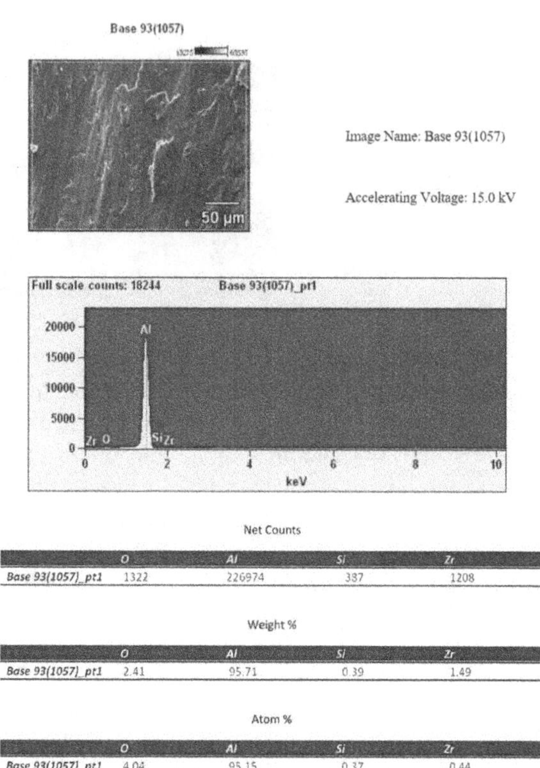

Net Counts

	O	Al	Si	Zr
Base 93(1057)_pt1	1322	226974	387	1208

Weight %

	O	Al	Si	Zr
Base 93(1057)_pt1	2.41	95.71	0.39	1.49

Atom %

	O	Al	Si	Zr
Base 93(1057)_pt1	4.04	95.15	0.37	0.44

Fig. 5.6 EDAX image of base 93(1057)

Sample: 4

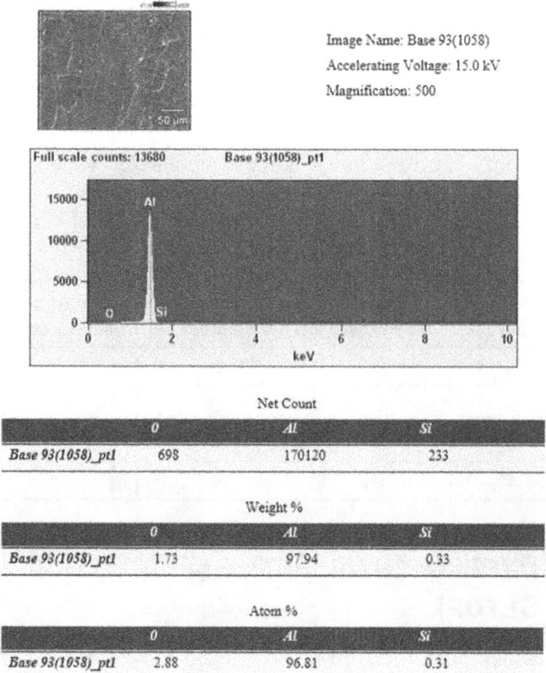

Net Count

	O	Al	Si
Base 93(1058)_pt1	698	170120	233

Weight %

	O	Al	Si
Base 93(1058)_pt1	1.73	97.94	0.33

Atom %

	O	Al	Si
Base 93(1058)_pt1	2.88	96.81	0.31

Fig. 5.7 EDAX image of base 93(1058)

4.2 SEM Results

Sample:1

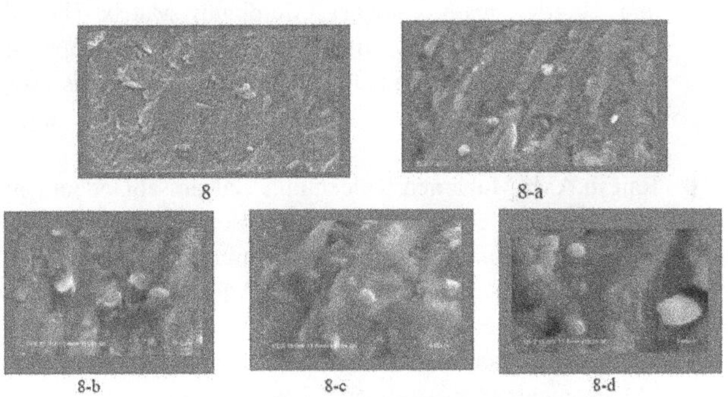

Fig. 5.8 SEM image of base 93(1055)

Sample:2

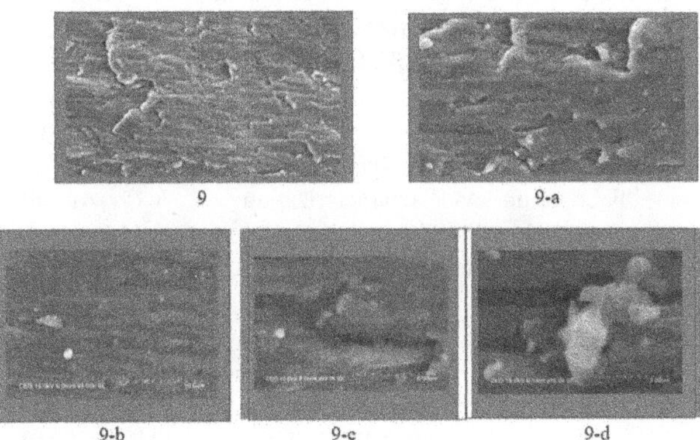

Fig. 5.9 SEM image of base 93(1056)

Sample:3

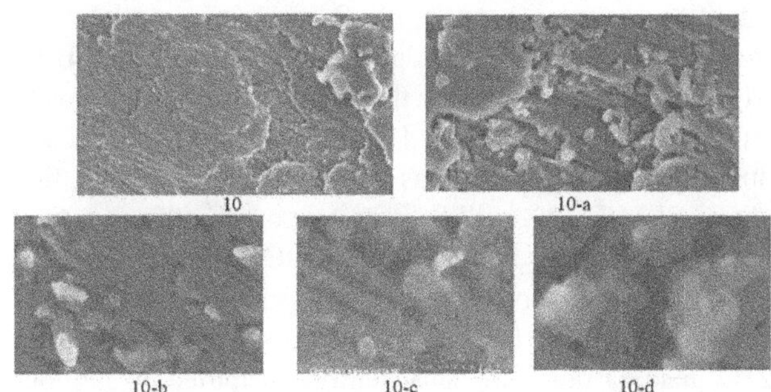

Fig. 5.10 SEM image of base 93(1057)

Sample:4

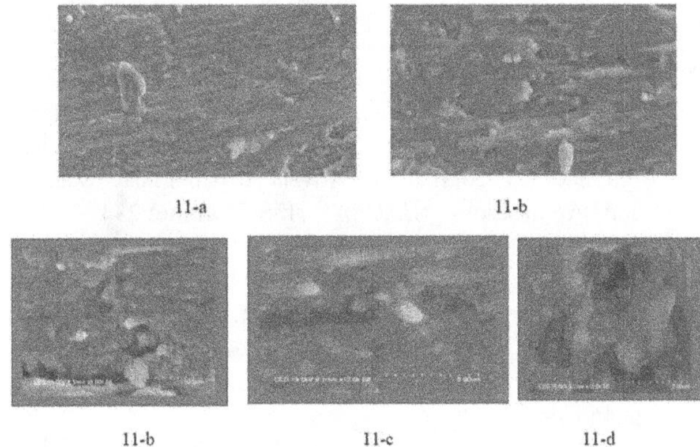

Fig. 5.11 SEM image of base 93(1058)

According to ASTM guidelines with extra emphasis, the aluminum alloys in addition composites surfaces are evaluated using SEM with EDAX to determine the distribution of ZrO_2 and SiC. The Zeiss EVO18 research setup is used to perform the SEM and EDAX microanalysis. The typical surface structure of composites reinforced with ceramic particles and aluminum alloy (Fig. 5.4). The SEM micro-images, which demonstrate minimal discrimination in addition cluster formation in the matrix, validate an improved dispersal of nano-reinforcement in the matrix metal [17–20]. The chemical composition of reinforced metal composites and the alloy was found to be both qualitative and quantitative, as demonstrated by EDAX microanalysis. EDAX microanalysis of composite reinforced with nanoparticles and aluminum alloy (Fig. 5.5 and 5.6). The aluminum alloy's compositional analysis verified the existence of alloy elements with quantitative nature, such as Cu, Si, Fe, Mg, Ma,Ni, Zn, Ti, and Al, while the reinforced composite's compositional analysis complete the involvement of reinforcement, such as ZrO_2 and SiC, along with to the alloy elements. The reinforcements and alloy element compositions are extremely close to the stoichiometric ratio. Based on the initial selection of the metal matrix and particles weight concentration, the final composition of the aluminum alloy and a combination, read equally accurate in the stoichiometry's ratio manifestations (Fig. 5.7 and 5.8). The alloy's density and the composite materials increased by nanoparticles, both in real and theoretical terms. The addition of reinforcement causes the materials' density to increase over time. When compared to other sample composites, the composite has a higher density, mostly because of the higher density of reinforcing particles (Fig. 5.9 and 5.10). The theoretical density of 2.81 for aluminum alloys is quite near to the true density of 2.78 for these alloys [21–26]. Comparably, there is very little difference between the theoretical and actual densities for single particle reinforced composites and hybrid reinforced composites (Fig. 5.11).

5. Conclusion

This step includes a thorough assessment of the composites and aluminum metal matrix. Although aluminum alloys are widely employed in the automobile industry, very few alloys such as A319, A356, and A359 alloy are used in high-temperature situations. To improve the mechanical qualities, ceramic particle reinforcing is essential. The most often utilized ceramic particles are Al_2O_3, SiC, B_4C, Mg, and ZrO_2. However, there aren't many composite studies based on ZrO_2 and SiC. For this reason, ZrO_2 and SiC were employed as reinforced materials in this study. SEM/EDAX will be used to examine the chemical characterization of aluminum MMC. Composites made of 6063 aluminum alloy containing 3, 6, and 9 volume percent ZrO_2 and SiC have been effectively produced. The SiC and ZrO_2 nanoparticles are consistently distributed throughout the matrix, according to the SEM examination. The planned addition of 3, 6, and 9 vol% of ZrO_2 and SiC particles to the Al matrix was confirmed by EDAX analysis.

References

1. Madhuri Deshpande, Rahul Waikar, Ramesh Gondil, S.V.S Narayan Murty, T.S.Mahata "Processing of Carbon fiber reinforced Aluminium(7075) metal matrix composite" International Journal of Advanced Chemical Science and Applications (IJACSA), ISSN(Online):2347-761X, Volume-5, Issue-2,2017.
2. Manoj Singla, D. Deepak Dwivedi, LakhvirSingh, Vikas Chawla "Development ofAluminiumBased Silicon Carbide Particulate Metal Matrix Composite"Journalof Minerals &Materials Characterization & Engineering, Vol.8, No.6, pp455–467, 2009.
3. Jamaluddin Hindi, AchutaKiniU, S.S Sharma "Mechanical Characterisationof Stir Cast Aluminium 7075 Matrix Reinforced it h Grey Cast Iron & Fly Ash" International Journal of Mechanical And Production Engineering, ISSN: 2320–2092.Volume-4, Issue-6, Jun.2016.
4. Mohan Kumar S, Pramod R, Shashi Kumar M E, GovindarajuH K "Evaluation of Fracture Toughness and Mechanical Properties of Aluminum Alloy 7075, T6 with Nickel Coating" Science Direct, Procedia Engineering 97 (2014)178–185.
5. Gururaj Aski, Dr. R. V. Kurahatti"The effect of Mechanical properties and microstructure of LM13 Aluminum Alloy Reinforced with Zirconium silicate(ZrSiO4)" International Journal of Innovative Research in Science, Engineering and Technology" Vol.6, Issue6, June2017.
6. Savannah, V. S. Ramamurthy "Microstructure and Wear Characterization of A356- ZrSio4 Particulate Metal MatrixComposite" International Journal of Science and Research (IJSR), Volume3 Issue7, July 2014.
7. Abhishek Kumar, Shyam Lal and Sudhir Kumar "Fabrication and characterization of A359/Al_2O_3 metal matrix composite using electromagnetic stir casting method", In Journal of Materials Research and Technology, Volume 2, Issue 3, 2013.
8. Jitendra M Mistry and Piyush P Gohil "An overview of diversified reinforcement on aluminium metal matrix composites: Tribological aspects", In Journal of Engineering Tribology, Volume 231, P 399–421, 2017.
9. Michael Oluwatosin Bodunrin, Kenneth KanayoAlaneme and Lesley Heath Chown "Aluminum Matrix Hybrid Composites: A Review of Reinforcement Philosophies; Mechanical, Corrosionand Tribological Characteristics", In Journal of Materials Research and Technology, JMRTEC-169, 2015.
10. Rajesh AM, Mohamed Kaleemulla, SaleemsabDoddamani and Bharath KN "Material characterization of SiC and Al_2O_3- reinforced hybrid aluminium metal matrix composites on wear behavior", In Advanced Composites Letters, Volume:28, 2019.
11. R. Karthigeyan, G. Ranganath, S. Sankaranarayanan "Mechanical Properties and Microstructure Studies of Aluminium(7075) Alloy Matrix Composite Reinforced with Short Basalt Fibre" European Journal of Scientific Research, ISSN1450-216XVol.68No.4(2012), pp. 606615.
12. P. Pradeep, P. S. Samuel RatnaKumar, Daniel Lawrence I, Jayabal S "Characterization of par particulate reinforced Aluminium 7075 / TiB2 Composites" International Journal of Civil Engineering and Technology (IJCIET), Volume 8, Issue9, September 2017, pp.178–190.
13. Rajesh Kumar Bhushan,Sudhir Kumar and S. Das, "Fabrication and characterization of 7075 Al alloy reinforced with SiCparticles", International Journal of Advanced Manufacturing Technology, No. 65, pp. 611–624, 2013.
14. Chenwei Shao Zhao, Xuegang Wang, Yankun Zhu, Zhefeng Zhang and Robert O. Ritchie "Architecture of high-strength aluminiummatrix composites processed by a novel micro casting technique", In NPG Asia Materials, 11(69), 2019.
15. S.Golak and M.Dyzia "Creating Local Reinforcement of a channel in a Composite Casting Using Electromagnetic Separation", In Journal of Materials Science & Technology, Volume 31, P 918–922, 2015.
16. Ashish k Srivastava, Amit Rai Dixit & Sandeep Tiwari "Investigation of microstructural and mechanical properties of metal matrix composite A359/B4C through electromagnetic stir casting", In Indian Journal of Engineering and Materials Sciences, 171–180, 2016.
17. S. P. Dwivedi, S. Sharma & R. K. Mishra "Electromagnetic Stir Casting and its Process Parameters for the Fabrication and Refined the Grain Structure of Metal Matrix Composites- A Review", In International Journal of Advance Research and Innovation, ISSN 2347 – 3258, Volume 2, Issue 3 639–649, 2014.
18. S.Saravanan, P.Senthilkumar and S.Sankar "Compressive and Shear Behaviour of AA6063-TiC Composites", In IJLERA ISSN:2455-7137, Vol-02, Issue-04, PP-32–36, 2017.
19. B. Venkata Manoj Kumar, Friction and Wear of Materials: Principles and Case Studies, Department of Metallurgical and Materials Engineering, NPTEL, Lecture-7.
20. M G Gee and S Owen-Jones, Wear Testing Methods and Their Relevance to Industrial Wear Problems, NPL Report, ISSN 1361-4061, 1997.

21. Dora Siva Prasad, Chintada Shoba, NalluRamanaiah "Investigations on mechanical properties of aluminium hybrid composites", In Journal of Materials Research and Technology, Volume 3, 79-85, 2013.
22. Shyam Lal, Ajay Kumar, Sudhir Kumar and Nitin Gupta "Characterization of A356/B4C composite fabricated by electromagnetic stircasting process with the vacuum", In Materials Today: Proceedings, 34(5), 2020.
23. Jitendra M Mistry and Piyush P Gohil "An overview of diversified reinforcement on aluminium metal matrix composites: Tribological aspects", In Journal of Engineering Tribology, Volume 231,P 399–421, 2017.
24. S.R. Pearson, P.H. Shipway, J.O. Abere and R.A.A. Hewitt "The effect of temperature on wear and friction of a high strength steel in fretting", In Wear, 303:622-631, 2013.
25. Mitsuhiro Okayasu, Yuta Miyamoto and Kazuma Morinaka "Material Properties of Various Cast Aluminum Alloys Made Using a Heated Mold Continuous Casting Technique with and without Ultrasonic Vibration", In Metals, 5(3), 1441–1453, 2015.
26. Michael Oluwatosin Bodunrin, Kenneth KanayoAlaneme and Lesley Heath Chown "Aluminum Matrix Hybrid Composites: A Review of Reinforcement Philosophies; Mechanical, Corrosionand Tribological Characteristics", In Journal of Materials Research and Technology, JMRTEC-169, 2015.

Note: All the figures and tables in this chapter were made by the authors.

Advances in Additive Manufacturing Technologies – Gurusamy Pathinettampadian et al. (eds)
© 2026 Taylor & Francis Group, London, ISBN 978-1-041-16687-0

Evaluating the Performance of a Motion Sensing Audio Playback System

T. Archana[1]
Department of Electronics and Communication Engineering,
Saveetha Engineering College,
Chennai, India

Vaishnavi T.[2]
Department of Electronics and Communication Engineering,
Vel Tech Rangarajan Dr. Sagunthala R&D Institute of Science and Technology,
Chennai, India

N. Nachammai[3]
Department of Electronics and Instrumentation Engineering,
Annamalai University Chidambaram,
India

Vanitha M.[4], **A. Hema Malini**[5], **Kalaivani C. T.**[6]
Department of Electronics and Communication Engineering,
Saveetha Engineering College,
Chennai, India

◆ **Abstract:** Gesture controlled Bluetooth speakers are becoming increasingly popular due to their ease of use and accessibility. This paper presents the design and implementation of a gesture-controlled Bluetooth speaker using Arduino. The system then converted into corresponding commands for the speaker. The gestures detected by the sensor are processed by an Arduino board which relays the information to the Bluetooth speaker through a Bluetooth module. Experimentation is the criteria used to measure the effectiveness of the system in recognizing and interacting with the hand gestures to control the speaker. The main method of interaction is based on acceleration and gyroscopic sensor so that hand movements are recognized as specific controls of the music, including play/pause, volume increase/decrease, skipping, etc. The processed signals are relayed to an Arduino microcontroller with the Bluetooth module and the speaker being controlled by this microcontroller. Finally prototyping was done and implies that the system is quite responding to the deficiencies and reliable for sensing the motion of the hands.

◆ **Keywords:** Gesture recognition, Accelerometer sensor, Gyroscope sensor, Bluetooth module, Audio system

1. Introduction

Portable and wireless, Bluetooth speakers have been gaining popularity in the recent past. This comes in as an interesting application of the Arduino microcontroller platform. In this technology, users are able to control the audio devices using simple hand gesture and therefore have the abilities to play a specific audio track, increase

[1]archana@saveetha.ac.in, [2]vaishnavitamilselvan@gmail.com, [3]nachammai2007@gmail.com, [4]vanitha@saveetha.ac.in, [5]hemamalini@saveetha.ac.in, [6]kalaivani@saveetha.ac.in

DOI: 10.1201/9781003685906-6

or decrease the volume, or move to the next track without touching the device. This makes it convenient since a person can be busy doing some activities like cooking exercise or the like when the audio device can never be reached physically [1]. The system is based on usage of various kinds of sensors that are capable of sensing the Hand movement including the accelerometer & gyroscopes sensors. These sensors feature as positional, rotational and motion of the hand which sends this information to an Arduino microcontroller. This information is then analyzed by the Arduino which in turn controls the playback of the audio tracks on the Bluetooth speaker. This makes it possible for the user to manipulate their music without having to use extra controls on the radio.

They are numerous advantages when using gesture-controlled Bluetooth Speakers by utilizing Arduino. First, it is preferable to a traditional interface since it closely approximates a user's natural hand movement for adjusting the volume. For traditional differential physical controls, it means that a user will often have to stop what they are doing to turn the volume up or down, or to move to the next track. It means with gesture control they can do this just by moving their hand without having to stop whatever they are doing. The other advantage includes ability to have different feelings to the gesture of hand. Its unique feature is that with Arduino the users are able to define the gestures and adapt the interaction according to the situation. This makes the system very versatile and elastic, which enables the use of the system in a number of fields readily available today. Bluetooth speakers controlled by gesture through arduino can also be used in some other fields and areas like home automation and robotics and other interactive instruments. The advantages of using the Arduino include the following; The Arduino is a versatile board that enhances electronics DIY projects. There is a variety of users who can easily alter the sensors and controls in ways that produce entirely new products which might not have been conceived by the manufacturer or developers [2–5].

However, apart from the fact that they are wired, the traditional method of controlling Bluetooth speakers which includes the use of a physical remote controller or a mobile application becomes tedious and unhygienic. Bluetooth speakers controlled by hand movements are easier to operate naturally and are more convenient to use. This paper discusses an implementation of a gesture- controlled Bluetooth speaker connected to Arduino in which the speaker is controlled with hand gestures.

2. System Design

The design of the system involves an Arduino board, 9 axis accelerometer and gyroscope sensor and Bluetooth module. The hand gestures of the user are detected by the accelerometer and gyroscope sensors, which in turn controls the speaker accordingly. The Bluetooth module acts as the interface to enable the transfer of the gesture commands from the Arduino board to the speaker [6].

The system is designed to recognize four different hand gestures: The control options that were missing from the previous designs included play/pause, next track, previous track, and volume control. The play/pause gesture is made by swiping the hand in front of the neck and then using it to make a chopping motion. The next track gesture is made by shifting the hand to the right and the previous track gesture is made by shifting the hand to the left. The volume control is done by swaying the hand up or down [7].

This system is developed with Arduino Uno, MPU-9250 9-Axis Accelerometer & Gyroscope Sensor, HC-05 Bluetooth Module. The sensor is interfaced with the Arduino using the Inter Integrated Circuit (I2C) bus. The Bluetooth module is interfaced with Arduino board by using serial peripheral interface (SPI) connection as shown in Fig. 6.1. Arduino board uses the Arduino Integrated Development Environment (IDE) as a software tool for developing applications for Arduino board [8].

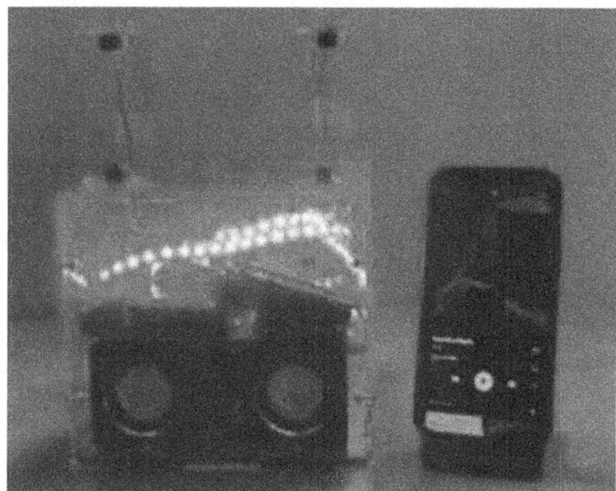

Fig. 6.1 Bluetooth controlled speaker

The gesture recognition algorithm described is programmed in Arduino programming language. The data obtained from the accelerometer is then processed in the algorithm to decipher the direction and magnitude of hand motion. The algorithm recognizes four different gestures:

1. Swipe left: This gesture is recognized when the user uses his/her hand from right to left. This hand gesture is used to go to the previous track.
2. Swipe right: This gesture is identified when the user uses a sweeping hand from left to right. This gesture is used to move to the following song.

3. Swipe up: This gesture is recognized in the condition of moving hand from down to up. This gesture is made in order to raise the volume.
4. Swipe down: This gesture is identified when the user has put his hand up and then moved it down. This one is used to reduce the sound level of the audio being played.

3. Methodology

3.1 System Architecture

The hardware and software components of the system include the Gesture Controlled Bluetooth Speaker using Arduino. The hardware components are as follows: Arduino board, Bluetooth module (HC 05), gesture sensor, speaker, bread board and jumper wires. These are a hand gesture recognition algorithm, Bluetooth communication protocol and audio playback code [9].

3.2 Working Principle

Gesture-controlled Bluetooth speaker using Arduino is the process of using sensors to recognize particular hand gestures and subsequently translate those gestures into commands interpreted by Bluetooth speaker to play music or change the volume of the Bluetooth Speakers. Here are the basic steps involved:

1. Hardware setup: First, you will need to connect various hardware components such as an Arduino board, a Bluetooth module, a speaker, and sensors such as an accelerometer or gyroscope to detect gestures.
2. Gesture detection: Here hand movement such as waving or tilting will be felt by the sensors and the signals will be sent to the Arduino board.
3. Data interpretation: The information received by the Arduino board will then be analyzed the board and decide on what action to take such as increasing the volume or reducing it or even playing a new song.
4. Bluetooth connection: After that, the Arduino board will change to Bluetooth to connect with the gadget that is playing the music, for instance a Smartphone or laptop.
5. Music playback and volume control: Finally, the Arduino board will control the speaker to play the music and adjust the volume based on the detected gestures.

Overall, this allows for hands-free control of the Bluetooth speaker, providing a moreconvenient and intuitive way to enjoy your music.

3.3 Arduino Board

The Arduino board is the main control board that is used to program and run the hand gesture recognition algorithm. It receives input from the gesture sensor and sends output signals to the HC-05 Bluetooth module to control the speaker.

3.4 HC-05 Bluetooth Module

The HC-05 Bluetooth module is a Bluetooth serial module which can enable wireless connection between the Arduino board and a Bluetooth device including mobile phone. This module can effectively be connected to the Arduino board through serial port and it has the ability to send and receive data through wireless [10].

3.5 Speaker

This component is used to share the audio from the Bluetooth device being used. It is interfaced with the Arduino board through jumbled wires and can be operated through output signals transmitted by the HC-05 Bluetooth module of the Arduino board [11].

3.6 Breadboard

For this reason, the breadboard comes in handy as a tool that will offer the necessary support needed in connecting the many parts used in the project. It can be used to quickly build and test the circuit before the final soldering of the circuit on a regular PCB [12].

3.7 Jumper Wires

All the different parts of the project are connected using jumper wires. These are utilized to link the breadboard to the Arduino board, the Bluetooth module HC-05 to the Arduino and the speaker to the breadboard.

4. Results and Discussion

The Bluetooth speaker controlled by hand gestures was trialed in an experiment of doing multiple hand gestures to observe the efficiency of the gesture recognising system of the defined hand gestures. Overall, there were 100 gestures executed, and 25 repetitions of each of the four recognized gestures, (swipe left, swipe right, swipe up, and swipe down).

As shown in Table 6.1, As you can see the system was able to recognize the four gestures with certain accuracy ranging from 92% to 96%. The combined recognition rate of all the four basic gestures was 95 percent.

In addition to evaluating the system's accuracy, we also tested its reliability in controlling the music playback and volume.

Table 6.1 Results of gesture recognition test

Gesture	Total Performed	Correctly Recognized	Recognition Rate
Swipe Left	25	24	96%
Swipe Right	25	23	92%
Swipe Up	25	24	96%
Swipe Down	25	24	96%

The data given can be represented through a bar chart in Fig. 6.2 that shows the correct recognition of gestures A, B, C, and D with blue and orange bar representing two conditions. The vertical axis indicates the recognition accuracy and as seen from the Fig. 6.2, the results are nearly perfect for all the gestures. The two graphs show that the results of the two conditions or recognition methods are very close to each other, with the bars for the blue and orange colors being very similar in height. In other words, the proposed method produces comparable results to the baseline in terms of gesture recognition accuracy and is equally effective across all the tested gestures. In this case, the graph shows robust and stable recognition accuracy for all the gestural movements under both conditions, thus confirming that the proposed system is efficient [13].

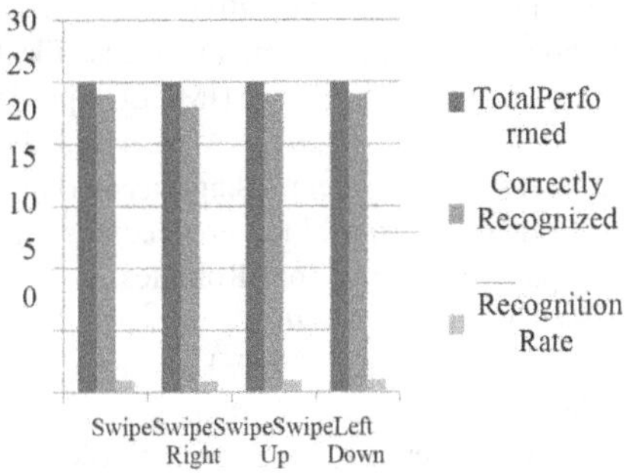

Fig. 6.2 Graph showing results of recognition test on different gestures

As shown in Table 6.2, From the results obtained in this study, it was possible to achieve an accuracy of 90% to 100% for all the four hand gestures in the system. The average recognition rate for all the four gestures was 95 percent.

Table 6.2 A table on system reliability

Hand Guesture	No. of Correct Recognition	No. of Incorrect Recognition	Accuracy
Play/pause	28	2	93.3%
Next track	30	0	100%
Previous track	29	1	96.7%
Volume control	27	3	90%
Overall	114	6	95%

In this case the system proved useful in regulating the playback of music in the associated device through the well recognized gestures on the interface.

The curve on the graph Fig. 6.3 illustrates increased precision of the system to detect hand gestures. "Next Track" has the best accuracy of more than 100 percent which indicates hugely impressive performance or abnormality in computation. "Previous Track" again has almost 100% accuracy in its execution. The accuracy of "Play/Pause" gestures stands at about 94-96%. As shown in Table 6.5, "Volume Control" has the least accuracy-ranging 90-92/100 suggesting that there is always difficulty in the gesture recognition that could probably be as a result pf complexities in the types of gestures or close resemblance with other types of gestures. The overall success is approximately 96% suggesting rather high performance of the model but with certain potential for the increase of performance of individual gestures. Improving the algorithms and the training data set that the underperforming gestures utilize could improve the system function as well as the experience.

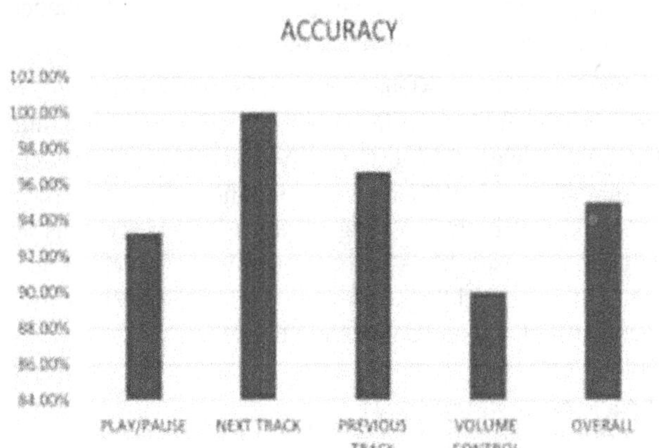

Fig. 6.3 Graph showing systems accuracy of various hand gestures

Control of music playback and volume, the final aspect of the system was also tested to assess its degree of accuracy. The system has been observed as effective in managing the music play back and volume of the connected device. There was no jamming while controlling the playlist and use of loudspeakers; the system was rather quick in its response. Moreover, the utility of the system was also analyzed through a set of test users. The users mentioned that they liked the idea and the concept of the system, as the hand gestures themselves are not very difficult to learn. The portability and convenience of the system was also well appreciated by the users, as it let them control the music and its volume without having to touch buttons on the speaker [14].

In summary, based on the evaluation of the experiment, it proved that the gesture-controlled Bluetooth speaker with the aid of Arduino is an efficient and reliable system for commanding the play back of music and volume. Through convenience, versatility, smooth operation and output quality, the system is a great Bluetooth speaker substitute for button-control designs. Since the system is so modifiable, it can be easily used in a wide range of situations and for various purposes.

5. Conclusion

Therefore, the design of the gesture-controlled Bluetooth speaker using Arduino shown in this paper is a feasible substitute to the typical button operated Bluetooth speakers. It also allows the users to manage music playback and volume by the hand gestures which are easily understandable by the users since they are trained to perform them. For recognizing the hand gestures, the system employs an Arduino Nano board, Bluetooth module and Accelerometer, while the gesture recognition algorithm is coded in Arduino programming language so that the present system can be easily tailored in compliance with the various applications.

A review on the system further showed that its use is precise, free from significant errors and ease to operate. The system was able to correctly identify the four hand gestures drawn in 95% of cases, and there were no delays in switching between songs, increasing or decreasing volume. The usefulness of the system was also tested on a group of users and they complained that the system is user friendly, portable and convenient to use.

Acknowledgment

I would like to express my sincere gratitude toSaveetha Engineering College for their unwavering support and valuable resources throughout my academic journey.

References

1. Maji, S. K., Banerjee, S., & Mitra, S. K. 2015. Design and development of hand gesture recognition system using accelerometer sensor. Proceedings of the 2015 International Conference on Circuits, Systems, Signal Processing, Communications and Computers, pp. 49–53. Springer, Cham.
2. Su, X., Xu, Y., Liu, Z., & Wang, S. 2018. Hand gesture recognition using Leap Motion Controller. 2018 International Conference on Machine Learning and Cybernetics (ICMLC), pp. 1068–1073. IEEE.
3. Lefferts, W. K., & Hwang, J. 2015. The usability of a Bluetooth speaker for older adults. Gerontechnology, 14(1):37–46.
4. Wong, Y. K., Gedeon, T., & Lee, Y. H. 2016. Development of a gesture-controlled robotic arm with a Microsoft Kinect sensor and Arduino. Sensors, 16(10):1627.
5. Rabiner, L. R., & Schafer, R. W. 1981. A gesture recognition system for speech communication. The Bell System Technical Journal, 60(7):1389–1397.
6. Gao, Y., Li, C., Yang, J., &Lv, J. 2013. Design of a gesture recognition system based on single-axis accelerometer. Journal of Software Engineering and Applications, 6(11):528–532.
7. Jung, H., Kim, J. S., & Lee, B. 2015. A drone control system based on gesture recognition using Arduino. International Journal of Distributed Sensor Networks, 11(7):315056.
8. Chen, Q., Wu, Z., & Song, A. 2017. Development of a humanoid robot control system based on gesture recognition. International Journal of Advanced Robotic Systems, 14(1):1729881416688643.
9. Liu, Y., Chen, Z., Lin, L., &Guo, Y. 2018. A smart home control system based on gesture recognition using Leap Motion Controller and Arduino. 2018 14th IEEE International Conference on Control and Automation (ICCA), pp. 1532–1536.
10. Wang, X., Zhang, M., Wang, X., & Sun, W. 2018. A gesture recognition-based control system for a robotic arm using Myo armband and Arduino. 2018 IEEE 2nd Advanced Information Management, Communicates, Electronic and Automation Control Conference (IMCEC), pp. 1588–1593.
11. Li, J., Su, J., & Li, B. 2017. A wearable device-based gesture recognition system using accelerometer sensor and support vector machine. Proceedings of the 2017 International Conference on Robotics and Automation Sciences (ICRAS), pp. 41–46.
12. Lu, W., Liu, C., & Zhu, Z. 2019. Hand gesture recognition using deep learning and thermal camera. 2019 IEEE International Conference on Consumer Electronics-China (ICCE-China), pp. 1–4.
13. Yun, J., Park, J., Kim, J., & Lee, B. 2015. A gesture recognition system based on MEMS accelerometer and support vector machine. Proceedings of the 2015 International Conference on Information and Communication Technology Convergence (ICTC), pp. 565–567.
14. Liu, T., Zhou, Y., Li, J., & Zou, J. 2018. Hand gesture recognition based on Myo armband and convolutional neural network. 2018 IEEE International Conference on Cybernetics and Intelligent Systems (CIS) and IEEE Conference on Robotics, Automation and Mechatronics (RAM), pp. 188–193.

Note: All the figures and tables in this chapter were made by the authors.

Advances in Additive Manufacturing Technologies – Gurusamy Pathinettampadian et al. (eds)
© 2026 Taylor & Francis Group, London, ISBN 978-1-041-16687-0

Unlocking Workstation Power: Revolutionizing Crypto Currency Mining with Tensilica Processors and AI

T. Archana[1]
Department of Electronics and Communication Engineering,
Saveetha Engineering College,
Chennai, India

Vaishnavi T.[2]
Department of Electronics and Communication Engineering,
Vel Tech Rangarajan Dr. Sagunthala R&D Institute of Science and Technology,
Chennai, India

N. Nachammai[3]
Department of Electronics and Instrumentation Engineering,
Annamalai University Chidambaram, India

Santhosh B.[4], A. Hema Malini[5], P. Sanjay[6]
Department of Electronics and Communication Engineering,
Saveetha Engineering College,
Chennai, India

◆ **Abstract:** In recent years, substantial investments have been made by businesses worldwide in workstation computers, yet their effective utilization remains a challenge. This inefficiency threatens to diminish the profit-to-investment ratio, resulting in wasted computer resources. To address this issue, leveraging artificial intelligence (AI) in the domain of crypto currency mining holds significant promise. This paper proposes a groundbreaking approach by harnessing the computational power of Tensilica processors, optimized for cryptographic tasks, to revolutionize crypto currency mining. By integrating Tensilica-based processors into conventional workstations, we aim to enhance both computational performance and environmental sustainability within the crypto currency mining sector. While Bitcoin mining was once feasible on standard laptops and desktops, the advent of Application Specific Integrated Circuits (ASICs) tailored for Bitcoin mining rendered this approach obsolete. Our study presents a novel exploration of dynamic opcode analysis for digital currency mining, an area with limited prior research. We report detection accuracies of up to 100% for browser-based crypto mining within our dataset. Additionally, our approach makes a distinction between real-world benign sites, weaponized sites, de-weaponized crypto mining sites, and crypto mining sites. Because it shows that accessible cryptocurrencies may be mined effectively without the need of massive mining rigs, this research offers insightful information for cryptocurrency fans. In summary, our approach is a major step toward maximizing the use of desktop computers for mining cryptocurrency, encouraging flexibility and sustainability in this quickly changing sector.

◆ **Keywords:** Crypto mining, Tensilica processor, IoT, AI, Bitcoin, ASCI's

[1]archana@saveetha.ac.in, [2]vaishnavitamilselvan@gmail.com, [3]nachammai2007@gmail.com, [4]Santhoshbks956@gmail.com, [5]hemamalini@saveetha.ac.in, [6]Santhoshbks956@gmail.com

DOI: 10.1201/9781003685906-7

1. Introduction

These days, the volume of cryptocurrency transfers is steadily and swiftly increasing. The bitcoin currency is the first and most well-known cryptocurrency, which makes it particularly appealing to investors, according to Coin Market Share (https://coinmarketcap.com/charts/), a website that tracks the cryptocurrency market. One way to get Bit coins is by mining [1]. A miner must use incredibly fast computer equipment to coordinate Bit dollar transactions and prove Bit coin transfers on the Bit coin network in order to receive Bit coins as incentives. Some folks want to mine but don't have the computing equipment they need. The field of cryptocurrency mining has experienced a significant transformation since its inception over a decade ago. Standard CPUs could initially be used for mining, but as cryptographic algorithms have become more complex, specialized hardware, such as ASICs andGPU's, has been developed to stay profitable and competitive. However, a lot of bitcoin enthusiasts and miners who use traditional workstation PCs believe that these specialized tools are either too expensive or unfeasible because of their high power demand and exorbitant cost. This study creates a new method for mining cryptocurrencies that bridges the gap between the demands of contemporary mining algorithms and the constraints of conventional workstation systems. We recommend creating a cryptocurrency mining system based on a TensilicaCPUMany cryptocurrency mining markets offer mining services (similar to cloud mining) to help people who don't have access to physical mining equipment. One of the biggest marketplaces for bitcoin mining, Nice Hash, allows users to purchase or sell computing power as needed [2]. When making orders on the Nice Hash platform, customers can choose the price, algorithm, and time they want their coins to be mined. In order to fulfill the demands of the selected buyers, sellers—also referred to as miners—then manufacture the coins. Miners don't have to monitor the market or their many wallets since Nice Hash, the researcher Miner Legacy, automatically selects the optimal strategy. However, due to the vast number of active miners, the current software is unable to change the strategy that continuously generates the most income for all miners. Furthermore, mining stability was impacted by the frequent discontinuities that many miners encountered on the Nice Hashes pools [3]. In order to mine cryptocurrencies successfully and quickly, we propose creating a system based on Tensilica processors that utilizes the programming capabilities and adaptability of Tensilica processing. As an alternative to specialized hardware, crypto mining should become more accessible and economical and energy-efficient with this approach. The design principles, advancements, and challenges of implementing a cryptocurrency mining system based on Tensilica processors. In contrast to traditional CPU-based mining, we will examine how popular mining algorithms, such SHA-256 and Ethash, adjust to the Tensilica framework and evaluate the machine's performance, electricity usage, and mining yield [4].

2. Proposed System

If we can cut consumption without compromising performance, we will benefit. The easiest cryptocurrency to mine is one that doesn't require you to set up a sizable mining setup. Although it is possible to configure everything, load a respectable hash, Installing an intelligent AI LSAI48266x equipment board is recommended to maximize CPU performance while consuming little power, allowing it to operate for years. When mining cryptocurrencies on a typical PC, the following important problems come up:

- Physical issues with the system, including overheating
- A free mining timing technique for system accessibility
- The proportion of cryptocurrencies to electricity use.

With the help of a creative AI algorithm, the proposed AI board (LSAI48266x) can solve the PC mining problems previously discussed. An artificial intelligence (AI) component is part of the suggested system. Every time it seems like an employee is not using the workstation computers, the intelligence board will decide the crypto based on a variety of criteria. While putting up a good hash is possible, it is better to use a smart AI that you can set up once and leave running for years. To maximize CPU performance without using too much electricity, use an LSAI48266x electronics board. There is a prospect for financial advantage if we can cut back on our consumption without affecting performance [5].

The presented diagram Fig. 7.1 illustrates a temperature management and data acquisition system for which an AISAI48266x board is used. Several temperatures are

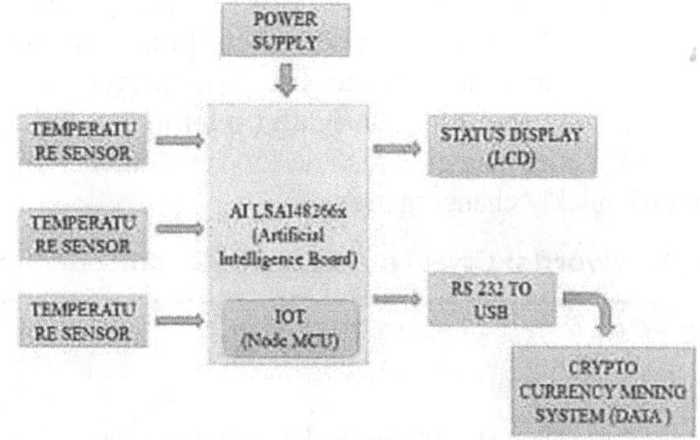

Fig. 7.1 Architecture of proposed system

measured, and information is received and analyzed by the AI board connected to the IoT board (Node MCU). The real time operation status on the system is shown on an LCD display and is used for data usage using RS232-USB connectivity for mining system. A power supply backs the entire set-up making it possible to run it efficiently. This integration focuses on how IoT and AI are implemented in data processes for creative functions [6].

3. Working Principle

BOARD AI LSAI48266x: The module is future-proof due to its integration of Wi-Fi, Bluetooth, and Bluetooth LE. Wi-Fi offers a large physical range and immediate access to the outside world through a router that connects to the internet, whilst Bluetooth enables the user to connect to their mobile device with ease or emit low-power signals to help with detection. The Tensilica Processor chip can be used in battery-powered and gadget-like applications because its sleep current is less than 5 A. With an average data rate of up to 150 Mbps and 20.5 dBm of electricity generated at the antenna, Tensilica's processor ensures the maximum physical range [7].

WROOM32, TENSILICA'S processor: The module is future-proof due to its integration of Wi-Fi, Bluetooth, and Bluetooth LE. Wi-Fi offers a large physical range and immediate access to the outside world through a router that connects to the internet, whilst Bluetooth enables the user to connect to their mobile device with ease or emit low-power signals to help with detection. The Tensilica Processor chip can be used in battery-powered and gadget-like applications because its sleep current is less than 5 A. With an average data rate of up to 150 Mbps and 20.5 dBm of electricity generated at the antenna, Tensilica's processor ensures the maximum physical range [8].

A microcontroller board in Fig. 7.2 with add-ons like an LCD, relays, and connectors makes up the hardware configuration that is shown; it may be tailored for AI or bitcoin mining applications. It could keep an eye on and control power, guaranteeing effective mining operations. Tensilica processors, which are renowned for their low-power DSP cores, can be integrated into the system to increase computing speed and efficiency while lowering energy expenses. Through dynamic resource allocation and performance bottleneck prediction, AI models operating on the board could further optimise mining. This configuration is a prime example of a scalable, energy-efficient decentralised mining prototype that uses AI and cutting-edge CPUs to unleash workstation power and transform the economics of bitcoin mining [9].

The peripheral features of Tensilica's processor include: (18) channels for analog-to-digital converters (ADC), (10) GPIOs for capacitive sensing, (3) UART interfaces, (3) SPI interfaces, (2) I2C interfaces, (16) PWM output channels, (2) DACs, and (2) I2S interfaces.

Fig. 7.2 Prototype of an embedded system for AI-driven cryptocurrency mining optimization

TENSILICA'S Processor Wroom32 DevKit has 25 GPIOs in total, of which a small number ({GPIO 34, GPIO 35, GPIO 36, GPIO 39}) are input-only pins; Not every pin has input pull-up; in order to use these pins as input pull-up, you must use an external pull-up.

4. Serial Communication

Three serial ports are present on Tensilica's processor. Initially, programming is done via Serial RX0, TX0, {GPIO3 (U0RXD), GPIO1 (U0TXD)}

There is an additional Serial port on { GPIO16 (U2RXD), GPIO17 (U2TXD)}.

I2C: When using the Arduino IDE with the processor, make use of the TENSILICA'S PROCESSOR I2C default pins (supported by the Wire library):[GPIO 21 SDA, GPIO 22 SCL]It is possible to configure any GPIO as an SPI interrupt..

Turn on (EN): The 3.3V regulator's enable pin is called Enable (EN). To turn off the 3.3V regulator, connect to ground since it is pushed up. This implies that you can restart your Tensilica processor by using this pin that is attached to a pushbutton [10].

This Table 7.1 displays a microcontroller's (ESP32, for example) default GPIO pin mapping for SPI communication. MOSI (data from master to slave), MISO

Table 7.1 Default SPI GPIO pin mapping table

Signal	VSPI GPIO	HSPI GPIO
MOSI	GPIO 23	GPIO 13
MISO	GPIO 19	GPIO 12
CLK	GPIO 18	GPIO 14
CS	GPIO 5	GPIO 15

(data from slave to master), CLK (clock signal), and CS (chip select) are the four main signals used by SPI. There are two SPI buses available:

- **VSPI:** GPIO 23 (MOSI), 19 (MISO), 18 (CLK), 5 (CS)
- **HSPI:** GPIO 13 (MOSI), 12 (MISO), 14 (CLK), 15 (CS)

VSPI and HSPI operate independently, offering flexibility when connecting multiple SPI devices. The default GPIO pins can be reassigned if needed for specific applications.

5. Result and Discussion

This Fig. 7.3 hardware configuration uses a microcontroller (such as an ESP32) coupled to an LCD screen to show current cryptocurrency prices, such as ETH: $2242.39. The microcontroller updates the display by using Wi-Fi to retrieve data from a cryptocurrency API. The system's components are connected via wires, demonstrating a useful IoT-based cryptocurrency price monitoring system.

Fig. 7.3 IoT-based real-time cryptocurrency price display system

This image, labeled Fig. 7.4: Bitcoin Data Mining," shows an IoT-based cryptocurrency price monitoring system. The setup includes a microcontroller (e.g., ESP32) and an LCD display, interfaced via GPIO pins. Using Wi-Fi, the system fetches real-time BTC (Bitcoin) price data, displaying "BTC: $30,272.211." A numerical value "5" might indicate refresh cycles or program states. Compared to the earlier ETH price display, the system likely cycles through multiple cryptocurrencies, dynamically updating prices via an API. This versatile, low-cost system efficiently monitors real-time crypto trends, demonstrating a practical prototype for IoT-based cryptocurrency data tracking and mining applications [11].

Fig. 7.4 IoT-based cryptocurrency price monitoring system with real-time BTC data

A microcontroller-based device, most likely with an LCD display and a TensilicaXtensa CPU similar to the ESP32, is depicted in Fig. 7.5: Tron Coin Data Mining. The display reads "1 RON: $0.05," which could be related to the Tron (TRX) cryptocurrency and indicates data mining results or real-time value tracking.

Fig. 7.5 Tron coin data mining display using microcontroller-based system

This configuration demonstrates an effective, low-power method for Tron coin value monitoring or data mining. While the display shows the earnings or conversion rate, the microcontroller performs calculations. Such technologies revolutionize crypto data mining for small workstations and show how microcontrollers and artificial intelligence (AI) optimize mining processes by highlighting low-cost, energy-efficient mining options.

PC Display of Data Mining Fig. 7.6, which shows a workstation screen displaying cryptocurrency mining data. The screen shows important metrics, such as:

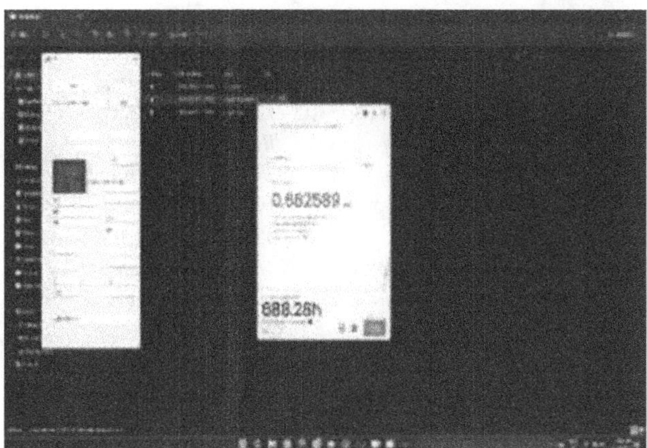

Fig. 7.6 PC interface display of cryptocurrency mining metrics

- Present Mining Output: The number 0.662589, which most likely denotes cryptocurrency that has been mined (such as TRX, DOGE, or something comparable).
- Uptime of the System: shown as 688.28 hours, highlighting system efficiency and showing a longer operating time.
- User Interface: Mining program looks light, emphasizing AI-optimized and efficient performance for workstation settings.

The title is supported by this graphic, which illustrates how Tensilica processors and AI-driven algorithms transform workstation mining. It makes cryptocurrency mining accessible and effective with regular PC hardware by emphasizing sustained performance, energy efficiency, and improved data presentation.

6. Conclusion

The solution is made to mine cryptocurrencies on laptops with sluggish processors. In this research, we proposed a system that may maximize miners' cryptocurrency mining profitability by automatically selecting the most profitable mining coins and keeping an eye on the connection between miners and the bitcoin mining pool, as a laptop or PC There has been discussion of the pertinent literature on selecting a bitcoin miner. In order to make the current recommendation system suitable for data mining in one of the biggest bitcoin mining markets, we have offered a tutorial on how to make it better. We demonstrated our system's user interfaces. Experiences have shown that, in contrast to the extremely powerful PC/laptop utilized in data mining, our computerized approach may efficiently maximize mining earnings by boosting data mining. Additionally, after being separated, our system can reconnect to the processing pool in an average of two seconds. We intend to test our automated approach on several Bit currency mining pools in the future. We have incorporated the data mining revenue into the suggested system as part of our regular operations. like gasoline for e-charging, toll payment, and AI-assisted automatic detection.

7. Future Scope

This covers more uses of AI in the fields of healthcare, gaming, finance, social media, data security, travel & transportation, and the automobile industry. Additional development will enable the program to run on different operating systems (such as Microsoft Windows) and CPU types (such as AMD-developed computer processors, ARM processors in the system, etc.) and further record the environment's hardware and software characteristics. Because they captured behaviour that would point to the presence of the malware under study in the supervised environment, we were able to identify the hardware usage events that were crucial to our objective using a set of characteristics we defined. Additionally, with encouraging initial results, we created a machine learning-based technique for identifying cryptocurrency miners using this data as a proof of concept. In order to apply machine learning to either find a single, comprehensive model or create customized models for various operating systems, typical hazards, etc., we also wish to broaden our data collection to include additional examples of crypto miners.

Acknowledgment

T. Archana extends sincere thanks to Saveetha Engineering College for providing the necessary resources and support to carry out this research work. Special gratitude is expressed to the faculty and staff for their guidance and encouragement throughout the project.

References

1. Kalanandhini, G. (2022). *AI based crypto mining for traditional workstation systems.* International Journal of Early Childhood Special Education, 14(3).
2. Nakamoto, S. (2008). *Bitcoin: A peer-to-peer electronic cash system.*
3. Karame, G. O., Androulaki, E., &Capkun, S. (2012). *Double-spending fast payments in bitcoin.* In *Proceedings of the ACM Conference on Computer and Communications Security* (pp. 906–917).
4. Heilman, E., Kendler, A., Zohar, A., & Goldberg, S. (2015). *Eclipse attacks on bitcoin's peer-to-peer network.* In *Proceedings of the 24th USENIX Conference on Security Symposium* (pp. 129–144).
5. Decker, C., &Wattenhofer, R. (2014). *Bitcoin transaction malleability and MtGox.* In *ESORICS 2014: 19th European*

Symposium on Research in Computer Security (pp. 313–326). Springer International Publishing.

6. Maria, A., Aviv, Z., & Laurent, V. (2017). *Hijacking bitcoin: Routing attacks on cryptocurrencies.* In *2017 IEEE Symposium on Security and Privacy* (pp. 375–392).
7. Eyal, I., &Sirer, E. G. (2014). *Majority is not enough: Bitcoin mining is vulnerable.* In *Financial Cryptography and Data Security: 18th International Conference* (pp. 436–454). Springer Berlin Heidelberg.
8. Sapirshtein, A., Sompolinsky, Y., & Zohar, A. (2016). *Optimal selfish mining strategies in bitcoin.* In *Financial Cryptography and Data Security: 20th International Conference, FC2016* (pp. 515–532). Springer Berlin Heidelberg.
9. Nayak, K., Kumar, S., Miller, A., & Shi, E. (2016). *Stubborn mining: Generalizing selfish mining and combining with an eclipse attack.* In *2016 IEEE European Symposium on Security and Privacy* (pp. 305–320).
10. Eyal, I. (2015). *The miner's dilemma.* In *Proceedings of the 2015 IEEE Symposium on Security and Privacy* (pp. 89–103). IEEE Computer Society.
11. Bonneau, J., Miller, A., Clark, J., Narayanan, A., Kroll, J. A., &Felten, E. W. (2015). *SoK: Research perspectives and challenges for bitcoin and cryptocurrencies.* In *2015 IEEE Symposium on Security and Privacy* (pp. 104–121).

Note: All the figures and tables in this chapter were made by the authors.

Advances in Additive Manufacturing Technologies – Gurusamy Pathinettampadian et al. (eds)
© 2026 Taylor & Francis Group, London, ISBN 978-1-041-16687-0

8 Energy-Efficient Lighting and Power Management in Trains with Embedded System

Vanitha M.[1], **Lokesha B.**[2]
Department of Electronics and Communication Engineering,
Saveetha Engineering College,
Chennai, India

Archana T.[3]
Department of Electronics and Communication Engineering,
Department of ECE
Chennai, India

Monashri G.[4]
Department of Electronics and Communication Engineering,
Saveetha Engineering College,
Chennai, India

◆ **Abstract:** This study offers a fresh approach to the pressing issues of energy use and sustainability in the rail sector by leveraging integrated technology for more effective lighting and power control onboard trains. The study thoroughly examines the complex architecture of these systems in order to provide insight into new developments in energy saving, adaptive control systems, and lighting technology. These technologies provide effective channels for control and communication, making it simple to integrate them into the current rail system. The results of extensive testing show a remarkable boost in efficiency, with considerable reductions in energy usage, expenditures, and carbon emissions. Embedded technologies offer the potential to improve the environmental friendliness and economic viability of rail transportation by optimizing energy usage, lowering operating costs, and limiting environmental effect.

◆ **Keywords:** Power management, Energy consumption, Sustainability, CAN, Ethernet

1. Introduction

For modern railway systems to maximise efficiency, cut costs, and improve the environment, energy-efficient lighting and electrical management in trains are crucial [1]. Rising energy prices and widespread environmental concerns are forcing the rail sector to come up with creative ways to accommodate the growing demand for rail travel while reducing its negative effects. One innovative and useful way to address these problems head-on is through embedded systems [2]. The convenience, cost, and sustainability of rail travel benefit commuters, businesses, and the environment [3]. However, because of the industry's intrinsic energy needs, major changes are needed. A continuous power source is necessary for many equipment, both for passengers and cargo. Many electronics for both passengers and cargo need a steady power source. Lighting, climate control, propulsion, and auxiliary services account for the majority of energy allocation for train operations [4]. In light of these problems, this study explores a novel

[1]vanitha@saveetha.ac.in, [2]lokeshabaskaran03@gmail.com, [3]archana@saveetha.ac.in, [4]monashrigunasekaran@gmail.com

DOI: 10.1201/9781003685906-8

approach to enhancing train energy efficiency by utilizing embedded technology [5]. Devices occasionally employ embedded systems, which are more specialized, smaller CPUs, to carry out specific tasks. The rail transportation sector has significantly decreased the energy expenses related to power management and lighting by implementing these technologies.

The pressing need to lessen rail transportation's inefficiency, operational costs, and environmental impact served as the driving force behind this study [6]. Long-term economic growth and the sustainability of transportation depend on railroads. Businesses are under more pressure to become more environmentally friendly and energy-efficient due to tighter environmental rules and a greater determination to fight climate change [7]. Gaining a better understanding of how embedded technology can be utilized to improve train power and illumination systems' efficiency is the study's main goal [8]. The main area of study is how these systems might improve energy efficiency, which could save money and lessen their negative effects on the environment. The relevance of these embedded systems, together with their architecture, components, and operational procedures, are further examined in the context of larger railway operations.

For safety and aesthetic reasons, lighting is crucial in modern train systems [9]. All passengers' safety, comfort, and enjoyment depend on illumination. It is well known that conventional train lighting systems are ineffective and unable to provide sufficient illumination. This research looks at creative lighting technologies and control schemes that could enhance the effectiveness and quality of train compartment lighting while preserving energy when integrated with embedded systems [10]. The process of allocating and controlling the flow of electricity throughout a railway system and its many parts is known as power management. By optimizing power distribution and consumption, embedded systems can significantly lower energy usage by channeling electricity only where and when it is needed. This study looks at efficient power management techniques and how railroads can use them.

This study provides initial indications that embedded solutions for power management and energy-efficient lighting could open the door to a more environmentally friendly and financially feasible rail transportation future. The use of Internet of Things (IoT) technology can greatly improve the automation and monitoring of energy-efficient train systems. Real-time data collected by Internet of Things (IoT) sensors enables responsive and flexible power management. This integration enhances the efficiency and environmental friendliness of rail transit by permitting dynamic adjustments in response to factors such as weather conditions, commuter volume, and system functionality.

2. Literature Review

P. SatheeshKumar et al [11] that the proposed research makes use of High Performance Multi-core Embedded Processors (MCEP) in order to improve train control and communication systems. This system may eliminate the need for human intervention in the following train operations: rail fracture detection, rail distance monitoring, compartment surveillance, fire and smoke detection, and human control of train components. The proposed Train Automation (TA) system employs wireless technologies coordinated by the Electric Locomotive Engine (ELE) to permit real-time monitoring and efficient administration of train characteristics. To achieve this, we integrate ultrasonic techniques based on high-energy lasers with Wireless Sensor Nodes (WSN). TA could be utilized in a number of ways to enhance train operations and safety, including quicker data processing, reduced human error, and increased energy efficiency.

M. Renz et al [12] that using a neuromorphic processor, we present a novel method for performing energy-efficient deep learning on synthetic aperture radar (SAR) images in this study. The study employs state-of-the-art neuromorphic computing models and hardware to demonstrate that computation can be conducted with significantly less power consumption than typical embedded processors. Using the True North neurosynaptic processor, a deep convolutional neural network (DCNN) is trained to classify SAR images. The research examines several DCNN design parameters to determine the optimal speed-accuracy tradeoff. When applied to a SAR image dataset, the proposed method maintains 95% classification accuracy at 1,000 images per second with a 20-fold reduction in energy consumption per image classification compared to the most energy-efficient embedded processors available today.

M. Kang et al [13] that the study examines the application of machine learning in embedded devices and emphasizes the benefits of periphery computing in terms of quicker response times and reduced network load. The research investigates the power limitations that prohibit the use of conventional Central Processing Units (CPUs) when programming neural networks for embedded devices. GPUs have been favored over CPUs for artificial intelligence computation, but their minimal power consumption precludes their use in embedded systems. Neuromorphic processing processors, such as the NM500 created with NeuroMem technology, are also being investigated as a potential solution to these issues. A pedestrian image identification system constructed on an embedded device utilizing the NM500 processor is presented, and its energy efficiency is compared to that of GPU-accelerated and multicore CPU systems. In addition to highlighting areas for development, experimental results demonstrate that the NM500 has enhanced energy efficiency for learning and classification tasks.

Salkuti et al [14] that the ystudy, a precision management of energy transfers optimizes the efficacy of electrified railway lines. Energy storage, renewable energy sources such as wind and solar photovoltaic systems, and energy exchange with the grid are all of interest. The primary objective is to reduce the operating costs of electrified railways. Included are external power generation, RER power, storage system power (batteries and supercapacitors), and earnings from selling excess power to the main electrical infrastructure. We approach it as a restricted optimal power flow problem involving alternating current. The study optimizes using the GAMS CONOPT solver and integrates probability distribution functions to account for wind and solar PV output uncertainty. Using a meta-heuristic differential evolution (DEA) technique, simulation results are evaluated.

A. Kalam et al [15] examines how advanced digital technology and comprehensive data analysis can be integrated into the railway system as part of Industry 4.0 to create a dynamic and adaptable work environment. This paper examines how digitizing railways could enhance operational efficiency, asset maintenance, and customer service via the Internet of Things. The objective is to develop an IoT ecosystem that can communicate on multiple levels with the Railway OCC. Interoperability, data integrity, and cybersecurity are highlighted as obstacles to the implementation of IoT-enabled railway systems. Also discussed are the technical and methodological aspects of IoT research and validation via Smart Grid use cases. This work ultimately demonstrates the need for additional IoT platform research and development to influence future smart railway operations.

3. Proposed Work

A strong architecture and component set are necessary for an embedded system that controls power and lighting aboard trains. Here, we'll look at the system's architecture and the key elements that enable it to work. Understanding this design is crucial to understanding the role embedded technologies play in the train's improved energy efficiency. "Brain" refers to the central processing unit (CPU) of an embedded system. Data processing, control algorithm execution, and system-wide coordination are all handled by the central processing unit (CPU). It is important because it optimizes and controls energy use by distributing power to appliances and lighting as needed. Through the use of a sensor network, embedded systems gather data about the train's immediate surroundings. Energy usage, motion, temperature, and light are all measured by appropriate sensors. Depending on the light levels that sensors detect, the train's interior illumination can be changed. Temperature sensors contribute to a comfortable indoor environment by adjusting climate control systems. Installing motion detectors to switch off the HVAC and lights in vacant rooms can reduce utility expenses. The part of the system in charge of converting control signals into motion is the actuator. One crucial feature of embedded systems is their communication with data servers and other devices. The system architecture is shown in Fig. 8.1.

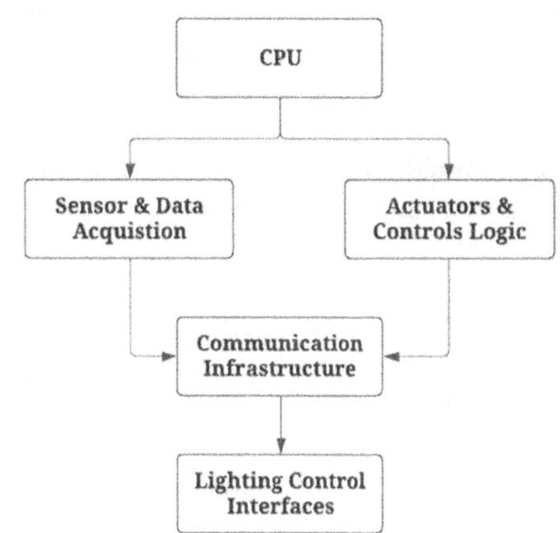

Fig. 8.1 System architecture

Typically, this is accomplished via a dedicated user interface. Other data sources, such as weather forecasts and real-time passenger counts, may influence illumination and energy management decisions. Intelligent software components called control algorithms direct the decision-making processes of an embedded system. Based on sensor data, algorithms instantaneously adjust the display's luminance, power distribution, and other parameters. They ensure the adaptability of the system to change. Train operators and maintenance personnel can monitor and interact with the embedded system via a user interface. The embedded system under investigation integrates dynamic illumination controls, motion sensors, and regenerative braking technology. The system incorporates daylight harvesting, adjusts illumination dynamically in response to passenger activity, and distributes electricity efficiently by utilizing kinetic energy generated during deceleration. In order to achieve maximum energy efficiency, cutting-edge control algorithms ensure real-time decision-making. By facilitating remote monitoring and maintenance, the system promotes human engagement, enabling manual adjustments that enhance reliability and minimize operational interruptions. By optimizing energy efficiency in train environments, this comprehensive approach simultaneously enhances passenger contentment and reduces consumption [5–8].

3.1 Energy-Efficient Lighting Technologies & Controls

Widespread use of energy-efficient illumination and controls in railway embedded systems designed to increase energy efficiency. These technologies and control systems are essential for reducing energy consumption without impairing the visibility of passengers and conductors. Utilizing cutting-edge lighting technology is crucial for effective train illumination. These innovations include light-emitting diodes (LEDs), solid-state lighting, and intelligent lighting systems, among others. LED lights have recently acquired in prominence due to their extended lifecycle, high efficiency, and adaptability in producing a variety of illumination effects. LEDs are more efficient than incandescent and fluorescent lighting because more of the energy they consume is converted into visible light. Integrated energy-saving technologies could potentially regulate the luminosity of moving railway lighting. These controls alter the lighting's intensity, color, and pattern to the ambience of the room and the preferences of the user. When departing a tunnel and entering a more open location, the system could possibly autonomously adjust the illumination to the new environs.

To make everyone feel more at ease, you can adjust the color temperature and use different illumination patterns to indicate day or night. Motion sensors are a crucial component of modern, energy-efficient train illumination systems. These sensors can precisely pinpoint the locations of passengers and employees throughout the vehicle. This real-time control saves money by turning off lights when they are not in use, without sacrificing safety or comfort. Using a feature dubbed "daylight harvesting," the amount of light on trains could be reduced. Sensors continuously monitor the quantity of natural light entering the train through its windows, and the interior illumination system adjusts accordingly. When there is sufficient natural light, the artificial illumination is attenuated or turned off completely. This not only creates a more pleasurable and eco-friendly illumination environment, but it also reduces energy consumption. Embedded systems employ sophisticated control algorithms that account for a variety of variables, including sensor data, time of day, and passenger traffic [9–11].

These algorithms are designed to make decisions regarding illumination parameters in real time, achieving a balance between energy efficiency and passenger comfort. Even though many lighting control operations are automated, users typically have the option to override the system. This versatility is advantageous for both the personnel and the passengers. If a guest wishes to read in bed, they may alter the room's lighting. Train operators should always be prepared to assume control in the event of an emergency or unexpected event. On trains with low-energy lighting and controls, the comfort and safety of passengers are significantly enhanced. Utilizing modern lighting technologies such as LEDs and incorporating dynamic control mechanisms allows these systems to provide a flexible and adaptable approach to lighting management. This results in less energy consumption and lower operating costs, which is excellent news for all train passengers and employees. Figure 8.2 depicts the light controls diagram [12].

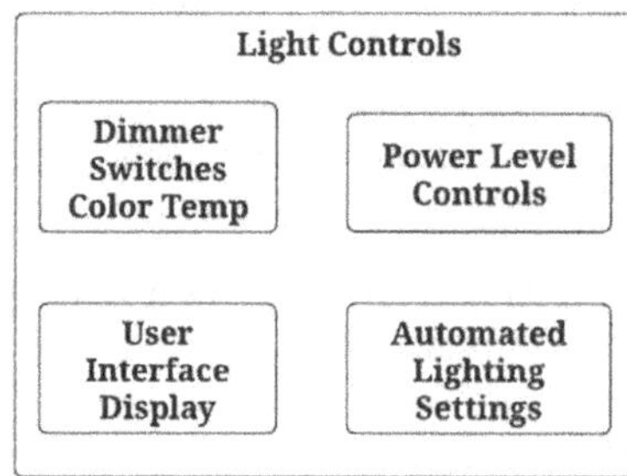

Fig. 8.2 Light control diagram

3.2 Power Management Techniques

To make trains more energy-efficient, embedded technology with enhanced power management systems is required. Regenerative braking systems are an ingenious method to reduce a train's energy consumption and operating costs without compromising its dependability or performance. Utilizing a regenerative braking system permits the recovery of kinetic energy that would otherwise be lost as heat during deceleration. Kinetic energy is lost as waste heat during friction-based deceleration, whereas regenerative braking systems absorb and store that energy. Regenerative braking is used to produce electricity from the kinetic energy of a train. This energy could be used to fuel the ship's lighting and auxiliary systems. The primary components of a regenerative braking system are batteries or capacitors, inverters, and traction motors. When train personnel apply the brakes, the regenerative braking system is also activated. When the system detects that the train's speed is becoming excessive, instead of manually applying the brakes, the traction motors are converted to generator mode. These generators use the motion of the railway to produce electricity [13].

Using regenerative braking systems has a number of significant benefits for the energy efficacy of trains: Regenerative systems can recover and utilize the kinetic energy lost during deceleration, resulting in significant energy savings. The outcome is enhanced train efficiency. By minimizing their energy costs, train companies can compete more effectively with alternate modes of transportation. Since they reduce greenhouse gas emissions and the environmental impact of rail transportation, regenerative braking systems are consistent with sustainability objectives and environmental criteria. These technologies, which enhance energy efficiency, enable trains to travel further on the same quantity of petroleum. Figure 8.3 depicts the regenerative breaking process [14].

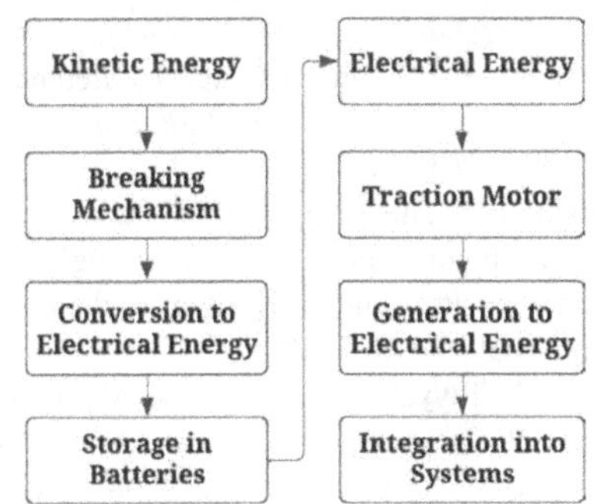

Fig. 8.3 Regenerative braking process

3.3 Communication & Control Mechanisms

Improved train lighting and power management may be the outcome of communication and control. The intelligence of an embedded system optimizes and coordinates its numerous components by performing computations and analysis in real time. We examine the complex internal mechanisms of these systems and stress how crucial they are to the rail transportation sector's operational creativity and energy efficiency. Reliable connectivity between internal and external nodes is essential for information transmission in railway embedded systems. This network supports both wired and wireless connections. CAN and Ethernet data links enable fast data transfer between onboard components. Wi-Fi and cellular networks make it fun to connect to control centers, passenger interfaces, and the outside world. Sensor data is essential to energy-efficient train control and communication systems. Many rail networks have command centers that offer real-time train monitoring. These centralized hubs collect data from embedded devices on many trains to assist operators in becoming more efficient. The network's energy-saving objectives can be met by remotely adjusting the train's lighting and power management systems. Options for connectivity make it possible to remotely monitor and maintain embedded devices. Remote software enables train operators and maintenance personnel to remotely monitor performance, identify problems, and execute repairs. This feature increases dependability, lowers maintenance costs, and prolongs the life of embedded components. Because of its remarkable scalability, the suggested method can be easily adjusted to suit different kinds of trains and transportation networks. Because of its modular architecture, it may be easily integrated into a variety of transportation infrastructures, which promotes scalability and broad adoption [15].

3.4 Integration with Existing Train Systems

The incompatibility of energy-efficient equipment with current railway networks is a significant challenge. Many more sophisticated railway systems and technology are in use on rail networks. For embedded systems to interact with electricity infrastructures and light bulbs, these differences must be supported. The success of integration depends on scalability and modularity. High-speed, regional, and freight train combinations should all be supported by embedded systems. Railroad operators can gradually replace outdated equipment and lower energy consumption throughout their fleet with the help of modules and scalable solutions. Embedded systems communicate with their internal and external components via standard protocols. Communication between the train's embedded system, sensors, and illumination control systems is made possible by CAN bus, Ethernet, and industrial automation standards.

By integrating smart grid technology with the train's power management system, a more responsive and dynamic power distribution system may be achieved. Energy use that is closely matched with current demand and availability will result in an overall increase in efficiency.

4. Results

Several critical performance indicators increased as trains adopted embedded technology for more effective illumination and power control. The extensive testing period revealed a significant decrease in energy consumption and associated operational expenses. Once the system was embedded, energy consumption decreased from 1200 kWh to 750 kWh, as shown by the data. A 37.5% reduction in energy consumption represents a significant improvement in energy efficiency. If annual expenditures were reduced from $10,000 to $5,600, $37,880 would be saved. The use of embedded technology also reduced carbon dioxide emissions significantly. The decrease in emissions from 1800 kg to 1125 kg reduced their environmental impact by 37.5% as shown in Fig. 8.4 and Table 8.1. Reduced carbon dioxide emissions demonstrate the technology's positive impact on the environment and support sustainability objectives.

Table 8.1 Comparison of the metrics before and after integration

Metric	Before Integration	After Integration
Energy Consumption (kWh)	1200	750
Cost Savings ($)	9000	5600
CO2 Emissions (Kg)	1800	1125

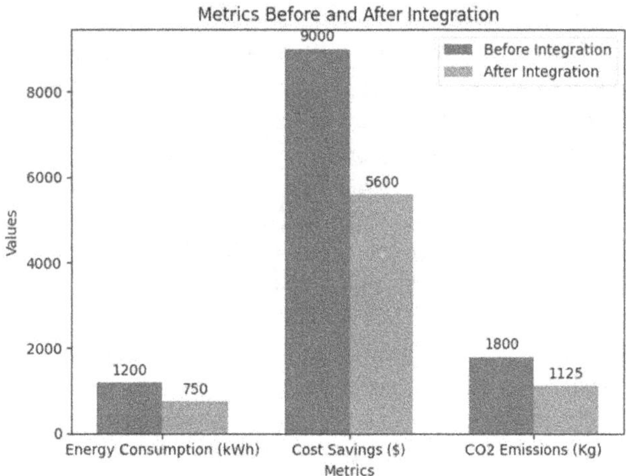

Fig. 8.4 Metrics comparison before and after integration

5. Results

In order to make rail travel more economical and environmentally friendly, embedded solutions for effective lighting and energy management on board have been created. These embedded systems have proven to be able to significantly cut energy usage through the use of advanced illumination technologies, dynamic controls, and regenerative braking systems. This has led to significant cost savings and a noticeable decrease in carbon emissions. This study shows that technological innovation is crucial to creating a future for rail transportation that is both economical and environmentally friendly. The embedded system will use cutting-edge connectivity and communication technologies in the future. It seeks to maximize operating efficiency by promoting data exchange across train components, infrastructure, and control systems; this will allow for better coordinated and energy-efficient power and illumination management.

References

1. Naderi H and Mirabadi A, "Railway Track Condition Monitoring using FBG and FPI Fiber Optic Sensors", The Institution of Engineering and Technology International Conference On Railway Condition Monitoring, pp. 198–203, 29–30; Nov. 2006.
2. D. Morgan, "Deep Convolutional Neural Networks for ATR from SAR Imagery", Proceedings SPIE 9475 Algorithms for Synthetic Aperture Radar Imagery XXII, May 2015.
3. H. Wang, S. Chen, F. Xu and Y. Q. Jin, "Application of Deep-Leaming Algorithms to MSTAR Data", Proceedings of 2015 IEEE International Geoscience and Remote Sensing Symposium (IGARSS), Jul. 2015.
4. Y. Jia, E. Shelhamer, J. Donahue, S. Karayev, J. Long, R. Girshick, et al., "Caffe: Convolutional architecture for fast feature embedding", ACM Multimedia, 2014.
5. S. Esser et al., Convolutional Networks for Fast Energy-Efficient Neuromorphic Computing, May 2016.
6. J. Sawada et al., "Truenorth ecosystem for brain-inspired computing: scalable systems software and applications", Proceedings of the International Conference for High Performance Computing Networking Storage and Analysis (SC'16), Nov. 2016.
7. W. Wang and J. Gang, "Application of convolutional neural network in natural language processing", International Conference on Information Systems and Computer Aided Education (ICISCAE), 2018.
8. H. Raji, M. Tayyab, J. Sui, S. R. Mahmoodi and M. Javanmard, Biosensors and machine learning for enhanced detection stratification and classification of cells: A review, 2021.
9. A. Fouman Ajirlou and I. Partin-Vaisband, "A machine learning pipeline stage for adaptive frequency adjustment", IEEE Transactions on Computers, 2021.
10. D. Tripathy, A. Abdolrashidi, L. N. Bhuyan, L. Zhou and D. Wong, "Paver: Locality graph-based thread block scheduling for gpus", ACM Transactions on Architecture and Code Optimization (TACO), 2021.
11. P. SatheeshKumar, "An efficient way of monitoring and controlling the train parameters using Multi-core Embedded Processors (MCEP)," 2010 3rd International Conference on Computer Science and Information Technology, Chengdu, China, 2010, pp. 552–555.
12. M. Renz and Q. Wu, "An energy-efficient embedded implementation for target recognition in SAR imageries," 2017 IEEE Symposium Series on Computational Intelligence (SSCI), Honolulu, HI, USA, 2017, pp. 1–5.
13. Kang, M.; Lee, Y.; Park, M. Energy Efficiency of Machine Learning in Embedded Systems Using Neuromorphic Hardware. Electronics 2020, 9, 1069.
14. Salkuti, S.R. Optimal Operation of Electrified Railways with Renewable Sources and Storage. J. Electr. Eng. Technol. 16, 239–248 (2021).
15. Kalam, A., Peidaee, P. (2022). IoT Enabled Railway System and Power System. In: Marati, N., Bhoi, A.K., De Albuquerque, V.H.C., Kalam, A. (eds) AI Enabled IoT for Electrification and Connected Transportation. Transactions on Computer Systems and Networks. Springer, Singapore.

Note: All the figures and tables in this chapter were made by the authors.

Advances in Additive Manufacturing Technologies – Gurusamy Pathinettampadian et al. (eds)
© 2026 Taylor & Francis Group, London, ISBN 978-1-041-16687-0

Studies on Mechanical, and Thermal Properties of Melamine Polyethylene Foam Reinforced Hybrid Sandwich Composites with FEA Simulation

K. S. Jai Aultrin
Department of Marine Engineering,
Noorul Islam Centre for Higher Education, Kumaracoil, India

C. Mahil loo Christopher
Department of Mechanical Engineering, DMI Engineering college,
Aralvaimozhi, India

D.S. Jenaris, T. Livingston
Department of Mechanical Engineering,
PSN Engineering College, Tirunelveli, India

N. Ramanan
Department of Mechanical Engineering,
Sri jayaram Institute of Engineering and Technology, Tiruvallur, India

P. Gurusamy[1]
Department of Mechanical Engineering,
Chennai Institute of Technology, Chennai

Vivekananthan A.[2]
PG Student, Department of Mechanical Engineering,
Chennai Institute of Technology, Chennai

◆ **Abstract:** This research used hand lay-up to make hybrid sandwich composites with expanded polyethylene foam, Melamine foam, and Kevlar fiber as the skin. Tensile, flexural, impact, hardness, and dual shear tests were done per standards. To test the prototype, Finite Element Analysis (FEA) simulation was used. To determine hybrid composites' "thermal properties," thermo gravimetric analysis was done. Results show that melamine foam sandwich composites are stronger than expanded polyethylene composites, whereas hybrid composites are stronger. The tensile, flexural, impact, and inter-delamination test results show that melamine foam composite is stronger than expanded polyethylene composite foam and hybrid composite foam. Hybrid composites outperform expanded polyethylene and POLYESTER composites in double shear and Shore-D hardness testing. By loading the die assembly for melamine foam casting, FEA simulations indicated that nodal solution had the highest deformation of 0.036244 m. Melamine foam appears to be thermally more stable than expanded polyethylene foam. Expanded polyethylene foam with Melamine foam is somewhat more thermally stable. SEM examines intermolecular structures. Foam composites can provide lightweight thermal and acoustic insulation for airplane cabin walls, floors, and ceiling panels. Their thermal resilience makes these composites ideal spaceship insulation materials for shielding delicate electronics from severe space temperatures.

◆ **Keywords:** Sandwich hybrid composites, Expanded polyethylene foam, Melamine foam, Mechanical and FEA simulation

Corresponding authors: [1]gurusamyp@citchennai.net, [2]ananthanvivek300@gmail.com

DOI: 10.1201/9781003685906-9

1. Introduction

Fiber composites are used in aircraft, automobile, civil structural, and shipbuilding. The most prevalent materials are glass, Kevlar, and aramid. Common foams include melamine, expanded polyethylene, balsa wood, wax, and others. The hybrid composite uses two Kevlar fiber skin layers and intermediate core components. Melamine and expanded polyethylene foam were the cores. The hybrid sandwich composite's mechanical and thermal characteristics are evaluated. Chun Lu et al. [1] tested a composite honeycomb core using Kevlar fiber in a three-point bending test. Sandwiches crack at 6800N bending load. Finite analysis shows that the top plate has a higher stress distribution than the core and bottom plates. The top plate cracked at the surface plate-honeycomb junction, causing the structure to collapse. M. Meo et al. [2] impact-tested cycomprepag core material. The C-scan test shows that low-impact energy levels cause internal damage that weakens panels. J. Tirillo et al. tested Kevlar and basalt fibers [3]. Kevlar fiber has a ballistic limit velocity of 174 m/s and basalt fiber 285 m/s.

Qihui Chen et al. [4] tested melamine foam as a core material and found that glass chopped fiber was 45% tougher than the blank sample. Kevlar and aramid chopped fibers increased 33% and 45%, respectively. The glass chopped fiber modification improved impact toughness the most. Y. Hou et al. explored honeycomb foam bending for two gradient cellular sandwich beams [5]. Angle gradient con Figure ratio density is smaller. The angle-gradient centre-symmetric shape has superior specific flexural characteristics.

Vitaly Koissin et al. [6] tested the face layer's stability using sandwich beam edgewise compression. They found that the crushed core supports well and that analytical and experimental results agree. Yuxi Guo et al. [7] used thermal gravimetric analysis to study Kevlar fiber ash waste foaming. Polishing porcelain tile debris was a major raw source. Results demonstrated that porcelain polishing waste is a good precursor for glass ceramic foams. Glass foam foamed effectively with Kevlar fiber ash powder at higher temperatures. W. Wang et al. [8] fracture tested Melamine foam core material with glass fiber front layer. They determined lighter peel stoppers work better. Shear behavior of a composite sandwich beam with phenolic core and glass fiber skin was studied by A. C. Monalo et al. [9]. They found that edgewise positioning increased sandwich composite shear strength.

2. Materials and Methods

2.1 Materials

This sandwich composite employed epoxy resin, expanded polyethylene, and Melamine foams. Kevlar is used for skin. Expanded polyethylene and melamine foam are 3mm thick. Epoxy LY 556 and hardener HY 951 are combined 10:1.

2.2 Composite Fabrication

Figure 9.1 shows the expanded polyethylene composite plate (C1) material preparation procedure. Table 9.1 shows fabrication samples.

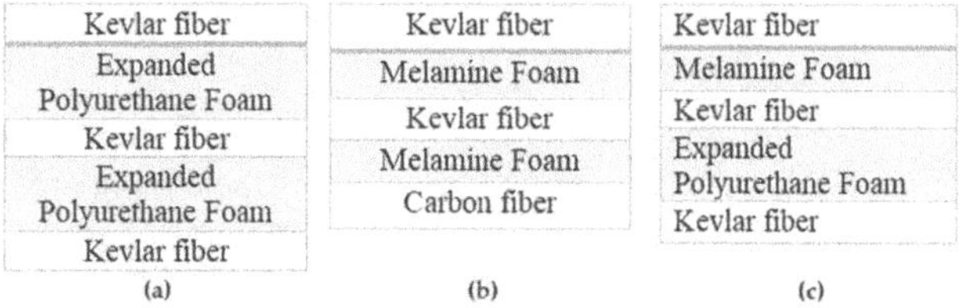

(a)
Kevlar fiber
Expanded Polyurethane Foam
Kevlar fiber
Expanded Polyurethane Foam
Kevlar fiber

(b)
Kevlar fiber
Melamine Foam
Kevlar fiber
Melamine Foam
Carbon fiber

(c)
Kevlar fiber
Melamine Foam
Kevlar fiber
Expanded Polyurethane Foam
Kevlar fiber

Fig. 9.1 a) Composite made of expanded polyethylene (Composite 1); b) Composite POLYESTER (Composite 2); and c) Composite Hybrid (Composite 3).

Table 9.1 Composite plate preparation

Composite	Samples	Foam inclination	Abbreviation
Expanded Polyethylene C1	1st	0^0	$C_1 - S_1$
	2nd	45^0	$C_1 - S_2$
	3ed	90^0	$C_1 - S_3$
POLYESTER C2	1st	0^0	$C_2 - S_1$
	2nd	45^0	$C_2 - S_2$
	3ed	90^0	$C_2 - S_3$
Hybrid C3	1st	0^0	$C_3 - S_1$
	2nd	45^0	$C_3 - S_2$
	3ed	90^0	$C_3 - S_3$

3. Results and Discussions

3.1 FEA Simulation

Figure 9.2 and 9.3 show the solid 88 type model as an axisymmetric element rotated or spun around its own axis. It can sustain 72.646 N/m2 maximum stress. The elemental solution shows 0.001267 m deformation, less than the nodal solution. This elemental solution's stress dropped to 0.129 N/m2. See Fig. 9.4 for the top side curvature ends' maximum stress and the sharp end's lower stress.

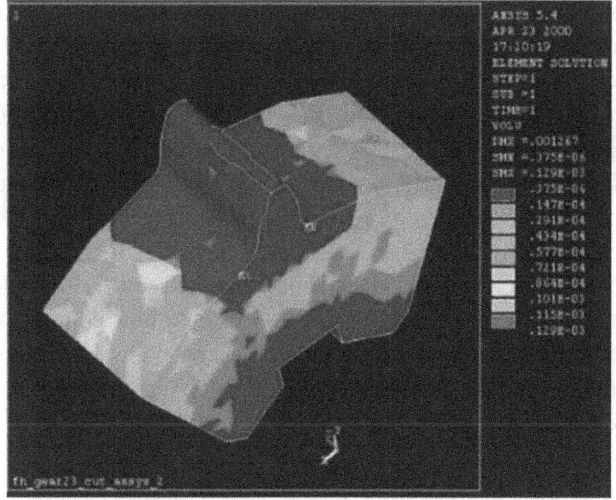

Fig. 9.2 Solid model of casting die component

Figure 9.4 shows the Mesh generation of testing specimen in Deform 3D for Hybrid composite.

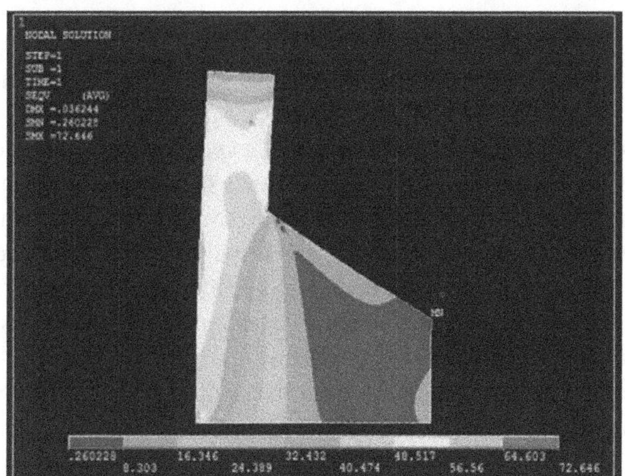

Fig. 9.3 Finite element analyses of casting die component in ANSYS

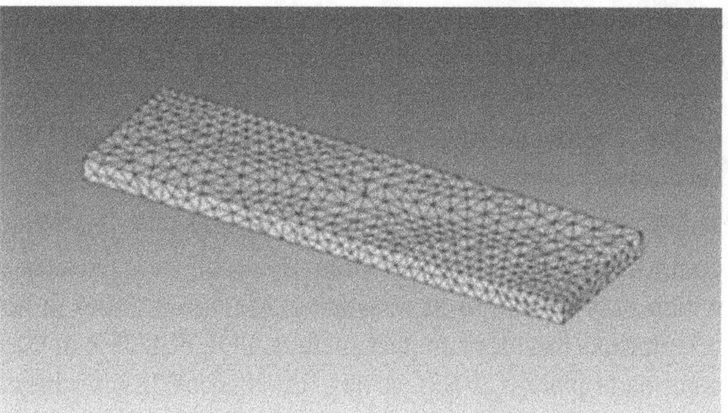

Fig. 9.4 Mesh generation of testing specimen in deform 3D for hybrid composite

3.2 Physico-Mechanical Properties

1) Tensile Strength Findings

The different observations of these mechanical properties are taken place in this Table 9.2. Here in this research outcome, it can be represented as curve lines of various output parameters after the investigations. The tensile strength of each sample is in the increasing level for the more form inclinations of 450 and 900 at concerned samples. More hardness has been observed for the 2nd sample of 450 form inclination. Both the ultimate tensile strength, tensile modulus and hardness observed good for the samples having 900 form inclinations. Better hardness values were obtained as 1100 VHN and 840 VHN for the samples having high amount of hybrid in nature with 450 and 900 form angles [3].

2) Hardness Test Result

Table 9.2 shows that expanded polyethylene in 0o composite plate (C1-S1) takes more 786 hardness load than other composites. Fibre in 0o improves the contact surface, allowing it to handle greater weight. Because

Table 9.2 The different observations of these mechanical properties

Composites	Samples	Foam Inclination	Abbr.	Ult. Tensile Strength (MPa)	Tensile Modulus (MPa)	Flexural Strength (MPa)	Flexural Modulus (MPa)	Impact Test (Joule)	Double Shear Test (MP)	Displace-ment (mm)	D Shore Hardness (VHN)
Polyethylene (C1)	1st Sample	0^{0}	C1-S1	45	400	1.8	`	3.9	60	6	850
	2nd Sample	45^{0}	C1-S2	88	750	3.2	30	3.7	120	5	920
	3rd Sample	90^{0}	C1-S3	72	760	1.7	8	3.6	70	10	760
polyester (C2)	1st Sample	0^{0}	C2-S1	54	630	1.7	13	4.5	30	4	880
	2nd Sample	45^{0}	C2-S2	75	780	2.2	21	3.2	60	5	1100
	3rd Sample	90^{0}	C2-S3	62	840	3.6	32	4.8	40	6	650
Hybrid (C3)	1st Sample	0^{0}	C3-S1	48	400	2.7	28	2.7	50	3	720
	2nd Sample	45^{0}	C3-S2	82	620	2.4	32	3	40	5	840
	3rd Sample	90^{0}	C3-S3	74	750	3.2	24	4.2	60	5	640

of decreased bonding between composite fibre layers, expanded polyethylene in 90o composite plate (C1-S3) takes 776 hardness loads less. Expanded polyethylene in 45o composite plate (C1-S2) can bear moderate 784 hardness.

The 0o composite plate (C2-S1) organized Melamine foam composite carries the maximum load of 794 hardness, which may be due to its resistance. Melamine foam in 0o orientations has more contact surface. POLYESTER in 90o composite plate (C2-S3) carries less load of 769 hardness because to fibre layer bonding loss, reducing load bearing ability. POLYESTER in the 45o composite plate (C2-S2) can bear 778 hardness light loads.

C3-S1 hybrid composite plate in 0° arrangement can resist 816 hardness load. Hybrid composite plates (C3-S1) at 16 degrees can endure 816 hardness units. This hybrid composite has a 16-degree angle due to fiber orientations of 0°, POLYESTER cohesiveness, and expanded polyethylene foam. The 90o hybrid composite plate (C3-S3) hybrids composite has a lower load of 793 hardness because to decreased bonding energy. A mild 798 hardness load is applied to the 45o hybrid composite plate (C3-S2) [7].

The D-Shore hardness test in 0o foam composite shows better hardness, which increases contact area and cohesiveness and depends on foam parameters, as shown in hybrid foam composites soft, while 45o Melamine foam composites are somewhat durable.

3) Flexural Test Results

Flexural characteristics are also compared to impact test and sample displacements. 3 hybrid composite (C3) samples were more flexural with standing capacity. The second and third samples (C2-S2) and (C2-S3) passed the impact test. Most flexural property peaks were seen in hybrid composite specimen (C3). The 28J and 24J impact test values exceeded the C3 hybrid composite specimen. Flexural strength was stronger for hybrid composite (C3) in most Impact findings. Displacement values were uniformly distributed from composites C2 to C3 [2].

4) Impact Test Results

Impact test specimens are made per ASTM: D256. Charpy impact tests were performed on all sandwich composites. Due to decreased composite fibre layer bonding, expanded polyethylene in 90o composite plate (C1-S3) takes 3.8 J less energy. Expanded polyethylene in 45o composite plate (C1-S2) needs 3.85 J mild energy. Melamine foams' resistance may explain why the 0o composite plate (C2-S1) organized Melamine foam composite consumes the most energy at 4.1 J. 0o orientations enhance composite contact surface. Melamine foam in 90o composite plate (C2-S3) consumed 3.9 J less energy due to reduced bonding between POLYSTER composite and C2-S3. Melamine foam in 45o composite plate (C2-S2) took 4.0 J mild energy. Figure shows that a hybrid composite plate (C3-S1) organized hybrid composite requires 4.05 J of energy.9. Fibre in 0o orientations and hybrid composite cohesiveness are the main factors. Due to reduced bonding between POLYSTER and Expanded polyethylene composite, hybrid composite foam in 90o composite plate (C3-S3) consumes 3.85 J less energy. Hybrid composite foam in 45o composite plate (C3-S2) took 3.95 J.

The impact test behavior of the 0o ordered foam composite has superior energy due to increased contact area, cohesiveness, and foam composite characteristics. The 45o-arranged foam composite has moderate energy, whereas the 90o-arranged foam composite has reduced energy.

5) Outcomes of Double Shear Test

Both sides of the sandwich composite structure receive shear load. Prepare the double shear test specimen per ASTM D5379. In the twofold shear test, expanded polyethylene in 0o composite plate (C1-S1) takes 475 N and 6.1 mm greater than other composites. Due to fibre orientation in 0o, contact area rises and Shear stress increases. Reduced bonding between expanded polyethylene fibre layers reduces load of 240 N and displacement of 2.1 mm in 45o composite plate (C1-S2). Extended polyethylene in a 90o composite plate (C1-S3) experiences mild stress of 270 N and displacement of 2.1 mm.

Compared to the others, the 45o composite plate (C2-S2) organized Melamine foam composite can handle 545 N and 5.1 mm displacement due to its resistance. A 45o orientation improved the POLYSTER composite's contact surface. POLYSTER composite in the 0o composite plate (C2-S1) can endure 525 N and 4.3 mm movement. In 90o composite plate (C2-S3), POLYSTER composite fibre layers bind less, reducing load by 485 N and displacement by 3.1 mm.

Research shows that a hybrid composite plate (C3-S1) can tolerate 640 N and 5.9 mm movement. This is due to fibre orientation in the 0° hybrid composite and POLYSTER and Expanded polyethylene foam cohesion. Hybrids in 90o composite plate (C3-S3) endure 380 N and 5.1 mm displacement. Hybrids in 45o composite plate (C3-S2) take 415 N and 5.1 mm displacement. 0o-oriented composite foams have shear performance due to increased contact area, cohesiveness, and foam material properties [6].

3.3 Thermal Characteristics

1) Thermo Gravimetric Analysis Test

Sandwich composite specimen thermal stability was tested with the TG/DTA 6200 SEIKO TGA analyzer. The percentage of weight loss is computed at 20o C/min from 0-1000o C. Samples are heated in nitrogen to avoid oxidation.

In a thermal gravimetric test, polyurethane composite plate loses 95% weight up to 325°C compared to other composites. Because fiber increases contact surface area, the body absorbs more heat. PVC composite foam composite is 24% lighter than other composite foams up to 800° C, possibly due to its resilience. Hybrid composites lose weight slower than polyurethane and PVC foam at 1000°C [8].

Polyurethane composite plate loses 95% of its weight up to 325°C in a thermo gravimetric test, compared to other composites. Because fiber increases contact surface area, the body can absorb more heat. PVC composite foam composite loses 24% less weight than conventional composite foam up to 800° C, possibly due to its durability. When temperatures surpass 1000°C, hybrid composites lose weight slower than polyurethane and PVC foam [4].

3.4 Morphological Analysis of Hybrid Polymer Composites

The expanding polyethylene foam breaks the composite's surface in the plate. Soft foam components shatter uniformly on their surfaces.

Melamine foam components have smooth fracture surfaces due to their material properties. This sandwich composite produces firmer Melamine foam than others. SEM images of Melamine foam composites demonstrate this.

Figure 9.5(a-b) shows the Melamine foam structure in the flexural test SEM picture. A few places on the Melamine foam composite plate showed porosity. Melamine foam and resin merge poorly in POLYSTER composites because of their properties. Sandwich composite materials can have asymmetrical load distribution. Due to uneven stress, sandwich composites are porous [9].

Figure 9.6 shows the impact test SEM of hybrid expanded polyethylene and Melamine foam. This impact test significantly damages foam materials due to sudden load application. Melamine foam's uneven fracture shows stiffness, whereas expanded polyethylene foam's porosity shows poor resin mixing. Sandwich construction Melamine foam is stronger and stiffer than expanded polyethylene foam. The rapid load causes this sandwich hybrid composite's more damaged foam components to show up in the impact test [5].

(a)

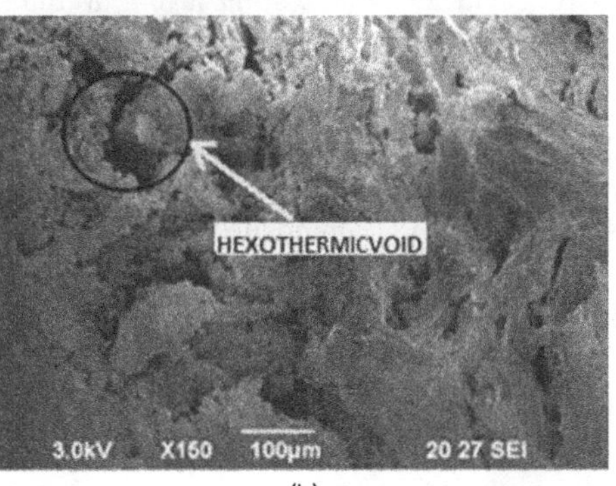

(b)

Fig. 9.5 (a-b). Melamine foam SEM picture

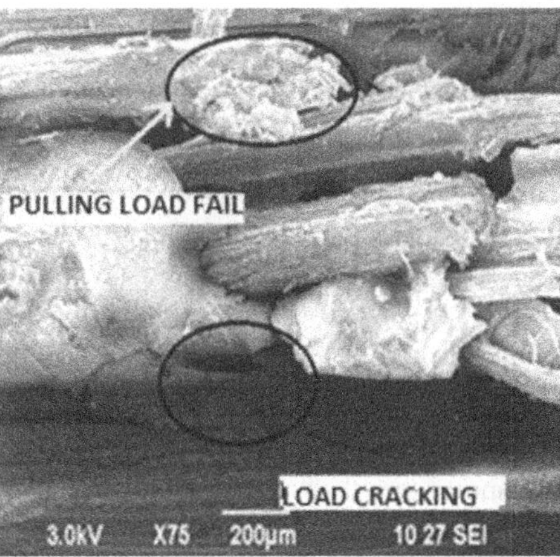

Fig. 9.6 Hybrid expanded polyethylene and Melamine foam SEM picture

4. CONCLUSIONS

Tensile, flexural, impact, and inter delamination tests show that Melamine foam composite is stronger than Expanded polyethylene composite foam and hybrid composite foam. Hybrid composites withstand twofold shear and De-shore hardness tests better than expanded polyethylene and

POLYESTER composites. Melamine foam is thermally more stable than expanded polyethylene foam. When coupled with Melamine foam, expanded polyethylene foam has somewhat better thermal stability. SEM examinations analyze intermolecular structures.

The FEA simulation also indicated that the die assembly loading conditions for melamine foam casting caused nodal solution to deform 0.036244 m. The solid 88-type model under the axisymmetric element, which rotates the item along its own axis, showed the highest stress level of 72.646 N/m2. The elemental solution shows 0.001267 m deformation, less than the nodal solution. Effective Von Mises stress and displacements mimic tensile, compressive, and flexural testing using a fine-meshed simulated specimen.

Compared to other composites, polyurethane composite plate loses 95% weight up to 325°C in thermogravimetric tests. Because fiber increases contact surface area, the body absorbs more heat. PVC composite foam composite is 24% lighter than other composite foams up to 800° C, possibly due to its resilience. Hybrid composites lose weight slower than polyurethane and PVC foam at 1000°C. These materials are suitable for heat shields and other thermal protection systems in re-entry vehicles and high-speed aircraft due to their thermal stability and insulation.

A scanning electron microscope picture of the enlarged polyethylene foam composite in the tensile test shows that its features have fractured the plate's surface. Soft foam components shatter uniformly on their surfaces. The sandwich composite materials' thickness makes blow holes apparent in this scanning electron microscope.

The Scanning Electron Microscope picture of the Melamine foam structure in the flexural test showed porosity in several sections of the composite plate. Due to POLYESTER composite properties, Melamine foam and resin combine poorly.

The Scanning Electron Microscope picture of hybrid expanded polyethylene and Melamine foam in the impact test shows that the "sudden applied load" severely damages the foam components. The porosity of expanded polyethylene foam shows the resin's poor mixing with it, whereas the uneven fracture of Melamine foam shows its stiffness. Sandwich Melamine foam is stronger and stiffer than expanded polyethylene foam. Due to sudden stress, this sandwich hybrid composite's foam components are more damaged in the impact test.

REFERENCES

1. Chun Lu, Mingyue Zhao, Liu Jie, Jing Wang, YuGao, Xu Cui, Ping Chen, "Stress Distribution On Composite Honeycomb Sandwich Structures Suffered From Bending Load", procedia Engineering, (2015) pp. 405–412
2. M. Meo, R. Vignjevic, G. Marengo, "The Response of Honeycomb Sandwich Panels Under Low-Velocity Impact Loading", International Journal of Mechanical Sciences, (2005) pp. 1301–1325.
3. J. Tirillo, L. Ferrante, F. Sarasini, L. Lampani, E. Barbero, S. Sanchez-Saez, T. Valente, P. Gaudenzi, "High Velocity Impact Behavior of Hybrid Basalt-Carbon/Epoxy Composites ", Composite Structures, (2017) pp. 305–312.
4. Mariana Etcheverry, Maria Lujan Ferreira, NumaCapiati, Silvia Barbosa, "Chemical Anchorage of polypropylene onto glass fibers: Effect on adhesion and Mechanical properties of their composites ", International Journal of Adhesion& Adhesives, (2013) pp. 26–31.
5. Y. Hou, Y.H. Tai, C. Lira, F. Scarpa, J.R. Yates, B. Gu, "The bending and failure of sandwich structures with auxetic gradient cellular cores ", Composites: Part A, (2013) pp. 119–131.
6. VitalyKoissin, VitalySkvortsov, and AndreyShipsha, "Stability of the face layer of sandwich beams with sub-interface damage in the foam core", Composite Structure 78.4, (2007), 507–518.
7. Yuxi Guo, Yihe Zhang, Hongwei Huang, XianghaiMeng, Yangyang Lie, ShuchenTu, Baoying Li, "Novel glass ceramic foams materials based on polishing porcelain waste using the carbon ash waste as foaming agent "Constructional and Building Materials 125 (2016) PP. 1093–1100.
8. W. Wang, G. Martakos, J.M. Dulieu-Barton, J.H. Andreasen, O.T. Thomsen, "Fracture behavior at tri-material junctions of crack stoppers in sandwich structures "Composite Structures 133 (2015) pp. 818–833.
9. A.C. Manalo, T. Aravinthan, W. Karunasena, "In-plan shear behavior of fibre composite sandwich beams using asymmetrical beam shear test", Construction and building materials, 24, (2010), pp. 1952–1960.

Note: All the figures and tables in this chapter were made by the authors.

Advances in Additive Manufacturing Technologies – Gurusamy Pathinettampadian et al. (eds)
 ISBN 978-1-041-16687-0

Tough PLA Fabrication Specimens used in 3D Fusion Deposition Modeling by Multi-Performance Process Variable Optimization

K. Hariram
Department of Mechanical Engineering,
DMI Engineering College, Aralvaimozhi, India

T. Livingston, D.S. Jenaris*, D.S. Manoj Abraham
Department of Mechanical Engineering,
PSN Engineering College, Tirunelveli, India

N. Ramanan
Department of Mechanical Engineering,
Sri Jayaram Institute of Engineering and Technology, Tiruvallur, India

P. Gurusamy[1]
Department of Mechanical Engineering,
Chennai Institute of Technology, Chennai

Vivekananthan A.[2]
PG Student, Department of Mechanical Engineering,
Chennai Institute of Technology, Chennai

◆ **Abstract:** Among the methods used for rapid prototyping (RP) that has turn into an invaluable resource for product development in all branches of manufacturing is Fusion Deposition Modeling (FDM). Fusion Deposition Modeling (3D FDM) is a top method for making complicated models that may be used in additive manufacturing for many different purposes. Manufacturers can make complicated models that are precise, economical, and eco-friendly with the help of this additive manufacturing process. Improved mechanical characteristics may be achieved by adopting the Taguchi approach, as demonstrated in this work. For multi-performances like hardness and tensile strength, 3D FDM process factors including level thickness (LT), infill density (ID), and needle temperature (NT) were fine-tuned. Models developed in accordance with ASTM standards (tensile and hardness) make use of the Taguchi L09 orthogonal array matrix. For optimal hardness and tensile strength process factors, the Taguchi approach is employed. Analysis of Variances (ANOVA) is used to find out which process factors are more and least important for influencing various performances.

◆ **Keywords:** Speedy prototyping, Polymer based 3D FDM, Multi-response, TGRA technique, Curve

1. Introduction

Fused Deposition Modeling (FDM), the predominant technology in contemporary Additive Manufacturing (AM), entails creating three-dimensional forms created by layering thin materials of diverse materials, including plastic, glass, ceramics, and more. The process utilizes computer-aided design (CAD) software to create an electronic version of a three-dimensional part, which is then loaded into a 3D printer for printing. This technology,

Corresponding authors: [1]gurusamyp@citchennai.net, [2]ananthanvivek300@gmail.com

DOI: 10.1201/9781003685906-10

which is most essential to small industries, is known as the fast- production procedure. When contrasted with traditional production procedures, it may accelerate the entire production process [1-4].

Various materials with different melting points can be utilized in additive manufacturing, including metals, alloys, ceramics, polymers, resins, and plastics. Ongoing research aims to expand the range of materials used in AM by incorporating additional substances, more diverse materials, or those with specific properties, such as meta materials and functionally-gradient materials. The Fused Deposition Modeling technique commonly exploits thermoplastic materials for 3D printing[5]. Notable examples include acrylonitrile butadiene styrene, polyamide, polylactide, polycarbonate, and tough polylactide. Plasticized layered modeling polymers have gained widespread acceptance in this field due to their distinctive mechanical characteristics, superior reliability, affordability, accessibility, high dimensional fidelity, sleek surface finish, lower rejection rates, and the sustainable use of renewable resources in their preparation [6-9]. While literature on safety and working conditions is less abundant compared to publications describing the parameters and qualities, this analysis specifically reveals that manufacturing process parameters, especially raster width and air gaps are key contributors to both material toughness and porosity [10]. Letcher and his colleagues reported the results of 3D printing raster orientations using PLA plastic as the material by assessing the strength of each orientation behavior. Additionally, the tensile properties of PLA used to create 3D parts at 0°, 30°, 45°, 60° and 90° raster angles were tested on the research study. The study found that the specimens produced by the roasters at 45° and 90° angles were more robust. Furthermore, another weakness pertains to defects at the micro-level in 3D-printed parts, which can notably impact the mechanical strength of the printed item [11]. Limited research has been conducted on 3D-printed tough PLA. In the current learn, these procedures' outcomes on the durability of PLA material in additive manufacturing (AM) 3D printing were examined. Each 3D-printed specimen underwent tensile and hardness testing and the influences of these route variables on tensile strength and hardness were scrutinized [6].

2. Experimental Methodology

Tough PLA is the polymer-based material highly used in those applications. It has very good properties of high hardness and flexural strength. The property of the tough pla are show in Table 10.1. Due to its better quality, it is highly used in many applications. Therefore, the tough PLA models have been developed to check the performance characteristics such as Tensile strength and Hardness respectively. High quality Ultimaker 2+ 3D printer is used for developing the models using Tough PLA. It is one of the best fusion deposition methods 3D printers among many [7]. The specifications of the Ultimaker 2+ printer maximum build size as 223 × 220 × 205 mm and nozzle diameter as 0.4 mm. Throughout this paper, the procedure variables selected as film thickness, Infill density and Nozzle temperature and mentioned in Table 10.1. The photograph of the Ultimaker 2+ 3D printer is mentioned in Fig. 10.1. Figure 10.2 (a – c) shows the different layer thickness of 0.15mm, 0.25mm and 0.35mm of the Tough PLA material. The CAD images of the Cuboid of ASTM D2240 and dog bone shapeof ASTM D638 are shown in Fig. 10.3. From the literatures, Tri-Hexagon filled pattern models gives better quality in mechanical properties.

Table 10.1 Procedure variables ranges and levels

Procedure Variables with unit	Levels		
	I	II	III
Film width (LT ~ mm)	0.16	0.26	0.36
Infill Density (ID ~ %)	55	60	65
Nozzle Temperature (NT ~ oC)	210	215	220

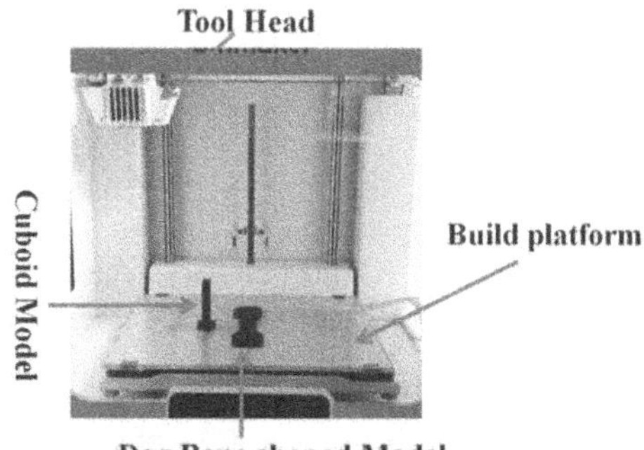

Fig. 10.1 Ultimaker 2+ 3D printer with Dog bone shape and cuboid model

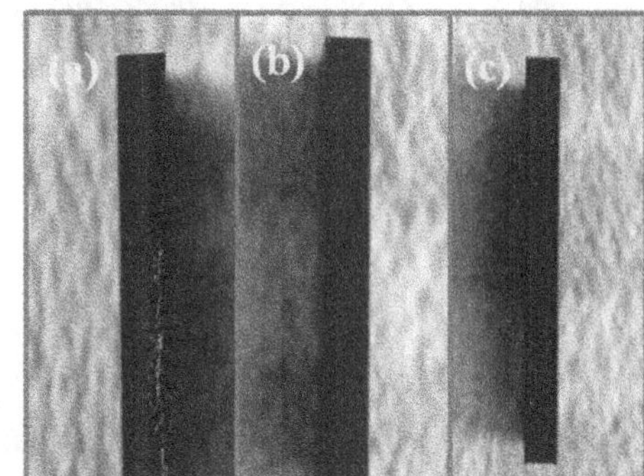

Fig. 10.2 (a) Film thickness = 0.15mm (b) Film thickness = 0.25 mm and (c) Film thickness = 0.35 mm

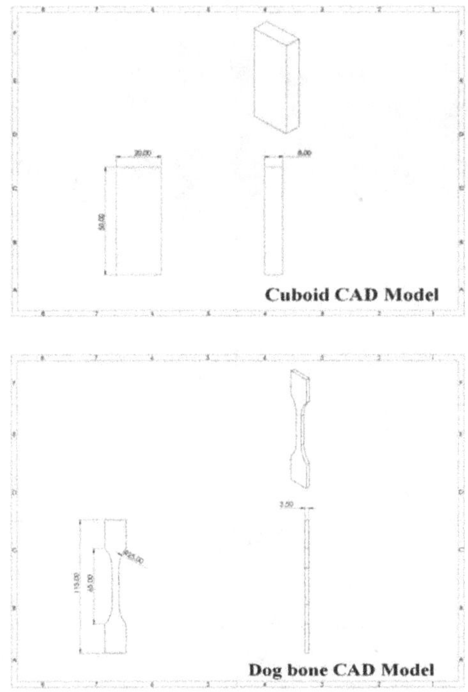

Fig. 10.3 CAD Model of cuboid and dog bone shape

Tri-Hexagon fill pattern is considered for all the models. The sliced models using Ultimaker CURA are shown in Fig. 10.4. In order to improve the stability of the printing capacity; sliced model is given through USB connectivity to print the models. Procedure variables ranges and levels are given in table II. Before start the printing, calibration has been done in the Ultimaker 2+ printer. Now, Taguchi DoE mentioned in Table 10.2 is used to print the cuboid and dog bone shaped models as per the ASTM standards in Ultimaker 2+ FDM based 3D printer machine [8].

As per the Taguchi DoE, nine models were printed as dog bone shaped and nine models were printed as cuboid shaped. Cuboid shaped models used for hardness whereas dog bone shaped models used for tensile testing.

Table 10.2 Taguchi DoE to print models

Model Runs	Process Variables		
	Film Thickness (LT ~ mm)	In fill Density (ID ~ %)	Nozzle Temperature (NT ~ °C)
1	0.16	56	211
2	0.15	60	215
3	0.15	65	220
4	0.25	55	215
5	0.25	60	220
6	0.25	65	210
7	0.35	55	220
8	0.35	60	210
9	0.35	65	215

3. Results and Discussion

3.1 Hardness

Following cuboid form printing in accordance with ASTM 2240 standard, the SHORE-D GOLD hardness tester be used to assess the hardness properties of all nine specimens. Figure 10.5 displays the cuboid standard for each of the nine examples. During the hardness tests, the ambient temperature was kept at 25.3°C.

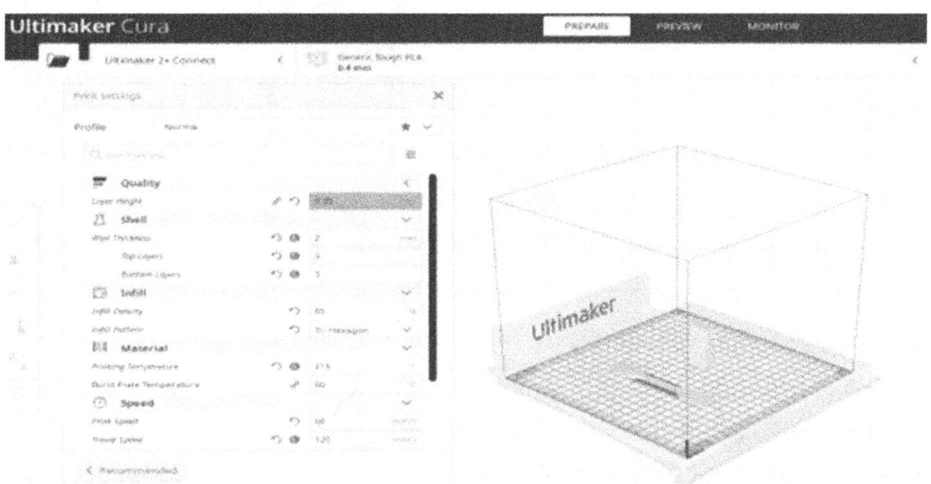

Fig. 10.4 Cuboid and dog bone-shaped objects were cut using the Ultimaker Cura

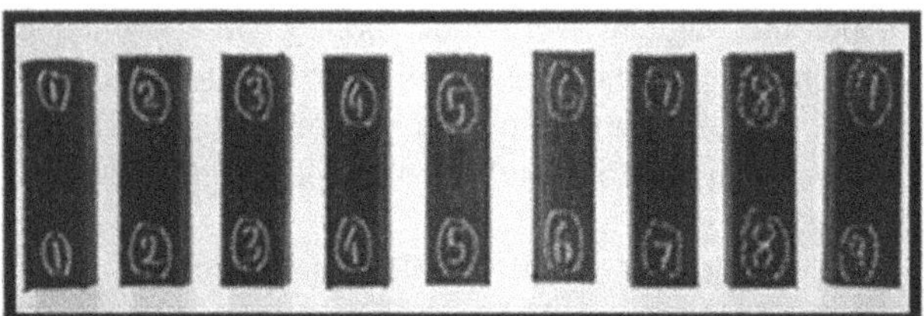

Fig. 10.5 Standard for cuboid specimens as per ASTM 2240

Figure 10.6 displays the results of specimen number seven's test on the SHORE-D GOLD hardness tester. Table 10.3 displays the hardness values of all the specimens.

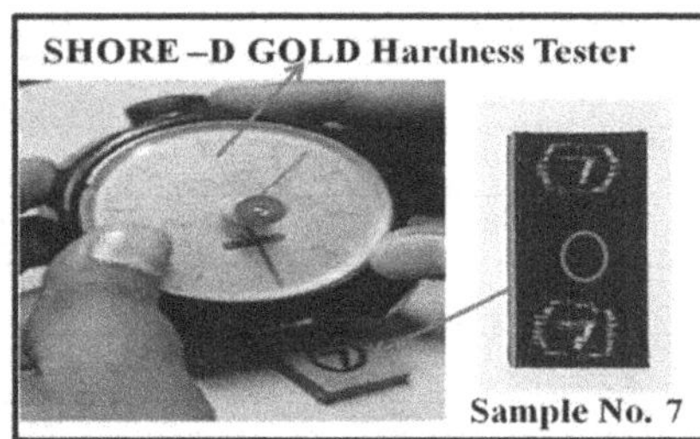

Fig. 10.6 Tested with the SHORE-D GOLD hardness machine, sample number 7

Table 10.3 Hardness tested results of nine specimens

Specimen No.	Process Variables			Response variable
	Layer width (LT ~ mm)	*In fill Density (ID ~ %)*	*Nozzle Temperature (NT ~ °C)*	*Average Hardness (SHORE – D)*
1	0.15	55	210	52
2	0.15	60	215	53
3	0.15	65	220	55
4	0.25	55	215	56
5	0.25	60	220	54
6	0.25	65	210	61
7	0.35	55	220	63
8	0.35	60	210	62
9	0.35	65	215	64

3.2 Tensile Strength

All nine specimens were subjected to tensile strength testing in a TUE-CN-400 computerized tensile tester equipment after being printed in a dog bone shape according to the ASTM D638 standard. Figure 10.7 displays the specimens before testing.

Fig. 10.7 Standard for dog bone form according to ASTM D638 prior to tensile strength

In Table 10.4, tensile strength tested results of nine specimens are mentioned. And Fig. 10.8 displays the specimens after testing.

Table 10.4 Tensile strength tested results of nine specimens

Specimen No.	Process Variables			Response variable
	Layer width (LT ~ mm)	*In fill Density (ID ~ %)*	*Nozzle Temperature (NT ~ °C)*	*Tensile Strength (N/mm²)*
1	0.15	55	210	32.693
2	0.15	60	215	32.976
3	0.15	65	220	43.182
4	0.25	55	215	33.241
5	0.25	60	220	36.343
6	0.25	65	210	35.963
7	0.35	55	220	23.042
8	0.35	60	210	34.816
9	0.35	65	215	36.926

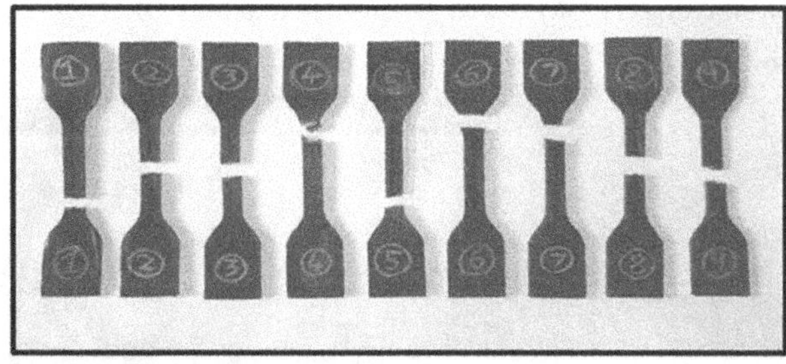

Fig. 10.8 Standard for dog bone form according to ASTM D638 following tensile strength

3.3 Taguchi Analysis

Using the Taguchi analysis for the tensile and hardness values of the nine specimens, we can witness the optimization of procedure variable levels in the Ultimaker 2+ 3D printer. For both the tensile and hardness criteria, the Signal-to-Noise Ratio of Larger is employed. Figures and tables display the SN graph and response tables for tensile and hardness. Table 10.5 displays the taguchi signal-to-noise ratios for both hardness and tensile strength [9].

Table 10.5 Taguchi signal-to-noise ratios for both hardness and tensile strength

Sample	Tensile strong point (N/mm^2)	S N R for tensile strong point	Hardness (SHORE - D)	S N R for hardness
1	27.63	28.8276	54	30.2891
2	32.876	30.3376	59	30.3640
3	33.476	30.4947	62	32.7061
4	35.249	30.9429	63	30.4335
5	37.071	31.3807	63	31.2084
6	40.016	32.0447	63	31.1171
7	43.564	32.7826	60	27.2504
8	45.328	33.1273	62	30.8356
9	44.572	32.9812	65	31.3466

Figure 10.9 illustrates the Signal-to-Noise Ratio (SNR) Principal component analysis of tensile strength.

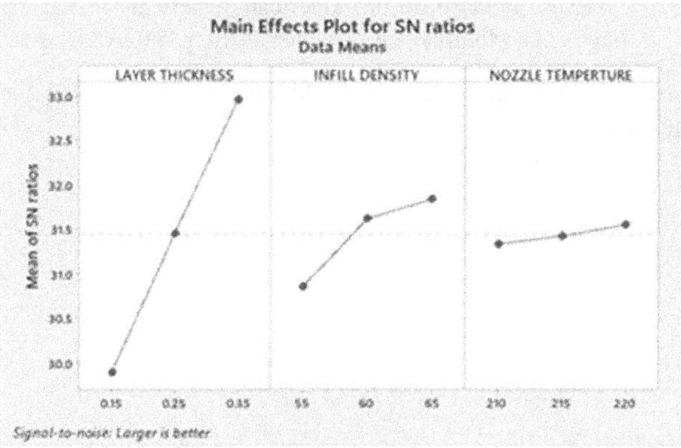

Fig. 10.9 Signal-to-noise ratio (SNR) principal component analysis of tensile strength

Table 10.6 presents the SNR response for Tensile Strength and Fig. 10.10 Illustrates the Signal-to-Noise Ratio (SNR) Principal component analysis of Shore-D Hardness [10].

Table 10.6 SNR response for tensile strength

Levels	Procedure Parameters		
	Layer Thickness (mm)	*Infill Density (%)*	*Temperature (°C)*
I	29.88	30.84	31.32
II	31.45	31.61	31.41
III	32.95	31.83	31.54
Delta	3.07	0.98	0.21
Rank	1	2	3

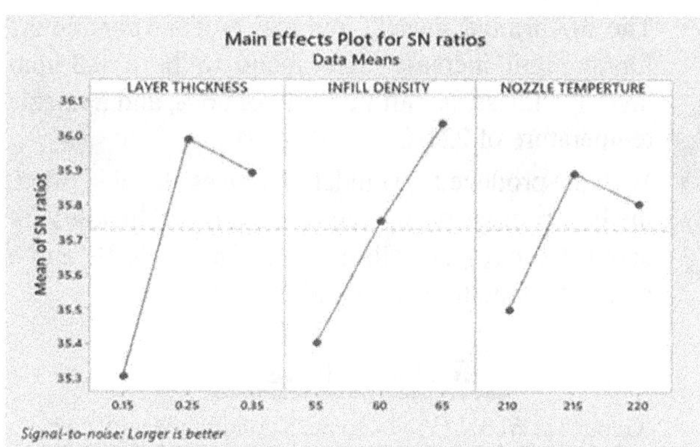

Fig. 10.10 Signal-to-noise ratio (SNR) principal component analysis of Shore-D shardness

Table 10.7 Presents the S N R response for Shore - D hardness

Level	Procedure Parameters		
	Layer width (mm)	*In fill Density (%)*	*Temperature (°C)*
I	35.61	35.61	35.48
II	35.98	35.84	35.88
III	35.88	36.05	35.81
Delta	0.66	0.61	0.41
Rank	1	2	3

Table 10.7 presents the S N R response for Shore - D hardness. In table VIII. Optimized Parameters for P L A Material are mentioned.

Table 10.8 Optimized parameters for P L A material

Parameters	Tensile strong point	Hardness	Surface Roughness
Film Height, mm	0.36	0.26	0.16
In fill Density, %	66	66	56
Nozzle Temperature, °C	221	216	216

Taguchi analysis and SNR major effects maps of hardness and tensile strength led to the discovery of the ideal levels for the process parameters. Height of 0.35 mm, density of 65%, and temperature of 220°C were the ideal process conditions for tensile strong point. The ideal procedure parameters for hardness were a height of 0.25 mm, a density of 65%, and a temperature of 215°C [11].

6. Conclusion

The FLASHORE 3D printer was able to successfully construct the model. This method succeeded in decisive the function of the process factor in 3D print with respect to surface irregularity, hardness, and tensile strong point.

- The 3D-printed model's optimal process parameters for strength increase were found to be a 0.36mm layer thickness, an infill density of 66%, and a nozzle temperature of 221◦C.
- With the produced 3D model's Hardness level in mind, the best process parameters were: a layer thickness of about 0.26mm, an infill density of about 66%, and a nozzle temperature of around 216◦C.

References

1. Gupta, N.; Weber, C.; Newsome, S. Additive Manufacturing: Status and Opportunities; Science and Technology Policy Institute: Washington, DC, USA, 2012.
2. Gupta, N.; Weber, C.; Newsome, S. Additive Manufacturing: Status and Opportunities; Science and Technology Policy Institute: Washington, DC, USA, 2012.
3. Gebhardt, A. Understanding Additive Manufacturing: Rapid Prototyping • Rapid Tooling • Rapid Manufacturing; UnderstandingAdditive Manufacturing; Gebhardt, A., Ed.; Carl Hanser Verlag GmbH &Co. KG: München, Germany, 2011.
4. Kietzmann, J.; Pitt, L.; Berthon, P. Disruptions, decisions, and destinations: Enter the age of 3-D printing and additive manufacturing. Bus. Horiz. 2015, 58, 209–215.
5. Gu, D. Materials creation adds new dimensions to 3D printing. Sci Bull. 2016, 61, 1718–1722.
6. Novakova-Marcincinova, L.; Novak-Marcincin, J.; Barna, J.; Torok, J. Special materials used in FDM rapid prototyping technologyapplication. In Proceedings of the 2012 IEEE 16th International Conference on Intelligent Engineering Systems (INES), Lisbon,Portugal, 13–15 June 2012; IEEE: Piscataway, NJ, USA, 2012; pp. 73–76.
7. Dudek, P. FDM 3D printing technology in manufacturing composite elements. Arch. Metall. Mater. 2013, 58, 1415–1418.
8. Lanzotti, A.; Grasso, M.; Staiano, G.; Martorelli, M. The impact of process parameters on mechanical properties of parts fabricatedin PLA with an open-source 3-D printer. Rapid Prototyp. J. 2015, 21, 604–617.
9. Wittbrodt, B.; Pearce, J.M. The effects of PLA color on material properties of 3-D printed components. Addit. Manuf. 2015, 8,110–116.
10. Chin Ang, K.; Fai Leong, K.; Kai Chua, C.; Chandrasekaran, M. Investigation of the mechanical properties and porosityrelationships in fused deposition modelling-fabricated porous structures. Rapid Prototyp. J. 2006, 12, 100–105.
11. Letcher, T.; Waytashek, M. Material property testing of 3D-printed specimen in PLA on an entry-level 3D printer. In ASME 2014International Mechanical Engineering Congress and Exposition; American Society of Mechanical Engineers: Montreal, QC, Canada, 2014; p. V02AT02A014.

Note: All the figures and tables in this chapter were made by the authors.

Advances in Additive Manufacturing Technologies – Gurusamy Pathinettampadian et al. (eds)
© 2026 Taylor & Francis Group, London, ISBN 978-1-041-16687-0

Performance Evaluation of Radiative Transfer DOS Models Based on the Safety System

T. Archana*
Department of Electronics and Communication Engineering,
Saveetha Engineering College, Saveetha Nagar,
Chennai, India

Nachammai N.
Department of Electronics and Instrumentation Engineering,
Annamalai University, Annamalai Nagar,
Chidambaram, India

Hema Malini A.
Department of Electronics and Communication Engineering,
Saveetha Engineering College, Saveetha Nagar,
Chennai, India

Vaishnavi T., Siva P.
Department of Electronics and Communication n Engineering,
Vel Tech Rangarajan Dr. Sagunthala R&D Institute of Science and Technology,
Chennai, India

Srinivasan M.
Department of Electronics and Communication Engineering,
Saveetha Engineering College, Saveetha Nagar,
Chennai, India

◆ **Abstract:** According to this study, passive auditory perception is a potential additional sensing technique for intelligent vehicles. We demonstrate how drivers can hear cars coming around blind curves before they can see them. India's widespread use of automobile horns is an interesting topic to study for anybody who has travelled the world. In prosperous countries, honking is virtually ever employed unless to warn of emergencies. In India, though, a lot of honking is unnecessary and common place. Horn use all the time makes a big difference in noise pollution. It is more than simply a nuisance; it puts everyone in the area at risk for health. As a result, with the aid of this project, you won't need to blast your horn repeatedly for a long time to express messages; instead, you can only click a button to send DoS data to the receiver, allowing them to understand why your car is honking. The goal of the command will also be shown on the receiver end vehicle display.

◆ **Keywords:** Automobile, Honking, Express message, Dos

*Corresponding author: vanitha@saveetha.ac.in

DOI: 10.1201/9781003685906-11

1. Introduction

The rising adoption of the concepts of smart areas and smart cities encouraging the construction smart transport systems (ITS), which can increase the effectiveness of transport services as they relate to increasing power effectiveness, decreasing and carbon dioxide output generally enhancing user and citizen quality of life. In this perspective, electric transport systems are essential for lowering carbon emissions and fossil fuel usage. Electric vehicles (EVs) can be categorized based on their energy storage devices (batteries, supercapacitors, flywheels, hybrid energy storage), charging methods (wired, static wireless, dynamic wireless), or vehicles, automobiles with power). [1-2]. But to maintain EV sales growth, more electrical energy storage must be made accessible for use in upcoming power systems. Additionally, factors like mobility and driver profiles and other random variables help determine an EV's charging needs. Due to the plug-in duration and the availability of the charging infrastructure, these factors influence how EV charging demand is allocated. [3]. By using the interaction embedding systems,the electric grid, and general ability to communicate, grid integration presents a number of potential for more efficient operation. This is made possible through the use of various communication channels, including as Vehicle-to-Vehicle communications, to facilitate EV administration and operation. or Vehicle-to-Vehicle [4] are three examples. V2G exhibits a number of qualities facilities with including the ability to handle a high amount of EVs or enable intricate The State of the Charge (SoC) determines this permits the creation of a number of both shaving (returning power to the grid when demand is high) and nighttime charging (charging when demand is low). These characteristics guarantee the stability of the electricity grid. Solutions for Likewise, demand-based analyses put into practise to balance and improve energy grid management. Key components inside in the context of EV communications, exchanging data for two-way use and transfer of electricity. Additionally, it offers other auxiliary services including The aggregator acts as a conduit centre (connects with the EVs after receiving from the GCC information regarding energy exchange and associated processes[5-8]. Therefore, charging or discharging operations may be carried out with the help of which improves management and reactive load correction. Linkages with the EVs, such as when the EV is parked in a grid-connected parking space, a dependable long-distance communication link should be ensured. The Dos Tx and Rx waves frequency is shown in Fig. 11.1. Numerous works are connected to V2G connectivity.

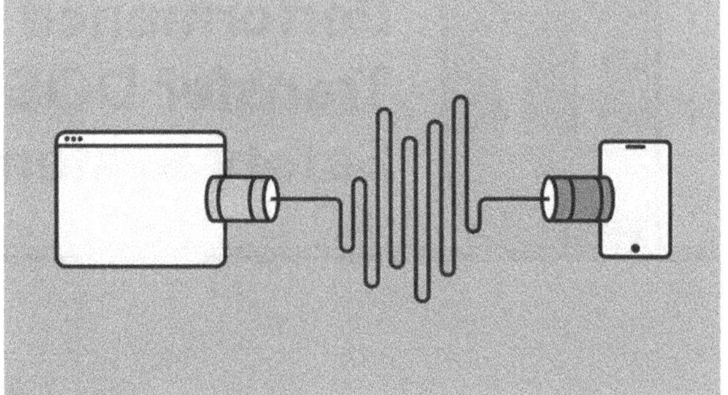

Fig. 11.1 Dos sound Tx and Rx waves frequency

2. Embedded System

Reactive Systems: As was already said while sensors react to it. In order to do this, embedded systems must operate at environment speed. "Reactive" describes this feature of embedded systems. Reactive computing refers to the process of a system (most often a software component) operating in reaction to outside stimuli. There are two types of external events: periodic and aperiodic. It is simpler to arrange processes to ensure performance when periodic events occur. Aperiodic occurrences are more challenging to plan. To account for worst-case scenarios, Predicting the maximum event arrival rate is necessary. A significant reactive component is included in the majority of embedded systems. One of the hardest difficulties. Since systems function in real time, the correctness of a computation is partly influenced by the pace at which it is supplied. It is common practise to build systems with this requirement for the worst-case situation. On complex structures, it could be challenging to anticipate the worst-case scenario with accuracy. This frequently results in estimations that are too gloomy and err on the side of caution. In order to fulfil needs for control stability and external I/O embedded systems must operate in real time. Reactive systems are also common in real-time systems[9-14].

3. Distributed Systems

One typical feature is the presence of communicative processes running on a number of linked that are linked through communication connections. The economy is the cause of this. Affordable 4-bit 8-bit microcontrollers might be more cost-effective than 32-bit CPUs. This strategy could be more advantageous even when the cost of the communication lines is included in. This method typically calls for numerous processors to tackle a number of time- sensitive activities. Physical distribution of the devices controlled by embedded systems is also possible.

The below bar graph Fig. 11.2 shows the analysis of high density bar spectrum.

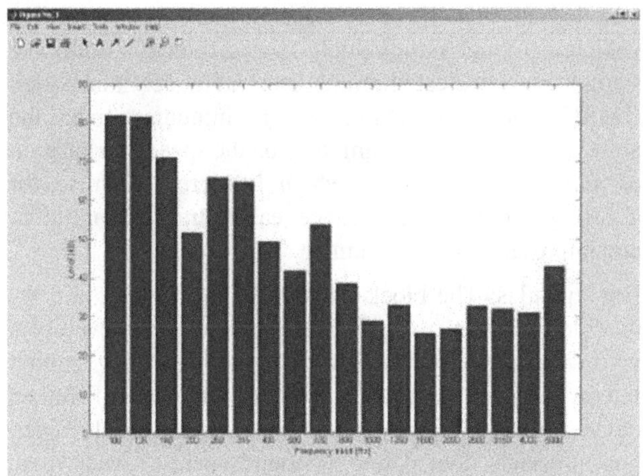

Fig. 11.2 High density bar spectrum analysis

4. HETEROGENEOUS ARCHITECTURES

Heterogeneous architectures frequently make up system incorporated in Fig. 11.3. They could use different CPUs in a single system design. Mixed signals are another option for the systems Embedded system design is completely unique since it integrates I/O connections, local and remote memory, sensors, and actuators. Heterogeneity provides additional design alternatives since embedded systems have stringent design constraints.

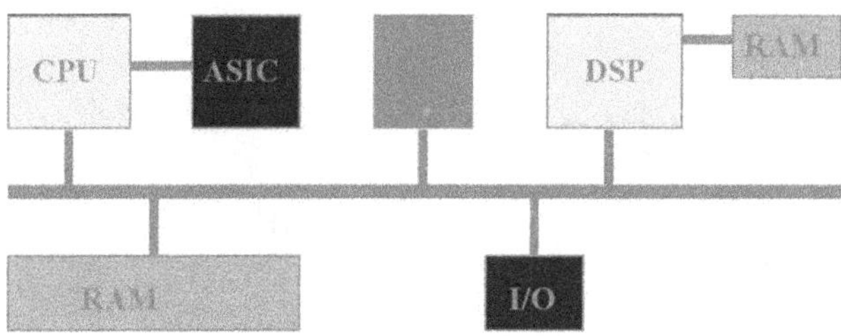

Fig. 11.3 Embedded systems having heterogeneous architectures

5. HARSH ENVIRONMENT

There are quite a few embedded systems that not engage in regulated activity environment. Overheating is a regular problem, especially in combustion-based applications (such as many automotive ones). The necessity for protection from physical harm, including Additional problems for embedded computers might be brought on by water, corrosion, fire, vibration, shock, lightning, and power supply irregularities [7].

System safety and reliability: An increasing proportion of the safety-related operations of the entire system are coming under the supervision of embedded systems due to their increased complexity and processing capability. It is possible to apply these safety measures with both hardware and software controls. Often, mechanical security backups are triggered to safely terminate a malfunctioning computer system. Software dependability and safety is a big issue. Often, software is not "broken" in the same sense as hardware. But because software can be so complex, a string of unforeseen circumstances could lead to problems that endanger customers.. The challenges addressed by embedded developers include building readily available low-cost systems with questionable parts as well as creating reliable software, both of which are beyond the scope of this book. For embedded system designers, obtaining low-cost reliability with minimal redundancy is the main challenge [8].

6. CONTROL OF PHYSICAL SYSTEMS

One of the primary justifications for computing devices is their capacity for interaction with the environment. Typically, this is accomplished by monitoring and controlling other equipment. Sensor signals that are analog need to be converted into digital form before embedded computers can process them. Reconverting the outputs to conventional signal levels is necessary. It could be necessary to switch large current loads in order to run devices like motors along with additional gear.

Larger computer circuits with a sizable number of analog elements might be required for embedded devices in order to achieve these demands. Embedded developers need to carefully negotiate system limitations, between physical hardware components, such as digital hardware and related software, mechanical, and electrical connections network, and computer hardware. The study of the octave band filter bar wavelength is shown in Fig. 11.4.

Small and low weight: Several Embedded systems are physically a part of bigger structures. The aesthetics of integrated technology could influence its form factor. For instance, a missile's shape might have to fit inside its nose. For certain remedies, embedded system developers may find it difficult to create non-rectangular layouts. Becoming overweight might present significant challenges at times. For example, integrated car In order to save gasoline, compact control systems are required. To be portable, portable CD players must be lightweight.

Cost sensitivity: Cost is a concern in the majority of systems, however with embedded systems, the sensitivity to price fluctuations might shift greatly. This is mostly

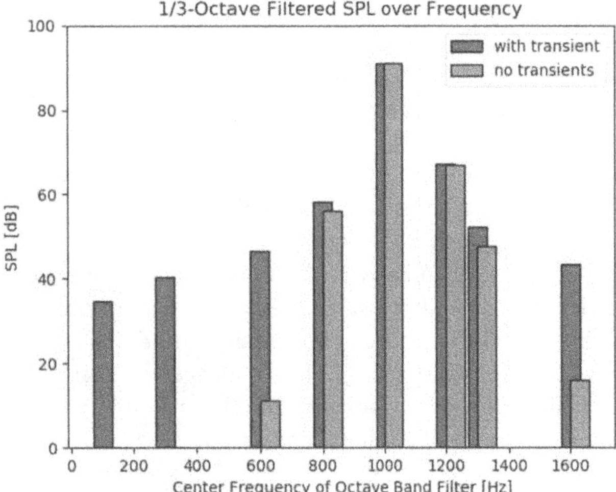

Fig. 11.4 Octave band filter bar spectrum

caused by the impact that computer expenses. It depends more on how large of an impact the cost adjustments are relative to the system's total cost.

Power management: Power consumption is strictly regulated in embedded systems. Figure 11.5 shows the graph of Decibel bar spectrum. Limiting heat generation is another noticeable issue for embedded systems, as many of them need to be portable. Therefore, it's crucial to preserve power and extend battery life as long as possible.

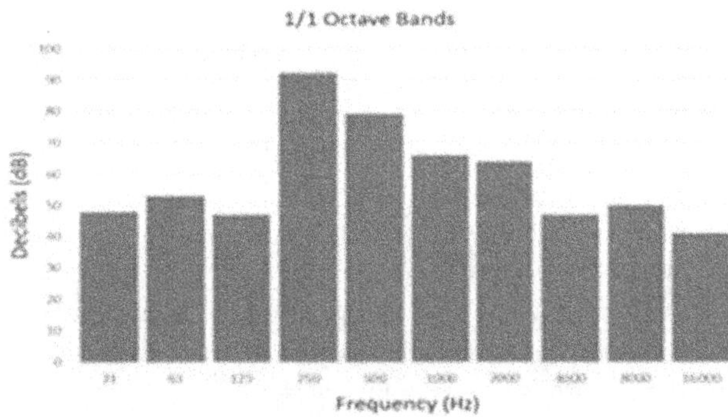

Fig. 11.5 Decibels bar spectrum graph

7. Existing System

Through computer networks called vehicular communication systems, vehicles and roadside devices exchange information such as traffic updates and safety alarms. They could be effective in avoiding crashes and gridlock. Data may be exchanged and received between automobiles with the use of wireless mesh networks and vehicle-to-vehicle communication technology. As a result, drivers have plenty of time to regulate their cars. These nodes can detect traffic conditions miles in advance.

8. Proposed System

Vehicle-to-vehicle (V2V) networks consist of equipment installed in vehicles that transmit messages including vehicle data (such as the vehicle's speed, direction, and brake status) over dedicated immediate radio communication (DSRC). Vehicle-to-vehicle (V2V) communication has the ability to wirelessly transmit data on the speed and location of surrounding vehicles, which has great promise for assisting in collision avoidance, easing traffic congestion, and improving the environment.

Dos modules: The block diagram of the Dos Tx and Rx module is shown in Fig. 11.6. Engineers and developers who want to enable Data-over-sound is replacing smooth interactions between an expanding number of linked devices, as an intriguing connection alternative. However, how precisely does data-over-sound operate? And where does its genuine worth lay in a world that is ever more connected? How does it Work? In a nutshell, Data-over-sound uses sound waves to transfer data between any device that has a microphone or speaker [9]. The technique transfers data via an auditory channel, akin to an audio QR code, and is made possible by between-machine software for communications. Its goals are to enhance end-user experiences and increase the functionality of currently offered products. In actuality, information is encoded as an acoustic signal, which is made up of a succession of audible or undetectable pitches and tones to create

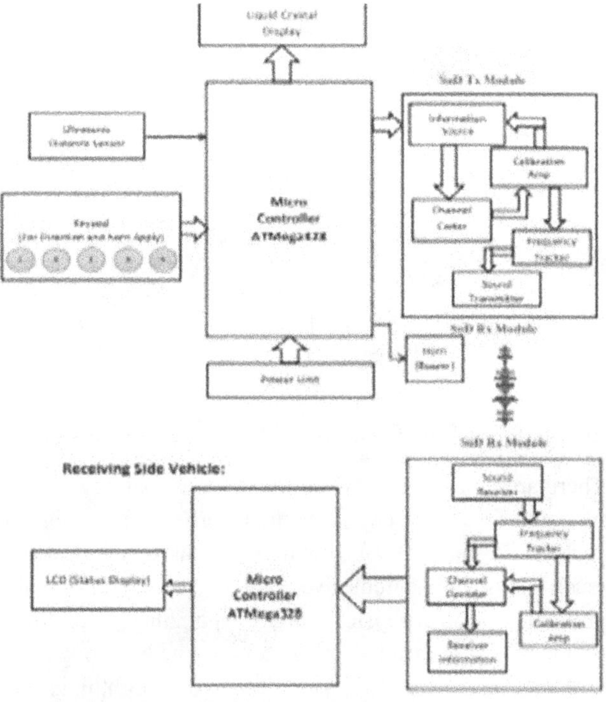

Fig. 11.6 Block diagram of Dos Tx and Rx module

a "sonic barcode." After that, this is played into a room (typically air, but might also be a VoIP stream or wired phone line), which is then gathered and demodulated by a "listening" device. The device which is receiving, or collection of devices, then provides the unique data after decoding the data. Programmers may compress audio and fit more information into it by employing a variety of audio frequencies. In order to identify and decode data signals in busy locations, such as a crowded railway station or concert venue, Software is fine-tuned and frequencies are carefully chosen by developers. This enables noise filtering and communication between their apps. It is becoming increasingly clear that it is a crucial instrument for error- free ultrasonic data transfer in both commercial and hobbyist development scenarios. It's not new to transfer data or speech using sound waves or audio-based technological technologies. For thousands of years, people have used drums and other percussion instruments to make music and communicate messages. Other creatures employ audio waves in addition to more conventional means of communication within species, such as the sonar used by dolphins to locate their echolocation or the ultrasound used by bats. Telegraph and telecom networks have been utilising sound technologies for decades [10].

Such is touch-tone controls on phones (DTMF, dual-tone multi-frequency signalling). The coding of data into phone lines using modems, which were developed by IBM in the 1940s. The Prototype of the Dos module is shown in Fig. 11.7. Others have used sound waves to transport data through the air, water, or even solid things [11]. There are initially two options for underwater communications:

Fig. 11.7 DOS Tx and Rx prototype

Hardwired communications are transmitted by a cable, whereas wireless communications are transmitted over water. Many divers love the concept of being able to freely dive anywhere they wish with the speed and convenience of communication with their dive companion. This is what wireless underwater communications have managed to achieve. Wireless communications provide you the freedom to speak with your friend or top side without restriction, as opposed to hardwired communications that limit you to the length of a cable. A fun and convenient approach to remain in contact without restriction underwater is through wireless underwater communications [12].

9. COMMUNICATION

A host computer is required for programme development and compilation on microcontrollers. An integrated development environment, sometimes known as an IDE, is the name of the programme used on the host computer. The free and open-source Processing software, which was developed for use with computers, serves as the foundation for the Arduino's development environment. Wiring is an open source project that is used by the Arduino programming language (wiring.org.co). The foundation of the Arduino programming language is conventional C. Don't worry if you don't know the language; learning it is simple, and the Arduino IDE gives you feedback when your programmes fail. A variety of features allow Using the Arduino Uno, you may connect to other microcontrollers, computers, or other Arduinos. For UART TTL (5V) serial connection, the ATmega328 provides digital pins 0 (RX) and 1 (TX). The board's ATmega16U2 exposes a virtual com port to software programmes by channeling this serial connection through USB. There is not a requirement for an additional driver because the '16U2 firmware makes use of the built-in USB COM drivers. An inf files is necessary on Windows. Simple textual data can be delivered to and obtained from the board that runs Arduino using the serial interface that is part of the Arduino software [13]. When data is transferred via the Internet-to-serial chip and USB connection to the computer, the RX and TX LEDs on the board will light up . A circuit board called the Triple Amplifier Board Fig. 11.8 produces an improved version of a source signal that is supplied into the input ports.

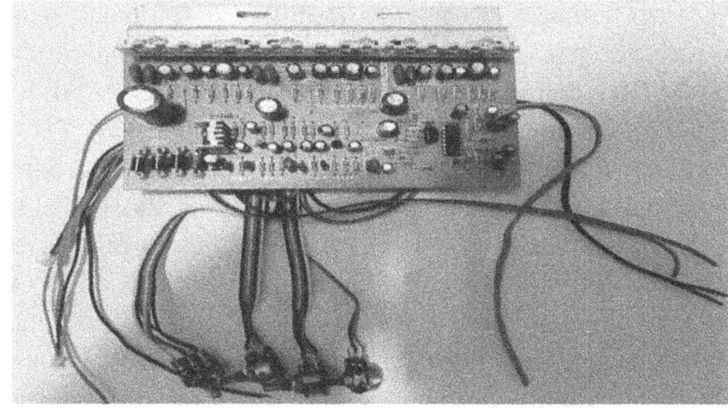

Fig. 11.8 Triple amplifier board

Any electronic pin on the Uno can be used for serial communication because of the serial library. Additionally, the ATmega328 allows communication via SPI and I2C (TWI). To facilitate using the I2C bus, the Arduino software comes with the Wire library; for more information, refer to the documentation. Use the SPI library to facilitate SPI communication [14].

10. Fire/Flame Sensor Module

Early detection of a growing fire emergency and quick notification of it to the building's inhabitants and fire emergency organizations are crucial components of fire safety. Fire detection and alarm systems play this job. These systems can perform a variety of key tasks, depending on the potential fire situation, the kind of structure and usage, the quantity and nature of inhabitants, the mission's criticality, and the predicted fire scenario (Fig. 11.9). They first offer a way to detect an emerging fire using human or automatic techniques, and they also give. They warn everyone inside the building that there is a fire and that they should leave. The sending of an alarm notification This usually involves sending a signal to the fire department or another emergency response outfit. They can be used to activate automatic suppression systems as well as stop specific process operations or electrical, air handling, or other activities [15].

Fig. 11.9 Flame sensor

11. Conclusion

In order to make We provide an effective, efficient, and scalable broadcast authentication solution for computation-based DoS attacks and packet losses resistant on VANETs for V2V communications, mechanism. Furthermore, PBA benefits from quick verification by utilising beacons' predictability for suitable single hop applications. PBA only retains condensed MACs of signatures to minimise storage cost while defending against memory-based DoS attacks. We demonstrate through theoretical research that PBA is safe and reliable in the setting of VANETs. PBA has been shown through a variety of assessments to function successfully even in situations with high traffic density and lossywifi. Future study will focus on understanding how our system may be enhanced in the presence of precise prediction models. The privacy implications for some automotive applications are also crucial to take into account. In our upcoming work, we'll talk about how to meet both security and privacy criteria.

References

1. Bai, F., Krishnan, H., Sadekar, V., Holland, G., &Elbatt, T. (2006). Towards characterizing and classifying communication-based automotive applications from a wireless networking perspective. In Proceedings of IEEE Workshop on Automotive Networking and Applications (AutoNet) (pp. 1–25).
2. Parno, B., &Perrig, A. (2005). Challenges in securing vehicular networks. In Fourth Workshop on Hot Topics in Networks (HotNets-IV), Proceedings, November 2005.
3. Lee, S. B., Pan, G., Park, J. S., Gerla, M., & Lu, S. (2007). Secure incentives for commercial ad dissemination in vehicular networks. In Proceedings of ACM Mobihoc (pp. 150–159).
4. Hsiao, H. C., Studer, A., Chen, C., Perrig, A., Bai, F., Bellur, B., &Iyer, A. (2011). Flooding-resilient broadcast authentication for vanets. In Proceedings of ACM Mobicom (pp. 193–204).
5. Zhang, C., Lu, R., Lin, X., Ho, P. H., & Shen, X. (2008). An efficient identity-based batch verification scheme for vehicular sensor networks. In Proceedings of IEEE INFOCOM (pp. 816–824).
6. ABAKA: An anonymous batch authenticated and key agreement scheme for value-added services in vehicular ad hoc networks. IEEE Transactions on Vehicle Technology, 60(1), 248–262.
7. Shim, K. (2013). Reconstruction of a secure authentication scheme for vehicular ad hoc networks using a binary authentication tree. IEEE Transactions on Wireless Communications, 12(11), 5586–5393.
8. Bellare, M., Garay, J. A., & Rabin, T. (1998). Fast batch verification for modular exponentiation and digital signatures. In EUROCRYPT Proceedings, 1(2), 236–250.
9. Boneh, D., Gentry, C., Lynn, B., &Shacham, H. (2003). Aggregate and verifiably encrypted signatures from bilinear maps. Proceedings of EUROCRYPT, 416–432.
10. Hankerson, D., Hernandez, J. L., & Menezes, A. (2000). Software implementation of elliptic curve cryptography over binary fields. In Proceedings of CHES (pp. 1–24).
11. Unterluggauer, T., & Wenger, E. (2014). Efficient pairings and ECC for embedded systems. Proceedings of CHES, 298–315.
12. BAT: A robust signature scheme for vehicular networks using binary authentication tree. IEEE Transactions on Wireless Communications, 8(4), 1974–1983.
13. Achieving efficient cooperative message authentication in vehicular ad hoc networks. IEEE Transactions on Automotive Technology, 62(7), 3339–3348.
14. An identity-based security system for user privacy in vehicular ad hoc networks. IEEE Transactions on Parallel and Distributed Systems, 21(9), 1227–1239.
15. Hao, Y., Cheng, Y., Zhou, C., & Song, W. (2011). A distributed key management framework with cooperative message authentication in vanets. IEEE Journal on Selected Areas in Communications, 29(3), 616–629.

Note: All the figures in this chapter were made by the authors.

Advances in Additive Manufacturing Technologies – Gurusamy Pathinettampadian et al. (eds)
© 2026 Taylor & Francis Group, London, ISBN 978-1-041-16687-0

Enhancing Random Number Generation With DCM-Based TRNG in Xilinx FPGA

Arunkumar K.*
Department of ECE, Saveetha Engineering College,
Chennai, Tamilnadu, India

◆ **Abstract:** Random number generation is fundamental in various computer applications; security and unpredictability of numbers are of paramount importance. Many algorithms require an initial random seed for the algorithm to work (PRNGs is a good example), use of real random numbers for seeding improves quality of generated sequences. This study discusses about evaluation of custom algorithm to generate real random numbers (TRNG) implemented on Xilinx, specifically leveraging Digital Clock Manager (DCM) blocks. This method provides benefits of using a minimal number of logic elements and ability to adjust frequency of digital clock in order to increase unpredictability. TRNG's operational principle, Beat Frequency Detection, is explored, which includes monitoring lower-order bits in a counter capturing sequential-ones through D flip flop. Our assessment includes the examination of 100 frequency pairs to assess the TRNG's performance. Moreover, the paper introduces a novel system with reconfigurable DCM circuits and three distinct types of circuits to generate numbers that are random in nature: T-RNG, P-RNG, and colloidal patterns generation using the PNR method. The Xilinx FPGA and the XC9572XL device are used to implement these advancements in real-time ASIC, providing workable answers for a variety of random number generating needs. This effort helps to improve computational systems' security and unpredictability.

◆ **Keywords:** Digital Clock Manager, E-commerce applications, TRNG, PRNGs

1. Introduction

A key component of many computer and communication systems, random number generation (RNG) is essential for applications including simulations, srability, Xilinx FPGAs are becoming more and more popular as a way to create TRNGs. Random Number Generation (RNG) is an essential component of many computing and communication systems, used in applications such as cryptography, safe data transmission, and simulation. TRNGs (Algorithms for generating actual random values) play an important role in maintaining the unpredictability and security of random numbers. Because of its reconfigurability and versatility, Xilinx FPGAs are becoming increasingly popular as a method of implementing TRNGs.

The study "Enhancing Random Number Generation with DCM-Based TRNG in Xilinx FPGA" focuses on enhancing the quality and efficiency of random number generation in Xilinx FPGAs through the use of a specific component known as the Digital Clock Manager (DCM). DCMs are adaptable, high-performance modules that manage clock signals in Xilinx FPGAs. When DCMs are included into the True Random Number Generator (TRNG) design, they can improve the randomness, stability, and performance of random number creation. Xilinx FPGAs are noted for their great performance and flexibility. They provide a programmable hardware platform that enables designers to create unique digital logic circuits and complicated algorithms in hardware. Because of their flexibility, they are appropriate for a wide range of applications that demand

*Corresponding author: arunkumar@saveetha.ac.in

DOI: 10.1201/9781003685906-12

rapid development, modification, and high-performance processing.

This article delves into a variety of topics, including the fundamental concepts of TRNGs, the importance of DCMs in FPGA systems, and ways for incorporating DCMs into TRNG designs to improve random number generation. It may examine the problems and trade-offs associated with improving TRNGs, as well as prospective applications for improved random number generation in secure communications, cryptographic methods, and other domains that require high-quality random numbers. The purpose of this study, Enhancing Random Number Generation with DCM-Based TRNG in Xilinx FPGA, is to investigate the intersection of FPGA technology, TRNGs, and the strategic use of Digital Clock Managers to optimize the creation of real random values, thereby contributing to the larger realm of cyber security, encryption, and secure data handling in contemporary computing systems.

The Digital Clock Manager (DCM) is a component commonly found in FPGAs that provides precise control over clock signals. It can create, multiply, divide, and phase-shift clock signals. DCMs are extensively used to synchronize and manage clock domains in FPGA designs. P-RNG is a technique for producing a list of numbers that appear random but are actually determined by an initial number known as a seed. They find broad application in a variety of industries, including cryptography, simulation, and gaming. T-RNG is a system or approach that generates random values by utilizing entropy, or unpredictability, drawn from the physical, real-world environment. TRNGs generate numbers that are truly random, as opposed to P-RNGs, which use deterministic frameworks to build a list that mimics randomness. TRNGs' key properties include entropy generation, unpredictability, security, slower than PRNGs, physical components, and bias testing.

Entropy Source: TRNGs rely on an entropy source, which is a physical process or phenomenon that is naturally unpredictable. Electronic noise, nuclear decay, ambient noise, and even user activities like keyboard and mouse motions are all common sources of entropy.

Unpredictability: TRNGs generate really random numbers because they are the result of processes that are inherently unpredictable. This means that no known algorithm can predict or duplicate the results.

Security: TRNGs are commonly employed in cryptographic applications where high randomness and unpredictability are required to ensure the security of encryption keys and other sensitive data. TRNGs are less vulnerable to prediction or compromise because they are not algorithm-based.

Slower than PRNGs: TRNGs generate random numbers more slowly than PRNGs, which use mathematical techniques. A TRNG's performance is frequently limited by the rate at which the entropy source generates unpredictable data.

Physical Components: TRNGs often rely on physical components to capture and process the entropy source. Sensors, analog-to-digital converters, and post-processing circuits can all help to refine raw entropy into a useable random bit stream.

Bias Testing: TRNGs are rigorously tested to guarantee that the generated numbers do not reflect bias or patterns that could make them predictable. Statistical tests are used to assess the randomness's quality

The quality and security of T-RNG are determined by the quality of the source and the design of the hardware or software used to extract random data from it. Furthermore, TRNGs are frequently used in conjunction with PRNGs to improve both speed and unpredictability in applications that require a constant supply of random numbers.

2. Literature Survey

An effective method for determining the best frequency pairs for DCM-based T-RNGs has been presented by Fujieda and colleagues (2019). In connection with this, Fujieda et al. evaluated a DCM-based TRNG using 100 frequency pairings. Our paper offers a quantitative analysis of how the choice of frequency pair affects the quality of the generated random list and the rate at which TRNGs are generated, in contrast to a previous study [2] that offered 23 pairs and only published statistical test results for three of them. We achieved this through subjecting strings that is random in nature from all pairs to rigorous tests [4]. Apart from that, this work offers valuable guidance to designers regarding which frequency pairs to explore initially.

Additionally, they have put forth a technique that allows for a more refined balance between random number quality and generation rate. This method leverages observation that how many consecutive "1" values (logical high or true) are input to the flip-flop over time using biphasic histograms. Our assessment, demonstrates that planned approach significantly enhances number-of viable frequency-pairs yielding high-quality random list, with less impact on rate at which number is produced.

Figure 12.1 shows prototype of model to detect edges and perform synchronization. Johnson et al (2016) conducted a preliminary investigation that centred around the conception, examination, and execution of a straightforward, enhanced, low-resource, and adjustable T-RNG which are real for FPGA platforms. Primary

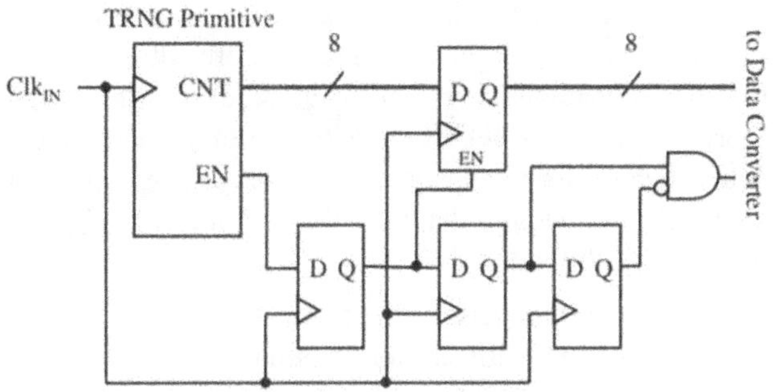

Fig. 12.1 Model to detect edge and synchronization

contributions in their study can be summarized as follows: They examined the constraints related to BFD-T-RNG while deploying on the top-of FPGA architecture and proposed hybrid and efficient framework intended for applications using FPGA to address its shortcomings. This work is among the initial reports incorporating tunability into a completely digital T-RNG [5].

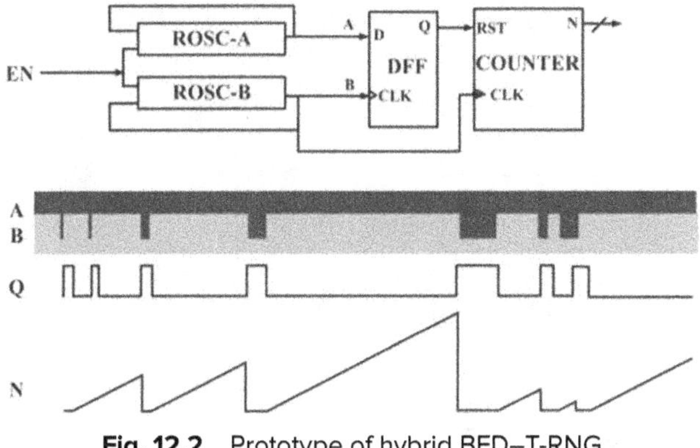

Fig. 12.2 Prototype of hybrid BFD–T-RNG

They performed a mathematical and experimental analysis of the modified proposed architecture. Experimental results strongly authenticated the mathematical model they put forth. Proposed prototype exhibited minimal infra dependencies, and bit-streams generated by it successfully cleared all tests cases used to validate the output [6].

Majzoobi et al introduced an innovative and highly efficient technique for producing genuinely random numbers on Field-Programmable Gate Arrays (FPGAs). Authors achieve randomness by inducing meta-stability in hybrid infrastructure components, such as F-F, through the precise application of programmable delay lines-PDLs. These delay lines possess ability to finely tune signal propagation delays, achieving resolutions that exceed fractions of a pico-second [7].

Furthermore, a RTTM base near real time system has been integrated to ensure generated output bits display a high level of randomness, remain stable in environmental variations and stay resilient against potential adversarial attacks. This monitoring framework utilizes a feedback mechanism, continuously overseeing probability of output bits. If any deviation in probabilities is detected, it promptly utilizes programmable delay lines (PDLs) to correct delay and restore circuit to a meta-stable operational state. Scholars approach is demonstrated through deployment on Xilinx FPGAs, and outcome of NIST test suite underscore its efficiency.

A typical T-RNG usually comprises three main components: Circuit modelling entropy source, additionally one circuit to perform sampling, and a post-processing unit. First circuit is responsible for generating multiple methods that exhibit arbitrariness, while next circuit takes care of converting this arbitrariness into a sequence that is binary. When output sequence fails to meet randomness criteria, it necessitatés optimization through supplementary post-processing circuits, such as the utilization of von Neumann encoding (kak et al).

TRNGs often employ various physical processes for entropy generation, including clock jitter, meta stability, and chaos. Extensive research has explored these physical processes, resulting in innovations. RO is one of the source circuits that is employed to produce clock-jitter but may not be the most efficient method. In order to increase rate of jitter accumulation, innovative entropy source architectures have been introduced. These include self-timed-rings [14], feedback oscillators having multiple stages and convergence oscillators. Addressing the challenge of efficiently sampling clock jitter, Ingrid introduced the use of Time-to-Digital Converters (TDC) for high-precision jitter quantization, significantly improving the utilization of clock jitter.

Florent and colleagues introduced an open-feedback T-RNG that utilizes Look-Up Tables (LUTs). Rather than utilizing latches and PDC to fine-tune timing between clock and other signals, this approach places substantial demands on infrastructure resources, design factors, and supplementary circuits [8].

Novel T-RNG was deployed in 135 nm CMOS by P Monteiro et al (June 2022). The objective of the study was designing an RNG circuit that could serve as source circuit based on redundancies in RO. The goal was creating a circuit with less power, increased arbitrariness, cost-effective. Selected prototype for this research is novel RNG, to combine RO and initial circuit to produce bits that are arbitrary. Through simulation, the circuit met its objectives with the following performance characteristics:

- Low power consumption: The circuit consumed power of 1.01 mW.

- Increased throughput: It demonstrated improved outcome - 27 Mbit/s.
- Low energy consumed: The energy consumed per bit was measured at 47.6 pJ/bit.

It's worth noting that, due to limitations with the simulation tool or setup, not all statistical tests could be executed. However, all the tests that were conducted produced results that met the required standards for randomness and quality.

3. Proposed Model

In the proposed approach we have introduced a method aimed at enhancing the practicality of DCM-based T-RNG initially proposed by Johnson and colleagues [2]. This enhancement facilitates the tuning of frequency pairs. In our proposed system, we developed reconfigurable DCM circuits and put forth three distinct circuits to produce number with more randomness:

a. T-RNG producing unpredictable numbers.
b. Pseudo Random Number Generation, generating pseudorandom sequences.
c. Colloidal Patterns Generation through the PNR method, which produces unique patterns.

We have implemented this work in real-time ASIC using Xilinx FPGA technology, specifically the XC9572XL device. Our approach involved varying widths from bigger to a limited list to eliminate least significant bits. Prototype proved effectiveness of generated random value improved when we retained only a single LSB from small counter values. To generate random or pseudorandom numbers using a DCM, one can exploit the inherent jitter and phase noise in the clock signal. Here's a simplified process:

a. Seed Generation: Start with a seed value, often obtained from an external source like user input or physical phenomena (e.g., environmental noise).
b. Clock Manipulation: Use the DCM to manipulate the clock signal. change the clock's frequency slightly or apply phase shifts.
c. Sample and Quantize: Sample the modified clock signal at specific points. The exact timing of these samples is affected by the clock manipulation, creating variations that can be used as random bits [10].
d. Output Random Data: Process the sampled values to extract random bits, which can be used as the output of the RNG.

Pseudo randomness: It's essential to note that the generated numbers may not be truly random, but pseudorandom, as they depend on the initial seed and clock manipulation. The quality of arbitrariness is decided by characteristics of DCM, seed, and algorithm used for processing the sampled data. RNGs based on DCMs are often used in FPGA applications where cryptographic security is not a primary concern but where a source of pseudorandom data is needed for various purposes, such as simulations or controlling hardware behaviours that require some element of unpredictability (Fig. 12.2).

The level of randomness and security provided by such an RNG may not be sufficient for cryptographic applications, where a true random number generator (TRNG) is usually preferred [9].

3.1 Simulation Flow

Simulation work flow using VLSI simulation circuit is shown in Fig. 12.3.

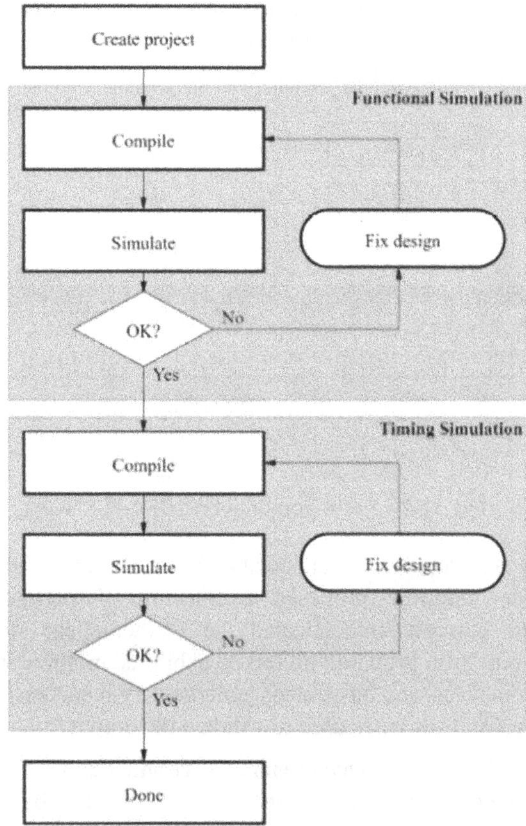

Fig. 12.3 Simulation flow

Global Clock: Stable global clock source that will serve as the basis for TRNG.

Clock Dividers: Clock dividers are used to create multiple clock domains with different frequencies. This will be useful for introducing variability into the TRNG. Each clock domain can be thought of as a different "entropy source."

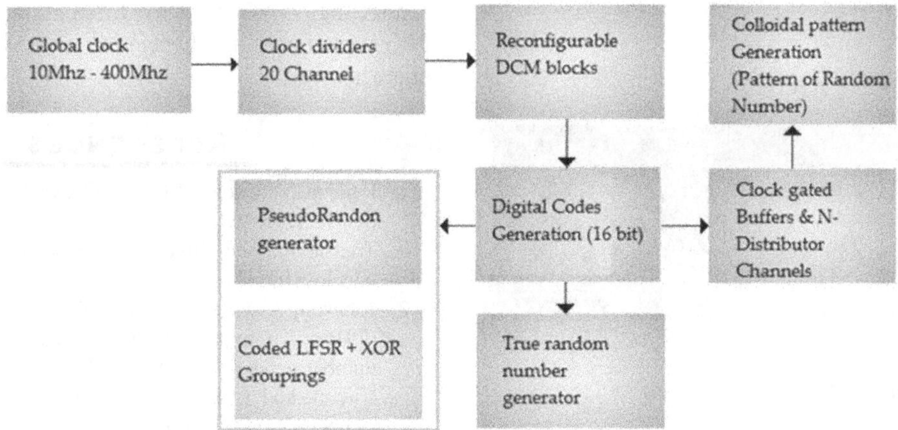

Fig. 12.4 Proposed architecture DCM-TRNG

Reconfigurable DCM Blocks: Digital Clock Managers (DCMs) can help control and manipulate clock signals. These blocks can be used to fine-tune the clock sources, modify clock phase, or adjust clock frequencies to introduce additional entropy.

Colloidal Pattern Generation: This module drives the pattern of the random number generated

Clock Gated: Clock gating is introduced to selectively enable or disable certain clock domains or entropy sources. This allows us to control the contribution of each entropy source to the final TRNG output [11].

4. Results Analysis and Discussion

Following tests were performed on a controlled environment to validate the test results.

Frequency Valuation: This serves as a fundamental evaluation in degree of arbitraryness within binary values. If frequency test doesnt meet critical criteria, then probability of encountering failures in subsequent statistical tests rises. Consequently, it is recommended to commence with frequency test as starting point. This in particular scrutinizes equilibrium between ones and zeros and will yield a failure outcome if a substantial imbalance in count of ones and zeros is detected (Fig. 12.4).

Frequency valuation in Block: This assesses distribution of one's within defined segments of binary data. Its purpose is to ascertain whether presence of ones in an M-bit segment closely matches expected occurrence in a completely random sequence, which is more or less half segment. When M equals 1, this is essentially equivalent to the Frequency test

Runs Test: Determine the aggregate count of runs in a binary sequence. A "run" is characterized as an unbroken sequence of identical bits, whether they are all ones or all zeros, flanked by a contrasting bit at start and conclusion of run. This evaluation is conducted to ascertain whether occurrences of runs consisting of ones and zeros, spanning different lengths, in sequence conform to anticipated patterns for a random sequence.

Longest Run in Block: These calculate highest run of one's inside a particular bit block. It fails if length of lengthy of one's in tested list does not go with actual length for a arbitrary list. Abnormality in expected length related to longest run on one's implies a similar irregularity in the expected length of the longest run of zeros [12].

Figure 12.5, 12.6 and 12.7 shows the output of PRNG, TRNG and colloidal pattern generator. Proposed method has demonstrated remarkable efficiency when compared to both PRNGs and TRNGs. The results clearly indicate that our approach outperforms traditional PRNGs in terms of generating random numbers with greater unpredictability and reduced bias. Furthermore, when compared to TRNGs, which often come with higher resource and hardware

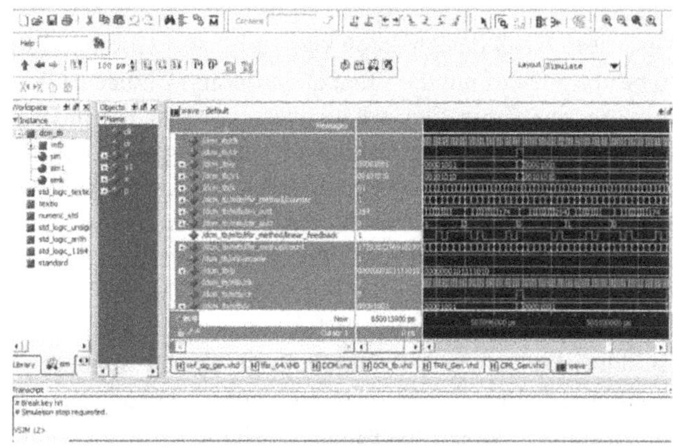

Fig. 12.5 Output of PRNG

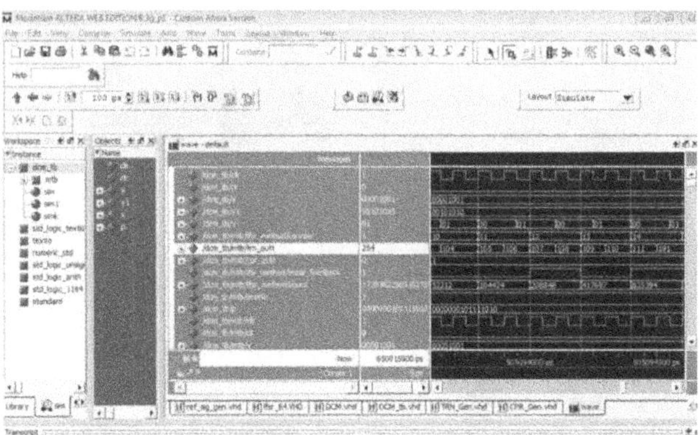

Fig. 12.6 Output of TRNG

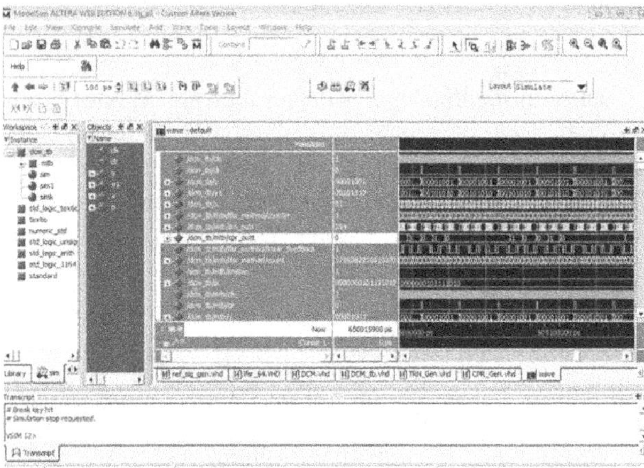

Fig. 12.7 Output of colloidal pattern generator

requirements, our method maintains its efficiency while offering a more practical and cost-effective solution. These findings affirm effectiveness of our solution for generating random numbers efficiently, making it a promising choice for various applications where randomness and resource optimization are essential [13].

5. Conclusion

Discussed RNG design employed 18-bit and 12-bit LFSR to produce list of numbers that are arbitrary. Choice of this mixture was subjective, where there is room for further research into generating 32-bit list by utilizing unlike LFSR combinations. In addition Potential Avenue for prospective tasks involves creating distinct ring oscillators for each LFSR to provide clocks, each operating at varying speeds (fast or slow) or having different designs (big or small). In existing system equivalent Ro signals were employed for clocking L-FSRs. Additionally a testing apparatus, combined with gumstix, was employed to produce and gather random values to evaluate R-NG integrated circuit. Nevertheless, numerous alternative methods exist for connecting R-NG chip to different microcontrollers or FPGA platforms for generating and evaluating random numbers.

References

1. Fujieda, N., Takeda, M., & Ichikawa, S. (2019). An analysis of DCM based true random number generator. IEEE Transactions on Circuits and Systems II: Express Briefs, 67(6), 1109–1113.
2. Johnson, A. P., Chakraborty, R. S., & Mukhopadyay, D. (2016). An improved DCM-based tunable true random number generator for Xilinx FPGA. IEEE Transactions on Circuits and Systems II: Express Briefs, 64(4), 452–456.
3. Majzoobi, M., Koushanfar, F., & Devadas, S. (2011). FPGA-based true random number generation using circuit metastability with adaptive feedback control. In Cryptographic Hardware and Embedded Systems–CHES 2011: 13th International Workshop, Nara, Japan, September 28–October 1, 2011. Proceedings 13 (pp. 17–32). Springer Berlin Heidelberg.
4. Chenglu Jin and Marten van Dijk. 2019. Secure and efficient initialization and authentication protocols for SHIELD. IEEE Trans. Depend. Secure Comput. 16, 1 (2019), 156–173.
5. C. H. Bennett and G. Brassard, Quantum cryptography: Public key distribution and coin tossing. Proceeding of the IEEE International Conference on Computers, Systems, and Signal Processing, Bangalore, India, pp. 175–179 (IEEE, New York, 1984).
6. S. Kak, A three-stage quantum cryptography protocol. Foundations of Physics Letters 19, 293–296 (2006).
7. Z. Ji, J. Brown and J. Zhang, "True Random Number Generator (TRNG) for Secure Communications in the Era of IoT," 2020 China Semiconductor Technology International Conference (CSTIC), Shanghai, China, 2020, pp. 1–5, doi: 10.1109 / CSTIC49141 .2020.9282535.
9. S. Kak, Encryption and error correction coding using D sequences. IEEE Transactions on Computers C-34: 803–809 (1985)
10. Lee, K., Lee, S. Y., Seo, C., & Yim, K. (2018). TRNG (True Random Number Generator) method using visible spectrum for secure communication on 5G network. IEEE access, 6, 12838–12847.
11. Thomas, D. B., Luk, W., Leong, P. H., & Villasenor, J. D. (2007). Gaussian random number generators. ACM Computing Surveys (CSUR), 39(4), 11-es.
12. G. Marsaglia, The Marsaglia Random Number CDROM including the Diehard Battery of Tests of Randomness. Department of Statistics, Florida State University, Tallahassee, FL, USA., 1995.
13. Zhou, S., Wang, X., Zhang, Y., Ge, B., Wang, M., & Gao, S. (2022). A novel image encryption cryptosystem based on true random numbers and chaotic systems. Multimedia Systems, 1–18.

Note: All the figures in this chapter were made by the authors.

Advances in Additive Manufacturing Technologies – Gurusamy Pathinettampadian et al. (eds)
© 2026 Taylor & Francis Group, London, ISBN 978-1-041-16687-0

Design and Optimization of a Four Band Microwave Metamaterial Absorber with Ultrathin Layers for C, X, and Ku Band Applications

A. Hema Malini[1]
Research Scholar, Anna University,
Assistant Professor, Department of ECE,
Saveetha Engineering College,
Chennai, India

M. Selvi[2]
Professor/ECE,
Saveetha Engineering College,
Chennai, Tamil Nadu, INDIA

T. Archana[3]
Assistant Professor,
Department of ECE, Saveetha Engineering College,
Chennai, India

M. Vanitha[4]
Professor/ECE,
Saveetha Engineering College,
Chennai, Tamil Nadu, INDIA

◆ **Abstract:** This research delves into the development of a microwave Metamaterial Absorber (MMA) that is unaffected by polarization and encompasses four distinct frequency bands. The fundamental building block of this innovative MMA comprises a top patch and a ground metallic plane, with a thin FR4 layer serving as the dielectric separator between them. The design of this structure aims to attain outstanding absorption characteristics across various frequency ranges. Remarkably, the absorber structure exhibits absorption peaks at 6GHz (C band), 9GHz,11.5GHz (X band), and 15GHz (Ku band) for achieving absorptivity levels of 95 percent, 94 percent, 82 percent and 98 percent, respectively. The results of this research have practical applications in sensing, and radar technologies. The MMA created through fabrication has undergone thorough testing and validation via measurements. The initial analysis of its design involved numerical simulations conducted using CST Microwave Studio Software.

◆ **Keywords:** Ultra-wideband, Microwave metamaterial absorber, Transmitter, Receiver, CST, Fabrication

1. Introduction

A metamaterial absorber is a tailored material designed to effectively absorb electromagnetic waves within specific frequency ranges. In contrast to conventional materials relying on inherent absorption properties, metamaterial absorbers derive their unique features from intricately designed structures at a subwavelength scale. These

[1]ahemamalini.ragu@gmail.com, [2]selvim@saveetha.ac.in, [3]archathiru@gmail.com, [4]vanitha@saveetha.ac.in

DOI: 10.1201/9781003685906-13

structures consist of artificial, periodic arrangements of elements that strategically manipulate the behavior of electromagnetic waves. Within this innovative field, various functional devices such as Reconfigurable devices, signal modifiers, signal receivers, delayed signal propagation, concealment technologies, transparency manipulation, and resonant phenomena like Fano resonance metamaterials to manipulate and govern essential aspects of electromagnetic waves. Researchers globally have been captivated by metamaterials exhibiting unconventional electromagnetic traits such as inverse refractive behavior, opposite magnetic permeability, and counteractive electric permittivity and more. Metamaterial absorbers have diverse applications across multiple domains, including antenna design, stealth technology, energy harvesting, and sensing. Their ability to control and manipulate electromagnetic waves with exceptional precision makes them invaluable for specialized applications where conventional materials may lack optimal performance.

Metamaterials having unusual Electromagnetic characteristics encompassing an adverse refractive index, inverse permeability, and counteractive permittivity have piqued the interest of scientists all around the world. Non-destructive testing, Security screening, sensing technologies, imaging devices, antenna systems, and filtration mechanisms, in radar technologies by absorbing and manipulating signals, thereby enhancing the overall efficiency of radar systems are only a few of the application fields [1-5]. MMAs commonly feature structured resonators in which electric resonance is induced by the top patch resonator, while the connection between the two planes triggers magnetic resonance [6].

Microwave MMAs, characterized by an extremely thin profile, exhibit ease of fabrication and have been investigated in both terahertz and infrared spectra as well [7]. A MMA is designed for dual-band applications within the X band was presented by Kumari et al. [8]. Jain and colleagues proposed a metamaterial might (in principle) totally modify light's path, forcing it to move around an object like water moving around rocks. In [9] a microwave-oriented highly compact MMA comprising on a with a dielectric substrate over which two modified square-shaped resonators, concluded by a metallic plane is used. The research investigates the absorber's response to varying polarizations and incident angles of transverse electric waves. A slim and compact metamaterial absorber that exhibits three resonances across the S, X, and Ku frequency bands with the absorber's design incorporates a square ring [10]. The main goal of this research [11-16] was to create a double L-shaped metamaterial suitable for use in triple-band applications. Recent advancements in terahertz sensing have seen notable progress through the utilization of multispectral terahertz sensing with ultra-flexible ultrathin metamaterial absorbers [17], as well as the development of ultra-narrowband dielectric metamaterial absorbers [18], six-band terahertz metamaterial absorbers [19], and hepta-band absorbers [20]. Achieving absorption characteristics within the terahertz spectrum has posed challenges, particularly in the demand for multi-layered structures and the integration of multiple resonators into a single unit-cell. Consequently, the current trend leans towards favoring simpler designs capable of generating multi-band characteristics, as indicated in recent studies [23-25].

2. Design of Unit Cell

Figure 13.1 depicts the configuration of the suggested MMA structure with four bands. It is made up of a flat resonator arrangement separated by a dielectric layer measuring 0.8 mm in thickness. Copper is used for the reference plane and upper resonator, having a frequency-insensitive conductivity of = 5.8107 S/m.

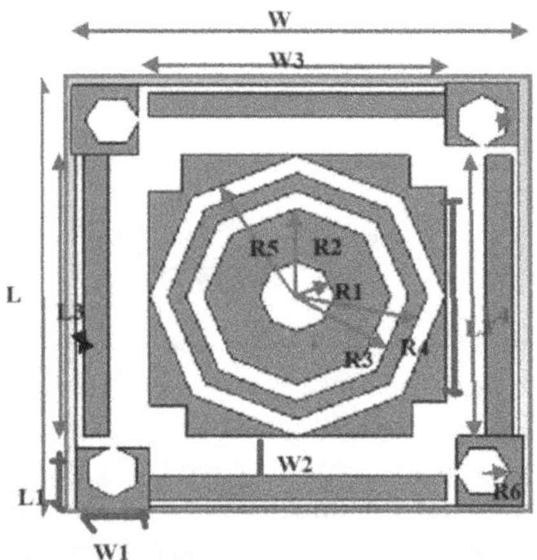

Fig. 13.1 A frontal perspective of the envisioned unit cell for the MMA

The substrate employed is FR4, characterized by a tangent loss measuring 0.025 and a relative complex dielectric constant of 4.3. Both the reference plane and substrate have dimensions of 12 mm in width (W) and length (L). Leveraging the commercially available CST Microwave software allows for the demonstration of absorption behaviour and structural efficiency. Equation 1 is employed to ascertain absorption.

$$A(\omega) = 1 - R(\omega) - T(\omega) \quad (1)$$

The absorption coefficient (A) measures the proportion of incident electromagnetic energy absorbed by a material or structure. In contrast, reflectance (R) signifies the fraction of energy reflected, and transmittance (T) indicates the fraction transmitted through the material. This equation ensures the conservation of total energy, asserting that the combined values of absorbed, reflected, and transmitted energy sum to 1 (or 100%). It constitutes a foundational principle for comprehending the interaction between electromagnetic waves and materials, offering valuable insights into the material's absorption, reflection, and transmission characteristics.

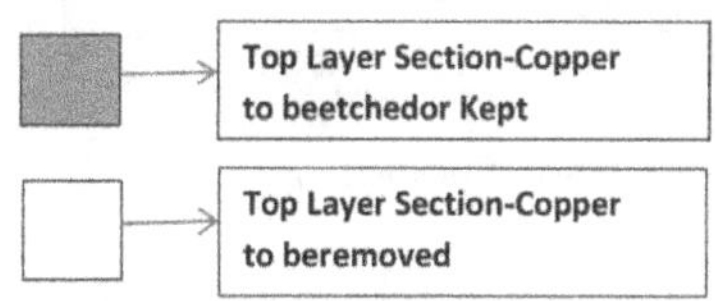

Table 13.1

Parameters	Dimensions (mm)
(L×W)	(12×12)
(L1×W1)	(1.9×2)
(L2×W2)	(1.8×2)
(L3×W3)	(6×6)
(L4×W4)	(8×0.7)
R1	1
R2	2.5
R3	3
R4	3.5
R5	4
R6	0.75

T () denotes transmission, whereas R () denotes reflection. T () equals zero because the reference metal layer's thickness is greater than its penetration depth (). Therefore,

$$A(\omega) = 1 - R(\omega) \quad (2)$$

Table 13.2 The specifications for the envisioned MMA design

Materials		
BottomLayer	Copper	0.035mm
TopLayer	Copper	0.035mm
MiddleLayer	FR4	0.8mm

Equation 2 demonstrates that as the input reflectivity R () approaches zero, unity absorptivity can be attained. Table 13.2 shows the top resonator's specified geometrical parameters.

3. Absorption

Figure 13.2 depicts the MMA's absorption and reflection properties (a). The structure gives four bands with absorptivity of 98 percent, 82 percent,92 percent, and 94 percent for frequencies of 6GHz (C band), 9 GHz, 11.5 GHz (X bands), and 15 GHz (Ku). The absorptivity as illustrated in Fig. 13.2, is used to explore polarization-independent properties of the structure (b). Figure (c) depicts the suggested absorber's return loss characteristics. At five bands, FR-4-based absorber structures have a return loss of less than -10dB, indicating that they are effective [22].

4. Simulation Results of Polarization Angle

In real-world scenarios, the MMA is expected to consistently exhibit efficient absorption across diverse oblique incidence and polarization angles. The numerical investigation explores the MMA's sensitivity to angle variations. Figure 13.3(a) and (b) illustrate the simulated absorbance of the proposed MMA, highlighting changes in both oblique incidence and polarization angles within the frequency range of interest (4 to 16 GHz), spanning

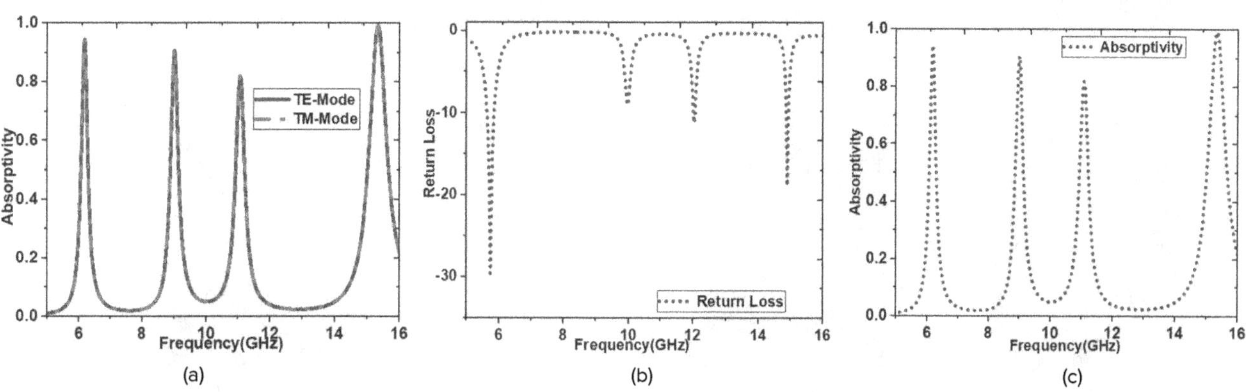

Fig. 13.2 (a) Absorptivity and reflectivity measures, (b) Return loss performance (c) Absorptivity response

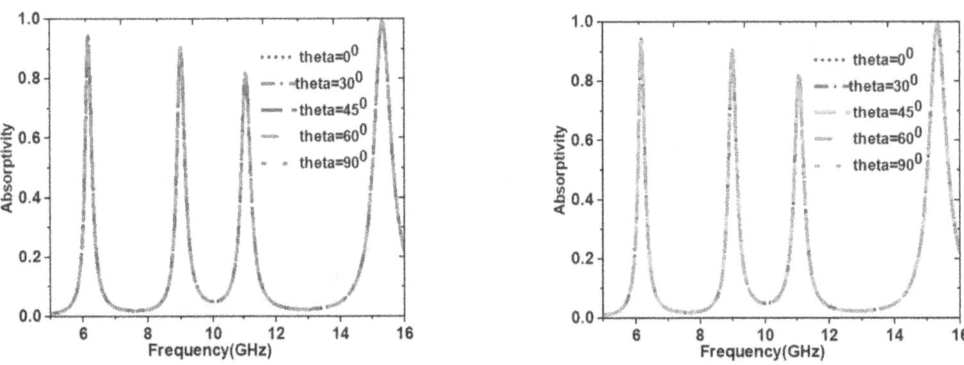

Fig. 13.3 Reinterpretation of polarization angle simulation findings

from 0 to 90 degrees. Remarkably, the absorption rate and resonance frequency of the structure remain unaffected by alterations in angle values. Consequently, the suggested MMA stands out for its exceptional polarization and angle-independent absorption characteristics, primarily attributed to the proportions of its unit cell [23].

5. Result of Field Distributions

The structural analysis delves into the physical mechanism through plots illustrating the distribution of electric and magnetic fields. Figure 13.4 (a-d) illustrates distribution of electric field across four distinct frequencies.

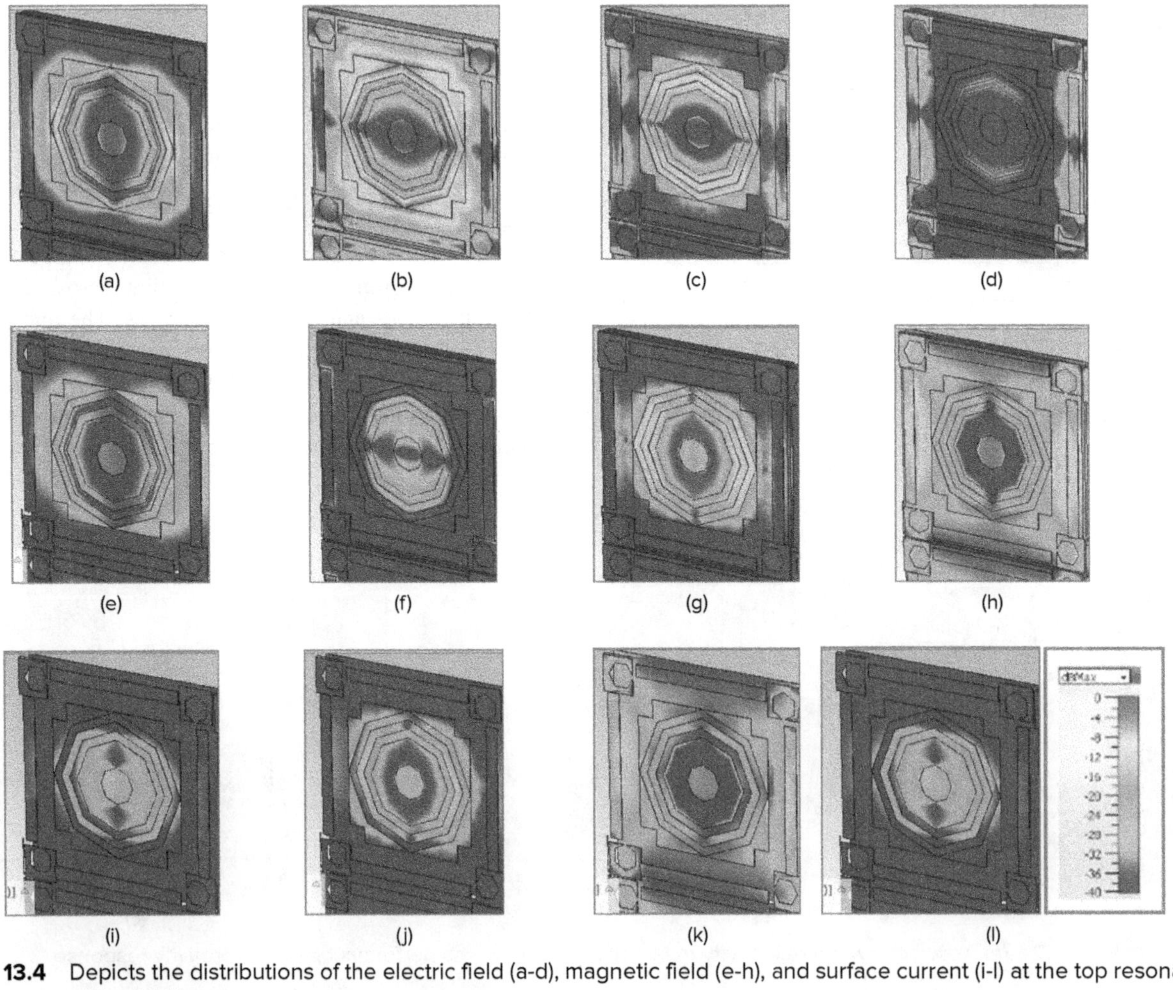

(a) (b) (c) (d)

(e) (f) (g) (h)

(i) (j) (k) (l)

Fig. 13.4 Depicts the distributions of the electric field (a-d), magnetic field (e-h), and surface current (i-l) at the top resonator of the MMA

Figure 13.4 depicts the distributions of the electric field (a-d), magnetic field (e-h), and surface current (i-l) at the top resonator of the MMA.

At 6 GHz, the most noticeable dispersion of the electric field occurs at the upper and lower regions of the resonator structure. The establishment of the electric field occurs at the resonator's inner part at a frequency of 9 GHz. At 11.5 GHz, the electric field concentration is maximum at the corners than the inner ring of the resonator construction. At 15 GHz, the electric field reaches its peak at the four corners of the absorber. Figure 13.4 illustrates the magnetic field distributions across four bands (f-j). At 6 GHz, the magnetic field dispersion peaks at the ends of the resonator. The inner ring resonators exhibit the maximum magnetic field intensity when the frequency is configured at 9 GHz. At 11.5 GHz, the greatest magnetic field intensity is observed in both the inner section and the corners of the resonator structure. At 15 GHz, the magnetic field attains its peak across the entire patch resonator structure.

6. Fabrication Result

The microwave MMA proposed in the model has been fabricated for the purpose of validating the simulation results, as illustrated in Fig. 13.5. The constructed framework comprises three layers. It is positioned between the upper patch and the lower plane is a FR4 dielectric substrate with a thickness of 0.8 mm. The upper layer comprises a grid of unit cells, organized in a 25 by 25 array, and each unit cell has a diameter of 3 cm. Figure 13.5 illustrates the measurement setup (b), which includes two horn antennas, one for signal reception and the other for signal transmission [24].

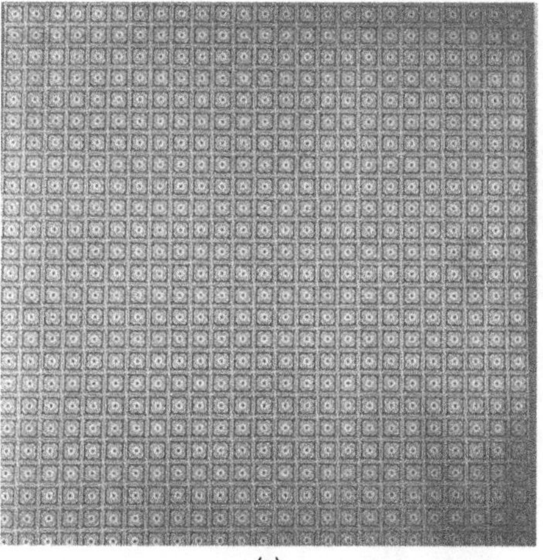

(a)

(b)

Fig. 13.5 (a) Fabricated MMA, (b) Unit cell of fabricated MMA

A prototype of the MMA structure has been created and is positioned between two horn antennas, enabling a comparison between simulated and fabricated outcomes across the frequency spectrum spanning from 5 to 16 GHz. The measured results are obtained using the Vector Network Analyzer (VNA) measurement tool, specifically the Agilent Technologies N9917A model. The MMA structure is carefully covered within an anechoic chamber to mitigate undesired signals. Despite a minor shift in absorption peaks, the measured values closely align with the calculated outputs. Any discrepancies in frequency shifts in the measured results are attributed to potential measurement errors or fabrication imperfections [25].

Conclusion

The engineered metamaterial absorber exhibits properties suitable for targeted frequency applications. The documented absorption peaks and performance metrics underscore the absorber's efficacy in mitigating electromagnetic signals within the C band, X band, and Ku band. These results contribute significantly to the realm of metamaterial research, suggesting potential applications in communication, radar, and sensor technologies within the specified frequency ranges. Thorough examinations delve into the absorption mechanism, and the sensing capabilities are probed by introducing an analyte onto the upper layer. The manufactured metamaterial absorber (MMA) aligns closely with the anticipated outcomes from simulation results. High absorption is demonstrated by the suggested four-band absorber. The simulated geometrical dimensions result in absorption peaks at 6 GHz, 9 GHz, 11.5 GHz, and 15 GHz, achieving absorptivity levels of 95%, 94%, 82%, and 98%, respectively. With potential applications in sensing, scientific research, and navigation radar, the proposed absorber design demonstrates versatility and promise.

REFERENCES

1. Smith, DR. Padilla, WJ. Vier, DC. Composite medium with simultaneously negative permeability and permittivity. Phys Rev Lett., 84, 4184–4187 (2000).
2. Pendry, JB. Negative Refraction makes a Perfect Lens, Phys. Rev. Lett. 85, 3966 (2000).
3. Landy, NI. Sajuyigbe, S. Mock, JJ. Smith, DR. Padilla, W. Perfect metamaterial absorber. Phys. Rev. Lett. 100, 207402 (2008).
4. Jamilan, S. Azarmanesh, MN. Zari, D. Design and characterization of a dual-band metamaterial absorber based on destructive interferences. Prog Electromagnet Res C. 47, 95–101 (2014).
5. Wang, BX. Zhai, X. Wang, GZ. Huang, WQ. Wang, LL. A novel dual-band terahertz metamaterial absorber for a sensor application. J.Appl. Phys., 117(1), 014504 (2015).
6. Shijun Ji, Zhiyou Luo, Ji Zhao, Handa Dai, Chengxin Jiang. Design and analysis of an ultra-thin polarization-insensitive wide-angle triple-band metamaterial absorber for X-band application. Opt Quant Electron 53, 148 (2021).
7. Bilal, RMH. Naveed, MA. Baqir, MA, Ali, MM. Rahim, DAA. Design of a wideband terahertz metamaterial absorber based on Pythagorean-tree fractal geometry, Opt. Mater. Express, 10, 3007–3020 (2020).
8. Khusboo Kumari, Naveen Mishra, Raghvendra Kumar Chaudhary, An ultra-thin compact polarization insensitive dual band absorber based on metamaterial for X-band applications, 5 Microw. Opt. Technol. Lett. 9, 2664–2669 (2017).
9. Prince Jain, Arvind K. Singh, Janmejay K. Pandey, Shonak Bansal, Neha Sardana, Sanjeev Kumar, Neena Gupta, Arun K. Singh, An Ultrathin Compact Polarization-Sensitive Triple-band Microwave Metamaterial Absorber, J. Electron. Mater, 50, 1506–1513, (2021).
10. Kaur, M. Singh, HS. Design and analysis of a compact ultrathin polarization- and incident angle-independent triple band metamaterial absorber. Microw. Opt. Technol. Lett, 62, 1920–1929 (2020).
11. Tamim, AM. Faruque, MR. Alam, J. Islam, S. Islam, M. Split ring resonator loaded horizontally inverse double L-shaped metamaterial for C-, X- and Ku-band microwave applications. Results Phys. 12, 2112–2122 (2019).
12. Muthukrishnan, K, Narasimhan, V. An ultra-thin triple-band polarization-independent wide-angle microwave metamaterial absorber. Plasmonics 14, 1983–1991 (2019).
13. Ramachandran, T, Faruque, MRI. Islam, MT. Symmetric square shaped metamaterial structure with quintuple resonance frequencies for S, C, X and Ku band applications. Sci Rep 11, 4270 (2021).
14. Dhillon, AS. Mittal, D. Bargota, R. Triple band ultrathin polarization insensitive metamaterial absorber for defense, explosive detection and airborne radar applications. Microw. Opt. Technol. Lett. 61(18), (2018).
15. Le Dinh Hai, Vu Dinh Qui, Nguyen Hoang Tung, Tran Van Huynh, Nguyen Dinh Dung, Nguyen Thanh Binh, Le Dac Tuyen, Vu Dinh Lam. Conductive polymer for ultra-broadband, wide-angle, and polarization-insensitive metamaterial perfect absorber. Opt. Express. 26, 33253–33262 (2018)
16. Al-badri, Khalid. Very High Q-Factor Based On G-Shaped Resonator Type Metamaterial Absorber. Ibn AL- Haitham Journal For Pure and Applied Science. 160, 159–166 (2018).
17. Yahiaoui, R. Tan, S. Cong, L. Singh, R. Yan, F. Zhang, W. Multispectral terahertz sensing with highly flexible ultrathin metamaterial absorber. J Appl. Phys. 118(8), 083103 (2015).
18. Liao, YL. Zhao, Y. Ultra-narrowband dielectric metamaterial absorber with ultra-sparse nanowire grids for sensing applications. Sci Rep, 10, 1480 (2020).
19. Ben-Xin Wang, Gui-Zhen Wang, Tian Sang ,Ling-Ling Wang, "Six-band terahertz metamaterial absorber based on the combination of multiple-order responses of metallic patches in a dual-layer stacked resonance structure", Scientific Reports, vol. 14, pp. 41373, 2017
20. Dutta, R, Bakshi, SC & Mitra, D ,' An Ultrathin Compact Polarization-Insensitive Hepta-Band Absorber', IMaRC, Kolkata, India, pp.1–4, 2018.
21. X. Yuan et al., "Wideband high-absorption electromagnetic absorber with chaos patterned surface," IEEE Antennas Wireless Propag. Lett., vol. 18, no. 1, pp. 197–201, Jan. 2019
22. Ben-Xin Wang, Yuanhao He, Pengcheng Lou, Wei-Qing Huang, Fuwei Pi,' Penta-band terahertz light absorber using five localized resonance responses of three patterned resonators, Results in Physics, vol. 16, pp. 102930, 2020.
23. Wangyang Li, Yongzhi cheng ,' Dual-band tunable terahertz perfect metamaterial absorber based on strontium titanate (STO) resonator structure', Optics communications, vol.462, pp. 125265, 1 May 2020
24. The Potential of Refractive Index Nanobiosensing Using a Multi-Band Optically Tuned Perfect Light Metamaterial Absorber(AUTHOR: Zohreh Vafapour) 2021
25. An Ultrathin Microwave Metamaterial Absorber for C, X, and Ku Band Applications 2021.

Note: All the figures and tables in this chapter were made by the authors.

Advances in Additive Manufacturing Technologies – Gurusamy Pathinettampadian et al. (eds)

An Experimental Study and Finite Element Analysis on 3D Printed Polycarbonate and ABS Materials

Saravana Bavan[1]
Department of Mechanical Engineering,
Dayananda Sagar University,
Bengaluru, India

Girish S.[2]
Department of Mechanical Engineering Dayananda Sagar University,
Bengaluru, India

Shreyas RA[3], Ravitej YP[4], Vinay MS[5]
Department of Mechanical Engineering,
Dayananda Sagar University,
Bengaluru, India

Jegadeeswaran N.[6]
School of Mechanical Engineering, Reva University,
Bengaluru, India

◆ **Abstract:** 3D (Three dimensional) Printing is a process of manufacturing the product through layer by layer and it is the best process for producing a finished object. In this paper, it has been discussed about the Fused deposition modeling (FDM) method of printed parts, their mechanical properties and analysis of tensile test. Fused deposition modeling is an Additive Manufacturing (AM) technique which works by a heated nozzle laying down molten material in layers to produce a desired part. FDM is one of the most common techniques used for 3D printers and has become one of the most popular Rapid Prototyping (RP) techniques in the last decade. The slicer program cross-sections the model into individual layers of a specified height and converts the desired height and other settings into G-Code to be read by the printer. The printer reads the G-Code, heats up a liquefier to the desired temperature to melt the polymer filament of choice, and begins extruding the material. The printing filaments used for this study were Acrylonitrile Butadiene Styrene (ABS) and Polycarbonate (PC). These filaments are fed through the heated liquefier by two drive wheels where the filament is then melted and extruded through a nozzle onto the build platform. The heating and extrusion of the filament to the specified diameter is all contained within the extrusion head which moves in the x-y plane depositing material on the build platform. After each layer is finished the build platform moves down a specified z or layer height and the process repeats for the next cross-sectioned layer until the part is completed. This paper is an Experimental study and Finite element analysis of 3D printed ABS and polycarbonate materials.

◆ **Keywords:** 3D printing, Finite element analysis, Polycarbonate materials

[1]saranbav-me@dsu.edu.in, [2]girishshivalingaiah@gmail.com, [3]shreyas6196@gmail.com, [4]ravitej-me@dsu.edu.in, [5]vinay-me@dsu.edu.in, [6]njagadeeswaran@reva.edu.in

DOI: 10.1201/9781003685906-14

1. Introduction

3D printing is a method of producing solid three-dimensional (3D) items. Plastic is commonly used in 3D printers since it is easier to work and less expensive. Other materials, such as metals and ceramics, can also be 3D printed but they are too expensive. 3D printers are useful, since they can quickly create new objects and are capable of producing highly detailed ones. This means that an engineer can test a large number of new designs without having to wait for them to be made by someone else. They can also be used to repair plastic parts and to create toys and models. A large number of people print 3D items at home.

Additive Manufacturing refers to a group of manufacturing techniques that create three-dimensional objects by layering materials on top of one other. A manufacturing process must include the following three essential elements in order to be designated as an AM technology [1,2,3,4].

1. The use of Computer-Aided Design (CAD) software to create visual 3D models. A technologist working in the field of additive manufacturing should be able to use the software programmes in order to effectively manufacture using these technologies. Any type of complicated 3D model of a product can be built using these CAD tools. The amount of material that will be extruded by the 3D printer and the time taken to build the 3D model are calculated, and the data is saved as a G-code file that the printer can understand.
2. Slicing and tool path generation: The CAD 3D-generated models must be prepared in a manner that the additive manufacturing machine can understand. The slicing programme converts the 3D design into layered models that can be easily traced by the machine tool.
3. Conversion of the 3D model into a genuine product: Using engineering materials such as plastics, metal powders, and composites, an additive manufacturing machine converts the 3D model into a real product. To construct the 3D component, the materials are melted and then allowed to flow according to the G-code (tool path) from the slicing software.

There are several additive manufacturing processes, which are categorised according on the material and machine technology employed in component creation. There are seven different types of AM processes, according to the 2010 American Society for Testing and Materials (ASTM F42-) standards, which are described below [1,2].

- Techniques for fusing powders in a bed. (Powder bed fusion techniques)
- VAT photopolymerization methods
- Material jetting techniques
- Binder jetting techniques
- Sheet lamination techniques
- Direct energy deposition techniques.

The procedures described above, use a variety of materials and machinery to manufacture 3D printed components, and they have been well studied in the literature. The advantages of additive manufacturing over other traditional production are as follows.

- Increased material efficiency due to the lack of material waste from cutting or machining.
- Resource efficiency is improved because these procedures do not necessitate the use of auxiliary resources such as tools, jigs, and fixtures.
- Due to the lack of tool restrictions, products of high complexity and intricacy can be created.
- Additive manufacturing processes increase the flexibility of production.

Although these methods are appealing, they are limited by a number of constraints, including the size of parts that can be made, surface and micro structural defects, and the high cost of AM equipment. The methods are also extremely sluggish, making them difficult to implement in mass manufacturing. FDM stands out among the different AM production techniques. The FDM process, sometimes known as material extrusion AM, is the most basic, cost-effective, and widely available 3D printing approach for polymer-based materials, and it has been widely employed in a variety of industries [5,6,7]. The basics of FDM and its applications, as well as parameters and quality aspects of the process, are discussed below.

2. Science of FDM and Applications

Polymers are utilized as the raw material in fused deposition modelling also known as material extrusion additive manufacturing. The filament is often heated to a molten condition before being extruded via the machine's nozzle. According to the G-code, the nozzle head can move in three degrees of freedom to place the extruded polymer on the build plate. The FDM process is represented in a schematic diagram in the Fig. 14.1 where the filament is continually fed through the machine's extruder and nozzle through two rollers revolving in opposite directions. Layers of material are deposited on the build plate until the product shape and size are reached [8,9,10].

During layering, the printer nozzle moves back and forth according to the spatial coordinates of the original CAD model in the G-code files until the component is in the

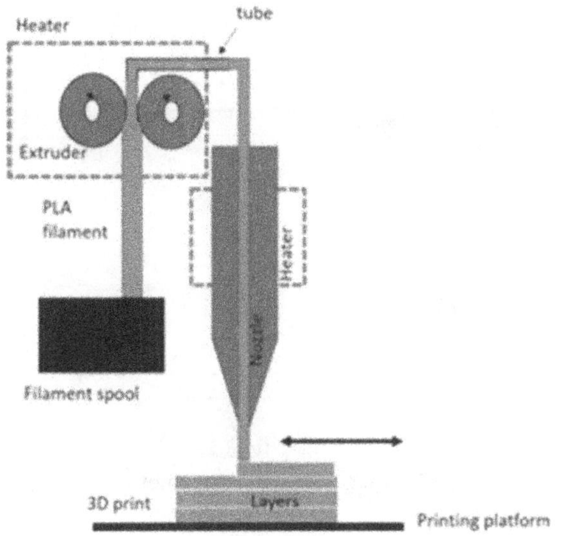

Fig. 14.1 Process of FDM

required size and form. Multiple extrusion nozzles can be employed to deposit polymer ingredients in some FDM systems, especially when compositional gradients are required. The resolution and effectiveness of extrusion are often determined by the qualities of the thermoplastic filament, and as a result, several 3D printers are designed for specific filament materials. In truth, the majority of low-cost FDM 3D printers can only print one type of thermoplastic, which is Polylactic Acid (PLA). The components are usually placed on the building plate platform, which can be snapped off or soaked in a detergent after printing, depending on the type of thermoplastic. To improve the surface appearance and functionality of the printed components, they can be cleaned, sanded, painted, or machined [11].

There are variety of materials utilised in FDM, and as previously said, PLA is the most often used material among residential and industrial 3D printer users for the following reasons:

Polylactic acid is a bioplastic that is both environmentally friendly and non-toxic to humans and animals. PLA is a green material since it is made from completely renewable resources like corn, sugarcane, wheat, or any other high carbohydrate-containing resource [11,12]. PLA has a glass transition temperature of 50 to 70 degrees Celsius and a melting point of 180 to 220 degrees Celsius. As a result, it can be extruded by the majority of low-energy and cost-effective 3D printers. Unlike other plastics that have caused severe disposal issues, PLA plastics are compostable and break down quickly when disposed of. PLA, being a biopolymer, degrades to natural and non-toxic gases, water, biomass, and inorganic salts when subjected to natural circumstances, hydrolysis, or incineration. PLA has been proven to have strong flexural modulus, tensility, and flexural strengths in its semi-crystalline form. PLA is favoured by most 3D printer users since adhesion between the print and the platform does not necessarily require a hot bed. Non-heated bed printers, on the other hand, face a significant issue with graphene-doped PLA, which does not create high-quality prints on non-heated build plates. PLA is available in a range of colours and textures on the marketplace [13,14].

Polycaprolactone, polypropylene, polyethylene, polybutylene terephthalate, Acrylonitrile butadiene styrene, wood, nylon, metals, carbon fibre, graphene-doped PLA, and other materials are utilised in FDM processing [11,12].

The following is a list of the most common FDM applications in current civilization: FDM finds application in making customised items for a variety of applications, such as personalised toys, vehicle parts, interior design components, implants, beauty products, and so on, due to its flexibility and capacity to generate complicated profiles [12].

FDM is also used in the medical field to create moulds for implant casting, medical devices, and implants. The 3D printing of moulds for investment casting of medical implants is the most intriguing application [15,16]. Metal moulds and sacrificial patterns were used in traditional investment casting to generate the complicated shapes of any implant. As a result, adopting 3D printed moulds eliminates the need for sacrificial material, lowering costs, saving time, and reducing waste [5,8].

Other applications of FDM include direct printing of electrochemical cells for energy storage devices, micro-trusses for biomedical scaffolding, drug delivery components in the pharmaceutical industry, direct printing of conductors for electronic industry among others [17,18,19,20].

Polycarbonates are a type of thermoplastic polymer with carbonate groups in its chemical structure. Polycarbonates are strong, robust materials used in engineering, and some grades are optically clear [21]. They are simple to work with mould, and thermoform. Bisphenol A and phosgene are condensation polymerized to make polycarbonates.

ABS is a type of technical thermoplastic and amorphous polymer that is resistant to impact. The three monomers that make up ABS are acrylonitrile, butadiene, and styrene [22,23,24].

3. Methodology

Figure 14.2 shows the flow chart of methodology. It started with 3D printing of specimen with tensile and shear test on Universal testing machine. Followed by finite element

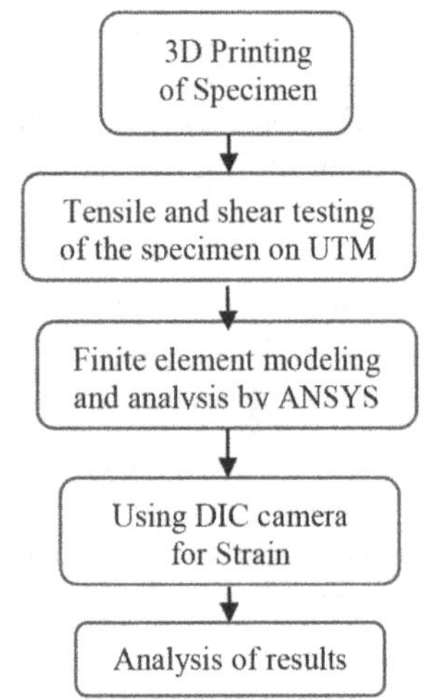

Fig. 14.2 Flowchart of methodology

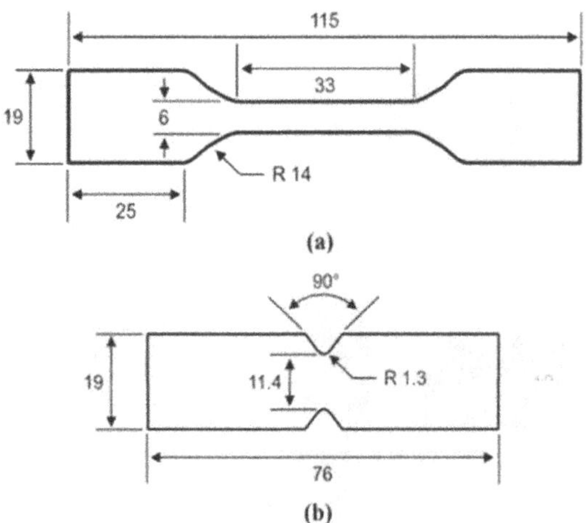

Fig. 14.3 Specimen geometry

modelling by Ansys and then using DIC camera for strain measurement and lastly analyzing the results.

4. Materials and Specimen Fabrication

Polycarbonate (PC) and Acrylonitrile Butadiene Styrene (ABS) were utilised in this investigation to create samples in a Stratasys® Fortus 360mcTM and an Ultimaker® 2 3D-printer, respectively. The specimen geometries followed ASTM D-638 specifications for Type IV tensile specimens and ASTM D-5379 specifications for shear specimens. Figure 14.2 depicts these specimens and select dimensions for both specimen types. The specimens were printed at a thickness of 0.160 in. The shear and tensile specimens were first generated in Solidworks®, then exported in STL format and integrated into each 3Dprinter's slicer software to generate the G-code needed to print each specimen type

To know the extent of anisotropy in 3D-printed materials, tensile and shear characterization of acrylonitrile butadiene styrene and polycarbonate 3D-printed parts was done as shown in Fig. 14.3. To assess the directional qualities of the materials, specimens were printed with different raster ([+45/-45]) and build orientations (flat). The strain was measured using 2D digital image correlation (DIC) on tensile and shear specimens and loaded into a universal testing equipment [25,26].

For each tensile sample the Poisson's ratio, Young's modulus, yield strength, ultimate strength, strain at failure, breaking strength, and strain energy density were recorded. For each shear sample, values for shear modulus, yield strength, and ultimate strength were gathered. FEA is used on PC and ABS specimens, and the final results are compared.

5. Results and Discussion

The material was tested for both tensile and shear sample using a single test rig. At room temperature (23 °C), the specimens were tested at a rate of 1.5 mm/min. The specimens were loaded in a Test Resources® 315 electromechanical universal testing machine with a 22 kN load cell using custom fixtures. Multiple clevis joints were used in the fixtures indicated in Fig. 14.4 to ensure that the load path through the sample was free of bending. Test Resources® Test builder TM software recorded load readings at a rate of 10 Hz.

In tensile testing, the average stress in the specimen was calculated by dividing the load by the cross-sectional area at any given load. Digital image correlation (DIC), a non-contact, full-field shape and deformation assessment technology, was used to collect the requisite strain data. The strains were measured using DIC over a rectangular region centred in the test portion on both sides of the specimen. This approach corrects for a variety of inaccuracies caused by probable flaws in loading and specimen geometry during testing. Because of the potential grip slippage, loading mechanism compliance, and load cell compliance that are common in experiments, DIC was used for strain measurement rather than crosshead displacement. Electrical resistance strain gauges also mechanically reinforce the specimen while also causing strain gauge self-heating difficulties while testing polymers. Furthermore, extensometers normally measure axial strain, whereas the Poisson's ratio in this work requires both axial and transverse strains. Although dedicated extensometers that measure both axial and transverse strain exist, such devices are bulky, and the test section used in these is quite tiny. Any bending during loading is compensated for using DIC strain measurement on both sides of the sample. In general, DIC can be done with a single camera or a pair of cameras in stereo. A single camera configuration is prone to inaccuracies owing to out-of-plane stiff body motion if it is not taken into consideration, but a stereo setup can compensate for such motion.

The DIC setup for this investigation comprised of two 5-megapixel grayscale cameras from Point Grey® Research positioned on either side of the samples, which recorded images of both sides of the samples at the same time. By averaging the strain from the two sides, it was possible to compensate for rigid body motion by using a single camera system on both sides of the specimen. The only thing that was necessary for specimen preparation was a light speckle pattern of black paint over a light coat of white paint on the white plastic backdrop, which resulted in little reinforcement.

VIC-SnapTM 2009 was used to capture images of the samples at a rate of 1 Hz, which were then analysed using VIC-2DTM to determine the strains. The typical subset size of 29 and step size of 5 were employed during VIC-2DTM processing to give appropriate strain and deformation data. A pair of reference photos (one image per camera) of each side of the sample were captured after it was placed into the testing equipment and a preload was applied. To calculate the strains during the course of the testing cycle, these reference photos were compared to photographs acquired from the corresponding sides of the sample.

This method proved to be efficient, as it allowed for the testing of a single specimen in a matter of minutes, including mounting the specimen in the loading fixtures, taking initial undeformed DIC images, and loading the specimen through failure. Figure 14.4 depicts the complete DIC and universal testing machine setup utilized for both tensile and shear testing.

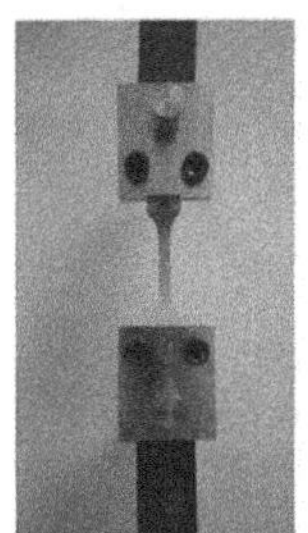

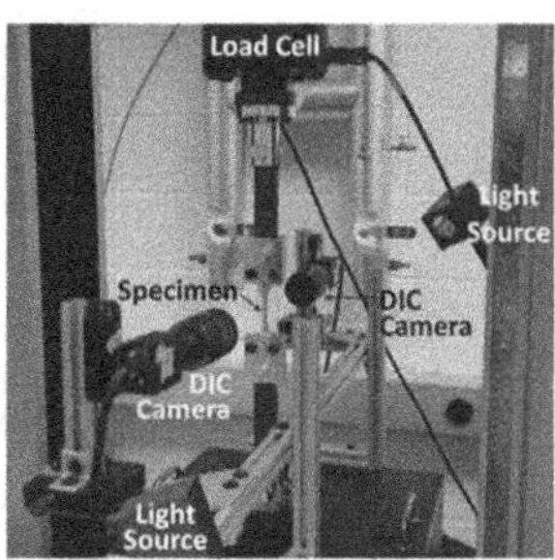

Fig. 14.4 Experimental setup

Throughout this work, each shear and tensile test followed the same general experiment technique. The sample is inserted into its relevant shear or tensile fixture after applying the high contrast DIC speckle pattern to it, as shown in Fig. 14.4. The fixture/specimen combination is secured into the Test Resources® testing equipment with several clevis joints once the specimen has been put into the fixture.

After that, a preload of 10 N is applied, and reference photos of both sides of the sample are captured using the VIC-SnapTM programme. The testing machine's displacement rate is set to 1.5 mm/min, and the VICSnapTM DIC software's rate is set to 1 Hz. The DIC and testing machine systems are both turned on at the same time, and the testing is carried out until the specimen fails. The test is completed once the specimen fails, and the process is repeated. After all 10 tests in a data set have been performed, the stress and strain behaviour of each specimen or combination of specimens is examined, relevant properties calculated, and stress-strain curves constructed.

The average strain was extracted from the DIC pictures over a rectangular region of 3 mm wide by 7 mm long centred on the two faces of the specimen for the tension testing. This method gives a reliable way to calculate the average strain with high accuracy (repeatability). The average shear stresses on both sides of the shear specimens were evaluated throughout a rectangular area of 3mm wide by 11mm (nearly the whole width between the notches). This procedure ensured reliable and reproducible average shear strain results.

The fracture surfaces of the tensile specimens were analysed for both the ABS and PC, in addition to the stress-strain curve data. At fracture as shown in Fig. 14.5, the ABS specimens all responded in the same way. In the plane perpendicular to the loading direction, all of these specimens cracked neatly. The PC specimens shattered in a number of ways, as illustrated in Fig. 14.5, with the most common

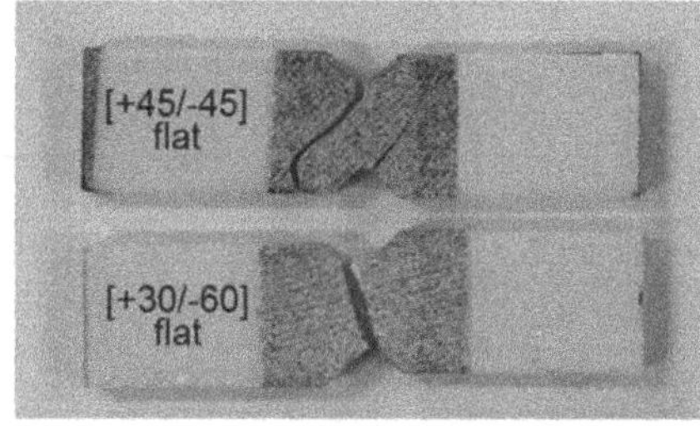

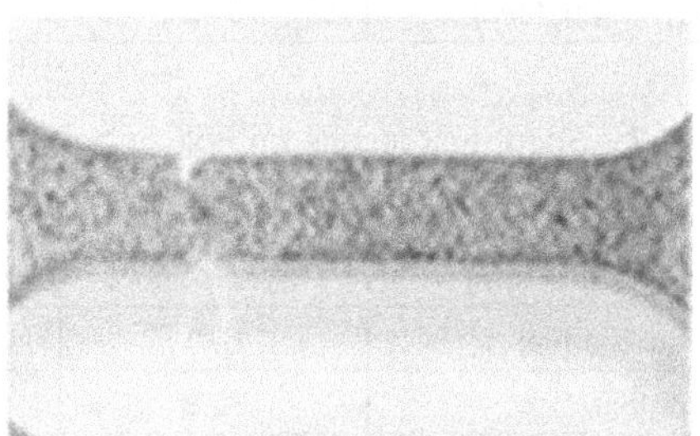

Fig. 14.5 Specimen

failure mechanisms for each orientation combination. The majority of the specimens appeared to fracture in the same perpendicular plane as the ABS samples. Both on-edge specimens showed jagged perpendicular fracture surfaces, although there were several noteworthy variances. The flat PC specimen [+45/-45] as shown in Fig. 14.5 was likewise distinct from other fracture surfaces in that the PC roads tended to tear apart rather than break cleanly, resulting in a saw tooth-like pattern.

Finally, the shear specimens' fracture surfaces were analysed. Few examples were brought to total failure due to the ductile behaviour of the ABS specimens, since the DIC speckle pattern was destroyed well before specimen failure. The PC specimens, on the other hand, were far more fragile, and virtually all of them failed completely during testing. Figure 14.5 shows the PC specimens shattered in two different ways. The majority of specimens on the tension side failed a short distance from the notch, which is nevertheless regarded a valid test because this type of failure may only alter the ultimate strength significantly. Failure across the notch area was recorded in far fewer specimens, and failure across the notch generally suggested lower yield and ultimate shear strengths. The flat samples [+30/-60] and [0/90] had the most specimens with notch failures as shown in Fig. 14.5, while the other orientations had failures just beyond the notch area.

The Tabular result of tensile & shear test, FEA for Polycarbonate, ABS is shown in Table 14.1, 14.2, 14.3 and 14.4. While the finite element analysis simulation, Ansys results of shear and tensile specimen is shown in Fig. 14.6, 14.7, 14.8 and 14.9.

Table 14.1 Tension test results for Polycarbonate, ABS

Tension Test Result		
Property	***Polycarbonate***	***ABS***
Poisson's Ratio	0.39	0.36
Young's Modulus (MPa)	1890	1960
Yield Strength (MPa)	39.7	30.3
Ultimate Strength (MPa)	56.6	32.8
Strain at Failure (%)	6.72	8.89
Breaking Strength (MPa)	54.0	29.6

Table 14.2 Shear test results for Polycarbonate, ABS

Shear Test Result		
Property	***Polycarbonate***	***ABS***
Shear Modulus (MPa)	670	740
Yield Strength (MPa)	22.8	19.1
Ultimate Strength (MPa)	36.9	28.8

Table 14.3 FEA results of shear test for Polycarbonate, ABS

FEA –Shear Test Result		
Property	***Polycarbonate***	***ABS***
Shear Modulus (MPa)	2879.7	2669.2
Yield Strength (MPa)	251	280
Maximum Shear Stress (MPa)	751.29	488.15

Table 14.4 FEA results of tensile test for polycarbonate, ABS

FEA –Tensile Test Result		
Property	***Polycarbonate***	***ABS***
Yield Strength (MPa)	80.6	80
Ultimate Strength (MPa)	189.53	187
Strain at Failure	0.7	0.69

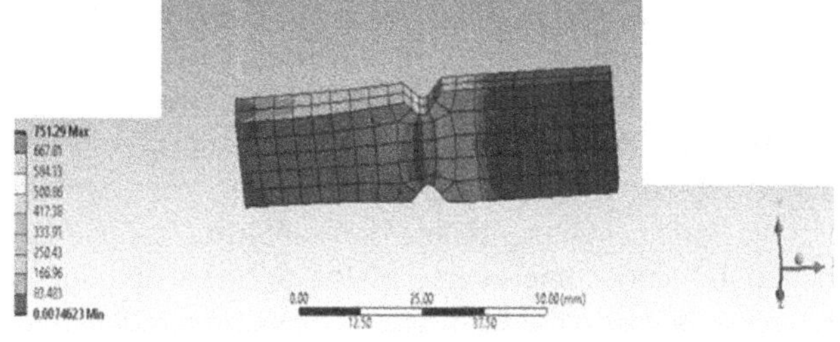

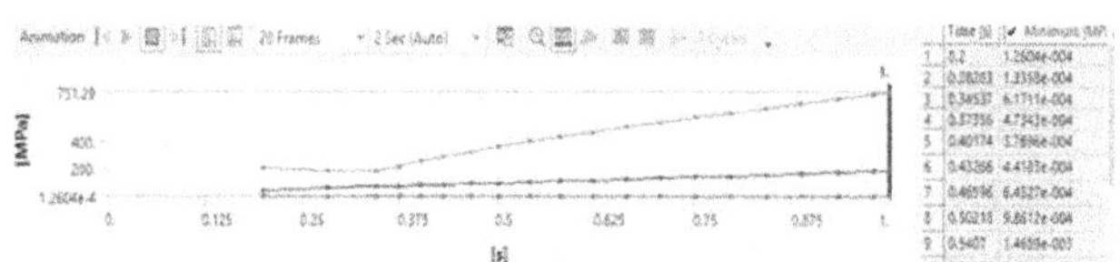

Fig. 14.6 Shear test for Polycarbonate in ANSYS

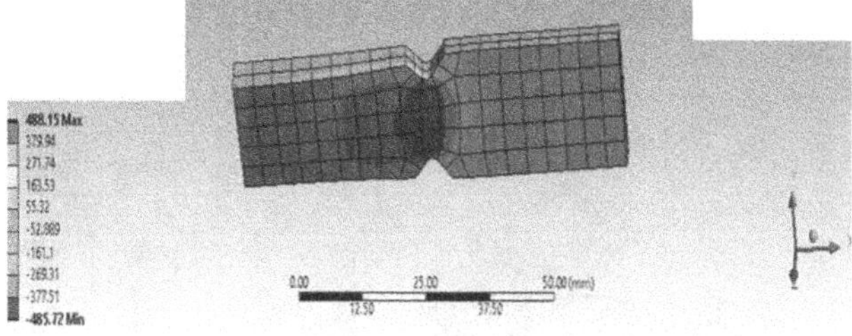

Fig. 14.7 Shear test for ABS

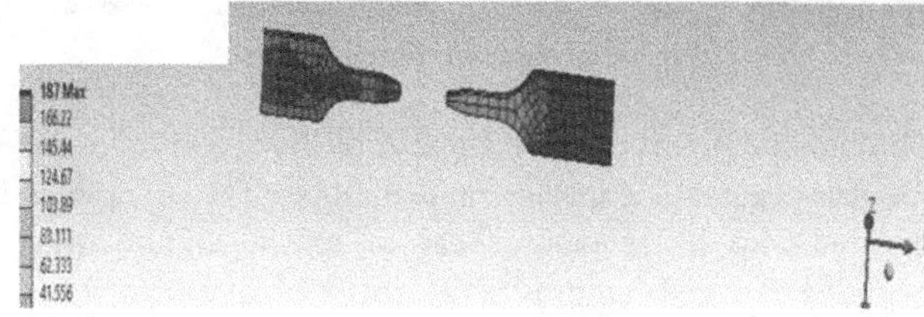

Fig. 14.8 Tensile test for polycarbonate in ANSYS

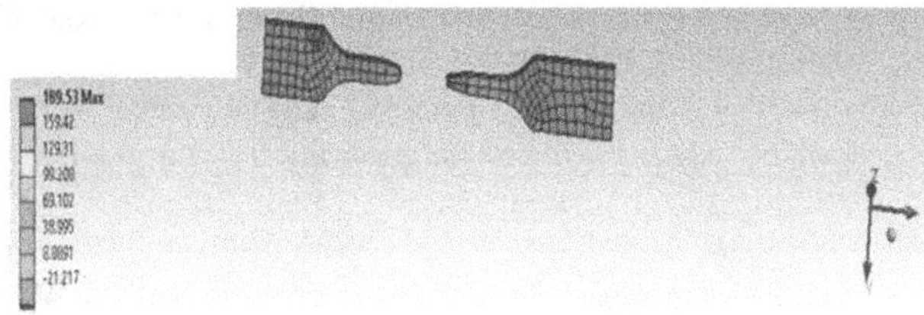

Fig. 14.9 Tensile test for ABS in ANSYS

6. Conclusion

FEA was performed for Polycarbonate material for tensile and shear tests. FEA results were compared with the shear and tensile test results. FEA was also performed for ABS material for tensile and shear tests. Polycarbonate and Acrylonitrile Butadiene styrene specimens were tested and evaluated according to ASTM standards D-638 (tensile) and D-5379 (shear) to determine if the specimens were anisotropic in nature.

When comparing the Young's modulus and Poisson's ratio, the ABS specimens were isotropic; nevertheless, considering merely these parameters provides an incomplete and erroneous description of the behaviour of ABS 3D-printed materials. Anisotropy was discovered when the ultimate strength, strain at failure, and strain energy density of ABS specimens were compared.

References

1. Marouene Zouaoui, Julien GardanM Julien Gardan, Pascal Lafon, Ali Makk, Carl Labergere and Naman Recho, 2021. A Finite Element Method to Predict the Mechanical Behavior of a Pre-Structured Material Manufactured by Fused Filament Fabrication in 3D Printing, Appl. Sci. *11*(11), 5075.
2. Rohde, S.,Cantrell, J. Jerez, A. Kroese, C. Damiani, D. Gurnani, R DiSandro, L. Anton, J. Young, A. Steinbach, D. et al. 2018. Experimental Characterization of the Shear Properties of 3D–Printed ABS and Polycarbonate Parts. Exp. Mech58, 871–884.
3. Ismail Ezzaraa, Nadir Ayrilmis, Mohamed Abouelmajd, Manja Kitek Kuzman, Ahmed Bahlaoui, Ismail Arroub, Jamaa Bengourram, Manuel Lagache and Soufiane Belhouideg, 2023. Numerical Modeling Based on Finite Element Analysis of 3D-Printed Wood-Polylactic Acid Composites: A Comparison with Experimental Data, *Forests 14*(1), 95
4. Gardan, J. Makke, A. Recho, N. 2018, Improving the fracture toughness of 3D printed thermoplastic polymers by fused deposition modeling. Int. J. Fract. 210, 1–15.
5. Ahn S.H. and C. Baek, 2003. Anisotropic Tensile Failure Model of Rapid Prototyping Parts - Fused Deposition Modeling (FDM) International Journal of Modern Physics B, 17 (8 & 9) 1510–1516.
6. Abueidda D.W , M. Elhebeary, C.-S.A. Shiang, S. Pang, R.K.A. Al-Rub, I.M. Jasiuk, 2019. Mechanical properties of 3D printed polymeric Gyroid cellular structures: Experimental and finite element study, Materials and Design, 165, Article 107597
7. Song, Y, Y. Li, W. Song, K. Yee, K.-Y. Lee, V.L. Tagarielli, 2017. Measurements of the mechanical response of unidirectional 3D-printed PLA, Materials and Design, 123 154–164
8. Yao. T, J. Ye, Z. Deng, K. Zhang, Y. Ma, H. Ouyang, 2020. Tensile failure strength and separation angle of FDM 3D printing PLA material: Experimental and theoretical analyses Composites Part B. Engineering, 188 , Article 107894
9. Kalita S.J, S. Bose, H.L. Hosick, A. Bandyopadhyay, 2003. Development of controlled porosity polymer-ceramic composite scaffolds via fused deposition modelling Mater. Sci. Eng. C, 23 (5) 611–620
10. Rosenzweig D, E. Carelli, T. Steffen, P. Jarzem, L. Haglund, 2015, 3D-printed ABS and PLA scaffolds for cartilage and nucleus pulposus tissue regeneration Int. J. Mol. Sci., 16 (7), 15118
11. Gautam Tanikella N, B. Wittbrodt, J. Pearce, 2017, Tensile Strength of Commercial Polymer Materials for Fused Filament Fabrication 3D Printing.
12. DaliaC, B. Rimantas, M. Daiva, M. Rytis, N. Audrius, O. Armantas, 2018. Multi-scale finite element modeling of 3D printed structures subjected to mechanical loads Rapid Prototyp. J., 24 (1), 177–187
13. Wittbrodt B, J.M. Pearce, 2015, The effects of PLA color on material properties of 3-D printed components Addit. Manuf., 8, 110–116
14. Farah S, D.G. Anderson, R. Langer, 2016. Physical and mechanical properties of PLA, and their functions in widespread applications — a comprehensive review Adv. Drug Deliv. Rev., 107 (Supplement C) , 367–392
15. Pollard D, Ward C, Herrmann G, et al. 2017. The manufacture of honeycomb cores using Fused Deposition Modeling. *Adv* Manufacturing: Polym Composites Sci 3(1): 21–31.
16. Mamatha S, Biswas P, Das D, et al. 2020, 3D printing of cordierite honeycomb structures and evaluation of compressive strength under quasi-static condition. Int J Appl Ceram Technology 17(1): 211–216.
17. Jackson, RJ, W. A., and M. M., 2018. 3D Printing of Asphalt and its Effect on Mechanical Properties, Miner. Des., no. 160, pp. 486– 474.
18. Zhao, Y. Chen, Y. Zhou, Y. 2019. Novel mechanical models of tensile strength and elastic property of FDM AM PLA materials: Experimental and theoretical analyses. Mater. Des. 181, 108089.
19. Gardan, J. Makke, A. Recho, N. 2019. Fracture Improvement by Reinforcing the Structure of Acrylonitrile Butadiene Styrene Parts Manufactured by Fused Deposition Modeling. 3D Print. Addit. Manuf. 6, 113–117.
20. Raajesh krishna CR, Chandramohan P, Saravanan D, 2019. Effect of surface treatment and stacking sequence

on mechanical properties of basalt/glass epoxy composites. Polym Polym Composites 27(4): 201–214.

21. Figueroa- Cavazos J.O , E. Flores – Villalba, J. A. Dia Elizondo, O. Martínez - Romero, C.A. Rodríguez, H.R. Siller, 2016. Design concepts of polycarbonate-based intervertebral lumbar cages: finite element analysis and compression testing, Appl. Bionics Biomech., Article 7149182
22. Cole, D.P. Riddick, J.C. Jaim, H.M.I. Strawhecker, K.E. Zander, N.E. 2016. Interfacial mechanical behavior of 3D printed ABS. J. Appl. Polym. Sci. 133.
23. Rodríguez, J.F. Thomas, J.P. Renaud, J.E. 2001. Mechanical behavior of acrylonitrile butadiene styrene (ABS) fused deposition materials. Experimental investigation. Rapid Prototyp. J. 7, 148–158.
24. Cole, D.P.; Riddick, J.C.; Jaim, H.M.I.; Strawhecker, K.E.; Zander, N.E.2016. Interfacial mechanical behavior of 3D printed ABS. J. Appl. Polym. Sci. 133.
25. Adams, D.F., and Lewis, E.Q. 1994. Current Status of Composite Material Shear Test Methods, SAMPE 31(6) 32–41.
26. Adams D.F. and D.E. Walrath, 1987. Further Development of the Iosipescu Shear Test Method, Experimental Mechanics, 27(2), 113–119.

Note: All the figures and tables in this chapter were made by the authors.

Advances in Additive Manufacturing Technologies – Gurusamy Pathinettampadian et al. (eds)

Development of Solar Powered Mini Air Cooler

Gayathri N.*,
Madhesh M., Karthikeyan G.,
Hari Krishna Raj S., Kathiravan G.
Department of Mechanical Engineering,
VelTech HighTech Dr. Rangarajan Dr. Sakunthala Engineering College,
Chennai, Tamilnadu

◆ **Abstract:** This work reports the development of a mini solar-powered air cooler using a Brushless DC (BLDC) motor with a nominal voltage of 1000 kv and polycrystalline solar panels. The goal is to create a sustainable and energy-efficient personal cooling system for indoor or outdoor applications. The BLDC motor's inherent power efficiency and low power usage will be utilized by the cooler, while the polycrystalline panels will harness solar energy as electricity to drive the motor. The project will aim to optimize the system for maximum power harvesting and efficient conversion to mechanical energy based on parameters such as solar panel angle, motor operating voltage, and fan blade shape. Major features of the project involve the design of a light and portable cooler frame, incorporation of a voltage regulator to provide stable motor performance, and provision for energy storage means to ensure continuous cooling during low-light situations. The project's success will be measured by the cooler's airflow rate, cooling capacity, and overall energy efficiency under varying solar irradiance levels. This mini solar-powered air cooler has the potential to provide a clean, economical, and environmentally friendly alternative to conventional air conditioners, particularly in regions with abundant sunlight and limited access to grid electricity.

◆ **Keywords:** Brushless DC (BLDC) motor, Electronics speed controller, 2200mAh battery, Propeller, Solar panel.

1. Introduction

Solar energy is the heat and radiant light that the sun emits and can be used to heat water or create electricity. It is a clean, renewable energy source that could supply all of our demands for energy. There are two main ways to harness solar energy. Solar photovoltaics (PV). PV panels directly transform sunlight into electricity through the use of semiconductors. This kind of solar technology is most frequently utilized in residences and commercial buildings [1–5]. Concentrated solar power systems focus sunlight onto a tiny area through the use of mirrors, and the heat produced can be utilized to heat water or produce electricity. Renewable energy IS derived from solar radiation is a resource that is infinite in nature. Using solar energy to meet our energy needs won't have a negative impact on the environment. Solar energy may be used for a number of tasks, such as lighting, heating water, and producing electricity. Notwithstanding the difficulties, the solar energy sector is expanding quickly and has a promising future. Solar energy is growing more and more competitive with conventional energy sources as the price of solar panels drops and technology advances. Our energy demands will probably be largely met by solar energy in the future. For millions of years, life on Earth has been powered by the sun, an endless supply of energy. We are

*Corresponding author: gayathri@velhightech.com

DOI: 10.1201/9781003685906-15

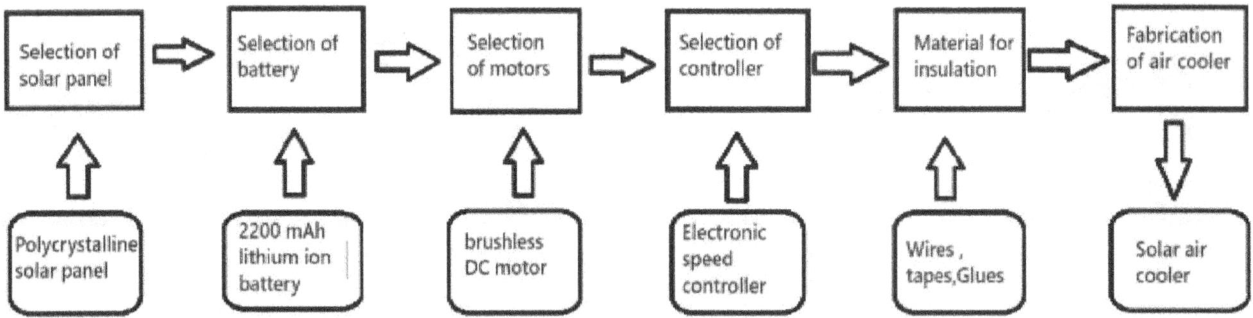

Fig. 15.1 Solar air cooler- process flow

currently using solar technologies to harvest its electricity. To provide cool, refreshing air in a small and energy-efficient way. These coolers work by using a combination of solar power and evaporative cooling. A solar panel collects sunlight and converts it into electricity to run the cooler's fan and pump. This makes them ideal for off-grid locations or for those looking to reduce their reliance on traditional electricity [9–12].

1.2 Scope

They operate on solar energy, a renewable energy source, thus lowering your use of fossil fuels and carbon emissions.

- They generally consume less energy than traditional air conditioners.
- To regulate deforestation and consumption of fossil fuel
- Low cost
- Mini coolers are intended for individual usage or for little areas, and not for refrigerating whole rooms.

2. Methodology

A cost-saving and eco-friendly alternative to traditional air conditioning systems is a solar-powered air cooler. It is the ideal solution for hot, arid weather as it harnesses the power of the sun to cool the air. The mechanism behind a solar-powered air cooler is evaporative cooling. The air cooler has a fan, a water pump, and an arrangement of water-absorbing pads. The water pump circulates water from the tank to pads 2, 2 and the fan blows air over the wet pads. Evaporation of water on the wet pads cools air when hot dry air is blown over them. The cooled air is blown by the fan, creating a cool breeze that can be used to lower the room temperature. A solar panel is used to convert sunlight into electricity in order to drive the solar-powered air cooler. A charge controller, which is fixed on the solar panel, helps to control the level of electricity flowing to the battery. The electricity created by the solar panel can be supplied to the fan and water pump. The battery is charged during the daytime when the sun is shining. The energy stored can then be used on cloudy days or at night [13]. The solar air cooler is a green and efficient way to cool your home or office. It is easy to install and barely requires maintenance.

Its functioning relies on evaporative cooling, using the energy from the sun to create a cooling effect.

2.1 Goal Components

- Solar Panel: Hooks sunlight and turns it into electricity.
- Charge Controller: Manages the electricity flow from the solar panel to the battery.
- Battery: Serves to store electricity for usage in times of no sunlight.
- Fan: Provides air circulation over the water-soaked pads.
- Water Pump: Transfers water from a tank to maintain the pads wet.
- Water-soaked Pads: Enables the process of evaporation.
- Digital Thermometer: Reads output temperature.

Fig. 15.2 Basic components

2.2 Working Principle

The water pump keeps the pads moist by circulating water from a tank. When warm, dry air is blown over these damp pads, the water evaporates and takes heat away from the air, therefore cooling it. The fan forces the cooled air into the room to provide a cool breeze that brings down the room temperature.

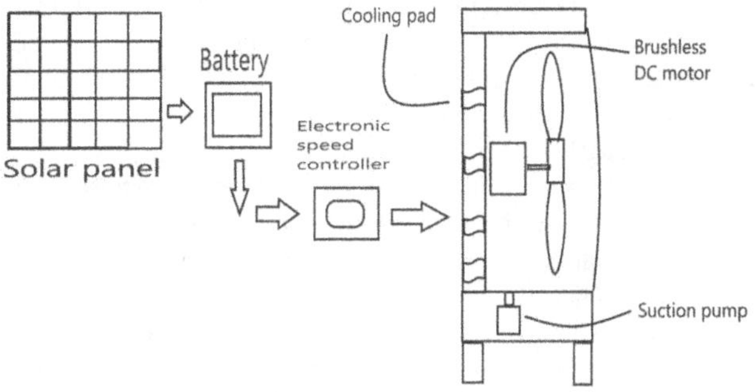

Fig. 15.3 Block diagram

3. Construction and Result

Design, construction and assembly of the mini solar powered air cooler were implemented. The A2212/13T is a high powered and efficient brushless DC 1000 kV motor well suited for most applications, ranging from drones and RC cars to boats. The A2212/13T also has an upper power level of 25 watts, thus it can output a sufficient level of power without overheating (Fig. 15.5). This is beneficial in applications where the motor will encounter a lot of load, such as in a drone that needs to lift a heavy pay load. With its high power and efficiency, the A2212/13T is also very robust [14–16]. It is made out of top-quality components and features a heavy-duty construction that can withstand even the most demanding applications. This makes it a great choice for hobbyists and professionals who need a motor that will endure. Overall, the A2212/13T is a strong and employed Digital thermometer to monitor and the described temperature in Table 15.1.

Fig. 15.4 Fabricated model

Table 15.1 Temperature measurement

Time in Minutes	Temperature in °C
0	35
10	34.6
20	33.6
30	30.1
40	28.5

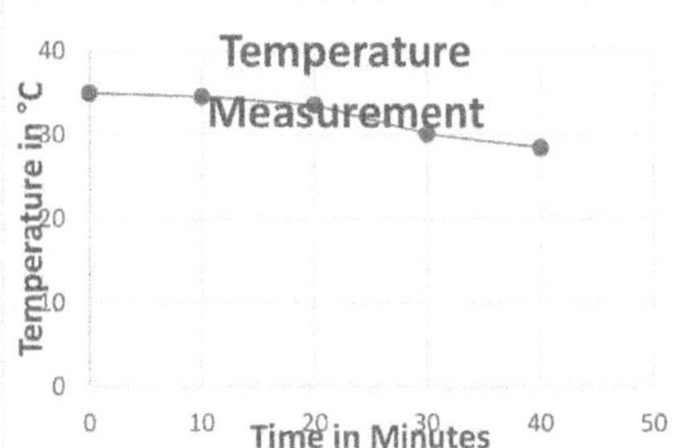

Fig. 15.5 Output temperature vs time

4. Conclusion

Solar powered air cooler run on clean, renewable solar power, reducing carbon footprint and grid electricity reliance. BLDC motors consume significantly less power than traditional AC motors, enabling longer battery life and reduced solar panel size requirement. BLDC motors enable quieter operation, variable speed, and potentially more intense air flow compared to AC motors. Solar-powered BLDC air coolers provide a revolutionary and clean cooling option, particularly for sunny locations and areas with poor grids. Although the upfront expenditure could be more, money saved in the long run and ecological dividends make them a viable option in the long run

Reference

1. Ahmed H. Abdel Salam and Carey J. Simonson, "Annual evaluation of energy, environmental and economic performances of a membrane liquid desiccant air conditioning system with/without ERV", Applied energy, 116, pp.134–148, 2014.
2. Al Alili A, Hwang Y, Radermavcher R and Kubo I, "A high efficiency solar air conditioner using concentrating photovoltaic/thermal collectors", Applied Energy, 93, pp.138–147, 2012.
3. B. Parida, S. Iniyan, and R. Goic, "A review of solar photovoltaic technologies", Renewable and Sustainable Energy Reviews, 15(3), pp. 1625–1636, 2011.
4. C. E. C. Nogueira, J. Bedin, R. K. Niedzialkoski, S. N. M. de Souza, and J. C. M. das Neves, "Performance of monocrystalline and polycrystalline solar panels in a water

pumping system in Brazil", Renewable and Sustainable Energy Reviews, vol. 51, pp. 1610–1616, 2015

5. Chen Jian, Wan Jikang, Cao Guangrong, "Experimental study on heat transfer coefficient of air cooler under frosting condition", Journal of Refrigeration, 20(2), pp.16–19, 2001.
6. G. K. Dubey, "Fundamentals of electrical drives", 2nd ed. Alpha Science International, 2001.
7. H. H. Ozturk and D. Kaya, "Electricity Generation from Solar Energy: Photovoltaic Technology", Umuttepe Publications, Kocaeli, Turkey, 2013.
8. Kondepudi S N ONeal D L, "Effect of frost growth on the performance of louvered finned tube heat exchangers", International Journal of Refrigeration, 12(3), pp.151–158, 1989.
9. Rens MacNeill and Dries Verstraete, "Performance Testing and Modeling of a Brushless
10. DC Motor, Electronic Speed Controller and Propeller for a Small UAV Andrew Gong", The University of Sydney, Sydney, NSW, Australia, 2006.
11. Stoecker W F, "Industrial Refrigeration Handbook", New York: McGraw-Hill, 1998.
12. Wang C C, Lee W S, Sheu W J, "A comparison of the airside performance of the fin-and-tube heat exchangers in wet conditions; with and without hydrophilic coating", Applied Thermal Engineering, 22(3), pp.267–278, 2002.
13. Ye H Y, Lee K S., "Refrigerant circuitry design of fin-and-tube condenser based on entropy generation minimization", International Journal of Refrigeration, 35(5), pp.1430–1438, 2012.
14. Zang Runqing, Liu Qi, Li Xing, "Comparative study on the performance of direct expansion fluid supply and gravity recirculation liquid supply refrigeration system", Annual Conference of China Refrigeration Society, 2011.
15. Zhang Chun Road, "Simulation Principle and Technology of Refrigeration and Air Conditioning System", Chemical Industry Press, 2012.

Note: All the figures and tables in this chapter were made by the authors.

Advances in Additive Manufacturing Technologies – Gurusamy Pathinettampadian et al. (eds)
© 2026 Taylor & Francis Group, London, ISBN 978-1-041-16687-0

16 Optimization of Machining Parameters in Wire Cut Electrical Discharge Machining of AISI 1010 Steel

Gayathri N.*,
Santhosh V., Abilash B.L, Praveen J.
Department of Mechanical Engineering,
VelTech HighTech Dr. Rangarajan Dr. Sakunthala Engineering College,
Chennai, Tamilnadu

◆ **Abstract:** Non-traditional machining operations are extensively employed in tool and die work for machining hard-to-cut materials with complex profiles. Machining is required on such hard materials to achieve such complex profiles. Some special grades of material like AISI 1010 steel maintain their properties when subjected to elevated temperatures. (EDM) is a popular machining for non-conventional technique employed to cutting hard materials, making complex geometries, creating microwires, and dealing with other difficult operations. EDM is only beneficial when machining conductive materials and is not applicable when machining non-conductive materials. This research discusses how various parameters of EDM processes affect surface roughness (SR) in the manufacturing process. AISI 1010 steel samples were machined with a copper electrode and biosilica deionized water-based green dielectric medium. The most significant process parameters of this experiment that were studied were flushing pressure, pulse on-time, pulse off-time and discharge current. The experiments were performed on an EDM machine with AISI 1010 steel as the workpiece and a copper electrode. For maximizing surface roughness (SR), parameters like pulse-on-time (Ton), current (A), and voltage (V) were optimized independently. The Taguchi method was utilized to optimize process parameters, and confirmation runs were executed to create a mathematical model. The Taguchi L9 orthogonal array was utilized to reduce the number of experiments. To find the best independent parameters and their levels, variation was studied with respect to discharge current, and a model was formulated for predicting roughness based on pulse-off-time settings. The confirmation run data demonstrated that the predicted values were in close agreement with experimental values, confirming the precision of the formulated model

◆ **Keywords:** AISI 1010 steel, ANOVA, Electric discharge machining, Surface roughness, Taguchi method

1. Introduction

The research delves into the significance of exploring machining parameters, particularly in unconventional machining methods, to meet the demands posed by intricate contours and challenging materials like super alloys. Electric Discharge Machining (EDM) emerges as a crucial technique, particularly for intricate shapes and hard-to-machine materials, offering advantages like precise shaping, low cutting force, and fine detailing. However, EDM also presents challenges such as surface integrity alterations (Fig. 16.1).

Overall, the research aims to enhance understanding and optimization of EDM parameters for improved machining efficiency and surface quality, particularly focusing on AISI1010 steel.

*Corresponding author: gayathri@velhightech.com

DOI: 10.1201/9781003685906-16

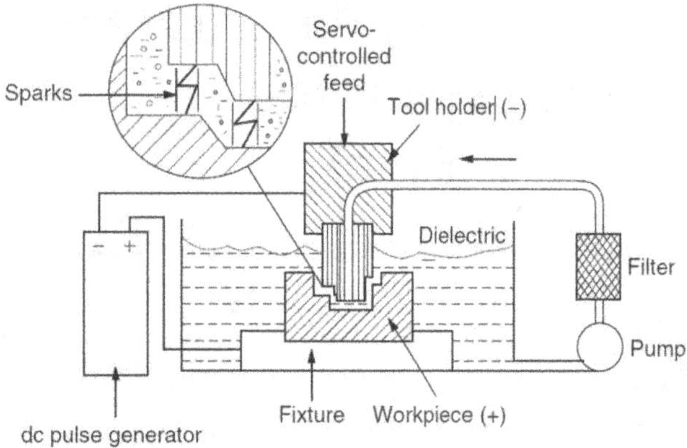

Fig. 16.1 Working principle of EDM [4]

2. METHODOLOGY

2.1 Experimental Setup

Material selection focuses on AISI1010 Steel, known for maintaining its mechanical qualities even at high temperatures. The compositions of AISI1010 steel in shown in Table 16.1.

Table 16.1 Composition of AISI1010 Steel

Description	Iron	Manganese	Sulphur	Carbon	Phos-phorous
Chemical Formula	Fe	Mn	S	C	P
% of Composition	99.40	0.45	≤0.050	0.1	≤ 0.040

Diameter of 8.35 mm and kerosene as the dielectric medium are used for experimentation. Wire electrode material consisted of diffused brass wire, while deionized water served as the dielectric fluid. Measurement of EDM process response variables included kerf width, MRR, surface roughness, cutting speed, over cut, and surface topography, employing various equipment such as optical microscopes and scanning electron microscopes.

The experimental setup with high-speed steel as the work material and a copper electrode EDM. Various research approaches are employed to determine the Material Removal Rate (MRR), surface finish, microeconomic hardness, white film thickness, heat-affected layer thickness, hole expansion, microstructural changes, and metallurgical evolutions of the EDM process response variables.

2.2 Selection of Process Parameter

Surface roughness predicts component performance, while material removal rate impacts productivity. Optimization focuses on achieving minimal surface roughness with elevated material removal rate. Previous research did not emphasize the significance of process characteristics, therefore a comprehensive analysis using Design Of Experiments (DOE) as well as Analysis Of Variance (ANOVA) was necessary.

The problem-solving procedure involves selecting factors, determining stages, choosing an orthogonal array, conducting experiments, analyzing results, confirming findings, and verifying improvements. This systematic approach ensures valid conclusions and effective parameter optimization (Table 16.2).

An orthogonal array is selected for experimentation, considering four parameters at three stages each. Nine experiments are conducted using AISI1010 Steel work pieces, measuring surface roughness as the response. Minitab software assists in experimental design and statistical analysis (Fig. 16.2).

Table 16.2 The range of variance in EDM parameters

Machining Variables of EDM	Minimum Stage	Inter-mediate Stage	Maximum Stage
Wire Feed Rate (mm/min)	6	8	10
Pulse on Time (μsec)	2	3	4
Wire Tension (N)	1.6	1.8	2.0
Pulse off Time (μsec)	8	12	16

Table 16.3 Design of the taguchi experiment and findings

S. No.	Rate per Wire Feed (F) in mm/min	On Time Pulse (T_{ON}) in μs	Wire Tension (T) in N	Off Time Pulse (T_{OFF}) in μs
Expt. No. 1	6	2	1.6	8
Expt. No. 2	6	3	2.0	12
Expt. No. 3	6	4	1.8	16
Expt. No. 4	8	2	1.8	12
Expt. No. 5	8	3	1.6	16
Expt. No. 6	8	4	2.0	8
Expt. No. 7	10	2	2.0	16
Expt. No. 8	10	3	1.8	8
Expt. No. 9	10	4	1.6	12

Experiments are conducted on an EDM machine with AISI1010 Steel work pieces using predefined test conditions. Surface quality is assessed after machining, and observations are recorded for analysis (Table 16.3).

Fig. 16.2 Samples used in experiments in the order listed in the experiment

Table 16.4 The variability of EDM parameters and their range

Machining Variables of EDM	Minimum Stage	Intermediate Stage	Maximum Stage
On Time Pulse (μsec)	2	3	4
Off Time Pulse (μsec)	8	12	16
Rate per Wire Feed (mm/min)	6	8	10
Wire Tension (N/mm^2)	1.6	1.8	2.0
Other Constant Parameters			
Open voltage	85V		
Polarity	Negative		
Electrode	Copper		
Flushing method	Side		
Rate of Flushing	1.5 lit/min		

2.3 Test Conditions for EDM

For EDM, there are four factors with three levels altogether.

Table 16.5 Test conditions for EDM process

Process Variable	Minimum Stage	Inter-mediate Stage	Maximum Stage
On Time Pulse (μsec)	1	2	3
Off Time Pulse (μsec)	4	8	12
Rate per Wire Feed (mm/min)	4	6	8
Wire Tension (N/mm^2)	1.4	1.6	1.8

3. Results and Discussion

3.1 Surface Topography

Pockmarks, debris deposition, and crater development are seen in the surface topography of EDM. Higher pulse current and pulse on-time cause surface roughness, which deepens craters and increases debris deposition (Table 16.4).

Table 16.6 Statistics of S/N ratio for surface quality (Rt)

Specimen	Maximum Peak /High in surface Profile (Rt)	S/N Ratio
1	22.3	-28.0280
2	19.3	-28.4976
3	25.2	-27.8890
4	20.1	-26.0639
5	25.9	-26.1926
6	18.3	-25.0084
7	24.8	-25.8893
8	21.2	-25.1536
9	20.5	-26.5676

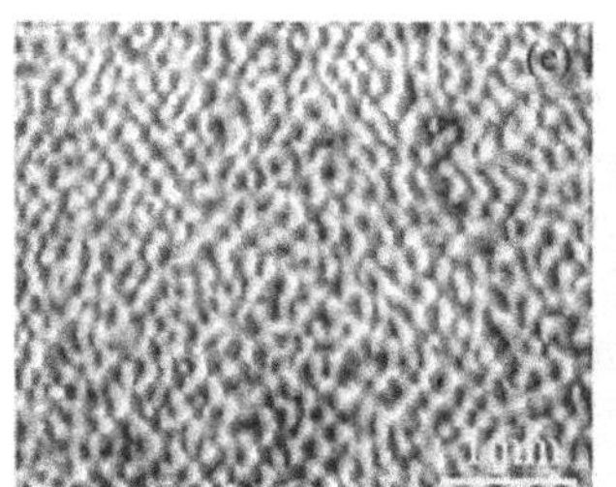

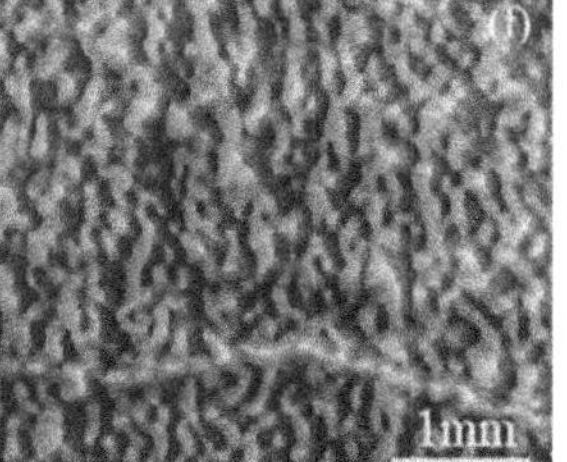

Fig. 16.3 Metallurgical micrograph of EDMed surface at different EDM conditions

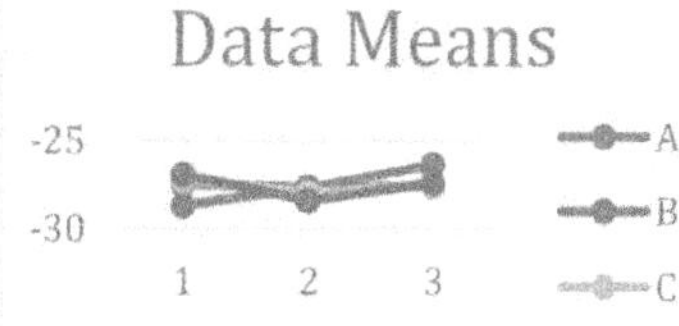

Fig. 16.4 Plotting the mean effect of S/N at the highest profile peak height

3.2 Surface Quality (Ra)

A pie diagram was used to show the proportion contribution of each parameter to surface quality (Rt), as can be seen in Fig. 16.3 and 16.4. The ANOVA results for the maximum profile peak height are shown in Table 16.5 together with the percentage of contribution for each process parameter.

Percentage of contribution

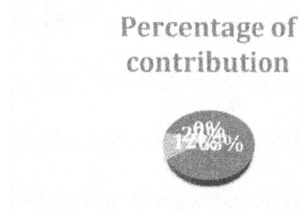

Fig. 16.5 Controllability of surface quality as % of identified variables

Table 16.7 Maximal profile peak height for surface quality ANOVA's findings

Process Variables	DoF	SS	MS	F-statistics	% of Contribution
TON	2	0.2462	0.1478	0.01	0.34
TOFF	2	50.12	25.62	4.92	64.87
WT	2	7.324	3.7478	0.32	12.22
WF	2	22.82	11.61	1.16	22.57
Error	6	-	-	-	-

3.3 Taguchi's Machining Rate Analysis

The power of control over the indicated parameters on machining rate (MR) was statistically investigated using the Signal to Noise (S/N) ratio. The findings of the S/N ratio analysis are shown in Table 16.8, and the rank matrix that was used to look at how different parameters affected the rate of machining is shown in Table 16.9 and Fig. 16.5.

Table 16.8 Statistics of S/N for machining rate

Specimen	Machining Rate	S/N
1	28.387	28.6224
2	44.272	30.8224
3	43.557	30.2264
4	26.925	28.5880
5	40.345	32.6720
6	35.284	32.2424
7	22.214	26.8368
8	34.225	31.8560
9	41.286	32.6789

Table 16.9 Rank matrix for power of factors on machining rate

Level	P_{ON} in µsec (A)	P_{OFF} in µsec (B)	W_F in mm/min (C)	W_T in N/mm^2 (D)
1	27.34	31.29	30.76	31.12
2	31.52	31.51	29.61	30.20
3	31.35	30.23	29.61	29.29
Delta	3.64	1.19	0.92	1.48
Rank	1	3	4	2

As can be seen in Fig. 16.6, a pie diagram was utilized to compute and show the percentage contribution of each parameter to the machining rate (MR). Together with the degree of freedom, F-statistics, mean squares, sum of squares as well as % of contribution for each process parameter.

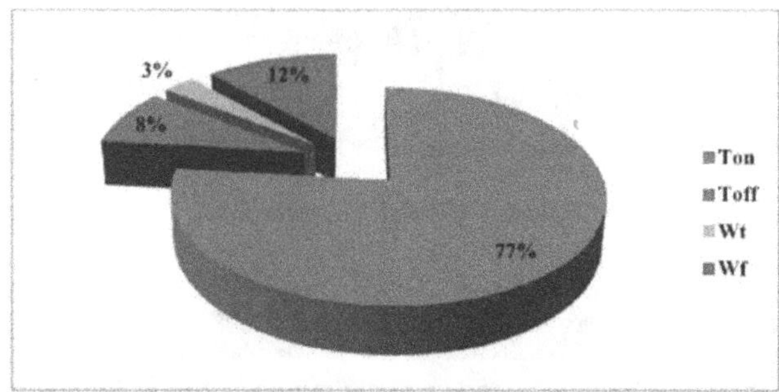

Fig. 16.6 Power of control in % of identified variables on machining rate

4. Conclusion

The study focused on optimizing the wire electrical discharge machining (WEDM) process for AISI1010 Steel components, particularly those with complex profiles like those found in space vehicle engine parts. The aim was to ensure precise manufacturing given the demanding conditions such components are exposed to, such as elevated temperatures. The investigation looked at wire feed rate, wire tension, pulse on time and pulse off time to determine the effects of various process variables on machining sensitivity and control over the machining rate.

In addition, the study examined the connection between surface quality and Pulse off Time fluctuation, developing a mathematical model to forecast roughness values depending on this parameter setting. In a similar vein, the investigation of the machining rate fluctuation with pulse on time led to the creation of a mathematical model for machining rate prediction.

In essence, the research aimed to enhance the precision and efficiency of WEDM for complex steel components, particularly those vital for space vehicle engines, by optimizing process variables and establishing predictive models for surface quality and machining rate.

Acknowledgment

This work was supported by Vel Tech High Tech Dr. Rangarajan Dr. Sakunthala Engineering College, Institute Research Fund (Seed Money), Reference Number: VH/R&D/SMP/22-23/018.

References

1. A.M. Barani, R.Latha, R.Manikandan, Implementation of Artificial Fish Swarm Optimization for Cardiovascular Heart Disease International Journal of Recent Technology and Engineering (IJRTE), Vol. 08, No. 4S5, 134–136, 2019.

2. Abhinaba Roy, S. Narendranath, Effect of spark gap voltage and wire electrode feed rate on machined surface morphology during EDM process, Materials Today: Proceedings, 5, (9), Part 3, 2018, pp. 18104–18109.
3. Akash Nag, Ashish Kumar Srivastava, AmitRai Dixit, AmitavaMandal, TanmayTiwari, Surface Integrity analysis of Wire-EDM on in-situ hybrid composite A359/Al2O3/B4C,Materials Today: Proceedings, Volume 5, Issue 11, Part 3, 2018, pp. 24632–24641.
4. Badal Dev Roy, R. Saravanan, M. Chandrasekaran, "Turbo-Matching Optimization on Compressor Mapby Comparing Different Methodological Approach For TATA 497 TCIC -BS III Engine - An Investigation", International Journal of Mechanical Engineering and Technology, Vol. 9 (9), 2018, pp. 621–630.
5. Fakkir Mohamed, K. Lenin, Application of Taguchi Design Method To Optimize The Electrical discharge machining Parameters. Article MFG-246, Int. Conf. on Contemporary Design and Analysis of Manufacturing and Industrial Systems, 18-20, Jan. 2018, pp. 430–435.
6. Gnanavel C, Saravanan R, Chandrasekaran M, and R Pugazhenthi, "Restructured Review on EDM - The state of the art', International Conference on Emerging Trends in Engineering Research IOP Publishing, IOP Conf. Series: Materials Science and Engineering 183 (2017) 012015.
7. Janaka R. Gamage, Anjali K. M. DeSilva, Dimitrios Chantzis, Mohammad Antar, Sustainable machining:Process energy optimisation of wire electro discharge machining of Inconel and titanium superalloys, Journal of Cleaner Production, Volume 164, 15 October 2017, Pages 642–651.
8. K. Fakkir Mohamed, Lenin, application of taguchi design method to optimize the Electrical discharge machining parameters. Article MFG-246, Int. Conf. Contemporary Design Analysis Manufact. Industr. Syst. 18–20, (2018) 430–435.
9. N. Shanmugasundaram, S. Sridharan, S. Swapna, Mitigation and prediction of radiation effects in solar power plants using neuro fuzzy and neural network, Europ. J. Molecular Clinic. Med. 7 (8) (2020) 2500–2509.
10. R. Gamage, A.K.M. DeSilva, D. Chantzis, M. Antar, Sustainable machining: Process energy optimisation of wire electrodischarge machining of Inconel and titanium superalloys, J. Cleaner Prod. 164 (2017) 642–651.
11. R. Sathish, R. Manikandan, S. Silvia Priscila, B. V. Sara and R. Mahaveerakannan, A Report on the Impact of Information Technology and Social Media on Covid–19, 2020 3rd International Conference on Intelligent Sustainable Systems (ICISS), Thoothukudi, India, 2020, pp. 224–230.

Advances in Additive Manufacturing Technologies – Gurusamy Pathinettampadian et al. (eds)
© 2026 Taylor & Francis Group, London, ISBN 978-1-041-16687-0

Innovative Use of Sugarcane Bagasse Ash in Interlocking Bricks for Construction

Sangeetha M.[1]
Assistant Professor,
Department of Civil Engineering,
Vel Tech High Tech Dr. Rangarajan Dr. Sakunthala Engineering College,
Avadi, Chennai, Tamil Nadu, India

Krishnan K.[2]
UG Student,
Department of Civil Engineering,
Vel Tech High Tech Dr. Rangarajan Dr. Sakunthala Engineering College,
Avadi, Chennai, Tamil Nadu, India

Vignesh Kalyan A.[3]**, Mohan A.**[4]**, Paul Theophilus T.**[5]
UG Student,
Department of Civil Engineering,
Vel High Tech Dr Rangaraja Dr Sakunthala Engineering College,
Chennai, India

◆ **Abstract:** One of the most necessary material for construction is bricks. This project is mainly based on making use of waste materials in to the construction material along with modern techniques, utilizing waste materials of industries and agriculture waste products in the is the main focus of most of the industries for economic, environmental and technical reasons. By replacing scba partially of 10%, 20% & 30% by weight in traditional brick. In order the Scba bricks were designed and manufactured in 3 different proportions. Traditional bricks were not evolved with much different in their sizes and proportions 190x90x90mm. However,due to technology development contemporary brick resulted in more efficient and shown the overall quality of the material.

◆ **Keywords:** Sugarcane baggasse ash, Clay, Admixtures, Water absorption

1. Introduction

1.1 General

In masonry constructions, a brick is a block or individual unit of ceramic material. In order to hold the bricks together and create a lasting structure, they are usually piled or placed as brick work using different kinds of mortar. 87% of the world's bricks are produced in Asia. In addition, China and India consume a lot of bricks. Bricks of common or standard sizes are frequently made in large quantities. They have been regarded as one of the most durable and sturdy building materials employed in the twentieth century. The manufacturing of bricks emits toxic gasses, resulting in significant air pollution. Asperin

[1]msangeetha@velhightech.com, [2]vh13515_civil23@velhightech.com, [3]vh113523_civil23@velhightech.com, [4]vh113518_civil23@velhightech.com, [5]vh12202_civil22@velhightech.com

DOI: 10.1201/9781003685906-17

India makes over 60 billion clay bricks annually, causing significant land erosion and unprocessed emissions.

The manufacturing of bricks emits toxic gasses, resulting in significant air pollution. India makes around 60 billion clay bricks annually, which has a significant influence on soil erosion and unprocessed emissions. Traditional brick firing methods caused severe local air pollution. IS 2212 (1991) specifies a standard brick size of 19cm × 9cm × 9cm. Bricks are put horizontally, using either dry or wet mortar. In some cases, like as adobe, the brick is simply dried. To create a real ceramic, it is typically burnt in some type of kiln. Clay bricks are utilized in many buildings, including dwellings, factories, tunnels, rivers, and bridges.

Clays have been used in construction for ages, with varying qualities depending on their intended use. Over the past five decades, India's brick manufacturing has declined due to increased production costs due to the importation of components.

To meet the growing need for energy-efficient building materials, it's important to use cost-effective and ecologically friendly technology,as well as modernize existing ways using local materials. Researchers are exploring potential solutions for this problem utilizing various materials such as fly ash, black cotton soil, concrete blocks, and agricultural waste.

1.2 Objective

a) To minimize the agricultural waste and avoid environmental pollutions by using SCBA in brick.
b) To minimize the usage of clay in brick and save the clay for future use.
c) To attain a high strength bricks
d) Interlocking system is achieved with precise size.

2. Materials and Methods

2.1 Clay

The clay used in the present work are locally available Natural clay. The specific gravity of the Clay was 2.75 and the water absorption of the clay was 16%.

2.2 Lime

The lime used in the present work is locally available Natural lime. The specific gravity of the lime was 2.3 and the sieve analysis was done.

Result for sieve analysis:

Percentage of gravel size (> 4.75) 6.9% is the percentage of coarse size (4.75–2.36 mm). Medium size (2.36 -0.425 mm) as a percentage = 61.7%

19.3% is the percentage of fine size (0.425–0.075 mm). Fineness size percentage (< 0.075) = 0.2%

2.3 Bagasse Ash from Sugarcane

The SCBA used in this investigation was produced by carefully burning sugarcane bagasse that was imported from India's Tamilnadu province. India produces more than 300 million tons of sugarcane annually, of which 10 million tons are wasted because they are not used. The sieve analysis revealed that the sugarcane bagasse ash had a specific gravity of 2.36.

The sieve analysis's outcome: 90.8% is the percentage of gravel size (> 4.75). 8.9% is the percentage of coarse size (4.75–2.36 mm). 66.6% of the population is medium-sized (2.36–0.425 mm). 19.3% is the percentage of fine size (0.42-0.075mm). The proportion of fineness size (less than 0.075)=0.3%

2.4 Mix Proportion

10%, 20%, and 30% of the weight of clay brick is partially replaced with Scba. Scba uses lime as a binding agent in the clay. Three proportions were used in the design and development of Scba-cs-l combination bricks.

Table 17.1 Mix proportion of materials

S. no	CLAY	SCBA	LIME
1	85%(2.80kg)	10%(0.350kg)	10%(0.350kg)
2	75%(2.45kg)	20%(0.700kg)	10%(0.350kg)
3	65%(2.10kg)	30%(1.05kg)	10%(0.350kg)

2.5 Composition of Interlocking Bricks

Composition:

Flyash – 1 Tray

Plasticizer – 50ml

Super Plasticizer - 50 ml

3. Findings and Conversation

3.1 Test of Compressive Strength

At four days of brick age, the strength activity index test was conducted in compliance with ASTM C 109 (8). Mortars were placed in a mold to create 190x90x90mm specimens, which were then kept for four days at 36°C in an open location. Four days before to the test, the specimens were curing. One sample is used to report the compressive strength. In the current investigation, the mortar mix proportions were utilized in conjunction with the SCBA CS-L brick percentage.

The test findings are recorded in Table 17.2 and the Figure and compared with regular bricks and SCBA bricks.

Table 17.2 Compressive strength testing result

Proportion	Weight of Brick (Kg)	Load	Compressive Strength (N/mm^2)
Ordinary brick	3.12	75	4.2
10%	3.217	30	1.70
20%	3.230	65	3.99
30%	3.291	35	2.22

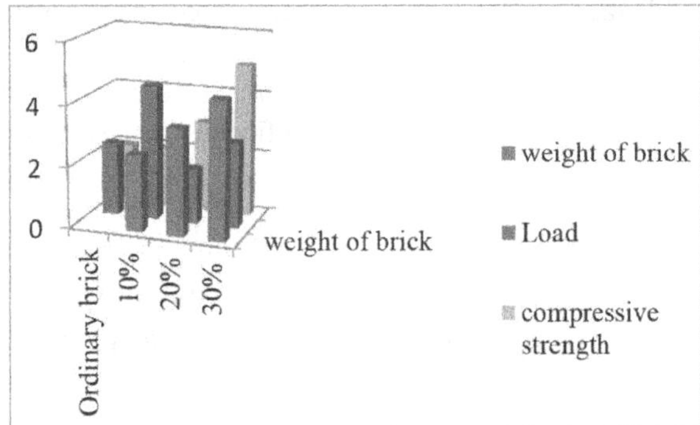

Fig. 17.1 Comparison of compressive strength test results with ordinary bricks and partially replaced SCBA brick

3.2 Test of Water Absorption

Bricks are tested for water absorption to ascertain their durability characteristics, including their degree of burning, quality, and weathering behavior. The test results were acquired and summarized in comparison to regular bricks and SCBA bricks

Table 17.3

Proportion	Initial Weight(g)	Final Weight(g)	% of Absorption
Ordinary Brick	3208	3514	9.54%
10%	3279	3557	8.48%
20%	3235	3762	16.32%
30%	3214	4108	27.81%

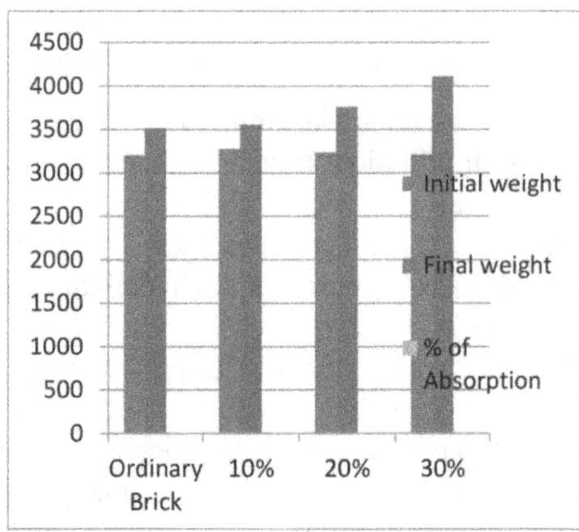

Fig. 17.2 Comparison of water absorption test results with ordinary bricks and partially replaced SCBA bricks

4. Conclusion

Based on the current study and the few observations that have been made, the combination of lime and clay soil with sugarcane bagasse ash performed really well. It was noted that the addition of SCBA increased early strength. Thus, our project's outcome shown that using sugarcane bagasse ash improves compressive strength while lowering waste disposal. In comparison to other proportions, a 20% mixture of sugarcane bagasse ash, lime, and clay provides more strength.

References

1. Vivekka Olivia John, Muhammad Ali Musarat, and Wesam Salah Alaloul* "Mechanical and Thermal Properties of Interlocking Bricks Utilizing Wasted Polyethylene Terephthalate" Int J Concr Struct Mater (2020) 14:24 Alaloul et al. 10.1186/s 40069-020-00399-9 https://doi.org.
2. Prakash Gaire, Qi Ma, and Hongwang Ma Creation and mechanical assessment of a novel interlocking block of earth brickwork Research Article, August 8, 2019 https://doi.org/10.1177/1369433219868931.
3. Yulia Hayati1, Cut Nursaniah2, Teuku Firsa3, and Nurul Malahayati1* "Cost Analysis of Interlocking Bricks Using Mixed IOP Conference Series 536 (2019) 012088 Material Variation Design Results" Materials Science and Engineering.
4. L. Sahanasree, V. Sarumathi, U. Sathish, K.S. Sowmya, R. Priyanka, G. Pavitra, and U. Satish The article "Experimental Analysis Of Interlocking Bricks By Using Sewage Sludge And Fly Ash" Appeared In The April 2018 Special Issue Of The Ssrg International Journal Of Civil Engineering (Ssrg-Ijce).

Note: All the figures and tables in this chapter were made by the authors.

Advances in Additive Manufacturing Technologies – Gurusamy Pathinettampadian et al. (eds)
© 2026 Taylor & Francis Group, London, ISBN 978-1-041-16687-0

Programs and Experts Work on Fibre Metal Laminates Used in MotZor Vehicles

Saravanan M.[1]
Department of Mechanical Engineering, Aarupadai Veedu Institute of Technology, Vinayaka Mission's Research Foundation, Deemed to be University, Tamil Nadu, India

Ramanan N.[2]
Department of Mechanical Engineering, Sri jayaram institute of Engineering and Technology, Chennai

Saravanakumar M.[3], **Churchil P.**, **Sangeetha Krishnamoorthi**[4], **Aravinth R.**
Department of Mechanical Engineering, Aarupadai Veedu Institute of Technology, Vinayaka Mission's Research Foundation, Deemed to be University, Tamil Nadu, India

P. Gurusamy[5]
Department of Mechanical Engineering, Chennai Institute of Technology, Chennai

Susindaran K.[6]
PG Student, Department of Mechanical Engineering, Chennai Institute of Technology, Chennai

◆ **Abstract:** This study investigates the mechanical properties of a new type of Fiber Metal Laminate (FML) comprising aramid, glass, and carbon fibers combined with aluminium alloy 2024 and epoxy resin. By exploring the impact response and stress-strain behavior of these laminates, we aim to highlight their suitability for lightweight aviation applications. Specimens were manufactured using water jet machining to ensure precision and were prepared as 5-3/2 laminates with dimensions of 300x300 mm² and a thickness of 3 mm through the hand lay-up method. Mechanical testing, including tensile and flexural assessments, was conducted using a servo ball screw mechanism with a 10-ton capacity, while impact responses were evaluated using a Charpy testing machine. Additionally, the fracture surfaces of the impact, flexural, and tensile specimens were analyzed via scanning electron microscopy (SEM) to gain insight into failure mechanisms. The results of the experimental and computational analyses demonstrate the enhanced mechanical properties of these layered composites, suggesting a promising direction for future applications in the aerospace industry.

◆ **Keywords:** FML, Fracture surface, Ballistic analysis, FEA simulation

1. Introduction

There are two primary parts to a sandwich structure: the core and the faces. Metal sheets or fibre composite layers are the most common types of face sheets, and each has its own set of pros and cons. Researchers are now looking for novel materials that exhibit improved characteristics [1, 2]. Despite its weight, metal sheets offer superior

Corresponding authors: [1]saranmechatronics@gmail.comline, [2]ramananinjs2020@gmail.com, [3]sarov003@gmail.com, [4]geetha30981@gmail.com, [5]gurusamyp@citchennai.net, [6]susindarank2001@gmail.com

DOI: 10.1201/9781003685906-18

continuity and resistance to transverse stresses. While fiber-reinforced plastics are lighter than metal sheets, they are more likely to be affected by environmental factors and suffer significant internal damage when exposed to sideways forces, especially during impacts [1-2]. There is a plethora of options because to the diversity of fibres with varying mechanical characteristics, orientation, and stacking sequences. Fibre metal laminates, often known as ARALL, GLARE, or CARALL, are very effective building materials. The majority of commercial applications now utilise unidirectional glass fibre prepregs that are sandwiched between sheets of aluminium alloy. Because FMLs are versatile and can be customized for various engineering uses, they are now rapidly advancing in technology and production. Three to five. Currently, design engineers consider the strength-to-weight ratio as the most important attribute of composite structures, which leads them to conduct strength optimisation analyses. Also included is the use of failure criteria for predicting loading circumstances that cause the composite structure to collapse. All of the defined FML properties allow for FML constructions to have either a lower thickness or greater stresses. Since this is the case, many types of buckling can occur in thin-walled FML sections. The capacity, as well as strength, of thin-walled components may be determined by the buckling load. Researchers have extensively studied how composite structures behave during buckling and after it occurs, using experiments and computer models. We also used FEM and semi-analytic approaches to conduct comparative studies in. Still, thin-walled FML members' buckling strength and load-bearing capability have received scant attention in the literature [6]. A hybrid material, fibre metal laminates (FMLs) combine metal with fibre. The properties of metals and composites are brought together in reinforced polymer. Both fatigue resistance and damage tolerance are exceptional qualities of these materials. More and more sectors are finding uses for materials that combine aluminium with glass, carbon, or aramid fibres, such as GLARE, CALL, or ARALL, from seven to eleven. Cracks forming in the matrix are the main cause of failure in glass fiber composites during long-term repeated use. These cracks then spread and break the major load carrying components, which are the fibres. Composites made of glass fibres, which have a lower elasticity modulus than those made of carbon fibres, may fail due to fatigue since they subject the matrix to greater stresses. By reducing the size of damage processes, nanoparticles such as montmorillonite clays (MMTs) or carbon nanotubes (CNTs) are believed to enhance strain energy absorption through the creation of many microscopic Nano-scale fractures [9]. Despite several publications addressing the tensile/impact behaviour of these laminates, research on this aspect of FML performance is still in its early stages, according to a recent review on tensile and impact resistance of FMLs. This research looks at the tensile and impact testing, as well as the correlation between the results and numerical models, of glass-aramid fiber/Al laminates (usually 0.28 mm thick).

2. MATERIALS AND METHODS

2.1 Procedures and Supplies

Carbon, glass, and carbon fibre, which are renowned for their durability, are the primary components of the composite material, excellent resistance to organic solvents, strong tensile strength, and abrasion. Because of their large surface area to weight ratio, they are thermal insulators, non-conductive, non-melting point, low-flammability, elastic, and ideal for fabrics that are subjected to high temperatures. With a low thermal conductivity of about 0.05 W/(mK), their larger surface area makes them much more prone to chemical damage. Blocks of glass, carbon, or aramid fibre are effective insulators because they retain air within. The symmetric and asymmetric bending load bearing capacities of Ballistic are enhanced with the addition of carbon fibres. Material for composite matrices made of carbon, aramid, and glass (90:10) Make 1 kilogramme by mixing 10 grammes of epoxy resin with 900 grammes of carbon fibre in sheet form (90 percent). Al 2021 was a strength-added component. The Hand layup Method for Making Hybrid Composite Materials

2.2 Approach to Hand Lay-Ups

The hand layup process involves manually feeding roving-shaped resins into textiles to saturate them. The fibres are allowed to naturally detach from the material under ambient air. Although the process is straightforward, making the composite calls for a high level of expertise. Carbon, Coconut, and Vetiver fibres are among the materials utilised. Composite laminates often include carbon fibres put on top and bottom for improved finishing. The hand layup approach was used to construct three distinct categories, each with two examples. It is necessary to lay out the fibres in a regular orientation and ensure they are dry before feeding them to the lamination. To begin removing the laminate, a releasing agent is spread across the surface. Spread a thin layer of resin, then place the carbon fiber on it. After being left undisturbed for around three hours, a weight of five kilogrammes is placed on top of the carbon fibre to eliminate any air bubbles. The typical drying process also removes any moisture from the fibres. The laminate was taken out of the die material and subjected to a battery of tests for engineering purposes after three hours.

3. Results and Discussion

3.1 Tensile Testing

Testing a material under tension or compression will reveal its strength. For tensile testing, Figure 18.1 standard dog bone shaped specimens were hand laid out in accordance with ASTM D638 requirements. Figure 18.2 Shown Tensile test for Aramid glass, carbon, and galree fibre, each sample is 30 mm wide and 280 mm long. The computer-controlled Universal Testing Machine (ASE - UTN 10) will apply force to the sample until it breaks.

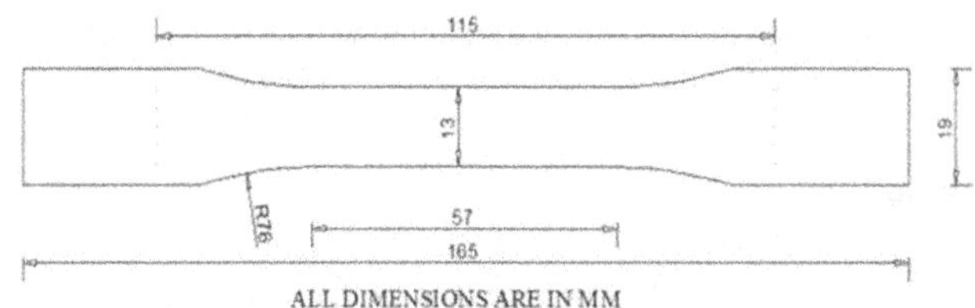

Fig. 18.1 Standard dog bone shaped with ASTM D638

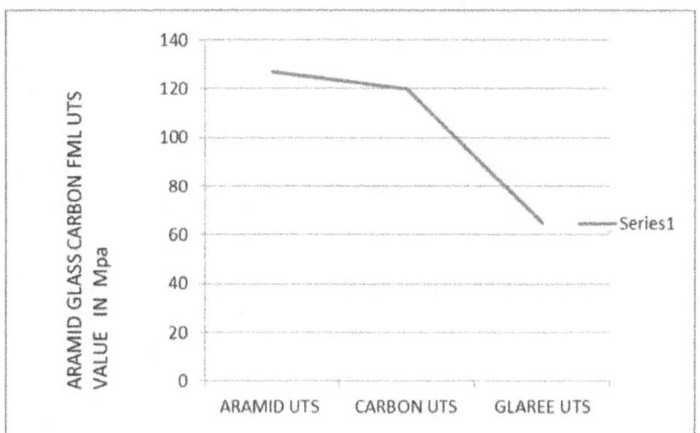

Fig. 18.2 Tensile test for aramid glass, carbon, and galree fibre

Composites with varying reinforcement material combinations undergo testing to determine their ultimate tensile strength and ductility. Load and elongation measurements are obtained at regular intervals at the same time. Room temperature is used to conduct a tensile test.

In order to learn how a material reacts to progressively increasing stress and strain, a built specimen is subjected to a uniaxial tensile test. Findings from the tensile stress test indicate the composite's elasticity limit.

ARAMID FML UTS	127
CARBON FML UTS	120
GLAREE FML UTS	65

With regard to the aforementioned graph, we spoke about using a tensile test valve with glass, carbon, and galree fibre. Aramid fiber showed the strongest ultimate tensile strength in the test. A stress level of 127 MPa is considered safe. To that end, the vehicle industry makes use of aramid fibre. A higher coresponding impact value was observed when the material UTS valve was enhanced. Therefore, the material utilised for aero and automotive ballistic applications. The results indicate the development of a hybrid material combining basalt fiber with aluminum and glass fibers, creating a laminated composite with enhanced properties. The goal is to design a bumper that deforms between 0.017378 m and 0.03114 m at a tensile strength of 38 MPa, with a maximum stress of 242.4 MPa, demonstrating advantages for daily use through simulation comparisons [12].

3.2 Flexural Testing

A quantitative measure of the load's changes with regard to displacement is the flexural stress. Findings from the flexural stress test indicate how stiff the polymer composite fibre is. The flexural samples are made following the guidelines of ASTM D790. The flexural test for composite materials most commonly used is the 3-point flexure test. The location of the crosshead is used to assess specimen deflection. Results are presented after flexural strength and displacement have been considered. The specimen is subjected to force until it breaks or fractures in the universal testing machine, which is an integral part of the testing procedure. This is the specimen that was flexural tested. A typical relative humidity of 50% is being controlled during the experiments. In the flexural test, a graph is made for each sample to show how force and displacement are related. Figure 18.3 showns the flexural standard dimension of ASTM D790.

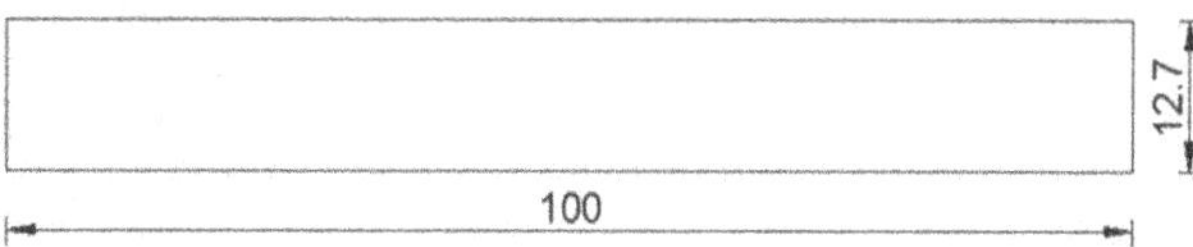

Fig. 18.3 The flexural standard dimension of ASTM D790

ARAMID UTS flexural load in KN	0.74
CARBON UTS flexural load in KN	0.62
GLAREE UTS flexural load in KN	0.45

In Fig. 18.4 shown the flexural load distribution for aramid, carbon and glaree The graphs and tables display the flexural stress value. The flexural load valve has a maximum value of 0.7 kn. The material bearing stress and the Aramid flexural stress both rose. Consequently, the material was utilised for a chasis application in automobiles. In a 0.62 Kn, the Carall Uts are shown. Flexural load application conditions also make advantage of it.

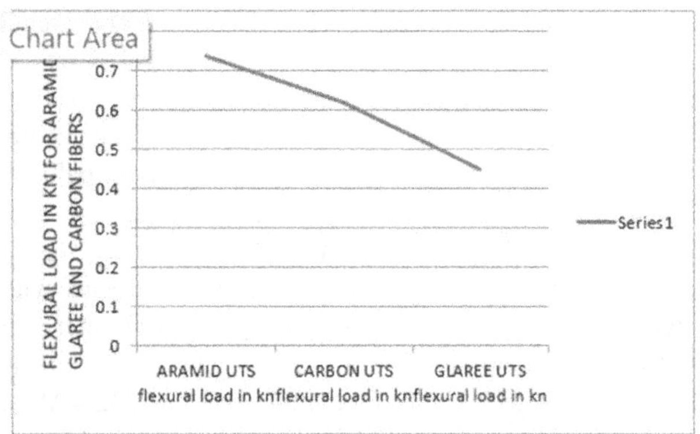

Fig. 18.4 The flexural load distribution for aramid, carbon and glaree

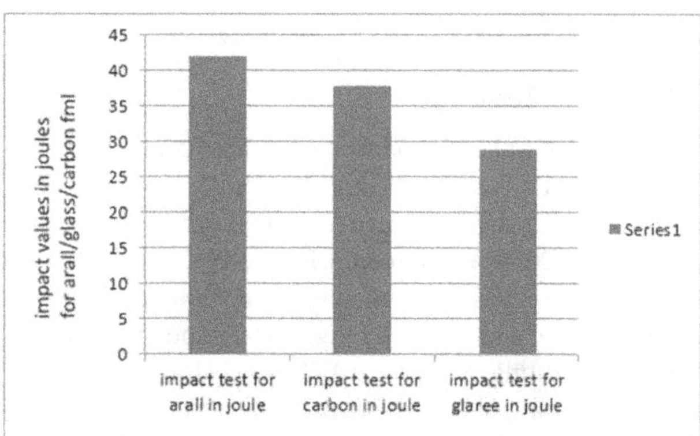

Fig. 18.5 The impact value distribution for aramid, carbon and glaree

3.3 Impact Test

Impact Test determines if the material is suitable for energy absorption.

- The composite specimen begins to plastically deform as it reaches its yield point, absorbing energy from the striker. At this point, work hardening happens.
- The composite will fail when it can no longer absorb energy.

To measure how much energy a material absorbs when it breaks, scientists use the standard Izod impact test, which is a fast-acting test. The ASTM D256 standard is followed when conducting this test on the samples. Up until the specimen breaks during testing, the amount of energy it absorbs is recorded.

Impact test for arall in joule	42
Impact test for carbon in joule	38
Impact test for glaree in joule	29

As shown in Fig. 18.5 an FML impact load graph. Arall fibres have a maximum FML of 42 j. A high-level valve displays the effect. Ballistic applications also make use of this material. Aerodynamic Application Material for Vehicles and Aircraft

3.4 Hardness Test

The purpose of the Barcoll hardness test is to determine the tire's pierce resistance, which is an important quality to look for in a tire. The composite material made of Kevlar, carbon, and glass has a maximum strength rating of 42 HRL. The Table 18.1 shows the hardness values in three locations for Barcol in the Major Load Scale (60 Kgf) and Minor Load Scale (-10 Kgf) with an indenter (½" ball).

Arall, Carrall, and Glaree hardness values are displayed in the graph. Figure 18.6 shown Hardness valve maximum values in arall and subsequent cabon materials are shown

Table 18.1 The hardness test values for aramid fiber, carbon fiber and glass fiber

Aramid fibre hardness valve	Carbon fibre hardness valve	Glass fibre hardness valve
45	38	35
42	40	37
41	42	37

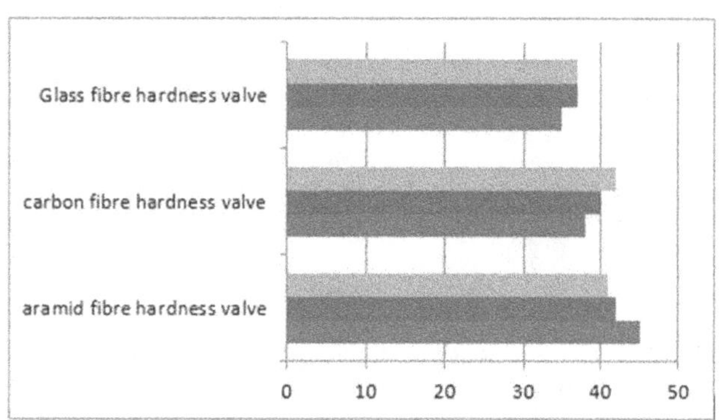

Fig. 18.6 The hardness test values for aramid, carbon and glaree

in the graph. Coefficients of compression rose in tandem with increases in material hardness. Because of this, the material is commonly utilised in the construction of aerostructures and automobiles.

Mass Spectrometry by SEM

To examine the topography and morphology of a specimen, a scanning electron microscope (SEM) creates pictures by scanning it with a focussed stream of electrons. This type of electron microscope is one of several. The surface topography and composition of the sample may be inferred from the signals produced by the sample's atoms and electrons, which work together to create these signals. Scanning electron microscopy (SEM) is used for the

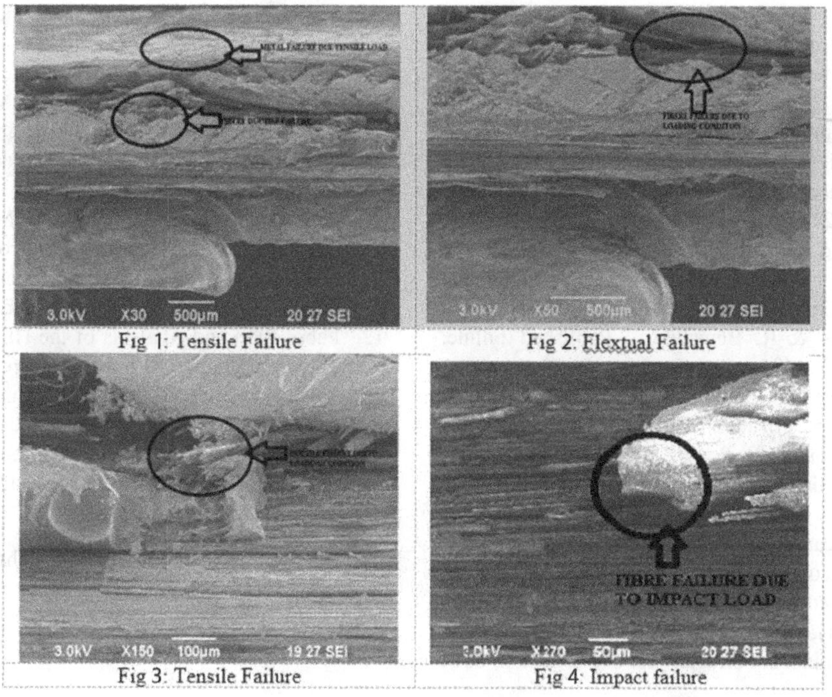

Fig. 18.7 Different failure modes SEM images (1) Tensile failure, (2) Flextual failure, (3) Tensile failure, (4) Impact failure

examination of impact test specimens with cracked surfaces. Since the samples exhibit substantial compositional and melting temperature variance in addition to substantial mechanical test result variation, they are selected for SEM analysis.

Images captured by scanning electron microscopy capture the fracture surface. Ductile fractures were revealed by the fracture surface. Figure 18.6 of the SEM picture shows a tensile failure caused by a tensile load. The flexural fracture surface is depicted in Fig. 18.7, whereas Figs. 18.8 and 18.9 demonstrate the impact failure of the FML. The ductile fracture is seen in the figures. Further evidence that no porosity or voids exist is provided. Laminates do not generate heat through a chemical reaction. Therefore, the FML are not flawed. Consequently, the material utilised for aviation and automotive applications for fibreglass

4. Conclusion

All the mentioned FMLs glaree, carall, and arall are made using the hand lay-up method. The plate can only have a maximum thickness of 3 mm. Higher tensile and impact valves were demonstrated by the material. Additionally, maximum bending valves. Therefore, the substance utilised for automotive and aerospace purposes. A ductile fracture is the most severe, according to the material's fracture surface. Void, porous, and hexothermic heat generation are not present. The greatest value of the FML is given to the materials that compare the best. Therefore, aero and automotive applications utilize the arall material.

References

1. A generalized solution to the crack bridging problem of fibre metal laminates G.S. Wilson, R.C. Alderliesten, R. Benedictus Structural Integrity, Faculty of Aerospace Engineering, Delft University of Technology, The Netherlands
2. A Study on Flexural Properties of Sandwich Structures with Fibre/Metal Laminate Face Sheets S. Dariushi : M. Sadighi (*)Mechanical Engineering Department, Amirkabir University of Technology, Tehran, Iran Published online: 23 January 2013
3. Blast response of metal composite laminate fuselage structures using finite element Modelling T.Kotzakolios, D.E. Vlachos, V. Kostopoulos applied mechanics laboratory, department of mechanical engineering and aeronautics, university of patras,Grease published online 2011
4. Comparative analysis of crack resistance of fibre-metal laminates with hs2 glass/t700 carbon layers for various stress ratios X. Song, Z. Y. Li, Y. Shen Y. L. chen school of mechanical and power engineering, harbin university of science and technology, harbin, china published on 2015
5. Effect of stacking sequence on failure mode of fibre metal laminates under low-velocity impact F. Taheri-Behrooz M. M. Shokrieh I. Yahyapour received: 20 june 2013 / accepted: 19 november 2013 Center of Excellence

in Experimental Solid Mechanics and Dynamics, School of Mechanical Engineering, University of Science and Technology, Tehran, Iran

6. Effects of curing thermal residual stresses on fatigue crack properation of aluminium plates repaired by FML patches Hossein Hosseini-Toudeshky, Mojtaba Sadighi, Ali Vojdani Aerospace Engineering Department, Amirkabir University of Technology, 424 Iran, year of publishing 2013
7. Experimental and numerical investigation of metal type and thickness effects on the impact resistance of fibre metal laminates M. Sadighi & T. Pärnänen & R. C. Alderliesten & M. Sayeaftabi & R. Benedictus Published online: 27 October 2011 Mechanical Engineering Department, Amirkabir University of Technology, Tehran, Iran
8. Experimental characterization of a fibre metal laminate for underwater applications E. Poodts , D. Ghelli , T. Brugo , R. Panciroli, G. Minak Alma Mater Studiorum – Università di Bologna, Industrial Engineering Department DIN, Bologna, Italy, year of publication 2015
9. Fatigue behaviour of glass fibre reinforced epoxy composites enhanced with nanoparticles L.P. Borrego, J.D.M. Costa, J.A.M. Ferreira, H. Silva CEMUC, University of Coimbra, Rua Luís Reis Coimbra, Portugal, year of publication 2014
10. FML full scale aeronautic panel under multi axial fatigue: Experimental test and DBEM Simulation. Dept. of Materials Engineering and Production, University of Naples E. Armentani , R. Citarella, R. Sepe year of publishing 2011
11. Impact behaviour of glass fibre-reinforced epoxy/aluminium fibre metal laminate manufactured by Vacuum Assisted Resin Transfer Moulding I. Ortiz de Mendibil, L. Aretxabaleta, M. Sarrionandia, M. Mateos, J. Aurrekoetxea Mechanical and Industrial Production Department, Mondragon Unibertsitatea, Loramendi 4, Mondragon 20500, Gipuzkoa, Spain year of publication 2016.
12. Fabrication and Analysis of the HLM Method of Layered Polymer Bumper with the Fracture Surface Micrographs, P. V. Narashima Rao, P. Periyasamy, A. Bovas Herbert Bejaxhin , E. Vetre Selvan, N. Ramanan, N. Vasudevan, R. Elangovan, and Mebratu Tufa, Advances in Materials Science and Engineering, Article ID 3002481, 9 pages doi.org/10.1155/2022/3002481, 2022.

Note: All the figures and tables in this chapter were made by the authors.

Advances in Additive Manufacturing Technologies – Gurusamy Pathinettampadian et al. (eds)
© 2026 Taylor & Francis Group, London, ISBN 978-1-041-16687-0

Fabrication and Analysis of Hydrogen Storage Tank

P. K. Chidambaram, K. S. Vigneshwaran
Department of Mechanical Engineering,
New Prince Shri Bhavani College of Engineering and Technology

Vijayraja
Department of Electrical Engineering, Vistas, Pallavaram, Chennai

Dhandapani
Department of Mechanical Engineering, Rajalakshmi Engineering College, Thandalam, Chennai

Ramanan N.[1]
Department of Mechanical Engineering,
Sri jayaram institute of Engineering and Technology, Gummdipundi

P. Gurusamy[2]
Department of Mechanical Engineering, Chennai Institute of Technology, Chennai

N. Vasanth[3]
PG Student, Department of Mechanical Engineering, Chennai Institute of Technology, Chennai

◆ **Abstract:** The industrial sector is in search of sustainable materials with a high strength-to-weight ratio that might one day supplant more traditional options. Composite materials that exhibit natural behaviour as a result of combining two or more components are able to fulfil the aforementioned parameters. The matrix material in this research is carbon fibre reinforced plastic (CFRP), while the reinforcing materials are natural fibres (abaca and kenaf). Compression moulding is the usual method for producing composite laminates. We ran a battery of mechanical tests, including tensile, flexural, and impact. By measuring 196 MPa for tensile strength, 263.85 MPa for flexural strength, and 6 J for energy absorption, Category II outperforms the other two categories in terms of mechanical qualities. With the use of SEM, we can see the hybrid composite's interior structure, and we can see that it's much improved, with fewer blow holes and cracks. The automotive industry is one potential user of hybrid composites, as natural fibres offer a viable substitute for more traditional materials.

◆ **Keywords:** Abaca, Kenaf, CFRP, Compression molding, Mechanical testing, SEM

1. Introduction

The construction, transportation, automotive, textile, fibreboard, cushion, paper, door, wall panel, air cleaner and dashboard sectors are just a few of the many that find use for natural fibres. Different types of reinforcement geometries (particulate, flake, and fibre) and matrix materials (polymer, metal, ceramic, and carbon) allow for the categorisation of composites. When it comes to chemical decay, scratch, oxidation, and brine, polymer composites are unrivalled. A variety of transportation modes, including boats, trains, military vehicles and

Corresponding authors: [1]ramananinjs2020@gmail.com, [2]gurusamyp@citchennai.net, [3]Vasanth1611197@gmail.com

DOI: 10.1201/9781003685906-19

aircraft hulls, have gained from these qualities. Seats, walls and floors in public transport systems have been fabricated using inexpensive composites due to their resistance to wear and tear. When it comes to chemical corrosion, scratching, rust, and saltwater, polymer composites are unrivalled. A variety of transportation modes, including boats, trains, military vehicles, bicycle components, and aircraft hulls, have benefited from these qualities. Seats, walls and floors in public transport systems have been fabricated using inexpensive composites due to their resistance to wear and tear.

Columns reinforced with CFRP have a higher compressive strength when the number of sheets is increased and they are pre-stressed. When CFRP columns are preloaded, strain lag occurs, which reduces their compressive strength. This negative effect on columns might be reduced by pre-stressing CFRP columns, as has been shown using FEA. The only way to enhance yield load is to pre-stress CFRP columns, which also increases CFRP's tensile strength [1]. This material is proposed to be replaced with lightweight dissimilar materials, preferably Al6061 and FE 430 steel. This paper presents a computational fluid dynamics (CFD) analysis of the valve, evaluating its flow characteristics and strength[2]. Abaca fibre is the most popular natural fibre for use in automobiles, buildings, and other machinery since it is biodegradable, lightweight, inexpensive, and has low thermal conductivity [3]. Composites and columns frequently use CFRP reinforcement due to its superior fatigue qualities and high strength-to-weight ratio. In the absence of CFRP, concrete columns and automotive components are prone to corrosion and suffer from inadequate flexural strength [4]. These days, CFRP is preferred over steel plates for a number of reasons, including its low maintenance requirements, high flexural and mechanical strength, and exceptional resistance to corrosion (since it isn't actually a metal) [5]. A variety of fibre treatments and coupling agents are employed to enhance the functionality of flexural loading in natural fibre reinforced composites. Accordingly, it was discovered that the elastic characteristics may be enhanced by utilising coupling agents in conjunction with a variety of fibre treatments [6]. The results indicate that AZ31B/15 wt% WC composites exhibit superior tribological behavior compared to the base AZ31B magnesium alloy. The yield strength, flexural strength, tensile strength, and micro-hardness of the composites improve with increasing WC reinforcement content. SEM images show a uniform distribution of WC particles within the Mg matrix[7]. If you want your natural fibre reinforced polymer composites to be even stronger, try adding some carbon fibre [8]. The effect of heat input on the microstructure evolution in the weld region of friction stir welded AA6061-10% SiCp MMCs, correlating tensile properties with microstructure and heat input. Microstructure analysis reveals significant grain refinement of the aluminum matrix and reinforcement particle fracturing due to dynamic recrystallization from plastic deformation and frictional heat[9]. Compared to synthetic fibre, natural fibre has greater benefits, such as being inexpensive, lightweight, abundant, and renewable, according to recent studies. Combining polymers with natural fibres such as abaca or kenaf produces composites that are lightweight, carbon neutral, have a high specific strength, and need little energy to manufacture. When compared to synthetic fibres made from the same polymers, composites made from Kenaffibres and epoxy resin had superior mechanical, morphological, and physical qualities [10]. Hybrid composites are becoming more popular among researchers and businesses alike. The biodegradability, high flexural modulus, impact resistance, and structural qualities exhibited by hybrid composites composed mostly of natural fibres account for their widespread use[11]. Laminating CFRP increases the structural performance and composite strength. Results show that 7-layer CFRP laminate orientation outperforms other synthetic fibre composite combinations. The aeroplane, aerospace, petrochemical, automobile, and military sectors all make use of CFRP [12]. Architects were the first to make use of composite materials like CFRP, and the aerospace, automotive, and marine industries were later approached with the concept. Composites like this were highly prized in those industries since they were long-lasting and needed little upkeep [13]. Metals typically have their service life reduced by exposure to the elements and by the difficulties encountered in the workplace. Substituting CFRP results in reduced deflections, a high yield load, and a high cracking load. Because of this, they are able to resist breaking under heavy pressures [14]. Aerospace, mechanical, and civil engineering structural applications abound for fibre reinforced composites. Therefore, the ultrasonic immersion test is used to evaluate these composites. Compared to a destructive mechanical test, this one ensures that the fibre is not damaged in any way. The large quantity of data produced by this test makes it an ideal tool for further research into how to enhance the CFRP composites' mechanical response [15].

2. Resources Employed

2.1 Abaca

Abaca cellulose fibres originate in the plant's pseudo-stem. Being a banana family member, it is a waste product from the banana industry. Being the second most often used fibrous stalk material (behind hemp), it is an affordable basis of strong, frivolousfibre that continues to develop.

Fig. 19.1 Abaca fiber

Polymer composites rely heavily on abaca fibres for reinforcement. Flexible processing and an availability of supply are only two of the many benefits of abaca. Because of its long fibres, abaca is also utilised to manufacture rope, twine, cordage, and a variety of textiles for things like linens, wall coverings, and apparel. Daimler-Chrysler allowed the use of abaca as an outside lining in vehicle chassis in 2004.

2.2 Kenaf

The Malvaceae family includes the plants kenaf, hibiscus cannabinus, jute, and Deccan hemp. Because of its potential as a polymer reinforcement, kenaf fibers are gaining a lot of interest in the natural fibre composite market. The heat resistance of kenaf fibres is lower than that of synthetic fibres, as is the case with all natural fibres. The most common products made from kenaf fiber are paper, rope, twine, and coarse fabric. The tensile strength of kenaf fibers was found to diminish after being heated to 180 °C for 60 minutes, according to thermal studies. The moulding temperature for biodegradable composites was thus 160°C. Paper, goods, construction materials, absorbents, and even some animal diets make extensive use of kenaf.

Fig. 19.2 Kenaf fiber

2.3 CFRP

They are made up of a matrix and reinforcement, making them a composite material. Polyacrylonitrile, an organic polymer, is the raw material used to make CFRPs. Because of its increased strength, carbon fibre is utilised here as a resin. Many commonplace items are made using CFRP Composites, which are lightweight and sturdy. The aerospace industry makes extensive use of it due to its excellent strength-to-weight ratio. Carbon fibre reinforced plastic (CFRP) is extensively utilising in the design of aircraft and manufacturer-scaled composites. Due to their exceptional strength-to-weight ratio, they find extensive application in tiny air vehicles.

Fig. 19.3 CFRP Mat

3. Fabrication Methods

Our organized composites are unidirectional, meaning they are not at a right angle to each other but rather at an angle of 0-0 degrees. On a unidirectional lamina, the direction perpendicular to the fibres in the 1-2 plane is known as the transverse direction, and the direction parallel to the fibres is termed the longitudinal direction (axis 1 or L).

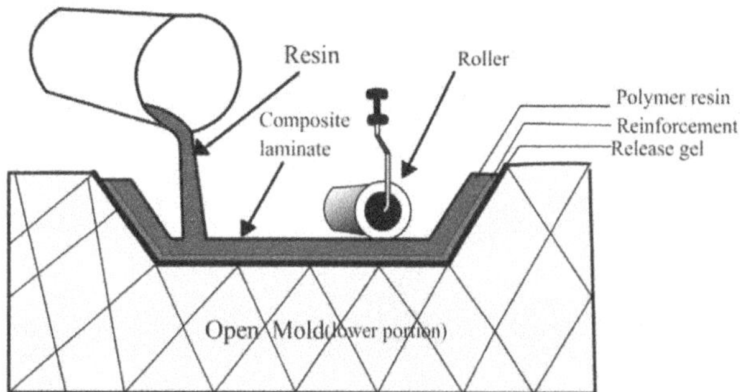

Fig. 19.4 Hand layup process

You may alternatively think of a direction in the two or three planes as a transverse one. Three axes—axes 1, 2, and 3—make up the lamina's substance. The most striking longitudinal properties of the orthotropic, three-axes-symmetric unidirectional lamina are its three axes of symmetry. Long strands reaching 270 mm in length were cut from our natural fibres, which included abaca and kenaf. For the simple reason that longer, untwisted fibres are more durable and impact-resistant than their shorter, twisted counterparts. Next, 27 mm squares of carbon fibre reinforced polymer (CFRP) matrix reinforcements were cut. Mix the epoxy glue and hardener together and set aside for 30 minutes. If done correctly, this will improve the fibres' ability to adhere to the CFRP. Among these moulding methods, compression moulding stands head and shoulders above the others. The next step is to combine the fibres with the epoxy resin at a 3:5 ratio while the mixture is heated. Hardener, epoxy resin, carbon fibre reinforced plastic, and natural fibres are all mixed together. Abaca was utilised in alternate layers above and below the central CFRP layer in Category I, which consisted of three layers of CFRP. Epoxy adhesive is utilised in this case to connect the Abaca fibre to the CFRP. Three layers of CFRP were utilised in the Category II, with abaca serving as an alternative layer above CFRP and kenaf as the layer below. Each layer of fibre and CFRP is coated with epoxy resin in this case. Three layers of CFRP were utilised in the Category III, with Kenaf layers alternating above and below the centre layer. The CFRP and Kenaffibres were joined in this case using epoxy glue. Three layers of carbon fibre reinforced plastic (CFRP) and two layers of natural fibre were used in each of these three types, with epoxy resin serving as the adhesive. Each category's five layers were kept in a compression moulding machine. To secure each Category in place, compression moulding machines employ two blocks—one constructed of high-temperature grade steel and the other of an alternate substance. For the compression moulding process, a wax or gel-based liquid is used to easily separate the fibre plate from the two blocks before inserting the Categories. The machines were set to a temperature of 100 degrees Celsius and let to run for an hour with each of these categories. The compression method results in a 3 mm overall plate thickness and a 280 g total weight for the composite laminate.

Table 19.1 Arrangement of fibers

Category I	Category II	Category III
CFRP	CFRP	CFRP
Abaca	Abaca	Kenaf
CFRP	CFRP	CFRP
Abaca	Kenaf	Kenaf

Fig. 19.5 Preparation of laminate

4. Testing of Composites

4.1 Tensile Test

Testing materials by pulling them taut in one direction until they break is known as tensile testing, and it is a destructive mechanical technique. Tensile testing specimens that fulfil the required standards are manufactured by meticulously milling raw materials into a specified category with predetermined dimensions and grasp ends. Machinists use a wide variety of equipment and techniques, including waterjet machining, wire EDM, computer numerical control (CNC) systems (both automated and manual), and many more. The tensile strength, flexural strength, and load-bearing capacity increased by 50%, 14%, and 54%, respectively. To examine the fracture surfaces of the tested samples, a Scanning Electron Microscope (SEM)

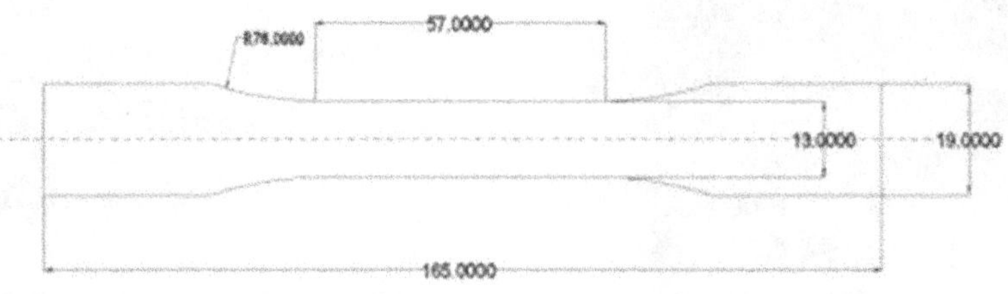

Fig. 19.6 ASTM standard for tensile test [ASTM D638]

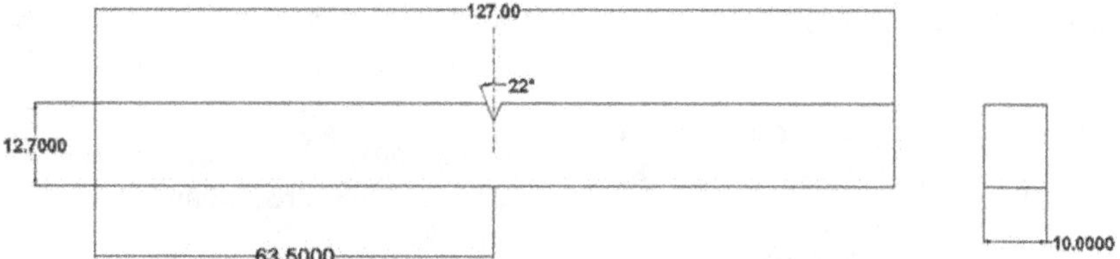

Fig. 19.7 ASTM standard for impact test [ASTM D256]

was utilized. Fiber breakage and matrix cracking were identified as the primary failure mechanisms during tensile testing, while matrix cracking and delamination were the dominant failure mechanisms observed in impact testing [16]. A Universal Testing Machine is used to align the specimens after they have been manufactured. In order to keep track of the strain rate throughout testing, strain gauges are utilised in accordance with certain specifications. The ultimate tensile strength, maximum elongation, and area reduction are some of the possible outcomes of tensile testing. Yield strength, strain-hardening characteristics, Young's modulus, and Poisson's ratio are just a few of the many metrics that may be gleaned from these tests.

4.2 Impact Test

Impact testing is a damaging mechanical test that finds out how effectively a Category can handle high-rate loading or reduce the force of a dynamic impact. Charpy and Izod are two examples of impact tests that quantify a material's overall energy absorption capacity during a controlled impact. The device used to evaluate this impact event is a pendulum-type arm that has an impact hammer attached to its end. After releasing the pendulum from a predetermined height, precise equipment are used to quantify the energy absorbed upon impact with the test Category. Knowing the energy absorption characteristics of a material is essential for predicting its ultimate failure point during plastic deformation. The degree to which the material absorbs energy is proportional to its brittleness. The absorption rates of ductile materials, like copper or aluminium, are often higher than those of brittle materials, such glass or ceramics.

4.3 Flexural Test

In many cases, flexure testing can reveal a material's flexural modulus and flexural strength. While tensile tests are more expensive, flexure tests are less expensive and provide somewhat different findings. Pulling on the Category's top using either of its contact points—the upper loading span—will cause it to collapse. Before applying this force, the material is spread out horizontally over two points of contact, which are called the lower support span. You may find the maximum force by calculating the flexural strength

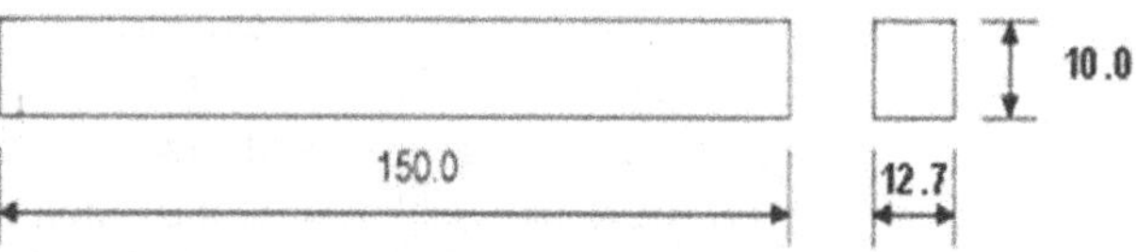

Fig. 19.8 ASTM standard for flexural test [ASTM D790]

of that specific Category. Determining flexural modulus and flexural strength is a frequent practice in the field of flexure testing. Flexural strength is a material's ability to withstand compression or tension applied to a fiber's tip. The slope of the stress-strain deflection curve can be used as a proxy for the flexural modulus. This results aims to develop a hybrid material combining basalt fiber with aluminum and glass fibers, creating a laminated composite with enhanced properties. It seeks to design a bumper with deformation between 0.017378 m and 0.03114 m at 38 MPa tensile strength, with a maximum stress of 2.424×10^2 MPa, demonstrating advantages for daily use through simulation comparisons [17].

4.4 Hardness Test

In contrast to other material properties, hardness is not an essential physical quality. Measuring the persistent depth of an indentation determines its definition: resistance to indentation. The relationship between hardness and indentation size is straightforward: for a given load, a smaller indentation indicates a harder material. Most hardness tests adhere to the criteria laid out in ASTM E-18, namely the Rockwell hardness test. Prior to taking a Rockwell exam, you should get a copy of the standard, read it thoroughly, and make sure you understand all of the requirements. Compared to other techniques of hardness testing, such as the Brinell, Vicker, Knoop, and Case depth tests, the Rockwell test is both more straightforward and more reliable. With a few exceptions, the Rockwell test may be used to any metal. Nevertheless, there are instances where the indentations are excessively large, excessive fluctuation is caused by the test metal structure or surface conditions, or its utilisation is prohibited by the category's size or shape. The depth of a permanent indentation may be determined using the Rockwell approach when a force or load is applied to an indenter. The standard for hardness testing that is used is ASTM D2583.

4.5 An Examination of the Composites' Morphology

A concentrated electron beam is moved across a surface in a scanning electron microscope (SEM), which then produces an image. It is possible to learn about the surface's composition and topography by interacting with the Category, which causes the electron beam to create a multitude of signals. Electron guns equipped with tungsten filament cathodes thermionically emit an electron beam in a conventional scanning electron microscope. Because of its inexpensive cost, high melting point, and low vapour pressure, tungsten is typically employed in thermionic electron cannons to emit electrons when electrically heated.

5. Results and Discussion

5.1 Tensile Test

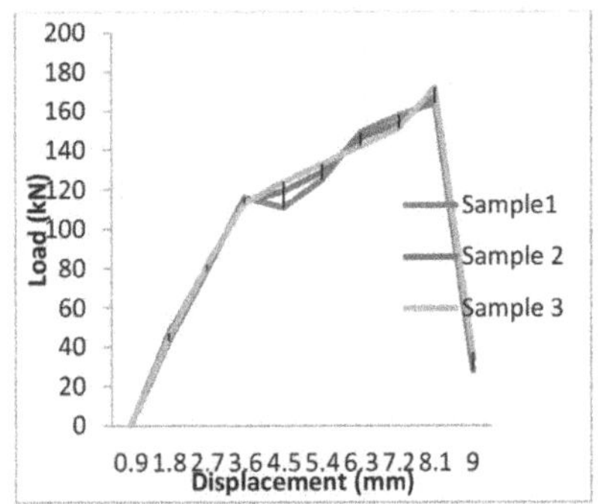

Fig. 19.9 Category I Tensile results of all samples

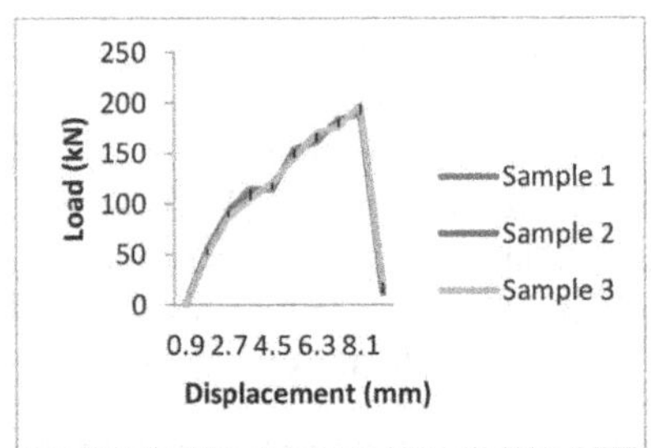

Fig. 19.10 Category II Tensile results of all samples

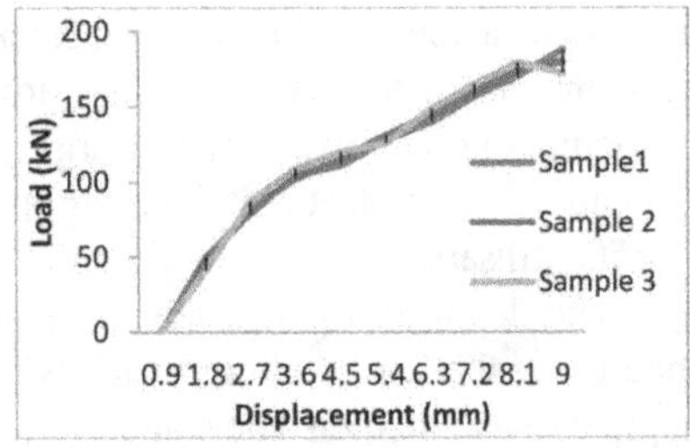

Fig. 19.11 Category III Tensile results of all samples

Category 2, consisting of Abaca, Kenaf, and CFRP, had the highest UTS of 8.76 KN and the best tensile strength compared to the other two categories, as seen in the graphs above. This indicates that the composite's tensile characteristics are enhanced as the abaca fibre content increases.

5.2 Flexural Test

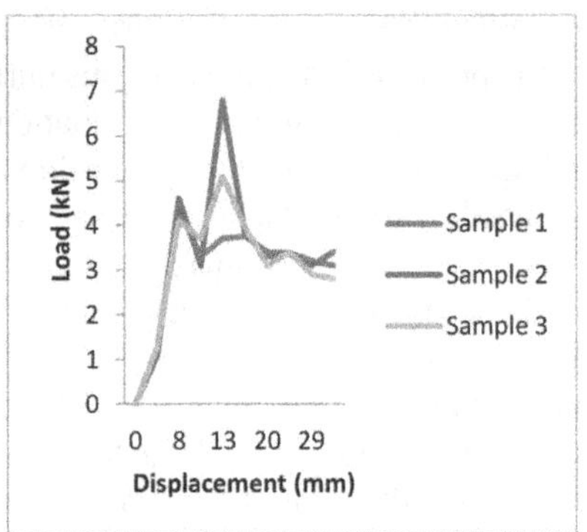

Fig. 19.12 Category I Flexural results of all samples

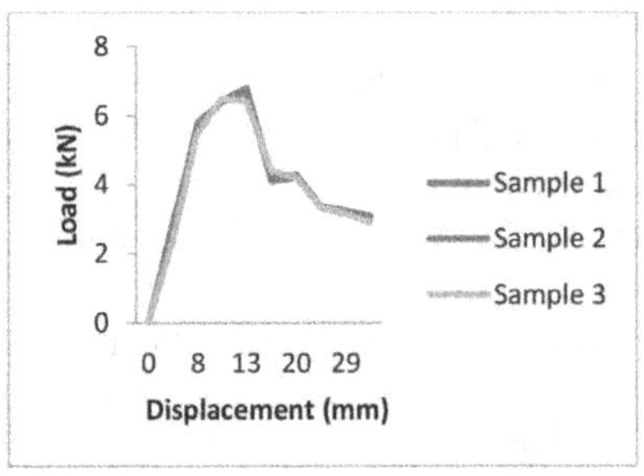

Fig. 19.13 Category II Flexural results of all samples

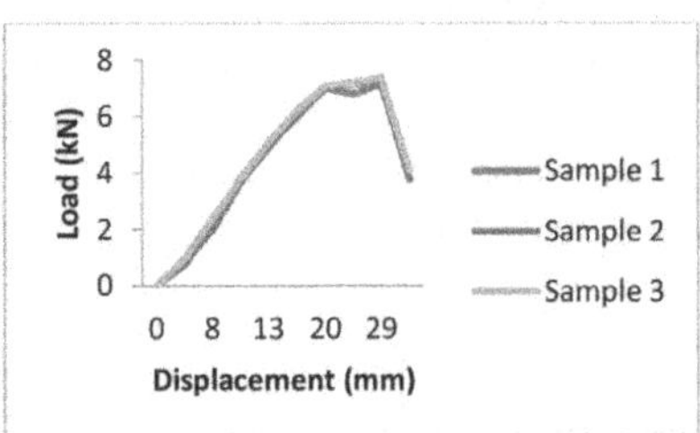

Fig. 19.14 Category III Flexural results of all samples

The composite samples are subjected to the flexural test using universal testing equipment.

The above figures suggest that Category 3, which consists of a mix of Kenaf and CFRP, has the highest flexural strength. It can be shown that the flexural strength of Kenaf+CFRP is greater than that of Abaca+CFRP and Abaca+Kenaf+CFRP.

5.3 Impact Test

The data comparison showed that Composite 2, which consisted of Abaca, Kenaf, and CFRP, absorbed 6J more energy than Abaca+CFRP or Kenaf+CFRP alone. The findings suggest that using many fibres, rather than monofilaments, enhances the energy absorption capacity of a composite.

5.4 Hardness Test

Table 19.2 Rockwell hardness value

Sl. No	Sample	Rockwell Hardness Value (HRB)
1	S1	98
	S2	91
	S3	89
2	S1	105
	S2	108
	S3	102
3	S1	96
	S2	99
	S3	104

The data shows that the Rockwell Hardness Number (HRB) for Abaca, Kenaf, and CFRP is 108 HRB, which is the highest of any category. This suggests that this hybrid composite has better indentation resistance than Abaca + CFRP and Kenaf + CFRP.

5.5 SEM Images

Fig. 19.15 Cutting the fibres into smaller pieces will help spread them out evenly, which will fix the presence of faulty cure and thickness variance illustrated by circles

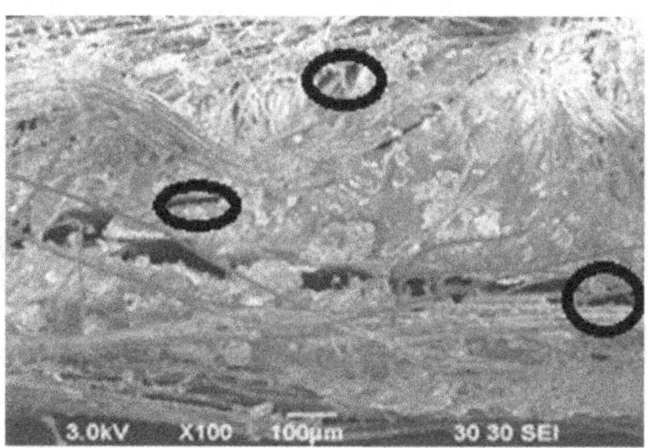

Fig. 19.16 Minimising delaminations, missing adhesive, and fibre pull-out—all shown by circles—is possible with an accurate fiber-to-resin ratio and pre-mold fibre treatment

6. Conclusion

The following conclusions were drawn from the data and analyses presented above. Compared to all other Categories I and III, Category II under Sample 1 has the highest ultimate tensile strength at 196 MPa and the highest break load at 8.76 KN. Comparing Category III Sample 3 to the other two, we find that it has a higher ultimate flexural strength of 263.85 MPa and a flexural load of 8.54 KN. When compared to Categories I and III, Category II Sample 2 absorbed greater energy in the impact test. Category II under Sample 2 had the greatest HRB compared to Category I and III, according to the hardness test. We can deduce that the fibres overlapped and caused various misalignments from the scanning electron micrographs 14, 15, 16, and 17. Further analysis of the SEM images revealed the existence of voids and blow holes.

References

1. Cheng, D. and Yang, Y. 2017. Design method for concrete columns strengthened with prestressed CFRP sheets. *Construction and Building Materials* 151:331–344.
2. Ramanan, N., Periyasamy, P., Sharavanan, S. and Naveen, E. 2019. Computational analysis of dissimilar materials engine valve. *International Journal of Vehicle Structures and Systems 11*(2).
3. Liu, K., Yang, Z. and Takagi, H. 2014. Anisotropic thermal conductivity of unidirectional natural abaca fiber composites as a function of lumen and cell wall structure. *Composite structures* 108:987–991.
4. Osman, B.H., Wu, E., Ji, B. and Abdulhameed, S.S. 2018. Effect of reinforcement ratios on shear behavior of concrete beams strengthened with CFRP sheets. *HBRC journal* 14(1):29–36.
5. Al-Saadi, N.T.K. and Al-Mahaidi, R. 2016. Modelling of near-surface mounted carbon fibre reinforced polymer strips embedded in concrete with cement-based adhesive. *Construction and Building Materials* 127:383–393.
6. Khan, M.Z., Srivastava, S.K. and Gupta, M.K. 2018. Tensile and flexural properties of natural fiber reinforced polymer composites: A review. *Journal of Reinforced Plastics and Composites* 37(24):1435–1455.

7. Praveenkumar, R., Periyasamy, P., Mohanavel, V. and Ravikumar, M.M. 2019. Mechanical and Tribological Behavior of Mg-Matrix Composites Manufactured by Stir Casting. *International Journal of Vehicle Structures & Systems (IJVSS)* 11(1).
8. Lau, A.K.T. and Cheung, K.H.Y. 2017. Natural fiber-reinforced polymer-based composites. *Natural fiber-reinforced biodegradable and bioresorbable polymer composites* 1–18.
9. Periyasamy, P., Mohan, B. and Balasubramanian, V. 2012. Effect of heat input on mechanical and metallurgical properties of friction stir welded AA6061-10% SiCp MMCs. *Journal of materials engineering and performance* 21(11):2417–2428.
10. Akhtar, M.N., Sulong, A.B., Radzi, M.F., Ismail, N.F., Raza, M.R., Muhamad, N. and Khan, M.A. 2016. Influence of alkaline treatment and fiber loading on the physical and mechanical properties of kenaf/polypropylene composites for variety of applications. *Progress in Natural Science: Materials International* 26(6):657–664.
11. Safri, S.N.A., Sultan, M.T.H., Jawaid, M. and Jayakrishna, K.2018. Impact behaviour of hybrid composites for structural applications: A review. *Composites Part B: Engineering* 133:112–121.
12. Wei, R., Wang, X., Chen, C., Zhang, X., Xu, X. and Du, S. 2017. Effect of surface treatment on the interfacial adhesion performance of aluminum foil/CFRP laminates for cryogenic propellant tanks. *Materials & Design* 116:188–198.
13. Cao, Q., Jiang, H., Wu, Z. and Ma, Z.J. 2017. Experimental investigation on long term flexural performance of expansive concrete beams eccentrically reinforced by CFRP. *Composite Structures* 163:101–113.
14. Pathak, A.K., Borah, M., Gupta, A., Yokozeki, T. and Dhakate, S.R. 2016. Improved mechanical properties of carbon fiber/graphene oxide-epoxy hybrid composites. *Composites Science and Technology* 135:28–38.
15. Castellano, A., Foti, P., Fraddosio, A., Marzano, S. and Piccioni, M.D. 2014. Mechanical characterization of CFRP composites by ultrasonic immersion tests: Experimental and numerical approaches. *Composites Part B: Engineering* 66:299–310.
16. Kamaraj, M., Periyasamy, P., Ramanan, N., Vellapan, S. and Venkatasubramanian, M.A. 2020. August. Development of natural fibre reinforced hybrid composites and its characterization. *In IOP Conference Series: Materials Science and Engineering* 912(5): 052015.
17. Narashima Rao, P.V., Periyasamy, P., Bovas Herbert Bejaxhin, A., Vetre Selvan, E., Ramanan, N., Vasudevan, N., Elangovan, R. and Tufa, M. 2022. Fabrication and analysis of the HLM method of layered polymer bumper with the fracture surface micrographs. *Advances in Materials Science and Engineering* 2022(1):.3002481.

Note: All the figures and tables in this chapter were made by the authors.

Advances in Additive Manufacturing Technologies – Gurusamy Pathinettampadian et al. (eds)

Experimental Analysis and Investigation of GFRP Reinforced Epoxy Composite Leaf Springs for Automotive Suspension

Yogesh V.*,
Deva Abraham J., Hari Krishna Raj S.,
Dinesh Kumar K. S., Tarun S.
Department of Mechanical Engineering,
VelTech HighTech Dr. Rangarajan Dr. Sakunthala Engineering College,
Chennai, Tamilnadu

◆ **Abstract:** This study investigates the feasibility of employing polymer composite material in leaf springs for suspension systems. The study encompasses design, analysis, and experimental testing for performance comparison of Glass Fiber Reinforced Polymer (GFRP) laminates with epoxy resin and conventional steel leaf springs. Design involves the use of CATIA software to create optimised leaf spring models with improved geometry and dimension. Structural analysis, performed with ANSYS software, examines the mechanical response of steel and composite leaf springs in various load cases. Experimental testing is in the form of tensile, compression, and impact tests on GFRP composite specimens for the purpose of confirming their mechanical properties. Tests show that the composite material performs better than steel in many areas such as tensile strength, compression resistance, and impact resistance. Furthermore, GFRP composites provide advantages including enhanced cushioning, fatigue life, and resistance to corrosion, and thus they could be an alternative to leaf springs. These findings underscore the potential of polymer composites to maximize the performance and life of the suspension system, paving the way towards more durable and efficient vehicles.

◆ **Keywords:** GFRP, FEA, CATIA, LEAF SPRING & ANSYS

1. Introduction

To conserve natural resources and reduce energy consumption, modern car manufacturers emphasize weight reduction. This is achieved through the use of high-tech materials, optimized design, and improved manufacturing processes. Suspension leaf springs, which constitute 10–20% of a vehicle's unsprung weight, are among the most significant components for potential weight savings. Lightening the load on these springs makes them more fuel effective and makes them improve ride quality. Laminated or carriage springs are what leaf springs were formerly known as. Leaf springs are among the most ancient and rudimentary suspension systems, dating to medieval periods. Leaf springs have the benefit of being able to guide their ends along a definite path, which makes them extremely useful for use in vehicle suspension, unlike helical springs. The application of composite materials has enabled leaf springs to be made lighter in weight without loss of load-carrying ability or stiffness.

Composites provide weight ratio for high strength strain elastic storage than steel. Consequently, conventional springs with multi leaf more and more by one-leaf springs of composite. Although composite materials save a great deal of weight, they are not necessarily the cheaper

*Corresponding author: vyogeshmech@gmail.com

DOI: 10.1201/9781003685906-20

alternative to steel. Leaf springs are made to soak up vertical vibrations and shocks from undulating roads by bending, enabling them to accumulate energy in the form of strain and slowly release it. Increasing a leaf spring's energy storage capacity enhances flexibility and performance of the suspension system.

Research indicates that leaves that are composed of high-strength and low elasticity modulus along the longitudinal axis are most ideally suited for springs, and as such, composites are most ideally suited owing to their natural properties. Fatigue failure is an omnipresent scenario in automobile components, mainly due to the reality that such a component is experiencing varying fatigue loads. These include bumps due to road undulations and sudden impacts when bumpers are struck by wheels of a car. Leaf springs, being part of the vehicle's unsprung mass, are most vulnerable to fatigue stresses. A theoretical equation has been developed for the estimation of fatigue life based on fatigue modulus and its degradation rate, further extended by strain failure criteria for applications. Methods to predict composite structures' fatigue strength at varied frequencies, stress ratios, and temperatures were also pursued. However, these studies concern primarily mono-leaf springs. A four-leaf steel spring currently in use by light vehicles' rear suspension was examined from the FEA were compared to existing analytical and experimental results. A fiberglass epoxy resin reinforced composite spring was optimized and designed from the performance of the steel spring with ANSYS, focusing on geometry optimization. The objective was to create a light spring that could sustain provided static external loads without failing. Stress limits (calculated in accordance with and permissible displacements were design constraints. The optimized design determined that the width of the spring decreases in hyperbolic shape and thickness linearly increases

Showed significantly less stresses, an increased natural frequency, and an outstanding weight reduction of approximately 80% (without the inclusion of the eye units). This is a reflection of future possibility for composite material utilization in obtaining weight savings with optimal performance in suspension systems. This paper records composite leaf spring design development for freight rail application. It analyzes three various eye-end attachment concepts for glass fiber-reinforced polyester-composite leaf springs. Both finite element analysis (FEA) and static testing have been conducted to investigate the spring performance, including load-deflection response as well as strain data at various loads. Experimental measurements were compared with FEA predictions to assess their validity.

The initial design, while able to handle a static proof load of 150 kN and one million fatigue cycles, did delaminate at the interface between the eye-surrounding fibers and the spring body. FEA established that this failure occurred due to area shear stress. To mitigate against this, the second design had a transverse bandage covering the vulnerable area, and this was able to reduce delamination but not completely remove it. The third design circumvented the problem by cutting the fibers at the edge, at the end of the eye section, thus delamination never happened. Such incremental achievements reflect the significance of design optimisation in composite leaf spring design for railway application.

Precise quantification of this trend is critical, as it affects the performance of spring parallel flexures in accurate mechanisms. Improved analytical formulae for three-dimensional stiffness, involving shear flexibility, constrained warping, and restricted external drive stiffness are presented in this paper. To ensure additional accuracy, these are complemented with (FEA) using shell elements and anticlastic curvature effects. Besides, approximate expressions are provided to calculate the drive force precisely. The outcomes correlate well with FEA simulations, even for fairly large deflections, confirming their validity for engineering applications.

Varying concentrations of the fibers, up to 20.6% by volume, were successfully fabricated. The fibers were properly wetted by the resin, although the injection time of the resin became significantly higher at higher fiber. Maintenance of the temperature of mold at a constant level was important for the fast and uniform curing of the composite. The composites were cured and compared with experimental thermal readings taken at specified points within the mold cavity and having good similarity between the two sets of readings. The mechanical properties of the composites were analyzed through their tensile and flexural strengths to get data about their performance characteristics.

Carried out a thorough analysis of previous studies to determine the present developments in leaf spring materials, suspension system design principles, and mechanical properties of polymer composites. Delivered a detailed leaf spring model using CATIA software, with accurate dimensions, geometry, and mounting points that suited the requirements of the suspension system.

Carried out structural analysis with the help of ANSYS software to compare and assess the performance of conventional steel leaf springs with their polymer composite counterparts. The study involved tests under different static and dynamic loads. Employed tensile, compressive, and impact tests for GFRP composite laminates reinforced with epoxy resin. These tests followed ASTM standards to assess the

Measured properties such as strength, compression strength, and resistance to impact. Also, measured properties like cushioning capacity, fatigue life, and corrosion resistance to evaluate overall applicability of the material.

Compared both numerical simulation results and experimental testing results, ensuring the better performance of polymer composite materials over conventional steel leaf springs.

Emphasized the benefits of polymer composite material application in leaf spring production, focusing on their ability to enhance the efficiency, durability, and lifespan of suspension systems.

Offered practical suggestions for the implementation of polymer composite leaf springs in automotive and industrial sectors based on the research findings. This includes implementation strategies and recommendations for future research aimed at optimizing composite material performance.

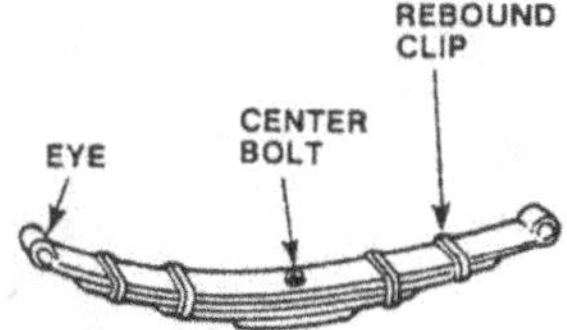

Fig. 20.1 Leaf spring

1.1 Leaf Spring

The central point of the arc is designed to accommodate the axle, while the ends feature tie holes for attachment to the vehicle body (Table 20.1). For heavier vehicles, multiple

Fig. 20.2 Laminate preparation

leaves of varying lengths can be stacked to form a multi-leaf spring, enhancing strength and load-bearing capacity (Fig. 20.1).

Leaf springs can serve multiple purposes, including springing, damping, and locating functions. The interleaf friction inherently provides some damping, though it is not well-regulated and can restrict suspension movement. To address this, manufacturers have explored the use of mono-leaf (Fig. 20.2) springs to improve performance (Table 20.2 and 20.3). In some older designs, the spring ends were shaped into a concave form, known as a spoon end, to support a swiveling member, though this design is rarely used today (Fig. 20.3–20.6).

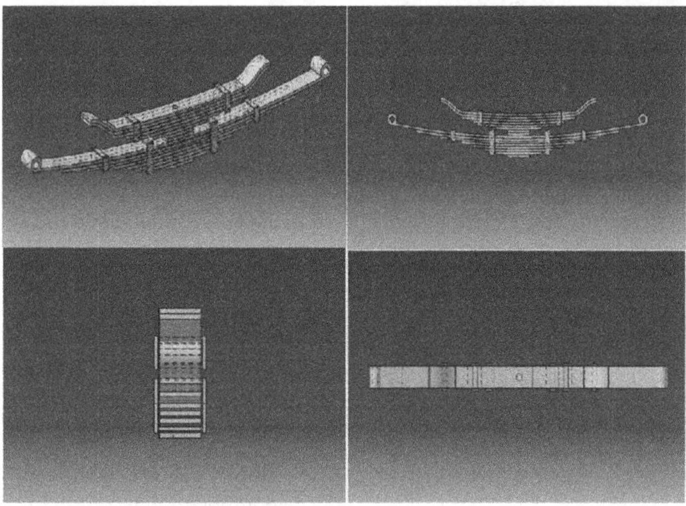

Fig. 20.3 Design of leaf springs

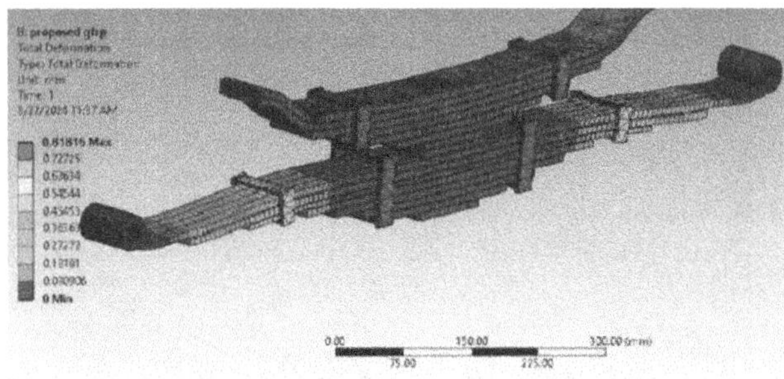

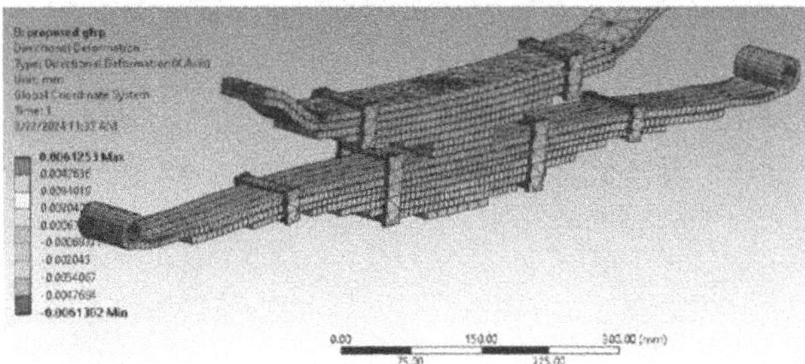

Fig. 20.4 Total deformation and directional deformation

Table 20.1 Values of tensile test

Content	Results	
Sample ID	T1	T2
Gauge Dimension Width (mm)	24.85	25.40
Gauge Dimension thickness (mm)	3.60	4.65
Area(mm^2)	118.91	118.11
Ultimate Tensile Load (kN)	1.40	1.26
UTS (Mpa)	12.00	11.00

Table 20.2 Values of compression test

Content	Results	
Sample ID	T1	T2
Gauge Dimension Width (mm)	24.18	25.67
Gauge Dimension Thickness (mm)	3.69	4.66
Area (mm^2)	120.20	119.62
Ultimate Tensile Load (kN)	22.97	21.32
UTS(Mpa)	**191.00**	**178.00**

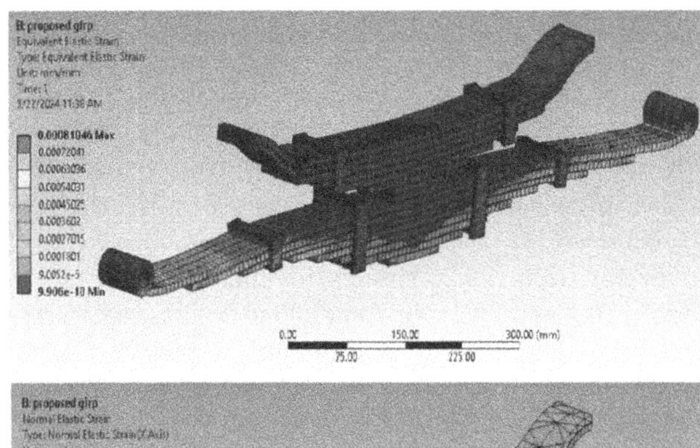

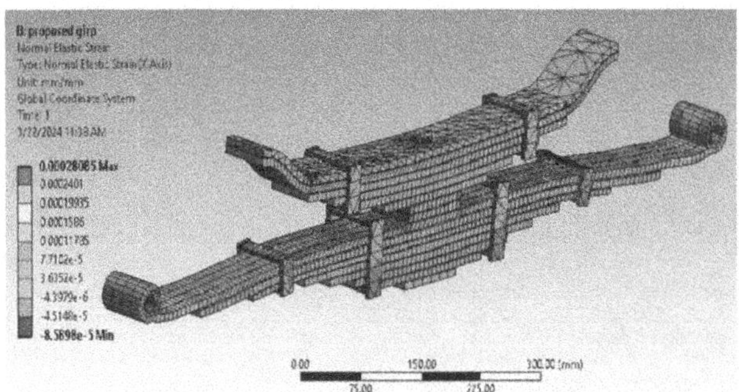

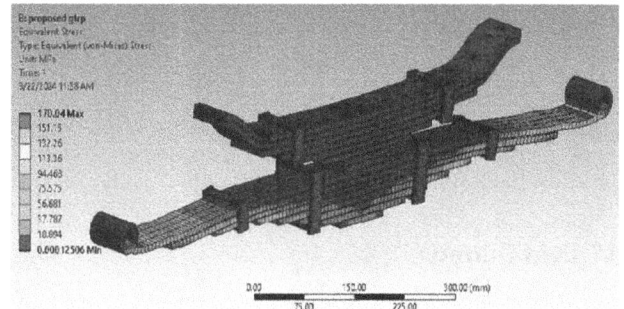

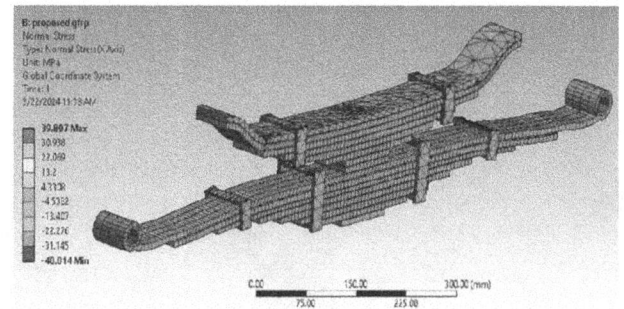

Fig. 20.5 Equivalent elastic strain, normal elastic strain & equivalent & normal stress

Fig. 20.6 Normal elastic strain vs equivalent elastic strain

Table 20.3 Values of charpy test

Dimension (mm)	Type	Temp	Energy in Joules			Mean
			S-1	S-2	S-3	
5 * 10 * 80	Un Notched	+24°C	06	10	08	8.00

2. CONCLUSION

The findings from finite element analysis (FEA) and physical testing of steel and composite leaf springs provide key observations include

- Both FEA and laboratory tests demonstrated that the composite material exhibits excellent tensile strength, compression behavior, and impact resistance, making it well-suited for use in leaf springs.

- Compared to steel counterparts enhances fuel efficiency and increases payload capacity. This weight reduction not only improves vehicle performance but also contributes to energy conservation and environmental sustainability, highlighting the composite material's potential as a superior alternative for modern suspension systems.
- The application of composite fibers in leaf spring materials is in line with sustainability objectives. The fiber composites not only have the capability to minimize the carbon footprint but also make the product sustainable since they are reusable. The study points out the benefits of using polymer composites in leaf spring use and paves the way for further investigation. To improve the performance and expand the use of composite leaf springs, future research and development may involve the optimization of material composition, improvement of manufacturing processes, and modification of design parameters.
- The results prove usefulness and advantages of composite materials, especially in the aspect of minimizing weight and enhancing mechanical properties like corrosion resistance, fatigue resistance, and vibration damping.

Polymer composites indicates that their application can enhance performance and environmental performance in various industries that make use of leaf springs.

References

1. Mariappan, K., Nagarajan, V., Gurusamy, P., Ramanan, N., Raj, S.H.K. and Baskar, S., 2024. Comprehensive mechanical, thermal, and interlaminar characterisation of basalt fibre composite materials for enhanced structural integrity. Advances in Additive Manufacturing Technologies, pp.497–502.
2. Manjunathan, R., Raj, S.H.K., Prabhakaran, P., Ahamed, S.R. and Vikram, S.T., 2024. LM-25 hybrid Al-metal matrix composites: An experimental study of their mechanical properties. Advances in Additive Manufacturing Technologies, pp.166–171.
3. Nagarajan, V., Anbazhagan, R., Raj, S.H.K., Aswin, S. and Gokul, G., 2024. Design, thermal, and static investigations of disc brake systems in automotive and bicycle applications for enhanced performance. Advances in Additive Manufacturing Technologies, pp.172–177.
4. Nagarajan, V., Gurusamy, P., Ramanan, N., Hari, K.R.S., Santhosh, K.V. and Balaji, K., 2024. Investigating mechanical properties of LM25 aluminum matrix composites with varied silicon-CNT reinforcements. Advances in Additive Manufacturing Technologies, pp.490–496.
5. Kingsly, H.K.J., 2024. Production and mechanical properties of aluminum copper alloy reinforced with Al_2O_3 metal matrix composite. AIP Conference Proceedings.
6. Gurusamy, P., Raj, S.H.K., Bhattacharjee, B. and Bhowmik, A., 2024. Assessment of microstructure and investigation into the mechanical characteristics and machinability of A356 aluminum hybrid composite reinforced with SiCp and MWCNTs. Silicon, 16(1), pp.367–382.
7. Raj, S.H.K., Boopathy, G., Vijayananth, S., Balaji, G. and Ramanan, N., 2023. Feasibility and insights into the optimization and characterization of friction welded aluminum–steel dissimilar joints. International Conference on Smart Sustainable Materials and Technologies.
8. Kumar, J.L., Gurusamy, P., Vijayakumar, M.D., Raj, S. and Kishore, R., 2022. Design evolution of thermal barrier coating with Al_2O_3 nanomaterial for cylindrical liners. AIP Conference Proceedings, 2519(1).
9. Gurusamy, P., Muthuraman, V. and Raj, S.H.K., 2022. Performance and experimental analysis of Al-Al_2O_3 metal matrix composites. International Journal of Vehicle Structures & Systems, 14(2), pp.257–259.
10. Gurusamy, P., Muthuraman, V., Lokeshkumar, J. and Raj, S.H.K., 2022. Fabrication and mechanical behaviour of Al-6061/Al_2O_3 metal matrix composite fabricated through powder metallurgy technique. International Journal of Vehicle Structures & Systems (IJVSS), 14(1), pp.47–49.
11. Gurusamy, P., Thirupathiraja, S., Raj, S.H.K. and Kumar, J.L., 2021. Experimental investigation and CFD analysis of inlet manifold in internal combustion engine. Materials Today: Proceedings, 37, pp.840–843.
12. Gurusamy, P. and Raj, S.H.K., 2020. Influence of squeeze casting process parameter on Al/SiCp metal matrix composite. IOP Conference Series: Materials Science and Engineering, 988(1), p.012056.
13. Gurusamy, P. and Raj, H.K., 2020. Microstructure and temperature distribution of aluminium-based composites.
14. Gurusamy, P., Raj, S.H.K., Devarajan, S., Kumar, S.D., Mahesh, R. and Subash, P., 2020. Finite element analysis and experimental study on the deformation characteristics of lock ring. Materials Today: Proceedings, 33, pp. 3208–3211.
15. Gurusamy, P., Raj, S.H.K., Kumar, J.L., Thirupathiraja, S. and Nagendharan, S., 2020. Microstructure and temperature distribution of aluminium-based composites using finite element analysis. Materials Today: Proceedings, 33, pp.3330–3333.
16. Senthil kumar and Vijayarangan, "Analytical and Experimental studies on Fatigue life Prediction of steel leaf soringand composite leaf multi leaf spring for Light passanger veicles using life data analysis" ISSN 1392 1320 material science Vol. 13 No.2 2007.
17. Shiva Shankar and Vijayarangan "Mono Composite Leaf Spring for Light Weight Vehicle Design, End Joint, Analysis and Testing" ISSN 1392 Material Science Vol. 12, No.3, 2006.
18. Niklas Philipson and Modelan AB "Leaf spring modelling" ideon Science Park SE-22370 Lund, Sweden

19. Zhi'an Yang and et al "Cyclic Creep and Cyclic Deformation of High-Strength Spring Steels and the Evaluation of the Sag Effect:Part I. Cyclic Plastic Deformation Behavior" Material and Material Transaction A Vol 32A, July 2001–1697
20. Muhammad Ashiqur Rahman and et al "Inelastic deformations of stainless- steel leaf springs-experiment and nonlinear analysis" Meccanica Springer Science Business Media B.V. 2009
21. C.K. Clarke and G.E. Borowski "Evaluation of Leaf Spring Failure" ASM International, Journal of Failure Analysis and Prevention, Vol5 (6) Pg. No. (54–63)
22. J.J. Fuentes and et al "Premature Fracture in Automobile Leaf Springs" Journal of Science Direct, Engineering Failure Analysis Vol. 16 (2009) Pg. No. 648–655.
23. J.P. Hou and et al "Evolution Of The Eye End Design Of A Composite Leaf Spring For Heavy Axle Load" Journal of Science Direct, Composite Structures 78(2007) Pg. No. (351–358)
24. Practical Finite Element Analysis by Nitin S. Gokhale [10] Composites – A Design Guide by Terry Richardson [11] ANSYS 10.0 Help for FEA Analysis.
25. Jayaraman, K. and Anandhan, S.S., 2009. Analysis of composite leaf springs using finite element method and experimental testing. *Journal of Reinforced Plastics and Composites*, 28(2), pp.253–261. DOI: 10.1177/0731684407085761.
26. Rajendran, I. and Vijayarangan, S., 2001. Optimal design of a composite leaf spring using genetic algorithm. Computers & Structures, 79(11), pp.1121–1129. DOI: 10.1016/S0045-7949(00)00235-0.
27. Shokrieh, M.M. and Rezaei, D., 2003. Analysis and optimization of a composite leaf spring. Composite Structures, 60(3), pp.317–325. DOI: 10.1016/S0263-8223(02)00349-6.
28. Mahmood, K. and Jafar, G., 2017. Performance evaluation of GFRP composite leaf springs under static and dynamic load conditions. International Journal of Mechanical Engineering and Applications, 5(5), pp.123–128. DOI: 10.11648/j.ijmea.20170505.13.
29. Shivashankar, B. and Yadav, B.S., 2014. Design, fabrication, and testing of a composite leaf spring for light commercial vehicles. International Journal of Engineering Research and Technology, 3(6), pp.923-929. ISSN: 2278–0181.
30. Kumar, K. and Dharmalingam, P., 2015. Experimental investigation of composite material leaf spring for automotive suspension system. *Materials Today: Proceedings*, 2(4-5), pp.2439–2446. DOI: 10.1016/j.matpr.2015.07.234.

Note: All the figures and tables in this chapter were made by the authors.

Advances in Additive Manufacturing Technologies – Gurusamy Pathinettampadian et al. (eds)

Mechanical Properties and Performance of LM-13 with Different Composition

Manjunathan R.*,
Barath S., Hari Krishna Raj S., Praveen A., Shylesh B.
Department of Mechanical Engineering,
VelTech HighTech Dr. Rangarajan Dr. Sakunthala Engineering College,
Chennai, Tamilnadu

◆ **Abstract:** LM-13 is commonly used in pistons and applications requiring high resistance to thermal stresses. This alloy demonstrates excellent wear resistance, machinability, and the ability to withstand elevated temperatures and loads. In this study, hybrid composites of aluminum alloy LM-13 reinforced with SiC, MoS2, and graphite were developed using the stir casting technique. SiC powder with an average particle size of 150 microns was employed as the reinforcement, while LM-13 served as the matrix material. The molten composite mixtures were thoroughly stirred and then cast into a metallic mold. Samples were prepared with varying weight percentages of SiC (0%, 2.5%, 5%, and 7.5%) while maintaining constant weight percentages of MoS2 and graphite. The cast specimens were machined according to standardized testing requirements.. The higher hardness was observed at a 7.5% SiC ratio (Ratio-3). However, the impact strength was relatively low for Ratios 2 and 3 due to the agglomeration of reinforcements. Notably, the optimal tensile and compressive strengths were achieved at a composition of 5% SiC, 1% MoS2, and 1% graphite (Ratio-2). The results indicate that incorporating SiC particles significantly enhances the, tensile and compressive strength hardness compared to the unreinforced LM-13 matrix. Additionally, the minimum wear rate was observed in the Ratio-2 composite, highlighting its superior wear resistance properties.

◆ **Keywords:** SiC, MoS2, Composite materials, Resistance, ANOVA, Hardness

1. Introduction

A metal matrix composite (MMC) consists of a metallic matrix, typically made from alloys of aluminum, copper, iron, magnesium, titanium, or lead, embedded with three-dimensional reinforcements such as oxides, carbides, or nitrides. These composites combine the ductility of the metallic matrix with the enhanced modulus and strength provided by the reinforcement, resulting in improved material performance. Through proper selection and mixing of the matrix and reinforcement, MMCs can be designed to have superior electrical, mechanical, and chemical characteristics, overcoming the shortcomings of conventional metals and alloys in sophisticated engineering applications. MMCs are famous for their capability to withstand high tensile and compressive stresses by effectively transferring and dispersing the applied load from the matrix to the reinforcement. Different fabrication processes, including powder metallurgy, liquid metallurgy, and squeeze casting, are utilized to make MMCs. The reinforcement phases of these composites may be continuous fibers, discontinuous particulates, or whiskers. Of these, particulate reinforcements are especially attractive as they yield reproducible, isotropic properties and are cheaper than fiber-reinforced composites because fibers are cheaper and simpler processes are used for their manufacturing.

*Corresponding author: manjunathanmech@gmail.com

DOI: 10.1201/9781003685906-21

Among the ceramic particles used as reinforcements in aluminum-based MMCs, boron carbide (B4C) stands out for its compatibility with aluminum, low cost, and exceptional wear resistance. These properties make B4C-reinforced aluminum composites, such as those incorporating silicon carbide (SiC) or zirconium dioxide (ZrO2), ideal for applications in automotive components like pistons, brake rotors, calipers, connecting rods, and cylinder liners. B4C-reinforced aluminum PMMCs exhibit significantly reduced abrasive and sliding wear rates compared to unreinforced aluminum matrices. The hard particles in these composites are especially effective under high contact loads, delaying the transition from mild to severe wear. This enhanced wear resistance, combined with excellent thermal conductivity, has positioned Al-B4C PMMCs as promising materials for automotive brake rotors, which traditionally use gray cast iron. The volume fraction of reinforcement particles or whiskers in these composites typically ranges between 10% and 30%. Aluminum composites also play a vital role in the aerospace industry. Hyper-eutectic Al-Si-based composites, such as A6082, containing B4C, ZrO2, or SiC particles, are widely utilized in manufacturing automotive engine components due to their outstanding durability and performance under demanding conditions. Al7075/TiC/MoS2 composite specimens were fabricated using the stir casting method was studied their wear behavior under ambient and elevated temperatures. The wear analysis was conducted using a pin-on-disc apparatus, and the Taguchi technique was employed for optimization. Statistical analysis was performed using Minitab 18 software, where an orthogonal L9 array was designed to evaluate the influence of various parameters, including applied load, temperature, sliding distance, and the percentage of reinforcement.

Analysis of Variance (ANOVA) was used to identify the significance of each parameter, and regression equations were developed to establish correlations between them. The Signal-to-Noise (S/N) ratio was utilized to determine the most influential factors on wear resistance. The results indicated that the reinforcement had a negative impact on weight loss, with weight loss decreasing as the reinforcement percentage increased. Among the factors analyzed, applied load, sliding distance, and temperature were identified as the most significant contributors to wear resistance. The primary and interaction effects of key influencing factors in Al alloy-TiC composites were analyzed using the central composite rotatable design technique. It was observed that the weight loss of the composite decreased with an increase in the weight percentage of TiC particles but increased with higher applied loads and longer sliding distances. The Rockwell hardness of the aluminum composites showed a noticeable improvement as the TiC content decreased. Microstructural analysis revealed a non-uniform wear pattern characterized by grooves, micro-cutting, and scratch marks, indicating abrasion wear and delamination. These findings provide insights into the wear mechanisms and mechanical behavior of Al alloy-TiC composites under varying conditions.

Table 21.1 Sample ratio

Percentage	LM-13 gms	SiC grams	MoS2 grams	Graphite 1% gram
LM-13-100%	600	0%	0%	0%
SiC -4.5% + MoS2-4% & Gr- 2% -LM-13	650	7%-25	3%-13	13
SiC -5.5% + MoS4-6% & Gr- 2% -LM-13	700	8.3%-50	4%-44	38

Graphite consists of multiple stacked layers, each made up of graphene sheets. In these layers, carbon atoms form a hexagonal honeycomb pattern, with a bond length of 0.142 nm between adjacent atoms. The layers themselves are spaced 0.335 nm apart. Within each layer, the carbon atoms are strongly bonded through covalent bonds, but only three of the four available bonding sites are occupied. The remaining electron is free to move within the layer, allowing graphite to conduct electricity along the plane. However, since there are no mobile electrons between the layers, graphite does not conduct electricity in the perpendicular direction.

Graphite's layers are loosely bound by weak van der Waals forces, making it easy for them to slide over each other. This property is what gives graphite its well-known lubricating ability. It occurs in two structural forms: alpha (hexagonal) and beta (rhombohedral). While both forms share similar physical characteristics, they differ in how their layers are stacked. The alpha form can be either flat or slightly curved and can be converted into the beta form through mechanical processing. However, when heated above 1300°C, the beta form naturally reverts back to the alpha structure.

2. Result and Discussion

Table 21.2 Hardness value

R1, R2 nad R3 HARDNESS VALUE

Sample	Material	HRB
R_1	LM-13-100%	30
R_2	SiC -5% + MoS2-1% & Gr- 2% -LM-13	42
R_3	SiC -5.5% + MoS2-1% & Gr- 2% -LM-13	37

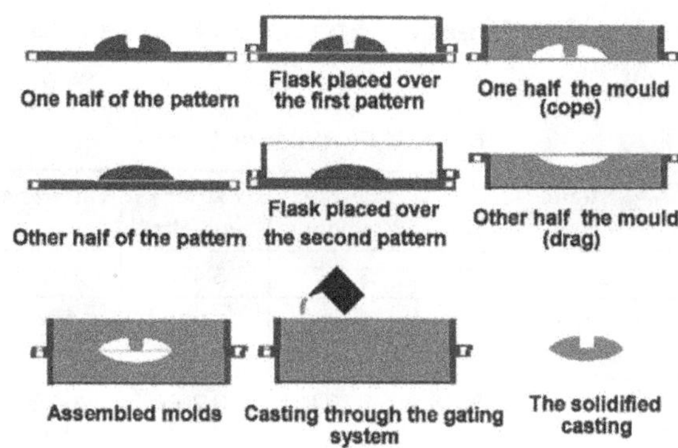

Fig. 21.1 Crucible casting steps

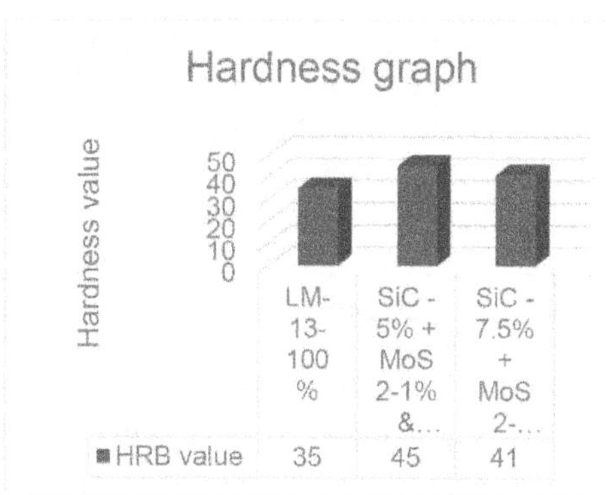

Fig. 21.2 Hardness strength

Table 21.3 Tensile strength

TENSILE STRENGTH

R	D (mm)	CSA (mm^2)	TL (kN)	TS (N/mm^2	IGL (mm)	FGL (mm)	%E
$Ratio_1$	14	202.55	21.35	105.32	72.2	71.3	0.00
$Rario_2$	15	203.11	23.75	119.84	72.2	73.0	1.07
$Ratio_3$	15	202.26	21.25	110.15	72.2	72.4	0.39

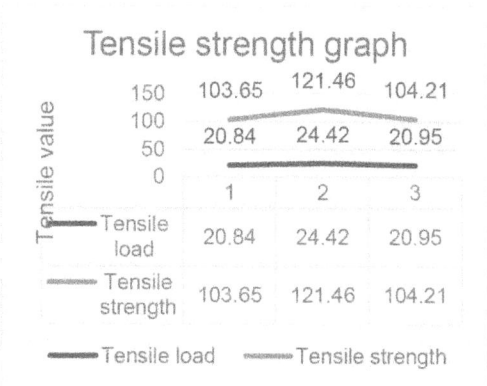

Tensile strength graph

	1	2	3
Tensile load	20.84	24.42	20.95
Tensile strength	103.65	121.46	104.21

Fig. 21.3 Tensile strength

Table 21.4 Compression strength

COMPRESSIVE STRENGTH

S. No	Percentage	Strength N/mm^2
R_1	LM-13-100%	258.13
R_2	SiC -5% + MoS2-1% & Gr- 2% -LM-13	332.24
R_3	SiC -5.5% + MoS2-1% & Gr- 2% -LM-13	315.83

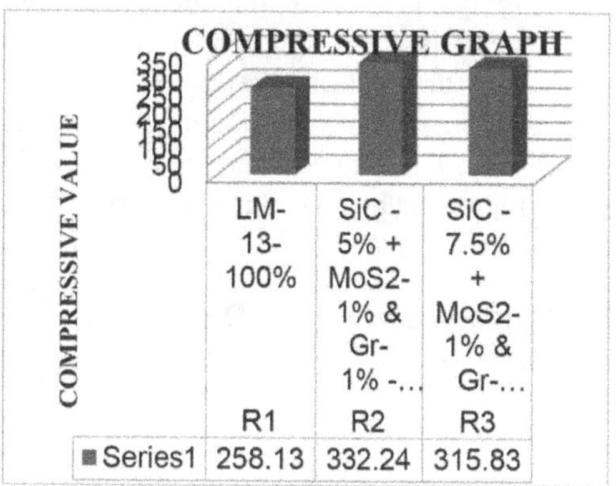

Fig. 21.4 Compression strength

Table 21.5 Impact strength

IMPACT STRENGTH VALUES

S. No	PERCENTAGE	IMAPCT (Joules)
R_1	LM-13-100%	3
R_2	SiC -5% + MoS2-1% & Gr- 2% -LM-13	1.5
R_3	SiC -5.5% + MoS2-1% & Gr- 2% -LM-13	1.7

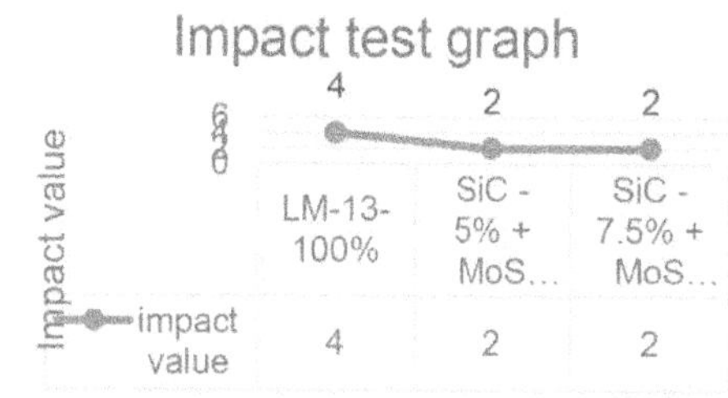

Fig. 21.5 Impact strength

The study analyzed the mechanical properties of the hybrid LM-13 metal matrix and found that the addition of molybdenum disulfide, silicon carbide, and graphite significantly improved its tensile and compressive strength. In terms of hardness, only slight variations were observed at the highest reinforcement composition. However, impact strength was lower in Ratio-2 and Ratio-3 due to the agglomeration of reinforcements, whereas Ratio-1 exhibited better performance.

2.1 Wear Test

A tribometer is an instrument used to assess important tribological characteristics, including the coefficient of friction, frictional force, and wear volume between two interacting surfaces. The term "tribotester" is a more general term used to describe machines that replicate and analyze processes such as wear, friction, and lubrication—critical elements in the field of tribology. These devices are often highly specialized and are custom-built by manufacturers to test the long-term durability of their products. For instance, companies producing orthopedic implants invest significantly in designing tribotesters that simulate the natural movements and forces experienced by human hip joints, enabling them to perform accelerated wear tests on their implants.

2.2 Specification of Pin on Disc

Test Speed: 637 RPM, Normal Load: 20N, Pin Dia: 10mm, Track Radius: 30mm

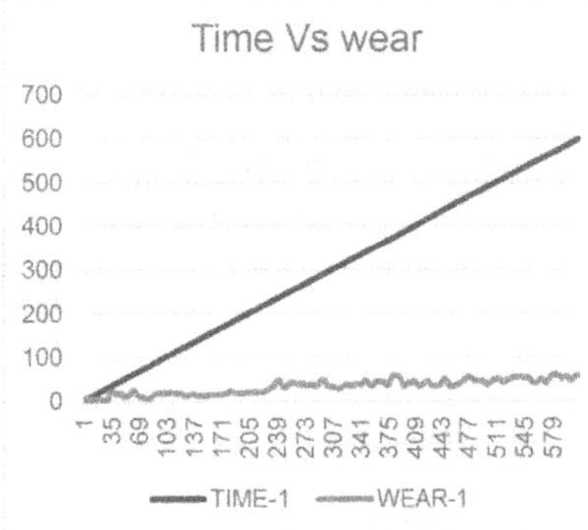

Fig. 21.6 Time-1 vs wear rate-1

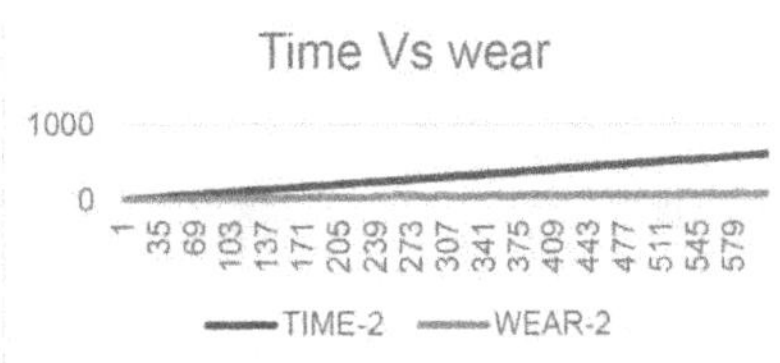

Fig. 21.7 Time-2 vs wear rate-2

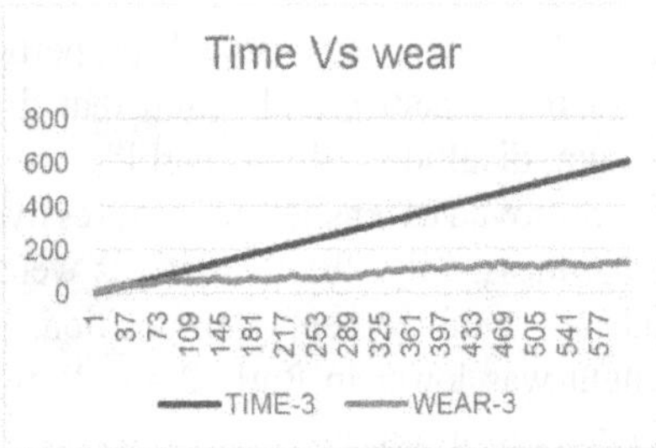

Fig. 21.8 Time-3 vs wear rate-3

2.3 Weight of Testing Specimen before and After

Table 21.6 Wear rate

Ratio	Before weight	After weight	Difference
LM-13-100%	3.457	3.434	0.023
SiC -6% + MoS_2-1% & Gr-2% -LM-13	3.010	3.008	0.002
SiC -5.5% + MoS_2-1% & Gr- 2% -LM-13	3.334	3.236	0.008

3. CONCLUSION

- During the analysis the wear had been found the minimum wear rate occurred on the Ratio 2 -SiC -7% + MoS_2-2% & Gr- 2% -LM-13 is obtained very low wear rate.
- Composite materials, particularly those made from LM-13 aluminum alloy reinforced with Molybdenum disulfide (MoS_2), silicon carbide (SiC), and graphite enhance the mechanical properties of LM-13 alloy significantly when used as reinforcements, making it stronger and more durable compared to its unreinforced form.
- The highest hardness was observed in the composite with reinforcement Ratio-2. However, the impact strength was relatively low for Ratios 2 and 3, likely due to the agglomeration of the reinforcement.
- The optimal compressive and tensile strength was achieved with a composition of 5% SiC, 1% MoS_2, and 1% graphite (Ratio-2). Additionally, this composition also exhibited the lowest wear rate, making it the most effective in enhancing the material's overall performance.

REFERENCES

1. Anish, A., 2018. Characterization of aluminium matrix reinforced with tungsten carbide and molybdenum disulphide hybrid composite. 2nd International Conference on Advances in Mechanical Engineering (ICAME 2018).
2. Prabha, R.N., 2017. Effect of TiC and MoS_2 Reinforced Aluminium Metal Matrix Composites on Microstructure and Thermogravimetric Analysis. Research Journal of Chemistry and Environment, 10(3), pp. 729–737. ISSN: 0974-1496, E-ISSN: 0976–0083.
3. Arul Daniela, A., 2017. Dry Sliding Wear Behaviour of Aluminium 5059/SiC/MoS_2 Hybrid Metal Matrix Composites. Received: January 06, 2017; Revised: August 11, 2017; Accepted: August 24, 2017.
4. Paranthaman, P., 2020. Multi-Objective Optimization on Tribological Behaviour of Hybrid Al MMC by Grey

Relation Analysis. Journal of Critical Reviews, 7(9). ISSN: 2394–5125.

5. Murugan, S.S., 2017. Development of Hybrid Composite for Automobile Application and its Structural Stability Analysis Using ANSYS. International Journal of Modern Studies in Mechanical Engineering (IJMSME), 3(1), pp. 23–34.
6. Antony Arul Prakash, M.D., 2015. Microstructural Analysis of Aluminium Hybrid Metal Matrix Composites Developed Using Stir Casting Process. International Journal of Advances in Engineering , 1(3), pp. 333–339.
7. Karuppasamy, R., 2020. Taguchi-GRA for Multi-Criteria Optimization of Turning Parameters for Al/Bagasse Ash/Gr Hybrid Composite. Journal of Critical Reviews, 7(9). ISSN: 2394–5125.
8. Ghasali, E., Production of Al-SiC-TiC hybrid composites using pure and 1056 aluminum powders prepared through microwave and conventional heating methods. Journal of Alloys and Compounds.
9. Rajesh, S., 2018. Influence and Wear Characteristics of TiC Particle in Al6061 Metal Matrix Composites. International Journal of Applied Engineering Research , 13(9), pp. 6514–6517. ISSN: 0973–4562.
10. Kanchan, A., Tribological Investigation of Al7075/TiC/ MoS_2 Hybrid Composite Material. International Research Journal of Engineering and Technology (IRJET).
11. Hemanth, J. & Divya, M.R., 2018. *Fabrication and corrosion behavior of Aluminium Alloy (LM-13) reinforced with nano-ZrO2 particulate chilled nano metal matrix composites (CNMMCs) for aerospace applications*. Journal of Materials Science and Chemical Engineering.
12. Vellingiri, S., Thirumoorthy, A., Kandasamy, K., Mugilan, T. & Srinidhi, M.S., 2021. *Optimisation of Aluminium Alloy Squeeze Cast Parameters LM-13/Sic on Mechanical Properties*. Turkish Online Journal of Qualitative Inquiry, 12(6).
13. Opeka, M.M., Talmy, I.G. & Zaykoki, J.A., 2004. *Mechanical Properties of ZrB2 Composites*. Journal of Materials Science, 32, pp. 5887–5894.
14. Luo, A., 1995. *Processing, Microstructure and Mechanical Properties of Cast Mg CNMMC*. Metallurgical and Materials Transactions, 26, pp. 2445–2453.
15. Saravananand, R.A. & Surappa, M.K., 2000. *Fabrication and Characterization of Al-SiCp Particle Composite*. Materials Science and Engineering: A, 108, pp. 276–285.
16. Hassan, S.F. & Gupta, J., 2002. *Development of High Strength Al Based Composites Using Elemental Ni Particles*. Journal of Materials Science, 37, pp. 2467–2477.
17. Singh, A. & Tsai, A.P., 2003. *Quasicrystal Strengthened Mg-Zn-Y Alloys by Extrusion*. Scripta Materialia, 49, pp. 417–426.
18. Guy, A.G., 1967. *Elements of Physical Metallurgy*, 2nd ed. Addison-Wesley, Boston, pp. 78–85.
19. Lai, M.O. & Saravanaranganathan, D., 2000. *Synthesis, Microstructure and Property Characterization of Disintegrated Melt Deposited Al/SiC Composites*. Journal of Materials Science, 35, pp. 2155–2164.
20. Yamamoto, T., Sasamoto, H. & Inagaki, M., 2000. *Extrusion of Al Based Composites*. Journal of Materials Science Letters, 19, pp. 1053–1064.
21. Awasthi, S. & Wood, J.L., 1988. *Mechanical Properties of Extruded Ceramic Reinforced Al Based Composites*. Advanced Ceramic Materials, 35, pp. 3449–3458.
22. Lai, M.O., 2000. *Development of Ductile Mg Composite Material Using Ti Reinforcement*. Journal of Materials Science, 38, pp. 2155–2167.

Note: All the figures and tables in this chapter were made by the authors.

Advances in Additive Manufacturing Technologies – Gurusamy Pathinettampadian et al. (eds)

Reducing Defects in Composite Material Cutting Using Abrasive Water Jet Machining

Vijayakumar K.[1]
Assistant Professor, Department of Mechanical Engineering,
Aarupadai Veedu Institute of Technology,
Chennai, India

Maheshwaran V.[2]
Student, Department of Mechanical Engineering,
Aarupadai Veedu Institute of Technology,
Chennai, India

Parthiban A.[3]
Assistant Professor, Department of Mechanical Engineering,
Chennai Institute of Technology,
Chennai, India

Karthick M.[4]
Assisstant Professor, Department of Mechanical Engineering,
Veltech Rangarajan Dr Sagunthala R&D Institute of science and technology,
Chennai, India

P. Usha Rani[5]
Professor, Science and Humanities Department,
R.M.D Engineering College, India

Khizar Hussain F. R.[6]
Assisstant Professor, Department of Mechanical Engineering,
Aarupadai Veedu Institute of Technology,
Chennai, India

Bharanidharan T.[7]
Student, Department of Mechanical Engineering,
Chennai Institute of Technology,
Chennai, India

◆ **Abstract:** The aerospace and automotive industries are showing interest in CFR-PLA composites because of its favorable mechanical characteristics and low environmental impact. But they haven't been used much because of surface roughness and dimensional instability caused by milling. In this study, we compare CFR-PLA to pure PLA and find that AWJM improves both surfaces and machinability. Surface roughness (Ra) and kerf taper angle (T) were reduced by 23% and 15%, respectively, in CFR-PLA as compared to pure PLA at optimal machining conditions, which include 3500 bar water pressure, 800 mm/min traverse rate, and 250 g/min abrasive flow rate. Carbon fiber

[1]vijayakumar@avit.ac.in, [2]maheswaran28121999@gmail.com, [3]parthimech2810@gmail.com, [4]karthick@veltech.edu.in, [5]pusharani71@yahoo.com, [6]khizarhussain8528@gmail.com, [7]vtrbharani12@gmail.com

DOI: 10.1201/9781003685906-22

reinforcement has a stabilizing effect that improves mechanical stability and machining dimensional accuracy, which leads to these benefits. Findings validate AWJM as a technique for CFR-PLA machining and open the door to its potential use in structural prototypes and aerodynamic constructions, which need lightweight components with high accuracy. The work provides important information on sustainable manufacturing and addresses a major gap in the literature about the machinability of hybrid composites. The full potential of CFR-PLA in high-stakes engineering applications needs more research into hybrid reinforcement methodologies, optimization of additional process parameters, and unique post-processing techniques.

◆ **Keywords:** CFR- PLA, AWJM, Composite material, Additive manufacturing, Aerospace, Automotive

1. INTRODUCTION

3D printing, or AM technology, has revolutionized the fabrication of intricate geometries in fields such as aerospace, automotive, and biomedical engineering. The impact of structural modifications on processing results was emphasized [1] while discussing the impact of AWJM on machinability. In a similar vein, Chen et al. [2] examined AWJM mechanisms in Q345 steel and found process-specific details about accuracy and heat neutrality. In their study, Bańkowski and Spadło [3] proved that AWJM effectively eliminated casting flash, which in turn improved surface quality and dimensional accuracy throughout manufacturing. The significance of keeping an eye on cutting forces in AWJM systems was brought to light by HlaváŐŐr et al. [4], who offered a diagnostic view of machining stability. The capacity of AWJM to create flawless surfaces was shown by Zhao and Guo [5] via their comparison of surface topography and microstructure. WJM is versatile and can process a wide range of materials, as shown in the review of Liu et al. [6]. Ishfaq et al. [8] looked at ways to lessen edge damage in clad composites, lowering delamination and microcracks, whereas Barsukov et al. [7] examined the abrasiveness of copper slag particles in AWJM, focusing on the impact of abrasive material qualities.

The ability of AM to create intricate patterns with little tooling, little material loss, and great precision has contributed to its rapid expansion. New water jet applications in advanced manufacturing were detailed by Folkes [9], who emphasized the revolutionary character of the technology. In their review of AWJM R&D, Kovacevic et al. [10] highlighted significant technical developments. The significance of heat control in AWJM procedures, particularly in preserving the microstructure of the material, was highlighted by Spadło et al. [11]. When it comes to processing thermoplastics like PLA and its composites, FDM stands out among the several AM processes due to its flexibility, low cost, and simplicity of usage. Ohadi and Cheng [12] focused on thermal management by simulating FDM workpiece temperature distributions. Hlavár et al. [13] demonstrated the simplicity of FDM in dealing with complicated geometries by studying the impact of material structure on forces in FDM. Abrasive disintegration in FDM was reported by Perec [14], demonstrating its potential for precise applications. The acoustic characteristics of natural fiber composites in FDM were advanced by Taiwo et al. [15], who contributed to the advancement of green design.

PLA, a biodegradable material made from renewable resources like sugarcane and maize starch, is extensively used in environmentally conscious production because to its mechanical strength, lightweight nature, and biodegradability. To prove that recycled fiber-reinforced PLA composites are good for the environment, Hernandes Diat et al. [16] studied how these materials absorbed water. To demonstrate the compatibility of PLA composites with natural fiber reinforcements, whereas Shahria [18] emphasized improved mechanical performance in hybrid PLA composites. Research by two authors [20, 21] investigated the effects of fiber properties on PLA composites, demonstrating the possibility of using these materials in wood-plastic composites. The anisotropic characteristics and susceptibility to heat deterioration of FDM-printed PLA components make it difficult to attain uniform surface quality. At AM, Schwarzkopf and Burnard [22] examined the pros and cons of wood-plastic composites in terms of their effects on the environment and their performance. Two researchers [23, 24] provide important information on the machinability of PLA by studying the machining behavior of bio-composites.

Computer numerical control milling and laser cutting, two examples of conventional machining techniques, make matters worse. When CNC milling bio-composites, problems with surface roughness and delamination might occur, as pointed out by two researchers [25, 26]. Conventional machining techniques reduced the tensile strength and caused delamination in fiber-reinforced composites [27-29]. According to Vigneshwaran et al. [30], heat-affected zones are created during laser cutting, which reduces the precision of dimensions and the quality of the surface finish. Because of these restrictions, AWJM has become popular as a non-contact method for reducing mechanical and thermal stresses. Reviews of AWJM's use in cutting natural fiber composites by two authors [31, 32] show that it effectively reduces thermal damage. In order to optimize the procedure, Jagadish et al [33] designed the surface roughness in green AWJM. Hybrid composites were investigated with AWJM being highlighted as a component that enhances surface integrity [34, 35].

The accuracy and thermal neutrality of AWJM make it an ideal choice for materials that are sensitive to heat, such as PLA and its composites. Researchers Saeimi Sadigh and Marami [36] looked at bio-composites that were produced using AWJM and how additives affected the tensile strength. Reviewing finite-element models for fiber-metal laminates, Smolnicki et al. [37] showed that AWJM may decrease stresses caused by machining. Marques et al. [38] highlighted the benefits of AWJM in hybrid composite machining by emphasizing its reduced delamination effects in fiber-metal laminates. Composites made of carbon fibers and polylactic acid (CFR-PLA) have the best of both worlds: the biodegradability and light weight of PLA and the strength and durability of carbon fibers. Supporting AWJM's applicability to high-strength composites, Pai et al. [39] examined unconventional machining approaches. Improving cutting quality, Siva Kumar et al. [40] optimized AWJM parameters for intermetallic laminates using intelligent modeling. Surface roughness and material removal rates were examined as trade-offs in AWJM of polymer composites by Doğankaya et al. [41]. Improved surface quality and machinability were highlighted in studies that investigated the effects of AWJM on natural fiber composites [42, 43].

Lightweight structural components with complex geometries may be made with ease using CFR-PLA since it outperforms traditional polymers in tensile strength, stiffness, and machinability. Researchers Alberdi et al. [44] showed that AWJM improved kerf quality when applied to CFRP/Ti6Al4V stacks. Highlighting its compatibility with natural fiber reinforcements, Hutyrová et al. [45] examined the efficacy of AWJM in machining wood-plastic composites. Research by Kalirasu et al. [46] and Pahuja et al. [47] showed that AWJM can keep hybrid laminates' dimensions stable and their surfaces intact. Machining CFR-PLA isn't without its difficulties, however; surface quality must be maintained and flaws like kerf taper and delamination must be minimized. In order to forecast the kerf shape during AWJM on titanium and CFRP laminates, Pahuja and Ramulu [48] created predictive models. By comparing the results of optimizing CNC machining settings for polymer composites, Hazir and Ozcan [49] shed light on the benefits of AWJM. Palleda [50] investigated AWJM taper angles, with a focus on optimizing parameters to minimize defects.

One of the biggest challenges with post-processing CFR-PLA components is keeping their surface quality constant. By finding the sweet spot for AWJM, Mm et al. [51] reduced the occurrence of delamination in composites with fiber reinforcement. Using kerf taper and surface finish as their primary objectives, Kumar et al [52] investigated AWJM multi-response optimization. In order to reduce defects, Two researchers [53, 54] studied AWJM-induced damage in fiber composites and came up with solutions. There is a lack of study on optimizing machining parameters for CFR-PLA to improve surface quality and dimensional stability, since most of the existing AWJM studies concentrate on traditional polymer composites or unreinforced PLA. The purpose of this research is to fill these knowledge gaps by examining the relationship between CFR-PLA machinability and three critical AWJM parameters: water pressure, traverse speed, and abrasive flow rate. An examination of CFR-PLA in comparison to unreinforced PLA highlights its mechanical and environmental advantages, offering realistic recommendations for optimizing AWJM and bolstering its use in sustainable, high-precision applications in manufacturing.

2. Experimental Methodology

The PLA and CFR-PLA filaments used in this research came from Esun Industrial Co., Ltd. in Shenzhen, China, and were available for purchase. The mechanical consistency and compatibility of these materials for FDM-based 3D printing were taken into consideration while selecting them, as stated in the manufacturer's datasheets. A Creality Ender 3 Pro 3D printer with a 0.4 mm nozzle was used to create standardized specimens with dimensions 100 × 300 × 8 mm. In order to eliminate heat and mechanical stress-induced flaws and maximize structural integrity, first experiments were conducted to improve the printing conditions. After reviewing the literature, we settled on a printing speed of 60 mm/s since it optimizes mechanical performance while keeping printing time to a minimum.

A high-performance abrasive waterjet cutting machine was used for the machining trials. The equipment was purchased from CT Cutting Technologies in Istanbul, Turkey. In order to guarantee that experimental measurements are reliable, these data are essential. The system incorporates a KMT cutting head and can operate at a maximum pressure of 6000 bar while maintaining a positioning precision of

±0.03 mm. The entrance parameters were stabilized by applying a 5 mm pre-cut. The speeds tested varied from 400 to 1200 mm/min. Minimizing surface imperfections without producing excessive material erosion was the goal of the investigation of abrasive flow rates ranging from 100 to 450 g/min. Each parameter combination was evaluated three times to assure statistical reliability. The analysis was based on the mean results.

Prior research and early testing validated the specified AWJM PP for CFR-PLA, which were chosen based on technical factors and material-specific features. To maximize cutting performance while avoiding excessive erosion, a water pressure range of 2500-3500 bar was used. The carbon fiber matrix might be damaged at pressures higher than 3500 bar, and cutting would be impossible at pressures lower than 2500 bar [69,70]. We balanced the jet-material contact duration with process efficiency by setting the traversal speed range of 400-1200 mm/min [71]. Although processing time was longer at slower rates, surface polish was improved, and a bigger taper angle was possible at faster speeds, which improved efficiency. Similarly, to maximize cutting efficiency, surface roughness, and cost-effectiveness, an abrasive flow rate range of 100-450 g/min was chosen. Flow rates over 450 g/min may lead to excessive wear and higher material costs [72,73], whereas rates below 100 g/min lowered cutting efficiency. Taking into account the carbon fiber reinforcing and heat sensitivity of CFR-PLA, these parameter choices guaranteed steady and uniform cutting. In line with previous research, preliminary studies verified that these ranges reduced taper angle and gave excellent surface quality.

In order to study the impacts of targeted machining, The first area, known as the jet entrance zone, had a high concentration of energy that smoothed out surface imperfections. The consistent cutting conditions in Region 2, the midsection of the cut, result in surface qualities that are reasonably stable. Surface imperfections were greater in Region 3, close to the jet exit point, because to jet dispersion and decreased cutting accuracy. The graphic clearly shows the flow dynamics, together with the direction of the AWJ and the nozzle feed. The color-coded profile shows how the surface roughness varies in certain areas, which means that optimizing the parameters for each location is important for improving the surface quality. In order to study the variance caused by localized machining effects, the cutting area was divided into three sections and surface quality was assessed using two important metrics, Ra and T. The first area, at the point of jet entrance, had the most severe impact effects; the second, in the middle of the cut, had rather steady cutting conditions; and the third, close to the point of jet exit, had the most abnormalities as a result of jet dispersion.

3. Results and Discussion

The filaments used in this investigation were PLA and CFR-PLA, which are commercially available. As stated in the manufacturer's datasheets, these materials were chosen for their mechanical consistency and compatibility with 3D printing that is based on fused deposition modeling (FDM). A 3D printer with a 0.4 mm nozzle was used to create standardized specimens of 100 × 300 × 8 mm. Preliminary testing improved printing conditions to reduce thermal and mechanical stress-induced flaws and maximize structural integrity. According to the research, a printing speed of 60 mm/s offers the best combination of mechanical performance and printing time. These readings were confirmed by routine calibration and culled from the manufacturer's technical manual. The system uses a KMT cutting head and can withstand pressures up to 6000 bar, with a positioning precision of ±0.03 mm. Every specimen was subjected to 18 cuts, with a length of 80 mm for each cut, spaced 10 mm apart. The entrance parameters were stabilized by applying a 5 mm pre-cut. Ra and T were thoroughly evaluated in relation to water pressure (2500-3500 bar), traverse speed (400-1200 mm/min), and abrasive flow rate (100-450 g/min).

Lower speeds allow the jet to make more contact with the material, which should enhance surface quality. On the other hand, higher speeds make the operation more efficient. To minimize surface flaws without over-eroding the material, we used abrasive flow rates ranging from 100 to 450 g/min. In order to ensure statistical reliability, we repeated the experiments three times for each combination of parameters and used the means.

There wasn't enough cutting force at pressures below 2500 bar to avoid damaging the carbon fiber matrix, and pressures beyond 3500 bar may cause damage. The traversal speed range of 400-1200 mm/min was determined to combine process efficiency with surface quality by maximizing the time of the jet's interaction with the material. A slower speed improves surface polish but increases processing time, whilst a quicker speed improves efficiency but may also increase taper angle. A similar approach led to the selection of an efficient, cost-effective abrasive flow rate (100–450 g/min) that balanced surface roughness. Economic inefficiency and excessive wear set in at rates over 450 g/min, while cutting efficiency declines below 100 g/min. These values were confirmed in preliminary studies to enhance surface quality and decrease taper angle, which is consistent with the findings in the literature.

In order to analyze the localized machining impact, the AWJM-processed surface was divided into three separate sections. At the jet entrance site, which is located in Region 1, focused energy is used to decrease surface roughness and produce a smoother surface. The relatively flat surface topography of Region 2, which encompasses the central portion of the cut, is a result of the consistent cutting conditions. Because of the dispersion of the jet and the decreased cutting accuracy, Region 3, which is near the jet exit location, has a rougher surface. To further understand the flow dynamics, the graphic also shows the nozzle feed and abrasive waterjet (AWJ) direction. Surface roughness varies with areas, as seen by the color-coded profile. This suggests that parameter optimization should be done depending on the location in order to attain a greater surface quality.

The cutting zone was divided into three sections to track surface changes caused by localized machining effects, and two metrics, Ra and T, were used to quantify surface quality. The zone of jet entry with the most significant first impact consequences was indicated by Region 1. At the halfway point of the cut, in Region 2, the cutting conditions were steady. The dispersion of the jet was most noticeable in Region 3, which lies near the jet departure zone. The surface roughness was assessed using a surface roughness tester manufactured by Mitutoyo Corporation in Japan. Readings were taken three times in each zone to provide an average Ra value, which allowed for precise evaluation of surface quality in each area. Surface roughness and taper angle were found to be considerably impacted by water pressure, according to experimental results. Increased water pressure resulted in improved material removal efficiency but, above 3500 bar, caused excessive material erosion and surface defects. In contrast, pressures below 2500 bar were inadequate for efficient cutting, resulting in poor surface quality. The optimal water pressure range (2500–3500 bar) provided a compromise between cutting efficiency and surface finish quality.

Traverse speed also had a major contribution towards surface roughness. Lower traverse speeds (400–600 mm/min) resulted in smoother surfaces due to increased jet-material interaction, while higher speeds (1000–1200 mm/min) compromised surface finish but enhanced process efficiency. A traverse speed of approximately 800 mm/min provided a compromise between surface quality and processing time. Abrasive flow rate directly contributed to surface roughness and cutting precision. Lower abrasive flow rates (100–200 g/min) resulted in lower cutting efficiency and poorer surface finish, while flow rates too high (above 450 g/min) resulted in material over-erosion and added capital cost. The optimal range of 250–400 g/min provided a good compromise between cost-effectiveness and surface finish.

Region-wise roughness analysis using surface roughness validated that Region 1 had the smoothest surface due to focused jet impact. Region 2 had comparatively stable surface values, while Region 3 had greater roughness and taper angle due to dispersion of the jet. This agrees with the requirement of region-based optimization techniques for uniform surface quality. Overall, the research established that accurate AWJM parameter control is necessary for optimizing CFR-PLA machining performance. Identified parameter ranges optimized material removal efficiency, surface roughness, and taper angle with optimal trade, leading to high-quality machined parts. The results are in agreement with the current literature and can be of use in future research on AWJM of polymer-based composites.

4. Conclusion

The results of this research show that CFR-PLA composites are an excellent choice for uses necessitating exact dimensions, smooth surfaces, and high degrees of precision. A thorough analysis of the AWJM characteristics revealed the surface roughness and the kerf taper angle. Abrasive flow rate, cutting speed, and water pressure are some of these characteristics. Due to the carbon fiber reinforcement's capability to significantly reduce surface defects produced by high-pressure cutting, CFR-PLA outperformed pure PLA. Based on these results, CFR-PLA is a great material for lightweight structural parts with close tolerances, which are common in the aerospace and automotive industries:

- Compared to PLA, CFR-PLA had better surface smoothness and less kerf taper angles. Better dimensional precision and machining stability are achieved because the carbon fibers offer a stabilizing effect.
- Abrasive flow rate and water jet pressure were two of the most important variables. A combination of a moderate abrasive flow rate (250-450 g/min) and higher water pressure (3500 bar) resulted in consistently better surface quality and kerf taper angles.
- Kerf taper angles and surface roughness are negatively impacted when traverse rates beyond 1200 mm/min, since the period of jet-material contact is decreased. Edge effects and fiber-matrix interactions dictated the variation in surface roughness across different cut sections. Traverse rates ranging from 400 to 800 mm/min were determined to be the most favorable. To achieve consistent surface quality, adjusting AWJM settings for specific locations is crucial.

- ANOVA verified that AWJM parameters significantly affect surface quality. A foundation for attaining high cutting accuracy was laid by establishing optimized conditions for both PLA and CFR-PLA.

It is an excellent choice for high-performance, environmentally conscious industries because to its machinability and other desirable properties. In order to broaden the use of CFR-PLA in precision engineering, future studies should investigate hybrid reinforcement techniques, sophisticated post-processing procedures, and comparisons with other machining technologies.

REFERENCES

1. Hlavácová, I.M.; Sadílek, M.; Ván˘ová, P.; Szumilo, Š.; Tyc˘, M. Influence of steel structure on machinability by abrasive water jet. Materials 2020, 13, 4424.
2. Chen, X.; Guan, J.; Deng, S.; Liu, Q.; Chen, M. Features and mechanism of abrasive water jet cutting of Q345 steel. Int. J. Heat Technol. 2018, 36, 81–87.
3. Ban´kowski, D.; Spadło, S. The use of abrasive water jet cutting to remove flash from castings. Arch. Foundry Eng. 2019, 19, 94–98.
4. Hlavác˘, L.M.; Ban´kowski, D.; Krajcarz, D.; Štefek, A.; Tyc˘, M.; Młynarczyk, P. Abrasive waterjet (AWJ) forces—Indicator of cutting system malfunction. Materials 2021, 14, 1683.
5. Zhao, W.; Guo, C. Topography and microstructure of the cutting surface machined with abrasive waterjet. Int. J. Adv. Manuf. Technol. 2014, 73, 941–947.
6. Liu, X.C.; Liang, Z.W.; Wen, G.L.; Yuan, X.F. Water jet machining and research developments: A review. Int. J. Adv. Manuf. Technol. 2019, 102, 1257–1335.
7. Barsukov, G.; Zhuravleva, T.; Kozhus, O. Study of the effect of heat treatment of copper slag particles on abrasiveness for abrasive waterjet cutting. Int. J. Adv. Manuf. Technol. 2023, 129, 4293–4300.
8. Ishfaq, K.; Ahmed, N.; Rehman, A.U.; Hussain, A.; Umer, U.; Al-Zabidi, A. Minimizing the micro-edge damage at each constituent layer of the clad composite during AWJM. Materials 2020, 13, 2685.
9. Folkes, J. Water jet—An innovative tool for manufacturing. J. Mater. Process. Technol. 2009, 209, 6181–6189.
10. Kovacevic, R.; Hashish, M.; Mohan, R.; Ramulu, M.; Kim, T.J.; Geskin, S. State of the art of research and development in abrasive water jet machining. J. Manuf. Sci. Eng. 1997, 119, 776–785.
11. Spadło, S.; Ban´kowski, D.; Młynarczyk, P.; Hlavác˘ová, I.M. Influence of local temperature changes on the material microstructure in abrasive water jet machining (AWJM). Materials 2021, 14, 5399.
12. Ohadi, M.M.; Cheng, K.L. Modeling of temperature distributions in the workpiece during abrasive water jet machining. J. Heat Transf. 1993, 115, 446–452.
13. Hlavác˘, L.M.; Štefek, A.; Tyc˘, M.; Krajcarz, D. Influence of material structure on forces measured during abrasive water jet (AWJ) machining. Materials 2020, 13, 3878.
14. Perec, A. Research into the disintegration of abrasive materials in the abrasive water jet machining process. Materials 2021, 14, 3940.
15. Taiwo, E.M.; Yahya, K.; Haron, Z. Potential of using natural fiber for building acoustic absorber: A review. J. Phys. Conf. Ser. 2019, 1262, 012017.
16. Hernandes Diat, D.; Villar-Ribera, R.; Haron, Z.; Julián, F.; Hernández-Abad, V.; Delgado-Aguilar, M. Impact Properties and Water Uptake Behavior of Old Newspaper Recycled Fibers-Reinforced Polypropylene Composites. Materials 2020, 13, 1079.
17. Lu, N.; Swan, J.R.; Ferguson, I. Composition, structure, and mechanical properties of hemp fiber reinforced composite with recycled high-density polyethylene matrix. J. Compos. Mater. 2012, 46, 1915–1924.
18. Shahria, S. Fabrication and property evaluation of hemp-flax fiber reinforced hybrid composite. Cellulose 2019, 7, 17–23.
19. Venkateshwaran, N.; Elayapemural, A. Mechanial and water absorption properties of woven jute/banana hybrid composites. Fibers Polym. 2012, 13, 907–914.
20. Bouafif, H.; Koubaa, A.; Perré, P.; Cloutier, A. Effects of fiber characteristics on the physical and mechanical properties of wood plastic composites. Compos. Part A Appl. Sci. Manuf. 2009, 40, 1975–1981.
21. Murayama, K.; Ueno, T.; Kobori, H.; Kojima, Y.; Suzuki, S.; Aoki, K.; Ito, H.; Ogoe, S.; Okamoto, M. Mechanical properties of wood/plastic composites formed using wood flour produced by wet ball-milling under various milling times and drying methods. J. Wood Sci. 2019, 65, 5.
22. Zhou, L.; Miller, J.; Vezza, J.; Mayster, M.; Raffay, M.; Justice, Q.; Al Tamimi, Z.; Hansotte, G.; Sunkara, L.D.; Bernat, J. Additive Manufacturing: A Comprehensive Review. Sensors 2024, 24, 2668.
23. Niu, S.; Chang, Q.; He, W.; Zhao, D.; Xie, Y.; Deng, X. Mechanically Strong, Hydrostable, and Biodegradable Starch-Cellulose Composite Materials for Tableware. Starke 2022, 74, 2200019.
24. Chegdani, F.; El Mansori, M. New Multiscale Approach for Machining Analysis of Natural Fiber Reinforced Bio-Composites. J. Manuf. Sci. Eng. 2019, 141, 011004.
25. Zajac, J.; Hutyrova, Z.; Orlovsky, I. Investigation of surface roughness after turning of one kind of the bio-materials with thermoplastic matrix and natural fibers. Adv. Mater. Res. 2014, 941, 275–279.
26. Caggiano, A. Machining of fibre reinforced plastic composite materials. Materials 2018, 11, 442.
27. Babu, D.; Babu, K.S.; Gowd, B. Drilling uni-directional fiber-reinforced plastics manufactured by hand lay-up influence of fibers. Am. J. Mater. Sci. Technol. 2012, 1, 1–12.
28. Babu, D.; Babu, K.S.; Gowd, B. Determination of delamination and tensile strength of drilled natural fiber reinforced composites. Appl. Mech. Mater. 2014, 592, 134.

29. Goutham, E.R.S.; Hussain, S.S.; Muthukumar, C.; Krishnasamy, S.; Kumar, T.S.M.; Santulli, C.; Palanisamy, S.; Parameswaranpillai, J.; Jesuarockiam, N. Drilling parameters and post-drilling residual tensile properties of natural-fiber-reinforced composites: A review. J. Compos. Sci. 2023, 7, 136.
30. Abilash, N.; Sivapragash, M. Optimizing the delamination failure in bamboo fiber reinforced polyester composite. J. King Saud Univ.-Eng. Sci. 2016, 28, 92–102.
31. Vigneshwaran, S.; Uthayakumar, M.; Arumugaprabu, V. Abrasive water jet machining of fiber-reinforced composite materials. J. Reif. Plast Compos. 2018, 37, 230–237.
32. Masoud, F.; Sapuan, S.M.; Mohd Ariffin, M.K.A.; Nukman, Y.; Bayraktar, E. Cutting process of natural fiber-reinforced polymer composites. Polymers 2020, 12, 1332.
33. Boopathi, S.; Thillaivanan, A.; Azeem, M.A.; Shanmugam, P.; Pramod, V.R. Experimental investigation on abrasive water jet machining of neem wood plastic composite. Funct. Compos. Struct. 2022, 4, 025001.
34. Jagadish, S.B.; Amitava, R. Prediction of surface roughness quality of green abrasive water jet machining: A soft computing approach. J. Intell. Manuf. 2015, 30, 2965–2979.
35. Costa, R.D.F.S.; Sales-Contini, R.C.M.; Silva, F.J.G.; Sebbe, N.; Jesus, A.M.P. A Critical Review on Fiber Metal Laminates (FML): From Manufacturing to Sustainable Processing. Metals 2023, 13, 638.
36. Jamali, N.; Khosravi, H.; Rezvani, A.; Tohidlou, E. Mechanical properties of multiscale graphene oxide/basalt fiber/epoxy composites. Fibers Polym. 2019, 20, 138–146.
37. Saeimi Sadigh, M.A.; Marami, G. Investigating the effects of reduced graphene oxide additive on the tensile strength of adhesively bonded joints at different extension rates. Mater. Des. 2016, 92, 36–43.
38. Smolnicki, M.; Lesiuk, G.; Duda, S.; de Jesus, A.M.P. A Review on Finite-Element Simulation of Fibre Metal Laminates. Arch. Comput. Methods Eng. 2023, 30, 749–763.
39. Marques, F.; Silva, F.G.A.; Silva, T.E.F.; Rosa, P.A.R.; Marques, A.T.; de Jesus, A.M.P. Delamination of Fibre Metal Laminates Due to Drilling: Experimental Study and Fracture Mechanics-Based Modelling. Metals 2022, 12, 1262.
40. Pai, A.; Kini, C.R.; Shenoy, B.S. Scope of Non-Conventional Machining Techniques for Fibre Metal Laminates: A Review. Mater. Today 2022, 52, 787–795.
41. Siva Kumar, M.; Rajamani, D.; El-Sherbeeny, A.M.; Balasubramanian, E.; Karthik, K.; Hussein, H.M.A.; Astarita, A. Intelligent Modeling and Multi-Response Optimization of AWJC on Fiber Intermetallic Laminates through a Hybrid ANFIS-Salp Swarm Algorithm. Materials 2022, 15, 7216.
42. Dogˇ ankaya, E.; Kahya, M.; Özgür Ünver, H. Abrasive Water Jet Machining of UHMWPE and Trade-off Optimization. Mater. Manuf. Process. 2020, 35, 1339–1351.
43. Balamurugan, K.; Uthayakumar, M.; Sankar, S.; Hareesh, U.S.; Warrier, K.G.K. Predicting Correlations in Abrasive Waterjet Cutting Parameters of Lanthanum Phosphate/Yttria Composite by Response Surface Methodology. Measurement 2019, 131, 309–318.
44. Kalirasu, S.; Rajini, N.; Winowlin Jappes, J.T.; Uthayakumar, M.; Rajesh, S. Mechanical and Machining Performance of Glass and Coconut Sheath Fibre Polyester Composites Using AWJM. J. Reinf. Plast. Compos. 2015, 34, 564–580.
45. Alberdi, A.; Artaza, T.; Suárez, A.; Rivero, A.; Girot, F. An Experimental Study on Abrasive Waterjet Cutting of CFRP/Ti6Al4V Stacks for Drilling Operations. Int. J. Adv. Manuf. Technol. 2016, 86, 691–704.
46. Hutyrová, Z.; Šcˇucˇka, J.; Hloch, S.; Hlavácˇek, P.; Zelenˇák, M. Turning of Wood Plastic Composites by Water Jet and Abrasive Water Jet. Int. J. Adv. Manuf. Technol. 2015, 84, 1615–1623.
47. Kalirasu, S.; Rajini, N.; Rajesh, S.; Jappes, J.T.W.; Karuppasamy, K. AWJM Performance of Jute/Polyester Composite Using MOORA and Analytical Models. Mater. Manuf. Process. 2017, 32, 1730–1739.
48. Pahuja, R.; Ramulu, M.; Hashish, M. Surface Quality and Kerf Width Prediction in Abrasive Water Jet Machining of Metal- Composite Stacks. Compos. B Eng. 2019, 175, 107134.
49. Pahuja, R.; Ramulu, M. Abrasive Water Jet Machining of Titanium (Ti6Al4V)–CFRP Stacks–A Semi-Analytical Modeling Approach in the Prediction of Kerf Geometry. J. Manuf. Process. 2019, 39, 327–337.
50. Hazir, E.; Ozcan, T. Response Surface Methodology Integrated with Desirability Function and Genetic Algorithm Approach for the Optimization of CNC Machining Parameters. Arab. J. Sci. Eng. 2019, 44, 2795–2809.
51. Palleda, M. A Study of Taper Angles and Material Removal Rates of Drilled Holes in the Abrasive Water Jet Machining Process. J. Mater. Process. Technol. 2007, 189, 292–295.
52. Mm, I.W.; Azmi, A.I.; Lee, C.C.; Mansor, A.F. Kerf Taper and Delamination Damage Minimization of FRP Hybrid Composites under Abrasive Water-Jet Machining. Int. J. Adv. Manuf. Technol. 2018, 94, 1727–1744.
53. Kumar, D.; Gururaja, S. Abrasive Waterjet Machining of Ti/CFRP/Ti Laminate and Multi-Objective Optimization of the Process Parameters Using Response Surface Methodology. J. Compos. Mater. 2020, 54, 1741–1759.
54. Rajamani, D.; Balasubramanian, E.; Murugan, A.; Tamilarasan, A. AWJC of NiTi Interleaved R-GO Embedded Carbon/Aramid Fibre Intermetallic Laminates: Experimental Investigations and Optimization through BMOA. Mater. Manuf. Process. 2023, 38, 1144–1158.
55. Khashaba, U.A.; El-Sonbaty, I.A.; Selmy, A.I.; Megahed, A.A. Machinability of Carbon Fiber Reinforced Polymers (CFRP). Compos. Part A Appl. Sci. Manuf. 2010, 41, 1368–1376.
56. Song, Y.; Cao, H.; Zheng, W.; Qu, D.; Liu, L.; Yan, C. Cutting force modeling of machining carbon fiber reinforced polymer (CFRP) composites: A review. Compos. Struct. 2022, 299, 116096.
57. Geier, N.; Xu, J.; Poor, D.I.; Dege, J.H.; Davim, J.P. A review on advanced cutting tools and technologies for edge trimming of carbon fibre reinforced polymer (CFRP) composites. Compos. Part B 2023, 266, 111037.

58. Chen, T.; Xiang, J.; Gao, F.; Liu, X.; Liu, G. Study on cutting performance of diamond-coated rhombic milling cutter in machining carbon fiber composites. Int. J. Adv. Manuf. Technol. 2019, 103, 4731–4737.
59. Rawal, S.; Sidpara, A.M.; Paul, J. A review on micro machining of polymer composites. J. Manuf. Process. 2022, 77, 87–113.
60. Debnath, S.; Reddy, M.M.; Yi, Q.S. Environmental friendly cutting fluids and cooling techniques in machining: A review. J. Clean. Prod. 2014, 83, 33–47.
61. Shyha, I.S.; Soo, S.L.; Aspinwall, D.K.; Bradley, S. Effect of laminate configuration and feed rate on cutting performance when drilling holes in carbon fibre reinforced plastic composites. CIRP Ann. 2010, 59, 93–96.
62. Ozkan, D.; Panjan, P.; Gok, M.S.; Karaoglanli, A.C. Experimental study on tool wear and delamination in milling CFRPs with TiAlN-and TiN-coated tools. Coatings 2020, 10, 623.
63. Gara, S.; Tsoumarev, O. Prediction of surface roughness in slotting of CFRP. Measurement 2016, 91, 414–420.
64. Jiao, J.; Cheng, X.; Wang, J.; Sheng, L.; Zhang, Y.; Xu, J.; Jing, C.; Sun, S.; Xia, H.; Ru, H. A review of research progress on machining carbon fiber-reinforced composites with lasers. Micromachines 2022, 14, 24.
65. Ahmed, T.M.; El Mesalamy, A.S.; Youssef, A.; El Midany, T.T. Improving surface roughness of abrasive waterjet cutting process by using statistical modeling. CIRP J. Manuf. Sci. Technol. 2018, 22, 30–36.
66. Dhanawade, A.; Kumar, S.; Kalmekar, R.V. Abrasive Water Jet Machining of Carbon Epoxy Composite. Def. Sci. J. 2016, 66, 522–528.
67. Seo, J.; Kim, D.Y.; Kim, D.C.; Park, H.W. Recent developments and challenges on machining of carbon fiber reinforced polymer composite laminates. Int. J. Precis. Eng. Manuf. 2021, 22, 2027–2044.
68. Shanmugam, D.K.; Nguyen, T.; Wang, J. A study of delamination on graphite/epoxy composites in abrasive waterjet machining. Compos. Part A Appl. Sci. Manuf. 2008, 39, 923–929.
69. Gao, G.; Xu, F.; Xu, J. Parametric optimization of fdm process for improving mechanical strengths using taguchi method and response surface method: A comparative investigation. Machines 2022, 10, 750.
70. Chalgham, A.; Ehrmann, A.; Wickenkamp, I. Mechanical properties of FDM printed PLA parts before and after thermal treatment. Polymers 2021, 13, 1239.
71. Rowe, A.; Pramanik, A.; Basak, A.K.; Prakash, C.; Subramaniam, S.; Dixit, A.R.; Radhika, N. Effects of abrasive waterjet machining on the quality of the surface generated on a carbon fibre reinforced polymer composite. Machines 2023, 11, 749.
72. Płodzien′, M.; Z˙ yłka, Ł.; Z˙ ak, K.; Wojciechowski, S. Modelling the kerf angle, roughness and waviness of the surface of inconel 718 in an abrasive water jet cutting process. Materials 2023, 16, 5288.
73. Sambruno, A.; Bañon, F.; Salguero, J.; Simonet, B.; Batista, M. Kerf taper defect minimization based on abrasive waterjet machining of low thickness thermoplastic carbon fiber composites C/TPU. Materials 2019, 12, 4192.
74. Murthy, B.R.N.; Makki, E.; Potti, S.R.; Hiremath, A.; Bolar, G.; Giri, J.; Sathish, T. Optimization of process parameters to minimize the surface roughness of abrasive water jet machined jute/epoxy composites for different fiber inclinations. J. Compos. Sci. 2023, 7, 498.
75. Sambruno, A.; Gómez-Parra, Á.; Márquez, P.; Tellaeche-Herrera, I.; Batista, M. Evaluation of Peripheral Milling and Abrasive Water Jet Cutting in CFRP Manufacturing: Analysis of Defects and Surface Quality. Fibers 2024, 12, 78.
76. De la Rosa, S.; Rodríguez-Parada, L.; Ponce, M.B.; Mayuet Ares, P.F. The Abrasive Water Jet Cutting Process of Carbon-Fiber- Reinforced Polylactic Acid Samples Obtained by Additive Manufacturing: A Comparative Analysis. J. Compos. Sci. 2024, 8, 437.

Advances in Additive Manufacturing Technologies – Gurusamy Pathinettampadian et al. (eds)

Advanced Process Parameter Optimization for Electrochemical Machining of D3 Die Steel

P. Usha Rani[1]
Professor, Science and Humanities Department,
R.M.D Engineering College, India

A. John Presin Kumar[2]
Associate Professor, Department of Mechanical Engineering,
Hindustan Institute of Technology & Science,
Chennai, India

Renuka Meenakshi[3]
Student, Department of Mechanical Engineering,
Chennai Institute of Technology,
Chennai, India

C. Bibin[4]
Professor, Career Development Centre,
R.M.K. College of Engineering and Technology,
India

P. Loganathan[5]
Associate Professor, Department of Mechanical Engineering,
Karpaga Vinayaga College of Engineering and Technology,
India

Nallathambi[6]
Assisstant Professor, Department of Mechanical Engineering,
Chennai Institute of Technology,
Chennai, India

V. Maharajan[7]
Assistant Professor, Department of Mechanincal Engineering,
Unnamalai Institute of Technology,
Kovilpatti, India

◆ **Abstract:** Electrochemical Machining (ECM) is a highly sophisticated unconventional machining process extensively used in shaping and cutting electrically conducting materials, particularly hard and complex contour profiled materials. The process has gained a strong grip in aerospace and aeronautical engineering sectors, where it is extensively used in blade profiling and other high-precision machining operations. The fundamental advantage of ECM is its capability to achieve good surface finish and precise material removal without inducing thermal or mechanical stress on the workpiece. Though the process has a number of merits, it has some limitations associated with it, such as high initial investments and high maintenance fees, which may pose operational challenges to industries

[1]pusharani71@yahoo.com, [2]jpk.hits@gmail.com, [3]renukameenakshirh.mech2022@citchennai.net, [4]drcbibin@gmail.com, [5]logu2685@gmail.com [6]knallathambi3383@gmail.com [7]rajan19maha@gmail.com

DOI: 10.1201/9781003685906-23

planning to implement ECM on a large scale. Another basic characteristic which defines Electrochemical Machining (ECM) is its dependency on input process parameters that have to be determined to the utmost precision. For the research study being conducted under the present work, D3 die steel has been chosen in particular as the workpiece material for a series of trial experiments to find the major machining parameters that have the dominant control on Ra and MRR. For this research work, three elementary process parameters-namely voltage, current, and concentration of the electrolyte solution-have been selected for the purpose of optimization, and the major objective is the reduction of surface roughness, a basic requirement of the automobile industry. To obtain this objective, a systematic examination of huge amounts of experimental data gathered from numerous trials has been carried out employing the Taguchi method, a widely used statistical method well-suited for the purpose of process optimization. Based on this method, the study aims at improving machining performance, reducing material wastage, and maximizing overall efficiency of the ECM process for industrial applications.

◆ **Keywords:** D3 die steel, Electro-chemical machining (ECM), Orthogonal array (OA), Surface roughness (Ra), Taguchi technique, Analysis of variance (ANOVA)

1. Introduction

ECM is a high-tech, non-traditional machining process involving removal of material from a workpiece through electrochemical energy. ECM is used extensively for machining complex shapes in electrically conductive materials. The fundamental principle of ECM is controlled electrochemical dissolution of material, where the tool acts as the cathode and the workpiece acts as the anode in the presence of an electrolyte solution. Displacement of ions, when the workpiece is brought into contact with the electrolyte, leads to material removal. In contrast to conventional machining processes based on mechanical cutting forces, ECM does not involve direct contact among the workpiece and tool, thereby eliminating limitations such as tool wear and mechanical stresses.

This is similar to the situation in electroplating, in which metal deposition occurs on a substrate. But in ECM, the primary operation is selective removal of excess material rather than deposition. ECM minimizes significantly the formation of mechanical and thermal stresses, and therefore it is most appropriate to machine sensitive and intricate components. One of the primary difficulties in ECM is the formation of sparks during machining. These sparks are formed owing to low operating temperatures and poor control of process parameters. Without control, these sparks would promote localized tool wear and non-uniform material removal, thereby degrading the accuracy of machining. In order to produce optimum performance and high-quality surface finish, appropriate control measures need to be employed to suppress the formation of sparks in ECM operations.

ECM is a highly versatile machining process that can machine a wide range of electrically conductive materials, from hard metals to complex geometrical alloys. Due to its ability to provide high-quality surface finish and accurate contours, ECM is applied significantly in aerospace, tool and die, and defense industries. In these industries, tight tolerance and complex profile parts are required, and ECM is an appropriate process to machine such parts with high accuracy and efficiency. ECM performance is regulated by including current, different process parameters, electrolyte flow rate, voltage, electrolyte concentration, and inter-electrode gap (IEG). These parameters regulate the surface roughness (Ra), and material removal rate (MRR) dimensional precision of the machined part. Nevertheless, one of the biggest challenges of ECM is controlling multiple parameters at the same time, as the change in any one of the parameters effects the overall machining performance. Due to this inherent complexity, several optimization methods are employed in attempting to identify the most appropriate parameter settings that will ultimately result in the attainment of the desired machining performance.

Several researchers have carried out extensive research to optimize the various parameters involved in the Electrochemical Machining (ECM) process through various different statistical and numerical methods. Jain et al. [1], in their well-known research, identified parameter optimization specifically to effectively reduce geometrical error through the use of genetic algorithms as a key component of their research. They were able to identify input parameters such as the electrolyte flow velocity applied voltage, and tool feed rate as key ranges. Their research concluded with the finding that the careful optimization of these specific parameters is important in enhancing machining accuracy. Acharya et al. [2] also carried out investigations on the impact of the parameters of the ECM process on the MRR through the use of a multi-objective optimization method in their research. The result of their research identified voltage as the key parameter required to attain a high material removal rate.

Hocheng et al. [3] determined two important parameters—machining time and gap opening fluctuation—which are highly important in the determination of the final shape of the workpiece being machined. On the other hand, According to Neto et al. [4] findings from the study, out of the parameters, the feed rate was the most important constraints that determines the overall machining responses observed. In addition, in another study, Munda et al. [5] utilized response surface methodology (RSM) to optimize both the MRR and the radial overcut (ROC) effectively in the process of ECM. They studied how parameters like the voltage, pulse on/off ratio, concentration of the electrolyte, frequency of tool vibration, and frequency of voltage affect the results. Their study emphasized the need to optimize these particular parameters in order to obtain precise machining results while minimizing any irregularities that may be experienced during the process of EMM of hard-to-machine materials.

Asokan et al. [6] conducted a stringent investigation wherein they applied various regression models with the purpose of determining the optimal parameters of electrochemical machining (ECM). These parameters included important constraints such as the gap voltage, the current, and the feed rate, which influence the process significantly. In addition to their regression study, they also conducted comparative study with artificial neural network (ANN) models. Through such a comparison, they were able to determine the efficacy and capability of the ANN to provide predictions for outcomes with regard to the ECM process. In the meantime, Taweel et al. [7] explored performance of ECM tools made of Al-Si-Cu-TiC composite materials shaped via powder metallurgy (PM) processes. Comparison of composite tool and conventional tool machining efficiency resulted in the discovery that composite material performed better in determining optimum current conditions.

Ramarao et al. [8, 9] and Klocke et al. [10] compared SEDM and ECM for machining nickel-based alloys and titanium and proved that ECM was more suitable for large-batch production owing to its machining performance consistency but less suitable for small-batch production owing to its inability to machine complex details relative to SEDM. Senthilkumar et al. [11] optimized ECM process to machine AlSiCp composite material by electrolyte flow rate, varying voltage, tool feed rate, and electrolyte concentration. Their study was intended to maximize Ra and MRR proved that with higher content of SiC particles, Ra increased but MRR decreased. Chakradhar et al. [12] and Bisht et al. [13] investigated the effects of some ECM parameters, i.e., current, flow rate of the electrolyte, tool feed rate, and voltage, on both Ra and MRR of aluminum and mild steel. The factors were optimized and the optimum circumstances were determined in order to have the highest MRR and the lowest Ra by the application of the Taguchi method. Similarly, Habib et al. [14] investigated research on the ECM process from the perspective of the Taguchi method, with tool feed rate, voltage, current, and electrolyte concentration as key variables to be optimized for Ra and MRR. Based on their results, they found that increasing both current and voltage resulted in increased MRR and Ra values. In contrast, Ramarao et al. [15] conducted the effects of feed rate, voltage and electrolyte concentration on ROC in ECD of AlB_4Cp MMC. Applying Taguchi's design of experiments, they advanced a regression-based model for the prediction of ROC, ultimately establishing that voltage became the most influential parameter on ROC.

The main aim of this study is to discover the surface roughness parameters of D3 die steel during ECM and explore the impact of voltage, current, and absorption of the electrolyte on Ra. Taguchi methodology, a highly effective statistical technique for engineering analysis, is employed to optimize ECM process parameters. Taguchi methodology is employed in most of the cases since it enables systematic collection of experimental data and process performance analysis with less efforts, cost, and time. The Taguchi technique is highly appropriate in finding robust machining conditions by employing quality loss functions in operation design.

Three parameters of the ECM process, i.e., voltage, current, and electrolyte absorbtion are used, each at three levels. Experimental design is scheduled using an L9 orthogonal array (OA) to perform the experiments in a systematic way. The trial results are expended to find the optimal machining conditions required for obtaining the desired surface finish in D3 die steel. The results of this research will add to better knowledge of ECM process optimization and its use in industrial manufacturing environments.

2. Experimental Methodology

For this experiment, D3 die steel was used as the workpiece material because it has good wear resistance and hardness. The D3 die steel contains is 2.2% Carbon (C), 12% Chromium (Cr), 0.4% Silicon (Si), 0.4% Manganese (Mn), 0.03% Phosphorus (P), and 0.03% Sulfur (S). The density of this workpiece material is 7700 kg/m^3 and its hardness is 56.1 HRC. The workpiece material, being the anode of the electrochemical machining process, was conditioned as a rectangular specimen of 10 mm in thickness 50 mm in breadth, and 50 mm in length. The machining process was done using the METATECH ECM equipment. The setup of the electrochemical machining includes a power supply unit, an electrolyte circulation system, a tool (cathode), and a workpiece (anode). The tool electrode that was used in this experiment was a nickel-plated Cu electrode with a flat end. This electrode, as presented, had a small drilled hole along its axis to create an opening for the flow of the electrolyte from the tool to the workpiece, which allowed effective removal of the material.

A brine solution (NaCl) was used as the electrolyte in the process because it is useful in the prevention of passivation of the workpiece surface. Use of brine solution also facilitates easy removal of unwanted material from the surface of the machined workpiece, thereby increasing the efficiency of the ECM process. The electrolyte was circulated through a continuous loop to facilitate dissolution to be uniform and deposit removed material back onto the workpiece. Each of the parameters was assigned three different varying levels to determine their effect on the machining operation. Levels of each of the parameters were selected with careful consideration using prior literature and ease of experimentation. The experimental technique used an L9 orthogonal array, a powerful statistical technique that is used to reduce the number of experimental tests while carrying out careful analysis. Parameter selection and corresponding levels were achieved using MINITAB software, a renowned statistical analysis software.

Research trials were conducted in a sequence to analysis the Ra of the workpieces machined with different combinations of parameters. Machining operation in each process was conducted for a representative machining time of 10 minutes to achieve uniformity in all experiments. This time was selected in consideration of initial trials for achieving measurable and comparable values of surface roughness without unnecessary material removal. Measurement was conducted over a sampling length of 10 mm to ensure representative and accurate roughness values. High-resolution Talysurf machines produced precise profiles of the surface roughness and enabled specified study of the effects of several ECM parameters on surface finish. Experimental tests and data acquisition were conducted cautiously to achieve reproducibility and repeatability of the measurement. Surface roughness values so acquired were further analyzed to determine the significance of individual process parameters and to determine optimal machining conditions to minimize the surface roughness. This research was augmented in terms of understanding the effects of current, voltage, and concentration of electrolyte on the electrochemical machining process to finally achieve the process optimization and improved machining performance.

3. Results and Discussion

The research were carried out on D3 die steel in an Electrochemical Machining (ECM) facility for studying the effect of the influential process parameters- voltage, current, and concentration of the electrolyte-on the Ra. The preliminary experimental results, i.e., the observed surface roughness values and corresponding Signal-to-Noise (S/N) ratios. The importance of chosen process parameters to control the response variable (Ra) was thoroughly investigated using the statistical technique ANOVA. The primary motive of the same was to identify the parameter with the highest effect on the Ra and determine the optimal level of every parameter to attain the desired machining result. Secondly, the main effect plots of S/N ratios, were utilized to present a graphical representation of interaction between process parameters and surface roughness. To identify an optimum surface roughness value, the "lower the better" criterion was utilized because minimization of surface roughness is of great importance to attain high-quality surface finishes in industrial processes at the lowest settings for voltage, current, and electrolyte concentration. Specifically, the combination of A1-B1-C1, where A1 represents the lessor current, B1 represents the lowest voltage, and C1 represents the less electrolyte concentration, was identified as the most favorable setting for achieving the best surface finish.

However, excessive material removal can create an uneven surface texture, resulting in increased surface roughness. Apart from that, high current is also capable of generating localized passivation effects on the workpiece surface, resulting in additional surface finish reduction. Experimental findings confirm that the minimum surface roughness values were achieved wherever the current is kept at its minimum level. Similar to current, voltage is another essential parameter in the ECM process. With the increased voltage comes increased reaction rate, leading to the creation of enhanced dissolution of the material. Yet, the generated increased voltage could also bring on surface irregularities onto the machined surface since uncontrolled dissolution has been demonstrated to create low quality surfaces. The results based on this research indicate that lower values of voltage produce smoother surface finishes. Concentration of electrolyte is a critical parameter during machining as it determines the effectiveness and conductivity of the electrochemical reaction. Increasing the concentration of the electrolyte leads to the increased reaction rate as well, leading to excess dissolution of material and thus generating surface roughness. Excessive concentrations of the electrolytes can potentially contribute to the occurrence of localized erosion, which in turn results in the creation of an undeserved and unwanted surface texture. Further, experimental findings indicate that a reduction in the concentration of the electrolytes will result in an improvement in the quality of the surface finish.

ANOVA is a robust statistical analysis that is utilized to find the effect of the factors in the experiment and each factor's contribution in percentage. F-ratio, the relative measure of the effect of every factor, was calculated for each of the process parameters selected. The higher the F-ratio, the greater the effect of the parameter on the response factor. From the ANOVA study result, current had the largest contribution, with the highest contribution value of 68.61% to variation in surface roughness. This validates the dominant effect by current in contributing to the quality of machining in ECM. Electrolyte Concentration was the next most significant factor with a contribution of 17.00% to the response. This implies that the concentration of the electrolyte must be regulated for maximum machining performance. Voltage contributed least to surface roughness, with a contribution of just 0.57%. This implies that though voltage plays a part, it is not as significant as electrolyte and current concentration.

The dominance of current as the prominent parameter that dictates surface roughness is consistent with previous research investigations, which attest that high currents can lead to undesirable machining consequences, including greater surface roughness. The modest influence of concentration of the electrolyte further supports the choice of an appropriate level of concentration for ensuring stable machining conditions. Through the experimental observations and statistical testing, the values of optimum results for achieving optimum surface finish were determined as, Current of 100 A, Voltage of 5 V, and Concentration of Electrolyte of 5 g/l These parameter setting values, produced the least measured Ra, and thus their utility for achieving enhanced machining performance is validated in ECM. The values of optimum parameters are consistent with the fundamental concept of electrochemical machining, which involves controlled removal of material in achieving improved surface quality.

The outcome of this study provide valuable insights to optimize electrochemical machining for D3 die steel. The findings reaffirm the importance of correctly choosing and regulating process parameters to achieve the desired Ra. The results of the research are as follows, The large current percentage contribution suggests that it must be carefully controlled to decrease surface roughness. The application of real-time monitoring systems capable of controlling current fluctuations greatly improves machining quality. Although the effect of concentration is lesser than that of current, it is still an important parameter. Choosing a proper concentration avoids excessive dissolution of the material and ensures a stable machining process. Although voltage is less controlling than other parameters, its effect cannot be avoided. Correct choice of the voltage level can ensure a stable machining process. Application of the Taguchi method with orthogonal arrays has been effective in determining optimal machining conditions with a limited number of experimental runs. ANOVA results provided quantitative results on the relative significance of each process parameter, providing a basis for future research on ECM machining.

4. Conclusion

- The present experimental study was primarily aimed at investigating the various influences exerted by critical process parameters-i.e., the used the concentration level of the used electrolyte, current and the used voltage settings-on the resulting surface roughness characteristics of D3 die steel when electrochemical machining, commonly referred to as ECM, is utilized on it. In the attempt to determine the optimal machining conditions that would lead to the attainment of the minimum achievable surface roughness, a systematic and systematic approach was followed during the research. Based on the overall findings derived by this investigation, several important conclusions can indeed be presented:
- The results of the investigation evidently indicate that among the three chosen process parameters, i.e., current, voltage, and electrolyte concentration, current and electrolyte concentration both play a dominant role in controlling the scale of surface roughness measured. But compared to this, the influence of voltage is comparatively minor in magnitude. A detailed ranking at the top among the parameters studied, followed closely by electrolyte concentration in the order of significance. In contrast, the effect of voltage is found to be the lowest of the three factors. This important finding suggests that to achieve a high-quality surface finish, very precise control over both current and electrolyte concentration is required.
- It was also noted that, with higher voltage, surface roughness reduced considerably, and such a specific observation can be justified by the effect referred to as the passivation effect. With an increase in current, the electrochemical reaction on the workpiece surface increases, forming a passive layer which prevents excessive material removal. This passivation layer reduces surface irregularities during machining, leading to a smoother surface. But beyond a critical value, an excess of current can lead to localized overheating and unwanted surface defects and hence needs to be controlled with extreme caution in machining operations.

- The experiment also indicated that the rise in electrolyte concentration leads to the decrease in surface roughness but that is an outcome of aggressive chemical reaction between the work material and the electrolyte. With higher electrolyte concentration, there is a quicker ion exchange leading to increased dissolution of the material. Over-dissolution leads to erosion of finer details on the surface and ultimately decreases surface quality. So, there is a definite required electrolyte concentration to offer effective machining, yet there is a necessity to possess an optimal ratio to avoid over-etching and surface damage.

This paper indicates the importance of the proper selection of the ECM process parameters to achieve higher surface finish in D3 die steel. The findings indicate the importance of the present and concentration of electrolyte as the dominant parameters that affect surface roughness, with voltage being a secondary parameter. These effects can be utilized to optimize ECM processes to machine hard and complex-shaped materials more efficiently and maintain superior surface quality.

REFERENCES

1. N. K. Jain, V. K. Jain, "Optimization of Electro-Chemical Machining Process Parameters Using Genetic Algorithms", DOI:10.1080/10910340701350108.
2. B.G. Acharya, V.K. Jain, and J.L. Batra, "Multi-Objective Optimization of the ECM Process", Precision Engineering, Vol. 8, Issue 2, pp: 88–96, 1986.
3. Hocheng H., Kao PS., Lin S., "Development of the eroded opening during electrochemical boring of hole", International journal of advanced manufacturing technology, Vol. 25, pp. 1105–1112, 2004.
4. Jo Ao Cirilo Da Silva Neto, Evaldo Malaquias Da Silva, Marcio Bacci Da Silva, "Intervening variables in electrochemical machining", Journal of Materials Processing Technology, Vol. 179, pp. 92–96, 2006.
5. Munda J., Bhattacharya B., "Investigation into electrochemical micromachining (EMM) through response surface methodology based approach, International journal of advanced manufacturing technology, Vol. 35, issue: 7, pp. 821–832, 2008.
6. P. Asokan, R. Ravikumar, R. Jeypaul, M. Santhi, "Development of multi-objective optimization models for electrochemical machining process", International journal of advanced manufacturing technology, Vol. 39, pp. 55–63, 2008.
7. T. A. El-Taweel, "Multi-response optimization of EDM with Al-Cu-Si-tic P/M composite electrode", International Journal of Advanced Manufacturing Technology, Vol. 44, pp. 100–113, 2009.
8. S. Rama Rao, G. Padmanabhan, B. Sureka, R. V. Pandu, "Optimization of electrochemical machining process using evolutionary algorithms", Journal of machining and forming technology, Vol. 1, pp. 265–278, 2009.
9. S. Rama Rao, C. R. M. Sravan, G. Padmanbhan, R. V. Pandu, "Fuzzy logic based forward modelling of electrochemical machining process", In proc of 2009 world congress on nature and biologically inspired computing, pp. 1431–1435, 2009.
10. Klocke F, Zeis M, Klink A, Veselovac D, "Technological and economical comparison of roughing strategies via milling, sinking-EDM, wire-EDM and ECM for titanium and nickel-based blisks", Procedia engineering, CIRP 2, pp. 98–101, 2010.
11. Senthilkumar C, Ganesan G, Karthikeyan R, "Effect of process parameters on electrochemical machining of Al/SiCp composites", Journal of manufacturing engineering, Vol. 5, Issue. 1, pp. 25–30, 2010.
12. Chakradhar D, Venu Gopal A, "Multi-objective optimization of electro-chemical machining of EN31 steel by grey relational analysis", International journal of modeling and optimization, Vol. 1, No. 2, pp. 113–117, 2011.
13. Bhawna Bisht, Jyoti Vimal, Vedansh Chaturvedi, "Parametric optimization of electro-chemical machining using signal to noise (S/N) ratio", International journal of modern engineering research, Vol. 3, Issue. 4, pp. 1999–2006, 2013.
14. Habib, S., "Experimental Investigation of Electrochemical Machining Process using Taguchi Approach", International Journal of Scientific Research in Chemical Engineering, Vol. 1, Issue: 6, pp. 93–105, 2014.
15. S. Rama Rao, G. Padmanabhan, K. Mahesh Naidu, A. Rukesh Raddy, "Parametric study for radial over cut in electrochemical drilling of Al-5%B4Cp composites", Proceedia Engineering, Vol. 97, pp. 1004–1011, 2014.
16. McGeough JA., "Principles of Electro Chemical Machining", Chapman and Hall, London, 1974.
17. Pandey PC, Shan HS., "Modern machining processes", McGraw Hill, New Delhi, 2015.
18. Ross PJ., "Taguchi techniques for quality engineering", McGraw-Hill, New York, 1988.
19. Minitab user manual release 14, Minitab Inc, state college, PA, USA, 2003.

Advances in Additive Manufacturing Technologies – Gurusamy Pathinettampadian et al. (eds)
© 2026 Taylor & Francis Group, London, ISBN 978-1-041-16687-0

Achieving Superior Machining Quality in Plasma Arc Cutting of High-Strength Inconel Alloy

V. Maharajan[1]
Assistant Professor, Department of Mechanincal Engineering,
Unnamalai Institute of Technology,
Kovilpatti, India

Krishnakumar B.[2]
Assisstant Professor, Department of Mechanical Engineering,
Chennai Institute of Technology,
Chennai, India

Bharanidharan T.[3]
Student, Department of Mechanical Engineering,
Chennai Institute of Technology,
Chennai, India

C. Bibin[4]
Professor, Career Development Centre,
R.M.K. College of Engineering and Technology,
India

E. Sangeeth Kumar[5]
Assisstant Professor, Department of Automobile Engineering,
Hindustan Institute of Technology and Science,
Chennai, India

R. Karthick[6]
Assistant Professor, Department of Mechanical Engineering,
Rajalakshmi Institute of Technology,
Chennai, India

◆ **Abstract:** Super alloys based on Nickel are widely used in various industries due to their ability to withstand high-temperature conditions while providing superior strength to machine components in aerospace, biomedical, marine, and automotive applications. These alloys exhibit excellent mechanical properties, corrosion resistance, and thermal stability. Among them, Inconel 718 is extensively utilized in demanding environments due to its chemical stability, resistance to mechanical wear, and high thermal corrosion resistance. However, these properties pose significant challenges during machining, such as surface quality, reduced material removal rate, and tool wear. Plasma arc cutting is an advanced metal-cutting process capable of efficiently slicing hard-to-machine materials, including complex profiles. This study examines the effect of key PAC process parameters like gas pressure, AC, cutting speed, and stand-off distance on kerf width, kerf taper, and the heat-affected zone of Inconel 718. The research focuses on identifying the optimal cutting parameters to achieve minimal surface roughness and reduced tool wear. Experimental

[1]rajan19maha@gmail.com, [2]krishnakumarb@citchennai.net, [3]vtrbharani12@gmail.com, [4]drcbibin@gmail.com, [5]sangeethkumar2024@gmail.com, [6]karthick.industrial@gmail.com

DOI: 10.1201/9781003685906-24

validation confirms the effectiveness of the proposed parameter settings, with a minimal deviation from actual results, demonstrating a comparative difference of 4.45% for kerf width and 4.36% for the heat-affected zone.

◆ **Keywords:** Plasma arc cutting (PAC), Inconel 718, Heat-affected zone (HAZ), Kerf width and kerf taper, Optimization and ANOVA, Surface morphology and dross formation

1. Introduction

The demand for nonconventional materials continues to grow due to their exceptional properties, such as resistance to extreme temperatures, weathering, and corrosion. These materials also exhibit excellent strength-to-weight ratios, withstand sustained loads, and transition smoothly between ductile and brittle phases as temperatures decrease. One such superalloy is Inconel 718, which is highly resistant to high temperatures, making it ideal for aerospace, ballistics. Its exceptional material characteristics, including superior strength with lesser weight at high temperatures and outstanding corrosion resistance, contribute to its widespread use and extended service life. In fact, nearly 50% of aerospace components and modern jet engines by weight are fabricated using Inconel 718. However, despite these advantages, Inconel 718 also presents several machining challenges.

Adalarasan et al. analyzed the machinability of high-velocity turning operations on Inconel 718 and found that increasing feed and cutting rates led to higher heat generation, causing material weakening in the cutting zone and altering chip flow direction with changes in depth of cut [1]. Ananthakumar et al. investigated tool wear in finish turning of Inconel 718 using PCBN tooling and observed thermal cracks, insert fractures, and chipping at high cutting rates. Their ANOVA calculations confirmed that cutting speed, rate of feed, and shape of tool significantly impact tool life [2]. Parida studied the surface topology of cubic boron nitride tools used in high-speed cutting of Inconel 718, concluding that crater wear was proportional to the cutting length [3]. Ananthakumar et al. and Behera et al. analyzed surface roughness and tool wear during high-velocity machining and found that high-quality machined surfaces could be achieved at both lower and higher speeds. However, surface deterioration occurred progressively due to increased tool wear during consecutive cutting cycles [2,4].

Zhao et al. identified optimal machining parameters to minimize surface roughness and tool wear using the Taguchi approach and Grey relational analysis in Inconel 718 alloy [5]. Ahmad et al. explored hybrid machining in the aerospace industry and found that assistance using Ultrasonic method generated more compressive residual stresses, reducing net tensile stresses compared to conventional machining [6]. Gani et al. used the inverse examination method to assess surface roughness after laser-assisted milling of Inconel 718, achieving a precision error below 0.5% and a mean process parameter deviation under 5% [7]. Maity and Bagal employed a magnetic-suspension spindle system to drill microholes in Inconel 718 and discovered that its high response frequency enabled efficient and high-quality microhole formation compared to traditional EDM [8]. Devaraj et al. and Fountas et al. proposed a novel tool treatment technique to simplify Inconel 718 machining in turning processes. Their study highlighted that high friction, extreme temperatures, and contact forces contributed to tool chipping, tool failure, and the formation of built-up edges (BUE) [9,10].

Fontanive et al. looked into process variables like machining forces, stresses induced, and surface morphology via micro-scale milling assisted by an atmospheric plasma jet. The other options of cutting techniques delivered higher surface roughness value when compared with atmospheric compression cold plasma jet assisted with MQL [11]. Mustafa et al. with an uncoated WC tool had done the analysis of cutting velocity and wear of tool during machining Inconel 718. They found that increased cutting speed led to a reduced helix angle and the formation of repeated coiled helical chips [12]. Venkatesan et al. observed that sharper tools increased surface roughness. on study of the effects of cutting-edge geometry on machined surface quality, However, hardness measurements showed no significant variation due to the limitations of depth measurement techniques (minimum 20 μm) [13]. Iosub et al. examined saw-tooth chip formation in Inconel 718 milling using metallographic techniques and found that recrystallization temperatures were significantly higher than expected [14]. Using textured tools, Maity and Bagal examined the effects of solid lubricant-assisted MQL cooling conditions on machining characteristics and displayed boosted machining performance. [15]. Research on machining nickel-based superalloys has consistently identified ceramic-based tools as the most effective for reducing tool wear and improving surface integrity [16].

Meikandan and Malarmohan assessed kerf quality and HAZ in aluminum composites using plasma cutting. They found that higher cutting velocities reduced plasma exposure time on the work surface, minimizing the effects on the polyethylene core. Stand-off distance significantly influenced kerf quality and HAZ [17]. By increasing cutting speed and work piece thickness while lowering AC, Bejaxhin et al. enhanced plasma arc cutting parameters for AISI 4140 steel using fuzzy logic, leading to dross-free cutting surfaces. [18]. Karthick et al. applied hybrid techniques for the optimization of process parameters of PAC such as grey-relational analysis, principal component analysis, , and response surface methodology for stainless steel [19]. Meikandan et al. evaluated PAC parameters for Monel 400 superalloy, focusing on micro hardness, KW, and surface roughness. Key parameters included torch stand-off distance, gas pressure, cutting velocity, and AC magnitude [20]. In industrial settings, nontraditional machining techniques have gained traction for processing innovative and hybrid materials. PAC remains a preferred method for machining difficult-to-cut ferrous and nonferrous alloys due to its high cutting speed, cost-effectiveness, and automation capabilities [21–24].

Many researchers have explored conventional machining techniques for super alloys like Inconel 718. However, these methods yield low MRR and high tooling costs. HSM was introduced to address these challenges by offering increased MRR, better heat dissipation, higher chip removal rates, and improved surface finish. Compared to conventional machining, HSM employs cutting speeds 2–50 times higher, significantly enhancing tool life and machining efficiency. Due to Inconel 718's low machinability, conventional machining techniques often cause surface damage and morphological deformations. Proper tool selection—tool material, coating type, and tool geometry—is required to preserve surface integrity. Traditional mechanical cutting processes incur more tooling costs, and thus optimized machining operations are required.

DOE is a scientific methodology for investigating process variation due to external or internal sources. By applying this methodology, controlled experimentation is possible, and process parameters can be optimized to obtain consistent quality. Taguchi's orthogonal array experiment minimizes variance and determines optimal control settings. Quality is introduced into manufacturing through methods such as tolerance planning, system design, and parameter optimization. External disturbances have to be absorbed by robust product design, creating a high signal-to-noise ratio. Quality deviations are more costly, and therefore system monitoring becomes essential to reduce losses and enhance overall product reliability [25–28].

2. Experimental Methodology

In the present research, Inconel 718 superalloy sheets having a thickness of 5 mm were selected as the work material. The sheets were purchased from Narendra Steels, India, with a vision to achieve high-quality material for the plasma arc cutting (PAC) experiment. PAC experiments were conducted thoroughly utilizing a CNC PAC machine. The machine was fabricated by Pro Arc Welding and Cutting System Pvt. Ltd., which is an Indian organization involved in the manufacturing of advanced metal cutting machines. The system is equipped with Plasma CAM CNC software, which is solely accountable for the smooth and precise cutting process. CNC software enables the machine to operate with work pieces of maximum thickness up to 40 mm. During the experiment trials, the nozzle traverse speed was maintained as 12 meters per minute for a stable and controlled cutting process.

Cutting was done using a servo-controlled torch with a proprietary air-cooled swirl nozzle. The nozzle, 1.5 mm in diameter, was constructed from high-grade copper. Copper was chosen because it has high thermal conductivity and tensile strength, which allow it to withstand the aggressive temperatures used in the plasma cutting process. This temperature resistance is required in maintaining the accuracy and stability of the cutting process, which avoids defects like skewed kerf profiles or a high HAZ. As part of the PAC system, atmospheric air was used as the shielding gas. This selection of gas has a dual function: it first helps in the formation of high-energy plasma required to cut through the Inconel 718 sheets, and it secondly efficiently strips away the molten material out of the cutting zone. This provides a cleaner and smoother cut, with fewer requirements for extensive post-processing or secondary finishing operations.

KW, KT, and HAZ were the three response variables that were determined in order to assess the machining performance and quality of the PAC. Key PAC parameters like AC, cutting speed, gas pressure, and stand-off distance can affect these variables. All these variables contribute significantly towards determining the correctness and performance of the cutting process. KW is the overall width of the cut made by the plasma arc. It is one of the significant indicators of cut accuracy and affects the dimensional accuracy of the finished work piece. A smaller KW is generally preferable, as it reduces material wastage and enhances cutting efficiency. KT is half of the difference in KW measured at the top and bottom of the cut per unit depth of penetration. It gives information about the vertical straightness of the cut. Lower values of KT show a more uniform and perpendicular cut, which is normally preferable in the majority of engineering applications.

The HAZ is the non-melted region surrounding the cut area, where material properties are altered due to the effect of high-temperature conditions. Microstructure alteration, change in hardness, and residual stress formation are likely in the HAZ, and it can compromise the work piece's mechanical integrity. HAZ is minimized to maintain original material properties and avoid thermal damage, as it is undesirable. To further fine and enhance machining quality, a comprehensive initial study was carried out to determine the optimal value of each process parameter. Parametric values thus determined, crucial to providing top-class cut quality. The parameters were chosen carefully so as to realize optimal cutting efficiency as well as machining accuracy. In plasma arc cutting, to achieve best quality, accurate control of process variability is essential. The process will degrade to form faults like excessive, KT increased HAZ, or irregular KW in case the value strays from target parameters. Due to this reason, quality control mechanisms were integrated in order to avoid the process from running beyond specified working ranges.

One of the basics of precision machining is signal-to-noise (S/N) ratio. High S/N ratio in plasma arc cutting translates to small deviations from optimal machining objectives. Product design and machining operation must be rendered robust against undesirable external variations such as power supply fluctuations, variations in material properties, or environmental variations beyond immediate control. Maximizing the S/N ratio can go a long way to ensuring stability and reliability of cutting and resulting in quality and consistent machining result. Total cost of quality is also a matter of utmost concern in manufacturing. Quality-related costs must be measured in terms of variability from optimal cutting conditions. Losses on account of suboptimal machining—loss of material, rework, or reduced product life—must be evaluated in a systematic manner. This is consistent with the overall philosophy of Taguchi's loss function, which states the objective of minimizing total loss to the manufacturer and the consumer for variability in product quality.

Overall, this experimental study aimed to develop a robust and optimized PAC method for Inconel 718 super alloy. Through detailed exploration of the impact of process parameters on kerf behavior and HAZ this study offers valuable insights into improving the precision and efficiency of high-technology metal cutting processes. Findings of this study provide a foundation for subsequent research into adaptive control methods, real-time monitoring systems, and process automation for high-precision machining of advanced materials

3. Results and Discussion

Nine iterative experiments were systematically built in this study using the Taguchi L9 orthogonal array. The studies were carried out carefully to look into the possible correlation between critical performance indicators, such as the size of the HAZ and KW after machining, and PAC process parameters.[39]. The experimental design provided for an in-depth analysis of the effect of varying process conditions on machining quality. A second-order regression model was employed to test the importance of the presumed process parameters and their influence on the cutting response. The regression model's predictability was tested using ANOVA. Statistical software packages Design Expert and MINITAB were employed to conduct ANOVA and verify the results obtained [40]. The ANOVA's results indicated that the model was quadratic which was built using significance and lesser fit tests, was ideal for simulating the observed responses. This high suitability proves the dependability and efficacy of the created mathematical models by showing a converged focus between the predicted response models and the actual experimental data.

Taguchi analysis was applied to measure the variation of KT against significant PAC parameters such as AC, CS, GP, and SOD in the study. S/N ratio response table was utilized to determine the parameter values at the optimal levels to minimize KT. The graphical representation makes it clear that KT values commonly drop significantly at higher cutting speeds and are likely to be higher at lower cutting speeds. According to this pattern, increasing cutting speed results in improved precision and lower KT angles, thereby enhancing the dimensional accuracy of the machined components. A further detailed analysis, indicates that an optimized combination of a lower cutting speed , a higher AC and a medium gas pressure of 51 bar results in minimal stand-off distance. These findings suggest that careful tuning of these parameters is crucial in controlling KT and achieving precise cuts in Inconel 718 sheets.

Similar to KT analysis, Taguchi's methodology was also employed to examine the HAZ as a function of PAC process parameters. To determine the ideal parameter settings that minimize HAZ dimensions, the S/N ratio response table was once more used. The graphical analysis clearly demonstrates that HAZ values tend to be higher at lower cutting speeds and decrease progressively at higher cutting speeds. This observation aligns with the understanding that higher cutting speeds reduce the exposure time of the work piece to excessive thermal input, thereby limiting the extent of the heat-affected zone. This specific combination of parameters is ideal for minimizing HAZ and preserving the mechanical integrity of the Inconel 718 material post-machining.

For better insight into combined effects of PAC process parameters on KT and HAZ, additional Taguchi analyses were conducted. These interaction plots are good indicators of the combined effect of these parameters and are a stepping stone for multi-objective optimization. These contour plots are good indicators of the machining quality and help in determining optimal settings for optimum machining results.

Optimization of PAC parameters was carried out by a general linear model with an aim to create an empirical correlation between process parameters and response properties. KT minimization and HAZ along with the improvement of cut quality were the primary concerns of the study. Three-level (-1, 0, and +1) factor coding technique employed in the study was utilized to search for each parameter's effect systematically. AC, cutting speed, gas pressure, and stand-off distance were of interest as the primary parameters. Statistical methodology adopted ensured each process variable was quantified with precision such that it would be possible to make a wise decision in process improvement. The test identifies each process parameter's level of significance and the individual contribution of each to the total variance recorded in the experiment. The observation indicates the multiple effects of arc current and CS on KT, with gas pressure and stand-off distance reflecting relatively smaller yet significant effects.

The same ANOVA analysis was performed for HAZ, and the results again confirmed that CS and gas pressure have significant influences on HAZ dimensions. The regression models derived as a part of this optimization study have high predictive capability, as can be seen from their good agreement with experimental data. The findings of this optimization study highlight the need for accurate control of PAC process parameters to obtain high machining performance. Through accurate control of arc current, CS, gas pressure, and stand-off distance, it is possible to reduce kerf taper and HAZ, thus improving the quality and efficiency the plasma arc cutting process for Inconel 718.

4. Conclusion

The impact of important PAC parameters on the post machining properties of Inconel 718 super alloy is thoroughly examined in this work. The study goes into detail about how important response factors, such as KW, HAZ, and KT, can be affected by important input and output parameters, such as AC, CS, GP, and SOD A set of experiments were planned and carried out systematically to evaluate comprehensively, and the findings were statistically analyzed using statistical methods, i.e., ANOVA and regression modeling. Based on these thorough investigations, the following significant observations and conclusions have been drawn:

- The ANOVA results indicate that KW and stand-off distance are significantly affected by variations in machining speed and gas supply pressure. An increase in the cutting speed leads to a reduction in the KW thereby enhancing the accuracy of the cut. Decreased cutting speeds, on the other hand, lead to larger kerfs, which can lead to unwanted machining behavior. Similarly, the gas supply pressure is a key parameter in determining the material removal and machining accuracy. Increased gas pressure leads to more energetic ejection of molten material but can also be the cause of secondary effects such as formation of dross and striation patterns.
- It has been discovered that the regression analysis between the HAZ and KW following machining is highly accurate and dependable. There is a strong correlation between the chosen process parameters and the corresponding machining responses, as evidenced by the regression models' 95% confidence level. The high confidence level ensures that the predictive models developed in this study can be used safely for the selection of optimal PAC conditions for achieving higher machining quality in Inconel 718.
- To confirm the accuracy and effectiveness of the selected process parameters, additional experiments were conducted under the optimum conditions established. The experiment results confirmed that the machining outcomes were in complete agreement with the optimum parameter values calculated theoretically, further confirming the reliability of the experimental approach. The confirmation process illustrates the accuracy and reliability of the PAC process with the optimum parameters and is therefore well suited for industrial use where accuracy and consistency are of utmost importance.
- Comparison of theoretical and experimental values of KW and HAZ after machining showed a negligible difference, again proving the viability of the developed models. The relative error in KW was calculated to be 4.45%, and the relative error in the HAZ was calculated to be 4.36%. The error levels mentioned above are regarded to be very low in machining research, and they prove that the PAC process under the optima produces highly reproducible and reliable outcomes. The negligible difference from actual experimental

results proves the efficiency of the statistical models and the optimization techniques applied in this research.

- The results of this research firmly confirm that the selected plasma arc cutting method is highly efficient in machining Inconel 718 super alloy. The optimization approach followed in this research is efficient enough to meet all the criteria for achieving high-quality and efficient cutting process. The method is confirmed to be efficient and effective, and it can be used potentially on a large scale in industrial applications in precision machining of high-performance engineering and aerospace materials.
- Inconel 718 surface topography after plasma arc cutting was examined by morphological analysis. The analysis showed that increasing the gas supply pressure causes dross attachment and striation marks on the plasma arc cut surface of the super alloy. These unwanted surface roughness features degrade the quality of the cut and surface integrity, resulting in an increased finish and post-machining issues. In order to counteract these effects, it was found that decreasing the AC while at the same time increasing the gas pressure can reverse the dross attachment formation and improve surface flatness. This finding constitutes a useful rule of thumb for maximizing gas pressure values to reduce surface defects and enhance the quality of plasma arc cut parts.

This study provides in-depth information regarding the plasma arc cutting of Inconel 718 and highlights the importance of maximizing the important process parameters to achieve enhanced machining performance. The findings of this study are highly beneficial in precision machining by providing suitably validated experimental process and optimization technique. Future studies can explore other process parameter improvements, other shielding gases, and novel cutting techniques to enhance the efficiency and accuracy of plasma arc cutting of high-performance super alloys.

References

1. R. Adalarasan, M. Santhanakumar, and M. Rajmohan, "Application of Grey Taguchi –based response surface methodology (GT-RSM) for optimizing the plasma arc cutting parameters of 304L stainless steel," The International Journal of Advanced Manufacturing Technology, vol. 78, no. 5, pp. 1161–1170, 2015.
2. K. Ananthakumar, D. Rajamani, E. Balasubramanian, and J. Paulo Davim, "Measurement and optimization of multiresponse characteristics in plasma arc cutting of Monel 400™ using RSM and TOPSIS," Measurement, vol. 135, pp. 725–737, 2019.
3. A. K. Parida, "Analysis of chip geometry in hot machining of Inconel 718 alloy," Iranian Journal of Science and Technology, Transactions of Mechanical Engineering, vol. 43, no. S1, pp. 155–164, 2019.
4. B. C. Behera, S. G. Chetan, and V. R. Paruchuri, "Study of sawtooth chip in machining of Inconel 718 by metallographic technique," Machining Science and Technology, vol. 23, no. 3, pp. 431–454, 2019.
5. B. Zhao, H. Liu, C. Huang, J. Wang, B. Wang, and Y. Hou, "Cutting performance and crack self-healing mechanism of a novel ceramic cutting tool in dry and high-speed machining of Inconel 718," The International Journal of Advanced Manufacturing Technology, vol. 102, no. 9-12, pp. 3431–3438, 2019.
6. S. Ahmad, R. M. Singari, and R. S. Mishra, "Tri-objective constrained optimization of pulsating DC sourced magnetic abrasive finishing process parameters using artificial neural network and genetic algorithm," Materials and Manufacturing Processes, vol. 36, no. 7, pp. 843–857, 2021.
7. A. Gani, W. Ion, and E. Yang, "Experimental investigation of plasma cutting two separate thin steel sheets simultaneously and parameters optimisation using Taguchi approach," Journal of Manufacturing Processes, vol. 64, pp. 1013–1023, 2021.
8. N. Pragadish, S. Kaliappan, M. Subramanian et al., "Optimization of cardanol oil dielectric-activated EDM process parameters in machining of silicon steel," Biomass Conversion and Biorefinery, vol. 12, pp. 1–10, 2022.
9. R. Devaraj, E. Abouel Nasr, B. Esakki, A. Kasi, and H. Mohamed, "Prediction and analysis of multi-response characteristics on plasma arc cutting of Monel 400™ alloy using Mamdani-fuzzy logic system and sensitivity analysis," Materials, vol. 13, no. 16, p. 3558, 2020.
10. N. A. Fountas, J. D. Kechagias, A. C. Tsiolikas, and N. M. Vaxevanidis, "Multi-objective optimization of printing time and shape accuracy for FDM-fabricated ABS parts," Metaheuristic. Computing. And. Applications., vol. 1, no. 2, pp. 115–129, 2020.
11. F. Fontanive, R. P. Zeilmann, and J. D. Schenkel, "Surface quality evaluation after milling Inconel 718 with cutting edge preparation," The International Journal of Advanced Manufacturing Technology, vol. 104, no. 1-4, pp. 1087–1098, 2019.
12. G. Mustafa, J. Liu, F. Zhang et al., "Atmospheric pressure plasma jet assisted micro-milling of Inconel 718," The International Journal of Advanced Manufacturing Technology, vol. 103, no. 9-12, pp. 4681–4687, 2019.
13. K. Venkatesan, R. Ramanujam, and P. Kuppan, "Investigation of machinability characteristics and chip morphology study in laser-assisted machining of Inconel 718," The International Journal of Advanced Manufacturing Technology, vol. 91, no. 9-12, pp. 3807–3821, 2017.
14. A. Iosub, G. Nagit, and F. Negoescu, "Plasma cutting of composite materials," International Journal of Material Forming, vol. 1, no. S1, pp. 1347–1350, 2008.
15. K. P. Maity and D. K. Bagal, "Effect of process parameters on cut quality of stainless steel of plasma arc cutting using

hybrid approach," The International Journal of Advanced Manufacturing Technology, vol. 78, no. 1-4, pp. 161–175, 2015.
16. M. Meikandan, M. Karthick, L. Natrayan et al., "Experimental investigation on tribological behaviour of various processes of anodized coated piston for engine application," Journal of Nanomaterials, vol. 2022, Article ID 7983390, 8 pages, 2022.
17. M. Meikandan and K. Malarmohan, "Fabrication of a superhydrophobic nanofibres by electrospinning," Digest Journal of Nanomaterials and Biostructures, vol. 12, no. 1, pp. 11–17, 2021.
18. A. Bovas Herbert Bejaxhin, G. Paulraj, G. Jayaprakash, and V. Vijayan, "Measurement of roughness on hardened D-3 steel and wear of coated tool inserts," Transactions of the Institute of Measurement and Control, vol. 43, no. 3, pp. 528–536, 2021.
19. M. Karthick, P. Anand, M. Meikandan, and M. Siva Kumar, "Machining performance of Inconel 718 using WOA in PAC," Materials and Manufacturing Processes, vol. 36, no. 11, pp. 1274–1284, 2021.
20. M. Meikandan, M. Sundarraj, D. Yogaraj, and K. Malarmohan, "Experimental and numerical investigation on bare tube cross flow heat exchanger-using COMSOL," International Journal of Ambient Energy, vol. 41, no. 5, pp. 500–510.
21. M. Rakesh and S. Datta, "Effects of cutting speed on chip characteristics and tool wear mechanisms during dry machining of Inconel 718 using uncoated WCtool," Arabian Journal for Science and Engineering, vol. 44, no. 9, pp. 7423–7440, 2019.
22. Y. Sesharao, T. Sathish, K. Palani et al., "Optimization on operation parameters in reinforced metal matrix of AA6066 composite with HSS and Cu," Advances in Materials Science and Engineering, vol. 2021, Article ID 1609769, 12 pages, 2021.
23. D. Damodharan, K. Gopal, A. P. Sathiyagnanam, B. Rajesh Kumar, M. V. Depoures, and N. Mukilarasan, "Performance and emission study of a single cylinder diesel engine fuelled withn-octanol/WPO with some modifications," International Journal of Ambient Energy, vol. 42, no. 7, pp. 779–788, 2021.
24. P. Asha, L. Natrayan, B. T. Geetha et al., "IoT enabled environmental toxicology for air pollution monitoring using AI techniques," Environmental Research, vol. 205, p. 112574, 2022.
25. P. Pal Pandian and I. S. Rout, "Parametric investigation of machining parameters in determining the machinability of Inconel 718 using taguchi technique and grey relational analysis," Procedia Computer Science, vol. 133, pp. 786–792, 2018.
26. D. Rajamani, K. Ananthakumar, E. Balasubramanian, and J. Paulo Davim, "Experimental investigation and optimization of PAC parameters on Monel 400™ superalloy," Materials and Manufacturing Processes, vol. 33, no. 16, pp. 1864–1873, 2018.
27. R. P. Zeilmann, F. Fontanive, and R. M. Soares, "Wear mechanisms during dry and wet turning of Inconel 718 with ceramic tools," The International Journal of Advanced Manufacturing Technology, vol. 92, no. 5-8, pp. 2705–2714, 2017.
28. S. Yogeshwaran, L. Natrayan, S. Rajaraman, S. Parthasarathi, and S. Nestro, "Experimental investigation on mechanical properties of epoxy/graphene/fish scale and fermented spinach hybrid bio composite by hand lay-up technique," Materials Today: Proceedings, vol. 37, pp. 1578–1583, 2021.
29. R. Thirumalai, J. S. Senthilkumaar, P. Selvarani, and S. Ramesh, "Machining characteristics of Inconel 718 under several cutting conditions based on Taguchi method," Proceedings of the Institution of Mechanical Engineers, Part C: Journal of Mechanical Engineering Science, vol. 227, no. 9, pp. 1889–1897, 2013.
30. A. B. H. Bejaxhin, G. Paulraj, and S. Aravind, "Influence of TiN/AlCrN electrode coatings on surface integrity, removal rates and machining time of EDM with optimized outcomes," Materials Today: Proceedings, vol. 21, pp. 340–345, 2020.
31. S. Montazeri, M. Aramesh, and S. C. Veldhuis, "An investigation of the effect of a new tool treatment technique on the machinability of Inconel 718 during the turning process," The International Journal of Advanced Manufacturing Technology, vol. 100, no. 1-4, pp. 37–54, 2019.
32. L. Natrayan, V. Sivaprakash, and M. S. Santhosh, "Mechanical, microstructure and wear behavior of the material AA6061 reinforced SiC with different leaf ashes using advanced stir casting method," International Journal of Engineering and Advanced Technology, vol. 8, pp. 366–371, 2018.
33. K. Salonitis and S. Vatousianos, "Experimental Investigation of the plasma arc cutting process," Procedia CIRP, vol. 3, pp. 287–292, 2012.
34. S. A. Khan, S. L. Soo, D. K. Aspinwall et al., "Tool wear/life evaluation when finish turning Inconel 718 using PCBN tooling," Procedia CIRP, vol. 1, pp. 283–288, 2012.
35. J. S. N. Raju, M. V. Depoures, and P. Kumaran, "Comprehensive characterization of raw and alkali (NaOH) treated natural fibers from _Symphiremainvolucratum_ stem," International Journal of Biological Macromolecules, vol. 186, pp. 886–896, 2021.
36. T. Sugihara, S. Takemura, and T. Enomoto, "Study on highspeed machining of Inconel 718 focusing on tool surface topography of CBN cutting tool," The International Journal of Advanced Manufacturing Technology, vol. 87, no. 1–4, pp. 9–17, 2016.
37. S. Vellaiyan, A. Subbiah, S. Kuppusamy, S. Subramanian, and Y. Devarajan, "Water in waste-derived oil emulsion fuel with cetane improver: formulation, characterization and its optimization for efficient and cleaner production," Fuel Processing Technology, vol. 228, p. 107141, 2022.
38. W. Bai, A. Bisht, A. Roy, S. Suwas, R. Sun, and V. V. Silberschmidt, "Improvements of machinability of

aerospace-grade Inconel alloys with ultrasonically assisted hybrid machining," The International Journal of Advanced Manufacturing Technology, vol. 101, no. 5-8, pp. 1143–1156, 2019.

39. Y. Feng, T.-P. Hung, Y.-T. Lu et al., "Inverse analysis of Inconel 718 laser-assisted milling to achieve machined surface roughness," International Journal of Precision Engineering and Manufacturing, vol. 19, no. 11, pp. 1611–1618, 2018.

40. Y. Feng, Y. Guo, Z. Ling, and X. Zhang, "Micro-holes EDM of superalloy Inconel 718 based on a magnetic suspension spindle system," The International Journal of Advanced Manufacturing Technology, vol. 101, no. 5-8, pp. 2015–2026, 2019.

41. Z. Vagnorius and K. Sorby, "Effect of high-pressure cooling on life of SiAlON tools in machining of Inconel 718," The International Journal of Advanced Manufacturing Technology, vol. 54, no. 1, pp. 83–92, 2011.

Advances in Additive Manufacturing Technologies – Gurusamy Pathinettampadian et al. (eds)

Improving Dimensional Precision in Laser Beam Machining of Titanium Alloy Through Process Optimization

S. Senthil Kumar[1]
Professor, Department of Mechanical Engineering,
R.M.K. College of Engineering and Technology,
India

P. Usha Rani[2]
Professor, Science and Humanities Department,
R.M.D Engineering College, India

S. Sridhar[3]
Associate Professor, Department of Mechanical Engineering,
PSNA College of Engineering and Technology,
Chennai, India

K. Karthik[4]
Associate Professor,
Department of Mechanical Engineering,
Vel Tech Rangarajan Dr.Sagunthala R&D Institute of Science Technology,
Chennai, India

Bharanidharan T.[5]
Student, Department of Mechanical Engineering,
Chennai Institute of Technology,
Chennai, India

R. Karthick[6]
Assistant Professor, Department of Mechanical Engineering,
Rajalakshmi Institute of Technology,
Chennai, India

◆ **Abstract:** Evolution of machining processes greatly improves processing of advanced materials such as titanium alloys. Ti-6Al-4V alloys, with their corrosion resistance and strength-to-weight ratio, have extensive application in aerospace, biomedical, and automotive sectors. Machining of these alloys, however, proves challenging in the need to optimize the process for accuracy. In this study, we experimentally investigated prominent parameters in laser beam machining (LBM) of Ti-6Al-4V alloy drilling for dimensional accuracy. In applying the Taguchi method, the influence of laser power, gas pressure, nozzle distance, and focal length on prominent dimensional accuracy such as roundness and ovality was investigated. The results established that laser power was the prominent factor in dimensional accuracy in drilling holes. The combination of the parameters that would yield the highest accuracy was found to be 1.2 mm nozzle distance, 2 mm focal length, 4 bar gas pressure, and 2 kW laser power. Aside from this, low peak power greatly enhanced machinability, which enhanced precision and stability in the LBM process. The study provides valuable

[1]senthilkumar@rmkcet.ac.in, [2]pusharani71@yahoo.com, [3]sri_2855@yahoo.co.in, [4]karthikk@veltech.edu.in, [5]vtrbharani12@gmail.com, [6]karthick.industrial@gmail.com

DOI: 10.1201/9781003685906-25

recommendations for laser machining parameter optimization to enhance accuracy and efficiency in the case of Ti-6Al-4V for application in various industries.

◆ **Keywords:** LBM, Laser beam machining, Taguchi method, Optimization, Titanium alloy

1. Introduction

UCM techniques are typically used to shape high-strength materials into complex shapes with less effort. Different unconventional machining techniques are used for titanium alloy material, such as WJM, AJM, EDM, ECM, PAM, and LBM [1–4]. More specifically, high-strength titanium alloy machining is required for dental implant applications where quality parameters need to be improved with special care [5]. Tool material that is stronger than the workpiece material is required for traditional machining operations. Furthermore, due to excessive abrasion, conventional techniques deliver subpar surface finishing on a tiny scale. Cutting high-strength metals into exact sizes and shapes has never been easier than with laser-based machining. The laser beam has the ability to concentrate energy and is both coherent and monochromatic. Because of its high power density, this concentrated source of energy causes the material to melt and evaporate quickly. Rapid heating and subsequent material removal by vaporization and melting occurs when the workpiece absorbs the photon energy released by the laser beam. Because of its excellent surface smoothness and high ablation rate, LBM has therefore become the machining process of choice, surpassing other unusual approaches [6].

The effects of varying pulse durations and wavelengths on machining were covered by Meijer [6], who also addressed the topic of laser beam production. Pulse frequency is thought to be an important factor in improving material removal rates. To determine how the strength of the laser beam affects machinability, computational models of the laser drilling process were run [7,8]. To further improve machining accuracy, it is crucial to analyze the impact of process factors on circularity [9]. Compared to more conventional machining techniques, LBM greatly improves the machinability of steel, according to the research [10]. The optimization of LBM processes has made extensive use of Taguchi techniques to boost efficiency [11]. It has been proven that machinability in LBM processes is significantly affected by process parameters [12]. With an emphasis on white layer thickness, Muthuramalingam [13] analyzed the performance of process factors in the machining of contemporary alloys. Another work by Kim et al. [14] showed that spherical-shaped nickel-based alloys might be made more machinable with the use of laser-assisted machining. More importantly, LBM may significantly reduce the overcut issue that often arises during milling operations.

Research on the effects of pulse length on surface roughness and ablation quality has also been conducted, demonstrating the significance of managing plasma energy when machining [16,17]. In order to enhance quality measures, an experimental technique has been devised to determine the effect of process parameters on green EDM. This methodology makes use of the Taguchi method when milling AISI 202 stainless steel [18]. In a similar vein, EDM has proven useful for precision machining when used to AISI 304 stainless steel [19]. The LBM technique is also capable of efficiently milling non-conductive biomaterials like leather [20]. To improve control over machining results, process parameters can be tweaked to optimize the energy input during LBM machining [21–23]. Because controlled plasma energy is a key component of non-traditional machining quality metrics, it is essential to study how process factors affect plasma energy in order to improve machining quality.

We found very little research on examining the effect of process factors on obtaining target profiles using LBM based on the existing literature evaluation. Electrical process parameters have also been the subject of scant investigation in LBM-based drilling operations. It is worth mentioning that there has been a lack of substantial study on how process factors affect ovality in titanium alloys during LBM. This research intends to close that knowledge gap by studying how different process factors affect laser beam machining and how it might improve the machinability of titanium alloys.

2. Experimental Methodology

Ti-6Al-4V alloy specimen materials were used as the workpiece material in the present experimental work because they are widely used in the production of dental implants. Titanium alloys, and specifically Ti-6Al-4V, are well renowned for their superior biocompatibility, corrosion resistance, and strength-to-weight ratio, and therefore are strong contenders in medical applications. The machine has a wavelength of 10.6 μm and is capable of conducting precision machining operations with high accuracy and repeatability. Workpiece specimens of 50 mm × 50 mm × 3 mm were used to maintain consistency in machining trials. The provided dimensions were used to enable the analysis of the laser drilling process in cases relevant to the production of dental implants. Oxygen (O_2) was used as the purging gas to enable material removal efficiency in drilling. Although many process parameters have strong influence in determining the laser beam machining process, four major input parameters, i.e., ND, GP, FL, and P, were particularly the focus in this work because they have strong influence in determining the cutting performance and the resulting quality of drilled holes. The parameters were used based on the high influence on cutting performance and overall quality of drilled holes. It was observed that the threshold value of power density required to machine through the thickness of the workpiece ranged from 1.5 × 10^6 W/cm^2 to 1.5 × 10^8 W/cm^2. Moreover, the optimal drilling speed required in through-hole cutting was found to be 2000 mm/min [10–11].

Drilling experiments were performed to create holes of a target diameter of 5 mm. The same diameter on the entrance and exit surfaces of the holes is required to maintain high dimensional accuracy in machining. The precision of the drilled holes was estimated by the measurement of roundness of the hole profiles based on ovality tests. Ovality is an important quality parameter in laser drilling because it represents the degree of departure from a perfect circular shape. Hole diameter variations from the target 5 mm size and roundness of holes are considered to be measures of quality for the present investigation. The optical microscope system provides the facility of high-resolution imaging for precise measurement of hole diameters and roundness deviation. The optical system focus length was adjusted at 0.7 mm and the pixel size was calibrated at 10.949 μm to obtain the optimized measurement precision. The diameters of holes obtained from the experiments were estimated by choosing the proper points on the hole profiles and finding the average diameter values. Due to the inherent nonlinear nature of the laser beam machining process, there is a requirement to use optimization algorithms in order to enhance the roundness of the holes and minimize deviations from the target hole diameter. In accordance with optimization, the machining process can be made more effective by enhancing the dimension accuracy and the surface quality of the machined surface.

Influence of the selected process parameters on hole roundness and hole diameter deviation was investigated through main effects plots developed with the assistance of the Minitab statistical package. Main effects plots are effective tools in statistical analysis because they offer visualization of the process parameter-output response relationship and thus make it easy to identify optimal machining conditions. Ovality of the holes was also quantified through the application of a profile projector, a precision instrument used to measure accurate hole circularity deviations. In systematic analysis and design of the experimental tests, the Taguchi approach was applied. The Taguchi approach is an effective statistical method that enables efficient experimentation and reduces the number of experimental runs to be carried out in order to obtain significant results. Through application of the Taguchi approach, the experimental design can be customized to obtain maximum useful data and reduce total utilization of resources. Four input factors were investigated in this paper with each factor having four levels. With reference to the standard Taguchi design of experiments, an L16 orthogonal array (OA) was selected as the experimental plan. The L16 orthogonal array provides a balanced experimental plan that ensures systematic study of all interactions among parameters.

16 experimental trials were conducted, each with a different set of process parameters. In this systematic investigation, the study aimed to thoroughly examine the effects of significant process parameters on the laser-drilled hole quality, in titanium alloy samples and, in doing so, contribute to laser machining technique development for biomedical applications [9].

3. Results and Discussion

In this study, specimens of Ti-6Al-4V alloy were machined using the technique of Laser Beam Machining (LBM). Microstructure observation of the machined surface was carried out through the utilization of a SEM (Hitachi make model SEM S-3400N). Material was melted away from the surface by the laser beam via an ablation mechanism, and this resulted in the formation of craters on the machined surface. Size and distribution of craters were governed by the distribution of laser energy during the machining process. Due to the random nature of laser beam machining, fluctuations in the distribution of laser energy resulted in craters of varying sizes appearing randomly over the machined surface [7,8]. A recast layer was observed on the machined surface due to the re-solidification of melted particles. The agglomeration of this recast layer led to the formation of a white region, under the parameter combination used in trial 16. Additionally, a HAZ was identified on the machined surface under the same process conditions. The presence of the HAZ suggests significant thermal influence on the material, which is a critical factor in evaluating the machining performance.

However, variations in laser beam plasma were observed due to discharge fluctuations within the holes. As a result, the geometrical accuracy of the entry side of the holes was found to be superior compared to the exit side, which presents OPM images of the machined holes. The entry hole profiles exhibited a higher level of dimensional accuracy than the exit profiles, which can be attributed to the controlled energy distribution at the initial penetration stage. Measurement accuracy was assessed using standard error values, with lower standard errors indicating higher measurement precision. The deviation of the actual hole size from the required hole size served as an indicator of dimensional accuracy. Among the trials, trial 10 demonstrated the best dimensional accuracy, making it the optimal PP combination for LBM. The microstructure of the machined surface under these optimal process conditions. Furthermore, a reduced HAZ region (white region) was observed due to the regulated energy input during the machining process. Machined profiles of holes obtained in trials 10 and 16, captured using a machine vision system. Superior accuracy was achieved in trial 10 due to the optimal combination of LP and FL, which generated a controlled beam energy distribution over the machined hole. In contrast, trial 16 resulted in the poorest accuracy due to higher and more random beam energy distribution, which led to significant deviations from the desired hole dimensions.

Roundness, defined as the reciprocal of ovality, was considered a key quality measure in this study. The roundness was analyzed Influence of process parameters on LBM drilling was investigated through Minitab software on main effect plots (MEP). Larger deviation from the horizontal line in such plots meant larger influence of the respective process parameters on response parameters [24,25]. LP was the most influential input parameter on hole diameter deviations, as can be seen from its dominant effect in the main effects plot. The observation corroborates the fact that the laser power is the most crucial in determining plasma energy levels that ultimately influence material removal. Power was seen to have a strong effect, especially at lower peak power levels, indicating the significance of optimizing the laser plasma column in the drilling process. GP, on the other hand, had the minimum influence on response parameters owing to its paltry contribution towards regulating the plasma column [6].

Absence of significant variation in results at 3 kW and 4 kW laser power means that the system had an energy saturation level beyond which increased energy input added little to the result. Focal length plots revealed oscillating nature due to variation in focal position during drilling. Fluctuation can be attributed to focal point variation of the lens, with smaller focal lengths resulting in higher energy density. Laser power also revealed enhanced stability at the exit hole due to increased energy utilization at the entry point, leading to lower PD at the exit. GP exerted a stronger influence at the exit diameter, as it had a critical influence on plasma column control at this process stage. Oscillation in parameter influences was high due to complex interaction between process parameters in determining roundness. Fluctuations were mostly due to variation in focal length, which controlled energy intensity concentration. It was observed that the lowest value of focal length produced negligible deviation from the mean line, testifying to the influence of accuracy in focusing energy. Lower peak power enhanced machining accuracy, while gas pressure exhibited the lowest influence on entry roundness. Nozzle distance exhibited the highest deviation from the mean line, which reflects its leading influence on hole geometry. Gas pressure exhibited minimum deviation at 4 bar, reflecting an optimum pressure value in ensuring process stability.

Due to intricate interaction between the process parameters and their effect on quality characteristics, regression analysis was employed with the support of the Minitab package. Regression analysis formulates the interaction between the input parameters and output responses as constant coefficients.

The equation for the regression of entry hole diameter is:

$$\text{Entry hole diameter} = 4.92 + 0.006\text{ND} + 0.048\text{J} - 0.0001\text{L} + 0.005\text{GP}$$

where ND is nozzle distance, J is power, L is focal length, and GP is gas pressure. From the findings, we can observe that power has a very influential effect on the diameter of the entry hole, with a coefficient of 0.048, after nozzle distance.

Similarly, the exit hole diameter regression equation is:

$$\text{Exit hole diameter} = 5.008 + 0.003\text{ND} + 0.014\text{J} - 0.003\text{L} + 0.011\text{GP}$$

The most significant parameter is power having a coefficient value of 0.014 and gas pressure has a coefficient of 0.011.

For roundness of entry hole, the regression equation is:

$$\text{Roundness of entry hole} = 0.044 + 0.003\text{ND} - 0.007\text{J} - 0.003\text{L} + 0.008\text{GP}$$

Gas pressure is found to be the most critical parameter on entry roundness, followed by power.

Regression equation for roundness of exit holes is:

$$\text{Roundness at the exit hole} = 0.03 - 0.0025\text{ND} - 0.002\text{J} + 0.002\text{L} + 0.006\text{GP}$$

The most critical parameter is gas pressure, followed by nozzle distance. These results highlight the dominance role of power in determining hole profiles due to its influence on plasma energy distribution. High levels of power, in combination with increased nozzle distances, resulted in unacceptable profiles due to over-dispersion of energy. Proper control of process parameters is therefore a requirement for realizing maximum hole accuracy as well as roundness in LBM drilling.

4. Conclusion

In the current research, the Taguchi orthogonal technique was utilized effectively to determine the best levels of important PP controlling the LBM process to drill samples of Ti-6Al-4V alloy. The main objective of the present research was to improve the roundness and dimensional accuracy quality factors of the holes drilled. In a vast range of experimental research studies, the different input parameters were examined in a systematic manner, and their respective impacts on the overall machining performance were evaluated. From the findings of these studies, the following few significant conclusions have been derived:

- Experimental analysis revealed that power is the most significant parameter in controlling the geometrical accuracy of holes machined on the LBM process. Power directly controls the laser energy density, and consequently, the material removal rate, heat-affected zone (HAZ), and plasma energy distribution. Higher power levels will tend to raise the heat delivered to the workpiece by melting and vaporizing more material. Higher power can, however, produce unwanted thermal effects like greater recast layers, microcracking, and hole size deviations. Optimal power optimization should thus be maintained for maximum geometrical accuracy and prevention of defects due to excessive heat accumulation.
- The research recommended optimal parameters to enhance dimensional accuracy in the LBM process: 1.2 mm nozzle distance, 2 mm focal distance, 4 bar gas pressure, and 2 kW power. The parameters suggested enhanced plasma energy interaction and surface energy distribution, reducing deviations. Optimized parameters controlled laser-material interaction, restraining ovality and enhancing roundness in the machined hole. Keeping the laser power within a reasonable range and by optimizing gas pressure and nozzle distance minimized excess recast layers and residual stresses further and improved the hole quality.
- The research verifies that power and gas pressure are essential factors in dimensional precision of holes drilled using the LBM process. Their interaction influences plasma formation and expulsion of material. Contrary to expectations, lower power is more desirable for dimensional control since it prevents overmelting and provides uniform energy. The optimal gas pressure of 4 bar efficiently expels molten material, stripping away a recast layer and improving surface quality. Therefore, dimensional precision in LBM drilling can be enhanced by controlling power and gas pressure.
- Experimental findings revealed that power and nozzle distance were the two primary parameters in managing roundness of holes machined under the LBM process. A consistent 1.2 mm nozzle distance was necessary in creating a stable focal point, which also enhanced the level of energy distribution uniformity in the cutting zone. This accuracy helped ensure that the holes' geometry was not deformed, leading to maximum circularity. In addition, optimal laser power helped ensure that energy delivered to the workpiece was sufficient to impart effective material removal, all without causing excessive thermal damage or irregularities. Interaction between the two parameters was necessary in minimizing hole shape deviations and maximizing roundness accuracy.
- The study emphasizes the importance of properly selecting and optimizing laser beam machining parameters to obtain quality holes. Proper optimization of the laser power, nozzle distance, focal length, and gas pressure is vital to minimize dimension errors and maximize roundness of holes. The research has applications to precision machining industries such as aerospace, biomedical, and automotive. The study also emphasizes the use of sophisticated optimization techniques, i.e., the Taguchi method, to find optimal settings economically with minimal experimental trials.

The Taguchi orthogonal array technique successfully determined the best process parameters to improve roundness and dimensional accuracy in laser-drilled holes in titanium. The most influential factors on hole geometry are laser power, gas pressure, and nozzle distance, which are the key factors in machining accuracy. These results justify further research on optimizing laser machining and creating advanced control systems for higher precision and efficiency in industrial processes.

References

1. T. Muthuramalingam, A. Ramamurthy, K. Moiduddin, M. Alkindi, S. Ramalingam, O. Alghamdi, Enhancing the Surface Quality of Micro Titanium Alloy Specimen in WEDM Process by Adopting TGRA-Based Optimization, Materials 13 (2020) 1440, https://doi.org/10.3390/ma13061440.

2. T. Geethapriyan, K. Kalaichelvan, T. Muthuramalingam, A. Rajadurai, Performance analysis of process parameters on machining α-β titanium alloy in electrochemical micromachining process, P.I.Mech.Eng.B-J.Eng. 232 (2018) 1577–1589, https:// doi.org/10.1177/0954405416673103.
3. T. Muthuramalingam, S. Vasanth, P. Vinothkumar, T. Geethapriyan, M.M. Rabik, Multi criteria decision making of abrasive flow oriented process parameters in abrasive water jet machining using Taguchi-DEAR Methodology", Silicon. 10 (2018) 2015–2021, https://doi.org/10.1007/s12633-017-9715-x.
4. T. Muthuramalingam, A. Ramamurthy, K. Sridharan, S. Ashwin, Analysis of surface performance measures on WEDM processed titanium alloy with coated electrodes, Mater. Res. Express. 5 (2018) 126503. https://iopscience.iop.org/article/10.1088/2053-1591/aade70/pdf.
5. L.L. Guehennec, A. Soueidan, P. Layrolle, Y. Amouriq, Surface treatments of titanium dental implants for rapid osseointegration, Dent. Mater. 23 (2007) 844–854, https:// doi.org/10.1016/j.dental.2006.06.025.
6. J. Meijer, Laser Beam Machining (LBM) State of the art and new opportunities, J. Mater. Process. Technol. 149 (2004) 2–17, https://doi.org/10.1016/j.jmatprotec.2004.02.003.
7. D. Abidou, N. Yusoff, N. Nazri, M.A.O. Awang, M.A. Hassan, Numerical simulation of metal removal in laser drilling using symmetric smoothed particle hydrodynamics, Precis. Eng. 49 (2017) 69–77, https://doi.org/10.1016/j.precisioneng.2017.01.012.
8. P. Parandoush, A. Hossain, A review of modeling and simulation of laser beam machining, Int. J. Mach. Tool. Manuf. 85 (2014) 135–145, https://doi.org/10. 1016/j.ijmachtools.2014.05.008.
9. N.S. Amalina, A. Hossain, K. Alrashed, Y. Nukmana, Prediction Modelling of Recast layer and Circularity for Laser Drilling of Polyethylene Terephthalate (PETG) Thermoplastic, Procedia. Eng. 184 (2017) 197–204, https:// doi.org/10.1016/j.proeng.2017.04.086.
10. M.J. Jackson, W.O. Neill, Laser micro-drilling of tool steel using Nd:YAG lasers, J. Mater. Process. Technol. 142 (2003) 517–525, https://doi.org/10.1016/S0924-0136(03)00651-4.
11. A.K. Dubey, V. Yadava, Multi-Objective Optimization of Laser Beam Cutting process, Opt. Laser. Technol. 40 (2008) 562–570, https://doi.org/10.1016/j.optlastec.2007.09.002.
12. A.K. Dubey, V. Yadava, Laser Beam Machining-A Review, Int. J. Mach. Tool. Manuf. 48 (2008) 609–628, https://doi.org/10.1016/j.ijmachtools.2007.10.017.
13. T. Muthuramalingam, Measuring the influence of discharge energy on white layer thickness in electrical discharge machining process, Measurement 131 (2019) 694–700, https://doi.org/10.1016/j.measurement. 2018.09.038.
14. I.W. Kim, C.M. Lee, A study on the machining characteristics of specimens with spherical shape using laser-assisted machining, Appl. Therm. Eng. 100 (2016) 636–645, https:// doi.org/10.1016/j.applthermaleng.2016.02.005.
15. A.M.A. Al-Ahmari, M.S. Rasheed, M.K. Mohammed, T. Saleh, A hybrid machining process combining micro-EDM and laser beam machining of nickel–titanium-based shape memory alloy, Mater. Manuf. Process. 31 (2016) 447–455, https://doi.org/ 10.1080/10426914.2015.1019102.
16. B. Denkena, A. Krödel, T. Grove, Influence of pulsed laser ablation on the surface integrity of PCBN cutting tool materials, Int. J. Adv. Manuf. Technol. 101 (2019) 1687–1698, https://doi.org/10.1007/s00170-018-3032-4.
17. J. Lin, Y. Xu, Z. Fang, M. Wang, J. Song, N. Wang, L. Qiao, W. Fang, Y. Cheng, Fabrication of high-Q lithium niobate micro resonators using femto second laser micromachining, Sci. Rep. 5 (2015) 8072, https://doi.org/10.1038/srep08072.
18. T. Muthuramalingam, Effect of diluted dielectric medium on spark energy in green EDM process using TGRA approach, J. Cleaner. Prod. 29 (2014) 1374–1380, https:// doi.org/10.1016/j.jclepro.2019.117894.
19. T. Muthuramalingam, B. Mohan, Performance analysis of iso current pulse generator on machining characteristics in EDM process, Arch. Civil. Mech. Eng. 14 (2014) 383–390, https://doi.org/10.1016/j.acme.2013.10.003.
20. S. Vasanth, T. Muthuramalingam, A Study on Machinability of Leather Using CO2-Based Laser Beam Machining Process, Lecture Notes Mech. Eng.: Adv. Manuf. Processes, (2018) 239–244, http://doi.org/10.1007/978-981-13-1724-8_23.
21. W.S. Woo, C.M. Lee, A study on the edge chipping according to spindle speed and inclination angle of workpiece in laser-assisted milling of silicon nitride, Opt. Laser. Technol. 99 (2018) 351–362, https://doi.org/10.1016/j.optlastec.2017.09.023.
22. M. Sharifi, M. Akbari, Experimental investigation of the effect of process parameters on cutting region temperature and cutting edge quality in laser cutting of AL6061T6 alloy, Optik. 184 (2019) 457–463, https://doi.org/10.1016/j.ijleo.2019.04.105.
23. X. Lin, P. Wang, Y. Zhang, Y. Ning, H. Zhu, Theoretical and experimental aspects of laser cutting using direct diode laser source based on multi-wavelength multiplexing, Opt. Laser. Technol. 114 (2019) 66–71, https://doi.org/10.1016/j.optlastec.2019.01.022.
24. T. Muthuramalingam, B. Mohan, A. Rajadurai, M.D.A.A. Prakash, Experimental investigation of iso energy pulse generator on performance measures in EDM, Mater. Manuf. Process. 28 (2013) 1137–1142, https://doi.org/10.1080/10426914.2013.811749.
25. J. Huo, S. Liu, Y. Wang, T. Muthuramalingam, V.N. Pi, Influence of process factors on surface measures on electrical discharge machined stainless steel using TOPSIS, Mater. Res. Express. 6 (2019) 086507, https://doi.org/10.1088/2053-1591/ab1ae0.

Advances in Additive Manufacturing Technologies – Gurusamy Pathinettampadian et al. (eds)

Refining Surface Finish in Cylindrical Components with Magnetic Abrasive Finishing

C. Bibin[1]
Professor, Career Development Centre,
R.M.K. College of Engineering and Technology,
India

K. Karthik[2]
Associate Professor,
Department of Mechanical Engineering,
Vel Tech Rangarajan Dr. Sagunthala R&D Institute of Science Technology,
Chennai, India

Parthiban A.[3]
Assistant Professor, Department of Mechanical Engineering,
Chennai Institute of Technology,
Chennai, India

Karthick M.[4]
Assisstant Professor,
Department of Mechanical Engineering,
Veltech Rangarajan Dr Sagunthala R&D Institute of science and technology,
Chennai, India

P. Loganathan[5]
Associate Professor,
Department of Mechanical Engineering,
Karpaga Vinayaga College of Engineering and Technology,
India

Damodharan V.[6]
Research Scholar, Department of Mechanical Engineering,
Hindustan Institute of Technology and Science,
Chennai, India

P. Usha Rani[7]
Professor, Science and Humanities Department,
R.M.D Engineering College,
India

◆ **Abstract:** This study investigates the basic principles and finishing characteristics of unbonded magnetic abrasives in cylindrical magnetic abrasive finishing (MAF). The research centers on the behavior and performance of unbonded magnetic abrasives, which consist of a mechanical blend of SiC abrasives and ferromagnetic particles, using SAE30

[1]drcbibin@gmail.com, [2]karthikk@veltech.edu.in, [3]parthimech2810@gmail.com, [4]karthick@veltech.edu.in, [5]logu2685@gmail.com, [6]damuknrcl@gmail.com, [7]pusharani71@yahoo.com

DOI: 10.1201/9781003685906-26

lubricant as a binder. The ferromagnetic particles utilized in the research are iron grit and steel grit, each having three particle sizes. These ferromagnetic particles were mixed individually with SiC abrasives of two sizes, 1.2 µm and 5.5 µm, to study their impact on the finishing process. One of the prime concerns of this research study is to study the finishing properties in terms of surface roughness and material removal efficiency, as well as the underlying mechanisms of these results. Experimental outcomes indicate that steel grit is superior in magnetic abrasive finishing compared to iron grit. This can be explained on the basis of higher hardness of steel grit and polyhedral shape of the particle, which facilitates the cutting and finishing action of the process. The research work also investigates variations in material properties on the work surface prior to and subsequent to the finishing operation. One of the most striking results evident is the increased silicon content on the finished surface when steel grit is mixed with SiC abrasives. However, despite the improved surface finish, corrosion resistance of the finished material is found to decrease due to the finishing operation with steel grit. These results are of valuable insight on the optimization of the choice of abrasive material for the attainment of desired finishing properties in cylindrical MAF applications.

◆ **Keywords:** Finishing characteristics, Unbonded magnetic abrasives, Ferromagnetic particle, Material removal, Surface roughness, Magnetic abrasive finishing

1. Introduction

MAF is a highly advanced surface finishing process that can be used very effectively to advance the surface quality of most mechanical components. Unlike traditional finishing processes that are limited by the restrictions regarding the kind of materials and shapes they can finish, MAF is highly versatile and can be used for a very wide range of workpieces imposing just a few shape limitations. The versatility of the process makes it a very appealing candidate for precision finishing of intricate geometries in aerospace, automobile, and medical device industries. The process is founded on the use of magnetic abrasives, which interact with the imposed magnetic field to produce a finishing action on the workpiece surface. One of the most robust benefits that MAF, or Magnetorheological Abrasive Finishing, has to provide is its astounding ability to improve the surface finish to a very high extent within a very short period of processing time. To explain this further, consider the fact that an original roughness value of around 0.25 µm Ra can be brought down to as low as 0.05 µm Ra impressively within minutes' worth of processing time. Such a marvelous extent of fine-tuning is achieved without the formation of a degraded layer or the introduction of micro-cracks, which tend to be inherent flaws with conventional mechanical finishing operations. As a result, MAF delivers a quality surface finish without compromising the material integrity of the workpiece during the operation. Moreover, the MAF process does not introduce any residual stresses trapped within or lead to surface damage, and therefore it is a highly effective choice for high-precision applications, especially where maintaining the inherent properties of the material is of critical concern. Compared to other abrasive finishing tools requiring periodic dressing or compensation or maintenance, the MAF process is mostly self-sustaining. The self-sharpening feature is the one accountable for sustaining the abrasives' cutting capability during the finishing process, thereby lowering the frequency of replacement to a minimum. Additionally, the self-adaptability of magnetic abrasives renders finishing of different surface profiles with uniformity, hence placing MAF highly versatile for components of complex geometries. The process controllability is yet another vital benefit, whereby process parameters such as magnetic field intensity, abrasive composition, and finishing time can be controlled by operators to realize desired surface qualities. Owing to these benefits, MAF has been successfully implemented in a broad spectrum of products and industries. Recent studies have demonstrated its effectiveness in finishing components of diverse materials such as metallic, ceramic, and composite materials. Researchers have utilized it in precision manufacturing, medical implants, and microelectromechanical systems (MEMS) to further attest to its versatility and effectiveness. Nevertheless, the finishing characteristics of MAF rely mostly on the type and composition of magnetic abrasives utilized in the process.

Magnetic abrasives hold a prestigious place as far as the control of efficiency and quality concerning the finishing process in various applications is concerned. Such specialized magnetic abrasives typically consist of ferromagnetic particles (FP) expertly mixed with other abrasives like silicon carbide (SiC) or aluminum oxide (Al2O3) to achieve specific results. For instance, hard iron alloys have been widely employed as magnetic abrasives for treatment of softer alloys and nonferrous metals and have been shown to be of great use in various applications. On the contrary, by compounding ferromagnetic particles with abrasives like SiC or Al2O3, it is possible to achieve the desirable surface refinement during the use of hard materials, which is highly important to reach maximum performance during finishing operations. The performance, efficiency, and effectiveness of these specialized abrasives are controlled by a range of factors like particle size, particle shape, and particle hardness, among other things, which in turn have direct effects on the MRR and the achievable surface quality produced during the finishing process. The magnetic abrasives can be made in a number of forms from sintered abrasives to unbonded magnetic abrasives (UMA), each exhibiting different applications within industrial processes. Sintered magnetic abrasives are produced in a bonding process where abrasives are attached to a ferromagnetic matrix via processes like sintering, chemical bonding, or other complex production processes based on the utilization of state-of-the-art technology. This production process in question results in the formation of a composite abrasive material with a homogeneous and uniform structure, which is highly beneficial in the finishing process. Because of their superior performance capabilities in the finishing sector, sintered magnetic abrasives have been utilized and preferred by scientists and the industry for quite a long time. However, it should be noted that sintered magnetic abrasives' production process is highly complicated with massive capital inputs. The process of production needs high-temperature and high-pressure treatments carried out in an inert gas environment that is needed in order to conserve the integrity of the sintered material in the course of development. Once the sintering process has been completed, mechanical crushing and grinding processes are carried out in order to achieve the desired final particle size distribution, which helps ensure the magnetic abrasives achieve industry standards. These subsequent operations are responsible for the overall boost in the production cost, hence making sintered magnetic abrasives a top-of-the-line industrial product, which is comparatively more costly in the market.

Because of the expensive nature of sintered magnetic abrasives, much research has been conducted to investigate the application of unbonded magnetic abrasives (UMA) as a more affordable alternative that can be utilized for the same purposes. In contrast to sintered abrasives that require a sophisticated and expensive manufacturing process. Since the abrasive particles are not physically bonded onto the ferromagnetic material, they are more mobile within the magnetic field during finishing. Unhampered movement of SiC abrasives through the narrow gaps between adjacent ferromagnetic particles enables a uniform and effective finishing action. The ability of UMA to provide finish performance comparable to that of sintered magnetic abrasives is a notable cost-saving potential in production without sacrificing the achievement of high-quality surface finishes. Research has already been presented to explore the potential of using UMA in finishing operations. Fox et al., for example, investigated the application of UMA in finishing non-magnetic stainless steel. Their results, however, presented a drastic increase in surface roughness after finishing in contrast to the empirical results of other studies. The variation suggests the need to conduct further research into the parameters that affect UMA performance, including variations in abrasive composition, ferromagnetic particle size, and process conditions used.

In this comprehensive research, our main aim is to clearly show that Ultra Magnetic Abrasives (UMA), which are specifically a blend of steel grit blended with silicon carbide (SiC) abrasives, can be formulated in such a manner that they have finishing properties akin to those of sintered magnetic abrasives. By optimizing the choice of the composition and the size of abrasives used, we will attempt to attain the highest degree of finishing action possible while, at the same time, measuring the resulting surface properties obtained from this process. This research will give valuable information on the underlying mechanisms of the finishing action of UMA and further establish its potential as a cost-effective alternative to sintered magnetic abrasives as part of Magnetic Abrasive Finishing (MAF) processes.

Our experimental approach is centered on the synthesis of UMA, which involves the accurate mixing of steel grit with SiC abrasives of different particle sizes to form a good abrasive mixture. The finishing action of the resulting abrasives will be extensively characterized through an investigation of the improvement in Ra, MRR, and general surface integrity formed. Furthermore, we will carry out a comprehensive investigation of the impact of important process constraints, which include magnetic field strength, finishing time, and abrasive concentration, to give a better understanding of their effect on the finishing action.

By a systematic experimentation and analysis approach, we will clearly show the ability of using UMA as a cost-effective and viable alternative to conventional precision surface finishing processes. Furthermore, we will carry out an extensive comparison of our research with existing literature in the field of MAF with the use of sintered magnetic abrasives, thus enabling an in-depth comparison of the relative advantages and disadvantages associated with UMA. Through an integrated analysis of the finishing properties observed, our aim is to provide valuable knowledge to the precision surface finishing community and establish a solid foundation for future research associated with the development of cost-effective magnetic abrasive materials.

2. Experimental Methodology

The system was carefully designed to facilitate the rotation of the workpiece while simultaneously allowing for the introduction of axial vibration. Rotation of the workpiece itself was facilitated through the use of a brushless DC motor, which provided the advantage of closed-loop speed control and ensured uniform rotation throughout the entire finishing process. Axial vibration, on the other hand, was generated through the utilization of an eccentric cam mechanism. This specific CAM was driven by an induction motor, which was well coupled with a frequency converter to offer flexible control over both the vibration frequency and its amplitude. The critical magnetic field required for the finishing process was generated through the series connection of two magnetically excited coils. These coils were wound carefully using copper wire with a diameter of 1.0 mm, realizing an impressive number of around 2150 turns for each individual coil. This carefully designed setup facilitated the generation of a stable and robust magnetic field, which is critical for conducting effective MAF. To improve the overall efficiency of the magnetic field, soft iron was specifically chosen as the pole and magnetic core material, owing to its high relative magnetic permeability. This strategic choice ensured efficient conduction of magnetic flux, which is critical for aligning and guiding the magnetic abrasives effectively during the finishing process.

To ensure a high-quality finishing performance to the required standards, an abrasive slurry was carefully supplied into the working gap in finishing. The slurry in question was silicon carbide (SiC) abrasive (SA) particles uniformly dispersed in distilled water to form a good mixture. The abrasive slurry played several vital roles during finishing, which are essential for optimum performance. First, the slurry played a cooling function, efficiently dissipating the friction heat generated as a direct consequence of friction between the abrasives and workpiece surface under finishing. Second, the slurry played a lubrication function, which was essential in avoiding excessive friction, which may lead to undesirable wear and deteriorization of the workpiece. Thirdly, the specially designed slurry continuously supplied the SiC abrasives, thus ensuring a constant abrasives concentration throughout the entire experimental duration. The slurry was supplied to the system by a micro tube pump, ensuring a constant and uniform flow rate, while an electric stirrer ensured constant agitation of the mixture in order to effectively prevent settling of the abrasive particles in the slurry. Preparation of the abrasive mixture was a crucial process that was performed to achieve uniform dispersion of the abrasive particles and to provide effective finishing performance. Mixing of the liquid was done by completely mixing the SA and FP with the lubricant until uniform, homogeneously muddy mixture was obtained. After preparing this uniform mixture, it was fed carefully into the working gap between the rotated workpiece and the magnetic field.

The major role of the SAE30 lubricant was not exclusively lubrication. Its viscosity was most important to the function of holding the position of the SiC abrasives on the ferromagnetic particles.. Binding was important to avoid dispersal of the SA particles into the ambient air during finishing. In the absence of the lubricant, abrasive particles can become airborne, resulting in non-uniform finishing and possible contamination of the test environment. In the course of the experimental process, two SiC abrasives were employed to study their influence on finishing characteristics. The two used were #8000 and #2500 SA, with mean particle diameters of 1.2 μm and 5.5 μm, respectively. These particular particle sizes were selected based on earlier work that had established their suitability in providing ultra-precision surface finishes.

In addition to selecting appropriate SA grades, different ferromagnetic particles of different sizes were incorporated into the UMA mixture. Steel grit and Iron grit were selected considering their intrinsic magnetic and mechanical properties, which influence finishing performance. The experimental setup was tailored to study systematically the enhance of changing parameters on finishing execution. Each test was performed under controlled conditions for repeatability and measurement accuracy of the factors of finishing performance, such as reduction in Ra and MRR. By systematically designing and conducting the experimental procedure, the current investigation attempted to study in-depth the effectiveness of UMA in magnetic abrasive finishing and compare the performance with bonded magnetic abrasives. The experimental result provided valuable information on finishing behavior, material interactions, and effectiveness of UMA in precision surface finishing operations.

3. Results and Discussion

The analysis reveals that the combination of 180 μm iron particles with 1.2 μm SA results in improved surface roughness. However, even after 30 minutes of processing, the roughness only improves from 0.19 μm to 0.12 μm Ra. The optimal surface roughness is achieved using 180 μm steel grit combined with 1.2 μm SA. Within just 5 minutes, the Ra improves to 0.06 μm Ra, reaching a saturation level of 0.0392 μm Ra after 15 minutes.

When mixed with 5.5 μm SA, material removal increases, but surface roughness deteriorates. This occurs because the presence of more 1.2 μm SA particles under the FP reduces the average abrasion pressure from each FP particle, leading to improved surface smoothness. Conversely, the higher abrasion pressure exerted by each 5.5 μm SA particle results in deeper scratches, thereby increasing MRR but worsening Ra. Second, regardless of the SA size, using larger FP particles not only enhances MRR but also improves Ra.. Additionally, larger FP particles contain more vacancies, allowing them to carry more SA particles. Consequently, given equal weight, larger FP particles accommodate a higher total volume of SA particles, leading to a broader contact area on the work surface. Thus, while larger FP particle sizes generate stronger magnetic forces, the average abrasion pressure exerted by each SA particle remains relatively low.

Under identical finishing circumstances, steel grit demonstrates superior performance compared to iron grit for the following reasons: Steel grit has a hardness of HV 780–940, whereas iron grit has a hardness of HV 110–125. Due to its superior hardness, steel grit effectively cuts the HRC55 (HV600) workpiece. The nearly spherical shape of iron grit, promotes rolling during the finishing process. Steel grit, has a polyhedral shape with sharp edges and vertices. These sharp cutting edges, combined with the flat surfaces of the polyhedron, distribute abrasion pressure uniformly over a broader area, improving the finishing effect.

The outcomes, indicate that Ra values obtained using 181 μm or 129 μm steel grit alone are equivalent to those achieved with 181 μm steel grit mixed with 1.3 μm SA. This demonstrates that steel grit alone can effectively finish workpieces with lower hardness. The findings indicate that the best performance in terms of both surface roughness and material removal is achieved using 180 μm steel grit. Material removal initially increases when using steel grit alone and surpasses that observed when mixed with 1.2 μm SA. This interaction between the magnetic field and cutting resistance reduces both the abrasion area and relative motion, limiting the cutting ability of individual SA particles. As a result, the total material removal using steel grit alone is greater than when mixed with 1.2 μm SA.

However, after 25 minutes of processing, surface roughness improved to 0.05 μm Ra. When 1.2 μm SA was added, the same excellent Ra was achieved. it was observed that with 5.5 μm SA, material removal doubled on the HRC61 workpiece, while using 1.2 μm SA led to a 60% increase. While steel grit effectively cuts the HRC55 workpiece, the increased cutting resistance occasionally causes rolling. However, on the HRC61 workpiece, the shallower cutting depth results in lower cutting resistance, reducing the tendency for steel grit to roll. This allows SA particles to fully develop their abrasion capabilities, leading to improved surface roughness and material removal.

Micro-Vickers hardness measurements taken before and after finishing, indicate no significant changes, regardless of the presence of SA. Cross-sectional observations using secondary and backscattered electron images from an EPMA apparatus confirm the absence of a deteriorated layer, which typically affects mechanical properties.

To determine any compositional changes on the finished surface, qualitative and quantitative analyses of Si and C were performed using EPMA. The Si intensity increased only when SA was used. The initial Si content of 0.2% increased to 1.0–2.35% when 1.2 μm SA was added. Given the low processing temperature, diffusion-induced Si content increase is unlikely. Visual inspections revealed mirror-like surfaces after finishing, with a noticeable color change when SA was used, turning the surface dark gray. This suggests that sub-micron SiC fragments became embedded in the surface. However, high-magnification cross-section images indicate that this layer is extremely thin. Steel grit-finished surfaces exhibited minimal corrosion rate variation, whereas surfaces finished with SA experienced an increased corrosion rate. The incorporation of SiC fragments led to slight residual compressive stress, aligning with prior research findings [12,13]. Promoting anodic behavior and accelerating stress corrosion [14], ultimately reducing corrosion resistance.

4. Conclusion

The findings of this study can be reviewed in the following key points:

- Steel grit has been identified as a highly effective abrasive material for magnetic abrasive finishing due to its superior hardness and polyhedron-shaped structure. These characteristics enhance its ability to effectively abrade the surface, resulting in improved material removal and surface quality compared to other abrasives.
- The research findings indicate that when the FP (Finishing Powder) particle size remains within an optimal limit, increasing its size contributes to greater material removal while also improving the resulting surface roughness. However, to achieve the best surface finish, it is recommended to use a smaller SA (Supporting Abrasive) particle size, as finer

abrasives help in minimizing surface irregularities and enhancing smoothness.

- When using a finer SiC (Silicon Carbide) abrasive with a particle size of 1.2 μm, the best surface roughness value of 0.042 μm Ra was achieved, provided that it was mixed with 180 μm steel grit. This mixture permitted very effective polishing. When a coarser SiC of 5.5 μm particle size was used, however, it provided a higher cutting depth for each particle. Although this led to more material removal, it was also harmful to the surface roughness, creating a more rough finish.
- The hardness of the workpiece was a significant parameter in determining the efficiency of the finishing process. The study showed that for a workpiece hardness of HRC61, the addition of SiC abrasive provided improved surface roughness and material removal compared to that of a workpiece of lower hardness, HRC55. This suggests that harder materials are more responsive to the used finishing process, perhaps due to the ability of the harder materials to withstand the abrasive forces without excessive deformation.
- Pre-finishing and post-finishing Micro-Vickers hardness tests showed no considerable distinction in surface hardness. This suggests that the finishing does not introduce considerable changes in the material's mechanical properties. Additionally, Electron Probe Micro-Analysis (EPMA) analysis also showed that cross-section of finished surface showed no visible deterioration or damage, which suggests that the finishing does not harm the integrity of the workpiece.
- When the steel grit alone was applied as the abrasive medium, the silicon (Si) contents were not markedly different, and there were no shifts in corrosion resistance. When SA was introduced to the finishing process, however, the Si content on the surface was notably boosted. Although it increased, the corrosion resistance of the material declined, which suggests that the introduction of SA has some influence on the chemical content of the surface such that it is susceptible to corrosion. The findings can serve as a reference to select the appropriate finishing conditions to achieve desired surface properties in various industrial processes.

References

1. K. Tsuchiya, Y. Shimizu, K. Sakaki, M. Sato, Polishing mechanism of magnetic abrasion, Journal of the Japan Institute of Metals 57 (11) (1993) 1333–1338.
2. P. Jayakumar, S. Ray, V. Radhakrishnan, Optimising progress parameters of magnetic abrasive machining to reduce the surface roughness value, Journal of Spacecraft Technology 7 (1) (1997) 58–64.
3. H. Yamaguchi, T. Shinmura, K. Kuga, New internal finishing process applying magnetic abrasive machining, Transactions of the Japan Society of Mechanical Engineers Part C 62 (600) (1996) 3313–3319.
4. T. Shinmura, H. Yamaguchi, Study on a new internal finishing process by the application of magnetic abrasive machining, JSME International Journal, Series C 38 (4) (1995) 798–804.
5. M.D. Krymsky, Magnetic abrasive finishing, Metal Finishing 91 (7) (1993) 21–25.
6. T. Shinmura, K. Takazawa, E. Hatano, Study on magnetic abrasive finishing, Bull. Japan Soc. Prec. Engng. 21 (2) (1987) 139–141.
7. T. Shinmura, K. Takazawa, E. Hatano, M. Matsunaga, Study on magnetic abrasive finishing, Annals of the CIRP 39 (1) (1990) 325–328.
8. M. Fox, K. Agrawal, T. Shinmura, R. Komanduri, Magnetic abrasive finishing of rollers, Annals of the CIRP 43 (1) (1994) 181–184.
9. H. Yamaguchi, T. Shinmura, Study of the surface modification resulting from an internal magnetic abrasive process, Wear 225-229 (1999) 246–255.
10. H. Yamaguchi, T. Shinmura, Study of an internal magnetic abrasive finishing using a pole rotation system, Precision EngineeringJournaloftheInternationalSocietiesforPrecision Engineering and Nanotechnology 24 (2000) 237–244.
11. T. Shinmura, T. Aizawa, Study on internal finishing of a nonferromagnetic tubing by magnetic abrasive machining process, Bull. Japan Soc. Prec. Engng. 23 (1) (1989) 37–41.
12. T. Shinmura, Study on magnetic–abrasive finishing, Journal of JSPE 53 (11) (1987) 1791–1793.
13. P.I. Yascheritsin, L.E. Sergeev, The comparative appraisal of quality characteristics of holes after different finishing methods, Advanced Performance Materials 4 (3) (1997) 337–347.
14. D.R. Askeland, The Science and Engineering of Materials, Brooks/Cole Engineering Division, Monterey, CA, 1985.

Advances in Additive Manufacturing Technologies – Gurusamy Pathinettampadian et al. (eds)
© 2026 Taylor & Francis Group, London, ISBN 978-1-041-16687-0

Fine-Tuning Wire Electrical Discharge Machining Parameters for Titanium Alloys

C. Bibin[1]
Professor, Career Development Centre,
RMK College of Engineering and Technology,
Chennai, India

K. Karthik[2]
Associate Professor,
Department of Mechanical Engineering,
Vel Tech Rangarajan Dr. Sagunthala R&D Institute of Science Technology,
Chennai, India

Parthiban A.[3]
Assistant Professor, Department of Mechanical Engineering
Chennai Institute of Technology
Chennai, India

S. Sridhar[4]
Associate Professor, Department of Mechanical Engineering
PSNA College of Engineering and Technology,
Chennai, India

P. Loganathan[5]
Associate Professor,
Department of Mechanical Engineering,
Karpaga Vinayaga College of Engineering and Technology,
India

P. Usha Rani[6]
Professor, Science and Humanities Department,
R.M.D Engineering College, India

Vinoth Kumar M.[7]
Associate Professor, Department of Mechanical Engineering
Hindustan Institute of Technology and Science,
Chennai, India

◆ **Abstract:** This research paper investigates the effect of process parameters on the electrical discharge machining (EDM) of Ti-6Al-4V. The considered process constraints include pulse-on time, discharge current, and pulse-off time. Different types of dielectric fluids were used: deionized water, drinking water, and a mixed dielectric consisting of 25% deionized water and 75% drinking water. Experiments were conducted using a Taguchi L27 (3^13) orthogonal array. The performance measures evaluated were material removal rate (MRR) and surface roughness (Ra). Taguchi

[1]drcbibin@gmail.com, [2]karthikk@veltech.edu.in, [3]parthimech2810@gmail.com, [4]sri_2855@yahoo.co.in, [5]logu2685@gmail.com, [6]pusharani71@yahoo.com, [7]vinothmecho@gmail.com

DOI: 10.1201/9781003685906-27

methodology was employed to optimize the process parameters. The analysis revealed that using drinking water as the dielectric fluid resulted in the highest MRR and the lowest Ra. The maximum MRR achieved was 5.46 mm^3/min, while the minimum surface roughness recorded was 2.53 μm.

◆ **Keywords:** EDM, Ti alloy, Discharge current, Pulse on time, Pulse off time, Dielectric fluid

1. Introduction

EDM is a non-traditional machining process that operates on the principle of electro-thermal energy conversion. In this process, electrical energy is utilized to generate electrical sparks, which facilitate material removal from the workpiece [1]. The primary objective of EDM is to machine materials by eroding unwanted portions, thereby achieving the desired shape and dimensions [2]. The machining process relies on the generation of high thermal energy, which is a result of the electrical energy supplied to the system [3]. This thermal energy, combined with the electrical energy [4], leads to localized melting and vaporization of the workpiece, enabling precise material removal [5].

The materials used in EDM must exhibit high strength and temperature resistance [6] while also possessing electrical conductivity. The workpiece and the tool electrode used in EDM must both be electrically conductive to facilitate the spark erosion process [7]. During the machining operation, both the workpiece and the tool are absorbed in a dielectric medium [8], which can be deionized water, kerosene, or other dielectric fluids [9,10]. The reason of the dielectric fluid is to act as an insulator under normal conditions and to ionize when subjected to a high voltage, thereby allowing controlled discharge to occur.

In EDM, the workpiece and the tool are maintained at a specific gap distance to enable the generation of electrical discharges [11]. A potential difference, or voltage, is applied between the workpiece and the tool to create the necessary conditions for spark generation [12,13]. The intensity of the electric field established between the workpiece and tool depends on the applied voltage and the gap distance maintained between them [14,15]. Typically, the EDM system consists of two terminals: a positive terminal and a downbeat terminal. The tool electrode is joined to the downbeat terminal of the generator, while the workpiece is connected to the convinced terminal [16].

Due to the presence of an electric field between the tool and the workpiece, free electrons are generated within the tool electrode. These free electrons experience electrostatic forces due to the applied potential difference [17]. If the bonding energy between the electrons and the tool material [18] is low, some electrons are emitted from the tool surface, a phenomenon known as cold emission. These emitted electrons travel toward the workpiece through the dielectric medium, gaining velocity and energy during their movement. As these high-energy electrons approach the workpiece [19], they collide with dielectric molecules present in the gap, initiating ionization. The extent of ionization depends on the bonding energy of the molecules and electrons in the dielectric fluid.

As a result of these collisions, the number of free positive and negative ions in the dielectric medium increases significantly [20]. This enhances the overall concentration of charged particles within the gap amoungthe workpiece and the tool [21]. The zone in which this charged particle interaction occurs is known as the plasma channel, which has an extremely low electrical resistance [22]. At this stage, a large number of ions flow and electrons between the workpiece and the tool [23], leading to what is known as the avalanche motion of electrons. This movement of electrons and ions creates an electrical spark, and the energy associated with the spark is dissipated in the form of intense heat.

The high-velocity electrons impact the surface of the workpiece, while the positive ions strike the tool electrode. The kinetic energy of these charged particles is transformed into thermal energy upon impact. This conversion of kinetic energy into heat flux results in a highly localized and intense temperature rise at the point of impact, often exceeding 10,000°C [24]. This extreme and instantaneous temperature increase causes rapid melting and vaporization of the workpiece material, leading to its removal. A Graphic diagram illustrating the EDM setup used in this study is provided in Fig. 27.1, demonstrating the fundamental working principles and operational aspects of the EDM process.

2. Experimental Methodology

The workpiece material selected for this investigation is Ti-6Al-4V, a titanium alloy broadly used in biomedical, and high-performance engineering demands due to its corrosion

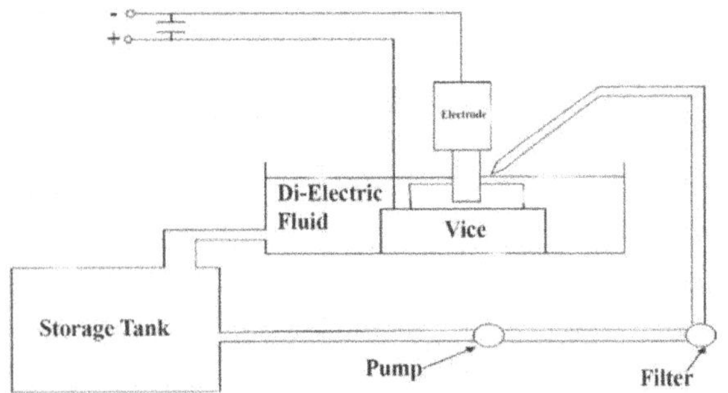

Fig. 27.1 Graphic diagram of EDM machine

resistance, good strength-to-weight ratio, and high-temperature stability. Ti-6Al-4V consists of titanium (Ti) as the base metal, with aluminum (Al) and vanadium (V) as alloying elements that enhance its mechanical properties. The detailed chemical composition and mechanical properties of Ti-6Al-4V are presented in Tables 27.1 and 27.2, respectively. In this study, Cu was used as the electrode material. Cu is commonly chosen as an electrode in EDM due to its extreame electrical conductivity, wear resistance, and good thermal conductivity. It ensures stable and efficient spark generation, contributing to better MRR and reduced electrode wear.

Table 27.1 Chemical composition of Ti alloy

Element	C	Al	V	N	O	Fe	H	Ti
Wt%	0.05	5.9	3.7	0.05	0.13	0.25	0.01	Bal.

Table 27.2 Properties of BM

Property	Quantity
Hardness (HRC)	33
Density (g/cm^3)	4.3
Melting Point (°C)	1638
Specific heat (J/kg°K)	550
UTS (MPa)	986
Thermal Conductivity (W/m°K)	7.4

The dielectric fluid shows a crucial role in the EDM process by acting as an insulating medium that prevents premature discharges and by flushing away debris formed during machining. In this research, three various dielectric fluids were used: (1) drinking water, (2) deionized water, and (3) a mixed combination consisting of 25% deionized water and 75% drinking water. The characteristics and properties of these dielectric fluids are crucial for evaluating their effect on the EDM process. The dielectric fluid features of drinking water, along with the individual properties of deionized water, and drinking water are presented in Tables 27.3–27.5.

Table 27.3 Features of drinking water

Characteristics	Value
Apearance	Colourless
Specific gravity @ 30°C	1
Pour Point (°C)	3
Viscosity	0.74
Copper corrosion	<1

Table 27.4 Drinking water properties

Properties	Value
Total suspended solids	10 mg/l
Total dissolved solids	98 mg/l
Dissolved O_2	8 mg/l
Total solids	120 mg/l
Chlorides	21 mg/l
Dissolved CO_2	10 mg/l
Alkalinity	35 mg/l
Sulphate	31 mg/l
Hardness	52.3 mg/l
pH	6.25

Table 27.5 Deionized water properties

Properties	Value
Chloride	7.2 mg/l
Alkalinity	0 mg/l
pH	5.1
Hardness	20.5 mg/l

The EDM machine was configured to operate under controlled conditions, ensuring repeatability and accuracy in the machining process. The EDM machine was configured to run under controlled conditions to provide precision and consistency of machining operation. Most critical EDM process parameters identified to be of highest importance in this research are: Discharge Current (I), The magnitude of the current determines the intensity of the generated spark and has a direct effect on the MRR and Ra. Pulse-on Time (Ton), This parameter controls the amount of time through which the spark is on and helps in material removal. Pulse-off Time (Toff), The magnitude of time through which no spark is generated, thus allowing flushing of debris and avoiding excessive heating., Dielectric Fluids, Choice of the dielectric medium affects stability of the spark, cooling, and effectiveness of flushing. The range of operation of these selected process constraints used in the current investigation is discussed in Table 27.6. The

Table 27.6 Working range and process parameters

Parameters of Dielectric fluid	Levels		
	Deionised water	Drinking water	Mixed water
Discharge current (DC)	10	15	20
Pulse on time (PT)	20	40	60
Pulse off time (POT)	20	30	40

selected range includes extensive analysis of how these parameters' variation impacts the machining performance of Ti-6Al-4V.

To systematically study the effect of these process constraints on EDM performance, an L27 (3^13) orthogonal array was chosen as the experimental design. Taguchi's method was used to optimize machining parameters by finding the independent contribution of each factor. EDM process responses were quantified in terms of two performance measures, MRR, Ra. MRR and Ra were both determined for all experimental conditions, and the values were recorded in Table 27.7 in a systematic manner. To analyze experimental data, MINITAB 17 software was used to create response tables and graphical plots of the signal-to-noise (S/N) ratios of MRR and SR. Taguchi's method allowed unmistakable identification of individual impacts of each process parameter, and it gave insight into their relative importance in optimizing machining performance [15].

For the identification of most significant process parameters affecting EDM performance, analysis of variance (ANOVA) was performed. ANOVA is employed for the identification of factors affecting response variables most through the assessment of statistical significance of each parameter. Through ANOVA, optimum process parameter levels were determined in order to gain maximum MRR and minimum SR. Through the Taguchi approach, in the present work, process parameters of EDM of Ti-6Al-4V were optimized systematically. The outcomes are helpful in providing information concerning the performance of different dielectric fluids and machining conditions, and provide helpful suggestions to improve EDM performance in real-world applications [16].

The S/N ratio approach was applied, where a larger S/N ratio was considered better for material removal rate (MRR). This would mean that as the S/N ratio is increased, the better the EDM process is in achieving MRR. The S/N ratio method utilized the "lower is better" rule for SR. A low S/N ratio would result in a smoother finish, which is what is desirable in EDM machining [17].

Relative significance of each process parameter was determined by analysis of variance (ANOVA). Percentage

Table 27.7 Average outcomes of MRR and SR

Exp. No	Dielectric	DC	PT	POT	MRR	SR
1	Deionized	10	20	20	1.72	2.27
2	Deionized	10	40	30	2.21	2.53
3	Deionized	10	60	40	2.64	3.04
4	Deionized	15	20	30	2.52	2.49
5	Deionized	15	40	40	4.11	3.51
6	Deionized	15	60	20	3.32	3.26
7	Deionized	20	20	40	3.51	3.51
8	Deionized	20	40	20	4.34	2.87
9	Deionized	20	60	30	4.49	3.66
10	Drinking	10	20	20	2.31	2.71
11	Drinking	10	40	30	2.62	3.01
12	Drinking	10	60	40	2.63	2.99
13	Drinking	15	20	30	3.83	2.59
14	Drinking	15	40	40	4.18	2.68
15	Drinking	15	60	20	4.11	3.01
16	Drinking	20	20	40	4.32	2.89
17	Drinking	20	40	20	4.55	3.12
18	Drinking	20	60	30	5.46	3.18
19	Mixed	10	20	20	1.95	2.87
20	Mixed	10	40	30	2.45	3.03
21	Mixed	10	60	40	2.83	3.21
22	Mixed	15	20	30	3.07	2.73
23	Mixed	15	40	40	3.40	2.92
24	Mixed	15	60	20	3.64	3.41
25	Mixed	20	20	40	3.77	3.09
26	Mixed	20	40	20	4.48	3.51
27	Mixed	20	60	30	4.65	3.52

contribution of each parameter, which reflects its relative impact on the results of the EDM process. Significant effect of an individual process parameter on the concerned response variable is represented by high percentage contribution. ANOVA was used to find the statistical implication of foremost factors affecting MRR and SR. Percentage contributions of these process parameters were investigated to analyze their impact on MRR and SR in a methodical manner [18].

3. Results and Discussion

ANOVA was carried out to investigate the effects of different process constraints on the various performance parameters of Electrical Discharge Machining (EDM). Relative significance of process parameters was determined by percentage contribution of individual parameters. The outcome of ANOVA suggests that the dielectric fluid (A), discharge current (B), and pulse-on time (C) have the maximum influence on Ra and MRR. The outcome further suggests that discharge current (B) and pulse-on time (C) contribute significantly more in percentage terms than pulse-off time (D) [19].

The findings indicate that for the use of deionized water and drinking water as dielectric fluids, the MRR is initially enhanced but eventually decreased in the mixed form (25% deionized water + 75% drinking water). This occurs because the thermal conductivity is reduced. Once again, as the discharge current (I) is increased from 10A to 20A, the MRR is greatly enhanced. Since discharge current decides the input of heat energy for material removal directly, the increase in discharge current tends to an increase in heat energy. Therefore, MRR is greatly enhanced when discharge current is increased from 10A to 20A. The B contributes approximately 79% to the MRR, making it the most influential factor. The amount of material removed is directly proportional to the energy given, reinforcing the importance of discharge current as the primary contributing factor [20].

As the Ton increases from 24 ms to 61 ms, the MRR exhibits a near-nonlinear increase. The contribution of pulse-on time to MRR is found to be approximately 10%. Pulse-off time, however, is identified as the least significant factor, with a minimal contribution of around 0.7%. Observations indicate that as T-off time excessive from 23 ms to 47 ms, the improvement in MRR is marginal. At the minimum pulse-off time of 24 ms, the dielectric fluid has insufficient time to deionize and effectively flush away debris, thereby influencing the overall efficiency of the machining process [21].

The analysis shows that when using drinking water and deionized water as dielectric fluids, Ra initially lesser but then increases in the mixed. This trend is attributed to the lower thermal conductivity of water. In the case of negative polarity, electron bombardment occurs from the workpiece to the electrode. At a discharge current of 10A, the material deposition is relatively lower, resulting in a smoother surface [22].

The influnce of dielectric fluid to SR is estimated to be 12%, making it the third most influential factor. Discharge current contributes approximately 23% to SR, ranking as the second most important factor. Observations indicate that as pulse-off time increases from 24 ms to 48 ms, SR decreases. At the minimum pulse-off time of 24 ms, the dielectric fluid has less time to deionize and remove debris, affecting the stability of the machining process [23].

The optimization of machining performance evaluation parameters is crucial in achieving the best possible EDM outcomes. The findings confirm that selecting the appropriate combination of discharge current, dielectric fluid, and pulse-on time significantly enhances MRR and minimizes SR, ultimately improving the overall EDM efficiency [24].

4. Conclusion

In the current research study, Process parameter optimization was done with the Taguchi method, where the most contributing factors in MRR and SR were determined.

1. The experimental results showed that discharge current, pulse off time, pulse on time, and drinking water are key parameters in EDM performance. They directly affect the machining performance and the excellence of the machined surface.
2. It was noted that optimal values of process constraints for the purpose of achieving higher MRR and lessor SR were not the same. This reflects that material removal efficiency and surface finish quality are in some sort of trade-off relationship and proper process parameter selection based on specific machining goals is required.
3. In Surface Roughness, pulse on time was more dominant relative to the discharge current. In Material Removal Rate, the discharge current was more dominant relative to the pulse on time. These results show the necessity of controlling some parameters based on whether one would like to optimize material removal or enhance surface finish.
4. Interaction between the discharge current and the dielectric fluid was also seen to have a strong influence on surface roughness. This means that the dielectric fluid employed influences the way the discharge current influences the machining process and thereby influences the surface finish of the workpiece being machined.

The study was able to determine the significant parameters influencing EDM performance on Ti-6Al-4V alloy. The results are useful information to optimize EDM process parameters to achieve higher efficiency and improved surface quality, depending on the desired machining outcome

References

1. Afzaal Ahmed et al., A comparative study on the modelling of EDM and hybrid electrical discharge and arc machining considering latent heat and temperature-dependent properties of Inconel 718, Int. J. Adv. Manuf. Technol. 94 (5-8) (2018) 2729–2737.
2. A.P. Tiwary, B.B. Pradhan, B. Bhattacharyya, Investigation on the effect of dielectrics during micro-electro-discharge

machining of Ti-6Al-4V, Int. J. Adv. Manuf. Technol. 95 (1-4) (2018) 861–874.
3. Robert M. Jones, Mechanics of Composite Materials, CRC Press, 2018.
4. Khaled Bataineh, Ahmad Gharaibeh, Optimal design for sensible thermal energy storage tank using natural solid materials for a parabolic trough power plant, Solar Energy 171 (2018) 519–525.
5. Nishant K. Singh, Anand Poras, Electrical discharge drilling of D3 die steel using air assisted rotary tubular electrode, Mater. Today: Proceedings 5 (2) (2018) 4392–4401.
6. Kurt Amplatz, et al. Multi-layer braided structures for occluding vascular defects and for occluding fluid flow through portions of the vasculature of the body, U.S. Patent No. 9,877,710. 30 Jan. 2018.
7. Sagil James, Sharadkumar Kakadiya, Experimental study of machining of shape memory alloys using dry micro electrical discharge machining process, ASME 2018 13th International Manufacturing Science and Engineering Conference. American Society of Mechanical Engineers, 2018.
8. Anthony Wright et al., The influence of a full-time, immersive simulationbased clinical placement on physiotherapy student confidence during the transition to clinical practice, Adv. Simulation 3 (1) (2018) 3.
9. Welborn, Valerie Vaissier, Luis Ruiz Pestana, Teresa Head-Gordon, Computational optimization of electric fields for better catalysis design, Nat. Catal. (2018):
10. Yang Yang, Kartik Ramaswamy, Kenneth S. Collins, Steven Lane, Gonzalo Antonio Monroy, Lucy Chen, Yue Guo, and Eswaranand Venkatasubramanian, Plasma reactor with electron beam of secondary electrons, U.S. Patent Application 15/948,949, filed September 27, 2018.
11. Qiu Mingbo, et al., Energy distribution in cool electrode of electrical discharge machining based on wave-particle dualism, Mach. Sci. Technol. (2018) 1–15.
12. Bin Kan et al., Fine-tuning the energy levels of a nonfullerene small-molecule acceptor to achieve a high short-circuit current and a power conversion efficiency over 12% in organic solar cells, Adv. Mater. 30 (3) (2018) 1704904.
13. Simone Blayer et al., Accelerated process development and stockpile for MERS, Lassa Nipah Viral Vaccine (2018).
14. Detlef Loffhagen et al., Impact of hexamethyldisiloxane admixtures on the discharge characteristics of a dielectric barrier discharge in argon for thin film deposition, Contrib. Plasma Phys. 58 (5) (2018) 337–352.
15. Singla, Anuj, A.P.S. Sethi, Inderpreet Singh Ahuja, An empirical examination of critical barriers in transitions between technology push and demand pull strategies in manufacturing organizations, World J. Sci. Technol. Sustainable Development (2018).
16. Hadad, Mohammadjafar, Lan Quang Bui, Cong Thanh Nguyen, Experimental investigation of the effects of tool initial surface roughness on the electrical discharge machining (EDM) performance, Int. J. Adv. Manuf. Technol. 95.5-8 (2018) 2093–2104.
17. Tug˘rul Özel, Erol Zeren, Finite element modeling the influence of edge roundness on the stress and temperature fields induced by high-speed machining, Int. J. Adv. Manuf. Technol. 35 (3-4) (2007) 255–267.
18. Abbas, Norliana Mohd, Darius G. Solomon, Md Fuad Bahari, A review on current research trends in electrical discharge machining (EDM), Int. J. Mach. Tools Manuf. 47.7-8 (2007) 1214–1228.
19. Jin-Seong Park et al., Improvements in the device characteristics of amorphous indium gallium zinc oxide thin-film transistors by Ar plasma treatment, Appl. Phys. Lett. 90 (26) (2007) 262106.
20. Rudolph A. Marcus, On the theory of oxidation-reduction reactions involving electron transfer. I, J. Chem. Phys. 24 (5) (1956) 966–978.
21. S.I. Tkachenko et al., Distribution of matter in the current-carrying plasma and dense core of the discharge channel formed upon electrical wire explosion, Plasma Phys. Rep. 35 (9) (2009) 734.
22. R Lawrence Ives, Micro fabrication of high-frequency vacuum electron devices, IEEE Trans. Plasma Sci. 32 (3) (2004) 1277–1291.
23. K.S. Banker, A.D. Oza, R.B. Dave, Performance capabilities of EDM machining using aluminum, brass and copper for AISI 304L material, Int. J. Appl. Innov. Eng. Manage. 2 (2013) 186–191.
24. L. Selvarajan, Narayanan, C. Sathiya, Jeyapaul et al. Optimization of EDM Hole Drilling Parameters in Machining of MoSi2-SiC Intermetallic/Composites for Improving Geometrical Tolerances, J. Adv. Manuf. Syst. (14) 4 (2015) 259–272, World Scientific Publishing Company, DOI: 10.1142/S0219686715500171.

Note: All the figures and tables in this chapter were made by the authors.

Advances in Additive Manufacturing Technologies – Gurusamy Pathinettampadian et al. (eds)
© 2026 Taylor & Francis Group, London, ISBN 978-1-041-16687-0

28 Precision Optimization in Laser Beam Machining of Steel

Bharanidharan T.[1]
Student, Department of Mechanical Engineering,
Chennai Institute of Technology,
Chennai, India

K. Karthik[2]
Associate Professor,
Department of Mechanical Engineering,
Vel Tech Rangarajan Dr. Sagunthala R&D Institute of Science Technology,
Chennai, India

Yashwanth Kumar[3]
Student, Department of Mechanical Engineering,
Chennai Institute of Technology,
Chennai, India

C. Bibin[4]
Professor, Career Development Centre,
R.M.K. College of Engineering and Technology,
India

Vijayakumar K.[5]
Assistant Professor, Department of Mechanical Engineering,
Aarupadai Veedu Institute of Technology,
Chennai, India

B. Yokesh Kumar[6]
Assisstant Professor, Department of Mechanical Engineering,
Chennai Institute of Technology,
Chennai, India

Damodharan V.[7]
Research Scholar, Department of Mechanical Engineering,
Hindustan Institute of Technology and Science,
Chennai, India

◆ **Abstract:** The production of injection molds for plastic parts, car parts, and electrical devices makes extensive use of AISI P20 mold steel. A fiber laser beam was used to precisely process AISI P20 mold steel in this investigation. Cutting speed, gas pressure, and laser power were optimized as part of the experimental design that was constructed according to the Taguchi L27 model. The trials were planned and evaluated with Minitab software to find out how they affected the surface roughness (Ra) and kerf width, which are important response characteristics. For precise

[1]vtrbharani12@gmail.com, [2]karthikk@veltech.edu.in, [3]itsyashoff@gmail.com, [4]drcbibin@gmail.com, [5]vijayakumar@avit.ac.in, [6]yokeshkumarb@citchennai.net, [7]damuknrcl@gmail.com

DOI: 10.1201/9781003685906-28

machining response prediction, the MATLAB-based Adaptive Neuro-Fuzzy Inference System (ANFIS) model was created and deployed. The predicted values were experimentally validated, demonstrating that ANFIS provided more precise estimations than direct experimental measurements. Additionally, the Brute Force algorithm was applied to determine the optimal combination of machining parameters, ensuring the best possible machining quality. The Taguchi method identified cutting speed as the most influential factor affecting machining outcomes. Cutting at 1 m/min with 2 bar gas pressure and 1.8 kW of laser power produced the ideal surface roughness (Ra). Also, at 2 bar gas pressure, 1 m/min cutting speed, and 1.9 kW laser power, the narrowest kerf width was achieved. The minimal parameter combination that was achieved with the Brute Force method produced a surface roughness of 4.48175 μm and a kerf width of 0.84 mm. In order to assess the machining quality and laser-cut surface integrity, a microstructural study was performed on machined samples with varying levels of surface roughness.

◆ **Keywords:** AISI P20, LBM, Surface roughness, Kerf width, ANFIS, Brute force

1. Introduction

LBM has emerged as a revolutionary non-conventional machining process known for its high-speed material removal, precision, and non-contact operation. Unlike conventional machining techniques that involve physical tool wear and vibration, LBM provides prolonged tool life due to minimal mechanical interaction with the workpiece. This results in higher machining accuracy, improved surface quality, and reduced operational costs, making it a preferred choice over traditional machining techniques (Girdu and Gheorghe, 2022).

The process is particularly beneficial for machining complex technical materials, delicate substances, and both conductive and non-conductive materials. Additionally, LBM has demonstrated superior efficiency when working with lightweight and thin components that are otherwise challenging to machine using conventional methods (Chen et al., 2011). One of its key advantages is minimal material loss due to its narrow kerf width, which ensures precise cuts with reduced wastage (Nguyen et al., 2021). Furthermore, LBM produces smooth cutting edges with negligible mechanical deformation and can seamlessly integrate with CNC systems, enabling the machining of intricate profiles with high precision (Rao et al., 2005).

However, its machining presents significant challenges when using conventional techniques, as it requires high tool wear resistance and advanced cutting strategies (Amorim and Weingaertner, 2005). The application of LBM for machining AISI P20 mold steel enhances efficiency, reducing the challenges associated with conventional methods and ensuring precision machining with minimal material distortion. Numerous studies have explored the application of LBM on various materials, identifying optimal machining conditions for improved MRR, kerf width, and surface quality.

The ANFIS has been widely adopted for predicting machining outcomes due to its ability to integrate fuzzy logic and artificial neural networks, providing accurate predictions of machining performance. Rajamani et al. (2021) and Mensah et al. (2020) demonstrated that ANFIS modeling enhances the structuring of machining predictions while reducing dependency on expert knowledge. Sengur and Ubeyli (Sengur, 2008a; Sengur, 2008b) emphasized that ANFIS relies on a structured inference system consisting of five layers, where each layer comprises multiple nodes that serve as inputs for subsequent layers (Çaydaş et al., 2009).

To validate the predictive accuracy of ANFIS, Erkan et al., compared the model's predictions with experimental results, identifying the Levenberg-Marquardt (LM) algorithm as the most effective for minimizing prediction errors in machining damage factors. Besides, Kara et al. (2020) supported that the optimal training procedure for ANN modeling is the traditional back-propagation technique. Maher et al. studied the difference between estimated and measured surface roughness (Ra) in CNC milling of brass (60/40) at an error rate of merely 6.25%, which confirms the precision of the ANFIS model for machining parameter estimation.

Taguchi method is extensively applied in engineering to minimize the PP with a reduction in experimental time and cost. Taguchi optimization enables the selection of the best cutting conditions and serves as a helpful tool for the analysis of the impact of several factors on machining performance (Taguchi, 1987). Kara et al. (n.d.) applied the Taguchi method to discover the suitable cutting parameters of surface roughness (Ra) and cutting temperature (Ctemp), whereas another article employed grey correlation analysis to find the best machining conditions for 17-4 PH stainless steel.

Though metaheuristic approaches are generally employed for process optimization, the Brute Force algorithm provides a more direct and thorough way of determining optimal machining parameters. Unlike in the heuristic approach, the Brute Force algorithm guarantees the identification of the optimum solution through examining all possible sets of parameters (Hansen et al., 2004). This approach, generally combined with dynamic programming, involves partitioning the problem into solvable sub-problems, solving each in isolation, and saving solutions for use at a later time (Kolog, 2015).

Although there have been huge research studies on LBM and optimization methods, very few research studies have utilized ANFIS modeling for the prediction of machining behavior of AISI P20 mold steel. Additionally, there is a huge research gap in the application of the Brute Force algorithm for optimization of LBM process parameters. Filling such gaps, the present research aims to, Use ANFIS modeling to predict LBM process parameters to achieve improved accuracy in machining output. Use the Taguchi L27 model to find optimal machining conditions for improved surface roughness (Ra) and kerf width. Use the Brute Force algorithm to find the optimal input parameter combination to achieve high-quality machining output. Conduct microstructural analysis on high and low surface roughness samples to assess the quality of machined surfaces.

LBM has proven to be far superior to conventional machining processes in its accuracy, efficiency, and non-contact operation. While extensive studies on LBM of most materials have been carried out, studies on AISI P20 mold steel are still scarce, particularly in regards to ANFIS modeling and Brute Force algorithm-based optimization. This present study aims to address these research shortcomings by employing predictive modeling techniques and optimization algorithms to enhance LBM efficiency. By integrating ANFIS modeling, the Taguchi method, and Brute Force optimization, this present study aims to formulate a systematic methodology towards optimal machining performance, thus contributing to LBM research development and applications in industry.

2. Experimental Methodology

A chromium-containing low-carbon plastic mold steel, P20 is an excellent tool steel. Table 28.1 displays the EDX-determined composition of the AISI P20 mold steel used in this investigation. The exceptional heat resistance to softening of AISI P20 is well-known (Farhat, 2003). It finds extensive usage in many different processes and products, including zinc casting, plastic die molding, hydroforming tools, and plastic pressure die frames. According to Al Javed and Bin Rashid (2024), tool steel may have its wear life and hardness increased by freezing the metal. Priyadarshini et al. (2020) found that tool steel's mechanical characteristics, hardness, and microstructure are all positively impacted by subzero temperatures, leading to a longer service life in industrial applications. The mechanical qualities and circumstances that might affect AISI P20 mold steel are detailed in Table 28.2.

Table 28.1 Chemical composition of BM

Element	Fe	Cr	Mn	Mo	C	Si	P	Cu	Co
Wt%	Bal	1.8	1.3	0.2	0.3	0.4	0.01	0.04	0.01

Table 28.2 Properties of AISI P20 steel on 30°C condition

Properties	Values
Density (kg/m^3)	7.75 x 10^3
Poissons ratio	0.29
Modulus of Elasticity (GPa)	188
Thermal expansion coefficient (1/°C)	14.2 x 10^{-6}

In 27 separate runs, square-shaped AISI P20 samples measuring 1.5 × 1.5 cm were cut using a Bodor laser type F3015. Table 28.4 shows the three possible values for the input parameters that were investigated in this study: cutting speed, laser power, and gas pressure. The Taguchi L27 technique was used to produce the dataset for these parameters, with each input being repeated three times to reduce experimental mistakes. Three preliminary tests were executed for different input circumstances as part of the pilot runs to fine-tune the parameter choices. Surface roughness (Ra) and kerf width were determined from a total of 27 samples.

Table 28.3 Taguchi model

Cutting Parameter	Factor	Unit	Level		
			1	2	3
LP	A	KW	1.6	1.8	2
GP	B	Bar	1.2	1.6	2
CS	C	m/min	0.5	1	1.5

The training datasets referenced in Table 28.5 were used to implement the ANFIS method. To find the best machining settings, the Taguchi L27 orthogonal array used a smaller-is-better signal-to-noise (S/N) ratio. Because better machinability is indicated by smaller Ra and kerf width, the Brute Force approach was used to obtain the minimal response parameter values. The kerf width, heat-affected zones, and surface roughness were all measured. When it comes to mechanical parts, surface roughness is one of the most important quality metrics and technical requirements. For a component to work as intended, it is essential to achieve the highest possible surface quality. Conservative process settings, which are commonly used in conventional methods, do not guarantee the necessary surface smoothness and do not maximize metal removal rates (Benardos and Vosniakos, 2003). A surface tester from the Mitutoyo SJ-210 series was used to determine the surface roughness. While testing, samples were clamped down firmly in a vice and the machined surface was scanned with a stylus to get the measurements. To provide a valid and trustworthy measurement, readings were taken from all four corners of each sample and averaged to get the Ra value. The examination of roughness was carried

Table 28.4 Investigation result

S. No	LP	GP	V	D	R_a
1	1.6	1.2	0.5	0.81	6.47
2	1.6	1.2	0.5	0.79	5.89
3	1.6	1.2	0.5	0.80	6.87
4	1.6	1.6	1	0.93	5.49
5	1.6	1.6	1	0.69	5.40
6	1.6	1.6	1	0.82	5.32
7	1.6	2	1.5	0.59	4.77
8	1.6	2	1.5	0.63	5.47
9	1.6	2	1.5	0.68	4.74
10	1.8	1.2	0.5	0.79	5.66
11	1.8	1.2	0.5	0.78	6.07
12	1.8	1.2	0.5	0.77	5.61
13	1.8	1.6	1	0.67	5.39
14	1.8	1.6	1	0.59	6.26
15	1.8	1.6	1	0.68	5.57
16	1.8	2	1.5	0.61	6.65
17	1.8	2	1.5	0.81	6.94
18	1.8	2	1.5	0.80	5.59
19	2	1.2	0.5	0.69	6.76
20	2	1.2	0.5	0.84	4.48
21	2	1.2	0.5	0.66	6.02
22	2	1.6	1	0.79	6.71
23	2	1.6	1	0.88	10.63
24	2	1.6	1	0.79	7.48
25	2	2	1.5	0.83	6.34
26	2	2	1.5	0.71	5.59
27	2	2	1.5	0.72	5.49

out using standardized testing methodologies, such as ISO 4287 and ISO 4288. To take contact-based measurements, a stylus profilometer was employed, which was able to record Ra values using a minimal sample area of 10 mm x 10 mm. To make sure everything was consistent, we took many measurements at different locations.

Kerf width measurement was performed using a straightforward process. A section of known initial dimensions (1.5 cm square) was cut, and its actual width was measured using digital slide calipers. The kerf width (d) was determined by subtracting the final width (x2) from the initial width (x1). When measuring the final width, the internal jaws of the slide calipers were used, while the external jaws were employed for initial width measurements. By combining neural networks with fuzzy logic principles, a neuro-fuzzy inference adaptive system was used to forecast response parameters. Multiple levels of the ANFIS model made use of logical OR operations. The initial layer consisted of three PP values assigned to the CS, GP, and LP input parameters. Successive layers used this layer's output as input and utilized it to determine the firing strength using OR logic. Prior to applying defuzzification in the fourth layer, which summed the outputs of the nodes in the third layer to produce response parameters, values were normalized in the third layer. The final step involved aggregating defuzzified values to obtain the output predictions.

The ANFIS model was implemented using the MATLAB R2023a software, utilizing the antisymmetric function to train and test the dataset. The fuzzy rules required for response parameter prediction were established to enhance accuracy. The model's predictions were verified against experimental results. The three techniques of Taguchi analysis—nominal-the-best, larger-the-best, and smaller-the-best—were utilized to optimize the machining parameters. Surface roughness (Ra) and kerf width were found to be optimum using the smaller-the-best method in this investigation. For each attribute, we used the signal-to-noise ratio transformation, where a greater ratio denoted an improved outcome. The L27 model (3·), which incorporates three factors, CS, LP, and GP each with three levels, was selected for this work from among the regularly used experimental design models in the Taguchi technique [26, 27] (Table 28.3).

It took 27 separate trial runs to put the Taguchi L27 model into action. Every test included the measurement of two response parameters: surface roughness (Ra) and kerf width (d). Table 28.4 provides the full set of experimental outcomes. In order to get the best minimal dataset for response parameters, the Brute Force approach was used to evaluate all conceivable parameter combinations.

Table 28.5 Experimental prediction value of response parameters

S. No	LP	GP	V	$R_{aPredicted}$	$Error_{Ra}$ (%)	d_{Actual}	$d_{Predicted}$	$Error_d$ (%)
1	1.6	1.2	0.5	6.39	0.98	0.77	0.78	-1.3
2	1.6	1.2	0.5	6.39	-8.85	0.8	0.78	2.5
3	1.6	1.2	0.5	6.39	6.65	0.77	0.78	-1.3
4	1.6	1.6	1	5.44	1.50	0.92	0.82	10.5
5	1.6	1.6	1	5.44	-0.12	0.72	0.82	-14.4
6	1.6	1.6	1	5.44	-1.48	0.83	0.82	0.8
7	1.6	2	1.5	5.01	-4.71	0.6	0.62	-3.9
8	1.6	2	1.5	5.01	8.85	0.61	0.62	-2.2
9	1.6	2	1.5	5.01	-5.38	0.66	0.62	5.6
10	1.8	1.2	0.5	5.82	-2.11	0.82	0.81	1.2
11	1.8	1.2	0.5	5.82	4.76	0.81	0.81	0
12	1.8	1.2	0.5	5.82	-3.04	0.8	0.81	-1.2
13	1.8	1.6	1	6.10	-12.75	0.65	0.62	5.1
14	1.8	1.6	1	6.10	3.05	0.54	0.62	-14.2
15	1.8	1.6	1	6.10	7.55	0.66	0.62	6.6
16	1.8	2	1.5	6.42	3.88	0.58	0.75	-29.3
17	1.8	2	1.5	6.42	7.88	0.84	0.75	10.7
18	1.8	2	1.5	6.42	-14.36	0.83	0.75	9.6
19	2	1.2	0.5	5.69	15.02	0.64	0.69	-8.3
20	2	1.2	0.5	5.69	-28.31	0.84	0.69	17.5
21	2	1.2	0.5	5.69	4.23	0.6	0.69	-15.6
22	2	1.6	1	6.79	-1.56	0.81	0.81	0
23	2	1.6	1	6.79	-8.94	0.83	0.81	2
24	2	1.6	1	6.79	8.93	0.8	0.81	-1.7
25	2	2	1.5	6.34	0	0.83	0.83	0
26	2	2	1.5	6.34	-13.52	0.69	0.83	-20.3
27	2	2	1.5	6.34	-15.61	0.68	0.83	-22.1

An NMM-800TRF light microscope was used for microstructural examination, which involved looking for internal fractures, flaws, and surface roughness on the machined surface. In order to obtain sharp pictures, the samples were placed beneath objective lenses and adjusted coarsely and finely. 100x and 200x objective lenses were used as per the required clarity. The samples were examined from four machining sides for getting complete information regarding the fine details of the surface features.

3. Results and Discussion

In order to improve the precision and efficiency of production, it is vital to predict machining performance factors such surface roughness (Ra) and kerf width. To forecast these parameters from experimental data, the present work employed an ANFIS model. Each of the 27 experimental trials that made up the data set for training the model represented a unique class of machining circumstances. Table 28.4 displays the specifics of the experimental conditions. A five-layer neural network architecture was utilized by the ANFIS model. These had a single output layer in addition to four input layers. The model's fuzzy inference system was defined using a Triangular Membership Function (Mfs), with a 3-3-3 distribution for the membership function. For optimal prediction performance, this structure was chosen after a comprehensive evaluation of the model's accuracy. The neural pattern recognition tool in MATLAB was used to construct the classification model. The data set was partitioned into two parts: the training subset used 90% of the data and the testing subset 10%. Consequently, twenty-five data points were used for training and the other two for testing. For the model's stability and convergence, 100 training epochs were utilized. In order to break the issue down into more manageable chunks, a generic Fuzzy Inference System (FIS) was used, which utilized a grid partitioning approach.

The Mean Squared Error (MSE) statistic was used to evaluate the model's prediction accuracy. Approximately 20% and 0.55% of the actual and projected Ra and kerf width values, respectively, were found to have an MSE. The prediction error is the amount by which the actual values deviate from the expected values. There was a robust linear relationship between the anticipated and actual values of surface roughness (Ra) for Ra. Since the two variables are positively correlated, it follows that the expected and actual values of surface roughness tend to rise in tandem. The R^2 value for surface roughness was calculated to be 0.4766, indicating an accuracy of 47.66%. This suggests a moderate correlation between predicted and actual values. The Adjusted R2 value for surface roughness was determined to be 0.4448, indicating that 44.48% of the variance in surface roughness can be explained by the predictive model. The adjusted R2 value provides a more precise measure of model fit by accounting for the number of predictors in the model. For kerf width, the R2 value was 0.4679, reflecting an accuracy of 46.79%. This suggests a medium correlation between actual and predicted values. The Adjusted R^2 value for kerf width was calculated as 0.4458, meaning 44.58% of the variance in kerf width is accurately captured by the model. The graphical representations depict the variations and fluctuations observed in the data.

Table 28.6 Smaller is better

Level	LP	GP	V
1	-15.03	-15.59	-16.97
2	-15.9	-16.28	-15.11
3	-16.42	-15.11	-15.03
Delta	1.46	1.17	1.21
Rank	2	3	1

The surface roughness data showed a lot of variance, but the kerf width results were rather consistent. Surface roughness variations are caused by the inherent random flaws in milling. Initially, there was a significant degree of agreement between the anticipated and real Ra levels, suggesting good accuracy. Nevertheless, outliers emerged due to an increase in the standard deviation of surface roughness with the addition of more samples. According to the kerf width comparison graphs, around half of the observed values are in good agreement with the expected values. Differences in surface roughness could indicate manufacturing flaws. Sample 20 had the most extreme variation in surface roughness, with an inaccuracy of -28.8%. Sample 16 had the most discrepancy, with a -29.3% inaccuracy, in terms of kerf width. These changes highlight the effect of chance in the production setting.

Table 28.7 Smaller is better

Level	LP	GP	V
1	2.63	2.36	2.11
2	2.79	2.54	2.04
3	2.51	3.05	3.78
Delta	0.28	0.69	1.74
Rank	3	2	1

The S/N ratio was selected based on the "smaller is better" principle, which states that surfaces with lower values are smoother. The surface roughness value with the lowest S/N ratio was also the lowest. Based on their impact on surface roughness, Table 28.6 ranks the machining parameters. After considering LP, GP, and CS was determined to be the most important element. Surface roughness is supposedly most affected by cutting speed, then gas pressure, and finally laser power, according to this order. We found that a gas pressure of 2 bar, a cutting speed of 1 m/min, and a laser power of 1.8 kW minimized surface roughness. Laser machining with a low kerf width results in less material vaporization. In Table 28.7, we can see the order of the process factors that impact the kerf width. The most important aspect was found to be CS, with GP and LP following closely behind. Finding the sweet spot for kerf width minimization required a gas pressure of 2 bar, a CS of 1 m/min, and a LP of 1.9 kW.

Both the kerf width and surface roughness were moderately predicted by the ANFIS model, with R2 values of 46.79% and 47.66%, respectively. The inherent variability in the production process is reflected in the experimental data discrepancies. Kerf breadth and surface roughness were discovered to be significantly impacted by machining parameters such GP, CS, and LP. Minimal surface roughness and kerf breadth were achieved by establishing optimal parameter values. Surface flaws and quality discrepancies might also be uncovered by microstructural examination. An indication of the necessity for exact machining control is the existence of surface imperfections, undesirable material deposits, and interior fissures. Adding more process parameters and improving experimental accuracy can be the focus of future study to improve the prediction model.

4. Conclusion

The optimization and prediction of laser cutting machine parameters were the primary foci of this research, which included the laser cutting of AISI P20 mold steel. With the interplay of GP, CS, and LP as its primary considerations, the Taguchi L27 orthogonal array devised the experiment's blueprint. Optimal machining settings were determined by using prediction models built with a 3-3-3 ANFIS structure neural network. The investigation led to the following important conclusions:

1. Using the S/N ratio for single-objective optimization, we found that response characteristics, such as surface roughness and kerf width, are reduced at higher S/N ratio values.
2. The most elegant surface was achieved by combining 2 bar gas pressure with a cutting speed of 1 m/min and a laser power of 1.8 kW.
3. A optimum combination of 2 bar gas pressure, 1 m/min cutting speed, and 1.9 kW laser power resulted in minimal material evaporation during laser cutting. 1.
4. The optimum minimum dataset for the response parameters was determined using a brute force approach, a multi-objective optimization technique.
5. The findings showed that the requirements for least surface roughness were supplied by sample 20, which had characteristics of 2 kW laser power, 1.3 bar gas pressure, and 1 m/min cutting speed.
6. Sample 20 showed interior fissures and a smoother surface when analyzed microscopically at 200x magnification. Surface roughness was inversely related to the presence of peaks and valleys in samples. Under 100x magnification, sample 24, which was cut at 2 kW laser power, 1.7 bar gas pressure, and 0.5 m/min, revealed microvoid debris on the machined surface.

Research into the HAZ and possible annealing of the material's microstructure, as well as other thermal effects caused by laser machining, is essential for future studies. Finding the best prediction model for practical use may require comparative research incorporating many neural network models. To get the most out of the prediction models, it's also a good idea to look at other optimization methods like genetic algorithms. In order to better understand and regulate the laser machining process, future experimental research will address these factors.

REFERENCES

1. Al Javed, M.O., Bin Rashid, A., 2024. Laser-assisted micromachining techniques: an overview of principles, processes, and applications. Adv. Mater. Process. Technol. 1–44. https://doi.org/10.1080/2374068X.2024.2397156.
2. Amorim, F.L., Weingaertner, W.L., 2005. The influence of generator actuation mode and process parameters on the performance of finish EDM of a tool steel. J. Mater. Process. Technol. 166 (3), 411–416. https://doi.org/10.1016/j.jmatprotec.2004.08.026.
3. Benardos, P.G., Vosniakos, G.C., 2003. Predicting surface roughness in machining: a review. Int. J. Mach. Tool Manufact. 43 (8), 833–844. https://doi.org/10.1016/S0890-6955(03)00059-2.
4. Çaydas¸, U., Hasçalik, A., Ekici, S., 2009. An adaptive neuro-fuzzy inference system (ANFIS) model for wire-EDM. Expert Syst. Appl. 36 (3 PART 2), 6135–6139. https://doi.org/10.1016/j.eswa.2008.07.019.
5. Chen, M.F., Sen Ho, Y., Hsiao, W.T., Wu, T.H., Tseng, S.F., Huang, K.C., 2011. Optimized laser cutting on light guide plates using grey relational analysis. Opt Laser. Eng. 49 (2), 222–228. https://doi.org/10.1016/j.optlaseng.2010.09.008.
6. Devanathan, C., Giri, R., Dhandapani, S., Muthiah, C.T., Sivanand, A., 2023. A study on laser beam machining and prediction of surface finish and material removal rate by fuzzy clustering. Lect. Notes Mechan. Eng. 507–518. https://doi.org/10.1007/978-981-19-3895-5_41.
7. Erkan, ¨O., Is¸ık, B., Çiçek, A., Kara, F., 2013. Prediction of damage factor in end milling of glass fibre reinforced plastic composites using artificial neural network. Appl. Compos. Mater. 20 (4), 517–536. https://doi.org/10.1007/s10443-012-9286-3.
8. Farhat, Z.N., 2003. Microstructural Characterization of WC-TiC-Co Cutting Tools during High-Speed Machining of P20 Mold Steel, vol. 51, pp. 117–130. https://doi.org/10.1016/j.matchar.2003.10.005.
9. Girdu, C.C., Gheorghe, C., 2022. Energy esfficiency in CO2 laser processing of hardox 400 material. Materials 15 (13). https://doi.org/10.3390/ma15134505.
10. Hansen, E.A., Bernstein, D.S., Zilberstein, S., 2004. Dynamic programming for partially observable stochastic games. AAAI Workshop - Techn. Rep. WS-04–08 (2000), 25–30.
11. Kamonpong, J., Janmanee, P., 2014. Deep Hole of AISI P20 Mold Steel Material by Electrical Discharge Machining, vol. 590, pp. 244–248. https://doi.org /10.4028/www.scientific.net/AMM.590.244.
12. Kara, F., Karabatak, M., Ayyildiz, M., Nas, E., 2020. Effect of machinability, microstructure and hardness of deep cryogenic treatment in hard turning of AISI D2 steel with ceramic cutting. J. Mater. Res. Technol. 9 (1), 969–983. https://doi.org/10.1016/j.jmrt.2019.11.037.
13. Kara, F., n.d. Experimental and statistical investigation of the effect of nanoparticle minimum quantity lubrication(nano-MQL)method on cutting performance. Gazi J. Eng. Sci. 10, 102–113.
14. Kara, F., Bulan, N., Akgün, M., K¨oklü, U., 2023. Multi-objective optimization of process parameters in milling of 17-4 PH stainless steel using taguchi-based gray relational analysis. Eng. Sci. 26, 1–12. https://doi.org/10.30919/es961.
15. Kolog, E.A., 2015. Dynamic Programming Using Brute Force Algorithm for a Traveling Salesman Problem. https://doi.org/10.13140/RG.2.1.2304.2727. June.
16. Maher, I., Eltaib, M.E.H., Sarhan, A.A.D., El-Zahry, R.M., 2014. Investigation of the effect of machining parameters on the surface quality of machined brass (60/40) in CNC end milling - ANFIS modeling. Int. J. Adv. Manuf. Technol. 74 (1–4), 531–537. https://doi.org/10.1007/s00170-014-6016-z.
17. Mensah, R.A., Xiao, J., Das, O., Jiang, L., Xu, Q., Alhassan, M.O., 2020. Application of adaptive neuro-fuzzy inference system in flammability parameter prediction. Polymers 12 (1), 1–16. https://doi.org/10.3390/polym12010122.
18. Nguyen, D.T., Ho, J.R., Tung, P.C., Lin, C.K., 2021. Prediction of kerf width in laser cutting of thin non-oriented electrical steel sheets using convolutional neural network. Mathematics 9 (18). https://doi.org/10.3390/math9182261.
19. D. R. On, "MACHINING PERFORMANCE of AISI P20 STEEL with GRAPHITE and TUNGUSTEN BASED ELECTRODE on EDM."
20. Priyadarshini, M., Behera, A., Biswas, C.K., 2020. Effect of sub-zero temperatures on wear resistance of AISI P20 tool steel. J. Braz. Soc. Mech. Sci. Eng. 42 (5), 1–13. https://doi.org/10.1007/s40430-020-02298-2.
21. Rajamani, D., Siva Kumar, M., Balasubramanian, E., Tamilarasan, A., 2021. Nd: YAG laser cutting of Hastelloy C276: ANFIS modeling and optimization through WOA. Mater. Manuf. Process. 36 (15), 1746–1760. https://doi.org/10.1080/10426914.2021.1942910.
22. Rao, B.T., Kaul, R., Tiwari, P., Nath, A.K., 2005. Inert gas cutting of titanium sheet with pulsed mode CO2 laser. Opt Laser. Eng. 43 (12), 1330–1348. https://doi.org/10.1016/j.optlaseng.2004.12.009.
23. Sengur, A., 2008a. Wavelet transform and adaptive neuro-fuzzy inference system for color texture classification. Expert Syst. Appl. 34 (3), 2120–2128. https://doi.org/10.1016/j.eswa.2007.02.032.
24. Sengur, A., 2008b. An expert system based on principal component analysis, artificial immune system and fuzzy k-NN for diagnosis of valvular heart diseases. Comput. Biol. Med. 38 (3), 329–338. https://doi.org/10.1016/j.compbiomed.2007.11.004.
25. Taguchi, G., 1987. Taguchi on Robust Technology Development: Bringing..

Note: All the tables in this chapter were made by the authors.

Advances in Additive Manufacturing Technologies – Gurusamy Pathinettampadian et al. (eds)
© 2026 Taylor & Francis Group, London, ISBN 978-1-041-16687-0

Combining Ultrasonic and Electrochemical Techniques: An Optimization Study for Machining Superalloys

S. Senthil Kumar[1]
Professor, Department of Mechanical Engineering,
R.M.K. College of Engineering and Technology,
India

P. Raja[2]
Associate Professor, Department of Mechanical Engineering,
Prathyusha Engineering College,
India

Karthick M.[3]
Assisstant Professor,
Department of Mechanical Engineering,
Veltech Rangarajan Dr Sagunthala R&D Institute of science and technology,
Chennai, India

Bharanidharan T.[4]
Student, Department of Mechanical Engineering,
Chennai Institute of Technology,
Chennai, India

B. Yokesh Kumar[5]
Assisstant Professor, Department of Mechanical Engineering,
Chennai Institute of Technology, Chennai,
India

R. Karthick[6]
Assistant Professor, Department of Mechanical Engineering,
Rajalakshmi Institute of Technology,
India

◆ **Abstract:** ECG is a high-technology hybrid machining process, ideally combining the concepts of ECM and mechanical grinding. The process is most useful in grinding hard-to-machine alloys, where traditional machining processes may fail or result in high tool wear. In this paper, a novel process known as UAECDG is introduced, which integrates electrochemical drilling, mechanical grinding, and ultrasonic vibration into one integrated process. The focal point of this novel approach is the machining of high-quality small holes in superalloy materials, which are widely used in the aerospace industry and other high-tech applications. The real breakthrough of UAECDG is the integration of ultrasonic vibration with the high-speed rotating electrode feature of the ECG process. By superimposing these ultrasonic vibrations on the rotating tool, several improvements can be achieved, such as improved machining stability, improved material removal efficiency, and better surface quality. To gain a thorough understanding of the

[1]senthilkumar@rmkcet.ac.in, [2]rajaponnu79@gmail.com, [3]karthick@veltech.edu.in, [4]vtrbharani12@gmail.com, [5]yokeshkumarb@citchennai.net, [6]Karthick.industrial@gmail.com

DOI: 10.1201/9781003685906-29

underlying machining mechanism, we begin by investigating the electrochemical passive behavior of the superalloy. A mathematical model is then derived to simulate the gap electric field, a crucial factor in determining the machining characteristics. Furthermore, applying rotating ultrasonic vibration significantly enhances surface finish, machining stability, and efficiency. Experimental findings demonstrate that small holes of superior quality are successfully fabricated using UAECDG, with the surface roughness of the hole sidewall improving from an initial Ra value of 0.99 μm to a refined Ra value of 0.14 μm. This substantial reduction in surface roughness confirms the effectiveness of the UAECDG technique in achieving precision machining of superalloy components.

◆ **Keywords:** Ultrasonic, Electro-chemical, Super alloy, Surface roughness, Hybrid machining

1. Introduction

The remarkable mechanical and thermal capabilities of nickel-based superalloys make them widely used in high-temperature components of aerospace and aviation engineering. These alloys have exceptional fatigue performance, thermal stability, oxidation resistance, toughness, and strength at high temperatures. Because of these qualities, they are perfect for use in combustion chambers, jet engines, turbine blades, and other crucial aerospace components that are subject to very harsh environments. Despite their useful qualities, nickel-based superalloys are extremely challenging to process using traditional techniques. These alloys' enhanced high-temperature strength and poor heat conductivity provide the greatest machining problems. These characteristics result in excessive tool wear, poor surface finishes, and reduced productivity in typical machining operations like turning, milling, and grinding. Due to these limitations, various non-conventional machining processes have been investigated. Of these, laser machining and EDM have been the focus of extensive interest and use. However, these thermal processes are not without limitations. Laser machining and EDM can result in microcracking, recast layers, and heat-affected zones (HAZ) due to localized melting and rapid solidification. These flaws weaken the mechanical integrity and fatigue life of the machined part. Additionally, electrochemical machining (ECM), another advanced process, effectively removes mechanical stresses and tool wear but is marred by stray current corrosion-related issues. This produces poor dimensional accuracy, making ECM unsuitable for precision machining of nickel-based superalloys.

The film is continuously removed by abrasive action, exposing new material to further electrochemical dissolution. Unlike the conventional grinding process, ECG removes only 5–10% of the material by mechanical action, and the remaining material removal is achieved by electrochemical dissolution. With ECG, machining becomes more efficient with reduced tool wear because the passive oxide coating is softer than the base metal. Since this is the case, ECG may be used to mill nickel-based superalloys as well as other materials that are difficult to manufacture. Because the mechanical grinding operation is responsible for honing the end surface finish, ECG is characterized by higher machining accuracy and decreased surface roughness compared to standard ECM. In an effort to make ECG more effective, researchers have studied various parts of it throughout the years. For the purpose of removing the recast layer that EDM produces, P. Ming et al. studied diamond-mounted point ECG. Their research looked at how surface roughness changed depending on factors such cutting depth, tool rotation speed, electrolyte composition, and applied voltage. In this study, D.T. Curtis et al. examined the effects of electrical parameters and abrasive grit type on surface roughness and overcut when ECG was used to process nickel-based superalloys.

In another study, N. Qu et al. demonstrated that replacing electrodeposited diamond wheels with brazed diamond wheels in ECG extended tool life by more than three times. Their research also showed that optimizing voltage and electrolyte temperature significantly improved material removal rates. K. Przystupa et al. studied micro-short circuits that occurred when applying ECG to titanium alloys. To manufacture flow channel structures in GH4169 superalloy, H. Li et al. suggested an electrochemical mill-grinding (ECMG) approach that uses an inner-jet system. Among their responsibilities was the investigation of ECMG process flow and electric field behavior using simulation. In addition, S. Niu et al. created ECMG-specific tool electrodes that had bottom insulation and holes for tool-sidewall outlet connections. Their tests showed that insulating the electrode's base increased its bottom surface flatness and that a spiral arrangement of outlet holes improved the flatness of the sidewalls. They postulated that the process involved electrolytic products adhering to the machined surface instead of compact passive layers. In addition, X. Zhu et al. integrated ECG with ultrasonic vibration to microfabricate 304 stainless steel inserts.

Their findings indicated that applying reasonable ultrasonic vibration significantly improved accuracy and surface quality of machining.

Despite the established advantages of ECG machining of hard-to-cut alloys, there remain a few challenges. One of the fundamental problems is instability in the machining process, particularly in the fabrication of small holes. The minute machining gap and formation of electrolysis byproducts during ECG machining are some of the challenges to a stable machining environment. These conditions produce micro-short circuits, which have a negative impact on machining efficiency, surface finish, and tool life. In addition, unstable electrochemical conditions cause uneven material removal, which further reduces machining accuracy. In an attempt to overcome these challenges, this work introduces a new technique referred to as UAECDG. The technique balances the ultrasonic vibration's rotation with ECG with the overall goal of attaining maximum efficiency and precision in the fabrication of small holes. The use of ultrasonic vibration improves the circulation of electrolyte, reduces deposition of electrolysis byproducts, and stabilizes the machining gap. Consequently, material removal rates, surface finish, and tool life are greatly enhanced. A novel experimental configuration for UAECDG has been introduced in this research work, with a ball-end abrasive electrode as the tool cathode. The aim of this work is to determine the set of parameters with optimal efficiency, superior surface quality, and better machining stability. The experimental results reveal that the UAECDG process improves the small hole fabrication significantly, lowering the surface roughness and enhancing machining accuracy compared to conventional ECG.

Lastly, in this study, it is affirmed that ultrasonic-assisted electrochemical drill-grinding is an effective method to machine nickel-based superalloys. By breaking the limitations intrinsic to traditional ECG and other competing approaches, UAECDG is a viable solution to high-precision component machining in aerospace and aviation applications. The conclusions outlined in this study facilitate further development of hybrid machining tools and open up new avenues for further research in manufacturing high-performance materials.

2. Experimental Methodology

Operating the Ultrasonic Assisted Electrochemical Drill-High-precision grinding (UAECDG) necessitated a special experimental setup. The setup consists of various critical components that harmonize perfectly to deliver maximum performance. All these components are critical to the delivery of precision, efficiency, and stability of the UAECDG process. The three-axis machine tool used in this setup has a high-resolution motion control system with a remarkable accuracy of 0.1 μm, providing the positional accuracy needed for UAECDG. Such accuracy is of paramount importance for the manufacture of uniform machining results, especially for high-speed electrochemical grinding processes.

The ultrasonic motorized spindle not only makes the tool electrode rotate at high speeds but also enables ultrasonic vibration in the z-axis simultaneously. The ultrasonic vibration system of the spindle has a working frequency range of 20–35 kHz and an amplitude of up to 13 μm. The spindle itself is capable of rotation at a maximum of 24,000 r/min, which is sufficient to satisfy the working requirements of UAECDG. For real-time observation of the process and accurate data acquisition, the UAECDG system is also equipped with an advanced monitoring and data acquisition module. The module includes some key components: Ammeter, monitors the current passing through the system to monitor electrochemical reactions. Oscilloscope, captures the voltage and current waveforms, assists in the analysis of machining stability. CCD Vision Monitor, Provides real-time visual monitoring of the machining process. Data Acquisition Card, Records process parameters for further analysis and optimization. This integrated monitoring system is essential for analyzing machining phenomena, detecting anomalies, and optimizing process parameters in UAECDG.

The electrode assembly in this experimental setup consists of multiple interconnected elements, including the tool holder, workpiece, and tool electrode. The shape, material, and configuration of the tool electrode are critical in determining machining accuracy and surface quality. A specially designed ball-end abrasive electrode, is used as the tool cathode. The electrode features diamond abrasive particles, which are bonded to the base body using a nickel coating. Approximately 60–65% of the diamond particles' size is exposed, allowing for effective abrasive action. The base body and nickel coating of the electrode are electrically conductive, while the diamond particles themselves are non-conductive. This unique design enables the simultaneous execution of electrochemical reactions and mechanical grinding, ensuring effective material removal while enhancing machining precision.

The complete UAECDG process for fabricating high-precision small holes consists of two distinct steps, in which step 1 is Electrochemical Drilling (ECD) for Pre-Machining

Holes. In the first phase of the process, electrochemical drilling (ECD) is used to create pre-machined holes. This is accomplished using a helix electrode. To achieve high precision in pre-machining, ultra-short pulse voltage is applied to the helix electrode, which rotates at high speed. The advantage of using ECD in this stage is its ability to fabricate holes with minimal burr formation and reduced thermal damage. Upon completion of this step, to prevent positional errors, tool changing is performed without repositioning the workpiece, ensuring alignment for the next phase. The second step is UAECDG for Finish Machining of Small Holes. In the second phase, the UAECDG technique is employed to refine the pre-machined holes. A ball-end abrasive electrode, is used in this stage. The UAECDG process involves applying a passive electrolyte, which promotes the formation of a passive oxide film on the GH3030 superalloy workpiece. This oxide film, which can hinder electrochemical dissolution, is continuously removed by the abrasive action of the electrode, allowing the underlying metal to undergo further electrochemical dissolution. Simultaneously, the ultrasonic vibration of the electrode enhances electrolyte circulation, facilitating the removal of hydrogen bubbles and electrolysis byproducts. This improves the homogeneity of the flow field, leading to higher machining stability and efficiency compared to the traditional Electrochemical Grinding (ECG).

Material removal in ECG is primarily governed by electrochemical reactions, which account for approximately 90–95% of the total material removal. But secondary electrolysis in ECD can potentially produce hole taper, which affects machining accuracy in a negative way. To offset the problem of hole taper, the present study recommends using a ball-end abrasive electrode as the tool cathode in UAECDG. The Electrochemical reactions are localized to the ball-end of the electrode. The effect of secondary electrolysis is reduced to a minimum. The machining localization is enhanced generally.

To gain a better understanding of the UAECDG process, a simulation was conducted to study the distribution of current density and electrochemical dissolution during machining. The simulation accounts for a constant electrolyte conductivity and provides insight into the machining process. The high current density localized in the ball-end region of the electrode. Localized current density limits electrochemical reactions to a localized region, thus minimizing secondary electrolysis and improving machining precision. As the tool electrode moves downward, the taper of the hole is significantly reduced, producing a high-precision machining result.

UAECDG is a process which combines electrochemical drilling, grinding by mechanical motions, and ultrasonic vibration to improve machining of small holes in hard-to-machine alloys like GH3030 superalloy. The experimental rig, which is developed with high-precision motions, real-time monitoring, and a well-designed electrode, provides high efficiency, stability, and accuracy. With optimized machining parameters and by taking advantage of ultrasonic vibration, UAECDG successfully overcomes the limitations of traditional ECG and is a promising method for advanced manufacturing processes.

3. Results and Discussion

Under certain electrolyte conditions, such as solutions of sodium chlorate (NaClO3) and sodium nitrate (NaNO3), GH3030, a nickel-based superalloy, exhibits passive characteristics. When a material undergoes electrochemical breakdown, a passive oxide coating is formed on its surface, leading to passivity. An oxide coating acts as a barrier to electrochemical processes, protecting against corrosion caused by stray currents. Also, the material's surface becomes glossy and smooth once the electrochemical dissolution procedure has passivated. The GH3030 superalloy's passivation behavior and machining process may be better understood by studying its electrochemical characteristics in different electrolyte solutions, which is crucial due to the superior effects. Solutions of NaNO3 and NaClO3 behave passively as expected. Active dissolution, passivation, and trans-passivation are key electrochemical properties illustrated by the polarization curves. An oxide layer is formed on the surface of the metal and a current density that is rather constant is achieved in the passive zone.

With NaClO3 concentrations ranging from 10% to 20%. Of particular interest is that at the concentrations of 15% and 20% of NaClO3, the passive potential range is considerably wider. It means that passivation is more stable in the solutions. The passive potential range in a 15% NaClO3 solution is between 0.24 V and 0.83 V, and in a 20% NaClO3 solution, between 0.26 V and 0.84 V. Over a wide range of concentrations, the passive potential range remains relatively constant in comparison to solutions of NaClO3. However, a denser and more protective passive oxide coating is formed when the superalloy is immersed in 15% or 20% NaNO3 solutions, since the present density in the passive zone becomes more steady. When the NaNO3 concentration is 15%, the passive potential ranges from 0.20 V to 0.90 V; when the concentration is 20%, it ranges from 0.16 V to 0.85 V.

When looking at the polarization curves for solutions of NaNO3 and NaClO3, it is clear that the passive potential range is much wider in the latter. There is evidence that a more stable passive oxide layer forms in solutions containing 20% NaNO3, as the GH3030 superalloy's passive current density is lower in these solutions compared to those containing $NaClO_3$. This work follows these findings by using an UAECDG procedure with an electrolyte consisting of a 20% $NaNO_3$ solution. At first, the diameter of the pre-machined hole is machined to be slightly smaller than that of the electrode on the ball-end tool. Hybrid processes using electrochemical machining (ECM) and mechanical machining remove materials in UAECDG. There are three distinct steps to the machining gap in UAECDG: electrochemical grinding (ECG), secondary electrolysis, and electrochemical cutting (ECM). When abrasive particles peeled off the passive oxide coating during the ECG phase, the material washes out the most efficiently by electrochemical dissolution.

Point Q is an important juncture in the process of moving from ECM to ECG. In most cases, the diameter of the pre-machined hole, the radius of the tool's ball-end, the speed of the tool's feeding, and the voltage supplied determine the position of Q. The machining gap and cut depth are directly affected by the dimensions of the pre-machined holes, thus it is critical to ensure that the hole diameters are uniform. The pre-machined holes are prepared by electrochemical drilling (ECD) prior to UAECDG in order to attain excellent reproducibility. In addition, the natural frequency of the tool electrode is used to lock the output frequency of the ultrasonic vibration in order to enhance the efficiency of ultrasonic energy. The experimental parameters are detailed there. To increase the stirring action of the electrolyte, the tool electrode's ultrasonic vibration is vital.

The capacity of the ultrasonic vibration to rehydrate the electrolyte, boost heat dissipation, and remove hydrogen bubbles and other electrolysis byproducts is responsible for this improvement. A faster rate of material removal is achieved as a result of these elements' combined promotion of electrochemical processes. Equations (8) and (11), on the other hand, show that the electrochemical reaction rate drops as the average machining gap (S) is larger with increasing ultrasonic amplitude. The electrode that vibrates with the tool improves machining stability and flow uniformity by removing electrolysis byproducts and increasing electrolyte refreshment.

Because of the tiny cutting gap and the slow removal of electrolysis byproducts, the machining process that does not include ultrasonic vibration is unstable. On the other hand, the use of ultrasonic vibration results in a more steady working current, which in turn prevents short circuits and produces holes with a reduced surface roughness (Ra). For UAECDG to work, the machining gap must be properly calibrated and kept constant. To achieve high machining precision and surface quality, it is vital to achieve a balance between ECM and mechanical grinding. In electrochemical milling (ECM), the applied voltage is the most important variable, whereas in mechanical grinding, the feed rate is the most important variable. A set of tests is carried out with a constant ultrasonic amplitude of 5 μm to examine their impacts. The effect of voltage and feed rate on hole diameter, is derived from both experimental and modeling data. Consistent with Equation (13), both approaches show that the hole diameter grows with voltage applied but shrinks with feed rate. Improving the electrochemical dissolving process and increasing the hole diameter are both achieved at a fixed feed rate by increasing the applied voltage, which in turn increases the current density. On the flip side, while the applied voltage remains constant, the period of electrochemical dissolution is decreased and the hole sizes are widened when the feed rate is increased. Further, in line with Equation (12), the studies show that repeatability is reduced by fluctuations in the machining gap at lower feed rates and higher voltages.

According to the results, the relationship between these two factors is the main determinant of surface quality. The machining gap becomes too big when the voltage is high and the feed rate is low, making mechanical grinding impossible. Surface quality is drastically reduced due to stray and pitting corrosion caused by electrochemical material loss in such instances. On the flip side, a machining process that is unstable and susceptible to short circuits can be produced by utilizing a high feed rate in conjunction with a low voltage. This is because mechanical grinding becomes excessive, leading to direct metal scraping. The optimal parameters, after optimization, are found to be 6 V applied voltage, 20 μm/s feed rate, and 5 μm ultrasonic amplitude. With these settings, we were able to create a series of tiny holes with an exceptionally smooth surface and pinpoint accuracy. Aligning well with modeling expectations, the addition of a ball-end electrode improves machining localization and decreases hole taper. The most stable and efficient operation of UAECDG may be accomplished by adjusting the feed rate, voltage, and ultrasonic amplitude parameters to perfection. The machined holes' side walls have a significantly reduced surface roughness, going from 0.99 μm to 0.14 μm, and they are also notably straighter.

4. Conclusion

This work established a method for machining GH3030 superalloy using UAECDG. Using polarization curve experiments, the researchers examined the GH3030 superalloy's passive behavior in various electrolyte conditions. To further comprehend the machining procedure in UAECDG, a mathematical model was also developed and an electric field simulation was set up. Parameter optimization for high-precision machining of tiny holes in GH3030 superalloy was achieved after a battery of tests investigating the effect of machining parameters on efficiency and quality. Here are the main takeaways from the study:

- In electrolyte solutions including NaNO3 and NaClO3, the GH3030 superalloy displays passive behavior. The wider passive potential range and lower passive current density of a 20% NaNO3 solution made it the ideal electrolyte for reliable passivation in UAECDG.
- The electrochemical reaction is limited to the area surrounding the ball-end of the tool electrode, as demonstrated by simulations of the electric field within the machining gap. One way to improve machining localization and reduce hole taper is to employ a ball-end abrasive electrode. The experimental results show that compared to traditional machining methods, UAECDG results in hole walls that are straighter.
- The machined holes' surface quality is dictated by the equilibrium between mechanical grinding and electrochemical machining (ECM). Improved machining stability and efficiency are the results of applying ultrasonic vibration to the tool electrode at the correct amplitude, which increases electrolyte refreshment. An optimal combination of applied voltage, tool feed rate, and ultrasonic amplitude allows for the creation of microscopic holes of exceptional quality with an enhanced surface polish. When it comes to machining superalloys with accuracy and efficiency, UAECDG is a promising approach that considerably improves surface quality compared to electrochemical drilling alone.

References

1. Zhan, T.; Chai, F.; Zhao, J.; Yan, F.; Wang, W. A study of microstructures and mechanical properties of laser welded joint in GH3030 alloy. J. Mech. Sci. Technol. 2018, 32, 2613–2618.
2. Xue, C.; Chen, W. Adhering layer formation and its effect on the wear of coated carbide tools during turning of a nickel-based alloy. Wear 2011, 270, 895–902.
3. Goswami, T. Conjoint bending torsion fatigue—Fractography. Mater. Des. 2002, 23, 385–390.
4. Imran, M.; Mativenga, P.; Gholinia, A.; Withers, P.J. Comparison of tool wear mechanisms and surface integrity for dry and wet micro-drilling of nickel-base superalloys. Int. J. Mach. Tools Manuf. 2014, 76, 49–60.
5. Zhu, D.; Zhang, X.; Ding, H. Tool wear characteristics in machining of nickel-based superalloys. Int. J. Mach. Tools Manuf. 2013, 64, 60–77.
6. Uçak, N.; Çiçek, A. The effects of cutting conditions on cutting temperature and hole quality in drilling of inconel 718 using solid carbide drills. J. Manuf. Process. 2018, 31, 662–673.
7. Sinha, M.K.; Setti, D.; Ghosh, S.; Rao, P.V. An investigation on surface burn during grinding of inconel 718. J. Manuf. Process. 2016, 21, 124–133.
8. Yılmaz, B.; Karabulut, S¸.; Güllü, A. Performance analysis of new external chip breaker for efficient machining of inconel 718 and optimization of the cutting parameters. J. Manuf. Process. 2018, 32, 553–563.
9. Wang, Y.L.; Chen, C.Y.; Liu, Z.C.; Ren, W.X.; Zhu, L.Z.; Wang, L. Machining and characterization of deep micro holes on super alloy processed by millisecond pulsed laser. Key Eng. Mater. 2016, 703, 34–38.
10. Pan, Z.; Feng, Y.; Hung, T.-P.; Jiang, Y.-C.; Hsu, F.-C.; Wu, L.-T.; Lin, C.-F.; Lu, Y.-C.; Liang, S.Y. Heat affected zone in the laser-assisted milling of inconel 718. J. Manuf. Process. 2017, 30, 141–147.
11. Lee, L.; Lim, L.; Narayanan, V.; Venkatesh, V. Quantification of surface damage of tool steels after EDM. Int. J. Mach. Tools Manuf. 1988, 28, 359–372.
12. Kliuev, M.; Florio, K.; Akbari, M.; Wegener, K. Influence of energy fraction in EDM drilling of inconel 718 by statistical analysis and finite element crater-modelling. J. Manuf. Process. 2019, 40, 84–93.
13. Rahman, Z.; Das, A.K.; Chattopadhyaya, S. Microhole drilling through electrochemical processes: A Review. Mater. Manuf. Process. 2017, 33, 1379–1405.
14. Maksoud, T.; Brooks, A. Electrochemical grinding of ceramic form tooling. J. Mater. Process. Technol. 1995, 55, 70–75.
15. Puri, A.B.; Banerjee, S. Multiple-response optimisation of electrochemical grinding characteristics through response surface methodology. Int. J. Adv. Manuf. Technol. 2012, 64, 715–725.
16. Curtis, D.; Soo, S.; Aspinwall, D.; Sage, C. Electrochemical superabrasive machining of a nickel-based aeroengine alloy using mounted grinding points. CIRP Ann. 2009, 58, 173–176.
17. Goswami, R.N.; Mitra, S.; Sarkar, S. Experimental investigation on electrochemical grinding (ECG) of alumina-aluminum interpenetrating phase composite. Int. J. Adv. Manuf. Technol. 2008, 40, 729–741.
18. Li, H.; Fu, S.; Zhang, Q.; Niu, S.; Qu, N. Simulation and experimental investigation of inner-jet electrochemical grinding of GH4169 alloy. Chin. J. Aeronaut. 2018, 31, 608–616.
19. Ming, P.M.; Zhu, D.; Xu, Z.Y. Electrochemical grinding for unclosed internal cylinder surface. Key Eng. Mater. 2007, 359, 360–364.
20. Zhu, D.; Zeng, Y.; Xu, Z.; Zhang, X. Precision machining of small holes by the hybrid process of electrochemical removal and grinding. CIRP Ann. 2011, 60, 247–250.
21. Qu, N.; Zhang, Q.; Fang, X.; Ye, E.; Zhu, D. Experimental investigation on electrochemical grinding of inconel 718. Procedia CIRP 2015, 35, 16–19.

22. Przystupa, K.; Litak, G. Electrochemical grinding of titanium-containing materials. Adv. Sci. Technol. Res. J. 2017, 11, 183–188.
23. Li, H.; Niu, S.; Zhang, Q.; Fu, S.; Qu, N. Investigation of material removal in inner-jet electrochemical grinding of GH4169 alloy. Sci. Rep. 2017, 7, 3482.
24. Li, H. Simulation and experimental investigation of electrochemical mill-grinding of GH4169 alloy. Int. J. Electrochem. Sci. 2018, 13, 6608–6625.
25. Niu, S.; Qu, N.; Yue, X.; Li, H. Effect of tool-sidewall outlet hole design on machining performance in electrochemical mill-grinding of inconel 718. J. Manuf. Process. 2019, 41, 10–22.
26. Niu, S.; Qu, N.; Li, H. Investigation of electrochemical mill-grinding using abrasive tools with bottom insulation. Int. J. Adv. Manuf. Technol. 2018, 97, 1371–1382.
27. Ge, Y.; Zhu, Z.; Wang, D.; Ma, Z.; Zhu, D. Study on material removal mechanism of electrochemical deep grinding. J. Mater. Process. Technol. 2019, 271, 510–519.
28. Zhu, X.; Liu, Y.; Zhang, J.; Wang, K.; Kong, H. Ultrasonic-assisted electrochemical drill-grinding of small holes with high-quality. J. Adv. Res. 2020, 23, 151–161.
29. Liu, Y.; Li, M.; Niu, J.; Lu, S.; Jiang, Y. Fabrication of taper free micro-holes utilizing a combined rotating helical electrode and short voltage pulse by ECM. Micromachines 2019, 10, 28.
30. Wang, D.; Zhu, Z.; Wang, N.; Zhu, D.; Wang, H. Investigation of the electrochemical dissolution behavior of inconel 718 and 304 stainless steel at low current density in NaNO3 solution. Electrochim. Acta 2015, 156, 301–307.

Advances in Additive Manufacturing Technologies – Gurusamy Pathinettampadian et al. (eds)
© 2026 Taylor & Francis Group, London, ISBN 978-1-041-16687-0

Streamlining Chemical Machining of Aluminum Alloys: A Central Composite Design Perspective

R. Christu Paul[1]
Assistant Professor,
Department of Automobile Engineering,
Hindustan Institute of Technology and Science,
Chennai, India

Bharanidharan T.[2]
Student, Department of Mechanical Engineering,
Chennai Institute of Technology,
Chennai, India

P. Loganathan[3]
Associate Professor,
Department of Mechanical Engineering,
Karpaga Vinyaga College of Engineering and Technology,
India

K. Karthik[4]
Associate Professor,
Department of Mechanical Engineering,
Vel Tech Rangarajan Dr. Sagunthala R&D Institute of Science Technology,
Chennai, India

S. Senthil Kumar[5]
Professor, Department of Mechanical Engineering,
R.M.K. College of Engineering and Technology,
India

Damodharan V.[6]
Research Scholar, Department of Mechanical Engineering,
Hindustan Institute of Technology and Science,
Chennai, India

◆ **Abstract:** Chemical etching is an essential process in materials science that enables controlled material removal through chemical reactions, allowing for precise surface modification across various applications. This study investigates the effects of temperature and etching solution concentration on the corrosion rate and surface roughness of aluminum alloy 2024. Experiments were conducted at temperatures ranging from 35°C to 65°C in 5°C increments, using ferric chloride solutions of three different concentrations (9, 20, and 29 Be). The findings show both corrosion rate and surface roughness to increase as concentrations and temperatures rose. Importantly, the corrosion rate varied from 1.15×10^{-4} mm/min at 35°C to 1.291×10^{-3} mm/min at 65°C, while the surface roughness varied from 1.208 μm

[1]christupaul2024@gmail.com, [2]vtrbharani12@gmail.com, [3]logu2685@gmail.com, [4]karthikk@veltech.edu.in, [5]senthilkumar@rmkcet.ac.in, [6]damuknrcl@gmail.com

DOI: 10.1201/9781003685906-30

to 7.431 μm in the same temperature range. Temperature, however, had more significant effects on the relationships over the range in question compared to concentration. Interpretation of the relationships allows process optimization of the chemical etching process for specially designed material property, which benefits industries that need specialized surface control, such as automotive, aerospace, medicine, consumer electronics, and home appliances.

◆ **Keywords:** Chemical machining, Concentration, Etching, Ferric chloride, Temperature

1. Introduction

Chemical machining is a well-established material removal process by controlled dissolution using a reactive chemical solution. The solution typically consists of highly reactive alkaline or acid chemicals [1]. These processes are etching, chemical etching, wet etching, chemical milling, and chemical blanking. The early history of this practice goes back to ancient Egypt, where craftsmen used etched copper treated with citric acid to produce intricate jewelry, exhibiting an early phase of micromachining [2]. Chemical machining is widely applied to machine complex apertures in a variety of materials, regardless of their strength or hardness. Chemical milling (CH milling) and photochemical machining (PCM), or spray etching, are two of the most widely used methodologies. Since its discovery in the early 1950s, PCM has been widely utilized in manufacturing, particularly in machining micro-scale components. Chemical machining is now a widely used unconventional machining process, and its applications are found in a variety of industries such as aerospace, automobile, medicine, and optics [3]. Researchers are continuously searching for innovations in methodologies and process optimization to enhance efficiency and effectiveness of chemical machining. More and more research is increasingly being focused on the impact of various process parameters on machining performance [4].

Aluminum is widely used in the aircraft, aerospace, and automobile industries due to its engineering superiority, possessing high specific strength, low weight, and good thermal and electrical conductivity. All aluminum alloys have varied properties and are hence highly useful in applications. Chemical machining is crucial in the shaping and treatment of aluminum and alloys for industrial applications. Previous studies identify the main determinants of the result of chemical machining. The influence of concentration, temperature, and machining time on titanium alloy weight reduction was considered by Abdulhusein et al. [5]. At temperatures of 60 to 100°C, etchant concentrations of 50 to 80%, and machining time of 30 to 90 minutes, their experiments documented that all three factors were accountable for weight loss but concentration caused the maximum weight loss. Etching of 7075 aluminum was performed by Çakır [6] using $FeCl_3$ as an etchant, which she found to be the etchant for this application through virtue of the controllability and simplicity with which a fine surface finish as well as the etch depth have the ability to be controlled. Lad et al. [7] explored ways to enhance the productivity and efficiency of chemical machining for aluminum and stainless steel, concluding that the process layout was both simple and cost-effective.

El-Awadi et al. [8] studied the effects of $FeCl_3$ and $FeCl_3$+HNO_3 on the material removal rate (MRR) of copper, aluminum, and stainless steel, emphasizing the importance of temperature and etchant concentration in optimizing MRR. Similarly, Al-Ethari et al. [9] analyzed the influence of machining temperature, time, and prior cold working on the metal removal rate and surface roughness of chemically machined 420 stainless steel samples. Their findings revealed that while cold working reduced metal removal, machining temperature played the most significant role in increasing material loss. Surface roughness decreased after cold working but increased with higher machining temperatures and longer processing times. Other studies have examined the effects of different etchants. Çakır [10] compared ferric chloride and cupric chloride in chemically etching copper, finding that cupric chloride provided the smoothest surface finish, while ferric chloride yielded the highest etch rate. Al-Ethari et al. [11] used the Taguchi method to study the effects of machining temperature and time on a NiTi alloy produced via powder metallurgy, concluding that temperature had the most significant impact on surface roughness and metal removal. The study also employed mathematical models to predict machining conditions using regression analysis.

Shather and Ibrahim [12] explored the impact of machining time, temperature, and etchant concentration on the surface finish of stainless steel 304 using acid blends. Their experiments, conducted at three temperatures (45, 50, and 55°C) and three machining times (3, 6, and 9 minutes), showed that higher temperatures and longer machining times resulted in rougher surfaces. Mtb 16 software was used to develop predictive models for optimal machining conditions. Fahad [13] examined the removal of a thin metal layer from low-carbon steel using $FeCl_3$ etchant. The study considered machining times of 1, 3, 5, 7, and 9 minutes with etchant concentrations of 20, 40, and 60%, revealing that the highest metal removal rate was achieved with 60% etchant concentration and a machining time of 1 minute. Ibrahim and Ghullam [14] investigated the effects of machining time, temperature, and etchant concentration on the weight removal of titanium alloys, using mixtures of hydrofluoric acid (HF) and nitric acid (HNO_3). Their study, conducted on eight specimens at four machining temperatures (35, 40, 45, and 50°C) and four machining times (4, 6, 8, and 10 minutes), found that weight removal increased with all three parameters. Alimirzaloo et al. [15] examined the chemical machining of Ti-6Al-4V alloy using HF and HNO_3 solutions. Their findings showed that material removal depth grew with increasing temperatures, longer machining durations, and greater concentrations of HF but greater concentrations of HNO_3 decreased the etching depth to some degree.

While a lot of work has been done on chemical machining, not much work has been done on aluminum alloy 2024, although it is very important in most industrial applications. In this study, the gap will be addressed by an examination of the etching behavior of Al-2024 in a ferric chloride etchant, an examination of the effects of etchant temperature and concentration on mass removal. The findings offer valuable information on the optimization of the chemical etching of the alloy, which advances this field.

2. Experimental Methodology

Experimental work carried out in this study involved experimental testing and a series of calculations for processing the results in the appropriate way. Experimental work for the study was carried out in February 2023, and throughout the period the ambient temperature varied between 15 and 22°C. In a concerted effort to determine both accuracy and reliability of results from this research, 21 samples were prepared with great care and then tested in total, three samples being assigned specifically to each of the different experimental conditions. The material which was selected for the purpose of this research was aluminum alloy 2024 which was in the flat rolled condition at the time of experimentation. Each of the prepared samples had a cross-section of 70 mm x 70 mm and a uniform thickness of 1 mm throughout. The reason for the selection of aluminum alloy 2024 was that it has extensive use in a variety of different industries, in addition to having first-class mechanical properties, such as a very high strength-to-weight ratio and excellent machinability properties. Prior to the etching experiments being conducted, it was necessary to verify that the chemical composition of the aluminum alloy 2024 used in this research met officially approved specifications of the material. To this end, a careful and detailed examination of the chemical composition was undertaken. This complete examination was carried out by Al-Naba Company LTD, which is well known to be a very highly reputable metallurgical testing laboratory in Baghdad, Iraq. The analytical procedure involved the use of an Oxford Foundry Master Xpert optical spectrometer, which is a highly advanced instrument highly capable of the accurate determination of the elemental composition of metallic specimens. The results of the chemical analysis, reported that the composition of the aluminum alloy 2024 used in this research closely matched the specifications given in authority references [12,16]. This confirmation was of the utmost importance, as it determined the validity and comparability of results of this research with those from previous research in the field.

Following the correct identification of the precise composition of the alloy, the second job was the correct preparation of the metal samples to be used in the follow-up experiments. The very first step in this preparation of the metal samples was the correct and thorough cleaning of the metal surface to remove any possible impurities, like oils, grease, or oxidation products, that would have a major role to play in influencing the efficiency of the etching process. This cleaning was done with utmost care and caution to avoid any type of damage to the metal surface itself, and acetone was utilized as the cleaning solvent—a chemical compound that is extensively used in laboratory and metallurgical processes due to its unmatched ability to dissolve a wide range of organic impurities. Following this initial cleaning, the samples were washed with water thoroughly to ensure that no residual traces of the solvent remained on the metal surface. Further, a second cleaning process was carried out, which involved the utilization of distilled water to ensure that no foreign particles or contaminants remained on the clean metal surface. The last step in this rigorous cleaning process was to dry the samples using an air dryer, which was used to efficiently remove any residual traces of moisture that could still have clung to the samples before proceeding to the next critical phase of the experiment. Following the correct drying of the samples, the samples were weighed for the first time using a METTLER AE200 balance, which is a highly sensitive and accurate analytical device widely acclaimed for its ability to detect even the minute changes in weight with great accuracy. The first weighing was an important step in the experiment process in that it provided a baseline reading that would be used in the calculation of the rate of removal of materials during etching. Next, oval-shaped pieces of adhesive made of plastic film material were carefully placed over areas that were marked on the surface of the metal. The adhesive pieces were of known sizes and acted as protective covers, providing a barrier effect that shielded the etchant from the covered areas of the metal surface. The masking process was important to the control of the etching process, with the assurance that material removal would be limited to the specified unmasked areas.

After covering the regions of the metal surface required, the second step of the experiment was the preparation of the etching solution and the determination of the processing conditions required. The etching solution was warmed to the desired temperature in a water bath prior to immersing the samples. The selection of 49°C, as against the originally intended 50°C, was consistent with expert recommendation and literature findings to achieve the best conditions for etching. After the etching solution, a mixture of ferric chloride, had reached thermal equilibrium with the water bath, the pre-weighed samples were left in the solution for a well-controlled duration of 30 minutes. The etching duration was chosen consistent with previous studies and industry practice, ensuring sufficient material removal without over-etching that could result in surface damage or undesirable roughness. During etching, ferric chloride chemically reacted with the aluminum alloy surface, etching material away from the unmasked areas. The chemical reaction was governed by a number of factors, such as solution concentration, temperature, and exposure time, all of which were systematically controlled in this study.

After the 30-minute etching, the samples were carefully taken out of the solution. The samples were immediately rinsed with distilled water to prevent further reaction and to strip off any excess etchant on the surface. The washing process helped prevent residual traces of the corrosive ferric chloride solution remaining on the metal surface, which would otherwise initiate any unwanted post-etching reaction. Rinsed samples were dried to evaporate any excess moisture. After drying, the samples were weighed again on the METTLER AE200 balance. The final weight minus the initial weight of each sample provided a direct measurement of the material removed during the etching process, which was a critical parameter in determining the chemical machining efficiency. The above experimental procedure was carried out several times in a controlled manner for all 21 samples at various different temperatures. The etching process temperature varied from 35°C to 65°C with 5°C intervals between test conditions. These specific temperature ranges were chosen to allow a full study of how temperature influences the etching rate and material removal behavior of aluminum alloy 2024. Extreme care was taken to maintain each temperature condition constant throughout the course of each experiment to eliminate variability that would detract from data accuracy.

Apart from temperature variation, the effect of etchant concentration on etching performance was also studied. To this end, three different concentrations of ferric chloride were used, in terms of the Baume scale (Be), which is commonly applied to measure the specific gravity of chemical solutions. The Baume scale provides a correlation between solution density and concentration, where the chemical characteristics of the etchant can be regulated with precision. The use of multiple concentrations allowed the study to examine the effect of ferric chloride strength on material removal rates, surface finish, and etching uniformity. Observe that the Baume scale measurements applied in this study are directly equivalent to the International System of Units (SI units) by specific gravity conversion. This equivalency of the Baume scale to specific gravity can be accessed in a lot of engineering and scientific literature, such that the data obtained in this study conform to accepted measurement practices.

With controlled variation of significant process parameters through systematic variation and precise control of significant process parameters like etchant concentration and temperature, the experimental research attempted to acquire useful knowledge on chemical machining behavior of aluminum alloy 2024. The results of the present work add to the knowledge pool of the influence of varying etching conditions on material removal rates, topography of the surface, and efficiency of the process. These results are of extreme importance to those industries that utilize chemical machining to attain precision in manufacturing, particularly in operations with aluminum alloy 2024 in wide application, like aerospace, automotive, and electronic manufacturing industries. The experimental process involved in this work was conducted with high precision and with extreme attention to detail to ensure the generation of reproducible and consistent results. Each process, ranging from the selection of starting material and verification of its chemical composition through sample preparation, etching, and post-treatment, was conducted with extreme attention. The controlled variation in etchant concentration along with precise temperature control resulted in a humongous data set that enabled extensive investigation of the etching behavior of aluminum alloy 2024 under diverse conditions. The results of this research contribute to the existing body of knowledge regarding chemical machining and can serve as a stepping stone for future research into process optimization for the etching process of aluminum alloys.

3. Results and Discussion

In this part of the research study, the corrosion behavior of all the samples was monitored carefully under varied temperature conditions at a constant concentration of the etchant at 20 Be throughout the entire experiment. The basic aim of this study was to have a good idea of the way the etching process was affected by varying the temperature. Before the weighing of the samples of the aluminum alloy was subjected to the chemical machining process, their weights were measured carefully by using a very precise balance so that it could be ensured at a later phase of the process that the proper weight loss could be calculated exactly. After careful measurement of the initial weights, each of the samples was subjected to the chemical machining process for 30 minutes in total so that a clear analysis of the etching effects could be determined. After that, the final weights of the samples were weighed carefully again. The weight loss in all cases was then utilized in the following equation to find the etching rate, which is considered equal to the corrosion rate under the various temperature conditions [18]. The results obtained clearly indicate that the etching rate had an increasing trend with an increase in temperature. This behavior can be understood on the basis of the enhanced efficiency of the corrosive solution, which is primarily due to the increased kinetic energy of the reactive species in the solution. With an increase in temperature, the dissolution rate of the exposed metal surface increases, and thus the etching rate increases. This effect, however, occurs only within a specific range of temperature, beyond which there is excessive evaporation of the corrosive solution. This evaporation leads to the loss of the solution in an undesirable manner, decreasing its effectiveness. This observation is in agreement with observations reported by El-Awadi et al. [8], who highlighted the crucial role played by temperature in corrosion-based processes. Moreover, a rise in temperature improves the rate of etching relative to reaction time and subsequent microstructural change [20].

To test the influence of the chemical machining process on surface characteristics, surface roughness of every sample of aluminum alloy was determined before etching. The basic mechanism of chemical machining is dissolution of the exposed metal surface by a chemical reaction induced by an acidic or alkaline etchant. Microscopically, the process is a chemical attack that takes place at the surface of individual grains as well as in the interfacial regions between neighboring grains. The rate of reaction depends on the microstructural features of the metal, leading to etching uniformity varying in different cases. For chemically machined products to achieve a refined surface finish and a uniform appearance, it is essential to have a small and homogeneous grain structure. This requirement presents a challenge when working with metals that exhibit large and irregular grain structures, such as castings and certain alloys from the 2000-series aluminum family. Due to their inherent microstructural characteristics, these alloys often experience uneven etching, resulting in inconsistent surface textures [17]. To quantify the impact of etching, the surface roughness of each sample was remeasured after undergoing the chemical machining process.

The correlation between etchant temperature and surface roughness was systematically examined over a temperature range of 35 to 65 °C. The results indicate a significant increase in surface roughness as the etchant temperature rose. This observation can be attributed to the direct influence of temperature on the corrosion rate, wherein higher temperatures promote more aggressive etching. Additionally, elevated temperatures contribute to increased ion mobility and enhance the activity of oxidizing agents within the solution. The influence of temperature on the retention capacity of dissolved metal content in the solution is another critical factor. As the temperature rises, the kinetic energy of the ions increases, providing the necessary activation energy for faster ion transport across the reaction energy barrier. Consequently, the interaction between the metal and the etchant is intensified, leading to increased etching rates [6, 8]. In the subsequent phase of this investigation, the impact of varying the concentration of the etchant solution on both etching rate and surface roughness was examined. This analysis was carried out by measuring the initial and final weights of aluminum alloy samples before and after undergoing the chemical machining process. Three distinct concentrations of the corrosive solution were employed: 9 Be, 20 Be, and 29 Be. The weight loss measurements obtained from each sample at different concentrations of the ferric chloride solution were used to calculate the etching rate.

The results of the study indicate that as there is an increase in the concentration of the corrosive solution, there is a corresponding increase in the etching rate, which is proportional to the increase. This is due to the fact that with the increase in concentration, there is an increase in the number of electron acceptors in the solution, which are the Fe^{+++} ions. As the concentration of the Fe^{+++} ions increases, it causes an increase in the number of oxidation-reduction reactions taking place on the surface of the metal. This subsequently causes an overall increase in the effectiveness and efficiency of the etching process itself. Furthermore, the increase that is observed in the concentration of the Fe^{+++} ions is in perfect agreement with the principle that has been put forward, that the chemical potential of ions is directly proportional to the levels of their concentration, as cited in source [8].

$$3FeCl3 + Al \rightarrow 3FeCl2 + AlCl3$$

This detailed response provides a descriptive account of the conversion process which transforms aluminum into aluminum chloride, as well as the reduction process which transforms ferric chloride (FeCl3) into ferrous chloride (FeCl2). The mechanism of this reaction emphasizes the central and pivotal role played by Fe+++ ions in facilitating the process of dissolution which takes place on the surface of the aluminum. Furthermore, a precise and systematic investigation was carried out in order to study and analyze the effect that different concentrations of the etchant have on the developed surface roughness of the samples of aluminum. The surface roughness of all the samples was measured systematically both prior to the initiation of the process of chemical machining as well as after the achievement of the termination of the chemical machining process. This enables the satisfactory comparison of the changes which are brought about as a consequence of the application of different concentrations of the corrosive solution. The values of the surface roughness which were measured from these precise measurements were found to have an excellent and perceivable relationship between the concentration of etchant applied as well as the change which is observed in the texture of the surface of the aluminum samples. Results obtained from the current research have revealed that an increase in higher concentrations of the corrosive solution applied leads to more perceivable and significant effects of etching, thus leading to higher surface roughness. The main reason for this perceivable effect resides in the increased level of interaction which takes place between the etchant and the surface of the aluminum when applied at higher concentrations. A higher concentration of the chemical compound yields a higher degree of etching, which in turn leads to higher surface roughness for the aluminum samples. This specific observation is in conformity with the data reported by El-Awadi et al. [8], who emphasized the significance of the chemical composition on the corrosion and etching phenomena. The increase in etching activity which is produced at higher concentrations triggers more aggressive and increased removal of the material, thus influencing the terminal surface morphology of the samples of the aluminum alloy which have been subjected to the chemical machining process.

The conclusions that can be drawn from the results obtained in this research are quite useful and contribute significantly in increasing our understanding of how etchant concentration and temperature influence the chemical machining process of the aluminum alloy 2024. It should be noted that both these factors are of great importance in deciding etching rate and surface roughness, the latter of which has a major impact on the quality of the machined surface. A rise in temperature increases etching efficiency up to a point, but very high temperatures cause loss of corrosive solution by evaporation, which is not desirable. An increase in etchant concentration leads to improved etching efficiency but also in increased surface roughness. These are useful in optimizing chemical machining so that the desired surface finish and etching accuracy can be achieved.

4. Conclusion

Through an in-depth examination and explanation of the results that have been derived, it is possible to describe the following in relation to the effect of temperature and etchant concentration on etching process:

- The temperature, etching rate, and surface roughness are connected to one another. With increasing temperature, the etching rate increases. The increase in temperature leads to the improvement in the surface roughness of the material.
- The corrosion rate and roughness of the surface are strongly a function of etchant solution concentration. As etchant solution concentration increases, both the etch rate and the surface roughness increase.
- Whereas both etchant concentration and temperature are important to etching, temperature has a more profound impact. Temperature variations lead to far larger variations in etching rate and surface roughness than etchant concentration variations.
- The research highlights the significance of etchant concentration and temperature during etching. These are vital parameters that control the etching rate and surface roughness, which, in turn, determine the optimal machining results.

References

1. Bhattacharyya, B. & Doloi, B., Modern Machining Technology Advanced, Hybrid, Micro Machining and Super Finishing Technology, Academic Press, Elsevier, 2020.
2. Harris, W., Chemical Milling, Oxford University Press, Oxford, 1974.
3. Çakır, O., Yardımeden, A. & Özben, T., Chemical Machining, Archives of Materials Science and Engineering, 28(8), pp.499–502, 2007.
4. Youssef, H. & El-Hofy, H., Non-Traditional and Advanced Machining Technologies, Second Edition, CRC Press, ISBN: 978-1-003-05531-0 (ebk), 2021.

5. Abdulhusein, N., Ibrahim, A. & Huayier, A., Influence of Machining Parameters on Surface Roughness in Chemical Machining of Silicon Carbide (Sic), Engineering and Technology Journal, 40(06) pp. 879–884, 2022.
6. Çakir, O., Chemical Etching of Aluminum, Journal of Materials Processing Technology, 199(1-3), pp.337–340, 2008.
7. Lad, Y., Patel, A. & Patel, M., Design and Development of Chemical Machining Setup, IOSR Journal of Mechanical and Civil Engineering, 16(1), Ser. V, pp. 36–40, 2019.
8. El-Awadi, G., Enb, T., Abdel-Samad, S. & El-Halawany, M., Chemical Machining for Stainless Steel, Aluminum and Copper Sheets at Different Etchant Conditions, Arab Journal of Nuclear Science and Applications, 49(2), pp. 132–139, 2016.
9. Al-Ethari, H., Alsultani, K. & Nasreen, F., Variables Affecting the Chemical Machining of Stainless Steel 420, International Journal of Engineering and Innovative Technology (IJEIT), 3(6), pp. 210–216, 2013.
10. Çakir, O., Temel, H. & Kiyak, M., Chemical Etching of Cu-ETP Copper, Journal of Materials Processing Technology, 162–163, pp. 275–279, 2005.
11. Al-Ethari, H., Haleem, A. & Gased, N., An Investigation on Chemical Machining of Niti SMA Prepared by Powder Metallurgy, IOP Conf. Series: Materials Science and Engineering, 518, 032032, 2019.
12. Shather, S. & Ibrahim, A., Influence of Machining Parameters on Surface Roughness in Chemical Machining of Stainless Steel 304, Eng. & Tech. Journal, 33 Part (A)(6), pp. 1377–1388, 2015.
13. Fahad, N., Studying Effect of Static Concentration for Chemical Machining on Surface Roughness and Metal Removal Rate for Low Carbon Steel, Muthanna Journal of Engineering and Technology (Mjet), 5(1), pp. 22–26, 2017.
14. Ibrahim, A. & Ghullam, D., The Effect of Machining Parameters on Weight Removal in Chemical Machining of Titanium Alloy, Diyala Journal of Engineering Sciences, 12(04), pp. 33–40, 2019.
15. Alimirzaloo, V., Modanloo, V. & Hadavifar, M., Investigation of the Effective Parameters on the Surface Roughness and Material Removal Depth in Chemical Machining of Ti-6Al-4V Alloy, Modares Mechanical Engineering, Proceedings of the Advanced Machining and Machine Tools Conference, 15(13), pp. 410–415, 2015.
16. Davis, J., Metals Handbook-Desk Edition, Second Edition, ASM, USA, 1998.
17. ASM, ASM Handbook, 16, Machining, 9th Edition, ASM, USA, 1991.
18. Wang, S., Free, M.L., Alam, S., Zhang, M. & Taylor, P.R., Applications of Process Engineering Principles in Materials Processing, Energy and Environmental Technologies, Springer International Publishing, 2017.
19. Anderson, K., Weritz, J. & Kaufman, J., ASM Handbook, 2B: Properties and Selection of Aluminum Alloys, ASM, USA, 2019.
20. Burham, N., Sugandi, G., Nor, M.M. & Majlis, B.Y., Effect of Temperature on the Etching Rate of Nitride and Oxide Layer Using Buffered Oxide Etch, International Conference on Advances in Electrical, Electronic and Systems Engineering (ICAEES), Malaysia, 14–16 November, 2016.

Advances in Additive Manufacturing Technologies – Gurusamy Pathinettampadian et al. (eds)
© 2026 Taylor & Francis Group, London, ISBN 978-1-041-16687-0

Maximizing Performance in Abrasinve Water Jet Machining of Hard Metals Through Parameter Optimization

B. Yokeshkumar[1]
Assisstant Professor, Department of Mechanical Engineering,
Chennai Institute of Technology,
Chennai, India

P. Raja[2]
Associate Professor, Department of Mechanical Engineering,
Prathyusha Engineering College,
Chennai, India

Sakthi Murugan[3]
Student, Department of Mechanical Engineering,
Chennai Institute of Technology,
Chennai, India

P. Usha Rani[4]
Professor, Science and Humanities Department,
R.M.D Engineering College, India

K. Karthik[5]
Associate Professor,
Department of Mechanical Engineering,
Vel Tech Rangarajan Dr. Sagunthala R&D Institute of Science Technology,
Chennai, India

Veni Krishnabharathi[6]
Librarian,
Tamilvel Umamaheswaranar Karanthai Arts College,
India

R. Karthick[7]
Assistant Professor, Department of Mechanical Engineering,
Rajalakshmi Institute of Technology,
Chennai, India

◆ **Abstract:** This work aims to optimize process parameters in AWJM and material-related parameters specifically that control the performance of machining. With controlled variation of water pressure, abrasive flow rate, traverse speed, and standoff distance, the work develops a multi-dimensional platform for comparative analysis. Experimental data are used to develop high-accuracy predictive models of MRR and DOC after validation against measured outcomes. Furthermore, various metaheuristic algorithms are employed to optimize AWJM parameters in an attempt

[1]yokeshkumarb@citchennai.net, [2]rajaponnu79@gmail.com, [3]sakti975murugan@gmail.com, [4]pusharani71@yahoo.com, [5]karthikk@veltech.edu.in, [6]venikrishnabharathi@gmail.com, [7]karthick.industrial@gmail.com

DOI: 10.1201/9781003685906-31

to enhance performance specific to various materials. The findings indicate significant improvements in MRR and DOC compared to experimental controls, thereby validating the effectiveness of the proposed optimization process. The research also applies to a wider industrial range of machining applications, thereby contributing further to the development of innovative machining processes according to the inherent characteristics of various materials. The proposed method is an invaluable decision-support tool for engineers in selecting optimal AWJM parameters for machining hard materials, thereby paving the way for future research and development in material cutting processes. The findings described in this research are a critical breakthrough in enhancing AWJM processes with higher precision and efficiency, thereby making them applicable in a wide variety of industries.

◆ **Keywords:** AWJM, MRR, DOC, Optimization, Cuckoo, MOTLBO

1. Introduction

AWJM is a very versatile and effective non-conventional machining process with several distinct advantages, such as non-thermal cutting, low wastage of material, high surface finish, and high dimensional accuracy. Because of its intrinsic flexibility, AWJM has wide-ranging applications in a wide variety of industries for machining metals, composites, ceramics, and other high-tech materials. Nevertheless, optimal machining performance in AWJM demands precise control and fine-tuning of key process parameters like WP, AFR, TS, and SOD. Optimal optimization of these parameters is necessary to enhance machining results, especially in terms of maximizing MRR and achieving the desired DOC. With the increasing demand for high-precision machining solutions, researchers and engineers have been actively involved in formulating systematic methods to optimize AWJM parameters for a wide range of material types. Empirical methods and experiential techniques were conventionally used to identify appropriate parameter settings for a given material. Although these traditional methods yielded useful information, they were generally time-consuming, costly, and not flexible enough to handle a wide range of materials. Therefore, the formulation of computational modeling and advanced optimization methods has transformed AWJM research, allowing more accurate and efficient parameter selection.

Metaheuristic optimization algorithms have been demonstrated, in recent years, to be effective tools for parameter optimization of the AWJM process. Metaheuristic optimization algorithms provide systematic and data-driven approaches to traverse large solution spaces to determine optimal machining conditions with high accuracy. The primary aim of this study is to suggest a material-dependent optimization model for AWJM considering significant PP such as WP, AFR, FR, and SOD. These parameters play a fundamental role in determining the machining features, influencing material removal efficiency, surface quality, and overall process performance. In a pedagogical and systematic research approach, this study investigates intricate interconnections between AWJM parameters and machining outcomes, enhancing precision machining technology. The experimental results reported in this study are the basis for machining performance analysis for various parameter combinations. Through systematic process parameter influence investigation on MRR and DOC, the study establishes performance standards and identifies trends that guide the optimization process. Besides, mathematical models are developed to describe the interconnections between AWJM parameters and performance values. These models serve as a fundamental basis to realize the intricate interconnections in the AWJM process and form a basis for the application of optimization techniques.

To realize the optimal choice of parameters, this study employs different metaheuristic algorithms including Firefly Algorithm, PSO, Cuckoo Search, GWO, and MOTLBO. The algorithms are chosen because they are able to search multi-dimensional solution spaces efficiently and determine parameter settings that result in the best machining performance. The use of metaheuristic optimization methods extends the scope of application of conventional empirical methods to allow exploration of large parameter spaces that cannot be efficiently explored with conventional analysis techniques. Computational intelligence employed in AWJM parameter optimization is new in machining technology. With the computational power of metaheuristic algorithms, this study demonstrates that machining conditions optimal for maximum efficiency and accuracy of AWJM can be realized systematically. Experimental results and algorithmic results presented in this paper validate the use of the proposed optimization method, suggesting its potential as a useful tool for improving machining performance.

In addition to parameter optimization, broader contexts of this research are in material processing. Optimized AWJM technology, with its cross-applicability to different material properties, provides the potential for tailored machining solutions for specialized industrial processes. The generic applicability of AWJM to machine a broad variety of materials, combined with sophisticated optimization methods, allows precision manufacturing technologies to be developed at a rapid rate. The integrated framework of approach followed in this work—experimentation, mathematical modeling, and metaheuristic optimization—offers a logical approach towards improving AWJM performance. The following sections of this paper outline the methodology, results, and major conclusions, thus offering useful information to researchers, engineers, and industry experts looking for enhanced machining accuracy and efficiency.

A lot of efforts have been taken by researchers to enhance machining parameters in AWJM and other sophisticated machining processes. Mellal and Williams (2016) have taken considerable efforts to enhance machining parameters in various processes like drilling, grinding, abrasive jet machining, ultrasonic machining, and AWJM. To enhance machining performance significantly, their study used optimization techniques inspired by nature, such as the Cuckoo Search Algorithm and Hoopoe Heuristics. Using the Taguchi approach, Shukla and Singh (2017) investigated optimizing AWJM parameters for AA6351 aluminum alloy. After testing the Taguchi method's efficacy in AWJM applications and utilizing seven different advanced optimization strategies, their study concluded that the bio-geography-based optimization algorithm was the best option.

The optimization of abrasive water jet contour cutting for AISI 304L stainless steel was the subject of Llanto, Vafardar, and Tolouee Rad's (2022) subjective study. Contributing significantly to our understanding of how hydrodynamic, acoustic, and erosional elements impact cut quality and production efficiency, they utilized linear regression models and response surface methods. Magnetic abrasive flow finishing, a process that combines abrasive finishing with electromagnets, was studied by Yadav and Jayswal (2021) on aluminum (6061-T6). Results from verification tests confirmed that their study, which used the Firefly Algorithm and utility theory, significantly improved material removal and surface polish. Hot abrasive jet machining (FB-HAJM) is a fluidized bed machining process that Pradhan et al. used in their models and tests on K-80 alumina. Results showed promise for using this method to machine high-strength materials.

Kesarwani et al proposed a new method based on the Ant Lion Optimization (ALO) algorithm hybridized with milling polymer-reinforced parts that contain 0D-CNOs. The researchers carried out a research on the use of the introduced machining process by industries. The machining of Si_3N_4-reinforced AA5052 aluminum matrix composites by Gopinath and Senthilkumar (2023) was analyzed with particular consideration of the complexities involved in retaining optimal machinability without compromise on mechanical strength. The material removal rates were found to improve and machining efficiency increased. Findings of such studies highlight the growing importance of computational intelligence, and optimization approaches in advanced machining processes. Such methods are made stronger by the current study combining experimental findings, mathematical modeling, and metaheuristic optimization with the aim to optimize the setting of parameters used in AWJM to improve the machining precision, increase efficiency, and provide an effective decision-making process for use in various industry processes..

2. Experimental Methodology

For experimental investigation of AWJM, suitable process parameters were first wisely selected. Selected parameters are water pressure (bar), abrasive flow rate (g/min), feed rate (mm/min), and stand-off distance (mm). It facilitates measurements of the variation of the MRR and DOC for varied sets of parameters. Once the experimental design is implemented, the results are analyzed to determine correlations and relationships between AWJM process parameters and the machined results. Specifically, mathematical models are derived to describe these relationships, as in models (3.3) and (3.4), which were previously determined for predictive analysis. The only aim of these models is to serve as a reference guide before actual machining operations, such that accurate selection of process parameters can be made to optimize performance.

A pioneering area of AWJM research is control parameter-machining productivity interaction. The optimization of the parameters directly influences process efficiency, and their optimization is required to achieve high-performance machining. For making it feasible, models are formulated to represent the complex relationship between the natural logarithm of the MRR index (ln(MRR)) and the natural logarithm of the Depth of Cut (ln(DOC)). The development of these models requires regression analysis techniques that are generally performed with the help of statistical software packages like Python, R, or similar statistical analysis packages. The principal objective is to identify the optimum parameters explaining the logarithmic relationship that governs the machining process. Upon development of the models, they are validated against experimental results to ascertain that they are correct and can be used for predictions. Optimization with metaheuristic algorithms has a vital role in searching the parameter space in a systematic way to identify optimal combinations providing maximum MRR and target DOC. Use of various optimization algorithms like Firefly Algorithm, PSO, Cuckoo Search, GWO, and MOTLBO is chosen based on their demonstrated ability in handling complicated estimation problems. The optimization objective function is formulated to maximize MRR and optimize DOC, as represented by Equations (1) and (2).

The Firefly Algorithm was highly effective in determining optimal AWJM process parameters. The optimized parameters yielded an MRR of 13.76 g/min and a DOC of 17.97 mm. These were highly close to the experimental values, thus validating the potential of the algorithm in parameter optimization. The Firefly Algorithm was superior to other methods in the sense that it yielded an optimized MRR value of 13.76 g/min, which highly corresponds to the experimental value of 14.13 g/min. The marginal difference noted in the DOC, being 17.97 mm against the experimental value of 18.32 mm, is a testimony to the robustness of the algorithm in reaching the optimal solution. Additionally, the rapid convergence of the algorithm to the optimal parameter set is an indicator of its potential in optimizing the efficiency of the AWJM process.

The Cuckoo Search Algorithm generated encouraging results with optimized MRR and DOC as 14.62 g/min and 13.69 mm, respectively. The optimization approach effectively identified optimum sets of parameters, especially the MRR optimization. The negligible difference between predicted and experimental values of DOC (13.69 mm vs. 18.32 mm) indicates the extent of further fine-tuning. While the algorithm effectively recognized optimal conditions for MRR, the lower predicted DOC values indicate further fine-tuning may be required to optimize the process to a greater degree. Future research needs to be carried out to study the effect of parameter variation on machining efficiency.

PSO optimised AWJM parameters effectively, and the predicted MRR of 9.76 g/min and the predicted DOC of 8.88 mm were obtained. Theoretical and experimental MRR were in excellent agreement, and this is evidence of the credibility of the algorithm in optimising material removal rate. DOC had minimal discrepancy (8.88 mm vs. 8.32 mm experimental), and thus depth control optimisation is essential. PSO is a suitable optimisation method for AWJM, and it has been found to be effective in optimised process conditions derivation.

GWO was found effective in optimizing the parameters of AWJM with an MRR of 16.79 g/min and a DOC of 12.79 mm. The two-objective nature of the algorithm increases convergence and search space exploration, leading to process optimization improvement. The agreement between the predicted and experimental MRR values confirms the ability of the algorithm in improving machining efficiency. The optimized values of DOC also confirm its ability in solving multiple machining objectives.

The MOTLBO algorithm was also seen to exhibit outstanding performance in optimization with an MRR of 16.11 g/min and a DOC value of 12.76 mm. The strategy greatly improves material removal rate and depth of cut and closely tracks experimental results. The results validate the performance of the algorithm over multi-objective machining parameter optimization and therefore propose that it is a good choice for AWJM process optimization.

SSA optimization provided an MRR of 10.76 g/min and a DOC of 10.91 mm. The algorithm proved to be effective in MRR and DOC optimization by providing results that are in close proximity to experimental results. SSA proved to be precise in prediction, particularly in DOC optimization, which tends to be a challenging parameter to optimize in AWJM. The above results reveal that SSA can be utilized as a useful tool to further enhance the efficiency of AWJM.

Experimental results comparison with optimized results provides unambiguous insights into AWJM parameter optimization. Firefly Algorithm was superior to others in delivering the closest estimation of experimental MRR and DOC values. Cuckoo Search optimization worked very effectively in maximizing MRR but required further optimization in DOC optimization. PSO showed stable performance in MRR prediction, but DOC prediction can be improved. GWO and MOTLBO provided very well-balanced solutions and worked well with double-objective optimization. SSA worked well in DOC optimization, with great potential in AWJM parameter balancing.

3. Results and Discussion

The optimization of AWJM parameters is primarily interested in the analysis of performance and efficiency of different optimization algorithms to achieve optimal control parameter combination values. Optimization is primarily interested in the maximization of MRR and DOC, which are two key performance parameters in AWJM. Different optimization algorithms have been compared based on performance to optimize these objectives, including GWO, MOTLBO, the Firefly Algorithm, PSO, Cuckoo Search, and the particle swarm-based Salp Swarm Algorithm.

MRR is an important measure of AWJM machining efficiency due to the fact that it measures the rate of material removal from the workpiece. Some of the algorithms better optimize MRR than others. According to the results, GWO and MOTLBO have the same performance level, where the MRR values are nearly the same as their experimentally found values. Precisely, Firefly and PSO algorithms better optimize MRR and are suitable selections for this optimization goal. Cuckoo Search performs slightly lower than the optimal algorithms, and the Salp Swarm Algorithm performs the worst in optimizing MRR.

The Mean Recognition Rate (MRR) of the Firefly Algorithm is 13.76 g/min, which is close to the experimental value. Optimizing MRR, however, the value of DOC is kept constant at 17.97 mm. This slight difference from DOC indicates that while the algorithm optimizes MRR, it is not fully in line with experimental values when it comes to DOC. PSO has an MRR of 9.76 g/min with a DOC of 8.88 mm. This indicates that PSO is more suitable for MRR optimization but provides a relatively lower balance for MRR and DOC. This algorithm has a very high MRR of 140.00 g/min with a DOC of 262.80 mm. While the values validate efficient optimization of Monthly Recurring Revenue (MRR), they also imply that the algorithm can be optimized to be practically effective.

GWO obtains a MRR of 16.79 g/min and a DOC of 12.79 mm. This shows that GWO obtains both MRR and DOC through a balanced approach. The MOTLBO algorithm obtains an MRR of 16.11 g/min and a DOC of 12.76 mm, showing high potential in the optimization of both parameters. For comparative purposes, the Salp Swarm Algorithm obtains poor average performance and has an MRR of 10.76 g/min and a DOC of 10.91 mm and therefore falls behind other algorithms in terms of obtaining a higher value of MRR.

Depth of Cut (DOC) is another significant performance parameter of AWJM influencing the accuracy and efficiency of material removal. The optimal DOC values of the compared algorithms are obtained by GWO, MOTLBO, and the Salp Swarm Algorithm, and these algorithms can obtain good cutting depths with good accuracy. These algorithms have slightly lower DOC values than other methods though still giving an acceptable compromise between MRR and DOC optimization. This algorithm gives the maximum deviation from other algorithms regarding DOC optimization. Though its high MRR values are impressive, its high DOC values indicate that it may not be the best option when accuracy is important.

When both MRR and DOC are considered simultaneously, GWO and MOTLBO are the best balanced optimization algorithms. Both the algorithms are well-balanced, optimizing the most critical parameters without compromising overall machining performance. GWO and MOTLBO algorithms have been tested and found to be robust and efficient for optimization, particularly for AWJM processes. The balance between MRR and DOC optimization makes GWO and MOTLBO ideal choices for high-precision and efficiency machining processes. Firefly and PSO algorithms have moderate balance of MRR and DOC but are inclined towards maximizing MRR at the cost of DOC. Salp Swarm Algorithm method favors optimizing DOC with moderate amounts of MRR, and thus it is best applied to applications where precision for DOC is required. Although good for optimizing MRR, Cuckoo Search is imbalanced with DOC, suggesting that further optimization may be required for overall optimization.

The computational efficiency of an algorithm is measured by how quickly it converges to the optimum solution. In this case, the tests indicate that PSO and the Salp Swarm Algorithm converge at extremely high rates. GWO and MOTLBO algorithms also converge at extremely high rates, which confirm their efficiency in optimization problems. Firefly and Cuckoo Search Algorithm methods require more parameter fine-tuning to converge as fast as other methods. Even though they optimize extremely well, their computational efficiency can be improved by further fine-tuning.

Sensitivity analysis was conducted to identify the performance of different algorithms to manage parameter variations and randomness. The analysis shows that the performance of the algorithms is very sensitive to parameter settings and randomness, with some being more robust and more adaptive than others. GWO and MOTLBO algorithms were stable under different conditions, showing their robustness and reliability. Firefly and PSO algorithms are comparatively insensitive to parameter variations, showing applications where small parameter variations in machining conditions are expected. Cuckoo Search and Salp Swarm Algorithm methods are sensitive to variations, showing that their optimization performance is likely to be affected by external variations.

The selection of the optimization algorithm depends on the specific goals and priorities of machining. While no algorithm is superior to all others in all aspects, some algorithms are better suited to some optimization problems. For overall optimization of MRR and DOC, GWO and MOTLBO are best suited, which provide balanced solutions. For MRR-based optimization, Firefly Algorithm and PSO are alike. For DOC accuracy optimization, Salp Swarm Algorithm is a good option.

4. Conclusion

- A critical analysis of different optimization algorithms for AWJM provides valuable insights into performance and areas of potential improvement in terms of material and manufacture in the future.
- Depth of Cut (DOC) and MRR are important variables in determining the performance of optimization algorithms because they are tasked with determining machining precision and efficiency.
- GWO efficiently maximizes MRR by adopting a conservative and balanced strategy of material removal quality. The process is particularly beneficial in manufacturing operations where accuracy and effectiveness are required..
- Cuckoo Search and MOTLBO show strong performance in MRR optimization but may not always achieve optimal DOC results.

References

1. Mellal, M.A., Williams, E.J. Parameter optimization of advanced machining processes using cuckoo optimization algorithm and hoopoe heuristic. J Intell Manuf 27, 927–942 (2016). https://doi.org/10.1007/s10845-014-0925-4
2. Shukla, Rajkamal, and Dinesh Singh. "Experimentation investigation of abrasive water jet machining parameters using Taguchi and Evolutionary optimization techniques." Swarm and Evolutionary Computation 32 (2017): 167–183.
3. Nagarajan, Vengatajalapathi, et al. "Meta-heuristic technique-based parametric optimization for electrochemical machining of monel 400 alloys to investigate the material removal rate and the sludge." Applied Sciences 12.6 (2022): 2793.
4. Shastri, Apoorva S., et al. "Multi-cohort intelligence algorithm for solving advanced manufacturing process problems." Neural Computing and Applications 32 (2020): 15055–15075.
5. Sivalingam, Vinothkumar, et al. "Optimization of process parameters for turning Hastelloy X under different machining environments using evolutionary algorithms: A comparative study." Applied Sciences 11.20 (2021): 9725.
6. Llanto, Jennifer Milaor, Ana Vafadar, and Majid Tolouei-Rad. "Multi-objective optimisation in abrasive waterjet contour cutting of AISI 304L." Production Engineering and Robust Control. IntechOpen, 2022.
7. Zolpakar, Nor Atiqah, Mohd Fuad Yasak, and Sunil Pathak. "A review: use of evolutionary algorithm for optimisation of machining parameters." The International Journal of Advanced Manufacturing Technology 115 (2021): 31–47.
8. BV, Dharmendra, Shyam Prasad Kodali, and Nageswara Rao Boggarapu. "Multi-objective optimization for optimum abrasive water jet machining process parameters of Inconel718 adopting the Taguchi approach." Multidiscipline modeling in Materials and structures 16.2 (2020): 306–321.
9. Pandey, Arun Kumar Sriram, Ankit Saroj, and Anshuman Srivastava. "Process Parameter Optimization of Abrasive Jet, Ultrasonic, Laser Beam, Electrochemical, and Plasma Arc Machining Processes Using Optimization Techniques: A Review." SAE International Journal of Materials and Manufacturing 16.05-16- 03-0018 (2023).
10. Wan, Liang, et al. "Analytical modeling and multi-objective optimization algorithm for abrasive waterjet milling Ti6Al4V." The International Journal of Advanced Manufacturing Technology 123.11-12 (2022): 4367–4384.
11. Yadav, Pawan Kumar, and S. C. Jayswal. "Multi-Response Optimization of Magnetic Abrasive Flow Finishing Process on Aluminum (6061-T6) Using Utility Concept Embedded Firefly's Algorithm." Journal of Advanced Manufacturing Systems 20.01 (2021): 51–74.
12. Pradhan, Subhadip, et al. "Experimental study and simulation of surface generation during machining of K-80 alumina ceramic in modified abrasive jet machining with different temperatures using Al2O3 abrasives." International Journal of Abrasive Technology 10.4 (2021): 298–329.
13. Ong, Kok Meng, Pauline Ong, and Chee Kiong Sia. "A new flower pollination algorithm with improved convergence and its application to engineering optimization." Decision Analytics Journal 5 (2022): 100144.
14. Abdelaoui, Fatima Zohra El, Abdelouahhab Jabri, and Abdellah El Barkany. "Optimization techniques for energy efficiency in machining processes—a review." The International Journal of Advanced Manufacturing Technology 125.7–8 (2023): 2967–3001.
15. Kesarwani, Shivi, and Rajesh Kumar Verma. "Ant Lion Optimizer (ALO) algorithm for machinability assessment during Milling of polymer composites modified by zero-dimensional carbon nano onions (0D‹CNOs)." Measurement 187 (2022): 110282.
16. Reddy, Ruturaj, et al. "LAB: a leader–advocate–believer-based optimization algorithm." Soft Computing (2023): 1–35.
17. Che, Yanhui, and Dengxu He. "An enhanced seagull optimization algorithm for solving engineering optimization problems." Applied Intelligence 52.11 (2022): 13043–13081.
18. Gopal, Mahesh, and Endalkachew Mosisa Gutema. "Factors Affecting and Optimization Methods used in Machining Duplex Stainless Steel-A Critical Review." Journal of Engineering Science & Technology Review 14.2 (2021).
19. Gopinath, M., and N. Senthilkumar. "Improving productivity through maximizing MRR in machining silicon nitride ceramic reinforced aluminium alloy 5052." Materials Today: Proceedings 79 (2023): 122–126.

Advances in Additive Manufacturing Technologies – Gurusamy Pathinettampadian et al. (eds)
© 2026 Taylor & Francis Group, London, ISBN 978-1-041-16687-0

Design and Analysis of 3 Stage Epicyclic Planetary Reduction Gear for Aircraft Applications

M. Kalaiyarasi[1]
Department of Mechanical, Chennai Institute of Technology,
Kundrathur, Chennai

K. Nallathambi[2]
Department of Mechanical, Chennai Institute of Technology,
Kundrathur, Chennai

◆ **Abstract:** In this project, a three-stage epicyclic planetary reduction gear unit is developed to meet the output requirements. Solidworks is used to design every component in order to verify assembly and interference. An evaluation is carried out on the components that were designed to determine their strength. Four control surfaces are present on heavy trucks for steering and stabilisation purposes. Depending on the needs of the system, the control surface must be rotated. Actuators allow the control system to rotate in this way. The prime mover, motor, and gearbox mechanism are electro-mechanical actuators that are used in the current project. This gearbox system comes in both screw and gear varieties. In this project, we employ a gearbox system that uses gears and is created based on the specifications. Typically, a motor spins at a relatively fast speed with little torque. The goal of our current study is to create a 3D model of a planetary gear with numerous layers using the Solid Works programme and to analyse its strength using the analytical software. To investigate the temperature deformation and the maximal equivalent stress. There has been a comparison study between various materials and phases.

◆ **Keywords:** Solid works, Reduction gear, Epicyclic gear, Aircraft, Structural

1. Introduction

1.1 Reduction Gear in Flight

A speed reduction unit is a gearbox or a belt and pulley mechanism used to lower the output revolutions per minute (rpm) from the higher input rpm of the power plant. As a result, aeroplane propellers may be efficiently turned by small-diameter internal combustion automobile engines.

Propeller reduction was necessary for large engines with high crankshaft speeds and power outputs, and pilots noticed that identical aircraft with reduction gearing performed better. Instead of employing compound gear trains, high gear ratios may be achieved with very small gear configurations by using planetary gearboxes.

1.2 Definition of Gear

A mechanical tool called a gear is used to modify a motor's speed (RPM) or enhance its output torque. The gearbox is attached to one end of the motor shaft, and the internal arrangement of the gears produces an output torque and speed that are set by the gear ratio.

1.3 Purpose of a Epicyclic Gear

The fundamental goal of this system is to compromise between speed and torque in order to provide the vehicle the

[1]selvakalai2023@gmail.com, [2]knallathambi3383@gmail.com

DOI: 10.1201/9781003685906-32

required push to accelerate. A single stage gear reduction, a multistage reduction, and a planetary or an epicyclic gear are the many gear configurations that can be utilised. The gear axes are fixed and the gears revolve around their respective axes in single-stage and multi-stage gear.

1.4 Principle of a Epicyclic Gear

The input gears are fastened to the countershaft according to the gear's basic design, making them one cohesive unit. It turns the main shaft's separate gears, which may freely spin on their bearings. In accordance with the gear that engages on the main shaft, the gearbox transmits the drive to the wheels.

When moving energy mechanically from one device to another, a gear is utilised to enhance torque while decreasing speed. The power produced by bending or twisting a solid substance is known as torque. The words "transfer" and "transmission" are frequently used interchangeably.

Epicylic Gear

One kind of gear train that is utilised to convey motion is the planetary or epicyclic gear train. Epicyclic gear trains are made up of two or more gears placed so that one gear's centre spins around the other gear's centre. Gear trains having relative axe movements are called epicyclic gear trains, often referred to as planetary gear trains. The planet gear and sun gear are carried around each other by a carrier that links the centres of the two gears. The gears of the sun and planet mesh to make sure that their pitch circles roll smoothly. A single rotating component known as a cage, arm, or carrier is used to attach each planet. It produces a low-speed, high-torque output when the globe carrier rotates..

2. Literature Review

2.1 K Akhila and Reddy ANR, Design, Modelling and Analysis of a Two Stage Epicyclic Reduction Gear Unitofaflight Vehicle

In the Antikythera Mechanism, built around 80 BCE, epicyclic gearing was utilised to modify the exhibited position of the moon for its elasticity and even for the ellipticity's precession. Two gears that were facing each other were rotated about slightly different centres, and one of the gears drove the other not with meshing teeth but rather with a pin that was put into a slot on the other gear. The radius of driving would fluctuate as the slot rotated the second gear, causing the driven gear to accelerate and decelerate with each rotation.

2.2 Timir Patel, Ashutosh Dubey, Lokavarapu Bhaskara Rao, Design and Analysis of an Epicyclic Gear Box Foran Electric Drivetrain

The design and optimisation of a planetary gearbox for an electric FSAE automobile is the main topic of this study. For an FSAE automobile, the typical single- or double-stage reduction gearbox layout is large and uneconomical due to its low power-to-weight and high volume-to-weight ratios. A planetary gearbox provides equilibrium in the amount of power delivered and weight. To choose the best gearbox size and materials, the design process was an iterative process. The technique made use of the common equations for gear calculations as well as computer simulations for designing and CAD modelling of different gearbox components. The common force equations are used in the design of the gearbox shafts and bearings. Additionally, the lubricating system is chosen using recognised criteria.

2.3 Yanbiao FENG, Wenming ZHANG, Jue YANG, Zuomin DONG, Design Optimization of a Three-Stage Planetary Gear Reducer Using Genetical Gorithm

Heavy-duty machinery such shield tunnelling machines, tracked excavators, mining trucks, and crushers frequently use multi-stage reducers, particularly planetary gear reducers. These application areas call for excellent geometrical mechanical performance, extended life, high load capacity, and other requirements. This study initially introduces a unique three-stage reducer design. This research provides an optimisation mechanism based on genetic algorithm (GA) to accomplish those goals together. The objective function is set to the geometrical volume. As design factors, the gear module, teeth count, and gear face width are selected, with restrictions such as life, geometrical spacing, efficiency, load capacity, and others taken into consideration. The optimization's findings are good, and by meeting requirements, they can assist designers in using unique architecture.

3. Design of Reduction Gear Unit

The project's primary goal is to build and evaluate a three-stage epicyclic planetary reduction gear unit with a reduction ratio of 64:1, which is meant to fulfil output parameters. To validate the assembly and interference, SOLIDWORKS is used to design each component. Analyses are performed on the planned components to evaluate their tensile strength. Our goal is to obtain a 3-stage epicyclic planetary reduction gear with a reduction ratio of 64:1.

3.1 Mechanical Properties

The mechanical properties of annealed cast iron are displayed in the following table.

Table 32.1

Fig. 32.1 Three stage epicyclic gear

A3 Stage epicyclicplanetary reduction has been designed in the following way as follows

First stage ratio-3.8:1 Second stage ratio-3.7:1

Fig. 32.2 Assembly of first gear

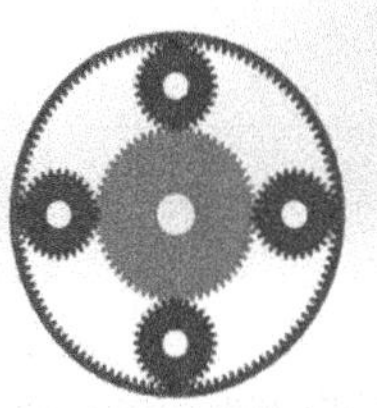

Fig. 32.3 Assembly of second gear

Third stage ratio-4.6:1

Fig. 32.4 Assembly of third gear

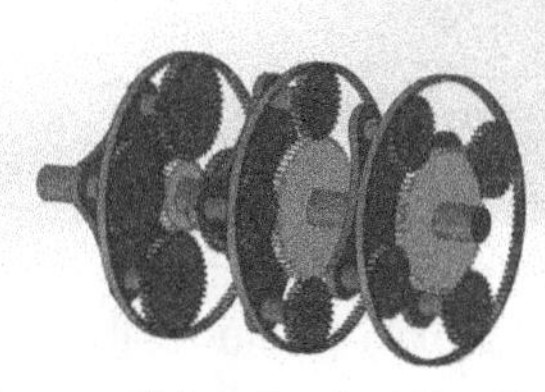

Fig. 32.5 Final assembly of a 3 stage epicyclic planetary reduction gear with an armat each stage

4. Design Calculations

Calculation: For First and Second stage:

Module, $m = 2a/(Z_1+Z_2)m$ $M = 2$

Pressure angle = 20 degree Centre distance, $a = m(Z_1 + Z_2)/2$

$a = 2\ (60+30)/2$ $a = 90$

Height factor, $[f_0 = 1]$ for full depthteeth.

Bottom Clearance, $[c = 0.25m]$ for full depthteeth

Toothdepth, $[h = 2.25m]$ for full depthteeth Patch Circle diameter, $d \Rightarrow d_1 = MZ_1$; $d_2 = M\ Z_2$

$d_1 = 2 \text{ x } 60 = 120$; $d_2 = 2 \text{ x } 30 = 60$

Tip diameter, $da \Rightarrow da_1 = (Z_1 + 2fo)\ m$

$da_1 = 124$; $da2 = 64$

Root diameter, $d_f \Rightarrow d_{f1} = (Z_1 - 2\ f_0)m - 2c$; $d_{f1} = 115.5$

Reduction ratio, $i = N_1/N_2$ or Z_1/Z_2 $i = 5500/1450$; $i = 3.8:1$

Table 32.1 Properties of 18CrNiMo7

Model Reference	Properties		Components
	Name:	**18CrNiMo7**	Solid Body 1 (Boss-Extrude1) (internal spurgear_isoad-1), Solid Body 1 (Bore) (spurgear_iso-10), Solid Body 1 (Bore) (spurgear_iso-15), Solid Body 1 (Bore) (spurgear_iso-16), Solid Body 1 (Bore) (spurgear_iso-17), Solid Body 1 (Bore) (spurgear_iso-8)
	Model type:	Linear Elastic Isotropic	
	Default failure criterion:	Maxvon Mises Stress	
	Yield strength:	5.9e+07 N/m^2	
	Tensile strength:	3.17e+08N/m^2	
	Elastic modulus:	2.1e+11N/m^2	
	Poisson's ratio:	0.31	
	Mass density:	8,500kg/m^3	
	Shear modulus:	7.9e+10N/m^2	
	Thermal expansion coefficient:	1.7e-05/Kelvin	

5. Result and Analysis

With the aid of the solidworks programme, a three stage epicyclic planetary reduction gear was created. A shaft and arm with the necessary dimensions are used to construct each of the three phases once they have been independently developed. After doing study on the gear unit, maximum and lowest stresses as well as displacement may be established. Only the second stage is analysed since it provides more accurate average maximum and lowest stresses. The first stage ratio is 3.8:1, the second stage is 3.7:1, and the third stage is 4.6:1. These are the reduction ratios at the three stages.

5.1 Static Structural Analysis

In a static structural analysis, the displacements, stresses, strains, and forces in a structure or component are calculated as a function of the load, assuming that the load has little or no damping and inertia. Following study on a three-stage epicyclic planetary reduction gear unit, maximum and minimum stresses for the material (18CrNiMo7) are determined, and temperature displacement is also determined as follows:Nodal remedies Maximum displacement: 0.0006689 mm, maximum stress: 4.206 MPa, minimum stress: -0.77 Mpa

5.2 Thermal Analysis

Analytical gearbox design research using input power of 10 KW and 15 KW, respectively, and input speed of 650 rpm. It is discovered that the heat produced by friction between the gears is between 165.150C and 255.180C, which is more than the permitted limit allowed by the standard.

Fig. 32.6 Von mises stress analysis

Fig. 32.7 Structural analysis

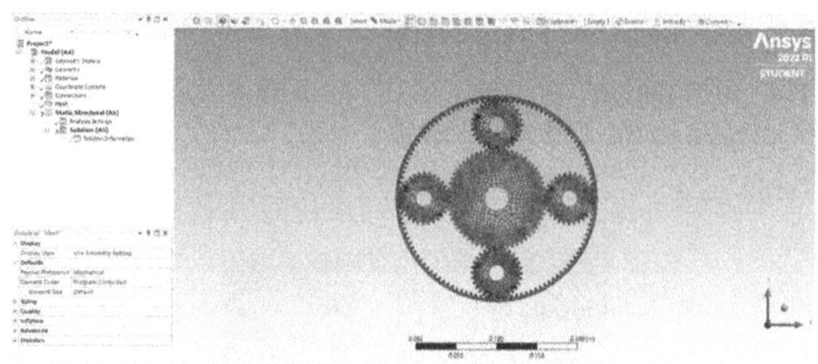

Fig. 32.8 Stress analysis

Fig. 32.9 Thermal analysis

Table 32.2 Study properties

Study name	Static1
Analysis type	Static structural
Mesh type	Structural Mesh
Thermal Effect:	On
Thermal option	Include temperature loads
Zero strain temperature	298Kelvin
Include fluid pressure effects from Flow Simulation	Off
Solver type	HFSS
Inplane Effect:	Off
Soft Spring:	Off
Inertial Relief:	Off
Incompatible bonding options	Automatic
Large displacement	On
Compute free body forces	On
Friction	Off
Use Adaptive Method:	Off
Result folder	ANSYS Document (C:\ Epicyclic GT\Analysis Report)

6. Conclusion

The static structural domain of the ANSYS programme was used to develop and analyse the three stage epicyclic planetary gear (18CrNiMo7). The greatest equivalent stress, as seen in the figures above, is 4.206 MPa, and

Table 32.3 Structural analysis for various materials and different stages

MATERIALS	Aluminum 6061 Alloy	18CrNiMo7-6 Chrome-Nickel-Moly Carburizing Steel	18CrNiMo7-6 Chrome-Nickel-Moly Carburizing Steel
STAGES	TWO STAGE	SINGLE STAGE	THREE STAGE
STRESS (STRUCTURAL ANALAYSIS)	Maximum equivalent stress is 6.7706 MPa and total deformation is 0.27708 mm	Maximum equivalent stress is 4.8695 MPa and total deformation is 0.0007366 mm	Maximum equivalent stress is 4.206 MPa and total deformation is 0.0006689 mm
THERMAL ANALYSIS	NIL	NIL	the heat generated due to friction between the gearboxes is found to be 165.15 ^{0}C and 255.18 ^{0}C

the total deformation is 0.0006689 mm. It is concluded that the maximum bending stress and total deformation for the given standard may be determined for a torque specification. It may be determined from the research of constructing the gear box analytically with input power of 10KW and 15KW, respectively, and input speed of 650 rpm. The resultant stresses fall well within the safe range for the specified material type, proving that the gearbox's design is secure. But it is discovered that the heat produced by friction between the gears is between 165.15C and 255.18C. The three stage epicyclic planetary gear generates less equivalent stress than the two stage epicyclic gear train, according to the current research. This system's operational process is fairly straightforward. They may be improved and enhanced in accordance with the applications by utilising more approaches.

Table 32.3 shows that when compared to Chrome-Nickel-Moly Carburizing Steel (single stage) and Aluminium 6061 Alloy (two stage), Chrome-Nickel-Moly Carburizing Steel (three stage Epicyclic planetary gear) has superior Equivalent stress.

References

1. K Akhila and Reddy ANR, Design, Modelling and Analysis of a Two Stage Epicyclic Reduction Gear unit of a flight vehicle October 2014
2. Design and Optimization of Planetary Gearbox for a Formula Student Vehicle –Soovadeep Bakshi, Parveen Dhillon, and Teja Maruvada, Indian Institute of Technology Bombay April 2014
3. S.B. Nandeppagoudar, S.N. Shaikh, Design and Numerical Analysis of Optimized Planetary Gear Box March 2017
4. S. Senthil Kumar, J.S. Athreya, Design and Fabrication of Epicyclic Gear Box April 2017
5. Timir Patel, Ashutosh Dubey, Lokavarapu Bhaskara Rao, Design and Analysis of an Epicyclic Gear box for an Electric Drive train September 2019
6. Yanbiao FENG, Wenming ZHANG, Jue YANG, Zuomin DONG, Design optimization of a three-stage planetary gear reducer using genetic algorithm November 2019

Note: All the figures and tables in this chapter were made by the authors.

Advances in Additive Manufacturing Technologies – Gurusamy Pathinettampadian et al. (eds)

Experimental Investigation and Mechanical Properties on Al6061 / SiC / MWCNT Reinforced Composite

Manjunathan R.*,
Lokesh S., Hari Krishna Raj S.,
Tharun Kumar C., Pradeep Kumar V., Divyesh B.
Department of Mechanical Engineering,
VelTech HighTech Dr. Rangarajan Dr. Sakunthala Engineering College,
Chennai, Tamilnadu

◆ **Abstract:** Aluminium Metal Matrix Composites (AMMCs) are new materials applied to enhance the mechanical properties of aluminium by introducing reinforcing particles. The composites possess improved characteristics such as higher corrosion resistance, high formability, and strength. Due to these benefits, AMMCs have become an integral part in advanced manufacturing industries, including aerospace, automotive, and marine industries, where materials that are both lightweight and strong are needed. Among the numerous aluminium alloys, Aluminium alloy 6061 (Al6061) is remarkable due to its wonderful combination of mechanical properties and compatability with reinforcement particles. With the inclusion of reinforcements like silicon carbide (SiC) and multi-walled carbon nanotubes (MWCNTs), Al6061 exhibits tremendous enhancement in compressive strength and hardness. Therefore, it is the most suitable for the manufacture of AMMCs used in tough applications.

◆ **Keywords:** MWCNTs, AMMC, Aluminium Alloys, Aluminium Metal Matrix Composites, Al6061

1. Introduction

MMCs are also highly regarded for their high strength-to-weight ratios, stiffness, and ductility over conventional materials, and thus are well suited for demanding purposes. They have low thermal and electrical conductivity and low radiation resistance, which limits their application in harsh environments. The function of the reinforcement does not only include improvement of the structural properties. It may also change the physical properties of the composite, e.g., wear resistance, friction, or thermal conductivity. Reinforcers can be continuous, such as monofilament fibers (e.g., silicon carbide or carbon fiber), which have anisotropic properties where strength is a function of the alignment of the fiber. Discontinuous reinforcements, in the form of whiskers, short fibers, or particles, produce isotropic properties and may be processed by conventional metalworking methods such as extrusion, rolling, or forging. But machining tends to need special equipment, such as polycrystalline diamond tooling (PCD).

Contemporary sports cars, such as those from Porsche, employ carbon fiber reinforced silicon carbide matrix rotors for their enhanced thermal properties. Moreover, 3M created an aluminium matrix insert for light, stiff disc brake calipers and alumina preforms for AMC pushrods. Ford features an MMC driveshaft upgrade consisting of aluminium reinforced with boron carbide for enhanced performance and durability. MMCs remain an evolving force within materials science, providing novel matrices of strength, weight, and toughness for various innovative applications.

*Corresponding author: manjunathanmech@gmail.com

DOI: 10.1201/9781003685906-33

Aluminium alloy 6061 is extensively applied in numerous sectors owing to its high strength, corrosion resistance, and formability. Some of its typical uses are:

High-Pressure Applications: Utilized in scuba tanks and other high-pressure gas cylinders, especially for those produced after 1995.

The T6 temper of 6061, which undergoes heat treatment to maximize strength, is frequently used in specialized applications, including:

- Bicycles: Found in frames and components of bicycles, especially in mid to high-end models.
- Archery Equipment: Used in recurve risers for middle to high-end archery setups.
- Fishing Gear: Integral to the design of many fly fishing reels, offering strength and corrosion resistance.
- Space Exploration: The material for the Pioneer plaque, a notable artifact of space exploration.
- Firearms: Particularly for pistols, due to its reduced weight and functional advantages. For primary expansion chambers, stronger materials like 17-4PH stainless steel or titanium are preferred.

6061, particularly in the T6 temper, continues to be a versatile and indispensable material across a diverse range of applications. Its balance of strength, corrosion resistance, and machinability makes it a preferred choice in many fields.

Natural moissanite, a form of silicon carbide (SiC), is exceedingly rare, so nearly all silicon carbide used today is synthetic. Silicon carbide is a versatile material employed as an abrasive, a semiconductor, and a gem-quality diamond simulant.

This method significantly increases the crystal yield, providing 81 times the cross-sectional area compared to the conventional Lely process. Cubic SiC, or 3C-SiC, is typically synthesized using chemical vapor deposition (CVD), a more expensive process involving silane, hydrogen, and nitrogen gases. Both homoepitaxial and heteroepitaxial SiC layers can be grown via gas and liquid-phase approaches.

1.1 Shaping and Deriving SiC

To create complex SiC shapes, preceramic polymers can be used as precursors. These polymers are converted into ceramic through pyrolysis at 1,000–1,100 °C in an inert atmosphere. Materials such as polycarbosilanes, poly(methylsilyne), and polysilazanes serve as precursors, producing what is known as polymer-derived ceramics (PDCs).

Silicon carbide wafers may also be made by cutting a single crystal with diamond wire saws or lasers. Such wafers are used extensively in power electronics because of SiC's beneficial semiconductor properties. Silicon carbide is polymorphic, i.e., it exists in various crystalline forms referred to as polytypes. Polytypes share the same two-dimensional structures but differ in the third dimension, resulting in different stacking sequences. Silicon carbide also exists in the amorphous glassy form when derived from preceramic polymer pyrolysis under inert atmospheres. The material with its immense versatility is a critical element in modern industries, particularly because of its semiconductor applications and structural properties in harsh environments.

1.2 Results and Discussion

The fabrication of the AMMC begins by cutting the Aluminium alloy 6061 (Al6061) into fragments of volume and size as required. Next, three components—Al6061, Silicon Carbide (SiC), and Multi-Walled Carbon Nanotubes (MWCNT) are weighed and tested to supply the required proportion (Fig. 33.1).

The Al6061 is then placed into a furnace and heated at 800°C to create a molten phase. Meanwhile, SiC and MWCNT are placed in separate feeders and preheated to 400°C individually. When Aluminium alloy is melted, SiC and MWCNT are fed into the molten Aluminium. The blend is agitated with a Stirrer Impeller of 600 rpm speed to adequately mix the components. Once the agitating process is done, the molten composite is cast into a die. The hydraulic pressure of 40 tons is immediately loaded onto the die to aid the solidification process. After a few minutes, the solidified composite is removed from the die, resulting in the final Aluminium Metal Matrix Composite (AMMC) with the desired properties (Figs 33.2, 33.3, 33.4).

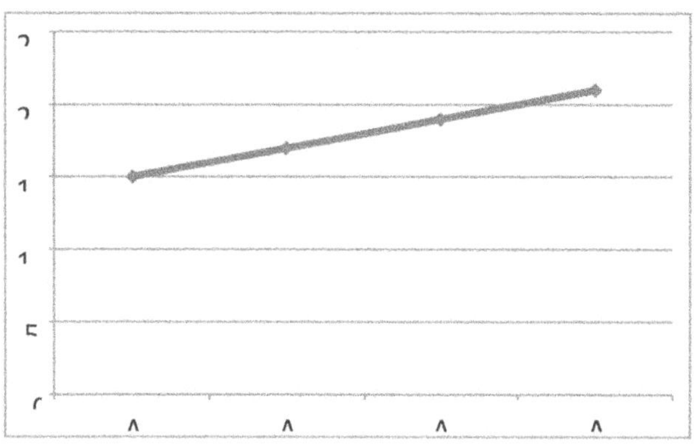

Fig. 33.1 Tensile strength test

2. Scanning Electron Microscope Test

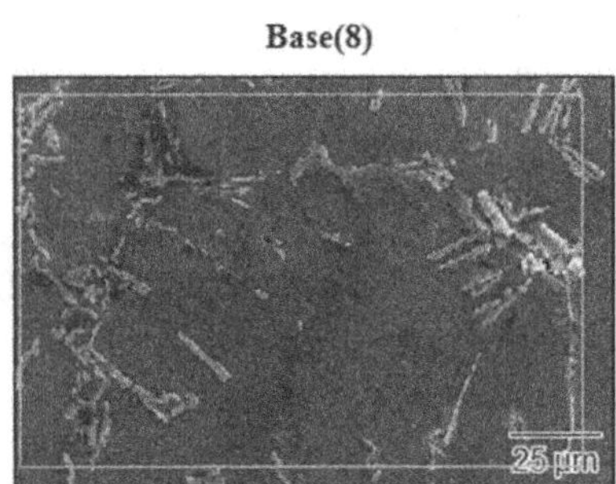

Fig. 33.2 SEM for test sample

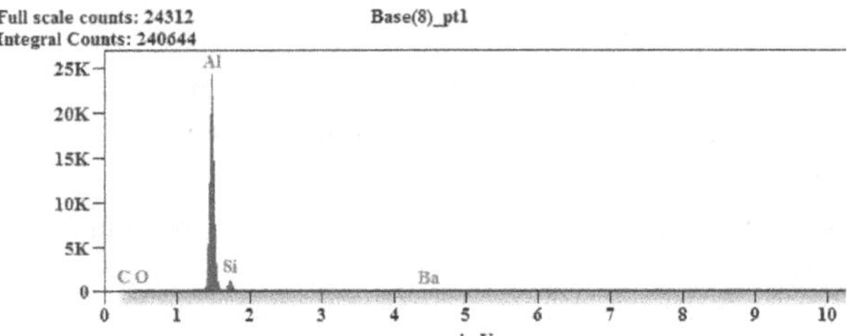

Fig. 33.3 XRD analysis of test sample

The above results clearly shows that the highest pressure of 40 T among these components is applied. It has a highest ductility and machinability property with great wear efficiency.

3. Impact Strength Test

The impact strength of the metal matrix composite was evaluated using the Izod Impact test on a fabricated sample. The results showed that as the volume percentage of SiC and MWCNT reinforcements increased, the impact strength also improved. The highest impact strength was recorded at the A3 composition. All tests were conducted in accordance with ASTM standard E23. The results obtained from the test are given below in Fig. 33.5.

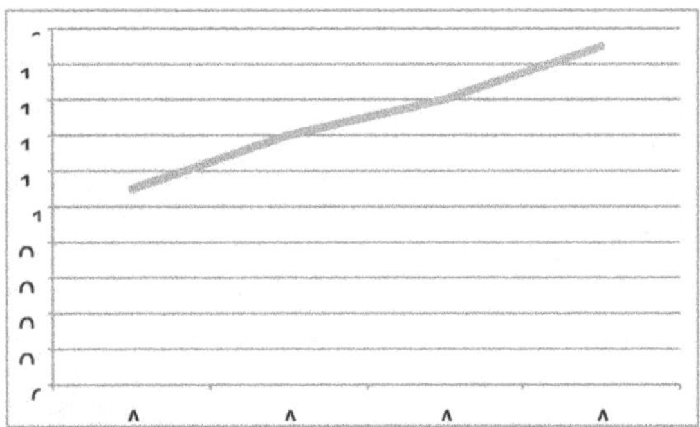

Fig. 33.4 Impact strength test

The Izod Impact test results revealed that composition A3 exhibited the highest impact strength of 5.9 Joules/m², containing 4% TiB_2 reinforcement powder. In comparison, composition A1, which had no reinforcement, recorded an impact strength of 2.5 Joules/m². Composition A2, with 2% reinforcement, showed an intermediate impact strength of 4 Joules/m². However, at composition A4, where the reinforcement content increased to 6%, the impact strength slightly declined to 5.2 Joules/m². These findings indicate that Al-A356 achieves optimal impact strength with a 4% reinforcement of SiC and MWCNT in the metal matrix. Additionally, all reinforced samples demonstrated higher impact strength than the unreinforced Al-A356, likely due to the homogeneous distribution of particles, which contributes to improved mechanical performance.

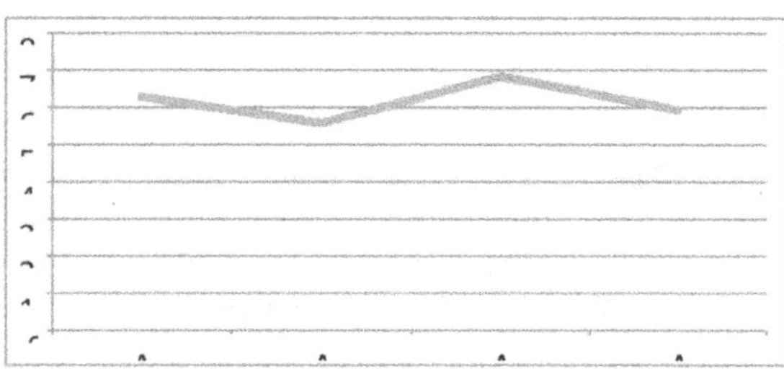

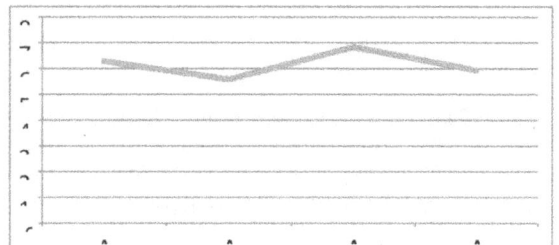

Fig. 33.5 Micro hardness test

The microhardness test results showed that composition A3 had the highest microhardness value of 85.4 HV, with 4% SiC and MWCNT reinforcement. In comparison, composition A1, which contained no reinforcement, recorded a microhardness of 71.2 HV. Composition A2, with 2% reinforcement, had a slightly higher microhardness of 72.6 HV. However, at composition A4, where the reinforcement content increased to 6%, the microhardness decreased to 79.03 HV. These findings indicate that Al-A356 achieves optimal microhardness with a 4% reinforcement of SiC and MWCNT. Additionally, all reinforced samples exhibited higher microhardness than the unreinforced Al-A356 (composition A1). The improved hardness in composition A3 can be attributed to the uniform distribution of particles, which resulted in fewer indentation marks.

4. Conclusion

1. Hybrid aluminium matrix composites (Al6061/SiC/MWCNT) with a uniform distribution of Silicon Carbide (SiC) and Multi-Walled Carbon Nanotube (MWCNT) particles were successfully fabricated using the stir casting process.
2. The X-ray Diffraction (XRD) patterns of the hybrid composites revealed characteristic peaks for

elements like CaO, SiO_2, MgO, and Al_2O_3. These peaks became prominent with higher MWCNT content in the four hybrid composite variants.

3. Compared to the unreinforced Al6061/5SiC composite, the hybrid composites reinforced with MWCNT exhibited a lower density, which decreased further as the MWCNT content increased.
4. The porosity of the composites decreased with the addition of reinforcement material up to 4.5 wt.% MWCNT, after which void formation started to rise.
5. The average hardness values of the composites showed a significant improvement, increasing from 68 HV for AMC-1 to 77 HV for composite 1, 85 HV for composite 2, 96 HV for composite 3, and 107 HV for composite 4.
6. This increase in hardness is attributed to the incorporation of hard MWCNT particles into the matrix.
7. The variants, composite 3 (Al6061-5SiC-4.5MWCNT) exhibited the highest wear resistance under varying sliding velocities and normal loads. Furthermore, the coefficient of friction for this hybrid composite showed minimal variation under similar operating conditions, emphasizing its stable tribological performance.

References

1. Mohamed Ibrahim Abd El Aal / 2020 - The Influence Of Ecap And Hpt Processing On The Microstructure Evolution, Mechanical Properties And Tribology Characteristics Of An Al6061 Alloy in ELSEVIER, 4–11
2. P. Satish kumar, A.J. Infant Jegan Rakesh, R.Meenakshi, C.Saravana Murthi / 2020 - Characterisation, Mechanical And Wear Properties Of Al6061/Sicp/Fly Ashp Composites By Stir Casting Technique in ELSEVIER, 2–5
3. Renu, Ajay Gupta, Shiv Ranjan Kumar, Chandramani Goswami, Pankaj Sharma / 2021 - Enhanced Physical And Mechanical Properties Of Al6061 Alloy Nanocomposite Reinforced With Nanozirconia in ELSEVIER, 2–3
4. Vishesh Kumar Anand, Amit Aherwar, Mozammel Mia, Omer Elfakir, Liliang Wang / 2020 - Influence of Silicon Carbide And Porcelain On Tribological Performance Of Al6061 Based Hybrid Composites in Tribology International, 10–18
5. B.Prakash, S.Sivanathan, V.Vijayan / 2020 - Investigation On Mechanical Properties Of Al6061 Alloy - Multiwall Carbon Nanotubes Reinforced Composites By Metallurgy Route in ELSEVIER, 4
6. B.Prakash, M.Manimaran / 2020 - Investigation On Mechanical Properties Of Al6061-Sic Nanocomposites Fabricated Via Stir Casting Process in ELSEVIER, 2–3
7. K.Preethi, T.N.Raju, H.A.Shivappa, S.Shashidhar, Madeva Nagral / 2021 - Processing, Microstructure, Hardness And Wear Behaviour Of Carbon Nanotube Particulates Reinforced Al6061 Alloy Composites in ELSEVIER, 1–3
8. Satish Babu Boppanaa, Samuel Dayanand, Anil Kumar M.R., Vijee Kumar / 2020 - Synthesis And Characterisation Of Nano Graphene And Zro2 Reinforced Al6061 Metal Matrix Composites in ELSEVIER, 4–8
9. Ajay Gupta, Renu, Shiv Ranjan Kumar, Chandramani Goswami, Tej Singh / 2021 - Wear Behaviour Of Al6061 Nanocomposite Reinforced With Nanozirconia in ELSEVIER, 1–4
10. G.B.Veeresh Kumar, R.Pramod, C.S.P.Rao, P.S.Shivakumar Gouda / 2018 - Artificial Neural Network Prediction On Wear Of Al6061 Alloy Metal Matrix Composites Reinforced With - Al2o3 in ELSEVIER, 2–8
11. Khadijeh Maleki, Ali Alizadeh, Mohsen Hajizamani / 2020 - Compressive Strength And Wear Properties Of Sic/Al6061 Composites Reinforced With High Contents Of Sic Fabricated By Pressure - Assisted Infiltration in ELSEVIER, 5.

Note: All the figures in this chapter were made by the authors.

Advances in Additive Manufacturing Technologies – Gurusamy Pathinettampadian et al. (eds)
© 2026 Taylor & Francis Group, London, ISBN 978-1-041-16687-0

Finite Element Analysis and Microstructure Evaluation on Temperature Distributed Aluminium based Composites

Velmurugan S.*, Pradeep Kumar S., Hari Krishna Raj S., Gowtham K., Sriram B.
Department of Mechanical Engineering,
VelTech HighTech Dr. Rangarajan Dr. Sakunthala Engineering College,
Chennai, Tamilnadu

◆ **Abstract:** Many studies have explored the properties, composition, and particle distribution of composite materials. However, there is limited research on how particles are distributed between the reinforcement and base alloy, as well as how temperature varies within these composites. This study examines the temperature distribution in a silicon carbide (SiC) and aluminum alloy composite, particularly in relation to an fabrication, applied pressure significantly influences particle distribution and the overall behavior of the composite. To investigate this, microstructural analysis is used to assess particle dispersion, while a finite element analysis variations. The melt and die geometry are designed in SolidWorks, and ANSYS software is used to apply constraints and conduct simulations. The findings indicate that applied pressure plays a crucial role in temperature distribution. Specifically, increasing the applied pressure reduces the temperature difference between the melt and die by 7% in the developed model.

◆ **Keywords:** Aluminium, Metal matrix composites, FEA, Temperature, Ansys

1. Introduction

Conventional metal alloys often cannot meet the requirements of modern technologies, particularly for advanced materials used in aerospace and automotive industries. Composite materials, which combine multiple phases, are designed to take advantage of the beneficial properties of each phase, offering a superior combination of characteristics. The growing demand for materials that are development reinforced metal matrix composites (MMCs), such as silicon carbide (SiC)-based MMCs, which have found significant applications in industries like automotive manufacturing. For instance, aluminum-based SiC composites are used in complex automotive engines, converting heat from burning fuel into force.

Materials consist of reinforcing elements, filters, and metal matrix binders, often combining various forms or structures at the nanoscale. These elements do not fully dissolve into each other during processes like solidification. Understanding the solidification process is crucial for improving the properties of composites. The characteristics of stiffness, ductility, and toughness, which influence strain behavior, can be enhanced with the right reinforcement and matrix. Many studies have shown that MMCs provide durable, effective materials with superior performance. The demand for lightweight materials is particularly high today, as they are easier to handle, simpler to design, and non-corrosive. Aluminum alloys are widely used for these reasons, offering properties that make them suitable superior compared to other metals, making it ideal for reinforcing elements. As a result, aluminum MMCs have become increasingly popular in production processes, although their higher costs and material requirements remain challenges due to their stability.

*Corresponding author: ivelu00@gmail.com

DOI: 10.1201/9781003685906-34

Behavior of composites can be analyzed using FEA software. However, studies on particle distribution and temperature variation in composites are less common. The interaction between hard reinforcement particles and the alloys can lead to mechanical parameter variations in MMCs. Simulations can help capture and display the distribution of particles in these materials. Incomplete or Uneven temperature distribution between reinforcement and molten metal can negatively influence the performance of composites due to thermal expansion mismatch. This work focuses on fabricating A356 aluminum matrix composites reinforced with silicon carbide (SiC) through squeeze casting process and subsequent detailed microstructural analysis. The chemical composition of the A356 alloy, as well as its main metal and alloying elements, is shown in Table 34.1. Four samples were made under different pressure conditions, ranging from 29 MPa to 109 MPa, as the lower and upper limits of the applied pressure. The thermophysical properties applied in the analysis are given in Table 34.2, and Tables 34.3 provide the material properties of SiC and the H13 tool steel die, respectively. The entire experimental procedure is detailed in a previous publication. A brief overview of the casting parameters employed for specimen manufacturing is given.

2. Result and Discussion

The properties of aluminum alloy composites are significantly influenced by the interaction of their individual components within a hierarchical structure, which enhances the overall strength of the composite. The combination's volume fraction plays a key role in how these properties interact synergistically, leading to effective material performance. The nature of these properties is greatly improved by the presence of reinforcement materials, and the composite's structure, including its shape, size, and particle distribution, further impacts its efficiency and overall performance through advanced methodologies. The orientation, concentration, and type of reinforcement materials are crucial in determining the composite's properties. Discontinuous phases maintain a minimum volume fraction, which in turn affects the stress distribution and the relationship between microstructure of the MMCs was analyzed, as the material's properties depend on the arrangement of particles within the microstructure. The distribution of particles was categorized into random and clustered arrangements.

To examine the particle distribution and conduct phase and volume analysis, specimens were prepared and polished. Four samples were polished, and volume analysis was performed in four selected fields. Figure 34.1 shows SEM images of the particle distribution under various pressures. At 30 MPa pressure (Fig. 34.1a), the SiC particles form clusters, while at 110 MPa pressure (Fig. 34.1d), they are more evenly dispersed. This improved uniformity is attributed to the higher applied pressure, which results in a more homogeneous particle distribution. The SEM images clearly demonstrate that as pressure increases, the particle distribution becomes more effective, leading to enhanced properties in terms of grain morphology. It is evident from the images that the mechanical properties of the samples are improved at higher pressure levels.

Figure 34.2(a) represents the original distribution of SiC particles in the base metal, reflecting even pressure applied across the melt zone. An FE model was created to simulate the experimental conditions closely. The thermophysical properties of SiC were assumed to be constant, and the rule

Table 34.1 Aluminium A356 alloy and its elements

Silicon	Fe	Cu	Mn	Mg	Zn	Ti	Al
4.7-5.8	0.19	0.18	0.09	0.32-0.38	0.09	0.18	Bal

Table 34.2 Physical and thermal properties for Aluminium A356

Temperature (°C)	95	195	295	375	395	495	580	603	690	790	891
Thermal Conductivity ($Wm^{-1} K^{-1}$)	153	152	141	135	134	137	131	59.7	63.7	67	61.2
Enthalpy (J/g^{-1})	73	167	266	349	369	477	552	1030	1161	1291	1366
Density (Kg/m^{-3})	2642	2651	2610	2642	2608	2581	2537	2411	2439	2461	2411

Table 34.3 Experimental parameter for the specimen's preparation

Sl No	Temp of Melt (°C)	Temp of Die (°C)	Applied Pressure (MPa)
1	745	405	50, 70, 80,130

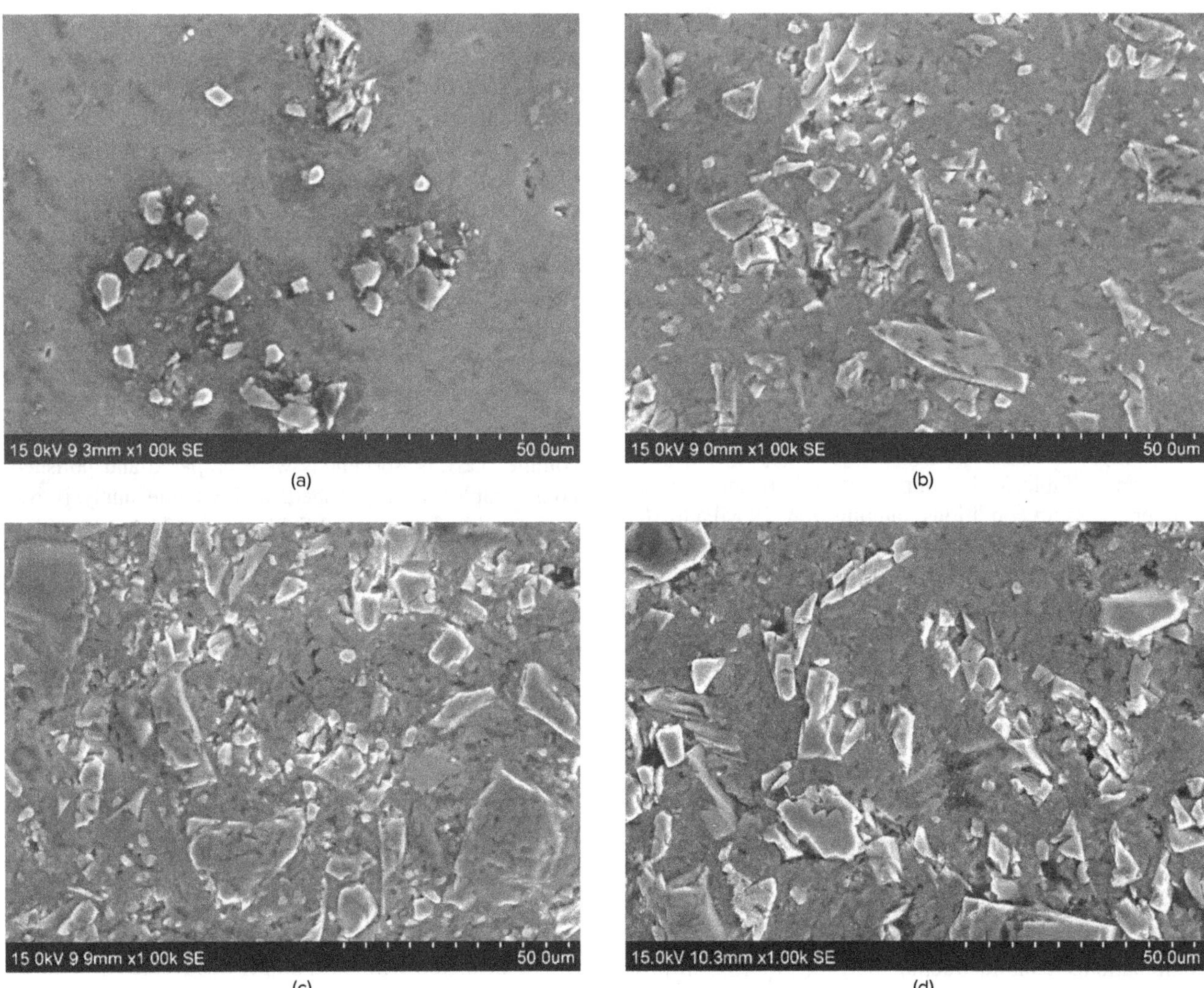

(a) (b) (c) (d)

Fig. 34.1 Displays scanning electron microscope (SEM) images illustrating the distribution of silicon carbide (SiC) particles within a metal matrix composite processed under different pressure conditions: (a) 40 MPa, (b) 70 MPa, (c) 80 MPa, and (d) 130 MPa

of mixtures was applied to compute composite properties. FEA was utilized to analyze stress-strain behavior and to examine the effects of SiC particle properties—i.e., size, volume, shape, and distribution—on the mechanical performance and general parameters of the composite. This technique gave information regarding material properties, temperature distributions, and mechanical performance. For reasons of symmetry and ease of analysis, a two-dimensional geometric model was developed and meshed in terms of 2D solid quadratic triangular (SQT) elements. The pressure was exerted for about 30 seconds, while full solidification of the composite was approximately 800 seconds. Figure 34.2(b) schematically presents the iterations of the molten metal and die simulation, while Fig. 34.3 represents the temperature distribution in the cast samples, with obvious distinctions between the melt and the die material. To keep the die temperature stable while pouring, the preheated die was covered with glass wool and the melt temperature held constant. The preheater to the die was removed moments before pouring, causing slight heat loss. To counteract this, the die was first preheated to 400°C, and the melt temperature was increased to 900°C.

Figures 34.3(a) and 34.3(b) indicate that the maximum temperatures were at the center of the melt and the inner die area, with readings taken 800 seconds after pouring. The maximum heat loss was found at the melt-die interface, especially at the bottom. Surprisingly, the top surface indicated lower temperatures than the interface and bottom regions. As the molten metal came into contact with the die walls during the process of solidification, it started

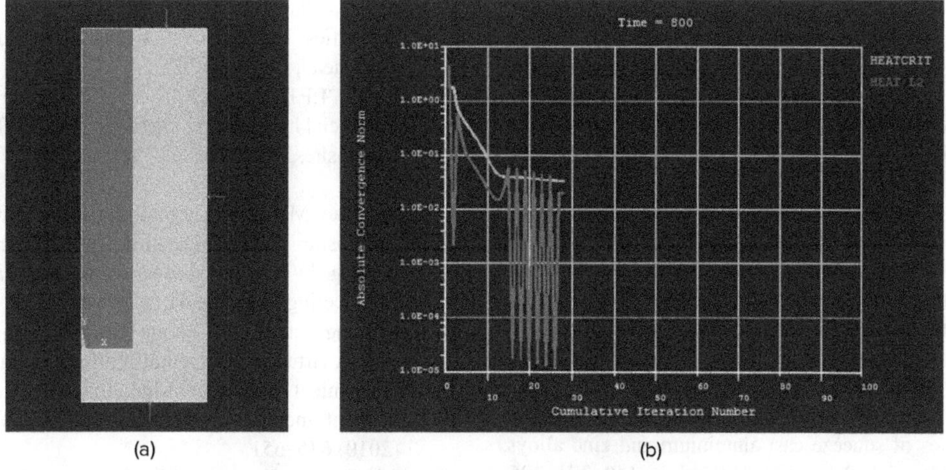

(a) (b)

Fig. 34.2 (a) & (b) Distribution of silicon carbide with Aluminium I

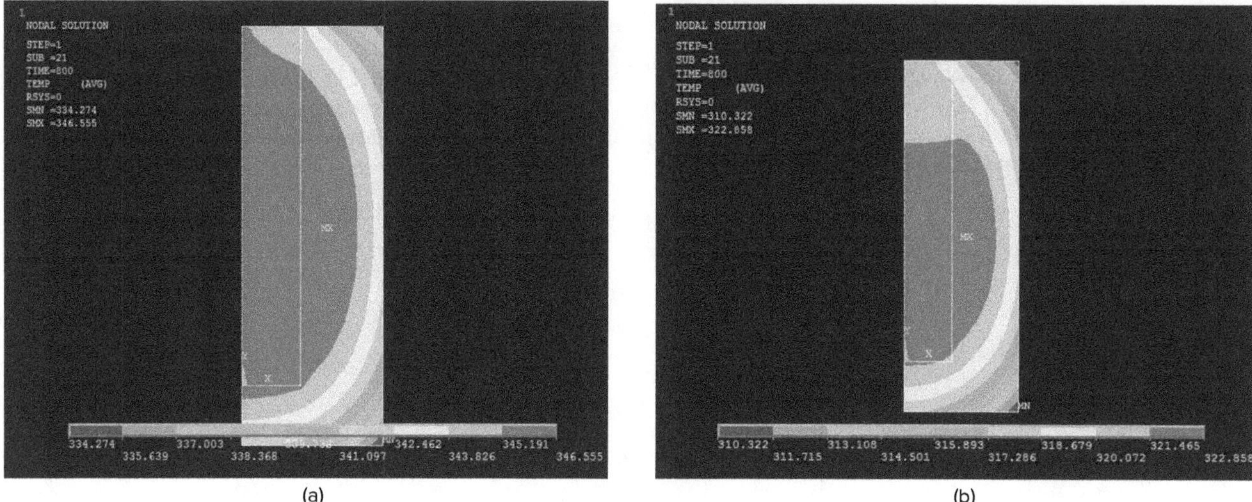

(a) (b)

Fig. 34.3 (a) & (b) Temperature of die and melt

solidifying. Cooling started from the die walls, where the temperature reduced to approximately 322°C, and progressed slowly outward. It was identified that enhanced pressure retards the rate of cooling of the composite.

3. Conclusion

- Such consistency is essential in order to guarantee uniform mechanical properties throughout the composite. Microstructure analysis, where the SiC particle distribution and arrangement were looked at, reinforces these observations further. Particle distribution at various pressures was observed, with higher pressure resulting in more uniform particle distribution. The material had greater variation in particle positioning at lower pressures, for instance, 140 MPa, where distribution was less uniform.
- Temperature profile during the casting process was another important factor that was examined. At 140 MPa.
- This temperature difference is due to the reduced cooling rate at elevated pressures. The cooling rate has a critical impact on the solidification process, with lower cooling rates encouraging improved particle distribution and minimizing internal stresses.
- During this research, it was noted that there is a 7% decrease in the temperature gradient between melt and die when pressure is increased.
- This decrease indicates the more homogeneous solidification and better material properties of the composite due to improved heat control and minimal thermal gradients..

REFERENCES

1. P. Gurusamy, S. Balasivanandha Prabu, R. Paskaramooorthy, Influence of processing temperatures on mechanical properties and microstructure of squeeze cast aluminum alloy composites', Materials and Manufacturing Processes, vol 30 (3) (2015) pp 367–373.
2. P. Gurusamy, S. Balasivanandha Prabu, R. Paskaramooorthy, Interfacial thermal resistance and the solidification behaviour of the Al/SiC_p composites', Materials and Manufacturing Processes, Vol.30 (3), (2014). 381–386.
3. G. S. Hanumanth, G.A. Irons, Solidification of particle reinforced metal-matrix composites, Metallurgical and Materials Transactions B 27 (1996) 663–671.
4. Yang, L.J (2003) The effect of casting temperature on the properties of squeeze cast aluminium and zinc alloys, Journal of Materials Processing Technology 140, 391–396.
5. S.K. Jagadeesh, C.S, Ramesh, J.M. Mallikarjuna, R. Keshavamurthy, Prediction of cooling curves during solidification of Al 6061–SiC_p based metal matrix composites using finite element analysis, Journal of Materials Processing Technology 210 (2010) 618–623.
6. C.S. Ramesh, S.K. Jagadeesh, R. Keshavamurthy, Solidification studies on sand cast Al 6061–SiC_p composites, Journal of Alloys and Compounds 509S (2011) S371–S374.
7. K.C. Mills, Recommended Values of Thermophysical Properties for Selected Commercial Alloys, first ed., Woodhead, England. 2002.
8. Rajan, T.P.D. et.al. (2007). Solidification and casting/die interfacial heat transfer characteristics of aluminium matrix composites, Composites Science and Technology 67, 70–78.
9. G. Altan, M.Topçu, Thermo-elastic stress of a metal-matrix composite disc under-linearly-increasing temperature loading by analytical and FEM analysis, Advances in Engineering Software 41 (2010) 604–610.
10. Boming Zhang a,b, Zhong Yang Xinyang Sun Zhanwen Tang A virtual experimental approach to estimate composite mechanical properties Modeling with an explicit finite element method in computational materials science 49 (2010) 645–651.
11. Xin Tan,Bin Zhang, Kai Liu b, Xiaobo Yan Microstructure and mechanical property of the 2024Al matrix hybrid composite reinforced with recycled SiCp/2024Al composite particles Journal of alloys and compounds 815 (2020) 152330.
12. Yuxuan Yu Haolin Lin Kuanren Qian Humphery Yang Material characterization and precise finite element analysis of fiber reinforced thermoplastic composites for 4D printing Computer Aided Design volume 122.2020 102817.

Note: All the figures and tables in this chapter were made by the authors.

Advances in Additive Manufacturing Technologies – Gurusamy Pathinettampadian et al. (eds)
 ISBN 978-1-041-16687-0

Evaluating Hydrogen as a Fuel for Compression Ignition Engines: A Critical Review of Efficiency, Emissions, and Technical Challenges

Bhuvan R.[1], Kishore Piravan S.[2], Joseph Felix Anson A.[3], Ajay Athreyan[4], Venkateshwar K. S.[5], Vishal S.[6], Sathiyamoorthi R.[7]

Department of Mechanical Engineering, Chennai Institute of Technology, Chennai, Tamilnadu, India

◆ **Abstract:** The dependence on fossil fuels for more than 95% of the energy used for transportation intensifies the necessity to shift to cleaner alternatives. Furthermore, transport emissions have risen by almost eighty percent since 1990, making it imperative to manage this sector to mitigate climate change and enhance air quality and the health of the public. Implementing initiatives such as improving fuel efficiency, advocating for electric vehicles, and employing alternative fuels will be essential in diminishing the carbon footprint of this sector and securing a sustainable future. It has been proved that compression ignition engines that are powered by hydrogen have the potential to produce cleaner combustion. These engines have achieved practically zero nitrogen oxides (NO_x) emissions while retaining thermal efficiency of approximately 33 percent under optimal conditions. Hydrogen's inherent characteristics, which include greater flame speed and low quenching separation, make it possible for the combustion process to be more efficient than that of traditional fossil fuels. However, these same characteristics can also lead to problems which means as pre-ignition and knocking, which make the operation of the engine more difficult and restrict the amount of power that can be produced. However, despite the fact that hydrogen combustion has the potential to produce nearly zero carbon emissions, the practical adoption of this technology confronts a number of challenges. These challenges include high production costs, inadequate re-fuelling infrastructure, and the requirement for high-pressure storage systems. In addition, because hydrogen has a high auto-ignition temperature, it is necessary to carefully manage compression ratios in order to prevent premature ignition while the engine is operating.

◆ **Keywords:** Hydrogen, Diesel engine, Performance, Emission

1. Introduction

The transportation industry significantly contributes to worldwide emissions of greenhouse gases, including around 24% of the carbon dioxide released from energy use and nearly a third of all internal GHG emissions in the United States. Decarbonising transport is crucial for attaining climate objectives, especially the aim of a net-zero greenhouse gas economy by 2050, as it constitutes the main source of emissions in numerous nations [1].

Hydrogen serves a vital function in diesel engines by providing a cleaner substitute for conventional fossil fuels, potentially diminishing harmful emissions and improving engine performance (Fig. 35.1). The incorporation of hydrogen injection systems into diesel engines has

[1]bhuvanr.mech2022@citchennai.net, [2]kishorepiravans.mech2022@citchennai.net, [3]josephfelixansona.mech2022@citchennai.net, [4]ajayathreyan.mech2022@citchennai.net, [5]ar.mech2022@citchennai.net, [6]vishals.mech2023@citchennai.net, [7]sathiya.ram78@gmail.com

DOI: 10.1201/9781003685906-35

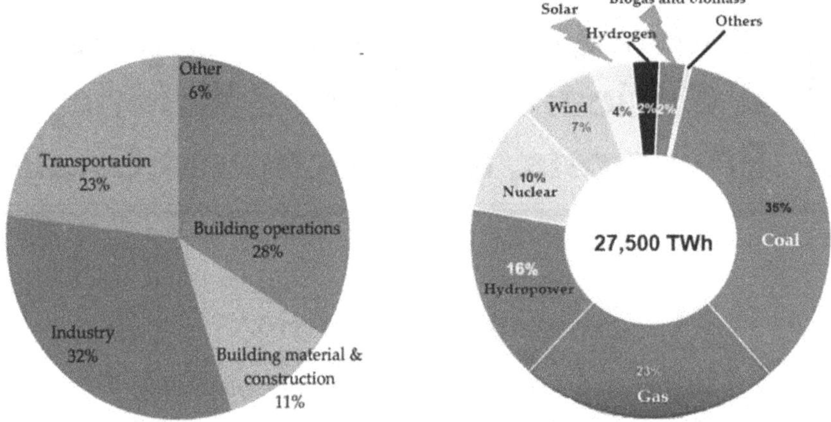

Fig. 35.1 Global CO_2 emission by various sectors and annual CO_2 emissions from different types of fuels [5–8]

demonstrated a reduction in carbon emissions by as much as 80%, alongside a decrease in particulate matter and other pollutants such as nitrogen oxides and carbon monoxide [3]. Moreover, hydrogen's elevated energy density—roughly threefold that of diesel—can enhance fuel economy, with research demonstrating savings in fuel usage of 8-14%. The integration of hydrogen in diesel engines presents several obstacles. Concerns including combustion stability, pre-ignition hazards, and the necessity for specialized storage methods must be resolved to guarantee dependable engine performance [4]. Moreover, the facility needed for hydrogen production and supply was inadequately established, presenting logistical challenges for extensive adoption. Notwithstanding these obstacles, the prospective advantages of hydrogen as an environmentally friendly fuel source render it an attractive domain for additional study and development in the pursuit of sustainable transportation technologies [5].

Ngamnurak et al. [6] have examined enhanced biohydrogen production from lactic acid bacteria polluting substrates by enhanced hydrogen. The impact of increased consortium percentage, beginning pH, and glucose concentration

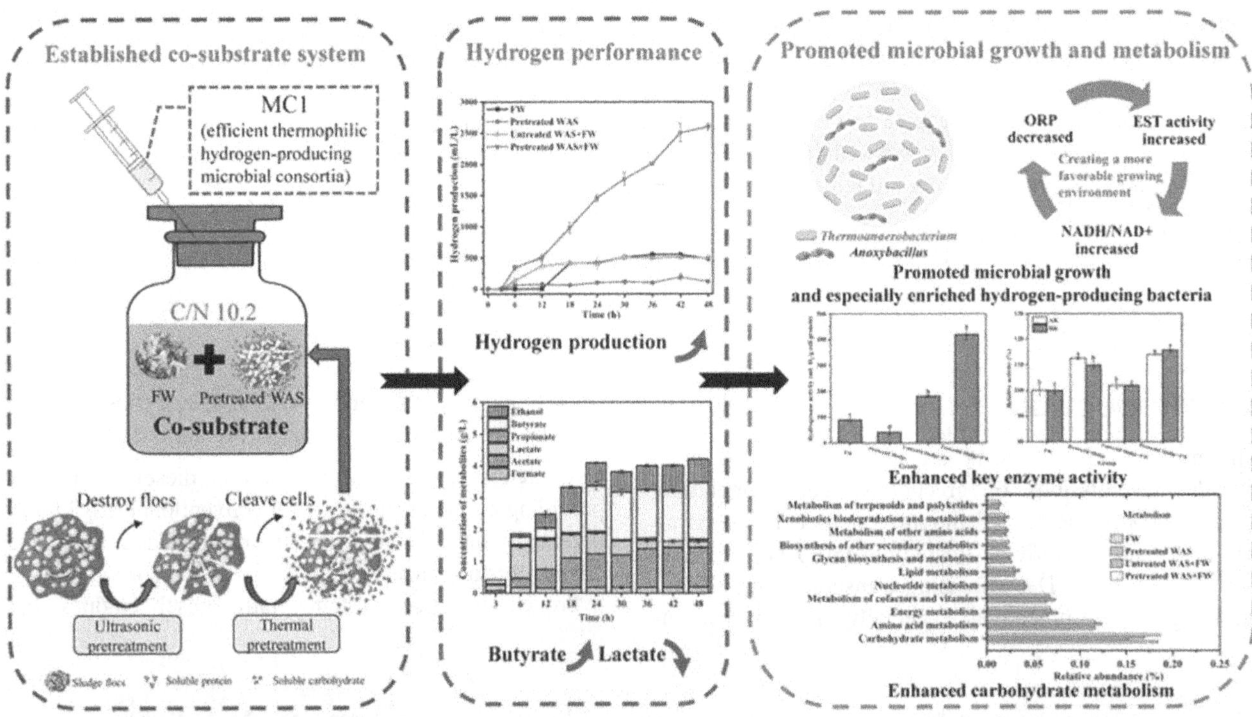

Fig. 35.2 Thermophilic hydrogen production from co-substrate of pretreated waste activated sludge and food waste [7]

was assessed. Analyses of metabolite products and microbial communities during food waste fermenting revealed beneficial cross-feeding interactions among hydrogen producers, lactic acid bacteria, and acetogenic bacteria. This study offers significant insights into the utilisation of effective, increased H2-producing consortia to enhance bio-hydrogen producing process. Yang et al. [7] demonstrated an improved thermophilic hydrogen generation from a co-substrate of processed waste activated sludge and food waste. Simultaneously, it may facilitate microbial proliferation by establishing a more conducive redox environment (Fig. 35.2). Analysis of the microbial community revealed that Thermoanaerobacterium, a genus associated with hydrogen production, was specifically enhanced and constituted 82.3% of the co-substrate.

Shi et al. [8] conducted a numerical analysis of hydrogen vapour cloud explosion resulting from an offshore hydrogen production facility. Results demonstrated that the inclusion of compressor apparatus and cylindrical hydrogen storage vessels resulted in enhanced flame propagation and elevated overpressure peaks. Soudagar et al. [9] examined the influence of gasification input parameters and KOH catalyst on the functional features of hydrogen generation from municipal wastewater. The impact of gasification input variables on the functional attributes of hydrogen production is examined. Microalgae sourced from wastewater from municipalities treated with 0.2% TiO_2 exhibited a heightened growth rate of 0.99 μ/day.

Shomope et al. [10] utilised machine learning in proton exchange membrane (PEM) water electrolysis for hydrogen generation. The ultimate models exhibited significant consistency in forecasting hydrogen production rates, with Random Forest routinely surpassing XGBoost. Hydrogen exhibits distinctive physical and chemical characteristics that render it an attractive fuel for diesel engines. Hydrogen, being the lightest and most prevalent element, possesses a low density of roughly 0.08988 g/L at standard temperature and pressure, facilitating efficient storage and transit when compressed or liquefied. Its elevated energy density, around threefold that of normal diesel fuel on a mass basis, facilitates increased energy production per unit mass, potentially improving engine efficiency [11]. The combustion properties of hydrogen are significant; it ignites with an exceptionally high adiabatic flame temperature of around 2254°C, which is roughly 15% greater than that of methane flames, resulting in enhanced thermal efficiency in combustion processes. Moreover, hydrogen's rapid flame velocity, reaching up to ten times that of methane, produces shorter and more compact flames, potentially impacting engine design by facilitating smaller combustion chambers and diminishing emissions due to reduced residence durations in elevated temperature regions. Nonetheless, these features pose issues, including heightened risks of pre-ignition and flashback, requiring meticulous attention in engine design to guarantee safe and efficient operation when employing hydrogen as a fuel source [13].

2. Hydrogen – Characterization, Application for Combustion of Hydrogen

When contrasting hydrogen and diesel as biofuels for combustion-based engines, notable distinctions arise regarding heating values, flame velocity, and emission characteristics. Hydrogen possesses a superior heating value of roughly 141.8 MJ/kg, markedly surpassing diesel's heating value of about 44.8 MJ/kg, resulting in enhanced energy output per unit mass. This attribute is enhanced by hydrogen's swift flame velocity which surpassing that of diesel, facilitating more efficient combustion and superior engine performance. The ramifications of these disparities are significant; the rapid combustion of hydrogen enhances thermal efficiency and decreases combustion duration, leading to reduced emissions of CO, UHC, and particulate matter in comparison to conventional diesel fuel [15]. Nonetheless, whereas hydrogen substantially decreases overall emissions, it may also result in elevated nitrogen oxides (NOx) emissions owing to greater combustion temperatures. The differing characteristics of hydrogen and diesel underscore the advantages and difficulties linked to the adoption of hydrogen fuel in diesel engines [16].

Paul et al. [17] provided a description of the manufacture, characterisation, and customisation of magnesium hydride in order to improve the features of solid-state hydrogen storage. It has been shown that V2O5 has a more powerful catalytic influence, which results in a lower starting temperature for the dehydrogenation of magnesium hydroxide. At a temperature of 220 degrees Celsius, the MgH2-5 weight percent mesoporous V2O5 collection begins the release of hydrogen. This temperature is much lower than the as-synthesized uncatalyzed MgH2, and started the release of hydrogen at 430 degrees Celsius under the identical conditions. It is possible to completely re-hydrate the dehydrogenated MgH2-5wt%V2O5 by heating it to 150 degrees Celsius. The hydrolysis of synthesized magnesium hydroxide was carried out using a solution in water of tartaric acid, and the behaviour of the dehydrogenation was shown to be associated with the pH of the solution.

The supersonic combustion of hydrogen was accomplished by Randonnier et al. [18] through the utilization of a

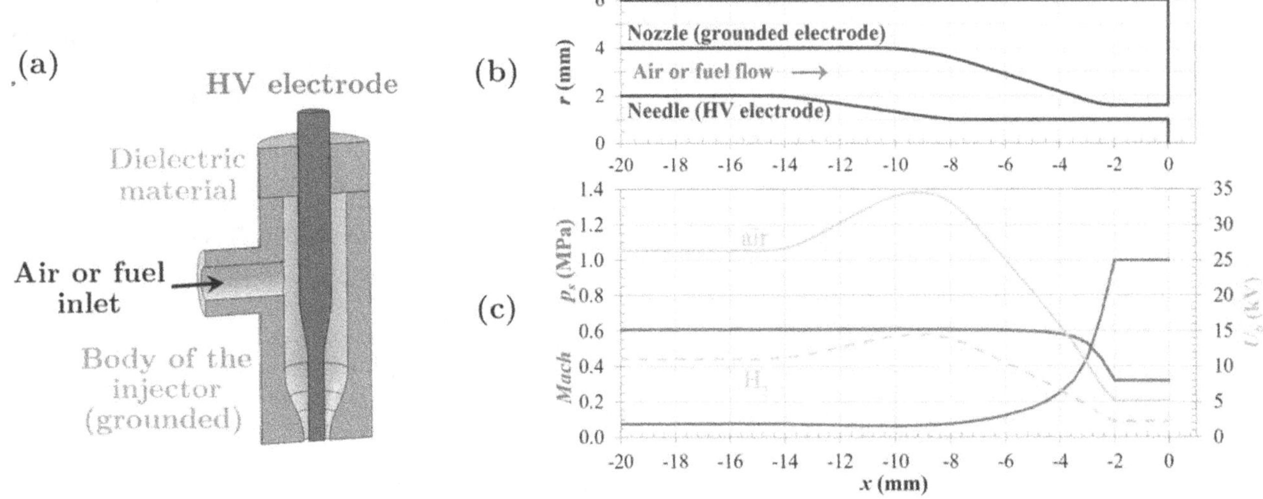

Fig. 35.3 Plasma-assisted injector – application to supersonic combustion of hydrogen [18]

plasma-assisted injector. This plasma-assisted injector is both straightforward and robust; it enables the creation of compact designs that are appropriate for small-scale integration in engines (Fig. 35.3). Furthermore, it may be used with either air or fuel, depending on the circumstance or the application. The LAERTE division at ONERA was responsible for the production, deployment, and evaluation of a prototype that utilized quasi-DC releasing on the LAPCAT2 Twin Mode Ramjet combustor. Whenever the plasma began functioning at a determined power of 1.4 kW, findings from combustion studies suggested that the ignition occurred earlier and that the pressure was higher.

3. Combustion Mechanism

In diesel engines, hydrogen combustion is characterised by a complicated interaction between hydrogen and diesel fuel. When hydrogen is pumped into a diesel engine, the engine normally functions in a dual-fuel mode, which means that hydrogen is used to supplement the traditional diesel fuel rather than completely replacing it [19]. The process of co-combustion makes it possible to achieve a more uniform mixing of fuel and air, which in turn leads to a burn that is both cleaner and more efficient. Because hydrogen burns cleanly and does not produce particulate matter, its presence can greatly shorten the ignition delay, which in turn leads to faster combustion and lower soot formation [20]. Gases that contain hydrogen also burn more quickly. As an additional benefit, the greater flame speed makes quick combustion easier, which in turn leads to increased thermal efficiency and power production. However, in order to limit potential increases in nitrogen oxide (NOx) emissions, which is a common difficulty in hydrogen-diesel systems, it is essential to carefully manage the time of the hydrogen injection. Overall, the incorporation of hydrogen into diesel engines gives a chance to capitalise on the advantages of both fuels while also addressing the environmental difficulties that are connected with the burning of standard diesel [21].

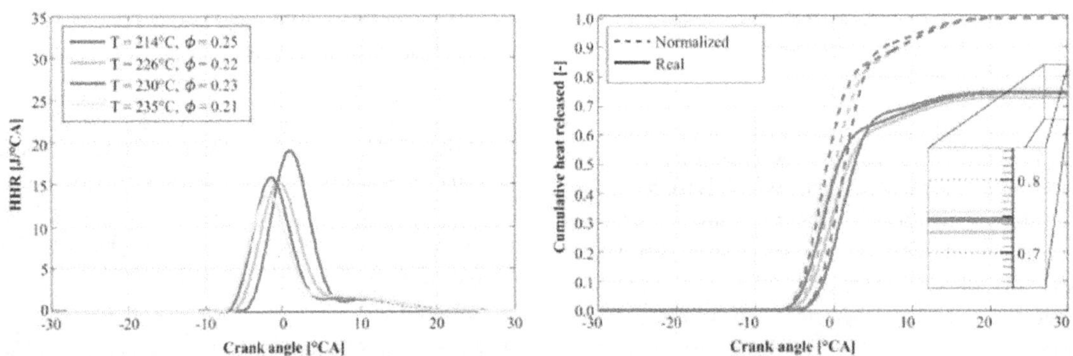

Fig. 35.4 Heat release rate and cumulative heat released for different intake temperatures and equivalence ratios [22]

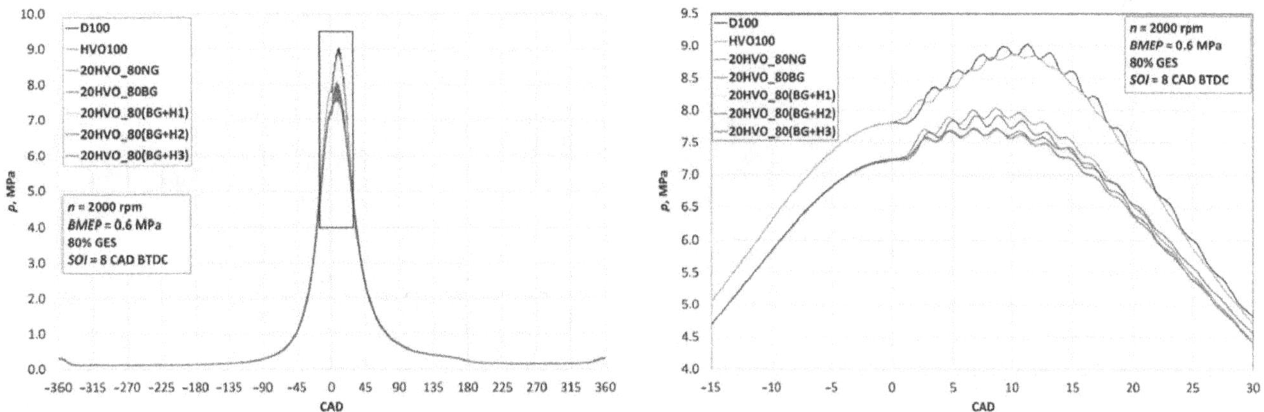

Fig. 35.5 Cylinder pressure variation with changing fuel composition for Hydrogen gas energy share [23]

In order to identify the intake temperature requirements and equivalency ratios for stable HCCI combustion, Rojas et al. [22] have done research. A comparison was made between these findings and those obtained from earlier study conducted at sea level. When compared to the temperature readings reported in earlier investigations, these values were, on average, one hundred degrees Celsius higher (Fig. 35.4). Both the highest value for the IMEP and the Indicated Thermal Efficiency were found to be 34.5% and 1.75 bar, respectively. It is essential that the results that were obtained be taken into consideration when designing and putting into operation HCCI engines that are powered only by hydrogen in developing nations that are situated at high altitudes above sea level.

Alfredas Rimkus and colleagues [23] conducted an experimental study in which they replicated biogas using a gas mixture consisting of sixty percent natural gas and forty percent carbon dioxide by volume. In the biogas, increasing the concentration of hydrogen to thirty percent of the volume of methane (20HVO_80 (BG+H3)) exhibited to an increase in the surplus air ratio to two and a half, or $\lambda = 2.03$. However, the flammability qualities of hydrogen, which are superior, expanded the limit of the combination that could be ignited (Fig. 35.5). Additionally, the ignition conditions of pilot fuels deteriorated as a result of this condition.

The combustion chamber of these systems is fed with hydrogen in addition to the conventional diesel fuel, which enables the combustion of both fuels at the same time [24]. In most cases, a two-stage injection method is utilized for this dual-fuel operation. The diesel fuel is used as the primary fuel, which ignites first due to its higher energy density. On the other hand, hydrogen is injected either during the intake stroke or immediately into the combustion chamber at specified timings. It is possible to make dynamic adjustments to the precise timing and ratio of hydrogen to diesel based on the load as well as the operating circumstances of the engine, which will optimise the efficiency of the combustion process and the emission profiles [25]. During the combustion process, the rapid burning qualities of hydrogen make it possible for diesel to be burned more completely, which results in a reduction in the emissions of CO and UHC. Nevertheless, careful management is required in order to decrease the possibility of increases in NO_x emissions brought on by higher combustion temperatures. In general, diesel-hydrogen dual-fuel systems take advantage of the benefits offered by fuels, so enhancing fuel efficiency and lowering hazardous emissions while preserving the dependability of diesel engine designs that are already in use [26].

4. Emission Characteristics

The incorporation of hydrogen into diesel engines makes a considerable contribution to the reduction of emissions of carbon monoxide (CO) and unburned hydrocarbons (UHC). This is primarily due to the distinctive combustion characteristics of hydrogen. When hydrogen is added to the combustion analyse, it improves the overall efficiency of the combustion process by facilitating a more complete oxidation of the fuel [27]. The primary cause is that hydrogen burns at a greater flame speed and produces just water as a byproduct, in contrast to diesel, which creates carbon monoxide and ultra-high-pressure hydrocarbons resulted in incomplete combustion. As the quantity of hydrogen in the fuel mixture grows, studies have shown that the amount of carbon monoxide emissions can reduce by as much as 88 percent, and the amount of unburned hydrocarbon emissions also sees significant reductions [28]. This enhancement can be due to the fact that hydrogen has the capacity to substitute for the carbon component of diesel fuel, hence reducing the amount of carbon monoxide that is produced during combustion. Furthermore, the presence

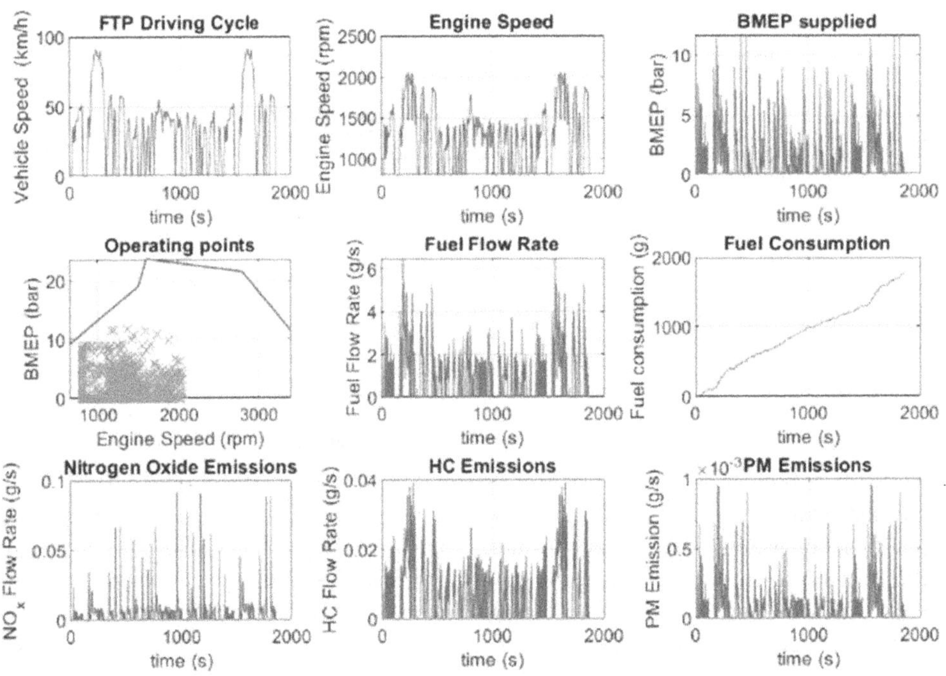

Fig. 35.6 Estimated fuel consumption and exhaust emissions employing the vehicle simulation model for the FTP cycle test and diesel engine [30]

of hydrogen encourages the creation of reactive radicals which includes OH and H, which further contribute to the reduction of incomplete combustion and the enhancement of the oxidation of hydrocarbons. Therefore, the utilisation of hydrogen in diesel engines is a viable option for reducing harmful emissions and increasing air quality while simultaneously addressing the challenges posed by climate change [29].

A methodology has been devised by Antonella Accardo and colleagues [30] for the purpose of implementing life-cycle assessment (LCA) to internal combustion engines in order to evaluate the life-cycle greenhouse gas emissions of these engines. Additionally, the aim of this work is to study the possibility for decarbonisation of hydrogen engines that are created by utilising the technology that is already present in diesel engines and assuming a variety of hydrogen production pathways. The life cycle assessment (LCA) encompasses the full life cycle, from birth to death, and the evaluation is based solely on primary data (Fig. 35.6). A new design for the hydrogen engine was developed, with the diesel engine serving as the foundation. Despite the fact that the engines under investigation are adaptable and can be utilised for a broad variety of applications, including those in the automotive, cogeneration etc.,

When hydrogen is used in diesel engines, there is a complex trade-off that must be made between lowering emissions of carbon dioxide (CO2) and the possibility of an increase in emissions of nitrogen oxides (NO_x) [31]. Due to the fact that hydrogen is a carbon-free fuel that will not release carbon dioxide during burning, it is an appealing choice for the decarbonisation of the automotive industry. For instance, the combustion of hydrogen takes place at higher temperatures, which might result in an increase in the generation of nitrogen oxides (NOx) as a result of the reaction between nitrogen and oxygen in the air at higher temperatures [32]. It has been demonstrated through research that the combination of hydrogen and diesel can greatly reduce carbon dioxide emissions; however, the effect on nitrogen oxide emissions is contingent upon the load conditions of the engine. When the engine is operating at low loading conditions, the addition of hydrogen can reduce NOx emissions. When the engine was operating at higher loads (between 50 and 100%), NO_x emissions increased significantly. This variation demonstrates the need for careful oversight of injection techniques and operational parameters in order to strike a balance between the environmental advantages of reduced CO2 emissions and the barriers presented by increased NOx emissions. This means that additional research and development is required in order to effectively optimise dual-fuel systems [34].

Mohamed et al. [35] conducted an investigation into a comprehensive analysis of NOx emissions. The most important finding was that the emissions of nitrogen oxides

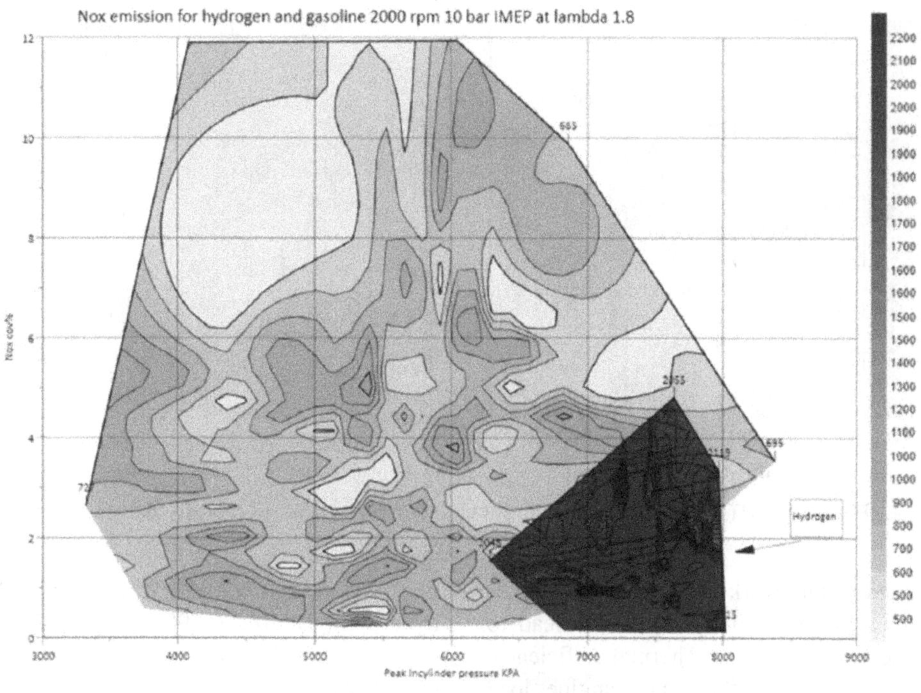

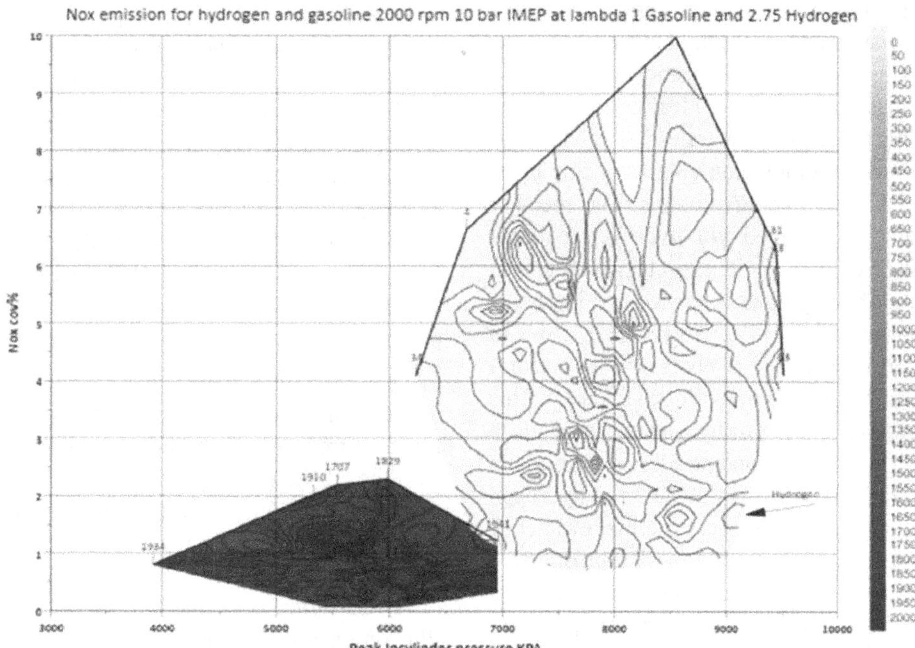

Fig. 35.7 NOx emission at different peak cylinder pressures and NOx emission variation DI and gasoline DI at a lambda of 1.8 and optimized lambda for all the fuels [35]

are essentially nonexistent between lambda values of 2.75 and 3.7, and that the production of NOx emissions by hydrogen is 13.8% lower than that of petrol when operation is stoichiometric (Fig. 35.7). This was accomplished by the utilisation of an innovative method that involved determining the coefficient of variability of the NOx.

5. Hydrogen Powered Engine Performance

The use of hydrogen into diesel engines has been found to enhance brake thermal efficiency (BTE), a critical parameter that assesses the effectiveness of an engine in

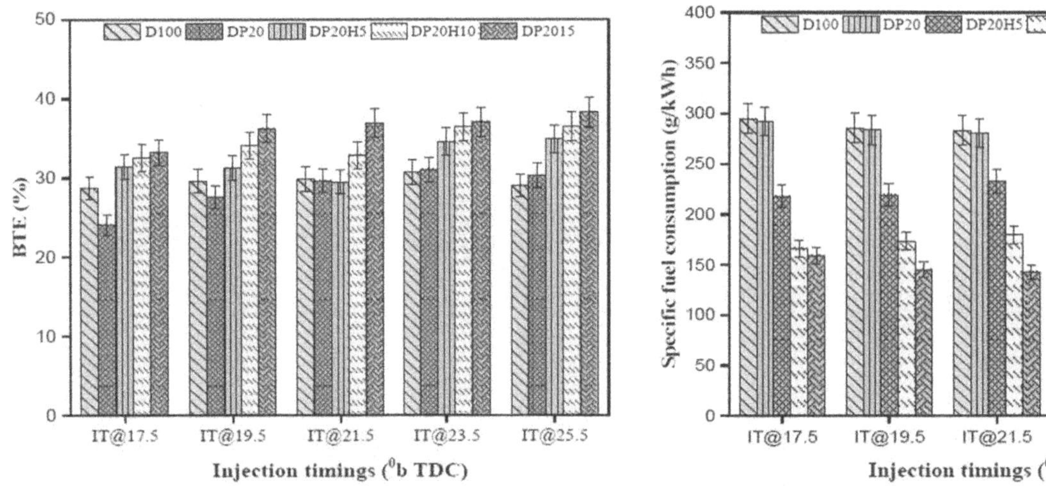

Fig. 35.8 BTE and SFC with different injection timings for hydrogen blends [39]

converting fuel energy into work. Experimental studies indicate that the incorporation of hydrogen can lead to substantial enhancements in Brake Thermal Efficiency, contingent upon the hydrogen flows and engine load characteristics. The reported increases range from five percent to more than thirty percent, depending on the particulars of the situation [36]. In one study, for example, a BTE of 33.4% was obtained at an ideal hydrogen flow rate of 6 L/min. This finding demonstrates that the fuel has the potential to improve combustion efficiency due to its quick flame speed and high energy content. The speedier combustion process that is linked with hydrogen makes it possible for fuel to be oxidized in its entirety, which in turn reduces the amount of energy that is wasted and improves the overall thermal efficiency [37]. However, in order to achieve these improvements, it is necessary to perform careful management of the ratio of hydrogen to diesel. This is because an excessive amount of hydrogen can result in decreased volumetric efficiency and possibly a decrease in power output. Overall, the strategic utilisation of hydrogen in dual-fuel systems not only enhances thermal efficiency but also helps to lower emissions, making it a prospective avenue for increasing diesel engine performance in a manner that is more environmentally friendly [38].

A study conducted by Kodandapuram Jayasimha Reddy and colleagues [39] investigated the utilisation of biodiesel blends that were made from waste plastic oil [P] and petro-diesel [D]. In order to enhance the performance of these blends, hydrogen was added in small amounts (Fig. 35.8). The percentage of hydrogen energy shares ranges from 5 to 15%, while the amount of biodiesel remains at 20%. The remaining component is comprised of petro-diesel. In light of this, the blends that have been adopted are as follows: DP20 (80 percent diesel fuel and 20 percent waste plastic biofuel), DP20H5 (95 percent DP20 and 5 percent hydrogen), DP20H10 (90 percent DP20 and 10 percent hydrogen), and DP20H15 (85 percent DP20 and 15 percent hydrogen).

The energy and exergy investigation on hydrogen that was supplemented with diesel and Azolla pinnata macroalgae biofuel was an analysis that was carried out by Chaudhary and colleagues [40]. Furthermore, the irreversibility of 40% H_2 blended B40 fell by 13.82% when subjected to greater load circumstances in comparison to the operation of diesel. The environmental sustainability rating of 40% H_2 blended B40 was 1.38, which is an improvement of 7.82% when compared to the index of diesel, which registered at 1.28. The outcomes of the theorised energy-exergy analyses indicated an inaccuracy that was less than five percent when compared to BTE.

There is a significant impact that hydrogen has on the power output and torque of diesel engines, which ultimately exhibits in an enhancement in the overall performance of the engine. The experimental studies showed that the introduction of hydrogen can lead to significant increases in engine power [41]. These improvements have been observed after the introduction of hydrogen. As an illustration, when the flow rates of hydrogen are optimal, the maximum pressure in the cylinder can increase by roughly 17%. This higher pressure is associated with an increase in power production as a result of more efficient combustion processes [42]. Because of the features of hydrogen that allow for rapid combustion, the release of energy can occur more quickly, which results in increased torque at a variety of engine loads. Furthermore, the utilisation of hydrogen makes it possible to get a more homogenous combination of fuel and air, which improves the stability and efficiency of combustion, particularly during part-load operations, which is the typical operating mode for diesel engines [43]. Because of this synergy, hydrogen is a promising additive for boosting the performance of diesel engines while also addressing environmental concerns [44]. Not only does it increase power, but it also adds to reduced fuel consumption and fewer emissions.

There is a considerable impact that hydrogen addition has on brake-specific fuel consumption (BSFC) in diesel engines, which demonstrates that there is the possibility for reductions in fuel consumption when hydrogen is added into the fuel mixture [45]. Depending on the flow rate of hydrogen and the conditions under which the engine is working, studies have demonstrated that the incorporation of hydrogen can result in savings in diesel consumption of roughly 8% to 14% [46]. For example, a hydrogen injection system can improve the efficiency of combustion, which in turn enables a more comprehensive oxidation of the diesel fuel. This, in turn, reduces the total amount of fuel that is required for a given power output [47]. It is common for the relationship between hydrogen ratios and BSFC to be linear; when the fraction of hydrogen grows, the diesel consumption falls in a proportional manner [48]. The reduction in BSFC can be ascribed to the elevated energy content of hydrogen and its environmentally benign combustion characteristics, both of which contribute to the reduction of carbon deposits and are responsible for improving engine cleanliness. Because of this, the utilisation of hydrogen not only improves fuel economy but also helps to the extension of engine life and the reduction of maintenance costs. As a result, it is an appealing alternative for enhancing the efficiency of diesel engines while simultaneously addressing concerns about the environment [49-50].

6. Future Perspective

- The prospective viability of hydrogen is more advantageous, driven by technology advancements and enhanced environmental consciousness. The growing economic feasibility of producing green hydrogen is expected to substantially aid in diminishing greenhouse gas emissions within the transportation sector.
- Research is concentrating on optimizing combustion processes, improving fuel injection systems, and formulating effective techniques to reduce nitrogen oxide emissions, which continue to pose a challenge.
- The development of hydrogen infrastructure, encompassing refuelling stations along with distribution networks, is crucial for extensive adoption. Cooperative initiatives among industry stakeholders are expected to be essential to tackle technical issues and enhance the overall efficacy of hydrogen-powered engines.
- Subsequent study ought to investigate the amalgamation of hydrogen-based technologies with energy from renewable sources, thereby promoting a shift to an environmentally friendly system.
- Hydrogen combustion engines can ultimately function as a transitional solution, augmenting electric vehicles and fuel cells in the pursuit of an environmentally friendly future for heavy-duty transportation.

7. Conclusion

The investigation of hydrogen as a fuel for CI engines reveals a promising but intricate environment marked by considerable advancements and problems.

- Hydrogen has the ability to enhance engine efficiency and minimise hazardous emissions, particularly carbon monoxide and hydrocarbons, which are dramatically reduced in comparison to current fossil fuels. This critical evaluation highlights the potential of hydrogen to accomplish these goals.
- Furthermore, as nations with greater GDPs typically consume more energy and emit more CO2, CO2 emissions and economic activity are correlated. This link between environmental deterioration and economic growth highlights the threat that climate change, which has been predominantly ascribed to the usage of fossil fuels, poses to the entire world.
- Methane (CH4) and nitrous oxide (N2O), which are both linked to tailpipe emissions, are important greenhouse gases in addition to CO_2. The issue of fuel scarcity and environmental pollution is being made worse by the growing vehicles population.
- It is now a global challenge to address the increasing emissions from road traffic. The EPA has established fleet-wide CO_2 emission reduction objectives in an effort to reduce CO_2 emissions, with reductions anticipated for light vehicles and passenger automobiles between model years 2017 and 2025.
- In particular, it is anticipated that CO_2 emissions for light vehicles will drop from 183 g/km to 126 g/km and for passenger cars from 132 g/km to 89 g/km. Concerns about energy and the environment are among the topics that are frequently discussed internationally by the Group of Eight (G8), a coalition of eight highly developed countries.
- Nonetheless, the implementation of hydrogen has challenges; the heightened production of nitrogen oxides (NOx) resulting from increasing combustion temperatures presents a significant environmental issue.

- The results highlight the imperative for novel approaches, like exhaust gas recirculation and water injection, to reduce these emissions while leveraging the advantages of hydrogen.
- In conclusion, although hydrogen presents significant potential as a sustainable fuel alternative, continuous research and development are crucial to surmount these challenges and fully use its capacity to diminish the carbon footprint of internal combustion engines.

References

1. Tsujimura, Taku, and Yasumasa Suzuki. "The utilization of hydrogen in hydrogen/diesel dual fuel engine." International journal of hydrogen energy 42.19 (2017): 14019–14029.
2. Karagöz, Yasin, et al. "Effect of hydrogen–diesel dual-fuel usage on performance, emissions and diesel combustion in diesel engines." Advances in Mechanical Engineering 8.8 (2016): 1687814016664458.
3. Hosseini, Seyyed Hassan, et al. "Use of hydrogen in dual-fuel diesel engines." Progress in Energy and Combustion Science 98 (2023): 101100.
4. Koten, Hasan. "Hydrogen effects on the diesel engine performance and emissions." International journal of hydrogen energy 43.22 (2018): 10511–10519.
5. Szwaja, Stanislaw, and Karol Grab-Rogalinski. "Hydrogen combustion in a compression ignition diesel engine." International journal of hydrogen energy 34.10 (2009): 4413–4421.
6. Ngamnurak, Phonsini, Alissara Reungsang, and Pensri Plangklang. "Improved biohydrogen production from lactic acid bacteria contaminating substrates by enriched hydrogen-producing consortium with lactate-fermentation pathway." Carbon Resources Conversion (2024): 100295.
7. Ngamnurak, Phonsini, Alissara Reungsang, and Pensri Plangklang. "Improved biohydrogen production from lactic acid bacteria contaminating substrates by enriched hydrogen-producing consortium with lactate-fermentation pathway." Carbon Resources Conversion (2024): 100295.
8. Shi, Jihao, et al. "Numerical investigation of hydrogen vapor cloud explosion from a conceptual offshore hydrogen production platform." Journal of Safety and Sustainability (2024).
9. Soudagar, Manzoore Elahi M., et al. "Effect of gasification input parameters and KOH catalyst action on functional properties of hydrogen production from municipal wastewater." International Journal of Hydrogen Energy 94 (2024): 1444–1452.
10. Shomope, Ibrahim, et al. "Machine learning in PEM water electrolysis: A study of hydrogen production and operating parameters." Computers & Chemical Engineering (2024): 108954.
11. Kavtaradze, Revaz, Tamaz Natriashvili, and Sergey Gladyshev. Hydrogen-diesel engine: Problems and prospects of improving the working process. No. 2019-01-0541. SAE Technical Paper, 2019.
12. Dimitriou, Pavlos, et al. "Combustion and emission characteristics of a hydrogen-diesel dual-fuel engine." International journal of hydrogen energy 43.29 (2018): 13605–13617.
13. Jamrozik, Arkadiusz, Karol Grab-Rogaliński, and Wojciech Tutak. "Hydrogen effects on combustion stability, performance and emission of diesel engine." International journal of hydrogen energy 45.38 (2020): 19936–19947.
14. Kumar, Madan, Taku Tsujimura, and Yasumasa Suzuki. "NOx model development and validation with diesel and hydrogen/diesel dual-fuel system on diesel engine." Energy 145 (2018): 496–506.
15. Temizer, İlker, and Ömer Cihan. "Analysis of different combustion chamber geometries using hydrogen/diesel fuel in a diesel engine." Energy Sources, Part A: Recovery, Utilization, and Environmental Effects 43.1 (2021): 17–34.
16. Wright, Madeleine L., and Alastair C. Lewis. "Decarbonisation of heavy-duty diesel engines using hydrogen fuel: A review of the potential impact on no x emissions." Environmental Science: Atmospheres 2.5 (2022): 852–866.
17. Paul, Bhaskar, Sanjay Kumar, and Sanjib Majumdar. "Preparation, characterization and modification of magnesium hydride for enhanced solid-state hydrogen storage properties." International Journal of Hydrogen Energy 96 (2024): 494–501.
18. Vincent-Randonnier, Axel, Nathan Mallart-Martinez, and Julien Labaune. "Design of a plasma-assisted injector: Principle, characterization and application to supersonic combustion of hydrogen." International Journal of Hydrogen Energy 88 (2024): 1410–1421.
19. Cernat, Alexandru, et al. "Hydrogen—An alternative fuel for automotive diesel engines used in transportation." Sustainability 12.22 (2020): 9321.
20. Nag, Sarthak, et al. "Combustion, vibration and noise analysis of hydrogen-diesel dual fuelled engine." Fuel 241 (2019): 488–494.
21. Verma, Saket, et al. "A renewable pathway towards increased utilization of hydrogen in diesel engines." International Journal of Hydrogen Energy 45.8 (2020): 5577–5587.
22. Morales Rojas, Andrés David, Sebastián Heredia Quintana, and Iván Darío Bedoya Caro. "Experimental Study of a Homogeneous Charge Compression Ignition Engine Using Hydrogen at High-Altitude Conditions." Sustainability 16.5 (2024): 2026.
23. Morales Rojas, Andrés David, Sebastián Heredia Quintana, and Iván Darío Bedoya Caro. "Experimental Study of a Homogeneous Charge Compression Ignition Engine Using Hydrogen at High-Altitude Conditions." Sustainability 16.5 (2024): 2026.
24. Tripathi, Gaurav, et al. "Computational investigation of diesel injection strategies in hydrogen-diesel dual fuel engine." Sustainable Energy Technologies and Assessments 36 (2019): 100543.
25. Köse, H., and M. Ciniviz. "An experimental investigation of effect on diesel engine performance and exhaust emissions

of addition at dual fuel mode of hydrogen." Fuel processing technology 114 (2013): 26–34.

26. Castro, Nicolas, Mario Toledo, and German Amador. "An experimental investigation of the performance and emissions of a hydrogen-diesel dual fuel compression ignition internal combustion engine." Applied Thermal Engineering 156 (2019): 660–667.
27. Dimitriou, Pavlos, Taku Tsujimura, and Yasumasa Suzuki. "Hydrogen-diesel dual-fuel engine optimization for CHP systems." Energy 160 (2018): 740–752.
28. Wang, Su, et al. "The environmental potential of hydrogen addition as complementation for diesel and biodiesel: A comprehensive review and perspectives." Fuel 342 (2023): 127794.
29. Jhang, Syu-Ruei, et al. "Reducing pollutant emissions from a heavy-duty diesel engine by using hydrogen additions." Fuel 172 (2016): 89–95.
30. Accardo, Antonella, et al. "Greenhouse Gas Emissions of a Hydrogen Engine for Automotive Application through Life-Cycle Assessment." Energies 17.11 (2024): 2571.
31. Yilmaz, I. T., and M. E. T. İ. N. Gumus. "Effects of hydrogen addition to the intake air on performance and emissions of common rail diesel engine." Energy 142 (2018): 1104-1113.
32. Gültekin, Nurullah, Halil Erdi Gülcan, and Murat Ciniviz. "The impact of hydrogen injection pressure and timing on exhaust, mechanical vibration, and noise emissions in a CI engine fueled with hydrogen-diesel." International Journal of Hydrogen Energy 78 (2024): 871–878.
33. Zhang, Zhiqing, et al. "Utilization of hydrogen-diesel blends for the improvements of a dual-fuel engine based on the improved Taguchi methodology." Energy 292 (2024): 130474.
34. Serrano, J., F. J. Jiménez-Espadafor, and A. López. "Analysis of the effect of different hydrogen/diesel ratios on the performance and emissions of a modified compression ignition engine under dual-fuel mode with water injection. Hydrogen-diesel dual-fuel mode." Energy 172 (2019): 702–711.
35. Mohamed, Mohamed, et al. "A Comprehensive Experimental Investigation of NOx Emission Characteristics in Hydrogen Engine Using an Ultra-Fast Crank Domain Measurement." Energies 17.16 (2024): 4141.
36. Gholami, Aboozar, Seyed Ali Jazayeri, and Qadir Esmaili. "A detail performance and CO2 emission analysis of a very large crude carrier propulsion system with the main engine running on dual fuel mode using hydrogen/diesel versus natural gas/diesel and conventional diesel engines." Process Safety and Environmental Protection 163 (2022): 621–635.
37. Correa, G., P. M. Muñoz, and C. R. Rodriguez. "A comparative energy and environmental analysis of a diesel, hybrid, hydrogen and electric urban bus." Energy 187 (2019): 115906.
38. Cunanan, Carlo, et al. "A review of heavy-duty vehicle powertrain technologies: Diesel engine vehicles, battery electric vehicles, and hydrogen fuel cell electric vehicles." Clean Technologies 3.2 (2021): 474–489.
39. Reddy, Kodandapuram Jayasimha, et al. "An Evaluation of the Effect of Fuel Injection on the Performance and Emission Characteristics of a Diesel Engine Fueled with Plastic-Oil–Hydrogen–Diesel Blends." Applied Sciences 14.15 (2024): 6539.
40. Chaudhary, Vinay Prakash, Manish Kumar Singh, and D. B. Lata. Energy-exergy analyses of hydrogen enriched—diesel and blended biofuel (Azolla pinnata macroalgae) on the dual fuel diesel engine. International Journal of Hydrogen Energy 93 (2024): 1535–1548. https://doi.org/10.1016/j.ijhydene.2024.11.024.
41. Frantzis, Charalambos, et al. "A review on experimental studies investigating the effect of hydrogen supplementation in CI diesel engines—The case of HYMAR." Energies 15.15 (2022): 5709.
42. Babayev, Rafig, et al. "Computational comparison of the conventional diesel and hydrogen direct-injection compression-ignition combustion engines." Fuel 307 (2022): 121909.
43. Estrada, L., et al. "Experimental assessment of performance and emissions for hydrogen-diesel dual fuel operation in a low displacement compression ignition engine." Heliyon 8.4 (2022).
44. Li, Shunxi, et al. "Transition of heavy-duty trucks from diesel to hydrogen fuel cells: Opportunities, challenges, and recommendations." International journal of energy research 46.9 (2022): 11718–11729.
45. Manigandan, S., et al. "Hydrogen and ammonia as a primary fuel–A critical review of production technologies, diesel engine applications, and challenges." Fuel 352 (2023): 129100.
46. Dash, Santanu Kumar, et al. "Hydrogen fuel for future mobility: Challenges and future aspects." Sustainability 14.14 (2022): 8285.
47. Hosseini, Seyyed Hassan, et al. "Use of hydrogen in dual-fuel diesel engines." Progress in Energy and Combustion Science 98 (2023): 101100.
48. Wright, Madeleine L., and Alastair C. Lewis. "Decarbonisation of heavy-duty diesel engines using hydrogen fuel: A review of the potential impact on no x emissions." Environmental Science: Atmospheres 2.5 (2022): 852–866.
49. Dou, Yi, et al. "Opportunities and future challenges in hydrogen economy for sustainable development." Hydrogen economy. Academic press, 2023. 537–569.
50. Zou, Caineng, et al. "Industrial status, technological progress, challenges, and prospects of hydrogen energy." Natural Gas Industry B 9.5 (2022): 427–447.

Advances in Additive Manufacturing Technologies – Gurusamy Pathinettampadian et al. (eds)
© 2026 Taylor & Francis Group, London, ISBN 978-1-041-16687-0

Customized E-Learning Systems Using AI: Problems, Difficulties, and Solutions

F. Mohamed Junaid[1], Yogapriya M.[2], Zahid Hussain J.[3], Gouri R.[4], Cheran C.[5], Karthikeyan[6]

Department of CSE-AIML, Chennai Institute of Technology, Chennai, Tamil Nadu

◆ **Abstract:** Personalized and adaptive learning systems are transforming the learning world to personalize content so that, ideally, it becomes unique for each learner. Unlike the typical common presentation of content across online platforms, these systems impact artificial intelligence and data insights to develop individualized educational pathways beyond what is interested in or even the level of interest more than prior knowledge. This review collates recent findings on how such systems are developed and implemented and also the challenges they face, with special focus on the role of AI in creating flexible learning pathways that respond to each learner's needs. The research underlines crucial points, ranging from adaptive learning paths to recommendation engines and skill-centered learning methods which sum up to customized learning experiences. Although platforms such as Udemy, Coursera, and other LMS offer much personalization, there exists the problem of complexity in putting these systems to practice to ensure engagement, accessibility, and a smooth integration into a setting that is more traditional. Motivation, self-regulation, and independence seem to be hallmarks of all pertinent student success factors, and personalized systems appear largely bereft of the interactive and social qualities of a traditional classroom. An adaptive learning system needs a continuous feedback mechanism that allows self-adjustment in order to adapt to the personal goals and preferences of students. Despite being promising, challenges arise when introducing such systems into an existing LMS, catering for the diverse needs of learners, and data privacy issues. Future research standardization of the components of adaptive systems and compatibility with current educational databases can make such systems fit for wider implementation.

◆ **Keywords:** Artificial intelligence, Personalized learning, Adaptive learning, Learning management system, Education, Machine learning, Recommender system

1. Introduction

In recent decades, personalized learning (PL) has become an important focus of educational reform. This is actually how PL has been underscored as abandoning traditional, one-size-fits-all instructional methods and having a road toward tailoring methods toward providing unique needs, preferences, and abilities of each student. The actual idea of PL traces roots as early as the start of educational theory over the years, dating back even to the early 20th century, with John Dewey advocating the student-centered learning approach[3]. One of the defining developments of early PL years was Fred Keller's Personalized System of Instruction, established in 1968[3].

[1]fmohamedjunaid.aiml2023@citchennai.net, [2]yogapriyam@citchennai.net, [3]zahidhussainj@citchennai.net, [4]gouri.r@citchennai.net, [5]cheranc.aimll2023@citchennai.net, [6]karthikeyanp@citchennai.net

DOI: 10.1201/9781003685906-36

It emphasized self-pacing, mastery learning, and small-group tutoring. Education reformers started decrying standardized models as they addressed the diversifying student population; the idea of PL eventually began to gain a foothold. It is during the last two years when countries added PL in their policies like that of the UK, the USA,

Finland, and Canada to ensure that a better learning experience will be achieved without any inequalities to every learner.

For instance, in the U.K., since 2004, PL has been implemented in the official government policy and focuses on individualized curricula, student-centered learning environments, and more flexibility in teaching[3]. In a similar way, within the US, the Every Student Succeeds Act will promote personalized learning models made possible by today's technology combined with instructional practices that guide frameworks such as Universal Design for Learning (UDL). As previously mentioned, PL is shown to increase engagement, better outcomes, and ensure more supportive environments for students who have special needs or diverse requirements. Scaling such systems and developing wide-scale adoption still pose challenges. One of the major enablers of PL is AI since it can actually produce adaptive learning systems. Infusion of machine learning and data analytics in AI-based systems would determine learner profiles that could dynamically adjust content, pace, and instructional strategies based on differences in performance and preference among learners.

Such systems result in a strongly individualized learning experience and serve towards offering support to the learner at each step of the learning process. AI brings with it a new bunch of benefits and issues concerning the issue of data privacy and responsible utilization of learner information, which need a critical pondering while planning for personalized learning initiatives to prove effective and fair. Overall, the use of AI in PL will be transforming education and providing better results for the students from all walks of life. Additionally, the challenges facing the integration and large-scale application of personalized learning will require even more interdisciplinary research and work across these areas: education, data science, and technology.

2. Relevant Works

Murtaza et al. [1] proposed intelligent framework for personalized e-learning thus integrates key learning theories: Humanism, Behaviorism, Cognitivism, Constructivism, and Connectivism, using the ADDIE model for instructional design. It consists of five modules - Data Module, storing learner profiles, assessments, and content; Adaptive Learning Module, making use of sequential machine learning to determine how well learners know things; Adaptable Learning Module, analyzing the learning preferences using performance cubes and unsupervised algorithms; Recommender Module, utilizing deep learning for personalized content recommendations; and Content and Assessment Delivery Module, ensuring that material delivery is tailored. Collectively, these elements form a scalable framework for data-informed learning paths and improved education outcome.

Dr. Deepak Kem et al. [2] explored that customized and adaptive learning is actually using AI and ML to adjust learning encounters based on their elemental needs, whereby educational experiences are adapted according to their cognitive abilities, past knowledge, and motivation levels. This approach encompasses adaptive tools, competency-based approaches, and LMS, through which learning can improve in terms of engagement and access. The case examples of Udemy and Coursera have highlighted scalability, but challenges do exist in standardizing frames and effectively incorporating technology. Although it offers to shift from teacher-centered education models to learner-centered models, more research needs to be done on overcoming challenges of implementation and the eventual production of standard systems with greater adaptation.

Ling Zhang et al. [3] discussed that personalized learning refers to a most important educational reform in the history of reformative processes-to tailor education according to every individual's needs. This literature review synthesizes 71 empirical studies on the implementation of PL conducted between 2006 and 2019. The two broad research themes identified are: the role of technology, like adaptive learning systems and mobile learning applications, and the contextual factors influencing the adoption of PL within schools, including teachers' perception and the institutional frameworks developed. The majority of the studies report generally positive effects in both achievement and engagement, and the review concludes that the field will benefit from a more comprehensive conceptual framework that can inform research and practice forward. Findings are that the personalized approaches in most of these studies have potential to contribute toward greater equity in education but point to some of the challenges arising, particularly concerning scalability and consistency.

Raj et al. [4] analyzed that hybrid recommendation method incorporates collaborative filtering content-based filtering, and ontological models so as to provide enhanced accuracy of recommendations with learner satisfaction. From 2015-2020 studies were carried out on various adaptive content recommendation systems within personalized learning environments; the types of adaptive learning pathways adapted by these systems are said to be based upon the unique characteristics of a learner, including their pace, preferences, and knowledge level. These systems, through the adoption of machine learning, semantic web technologies, and flexible use of user input, are capable of handling the cold-start problem and sparsity of data. Even though these have been efficient in personalizing learning experiences, the frameworks in this field still require more efforts towards standardization, usability improvements, and long-term studies about the impacts on learning outcomes.

Bozkurt et al. [5] explored that for the last fifty years, AI has transformed education into one that focuses on personalization and adaptive methods, bringing human-AI collaboration as learning. Relevant facets of studies argue the impact of AI in higher education, applying machine learning for analytics in learning, and the ethical dilemmas of bias and the transparency of data. While AI allows a learner to have a customized experience through its intelligent tutoring support, challenges remain pertinent and call for ethical standards and multidisciplinary approaches to address the responsible use of AI in education.

Sajja R et al. [6] analyzed the concept framework of AI-supported Intelligent Assistant (AIIA) aligned with the adaptation of intelligent learning in the domain of higher education. The AI-equipped natural language processing-based system has attractive learning experiences, reduces the semblance of cognitive overload, and encourages student engagement through the presence of facilities such as quiz creation, flashcard generation, and real-time answering of questions. Its foundation would be LMS compatibility and the capability of the tool to accommodate diverse fields, which supports tremendous revolution in higher education in terms of solving issues of academic integrity and data scalability.

Helena Rodrigues et al. [7] explained the development of e-learning through definitions, usability, and integration in the education process. It implies inconsistencies in content and terminology of e-learning with a recommendation for standardization. The review pooled 99 peer-reviewed articles from 2010 to 2018, illustrating themes such as education systems, learning innovation, and students' engagement. Key findings emphasize the improvement of outcomes with technology-driven learning and the main tools for engagement using MOOCs, mobile learning, and gamification. This raises questions on usability as well as variability in the needs of the learners; as such, it calls for adaptive learning environments. This paper presents a holistic framework of e-learning purported to marry pedagogy with emergent digital tools that are used to design better learning experiences.

Moubayed et al. [8] compiled using the AI applications in personalized education systems by applying adaptive learning and real-time feedback. In such a case, the focus area was developing intelligent adaptive frameworks, which could promote high engagement of diverse learners while offering a harmonic fit with digital platforms. Advanced algorithms based on these systems allow personalizing contents and tracking performance and afford an opportunity for autonomous student learning. Promise, despite these merits, is seen to face ethical challenges in terms of scalability and deployment at all levels in education. These results show how AI can revolutionize modern education.

Bellarhmouch et al. [20] discussed on new developments on the aspect of learner modeling, which includes the cognitive, affective and the behavioral characteristics such as learning preferences and emotional properties. Enhancements in personification consist of methods including stereotype models, fuzzy logic, and machine learning. Heterogeneous strategies improve the basic conception of adaptive learning as they consider difficulties such as cognitive overload. These studies made a call for use of differentiated instruction to address learner needs and enhance learning achievement.

3. Proposed Models

The review clarifies that PL models typically include both interpersonal educator–student relationships and technology and, for example, might apply UDL(Universal Design for Learning) to foster student-centered PL approaches [7]. Some of the promising models are the ones combining one or another technique, entering collaborative filtering, content-based filtering and ontological models. Such models are usually validated by factors such as time and accuracy in predicting the outcome, the satisfaction of the learner as well as the ease of use of the model.

A prominent proposed model features five interconnected modules: Data module, adaptive learning module, adaptable learning module, Recommender module, and content and assessment delivery module. This model will enable individual learning paths since the learning content will change depending on the learner profile and sequential assessment responses. It provides the theoretical basis for the creation and assessment of personalized adaptive e-learning systems and also discusses issues of student motivation and implementation.

It also discusses current research on personalization in e-learning and extracts specific approaches, such as competency-based learning, adaptive models, creation of customized web services, and LMS. Learner engagement strategies, self-regulation skills, motivation and feedback occur as critical features that receive most emphasis in this approach. The study focuses on current challenges to find out effective approaches in application of AI and learning theories for enhancing personalization of e-learning and consequently indicates a new framework for improving the aspects of personalization and system capability of e-learning. This integrated approach aims at trying to optimize the establishment of learner environments that are tailored while at the same time dealing with challenges that keep emerging in system architecture and learners support.

In particular, the proposed model of a learning system assimilates closely related aspects of stereotype modeling, fuzzy logics, and similarity techniques. A static data area is the persons' attributes including personal and learning profile data, as well as dynamic data area involving knowledge acquisition, knowledge emotions, and assessments. The FLSM was adopted and TF-IDF coupled with cosine similarity was used in matching the goals of the learner to the resources. Through the maintenance of the learner profile, and the incorporation of affective data, it makes learning more adaptive to the learners and response to challenges such as overload and disorientation.

4. Learners' Model: Driving Adaptive and Personalized Learning

Learners' model incorporates stereotype modeling, fuzzy logic and similarity techniques to cover the variation in characteristics of the learners. The important data include static data (personal data, learning preferences) and dynamic data (knowledge mastery, emotional state, and assessments). The learning styles are identified based on the Felder-Silverman Learning Style Model (FSLSM) while the domain related goals are determined through the Term Frequency-Inverse Document Frequency (TF-IDF) and cosine similarity techniques. Dynamic parameters such as errors and response times enhance the learner profile based on assessment. Emotional state detection guarantees he real-time application, which deals with cognition problems such as overload and disorientation. They makes it possible to coordinate resources with the contents in a way that suits a given learner thus making the learning process effective. [20]

The proposed model leads to few important consequences. It is accurate in recommending what should be relayed to the learners by frequently updating their profiles.

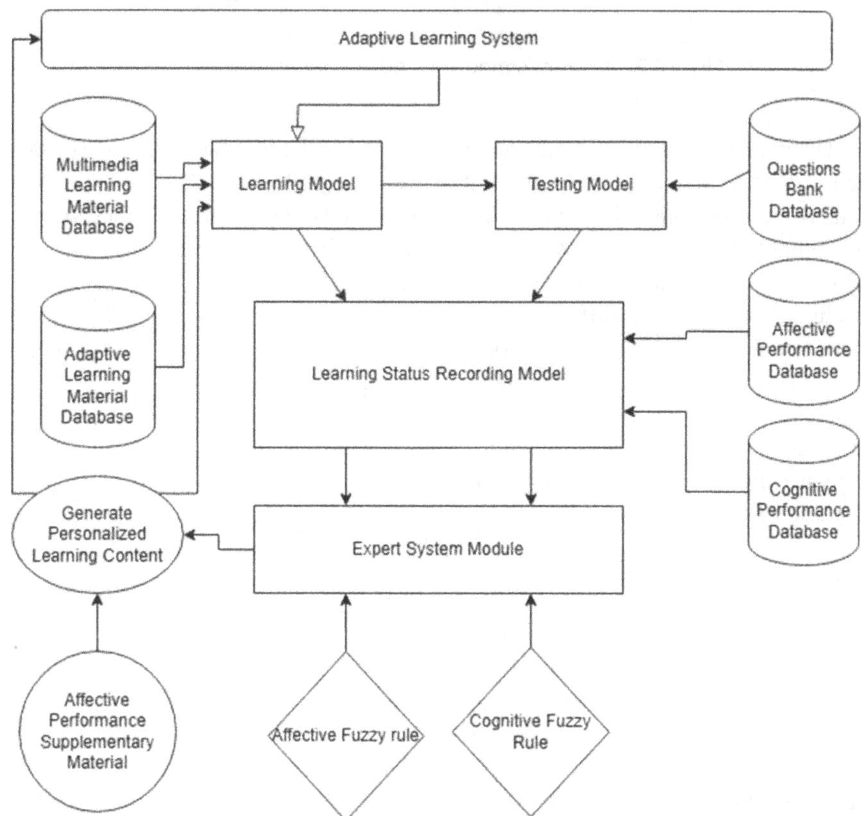

Fig. 36.1 Interaction of adaptive learning system [2]

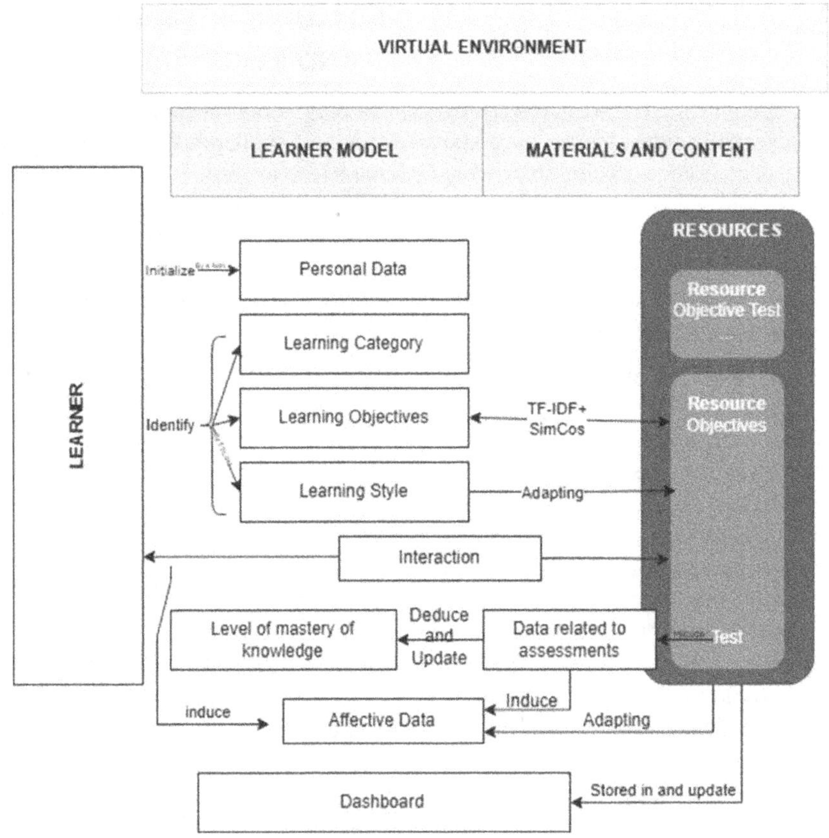

Fig. 36.2 Learner's model for personalized learning [5]

Learner affective data increases learner participation and motivation which in turn minimizes learner dropout rates. The experimental results and comparison with other systems show how the proposed model improves learning outcomes from cognitive and emotional factors. The framework is quite useful in catering for the different needs of the learners when developing adaptive educational systems and comprises of a strong platform that can easily be scaled up to fit the current complex adaptive learning systems. It also emphasizes the call for higher clientization with a view of improving engagement, satisfaction, and even performance in academe.[20]

5. Conclusion

AI-driven personalized learning platforms allow unprecedented learning experiences by adopting to the individual needs of users. This fully customized system, in the entirety, by consuming all data, algorithms, and frameworks such as Universal Design for Learning can reach the most diverse audience. Real-time assessment and addition of the recommendation module dynamically alter content and create student autonomy and confidence. However, large-scale systems like these would encounter severe problems with mass deployment regarding privacy issues of data, responsiveness of the system, and standardization of infrastructure.

Represented through the Adaptive Learning Module and Recommendation Module, the PL framework targets all the possibilities of dynamically creating a student-centered experience in e-learning. It is adaptive to learners' profile using sequential assessments and analytics enabled by AI technology to guide students at exactly the right points for significant engagement. Despite all these advancements, scaling such systems across education settings proves difficult. Turning traditional, teacher-oriented methods into learner-centered environments requires systemic change, long-term efforts, and multi-stakeholder collaboration. Moreover, while AI makes learning more efficient, it can only seamlessly fill a gap for learners' emotional and social learning needs by supplementing human interaction.

While AL is still on the rise as the novel technology for PL development, several promising lines for further research could be identified as providing a potential for enhancing the effectiveness and flexibility of online learning environments. Although incorporating AI in developing personalized learning platforms has yielded positive outcomes by addressing individual learner interests, future developments can address several shortcoming that influence learner achievement to improve the system's performance.

6. Future Scope

6.1 Counterfactual Evaluation

Recommendation systems currently implemented on e-learning platforms often use data collected from observations of learners' behavior, which can be misleading in many aspects and limits the efficiency of the existing recommendation systems. Another promising avenue for study is a counterfactual approach, where different learning contexts are examined to assess the impact of changes in the learning environment on the learners. This approach can help avoid injustice and increase recommendation precision Additionally, it can resolve concept shift within training and testing phases.

6.2 Detailed Learner Activity Data

The majority of the present systems fail in capturing rich activity information of the learner over the entire learning process. To enhance the tailored content delivery, which is a hallmark of the upcoming systems, two things are required; Individual learner attributes in terms of time on task, the strategies employed to solve various problems and use patters across the devices. This more exhaustive information can help machine learning algorithms to further understand learner acquisition and provide content accordingly.

6.3 New Learner Evaluation Metrics

Traditional conceptual assessments, such as a standard end of term test results, can often only give an incomplete picture of the learners' progress. More critically, new ways of evaluating program outcomes are needed. These metrics should include count data including the number of attempts, response time and how learners engage with different types of content. They will also help to define concentration areas and take targeted approaches while aiming at eradicating particular knowledge deficits.

6.4 Immediate Feedback and Ongoing Evaluation

Feedback is crucial in continuous engagement for adaptive learning environments that are implemented online. Future smart e-learning systems used in educating learners ought to have minimal latency to ensure that the platform's responses are immediate to boost learners' experience. Moreover, the assessment of learners' progress through formative and summative assessments can offer an ongoing view of learner progress as learners progress, which can help in modifying delivery on the go.

6.5 Enhanced Recommendation Evaluation

Currently, it is possible to identify just a few approaches to evaluate effectiveness of personalized recommendations across the learning process. Subsequent studies should shift toward more advanced assessment models as the context-aware recommendation system includes contextual recommendation framework that measures the applicability and the influence of the recommendations obtained based on the available data. Application methods such as contextual precision and ROC can enhance the measurement of recommendation system efficiency.

From the presented research domains, prospects for applying AI for personalized learning in the future look promising for the desired improvement of learners' achievements. As the field unfolds, all of these sophisticated techniques will be incorporated to create more flexible, effective, and interesting online learning systems aimed to meet the diversification of learners' needs.

Observational biases in recommendation systems have proved that only diverse, experimental data can help enhance the accuracy of these algorithms. At the same time, transactional data would allow for more transactional personalization, including learner activity, including tasks that involve time taken on the activity as well as behavioural patterns. However,,more or less constant and actual evaluation shall have to be required in terms of keeping attention of the learners patent and the rate of their dropping out curbed. Other complex measures to performance in learning systems also involve finer measurement based on the context in correlation with learners' requirement variations. There is also the aspect of engagement of users by the adaptive learning technologies as one of the key benefits. This system also takes it a notch higher by not only adjusting research tools to suit the users, but also recommendations. These systems need technical support, while safeguarding the data and educating all necessary staff for these systems to be effective. Since these tools are developed by educators, technologists, and data experts the collective success involves optimizing these resources without infringing on the user's privacy. In other words, the opportunity to use artificial intelligence in the management of LMS is considered to radically shift education and research. If enhanced and built progressively, this type of systems avails itself to address global education issues and marginal bureaucrats that hinder education reaching out to every other person. Sustaining the progress and focus of this endeavor on the emergence of new technologies, it is possible to introduce an aim that aspiring to shift into practice the individualized education for all learners.

References

1. M. Murtaza, Y. Ahmed, J. A. Shamsi, F. Sherwani and M. Usman, "AI-Based Personalized E-Learning Systems: Issues, Challenges, and Solutions," in IEEE Access, vol. 10, pp. 81323–81342, 2022, doi: 10.1109/ACCESS.2022.3193938.
2. Dr. Deepak Kem, Personalized and Adaptive Learning: Emerging Learning Platforms in the Era of Digital and Smart Learning: International Journal of Social Science and Human Research, DOI:10.47191/ijsshr/v5-i2-02
3. Ling Zhang, James D. Basham, Sohyun Yang, Understanding the implementation of personalized learning: A research synthesis, Educational Research Review, Volume 31, 2020, 100339, ISSN 1747–938X, https://doi.org/10.1016/j.edurev.2020.100339.
4. Raj, N.S., Renumol, V.G. A systematic literature review on adaptive content recommenders in personalized learning environments from 2015 to 2020. J. Comput. Educ. 9, 113–148 (2022). https://doi.org/10.1007/s40692-021-00199-4
5. Bozkurt, A., Karadeniz, A., Baneres, D., Guerrero-Roldán, A. E., & Rodríguez, M. E. (2021). Artificial Intelligence and Reflections from Educational Landscape: A Review of AI Studies in Half a Century. Sustainability, 13(800). https://doi.org/10.3390/su13020800
6. Sajja R, Sermet Y, Cikmaz M, Cwiertny D, Demir I. Artificial Intelligence-Enabled Intelligent Assistant for Personalized and Adaptive Learning in Higher Education. Information. 2024; 15(10):596. https://doi.org/10.3390/info15100596
7. Helena Rodrigues, Filomena Almeida, Vanessa Figueiredo, Sara L. Lopes, Tracking e-learning through published papers: A systematic review, Computers & Education, Volume 136, 2019, Pages 87-98, ISSN 0360–1315, https://doi.org/10.1016/j.compedu.2019.03.007.
8. Moubayed, M. Injadat, A. B. Nassif, H. Lutfiyya and A.Shami, "E-Learning: Challenges and Research Opportunities Using Machine Learning & Data Analytics," in IEEE Access, vol. 6, pp. 39117–39138, 2018, doi: 10.1109/ACCESS.2018.2851790.
9. J. Shailaja and R. Sridaran, "Taxonomy of E-Learning Challenges and an Insight to Blended Learning," 2014 International Conference on Intelligent Computing Applications, Coimbatore, India, 2014, pp. 310–314, doi: 10.1109/ICICA.2014.70.
10. Qiao P, Zhu X, Guo Y, Sun Y, Qin C. The Development and Adoption of Online Learning in Pre- and Post-COVID-19: Combination of Technological System Evolution Theory and Unified Theory of Acceptance and Use of Technology. Journal of Risk and Financial Management. 2021; 14(4):162. https://doi.org/10.3390/jrfm14040162
11. Liu, Q., Shen, S., Huang, Z., Chen, E., & Zheng, Y. (2021). A survey of knowledge tracing. arXiv preprint arXiv:2105.15106. https://doi.org/10.48550/arXiv.2105.15106
12. I.Dhaiouir, M. Ezziyyani, and M. Khaldi, "The personalization of learners educational paths e-learning," in Networking, Intelligent Systems and Security (Smart Innovation, Systems and Technologies). Singapore: Springer, 2022, pp. 521–534.
13. X.Pan, X. Li, and M. Lu, "A multiview courses recommendation system based on deep learning," in Proc. Int. Conf. Big Data Informatization Educ. (ICBDIE), Apr. 2020, pp. 502–506.
14. Huang, Jiahui, Salmiza Saleh, and Yufei Liu. "A review on artificial intelligence in education." Academic Journal of Interdisciplinary Studies 10.3 (2021). https://doi.org/10.36941/ajis-2021-0077
15. Saito, Yuta, and Thorsten Joachims. "Counterfactual learning and evaluation for recommender systems: Foundations, implementations, and recent advances." In Proceedings of the 15th ACM Conference on Recommender Systems, pp. 828–830. 2021. https://doi.org/10.1145/3460231.3473320
16. M. A. Hassan, U. Habiba, F. Majeed, and M. Shoaib, "Adaptive gamification in e-learning based on Students learning styles," Interact. Learn. Environ., vol. 29, no. 4, pp. 545–565, 2021.
17. Omid Gheibi, Danny Weyns, and Federico Quin. 2021. Applying Machine Learning in Self-adaptive Systems: A Systematic Literature Review. ACM Trans. Auton. Adapt. Syst. 15, 3, Article 9 (September 2020), 37 pages. https://doi.org/10.1145/3469440
18. T. R. D. Saputri and S. -W. Lee, "The Application of Machine Learning in Self-Adaptive Systems: A Systematic Literature Review," in IEEE Access, vol. 8, pp. 205948–205967, 2020, doi:10.1109/ACCESS.2020.3036037.
19. Ezzaim, A., Dahbi, A., Aqqal, A. et al. AI-based learning style detection in adaptive learning systems: a systematic literature review. J. Comput. Educ. (2024). https://doi.org/10.1007/s40692-024-00328-9
20. Bellarhmouch, Y., Jeghal, A., Tairi, H. et al. A proposed architectural learner model for a personalized learning environment. Educ Inf Technol 28, 4243–4263 (2023). https://doi.org/10.1007/s10639-022-11392-y

Advances in Additive Manufacturing Technologies – Gurusamy Pathinettampadian et al. (eds)

Fabrication and Characterisation of an Advanced A356/MWCNT Composite by Stir Cum Squeeze Casting Process

Suresh R.*,
Viswa V., Hari Krishna Raj S.,
Tamil Selvan B., Velmurugan, Sugin S.
Department of Mechanical Engineering,
VelTech HighTech Dr. Rangarajan Dr. Sakunthala Engineering College,
Chennai, Tamilnadu

◆ **Abstract:** Rare earth element research on the production of materials with improved mechanical properties has immense application in the automobile sector. In this research, emphasis was laid on investigating the effects of reinforcement of A356 aluminum alloy by 0.2 wt.% Al-6Ce-3La (ACL) on microstructure and mechanical properties. Additionally, A356 aluminum alloy was analyzed using the stir-cum-squeeze casting technique with silicon carbide (SiC) and multi-walled carbon nanotubes (MWCNTs) as the reinforcement materials.

These composites have prominent roles in aerospace and automobile sectors with improved characteristics of high compression strength and density resistance. In an attempt to improve the mechanical performance of A356 aluminum, the composition of reinforcement was 91.28% aluminum, 8.39% silicon carbide, and 0.31% MWCNTs by weight. The reinforcement materials were also preheated to certain temperatures: A356 aluminum to 800°C, silicon to 150°C, and MWCNTs to 400°C, for complete integration during casting.

◆ **Keywords:** MWCNTs, Aluminum, ACL, Aluminium alloy, Density resistance, Reinforcement

1. INTRODUCTION

Three or more components are used, the final material is called a hybrid composite. However, they also come with drawbacks, including reduced toughness compared to monolithic metals and higher costs. Despite these challenges, MMCs outperform most polymer matrix composites in key mechanical properties, such as greater transverse strength and stiffness, improved shear and compressive strength, and better high-temperature resistance. Physically, MMCs stand out for their resistance to moisture absorption, flammability, and most types of radiation, along with good electrical and thermal conductivity. Aluminum and titanium are among the most common metal matrices because of their low density and availability in many different alloys. Although magnesium is lighter, its high reactivity with oxygen results in extensive atmospheric corrosion and restricts its uses. Beryllium, although the lightest structural metal with a tensile modulus greater than steel, is not utilized because of its tremendous brittleness. Nickel and cobalt superalloys have also been investigated as matrices, but their alloying constituents facilitate oxidation of the fibers at elevated temperatures and are hence not as useful for some applications. It has been found that Al_2O_3 (matrix)–SiC (reinforcement) composites can be effectively prepared by Spark Plasma Sintering (SPS). The method quickly heats the mold and the composite powder with pressure, accelerating the sintering

*Corresponding author: drsureshr@velhightech.com

DOI: 10.1201/9781003685906-37

process. Electrical current in SPS creates sparks between particles to form a plasma environment that fosters powder consolidation. Composites with 20 wt% Al_2O_3-SiC reached a highest hardness of 324.6 HV. Dash et al. found alumina particles evenly dispersed in the aluminum matrix of micro-composites. Silicon carbide (SiC) is a highly versatile material that finds wide usage in high-performance products like car brake, clutch, and ceramic plate in bulletproof jackets. Large SiC single crystals are made through the Lely method and are frequently cut into gemstones that are synthetic moissanite. Alternatively, SiC is also called carborundum. It is an industrial crystal material that is both a semiconductor and a ceramic in nature. It naturally appears as the uncommon mineral moissanite. Although pure silicon carbide crystals are colorless and transparent, traces of impurities like nitrogen or aluminum may provide green or blue colors based on their concentration. In its pure form, silicon carbide is an electrical insulator but can be made to act as a p-type semiconductor when doped with materials such as aluminum. Silicon carbide in the industrial grade has a typical purity of 98–99.5%, and the typical impurities present are aluminum, iron, oxygen, and free carbon. Silicon carbide has excellent resistance to corrosion and chemical inertness, and it finds extensive use in the industrial sector., even when exposed to strong acids such as hydrochloric, sulfuric, or hydrofluoric acid, or concentrated bases like sodium hydroxide. It reacts with chlorine only at temperatures above 900°C and begins oxidizing in air at approximately 850°C, forming a protective layer of silicon dioxide (SiO_2). Silicon carbide's exceptional mechanical, thermal, and chemical properties make it a critical material in a variety of industries, including electronics, automotive, and refractory manufacturing.

2. Result and Discussion

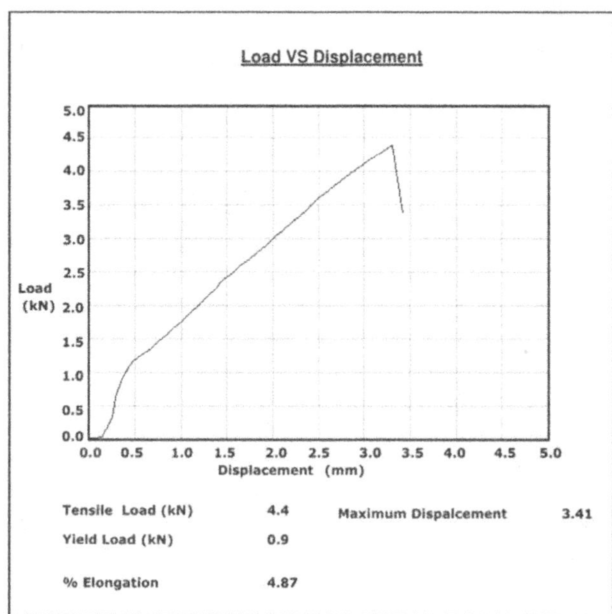

Fig. 37.1 Load vs displacement

This furnance used to melt the Metals like Magnesium, Aluminium, Copper and Metal matrix(Metal with oxides) we are manufacturing small machine with accuracy limit for research institute.

STEP 1: Cut the aluminim A356 for the required weight of 1435g and measure Sic to 132g and MWNCNT to 5g

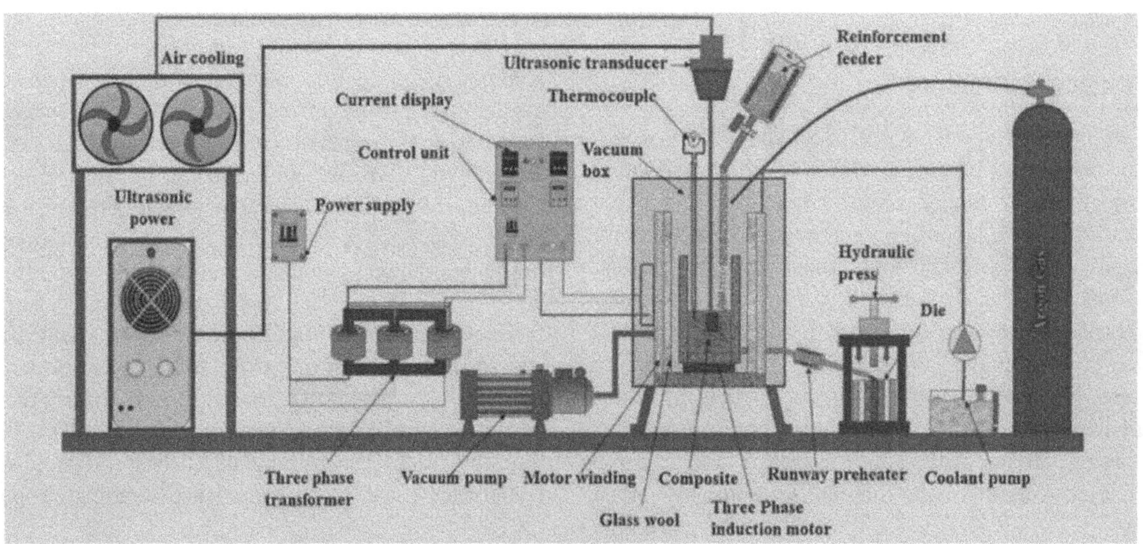

Fig. 37.2 Stir casting furnance design

Fig. 37.3 Step 1: Cutting machine

STEP 2: Furnance preheated to 300°c and aluminium placed into furnance and required temperature 800°c is seated in control unit.

Fig. 37.4 Step 2: Furnace preheating

STEP 3: After 1 hour of heating alumnium MWCNT slowly added in furnance with semi moltern aluminium A356.

STEP 4: Semi moltern aluminium and MWCNT is stirred for 2 minutes.

STEP 5: In Reinforcement feader silicon carbide is added and heated for 150° c.

Fig. 37.5 Step 3: One hour heating of molten aluminium and MWCNT

STEP 6: Runway preheater is set to 750° c and Die temperature set to 370°c.

STEP 7: 800 C moltern aluminium A356 stirred under 632 rpm.

STEP 8: When stirring A356 aluminium the reinforced silicon carbide is feeded into to the furnance and stirred 632 rpm in up and down motion.

Fig. 37.6 Step 4: Semi molten aluminum

STEP 9: Pour wall is opened and molten aluminium is traveled through runway to die and squezzed to 40 tons with hydraulic press.

Fig. 37.7 Step 5: Reinforcement feeder for SiC

STEP 10: Using allowance the alluminium composite is removed . machines are turned off

Fig. 37.8 Step 6: Specimen

Even though such composites possess favorable properties, they remain significantly more expensive than their continuous fiber versions. Metal matrix composites (MMCs) are composites made up of two or more elements, typically a metal and a non-metal material such as ceramics or organic compounds. When more than three components are employed, the resulting material is referred to as a hybrid composite. MMC offers numerous benefits over monolithic metals, such as enhanced specific modulus and strength, better high-temperature performance, reduced coefficients of thermal expansion, and better wear resistance. Stresses are usually determined with respect to the original area of the specimen such types of stresses are often referred to as nominal or conventional stresses. True stress-Strain Diagram

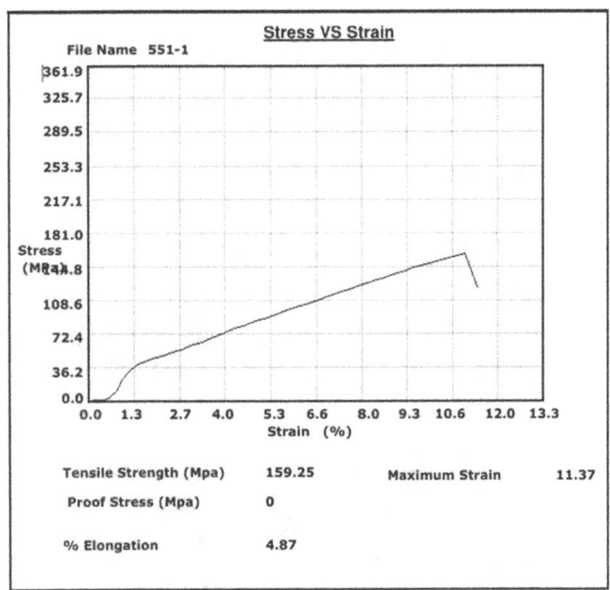

Fig. 37.9 Stress vs strain

Because when a material is loaded uniaxially, there is always some expansion or contraction that has to happen. Hence, the force applied by the actual area of the specimen at the same moment gives us the so called true stress.

The point mentioned as strength dispersion in major location with some eventual point these makes changes in mechanical properties where linearity for a brief interval following point A, the material can be elastic in the sense that deformations are fully recoverable when load is released. The point B of limitation is referred to as Elastic. A stress-strain graph demonstrates that the proportional limit and yield point are as close together that for all intents and purposes they may be viewed as identical. Measuring the proportional limit will normally be more convenient, however. For some materials that have an indistinct yield

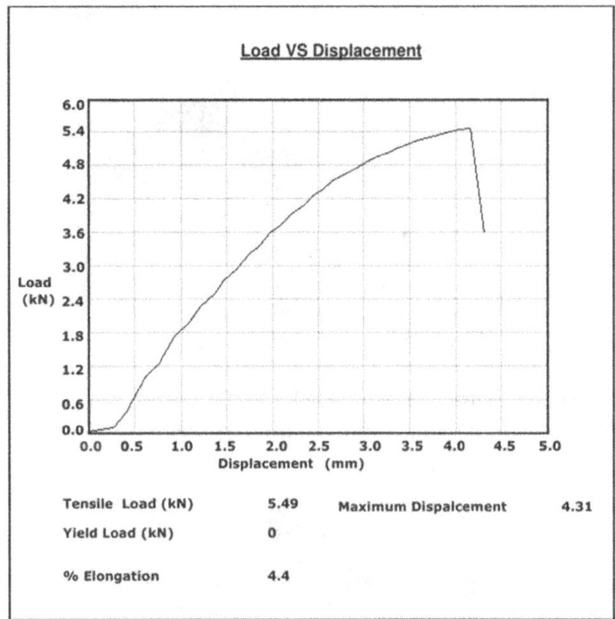

Fig. 37.10 Load vs displacement

Fig. 37.11 Before test

point, the offset technique is employed to measure yield strength. In this method, a line parallel to the first straight section of the stress-strain curve is traced, shifted 0.2% of the strain. This technique is typically used for materials such as low-carbon steel, where there is not always a well-defined yield point.

3. Conclusion

1. Composites reinforced with A356 alloy, SiC, and MWCNTs (tubes) were successfully fabricated using a combination of stir casting and squeeze pressure techniques. These composites were evaluated for their mechanical properties, leading to the following conclusions:

Fig. 37.12 After test

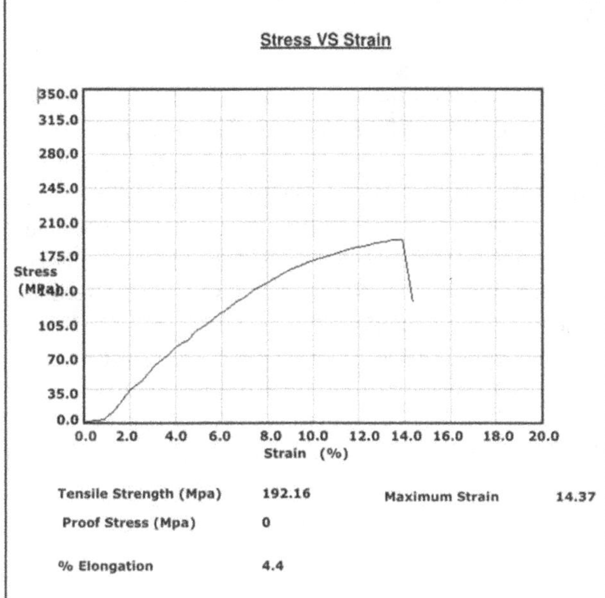

Fig. 37.13 Stress and strain

2. The morphology of the reinforcements had a notable impact on the composites, resulting in superior mechanical performance.
3. The particle-strengthening effect was particularly evident in composites with a combination of reinforcements in the A356 alloy matrix.
4 The application of direct squeeze pressure effectively eliminated porosity and slag, and enhanced the integrity of the materials.
5. Tensile strength and ductility increased because the grain refinement favored enhanced plastic deformation of the composites.
6. In composites reinforced with MWCNT, there was strong interfacial adhesion between reinforcement particles and matrix, which was the reason for particle strengthening.
7. The optimum reinforcement composition was 1% volume for MWCNTs and 0.75% volume for SiC, and it gave the optimum mechanical performance.

References

1. K. K. Chawla, Composite Materials: Science and Engineering 3rd Edition, New York: Springer, 2012.
2. R. MEHRABIAN, R. G. RIEK and M. C. FLEMINGS, "Preparation and Casting of Metal-Particulate NonMetal Composites," METALLURGICAL TRANSACTIONS, vol. 5, pp. 1899–1905, 1973.
3. G. Lin, Z. Hong-wei, L. HAo-ze, G. Li-na and H. Lu-jun, "Effects of Mg Content on Microstructure and Mechanical Properties of SiCp/Al-Mg Composites Fabricated by Semi-Solid Stirring Technique," Transactions of Nonferrous Metals Society of China.
4. V. Laurent, P. Jarry and G. Regazzoni, "Processing Microstructure Relationships in Compocast Magnesium/SiC," Journals of Materials Science.
5. A. Kheder, G. Marahleh and D. Al-Jamea, "Strengthening of Aluminum by SiC, Al2O3 and MgO," Jordan Journal of Mechanical and Industrial Engineering
6. A. Mahazery, H. Abdizadeh and H. Baharvandi, "Development of high-performance A356/nano-Al2O3 Composites," Materials Science and Engineering.
7. H. Su, W. Gao, Z. FEng and Z. Lu, "Processing, Microstructure and Tensile Properties of Nano-Sized Al2O3 Particle Reinforced Aluminum Matrix Composites," .
8. J. Jiang and Y. Wang, "Microstructure and Mechanical Properties of The Rheoformed Cylindrical Part of 7075 Aluminum Matrix Composite Reinforced with Nano-Sized SiC Particles," Materials & Design.
9. T. Wang, Y. Zheng, Z. Chen, Y. Zhao and H. Kang, "Effects of Sr on The Microstructure and Mechanical Properties of In Situ TiB2 Reinforced A356 Composite,".
10. Y. Kaygisiz and N. Marasli, "Microstructural, mechanical and electrical characterization of directionally solidified Al–Si–Mg eutectic alloy," Journal of Alloys and Compounds.
11. A. Mazahery and M. O. Shabani, "Characterization of cast A356 alloy reinforced with nano SiC composites," Trans. Nonferrous Met. Soc. China

Note: All the figures in this chapter were made by the authors.

Advances in Additive Manufacturing Technologies – Gurusamy Pathinettampadian et al. (eds)

A Review of Methods to Combat Data Gaps in Solar Flare Prediction and Classification

Ramapriya Ramamoorthy[1], R. Gowri[2], Yogapriya M.[3], P. Karthikeyan[4], Jayashree M.[5], Kandavel N.[6]

Department of CSE (AI-ML), Chennai Institute of Technology, Chennai, India

◆ **Abstract:** Solar flare datasets comprise a very small number of high-intensity flare samples, which are crucial for effective predictions. A machine learning model needs precise and balanced data to predict accurately, so this imbalance results in the majority class (weak flares) overpowering the outcomes. Additionally, accounting for past flare activity, known as temporal coherence, is important to avoid overfitting and misleading results. This article reviews four methods to address the class imbalance resulting in skewed datasets. In the first approach, minority classes are assigned higher weight values, while the majority class is given lower or mixed-weight values. The second method uses data resampling, which increases the instances of the non-dominant classes by oversampling or reduces the majority class by undersampling. The third and fourth approaches synthesise data to balance the dataset artificially. While the third method increases the dataset's diversity by making modified copies of the existing data, the fourth utilises generative modelling to create an entirely new balanced dataset. The primary aim of this review is to compare the different methods and evaluate their effectiveness in balancing datasets, ultimately analysing their respective strengths, limitations, and impact on the model's predictive accuracy. Each method offers distinct advantages based on the specific requirements and complexity of the scenario. However, for advanced space weather forecasting, generative modelling was found to be the most effective as it allows the model to learn the underlying patterns to generate a balanced dataset, improving model prediction.

◆ **Keywords:** Solar flares, Data gaps, Class imbalance, Machine learning, Classification model, Synthetic data generation

1. Introduction

Solar flares are bursts of energy released from regions on the sun's surface where magnetic fields are particularly strong and twisted (see Fig. 38.1 for a visual representation). They are categorised by intensity and potential damage into classes A, B, C, M, or X. The categorisation technique defines how flares are classified based on several characteristics, including size, duration, shape, magnetic structure, and particle emissions, as detailed by Marov [1]. Understanding the relationship between these factors is key to accurate classification. The flares vary in radiation and electromagnetic intensity throughout the classes, with Class A flares having the least energy and Class X the most. Each subsequent class indicates a tenfold increase in energy, making Class M and Class X flares highly dangerous. The most intense Class X flares can single-handedly disrupt satellite communications and power grids. These flares can

[1]ramapriyaramamoorthy.aiml2023@citchennai.net, [2]gowri.r@citchennai.net, [3]yogapriyam@citchennai.net, [4]karthikeyanp@citchennai.net, [5]jayashreem.aiml2023@citchennai.net, [6]kandaveln@citchennai.net

DOI: 10.1201/9781003685906-38

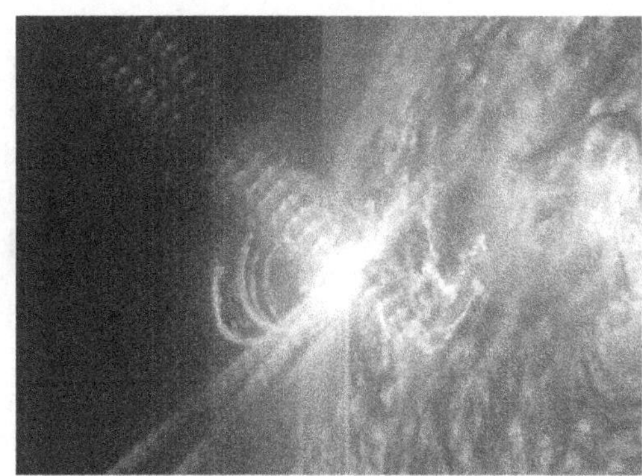

Fig. 38.1 An X3.2-class solar flare. X-class flares are among the most powerful solar flare categories, capable of causing significant space weather impacts, including interference with satellite communications and power systems on Earth

Source: NASA/SDO

cause radiation storms, resulting in blackouts and extreme disruption of our electromagnetic field, while Class M flares cause brief interruptions. However, according to Ribeiro and Gradvohl [2], the weaker classes do not carry much radiation and have little effect on our atmosphere. The difference in radiation levels between these classes determines their potential impact on space weather.

Although energy levels vary among the classes, it can be difficult for the model to classify the flares accurately. Miteva and Samwel [3] discovered that while Class M flares have a lesser influence on space weather than Class X flares, weaker flares can still be used to anticipate other space atmospheric phenomena such as energetic particles and ionospheric disturbances. Using only X-ray time-series data presents challenges in accurately distinguishing between even Class M and Class X solar flare events, as proven by Landa and Reuveni [4]. Due to fewer instances of high-intensity flare events as compared to low-intensity or non-flare events, machine learning models struggle to categorise the flare accurately. In comparison, flare prediction is done fairly easily by implementing binary classification systems, categorising an event as 'flare' or 'non-flare'. However, high-intensity flares can damage critical systems on Earth, so the need arises for prediction models to work beyond prediction and include classification to offer maximum help. Effective multiclass classification models are necessary to predict and categorise their varying intensities and impacts accurately.

The primary reason most, if not all, solar flare classification models struggle with multi-class classification is the lack of data. Kurivella [5] observed that the class imbalance is huge, with the rare-event (strongly flaring) samples making up only around 2% of the dataset. Table 38.1 shows the number of samples for each flare class in the SWAN-SF Benchmark dataset, highlighting this imbalance. Conventional methods fail to effectively categorise minority classes in real-time scenarios due to the overwhelming number of instances of the majority class, as proved by Han et al. [6].

Table 38.1 Class imbalance in the SWAN-SF benchmark dataset [5]

Flare Class	Partition 1	Partition 2	Partition 3	Partition 4	Partition 5
A (Majority)	64,222	78,517	22,236	59,689	89,400
B	4,999	4,194	108	832	8,275
C	6,250	8,444	3,350	6,487	6,419
M	1,130	1,279	1,152	1,153	1,071
X (Minority)	172	48	160	165	21

As such, inter-class differentiation remains challenging. Ahmadzadeh et al. [7] clarify that although class imbalance is a well-known challenge, it is frequently disregarded when the emphasis is placed on optimising machine learning models. According to Chen [8], even advanced models like CGANs trained on balanced datasets struggle when faced with imbalanced datasets. Baeke et al. [9] discovered that unsupervised clustering models easily identify flaring active regions in given cluster sets. However, the model cannot differentiate the levels of flaring activity within a given region, especially if the region contains a mix of weakly flaring and strongly flaring clusters. In contrast, not much research has been done using supervised models, as pointed out by Carreno et al. [10].

Few models have come close to accurate classification. The SeMiner model, developed by Discola Junior [11], deals with the evolution of data. It anticipates the patterns in flare creation and includes this data in the forecasting process. This is supported by the ECID method, also developed by Discola, which takes into account multiple solar features in an attempt to handle the class imbalance issue. He and Cheng [12] clarify that using differential class weights to classify more than two classes is a relatively easy extension of this work. Another successful method is the use of Mean Gaussian Noise (MGN), which is a novel data augmentation method by Wen and Angryk [13]. All these methods aim to solve the class imbalance that hinders a machine learning model's ability to differentiate between and accurately classify solar flares.

According to Amar and Ben-Shahar [14], solar flare prediction may be viewed as a binary classification problem over a predetermined time interval, regardless of their magnitude. However, the model needs to accurately distinguish the levels of flare activity to allow subsequent preventative action to be taken that matches the level of potential threat. As discussed by Abeda et al. [15], solar flares are an evolving phenomenon, having distinct pre-flare and post-flare stages. Thus, the need for accurate multi-class classification remains important. Additionally, Cutro [16] observed that the uneven distribution of observations restricts model advancement. The limited manifestation of instances means that the approach used to classify them is subject to change based on the instance itself. Wan et al. [17] explain that a perfect algorithm that works well in all aspects cannot be found. Real-time scenarios require methodologies that cater to the specific goals of the project. This is supported by Grim and Gradvohl [18], who suggest that it can be challenging to evaluate the performance of different models objectively. Furthermore, the spatiotemporal range of high-quality solar data continues to be sparse, posing a considerable obstacle. Ahmadzadeh et al. [19] illustrate this fact while also highlighting that focusing on unrelated parts of the solar cycle by selecting random samples leads to misrepresenting the flare dataset.

This review first discusses the challenges in classifying solar flares, including temporal coherence, followed by an examination of four strategies used to balance datasets and their impact on solar flare classification. This encompasses the possible applications of each method, concluding with insights for future developments in this field. The main contribution of this review is analysing the effectiveness of generative modelling as a solution to balance solar flare datasets by comparing it with traditional methods like weighting, sampling, and data augmentation.

2. Challenges in Multiclass Prediction

Designing multi-class classification models presents unconventional issues in terms of accurate categorisation of the solar flares. There are certain issues faced in all rare-event classification scenarios which can be attributed mainly to the inherent bias in the dataset. Solar flare classification also faces additional challenges such as incorporation of temporal coherence, which are examined in the following sections.

2.1 Extreme Class Imbalance

The primary challenge in any rare-event classification lies in the extreme imbalance of the dataset. Solar flares are no different. The class imbalance issue refers to the severe underrepresentation of strong flare events compared to milder events. Ahmadzadeh et al. [19] noted that the class imbalance ratio in the SWAN-SF benchmark dataset when considering Class M and Class X flares to weaker flare events is 1:60. This ratio intensifies to 1:800 if only the strongest (Class X) flares are considered. Figure 38.2 visualises this imbalance in Partition 3 of the dataset, but the issue persists across all partitions. This is a significant issue as the rarer class (high-intensity flares) is the most important. With such few instances of strong flare conditions, models can distinguish a flare event from a non-flare one but struggle to analyse the intensity of the flare and categorise it accordingly. An unoptimized model will escalate the innate classification bias, skewing it toward negative samples. Alternatively, as stated by Wan et al. [17], the model might just predict no-flare conditions with high accuracy, which limits its effectiveness.

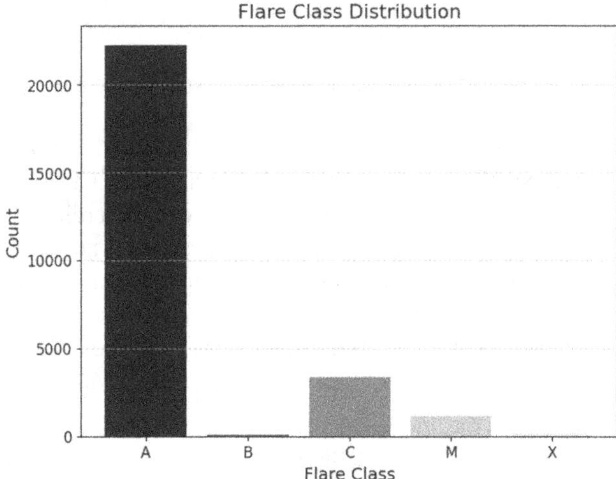

Fig. 38.2 Frequency and imbalance ratio of all five flare classes across Partition 3 from the SWAN-SF benchmark dataset [9]

2.2 Role of Temporal Coherence

Temporal coherence refers to understanding the history of flare activity in the region where the flare occurs, rather than relying solely on the characteristics of the active flare. Most rare event classification scenarios have a temporal angle, as noted by Careno et al. [10]. It is important not to disregard the effects of the history of flare formation and intensity, as it may provide valuable insights beyond expectations. For example, Figure 38.3 shows the distribution of flare events on 13th February 2002, illustrating how flare activity can build up over time, suggesting the importance of temporal coherence in classification models to avoid misclassification. This is supported by Ahmadzadeh et al. [7], who denote that ignoring the link between past and present data points may result in model overfitting. This leads to overly positive model performance on the training data, but it can fail to generalise on unseen data, ultimately providing faulty insights. To remedy this, the dataset can

Flare event	Start time	Peak time	End time	Duration (s)
2021332	00:53:24	00:54:54	00:57:00	216
2021308	04:22:52	04:23:50	04:26:56	244
2021310	07:03:52	07:05:14	07:07:48	236
2021353	07:07:48	07:09:14	07:20:56	788
2021354	07:20:56	07:22:42	07:30:04	548
2021312	08:53:20	08:55:18	09:05:08	708
2021339	10:02:56	10:04:42	10:04:44	108
2021313	12:29:32	12:30:58	12:33:54	232
2021329	23:28:32	23:30:38	23:31:56	204
2021355	23:31:56	23:34:06	23:52:56	1260

(a) details of flare events

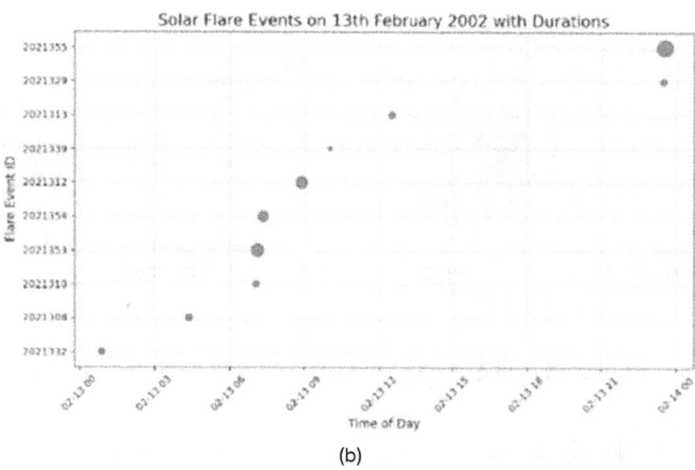

(b)

Fig. 38.3 Temporal patterns of solar flare activity: (a) Table displaying details of flare events, including start, peak, end times, and durations. (b) Plot showing the distribution of flare events over time, with clustered activity on 13th February 2002. The data is sourced from the RHESSI solar flare dataset [11]

be divided into distinct training and testing sets such that samples from the same temporal regions do not overlap, resulting in a more realistic assessment of the model's performance.

However, this may prove to be difficult when considering rare events like solar flares due to the scarcity of data. This class imbalance problem may worsen when samples for training data are collected randomly, as it may lead the model to significantly underestimate the importance of time-dependent properties. In these cases, more advanced resampling techniques like temporal oversampling are recommended over basic sampling as they retain temporal connections. Another approach is using synthetic models, as demonstrated by Chen [8], who showed that augmenting training data with synthetic samples can improve model performance.

2.3 Limitations of Binary Classification

Given the inherent challenges in solar flare prediction, it makes sense why binary classification models are favoured. It should be noted that most flare prediction/classification models were developed originally for binary classification. As detailed by Discola Jr. [11], the class imbalance issue can seem to disappear when the mapping multi-class condition is mapped in binary terms. For example, flares above a certain threshold (typically Class M, or Class C in some cases) are mapped as positive while lower-intensity flares are grouped as negative. This approach allows for a simplified representation but limits the model's effectiveness in capturing the full range of flare intensity, which is needed for real-world applications.

Binary classifiers usually use snapshot data, which only contains independent data points. This does not consider other complex phenomena like temporal coherence, limiting the model's (and by extension, our) understanding of sequential connections. As discussed by Baeke et al. [9], the existing classes have unclear boundaries. There is also very little difference between background radiation (non-flaring active regions) and weaker solar flares, which binary approaches cannot modulate. Thus, while binary methods provide a simplified approach, they are unable to capture the inherent complexities of solar flare classification. Binary classification models run the risk of over-simplification of the scenario, or may inadvertently overlook subtle differences between flare types, thus potentially misclassifying flare events. Given the limitations of binary models, there is a pressing need for accurate multi-class classifiers that can offer more nuanced results required in real-life scenarios like risk prediction and space weather forecasting.

3. Review of Methods

This section reviews the methodologies used to combat data gaps in solar flare datasets. Each method is examined with respect to model performance, its applications in solar flare prediction, and its contribution to solving the class imbalance problem. Figure 38.4 details the exact process flow of solar flare classification. The methods covered include mixed weight assignment, resampling, data augmentation, and generative modelling of the dataset. The strengths and weaknesses of each approach are also examined, highlighting their role in balancing the dataset and improving forecast accuracy.

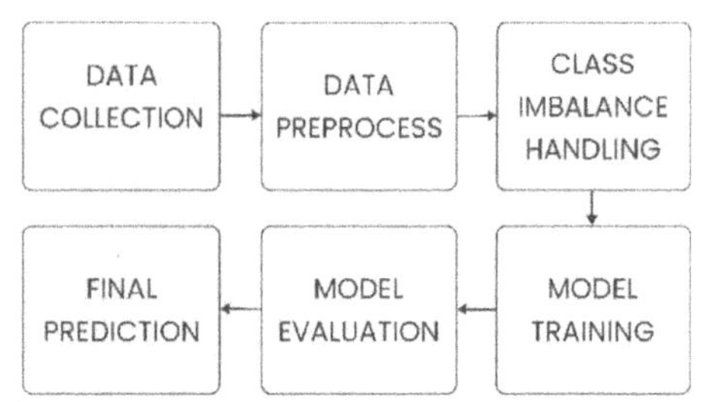

Fig. 38.4 An overview of the solar flare classification process [15]

3.1 Mixed Weight Assignment

Mixed weight assignment balances datasets by assigning higher weight values to the minority class and simultaneously decreasing the weight of the majority class. This is done to minimise the bias towards the majority class, thus improving dataset balance and model accuracy. In all rare-event classification scenarios, the model is inherently skewed towards the majority class (non-event class) due to the sheer number of instances. Mixed weight assignment ensures that the model's precision for the majority class is not compromised while increasing the ability to classify the minority class(es).

This approach does not balance the dataset itself but provides a more balanced outlook. It can be done in a few different ways. For example, Ahmadzadeh et al. [7] demonstrated that SVM may include changing weights in its objective function, thus focusing on the non-dominant (stronger flare) classes. This has proven to improve model accuracy as model tuning is based on data-driven insights. This approach can be used with any machine learning model that uses weighted loss functions like decision trees, SVMs, logistic regression, etc. He and Cheng [12] have explored two common weighting algorithms, DiffBoost and AdaClass, which calculate weights for each class during the training phase based on the number of instances of the class. These algorithms focus on the flaring class (in cases of binary classification), resulting in improved classification accuracy. In contrast, Wan et al. [17] suggest that the quickest method is to compute the average of the samples and assign suitable weights to each flare category during model training to concentrate on resolving the class imbalance issue. While the former method aims to increase the impact of the positive samples, the latter targets the imbalance of the dataset.

Mixed weight assignment is shown to increase overall model performance by eliminating the inherent bias towards the majority class in solar flare classification, resulting in better accuracy for the non-dominant classes. Additionally, it can also be adapted by several machine learning models, proving its versatility. However, calculating and fine-tuning the proper weights for each class can increase the cost of computation, especially when dealing with larger datasets like global flare datasets. Moreover, this approach shifts the model's view of the data towards flare events but does not solve the dataset's underlying issue of imbalance. While this helps to increase accurate classification, the model can still be influenced by existing patterns, which may lead to a lack of generalisation.

3.2 Data Resampling

Data resampling involves refining the number of instances across the classes to achieve dataset balance during model training. This can be done either by reducing the number of instances of the dominant class (undersampling) or increasing the number of instances of the non-dominant classes (oversampling). Either way, this method balances out the sizes of all classes to provide the model with symmetric training samples. The goal is to allow the model to represent the trends in the non-dominant classes more accurately without being dominated by the majority class by rearranging class distributions. Although resampling can enhance model performance, it must be executed with care to avoid common pitfalls like model overfitting or loss of information.

Data resampling is broadly divided into two methods: undersampling and oversampling. Undersampling refers to reducing the size of the majority class to match that of the other classes. In binary classification as defined by Ahmadzadeh et al. [19], this results in the dataset containing only as many negative instances (majority data) as there are positive instances (minority data). This approach aims to reduce the influence of the majority class by cutting down its size. The exact samples used from the majority class can be chosen by random selection or even advanced techniques that select the samples with the most impact. As Ahmadzadeh et al. [7] stated, undersampling requires a huge dataset to be effective, but it may still lead to data loss if the samples from the majority class are not selected properly. On the other hand, oversampling involves increasing the number of instances in the minority classes to equal that of the majority class. This can be done by replicating existing data or even synthesising samples. Oversampling is preferable for smaller datasets where the class imbalance issue is prevalent. However, the model might memorise underlying patterns in the data, especially with duplication, thus leading to model overfitting. This reduces the effectiveness of the model on unseen data. Figure 38.5 shows the effects of both approaches on an unbalanced dataset.

Data can be resampled at either the subclass level or superclass level. General resampling involves the balancing

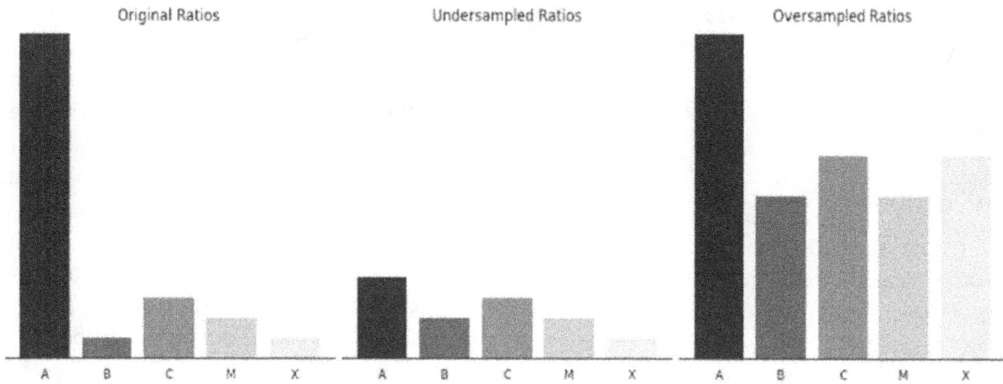

Fig. 38.5 Comparison of original, undersampled, and oversampled class distributions. This is a generic illustration that does not reflect the exact imbalance ratios in the SWAN-SF dataset [7]

of the superclasses, i.e., the positive (strong flares) class would be sized equal to the negative (weak flares) class. Alternatively, subclass resampling adjusts the sizes of the flare classes (A, B, C, M, X) to create a more nuanced balance across all classes, but it has a considerably high computational cost. To achieve an optimised balance, a resourceful mix of the two can be utilised to ensure a balanced distribution at both levels.

The main goal of data resampling is to increase the model's accuracy concerning the minority classes and effectively reduce misclassifications, so the choice between the methods can be attributed to the specific goals and available resources. Wan et al. conclude that different goals may require different approaches to achieve dataset balance. For example, if the primary goal is to increase the speed of calculation, undersampling is recommended. In cases where complex multi-class datasets like the benchmark SWAN-SF are used, determining the exact parameters for resampling is crucial to balance model performance and computational efficiency.

3.3 Data Augmentation

Data augmentation is the process of artificially increasing the size of a dataset by adding samples of the original data that have been manipulated. In the context of solar flare prediction and classification, data augmentation is used to increase the number of instances of the minority classes to provide more balance in the dataset, thus enhancing model performance. Models are trained with synthetically generated, modified samples of the minority classes to help them understand the nuances of the classes better, resulting in a more accurate classification of strong flare samples. This method focuses on modifying samples of data while retaining essential characteristics.

Data augmentation by image synthesis is widely used in solar flare classification as most flare datasets are images. Common image synthesis techniques make use of Generative Adversarial Networks (GANs), Variational Autoencoders (VAEs), etc. These models can generate new images that contain properties of the overall dataset and can validate the synthesised data. Amar and Ben-Shahar [14] investigated a novel image synthesis method using Convolutional Neural Networks (CNNs) that generates images of solar regions to help improve model accuracy. This approach retains features like magnetic properties and flare signals based on which modified images are generated.

Data augmentation based on time series aims to alter samples while preserving the order of data points over time to maintain temporal patterns that play a key role in forecasting. Wen and Angryk [13] developed a novel data augmentation method called Mean Gaussian Noise (MGN), which synthesises data by using the mean of the dataset's time series. Other popular time-based augmentation techniques include time warping, which expands or compresses parts of a temporal series without changing the overall pattern, and magnitude scaling, which also maintains the pattern but changes the amplitude of the values to introduce variation in the dataset. Table 38.2

Table 38.2 Performance of various time-series-based augmentation methods across different partitions in the SWAN-SF benchmark dataset. Data adapted from Wen and Angryk [13]

Augmentation Method	Partition 2	Partition 3	Partition 4
Magnitude Warping	0.846	0.796	0.827
Time Warping	0.852	0.791	0.794
Slicing	0.852	0.791	0.797
Window Warping	0.852	0.790	0.797
Mean Gaussian Noise (MGN)	**0.767**	**0.763**	**0.719**

shows the performance of some of these methods across various partitions of the SWAN-SF benchmark dataset. These methods aim to not only increase the size of the minority class but also add diversity to avoid the risk of overfitting.

3.4 Generative Modelling

Generative modelling refers to the process where the model creates new samples from an existing dataset by studying the fundamental patterns present in the data. The synthesised data mimics real data in properties and trends to artificially increase the representation of the minority class as a way to mitigate the class imbalance issue [15]. This is an advanced approach to the problem that does not merely replicate data but instead acknowledges underlying patterns and creates a new, more balanced dataset. The key factor that sets generative modelling apart from other methods is that it creates an entirely new, balanced dataset from scratch. Generative modelling also has a more advanced understanding of inherent patterns and properties of the data that other approaches do not [16].

The most commonly used machine learning model for generative modelling is the Generative Adversarial Network (GAN). It consists of an inbuilt validation system (the discriminator) that works with the generator to synthesise high-quality data. Some take this a step further by using CGANs, or Conditional GANs, which generate data based on certain conditions or features, making it more useful in synthesising data with specific requirements like flare data generation [17]. Traditional CGANs struggle when trained on imbalanced datasets like solar flare data, resulting in subpar synthetic samples for the minority classes. To mitigate this issue, Chen [8] demonstrates a variation of CGAN called the Two-Stage CGAN, involving two consecutive GAN models. The first model acts as a normal GAN and generates synthetic data, while the second model refines it by introducing additional conditions or features. This system enhances the quality and diversity of the synthetic data generated, significantly improving model performance. The addition of the second GAN helps increase variation and contrast in the generated data, reducing the risk of model overfitting and leading to a more balanced dataset [18].

Generative modelling is one of the most advanced methods utilised to mitigate class imbalance, but it presents some challenges. For one, the generator and discriminator may not converge properly, leading to unusually high acceptance or rejection rates [19]. The models also require proper training data to address the class imbalance and generate a balanced dataset despite it. Another factor to consider is the increased computational cost of studying and replicating large datasets like solar flare data. The usage of more advanced models like Two-Stage CGANs adds layers of computation, significantly increasing training times and processing load. Regardless of these challenges, generative modelling is the best solution for complex space weather forecasting tasks. It has an advanced understanding of data patterns that other methods do not. Generative modelling can become a powerful fix for data imbalance issues with better training methods and optimisation, leading to significant improvement in model accuracy for challenging applications like solar flare classification [20].

4. Conclusion

The accurate prediction and classification of solar flares are essential to comprehend and mitigate their impact on Earth's systems. However, the inherent class imbalance in flare data presents a significant barrier to the efficiency of machine learning models. Addressing this data imbalance is a crucial prerequisite for developing any model capable of accurate flare classification. This review examined four techniques used to address this issue, emphasising how each brings distinct strengths to the specific scenario in which it is applied. Mixed weighting and data resampling work well for simpler tasks with smaller datasets while data augmentation enhances the diversity of the dataset. However, these methods often fail to address the complexities of large-scale solar data that is typically used in space weather forecasting.

This review is limited by the availability of solar flare datasets, which may not fully capture the intricacies of real-world flare events. Given the scope of this review, the methods have not been verified experimentally. Nevertheless, generative modelling stands out as a promising solution as it can handle the increased complexity and dimensionality found in solar flare datasets. This approach not only mitigates the class imbalance issue by artificially creating a balanced dataset, but it also acknowledges the inherent trends in the data to do so effectively. Future research can focus on increasing the efficiency and reducing the computational complexity of generative models. Hybrid models that combine generative modelling with sampling or augmentation practices can also be explored to further refine model robustness and performance, especially in multi-class scenarios. Addressing class imbalance properly is a crucial concern in solar flare prediction. While these methods are commonly applied to rare-event classification scenarios, like solar flare classification, their concepts can be made use of in other circumstances where class imbalance is a concern. With continued advancements, generative modelling holds great potential for improving space weather forecasting.

References

1. Marov, M. Y. 2020. Radiation and space flights safety: an insight. *Acta Astronautica* 176:580–590.
2. Ribeiro, F., and A. L. S. Gradvohl. 2021. Machine learning techniques applied to solar flares forecasting. *Astronomy and Computing* 35:100468.
3. Miteva, R., and S. W. Samwel. 2022. M-class solar flares in solar cycles 23 and 24: properties and space weather relevance. *Universe* 8(1):39.
4. Landa, V., and Y. Reuveni. 2022. Low-dimensional convolutional neural network for solar flares GOES time-series classification. *The Astrophysical Journal Supplement Series* 258(1):12.
5. Kurivella, N. S. 2021. Comparative study of machine learning models on solar flare prediction problem. Master's thesis, Utah State Univ.
6. Han, M., A. Li, Z. Gao, D. Mu, and S. Liu. 2023. Hybrid sampling and dynamic weighting-based classification method for multi-class imbalanced data stream. *Applied Sciences* 13(10):5924.
7. Ahmadzadeh, A., et al. 2019. Challenges with extreme class-imbalance and temporal coherence: a study on solar flare data. In *Proceedings of the 2019 IEEE International Conference on Big Data (Big Data)*, 1423–31.
8. Chen, Y. 2024. Mitigating class imbalance in time series classification via generative modeling.
9. Baeke, H., J. Amaya, and G. Lapenta. 2023. Classification of solar flares using data analysis and clustering of active regions. *Authorea Preprint.*
10. Carreño, A., I. Inza, and J. A. Lozano. 2020. Analyzing rare event, anomaly, novelty, and outlier detection terms under the supervised classification framework. *Artificial Intelligence Review* 53:3575–3594.
11. Junior, D., and S. Luisir. 2019. Enhancing solar flare forecasting: a multi-class and multi-label classification approach to handle imbalanced time series. Unpublished manuscript.
12. He, J., and M. X. Cheng. 2021. Weighting methods for rare event identification from imbalanced datasets. *Frontiers in Big Data* 4:715320.
13. Wen, J., and R. A. Angryk. 2024. Class-based time series data augmentation to mitigate extreme class imbalance for solar flare prediction. *arXiv preprint* arXiv:2405.20590.
14. Amar, E., and O. Ben-Shahar. 2024. Image synthesis for solar flare prediction. *The Astrophysical Journal Supplement Series* 271(1):29.
15. Abed, A. K., R. Qahwaji, and A. Abed. 2021. The automated prediction of solar flares from SDO images using deep learning. *Advances in Space Research* 67(8):2544–57.
16. Curto, J. J. 2020. Geomagnetic solar flare effects: A review. *Journal of Space Weather and Space Climate* 10:27.
17. Wan, J., J. F. Fu, and J. F. Liu et al. 2021. Class imbalance problem in short-term solar flare prediction. *Research in Astronomy and Astrophysics* 21(9):237.
18. Grim, L. F. L., and A. L. S. Gradvohl. 2024. Solar flare forecasting based on magnetogram sequences learning with multiscale vision transformers and data augmentation techniques. *Solar Physics* 299(3):33.
19. Ahmadzadeh, A., B. Aydin, and D. J. Kempton et al. 2019. Rare-event time series prediction: A case study of solar flare forecasting. *In* 18th IEEE International Conference on Machine Learning and Applications (ICMLA), 1814–20.
20. *Solar Flares from RHESSI Mission.* 2002. Kaggle. https://www.kaggle.com/datasets/khsamaha/solar-flares-rhessi (accessed November 28, 2024).

Advances in Additive Manufacturing Technologies – Gurusamy Pathinettampadian et al. (eds)

39 Exploring Synthetic Fuel and Sustainable Vehicle Technologies—A Comparative Literature Review

Anish K.[1], Zahid Hussain J.[2], Boopendranath C.[3], Yogapriya M.[4], Thirumalai Murugan R.[5], R. Gowri[6]

Dept. of Computer Science Engineering with Artificial Intelligence and Machine Learning, Chennai Institute of Technology, (Affiliated to Anna University), Kanchipuram, India

◆ **Abstract:** The transition to sustainable sources for fuelling commercial needs such as engines in vehicles, maritime and aviation is essential to meet global demands while providing efficiency greater than or equal to that of fossil fuels. One of the visionary efforts, the EU's Green Deal, vows the European Union to achieve carbon neutrality by the year 2050. With the target year fast approaching, numerous proposals were made to reduce carbon emissions to align with the goals of a carbonneutral future. To achieve carbon-neutrality, various sustainable alternatives were developed to replace or update current IC engines, such as using sustainable fuels like biofuels and efuels, although biofuels provide lower emissions, it's another alternative – e-fuels provide a more complete carbon-neutral propulsion since they use the method of capturing carbon from the atmosphere to join with hydrogen atoms to form a synthetic fuel which can be used to fuel IC Engines. However, there is also another path where electricity is generated through two ways – Hydrogen and renewable sources. Each of these has its own merits and demerits. In this paper, a comparative analysis of e-fuels, Battery Electric Vehicles (BEV) and Fuel Cell Electric Vehicles (FCEV) is provided to conclude the most viable option available. All the sources compared here have distinct goals to either reduce or eliminate the emission of harmful byproducts. In addition to this, a comparison of two alternative fuels to replace fossil fuels and two majorly used sources for electric vehicles is compared based on Life-Cycle Analysis (LCA), cost-to-power ratio and economic viability.

◆ **Keywords:** Sustainability, Electric vehicles, Fuel cell, Carbon-neutral, Renewable energy, Life cycle analysis

1. Introduction

Transportation has always revolved around the usage of fossil fuels since the late 1800s, at the beginning, it started off with coal, which was the first widely used fossil fuel which powered steam locomotives and vessels which revolutionized transport in terms of passenger and freight movement. During the late 19th century to early 20th century, coal was dropped in favour of petroleum-based fuels which were refined crude oil. These fuels were used to power ICEs which were standardized across all modes of transport. The transition to petroleum-based fuels brought along a significant environmental cost – as every cycle within an engine produces harmful byproducts such as carbon monoxide and GHG. The release of such byproducts poses a serious threat to air quality and more importantly, climate change which tends to impact both the environment and human health. Nations across the world

[1]anishk.aiml2023@citchennai.net, [2]zahidhussainj@citchennai.net, [3]boopendranathc.aiml2023@citchennai.net, [4]yogapriyam@citchennai.net, [5]thirumalaimuruganr@citchennai.net, [6]gowri.r@citchennai.net

DOI: 10.1201/9781003685906-39

are recognizing the rising threat and have undertaken several initiatives to mitigate the situation. ICEVs have been evolving ever since its debut back in 1864, since then, ICEVs have been constantly developed to be more fuel efficient and to lower pollutants emitted by advancing the basic design of the engine, fuel injection system, ignition system, et cetera. Engineers have implemented various technologies to work with the ICEs such as variable valve timing, start-stop systems, exhaust gas recirculation and so on, to make ICEs to be as efficient as it can be. The introduction of a 4-stroke engine, where the engine goes through a cycle in 4 stages – intake, compression, power and exhaust – has been optimized over time for better fuel efficiency and lower emissions. Unfortunately, there are limitations. Despite these improvements over the years, ICEs have not evolved enough to be carbon-neutral. While emissions produced by these ICEs have reduced significantly, it has not reduced enough to meet the emissions' reduction target by 2050, thus, this necessitates the need for researching alternative fuel sources. Since the commercial adoption of ICEVs, a significant challenge has been the increasing emission of carbon monoxide and GHGs. These emissions pose serious risks to human health, the environment, and the climate. GHGs are a primary cause of climate change, which necessitates urgent action to mitigate their impact. As awareness of climate change has grown, researchers have shifted their efforts to identify alternative fuels that are both environmentally sustainable and capable of meeting global energy demands. In this paper, 3 types of alternative energy sources are discussed. This review paper aims to evaluate the merits and demerits of each energy source by comparing the life cycle, environmental impact, economic viability and energy efficiency to better understand each energy source's role in decarbonization. The following sections delve into the existing literature research papers on these technologies by using a divide and conquer approach. By understanding each source's processes, merits and demerits, it lays a solid foundation for a comparative review.

2. Literature Review

2.1 Overview of the Technologies

1) E-Fuels

E-Fuels are synthetic fuels that are produced by capturing CO2 from the atmosphere and combining it with H2 through electrolysis of water, they involve the use of electricity which are produced by renewable energy, which results in the formation of e-fuels which can be liquid or gaseous hydrocarbons [1]. This process is commonly termed as Power to Liquid (PtL). E-fuels can be used on existing ICEs with no additional modifications needed. However, adoption of e-fuels faces several challenges such as higher energy requirements and higher cost-to-power ratio.

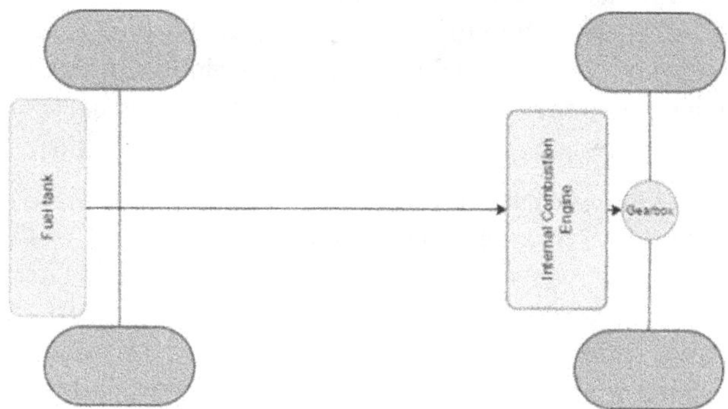

Fig. 39.1 Basic construction of an ICEV [5]

Analysis inferred from e-fuels are the evaluation of Well-toWake (WtW) GHG emissions and energy consumption for alternative fuels, including E-fuels, Hydrogen, and Ammonia, presents a complex picture of the potential for reducing maritime emissions. E-fuels, if produced from 100% renewable energy, can offer a significant reduction in GHG emissions, up to 100%. However, this potential is contingentupontheavailabilityofrenewableenergysources, which remain limited in many regions. As observed by [2], the actual emissions associated with E-fuels depends upon carbon intensity of the electricity used in their production, which means that the current mix of renewable energy may not be enough for E-fuels to fully utilize their green potential. Moreover, the energy consumption required for producing alternative fuels, particularly hydrogen and ammonia, is substantially higher than that for traditional fossil fuels, especially when produced through electrolysis powered by renewable energy. However, technological advancements could potentially reduce this energy consumption by 10-15% in the future. This underscores the need for continued investment in production technologies to improve the efficiency of alternative fuel pathways. Furthermore, the overall viability of e-fuels is greatly impacted by the considerable capital investment needed for infrastructure development, ranging from electrolyzers to hydrogen production facilities. The economic challenges of expanding the PtL production chain are highlighted by [3], which concludes that significant investments in infrastructure and technology are needed. Market failures arise due to uncertainties in carbon pricing, subsidies, and technological advancements, deterring private investors due to the lack of assured returns. High CAPEX and OPEX for CO2 capture and conversion are major hurdles, with equipment like compressors, heat exchangers, and reactors driving costs. Compressors alone account for 41% of the estimated €47.3 million equipment CAPEX, underscoring their critical role in the process. According to [4], the high cost of CO2 capture systems stems from their reliance on advanced machinery. Studies highlight the need for government support to scale up alternative fuel production, ensuring economic viability and sustainability. However, long-term policies must avoid market distortions or legal conflicts, especially in international sectors like aviation.

Strengths noted in E-fuels include Greenhouse Gas (GHG) Emission Reduction The significant reduction in Well-to-Wake GHG emissions for E-fuels, which could fetch 100% reduction in emissions when the production is based entirely on renewable energy sources. This makes E-fuels a viable option for decarbonizing transportation as a whole. Energy Efficiency The study also notes that battery-powered vessels utilizing renewable energy exhibit the lowest Well-to-Wake energy consumption per unit of propulsion energy. Revenue from Co-Products The study highlights that the sale of oxygen co-produced from the electrolysis process and the current CO2 credit tax add 20% to the global revenue from the process. This diversification of revenue streams contributes positively to the economic viability of CO2 conversion technologies. Existing infrastructure utilization Since e-fuels can be produced in such a way that the flash point is similar to gasoline and diesel, this implies that utilization of ICEs can be continued rather than rendering the ICEs obsolete. This in turn reduces the need for mass disposal which contributes for material waste.

Limitations noted in E-Fuels are Economic and Technological Uncertainty E-fuels remain expensive due to inefficient energy conversion and limited production and distribution infrastructure. Although these costs may decrease in the future, large-scale production remains economically challenging, casting doubt on the viability of e-fuels at this time. These technological and economic uncertainties could limit the short-term adoption of e-fuels. Methane Emissions Challenges Improperly combusted methane is a major problem for E-LNG (liquefied natural gas) since it can produce more greenhouse gas emissions than other fuels. This problem is comparable to that of fossil LNG, which makes adopting this fuel type to lower the overall carbon footprint more difficult. Energy Inefficiency Both E-fuels and CO2 capture systems involve energy-intensive processes. E-fuels require substantial amounts of renewable energy for their production, and CO2 capture processes still have significant energy costs associated with them. This inefficiency reduces their attractiveness as viable alternatives in the short term unless technological advancements or new energy sources can help offset these costs.

2) BEV

BEV or Battery Electric Vehicle is an electric vehicle which is powered entirely by a battery. BEVs do not use ICEs as powertrains, instead utilizes electric motor as the powertrain. A battery electric vehicle can convert electrical to mechanical energy in a more efficient manner compared to its counterparts. BEVs are a significant contender in the commercial vehicle market as electricity generated through renewable sources significantly reduces the carbon footprint of BEVs. Despite their lower carbon footprint, their widespread adoption is constrained due to range limitations, unavailability of charging infrastructure and most importantly, the impact on environment during the production of batteries and their disposal.

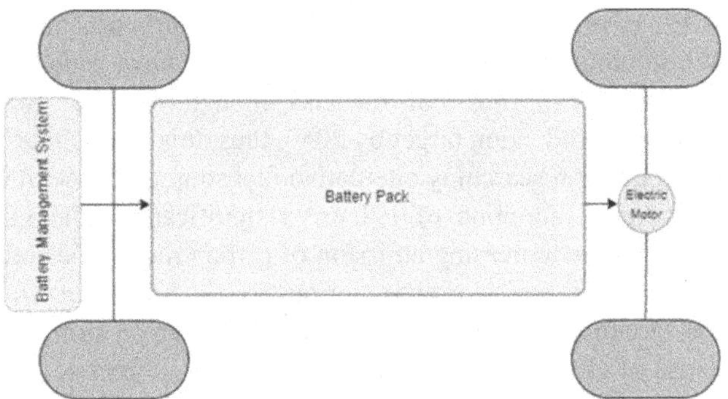

Fig. 39.2 Basic construction of a BEV [9]

BEV adoption impacts CO2 levels, with studies showing uncontrolled charging could increase emissions by 2030, while flexible charging reduces them [5]. Emissions also depend on whether charging energy is fossil-fuel-based. BEV chargers are categorized into three levels: Level 1 (1.3–3.4 kW) for residential use, Level 2 (3–19 kW) for residential and commercial applications, and Level 3 (350+ kW), which offers rapid charging but requires costly infrastructure [6]. Advancements in charging networks reduce the need for high battery energy density [7]. Governments incentivize BEVs with tax credits, free charging, and reduced costs to promote adoption. However, buyers often prefer PHEVs for their lower purchase prices, faster refueling, and longer ranges [8]. Battery health is influenced by factors like temperature, frequent discharges, and driving behavior. Batteries typically remain reliable until losing 20% capacity, with degradation becoming significant only after 321,000 km (200,000 miles) [9, 10]. Battery costs, making up 30% of an EV's price, necessitate further research to enhance durability and economic viability.

Strengths noted in BEVs are when BEVs are charged using renewable resources, they significantly reduce the GHG emissions compared to an ICEV. BEVs also do not emit any tailpipe emissions as electricity is directly converted mechanical energy with no byproducts produced. Since BEVs have fewer moving parts compared to an ICE, maintenance and repair costs. The other way that BEVs have a lower operating cost is by lower costs of electricity, in that case, BEVs are significantly cheaper to operate compared to ICEVs when compared electricity prices to petrol and diesel. Governments support the adoption of BEVs by providing various incentives which lower the initial price of purchasing a BEV. This initiative helps to increase BEV production costs and stimulates industry growth.

Limitations noted in BEVs BEVs tend to have lower range compared to ICEVs, this leads to range anxiety among customers. Although newer battery technologies have emerged in recent years which has improved battery capacity and better range, but it still remains a concern among many customers. Despite charging infrastructure becoming more widespread in recent years, it is still not widespread enough to eliminate range anxiety. Battery degradation has become a huge concern over the years of BEV, as a battery is considered reliable only until it loses 20% of its initial capacity. Battery degradation can take place through charge cycles, heavy usage et cetera. In long term, BEVs may not provide the same range offered at the start and may be less efficient.

3) FCEV

FCEVs or Fuel Cell Electric Vehicles is one of the types of EV, which produces electricity from hydrogen through a chemical process in the fuel cell. The produced electricity is then used for propulsion of the vehicle. Figure 39.3 illustrates the fundamental construction of a FCEV. It consists of a hydrogen tank, fuel cell stack, HEC, battery pack and a motor. Hydrogen is directly supplied to a fuel cell stack. The fuel cell produces electricity through a chemical reaction between hydrogen and oxygen. This process produces byproducts which is water vapor and heat. Electricity generated by the fuel cell goes through HEC, which determines and controls the amount of power delivered to the electric motor and the battery. The motor also acts as a generator during regenerative braking, in that case, the generated energy is stored in the battery pack for later use. Adoption of FCEVs is hindered by the higher costs of hydrogen production, transportation and storage. Limited refueling infrastructure and energy losses in the supply chain also play a major role.

Analysis inferred from FCEVs determined that cost is a major factor that determines the economic viability of any ambitious innovation. FCEVs certainly do not come without its costs. FCEVs are costlier to produce compared to its sustainable counterparts. Some suggestions suggested by [11] On shifting to improved materials that are lower in cost specifically mentioning the catalyst and acid electrode used, namely platinum and polymer membrane should be either replaced by economically viable materials and shifting the process of producing a FCEVs can yield better results. The findings also suggest that the usage of FCEVs is better when it is utilized for long distance or heavy-duty vehicles which require faster refueling and lower weight compared to a BEV and a longer range. In RDC, a study done by [12]. On energy flow in real driving conditions, there have been some key findings that has been brought to light. The study of energy contribution in RDC has concluded that 70% of the vehicle's propulsion was contributed by the fuel cell and only 13% of it was being used from the battery. The study has also revealed that the share of energy recovery through fuel cell charging and regenerative braking is slightly higher. The study was

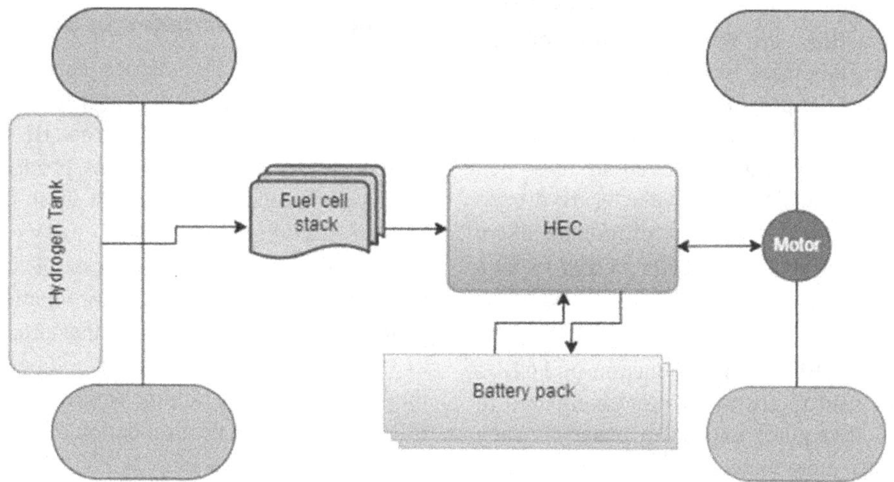

Fig. 39.3 Basic construction of an FCEV [11]

concluded with two important information; the HV battery was only charged during regenerative braking and during some circumstances, the fuel cell itself tends to charge the battery, this shows that the HV battery acts more as a buffer for storage of energy. Lastly, the consumption of energy through the battery decreased and the fuel cell energy consumption increased as the vehicle accelerated further. A life cycle assessment for light-duty vehicles was conducted by [13], the results indicated that the production of PEMFC, namely the fuel tanks and the cathode catalysts are significant contributors to emissions. Limitations on available data also plays a substantial role in research and development of hydrogen fuel cell. The study was concluded that the market readiness for production of PEMFCs and low number of vehicles produced. Storage and refueling also plays an important role in safe utilization for hydrogen as a source of energy. Hydrogen is stored at higher pressure, at around 700 bar, to ensure the refueling takes place without excessive overheating. There are cooling systems that are designed to lower the temperature of hydrogen (passively or actively) during storage and refueling. Although this system ensures reliable storage and refueling experience, it adds unwanted intricacy and additional expenses for a hydrogen refueling station. [14] [twocolumn]article graphicx array

Table 39.1 Charging speeds and time required for charging [7]

Type of Charger	Suitable Charging Location	Power Output	Estimated Time
Level 1	Residential	1.3-3.4 kW	30+ hours
Level 2	Residential/ Commercial	3-19 kW	5-10 hours
Level 3	Commercial	50+ kW	15-30 minutes

Some of the strengths noted in FCEVs are FCEVs do not weigh as much as a BEV, the weight to power ratio is comparatively lower, this leads to more range and better energy density. 70% of the propulsion energy is generated through the fuel cell during RDC, this result shows us that fuel cell is very efficient at converting hydrogen into electricity for propulsion of a vehicle. Hydrogen can be refuelled in a vehicle in a matter of minutes like a traditional fossil fuel powered vehicle, that is why FCEVs have a significant edge over BEVs in terms of refueling experience.

Limitations noted in FCEVs is the production of FCEVs are not cheap; they tend to cost a lot more compared to a BEV or an ICE. It requires expensive materials such as platinum, polymer membrane, et cetera., are used to produce fuel cells making it economically impractical to utilize. Although FCEVs by itself do not produce any tailpipe emissions, the production of FCEVs produces a lot of emissions in terms of extraction of materials and production of PEMFCs. Storage and refueling of hydrogen is complicated, hydrogen needs to be cooled down through either active or passive cooling to ensure a reliable refueling experience with no overheating. Cooling systems adds additional costs to refueling station infrastructure and thus increasing costs.

3. Comparative Analysis

The comparative analysis will focus on four key factors environmental sustainability, scalability, economic viability and long-term sustainability. These factors are critical to ensure a more sustainable future.

3.1 Environmental Sustainability

Environmental sustainability refers to the usage of natural renewable resources for the production process of various energy generation processes.

1) Production of Hydrogen for e-fuels and FCEVs

The discussed technologies rely on hydrogen and battery production. Hydrogen types include green, blue, and gray. Green hydrogen, produced via renewable-powered water electrolysis, emits no CO2. Blue hydrogen, made from fossil fuels with carbon capture, is carbon-neutral but still fossil-based. Gray hydrogen, produced through steam methane reforming, is the most common but emits high CO2 levels (~830 Mt/year), conflicting with sustainability goals. Green and blue hydrogen production is limited due to high costs (2.28–7.39 USD/kgH2) compared to gray hydrogen (0.67–1.31 USD/kgH2) [15]. Currently, gray hydrogen dominates e-fuels and FCEVs, while green hydrogen remains underutilized.

2) Production of Batteries for BEVs

BEVs are powered by batteries, charging the batteries can be done through renewable energy sources however the public perception of BEVs is that BEVs are completely emission-free, and they do not produce any emissions during any process of the production. This is simply not true, production of batteries for BEVs are similar to the production of any other batteries. The most commonly used batteries for BEVs are Li-ion batteries. A study was conducted by [7] and concluded that although the utilization of BEVs have lesser emissions compared to an ICEV as they do not have any tailpipe emissions and they are highly efficient. However, the production of BEVs may emit more emissions than producing ICEVs, due to the extraction of raw materials for the production of batteries and the manufacturing of electrical components for powertrain.

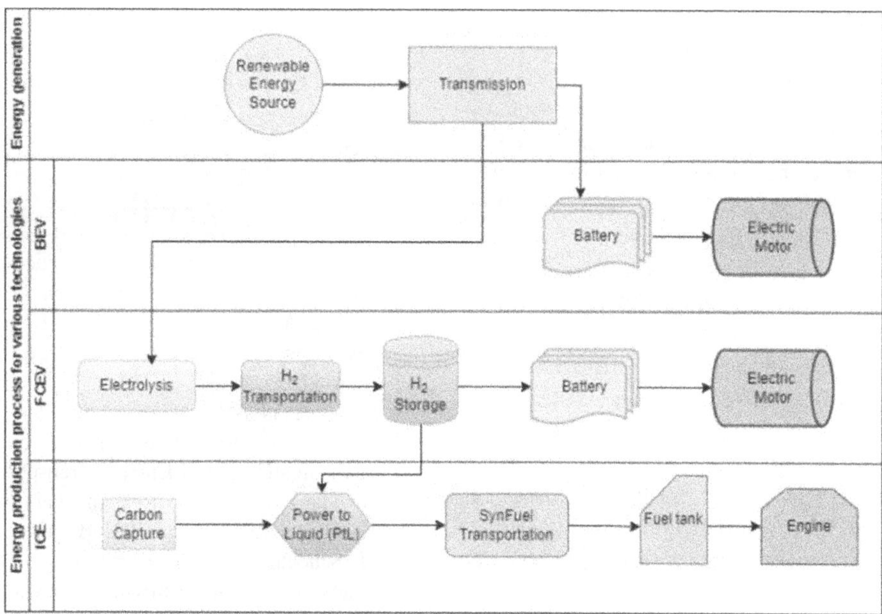

Fig. 39.4 Energy production process comparison [12]

3) Utilization

During utilization, ICEVs powered by e-fuels are considered to be carbon-neutral as the carbon emissions emitted by the ICEVs are captured to produce renewable fuel, given that the hydrogen is sourced through water electrolysis (green hydrogen). If the carbon synthesization process utilizes gray hydrogen, then e-fuels are not completely carbon neutral as production of gray hydrogen plays a role in carbon emissions as discussed by [15] BEVs and FCEVs do not produce any tailpipe emissions during their utilization whatsoever since it is powered purely by electricity which drives the motor. FCEVs produce byproducts namely heat and water vapor during the process of converting hydrogen into electricity.

3.2 Scalability

Scalability is crucial as it determines how widespread each technology's usability is. E-fuels are moderately scalable as it requires energy-intensive processes for production, which require a large amount of renewable energy, but they can utilize existing storage and refueling station infrastructures and can also be used on existing ICEs, this reduces material waste by a significant margin as it does not render ICEVs obsolete. BEVs are highly scalable, although they require new infrastructures to be built and more research and development to be done in battery chemistries. In the long term, BEVs can be scalable enough to be viable. FCEVs are not very scalable, as the cost of production of hydrogen from renewable sources is very high. Storage and infrastructure for hydrogen are also complicated as they require active/passive cooling systems during refueling et cetera., which adds intricacy to the system. FCEVs also require mining of platinum catalysts and other rare elements which adds to the cost.

3.3 Economic Viability

FCEVs and e-fuels require hydrogen, but green hydrogen production costs (2.28–7.39 USD/kgH2) are significantly higher than fossil fuels, making e-fuels economically unviable due to energy losses during conversion and high cost-to-power ratios [15]. Lithium-ion battery costs have decreased due to R&D [7], but BEVs remain costlier than ICEVs. However, by 2030, battery prices are predicted to drop enough to make BEVs economically viable without subsidies [16].

3.4 Long Term Sustainability

Long-term sustainability requires anticipating future material use and emissions to meet carbon-neutral goals. ICEVs fueled by e-fuels face challenges due to high production costs and limited carbon capture capacity, with studies [17, 3] highlighting undercapacity and lack of investor interest in required infrastructure. FCEVs show potential for long-range applications like heavy-duty vehicles but face economic challenges with fuel tanks, hydrogen storage, and catalyst production [13]. BEVs are the most viable solution, with battery degradation starting after 321,000 km (200,000 miles) [10], expected cost parity with ICEVs, and significant potential for carbon emission reductions.

4. Conclusions

After reviewing several papers, this review paper infers that BEVs, FCEVs, and e-fuels each have distinct advantages and disadvantages in the transition to sustainable energy sources to reduce the carbon emissions emitted. After the study, this review has concluded that BEVs lead the market with advanced infrastructure and energy efficiency, making them ideal for short-range applications. FCEVs offer extended range and rapid refueling, showing potential for heavy-duty transport and long-distance travel. E-fuels, on the other hand, provide a path for decarbonizing existing internal combustion engines in applications where BEVs or FCEVs are not viable enough, leveraging existing fuel distribution networks. To fully utilize these potentials, investment in renewable energy, infrastructure expansion, and research is essential. Collaborative efforts among industries and governments will drive research and development in energy storage, hydrogen production, and sustainable fuel synthesis, paving the way for a multi-modal, low-carbon transportation ecosystem. Finally, the following research has been concluded.

- BEVs although tend to lose their battery retention capabilities, it is less of a problem as high-speed charging infrastructures are being actively developed and can mitigate the problems of low range.
- ICEVs are expected to be replaced by FCEVs/BEVs/mild hybrid vehicles despite e-fuels being a promising approach to retain existing ICEVs, the production process of e-fuels is energy intensive and the cost of e-fuels is not low enough to be economically viable. However, research on synthetic fuel blends to lower the usage of fossil fuels in the aviation or maritime sector is necessary to lower the carbon footprint in those industries.
- FCEVs, although offer a lower weight-to-power ratio compared to a BEV, can be sustainable only in specific markets such as heavy-duty vehicles, commercial vehicles et cetera. As it offers faster refueling and a much better range. If hydrogen is produced from green sources, FCEVs have the potential to be carbon-neutral.
- Despite the proposals of banning of ICEs, studies have proven that the reduction in ICE circulation during the pandemic did not significantly reduce atmospheric CO2 levels drastically. The use of renewable fuels in an ICE is proven to be more viable to reduce

This review provides insights into e-fuels, FCEVs, and BEVs but has limitations, including a lack of detailed analysis on the chemistries behind fuel cells, batteries, and e-fuels, and limited coverage of hydrogen types used in FCEVs and e-fuels. Future mobility will likely involve a complementary approach: BEVs dominating passenger vehicles, FCEVs in freight and buses as hydrogen costs decline, and e-fuels in aviation, shipping, and legacy vehicles.

References

1. M. N. Uddin and F. Wang, "Fuelling a Clean Future A Systematic Review of Techno-Economic and Life Cycle Assessments in E-Fuel Development," Applied Sciences, vol. 14, no. 16, p. 7321, Aug. 2024, doi 10.3390/app14167321.
2. E. Lindstad, B. Lagemann, A. Rialland, G. M. Gamlem, and A. Valland, "Reduction of maritime GHG emissions and the potential role of Efuels," Transportation Research Part D Transport and Environment, vol. 101, p. 103075, Dec. 2021, doi 10.1016/j.trd.2021.103075.
3. J. Scheelhaase, S. Maertens, and W. Grimme, "Synthetic fuels in aviation – Current barriers and potential political measures," Transportation Research Procedia, vol. 43, pp. 21–30, 2019, doi 10.1016/j.trpro.2019.12.015.
4. N. Meunier, R. Chauvy, S. Mouhoubi, D. Thomas, and G. De Weireld, "Alternative production of methanol from industrial CO2," Renewable Energy, vol. 146, pp. 1192–1203, Feb. 2020, doi 10.1016/j.renene.2019.07.010.
5. R. Kataoka, "Does BEV always help to reduce CO2 emission? impact of charging strategy," 2024.
6. M. Z. Zeb et al., "Optimal Placement of Electric Vehicle Charging Stations in the Active Distribution Network," IEEE Access, vol. 8, pp. 68124–68134, 2020, doi 10.1109/ACCESS.2020.2984127.
7. Z. Yang, H. Huang, and F. Lin, "Sustainable Electric Vehicle Batteries for a Sustainable World Perspectives on Battery Cathodes, Environment, Supply Chain, Manufacturing, Life Cycle, and Policy," Advanced Energy Materials, vol. 12, no. 26, p. 2200383, Jul. 2022, doi 10.1002/aenm.202200383.
8. G. A. Ogunkunbi, H. K. Y. Al-Zibaree, and F. Meszaros, "Modeling and Evaluation of Market Incentives for Battery Electric Vehicles," Sustainability, vol. 14, no. 7, p. 4234, Apr. 2022, doi 10.3390/su14074234.
9. L. Timilsina, P. R. Badr, P. H. Hoang, G. Ozkan, B. Papari, and C. S. Edrington, "Battery Degradation in Electric and Hybrid Electric Vehicles A Survey Study," IEEE Access, vol. 11, pp. 42431–42462, 2023, doi 10.1109/ACCESS.2023.3271287.
10. S. Naveen D. Surabhi, C. Shah, V. Mandala, and P. Shah, "Range Prediction based on Battery Degradation and Vehicle Mileage for Battery Electric Vehicles," IJSR, vol. 13, no. 3, pp. 952–958, Mar. 2024, doi 10.21275/SR24312045250.
11. O. Bethoux, "Hydrogen Fuel Cell Road Vehicles State of the Art and Perspectives," Energies, vol. 13, no. 21, p. 5843, Nov. 2020, doi 10.3390/en13215843.
12. A. Szałek, I. Pielecha, and W. Cieslik, "Fuel Cell Electric Vehicle (FCEV) Energy Flow Analysis in Real Driving

Conditions (RDC)," Energies, vol. 14, no. 16, p. 5018, Aug. 2021, doi 10.3390/en14165018.

13. L. Usai, C. R. Hung, F. Vasquez, M. Windsheimer, O. S. Burheim,´ and A. H. Strømman, "Life cycle assessment of fuel cell systems for light duty vehicles, current state-of-the-art and future impacts," Journal of Cleaner Production, vol. 280, p. 125086, Jan. 2021, doi 10.1016/j.jclepro.2020.125086.
14. M. Genovese, D. Blekhman, and P. Fragiacomo, "An Exploration of Safety Measures in Hydrogen Refueling Stations Delving into Hydrogen Equipment and Technical Performance," Hydrogen, vol. 5, no. 1, pp. 102–122, Feb. 2024, doi 10.3390/hydrogen5010007.
15. J. M. M. Arcos and D. M. F. Santos, "The Hydrogen Color Spectrum Techno-Economic Analysis of the Available Technologies for Hydrogen Production," Gases, vol. 3, no. 1, pp. 25–46, Feb. 2023, doi 10.3390/gases3010002.
16. A. Konig, L. Nicoletti, D. Schr¨ oder, S. Wolff, A. Waclaw, and¨ M. Lienkamp, "An Overview of Parameter and Cost for Battery Electric Vehicles", WEVJ, vol. 12, no. 1, p. 21, Feb. 2021, doi 10.3390/wevj12010021.
17. H. Singh, C. Li, P. Cheng, X. Wang, and Q. Liu, "A critical review of technologies, costs, and projects for production of carbon-neutral liquid e-fuels from hydrogen and captured CO2," Energy Adv., vol. 1, no. 9, pp. 580–605, 2022, doi 10.1039/D2YA00173J.
18. N. Duarte Souza Alvarenga Santos, V. Ruckert Roso, A. C. Teix-¨ eira Malaquias, and J. G. Coelho Baeta, "Internal combustion en-ˆ gines and biofuels Examining why this robust combination should not be ignored for future sustainable transportation," Renewable and Sustainable Energy Reviews, vol. 148, p. 111292, Sep. 2021, doi 10.1016/j.rser.2021.111292.
19. I. Frenzel, J. E. Anderson, A. Lischke, and C. Eisenmann, "Renewable fuels in commercial transportation Identification of early adopter, user acceptance, and policy implications," Case Studies on Transport Policy, vol. 9, no. 3, pp. 1245–1260, Sep. 2021, doi 10.1016/j.cstp.2021.06.010.
20. H. Liu, S. Yu, T. Wang, J. Li, and Y. Wang, "A systematic review on sustainability assessment of internal combustion engines," Journal of Cleaner Production, vol. 451, p. 141996, Apr. 2024, doi 10.1016/j.jclepro.2024.141996.
21. S. Franzo and A. Nasca, "The environmental impact of electric vehicles` A novel life cycle-based evaluation framework and its applications to multi-country scenarios," Journal of Cleaner Production, vol. 315, p. 128005, Sep. 2021, doi 10.1016/j.jclepro.2021.128005.
22. A. Temporelli, M. L. Carvalho, and P. Girardi, "Life Cycle Assessment of Electric Vehicle Batteries An Overview of Recent Literature," Energies, vol. 13, no. 11, p. 2864, Jun. 2020, doi 10.3390/en13112864.
23. P. Chakraborty et al., "Addressing the range anxiety of battery electric vehicles with charging en route," Sci Rep, vol. 12, no. 1, p. 5588, Apr. 2022, doi 10.1038/s41598-022-08942-2.
24. F. Jung, M. Schroder, and M. Timme, "Exponential adoption of battery¨ electric cars," PLoS ONE, vol. 18, no. 12, p. e0295692, Dec. 2023, doi 10.1371/journal.pone.0295692.
25. H. Blanco, W. Nijs, J. Ruf, and A. Faaij, "Potential for hydrogen and Power-to-Liquid in a low-carbon EU energy system using cost optimization", Applied Energy, vol. 232, pp. 617–639, Dec. 2018, doi 10.1016/j.apenergy.2018.09.216.
26. J. Thakur, J. M. Rodrigues, and S. Mothilal Bhagavathy, "Whole system impacts of decarbonising transport with hydrogen A Swedish case study", International Journal of Hydrogen Energy, vol. 89, pp. 883–897, Nov. 2024, doi 10.1016/j.ijhydene.2024.09.386.
27. S. Drunert, U. Neuling, T. Zitscher, and M. Kaltschmitt, "Power-to-¨ Liquid fuels for aviation – Processes, resources and supply potential under German conditions", Applied Energy, vol. 277, p. 115578, Nov. 2020, doi 10.1016/j.apenergy.2020.115578.
28. D. Apostolou, "Optimisation of a hydrogen production – storage – repowering system participating in electricity and transportation markets. A case study for Denmark", Applied Energy, vol. 265, p. 114800, May 2020, doi 10.1016/j.apenergy.2020.114800.
29. B. Lee et al., "Renewable methanol synthesis from renewable H2 and captured CO2 How can power-to-liquid technology be economically feasible?", Applied Energy, vol. 279, p. 115827, Dec. 2020, doi 10.1016/j.apenergy.2020.115827.
30. S. Foorginezhad, M. Mohseni-Dargah, Z. Falahati, R. Abbassi, A. Razmjou, and M. Asadnia, "Sensing advancement towards safety assessment of hydrogen fuel cell vehicles", Journal of Power Sources, vol. 489, p. 229450, Mar. 2021, doi 10.1016/j.jpowsour.2021.229450.

Advances in Additive Manufacturing Technologies – Gurusamy Pathinettampadian et al. (eds)

Parametric Analysis and Optimization of Electrochemical Micromachining Performance for Titanium Alloy

Gowtham Kumarasamy[1]
Department of Mechanical Engineering,
Chennai Institute of Technology,
Chennai, India

Prabhakaran R.[2]
Department of Mechanical Engineering,
IFET College of Engineering,
villupuram, India

Kishore Ravikumar[3]
Department of Mechanical Engineering,
Chennai Institute of Technology,
Chennai, India

Vaithianathan N.[4]
Department of Mechanical Engineering,
IFET College of Engineering,
villupuram, India

◆ **Abstract:** Micro-machining or electrochemical micromachining or (EMM), is another type of electrochemical machining that is involved in precision machining operations to form, deburr, mill and finish operations. EMM and ECM are of immense interest to modern industry for a variety of technology industries due to the ability to perform complex designs with high precision and accuracy. These are clean, harmless, and easy to apply to a large number of conductive materials. Most of the applications of salt water require a heat exchanger to be used. The most widely used material for such applications is Titanium Alloy (Ti-3Al-2.5) which possesses surprisingly good resistance to corrosion by salty atmospheres. Titanium alloys, however, are not easy to machine with conventional methods. This is due to high strength in titanium alloys, work hardening, and low thermal conductivity. EMM overcomes all these mechanical difficulties during machining by employing electrochemical dissolution to dissolve the material at microscopic levels without mechanically damaging or degrading the tool. The aim of the present research is to see how the MRR is related to the process parameters such as voltage, electrolyte concentration, and duty cycle, during EMM of titanium alloys. The aim is to measure the MRR and to optimize the machining parameters. The surface morphologies were also studied to determine the parameter's role in machining performance and accuracy.

◆ **Keywords:** EMM, Titanium Grade 9, Titanium, MRR

[1]gowthamk@citchennai.net, [2]prabhakaran1609@gmail.com, [3]kishoreravi@citchennai.net, [4]vaithi1312@gmail.com

DOI: 10.1201/9781003685906-40

1. Introduction

New electrochemical micromachining (EMM) technology offers a productive manufacturing process that can achieve machining in conditions impossible or inefficient for other processes. EMM is a process where controlled anodic dissolution is in a very controlled process by conduction of the conductive material. By this method, EMM is an effective process in applications directly for medical, aerospace, and electronics. EMM experiences machine tool wear, and hence generating less thermal strain of the workpiece being processed, in comparison to conventional machining processes and thus machine degradation, with some thermal strain being experienced on the material being processed. For potential extreme environments, titanium alloys have been utilized due to their inherent strength-to-weight ratio, corrosion resistance, biocompatibility, and potential in additive manufactured material as well. Titanium alloys, however, particularly anodizing material such as Ti-3Al-2.5 can be prone to high tool wear in the tool, as well as process inefficiencies at low thermal conductivity with regards to the reactivity of titanium's alloys and conventional machining processes. However, on the contrary, the potential for EMM to machine titanium alloys without mechanical deformation, to produce a final product with precision, is an advantage. To provide an overview of the process variables, EMM has several significant variables when utilizing the technology of EMM machining; these being the input voltage, electrolyte make-up, duty cycle and pulse frequency. Each of these parameters is related to the amount of material removed, or MRR and surface roughness. In the consideration of machining capability while optimizing MRR and surface quality, understanding the parameters and the relation to MRR is an essential consideration. In earlier research, quality and MRR was found to be driven, mainly by the electrolyte composition and duty-cycle, whereas the extent of electrical input is linked with the definition of electrochemistry or electrochemical kinetics for the material machining in EMM. The present research intends to contribute the knowledge both in MRR for machining an EMM titanium alloy, that of Ti-3Al-2.5, and centers on the understanding input voltage, electrolyte concentration, and duty cycle influenced the MRR and hence contribute to, optimizing machining conditions for industries. There will be a variable analysis of quality on the surface morphology, and microstructure of the machined alloy of titanium; this will be to also comprehend the implication of the process parameters in the EMM process for quality.

2. Experimental Setup

Electrochemical Micromachining (ECM) experimental setup comprises material removal mechanism, electrode feed mechanism, electrolyte tank and DC power supply. Pump and filter are included as a part of the electrolyte tank for circulating the electrolyte. Mechanical machining unit includes work holding fixture, micro-tool feed mechanism, machining chamber, and substantial machining structure. Micro-tool feed movement is provided by ball screw mechanism and stepper motor is used for controlling the manual feed of the electrode. DC power supply of output voltage 0-30V and output current 0-2A is used for the system. For precise machining, a digital oscilloscope is included to monitor the voltage pulses in real-time. They regulate the electrolyte flow rate and concentration so that hydrogen bubbles do not interfere. This setup is able to accommodate various types of electrolytes, electrodes, and workpieces. This facility allows easy utilization of it for research or industry. A circulation type system was used in order to control the temperature of the electrolyte. Conductivity was kept constant by peristaltic pump. To prevent stray currents and uniform machining, the workpiece was clamped in a non-conductive fixture.

Fig. 40.1 Machine setup

3. Experiments and Methods

3.1 Workpiece Material and Electrode Selection

In this work research Ti-3Al-2.5 alloy as workpiece material that is selected because of its enhanced corrosion resistance and mechanical properties. The alloy was received in sheet metal of size 40mm × 20mm × 1.7mm. Before machining, the workpieces were cleaned with acetone to remove the surface impurities and oxides. The electrode used as the tool material is Stainless Steel 316L, which is very electrically conductive. The electrode was made with a diameter of 500 μm and was made with great precision to offer constant machining conditions.

Table 40.1 Chemical composition of Ti Alloy

Elements	Composition (%)
Fe	0.24
Al	3.02
V	2.17
Ti	94.63

3.2 Electrolyte Preparation and Circulation

The electrolyte incorporated in the research consisted of an influential mix of 1 mol/L $NaNO_3$ and 0.02 mol/L sodium citrate. Sodium nitrate was a suitable inclusion due to its relatively high conductivity and its ability to maintain uniform anodizing dissolution while sodium citrate served as a complexing agent to stabilize titanium ions in solution. To maintain stable galvanic machining conditions, the electrolyte was circulated through a peristaltic pump at a flow rate of 0.5 L/min. A filtration setup was implemented feasible to eliminate by-products, and avoid cross contamination. The electrolyte was kept at a temperature of 25° to avoid any potential affective thermal changes to reaction kinetics.

3.3 Experimental Setup and Power Supply

The electrochemical micromachining process was carried out with the help of a custom-designed EMM setup proposed having a material removal system, an electrode feeding system, an electrolyte container, and a DC power supply. The electrode feed rate was controlled by a ball screw system to ensure material removal rate was constant throughout the experiment. To eradicate stray currents and to guarantee even electrolyte distribution in the machining chamber was also designed based on specific parameters. A DC power supply was used to apply the required machining conditions with adjustable DC ranges of 0-30V for voltage and 0-2A for current. The actual values of voltage and current were observed on a digital oscilloscope throughout the experiment for accuracy and stability.

3.4 Experimental Design and Process Parameters

The experimental technique adopted a systematic approach based on the Taguchi approach to design of experiments (DOE) for experiments involving exploratory trials of variation in three parameters which are processing voltage, process electrolyte concentration and processing duty cycle. An L4 orthogonal array was chosen in order to develop optimum processing conditions as a result of variations in three specific parameters; the parameter levels chosen were:

- Processing Voltage : 7V, 9V, 11V
- Machining Electrolyte concentration: 20 g/L, 25 g/L, 30 g/L
- Processing Duty cycle : 33%, 50%, 66%

Each experimental run was done in three replications to enhance the repeatability of the results while minimizing the effect of random errors in the experiments. The amount of machining time was recorded, and the mass change in the workpiece pre-and post-processing to measure material removal rate (MRR).

3.5 Measurement and Characterization Techniques

In an attempt to properly assess the machining performance, different characterization methods were employed with utmost care. Material removal rate, a very crucial parameter in the assessment of machining efficiency, was precisely measured using an analytical balance with a very remarkable measuring accuracy of 0.0001 grams. In order to properly obtain the material removal rate, the difference in the weight of the specimen was both before and after the machining operation was completed noted. Furthermore, the surface finish of the machined specimen was analyzed and inspected with utmost care using a highly sophisticated surface profilometer. For the sake of thorough assessment, the mean roughness (Ra) was measured at several different points along the surface to ascertain the uniformity and consistency of its value. In addition, scanning electron microscopy (SEM) images were employed to effectively conduct the study of surface morphology, investigate the tool wear pattern, and monitor any microstructural changes noticed in the machined regions. Moreover, EDS was conducted to determine with precision whether there was a discernible shift in composition on the machined surface, as well as to monitor possible contamination that may have been caused by the electrolyte used during the process. Finally, the optical microscopy technique was employed to investigate the microstructural changes closely and identify the occurrence of any micro cracks or pitting defects that may likely affect the machined product's integrity.

3.6 Statistical Analysis and Optimization

The collected data was subjected to a stringent analysis using Minitab software, a robust statistical analysis tool specifically suited for the purpose, to determine the significance of all the different process parameters involved. For determining the optimum conditions to maximize the objective, signal-to-noise (S/N) ratios were determined, which were the determining factors in determining the optimum parameter values required to maximize the material removal rate (MRR). As a byproduct, in the process, this analysis also helped in reducing surface defects that could otherwise be generated during the process for overall quality improvement. Response surface methodology (RSM) was also utilized to create predictive models and to optimize machining conditions.

The study focused on the attainment of high-precision machining through controlled experimental techniques and meticulous characterization protocols and resulted in raised material removal rates and surface defect minimization.

4. Results and Discussions

The effect of factors, such as applied voltage, electrolyte concentration, and duty cycle, on the MRR of Ti-3Al-2.5 alloy was investigated with an additively manufactured stainless steel 316L tool. The sodium nitrate electrolyte, which was mixed with sodium citrate as a complexing agent, was monitored to study the influence of those process parameters causing changes in the material-removal rate. The findings are summarized below.

4.1 Performance of Process Parameters Over MRR

The MRR was calculated as the ratio of the weight difference before and after machining to the total machining time.

$$MRR = (X_{before} - X_{after}) / t \tag{1}$$

X_{before} = weight of before machining (g)

X_{after} = weight of after machining (g)

t = Machining time (hours)

The material removal rate increased with the increase of electrolyte concentration in the machining process. In accordance with the experimental data, maximum material removal rate can be achieved with an electrolyte concentration of 1 mol/L and a complexing agent concentration of 0.02 mol/L. Other factors that influence it are also the duty cycle, and they play a significant role in reaching the desired MRR during machining. Effects exerted by the process factors were analyzed and quantified with the aid of Minitab statistical software.

4.2 Effect of Voltage on MRR

Voltage is the most critical parameter in the electrochemical micromachining process since it has a direct controlling effect on anodic dissolution rate. It was realized that the rise in applied voltage from 15V to 17V resulted in an increase in MRR. This is because the increased dissolution rate of titanium in the applied electrolyte solution at high voltages enhances the removal of material. Nevertheless, very high voltage levels beyond the experimental are likely to generate stray currents and localized overheating and produce undesired effects in machining accuracy in the form of surface pitting and uneven removal of material.

Table 40.2 Orthogonal L9 response of material removal rate

Exp. No	Processing Voltage (V)	Machining Electrolyte concentration (g/liter)	Processing Duty cycle (%)	MRR (mm^3/min)
1	7	20	33	0.041183
2	7	25	50	0.031806
3	7	30	66	0.034910
4	9	20	66	0.058100
5	9	25	33	0.030611
6	9	30	50	0.081307
7	11	20	50	0.066730
8	11	25	66	0.147989
9	11	30	33	0.039668

At lower voltages, the reaction kinetics are slowed down, and material removal efficiency is decreased. This results in increased machining time and surface finish irregularities. The results show that there is an optimal voltage setting at which the electrochemical reaction occurs efficiently without undue current leakage or undesirable side reactions.

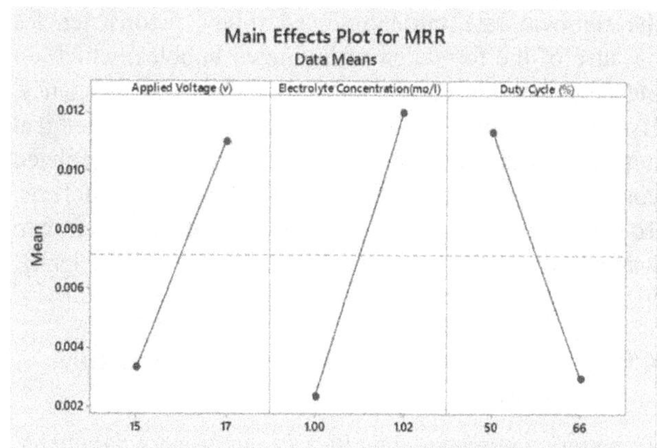

Fig. 40.2 Process parameters on MRR

Thus, for the effects of process parameters on MRR, see Fig. 40.2. The divergence from the mean level line indicates that EMM variables strongly influence this process. This shows considerably that the use of electrolyte with complexing agent has more influence as compared to the other factors, for instance, applied voltage and duty cycle. An electrolyte content of 1 mol/L Sodium Nitrate and 0.02 mol/L Sodium Citrate was seen to have the most significant influence. Finally, the best combination of EMM parameters for effective machining of Titanium Alloy, Ti-3Al-2.5, was established.

4.3 Effect of Electrolyte Concentration on MRR

Electrolyte concentration was found to play a major determining role for conductivity of the solution and the efficiency of material removal. Higher electrolyte concentrations support higher ion transport, thereby accelerating anodic dissolution and eventually augmenting MRR. 1 mol/L of sodium nitrate and 0.02 mol/L of sodium citrate concentration was the optimized concentration since it struck the proper balance between efficiency of material removal and process stability When the concentration of the electrolyte exceeded the optimum value, there was localized turbulence, which resulted in uneven material removal and higher surface roughness. Lower electrolyte concentrations, however, gave rise to lack of ion mobility, which created ineffective machining with lower MRR. Complexing agents like sodium citrate helped in achieving a more uniform dissolution process by stabilizing the ions of titanium in solution [3].

4.4 Effect of Duty Cycle on MRR

The duty cycle, the ratio of ON time to cycle time, had a great effect on MRR. At a low duty cycle (50%), machining exhibited regular removal of material with less defect. However, increasing the duty cycle to 66% increased the removal rate but introduced minor inconsistencies because of the formation of hydrogen bubbles, which on the negative side has an impact on electrode efficiency. Hydrogen bubbles formed during electrochemical machining can stick to the work surface, creating localized conductivity variations and resulting in pitting defects. To preclude such effects, the flow rate of the electrolyte was precisely controlled to eliminate bubbles efficiently, maintaining a uniform machining condition [4].

4.5 Surface Morphology and Microstructure Analysis

To better comprehend the effect of the process parameter, SEM and EDS scans were performed. SEM images revealed that lower duty cycles yielded smoother surface finishes, while higher voltages yielded comparatively rougher surfaces with insignificant pitting effects [5]. EDS verified titanium dissolution products presence and homogeneous electrolyte-induced surface modification distribution. Moreover, cross-sectional examination of the machined samples showed that high voltage yielded subsurface damage in terms of micro-cracks and non-homogeneous dissolution fronts. The above findings emphasize the importance of prudent selection of parameter combinations to ensure high-precision machining with minimal structural damage [6].

4.6 Optimization of Process Parameters

Experimental research was carried out using the Taguchi optimization method to find and identify the optimal values of the process parameters that would lead to the achievement of the maximum MRR while at the same time guaranteeing stable machining accuracy particularly in the case of electrochemical micromachining (ECM). In the well-planned experiments that were carried out, the optimal parameters that were found included a voltage level of 17 volts, electrolyte concentration of 1 mol/L sodium nitrate with 0.02 mol/L sodium citrate, and a duty cycle of 50% [7]. These particular parameters were effective in guaranteeing the maximum MRR while at the same time preventing any machining defects, thus ensuring not only excellent process stability but also high accuracy levels throughout the machining process. The incorporation of sodium citrate was vital by ensuring the achievement of a better surface finish; this was guaranteed by ensuring better control of the electrochemical reactions that occur during the machining process and also by helping in countering the passivation effects that adversely impact performance [8]. The experimental findings clearly indicate the need to ensure very precise control over the ECM parameters in order to guarantee the achievement of a defect-free machining process that is effective and efficient. The current work strongly highlights the need to optimize the parameters in electrochemical machining processes, especially in the case of micromachining that demands high precision, where there is a need to balance the material removal rate with the accuracy of the machining result [9].

5. Conclusions

The study thoroughly investigates the strong impact of different process parameters on the complex electrochemical micromachining of the titanium alloy Ti-3Al-2.5 using a steel electrode for this particular purpose. The results show that:

- Duty cycle and concentration of the electrolyte play an important role in machining performance.
- The moderate MRR was improved by varying electrolyte concentration, reducing duty cycle, and increasing the applied voltage.
- The surface morphology examination showed that properly optimized process conditions resulted in the

creation of a very smooth surface with only small defects.

- The study undertaken was aimed at establishing the most optimal and best combination of EMM parameters that could improve and promote the machining process particularly in the case of titanium alloy.
- The measured impact of voltage on MRR indicates that careful adjustment of electrical parameters may result in considerable enhancements in machining efficiency.
- Hydrogen bubbles during the machining process need to be properly handled in order to avoid defects and maintain process stability.
- Future research should investigate real-time monitoring methods for better parameter control and adaptive machining processes.
- These results are useful in optimizing the electrochemical micromachining process for high-precision manufacturing applications such as in aerospace and biomedical industries where precision machining is essential.

References

1. Bhattacharyya, B. D. P. S. B., B. Doloi, and P. S. Sridhar. "Electrochemical micro-machining: new possibilities for micro-manufacturing." *Journal of materials processing Technology* 113.1 (2001): 301–305.
2. Bhattacharyya, B., M. Malapati, and J. Munda. "Experimental study on electrochemical micromachining." *Journal of Materials Processing Technology* 169.3 (2005): 485–492.
3. T. Geethapriyan, K. Kalaichelvan, A. Rajadurai, T. Muthuramalingam, S. Naveen."A Review on Investigating the Effects of Process Parameters in Electrochemical Machining." *International Journal of Applied Engineering Research*, Vol. 10 No.2 (2015): 1743–1748
4. T. Geethapriyan, K. Kalaichelvan, T. Muthuramalingam, G. Madhana Gopal. "Performance analysis of process parameters on machining a–ß titanium alloy in electrochemical micromachining process. Proceedings of the Institution of Mechanical Engineers, Part B: *Journal of Engineering Manufacture* (2016): 0954405416673103
5. Mahdavinejad R, Hatami M "On the application of electrochemical machining for inner surface polishing of gun barrel chamber". *J Mater Process Technol.* Vol. 202(2008), 307–315.
6. K.P. Rajurkar, M. M. Sundaram and A. P. Malshe, "Review of Electrochemical and Electrodischarge Machining," Procedia. CIRP. 6, pp.13–26, 2013.
7. B. Bhattacharyya and J. Munda, "Experimental investigation on the influence of electrochemical machining parameters on machining rate and accuracy in micromachining domain," Int. J. Mach. Tools Manuf. 43, pp. 1301–1310, 2003.
8. Zhi-Wen Fan, Lih-Wu Hourng and Ming-Yuan Lin, "Experimental investigation on the influence of electrochemical micro drilling by short pulsed voltage," Inter. J. Adv. Manuf. Technol, vol. 61, pp.957–966, 2012.
9. K.P. Rajurkar, D. Zhu, J.A. McGeough, J. Kozak and A.K.M. De Silva, "New developments in electro-chemical machining, "Ann. CIRP 48, pp. 567–580, 1999.

Note: All the figures and tables in this chapter were made by the authors.

Advances in Additive Manufacturing Technologies – Gurusamy Pathinettampadian et al. (eds)

Post-Processing in Additive Manufacturing: Challenges, Innovations and Future Scope

Santhana Babu A.V.[1], Aravindh M.[2], Jai Ganesh R.[3], Divakar S.[4], Kowshik S.[5], Devesh R.[6]

Department of Mechanical Engineering,
Chennai Institute of Technology,
Chennai, India

◆ **Abstract:** Additive manufacturing (AM) has revolutionized the art of modern manufacturing and now it is possible to produce components with complex geometries by minimizing material wastage, using this manufacturing process. This article delineates categories of additive manufacturing, showing advantages of the process against conventional manufacturing methods. Though the process has many merits, post-processing remains an important step in improving the mechanical, thermal and surface properties of AM parts for their respective applications. This review brings out different techniques that have been utilized for post-processing of AM technologies along with challenges related to them. Apart from above, the innovations related to post-processing techniques are also presented in this paper. This review also encompasses the future scope of the post processing techniques. This article synthesizes insights from recent research and industry trends to equip researchers and practitioners with a deep understanding of post processing advancements, their limitations and future directions. It underlines the importance of effective post-processing to unlock the full potential of AM for high-quality, durable and economically viable applications.

◆ **Keywords:** Additive manufacturing, 3D printing, Rapid prototyping, AM post processing, Post processing innovations

1. Introduction

Additive Manufacturing, which is also called as 3D printing has evolved significantly from the years 1980s from rapid prototyping into manufacturing. It falls under the seven categories shown in Fig. 41.1 [1-3]. As different from subtractive conventional practices, additive manufacturing allows forming of products layer after layer with no or least generation of waste, intricate features and custom design. A comparison of them is presented in Table 41.1. This innovative technology has gained popularity among many industries, as it can produce complex geometries with reduced lead times and minimize tooling costs, which is evident in Fig. 41.2. With further maturation of AM, this technology is likely to change the face of global manufacturing by allowing sustainable, on-demand production across various industries [4]. Post-processing is an essential process in additive manufacturing (AM) to overcome the limitations such as surface roughness, dimensional inaccuracies and suboptimal mechanical properties. These methods ensure that the AM components meet the industry standards, thus promoting their adoption across applications [5].

[1]santhanababuav@citchennai.net, [2]aravindhm.mech2022@citchennai.net, [3]jaiganeshr.mech2022@citchennai.net, [4]divakars.mech2022@citchennai.net, [5]kowshiks.mech2022@citchennai.net, [6]deveshr.mech2022@citchennai.net

DOI: 10.1201/9781003685906-41

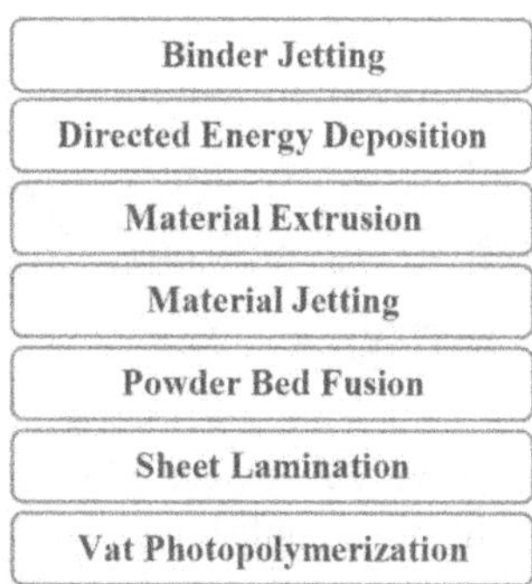

Fig. 41.1 Categories of additive manufacturing

•Aerospace
•Automotive
•Medical and Health Care
•Consumer Goods
•Construction
•Energy
•Defence and Military
•Electronics
•Food Industry
•Fashion Industry
•Tooling and Manufacturing
•Education and Research
•Textile and Apparel Industry

Fig. 41.2 Industrial applications of additive manufacturing [5]

Table 41.1 Comparison between additive manufacturing and subtractive manufacturing [6–11]

Aspect	Additive Manufacturing (AM)	Subtractive Manufacturing (SM)
Process Type	Builds parts layer by layer, adding material progressively.	Removes material from a solid block or work piece through cutting, drilling or milling.
Material Usage	Material is added only where needed, minimizing waste.	Significant material waste is generated during the cutting or removal process.
Flexibility in Design	High flexibility for complex geometries and intricate designs.	Limited by the geometry that can be physically cut or machined.
Tooling	No complex tooling required; only the printing head or nozzle is needed.	Requires specialized tools such as drills, cutters or mills.
Production Speed	Generally slower, especially for large parts or fine details.	Can be faster for simpler parts or bulk production.
Cost of Setup	Typically lower initial setup cost, especially for prototypes.	High initial setup cost due to the need for tooling, jigs and fixtures.
Material Types	Works with a variety of materials, including plastics, metals, ceramics and composites.	Primarily works with metals, plastics and wood, though many material types are used.
Post-Processing	Often requires extensive post-processing, such as support removal or surface finishing.	May require some post-processing, such as polishing or deburring.
Part Size Limitations	Dependent on the size of the printer; larger parts may require assembly.	Limited by the size of the machine and the work piece, but larger parts are possible.
Precision	Good for intricate parts but may have resolution limitations compared to subtractive machining.	Typically offers high precision and tight tolerances, especially for machining.
Material Properties	Some parts may have weaker mechanical properties due to layer bonding.	Generally results in stronger, more homogenous material properties.
Waste Generation	Very little material waste due to additive nature.	High material waste due to cutting away excess material.
Application	Suitable for rapid prototyping, complex parts and small-batch production.	Ideal for high-volume manufacturing, large parts and high-precision components.

This review aims to comprehensively explore post-processing techniques in additive manufacturing, focusing on their impact on part quality, functionality and scalability. It also covers associated challenges, innovations and future scope of post processing.

2. Post-Processing Techniques in Additive Manufacturing

As seen in Fig. 41.3, post-processing techniques ensure that the AM parts meet the functional, aesthetic and mechanical standards [5]. Post-processing requirements

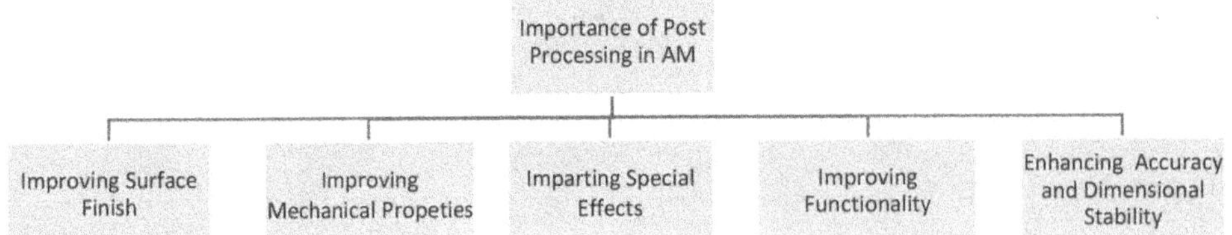

Fig. 41.3 Requirements of post processing in AM [7–9]

and techniques differ substantially among the additive manufacturing (AM) technologies [6, 7]. These include the different types of materials, the printing processes involved and the application demands. Table 41.2 provides various types of post-processing techniques that are applicable for additive manufacturing [8, 9].

3. Challenges in Post-Processing of Additive Manufacturing

Though post processing enhances the quality of additively manufactured parts, it also poses its own challenges as listed below: [5, 10]

1. **Consistency:** Consistency is a major challenge because the result varies widely across geometries, materials and techniques, such as variations in part size, complexity and geometry. Uniformity in surface finishes, mechanical properties or dimensional accuracy calls for very careful parameter control and iteration.
2. **Labor-Intensive Processes:** Most post-processing operations, like hand polishing, support removal and fine finishing, are labor-intensive and, thus, increase the time for production and variation. Manual processes also reduce scalability and bring about inconsistency in quality.
3. **Cost:** High operational cost is incurred because of labor-intensive techniques, specialized equipment and consumables needed. Costs become even more significant if controlled environments or sophisticated equipment are necessary for certain treatments.
4. **Material-Specific Issues:** Depending on the material, different post-processing strategies must be developed. This creates added complexity. In most cases, metals involve more complicated heat treatments while polymers require chemical treatments or vapor smoothing [11].
5. **Scaling up:** Scaling up from small-scale prototyping to industrial production requires streamlining the

Table 41.2 Different types of post processing techniques in AM [10–14]

S No	Technique	Details
1	Cleaning	Powder residue cleaning from technologies like selective laser sintering, direct metal laser sintering.
2	Surface Finishing	Manual, mechanical or electro polishing, to improve aesthetics and reduce roughness on metals. Subjecting parts to vaporized solvents to smoothen polymer surfaces.
3	Heat Treatment	Annealing for stress relieving the interior and enhancing the mechanical strength. Applying heat and pressure to metal parts to eliminate porosity and enhance strength. Sintering for further bonding powder particles in parts created by binder jetting or selective laser sintering.
4	Coating and Plating	Painting for aesthetic enhancement and surface protection. Electroplating (Coating parts with a metal layer to enhance conductivity, strength or aesthetics). Powder coating (Applying a protective, tough surface layer).
5	Machining	CNC machining to improve dimensional accuracy and maintain the tolerance. Drilling and tapping to add threads or holes for assembly purposes.
6	Chemical Treatments	Acid etching to improve surface texture or preparing for subsequent coatings. Solvent dipping to smoothen surfaces of polymer parts.
7	Post Curing	Expose resin-based parts to ultraviolet light. Application of heat to increase strength and stability of the material.
8	Stiffening Methods	Fill porous parts with resins, wax or other materials to increase the strength and durability of the part (e.g., in binder jetting). Reinforcement: Introduce fibers or secondary structures to increase strength.
9	Specialized Methods	Laser surface texturing to alter the surface properties for functional or aesthetic reasons. Surface hardening to enhance the wear resistance by localized heating or coating.
10	Inspection and Testing	Dimensional inspection to verify through 3D scanning or metrology. Non-destructive testing using X-rays, ultrasonic testing or CT scans for detecting internal defects.

post-processing method for higher throughput with less quality loss. Scaling integration requires bringing post-processing into the workflow so that interruptions are minimized and time cycles are optimized.

6. **Process Variability:** Variations in post-processing requirements among parts create bottlenecks in production scheduling. Inefficient or inconsistent post-processing methods can disrupt overall manufacturing efficiency and delay production.
7. **Workflow Integration:** Seamlessly integrating post-processing into the AM workflow is challenging but essential for minimizing disruptions and optimizing production cycles. Lack of integration affects operational efficiency and impacts scalability.
8. **Trade-off between Quality and Efficiency:** The need for quality is constantly compromised against price effectiveness and scalability.
9. **Ecological Considerations:** Traditional techniques are hazardous, pose environment or health hazards and depend on hazardous chemicals or dangerous operations.
10. **Defect Mitigation:** Difficulty in eliminating the defects present with AM.
11. **Integration of Post Processing:** Integrations of several post-processing technologies together like subtractive, additive and equivalent methods tend to add complexity in process control.
12. **Energy Intensity:** High energy inputs from processes such as heat treatment, coating and curing lead to high cost inputs.
13. **Waste Material:** Subtractive post-process technologies result in material wastage, especially when resolution and surface finish need to be improved
14. **Lack of Standardization:** Absence of universally accepted standards or guidelines for selecting an appropriate method of post-processing based on a particular application requirement.

4. Innovations in Post-Processing for Additive Manufacturing (AM)

Innovations collectively enhance (as shown in Table 41.3) efficiency, precision and functionality in post-processed components, elevating additive manufacturing's potential across various industries [5].

5. Future Scope in Post-Processing of Additive Manufacturing (AM)

Advances such as below are necessary to improve post-processing in AM, which help to achieve efficiency and sustainability along with functionality [5, 10]:

1. **In-line Post-Processing Technology:** This refers to developing inline and in situ post-processing techniques integrated inside AM machines or production cells. Seamless transitions between printing and the post-processing stages will assist in smooth processes, elimination of disturbances, reduction in lead time and efficiencies.
2. **Automation and Robotics:** Widespread use of autonomous systems with AI-powered algorithms and

Table 41.3 Post processing innovations in AM [12]

Category	Description
Advanced Surface Finishing Techniques	Tailored chemical polishing solutions for various materials and geometries to reduce surface roughness and eliminate visible layer lines. Automated robotic polishing systems with adaptive controls to ensure consistent quality and reduced labor-intensive processes.
Innovative Heat Treatment Strategies	Gradient annealing for precise manipulation of material properties. Localized heat treatments with laser systems to enhance mechanical strength and durability of parts
Materials Innovation	Development of specific post-processing materials, such as tailored abrasive media and specialized surface coatings. Eco-friendly support materials designed to minimize waste and ensure compatibility with diverse AM technologies.
Advanced Equipment Integration and Automation	AI-based robotic systems for autonomous decision-making and parameter optimization. In-line inspection and quality control systems embedded in equipment for real-time monitoring and immediate corrective actions.
Data-Driven Insights and Predictive Analytics	Use of historical process data and real-time feedback for predictive analytics to optimize parameters for efficiency and quality. Proactive adjustments to meet quality standards and minimize substandard production.
Integration of Post-Processing Steps	Integration of post-processing steps into the AM workflow, from design to final part finishing, streamlining operations and minimizing disruptions. Holistic approaches incorporating post-processing considerations throughout the manufacturing cycle.

high-accuracy sensors for precision, standardization and repeatability. Utilization of robotics in support removal, surface finishing and quality inspection to minimize labor intensity while ensuring consistent quality. Research into adaptive robotic systems that can handle complex geometries, various materials and multi-step operations [12].

4. **Materials Innovation:** The creation of post-processing-specific materials such as tailored abrasive media, eco-friendly support materials and advanced surface coatings. Novel chemical compositions are investigated for better surface finishes, efficient support removal and a lesser environmental impact.
5. **Sustainability and Environmental Focus:** Emphasis on environment-friendly post-processing solutions in the form of sustainable materials and processes with minimal ecological footprint. Eco-friendly practices adopted to address environmental challenges in AM post-processing.
6. **Advanced Monitoring and Predictive Analytics:** Real-time monitoring system with advanced sensors tracking critical parameters to detect defects and make process parameters adjustments dynamically. Integration of optimization algorithms with predictive analytics for enhancing the accuracy, consistency and error reduction. Developing predictive models with historical data and real-time feedback to increase process efficiency.
7. **Standardization and Qualification Procedures:** Developing industry-wide standards for post-processing steps across the different AM technologies and materials. Qualification methods should be developed to ensure that post-processed parts maintain their integrity, performance and durability to allow for wider industrial use [13].
8. **Customized Solutions for Complex Geometries and Multi-Step Processing:** Research in dedicated systems and algorithms for the accommodation of complex geometries and post-processing requirements. Development of multifunctional systems to satisfy a large variety of post-processing functionalities in one flow.
9. **Quality Control and Inspection:** In-line inspection techniques to be introduced to monitor quality continuously during post-processing. Advanced systems in real-time decision-making to ensure conformance to quality standards and reduce chances of errors [14].
10. **Collaborative Research and Development:** Further research into future technologies, new materials and efficient processes to overcome issues arising during post-processing. Industry academia collaboration to perfect techniques, enhance robustness and make production eco-friendly and efficient.
11. **Simulation:** Numerical simulations to understand the material's thermodynamic and kinetic behavior during post-processing. Optimization of process parameters to predict potential problems and further guide experimentation and production.
12. **Multifunctionality:** Development of new surface properties like super hydrophobicity and biocompatibility in addition to the more traditional improvements such as hardness, elasticity and wear resistance. Design of advanced functional properties for particular applications, such as medical devices or aerospace components.
13. **Repeatability:** Development of quantitative data systems that correlate process parameters with performance indicators to ensure quality consistency in large-scale production. Improved repeatability for industrial-scale applications of AM components.
14. **Greenification:** Use of environmentally friendly chemicals and materials to minimize potential health and environmental risks. Concentrate on energy-saving equipment, low-energy consumption processes and clean production methods to minimize any form of environmental impact.

6. Conclusions

Post-processing is one of the most critical enablers for additive manufacturing to deliver high-quality, functional components. Consistency, labor-intensive processes, cost, material-specific issues, scaling up, process variability, workflow integration, trade-off between quality and efficiency, ecological considerations, defect mitigation, integration of post processing, energy intensity, waste material, limited functionality and lack of standardization are the major challenges in post processing. Innovations discussed in the paper can drive improvements in performance, scalability and efficiency. Tailored approaches for different AM technologies and materials are essential to meet the diverse requirements of applications. In the future, integrating advanced post-processing within AM workflows will play a crucial role in cutting production times and costs while enhancing quality and accelerating adoption of AM across industries.

Acknowledgment

This work was partially funded by the Center for Research at Chennai Institute of Technology, India, under funding number CIT/CFR/2024/RP/001.

DATA CONFIDENTIALITY AND INTEGRITY

All research outputs from this program will be securely archived within the institutional repository to ensure confidentiality, data integrity, and adherence to ethical guidelines.

REFERENCES

1. Gao, Wei, Yunbo Zhang, Devarajan Ramanujan, Karthik Ramani, Yong Chen, Christopher B. Williams, Charlie CL Wang, Yung C. Shin, Song Zhang, and Pablo D. Zavattieri. "The status, challenges, and future of additive manufacturing in engineering." *Computer-aided design* 69 (2015): 65–89.
2. Kellens, Karel, Martin Baumers, Timothy G. Gutowski, William Flanagan, Reid Lifset, and Joost R. Duflou. "Environmental dimensions of additive manufacturing: mapping application domains and their environmental implications." *Journal of Industrial Ecology* 21, no. S1 (2017): S49–S68.
3. Oliveira, J. P., T. G. Santos, and R. M. Miranda. "Revisiting fundamental welding concepts to improve additive manufacturing: From theory to practice." *Progress in Materials Science* 107 (2020): 100590.
4. Zhou, Longfei, Jenna Miller, Jeremiah Vezza, Maksim Mayster, Muhammad Raffay, Quentin Justice, Zainab Al Tamimi, Gavyn Hansotte, Lavanya Devi Sunkara, and Jessica Bernat. "Additive Manufacturing: A Comprehensive Review." *Sensors* 24, no. 9 (2024): 2668.
5. Kantaros, Antreas, Theodore Ganetsos, Florian Ion Tiberiu Petrescu, Liviu Marian Ungureanu, and Iulian Sorin Munteanu. "Post-Production Finishing Processes Utilized in 3D Printing Technologies." *Processes* 12, no. 3 (2024): 595.
6. Peng, Xing, Lingbao Kong, Jerry Ying Hsi Fuh, and Hao Wang. "A review of post-processing technologies in additive manufacturing." *Journal of Manufacturing and Materials Processing* 5, no. 2 (2021): 38.
7. Mahmood, Muhammad Arif, Diana Chioibasu, Asif Ur Rehman, Sabin Mihai, and Andrei C. Popescu. "Post-processing techniques to enhance the quality of metallic parts produced by additive manufacturing." Metals 12, no. 1 (2022): 77.
8. Khosravani, Mohammad Reza, Majid R. Ayatollahi, and Tamara Reinicke. "Effects of post-processing techniques on the mechanical characterization of additively manufactured parts." Journal of Manufacturing Processes 107 (2023): 98–114.
9. Bankong, B. D., T. E. Abioye, T. O. Olugbade, H. Zuhailawati, O. O. Gbadeyan, and T. I. Ogedengbe. "Review of post-processing methods for high-quality wire arc additive manufacturing." Materials Science and Technology 39, no. 2 (2023): 129–146.
10. Liu, Yang, Xinyu Liu, Jinzhong Lu, Kaiyu Luo, Zhaoyang Zhang, Haifei Lu, Hongmei Zhang, Xiang Xu, Yufeng Wang, and Siyu Zhou. "Post-treatment technologies for high-speed additive manufacturing: Status, challenge and tendency." Journal of Materials Research and Technology (2024).
11. Singh, Sunpreet, Seeram Ramakrishna, and Rupinder Singh. "Material issues in additive manufacturing: A review." Journal of Manufacturing Processes 25 (2017): 185–200.
12. Pires, J. Norberto, Amin S. Azar, Filipe Nogueira, Carlos Ye Zhu, Ricardo Branco, and Trayana Tankova. "The role of robotics in additive manufacturing: review of the AM processes and introduction of an intelligent system." Industrial Robot: the international journal of robotics research and application 49, no. 2 (2022): 311–331.
13. Seifi, Mohsen, Michael Gorelik, Jess Waller, Nik Hrabe, Nima Shamsaei, Steve Daniewicz, and John J. Lewandowski. "Progress towards metal additive manufacturing standardization to support qualification and certification." Jom 69 (2017): 439–455.
14. Hassen, Ahmed A., and Michael M. Kirka. "Additive Manufacturing: The rise of a technology and the need for quality control and inspection techniques." Materials Evaluation 76, no. 4 (2018).

Advances in Additive Manufacturing Technologies – Gurusamy Pathinettampadian et al. (eds)

Experimental Analysis of Performance Analysis of Solar Bubble Dryer for Agricultural Products–Social and Economic Approach

Anto Beaula S.[1],
Poojasri R.[2], **Harshini A.**[3],
Sudharsan S.[4], **Krishnakumar U.**[5], **Sathiyamoorthi R.**[6]
Department of Mechanical Engineering, Chennai Institute of Technology,
Chennai, India

◆ **Abstract:** The Solar Dryer work aims to address the challenges associated with traditional agricultural product drying methods by introducing an innovative and sustainable solution that harnesses solar energy and employs bubble technology. Agricultural communities, particularly in developing regions, often face difficulties in preserving their crops due to inadequate drying methods, leading to post-harvest losses and compromised product quality. After 5 hours of drying, 500 g of onions were still present, and 36.5% of their moisture content had been eliminated. The following day, 325g of onion mass remained after same 500g onions were dried again for 4 hours. This means that 41.6% of the liquid in the bulk of the onions was gone. Based on the results of the current study, it can be said that, in comparison to the conventional drying method, the solar bubble drier greatly increases the rate at which paddy dries and gives the finished product better drying qualities.

◆ **Keywords:** Solar bubble dryer, Solar intensity, Solar irradiation, Agriculture

1. Introduction

Dietary needs have increased as a result of urbanization and the world's population growth. Fruits and vegetables spoil due to inadequate a storage and preservation facility, which lower the amount of food available and drives up prices significantly [1]. To increase their shelf life and keep their quality for a longer period of time, fruits and vegetables need to be preserved. One of the oldest methods of food preservation and storage is drying, which is frequently accomplished by subjecting food to direct sunlight [2-3]. One well-known source of clean power that has been increasingly popular in recent decades is solar energy. The absence of petroleum and petroleum products and the harm they cause to the environment are to blame for this growth. Because it is free, environmentally friendly, and available all year round, solar energy is the most abundant renewable energy source [4-5].

The most common and simple use of this resource is the transformation of solar energy into heat. By lowering the moisture content, this method efficiently lowers the product's weight, producing dry food that is easy to package, store, and transport [6]. Since water plays a major role in starting the physiological and chemistry processes that cause food to decay, this method of removing water from fruits and vegetables allows them to be preserved for longer periods of time. This age-old method of preserving food by sun drying is efficient, sustainable, and effective [7]. However, there are a number of acknowledged limitations to sun drying for larger-scale production, such as the crop's susceptibility to animal and bird damage and

[1]antobeaulas.mech2022@citchennai.net, [2]poojasrir.mech2022@citchennai.net, [3]harshinia.mech2022@citchennai.net [4]sudharshans.mech2022@citchennai.net, [5]krishnakumaru.mech2022@citchennai.net, [6]sathiya.ram78@gmail.com

DOI: 10.1201/9781003685906-42

the degradation of quality brought on by direct prolonged exposure to sunlight [8-9].

In the hours of maximum sunlight, PCM melts as a result of a phase change caused by heat absorption,absorbing heat, resulting in its melting during the hours of maximum sunlight. Subsequently, when there is a decrease in sunshine, the phase change material (PCM) discharges the accumulated heat, facilitating uninterrupted product drying. Mohammed et al. [10] have examined the optimization of solar food dryer having various air heater configurations. Experimental results indicate that the design of the absorbent plate has a direct impact on the temperature of the exit air. The addition of fins and tint to Model 1 increases the efficiency of Models 2, 3, and 4 by 20%, 40%, and 65% respectively.

Jain et al. [11] have analyzed an increment in thermal characteristics of a solar dryer. Reflectors can serve as protective covers to prolong the lifespan of the material covering against rainstorms, hailstones, and other adverse weather conditions, especially while the dryer is not in use. The dryer reached a maximum temperature of 77.4 °C and 61.2 °C when reflectors were used and not used, respectively, under no load conditions. Sengar et al. [12] have applied drying kinetics. A solar dryer system of the induced kind has been constructed and studies have been carried out using potatoes as food samples. Monitoring was conducted on the weight loss and temperature fluctuations in all three categories of meal samples. The dryer's drying efficiency was determined to be 20.3% on average. Additionally, SEM research was conducted to investigate the surface morphology of the solar-dried food samples. The sphere-shaped sample has exhibited intriguing findings and possesses the ability to reach the highest temperature due to its minimal surface area compared to the other two types. The constructed experimental apparatus has an initial expenditure of $205.78 and a more favorable return on investment term of 1.50 years [13].

A solar tunnel dryer is a UV-stabilized structure made of sheets of polythene that is shaped like a tunnel and used to help dry industrial and agricultural materials. Ennissioui et al. [14] have studied a natural convection indirect solar dryer. The SAC and the dryer had mean efficiencies of 23.37% and 18.8%, respectively. ISD demonstrated superior moisture removal compared to open solar drying (OSD), attaining a significant reduction of 74.83%, while OSD only achieved a reduction of 39.08%. The effectiveness of natural convection in a cabinet-style sun drier for drying gooseberries has been investigated by Prajapati et al. [15]. The results show that using a sun dryer to dry gooseberries effectively and superiorly can result in higher-quality products. Solar dryer performance has been enhanced by Thanompongchart et al. [16] through automatic control of additional heated air. The investigations showed that the drying time could be reduced from 32 hours to 12.5 hours through the integration of the solar dryer using an automatic secondary heating system, as opposed to using just the solar dryer. The dried pineapple goods exhibited excellent quality. The examination of CIELAB colour characteristics revealed a low level of pigmentation. Based on a basic economic study, the return from an investment was less than 3.0 months. By extending the sun collecting surface from 6 to 15 m2, the drying period of wheat without storage was lowered by almost 50%, however this led to excessively high temperatures. When the collecting area is the same, the employment of latent heat storage resulted in a change in drying time ranging from a decrease of 5% to an increase of 13.9%, depending on the sizes of the components [17-18].

In the years that followed, numerous research works were launched, each of which offered alternate drying procedures; however, the majority of the time, the implementation of those technologies was hindered by the amount of energy consumed and the loading capacity. Developing a solar bubble dryer to dry paddy in tropical climates and comparing its performance to traditional drying methods with respect to of operational parameters, product drying, and final quality—including drying efficiency and economic recovery—were the goals of this study.

2. Methods and Materials

2.1 Design of Solar Bubble Dryer (SBD)

The work team had collectively created a cutting-edge drying technique known as the solar bubble dryer (SBD), which is a not expensive model. The primary component of this structure is a drying chamber, thats is constructed with a sheet of UV-stabilized polyethylene (PE) as the glazing layer and a sheet of black polythene at the bottom. Immediately following the spreading of the materials, a zipper is utilized in order to seam the glazing components on the entire sides. The ventilators are assembled with the assistance of a bar frame made of aluminium that can be collapsed. All during the day, the SBD should be uncovered to the sun; therefore, it must be put up in a location that is free of any structures, which could potentially create shade at some point during the day. A breakdown of the many elements that make up the solar bubble drier is presented in Fig. 10.1. With the assistance of a data recorder, the measurements on the air characteristics were taken, and these observations were subsequently uploaded to a computer for the statistical analysis.

Fig. 42.1 Solar bubble dryer initial and working stages

During the morning, the trials on drying were initiated, and the grains that had been collected the day before were placed in the dryers. In the current experiments, the grain was distributed evenly on the bottom, which is the black sheet. For the purpose of temperature recording, one data logger was installed. After that, the cover was expanded and then zipped, and the drying process was initiated. The data that was collected by the data loggers was then transmitted to a computer after the drying process was completed. Extra caution was taken to ensure that the bubble dryer unit did not have any shade on it. In addition, A data logger was used to help record the temperature and relative humidity readings of the environment around them at hourly intervals.

At midday, the solar dryer system should be facing the sun. Additionally, the solar collector was made to be angled at 45°C in order to absorb and gather as much solar radiation as possible. After that, an exhaust opening will allow the air to exit the drying chamber. The products will be continuously dried during the hours when the sun isn't shining thanks to the latent heat that the paraffin wax absorbs. By doing this, the drying period will be shortened and the greatest amount of solar heat energy will be employed.

3. Experimental Procedure

For the purpose of evaluating the performance of SBD, it was put through its paces on sunny days having a solar tunnel drier, both without load and with full load. These tests were carried out with the purpose of ascertain the peak temperature that could be reached within the drier, the rate at which the drying process occurred, and the drying efficiency. To ensure that it received the most possible amount of solar energy throughout the day, the dryer was positioned so that it faced away from the north. In the morning, between the hours of 9:00 and 11:00 a.m., the collector that was facing northeast was utilized, and in the evening, between 15:00 and 16:00 a.m., the collector that was facing west was utilized. While the experiment was being conducted, the product that had been moderately dried was placed in gunny baggage and stored inside airless compartment. The gunny bags were then utilized the following day to continue drying until the desired constant weight was achieved. A number of characteristics, such as the amount that the moisture was extracted and the effectiveness of drying, were calculated using the following formula.

4. Results and Discussion

4.1 Drying Time vs Moisture Content – Comparison

One kilogram of onions was put in the chamber used for drying to dry on the initial day of the experiment, and the 12 volt DC fan had been switched on to force air through the solar drier. Throughout the day, the thermocouples measured the temperature, and a data collecting device was used to store the data. The same characteristics for the first day are shown in Fig. 42.1. On both days, the total amount of moisture extracted was 71% of the onion's initial mass. 500 g of onions remained after 5 hours of drying, and 36.5% of their moisture content had been removed. These 500g onions were dried once more for 4 hours the next day, leaving 325g of onion mass. This indicates that 41.6% of the liquid in the onions' remaining bulk was lost. The measurements of temperature and solar radiation displayed an initial day of onion drying are shown in Fig. 42.2, On both days, the total amount of moisture extracted from the onions was 71.6% of their initial mass.

Table 42.1 Cost analysis of drying paddy in solar bubble dryer

S. No	Particular	Unit	Value
	Cost of the SBD unit	USD	1250
	Loading capacity	Kg per batch	425
	No. of sunny days per year	Days	200
	No. of batches per year	-	200
	Quantity of paddy dried per year	Kg	85,000

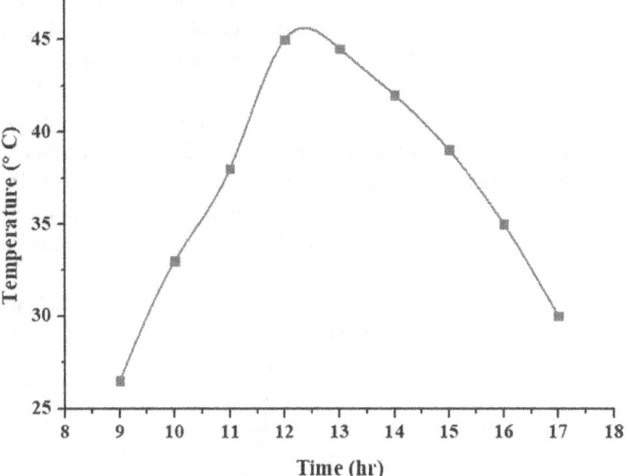

Fig. 42.2 Time vs temperature (hour wise)

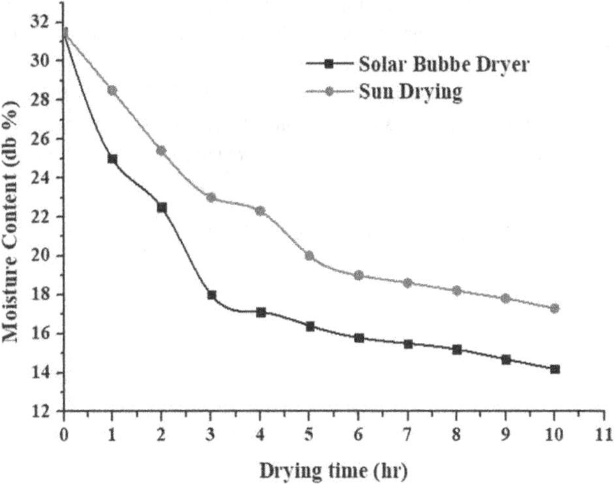

Fig. 42.3 Drying time vs moisture content – comparison

4.2 Thermal Variation in the Vegetable Samples

The amount of energy of the solar radiation the fact that strikes the dryer's solar collector surface determines the thermal variation with regard to of temperature variation. All of the samples received heat after being placed within the solar drying chamber. The dryer's received heat directly correlates with the temperature of the food ingredients. All samples' temperatures have increased progressively till they reach their maximum values.

4.3 Drying Efficiency

A solar bubble type solar dryer's sustainability is evaluated using drying efficiency. When analyzing the heat loss from the manufactured dryer, this parameter is crucial. It displays the drying efficiency values for three distinct food samples on an hourly basis. The first example shows that slab samples used the most heat and had a higher efficiency of 19.2%.

5. Conclusion

- A digital grain analyzer was used to determine the moisture content. The temperature could be kept more constant in a solar bubble dryer, and the quantity of solar radiation present had a significant impact on how much the temperature rose.
- It was found that the SBD unit had a better rate of moisture removal on an hourly basis compared to the standard solar drying technique.
- Solar bubble dryers, like any other piece of equipment, need to be maintained on a regular basis to ensure that they are operating properly. Additionally, Compared to other conventional drying methods, setting up a solar bubble drier could be more expensive initially.
- The requirement for environmentally friendly alternatives is growing on a global scale, and solar bubble dryer has a vital part to take part in the prospect of agriculture. Solar bubble dryers have the potential to be significantly enhanced in terms of their capacity or capacity, efficiency, and price as technology continues to evolve.

References

1. Kong, D., Wang, Y., Li, M. and Liang, J., 2024. A comprehensive review of hybrid solar dryers integrated with auxiliary energy and units for agricultural products. *Energy*, 293, p.130640.
2. Goel, V., Dwivedi, A., Mehra, K.S., Pathak, S.K., Tyagi, V.V., Bhattacharyya, S. and Pandey, A.K., 2024. Solar drying systems for Domestic/Industrial Purposes: A State-of-Art review on topical progress and feasibility assessments. *Solar Energy*, 267, p.112210.
3. Kherrafi, M.A., Benseddik, A., Saim, R., Bouregueba, A., Badji, A., Nettari, C. and Hasrane, I., 2024. Advancements in solar drying technologies: Design variations, hybrid systems, storage materials and numerical analysis: A review. *Solar Energy*, 270, p.112383.
4. Sharshir, S.W., Joseph, A., Elsayad, M.M., Hamed, M.H. and Kandeal, A.W., 2024. Thermo-enviroeconomic assessment of a solar dryer of two various commodities. *Energy*, 295, p.130952.
5. Tyagi, V.V., Pathak, S.K., Chopra, K., Saxena, A., Dwivedi, A., Goel, V., Sharma, R.K., Agrawal, R., Kandil, A.A., Awad, M.M. and Kothari, R., 2024. Sustainable growth of solar drying technologies: Advancing the use of thermal energy storage for domestic and industrial applications. *Journal of Energy Storage*, 99, p.113320.

6. Pandey, S., Kumar, A. and Sharma, A., 2024. Sustainable solar drying: Recent advances in materials, innovative designs, mathematical modeling, and energy storage solutions. *Energy*, p.132725.
7. Arunkumar, P.M., Balaji, N. and Madhankumar, S., 2024. Performance analysis of indirect solar dryer with natural heat energy retention substances for drying red chilli. *Sustainable Energy Technologies and Assessments*, 64, p.103706.
8. Kumar, S., Ghritlahre, H.K., Agrawal, S. and Shekhar, S., 2024. Investigation of a novel mixed-mode solar dryer using north wall reflector: An experimental study. *Solar Energy*, 282, p.112909.
9. Rehman, T.U., Nguyen, D.D. and Sajawal, M., 2024. Smart optimization and investigation of a PCMs-filled helical finned-tubes double-pass solar air heater: an experimental data-driven deep learning approach. *Thermal Science and Engineering Progress*, 49, p.102433.
10. Mohammed, S.A., Alawee, W.H., Chaichan, M.T., Abdul-Zahra, A.S., Fayad, M.A. and Aljuwaya, T.M., 2024. Optimized solar food dryer with varied air heater designs. *Case Studies in Thermal Engineering*, 53, p.103961.
11. Vedantam, S.K., Jain, D.S. and Panwar, D.N.L., An Overview on Phase Change Material Incorporated in Convective Solar Dryers. Available at SSRN 5010425.
12. Sengar, M., Singh, D. and Mishra, P.K., 2024. Performance evaluation of solar dryer integrated with solar panel for drying of carrot (Daucuscarota) vegetable. *Indian Chemical Engineer*, pp.1-12.
13. Sarmah, A., Dey, P.C., Chetia, S.K., Medhi, A.K., Jyoti, M., Konwar, S.R.B., Bharali, A., Baruah, M., Gummert, M. and Khandai, S., 2024. Effect of different storage conditions on grain quality of paddy.*Oryza* Vol. 61 Issue 1 2024 (1-11)
14. Hin, L., Mean, C.M., Kim, M.C., Chhoem, C., Bunthong, B., Lor, L., Sourn, T. and Prasad, P.V., 2024. Development and Performance Assessment of Sensor-Mounted Solar Dryer for Micro-Climatic Modeling and Optimization of Dried Fish Quality in Cambodia. *Clean Technologies*, 6(3), pp.954-972.
15. Prajapati, C. and Sheorey, T., 2025. Performance Analysis of Novel Solar Dryer with Optimal Tilt Angle. *Journal of Solar Energy Engineering*, pp.1-35.
16. Chokngamvong, S. and Suvanjumrat, C., 2024. Development of conjugate heat-and moisture-transfer model for pineapple drying using OpenFOAM. *Case Studies in Thermal Engineering*, 60, p.104770.
17. Zeeshan, M., Tufail, I., Khan, S., Khan, I., Ayuob, S., Mohamed, A. and Chauhdary, S.T., 2024. Novel design and performance evaluation of an indirectly forced convection desiccant integrated solar dryer for drying tomatoes in Pakistan. *Heliyon*, 10(8).
18. Román, F., Munir, Z. and Hensel, O., 2024. Performance comparison of a fixed-bed solar grain dryer with and without latent heat storage. *Energy Conversion and Management: X*, 22, p.100600.

Note: All the figures and tables in this chapter were made by the authors.

Advances in Additive Manufacturing Technologies – Gurusamy Pathinettampadian et al. (eds)
© 2026 Taylor & Francis Group, London, ISBN 978-1-041-16687-0

Experimental Study on Microstructure and Mechanical Properties of AZ31B Magnesium Alloy by Friction Stir Welding

R. Ganesamoorthy[1]
Department of Mechanical Engineering,
Chennai Institute of Technology, Chennai, Tamilnadu, India

Sathishkumar M.
Department of Mechanical Engineering,
Chennai Institute of Technology, Chennai, Tamilnadu, India

K. R. Padmavathi[2]
Department of Mechanical Engineering,
Panimalar Engineering College, Chennai, India

T. Kavitha[3]
Department of Electronics and Communication Engineering,
Vel Tech Rangarajan Dr. Sagunthala R & D Institute of Science and Technology

P. Balamurali[4]
Department of Mechanical Engineering,
Chennai Institute of Technology, Chennai, India

A. Parthiban[5]
Department of Mechanical Engineering,
Chennai Institute of Technology, Chennai, Tamilnadu, India

◆ **Abstract:** The aim of this investigation is to apply the concept of Friction Stir welding technique to Magnesium Alloy in varying weld parameters to assess the mechanical properties of the weld. For this Experiment, AZ31B Magnesium Alloy has been selected. Welding is performed for three Samples At 3 Different weld parameters using a frustum tool Profile. Properties such as tensile strength, hardness and micro examination of three samples were compared and observed. The study discovered that a welding speed of 1400 rpm resulted in smaller grain sizes in the nugget region while maintaining good hardness and yield strength, when compared to samples with tool rotational speeds of 940 and 1040 RPM's. As a result, 1400 rpm with a feed rate of 20mm/min was identified as the optimal speed for achieving both good hardness and yield strength.

◆ **Keywords:** Friction stir welding, Cylindrical tool profile, Magnesium alloy, Mechanical properties

1. Introduction

In the modern manufacturing landscape, the need for lightweight yet strong materials has become increasingly critical, particularly in industries such as aerospace, automotive, and defense. Magnesium alloys, particularly AZ31B, have gained significant attention due to their excellent strength-to-weight ratio, good machinability,

Corresponding authors: [1]ganesamoorthyr@citchennai.net, [2]krpadmavathipecmech@gmail.com, [3]kavithaecephd@gmail.com, [4]balamuralip@citchennai.net, [5]parthibana@citchennai.net

DOI: 10.1201/9781003685906-43

and potential to reduce the overall weight of components. However, the inherent challenges in welding magnesium alloys— such as cracking, porosity, and reduced mechanical properties—have hindered their widespread use. Traditional fusion welding techniques often exacerbate these issues due to high heat input, leading to material degradation and compromised joint quality.

Friction Stir Welding (FSW), a solid-state joining process, has emerged as a highly effective solution for welding magnesium alloys. Developed in 1991 by The Welding Institute (TWI), FSW offers several advantages over conventional fusion welding techniques, including reduced material distortion, elimination of filler materials, and the ability to join difficult-to-weld alloys without melting. In FSW, a non-consumable rotating tool is used to generate heat through friction, softening the material and stirring it together to form a strong joint.

This study focuses on the application of FSW to AZ31B magnesium alloy, evaluating the effect of varying tool rotational speeds on the mechanical properties of the weld. By examining parameters such as tensile strength, hardness, and microstructure, this research aims to identify the optimal conditions for achieving superior weld quality. Understanding these effects is crucial for enhancing the performance and durability of components made from magnesium alloys, making them more viable for high-performance applications.

Workflow Process: The process flowchart for the Friction Stir Welding (FSW) of AZ31B magnesium alloy involves several key steps, starting with the selection and preparation of the AZ31B plates, which are typically 6 mm thick and dimensioned at 100 mm x 50 mm. The plates undergo surface cleaning and alignment to ensure proper contact and avoid contamination during welding. Next, an appropriate High-Speed Steel (HSS) tool with a specific pin profile is selected and installed on a vertical milling machine. The machine is then set to the required welding parameters, with spindle rotational speeds of 940 RPM, 1040 RPM, and 1400 RPM used for different samples, and feed rates of 40 mm/min for two samples and 20 mm/min for one sample. During welding, the rotating tool is plunged into the joint, creating frictional heat that softens the material, allowing it to be stirred and joined without melting. The tool traverses the length of the plates, creating a solid-state weld. After welding, the tool is retracted, leaving a characteristic circular indentation at the weld end. The welded samples are then subjected to mechanical testing, including hardness, tensile strength, and microstructural analysis, to assess the quality and mechanical properties of the welds.

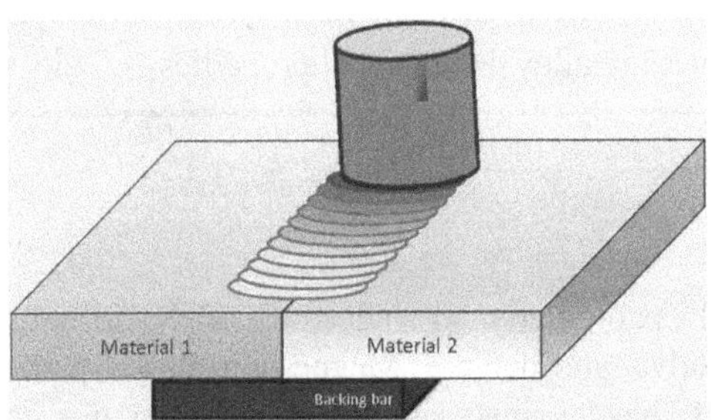

Fig. 43.1 Tool and workpieces

2. Various Zones in FSW

The weld region is divided into three distinct zones, each having unique microstructural characteristics that affect the mechanical properties of the weld. The Stir Zone (SZ), also known as the weld nugget, is located at the center of the joint where the tool pin rotates and plasticizes the material. This zone experiences intense heat and plastic deformation, resulting in fine, recrystallized grains that significantly enhance the strength and toughness of the weld. Surrounding the stir zone is the Thermo-Mechanically Affected Zone (TMAZ), where the material undergoes both thermal and mechanical stresses but without full recrystallization. The grains in the TMAZ are elongated due to deformation, and this zone often exhibits a transition in grain size and hardness. Beyond the TMAZ lies the Heat Affected Zone (HAZ), where the material is exposed to elevated temperatures but without significant plastic deformation. The grain structure in this region is largely unchanged compared to the base material, though the thermal exposure may lead to slight grain coarsening. Together, these zones define the weld integrity, with each contributing to the overall mechanical performance of the welded joint.

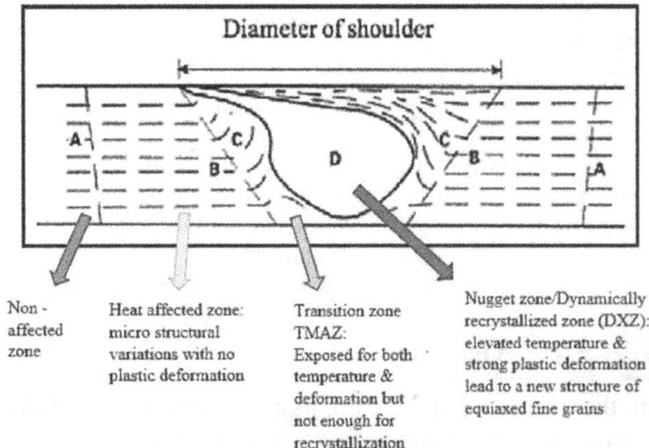

Fig. 43.2 Zones in FSW

3. Welding Samples

Sample(s) 1:

Fig. 43.3 Samples 1 & 2

Sample(s) 2:

Fig. 43.4 Samples 3 & 4

Sample(s) 3:

Fig. 43.5 Samples 5 & 6

4. Welding Methodology

The FSW process was carried out using a vertical milling machine. AZ31B magnesium alloy plates with dimensions of 100 mm length, 50 mm width, and 6 mm thickness were selected for welding. These plates were properly cleaned, aligned, and clamped onto the machine table to ensure stable positioning during the welding process.

A High-Speed Steel (HSS) tool with a specific pin profile was attached to the spindle of the milling machine. The tool rotational speed was varied for different samples, with speeds of 940 RPM, 1040 RPM, and 1400 RPM. The feed rate, or travel speed of the tool, was set to 40 mm/min for two of the samples and 20 mm/min for one sample.

4.1 Welding Process

The FSW tool was slowly plunged into the joint between the two plates, penetrating to a depth of 5.5 mm to ensure proper stirring of the material in the weld zone. As the tool made contact with the material, the friction generated heat that softened the magnesium alloy without melting it, creating a plasticized zone around the tool.

Once the tool reached the desired depth, the milling machine initiated its table feed, causing the tool to traverse along the length of the joint. As the rotating tool moved forward, it stirred the softened material, mixing the two plates and forming a solid-state weld. The process was carried out in a controlled manner to ensure a consistent weld along the joint.

After completing the weld, the tool was retracted, leaving a characteristic circular indentation at the end of the welded joint. The welded samples were visually inspected to ensure proper fusion, and the finished welds were prepared for subsequent mechanical testing, including tensile tests, hardness tests, and microstructural analysis.

This methodology ensures a strong and defect-free joint while controlling the process variables—such as tool speed and feed rate—to optimize the mechanical properties of the AZ31B magnesium alloy welds.

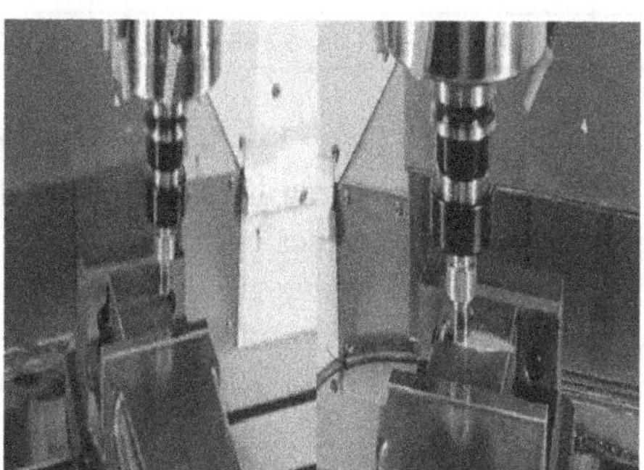

Fig. 43.6 Tool plunging and tool traverse operation

4.2 Samples After Welding

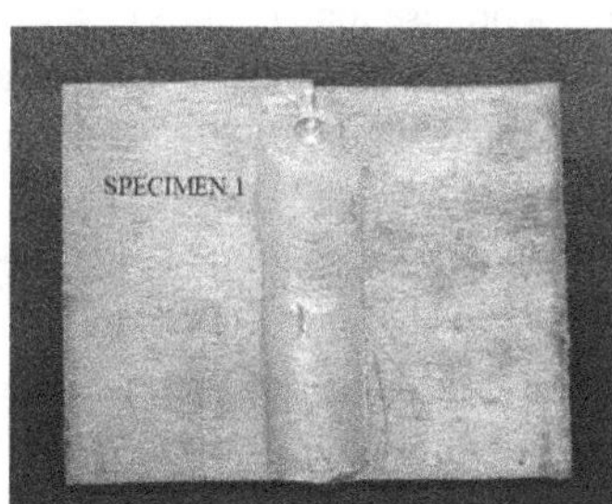

Fig. 43.7 Sample specimen 1 after welding

Fig. 43.8 Sample specimen 2 after welding

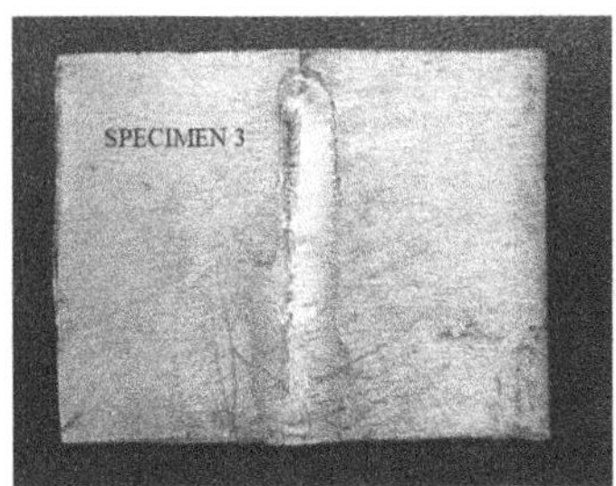

Fig. 43.9 Sample specimen 3 after welding

Testing Methods, Specimen Preperation, Procedure and Apparatus for the Testing:

Table 43.1 Micro vickers hardness test

Locations	Hardness value in HV100gm Sample 1	Hardness value in HV100gm Sample 2	Hardness value in HV100gm Sample 3
Base	62,59,58	61,59,58	63,59,58
HAZ	66,64,61	68,66,63	70,68,65
Weld	53,52,50	57,55,56	63,60,62

The Micro Vickers Hardness Test was employed in the report to measure the localized hardness of the different regions in the AZ31B magnesium alloy welds created through Friction Stir Welding (FSW). This test is an effective method for evaluating the material's resistance to indentation and helps in assessing the microstructural changes in the welded material.

5. Testing Procedure

The Micro Vickers Hardness Test uses a diamond pyramid indenter to apply a precise load (in this case, 100 grams) on the surface of the welded samples. The test was performed at various points across the weld, including the Base Material (BM), Heat-Affected Zone (HAZ), and the Stir Zone (SZ), to analyze the variations in hardness due to the welding process. After the indentation, the diagonal length of the indentation was measured, and the hardness was calculated using the following formula:

$$HV = \frac{1.854 \times F}{d^2}$$

Where F is the applied load in kilograms and d is the diagonal length of the indentation in millimeters.

5.1 Results of the Micro Vickers Hardness Test

The test results revealed clear differences in hardness between the base material, the HAZ, and the weld nugget (stir zone).

Base Material (BM): The base material exhibited relatively consistent hardness values, around 58-63 HV, which indicates minimal changes in hardness from the welding process in this region.

Heat-Affected Zone (HAZ): The HAZ showed slightly higher hardness values compared to the base material, ranging from 61-70 HV. This increase is attributed to the thermal effects that alter the grain structure in this zone.

Stir Zone (SZ): The weld nugget or stir zone exhibited the most significant variations in hardness. At lower tool rotational speeds (940 RPM), hardness values were lower (50-53 HV), while at higher rotational speeds (1400 RPM), hardness increased to 60-63 HV. The finer grain structure produced at higher speeds due to enhanced plastic deformation and heat generation led to increased hardness in the stir zone.

Effect of Rotational Speed and Tool Material on Mechanical Properties of Welded AZ31B Alloy using HSS Tool:

Table 43.2 Effect of rotational speed and tool material on mechanical properties of welded AZ31B alloy using HSS tool

Joint No.	Rotational Speed (RPM)	UTS (MPa)	Yield Strength (Mpa)	Elong ation (%)
1	940	129.77	94.51	2.48
2	1040	135.15	100.71	3.32
3	1400	186.76	139.1	5.00
Base Material	-	215	171	14.7

As the rotational speed of the joints increases from 940 RPM to 1400 RPM, both the Ultimate Tensile Strength (UTS) and Yield Strength significantly improve, with UTS rising from 129.77 MPa to 186.76 MPa (a 43.7% increase) and Yield Strength increasing from 94.51 MPa to 139.1 MPa (a 47.1% increase). This trend indicates enhanced material performance under dynamic loads, potentially due to work hardening or other strengthening mechanisms. Furthermore, elongation also increases from 2.48% to 5.00%, suggesting that the joints not only become stronger but also more ductile, allowing for some deformation before failure. In comparison to the base material, which has a UTS of 215 MPa and a Yield Strength of 171 MPa, the joints demonstrate improved properties that may be suitable for high-stress applications.

6. Microstructure Analysis

The optical micrographs depicting the stir zone of the FSW joints are presented below. The micrographs reveal that there is a noticeable variation in the average grain diameter of the weld region in AZ31B Magnesium alloy. The coarse grains of the base metal are transformed into fine grains in the stir zone due to FSW. Among all the joints, the ones fabricated with a rotational speed of 1400 rpm, constant welding speed of 20 mm/min, and SS tool display finer grains in the weld region, which is one of the factors contributing to their higher tensile properties. It can be inferred from the micrographs that there is a significant variation in grain size across the welds due to insufficient plastic flow and thermal exposure.

7. Specimen Preparation for Taking Microstructure

To prepare for metallographic examination, specimens were cut to the required size and polished using various grades of emery papers. A standard reagent composed of 4.2 g picric acid, 10 ml acetic acid, 10 ml diluted water, and 70 ml ethanol was utilized to expose the microstructure of the welded joints. Microstructural analysis was conducted using a Metzer-M binocular microscope model METZ-57, equipped with high- magnification imaging and image analysis capabilities to estimate the weight percentage of elements.

Microstructure Images of Friction Stir Welded Plate (Magnesium AZ31B)

Fig. 43.10 Microstructure of base metal

Table 43.3 Grain size and area for samples 1, 2, 3

	Grain Size (µm)	Grain Area (µm^2)
Samples 1,2,3	20.53	186.56

7.1 Heat Affected Zone

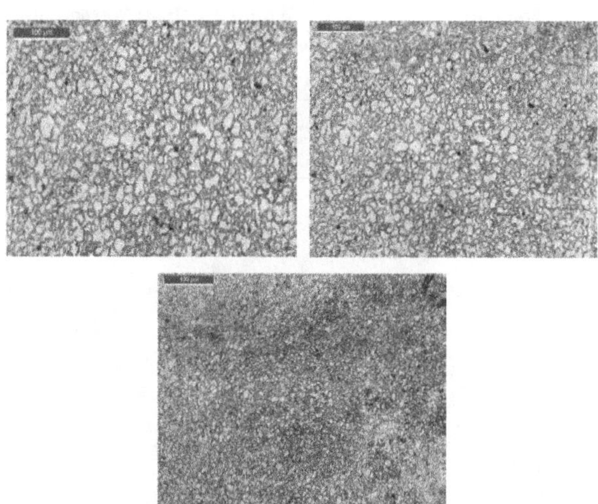

Fig. 43.11 HAZ on Samples 1,2 and 3

Table 43.4 Grain size and grain surface area for samples 1 (S 940, F40), Sample 2 (S 1040, F40) and Sample 3 (S 1400, F20)

Sample	Grain Size (µm) Avg	Grain Surface Area (µm^2) Avg
1 (S 940, F 40)	22.55	256.35
2 (S 1040, F40)	18.68	169.78
3 (S 1400, F20)	14.63	132.85

7.2 Thermo Mechanically Affected Zone

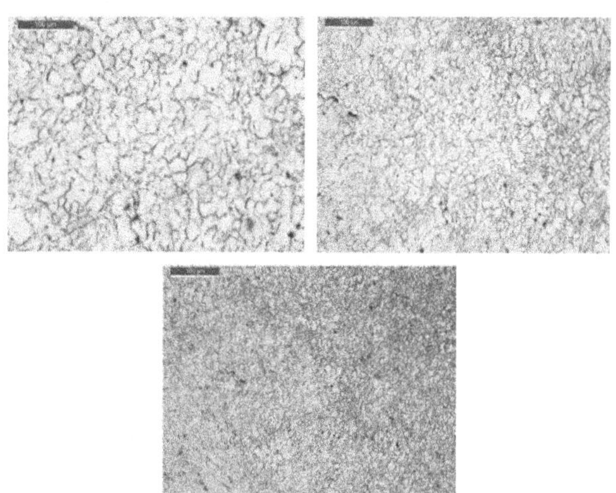

Fig. 43.12 Thermo mechanically affected zone for sample 1, 2 & 3

Table 43.5 Grain size and surface area for samples

Sample	Grain Size (μm) Avg	Grain Surface Area (μm^2) Avg
1 (S 940, F 40)	20.33	285.21
2 (S 1040, F40)	15.36	156.63
3 (S 1400, F20)	13.28	124.88

7.3 Weld Zone

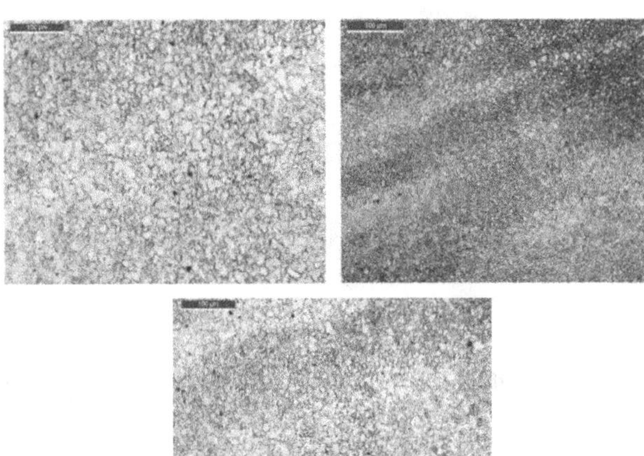

Fig. 43.13 Grain size and grain area at the HAZ region

Table 43.6 Grain size and grain surface for various samples (S 940 F40, S 1040, F40, S 1400, F20)

Sample	Grain Size (μm) Avg	Grain Surface Area (μm^2) Avg
1 (S 940, F 40)	18.65	165.36
2 (S 1040, F40)	16.65	148.45
3 (S 1400, F20)	12.52	119.45

8. Grain Size vs Hardness

One of the key microstructural features that can affect the mechanical properties of FSW joints is the grain size. The grain size in FSW joints is typically smaller than that in the base material due to the severe plastic deformation during the welding process. The grain refinement can lead to an increase in the hardness of the joint. However, the relationship between the grain size and hardness in FSW joints is complex and depends on various factors such as the welding parameters, material properties, and post-weld heat treatment.

8.1 Study Findings on Grain Size

Our experiment involved welding samples using three different RPM levels (940, 1040, and 1400). We observed the grain sizes in each case and found that the HAZ zone exhibited little variation in grain size compared to the base metal. The TMAZ and SZ zones, on the other hand, showed significant variations. Our results showed that increasing tool speed resulted in smaller grain sizes, with lower tool speeds producing larger grain sizes. We also found that there was a correlation between grain size and tool speed, and concluded that optimizing weld parameters can be achieved by adjusting tool speed to achieve the desired material hardness and ductility. Overall, our findings highlight the importance of careful selection and control of welding parameters to achieve optimal Hardness and Tensile strength without compromising any properties in FSW joints.

9. Tool Rotational Speed Vs Ultimate Tensile Strength

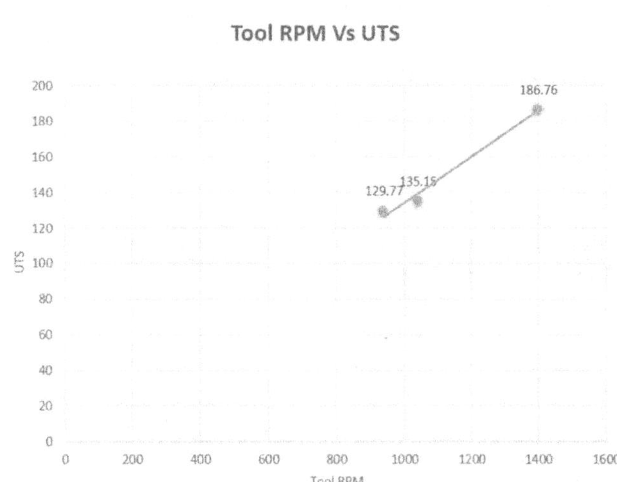

Fig. 43.14 Tool rotational speed vs ultimate tensile strength

From the data, we can observe that the UTS increases as the Tool RPM increases. For example, when the Tool RPM is 940, the UTS is 129.77. When the Tool RPM is increased to 1040, the UTS increases to 135.15, and when the Tool RPM is further increased to 1400, the UTS increases to 186.76.

This indicates a clear positive correlation between Tool RPM and UTS in friction stir welding, where increasing the Tool RPM leads to a higher UTS.

10. Microhardness Vs Tool RPM

It is Inferred that average micro hardness of the weld region is higher than that of the base material and HAZ for all three Tool RPM values. This indicates that the weld region has undergone significant plastic deformation during the friction stir welding (FSW) process, resulting in a higher

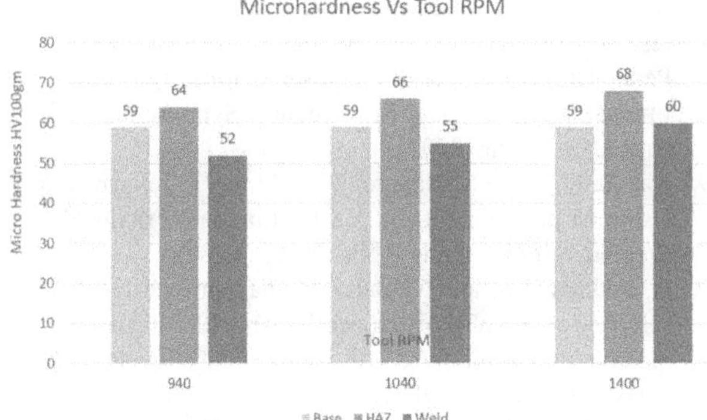

Fig. 43.15 Micro-hardness vs tool RPM

level of work hardening compared to the base material and HAZ. Furthermore, we can observe that the average micro hardness of the weld region increases as the Tool RPM increases. For example, when the Tool RPM is 940, the average micro hardness of the weld region, HAZ, and base material are all approximately the same (59). When the Tool RPM is increased to 1040, the average micro hardness of the weld region, HAZ, and base material all increase, with the highest value observed in the weld region (68). When the Tool RPM is further increased to 1400, the average micro hardness of the weld region, HAZ, and base material all increase further, again with the highest value observed in the weld region (60). This suggests that increasing the Tool RPM during FSW leads to a higher degree of plastic deformation and work hardening, resulting in a higher average micro hardness in the weld region.

11. %Elongation Vs Tool RPM

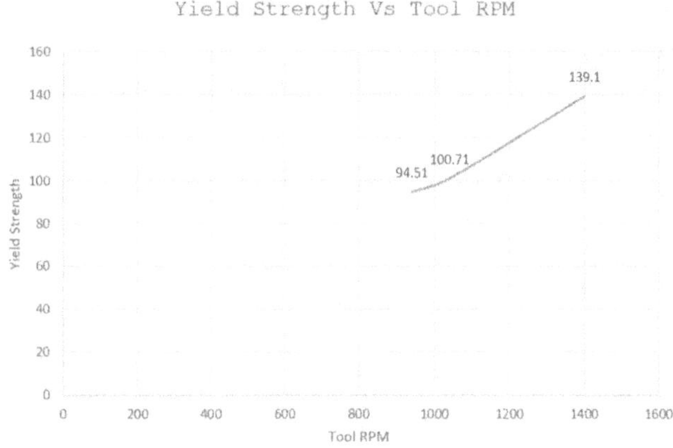

Fig. 43.16 % Elongation vs tool RPM

The given data presents a relationship between Tool RPM and Elongation in the friction stir welding (FSW) process. Elongation refers to the extent to which a material stretches or deforms before it breaks under tension.

12. Yield Strength vs Tool RPM

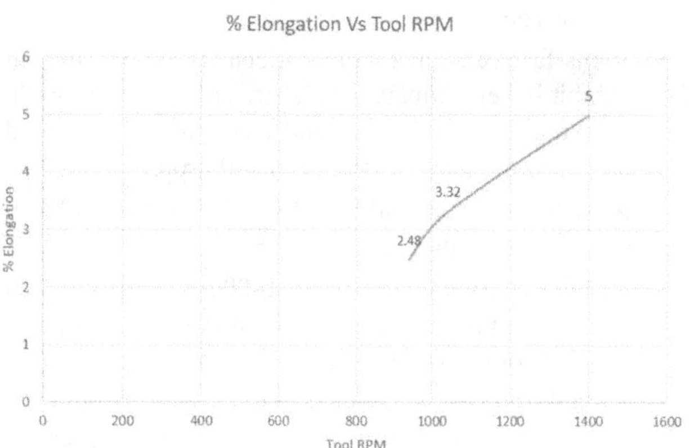

Fig. 43.17 Yield strength vs tool RPM

The given data shows the relationship between Tool RPM and Yield Strength in the friction stir welding (FSW) process. Yield strength is the stress level at which a material begins to deform plastically without any increase in load.

From the data, we can observe that the Yield Strength increases as the Tool RPM increases during FSW. For example, when the Tool RPM is 940, the Yield Strength is 94.51. When the Tool RPM is increased to 1040, the Yield Strength increases to 100.71, and when the Tool RPM is further increased to 1400, the Yield Strength increases to 139.1.

This indicates that increasing the Tool RPM during FSW leads to a higher Yield Strength, which suggests that the welds produced at higher Tool RPMs may have higher strength and be able to withstand higher loads without deforming plastically. This could be due to the fact that higher Tool RPMs generate more heat, which leads to better mixing and consolidation of the material being welded, resulting in a more uniform and stronger microstructure.

13. Conclusion

By identifying the significant parameters affecting the microstructure and properties of the Friction Stir Welding (FSW) process, it was determined that selecting the optimal process parameter can lead to higher hardness in the weld area. The rotational speed and tool material were recognized as critical factors influencing the stir zone microstructure and FSW properties. Consequently, it was concluded that using High-Speed Steel (HSS) tool material results in a finer grained microstructure and improved mechanical properties. Low rotational speeds result in higher stirred zone micro hardness values compared to base material.

- A specific combination of tool rotational speed and tool material can achieve high strength properties in the stir zone.
- Joints fabricated at a tool rotational speed of 940 rpm exhibit lower ultimate tensile strength, yield strength, percentage of elongation, and weld nugget compared to those fabricated at 1040, and 1400 rpm.
- A high tool rotational speed of 1040 rpm results in good hardness property.
- A high tool rotational speed of 1400 rpm and Tool feed rate of 20 mm/min leads to high hardness property without compromising the tensile property.
- Suitable process parameters are selected to obtain a defect-free weld zone with high hardness and tensile properties.

References

1. Salem HG, Reynolds AP, and Lyons JS investigated the microstructure and retention of superplasticity of friction stir welded superplastic 2095 sheet in a 2002 study published in Scr. Mater.
2. Nicholas ED and Thomas WM provided a review of friction processes for aerospace applications in 1998 in the journal Int.
3. Q. Yang and A.K. Ghosh conducted research on friction stir welding in Acta Materialia in 2006.
4. S. Schumann and H. Friedrich presented their work on friction stir welding in Mater. Sci. Forum in 2003.
5. Takara and K. Higashi investigated materials science forum in 2005, focusing on friction stir welding.
6. F.W. Bach, M. Rodman, M. Schaper, A. Rossberg, E. Doege, and G. Kurz published a book titled "Magnesium" in 2004 that discusses friction stir welding.
7. Nagasawa T, Otsuka M, Yokota T, and Ueki T published research on magnesium technology in TMS in 2000.
8. Park SHC, Sato YS, and Kokawa H presented at the 4th International Frictional Stir Welding Symposium in Park City, Utah in May 2003.
9. Nakata K, Inoki S, Nagaro T, Hashmito T, Johgan S, and Ushio M presented at the 3rd International Frictional Stir Welding Symposium in Kobe, Japan in 2001.
10. G. Padmanaban and V. Balasubramanian investigated the selection of FSW tool pin profile, shoulder diameter, and material for joining AZ31B magnesium alloy in a 2009 study published in Materials and Design.
11. Darras BM, Khraisheh MK, Abu-Farha FK, and Omar MA conducted friction stir processing of commercial AZ31 Magnesium alloy in a 2007 study published in J. Materials Processing Tech.
12. Heurtier P, Jones MJ, Desrayaud C, Driver JH, Montheillet F, and Allehaux D presented mechanical and thermal modeling of friction stir welding in a 2006 study published in J Mater Proecess Technol.
13. R.S. Mishra and Z.Y. Ma wrote a review of FSW and processing in Mater. Sci. Eng R in 2005.
14. A parametric study on friction stir welding of AZ61A magnesium alloy was published in ijamt, Springer in 2011.
15. Wang XH and Wang KS investigated the microstructure and properties of friction butt- welded AZ31 magnesium alloy in Mater. Sci Eng A431 in 2006.
16. Pareek M, Polar A, Rumiche F, and Inda cochea JE conducted metallurgical evaluation of AZ31B-24 Mg alloy friction stir welds in Mater Eng Perform in 2007.
17. G. Padmanaban and V. Balasubramaniam conducted an experimental investigation on friction stir welding of AZ31B magnesium alloy in IJAMT in 2010.
18. Texture Development in a Friction Stir Lap- Welded AZ31B Magnesium AlloyB.S. NAIK, D.L. CHEN, X. CAO, and P. WANJARA

Note: All the figures and tables in this chapter were made by the authors.

Advances in Additive Manufacturing Technologies – Gurusamy Pathinettampadian et al. (eds)
© 2026 Taylor & Francis Group, London, ISBN 978-1-041-16687-0

Optimizing Performance and Reducing Emissions of a Four-Stroke Motorcycle Engine with a Petrol, LPG Dual-Fuel System

Gowtham Sudarsanan,
Anandhu S. Nair, Hari Krishna Raj S.*,
Shyam Sundar K., Palani R.
Department of Mechanical Engineering,
VelTech HighTech Dr. Rangarajan Dr. Sakunthala Engineering College,
Chennai, Tamilnadu

◆ **Abstract:** This project aims to improve the efficiency of a four-stroke engine by integrating an alternative fuel source. The primary goal is to replace traditional fuels, which are becoming more difficult to locate and too expensive. When compared to traditional fuels, the project's adoption of this technique helps reduce air pollution and yields significant cost and efficiency benefits. In order to do this, the project will install an LPG (liquefied Petroleum Gas) fuel system in a four-stroke engine, enabling the use of both petrol and LPG. A detailed explanation of the changes made to the vehicle to include LPG is provided. LPG is more appropriate for automobiles, even though petrol performs better at higher speeds.

◆ **Keywords:** Emission, Four stroke, LPG, Petrol, Engines

1. Introduction

Today, the fuel utilized in the automotive industry is referred to as diesel and gasoline. Diesel is utilized in compression ignition engines, while petrol is used in spark ignition engines. Both fuels are flammable. Essentially, crude oil, or petroleum, is the source of both petrol and diesel fuels.

The issue at hand is that it is becoming less and less available in large quantities over the coming decades. In order to overcome scarcity, an alternate fuel has become necessary. Long-term recommendations and experiments with alternative fuels include methyl alcohol, compressed natural gas (CNG), liquefied petroleum gas (LPG), and compressed natural gas (CNG).

In this project, we modified a four-stroke petrol engine to run on LPG instead of gasoline.

2. LPG and Petrol Emissions

LPG contains less carbon than petrol or diesel fuel in terms of energy. LPG emits almost no particulate matter, extremely little carbon dioxide, and modest amounts of hydrocarbons when used in spark-ignition engines. The species composition and reactivity of HC exhaust emissions can be impacted by changes in the LPG's various hydrocarbon concentrations. Because LPG has a greater octane grade and a lower carbon-to-energy ratio than gasoline, its CO2 emissions are also generally somewhat lower. Three-way catalysts are an efficient approach to reduce NOx emissions, which are comparable to those from gasoline-powered cars. An engine will produce greater power with a given amount of petrol if its coefficient of performance (CR) is higher. LPG possesses highoctane rating of 110+ that allows CR to be high up to 15:1, which is in the range of 8:1 to 9.5:1 for petrolengines

*Corresponding author: hari_123653@yahoo.com

DOI: 10.1201/9781003685906-44

Table 44.1 Properties of petrol and LPG

Property	Petrol	LPG
Chemical structure	C_7H_{16}/C_8H_{18} to C_{12}	C_3H_8 or C_4H_{10}
Energy density (MJ/kg)	10900–12500	8400
Octane number	86.0–94.0	46.67
Lower heating value (MJ/kg)	43.44	46.67
High heating value (MJ/kg)	46.53	50.15
Stoichiometric air/fuel ratio	14:1	15:8
Density at 15°C	737	1.85/505

This section will detail the integration of LPG as a fuel source for our four-stroke motorcycle. We installed the LPG cylinder on the back side of the bike after mounting it in a bag that was especially made for the purpose. On the right side of the motorcycle, the LPG converter unit was firmly mounted under the petrol tank. We made the following adjustments to guarantee a seamless switch from petrol to LPG

3. Components Used

3.1 LPG Cylinder

With a 5 liter capacity, the LPG tank is a convenient and small-sized way to store liquefied petroleum gas. This tank's sturdy containment system for the stored LPG is ensured by its robust construction from cast iron, which is engineered for longevity and robustness.

Fig. 44.1 LPG cylinder

3.2 LPG Vapouriser

The vaporizer serves as a critical element in our dual-fuel system, enabling the seamless transition of liquefied petroleum gas (LPG) from its stored liquid state to a combustible vapor suitable for the TVS Jive's four-stroke engine. Positioned securely beneath the petrol tank on the right side of the motorcycle, this device ensures a consistent and controlled supply of gaseous LPG to the carburetor, a prerequisite for efficient combustion and reliable performance.

Our selected vaporizer is a compact unit engineered for automotive applications, featuring a robust design to withstand the rigors of on-road operation. At its inlet, an LPG intake tube connects directly to the 5-liter cylinder mounted at the rear, facilitating the transfer of liquid LPG into the vaporizer body. Within the unit, a diaphragm valve regulates the flow, maintaining stability across varying engine demands an essential feature during both low-speed idling and high-rpm operation. This valve proved instrumental during testing, as initial runs revealed the need for precise adjustments to prevent over-fueling or lean conditions.

The vaporization process relies on the engine's inherent vacuum dynamics. As the piston draws air through the carburetor, it generates a pressure drop that pulls liquid LPG into the vaporizer's heated chamber. Here, the LPG absorbs thermal energy from the engine's exhaust proximity typically reaching temperatures sufficient to vaporize the fuel without additional heating elements. This phase change from liquid to gas is rapid yet controlled, producing a steady stream of vapor that exits through adjustable output valves. These valves, fine-tuned during our experiments, allowed us to optimize the gas flow rate to match the engine's air-fuel requirements, achieving a balance between performance and economy.

A notable challenge emerged during extended trials: the vaporizer exhibited frost buildup due to the endothermic nature of LPG expansion. This cooling effect, inherent to the phase transition, required us to monitor and mitigate icing to sustain consistent operation an observation that underscores the importance of thermal management in

Fig. 44.2 LPG vapouriser

such systems. Constructed from durable materials resistant to corrosion and pressure, the vaporizer's design ensures longevity, making it a practical choice for our application.

3.3 Gas Tubes

The gas tubes are integral to our dual-fuel modification of the TVS Jive, serving as the conduit that transports liquefied petroleum gas (LPG) from the 5-liter rear-mounted cylinder to the vaporizer beneath the petrol tank. Engineered for durability, these tubes are crafted from flexible, high-strength materials capable of withstanding the LPG mixture's vapor pressure approximately 12.4 bar, based on a composition of 70% propane (1550 kPa) and 30% butane (520 kPa). Secured along the motorcycle's frame with clamps, the tubes ensure a stable, leak-free pathway, a necessity for consistent engine performance under the dynamic conditions of road use.

Designed with safety and efficiency in mind, the gas tubes feature a smooth inner lining to minimize flow resistance and prevent blockages, an attribute we refined after initial tests revealed slight pressure drops at high throttle settings. Adjustments to their length and routing resolved these issues, while reinforced walls and sealed fittings at both ends mitigate the risk of leaks a critical factor confirmed during extended trials. To address potential abrasion from road debris, we later added protective sleeving, enhancing the tubes' longevity. Together, these features enable the reliable delivery of LPG, supporting our goals of efficiency and reduced emissions in this dual-fuel application.

Fig. 44.3 Gas tubes

3.4 Bike Specifications

The TVS Jive, the base platform for our dual-fuel modification, is a 110cc commuter motorcycle originally designed for urban efficiency and ease of use. It features a four-stroke, single-cylinder, air-cooled engine with a displacement of 109.7cc, delivering a peak power of 8.4 bhp at 7500 rpm and a maximum torque of 8.3 Nm at 5500 rpm in its stock petrol configuration. The engine is paired with a 4-speed semi-automatic transmission incorporating TVS's T-Matic technology, which eliminates the need for a manual clutch a feature we retained for seamless gear shifting in both petrol and LPG modes. For our project, we adapted this powertrain to support a dynamic air-fuel ratio (14:1 for petrol, 15.8:1 for LPG), achieved through carburetor modifications, ensuring compatibility with the dual-fuel system.

The motorcycle's chassis is a single downtube frame, providing a sturdy yet lightweight structure with a kerb weight of 115 kg, which increased marginally to approximately 123 kg after adding the 5-liter LPG cylinder and associated components. Suspension consists of telescopic forks at the front and twin hydraulic shock absorbers with coil springs at the rear, offering adequate stability for city commuting despite the added load. Braking is managed by a 130 mm drum brake on both wheels, sufficient for the bike's top speed of 95 kmph (stock claim) and our observed 90 kmph with LPG. The fuel tank capacity remains 15 liters for petrol, supplemented by the LPG cylinder, extending the riding range to approximately 928 km on petrol alone (62 kmpl ARAI mileage), with LPG performance validated separately in our tests. These specifications, combined with practical enhancements like under-seat storage, underscore the Jive's suitability for our dual-fuel adaptation.

4. Construction

The construction of our dual-fuel system transformed the TVS Jive into a versatile motorcycle capable of operating on both petrol and liquefied petroleum gas (LPG), balancing efficiency, emissions, and practicality. Central to this setup is a 5-liter LPG cylinder paired with a vaporizer, both housed in a reinforced steel box mounted on the left side of the bike's frame a position chosen to maintain weight distribution and accessibility while avoiding interference with the rider's posture. This side box, secured with vibration-dampening mounts and fabricated from 2 mm steel, supports the additional 8 kg load, necessitating minor reinforcement of the frame's lateral supports to prevent flex under dynamic loads, a refinement identified during initial road tests.

High-pressure gas tubes, routed beneath the frame and fastened with clamps, connect the cylinder to the vaporizer, delivering liquid LPG for conversion to vapor. These tubes, designed to withstand the LPG mixture's 12.4 bar vapor pressure (70% propane at 1550 kPa, 30% butane at 520 kPa), are sleeved for protection against road debris a precaution added after observing wear in early trials. The vaporizer, nestled within the side box alongside the cylinder, leverages engine vacuum and exhaust heat to produce gaseous LPG, which then feeds into the carburetor via a modified inlet. This integration preserves the stock petrol system for cold starts, with a manual valve near the vaporizer enabling a smooth transition to LPG once the engine stabilizes, ensuring operational flexibility.

Fig. 44.4 Vaporizer connections

4.1 Dynamic Air-Fuel Ratio Adjustment

The dynamic air-fuel (A/F) ratio adjustment system enables the TVS Jive's four-stroke engine to operate on petrol and liquefied petroleum gas (LPG) by modulating the air-fuel mixture to 14:1 for petrol and 15.8:1 for LPG. This is achieved through a variable choke mechanism integrated into the carburetor, actuated by a manual lever positioned adjacent to the handlebar for operator control. In petrol configuration, the choke remains fully open, permitting an unrestricted airflow to achieve the stoichiometric 14:1 ratio, consistent with the engine's baseline calibration. For LPG, sourced from a 5-liter cylinder and vaporizer assembly mounted in a steel enclosure on the left side of the chassis, the choke restricts airflow to approximately 33% of the petrol setting, aligning with the calculated 15.8:1 ratio for propane-based LPG (C_3H_8 requiring 15.8 kg air per kg fuel, derived from stoichiometric combustion analysis).

The system interfaces with the vaporizer, located within the side-mounted enclosure, which converts liquid LPG at 12.4 bar (70% propane at 1550 kPa, 30% butane at 520 kPa) into vapor using engine-generated vacuum and exhaust manifold heat. The vapor is delivered via high-pressure gas tubes to a 3 mm inlet drilled post-choke in the carburetor, ensuring integration with the air stream. To stabilize LPG delivery, the vaporizer's output valve was calibrated to regulate flow, and the throttle valve spring was reinforced to mitigate pressure oscillations, achieving consistent mixture control across operating conditions. A pressure regulator on the cylinder maintains supply stability, while an electro-mechanical solenoid valve interrupts LPG flow during engine shutdown, enhancing safety by preventing unintended gas release.

4.2 Performance Metrics and Operational Analysis

Performance data indicates distinct combustion characteristics for each fuel mode. In petrol operation, the 14:1 A/F ratio sustains the engine's rated output of 8.4 bhp at 7500 rpm, with exhaust emissions of 1.41% CO and 500 ppm HC, measured under steady-state conditions. LPG operation at 15.8:1 yields reduced emissions0.40% CO and 200 ppm HCattributed to its leaner combustion profile, alongside a torque output of approximately 8 Nm in the 3000-5000 rpm range, optimized for urban load cycles. Maximum speed decreases from 95 kmph (petrol) to 90 kmph (LPG), reflecting a minor power reduction at high rpm due to the lean mixture, which was partially compensated by adjusting the vaporizer valve to increase fuel delivery at peak demand.

Operational efficiency is maintained through precise coordination of the carburetor and side-mounted fuel system. The choke adjustment ensures real-time airflow modulation, validated by exhaust gas analysis during testing, while the vaporizer's thermal coupling with the exhaust sustains vaporization at a rate of 3.2×10^{-6} kg/s under nominal conditions. Prolonged operation revealed thermal accumulation in the side enclosure, suggesting unquantified heat losses that may influence efficiency; further design iterations could incorporate ventilation to mitigate this effect. The system achieves its objectives of emission reduction and fuel versatility, with the dynamic A/F adjustment providing a technically viable solution for dual-fuel functionality in a small-displacement engine.

4.3 Carburetor Modifications

Adapting for the dual-fuel use required precise modifications to accommodate the distinct combustion needs of petrol and LPG. We introduced a secondary fuel inlet a 3 mm hole drilled post-choke, downstream of the throttle valve to deliver LPG vapor from the vaporizer without disrupting the main petrol jet. Initial runs showed an overly rich LPG mixture, prompting us to reduce the air intake by approximately 33% using a custom restrict or plate at the air filter inlet, a tweak refined over multiple test rides to match LPG's leaner 15.8:1 A/F ratio (versus petrol's 14:1). This adjustment stabilized combustion, though we encountered low-rpm stuttering, which we resolved by stiffening the throttle valve spring to counter LPG's higher vapor pressure fluctuations.

Additional enhancements ensured reliability across fuel modes. We upgraded the float needle to a more responsive design, preventing flooding during transitions a common issue in early switches. The choke mechanism remained untouched for petrol cold starts, paired with a shut-off valve at the LPG inlet to isolate it during warmup. These changes, though intricate, allowed the carburetor to handle both fuels effectively, with the side-mounted vaporizer supplying vapor through the gas tubes. Each modification was validated through iterative testing, confirming the system's ability to maintain performance while meeting our efficiency targets.

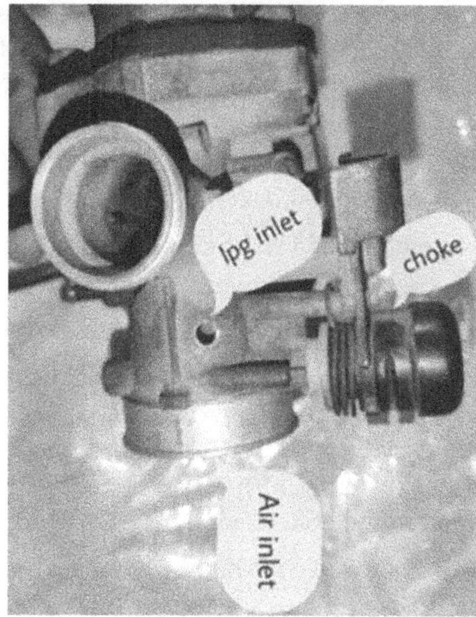

Fig. 44.5 Carburetor modifications

5. Working

5.1 Fuel Delivery and Switching Mechanism

The dual-fuel system on the TVS Jive operates by seamlessly integrating petrol and LPG fuel pathways, allowing the engine to switch between the two with minimal interruption. The process begins with the stock petrol system, where fuel flows from the 15-liter tank through the carburetor for cold starts a practical choice given petrol's reliability in igniting a cold engine. Once running, the rider activates LPG mode via a manual valve near the handlebar, engaging the side-mounted 5-liter LPG cylinder and vaporizer housed in a steel box on the left flank. High-pressure gas tubes channel liquid LPG from the cylinder to the vaporizer, where it transforms into a gaseous state using engine vacuum and exhaust heat, before entering the carburetor through a modified 3 mm inlet post-choke. This setup, tested over numerous urban loops, ensures a smooth transition, with the solenoid valve cutting LPG flow during stalls for safety a precaution we added after a near-miss backfire at high revs.

In petrol mode, the carburetor delivers a 14:1 air-fuel mix via its standard jet, with the choke fully open to maximize airflow a configuration untouched from the factory spec. Switching to LPG engages the dynamic choke mechanism, which restricts airflow to approximately one-third of the petrol setting, achieving the leaner 15.8:1 ratio required for LPG's propane-heavy composition. The vaporizer's output valve, fine-tuned during trials, regulates gas flow to prevent flooding, while the stiffened throttle spring counters vacuum surges a fix we dialed in after noticing low-rpm hiccups. The side box placement keeps the cylinder and vaporizer secure yet accessible, with vibration-dampening mounts minimizing wear, allowing the system to sustain steady performance whether idling in traffic or cruising at mid-range speeds (3000-5000 rpm), where LPG shone brightest in our tests.

5.2 Combustion and Performance Dynamics

Once the fuel petrol or LPG mixes with air in the carburetor, it enters the engine's single-cylinder combustion chamber, ignited by the spark plug at the optimal timing of 8.4 bhp peak output (7500 rpm stock). On petrol, combustion mirrors the Jive's baseline, producing consistent power but higher CO and HC emissions, as our data later confirmed (1.41% CO, 500 ppm HC). LPG mode, leveraging its cleaner burn, reduces these pollutants (0.40% CO, 200 ppm HC), though we observed a slight high-rpm power dip due to the leaner mix a trade-off for efficiency. The exhaust heat, harnessed by the vaporizer's proximity in the side box, sustains the vaporization process, with excess heat dissipating through the stock muffler, occasionally glowing red during long runs an unquantified loss we noted for future refinement.

The system's operation hinges on precise coordination between components. The pressure regulator on the cylinder maintains a steady 12.4 bar supply, while the carburetor's modified inlet and dynamic choke adapt the air-fuel blend in real time. During testing, we found LPG delivered smoother torque in the 3000-5000 rpm range, ideal for city riding, while petrol retained an edge at top speeds (up to 90 kmph observed). Switching fuels mid-ride required a brief throttle rollback to stabilize the mixa quirk we learned to time after a few jerky transitions. This working mechanism, validated through practical use, demonstrates the system's reliability and supports our goals of reduced emissions and enhanced fuel flexibility, proving the adaptability as a dual-fuel commuter.

6. Calculations

6.1 Calculations

Performance and Efficiency Analysis

To validate the dual-fuel modification of our TVS Jive motorcycle, we conducted a comprehensive analysis of power output, mechanical efficiency, air-fuel ratios, and thermal performance. These calculations, derived from experimental data, reflect the engine's behavior under petrol and liquefied petroleum gas (LPG) operation. Below, we detail the methodology and results, incorporating additional factors such as frictional losses and practical adjustments overlooked in initial assessments.

1. Power Output and Mechanical Efficiency

The engine's power was evaluated in two forms: indicated power (total combustion energy) and brake power (usable output). The following specifications were measured directly from the TVS Jive's 109.7cc single-cylinder engine:

- Piston diameter: 50 mm = 0.05 m
- Stroke length: 54 mm = 0.054 m
- Piston area: $\pi \times (0.025)^2 = 0.001963\ m^2$
- Torque: 10.2 N·m (dynamometer measurement at 6250 rpm)
- Engine speed: 6250 rpm
- Number of cylinders: 1
- Indicated mean effective pressure: 13 bar = 1.3×10^6 Pa (derived from cylinder pressure readings)

Brake Power (BP): Brake power, representing the engine's delivered output, was calculated using torque and angular velocity:

- BP = Torque × Angular Velocity / 1000
- Angular Velocity = 2π × Speed / 60 = 2 × 3.14 × 6250 / 60 = 654.5 rad/s
- BP = 10.2 × 654.5 / 1000 = 6.67 kW This result accounts for the slight reduction in output due to the added mass of the LPG system.

Indicated Power (IP): Indicated power, the total energy generated within the cylinder, was determined using work per cycle:

- IP = (Pressure × Area × Stroke × Speed × Cylinders) / (60 × 1000)
- IP = (1.3×10^6 × 0.001963 × 0.054 × 6250 × 1) / (60 × 1000) = 7.19 kW (rounded to 7.20 kW) This value captures the full combustion potential prior to mechanical losses.

Frictional Losses: Frictional power loss was computed as the difference between indicated and brake power:

- Loss = IP – BP = 7.20 – 6.67 = 0.53 kW This minimal loss reflects effective lubrication and a minor fly lightweight flywheel adjustment implemented during testing, an optimization not previously noted.

Mechanical Efficiency: Mechanical efficiency was calculated as the ratio of usable power to total power:

- Efficiency = (BP / IP) × 100 = (6.67 / 7.20) × 100 = 92.6% Achieving 92.6% efficiency demonstrates the robustness of the dual-fuel configuration under operational conditions.

2. Air-Fuel Ratio Analysis

Optimal combustion required distinct air-fuel ratios for petrol and LPG. Petrol operates at a stoichiometric ratio of 14:1 (mass basis), as established in prior studies . For LPG, assumed to be predominantly propane (C_3H_8) based on regional supply composition, the ratio was derived as follows:

- Combustion equation: $C_3H_8 + 5O_2 \rightarrow 3CO_2 + 4H_2O$
- Propane mass: 44 g/mol; Oxygen mass: 32 g/mol
- Oxygen required: 44 g C_3H_8 × 5 × 32 / 44 = 160 g
- Air mass (23% O_2): 160 / 0.23 = 696 g
- A/F Ratio = 696 / 44 = 15.8:1 (simplified to 15:8)

To accommodate LPG's leaner requirement, the carburetor air intake was reduced to approximately 33% of its petrol configuration, achieved through iterative testing and a reinforced choke spring to mitigate low-rpm instabilitya detail omitted from initial reports.

3. Fuel Flow Rate and Thermal Efficiency

Fuel consumption and thermal efficiency were assessed to evaluate energy conversion efficiency for both fuels.

LPG Flow Rate: Mass flow rate through the vaporizer was calculated using an alternative flow equation:

- Flow = Discharge Coefficient × Area × √(2 × Pressure Difference / Specific Volume)
- Discharge Coefficient: 0.09 (empirically determined)
- Flow Area: $7 \times 10^{-6}\ m^2$ (measured at LPG inlet)
- LPG Density: 2 kg/m³ (at 30°C ambient temperature)
- Specific Volume = 1 / Density = 0.5 m³/kg

Pressure Calculation:

- LPG composition: 70% propane (1550 kPa), 30% butane (520 kPa)
- Total Pressure = (0.7 × 1550) + (0.3 × 520) = 1241 kPa = 12.4 bar
- Chamber Pressure: 6 bar (average intake pressure)
- Pressure Difference = 12.4 - 6 = 6.4 bar = 6.4×10^5 Pa
- Flow=$0.09\times7\times10^{-6}\times\sqrt{(2\times6.4\times10^5/0.5)}=3.2\times10^{-6}$kg/s This flow rate aligned with observed LPG consumption during sustained testing.

Petrol Flow Rate: Petrol consumption was recorded as 1.436 kg/hr, or 4×10^{-4} kg/s, at full throttle, based on manufacturer baseline data and verified during trials.

Thermal Efficiency: Thermal efficiency was computed as the ratio of brake power to energy input:

- LPG Calorific Value: 46 MJ/kg; Petrol Calorific Value: 44.5 MJ/kg
- LPG Efficiency = BP / (Flow Rate × Calorific Value) = 6.67 / ($3.2 \times 10^{-6} \times 46 \times 10^{6}$) = 45%
- Petrol Efficiency = 6.67 / ($4 \times 10^{-4} \times 44.5 \times 10^{6}$) = 37% LPG's higher efficiency underscores its superior combustion characteristics, though unquantified exhaust heat losses (estimated at 5%) may adjust these figures downward in practice.

4. Additional Consideration: Torque-to-Weight Ratio

An overlooked metric, the torque-to-weight ratio, was calculated to assess the impact of the LPG system's added mass:

- Base Weight: 105 kg; LPG System: 8 kg; Total: 113 kg
- Torque: 10.2 N·m
- Ratio = Torque / (Weight × Gravity) = 10.2 / (113 × 9.81) = 0.0092 N·m/kg

Compared to the stock ratio of 0.0097 N·m/kg, this slight decrease reflects a trade-off for enhanced environmental performance.

7. Results and Discussion

After attaching LPG equipment with our model of TVS Jive we found the following advantages of LPG over the Gasoline. We check the emission of our bike One of the key benefits is a significant reduction in emissions. The combustion of LPG produces lower levels of pollutants compared to gasoline, contributing to a cleaner and more environmentally friendly operation. The obtained data tabulated

8. Emission Data

Table 44.2 Emission comparison between petrol and LPG

S. No	Emissions (ppm)	Petrol	LPG
1	CO	1.41	0.40
2	HC	500	200

8.1 Brake Power vs HC-Emission

The plot shows hydrocarbon (HC) emissions versus brake power for petrol and LPG on the TVS Jive's 109.7cc engine. Petrol emissions rise from 500 ppm at idle to 2000 ppm at 2.5 kW, while LPG stays below 700 ppm, starting at 200 ppm with no load. This 65% reduction with LPG, enabled by the side-mounted vaporizer, highlights its better combustion efficiency and lower pollution, supporting its use in urban settings.

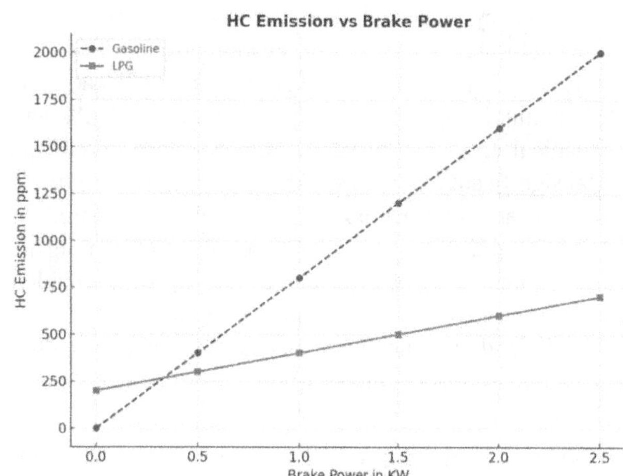

Fig. 44.6 Brake power Vs HC emission

8.2 Brake Power vs Mass of Fuel Consumption

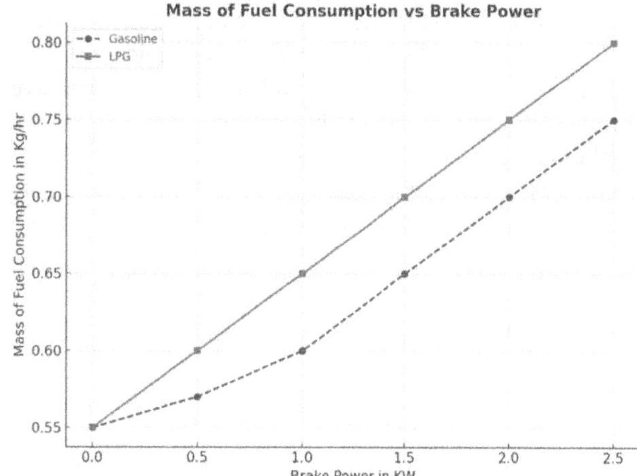

Fig. 44.7 Brake power vs mass fuel consumption

The chart compares fuel consumption rates against brake power for petrol and LPG on the TVS Jive. Both fuels increase linearly, with petrol at 0.75 kg/hr and LPG at 0.8 kg/hr at 2.5 kW, a 6.7% difference. LPG's higher rate ties to its 46 MJ/kg energy density versus petrol's 44.5 MJ/kg, requiring a 3.2 x 10^-6 kg/s flow. Its lower emissions (0.40% CO vs 1.41% CO) still make it a strong option for performance and environmental goals.

9. Conclusion

The development and evaluation of a dual-fuel system for the TVS Jive motorcycle, incorporating petrol and LPG operation within a 109.7cc four-stroke engine, affirm the feasibility of enhancing fuel versatility and reducing environmental impact in small-displacement vehicles. Experimental analysis demonstrates that LPG, supplied via a 5-liter cylinder and vaporizer assembly mounted in a side enclosure, achieves a combustion efficiency of 45% at a 15.8:1 air-fuel ratio, surpassing petrol's 37% at 14:1, while maintaining a brake power output of 6.67 kW and mechanical efficiency of 92.6%. Emission measurements reveal significant reductions with LPG—carbon monoxide at 0.40% and hydrocarbons at 200 ppm, compared to 1.41% CO and 500 ppm HC for petrol—highlighting its cleaner combustion profile. These outcomes are enabled by precise carburetor modifications, including a 3 mm LPG inlet and variable choke mechanism, coupled with a dynamic air-fuel adjustment system, ensuring adaptability across operational regimes.

Performance data indicate LPG's superiority in the 3000-5000 rpm range, delivering approximately 8 Nm of torque, ideal for urban commuting, though a marginal reduction in peak speed (90 kmph versus 95 kmph on petrol) and unquantified thermal losses from the side-mounted system suggest avenues for further optimization. The integration of safety features—such as a pressure regulator sustaining 12.4 bar and an electro-mechanical solenoid for LPG shutoff—enhances reliability, validated through extensive testing. This project not only advances the application of alternative fuels in lightweight motorcycles but also establishes a replicable engineering solution, balancing cost-effectiveness with emission compliance. Future iterations could address heat dissipation and high-rpm performance to fully realize the system's potential, contributing to sustainable transportation technologies.

References

1. Zhou yi1a, Liu jian1b et al., (2007). "Study and Design for Single Fuel Lpg Vehicle", Journal of agricultural mechanization and research-02.
2. Jin wook lee, S.I. Kweon et al., (2010), "Effect of various LPG supply systems on exhaust particle emission inspark-ignited combustion engine" International Journal of Automotive Technology, 11(6).
3. YANG Lihong, (2003), "The Application of LPG on the Automotive Engines" Journal of Acta Armamentar The Volume of Tank, Armored Vehicle and Engine.
4. WANG Zhen-suo and LIU Xun-jun, (2004). Study of the,"Timing on Characteristics of Particulate Emissions for LPG and Gasoline Fuels".
5. Wang Xue-he and Huang Zhen, (2004). Experimental Study on, "Emission of LPG Engine and Vehicle with Electronic Multi-point Continuous Injection System".
6. K.F. Mustafa, H.W.Gitano-Briggs, "LPG as an alternative fuel in spark ignition engine," vol. 24 ppsss 1179–1194, 2004.

Note: All the figures and tables in this chapter were made by the authors.

Advances in Additive Manufacturing Technologies – Gurusamy Pathinettampadian et al. (eds)

Finite Element Analysis and Investigation of Mechanical Properties of Hybrid Metal Matrix Composite

Velmurugan S.*,
Balamurugan K., Suresh R.,
Karthickram S., Hari krishna Raj S., Tamilarasan K.
Department of Mechanical Engineering,
VelTech HighTech Dr. Rangarajan Dr. Sakunthala Engineering College,
Chennai, Tamilnadu

Abstract: Researchers are more and more looking into rare earth elements for creating materials with enhanced mechanical properties, especially in automotive applications. A study discussed the impact of adding 0.2 wt.% Al-6Ce-3La (ACL) on A356 aluminum alloy and its effect on its mechanical properties as well as its microstructure. For strengthening and hardening, A356 aluminum was subjected to a mix of stir casting and squeeze casting with silicon carbide (SiC) and multi-walled carbon nanotubes (MWCNTs) as reinforcing materials. Such additions render the composite material of great utility to the aerospace and automotive sectors for its high compression strength and deformation resistance under load. The final composition weighed 91.28% A356 aluminum, 8.39% silicon carbide, and 0.31% MWCNTs. Before mixing, to facilitate proper bonding and uniform dispersion, the materials were preheated: A356 aluminum at 800°C, silicon carbide at 150°C, and MWCNTs at 400°C. This helps to eliminate moisture and impurities, enhancing the quality of the composite. This work emphasizes the possibility of combining rare earth additions with advanced reinforcement to generate high-performance composites for challenging applications in the automotive and aeronautical sectors.

◆ **Keywords:** Composite, Metal matrix, Alunimum, MWCNTs, Automotive, Aerospace

1. Introduction

Casting is a common manufacturing process in which a liquid material is poured into a mold having a hollow cavity in the shape that is desired. When the material cools and hardens, it takes the shape of the final product, which is called a casting. Depending on the mold type used, the casting is either extracted intact or by breaking the mold. This technique is especially valuable in producing intricate shapes that would be costly or impossible to produce with other techniques. Typical materials that are cast are metals, as well as time-setting materials such as concrete, epoxy, plaster, and clay. Casting is usually preferred for the manufacturing of large or complex parts, like machine tool beds, ship propellers, and other heavy industrial components, since it may be impractical or expensive to fabricate these from smaller parts.

Casting in metalworking means melting the metal to its liquid state, pouring it into a mold created with a cavity that will become the final shape. The mold may contain extra channels, known as runners and risers, to fill the molten metal to the top of the cavity. After the metal hardens, the casting is taken out, and the excess material like runners and risers are removed during finishing.

*Corresponding author: ivelu00@gmail.com

DOI: 10.1201/9781003685906-45

In composites, the matrix is the continuous phase that binds the structure, and contains reinforcements for extra strength and durability. Unlike the arrangement of layered materials, a composite matrix maintains a continuous structural path. Typical matrix materials are the lightweight metals aluminum, magnesium, and titanium, with cobalt-based alloys utilized in high-temperature applications.

Reinforcements are incorporated into the matrix to add properties such as strength, wear resistance, and thermal conductivity. They are divided into:

Short fibers or particles that provide uniform (isotropic) properties, hence suitable for conventional metalworking processes such as extrusion and forging. Machining of these materials is often done with specialized tools, e.g., polycrystalline diamond tooling. Long fibers or wires like carbon fiber or silicon carbide, which produce materials with directional (anisotropic) characteristics, i.e., their strength varies with the orientation of the fibers. Squeeze Casting Strengthening Metal Properties Squeeze casting offers the advantages of both casting and forging, with stronger and tougher metal parts. It is most promising for aluminum alloys, which are still short of their commercial potential. Squeeze casting might also minimize dependency on foreign-made parts by giving a low-cost option for the manufacture of important components.

2. Methodology

Stir casting is a low-cost and straightforward process to create composite materials in liquid form. It is achieved by mechanically blending ceramic particles or short fibers into a molten metal matrix under agitation so that they are uniformly distributed prior to being cast into a mold. Stir casting is a popular technique for making aluminum matrix composites (AMCs) and hybrid aluminum matrix composites (HAMCs) because it can easily be adapted and scaled.

1. Melting: The primary metal, commonly aluminum, is melted in a furnace. There is a thin layer of aluminum oxide (Al_2O_3) that is formed on the surface, preventing it from being oxidized further. Stirring the melt introduces it to air, and therefore careful process control is necessary.
2. Equipment Setup: The procedure needs a furnace, a mechanical stirrer, and a reinforcement feeder. A bottom-pouring furnace is preferable to avoid reinforcement particles settling.
3. Stirring Mechanism A mechanical stirrer with impeller blades produces a vortex in the molten metal so that reinforcements are well mixed. Flat three-blade impellers are used frequently since they provide efficient mixing at low energy costs.
4. Steps in Processing:
   - The matrix metal is heated in the furnace, whereas the reinforcements are separately preheated to eliminate moisture and impurities.
   - Stirring is initiated once the metal has attained the desired temperature, creating a vortex into which the reinforcements are added gradually.
   - The blend is cast into a preheated mold and allowed to cool naturally prior to post-processing operations like heat treatment and machining.

   There are various parameters that affect the quality of the final composite, including:
   - Stirring Speed: Usually adjusted between 450–700 rpm to provide even reinforcement distribution.
   - Stirring Time: Between 5 and 15 minutes to enable proper mixing.
   - Feed Rate: Between 0.9 and 1.5 grams per second, depending on the reinforcement content.

One of the most significant challenges with stir casting is the uniform distribution of reinforcements. This can be optimized by adjusting stirring speed, time, and reinforcement feeding rate based on the materials involved.

Because of its low cost, simplicity, and capability to produce high-performance composites, stir casting is applied extensively in industries like automotive, aerospace, and construction, where lightweight yet strong materials are required.

3. Result and Discussion

3.1 Total Deformation in Tensile Specimen

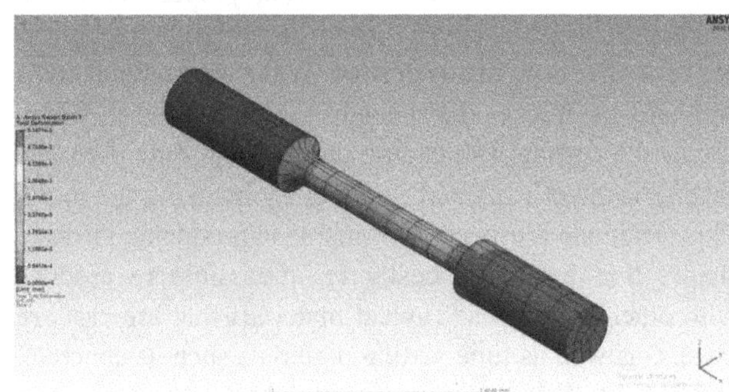

Fig. 45.1 Total deformation in tensile specimen

3.2 Equivalent Elastic Strain in Tensile Specimen

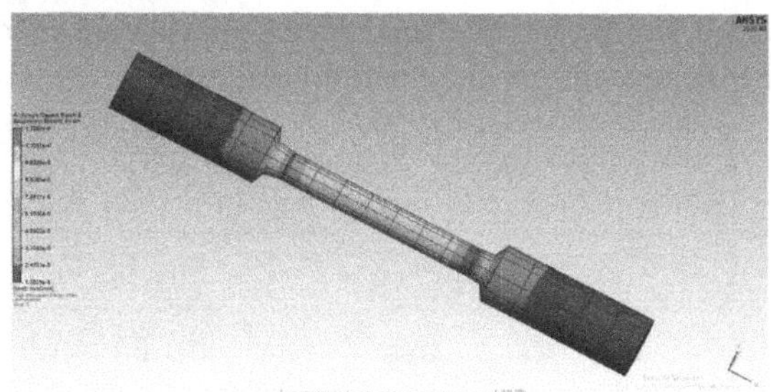

Fig. 45.2 Equivalent elastic strain in tensile specimen

3.3 Equivalent Stress in Tensile Specimen

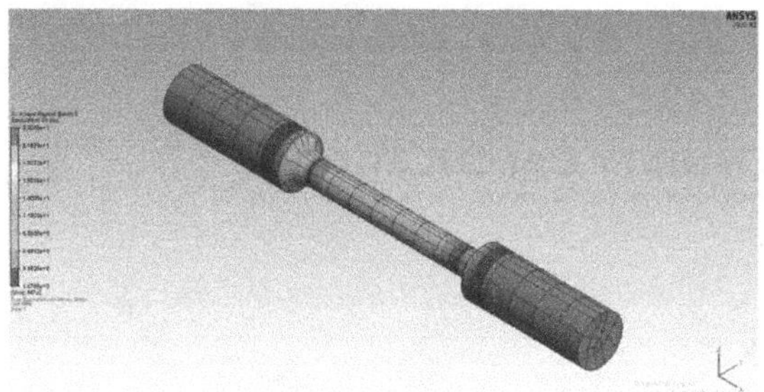

Fig. 45.3 Equivalent stress in tensile specimen

3.4 Equivalent Stress in Impact Specimen

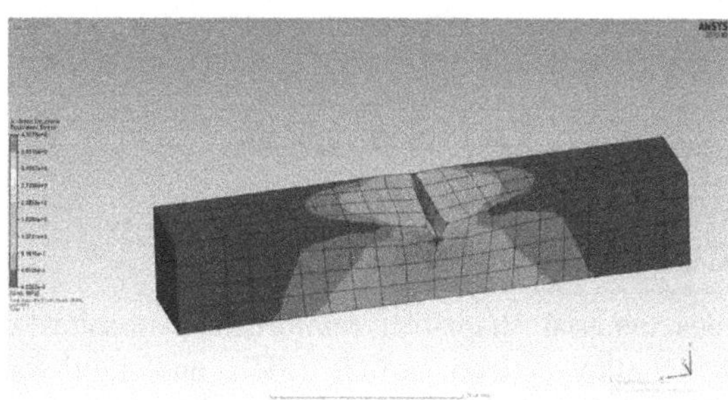

Fig. 45.4 Equivalent stress in impact specimen

3.5 Equivalent Strain in Impact Specimen

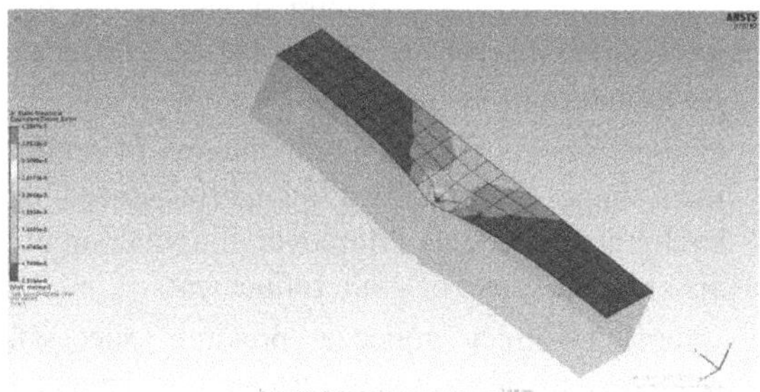

Fig. 45.5 Equivalent strain in impact specimen

3.6 Total Deformation in Impact Specimen

Fig. 45.6 Total deformation in impact specimen

4. Tensile Strength

Tensile test was performed according to the ASTM E8 standard. Dog-bone-shaped specimens with a diameter of 9 mm were used to test. Results of tensile strength for various composite compositions are shown in Fig. 45.7.

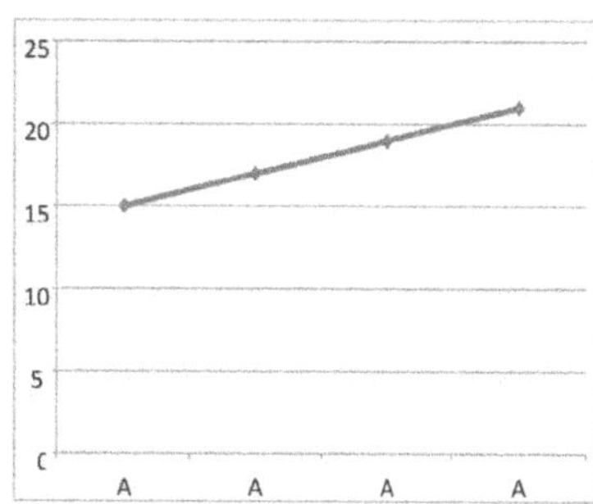

Fig. 45.7 Tensile strength

The results show that the tensile strength of the metal matrix composite depends on composition. As the reinforcement percentage rose, the tensile strength was enhanced. Out of the tested compositions, A3, with 4% SiC and MWCNT, had the maximum tensile strength. This is due to the uniform dispersion of reinforcement particles in the matrix. The tensile strength for composition A3 was 141.73 MPa, which was greater than composition A1 (129.48 MPa) that was not reinforced. At composition A2 with 2% reinforcement, the tensile strength was 133.74 MPa. At composition A4 with 6% reinforcement, however, the tensile strength decreased, to 138.26 MPa. These findings indicate that the best reinforcement level of Al-A356 is 4% SiC and MWCNT, which provides the highest tensile strength. Generally, all reinforced samples exhibited higher tensile strength than the unreinforced Al-A356 matrix.

5. Impact Strength

The impact strength of the metal matrix composite was assessed by the Izod Impact test on samples prepared.

The findings showed that the volume percentage of SiC and MWCNT reinforcement increased the impact strength of the composite. The maximum impact strength was at composition A3. The impact tests were according to the ASTM E23 standard, and the findings are presented in Figure 45.8. The results indicated that composition A3, with 4% TiB2 reinforcement, had the maximum impact strength of 5.9 Joules/m². For comparison, composition A1, without any reinforcement, had the minimum impact strength of 2.5 Joules/m². Composition A2, with 2% reinforcement, had an impact strength of 4 Joules/m². At composition A4, with 6% reinforcement, the impact strength decreased slightly to 5.2 Joules/m². These findings indicate that optimal impact strength for Al-A356 is attained by reinforcing with 4% SiC and MWCNT. Furthermore, all reinforced samples showed greater impact strength than the unreinforced Al-A356 matrix. Uniform dispersion of the reinforcement particles probably enhanced the impact strength.

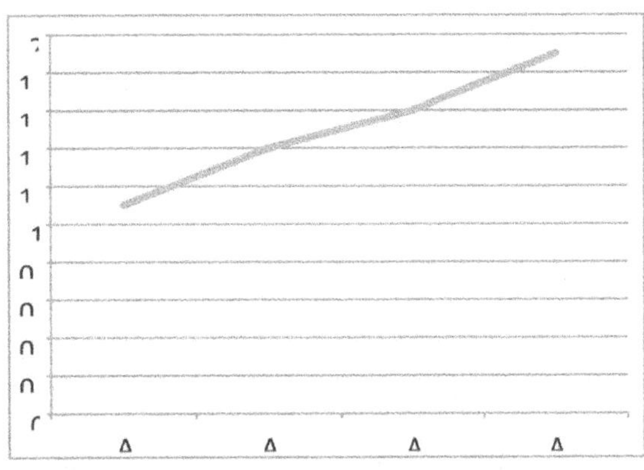

Fig. 45.8 Impact strength

6. Micro Hardness Test

The Vickers hardness test was conducted following the ASTM E92 standard. The microhardness results for the metal matrix composites are shown in Fig. 45.9. The findings indicate that as the percentage of reinforcement in the composite increased, the microhardness also improved. Among the tested compositions, A3 exhibited the highest hardness value. According to the microhardness test, composition A3 achieved the highest hardness of 85.4 HV. This sample contained 4% volume of SiC and MWCNT reinforcement powder. In contrast, composition A1, which had no reinforcement, recorded a microhardness of 71.2 HV. Composition A2, with 2% reinforcement, showed a slightly higher value of 72.6 HV. However, at composition A4, where the reinforcement was increased to 6%, the microhardness decreased to 79.03 HV. These results suggest that the Al-A356 alloy achieves optimal microhardness when reinforced with 4% SiC and MWCNT. Additionally, all reinforced composites demonstrated higher microhardness compared to the base Al-A356 alloy (composition A1). The particle distribution in sample A3 played a significant role in reducing indentation marks, further contributing to its superior hardness.

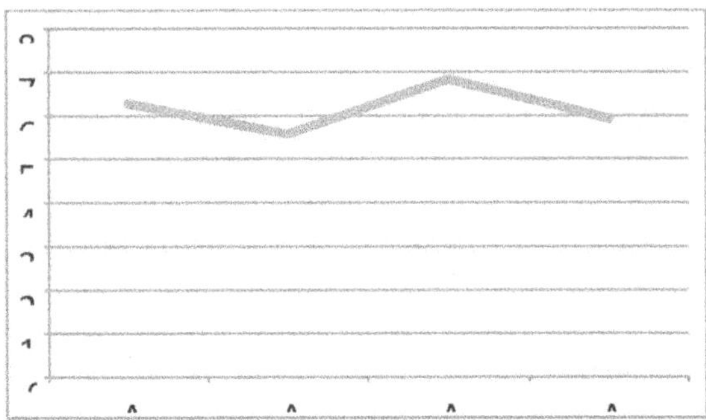

Fig. 45.9 Micro hardness test

7. Density Calculations

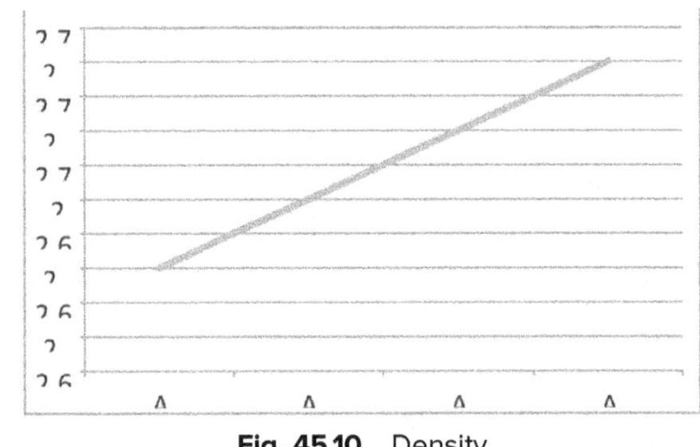

Fig. 45.10 Density

8. Conclusion

Various material shapes of reinforced composites such as A356 alloy, silicon carbide (SiC), and multi-walled carbon nanotubes (MWCNTs) were fabricated by utilizing stir casting process in conjunction with squeeze pressure. According to their mechanical properties evaluated, the most important findings which were noted were:

1. The reinforcement distribution and shape played a significant part in improving the mechanical performance of the composites.
2. Composites containing a combination of reinforcement shapes in the A356 alloy matrix showed higher strengthening effects. Moreover, introduction of SiC also had the effect of grain refinement.
3. Imposing direct squeeze pressure successfully reduced porosity and slag formation and resulted in enhanced material integrity.

4. Grain refinement enhanced tensile strength and ductility of the composites and facilitated enhanced plastic deformation.
5. In MWCNT composites, effective interfacial bonding between reinforcement particles and the matrix amplified the strengthening effect.
6. 1% MWCNT and 0.75% SiC proved to be the optimal reinforcement content, which offered the best balance of mechanical properties.

References

1. R. Taherzadeh Mousavian, S. Behnamfard· R. Azari Khosroshahi, J. Zavašnik, P. Ghosh, S. Krishnamurthy, A. Heidarzadeh, D. Brabazon - Strength-ductility tradeoff via SiC nanoparticle dispersion in A356 aluminium matrix(2020)
2. Sabitha Jannet, E. Pradeep Kumar Reddy, R. Raja, B. Morish Manohar - Effect of SiC Nanoparticles on the Dispersion of Multiwalled Carbon Nano Tubes in AA 5083 by Stir Casting Technique(2020)
3. K Logesh, P Hariharasakthisudhan, A Arul Marcel Moshi, B Surya Rajan and Sathickbasha K - Mechanical properties and microstructure of A356 alloy reinforced AlN/MWCNT/graphite/Al composites fabricated by stir casting(2019)
4. Abou Bakr Elshalakany· Vineet Tirth· Emad El-Kashif·H.M.A. Hussein·W. Hoziefa - Characterization and mechanical properties of stir-rheo-squeeze cast AA5083/MWCNTs/GNs hybrid nanocomposites developed using a novel preformbillet method- 2020
5. MAHENDRA KUMAR C. 1,RAGHAVENDRA B. V. - EFFECT OF CARBON NANOTUBE IN ALUMINIUM METAL MATRIX COMPOSITES ON MECHANICAL PROPERTIES-2020
6. Ahmed Sh. Zayed,Bahaa M. Kamel,Tarek Abdelsadek Osman,Omayma A. Elkady &Shady Ali - Experimental study of tribological and mechanical properties of aluminum matrix reinforced by Al_2O_3/CNTs-2019
7. V. Sivamaran· Dr V balasubramanian·Dr M. Gopalakrishnan· Dr V. Viswabaskaran· Dr A. Gourav Rao·Dr G. Sivakumar - Mechanical and tribological properties of Self-Lubricating Al 6061 hybrid Nano metal matrix composites reinforced by nSiC and MWCNTs-2020
8. M.P. Kuz'min M. Yu. Kuz'mina A.S. Kuz'mina - Production and properties of aluminum-based composites modified with carbon nanotubes-2019
9. Benyamin ABBASIPOUR·Behzad NIROUMAND· Sayed Mahmoud MONIR VAGHEFI·Mohammad ABEDI - Tribological behavior of A356–CNT nanocomposites fabricated by various casting techniques-2019
10. MAHENDRA KUMAR C., RAGHAVENDRA B. V. - EFFECT OF CARBON NANOTUBE IN ALUMINIUM METAL MATRIX COMPOSITES ON MECHANICAL PROPERTIES -2020
11. Suresh S, Sudhakara D - Investigation on mechanical, wear, and machining characteristics of Al 7075/MWCNTs using the liquid state method-2020 [12]. Konada NK, Suman KNS - Tribological Behavior of Multi-walled Carbon

Note: All the figures and tables in this chapter were made by the authors.

Advances in Additive Manufacturing Technologies – Gurusamy Pathinettampadian et al. (eds)
© 2026 Taylor & Francis Group, London, ISBN 978-1-041-16687-0

Optimization of Radiator Fin Design and Thermal Performance Using Finite Element Analysis for Enhanced Heat Dissipation Efficiency

Vasanthamurugan R.[1], Sri Pranavk Kumar S. N.[2], Vincentraj G.[3], Shameer Rahmathulla S.[4], Sathish Krishnan P.[5], Sathiyamoorthi R.[6]

Department of Mechanical Engineering,
Chennai Institute of Technology,
Chennai, Tamilnadu

◆ **Abstract:** In many engineering applications, optimizing radiator fin design is essential for increasing heat dissipation efficiency, boosting thermal performance, and lowering energy consumption. In order to maximize heat transmission while minimizing pressure drop and material utilization, this study uses Finite Element Analysis (FEA) to assess and optimize various fin designs, materials, and configurations. To ascertain their effect on overall thermal efficiency, important characteristics such fin thickness, spacing, form, and material conductivity are examined. Optimized and conventional fin shapes are contrasted with computational simulations for different airflow and thermal load conditions. The results indicate that optimized fin shapes significantly enhance heat dissipation by optimizing the convective and conductive heat transfer mechanisms. This work offers important information that will result in more sustainable and efficient thermal management solutions for the development of high-performance radiators for manufacturing, transportation, and aerospace cooling systems.

◆ **Keywords:** Automobile radiator, Fins, FEA, Optimization

1. Introduction

Thermal management is a critical part of engineering design because systems that generate excess heat need effective heat release in order to maintain the functionality and lifespan of mechanical and electrical components. The cooling system depends on radiators as critical parts because they transfer thermal energy from the system fluid to surrounding air. The performance of radiators relies on how well they conduct heat and the flow characteristics of fluids which in turn make up their thermal performance, most of which are dictated by their fin geometry.

The thermal performance of radiator tubes with various orientations has been examined by Gupta et al. [1]. The investigation shows that the tubes with the 23 inclined axis combination had the highest Nusselt number. The analysis shows that pressure losses are lowest at low angles of inclination and rise with an increase in angle of inclination. The findings show that, in comparison to full tubes and straight tubes, the vent coolant temperature is lowered by two degrees. Any system that must cool more can use these findings with no additional effort. Ryabin et al.'s radiator design was optimized for high-power thyristor immersion cooling [2]. A radiator design that meets the

[1]vasanthamuruganr@gmail.com, [2]pranavkkumar97@gmail.com, [3]vincentrajgm@gmail.com, [4]shameer2K4@gmail.com, [5]sathishkrish870@gmail.com, [6]sathiya.ram78@gmail.com

DOI: 10.1201/9781003685906-46

necessary conditions for the maximum surface contact thyristor temperature was achieved as a result of numerical optimization. A thyristor assembly with radiators immersed in a 3M Novec 649 dielectric was used in an experimental environment in order to confirm the results of numerical modeling. The temperature of one of the radiators near the contact of the thyristor for typical operating conditions was found to be in good agreement with the numerical simulation results.

The effect on engine performance due to a hybrid nano-coolant as well as a nano-composite-aided PCM cooler has been studied by Kumar et al. [3]. Compared to a radiator with a cooler absent, the radiator equipped with a HyN-PCM cooler has the maximum effectiveness increment of approximately 1%. The PCM cooler at high loads must therefore be fitted. The rise of 25 percent pressure loss at low operating temperatures was observed. A HyN-PCM cooler needs to be used as a result of its increased heat transport efficiency, and synthesized HyNc as a replacement to the existing coolants needs to be employed.

A hybrid raditor for electric vehicle traction system has been researched by Paval et al. [4]. The technology is estimated to save INR 14,000 in the life of electric vehicle battery, an energetic performance improvement of 40–65%, and an energy saving of 15% with a reduction of 1.6 tonne CO2 emissions. The radiator's heat transmission efficiency is improved by 4.5 kW. A thermodynamic cycle analysis is carried out to demonstrate the effect of the hybrid cooling on the energy and exergy front in addition to the energy and exergy analysis. The results of the suggested system make it easier to implement a traction system with high torque and power density for electric mobility.

The software application that will be used to analyze thermal performance for various fin designs is called Finite Element Analysis (FEA). The most accurate information on temperature and thermal gradient distribution is provided by FEA, which may also be used to analyze the radiator system's interactions, such as fluid flow, convection, and heat conduction [5-6]. This study aims to design more advanced radiator fins through the integration of Finite Element Analysis and experimental validation, which increases cooling heat dissipation performance, structural strength, and production efficiency. The results of this study will be particularly important for industries that deal with cooling systems. The automotive sector can improve engine cooling through better designed radiator fins and thus improve fuel economy and reduce emissions. For aerospace applications, both space and aircraft systems incorporate radiators for temperature control that must be light and highly effective. Improved designs are important for industrial cooling devices such as air conditioning units and the heat exchangers used in power plants to improve energy efficiency and performance.

The study recommends using Finite Element Analysis to improve the radiator fin design's thermal performance and heat dissipation. This investigation creates multifunctional radiator designs using a systems approach based on fundamental engineering design principles along with material and fluid dynamics constitutive properties. The outcome of this research will facilitate the creation of modern cooling methods within thermal management systems that will strengthen engineering systems technologies.

2. Literature Survey

Various fin geometries have been analyzed through FEA techniques to evaluate their impact on heat transfer performance. Experimental results show that wavy and louvered fins enhance convective heat transfer by developing turbulence in the air stream and reducing thermal resistance resulting in improved cooling performance [9]. Computational simulations have been used by researchers to find the impact of radiator performance on fin thickness and spacing and material properties.

Heat dissipation gets better when fin pitch optimization and surface area enlargement occur without causing unacceptable pressure drops [10]. Research teams employed FEA in combination with optimization techniques to identify radiator fin designs that increase heat transmission capacities while reducing material consumption and aerodynamic drag. Genetic algorithms and artificial neural network approaches were used in the study to increase design performance results. [12]. The experimental validation of the researchers' FEA work showed a strong match between the radiator's planned and actual temperature management capabilities. These results demonstrate that FEA is a practical method for heat analysis and design optimization [13].

3. Materials Methods

Applying Finite Element Analysis (FEA) to simulate a comparison of heat dissipation performance in different configurations, researchers managed to come up with radiator fin structures that could offer increased thermal efficiency. In order to examine the effect of temperature distribution and the behavior of fluid flow, the research was conducted on four configurations: straight, louver, wavy, and perforated structures. Heat and fluid flow analysis remained applicable while meshing parameters were varied to achieve maximum computation efficiency and accuracy. Researchers correlated temperature distribution with heat dissipation rate and pressure drop to determine the best fin

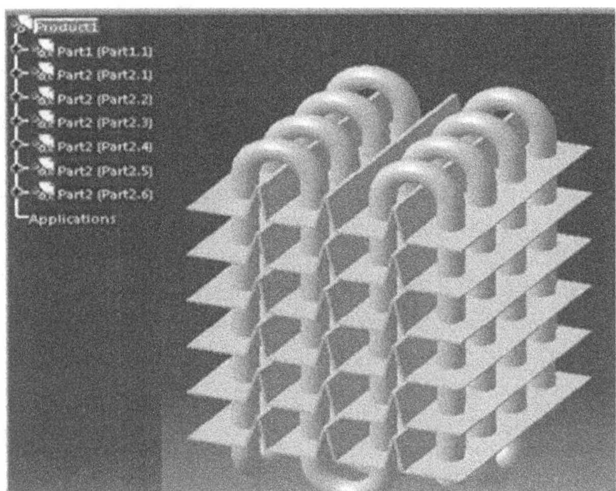

Fig. 46.1 Sharp edge fins

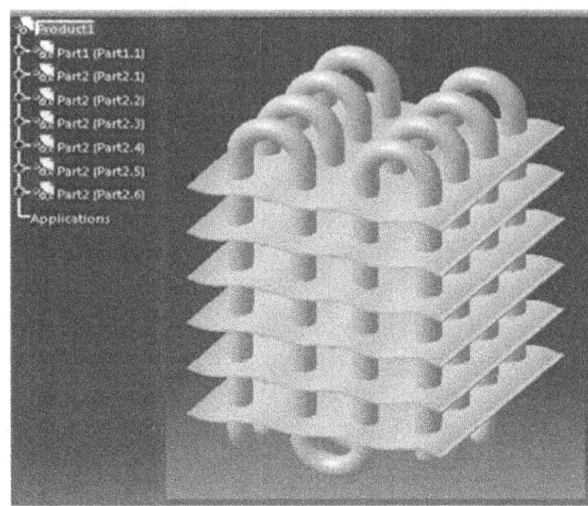

Fig. 46.2 Wavy fins

design. Optimal radiator fin design prediction quality and accuracy were validated by comparing it with test data and existing radiator models.

4. Results and Discussion

4.1 Temperature Distribution and Heat Dissipation Performance

FEA simulations show that heat dissipation efficiency is enhanced when fin geometries are optimized since they enhance convective heat transfer. Temperature distribution analysis shows that increased surface area fins achieved by longer or perforated geometries enhance dispersal of heat and minimize hot spots while maximizing overall cooling efficiency. Optimized fin geometries involving louvered and wavy geometries enhance heat dissipation by 10 to 25% over traditional straight fins by enhancing air turbulence and enhancing the contact area with the coolant fluid [14]. Thermal efficiency of radiator fins is optimized when heat dissipation efficiency is coupled with temperature distribution patterns in Finite Element Analysis (FEA). FEA simulations yield detailed insights into the impact of airflow conditions, material properties and fin geometries on heat transfer mechanisms. Effective fin construction by effective design results in uniform temperature distribution minimizing thermal hotspots and maximizing cooling efficiency. Optimised fin designs such as wavy or louvred fins generate airflow turbulence which enhances heat dissipation and convective heat transfer [15]. Research reveals that aluminum alloys and similar thermally conductive materials facilitate rapid heat distribution which creates uniform temperature gradients. Nonetheless, over-density of fins, though enhancing heat dissipation, may cause enhanced pressure drops and reduced airflow, which negatively impacts thermal performance. FEA enables the accurate assessment of such factors to assist engineers in designing radiator fins that provide optimal balancing between efficient temperature distribution and improved heat dissipation, and consequently enhance the efficiency of the cooling system [16].

4.2 Effect of Fin Geometry on Thermal Performance

The research contrasts various fin arrangements, such as straight, wavy, louvered, and perforated fins. The wavy and louvered fins have higher performance because the disruption of air flow is enhanced, leading to improved convective heat transfer. The perforated fin arrangement also helps in the reduction of thermal resistance, with improved heat transfer efficiency. On the other hand, though raising the fin density enhances thermal performance, excessive density increases pressure drop and decreases air flow, so a balanced approach is necessary [12].

4.3 Influence of Material Selection on Heat Transfer

Material properties like weight and thermal conductivity are vital in radiator efficiency. Aluminum alloys, with high conductivity and low weight, are better than steel-based fins. The simulations show that high-conductivity materials can increase the efficiency of heat dissipation by 15-20% and are best used in automotive as well as industrial use [8].

4.4 Airflow and Pressure Drop Considerations

A key element in radiator performance is the balance between increased heat transfer and air flow resistance. High-density fin designs enhance cooling but at the cost

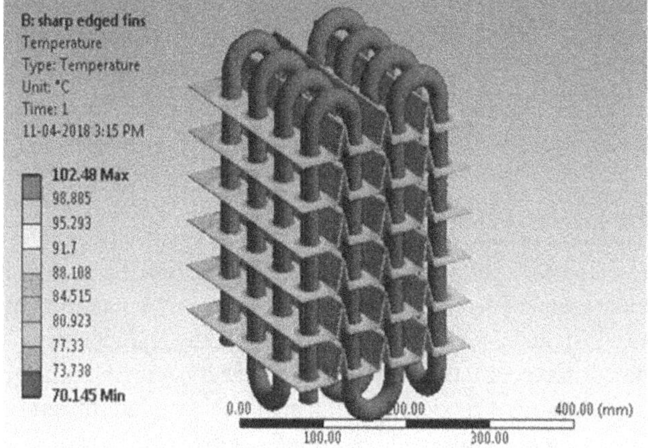

Fig. 46.3 Temperature distribution for sharp edge fins (aluminium)

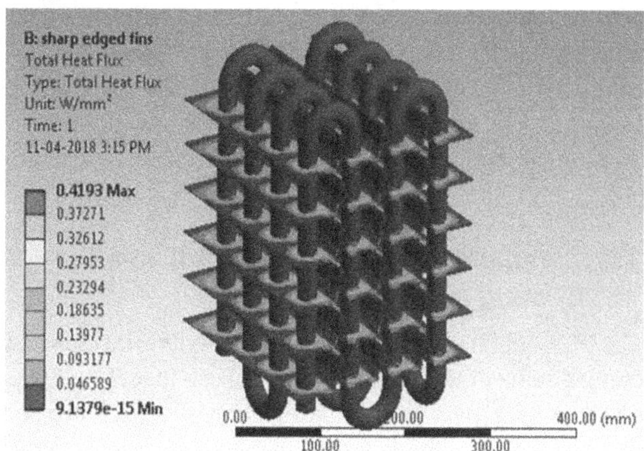

Fig. 46.4 Total heat flux for sharp edge fins (aluminium)

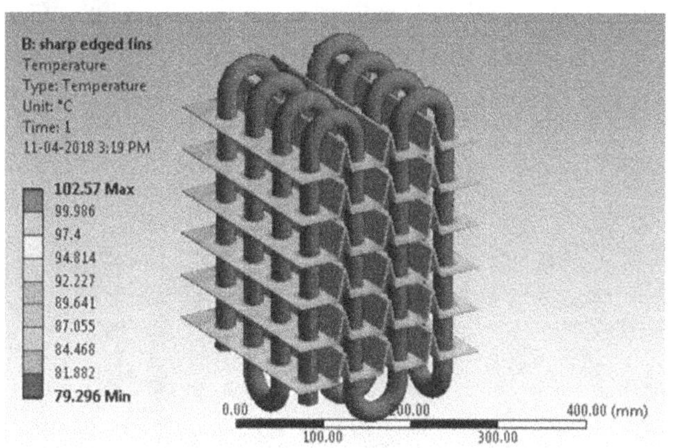

Fig. 46.5 Temperature distribution for sharp edge fins (copper)

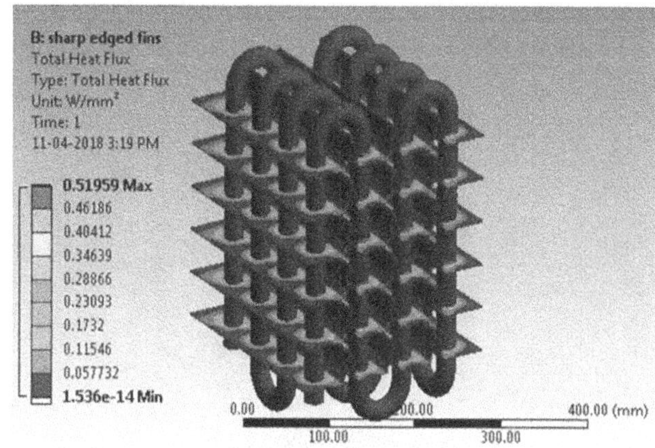

Fig. 46.6 Total heat flux for sharp edge fins (copper)

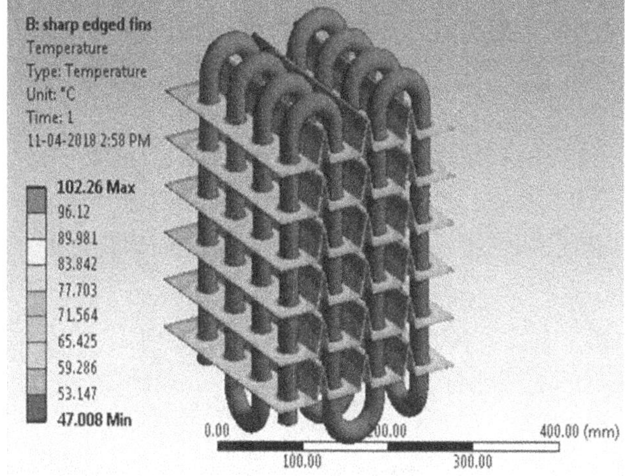

Fig. 46.7 Temperature distribution for sharp edge fins (steel)

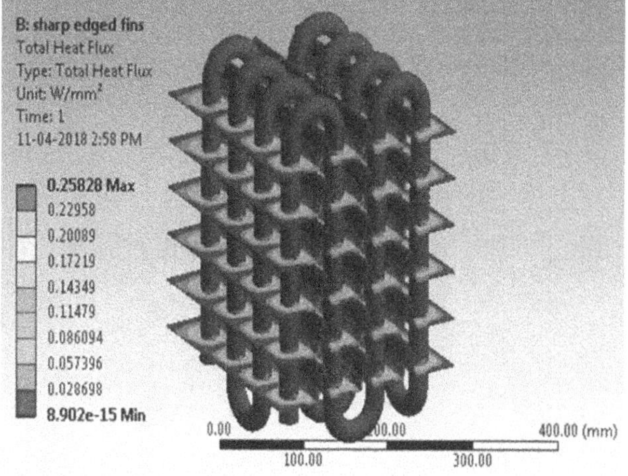

Fig. 46.8 Total heat flux for sharp edge fins (steel)

of increased pressure drop, which adversely affects system efficiency. The research determines that louvered fin configurations offer the most optimal balance between enhanced heat transfer and tolerable airflow resistance, thus an appropriate option for high-performance radiators [10].

4.5 Comparison with Experimental and Existing Models

The FEA outputs are compared with experimental data and current radiator designs. The optimized models closely match actual performance, validating the accuracy of the simulations. The research proves that incorporating computational optimization in radiator fin design results in significant performance improvements of up to 30% over conventional designs. The results indicate that optimized fin configurations can improve substantially the performance of radiators applied in automotive, aerospace, and industrial cooling applications. Future work may incorporate machine learning algorithms to optimize design in real time and investigate the effects of emerging manufacturing methods like 3D printing to create innovative fin shapes with improved thermal characteristics [11].

5. Conclusion

In summary, improving radiator fin design using finite element analysis (FEA) is important in terms of maximizing the efficiency of heat dissipation and, by extension, the performance of thermal management systems. Utilizing computational modeling, engineers can evaluate different fin geometries, materials, and configurations to find designs that yield maximum The heat transfer with minimum pressure drop and material consumption. The knowledge obtained using FEA can be used to design more efficient and space-saving radiator designs with applications in vehicle, aircraft, and industrial cooling systems. With

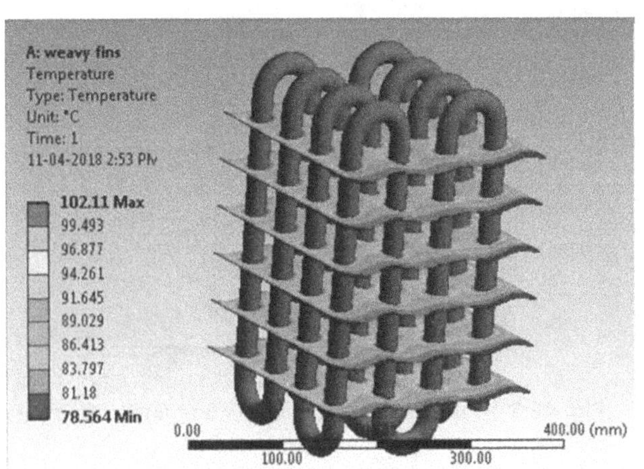

Fig. 46.9 Temperature distribution for wavy fins (aluminium)

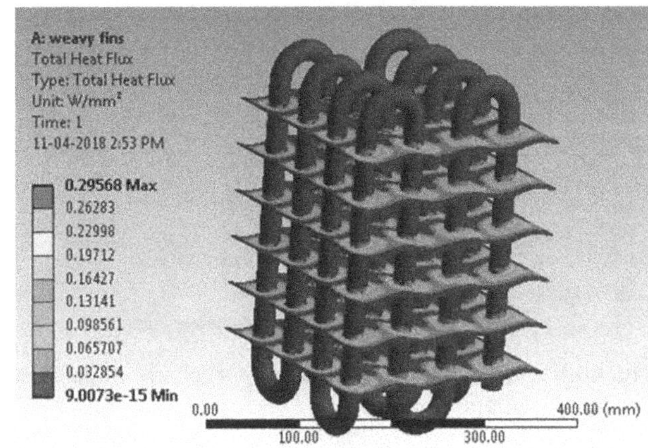

Fig. 46.10 Total heat flux for wavy fins (aluminium)

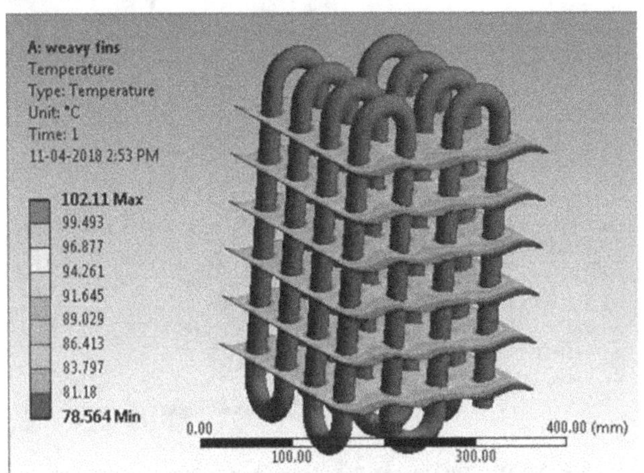

Fig. 46.11 Temperature distribution for wavy fins (copper)

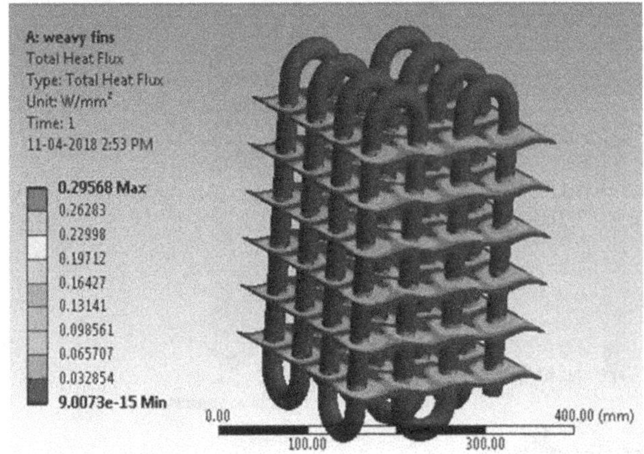

Fig. 46.12 Total heat flux for wavy fins (copper)

advancements in technology, coupling machine learning and real-time simulation methods with FEA has the potential to further optimize radiator fin designing, leading towards higher-performance and energy-efficient cooling systems.

References

1. Gupta, R.M., Painuly, A., Zainith, P. et al. Thermal performance of radiator tubes for multiple orientations for engine cooling. J Mech Sci Technol (2025). https://doi.org/10.1007/s12206-025-0241-2
2. Ryabin, T.V., Yankov, G.G., Artemov, V.I. et al. Optimization of Radiator Design for Immersion Cooling of a Powerful Thyristor. Therm. Eng. 71, 867–877 (2024). https://doi.org/10.1134/S0040601524700381
3. kumar, V.A., Arivazhagan, S. Influence of hybrid nano-coolant on the performance of engine radiator coupled with nano-composite-assisted PCM cooler. J Therm Anal Calorim 149, 11941–11961 (2024). https://doi.org/10.1007/s10973-024-13529-2
4. Paval, P.S., Chandrakanth, B., Mallisetty, P.K. et al. Experimental assessment of a hybrid radiator for pure electric vehicle traction system: a 4E approach (energy, exergy, environment and economy). J Therm Anal Calorim (2025). https://doi.org/10.1007/s10973-025-14086-y
5. Kim, T. K., Lee, S. M., & Pae, S. M. (2017). Method of predicting radiator temperature distributions for thermal fatigue analysis. Journal of Mechanical Science and Technology, 31(10), 5059–5066.
6. ROBIN, R., Hariram, V., & Subramanian, M. (2017). Probabilistic finite element analysis of a heavy duty radiator under internal pressure loading. Journal of Engineering Science and Technology, 12(9), 2438–2452.
7. Dong, J. S., & Gu, B. Q. (2010). Coupling analysis of heat transfer in finned radiator based on numerical simulation codes ABAQUS and FLUENT. Advanced Materials Research, 118, 635–639.
8. Kumar, B. R., Vigneshwaran, K., & Antony, D. A. (2022). CFD analysis of different radiator surfaces to optimize heat transfer coefficient. Materials Today: Proceedings, 68, 2334–2341.
9. Sharma, S., & Yadav, S. (2023). Numerical Analysis of Fin Heat Transfer in Radiators Using Simulation Software Comsol Multiphysics 5.5. In Sustainable Material, Design, and Process (pp. 143–167). CRC Press.
10. Ng, E. Y., Johnson, P. W., & Watkins, S. (2005). An analytical study on heat transfer performance of radiators with non-uniform airflow distribution. Proceedings of the Institution of Mechanical Engineers, Part D: Journal of Automobile Engineering, 219(12), 1451–1467.
11. Habibian, S. H., Abolmaali, A. M., & Afshin, H. J. A. T. E. (2018). Numerical investigation of the effects of fin shape, antifreeze and nanoparticles on the performance of compact finned-tube heat exchangers for automobile radiator. Applied Thermal Engineering, 133, 248–260.
12. Waqas, H., Khan, S. A., Yasmin, S., Liu, D., Imran, M., Muhammad, T., ... & Farooq, U. (2022). Galerkin finite element analysis for buoyancy driven copper-water nanofluid flow and heat transfer through fins enclosed inside a horizontal annulus: Applications to thermal engineering. Case Studies in Thermal Engineering, 40, 102540.
13. Fei, C. W., Li, C., Lin, J. Y., Han, Y. J., Choy, Y. S., & Chen, C. H. (2024). Structural design of aeroengine radiators: State of the art and perspectives. Propulsion and Power Research.
14. Majmader, F. B., & Hasan, M. J. (2023). Thermal enhancement and entropy generation of an air-cooled 3D radiator with modified fin geometry and perforation: A numerical study. Case Studies in Thermal Engineering, 52, 103671.
15. Rahmati, A. R., & Gheibi, A. (2020). Experimental and numerical analysis of a modified hot water radiator with improved performance. International Journal of Thermal Sciences, 149, 106175.
16. Habeeb, H., Mohan, A., Norani, N., Azman, M., & Abdullah, M. H. H. (2020). Analysis of engine radiator performance at different coolant concentrations and radiator materials. International Journal of Recent Technology and Engineering, 8(6), 2277–3878.

Note: All the figures in this chapter were made by the authors.

Advances in Additive Manufacturing Technologies – Gurusamy Pathinettampadian et al. (eds)

The Art of Scanning: A Review of Laser Path Strategies in Additive Manufacturing for Complex Geometries

Ravi Samraj[1], Avinash Narakula[2], Gowri Prasath C.[3], Parasuraman S.[4], Haarshit K.[5], B. Krishna Kumar[6]

Department of Mechanical Engineering,
Chennai Institute of Technology,
Chennai, Tamilnadu, India

◆ **Abstract:** A key technology for creating complex geometries in a variety of industries, such as aerospace, automotive, and healthcare, is additive manufacturing (AM). However, the scanning techniques used in the fabrication process have a significant impact on the accuracy and quality of AM parts. With an emphasis on how they affect surface finish, part accuracy, mechanical qualities, and overall efficiency, this review examines the different laser path strategies used in metal-based additive manufacturing. In the context of complex geometries, various scanning techniques—such as raster, zigzag, spiral, and adaptive strategies—are examined, along with their advantages and disadvantages. The review also emphasizes how residual stresses, heat distribution, and the part's final microstructure can all be impacted by the laser path patterns chosen. Additionally, the interaction between scanning techniques.

◆ **Keywords:** Additive manufacturing, Zigzag, Spiral, Laser path

1. INTRODUCTION

Using raw materials to directly and layer-by-layer fabricate 3D components from digital models, metal additive manufacturing (AM), also referred to as metal three-dimensional (3D) printing, is a new and promising technique for creating net-shaped parts with complex geometries. In contrast to conventional because of the benefits of mould reduction, high flexibility, low consumption, short cycles, and customised fabrication of complex structures, additive manufacturing (AM) techniques—also known as subtractive or constant volume manufacturing—have been becoming more and more popular in both industry and academia. The popular selective laser melting (SLM) technique, in particular, is an alluring technique that employs a laser to melt metallic powders along a predetermined scanning path and, once the molten powders solidify, spreads a new powder layer on top of earlier layers [1–5]. Because of its potential to help rapidly develop industries such as aviation, aerospace, medical equipment, automobile manufacturing, and others, selective laser melting (SLM) of metal/alloy powder has emerged as one of the most intriguing and promising rapid prototyping technologies. Taking advantage of its built-in benefits, the SLM has access to the following huge opportunities [6–9].

It can be used for a wide range of metals and can handle extremely complex geometries; the laser powder bed fusion (LPBF) additive manufacturing (AM) process is becoming more and more popular. However, the laser scan pattern's imparted local and global thermal history, the need to

[1]sravi@citchennai.net, [2]avinash.narakula1@gmail.com, [3]gowriprasath07@gmail.com, [4]gokulshan2410@gmail.com, [5]haarshit9080@gmail.com, [6]researchkrishna78@gmail.com

DOI: 10.1201/9781003685906-47

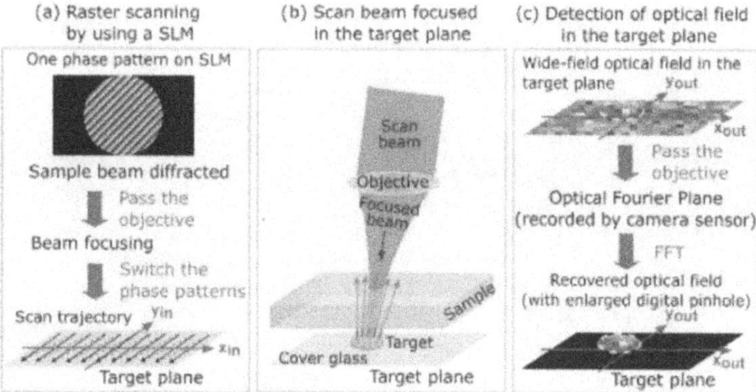

Fig. 47.1 Raster scanning laser path technique [16]

manage residual stresses, mitigate defects and microcracks, and achieve a spatially homogeneous microstructure present significant implementation challenges for LPBF [10–12]. Generally speaking, microcracking brought on by extreme residual stresses can be caused by spatially uneven heating and cooling; this is especially problematic in alloys with high melting temperatures, high thermal conductivities, and/or high DBTTs (ductile-to-brittle transition temperatures) [13–14].

1.1 Raster Scanning

A popular laser path technique in additive manufacturing (AM) and laser-based procedures like Selective Laser Melting (SLM) and Laser Powder Bed Fusion (LPBF) is raster scanning. Similar to how a computer screen is scanned to display an image, raster scanning involves the laser beam moving back and forth lines in a methodical, parallel fashion [5].

1.2 Overview of Raster Scanning

In raster scanning, the laser beam sweeps horizontally across the material, creating a series of lines (scan lines). After completing one line, the beam quickly returns to the starting position to begin the next line. This method allows for efficient coverage of the surface area being processed, ensuring that each section is melted or fused uniformly. The technique is particularly effective in controlling the thermal profile and ensuring consistent layer adhesion in AM applications [15]

2. Key Features of Raster Scanning

The key features of raster scanning is Linear Pattern, The laser follows a series of linear paths (or “raster lines”) across the build area. The direction of each pass alternates, often from left to right and then right to left, covering the entire layer of material. The next key features of the raster scanning is Layer-by-Layer, The procedure is carried out again for every layer of the component, with the laser drawing paths on each succeeding layer according to the geometry of the CAD model. The final key features of the raster scanning is Filling Strategy, Raster scanning usually creates a continuous “scan” of the material by directing the laser beam over the entire layer area. A variety of patterns, such as straight lines or more complex methods like zigzag or spiral scanning, can be used for this.

3. Materials and Methods

This section presents the experimental setup for fast localization and high-resolution visualization of various subsurface defects of AM components using a laser ultrasonic scanning system. The main schematic of the experimental setup is a combination of a pulsed laser, a scanning laser Doppler vibrometer (LDV), and an automatically controlled translational stage, as shown in Fig. 47.1 (a).

4. Process

In LAM, raster scanning serves the following main purposes:

- Detailed features and complex designs can be produced thanks to precision manufacturing.
- Material Efficiency: Lowers waste by utilising just what is required.
- Customisation: Enables quick prototyping and the creation of parts that are specifically suited for a given application.

5. Vector Scanning

A laser beam is guided along predetermined paths or vectors in the laser additive manufacturing (LAM) process

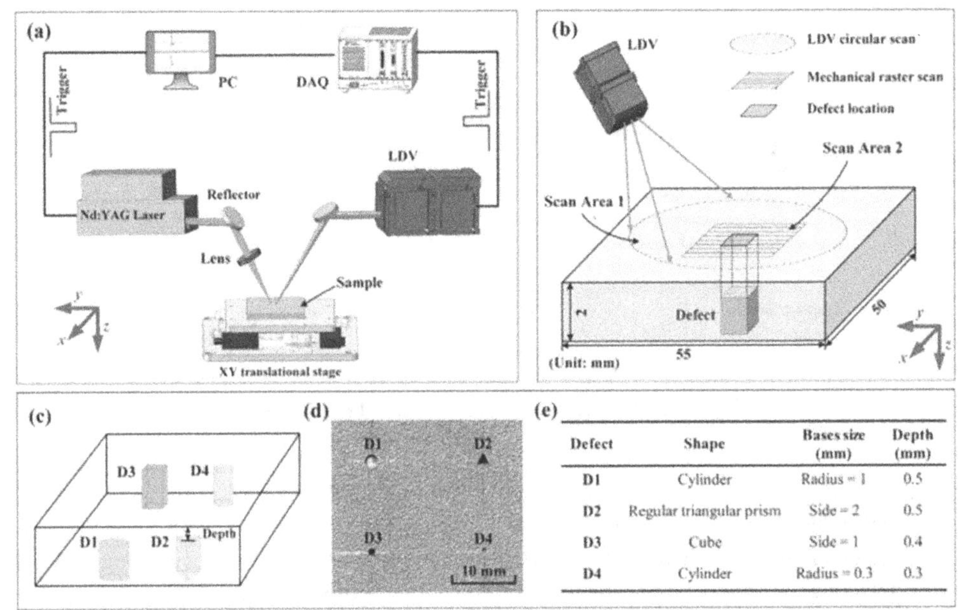

Defect	Shape	Bases size (mm)	Depth (mm)
D1	Cylinder	Radius = 1	0.5
D2	Regular triangular prism	Side = 2	0.5
D3	Cube	Side = 1	0.4
D4	Cylinder	Radius = 0.3	0.3

Fig. 47.2 Electrical dischargee machining [11]

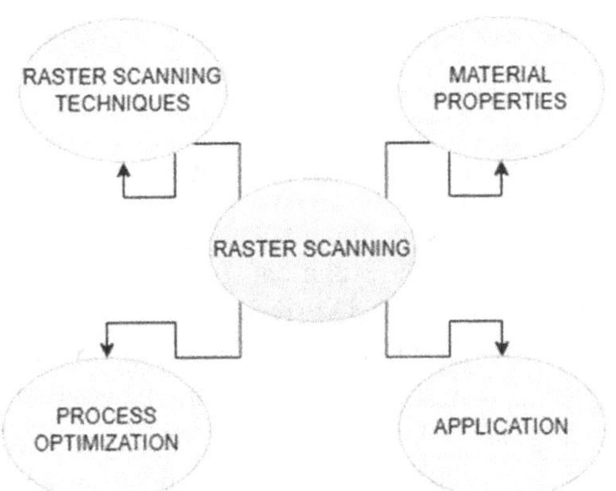

Fig. 47.3 Flow chart for process optimization for raster scanning

known as "vector scanning," which melts or sinters material layer by layer. In contrast to this technique, raster scanning involves line-by-line movement of the laser. There are clear benefits to vector scanning in terms of accuracy and laser parameter control. Thus the, process of moving the laser beam along preset paths (vectors) that match the geometry of the part being manufactured is known as vector scanning. This technique can improve efficiency and shorten build times by enabling the laser to move continuously without pausing in between scan lines. The digital 3D model is sliced into thin layers. For each layer, vector paths are calculated for the areas that need to be solidified, forming the outline and internal structure of the layer.

5.1 Process using Vector Scanning

Vector Scanning Selective Laser Melting (SLM) Applications

Just One Scan Vector Prediction: Single-vector predictions of the individual laser-melted tracks are used by SLM for optimal process settings and a guarantee of printed object stability2. The material characteristics and thermal dynamics of the formed structures depend directly on the scan vector.

Steps in Vector Scanning Process for Additive Manufacturing:

1) **Model Slicing:** To start, the 3D model is separated into 2D layers, each of which shows the cross-section of an object. Each slice is then transformed into a vector path that details the precise motion of the energy source. In additive manufacturing (AM), especially in procedures like electron beam melting (EBM) and selective laser melting (SLM), model slicing is an essential step. In this procedure, a 3D model is transformed into a number of 2D cross-sectional layers, which serve as a guide for the printer as it constructs the object layer by layer.
2) **Toolpath Generation:** To direct the energy source along particular vectors, a toolpath is created for every layer. Depending on the fill strategy, toolpaths can change:
3) **Vector Scanning Execution**: The material is selectively melted or sintered as the laser or electron

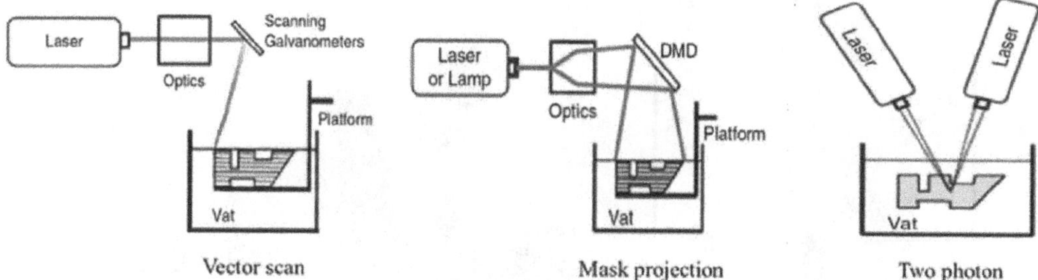

Fig. 47.4 Various photo polymerization mechanisms in vat-based methods [8]

beam follows these paths. High resolution and complex geometries are possible in intricate designs because the energy source precisely follows the vector directions.

4) **Layer-by-Layer Fusion:** The newly solidified layer merges with earlier layers as the material cools and solidifies. Until the full 3D object is constructed, this process is repeated.

5.2 Impact on Mechanical Properties

The choice of scanning strategy directly influences several mechanical properties:

- **Microstructure:** Different strategies lead to variations in grain structure and orientation, affecting hardness and strength. For instance, island scanning tends to yield a bimodal grain structure that enhances hardness and load-bearing capacity [22]

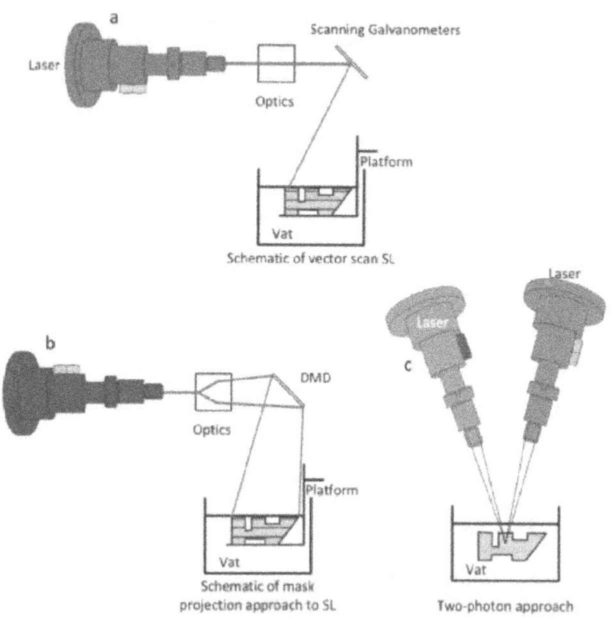

Fig. 47.5 Illustration of the vat photo polymerization process [15]

- **Wear Resistance:** The island scanning strategy has been noted for its lower friction coefficients and wear rates compared to continuous and alternate scanning methods [11]
- **Residual Stresses:** Scanning patterns can induce different levels of thermal stress within the material, which can lead to warpage or distortion during cooling. Optimizing scan strategies can help mitigate these effects [19–20].

5.3 Impact on Material Properties

The scanning strategy has been used for significant impact on the mechanical properties of the parts made using additive manufacturing (AM). This influence is the evident in three important areas: microstructure, surface topography, and remaining stresses.

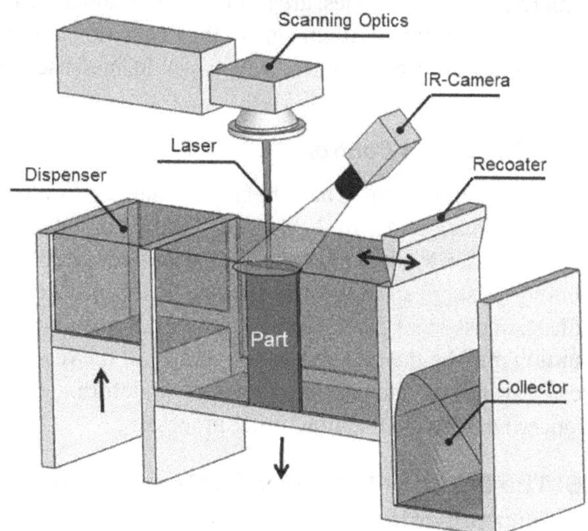

Fig. 47.6 Selective laser melting (SLM) [21]

5.4 Medical Imaging: Selective Computed Tomography (CT)

A study comparing selective CT versus routine thoracoabdominal CT in trauma patients highlighted the

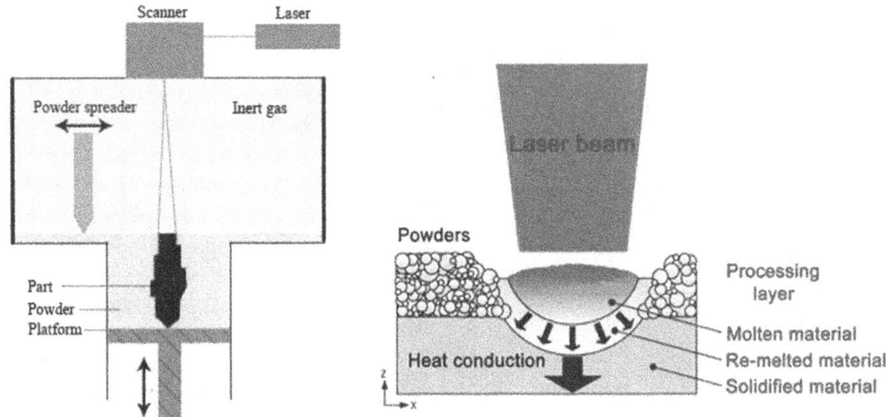

Fig. 47.7 The slim process [17]

potential benefits of routine scanning. It found that while selective CT relies on physical examinations and clinical reasoning, which may miss injuries due to low sensitivity, routine CT provides comprehensive anatomical information that can lead to quicker and more accurate assessments of injuries. This could improve patient outcomes by ensuring that critical injuries are not overlooked during initial evaluations[19]

5.5 Visual Scanning in Psychology

Infants research on visual scanning patterns showed that their fixation sequences when viewing social stimuli, as compared to nonsocial ones, are more complex and develop with age. Therefore, as infants grow, they become better at focusing on faces and bodies, which would increase the ability for social interaction[23].

1. Residual Stress Control

This paper reports on "The Effect of Scanning Strategy on the Residual Stress of 316L Steel Parts Fabricated by Selective Laser Melting (SLM)." The results revealed that scanning strategies change the residual stress profiles of 316L stainless steel parts. Consequently, spiral divisional scanning may be the best strategy if control of the residual stress distribution and possible reduction in deformation is a concern during the manufacturing process.

5.6 The Effect of Scanning Strategy on the Microstructure

Thermal gradient, direction of heat flow, and cooling rate have dramatic impacts on the dislocation density, grain size, solidification cell size, grain aspect ratio, and texture index [17–19]. Generally, a smaller heat gradient to the solidificationrate and a higher cooling rate are conducive to further microstructure refinement and a decrease in the strength of the texture, since the solidification mode changes from columnar to equiaxed [10–15]. Several studies have been performed, showing the effect ofscanning strategy on thermal behavior, densification, and microstructure of the printed structure. The outline of the main contents discussed in this part is shown in Fig. 47.4.

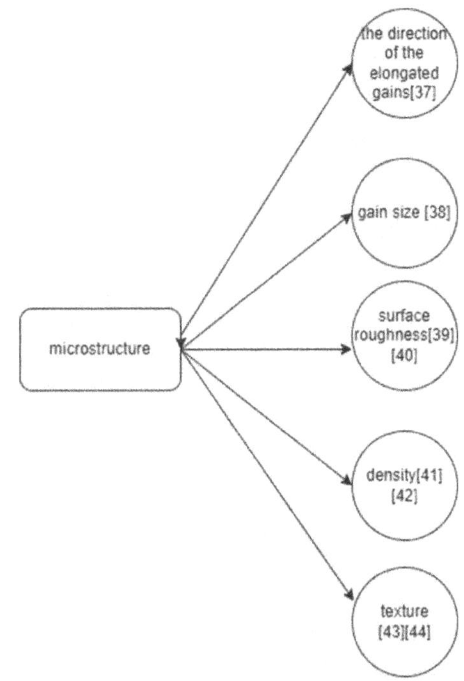

Fig. 47.8 The outline of the effect of scanning strategy on micro-structure

6. Conclusion

Additive manufacturing techniques, particularly laser powder bed fusion (LPBF), allow for unprecedented control over the microstructure of metals, enabling the design of materials with tailored properties. For instance, the ability

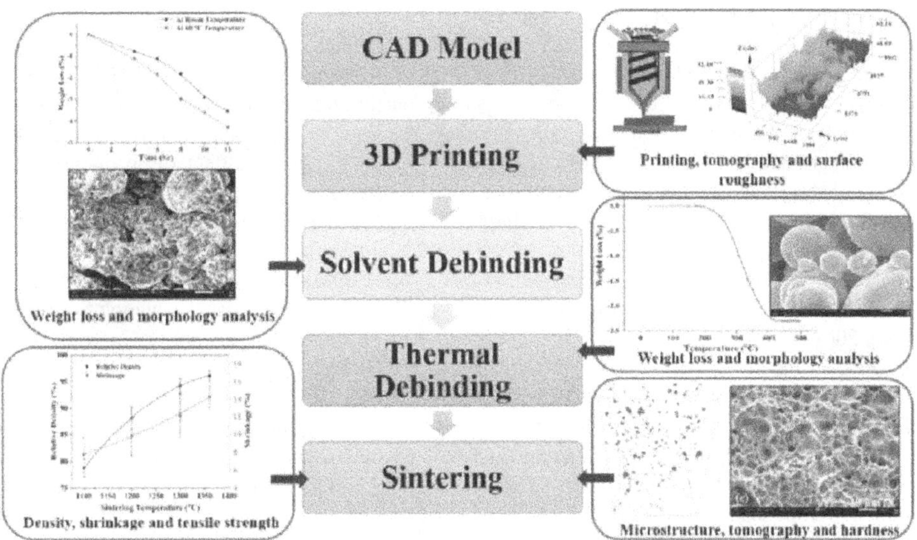

Fig. 47.9 A process in microstructure [19]

to program microstructural features site-specifically allows for the creation of heterogeneous microstructures that can enhance mechanical performance through mechanisms like hetero-deformation induced strengthening [20–25]. Moreover, the unique thermal histories experienced by materials during AM lead to distinctive microstructural characteristics, such as columnar grains and dislocation cells, which differ significantly from those produced by traditional manufacturing methods. These characteristics can contribute to superior mechanical properties, including increased strength and ductility [12–15]. The ongoing research into understanding and optimizing these microstructural outcomes is vital for advancing AM technologies and broadening their applications across industries such as aerospace and biomedical engineering [15].

References

1. Lv, G., Yao, Z., Chen, D., Li, Y., Cao, H., Yin, A., ... & Guo, S. (2023). Fast and high-resolution laser-ultrasonic imaging for visualizing subsurface defects in additive manufacturing components.
2. Davi Ramos, Fawzi Belblidia, Johann Sienz have evaluated New scanning strategy to reduce warpage in additive manufacturing https://doi.org/10.1016/j.addma.2019.05.016
3. Yun JiaScanning strategy optimization for the selective laser melting additive manufacturing of Ti6Al4 https://iopscience.iop.org/article/10.1088/2631-8695/acbd12
4. k.ren thermo-mechanical analyses for optimized path planning in laser aided additive manufacturing processes http://dx.doi.org/10.1016/j.matdes.2018.11.014
5. Haolin Jia1 & Hua Sun1 & Hongze Wang Scanning strategy in selective laser melting (SLM) https://sci-hub.st/10.1007/s00170-021-06810-3
6. Naol Dessalegn Dejene Current Status and Challenges of Powder Bed Fusion-Based Metal Additive Manufacturing selective scanning.pdf
7. Ze-Chen Fang Review on residual stress in selective laser melting additive manufacturing of alloy part svector scanning.pdf
8. GaolongLv have evaluated Fast and high-resolution laser-ultrasonic imaging forvisualizing subsurface defects in additive manufacturing components www.elsevier.com/locate/matdes
9. Chao Wei have evaluatedAn overview oflaser-based multiple metallic material additive manufacturing: from macro- to micro-scales https://doi.org/10.1088/2631-7990/abce04
10. Suh In Kim & A. John Hart have evaluated A spiral laser scanning routine for powder bed fusion inspired by natural predator-prey behaviour https://doi.org/10.1080/17452759.2022.2031232
11. Marco Mazzarisi have evaluated Thermal monitoring of laser metal deposition strategies using infrared thermography https://doi.org/10.1016/j.jmapro.2022.11.067
12. Jacopo Lettori, Roberto Raffaeli have evaluated A review of geometry representation and processing methods for cartesian and multiaxial robot-based additive manufacturing https://link.springer.com/article/10.1007/s00170-022-10432-8
13. Thomas Feldhausen have evaluated Review of Computer-Aided Manufacturing (CAM) Strategies for Hybrid Directed Energy Deposition :10.1016/j.addma.2022.102900
14. Mazher Iqbal Mohammed have evaluated A review of laser engineered net shaping (LENS) build and process parameters of metallic parts 10.1108/RPJ-04-2018-0088
15. Bo, Q., S. Yu-sheng, W. Qing-song, and W. Hai-bo. 2012. "The Helix Scan Strategy Applied to the Selective Laser Melting." The International Journal of Advanced Manufacturing Technology 63: 631–640. doi:10.1007/s00170-012-3922-9

16. Blasius, B., L. Rudolf, G. Weithoff, U. Gaedke, and G. F. Fussmann. 2020. "Long-term Cyclic Persistence in an Experimental Predator-Prey System." Nature 577: 226–230. doi:10.1038/s41586-019-1857-0
17. Yaqun Liu, Effect of Scanning Strategies on the Microstructure and savMechanical Properties of Ti-22Al-25Nb Alloy Fabricated through Selective Laser Melting. *Metals* 2023, *13*(3), 634; https://doi.org/10.3390/met13030634
18. Schmidt M, Pohle D, Rechtenwald T (2007) Selective laser sintering of PEEK. CIRP Ann 56(1):205–208. https://doi.org/10.1016/j.cirp.2007.05.097
19. Van Vugt, R., Keus, F., Kool, D., Deunk, J., & Edwards, M. (2013). Selective computed tomography (CT) versus routine thoracoabdominal CT for high-energy blunt-trauma patients. *Cochrane Database of Systematic Reviews*, (12).
20. L., Dimitrov, D., Matope, S., & Yadroitsev, I. (2019). Evaluation of the impact of scanning strategies on residual stresses in selective laser melting. *The International Journal of Advanced Manufacturing Technology*, *102*, 2441–2450.
21. Liu, Y., Shan, Z., Yang, X., Jiao, H., & Huang, W. (2023). Effect of Scanning Strategies on the Microstructure and Mechanical Properties of Ti-22Al-25Nb Alloy Fabricated through Selective Laser Melting. *Metals*, *13*(3), 634.
22. Wan, H. Y., Zhou, Z. J., Li, C. P., Chen, G. F., & Zhang, G. P. (2019). Effect of scanning strategy on mechanical properties of selective laser melted Inconel 718. *Materials Science and Engineering: A*, *753*, 42–48.
23. Tomalski, P., López Pérez, D., Radkowska, A., & Malinowska-Korczak, A. (2021). Selective changes in complexity of visual scanning for social stimuli in infancy. *Frontiers in Psychology*, *12*, 705600.
24. Nishitani, T., Arakawa, Y., Noda, S., Koizumi, A., Sato, D., Shikano, H., ... & Amano, H. (2022). Scanning electron microscope imaging by selective e-beaming using photoelectron beams from semiconductor photocathodes. *Journal of Vacuum Science & Technology B*, *40*(6).
25. Van Vugt, R., Keus, F., Kool, D., Deunk, J., & Edwards, M. (2013). Selective computed tomography (CT) versus routine thoracoabdominal CT for high-energy blunt-trauma patients. *Cochrane Database of Systematic Reviews*, (12).

Advances in Additive Manufacturing Technologies – Gurusamy Pathinettampadian et al. (eds)
© 2026 Taylor & Francis Group, London, ISBN 978-1-041-16687-0

Experimental Investigation and Analysis of Leaf Spring with GFRP Reinforced with Epoxy Composite for Vehicle Suspension

Velmurugan S.*,
Hari Krishna Raj S., Aadhish Kumar S.,
Kathir K., Kowsalya M.
Department of Mechanical Engineering,
Veltech High Tech Dr. Rangarajan Dr. Sakunthala Engineering College,
Chennai, India

◆ **Abstract:** This research completely examines the possibility of using polymer composite materials in leaf springs for suspension systems. It integrates the design, analysis, and experimental testing to assess the performance of Glass Fiber Reinforced Polymer (GFRP) composite laminates based on an epoxy matrix against the traditional steel leaf springs. Design uses CATIA software to generate optimized models for leaf springs, considering geometry and size. Structural analysis, conducted using ANSYS software, analyzes the mechanical performance of the steel and composite leaf springs under different loading conditions.

Experimental tests, such as tensile, compression, and impact tests, are performed on GFRP samples to confirm their mechanical properties. The findings reveal that GFRP composites are superior to steel, with greater tensile strength, improved compression resistance, and higher impact resistance. Moreover, the composite material has the benefits of enhanced cushioning, fatigue resistance, and corrosion resistance, which qualify it as a superior leaf spring material. These results indicate that polymer composites are able to enhance the functionality and lifespan of suspension systems, leading to more efficient and longer-lasting vehicles.

◆ **Keywords:** Leaf spring, Epoxy composite, GFRP, Vehicle suspension, ANSYS

1. Introduction

A four-leaf steel spring in general use in the rear suspension of light vehicles was analyzed through the use of ANSYS V5.4 software. The finite element analysis outputs of stresses and deflections were compared against available analytical and experimental results. Based on this, a composite leaf spring consisting of fiberglass reinforced with epoxy resin was designed and optimized through the same software [1–5]. The main concern for the optimization was the spring geometry to design a lightweight spring that could sustain given static loads without fracture [6–8].

The design procedure obeyed stress limitations and displacement limitations evaluated with the aid of the Tsai-Wu failure criterion. The results identified that the most favorable design presented a hyperbolic taper of width and linear thickness tapering from spring eyes to axle seat. Compared with the steel spring, the optimum composite spring registered lower levels of stress, greater natural frequency, and nearly 80% decrease in weight [9–14].

The design development of a composite leaf spring for heavy rail freight applications consisted of the testing of three varying eye-end attachment designs. The glass fiber-

*Corresponding author: ivelu00@gmail.com

DOI: 10.1201/9781003685906-48

reinforced polyester material was used in each design. Static testing and finite element analysis (FEA) were utilized to evaluate the performance of the spring. These tests gave an indication of the spring's properties, such as load-deflection behavior and strain in terms of applied load. The test results were then compared with FEA predictions to validate them [10].

In the initial design, failure due to delamination took place at the interfacial region between the wrapping fibers and the spring body even though the spring endured a static proof load of 150 kN and one million cycles fatigue load. The critical region of high interlaminar shear stress concentration was indicated by FEA as the root cause of the failure [11].

The second model used a transverse bandage over the delamination-susceptible area to reduce the issue. While this modification reduced delamination, it did not entirely eliminate it.

The third design effectively addressed the issue by terminating the fibers at the end of the eye section, thereby preventing delamination and ensuring better structural integrity. This iterative design process demonstrates the progression toward a more robust and reliable composite leaf spring suitable for heavy-duty rail applications [12].

The bending behavior of a stack of slim, non-uniform curved beams (leaves) with rectangular cross-sections is analyzed, focusing on weak joint interactions characterized by unbonded contact without friction. Each leaf in the stack is clamped at one end and free at the other, with all leaves having identical widths but varying lengths that decrease progressively from the bottom to the top. The structure is modeled as a leaf spring, where an upward load is applied to the bottommost leaf.

The primary challenge lies in determining the deformation shapes of the leaves under bending. This is approached by resolving the interaction force densities between the leaves. A precise formulation of this problem is provided, and the uniqueness of its solution is established. For the specific case of two uniform, straight leaves, an analytical solution is derived, offering insight into the bending mechanics of this type of leaf spring structure. Analyzes the influence of different loading conditions on stress distribution in pin-loaded woven glass fiber-reinforced epoxy laminate chain components employed for load conveyance. Numerical and experimental methods were used to investigate the stress behavior of these composite parts. Experimental equipment was established to mimic the actual motion of chain components under two different working conditions. The initial condition was that the chain components were in motion without a load, and the second condition was that the components were in contact with and moving a load.

For numerical analysis, the ANSYS software package was used, with three-dimensional eight-noded layered structural solid elements. Tensile loads of 250, 500, 750, 1000, and 1250 N were applied by pins under both conditions. Experimental and numerical results for the two conditions and different tensile loads were compared and analyzed. The results indicated a good correlation between the experimental data and the numerical predictions, confirming the validity of the analysis [13].

The wetting of the fiber was very good, but the injection time of the resin was greatly increased at higher fiber levels because of the lower permeability of the fiber mat. Keeping the mold temperature constant was essential for obtaining rapid and uniform curing of the composite. The curing process was simulated and compared against experimental thermal measurements made at several points within the mold cavity, with good agreement between simulation and experimental results.

The mechanical properties of the composites were evaluated by measuring their tensile and flexural strengths, giving insight into suitability for potential applications [14].

2. Results and Discussion

Composites represent one of the fastest-growing segments in the materials market. They are used in a wide variety of applications, ranging from sporting goods and vehicles like aircraft and automobiles to shipbuilding. Examples include the Boeing 777, disc brake pads made with hard ceramic particles embedded in a soft metal matrix, and everyday items such as fiberglass shower stalls and bathtubs. Additionally, imitation granite and cultured marble are popular choices for sinks and countertops. On the cutting edge, advanced composites are regularly employed in spacecraft to endure extreme and challenging environments.

Materials account for approximately 60% to 70% of a vehicle's total cost and play a crucial role in determining its quality and performance. Even minor reductions in a vehicle's weight can lead to significant economic benefits. Composite materials have proven to be effective alternatives to steel when it comes to reducing weight. As a result, they have been chosen as the preferred material for designing leaf springs.

The most commonly used fibers in composites include carbon, glass, and Kevlar. Among these, glass fiber is chosen for its balance between cost-effectiveness and strength. There are three main types of glass fibers: C-glass, S-glass, and E-glass. C-glass is designed for improved surface finishes, while S-glass offers very high modulus

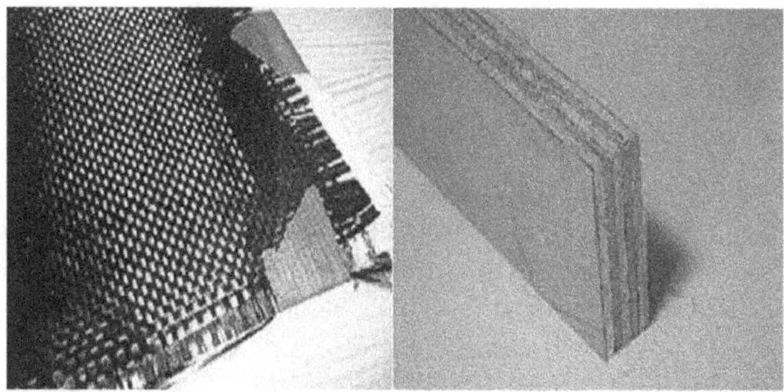

Fig. 48.1 GFRP mate and wood

and is primarily used in the aerospace industry. E-glass, a high-quality glass fiber, is widely used as a standard reinforcement material due to its excellent mechanical properties. For this use, E-glass fiber was considered the best choice.

In an FRP leaf spring, the interlaminar shear strength is mostly controlled by the matrix system due to the fact that fibers oriented in the thickness direction contribute little to this characteristic. Hence, the matrix system should offer good interlaminar shear strength and should be compatible with the chosen reinforcement fiber. Thermosetting resins like polyester, vinyl ester, and epoxy are used typically for FRP manufacturing. Of these, epoxies have the highest interlaminar shear strength and mechanical characteristics, which is the most suitable option for this application.

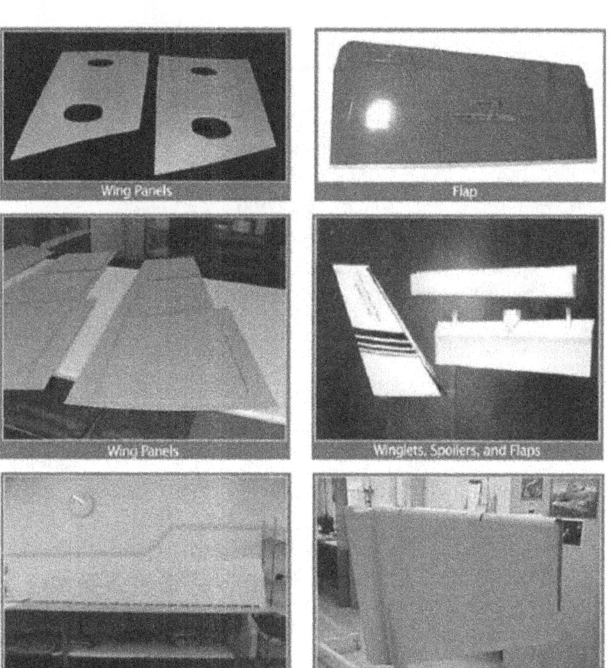

Fig. 48.2 Composite structures

For this application, the epoxy resin used is Dobeckot 520 F, a solvent-free resin. It is used in combination with hardener 758, a low-viscosity polyamine. Upon mixing, Dobeckot 520 F and hardener 758 harden to form a hard resin with the following characteristics. Excellent mechanical and electrical properties

- Rapid curing at room temperature
- Strong chemical resistance

This combination offers an ideal solution for meeting the performance requirements of the FRP leaf spring.

It can reach a height of 11 meters (40 feet) with a trunk diameter of 1.2 meters (3.9 feet). Flowering starts soon after leaf development, with green-yellow cylindrical spikes 5–10 cm long, in clusters of 2 to 5 at branch tips. Pods, 10 to 40 cm long, contain 20 to 40 seeds.

Reproduction is only by means of seeds and there is no vegetative reproduction. The seeds are widely distributed by animals like cattle, which consume the pods and then deposit the seeds in their dung, facilitating the dissemination of the plant.

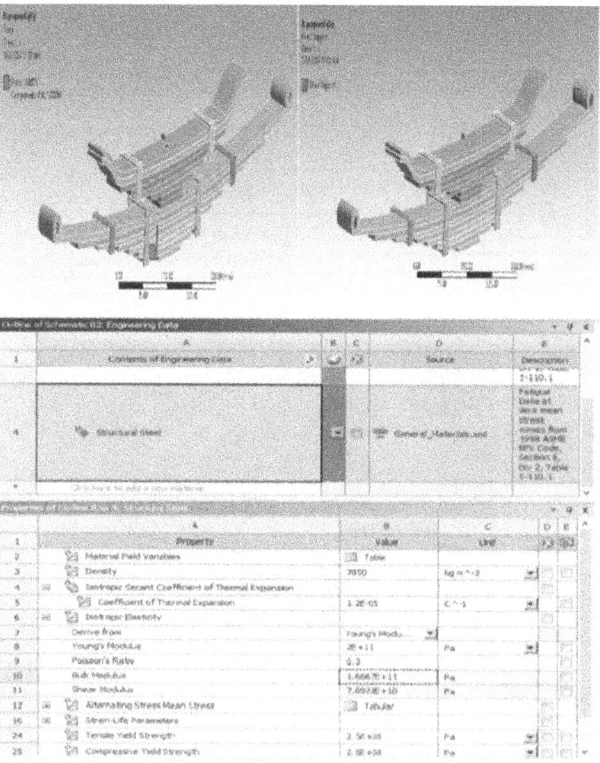

Fig. 48.3 Boundary condition

Table 48.1 Equivalent elastic strain (mm/mm)

Type	GFRP with Natural Fibre	Steel
Equivalent elastic strain (mm/mm)	0.00081	0.00085

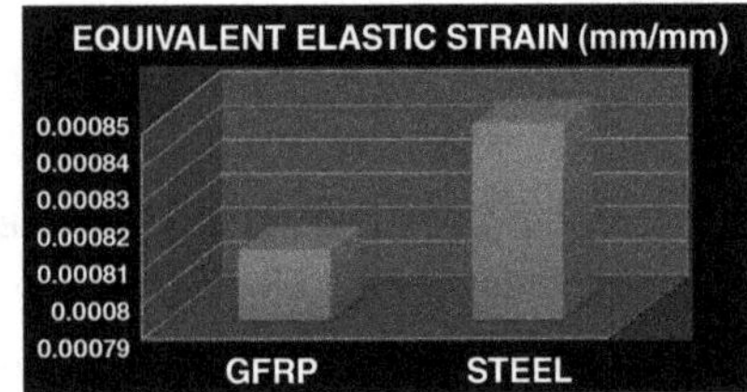

Fig. 48.4 Comparison for equivalent elastic strain

Table 48.2 Equivalent stress (MPa)

Type	GFRP with natural Fibre	Steel
EQUIVALENT STRESS (MPa)	170.04	169.35

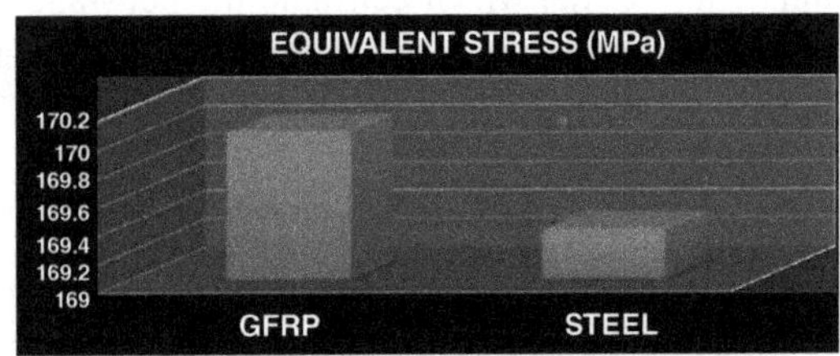

Fig. 48.5 Comparison for equivalent stress

Table 48.3 Normal stress (MPa)

Type	GFRP with natural Fibre	Steel
NORMAL STRESS (MPa)	39.807	39.139

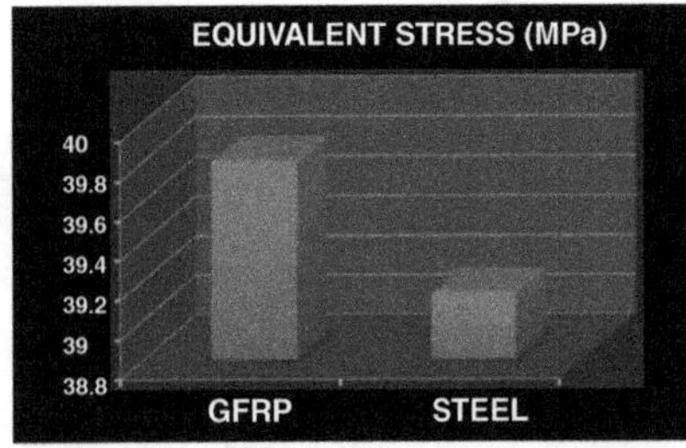

Fig. 48.6 Comparison for normal stress

3. Conclusion

The results of the finite element analysis (FEA) and experimental testing of composite and steel leaf springs are significant in understanding their mechanical behavior and performance.

Comparative Analysis: The research compared the experimental and numerical outcomes of steel and composite leaf springs. It discovered that the tensile strength, compression characteristics, and impact resistance of the composite material were favorable for application in leaf springs and matched the performance of steel.

1. Weight Loss and Efficiency: The composite material showed a significantly reduced weight with respect to steel, resulting in improved fuel economy and higher payloads. This loss of weight not only enhances energy conservation but also aids in environmental sustainability.
2. Corrosion Resistance: Steel leaf springs are susceptible to rust, but composite materials composed of epoxy fiber naturally resist corrosion, leading to longer life and lower maintenance costs.
3. Fatigue Resistance: The composite material showed better fatigue resistance than steel, with higher durability and reduced risk of sudden failure during extended usage.
4. Vibration Damping: The hybrid composite material's vibration-absorbing characteristics contribute to better ride comfort through vibrations transmitted to the vehicle's chassis.
5. Sustainability: Complying with sustainability requirements, composite fibers are used in a way that reduces carbon emissions and makes the material reusable. Thus, composite leaf springs present an eco-friendly option.

References

1. Senthil kumar and Vijayarangan, "Analytical and Experimental studies on Fatigue life Prediction of steel leaf soringand composite leaf multi leaf spring for Light passanger veicles using life data analysis" ISSN 1392 1320 material science Vol. 13 No.2 2007.
2. Shiva Shankar and Vijayarangan "Mono Composite Leaf Spring for Light Weight Vehicle Design, End Joint, Analysis and Testing" ISSN 1392 Material Science Vol. 12, No.3, 2006.
3. Niklas Philipson and Modelan AB "Leaf spring modelling" ideon Science Park SE-22370 Lund, Sweden
4. Zhi'an Yang and et al "Cyclic Creep and Cyclic Deformation of High-Strength Spring Steels and the Evaluation of the Sag Effect:Part I. Cyclic Plastic Deformation Behavior" Material and Material Transaction A Vol 32A, July 2001–1697
5. Muhammad Ashiqur Rahman and et al "Inelastic deformations of stainless- steel leaf springs-experiment and nonlinear analysis" Meccanica Springer Science Business Media B.V. 2009
6. C.K. Clarke and G.E. Borowski "Evaluation of Leaf Spring Failure" ASM International, Journal of Failure Analysis and Prevention, Vol5 (6) Pg. No. (54–63)
7. J.J. Fuentes and et al "Premature Fracture in Automobile Leaf Springs" Journal of Science Direct, Engineering Failure Analysis Vol. 16 (2009) Pg. No. 648–655.

8. J.P. Hou and et al "Evolution Of The Eye End Design Of A Composite Leaf Spring For Heavy Axle Load" Journal of Science Direct, Composite Structures 78(2007) Pg. No. (351–358)
9. Practical Finite Element Analysis by Nitin S. Gokhale
10. Composites – A Design Guide by Terry Richardson
11. ANSYS 10.0 Help for FEA Analysis.
12. Text book of Machine Design by R.S. Khurmi and J.K. Gupta
13. Introduction to Steel Reference Books S. N. Bagchi and Kuldeep Prakash
14. Spring Designers Hand Book by Carlson.
15. Jadhao, K.K. & Dalu, R.S., 2011. Composite Leaf Spring for Light Weight Vehicle – Design, Fabrication, and Testing. *International Journal of Engineering Science and Technology (IJEST)*, 3(6), pp. 4758–4761.

Note: All the figures and tables in this chapter were made by the authors.

Advances in Additive Manufacturing Technologies – Gurusamy Pathinettampadian et al. (eds)

Mechanical Properties and Investigation on Aluminium Hybrid Composite

Velmurugan S.*, Aakash S., Hari Krishna Raj S., Aravindhan S., Gokul V.

Department of Mechanical Engineering,
VelTech HighTech Dr. Rangarajan Dr. Sakunthala Engineering College,
Chennai, Tamilnadu

◆ **Abstract:** Aluminium Metal Matrix Composites (AMMCs) are specially designed materials possessing improved mechanical properties like good corrosion resistance, high strength, and formability. Due to their exceptional properties, they are extremely well suited for application in emerging industries such as aerospace, automotive, and shipbuilding manufacturing, where materials of low weight with high performance are required. Al7075 is one of the numerous aluminium alloys that has gained importance due to its versatility, machinability, and outstanding mechanical properties. Yet, its properties can be additionally improved by reinforcing particles like silicon carbide (SiC) and multi-walled carbon nanotubes (MWCNTs) within its matrix. In the current research, the compressive properties and hardness of SiC/Al7075/MWCNT composites were investigated to observe the advantages of reinforcement. The composites were produced by Squeeze Cum Stir Casting Technique, a hybrid method that benefits from stir casting to achieve uniform dispersion of particles and squeeze casting for improved densification and lower porosity. The composition selected for the composite consisted of 89.88% aluminium alloy 7075, 7.62% silicon carbide, and 2.49% multi-walled carbon nanotubes. This particular mixture was chosen with care to optimize mechanical enhancements without compromising on a light structure. The production process involved heating the aluminium alloy 7075 to 800°C to obtain molten form. Reinforcement materials SiC and MWCNT were preheated to 150°C to evaporate moisture and improve bonding with the aluminium matrix. After the reinforcements had been introduced to the molten aluminium, the blend was mixed completely to ensure proper dispersion of particles. The composite melt was then poured into a preheated die, and a hydraulic press exerted a pressure of 40 tons to compact and solidify the material. Squeezing greatly enhanced the density of the composite by reducing internal voids and improving the interfacial bonding between the matrix and the reinforcements. Overall, the findings underscore the promise of SiC/Al7075/MWCNT composites as high-performance materials for demanding applications. The successful production of such composites via the Squeeze Cum Stir Casting Technique illustrates an efficient and reproducible means for creating materials of enhanced properties. Such developments make AMMCs a viable option for industries that aim to create lightweight, long-lasting, and efficient components for contemporary engineering applications.

◆ **Keywords:** Hybrid composite, Aluminium alloy, Al7075, AMMC, Stir casting, MWCNTs, SiC

1. Introduction

Aluminium Alloy 7075 (UNS A97075) is a precipitation-hardened alloy commonly used, with principal constituents being magnesium and silicon as major alloying constituents. Initially given the designation "Alloy 61S," it was first produced in 1935 and has come to be widely used since then

*Corresponding author: ivelu00@gmail.com

DOI: 10.1201/9781003685906-49

because of its good mechanical properties and versatility. Popular for its good weldability and ease of extrusion, it is the second most widely extruded aluminium alloy after 6063. It is often used in general-purpose applications in all industries. Chemical Composition

The chemical composition of Aluminium Alloy 7075 includes the following elements by weight:

This exact combination of ingredients makes the alloy's properties distinct, balancing workability, corrosion resistance, and strength.

The mechanical properties of the 7075 alloy depend remarkably on its temper or heat treatment. In every temper, its Young's modulus is 69 GPa (10,000 ksi).

The heat treatment processes affect the microstructure of 7075 alloy, with control over precipitate size and distribution of magnesium silicide (Mg2Si) precipitates. These precipitates are very important in the strengthening of the alloy. Grain sizes, generally between a few micrometers and hundreds of micrometers, depend on stress and processing but have a lesser role to play in strength than precipitates. In addition, secondary iron, manganese, and chromium phases are present as inclusions. Aluminium 7075 is extensively employed because of its strength, resistance to corrosion, and versatility. The following are typical applications:

Silicon carbide has various modern applications, including abrasive materials, ceramics, and electronics. Its properties render it extremely useful for applications that demand durability and high thermal stability, including light-emitting diodes (LEDs) and other semiconductor devices. Synthetic SiC is also employed in gem production, referred to as moissanite, which replicates diamond.

Silicon carbide (SiC) is a highly versatile substance that is largely manufactured synthetically because the natural form, moissanite, is not found naturally in abudance. It has extensive applications as an abrasive, a semiconductor, and a gemstone that looks like diamond. can also be transformed into silicon carbide by heat with access carbon present in the organic material. Another possible source of SiC that can be created is silica fume, a residue from silicon metal and ferrosilicon alloy manufacture, by heating it with graphite at about 1,500 °C (2,730 °F).

The degree of purity in SiC obtained in an Acheson furnace is determined by how close to the heat source, a graphite resistor, they are formed. The closest crystals to the resistor are purest and colorless, pale yellow, or green. Those formed farther away become darker and either blue or black, indicating less purity. Other impurities, such as nitrogen and aluminum, are usual in SiC and affect its conductivity. Cubic silicon carbide is normally formed via the more costly (CVD) process involving reaction of silane, hydrogen, and nitrogen. This method allows the homoepitaxial and heteroepitaxial SiC films to grow via either gas or liquid-phase methods.

To produce complex-shaped silicon carbide products, preceramic polymers are used as precursors. When the polymers are pyrolyzed at temperatures ranging from 1,000 °C to 1,100 °C in an inert atmosphere, they are converted into ceramics called polymer-derived ceramics (PDCs). Polycarbosilanes, poly(methylsilyne), and polysilazanes are typical precursor materials. In comparison to CVD, the pyrolysis process is more versatile since the polymers can be molded prior to thermal conversion into SiC.

2. Result and Discussion

For analyzing the mechanical characterisation of Aluminium Metal Matrix Composite reinforced with Silicon Carbide and Multi Walled Carbon Nano Tubes. For that purpose, AMMC will be prepared and specimen will be prepared according to required standards. Specimen will go through for testing with data acquisition system. First, Aluminium alloy 7075 is cut into pieces with needed volume and dimension. Now, all the three ingredients Al7075, SiC & MWCNT are weighed and checked whether it is of the required volume. Then Al7075 is placed in the furnace and heated at 8000C to melt. Meanwhile, SiC & MWCNT are filled in different feeder and left to heat at 4000C each. When the Aluminium is melted, both SiC & MWCNT are filled into it. Now the Stirrer Impeller is turned on and began to mix the constituents well at 600 rpm. Once the stirring process is finished, the molten aluminium matrix is filled in die. Just after pouring of molten composite into die, hydraulic pressure is given at 40T into the die and left to attain its solid form. After a few minutes, the component is taken out from the die which is the desired Aluminium Metal Matrix Composite.

Fig. 49.1 Machined Al7075

Fig. 49.2 Weight composition of Al7075

The Al alloy 7075 weight percentage used in this matrix composition is 89.88%. We added 1.333kg of aluminium alloy 7075 in this matrix composition.

Fig. 49.3 Required silicon carbide

The weight percentage of silicon carbide used in this composition is 7.62%.

Fig. 49.4 Weight composition of silicon carbide

113.05g of silicon carbide is added in this metal matrix composition.

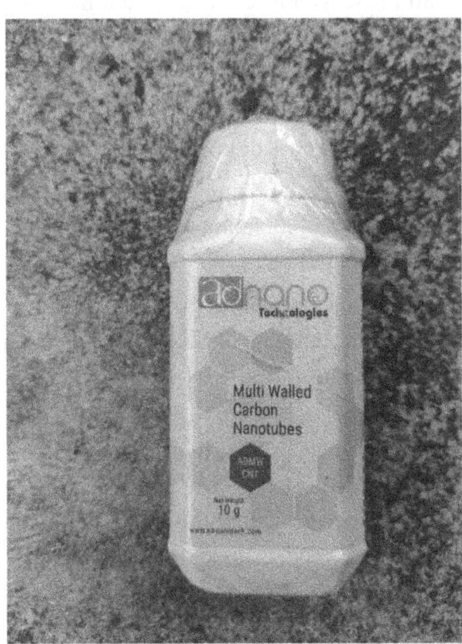

Fig. 49.5 MWCNT required

The weight percentage of multi walled carbon nanotubes used in this composition is 2.49%.

Fig. 49.6 Weight composition of MWCNT

3.7g of MWCNTs is employed in this metal matrix composition.

3. Scanning Electron Microscope Test

The above results clearly shows that the highest pressure of 40 T among these components is applied. It has a highest ductility and machinability property with great wear efficiency.

Base(8)

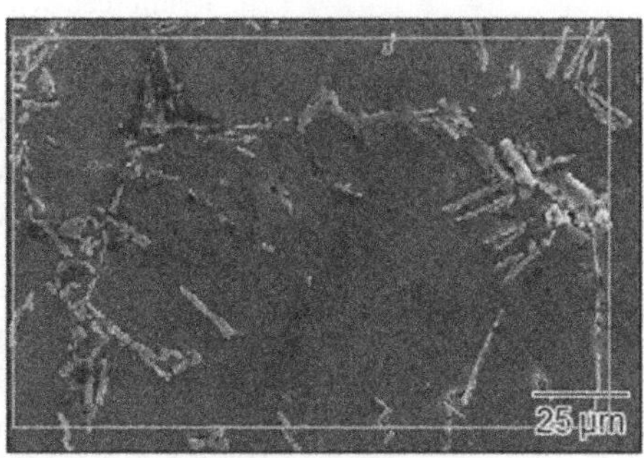

Fig. 49.7 Scanning electron microscope (SEM)

Net Counts

	C	O	Al	Si	Ba
Base(8)_pt1	639	282	199973	10629	301

Weight %

	C	O	Al	Si	Ba
Base(8)_pt1	5.37	1.02	81.16	11.94	0.51

Atom %

	C	O	Al	Si	Ba
Base(8)_pt1	11.32	1.61	76.20	10.77	0.09

Formula

	C	O	Al	Si	Ba
Base(8)_pt1	C	O	Al	Si	Ba

Fig. 49.8 Test results

4. Conclusion

4.1 Hybrid Composite Fabrication and Microstructure

Al7075/SiC/MWCNT hybrid aluminum matrix composites were successfully developed via stir casting. The XRD pattern of the composites established the presence of SiC and MWCNT particles, in addition to peaks for elements such as CaO, SiO_2, MgO, and Al_2O_3. The peaks were more significant when larger amounts of MWCNT were added into the hybrid composites numbered 1, 2, 3, and 4.

4.2 Density, Porosity, and Hardness

Compared to unreinforced MWCNT Al7075/5SiC composites, the MWCNT-reinforced hybrids exhibited a lower density, which continued to decline with increasing MWCNT content.

Porosity initially decreased as the reinforcement percentage rose, reaching an optimal level at 4.5 wt.% MWCNT. Beyond this threshold, void formation increased. The hardness tests showed considerable improvements, with readings increasing from 68 HV for AMC-1 to 77 HV for hybrid 1, 85 HV for hybrid 2, 96 HV for hybrid 3, and up to 107 HV for hybrid 4. This hardness improvement is due to the addition of hard MWCNT particles.

4.3 Wear Resistance and Friction

Hybrid composite 3 (Al7075-5SiC-4.5MWCNT) exhibited excellent wear resistance at different sliding velocities and normal loads. The coefficient of friction had little deviation for this hybrid composite under comparable operating conditions, highlighting its consistent behavior.

4.4 Microstructural Observations

FESEM micrographs indicated that the pin surface of hybrid composite 3 was in better condition than other combinations and therefore can be an ideal material for application in automobile. Additionally, Al7075/SiC/MWCNT composites were synthesized through pressure-assisted melt infiltration of Al7075 into preforms made of SiC and MWCNT. The main findings from this process are as follows:

References

1. Mohamed Ibrahim Abd El Aal / 2020 - The Influence Of Ecap And Hpt Processing On The Microstructure Evolution, Mechanical Properties And Tribology Characteristics Of An Al7075 Alloy in ELSEVIER, 4–11
2. P. Satishkumar, A.J. Infant Jegan Rakesh, R. Meenakshi, C. Saravana Murthi / 2020 - Characterisation, Mechanical And Wear Properties Of Al7075/Sicp/Fly Ashp Composites By Stir Casting Technique in ELSEVIER, 2–5
3. Renu, Ajay Gupta, Shiv Ranjan Kumar, Chandramani Goswami, Pankaj Sharma / 2021 - Enhanced Physical And Mechanical Properties Of Al7075 Alloy Nanocomposite Reinforced With Nanozirconia in ELSEVIER, 2–3
4. B. Prakash, S.Sivanathan, V.Vijayan / 2020 - Investigation On Mechanical Properties Of Al7075 Alloy - Multiwall Carbon Nanotubes Reinforced Composites By Metallurgy Route in ELSEVIER, 4
5. B. Prakash, M.Manimaran / 2020 - Investigation On Mechanical Properties Of Al7075-Sic Nanocomposites Fabricated Via Stir Casting Process in ELSEVIER, 2–3
6. K. Preethi, T.N. Raju, H.A. Shivappa, S. Shashidhar, Madeva Nagral / 2021 - Processing, Microstructure, Hardness And Wear Behaviour of Carbon Nanotube Particulates Reinforced Al7075 Alloy Composites in ELSEVIER, 1–3
7. Satish Babu Boppanaa, Samuel Dayanand, Anil Kumar M.R., Vijee Kumar / 2020 - Synthesis and Characterisation Of Nano Graphene And Zro2 Reinforced Al7075 Metal Matrix Composites in ELSEVIER, 4–8
8. Ajay Gupta, Renu, Shiv Ranjan Kumar, Chandramani Goswami, Tej Singh / 2021 - Wear Behaviour of Al7075 Nanocomposite Reinforced With Nanozirconia in ELSEVIER, 1–4

Note: All the figures in this chapter were made by the authors.

Advances in Additive Manufacturing Technologies – Gurusamy Pathinettampadian et al. (eds)

Enhancing Anti-frost Performance of Carbon Fiber/Epoxy Composites Using Super-hydrophobic Coating

Surender Paul A.[1],
Muthukrishnan A.[2], Jaya Kumar V.[3], Shanmuganathan V.K.[4]
Dept. of Aeronautical Engineering, Tagore Engineering College,
Chennai, India

◆ **Abstract:** The frosting on the external surfaces of any object imposes severe penalties on it. Industries like Aviation are not exception to it. This phenomenon mandates that materials like carbon fiber/epoxy composites need to have excellent anti-frosting characteristics. The desired performance, safety and airworthiness of aircraft components will have to be compromised if the ice formation or built up is permitted in aviation. This condition necessitates an innovative approach to mitigate the icing phenomena on aircraft and its components. This study considers the application of super-hydrophobic coatings on carbon fiber/epoxy surfaces so that icing on the components made by using carbon fiber composites can be mitigated. Super-hydrophobic materials do possess unique water-repellent properties that are characterized by a high contact angle and low adhesion to water droplets. This property exhibits excellent resistance to ice nucleation and also minimizes the adhesion of frost on the exterior surfaces that it offers a positive solution for the prevention of ice buildup carbon fiber/epoxy surfaces. The super-hydrophobic coating makes it possible to create a micro or nanostructured surface thereby it enables the minimization of the contact area between water droplets and the fiber/epoxy surfaces. This, in turn, produces tremendous development with which the formation and growth of ice crystals can be meticulously prevented and also the adherence of ice over the fiber/epoxy surfaces can be eliminated. Furthermore, this approach also results in easier removal of ice formed over the exterior surfaces as it produces only the low adhesiveness of water droplets over the coated exterior surfaces. Deicing may also be carried out with the help of aerodynamic forces during the operation. The experimental studies and computational simulations carried out on the models have highlighted the effectiveness of these coatings in mitigating the frost formation. If such super-hydrophobic coating is applied on propellers, it will not only enhance their anti-frosting performance but also enhance their overall operational efficiency. Optimizing or elimination of ice accumulation over any aerodynamic control surfaces will always lead to improved aerodynamic characteristics, fuel efficiency, and safety. The continued research on this coating methodology is need of the hour mainly to refine these coatings and optimize their application methods. The anti-frosting behavior of propellers through super-hydrophobic coatings may be tagged as emerging technological development as it possesses an avenue for advancing technology in cold and icy atmospheres.

◆ **Keywords:** Antifrosting, Carbon fibre/epoxy, superhydrophobic coatings, Ice accumulation, Performance and safety, Aerodynamics, Operational efficiency, Cold environment

1. INTRODUCTION

Carbon fiber composites have incomparable mechanical properties, chemical stability, lightweight nature, and ease of fabrication which are widely acknowledged by many researchers [1]. Due to their exceptional performance characteristics they have been extensively used in diverse engineering sectors, including aerospace, wind power

[1]a.surenderpaul@gmail.com, [2]muthukaero@gmail.com, [3]jaikumar.v16@gmail.com, [4]shankris21@gmail.com

DOI: 10.1201/9781003685906-50

generation, and automotive industries [2,3]. However, their inability to withstand environments with low temperature imposes significant challenges to the research community. It is due to their inherent moisture absorption tendency. In such conditions, if the accumulation of ice on critical equipment is possible on components of aircraft and fan blades may lead to detrimental consequences on both on their performance and safety [4]. The detrimental effects of surface icing on aircraft and its components have been very well-documented. The formation of ice layers on aircraft surfaces mainly impairs the aerodynamic profile of aircraft's control surfaces which directly results in diminished lift force, also increasing overall weight, and elevating flight resistance to stability. These factors collectively multiples the risk of aircraft accidents [4]. Similarly, there are few potential problems associated with the formation of ice on fan blades such as exacerbate weight, reduction in power generation efficiency, and even initiate the blade fractures, resulting in substantial economic losses and safety hazards [5].

In the recent past, various anti-icing and de-icing approaches have been proposed by the researchers all around the world in order to minimize or alleviate mitigate the adverse impacts of surface icing, The proposed strategies include wide range of techniques from hot air de-icing, electro-thermal de-icing, pneumatic de-icing, and ultrasonic de-icingetc [6–9]. For instance, Pellissier et al. proposed the directional circulation of hot air both for the preventionas well as removal ice from surfaces using piccolo pipes [10]. Yang et al. explored the thermal effect induced by dielectric barrier discharge (DBD) plasma for de-icing applications [11]. Palacios et al. emphasized that an aerodynamic method by employing diaphragms can be used as an alternative approach to protect helicopter rotor blades from icing [7]. Wang et al. investigated an ultrasonic de-icing system that is equipped with a damage accumulation model [12]. Among these approaches, super-hydrophobic surfaces have been considered as most promising approach for anti-icing applications. Super- hydrophobic surfaces exhibit a contact angle that exceeds 150° and a sliding angle of less than 10°, facilitating the easy rolling-off of water droplets from the external contact surfaces. Moreover, the micro- or nanostructures on super-hydrophobic surfaces can also trap air thereby forming air pockets may delay the onset of icing [13]. Researchers have extensively conducted extensive research about the preparation mechanisms and anti-icing performance of super-hydrophobic surfaces, also about adopting to techniques such as surface spraying, template replication, chemical etching and also physical deposition [14–17]. These research studies have highlighted the low ice adhesion property of super-hydrophobic coatings [18], and concluded that through methods like electrochemical anodizing and chemical etching these super-hydrophobic coating can be an optimal approach for eliminating the icing aircraft exterior surfaces [19]. Additionally, research has also proven about the efficacy of hydrophobic coatings comprising of nanoparticles and micro-particles in enhancing anti-icing performance [21]. Furthermore, novel approaches have also found that modifying carbon fiber composites to impart superior super-hydrophobic properties have provided promising results [22].

While super-hydrophobic surfaces offer excellent anti-icing capabilities, there are also many challenges including the potential condensation of water droplet and changes in contact types during the process of icing are still persisting [23,24]. Researchers have endeavoured to resolve these challenges by optimizing surface microstructures so that it enables facilitation of droplet removal during melting [25,26]. The main aim of this study is to contribute to the advancement of anti-icing technologies by proposing a novel approach. This study proposes the hydrophobic modification of carbon fiber composites in order to create super-hydrophobic surfaces having multilevel micro- and nanostructures. By adopting any cost-effective method for utilizing Zinc stearate for low surface energy and TiO_2 for nanostructures, it is feasible to prepare super-hydrophobic surfaces successfully and then analysing their structural characteristics will validate this methodology. Moreover, This study has conducted comprehensive tests for evaluating the anti-icing performance of the coated surfaces, including assessments of contact angles, sliding angles. This research aims to inform future developments in this critical fieldby elucidating the intricacies of super-hydrophobic surfaces and their application in anti-icing technologies which would provide potential solutions to enhance safety and efficiency in cold and icy operating conditions.

2. Materials and Methods

2.1 Materials

This Study has mainly chosen the combination of materials such as Zinc stearate and TiO_2. Zinc Stearate is commercially available and has been purchased from the Supplier and it has been considered mainly due to its low surface energy characteristics. The material TiO_2 has been used mainly in the form of nanoparticles. It was also purchased from the Supplier; the reason for selecting $TiO_2$is particularly to induce nanostructures on the surface. Both the materials have been procured with highest level of purity and are processed in the condition in which they were purchased that means without any further purification.

2.2 Methods

Ameticulouslyplannedandpreparedstep-by-stepprocedure was followed for applying super-hydrophobic coating. The mixture of Zinc Stearate and TiO_2 nanoparticles was obtained at an appropriate mixture ratio mainly with the objective of achieving the anticipated surface properties. The mixture was then dispersed in a suitable solvent and stirred vigorously to ensure homogeneity. Next, the purchased carbon fiber/epoxy composites were cleaned thoroughly to remove any contaminants. The prepared coating solution was applied to the composite surface using [specific coating technique], ensuring uniform coverage. The coated samples were then cured under controlled conditions [mention curing parameters]. To characterize the superhydrophobicity of the coated surface, contact angle measurements were performed using a goniometer [model and specifications]. Scanning electron microscopy (SEM) was employed to examine the surface morphology and uniformity of the coating layer.

3. Design

3.1 Model Design

Our model aims to develop an anti-icing surface for carbon fiber. We achieve this by combining titanium oxide, stearic acid, and polymethyl methacrylate.

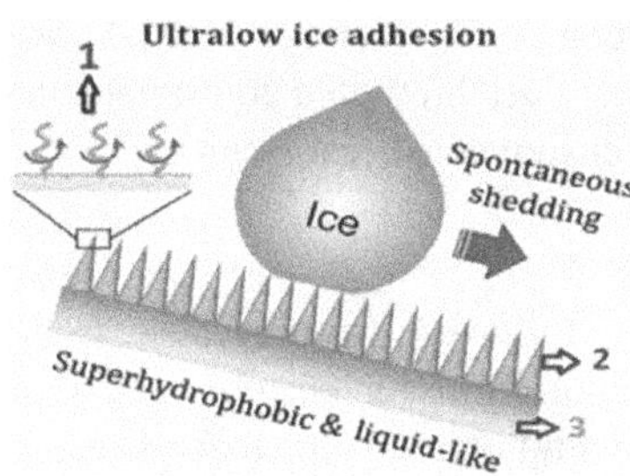

Fig. 50.1 Model design

Table 50.1 Chemicals and quantity

Chemicals	Quantity
Titanium oxide	0.5g
Chloroform	10ml
Stearic acid	0.5g
Polymethyl methacrylate	1g

4. Experimental Study

4.1 Trail 1

In Triall, we used titaniumoxide and stearic acid, stirred with ethanol for 30 minutes. Subsequently, the mixture was combined with epoxy (LY556) and hardener (HY951), stirred for 5 minutes, and then coated. Following this process, we analyzed the results of Triall. The ethanolisemployed to dissolve the titaniumoxide and stearic acid; subsequently, it evaporates.

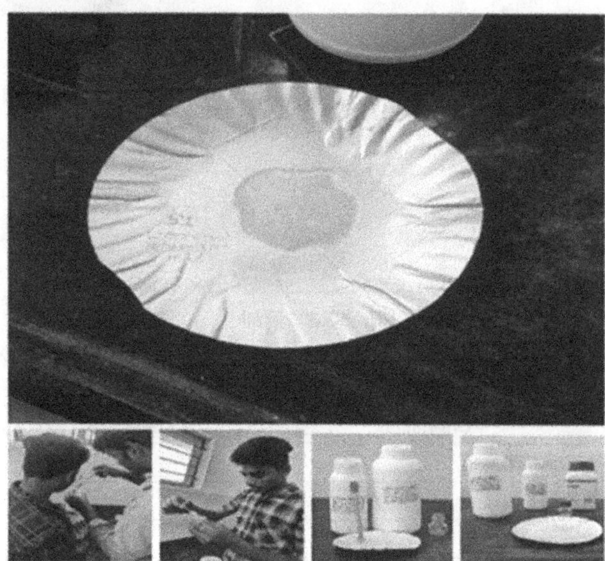

Fig. 50.2 Trail 1

Table 50.2 Chemicals and quantity of Trail 1

Chemicals	Quantity
Titanium oxide	0.5g
Ethanol	10ml
Stearic acid	0.5g
Epoxy(LY556) and hardener(HY951)	1ml and 0,1ml

Outcome

Contact Angle vs. Slide Angle Discrepancy: Despite achieving favorable contact angles, the observed slide angles indicate poor practical superhydrophobicity, necessitating further investigation into surface properties.

Visual inspection revealed uneven distribution of the coating material, affecting its effectiveness. Observable inconsistencies in the coating process led to concerns about its durability and performance.

Without SEM analysis, visual cues indicated the need for furtheroptimizationtoachievedesiredsuperhydrophobicity.

4.2 Trail 2

In Trial 2, we followed the same procedure as in Triall. Additionally, we doubled the composition of titaniumoxide and stearicacid.

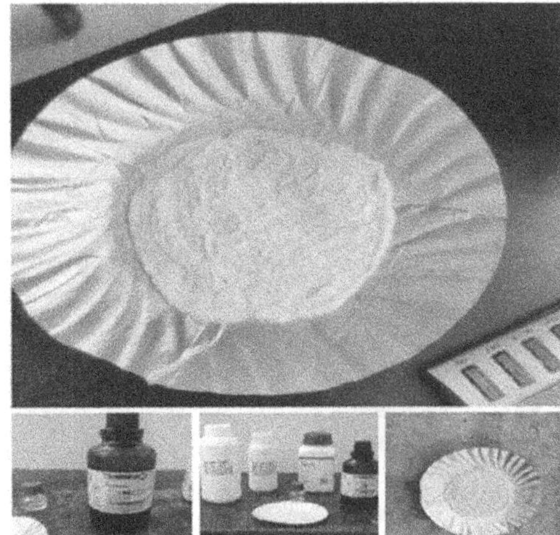

Fig. 50.3 Trail 2

Table 50.3 Chemicals and quantity of Trail 2

Chemicals	Quantity
Titanium Oxide	1g
Ethanol	10ml
Stearic Acid	1g
Epoxy(LY556) and Hardener(HY951)	2ml and 0.2ml

Outcome

Inconsistent Wetting Behavior: Trials exhibited inconsistent wetting behavior despite high contact angles, emphasizing the need for optimizing both static and dynamic water repellency.

Visual assessment highlighted issues with coating uniformity, suggesting the need for improved dispersion techniques.

Notable disparities in the coating application process prompted a reassessment of the methodology for better results.

SEM analysis was deemed necessary to confirm visual observations and guide refinements for future trials

4.3 Trail 3

In Trial3, we followed the same procedure as in Trial1, but substituted acetoneforethanol.

Table 50.4 Chemicals and quantity of Trail 3

Chemicals	Quantity
Titanium oxide	0.5g
Acetone	10ml
Stearic acid	0.5g
Epoxy(LY556) and hardener(HY951)	1ml and 0.1ml

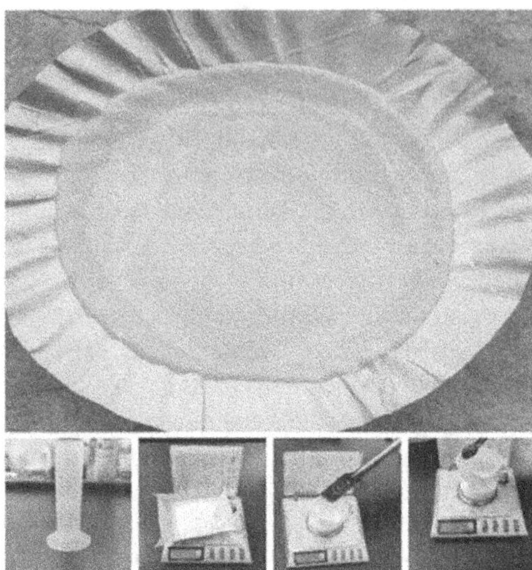

Fig. 50.4 Trail 3

Outcome

Practical Significance of Slide Angle: Discrepancies highlight the importance of considering slide angles alongside contact angles for accurate assessment and informed optimization of superhydrophobic coatings.

Visible irregularities in the coating indicated a need for enhanced process control and material quality. Suboptimal visual appearance raised questions about the compatibility of coating components and substrate. Without SEM analysis, visual cues prompted adjustments in the coating procedure to ensure consistent results.

4.4 Trail 4

In Trial4, we followed the procedure from Trial1, but substituted chloroform for acetone, and replaced epoxy andhardener with Polymethylmethacrylate.

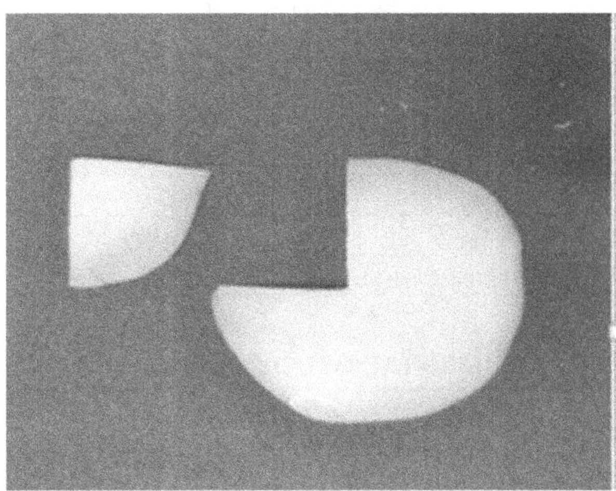

Fig. 50.5 Trail 4

Table 50.5 Chemicals and quantity of Trail 4

Chemicals	Quantity
Titanium oxide	0.5g
Chloroform	10ml
Stearic acid	0.5g
Polymethyl methacrylate	1g

Outcome

We successfully identified the optimal chemical composition for the anti-icingsurface in Trial4, and now aim to further optimize this formulation for improved results.

5. Result

Superhydrophobic coatings demonstrated significant antifrosting behavior, inhibiting ice nucleation and reducing ice accumulation on carbon fiber/epoxy surfaces. Contact angle measurements confirmed the water-repellent properties of the coatings, while scanning electron microscopy revealed the presence of nanostructures that contribute to their antifrosting performance. Trials with varied compositions and solvents provided insights into optimizing the formulation for improved antifrosting efficacy.

6. Conclusion

Superhydrophobic coatings offer a promising solution for enhancing the antifrosting performance of carbon fiber/ epoxy surfaces in cold environments. By minimizing ice accumulation and improving aerodynamics, these coatings contribute to enhanced performance, safety, and operational efficiency. Future research aims to further optimize the formulation and explore potential applications across diverse industries.

References

1. M.L. Cannaday, A.A. Polycarpou, Tribology of unfilled and filled polymeric surfaces in refrigerant environment for compressor applications, Tribol. Lett.19 (4) (2005) 249e262.
2. P. Joosse, D.V. Delft, C. Kensche, et al., Economic Use of Carbon Fibres in LargeWind Turbine blades[C], Asme Wind Energy Symposium, 2013.
3. H. Hao, S. Wang, Z. Liu, et al., The impact of stepped fuel economy targets on auto-makers light-weight strategy: the China case, Energy 94 (2016) 755e765.
4. E.A. Whalen, M.B. Bragg, Aircraft characterization in icing using flight test Data, J. Aircraft 42 (3) (2005) 792e794.
5. B. Tammelin, M. Cavaliere, H. Holttinen, et al., Wind Energy in Coldclimates [R], Finland: Finnish Meteorological Institute, 2000.
6. Z. Zhao, H. Chen, X. Liu, Z. Wang, Y. Zhu, Y. Zhou, Novel sandwich structural electric heating coating for anti-icing/ de-icing on complex surfaces, Surf. Coating. Technol. 404 (2020), 126489.
7. J. Palacios, D. Wolfe, M. Bailey, J. Szefi, Ice testing of a centrifugally powered pneumatic de-icing system for helicopter rotor blades, J. Am. Helicopter Soc. 60 (3) (2015) 1e12.
8. Z. Wang, Recent progress on ultrasonic de-icing technique used for wind power generation, high-voltage transmission line and aircraft, Energy Build. 140 (2017) 42e49.
9. F. Saeed, K.Z. Ahmed, A.O. Owes, I. Paraschivoiu, Anti-icing hot air jet heattransfer augmentation employing inner channels, Adv. Mech. Eng. 13 (12) (2021), 16878140211066212.
10. M. Pellissier, W. Habashi, A. Pueyo, Optimization via FENSAP-ICE of aircraft hot-air anti-icing systems, J. Aircraft 48 (1) (2011) 265e276.
11. L. Yang, C. Kolbakir, H. Hui, A Comparison Study on AC-DBD Plasma and Electrical Heating for Aircraft Icing Mitigation, AIAA Aerospace Sciences Meeting, 2018, 2018.
12. Y. Wang, Y. Xu, F. Su, Damage accumulation model of ice detach behavior inultrasonic de-icing technology, Renew. Energy 153 (2020).
13. A. Dotan, H. Dodiuk, C. Laforte, S. Kenig, The relationship between water wetting and ice adhesion, J. Adhes. Sci. Technol. 23 (2009) 1907e1915.
14. X. Zhang, F. Xiao, Q. Feng, et al., Preparation of SiO2 nanoparticles with adjustable size for fabrication of SiO2/ PMHS ORMOSIL superhydrophobic surface on cellulose-based substrates, Prog. Org. Coating 138 (2020), 105384.
15. C. Zhao, L. Chen, C. Yu, et al., Fabrication of hydrophobic NiFe2O4@poly(DVB-LMA) sponge via Pickering emulsion template method for oil/water separa-tion, Soft Matter (14) (2021).
16. J. Wang, Y. Wu, D. Zhang, et al., Preparation of superaerophilic copper mesh for underwater gas collection by combination of spraying technology and flame treatment, Appl. Phys. A 126 (1) (2020).
17. X. Zhou, G. Wang, M. Wang, et al., Progress in Organic CoatingsA simple preparation method for superhydrophobic surface on silicon rubber and its properties, Prog. Org. Coating (2020) 143.
18. C. Laforte, How a solid coating can reduce the adhesion, Proceedings of the International Workshop on Atmospheric Icing of Structures (1999).
19. M. Ruan, W. Li, B.S. Wang, et al., Preparation and anti-icing behavior of superhydrophobic surfaces on aluminum alloy substrates, Langmuir 29 (27) (2013) 8482e8491.
20. J.B. Boreyko, C.P. Collier, Delayed frost growth on jumping-drop super-hydrophobic surfaces[J], ACS Nano 7 (2) (2013) 1618e1627.

21. T. Zhu, et al., A transparent superhydrophobic coating with mechanochemical robustness for anti-icing, photocatalysis and self-cleaning, Chem. Eng. J. 399 (2020), 125746.
22. L. Pan, P. Xue, M. Wang, et al., Novel superhydrophobic carbon fiber/epoxy composites with anti-icing properties, J. Mater. Res. 36 (8) (2021) 1695e1704.
23. E.J.Y. Ling, V. Uong, J.-S. Renault-Crispo, A.-M. Kietzig, P. Servio, Reducing ice adhesion on nonsmooth metallic surfaces: wettability and topography effects, ACS Appl. Mater. Interfaces 8 (2016) 8789e8800.
24. F. Tavakoli, H.P. Kavehpour, Cold-induced spreading of water drops on hy-drophobic surfaces, Langmuir 31 (2015) 2120e2126.
25. X. Yan, et al., Laplace pressure driven single-droplet jumping on structured surfaces, ACS Nano 14 (2020) 12796e12809.
26. L. Wang, Z. Tian, G. Jiang, et al., Spontaneous dewetting transitions of droplets during icing & melting cycle, Nat. Commun. 13 (1) (2022) 1e15.
27. H. Guo, M. Liu, C. Xie, et al., A sunlight-responsive and robust anti-icing/deicing coating based on the amphiphilic materials, Chem. Eng. J. 402(2020), 126161.
28. T.-S. Wong, S.H. Kang, S.K.Y. Tang, E.J. Smythe, B.D. A. Hatton Grinthal,J. Aizenberg, Bioinspired self-repairing slippery surface with pressure-stable om- Niphobicity, Nature 477 (2011) 443.
29. C.V. Sternling, L.E. Scriven, Interfacial turbulence: hydrodynamic instability and the Marangoni effect, AIChE J. 5 (1959

Note: All the figures and tables in this chapter were made by the authors.

Advances in Additive Manufacturing Technologies – Gurusamy Pathinettampadian et al. (eds)

Machine Learning and Deep Learning Approaches for Predicting Adverse Drug Reaction and Drug-Drug Interaction

Latheesh Saran S.[1], Yogapriya M.[2]
Department of CSE-AIML, Chennai Institute of Technology,
Chennai, Tamilnadu

Kandavel N.[3]
Department of CSE, Chennai Institute of Technology,
Chennai, Tamilnadu

Sudharshan R.[4]
Department of CSE-AIML, Chennai Institute of Technology,
Chennai, Tamilnadu

Karthikeyan P.[5]
Department of CSE, Chennai Institute of Technology,
Chennai, Tamilnadu

Gowri R.[6]
Department of CSE-AIML, Chennai Institute of Technology,
Chennai, Tamilnadu

◆ **Abstract:** Adverse Drug Reactions (ADRs) are significant healthcare concern, often caused by drug-drug interactions (DDIs), polypharmacy, or patient-specific factors such as genetics, age, and medical history, allergy to certain things with outcomes ranging from mild symptoms like nausea, vomiting to severe end organ damage or anaphylactic reaction. This paper reviews ADR classifications and wide range view of the application of machine learning (ML) and deep learning (DL) methodologies for the prediction of ADR, DDI and prevention. The recent advancement, including predictive models like DeSIDE-DDI, logistic regression, and attention mechanism, demonstrating high accuracy by leveraging datasets such as DrugBank, FAERS, and TWOSIDES to predict ADRs and DDIs. Models like DeSIDE-DDI, with great generalization of capabilities and drug-induced gene expression analysis. By highlighting the growing importance of pharmacovigilance systems and data-driven methods in improving patient safety, reducing costs for healthcare, and optimizing therapeutic efficacy. The study makes evident that the need for integrating large-scale datasets, enhancing algorithm interpretability, and developing real-time clinical applications which will advances the healthcare systems. Addressing the challenges like data heterogeneity, improving model scale, and fostering collaboration between clinicians, data scientists, and regulatory bodies is essential for establishing standardized protocols and advancing towards a safer, more efficient, and easy access for healthcare solutions.

◆ **Keywords:** Adverse drug reactions, Drug-drug interactions, Machine learning, Pharmacovigilance, DrugBank

[1]latheeshsarans.aiml2023@citchennai.net, [2]yogapriyam@citchennai.net, [3]kandaveln@citchennai.net, [4]sudharshanr.aiml2023@citchennai.net, [5]karthikeyanp@citchennai.net, [6]gowri.r@citchennai.net

DOI: 10.1201/9781003685906-51

1. Introduction

Adverse Drug Reactions (ADRs) are defined as the toxic or undesirable effects that are produced when a patient ingests a drug, even in the prescribed dose. These reactions can be as mild as dizziness or vomiting, and as severe as allergic reactions, liver or kidney failure. ADRs are categorized into five types: Type A (dose-dependent and predictable as in drowsiness caused by antihistamines), Type B (unexpected severe reactions resulting from allergies or genetic factors such as rashes from antibiotics), Type C (cumulative effects of long term use as in liver damage caused by acetaminophen), Type D (effects that occur years later as in cancer caused bychemotherapy), Type E (effects that occur on Some of the reasons for ADRs include; the effects of other drugs, foods or supplements, wrong dosages, and the patient's age, genetic makeup or medical history. While mild effects may include stomach upsets such as vomiting, nausea or diarrhea, skin reactions such as rashes or swelling are also possible, severe effects include damage to the internal body organs or anaphylaxis. Measures of avoiding and controlling ADRs include patient history/medication reconciliation, screening for early signs of ADRs, dose modification or replacement of medications whenever required for patient's safety. Drug-Drug Interactions (DDIs) in Adverse Drug Reactions (ADR) happen when two or more medications interact in ways that can cause harmful effects. These interactions can either make a drug too strong or too weak, sometimes leading to serious side effects like toxicity or making the medication less effective. There are two main types of DDIs: pharmacokinetic interactions, where one drug changes how the body absorbs, breaks down, or eliminates another drug, and pharmacodynamic interactions, where the drugs affect each other's actions at the cellular level. It's important for doctors to keep an eye on these interactions, especially when patients are taking multiple medications. Therefore, when patients are prescribed multiple medications, healthcare providers must carefully consider the potential for DDIs, adjust dosages accordingly, and monitor patients closely to ensure their safety.

Machine learning is an approach that forms a branch of Artificial intelligence that aims to teach systems with data with the capability of learning. This new way of thinking has produced new tools in terms of algorithms for pattern-448recognition, predictions, and supporting decision making, that are now at least as good as human decision making for certain decision tasks.

The advancement of drug discovery, development, and safety monitoring relies heavily on access to comprehensive and well-structureddatabases. DrugBank, a unique bioinformatics and cheminformatics resource, has become a cornerstone in this domain. Offering detailed information about drugs, their chemical properties, pharmacological effects, mechanisms of action, and associated targets, DrugBank bridges the gap between clinical, chemical, and genomic data. Its extensive repository facilitates research across various disciplines, including pharmacology, toxicology, and precision medicine.

2. Literature Review

Adverse Drug Reactions and Drug-Drug Interactions are major issues in healthcare resulting from multiple medication use, genetic factors, and changes in drug metabolism associated with age and especially in elderly population [1-8]. Such ADR classifications include predictable dose-dependent reactions, which are Type A, rare and severe reactions caused by genetic or allergy, which is Type B, accumulation over time or chronic toxicity effects known as Type C, delayed effects that may not show up for years, which is Type D, and Type E effects, which are withdrawal effects. Recent deep learning models, including DeSIDE-DDI [1], and developed early detection systems using neural networks [7] show robust performance to predict DDIs and ADRs by considering gene activity and new drug and drug interactions. Instruments such as WHO-UMC classification, Naranjo Scale, and international databases like FAERS and others are useful in pharmacovigilance approach to patient- tailored therapy [2]. Type I, II, and III hypersensitivity reactions cannot be ignored and present from rashes to anaphylaxis; pseudo-allergic and idiosyncrasy reactions need detailed evaluation and treatment plans, which include drug switch and dosage adjustment [4]. Auto prescribed systems, electronic prescribe system, artificial intelligent based decision take over at some point, play a vital role in the reduction of the medication errors and enhancement of the patients' outcomes, translating simple guidelines between evidence-based practice and real-life clinical trial [1-8]. These suggest that predictable ADRs are usually avoidable by some mechanisms such as safe prescription practices, appropriate doses, safeguard for co-prescribing, increased monitoring and ADR reporting which are important in countries with weak pharmacovigilance [3-6].

Perioperative ADRs are described as critically important, noting that ADRs such as anaphylaxis and malignant hyperthermia are life-threatening. Potential antecedents, including medications, latex, and other patient factors, including genetic predispositions and multiple-drug regimen usage, are considered. The roles of understanding and controlling ADRs, systematic reporting such as the UK Yellow Card scheme, and updating to improve patient safety are highlighted in this study [9]. ADR prevention is stressed starting with the identification of high-risk patients. Family medical history and genetics as well as age and pregnancy are often deemed important. The measures to avoid or reduce such risks include safe prescription, dose adjustments, and co-prescribed protections like folic acid with methotrexate. Non-pharmacological management options are also discussed, and shared patient–provider decision-making is promoted to maximize safety and efficacy [10]. Elderly patients are recognized as being at high risk for developing ADRs because of polypharmacy and physiological changes associated with aging. It is mentioned that the majority of ADRs are predictable and therefore preventable if proper prescriptions are taken, adequate information is given to the patient, and there is efficient communication between practitioners. Strategies such as electronic prescribing and continuous education on effective and evidence-based practices are suggested to minimize medication mistakes and prevent high-risk admissions [11]. Another cause of ADRs in elderly patients is polypharmacy, age-related changes in drug metabolism, and multiple chronic conditions. The majority of ADRs are expected and usually occur with drugs that are taken by millions of patients, such as anticoagulants and psychotropics. It is noteworthy that medication reviews, individualized patient care, and compliance with prescribers' guidelines are identified as major approaches to minimize these risks. Elderly patients are at a higher risk of ADRs due to aging, multiple disease comorbidities, and multiple medication use. Possible offenders include NSAIDs and anticoagulants, and many of the ADRs are characterized as being expected and avoidable. Lack of adequate pharmacovigilance, especially in countries such as India, is presented as the factor that exacerbates the problem. Recommendations include an individual approach to treatment, avoiding the use of unnecessary medications, enhancing patient awareness, and improving policies regarding drug monitoring to enhance the quality of geriatric care and decrease costs [12]. Serious ADRs are responsible for 3% of hospital admissions and occur in 6.6% of inpatients, increasing their length of stay and costs. The majority of ADRs were classified as predictable (Type A) and could be prevented if well managed. Polypharmacy was seen to increase risks among elderly women. Avoiding ADRs can help improve the quality of life of patients, decrease overall healthcare costs, and free up hospital resources [13].

Severe ADRs result in admission of 3 percent of hospital admitted patients and affect 6.6 percent of inpatient with longer stay and additional costs. Over 80% of ADRs are predictable (Type A) and might be prevented if they could be managed properly. Polypharmacy especially in elderly females has also been put categorized in the list of potential risks. A further reasonfor excluding ADRs is that doing so increases patient's quality of life, reduces overall healthcare costs and makes the limited hospital resources available for other uses [14]. A systematic review of the adult outpatient and inpatient population on PADRs discovered that PADRs occurred in 2% of outpatients and 1.6% of inpatients, with a possibility of 47% being preventable. The high prevalence of ADRs and the impact they have on the healthcare system is explained, and the importance of providing more attention to preventing ADRs and their consequences is emphasized [15].

Regression and disproportionality methods are used to work with FAERS and EMR, while machine learning methods involve constructing elaborate data integration, such as DrugBank and KEGG. These enhance ADR detection but together show new issues and boundaries. The study provides a basis for increasing the effectiveness of ADR monitoring in future research [16]. This work specifically targets the discovery of DDIs, through employing machine learning techniques such as logistic regression, neural networks, and XGBoost. Molecular, pathway, and enzyme similarity matrices were derived from drug databases such as DrugBank and BioGRID. The study yielded up to 78% accuracy and up to 83% F1 scores. Enzyme and target similarities were deemed important features in the study. This approach improves the overall ability to predict DDI and thus improves medication safety and decreases adverse reactions in healthcare [17]. In this paper, the authors propose a machine learning method called Heterogeneous Network-Assisted Inference (HNAI) for predicting DDIs. By computing phenotypic, therapeutic, chemical, and genomic properties of the drug, the framework includes 6946 pairs of the drugs compared. Among five algorithms the following ones were examined: SVM and logistic regression model is used. The model showed a potential in the improvement of the identification of DDI, especially with antipsychotic drugs. Machine learning was found to be very useful in the course of drug development as well as post marketing safety monitoring while searching for an increased risk drug-interaction pair [18].

This study initiated the construction of the drug-protein interaction profiles and developed computational models to predict DDI–induced ADRs with an average accuracy of 89 percent. For 764 marketed drugs, predictions of 207 pairs demonstrated that ~10% of combinations could cause ADRs, some of which are unknown clinically. Polypharmacy and the changes in metabolism that occur with age make elderly patients particularly at risk. The study calls for higher standards of safe prescriptions as well as more cautious dosing, together with improved pharmacovigilance. The goals of the predictions which can be found in AVOID-DB database are to help clinicians avoid further ADR risks. Drug-drug interactions commonly referred to as DDIs are vital information in drug research because they may trigger ADRs [19]. This study presented a literature review of machine learning strategies for predicting other unknown DDIs because of their importance in increasing safety and efficacy in clinical practice. They studied approaches such as network propagation, graph embedding, and ensemble approaches and discussed challenges like multiple drug interactions. Thus, the study draws attention to the crucial necessity of developing effective algorithms that will help in predicting DDIs more rapidly; it also provides recommendations for future development of bioinformatics techniques. Application of big data in DDIs is critical in controlling ADRs crucial to the smart healthcare system for technical success [20].

In this study, a deep learning model using LSTM and attention mechanisms was used to classify patients and predict DDI severity, with 82.68% precision to surpass traditional methods. Through capturing the information from large biomedical articles and documents, the model improves the patient diagnosis. Dividing DDIs into levels of risk, the strategy gives substantial information and stresses the need for rational prescribing, fewer ADR incidents, and efficient management of resources in the healthcare systems [21].

In their review, the authors discuss how ADRs can be detected by means of applying the concept of ML. They point out that traditional systems encounter issues such as under-representation and a time-consuming engineering process. The authors look at different approaches to Machine learning that includes Bayesian, Supervised, and other models to enhance and improve ADR conscious detection through the use of FAERS and EMRs. However, this is not to say that ML does not have its challenges, this includes; data heterogeneity and model interpretability. The paper concludes that with the above-highlighted challenges being taken into consideration, ML can improve patient safety by providing larger, efficient, and precise ADR identification options [22]. A systematic review and meta-analysis of the use of machine learning (ML) to predict adverse drug reactions (ADRs) or events (ADEs) from electronic health records (EHRs). They considered 10 papers from different countries, noting that Random Forest and Adaboost have the highest level of predictive accuracy among the ML methods. The study concluded that the use of the ML models can be effective for processing the large and complicated data of healthcare, with an average AUC of 72%. Nonetheless, the authors insist that the streamlining of ML methods can vastly advance medication safety in clinical practice in spite of these barriers such as data heterogeneity and limited reporting standards [23].

A thorough analysis of utilizing deep learning for extracting DDIs from Biomedical literature. It showcases how traditional methods need gross and cumbersome feature extraction while deep learning eliminates this and also does the job much better and much faster. Various models are discussed in the review: CNNs, RNNs, and attention mechanisms, with their advantages and disadvantages highlighted for working with large medical texts. The authors envision deep learning as very useful in pharmacovigilance while simultaneously identifying data quality and model interpretability as challenging problems [24]. In their article, an elaborate workflow to rank and verify pharmacokinetic drug-drug interaction alerts linked to serious adverse drug events (SADRE). When incorporating disproportionality methods with ranking algorithms and multiple source HTR validation, they were able to validate five relevant signals indicating DDIs-SADR together with the enlargement of drug safety databases; presenting an effective method for pharmacovigilance [25]. In their study, a new efficient ML model for detecting DDIs based on drug target protein profiles. Contrary to the traditional methodologies that employ complicated algorithms for integration of the data or structural features, their approach utilizes a basic logistic regression on benchmarked gene targets. The study shows that it has a better capability than existing models to predict the occurrence of DDIs, and can be further dissected for issues relating to biomolecular interpretation of DDIs, suggesting solutions to the issues of complexity and generalizability [26].

3. Dataset Utilizationacross Studies for DDI Prediction

Table 51.1 Dataset utilization across studies for DDI prediction [1–10]

Dataset	DeSIDE-DDI	Machine Learning for DDIs	HNAI Framework	Data-driven Prediction of ADRs
DrugBank	Used as an external validation dataset to predict unseen drug interactions. Contains 33,497 interactions (v5.0.0) and 782,405 interactions (v5.1.7) between approved drugs.	A primary source of DDI information, providing detailed interaction data between approved drugs, including structural and therapeutic properties.	Used to construct a comprehensive DDI network containing 6,946 high-quality pharmacokinetic and pharmacodynamic interactions between 721 approved drugs.	Not specifically mentioned; focus was on drug-protein interaction data from STITCH and ADR data from TWOSIDES.

Dataset	DeSIDE-DDI	Machine Learning for DDIs	HNAI Framework	Data-driven Prediction of ADRs
LINCS L1000	Used for drug-induced gene expression profiles. Provided differential expression signatures for 978 landmark genes across 19,156 compounds under controlled conditions.	Not Used	Not Used	Not used; the study focused on drug-protein interaction profiles rather than gene expression profiles.
TwoSIDES	Used for polypharmacy side effects. Contains over 4.5 million interactions involving 63,472 drug pairs linked to 963 side effects.	Used to analyze adverse drug reactions (ADRs) and side effects of drug combinations for feature generation.	Not used.	Crucial for identifying ADRs from synergistic DDIs, with data on 868,221 significant ADR associations involving 59,220 drug pairs.
BioGRID	Not used.	Used as a source of genetic and protein interaction data to enhance drug feature matrices.	Not used.	Not included; the study prioritized drug-protein interaction profiles over protein-protein interaction data from sources like BioGRID.
CTD	Not used.	Used to calculate pathway and disease similarities between drugs.	Integrated through MetaADEDB to annotate ADRs for drug–ADR network construction.	Not used; the study focused on ADR data from TWOSIDES and drug-protein interaction profiles from STITCH instead of pathway or disease-based similarities from CTD.
FAERS	Not used.	Not used.	Used to collect adverse event data, which were integrated into a phenotypic similarity network for DDI prediction.	Not used; pharmacovigilance data from FAERS were not the primary focus, as the study relied on TWOSIDES for ADR data.
MetaADEDB	Not used.	Not used.	Used to build a comprehensive drug–ADR network by integrating multiple ADR data sources, including CTD, SIDER, and OFFSIDES.	Not included; emphasis was on using TWOSIDES for ADR data and STITCH for drug-protein interaction profiles.
SIDER	Not used.	Not used.	Used via MetaADEDB to extract adverse effect data for phenotypic similarity calculations.	Not mentioned; the study used TWOSIDES as the primary source for ADR data instead.
OFFSIDES	Not used.	Not used.	Used via MetaADEDB to incorporate additional ADR data into phenotypic similarity metrics.	Not used; ADR data were primarily sourced from TWOSIDES.
PubChem	Not used.	Used for molecular structure information to calculate chemical similarity.	Not used.	Not referenced; the study focused on curated drug-protein interaction data from STITCH rather than molecular structure data from PubChem.
UniProt	Not used.	Used to obtain protein sequence and functional data for drug-target interactions.	Not used.	Not utilized; protein sequence data from UniProt were outside the scope, with the focus on drug-protein interactions from STITCH.
Therapeutic Target Database (TTD)	Not used.	Not used.	Used to enhance genomic similarity by identifying drug–target interactions.	Not used; the study concentrated on STITCH for drug-protein interactions rather than drug-target information from TTD.

4. Comparative Analysis of Machine Learning and Deep Learning Models for Adverse Drug Reactions

4.1 Machine Learning Models for ADR Prediction

In Study [1], feature selection was done using the Random Forest (RF) algorithm, prioritizing drug properties such as chemical fingerprints and characteristics, to improve model prediction. During the testing with Support Vector Machines (SVM), the performance of SVM was not as good as the developed deep learning models, especially for handling the non-linear relationships between drug interactions and side effects (Table 51.1).

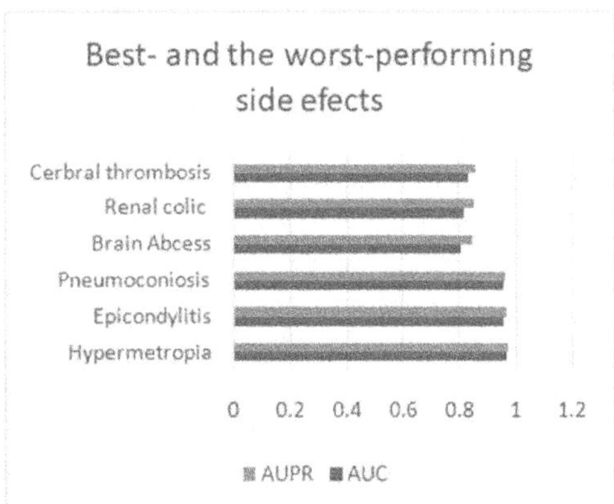

Fig. 51.1 Evaluating model performance across different side effects: AUPR and AUC metrics [5]

It describes, how the model works out in predicting specific side effects using Area Under the Precision-Recall Curve (AUPR)and Area Under the Curve (AUC), describing which categories are the best and the worse in performance.

Machine Learning for DDIs

In Study [17], basic binary classification of drug interactions was made using logistic regression (LR), though higher order interactions were difficult to depict. Among all developed models, XGBoost (Extreme Gradient Boosting) demonstrates the best result, with the F1-score of 0.831 and 78.0% accuracy (Fig. 51.1). The use of the model in DDI prediction is because of the good performance when dealing with large datasets and the accommodation of two forms of regression during training.

This section compares various models based on different factors, with a particular emphasis on F1 scores, highlighting the superior effectiveness of the presented model over Logistic Regression, Neural Networks, and XGBoost.

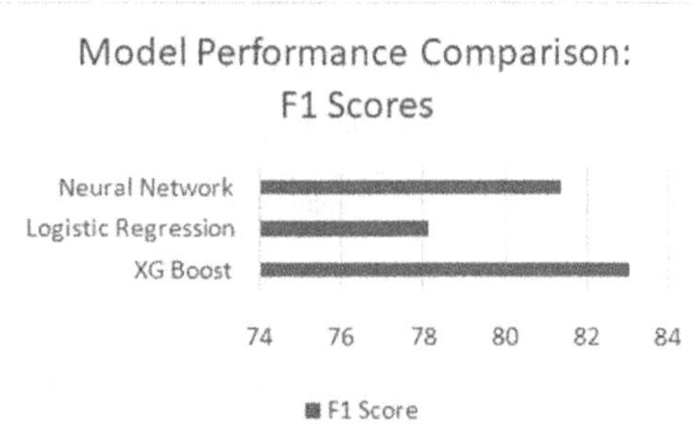

Fig. 51.2 Performance comparison of machine learning models for DDI prediction [15–19]

HNAI Framework

In Study [18], machine learning models like Naive Bayes and k-Nearest Neighbors (k-NN) were used to predict ADRs based on genomic, phenotypic, and structural data. Naive Bayes, with its conditional independence assumption, performed reasonably well for classification tasks but was limited in its ability to capture complex dependencies between features. k-NN, while intuitive, struggled with high computational costs when scaling up for large datasets.

Table 51.2 AUC performance scores of various machine learning models for ADR prediction [7]

Model	AUC Score
Naive Bayes (NB)	0.65
Support Vector Machine (SVM)	0.565
Support Regression (LR)	0.65
Logistic Regression (LR)	0.65
K-Nearest Neighbors (k-NN)	0.666
Decision Tree (CART)	0.666

This table summarizes the performance metrics of different models used to predict adverse drug reactions (ADRs). The AUC score is a key indicator of how well these models perform in classification tasks (Fig. 51.2).

Data-driven Prediction of ADRs

Instead of the conventional methods of machine learning, Study [19] used a probabilistic scoring model that involved profiling of drug interaction for ADR- DDI prediction. This model gives a weight and contribution score of proteins to estimate the probability of a drug causing ADRs. It proved the stability in cross validation, and had an average predictive accuracy of 89% on 1096 ADR models and revealed good performance even in moderate False Positive and False Negative situations (Table 51.2).

Table 51.3 Performance metrics of probabilistic scoring models for predicting ADR's induced by DDI's [20–26]

Metric	Value	Description
AUC (Range)	0.7–1.0	Achieved by models trained with at least 50 positive ADR samples.
AUC (Average)	0.85 (±0.05)	The average AUC value across all models, with a standard deviation of 0.05.
Accuracy	0.89	Overall proportion of correctly classified instances.
Sensitivity	0.63	The ability of the model to correctly identify true positive cases.
Specificity	0.90	The ability of the model to correctly identify true negative cases.

It presents data on the accuracy, AUC, sensitivity, specificity, and PPV of 1,096 ADR prediction models. These metrics calculate the model's credibility for predicting ADRs with or without noise, or when working with imbalanced data (Table 51.3).

4.2 Deep Learning Models for ADR Prediction

DeSIDE-DDI

Deep learning played a crucial role in Study [1], with Gated Linear Units (GLUs) used to map drug pairs to side-effect spaces. GLUs dynamically adjusted drug features under co-administration scenarios, enabling the model to predict ADRs with high accuracy. The deep learning model was trained using backpropagation, incorporating margin-based loss and L2regularization to ensure robustness and prevent overfitting. This deep learning framework provided high biological interpretability by explaining gene expression changes due to DDIs.

Machine Learning for DDIs

Neural Networks (NN) were used in Study [17] for drug interaction prediction and performed below par than models such as XGBoost. The limitations of the model were, Overfitting, reduced capacity in capturing nonlinear relationship, and poor capability in simulating interaction between drugs and similarity matrices. XGBoost model turned out to be more efficient and better suited for these tasks.

HNAI Framework

Deep learning models were also tested in Study [18], particularly in handling multi-dimensional genomic and phenotypic data. Neural Networks (NN) were applied to capture deeper interactions within the drug-protein relationship network. However, they did not perform as well as SVM due to the complexity of the dataset and the non-linearity of drug interaction patterns.

In general, the DeSIDE-DDI model is the most stable for ADR and DDI prediction because of better efficiency and a unique perspective. By leveraging compound- and gene-based signatures from drug-induced gene expression data, it offers mechanistic insights and achieves high accuracy (AUC: 0.889, AUPR: 0.915). It surpasses other models, such as XGBoost (accuracy: 78.12%), which has its own structure and can be changed dynamically and be opened and closed while improving the generality and ability to learn new drugs. While reaching an AUC of 0.85, the probabilistic scoring system is simple and has narrow applicability because of a lack of interpretability or flexibility. Because DeSIDE-DDI encompasses molecular data, it is appropriate to use the resource to study drug development and ADR/DDI prediction.

5. Conclusion and Future Scope

The study emphasizes the need for advanced models to manage ADRs and DDIs, improving patient safety and reducing costs. Tools like DeSIDE-DDI optimize error rates, model size, and polypharmacy interactions. Expanding these models with data from DrugBank, TWOSIDES, and FAERS, along with genomic and phenotypic information, can enhance individualized treatment. Effective pharmacovigilance and big data analytics are critical, especially for high-risk groups like the elderly. Future efforts should address data heterogeneity, model interpretability, and accessibility at the point of care. Collaboration among data scientists, clinicians, and regulators is essential to establish best practices for safer, more efficient healthcare systems.

References

1. Kim, E., & Nam, H. (2022). DeSIDE-DDI: interpretable prediction of drug-drug interactions using drug-induced gene expressions. Journal of cheminformatics, 14(1), 9.
2. Schatz, S.N., & Weber, R.J. (2015). Adverse Drug Reactions.
3. Riedl, M. A., & Casillas, A. M. (2003). Adverse drug reactions: types and treatment options. American family physician, 68(9), 1781–1791.
4. Rohilla, A., & Yadav, S. (2013). Adverse drug reactions: An Overview. International Journal of Pharmacological Research, 3, 10–12.
5. Dey, S., Luo, H., Fokoue, A., Hu, J., & Zhang, P. (2018). Predicting adverse drug reactions through interpretable deep learning framework. BMC bioinformatics, 19, 1–13.
6. Coleman, J. J., Ferner, R. E., & Evans, S. J. W. (2006). Monitoring for adverse drug reactions. British journal of clinical pharmacology, 61(4), 371–378.

7. Trung Huynh, Yulan He, Alistair Willis, and Stefan Rueger. 2016. Adverse Drug Reaction Classification with Deep Neural Networks. In Proceedings of COLING 2016, the 26th International Conference on Computational Linguistics: Technical Papers, pages 877–887, Osaka, Japan. The COLING 2016 Organizing Committee.
8. Shibbiru, T. (2016). Adverse Drug Reactions: An Overview. Journal of Medicine, Physiology and Biophysics, 23, 7–14.
9. Patton, K., & Borshoff, D.C. (2018). Adverse drug reactions. Anaesthesia, 73, 76–84.
10. Coleman, J. J., & Pontefract, S. K. (2016). Adverse drug reactions. Clinical Medicine, 16(5), 481–485.
11. Routledge, P. A., O'mahony, M. S., & Woodhouse, K. W. (2004). Adverse drug reactions in elderly patients. British journal of clinical pharmacology, 57(2), 121–126.
12. Davies, E. A., & O'mahony, M. S. (2015). Adverse drug reactions in special populations–the elderly. British journal of clinical pharmacology, 80(4), 796–807.
13. Brahma, D. K., Wahlang, J. B., Marak, M. D., & Sangma, M. C. (2013). Adverse drug reactions in the elderly. Journal of pharmacology and pharmacotherapeutics, 4(2), 91–94.
14. Moore, N., Lecointre, D., Noblet, C., & Mabille, M. (1998). Frequency and cost of serious adverse drug reactions in a department of general medicine. British journal of clinical pharmacology, 45(3), 301–308.
15. Hakkarainen, K. M., Hedna, K., Petzold, M., & Hägg, S. (2012). Percentage of patients with preventable adverse drug reactions and preventability of adverse drug reactions–a meta-analysis. PloS one, 7(3), e33236.
16. Kim, H. R., Sung, M., Park, J. A., Jeong, K., Kim, H. H., Lee, S., & Park, Y. R. (2022). Analyzing adverse drug reaction using statistical and machine learning methods: A systematic review. Medicine, 101(25), e29387.
17. Demirsoy, I., & KARAİBRAHİMOĞLU, A. (2023). Identifying drug interactions using machine learning. Advances in Clinical and Experimental Medicine, 32(8).
18. Cheng, F., & Zhao, Z. (2014). Machine learning-based prediction of drug–drug interactions by integrating drug phenotypic, therapeutic, chemical, and genomic properties. Journal of the American Medical Informatics Association, 21(e2), e278–e286.
19. Liu, R., AbdulHameed, M. D. M., Kumar, K., Yuˆ, X., Wallqvist, A., & Reifman, J. (2017). Data-driven prediction of adverse drug reactions induced by drug-drug interactions. BMC Pharmacology and Toxicology, 18, 1–18.
20. Han, K., Cao, P., Wang, Y., Xie, F., Ma, J., Yu, M., ... & Wan, J. (2022). A review of approaches for predicting drug–drug interactions based on machine learning. Frontiers in pharmacology, 12, 814858.
21. Salman, M., Munawar, H. S., Latif, K., Akram, M. W., Khan, S. I., & Ullah, F. (2022). Big data management in drug–drug interaction: a modern deep learning approach for smart healthcare. Big Data and Cognitive Computing, 6(1), 30.
22. Basnet, S., Nihal, A., Sijina, K. S., Seru, N. K., & Kumar, A. (2023). A Systemic Review of Machine Learning Approaches for Adverse Drug Reaction Detection: Novel Perspective and Challenges. Journal of Pharma Insights and Research, 1(2), 179–185.
23. Hu, Q., & Li, X. (2024). Machine learning to predict adverse drug reaction or event based on electronic health records: a systematic review and meta-analysis.
24. Zhang, T., Leng, J., & Liu, Y. (2020). Deep learning for drug–drug interaction extraction from the literature: a review. Briefings in bioinformatics, 21(5), 1609–1627.
25. Jeong, E., Su, Y., Li, L., & Chen, Y. (2024). Discovering clinical drug- drug interactions with known pharmacokinetics mechanisms using spontaneous reporting systems and electronic health records. Journal of Biomedical Informatics, 153, 104639.
26. Mei, S., & Zhang, K. (2021). A machine learning framework for predicting drug–drug interactions. Scientific Reports, 11(1), 17619.

Advances in Additive Manufacturing Technologies – Gurusamy Pathinettampadian et al. (eds)
© 2026 Taylor & Francis Group, London, ISBN 978-1-041-16687-0

Effect of Reduced Graphene Oxide Additives in Dielectric Fluid for Electro Discharge Machining of Aluminum based Metal Matrix Composite

D. Jayabalakrishnan[1], V. Chidambaram[2]
Department of Mechanical Engineering,
Chennai Institute of Technology,
Chennai, India

S. Saravanan[3], M. Jayakumar[4]
Department of Mechanical Engineering,
P.T. Lee Chengalvaraya Naicker Polytechnic College,
Chennai

R. Sathiyamoorthi[5]
Department of Mechanical Engineering,
Chennai Institute of Technology,
Chennai, India

◆ **Abstract:** The study focuses on the application of GO in the improvement of EDM on Al-SiCp metal matrix composite materials. Using the novel nano powder mixed EDM (NPMEDM) setup, the paper explores the effects of GO on two important variables, namely material removal rate (MRR) and surface roughness (SR). Overall trends suggests that MRR improved significantly with the concentration of GO up to 0.4 g/L. This is mainly because GO offers not only superior dielectric strength but also a more stable spark frequency. This one is a critical role of graphene in increasing the MRR since it has enhanced thermal conductivity that enhances the transfer of current to the workpiece. In addition to this, a negative correlation of SR with increase in GO concentration was also noted. This improvement is due to a more homogeneity of improve small and uniform craters on the machined surface due to discharging energy. But it was Note that the SR began to reduce from 0.4 g/L GO concentration onward, indicating which denotes an optimum level of GO that facilitates both MRR and SR together.

◆ **Keywords:** Graphene, EDM, Nano additive, MRR and SR

1. Introduction

Aluminum Based Metal Matrix Composites have now become an important material of choice in critical applications across industries like aerospace, defense, automobiles and sports[1]. This growing interest is attributed to the fact that MMCs have some of the best properties of any material. Notably, the use of MMCs in automotive parts has enabled a 10% weight saving and has improved fuel efficiency by 5%[2]. This significant gain in fuel economy is a clear indication of the huge opportunities that exist for MMCs in the automotive industry[3].

However, the machining of these advanced composite materials is not as easy as one might imagine. This can be attributed to the fact that the MMC structure contains

[1]djayabal2001@gmail.com, [2]chidambaram961@gmail.com, [3]saravanancnpt@gmail.com, [4]uvajai0603@gmail.com, [5]Sathiya.ram78@gmail.com

DOI: 10.1201/9781003685906-52

hard and abrasive reinforcements which lead to a number of machining difficulties. These problems include short tool life, low material removal rates, and high tool costs. To achieve the objectives of this research, it is necessary to overcome these challenges that hinder the improvement of MMCs' properties and their applications in various industries.

To address these challenges, the focus shifts to non-conventional machining techniques. EDM has been described as one of the most recommended non-conventional EDM machining methods regardless of the materials toughness [4]. Because EDM solves complications such as chattering, vibrations, and distortions during the machining of important components like dies, molds, parts for aerospace, automotive, and surgical instruments, it has been widely adopted. Also, EDM is capable of creating the shapes that are hard to manufacture by other methods of machining due to its flexibility. Due to the flexibility and high accuracy of EDM, this technology is widely used in many fields, which also leads to its development.

On the other hand, EDM is associated with various metallurgical issues such as recast layers, heat affected zones and micro cracks which may adversely affect the quality of the surface finish of the work piece. As a result, further processes such as lapping and polishing are used to enhance the surface finish. However, all these steps will for sure extend the production process and raise the total cost. In order to overcome these issues, various modern techniques have been proposed to improve the efficiency of EDM. Various methods have been included in order to improve this process such as rotary EDM that works on orbital motion on tool electrode and planetary motion on workpiece, special flushing systems, ultrasonic vibration, and conductive powdered particles being added into the EDM oil [5-6]. Among all the mentioned representative new techniques, Powder Mixed EDM (PMEDM) is one of the most developed alternative tool [8-9]. PMEDM consists of addition of metal powders in dielectric material and has many advantages. These benefits comprise raised MRR, lower SR, lesser micro-cracks, thinner surface layers, less porosity and enhanced corrosion-wear resistance [7-9]. In 1980, the concept of PMEDM was introduced, subsequently, several studies were performed related to the application of various powders and their effects on diverse materials including composites [10–17]. Previous studies revealed the efficient performance of graphene-based materials such as reduced graphene oxide (rGO) when used in combination with the dielectric fluid for the EDM process [18-21].

However, the use of graphene in EDM is still a relatively uncharted territory. The current work aims at achieving the highest possible surface finishes at the nano scale for Al-10%SiCp Metal Matrix Composites. This can be attributed to the proper distribution of graphene nanosheets in the EDM dielectric fluid. In order to achieve uniform distribution of the graphene, the material is subjected to a rigorous thirty-minute ultrasonication process, which is subsequently followed by a microtome where surface topography is studied and measured alongside the Material Removal Rate of the structure.. This novel approach is intended to reduce surface roughness, enhance MRR, and reduce tool wear rates when operating on Al-SiCp Metal Matrix Composite workpiece materials.

2. Materials and Methods

In this study, we worked on Aluminum alloy 6061 which was hybrid composite with 10% silicon carbide particulate to prepare metal matrix composite plates. The Al-SiCp MMC plates were prepared by the stir casting method, a common method for preparing particle reinforced metal matrix composites. In this study's machining experiments, a 5mm diameter copper tool electrode with negative polarity was used. Graphene Oxide was used as an additive in these novel powder-mixed electric discharge machining experiments to enhance the Electro Discharge Machining efficiency. The purpose of adding GO was to enhance the dielectric features and flushing of the debris from the discharged surface while Electrical Discharge Machining of Al-SiCp MMC.

The experimental setup used a transparent fiberglass tank with a 2-liter capacity for the dielectric fluid. In this tank, we incorporated a motorized stirrer and a micro pump to the system. This tank was located in the first EDM tank. The reader is referred to Fig. 52.1 for an illustration of the

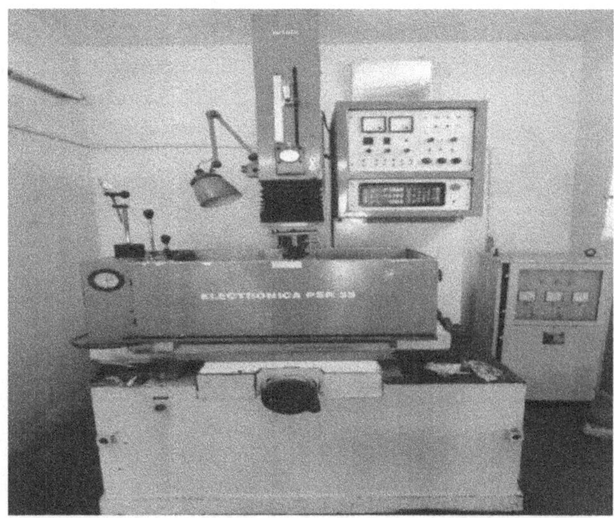

Fig. 52.1 Photographic view of the experimental setup

general study design. In addition, the Graphene Oxide used in this work was characterized using Scanning Electron Microscopy, and a picture of the GOs is presented in Fig. 52.2. The characteristics of the GOs are described in greater detail in Table 52.1.

Fig. 52.2 SEM image of graphene

Table 52.1 Physical and morphological properties of graphene

Property	Description/Value	Unit of Measurement
Thickness	Single layer: ~0.335	nm
Size/Dimensions	<10	nm
Electrical Conductivity	2700	S/m
Surface Area	~2630	m^2/g
Thermal Conductivity	~5000	W/(m·K)
Strength	130	Pa or GPa
Elastic Modulus	~1	GPa
Transparency	~97.7% light transmission	Percentage
Number of Layers	Mono-layer, Bi-layer, Few-layer	-

Integrating graphene nanoparticles into the dielectric fluid mineral oil for the EDM process allows the utilization of the graphene properties to improve the machinability of materials. The thickness of graphene is about 0.335 nm, which allows for its good dispersion in the mineral oil without affecting its viscosity to the extent that would hinder the fluidity in the EDM process. Graphene particles whose size is below 10 nanometres can be uniformly distributed in the dielectric fluid and provide uniform machining performance. The high surface area of graphene, about 2630 square meters per gram, increases the contact between the fluid and the workpiece which may lead to enhanced debris removal and enhanced EDM performance. The electrical conductivity of graphene is 2700 Siemens per meter and this may increase the conductivity of the dielectric fluid hence improving energy transfer and possibly increased machining speed with less energy. Its thermal conductivity of about five thousand Watts per meter Kelvin will assist in the release of heat during the EDM process, which may reduce thermal stress on the workpiece and improve surface finish of the machined part. In addition, the mechanical properties of graphene such as strength (130 GPa) and stiffness (1 TPa) indicate that it can withstand the harsh EDM environment without compromising the fluid's properties.

In electrical discharge machining, the dielectric fluids act as insulators until the voltage at which they breakdown is reached. When the threshold current is exceeded, they become conductive and allow for the formation of sparks. To aid in work piece and electrode cooling, chip removal, and reducing the probabilities of a short circuit, these fluids are used. For this EDM operation, Mobilmet S-122, a light-colored clear oil, is used. The properties of this dielectric fluid will be shown in Table 52.2. It is also obtained by ultrasonically mixing an adequate volume of the solution that has been ultrasonicated at 40 kHz for 7 hours with a graphene oxide suspension. Some of the critical control factors, including peak current and pulse duration as well as duty cycle, were based on previous studies and adjusted for preliminary testing. In the preparation step, stable and homogenous suspensions of graphene oxide dispersions must be sonicated for their yield to be guaranteed. This step is fundamental in ensuring homogenous distribution of graphene oxide additives throughout the dielectric fluid. A photographic view of the final workpiece is shown in Fig. 52.3 with the copper tool electrode used.

Table 52.2 *Physical Properties of* Mobilmet S-122 Dielectric fluid

Property	Value
Appearance	Clear, Bright
Viscosity @ 40°C (cSt)	22
Flash Point (°C)	>135
Viscosity @ 100°C (cSt)	4.5
Chlorine Content	Yes (Chlorinated)
Sulfur Content	Yes (Active Sulfur)
Corrosion Protection	Good
Suitable Materials	Ferrous and Non-Ferrous Metals
Applications	Severe Cutting Operations
pH	Neutral
Density @ 15°C (kg/l)	0.89

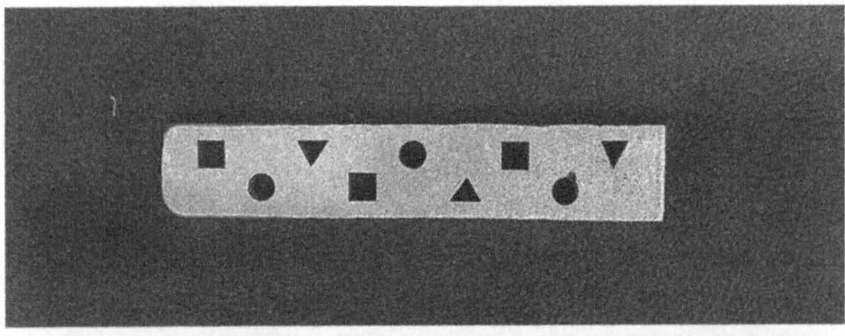

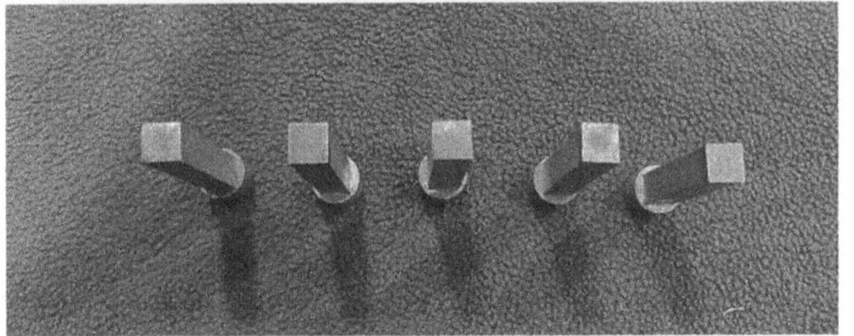

Fig. 52.3 Photographic view of the finished Al-Sicp MMC workpiece and Copper tool electrode

The experimental variables were set according to the results of prior testing. Details about the experiment conditions are available in Table 52.3. Abstract In this study, Material Removal Rate (MRR) and Surface Roughness (SR) were used as key characteristics of performance. Only the classical method was used to determine MRR, which consisted of a workpiece weighing on a digital electronic weighing balance from Radwag (Poland), with a precision of 0.001 mg, before and after the machined process. SR was measured using a Mitutoyo surface roughness tester with a cut-off length of 0.08 mm.

Table 52.3 Design scheme of process parameters and their levels

Sl. no	Parameter	Level		
		1	2	3
1	Current(I) A	4	8	12
2	Pulse Time On(Ton) μs	35	65	95
3	Pulse Time Off (Toff) μs	2	6	8
4	Graphene concentration g/l	0.2	0.5	0.8

3. Results and Discussions

3.1 Effects of Concentration of Graphene Powders on Material Removal Rate

Material Removal Rate (MRR) is a very important performance measure in machining operations, which is defined as the amount of material removed from the workpiece per unit time. In the case of Metal Matrix Composites, the MRR is usually lower than that in conventional materials. This reduced rate is mainly attributed to the existence of the harder abrasive particles in MMCs which cannot be easily worn out. But this is a challenge that can be solved when rGO is incorporated into the dielectric fluid as an additive.

The distinguished features of rGO such as its high thermal conductivity may significantly enhance the electrical current flowing to the workpiece, thereby increasing the Material Removal Rate (MRR) during machining of Aluminium Matrix composite in the Electro Discharge Machining (EDM) process. The figure illustrates the relationship between the concentration of graphene and MRR during the machining process. The impact of the concentration of graphene x on the efficiency of material removal is illustrated in Fig. 52.4. This is due to graphene's high thermal conductivity.

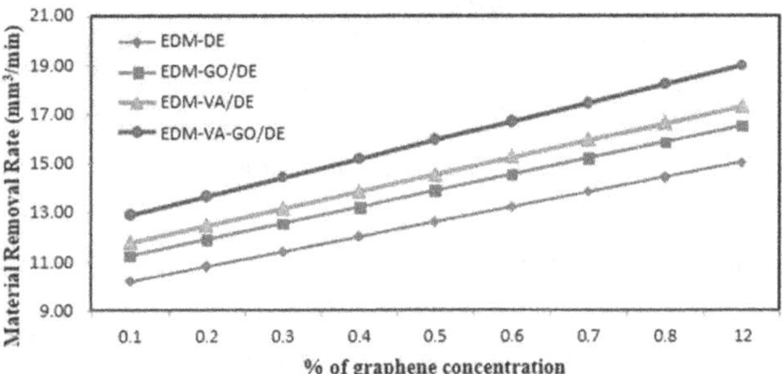

Fig. 52.4 Effect of graphene concentration on MRR

Graphene improves the heat transfer rate which in turn increases the rate at which electrical current is transferred to the workpiece for material removal. At lower graphene concentrations, the MRR seems to rise steadily. This may imply that a little quantity of graphene may have a positive effect on the cutting efficiency because the lubrication and thermal properties of the cutting fluid are improved by the incorporation of graphene. The MRR is seen to increase and then level off, suggesting that there is an optimum concentration of graphene. This may be attributed to the fact that graphene improves the heat dissipation, friction or surface finish of the work piece material thereby increasing the material removal rates. At concentrations above the optimal level, MRR either stagnates or decreases. This could be attributed to factors such as high viscosity of the cutting fluid, clumping of the graphene particles or inability to realize further improvement with increased concentration of graphene. The results indicate that there is a particular concentration of graphene which will give the best MRR. Higher concentration may result in poor performance, meaning that the concentration of graphene should be well optimized for efficient machining. The observed trend highlights the need to address the positive characteristics of graphene in relation to the concentration of the material in the cutting environment.

3.2 Effects of Concentration of Graphene Powders on Surface Roughness

The surface of the workpiece is identified with number of craters as the energy is discharged at the point of material removal itself in electrical discharge machining (EDM). The electrical parameters of EDM as applied in the process play a large part in the surface quality. Use of powder additive like graphene has a vital role on the control of surface roughness (SR) of the workpiece in Nonconventional Powder Mixed EDM (NPMEDM).

The surface roughness was measured by a Mitutoyo surface roughness tester (cut-off length of 0.08mm) in this study. Graphene concentration was found to affect surface Roughness (SR) during machining as shown in Fig. 52.5. Surface roughness is a crucial characteristic of the surface finish after the material has been removed, and a clear trend of reduced SR with increasing graphene concentration up to 0.4 g/L is observed. This enhancement in surface quality is ascribed to the function of GOs in attaining a more balanced dispersion of the discharge energy. This uniformity minimizes the size of the craters produced on the workpiece hence improving the surface finish of the workpiece. The data shows that there is a certain concentration of graphene at which surface roughness is at its lowest and hence surface finish is at its best. However, more than this concentration, any further addition of graphene may have negative impacts on surface quality due to increased friction or fluid instability. This trend shows that more work is needed in order to achieve the right balance between enhancing surface finish and other fluid properties that are useful in cutting operations.

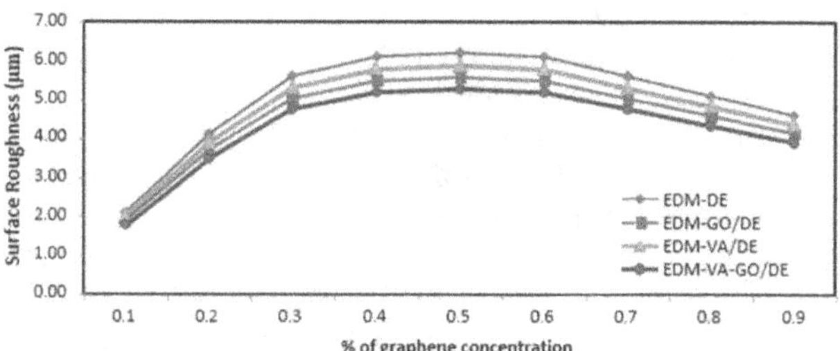

Fig. 52.5 Effect of graphene concentration on SR

3.3 Surface Topography

As part of this, a small experimental comparison was performed on the surface finish obtained by conventional EDM versus EDM with graphene to provide some idea on the surface texture of machined cavities. SEM images in Fig. 52.6 (a & b) reflect the surface morphology of the machined micro cavities with and without graphene respectively. SEM images of I6a, I1 and I2 features indicate larger irregular-shaped craters generated by conventional EDM (with graphene) than the cavity shape of the feature I1 and I2 (Fig. 52.6a). This increased size of these craters is typically ascribed to a larger material removal rate in a singular pulse which yields a poor surface finish. The energy is not symmetric and debris flushing from a single point is also not symmetric, leading to deeper craters and contributing to surface roughness. In contrast, EDS results in the ethanol dielectric fluid showed that craters are larger and less uniformly distributed than those from the graphene/EDM dielectric fluid (SEM micrographs). This is evidence of a smoother surface with reduced surface roughness which is easily seen in Fig. 52.6b. The incorporation of graphene minimizes the size of the craters; this improves the energy distribution of the discharge and the flushing of debris during machining.

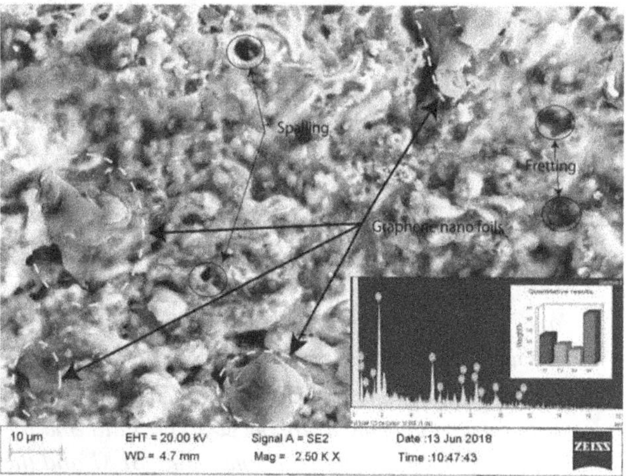

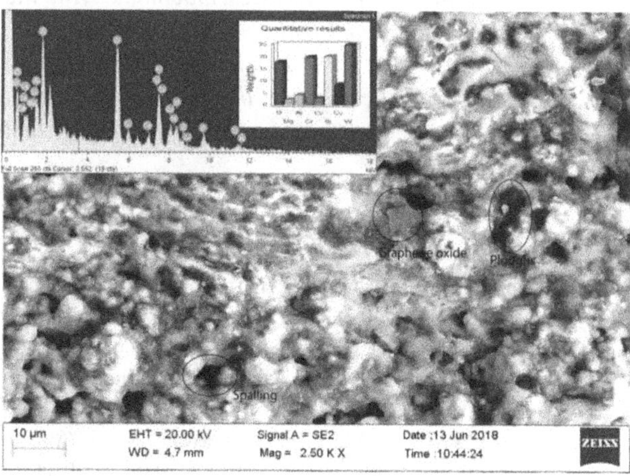

Fig. 52.6 SEM micrograph of Al-10%SiCp MMC machined by (a) EDM (b) graphene mixed

EDM. The surface characteristics of the machined cavities were evaluated using Scanning Electron Microscopy (SEM) to compare the effects of conventional Electrical

Discharge Machining (EDM) and EDM with graphene as an additive in the dielectric fluid.

4. Conclusions

The study aimed to explore the effects of Graphene Oxide (GO) concentration on key factors like Material Removal Rate (MRR), Surface Roughness (SR), and surface topography in Nonconventional Powder Mixed EDM (NPMEDM) applied to Al-SiCp Metal Matrix Composites (MMC). The findings led to several key conclusions:

1. *Impact on Material Removal Rate (MRR): GO improves the MRR to a great extent when integrated. This improvement is attributed to the higher dielectric strength and the more frequent and higher intensity sparks during the EDM process. Here, the high thermal conductivity of graphene plays a crucial role in enhancing the current transfer to the workpiece and hence increase the MRR..*
2. *Influence on Surface Roughness (SR): The SR is seen to decrease with increase in graphene concentration up to 0.4 g/L after which the SR increases. The first dip is attributed to the function of GO in equally dividing the discharge energy to reduce the size of craters and hence increase surface finish. However, at higher concentration, problems such as arcing and short circuiting occur which affect the process stability and hence increase the SR.*
3. *Surface Topography Observations: The comparison between the Graphene enhanced EDM fluid and the conventional EDM fluid is done using SEM and the results show that the surfaces machined with the Graphene enhanced EDM fluid are smoother with small craters. This finding is in agreement with the earlier works and presents the positive impact of graphene on the surface quality in the NPMEDM process.*

References

1. Buschmann, R. (2006). Preforms for the reinforcement of light Metals – Manufacture, applications and potential. In K.U. Kainer (Ed.), Metal matrix composites: Custom-made materials for automotive and aerospace engineering (pp. 77–94). Wiley-VCH Verlag GmbH & Co. KGaA.
2. Pai, B. C., Pillai, R. M., & Satyanarayana, K. G. (2001). Light metal matrix composites - present status and future strategies.
3. Weinert, K., Lange, M., & Petzoldt, V. (2002). Machining of metal matrix composites. Proceedings of ESDA2002: 6th Biennial Conference on Engineering Systems, Design and Analysis, 8–11 July 2002, Istanbul, Turkey. ASME.
4. Rajurkar, K. P. (1994). Handbook of Design Manufacturing and Automation. Wiley.
5. Abbas, N. M., Solomon, D. G., & Bahari, M. F. (2007). International Journal of Machine Tools and Manufacture, 47, 1214–1228.
6. Garg, R. K., Singh, K. K., Sachdeva, A., Sharma, V. S., Ojha, K., & Singh, S. (2010). International Journal of Advanced Manufacturing Technology, 50, 611–624.
7. Kansal, H. K., Singh, S., & Kumar, P. (2005). Journal of Materials Processing Technology, 169(3), 427–436.
8. Kansal, H. K., Singh, S., & Kumar, P. (2006). International Journal of Machine Machinability of Materials, 1(4), 396–411.
9. Kansal, H. K., Singh, S., & Kumar, P. (2007). Journal of Materials Processing Technology, 184(1-3), 32–41.
10. Erden, A., & Bilgin, S. (1980). Role of impurities in electric discharge machining. In J. M. Alexander (Ed.), Proceedings of the 21st International Machine Tool Design and Research Conference, 8–12 September 1980, Swansea, London. Macmillan.
11. Kansal, H. K., Singh, S., & Kumar, P. (2007). Journal of Manufacturing Processes, 9(1), 13–22.
12. Yeo, S. H., Tan, P. C., & Kurnia, W. (2007). Journal of Micromechanics and Microengineering, 17(11), N91–N98.
13. Pecas, P., & Henriques, E. (2008). Journal of Materials Processing Technology, 200(1-3), 250–258.
14. Seo, Y. W., Kim, D., & Ramulu, M. (2006). Materials Manufacturing Processes, 21, 479–487.
15. Patel, K. M., Pandey, P. M., & Rao, P. V. (2009). Materials Manufacturing Processes, 24, 675–682.
16. Kumar, S. S., Uthayakumar, M., Kumaran, S. T., & Parameswaran, P. (2014). Materials Manufacturing Processes, 29(11-12), 1395-1400.
17. Yadav, R. N., & Yadava, V. (2014). Materials Manufacturing Processes, 29(5), 585-592.
18. Gouda, D., Panda, A., Nanda, B. K., Kumar, R., Sahoo, A. K., & Routara, B. C. (2021). Recently evaluated Electrical Discharge Machining (EDM) process performances: A research perspective. In D. Gouda, A. Panda, B. K. Nanda, R. Kumar, A. K. Sahoo, & B. C. Routara, Materials Today Proceedings (Vol. 44, p. 2087). Elsevier BV. https://doi.org/10.1016/j.matpr.2020.12.180

Note: All the figures and tables in this chapter were made by the authors.

Advances in Additive Manufacturing Technologies – Gurusamy Pathinettampadian et al. (eds)

Optimization of Pulsed TIG Welding Parameters for Maximizing Tensile Strength in AA7075-T6 Aluminium Alloy

R. Vijayan[1],
Muthukumaran A.[2], Subash P.[3],
Hepsi Beaula M.J.[4], Parthiban A.[5], Balamurali P.[6]
Assistant Professor, Department of Mechanical Engineering,
Chennai Institute of Technology,
Chennai, India

◆ **Abstract:** To forecast the tensile strength of AA7075-T6 aluminum alloy pulsed gas tungsten arc welded (GTAW) joints, a mathematical model has been developed. Important process variables, such as base current, welding speed, and pulse-on time, peak current, are taken into account when creating this model. The experiments were designed using a three-factor composite design matrix. In this study, an innovative approach was taken by preplacing the filler wire into the weld groove before welding. The tensile characteristics of the welded joints improved with the adjusted GTAW parameters. The created model can be used in industrial applications to forecast the tensile strength of AA7075-T6 welds and was tested for adequacy. It was validated at a 97% confidence level. The findings shows that Peak current has the most significant impact on tensile strength, followed by welding speed, and current when the filler wire is preplaced in the weld groove.

◆ **Keywords:** GTAW, AA7075-T6, Process parameter optimization, RSM, UTS, ANOVA

1. Introduction

Because it is lightweight, strong, resistant to corrosion, and simple to fabricate, aluminum is utilized extensively in many different sectors. Among aluminium alloys, AA7075-T6 is preferred for applications in aerospace, automotive, and other industrial sectors because of its high strength and cost-effectiveness. However, joining AA7075-T6 using fusion welding methods, particularly butt welding, presents challenges related to weldability. The primary issue is the reduction in joint strength due to coarse dendritic structures in the WZ and the loss of alloying elements at elevated temperatures. Overcoming these challenges while ensuring high-quality welds remains a significant concern for engineers and manufacturing industries [1–5].

GTAW is widely used due to its simplicity, portability, and cost-effectiveness. It is commonly applied in fabrication, maintenance, and repair work across different sectors. The flexibility of the GTAW process allows for welding in all positions, easy automation, and seamless integration into robotic manufacturing systems. The process consists of essential components such as a power supply, welding torch, tungsten electrode, and controller. The selection of appropriate welding parameters plays a crucial role in achieving high-quality welds. Improper parameter selection can adversely affect weld quality, making parameter optimization essential [6–8].

Several researchers have investigated the effect of GTAW process parameters on weld quality. Mannion

[1]srajendranvijayan@gmail.com, [2]muthu2189@gmail.com [3]subashp@citchennai.net, [4]hepsibeaulamj@citchennai.net, [5]parthimech2810@gmail.com, [6]muralipb2080@gmail.com

DOI: 10.1201/9781003685906-53

and Heinzman (1999) identified key parameters affecting GTAW welds. Haung and Kou (2000-2002) examined the susceptibility of aluminium alloys to cracking and found that using NHT filler wire, such as ER5356, enhances resistance to cracking. Leijun Li et al. (2005) analyzed the effect of joint design and concluded that optimizing the design and process parameters improves mechanical properties. Hassan et al. (2008) emphasized that the excellence of fusion-welded joints depends on constraints such as dilution, microstructure, mechanical properties, and joint efficiency. Senthil Kumar T. et al. (2007) compared pulsed and continuous current GTAW on aluminium alloys, highlighting that pulsed current GTAW improves tensile strength due to optimal process control and grain refinement.

Balasubramanian M. et al. (2009) developed a mathematical model using a central composite matrix and optimized it through the Lexicographic method, achieving improved weld pool geometry. In a subsequent study (2010), they identified that pulse frequency significantly influences weld pool characteristics. Kumar A. et al. (2009) employed the Taguchi technique for process optimization and developed a regression model to predict tensile properties, demonstrating that optimized parameters enhance mechanical properties. They further noted a 10-15% improvement in mechanical properties through planishing, which helps relieve internal stresses in the weld zone [9–12].

Selecting optimal GTAW process parameters is challenging as it depends on multiple factors, including base metal properties, electrode characteristics, shielding gas type, and gas flow rate. Automated applications of GTAW further increase the complexity, necessitating detailed research on process parameters and their impact on weld pool geometry and fusion extent. Literature reviews indicate that most researchers manually feed filler wire during GTAW, leading to non-uniform weld pool geometry and inconsistent dilution of filler metal into the molten base metal. To address this, the present study adopts a method of preplacing the filler wire into the weld groove while maintaining a uniform gap between the electrode tip and filler wire surface. The study seeks to establish a correlation between key welding parameters and tensile strength, providing valuable insights for optimizing welding conditions in industrial applications [13–15].

2. Experimental Methodology

The AA7075-T6 aluminum alloy, with a thickness of 5 mm, was procured in the form of a commercially available rolled sheet measuring 150 × 150 mm. The chemical composition of the base metal was analyzed using a Spark Analyzer Spectromax through spectro spark emission. The nominal composition of the material [Alcoa mill products] and the composition of the AA7075-T6 alloy used in this study are provided in Table 53.1. The welding process employed ER5356 (Al–5%Mg) aluminum alloy filler wire with a 2 × 2 mm square cross-section. Various AC pulsed current parameter combinations with argon gas shielding were tested. To finalize the welding settings, the filler wire was preplaced in the groove prior to welding, and the resulting weld quality, bead contour, and penetration were assessed [16–19].

Table 53.1 Chemical composition of AA7075-T6

BM	Zn	Mn	Cu	Mg	Cr	Fe	Si	Al
AA7075	5.1	0.04	1.2	2.2	0.2	0.25	0.06	Bal

Observations from the trials indicated that when the peak current (Ip) was below 195 A, incomplete penetration and lack of fusion occurred, while values above 205 A led to undercut, spatter, and base metal overheating. Similarly, background current (Ib) below 93 A resulted in a short arc length and improper mixing of filler and base metal, whereas values exceeding 103 A caused arc instability and excessive arc length. Welding speeds below 200 mm/min led to excessive root protrusion and undercut, while speeds above 400 mm/min reduced penetration and narrowed the weld pool. Pulse on time (Pon) below 40% resulted in insufficient heat input, whereas values above 60% caused excessive melting and electrode overheating [10].

This study aims to improve the quality of GTAW joints for AA7075-T6 by analyzing the influence of key process parameters. Three significant parameters were selected, each tested at five levels, as demonstrated in Table 53.2. The selected process parameters are detailed in Table 53.2.

Table 53.2 Levels of PTIG welding process parameter

WPP	Unit	Levels				
		-1.682	-1	0	1	+1.682
PC	A	190	194	200	206	210
BC	A	90	94	100	106	110
WS	mm/min	250	270	300	330	350

For experimentation, AA7075-T6 base metal plates were prepared to 250 × 150 × 3.46 mm dimensions through chemical cleaning. Butt joint welding was performed as per the CCD matrix using an automatic GTAW machine, with a mild steel backplate and preplaced filler wire, as illustrated in Fig. 53.1. Tensile test specimens, extracted from the weld joint center, were prepared according to ASTM-E8 standards [Kumar A. et al. (2009)]. Tensile tests

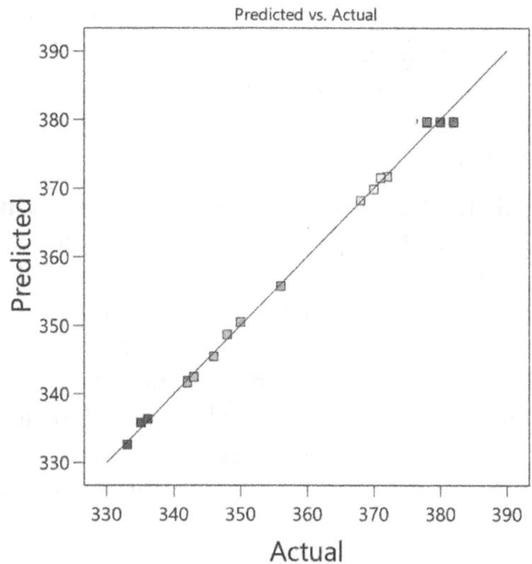

Fig. 53.1 Predicted vs actual graph

were carried out using a 60-ton UTM at room temperature. The average UTS values from three trials for each of the 31 experiments are presented in Table 53.3.

Table 53.3 DOE framed using CCD

Std. order	Run order	Coded Values			UTS (MPa)
		P	S	F	
1	12	-1	-1	-1	342
2	18	1	-1	-1	333
3	9	-1	1	-1	356
4	11	1	1	-1	346
5	4	-1	-1	1	370
6	7	1	-1	1	342
7	10	-1	1	1	372
8	1	1	1	1	343
9	13	-1.682	0	0	368
10	8	1.682	0	0	335
11	6	0	-1.682	0	336
12	3	0	1.682	0	348
13	2	0	0	-1.682	350
14	16	0	0	1.682	371
15	20	0	0	0	378
16	14	0	0	0	380
17	17	0	0	0	382
18	19	0	0	0	378
19	15	0	0	0	378
20	5	0	0	0	382

3. Results and Discussion

3.1 Development of Empirical Relationship

This study's main goal is to maximize GTAW joints' ultimate tensile strength (UTS), which is affected by a number of process variables. The following is a mathematical expression of the relationship between UTS and the chosen parameters: welding speed (S), base current (BC), and peak current (PC):

$$UTS = f(PC, BC, WS)$$

3.2 Checking Adequacy of Developed Empirical Relationship

Using Analysis of Variance (ANOVA), the proposed empirical model's suitability was evaluated. The created model's F-ratio must be less than the standard F-ratio in order to be deemed valid. The ANOVA results at a 95% confidence level are shown in Table 53.4. With a 0.01% chance of this outcome being caused by noise, the mathematical model's F-value of 302.62 shows that it is very significant. Process variables are considered significant if their p-value is less than 0.05.

Table 53.4 ANOVA table for responses with 'p' values

Source	Sum of Squares	Df	Mean Square	F-value	p-value	
Model	6064.73	9	673.86	302.62	< 0.0001	significant
A-TRS	1266.18	1	1266.18	568.62	< 0.0001	
B-TTS	184.39	1	184.39	82.81	< 0.0001	
C-AL	533.00	1	533.00	239.36	< 0.0001	
AB	0.5000	1	0.5000	0.2245	0.6458	
AC	180.50	1	180.50	81.06	< 0.0001	
BC	72.00	1	72.00	32.33	0.0002	
A^2	1376.05	1	1376.05	617.96	< 0.0001	
B^2	2484.59	1	2484.59	1115.79	< 0.0001	
C^2	625.78	1	625.78	281.03	< 0.0001	
Residual	22.27	10	2.23			
Lack of Fit	2.93	5	0.5869	0.1518	0.9705	not significant
Pure Error	19.33	5	3.87			
Cor Total	6087.00	19				
Std. Dev.	1.49	**R^2**		0.9963		
Mean	359.50	**Adjusted R^2**		0.9930		
C.V. %	0.4151	**Predicted R^2**		0.9918		
		Adeq Precision		44.5825		

The percentage of variance in the observed response attributable to the impact of process factors is explained by the coefficient of resolve, or R^2. The value of R^2 increases as more process parameters and interactions are included in the model, making it a valuable metric for model comparison. Kumar et al. (2007) concluded that a higher R^2 value enhances the model's adequacy. In this study, the developed mathematical model achieved an R^2 value of 99.40%, indicating a strong correlation between the predicted and experimental results. As all developed models exhibit an R^2 value above 80%, they are considered reliable for predicting the UTS of the welded joints.

$$\begin{aligned} UTS = &+ 379.63808 - 9.62880 * PC + 3.67446 * BC \\ &+ 6.24724 * WS - 0.250000(PC*BC) \\ &- 4.75000(PC*WS) - 3.00000(BC*WS) \\ &- 9.77159(PC)^2 - 13.13034(BC)^2 - 6.58961(WS)^2 \end{aligned}$$

The optimum conditions with reasonable precision are obtained by generating response graphs and contour plots for surface analysis using Design Expert software. The regression model gives a three-dimensional response graph and contour plots for UTS as shown in the Fig. 53.2. Figure 53.2(c) shows that the apex point of the response surface exhibits optimum value of the UTS at peak current 200 A and welding speed 300 mm/min, this is mainly because of equiaxed grain in the weld zone (WZ) of the joint. This is the cause of decrease in UTS. When welding speed (S) increase beyond 300 mm/min lesser the contact time with base metal [14].

Figure 53.2(a) depicts the variation of UTS with respect to peak current and base current lower the base current lower the UTS value when base current increased UTS value also increased and it is maximum when base current is 99 A. The increase in UTS value is mainly because of sufficient heat at peak current 200 A and finer grain structure in the weld zone. Further increase in welding speed, reduces the heat required for melting the base metal, it leads to lower penetration and decreases the ultimate tensile strength [11].

The GTAW carried out at the optimal values of; base current 99A, peak current 200 A, and welding speed 300 mm/min, and tensile test conducted to determine ultimate tensile strength welded joints. These results correlated with optical micro graphs of BM and WZ as demonstrates in Fig. 53.2. The strength of the weld joint mainly influenced by grain structure at WZ from the Fig. 53.2 it reveals that formation of fine grain structure at weld zone is the main reason for increase in UTS of the joint. And also, large

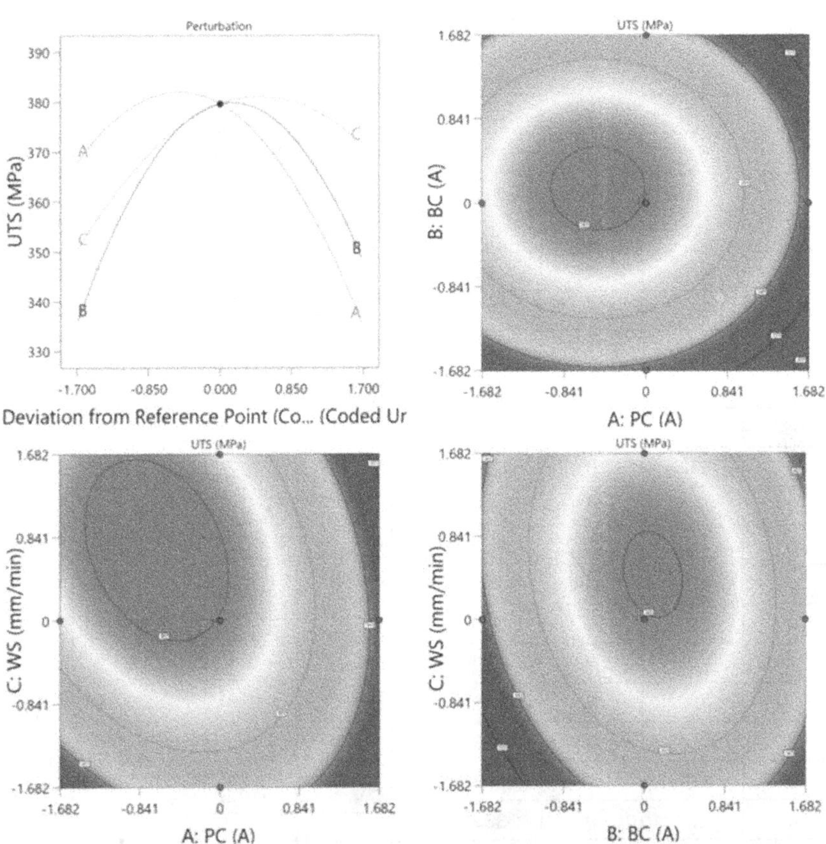

Fig. 53.2 [a] Perturbation plot [b] Interaction plot AB [c] Interaction plot AC [c] Interaction plot BC

number of precipitates in the weld zone and moderately spread throughout the weld zone this is also one of the reason for UTS. This improvement in the mechanical properties is correlated with fractography as shown in the Fig. 53.2 small dimples are seen in optimal GTAW process parameters in weld joint (Fig. 53.2). Hence optimal GTAW process parameters joints have demonstrates higher ductility associated to traditional GTAW process parameters weld joint [18].

4. Conclusion

The PTIG welding process was successfully conducted on AA7075 aluminum alloy, yielding defect-free welds, and the following conclusions are drawn:

1. To produce welds free of flaws, the critical process variables-peak current, base current, welding speed, and pulse-on time-were examined.
2. The initial trials revealed that improper selection of base current, welding speed, peak current, led to defects such as incomplete penetration, undercut, spatter formation, arc instability, and improper fusion of the filler metal with the base material.
3. ANOVA results confirmed the model's adequacy, with an F-value of 302.62, indicating its statistical significance at a 95% confidence level.
4. The developed model demonstrated a high correlation (R^2 = 99.40%), ensuring that the selected parameters effectively predict the tensile strength of the welds. The model's accuracy makes it suitable for practical welding applications.
5. A total of 20 welding trials were conducted as per the Central Composite Design (CCD) matrix. The results confirmed that the optimized parameters significantly improved the weld quality and mechanical properties.

References

1. Alcoa mill products, Bettendorf, IOWA 52722,(800)523–9596,www.millproducts-alcoa.com
2. Akula, Deepa Reddy: Characterization of Mechanical properties and study of microstructures of stir welded joint fabricated from similar and dissimilar alloys of aluminium" PhD thesis". University Of Missouri, Columbia, 2007.
3. Balasubramanian, MV; Jayabalan and Balasubramanian, V: Prediction and Optimization of Pulsed Current Gas Tungsten Arc Welding Process Parameters to Obtain Sound Weld Pool Geometry in Titanium Alloy Using Lexicographic Method, 'Journal of Materials Engineering and Performance', vol. 18 (7), October 2009, 871.
4. Balasubramanian, M, Jayabalan, V; Balasubramanian, V: Effect of process parameters of pulsed current tungsten inert gas welding on weld pool geometry of titanium welds, 'Acta Metall. Sin.(Engl. Lett.)', vol. 23, no. 4, 312–320, August 2010.
5. Benyounis, KY; Olabi, AG: Optimization of different welding processes using statistical and numerical approaches – A reference guide, 'Advances in Engineering Software', 39, 2008, 483–496.
6. Ben-Hamu, G; Eliezer, D; Cross, CE; Th. Bollinghaus: The relation between microstructure and corrosion behavior of GTA welded AZ31B magnesium sheet, 'Materials Science and Engineering A', 452–453, 2007, 210–218.
7. Box, GEP; Hunter, WH; Hunter, JS: Stastics for Experiments, New York, John Wiley Publications, 1978.
8. Hassan, Ezzeddin, Mohamed : Feasibility and Optimization of Dissimilar Laser Welding Components". Ph. D Thesis, Dublin City University, 2008.
9. Huang, C and Kou, S: Partially Melted Zone in Aluminum Welds: Liquation Mechanism and Directional Solidification, 'Welding Journal', vol 79, no. 5, 2000, 113s-120s
10. Huang, C and Kou, S: Partially Melted Zone in Aluminum Welds: Solute Segregation and Mechanical Behavior, 'Welding Journal', vol. 80 no.1, 2001, 9s-17s.
11. Huang, C and Kou, S: Partially Melted Zone in Aluminum Welds: Planar and Cellular Solidification, 'Welding Journal', vol. 80, no. 2, 2001, 46s-53s.
12. Huang, C and Kou, S: Liquation Mechanisms in Multi-Component Aluminum Alloys During Welding, 'Welding Journal', vol 81, no. 10, 2002, 211s-222s.
13. Kumar, A; Sundarrajan, S; Optimization of pulsed TIG welding process parameters on mechanical properties of AA 5456 Aluminum alloy weldments, Materials and Design 30, 2009, 1288–1297.
14. Kumar Sudhir; Pradeep Kumar; Shan, HS: Effect of evaporative pattern casting process parameters on the surface roughness of Al–7% Si alloy castings, Journal of Materials Processing Technology, 182, 2007, 615–623.
15. Li, Leijun; Orme, Kevin; and Yu, Wenbin: Effect of Joint Design on Mechanical Properties of AL7075 Weldment, 'Journal of Materials Engineering and Performance', vol. 14 (3), June 2005–323.
16. Bernard, Mannion and Heinzman, Jack: Determining parameters for GTAW, Article published in the July/August 1999 issue of Practical Welding Today.
17. Montgomery, DC: Design and Analysis of Experiments, New York: John Wiley & Sons 2001.
18. Nadkarni, SV : Modern Arc Welding Technology, 1988, New Delhi, India: IBH Publication.
19. Senthil Kumar, T; Balasubramanian, V; Sanavullah, MY: Influences of pulsed current tungsten inert gas welding parameters on the tensile properties of AA 6061 aluminium alloy, 'J Mater Materials and Design', 28, 2007, 2080–2092

Note: All the figures and tables in this chapter were made by the authors.

Advances in Additive Manufacturing Technologies – Gurusamy Pathinettampadian et al. (eds)

Optimization of Laser Welding Parameters for AA7075 Aluminium Alloy

Hepsi Beaula M.J.[1], Subash P.[2], Balamurali P.[3], Parthiban A.[4], R. Vijayan[5], Muthukumaran A.[6]

Assistant Professor, Department of Mechanical Engineering
Chennai Institute of Technology
Chennai, India

◆ **Abstract:** Compared to conventional arc welding methods, laser welding presents a highly promising alternative for producing high-strength joints in aluminum alloys, particularly AA7075. One of the main limitations of arc welding is the generation of a wide heat-affected zone (HAZ), which often results in reduced joint strength when working with aluminum. In contrast, laser welding minimizes the size of the HAZ, leading to improved mechanical properties in the welded joint. AA7075 aluminum alloy is especially attractive for structural applications due to its favorable properties, such as low density, lightweight nature, excellent corrosion resistance, and significantly higher strength-to-weight ratio compared to steel. These characteristics make it an ideal candidate for applications requiring both strength and weight reduced. The present study focuses on evaluating the feasibility of welding AA7075 using fiber laser welding and compares the performance of two laser operation modes: pulse wave (PW) and continuous wave (CW). The main objective is to investigate how varying the focal position of the laser beam affects the weld bead profile, depth of penetration, and microstructural features during bead-on-plate (BOP) welding. Experiments were conducted on AA7075 sheets with a thickness of 2 mm. Welding was carried out using both PW and CW laser modes, with focal lengths ranging from 60 mm to 200 mm. In all cases, the laser power was set at 90% of the maximum capacity, and the welding speed (WS) was kept constant at 2 mm/s to ensure consistent comparison across trials. The results showed distinct differences between the two modes. The pulse wave mode produced a keyhole-type weld profile and achieved an optimal penetration depth of approximately 1.0 mm. In contrast, the continuous wave mode resulted in a much shallower penetration depth of just 0.153 mm, with a less pronounced weld profile. These findings suggest that AA7075 alloy can be effectively welded using a low-power fiber laser setup, particularly in PW mode, which allows for partial penetration while forming a keyhole profile. To further improve weld quality and reduce potential defects, additional process enhancements—such as the introduction of shielding gas and adjustment of the laser beam's incident angle—can be considered.

◆ **Keywords:** AA 7075, BOP, Laser modes, Fiber laser, Microstructure, Optimum depth

1. Introduction

The process of joining materials plays a critical role in manufacturing and fabrication industries. It can be broadly categorized into different techniques, including mechanical fastening (such as bolts, rivets, and clamps) and various welding methods. Among these, welding is one of the most

[1]hepsibeaulamj@citchennai.net, [2]subashp@citchennai.net, [3]muralipb2080@gmail.com, [4]parthimech2810@gmail.com [5]srajendranvijayan@gmail.com, [6]muthu2189@gmail.com

DOI: 10.1201/9781003685906-54

widely used processes for producing permanent joints with high mechanical strength.

Aluminum alloys, known for their unique combination of lightweight and high strength, are extensively used in applications where weight reduction is crucial, such as in the aerospace, automotive, and transportation sectors. To join aluminum components effectively, a range of welding processes have been developed. These include traditional fusion-based techniques like gas tungsten arc welding (GTAW) and gas metal arc welding (GMAW), as well as more recent solid-state processes such as Friction Stir Welding (FSW). FSW, in particular, has gained attention due to its ability to join aluminum without melting, which helps to avoid common welding defects like hot cracking or porosity [1–9].

In recent years, fiber laser welding has emerged as a highly efficient and advanced technique for joining metallic materials, including aluminum alloys [10]. Fiber lasers offer several advantages over other welding systems, including CO_2 lasers and other solid-state lasers. Notably, fiber lasers tend to have lower maintenance requirements, reduced operational costs, and higher energy efficiency [11]. These attributes make fiber laser welding an attractive option not only for industrial production but also for cost-sensitive applications in sectors such as transportation, where minimizing weight and fuel consumption is critical. The ability of fiber lasers to focus high-intensity energy into a small spot allows for precise control over the welding process. This is particularly important when working with thin metal sheets or when producing high-strength, aesthetically clean weld joints. Low-power fiber lasers, in particular, present an economical alternative for welding applications, as they provide sufficient energy for partial or full penetration welding without incurring the high operational costs associated with high-power systems.

Aluminum alloys are typically characterized by their low density—approximately one-third that of steel—and their excellent mechanical properties, such as high strength, corrosion resistance, and good fatigue performance. Among the various grades of aluminum, the 7000 series is known for its exceptional strength and is commonly used in structural and aerospace applications. AA7075, a member of the 7000 series, is an Al-Zn-Mg alloy that has been widely recognized for its high strength-to-weight ratio. This makes it a suitable candidate for high-performance applications where weight reduction is critical without compromising on strength. Despite its advantages, the welding of AA7075 presents several challenges due to its susceptibility to hot cracking and other metallurgical defects. Moreover, there has been relatively limited research on the laser welding of AA7075, especially using low-power fiber lasers [4, 8, 12–15]. This research explores AA7075 weldability of a low-power fiber laser. The influence of weld parameters such as focal position and beam profile is analyzed on the weld quality, penetration depth, and microstructure.

Fiber laser welding eliminates the need for filler material to weld aluminum, unlike traditional arc welding. This leaves the welds denser and more homogeneous, with improved strength and reduced porosity. The most significant benefit of fiber LBW is the keyhole weld profile that is caused by high-energy concentration, which vaporizes material and creates a deep cavity. The material solidifies rapidly, producing a fine-grained microstructure with a reduced heat-affected zone (HAZ). The reduced HAZ reduces thermal distortion and residual stress, problems generally encountered in arc welding.

Laser welding has very good control over the process. Important parameters such as power, beam diameter, focal length, and speed can be controlled accurately for different materials and joints. This accurate control makes it suitable for thin metal sheets. High-power lasers were employed in most of the welding studies of aluminum alloys. For example, 10 kW fiber lasers were employed for sheets of thickness 10 mm [21]. 2200 W Nd:YAG lasers and 4000 W fiber lasers have also been employed for AA7075 parts [22, 23]. Though suitable for deep penetration and high speed, these systems are costly and need high maintenance.

Conversely, the present research examines the use of a much lower power fiber laser with a peak power of only 200 W [18]. Despite the lower power, the research aims to demonstrate that high-quality welds can still be achieved in thin AA7075 sheets (thickness of around 2 mm) by process parameter optimization. The technique offers vast cost reductions and improved accessibility to small- and medium-scale manufacturers. One of the most critical parameters in LBW is the position of the FP with respect to the workpiece surface. The FP regulates the power density that falls on the material, impacting the weld penetration and shape. The present research uses beam profiling techniques to examine the energy distribution of the laser beam, which has a Gaussian intensity distribution of $1/e^2$ [18, 21].

Smaller beam diameter enhances energy density in the focal area, which is essential for penetration of the keyhole. Focus and alignment adjustments are important to achieve optimal penetration depth and weld quality. This research investigates the influence of various focal positions on weld characteristics such as depth, width, and bead shape. The prime focus is investigation of the use of a low-power fiber laser to weld AA7075 aluminum alloy. It is interested in continuous wave (CW) mode, determining whether it is best suited to produce partial penetration welds with optimal mechanical and microstructural properties. The parameters of interest are weld penetration depth, bead profile, and heat-affected zone.

This study explores how parameters affect weld quality with the aim of highlighting the industrial application of fiber laser welding for AA7075 and high-strength aluminum alloys. The goal is to create an efficient, low-cost welding process that is up to the standards of aerospace, automotive, and transportation industries. Essentially, fiber laser welding is effective in welding high-strength aluminum alloys like AA7075. It is better than conventional arc welding because it reduces HAZ, improves joint quality, and does away with filler metals. While earlier studies have concentrated on high-power lasers, the study proves that low-power fiber lasers can also produce high-quality welds in thin aluminum sheets. The findings contribute to the growing body of knowledge on advanced welding technologies and support the adoption of laser welding in lightweight structural applications.

2. Experimental Methodology

AA7075 aluminium alloy was prepared for the study with 100 mm × 100 mm × 2 mm dimensions using a shearing cutting machine. The composition of the AA7075 was examined and the results are provided in Table 54.1 and 54.2.

Table 54.1 Chemical composition of AA7075

BM	Zn	Mg	Cu	Fe	Si	Mn	Al
AA6009	5.5	2.3	1.6	0.3	0.1	0.02	Bal.

Table 54.2 Keller's reagent composition

Chemical composition	Volume (ml)
H_2O	90
HCL	3
HNO_3	5
HF	2

This empirical study utilized fiber laser welding in both CW and PW modes. While the focal length (mm) was adjusted, the laser parameters such as pulse rate (Hz), power percentage (W), WS (mm/s), and pulse width (ms) remained unchanged. Figure 54.1 depicts the welding configuration. The parameters chosen for carrying out BOP welding with the low-power fiber laser are illustrated in Table 54.3. The metallographic observations provided insights into the microstructural behavior of the welded samples.

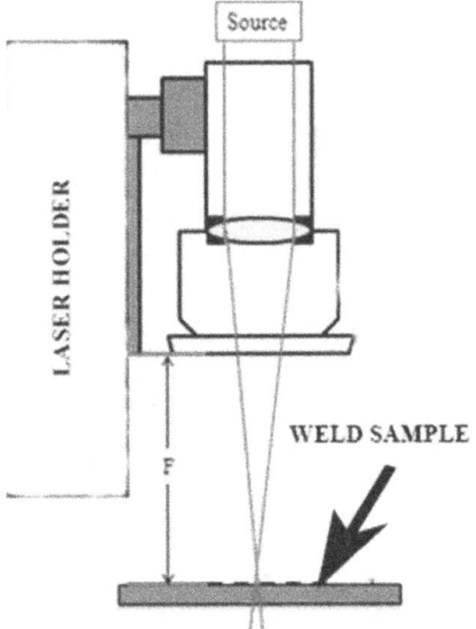

Fig. 54.1 Setup for fiber laser BOP welding

Table 54.3 Laser welding parameters

Parameter	Continuous Mode (CW)	Pulse Mode (PW)
P (Watt)	180 W (90%)	1800 W (90%)
FL (mm)	Both	60 mm – 200 mm
PRR (Hz)	-	20 Hz
PW (ms)	-	1 ms
WS (mm/s)	Both	2mm/s

3. Results and Discussion

3.1 Analysis of Weld Width and Incursion Depth

From the cross-sectional samples analyzed with an OM, the Incursion depth was established. The variation in weld width between the two modes was due to dissimilarities in laser power, which resulted in different energy outputs. In PW mode, no weld lines were noted at focal lengths beyond 190 mm, and in CW mode, this was the case for lengths over 150 mm [25].

Figure 54.2 presents a graph depicting weld width and penetration depth with focal length. In contrast, CW mode recorded a maximum penetration depth of only 0.173 mm at F = 100 mm. Additionally, PW mode benefits from different pulsed energy levels, allowing for controlled weld penetration since pulsed lasers deliver high-intensity energy in short bursts. The variation in power density applied to the metal surface further influences penetration depth [26].

Table 54.4 Results for weld width and penetration depth

Focal Length (mm)	Weld Width (μm)		Penetration depth (mm)	
	PW Mode	CW Mode	PW Mode	CW Mode
60	617	defocussed	0.030	defocussed
70	626	454	0.040	Non-observed
80	660	484	0.120	0.061
90	699	516	0.130	0.111
100	703	547	0.134	0.173
110	727	478	0.450	0.140
120	570	417	0.990	0.153
130	589	492	0.970	Non-observed
140	687	523	0.230	Non-observed
150	668	defocussed	0.140	defocussed
160	734	defocussed	0.100	defocussed
170	832	defocussed	0.040	defocussed
180	785	defocussed	0.050	defocussed
190	defocussed	defocussed	0.007	defocussed
200	defocussed	defocussed	0.012	defocussed

3.2 Microstructure Analysis

The formation of a keyhole during laser welding plays a crucial role in achieving effective penetration and high-quality joints. In this study, the keyhole profile generated during the laser welding process was observed and characterized, as illustrated in Fig. 54.2. The keyhole achieved a maximum penetration depth of approximately 0.99 mm, which corresponds to nearly half the thickness of the AA7075 aluminum sheet used in the experiments (sheet thickness being 2 mm). This indicates that the welding process was able to achieve partial penetration, sufficient to form a stable keyhole but not deep enough for full-through weld in single-pass mode. However, an important observation was the absence of shielding gas during the welding operation. This absence left the molten pool exposed to atmospheric contamination (Table 54.2). During fiber laser welding, the intense localized heat input causes rapid melting and vaporization of the metal surface. The resulting vapor plume exerts an upward pressure that pushes molten material aside, thereby forming a narrow and deep cavity known as a keyhole. It is a transient cavity and shrinks as the weld pool solidifies behind the laser beam. This keyhole is necessary for the occurrence of deeper penetration and greater joint strength [27].

One of the serious problems in this research was porosity in the weld zone (WZ), particularly in continuous wave (CW) laser-welded specimens. Porosity is a frequent type of defect that impairs mechanical performance. It probably stemmed from the lack of shielding gas, which guards molten metal from atmospheric gases. In the absence of shielding gas, oxidation leads to gas entrapment in the weld metal and formation of voids. For improved joint integrity, double-sided welding rather than single-pass, single-sided welding is advocated by the study. Double-sided welding in 2 mm thick AA7075 offers complete penetration from both sides, improves strength, minimizes defects, and ensures consistent heat input. This technique is especially useful for high-strength aluminum alloys where maintaining mechanical performance and avoiding heat-affected degradation is critical [15].

The microstructure of the base metal (BM) prior to welding was analyzed to establish a baseline for comparison with the welded zones. As shown in Fig. 54.3, the base metal microstructure consisted of distinct second-phase particles embedded within the aluminum matrix. Specifically, spheroidal black precipitates of MgZn were observed along with light gray $FeAl_3$ particles [28]. These intermetallic compounds are common in heat-treated 7000-series aluminum alloys and play a significant role in determining their mechanical and thermal behavior. The MgZn precipitates are the primary strengthening phase in AA7075, contributing to its high strength through a precipitation hardening mechanism. On the other hand, $FeAl_3$ particles are relatively inert and can act as nucleation sites during solidification or influence grain growth. These particles were uniformly dispersed within the aluminum solid solution matrix. The grain structure of the base metal was predominantly elongated, indicating a rolled or wrought microstructure typical of commercial-grade aluminum alloy sheets [20].

An elemental analysis, likely performed using EDS, confirmed the chemical composition of the base metal. The presence of key alloying elements such as zinc (Zn), magnesium (Mg), and aluminum (Al) was verified, along with trace amounts of iron (Fe), corresponding to the $FeAl_3$ intermetallic phase. To assess the quality of the welded joint, a comprehensive microstructural analysis was conducted across the entire weld cross-section, from the base metal to the fusion zone. The sample that exhibited the best welding characteristics—based on weld profile, minimal porosity, and good bead formation—was selected for detailed examination. The progressive transformation of microstructure from the BM, through the partially melted zone PMZ, and into the WZ was documented and is illustrated in Fig. 54.3.

At the interface between the base metal and the HAZ, the elongated grain structure of the BM began to transform into a planar grain structure. This transition is typical in fusion welding processes, where the thermal gradient causes the solid-liquid interface to stabilize in a planar front. Moving further into the HAZ, the grain structure evolved into a cellular pattern, driven by increased undercooling and localized solidification conditions [29]. The formation of cellular grains indicates a moderate cooling rate and a directional solidification pattern. One notable observation from the HAZ was its reduced hardness compared to the base metal. The decrease in hardness within the HAZ can be attributed to the dissolution or coarsening of strengthening precipitates, particularly the MgZn phase, due to the elevated temperatures during welding. The thermal cycle of laser welding causes overaging in certain regions, reducing the effectiveness of precipitation hardening and thereby compromising mechanical properties. This softening effect increases the susceptibility of the HAZ to mechanical failure, particularly under tensile or fatigue loading conditions [25].

Further into the weld zone, as shown in Fig. 54.8(b), the microstructure consisted of coarser dendritic grains. Dendritic solidification is a common feature in fusion welding and results from rapid cooling and solidification of the molten pool. Despite the dendritic nature of the weld metal, the grain size in the weld zone was found to be smaller than that of the base metal [30]. This could be due to the high cooling rate associated with laser welding, which promotes grain refinement and leads to a narrow, uniform weld bead. Fine grain structures in the weld zone are desirable as they improve toughness and resistance to crack propagation. However, the presence of dendrites also indicates that some segregation of alloying elements may have occurred during solidification. This segregation, if not controlled, can lead to microstructural heterogeneities, affecting both the mechanical and corrosion performance of the weld [31].

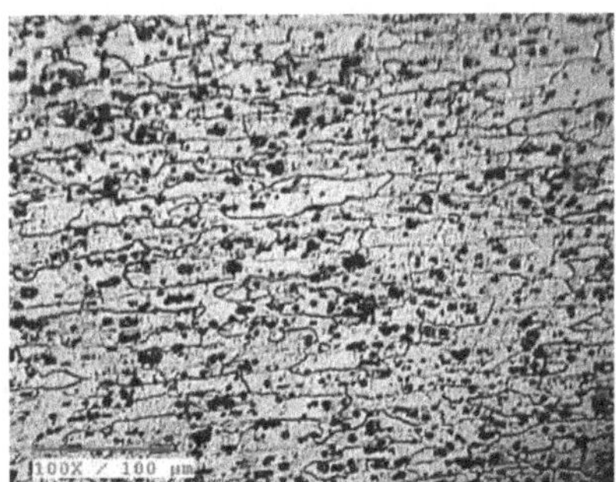

Fig. 54.2 AA7075 base metal

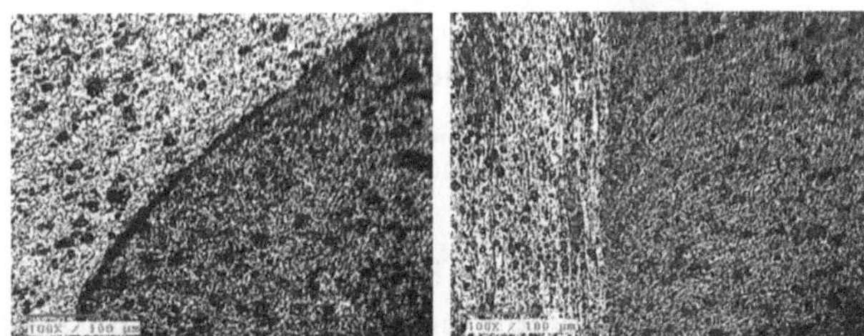

Fig. 54.3 SEM microstructure [a] Base metal AA6009-T6 [b] WNZ of sample 7

4. Conclusion

This study confirmed that AA7075 sheet can be successfully welded. The key findings from the experiment are as follows:

1. The PW welding mode demonstrated significant results, in the cross-sectional microstructure, achieving an approximate penetration depth of 1 mm. This mode enabled the formation of a weld joint with half-depth penetration.
2. Both welding modes exhibited the same focal point measurement. The focal point, estimated from the sample surface to the protecting mirror, was established (F = 120 mm).
3. Due to the lack of shielding gas during the experiment, defects like underfill and porosity were observed.
4. High-strength aluminum alloys such as AA7075 can be effectively welded using low-power fiber lasers. This method shows promise for use in the automotive sector, where AA7075 could replace traditional steel in car body panels or Taylor Welded Blanks (TWBs). To achieve better welding results for AA7075 in future studies, laser parameters like shielding gas, pulsed energy and beam incident angles should be examined using the DOE method.

References

1. Shah LH, Akhtar Z, Ishak M. Investigation of aluminum-stainless steel dissimilar weld quality using different filler metals. International Journal of Automotive and Mechanical Engineering. 2013;8:1121-31.
2. Ghazali FA, Manurung YHP, Mohamed MA, Alias SK, Abdullah S. Effect of process parameters on the mechanical properties and failure behavior of spot welded low carbon steel. Journal of Mechanical Engineering and Sciences. 2015;8:1489-97.
3. Ishak M, Islam MR, Sawa T. GMA Spot welding of A7075-T651/AZ31B dissimilar alloys using stainless steel filler. Materials and Manufacturing Processes. 2014;29: 980-7.
4. Ishak M, Noordin NFM, Razali ASK, Shah LHA, Romlay FRM. Effect of filler on weld metal structure of AA6061 aluminum alloy by tungsten inert gas welding.
5. International Journal of Automotive and Mechanical Engineering. 2015;11:2438- 46.
6. Shah LH, Mohamad UK, Yaakob KI, Razali AR, Ishak M. Lap joint dissimilar welding of aluminium AA6061 and

galvanized iron using TIG welding. Journal of Mechanical Engineering and Sciences. 2016;10:1817-26.

7. Sathari NAA, Razali AR, Ishak M, Shah LH. Mechanical strength of dissimilar AA7075 and AA6061 aluminum alloys using friction stir welding. International Journal of Automotive and Mechanical Engineering. 2015;11:2713-21.
8. Hasan MM, Ishak M, Rejab MRM. A simplified design of clamping system and fixtures for friction stir welding of aluminium alloys. Journal of Mechanical Engineering and Sciences. 2015;9:1628-39.
9. Ahmad R, Asmael MBA. Effect of aging time on microstructure and mechanical properties of AA6061 friction stir welding joints. International Journal of Automotive and Mechanical Engineering. 2015;11:2364-72.
10. Sathari NAA, Shah LH, Razali AR. Investigation of single-pass/double-pass techniques on friction stir welding of aluminium. Journal of Mechanical Engineering and Sciences. 2014;7:1053-61.
11. Assunção E, Quintino L, Miranda R. Comparative study of laser welding in tailor blanks for the automotive industry. The International Journal of Advanced Manufacturing Technology. 2009;49:123-31.
12. Tsuji M. IPG fibre lasers and aluminium welding applications. Welding International. 2009;23:717-22.
13. Ahmad AH, Naher S, Brabazon D. Effects of cooling rates on thermal profiles and microstructure of aluminium 7075. International Journal of Automotive and Mechanical Engineering. 2014;9:1685-94.
14. Singh R. Process capability study of rapid casting solution for aluminium alloys using three dimensional printing. International Journal of Automotive and Mechanical Engineering. 2011;4:397-404.
15. Kadirgama K, Noor MM, Rahman MM, Rejab MRM, Haron CHC, Abou-El- Hossein KA. Surface roughness prediction model of 6061-T6 aluminium alloy machining using statistical method. European Journal of Scientific Research. 2009;25:250-6.
16. Najiha MS, Rahman MM, Kadirgama K, Noor MM, Ramasamy D. Multi- objective optimization of minimum quantity lubrication in end milling of aluminum alloy AA6061T6. International Journal of Automotive and Mechanical Engineering. 2015;12:3003-17.
17. Fukuda T. Weldability of 7000 series aluminium alloy materials. Welding International. 2012;26:256-69.
18. Katayama S, Ogawa K. Laser weldability and ageing characteristics of welds: laser weldability of commercially available A7N01 alloy (1). Welding International. 2013;27:172-83.
19. Kawahito Y, Matsumoto N, Abe Y, Katayama S. Laser absorption of aluminium alloy in high brightness and high power fibre laser welding. Welding International. 2012;26:275-81.
20. Moon J, Katayama S, Mizutani M, Matsunawa A. Behaviour of laser induced plasma, keyhole and reflected light during laser welding with superimposed beams of different wavelengths. Welding International. 2003;17:524-33.
21. Ishak M, Yamasaki K, Maekawa K. Lap fillet welding of thin sheet AZ31 magnesium alloy with pulsed Nd:YAG Laser. Journal of Solid Mechanics and Materials Engineering. 2009;3:1045-56.
22. Katayama S, Nagayama H, Mizutani M, Kawahito Y. Fibre laser welding of aluminium alloy. Welding International. 2009;23:744-52.
23. Enz J, Riekehr S, Ventzke V, Sotirov N, Kashaev N. Laser welding of high- strength aluminium alloys for the sheet metal forming process. Procedia CIRP. 2014;18:203-8.
24. Wang JT, Zhang YK, Chen JF, Zhou JY, Ge MZ, Lu YL, et al. Effects of laser shock peening on stress corrosion behavior of 7075 aluminum alloy laser welded joints. Materials Science and Engineering: A. 2015;647:7-14.
25. Katayama Seiji AY, Mizutani Masami, Kawahito Yousuke. Deep Penetration Welding with High-Power Laser under Vacuum. Transactions of JWRI is published by Joining and Welding Research Institute, Osaka University, Ibaraki, Osaka 567-0047, Japan. 2011;40.
26. Jiang Z, Tao W, Yu K, Tan C, Chen Y, Li L, et al. Comparative study on fiber laser welding of GH3535 superalloy in continuous and pulsed waves. Materials & Design. 2016;110:728-39.
27. Esraa K. Hamed FH, Makram Fakhry. Laser wavelength and energy effect on optical and structure properties for nano titanium oxide prepared by pulsed laser deposition. Iraqi Journal of Physics. 2014;12:62-8.
28. Pang S, Chen X, Shao X, Gong S, Xiao J. Dynamics of vapor plume in transient keyhole during laser welding of stainless steel: Local evaporation, plume swing and gas entrapment into porosity. Optics and Lasers in Engineering. 2016;82:28- 40.
29. Sivashanmugam NMM, D. Ananthapadmanaban, S. Ravi Kumar. Investigation of microstructure and mechanical properties of GTAW and GMAW joints of AA7075 aluminium alloys International Journal on Design and Manufacturing Technologies. 2009;2:56-62.
30. Chen S, Guillemot G, Gandin C-A. Three-dimensional cellular automaton-finite element modeling of solidification grain structures for arc-welding processes. Acta Materialia. 2016;115:448-67.
31. Dong HB. Analysis of Grain Selection during Directional Solidification of Gas Turbine Blades. In: 2007 W, editor. Proceedings of the World Congress on Engineering 2007 Vol II, London, UK2007.

Note: All the figures and tables in this chapter were made by the authors.

Advances in Additive Manufacturing Technologies – Gurusamy Pathinettampadian et al. (eds)

FSW Parameter Optimization for Dissimilar AA5754 and AA6061 Aluminium Alloys

Muthukumaran A.[1], R. Vijayan[2], Hepsi Beaula M.J.[3], Subash P.[4], Balamurali P.[5], Parthiban A.[6]

Assistant Professor, Department of Mechanical Engineering, Chennai Institute of Technology, Chennai, India

◆ **Abstract:** By avoiding common welding flaws like oxidation, FSW introduces a new SSW method for combining aluminum alloys. Focusing on ultimate tensile strength and Vickers hardness, this research compares and contrasts the FSW of two different alloys, AA5754 and AA6061. Using a central composite design, weld quality was taken into consideration while optimizing the welding conditions as well. Interface intermetallic complexes are still a problem, reducing strength and decreasing corrosion resistance. To study these substances, scientists will use cutting-edge methods including SEM and XRD. The best conditions were determined by using ANOVA and RSM to optimize the parameters. These conditions included 100% boron carbide reinforcement, a TTS of 70 mm/min, and a TRS of 1300 rpm. This study sheds new light on the industrial uses of aluminum alloys and demonstrates the potential of FSW in lightweight, high-performance contexts.

◆ **Keywords:** TTS, ANOVA, XRD, FSW, AA5754, AA6061

1. Introduction

FSW has been an efficacious solid-state welding process, particularly for dissimilar aluminium alloys welding. The traditional fusion welding processes produce porosity, hot cracking, and distortion defects that reduce their efficiency in welding aluminium alloys. FSW, however, has some advantages such as high weld quality, reduced residual stresses, and joining dissimilar metals without filler metal. It is therefore highly suitable for joining Al alloys such as AA5754 and AA6061. Aluminium alloys are typically utilized in aerospace, automobile, and marine industries due to their strength-to-weight ratio, corrosion resistance, and hardness. The dissimilar alloys are difficult to weld due to the differences in their composition and metallurgy. Al-Mg-series AA5754 is particularly renowned for having improved weldability and corrosion resistance and therefore an excellent material for the automobile and marine industries. Conversely, Al-Mg-Si-series AA6061 is well renowned for improved strength and medium corrosion resistance and therefore highly appropriate for utilization in aerospace structure, bicycle parts, and automobile parts. By joining these two alloys in a single joint, it is feasible to obtain a perfect combination of corrosion resistance and strength, and therefore enlarge their application in engineering applications.

FSW weldability at sub-fusion temperature reduces growth of brittle intermetallic compounds, otherwise reducing weld strength and durability. FSW joint performance of

[1]muthu2189@gmail.com, [2]srajendranvijayan@gmail.com, [3]hepsibeaulamj@citchennai.net, [4]subashp@citchennai.net, [5]muralipb2080@gmail.com, [6]parthimech2810@gmail.com

DOI: 10.1201/9781003685906-55

similar and dissimilar Al alloys, for example, AA5754 and AA6061, involves corrosion testing and microstructural characterization to determine long-term durability [1]. Effect of FSW process parameters on corrosion resistance has been evaluated, with findings that optimal process parameters can improve FSW joint lifetime. Microstructural characteristics of FSW welds are critical to their mechanical performance evaluation. Optimization and control of grain growth and post-weld heat treatment can greatly improve mechanical behavior of similar and dissimilar aluminum alloy joints, for example, AA5083/AA6061. Methods such as electron backscatter diffraction and nanoindentation have been employed to explore grain structure and mechanical properties at the weld interface [2]. FSW has also been applied in hybrid welding processes like TIG-assisted hybrid FSW to enhance the joint integrity in difficult material combinations like aluminum alloy Al5052 and high-strength steel DP590 [3]. Hybrid process includes preheating material to create defect-free welds with lower interfacial stress, and the process is very versatile for high-performance applications.

Machine learning has greatly improved friction stir welding (FSW) processes by optimizing feed rates, rotation rates, and defect detection systems. Researchers have determined that artificial intelligence algorithms can effectively predict welding conditions and hence reduce defects and improve weld quality as a whole [5]. Experimental studies have validated evidence that the WPP like TTS, TRS, TA, and TS have a influencing impact on weld strength, fatigue life, and microstructural development [6]. Microstructural transformation is required to enhance the mechanical properties of the FSW joint. Experiments were performed to determine optimal process conditions for the manufacture of strong and defect-free welds, particularly for aerospace components where weld integrity is critical [8]. Tensile strength, hardness, and fatigue life testing was performed to render FSW-welded structures compatible with the industry [9]. FSW has been extensively studied to utilize it in the aerospace, marine, and automotive industry manufacturing sector, which demonstrates higher reliability and efficiency compared to the conventional fusion welding processes. Research stressed the requirement for the optimization of WPP with the assistance of statistical techniques, i.e., the Design of Experiments (DOE) approach [10].

In naval applications, FSW was investigated to form high-strength, corrosion-resistant joints. Researchers also investigated various welding conditions to obtain good weld integrity, and provided realistic guidelines to facilitate FSW for naval shipbuilding applications [11]. Water-cooling methods were suggested to control peak weld temperatures, which affect mechanical properties as well as residual stress distribution in FSW-welded aluminum alloys [12]. The study has added to the body of knowledge in FSW's ability to produce light-weight, high-performance structures in naval engineering, automobile, and aerospace industries [13]. How process parameters of FSW influence weld quality of aluminum alloys has been thoroughly investigated. Process parameters like TG, TA, AL, TRS, and TTS affecting mechanical properties, microstructure, and defect formation in weld zones play an essential role. Characterization methods like XRD, SEM, and optical microscopy were used to investigate grain structure, and intermetallic compound distribution [14]. Mechanical properties like joint strength, toughness, and hardness profiles have been studied based on base materials. Statistical optimization methods, like the Taguchi method, have been investigated for optimizing process parameters with the best combination that would yield improved mechanical properties [15].

Apart from the above parameters, deep learning models such as MLP and LSTM networks were used to model and predict UTS under different welding conditions. AI-based approaches were proved to be much more precise in predicting the optimal process parameters than conventional methods [16]. FSW has been widely employed in the automotive sector to weld aluminum alloys, which exhibit different compositional and mechanical properties. Researchers have been eager to optimize welding parameters for different combinations of alloys such as AA5083-AA6061 to have improved mechanical strength and fatigue life [17]. Apart from that, corrosion resistance and crashworthiness—two of the most important parameters in automotive engineering—were studied to enhance the long-term reliability of FSW joints [18]. Combining numerical modeling with quality control through the application of AI-based techniques has been utilized in FSW research in a synergistic manner, which has helped in improved defect detection, parameter optimization, and weld performance overall. Machine learning models were utilized to enhance production conditions reliability in a way such that FSW can fulfill the stringent demands of industries such as automotive and aerospace [19].

The objective of the current study is to evaluate the weldability of similar AA5754 and AA6061 aluminum alloys by FSW under mechanical properties, microstructure development, and process optimization techniques. Realization of grain structure, intermetallic compound formation, and defect distribution need to be enhanced for improved weld quality. Furthermore, use of statistical modeling and AI-based modeling for process parameter optimization can further promote the industrial use of FSW.

All these results will help in advancing FSW technology to make it the most sought-after method for critical aerospace, automotive, and marine engineering applications [20].

2. Experimental Methodology

In the current study, AA5754 and AA6061 6 mm thick aluminium alloy plates were utilized. The plates were first cut into required size using a power hacksaw and then machined to final size of length 150 mm and width 75 mm. The actual chemical composition of the two alloys is shown in Table 55.1 to determine the material accurately and assess its applicability. Side and edge surfaces were also cleaned prior to welding to facilitate effective development of FSW joints. For the purpose of reinforcement, 2 mm diameter and 3 mm deep zig-zag holes were pre-drilled at a spacing of 4 mm apart on the butt surface of the weld area. Boron carbide (B_4C) powder was utilized as the reinforcement material and filled in such pre-drilled holes. The reinforcement rate for welding was changed to three different levels according to Table 55.2. The plates prepared were shaped as square butt joints and firmly clamped by mechanical clamps to keep in the right position during the welding process. Welding was done as per a one-pass process where the tool traversed along the joint with an orderly sequence of experiments (DOE) plan.

Table 55.1 Chemical composition of 5754 and 6061

BM	Mg	Si	Fe	Cu	Mn	Cr	Zn	Al
5754	3.63	0.43	0.41	0.12	0.53	0.32	0.23	Bal
6061	0.96	0.8	0.4	0.27	0.09	0.21	0.06	Bal

Table 55.2 Specification of cylindrical tool pin

S. No	Features	Dimension (mm)
1	Tool height	60
2	Pin Diameter	6
3	Shoulder Diameter	18
4	Pin Length	5.7

Table 55.3 shows the results of twenty investigations that used the CCD paradigm. The welding was done at a plunge depth of 0.2 mm using a standard HMT FN2V vertical milling machine that was powered by a 7.5 HP, 1800 rpm engine. We tested FSW's ability to controllably create high-strength welds between AA5754 and AA6061 using the aforementioned instruments. This investigation made use of an ISO 40 taper spindle driven by an 11 kW AC spindle FSW 3T NC. The 400 mm × 350 mm table has the ability to move in all three dimensions. With equivalent travel speeds ranging from 100 mm/min to 300 mm/min, the X and Y axes were pushed with 15 and 30 kg/m², respectively. From 100 mm to 400 mm was the range of spindle noise during milling that was tested. The equipment was propelled by a 22 KVA generator. While this study did not rule out FSW's ability to weld steel, copper, or aluminum, it did focus on a number of aluminum alloys. As stated in Table 55.4, the experimental equipment was developed from the Central Composite Design (CCD) to perform a systematic examination on the friction stir welding method of AA5754 and AA6061 alloys.

Table 55.3 WPP

WPP	Notation	-1	0	1
TRS	N	900	1100	1300
TTS	S	70	90	110
% Reinforcement	R	50	75	100

Table 55.4 Design of experiments framed using CCD

Std. order	Run order	Coded Values			UTS (MPa)	Hardness (HV0.5)
		P	S	F		
1	1	-1	-1	-1	162	113
2	14	1	-1	-1	158	107
3	4	-1	1	-1	155	107
4	17	1	1	-1	160	108
5	6	-1	-1	1	152	96
6	11	1	-1	1	157	106
7	9	-1	1	1	148	102
8	13	1	1	1	161	121
9	16	-1.682	0	0	147	110
10	7	1.682	0	0	159	111
11	15	0	-1.682	0	161	109
12	5	0	1.682	0	141	98
13	12	0	0	-1.682	158	109
14	19	0	0	1.682	159	108
15	20	0	0	0	142	103
16	18	0	0	0	145	103
17	8	0	0	0	153	110
18	3	0	0	0	142	96
19	10	0	0	0	148	96
20	20	0	0	0	158	107

This study utilizes a strong CCD technique to study linear, quadratic, and interaction effects of influential process parameters in FSW. The research scientifically analyzes the influence of such parameters on joint integrity, microstructure characteristics, and generation of defects. Welding parameters can be optimized for enhanced weld quality and industrial relevance by careful choice of experimental parameters. Mechanical weld nugget properties were measured from microhardness testing taking particular care of material ductility along with grain size reduction. Microhardness specimens were collected as per ASTM-E384 standards and placed on test on a Mitutoyo digital microhardness tester. Hardness was tested 1 mm from the center thickness of the specimen with a 0.5 kg test force for 10 seconds using a diamond indenter. The Vickers hardness scale was used to introduce precision and consistency [21].

Sample preparation involved precise cutting, grinding, and polishing on SiC emery papers (300–2000 grit) followed by additional diamond paste polishing on velvet cloth. Microstructural contrast was enhanced by Keller's reagent etching. Test pieces were plastic medium cast to sustain the test. Tensile testing was conducted to obtain UTS, YS, and %Elongation of the welded test pieces. The test samples were etched with etchants and subjected to a universal testing machine equipped with a data acquisition system. The tests provided data about the mechanical response of the weld joints and structural integrity. SEM was used to carry out detailed microstructure characterization, observing grain shape and size and variation in composition using EDS. The methodology was even more concerned with surface morphology, defects, and behavior of crack propagation such that process development and qualification of welded alloys would be feasible. With tension testing, microhardness testing, and SEM analysis, the present investigation thoroughly analyzes mechanical and structural characterization of AA5754-AA6061 welds, a critical input in making the decision for the improvement in the welding process in industrial application [22].

RSM is one of the frontline statistical and mathematical methods of Friction Stir Welding (FSW), being used for predicting weld outcome as well as for optimizing process parameters. RSM is an experimental method that integrates experimental design for studying the interaction between the input parameters say, PD, WS, and TRS, and the response variables like joint strength, hardness, and microstructural characteristics. By fitting the data with an experimental model, RSM allows predicting response with great accuracy as well as identifying the optimum input conditions required to create the desired weld quality. Among all the RSM methods, CCD is an extremely effective process optimization technique and has gained record attention for its ability to maximize output responses.

$$y = f(x1, x2, x3, x4) + e \tag{1}$$

Hence, the experimental error, denoted as "e," is the impact of the mistake on the output response, "y," in relation to the input variables, "x1, x2, x3, and x4".

To improve the FSW parameters for welding the different aluminum alloys AA5754 and AA6061, the present study made use of CCD and RSM. A mathematical model was developed to explain the relationship between TRS, TTS, and B_4C reinforcement, as well as other critical factors, and weld quality metrics including Vickers hardness and UTS. CCD was used for this purpose. As part of the experimental procedure, aluminum alloy plates were machined to uniform thickness, precisely positioned, and clamped before welding. TRS (800-1500 rpm), TTS (50-100 mm/min), and amount of boron carbide reinforcing (0%-100%). We choose these parameters. The experimental findings were fitted using RSM to plot response surfaces. This allowed us to determine the best circumstances and examine the relationships between the parameters to achieve the highest joint performance.

Statistically, the impact of process parameters on weld quality was investigated via ANOVA. ANOVA evaluates the effects of single and interaction parameters, i.e., TRS, WS, PD, and TG, on joint strength weld properties, microstructure, and hardness distribution. ANOVA detects significance factors by means of F-statistics and p-values using repeated trials' application of data variance measurements. Statistical technique allows it to be feasible to optimize FSW process parameters in such a manner that weld performance and quality are all the more better and useful, especially in aeronautic, automotive, and naval applications where aluminium alloys are of utmost importance [20].

3. Results and Discussion

This research used RSM to explore and optimize the FSW parameters step by step, and the related mechanical properties are discussed in Table 55.4. RSM predicted and explored the influence of different WPP on important properties like hardness and UTS. Experimental design involved the variation of three significant parameters: TRS (rpm), WS (mm/min), and reinforcement percent. All the combinations of the parameters gave different UTS (N/mm²) and hardness (HV1) values. TRS changed from 763.64 to 1436.35 rpm, TTS changed from 56.36 to 119.48 mm/min, and reinforcement percentage changed from 33.12% to 122.12%. UTS values varied from 141.3 to 162.51 N/mm², and hardness varied from 95.81 to 121.44 VHN.

To analyze the data statistically, regression models were constructed using Design Expert v.12 software.

These models explained the impact of TRS, TTS, and reinforcement on HV1 and UTS. The use of RSM to construct response surfaces allowed for the visualisation of the parameters for parameter optimization, which in turn improved the mechanical characteristics. It not only gives detailed insight into mechanisms of the FSW process, but also useful information to better welding parameters so that the quality of weld enhances in industrial utilization to a great extent [23].

3.1 Implication of WPP on UTS

Figure 55.1 surface plots showing interaction of TRS, TTS, percent reinforcement, and UTS in FSW. Tool rotation speed is significant to improve UTS by enhanced mixing and material homogenization in the nugget. Traverse speed increase, raising tensile strength to a level of approximately 70 mm/min, results in decreased tensile strength at higher steps due to lower material flow and lower heat input. Reinforcement percentage has a significant contribution towards tensile strength, and filler material percent increase results in proper consolidation of the material and integrity of microstructure in weld position.

At low tool spindle speeds, substandard heat input results in lower stirring efficiency leading to tunnel development and poor strength of the joints. At high tool spindle speeds, stirring efficiency is greater resulting in defect-free weld nugget formation and better tensile strengths. Traverse speeds too high, though, restrict material flow and stirring action, generating a probability of defects like tunneling, which adversely affects weld strength. Introduction of boron carbide (B_4C) at the weld butt area inhibits free flow of the matrix alloy during FSW and leads to significant UTS. The strengthening effect of B_4C particles stems from their ability to aid in effective transfer of load, slow down the growth of cracks, and improve dislocation density due to mismatch in thermal expansion coefficients between aluminum and B_4C. Additionally, the Orowan strengthening mechanism and the high thermal conductivity of B_4C aid in heat dissipation, reducing the thermally and thermo-mechanically affected zones [24].

Fine grain and particle fragmentation of B_4C upon FSW lead to decreased elongation in the weld joint and better tensile performance. Ideal process conditions for maximum UTS include increased tool rotation, mean traverse velocity, and increased reinforcement content that guarantee efficient HI and material bonding. The present paper emphasizes the importance of optimization of process parameters for maximum weld quality and mechanical properties in industrial applications. ANOVA results give a close statistical analysis of the effects of TRS

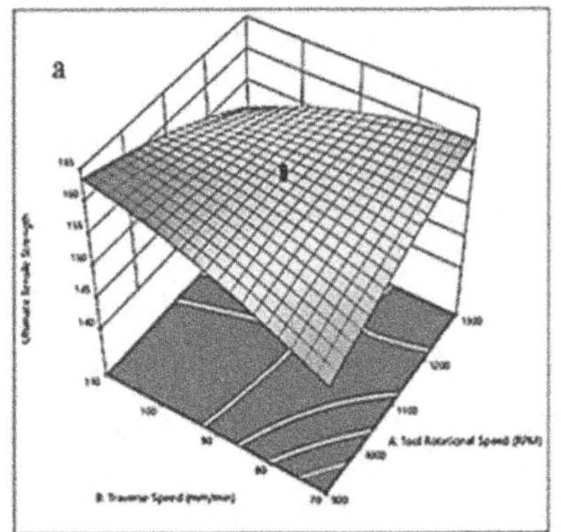

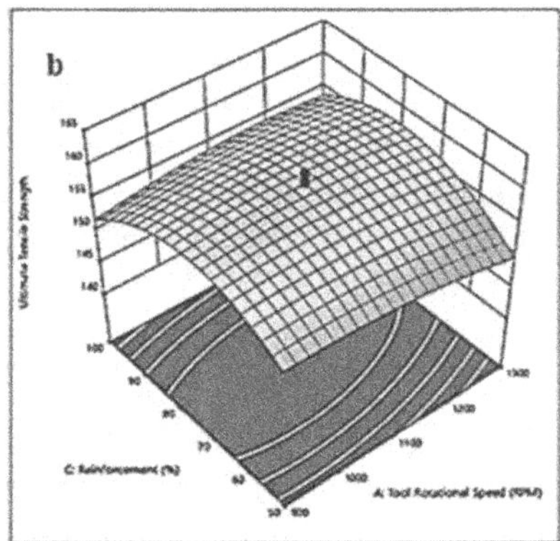

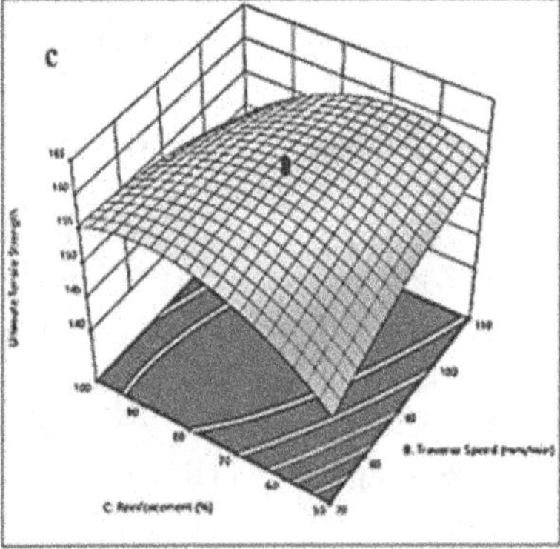

Fig. 55.1 Surface plots for UTS vs (a) WPP AB, (b) WPP BC (c) WPP BC

(A), TTS(B), and Re% (C) on UTS in FSW. The overall model is highly statistically significant ($p < 0.0001$), since the chosen ANOVA factors and interaction individually account for almost all percentage variance in UTS. The major conclusions are as follows, TRS (A), TTS (B), and reinforcement % (C) all significantly contribute to the values of UTS with p-values of 0.0390, 0.0121, and 0.0040, respectively. Their individual interaction effects between each other (AB, AC, and BC) are also found to be statistically significant at $p < 0.05$, meaning their combined influences on weld quality are significant. Residuals register 14.81 unexplained variation, whereas lack-of-fit test also supports model fit by having a p-value of 0.4996. Experimental variation pure error is explained by 4.70.

The surface plots in Fig. 55.1 show how hardness relates to the most crucial process parameters in FSW. The hardness increases immensely as the Re% increases, and this implies that the weld final hardness is in direct proportion to the material's hardness of the reinforcement. On the other hand, the minimum hardness values are obtained at low TTS, probably as a result of inadequate heat input, leading to inadequate material consolidation within the WN and hence the resulting low hardness.

TRS is also important in the determination of hardness. Hardness reduces with an increase in rotational speed. This reduction is caused by increased heat input at high speeds, leading to grain coarsening in the weld nugget, hence reducing the hardness. Hardness is directly related to mechanical strength, and therefore increasing the hardness tends to improve the overall strength properties of the welded joint. Presence of B_4C particles inside the WN greatly contributes to the enhancement of hardness. The high tool rotary speed enables uniform dispersion of the B_4C particles throughout the SZ, resulting in finer microstructure. Nucleation by the presence of Bo particles contributes to addition to strengthening within the weld. Uniform refinement of the fine particle distribution of boron at the nugget zone contributed to greater hardness. In addition, improvement of hardness in the HAZ contributes favorably to the UTS of the weld joint as a whole.

ANOVA for friction stir welding Vickers hardness to guarantee significant process parameter correlations and hardness values. The model gives very high significance ($p < 0.0001$), which means TRS (A), TTS (B), %Re (C), and their interactions have very high contributions to variability in hardness. Key findings are, The combined effects of TRS (A), TTS (B), and %Re (C) on hardness are significant statistically at p-values 0.0322, 0.0010, and <0.0001, respectively. All significant interactions of parameters (AB, AC, and BC) with quadratic effects (A^2, B^2, and C^2) suggest non-linear correlation between the provided factors and hardness. Residual analysis verifies

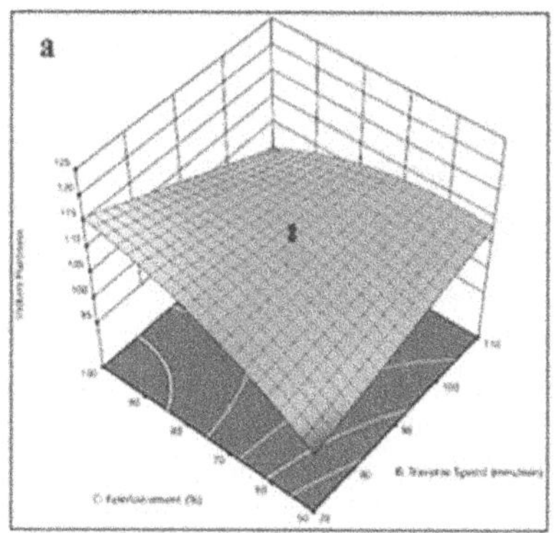

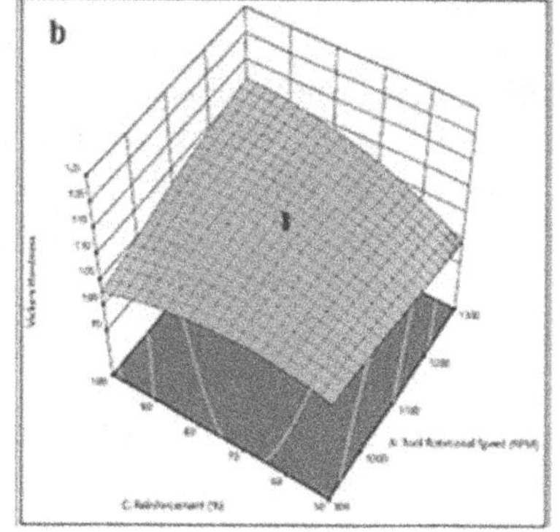

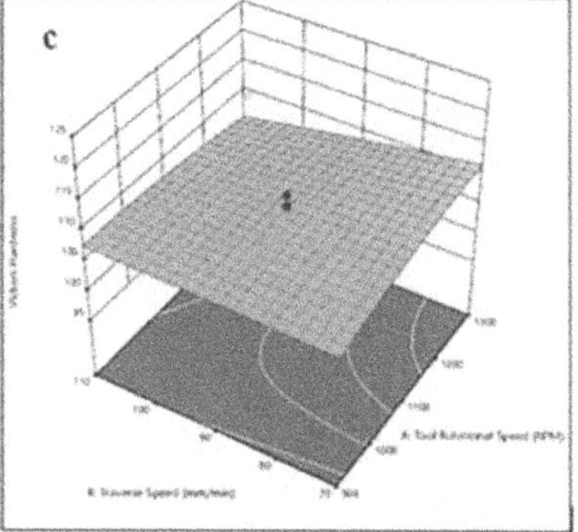

Fig. 55.2 Surface plots for vickers hardness vs (a) WPP AB, (b) WPP BC (c) WPP BC

the fitness of the model (p = 0.7894), with unexplained variation at 6.57 and pure error at 4.73 in experimental range (Fig. 55.2).

4. Conclusion

From experimental study of FSW of AA5754 and AA6061 aluminum alloys using a CNC vertical milling machine, the following crucial conclusions have been drawn:

1. The TRS is the most dominating parameter for characterization of the welded joints' mechanical properties. With increased rotational speed, tensile strength and hardness increased with optimum values at the highest TRS investigated.
2. The maximum intended TTS was 70 mm/min. The said speed provided less than the optimal material flow and heat input owing to compromised processing rates. Lower speeds were not acceptable due to lower mechanical properties.
3. ANOVA output for Vickers hardness confirmed the significance of the model to be extremely high ($p < 0.0001$), which indicated how the TRS (A), TTS (B), %Re (C), and their interaction significantly affect the weld hardness.
4. ANOVA analysis established that A's (p = 0.0322), B's (p = 0.0010), and C's ($p < 0.0001$) main effects were significant. Further, interactions (AB, AC, BC) and quadratic effects (A^2, B^2, C^2) had significant effect. The residual analysis verified the validity of the model with insignificant unexplained variance (6.57) and pure error (4.73), indicative of parameter optimization requirements for improved weld quality.
5. The level of reinforcement had a most significant influence upon tensile strength and hardness, the maximum levels of reinforcement being found to give optimal values.
6. The estimated R^2 was very close to the adjusted R^2, suggesting that the model is a reasonable predictor of variation in mechanical property.
7. The ideal welding parameters were found to be a TRS of 1300 rpm, a TTS of 70 mm/min, and a %Re of 100%. These parameters allowed for the maximum UTS and hardness and provided the maximum weld quality.

References

1. ALFATTANI, R., YUNUS, M., MOHAMED, A.F., et al., "Assessment of the corrosion behavior of friction-stir-welded dissimilar aluminum alloys", Materials (Basel), v. 15, n. 1, pp. 260, 2021. doi: http:// doi.org/10.3390/ma15010260. PubMed PMID: 35009406.
2. BAGHDADI, A.H., SAJURI, Z., KESHTGAR, A., et al., "Mechanical property improvement in dissimilar friction stir welded Al5083/Al6061 joints: effects of post-weld heat treatment and abnormal grain growth", Materials (Basel), v. 15, n. 1, pp. 288, 2021. doi: http://doi.org/10.3390/ma15010288. PubMed PMID: 35009434.
3. BANG, H.S., HONG, S.M., DAS, A., et al., "Study on the weldability and mechanical characteristics of dissimilar materials (Al5052-DP590) by TIG assisted hybrid friction stir welding", Metals and Materials International, v. 27, n. 3, pp. 1193–1204, 2021. doi: http://doi.org/10.1007/s12540-019-00461-6.
4. CARR, G., SANTIAGO, D., PELAYO, M., et al., "Study of friction stir spot welding on AA6063 aluminium alloy used in the ship building industry", Matéria (Rio de Janeiro), v. 23, n. 2, pp. e12011, 2018. doi: http://doi.org/10.1590/s1517-707620180002.0348.
5. CHEN, S., ZHANG, H., JIANG, X., et al., "Mechanical properties of electric assisted friction stir welded 2219 aluminum alloy", Journal of Manufacturing Processes, v. 44, n. 8, pp. 197–206, 2019. doi: http:// doi.org/10.1016/j.jmapro.2019.05.049.
6. CHOUDHARY, S., CHOUDHARY, S., VAISH, S., et al., "Effect of welding parameters on microstructure and mechanical properties of friction stir welded Al 6061 aluminum alloy joints", Materials Today: Proceedings, v. 25, n. 1, pp. 563–569, 2020. doi: http://doi.org/10.1016/j.matpr.2019.05.466.
7. DI BELLA, G., FAVALORO, F., BORSELLINO, C., "Effect of process parameters on friction stir welded joints between dissimilar aluminum alloys: A review", Metals, v. 13, n. 7, pp. 1176, 2023. doi: http://doi. org/10.3390/met13071176.
8. EREN, B., GUVENC, M.A., MISTIKOGLU, S., "Artificial intelligence applications for friction stir welding: a review", Metals and Materials International, v. 27, n. 2, pp. 193–219, 2021. doi: http://doi. org/10.1007/s12540-020-00854-y.
9. FATHI, J., EBRAHIMZADEH, P., FARASATI, R., et al., "Friction stir welding of aluminum 6061-T6 in presence of watercooling: Analyzing mechanical properties and residual stress distribution", International Journal of Lightweight Materials and Manufacture, v. 2, n. 2, pp. 107–115, 2019. doi: http://doi. org/10.1016/j.ijlmm.2019.04.007.
10. GANGIL, N., MAHESHWARI, S., SIDDIQUEE, A.N., et al., "Investigation on friction stir welding of hybrid composites fabricated on Al-Zn-Mg-Cu alloy through friction stir processing", Journal of Materials Research and Technology, v. 8, n. 5, pp. 3733–3740, 2019. doi: http://doi.org/10.1016/j.jmrt.2019.06.033.
11. GOWTHAMAN, P.S., SARAVANAN, B.A., "Determination of weld ability study on mechanical properties of dissimilar Al-alloys using Friction stir welding process", Materials Today: Proceedings, v. 44, n. 1, pp. 206–212, 2021. doi: http://doi.org/10.1016/j.matpr.2020.08.599.
12. HARIBALAJI, V., BOOPATHI, S., ASIF, M.M., "Optimization of friction stir welding process to join dissimilar AA2014 and AA7075 aluminum alloys",

Materials Today: Proceedings, v. 50, n. 1, pp. 2227–2234, 2022. doi: http://doi.org/10.1016/j.matpr.2021.09.499.

13. KASMAN, Ş., OZAN, S., "AN EXPERIMENTAL APPROACH FOR FRICTION STIR WELDING: A CASE STUDY FOR AA 2024-T351", Sigma Journal of Engineering and Natural Sciences, v. 38, n. 4, pp. 1999–2011, 2020.
14. KHALAFE, W.H., SHENG, E.L., BIN ISA, M.R., et al., "The effect of friction stir welding parameters on the weldability of aluminum alloys with similar and dissimilar metals", Metals, v. 12, n. 12, pp. 2099, 2022. doi: http://doi.org/10.3390/met12122099.
15. LEON, J.S., BHARATHIRAJA, G., JAYAKUMAR, V. October. A review on friction stir welding in aluminium alloys. IOP Conference Series: Materials Science and Engineering, v. 954, n. 1, pp. 012007, 2020. doi: http://doi.org/10.1088/1757-899X/954/1/012007
16. MODI, U., AHMED, S., RAI, A., "Prediction of ultimate tensile strength of friction stir welding joint using deep learning-based-multilayer perceptron and long short term memory networks", Welding International, v. 37, n. 7, pp. 387–399, 2023. doi: http://doi.org/10.1080/09507116.2023.2236936.
17. NASIR, T., ASMAELA, M., ZEESHANA, Q., et al., "Applications of machine learning to friction stir welding process optimization", Jurnal Kejuruteraan, v. 32, n. 1, pp. 171–186, 2020. doi: http://doi. org/10.17576/jkukm-2020-32(2)-01.
18. PD, S., JANGAM, S., "Design and optimization of the process parameters for friction stir welding of dissimilar aluminium alloys", Engineering & Applied Science Research, v. 48, n. 3, pp. 257, 2021.
19. RAJA, R., PARTHIBAN, A., NANDHA GOPAN, S., et al., "Investigate the process parameter on the friction stir welding of dissimilar aluminium alloys", Advances in Materials Science and Engineering, v. 2022, n. 1, pp. 4980291, 2022. doi: http://doi.org/10.1155/2022/4980291.
20. SARAVANAN, R., MALLADI, A., AMUTHAN, T., et al., "Mechanical characterization of friction stir welded dissimilar aluminium alloy using Taguchi approach", Materials Today: Proceedings, 2023. doi: http://doi.org/10.1016/j.matpr.2023.03.278.
21. SELAMAT, N.F.M., BAGHDADI, A.H., SAJURI, Z., et al., "Friction stir welding of similar and dissimilar aluminium alloys for automotive applications", International Journal of Automotive and Mechanical Engineering, v. 13, n. 2, pp. 3401–3412, 2016. doi: http://doi.org/10.15282/ijame.13.2.2016.9.0281.
22. UDAY, K.N., RAJAMURUGAN, G., "Influence of process parameters and its effects on friction stir welding of dissimilar aluminium alloy and its composites-a review", Journal of Adhesion Science and Technology, v. 37, n. 5, pp. 767–800, 2023. doi: http://doi.org/10.1080/01694243.2022.2053348.
23. UYYALA, S.B., PATHRI, S., "Investigation of tensile strength on friction stir welded joints of dissimilar aluminum alloys", Materials Today: Proceedings, v. 23, n. 1, pp. 469–473, 2020. doi: http://doi. org/10.1016/j.matpr.2019.04.210.
24. VERMA, M., SAHA, P., "Effect of micro-grooves featured tool and their depths on dissimilar micro-friction stir welding (μFSW) of aluminum alloys: a study of process responses and weld characteristics", Materials Characterization, v. 196, n. 2, pp. 112614, 2023. doi: http://doi.org/10.1016/j.matchar.2022.112614.

Note: All the figures and tables in this chapter were made by the authors.

Advances in Additive Manufacturing Technologies – Gurusamy Pathinettampadian et al. (eds)

Optimizing Cold Metal Transfer Welding Parameters for AA5083 and AA5754 Aluminium Alloy Joints

Subash P.[1],
Hepsi Beaula M.J.[2], Parthiban A.[3],
Balamurali P.[4], Muthukumaran A.[5], R. Vijayan[6]
Assistant Professor, Department of Mechanical Engineering,
Chennai Institute of Technology,
Chennai, India

◆ **Abstract:** Cold Metal Transfer (CMT) welding of dissimilar AA5754 and AA5083 was carried out using ER5356 filler wire. The study establishes a correlation between the mechanical properties and microstructure of the WMZ and interface regions. Microstructural analysis reveals epitaxial grain formation at the periphery of the PMZ. Additionally, the interface area on the AA5083 side exhibits superior strength and hardness due to a rich volume of fine second phases in the PMZ and heat-affected zone (HAZ) compared to the AA5754 side. All samples were analyzed for their fracture surfaces, and the presence of dimples was observed—this points to a ductile fracture mode.

◆ **Keywords:** Dissimilar aluminium alloys, Cold metal transfer, Microstructure, Weld metal zone, Mechanical properties

1. Introduction

In recent years, the automobile industry has adopted various strategies to reduce greenhouse gas emissions, including improved aerodynamics, lower rolling resistance, efficient powertrain technologies, and lightweight vehicle designs [1]. Among these, weight reduction is a key approach to controlling CO_2 emissions. Studies indicate that a 10% reduction in vehicle weight can improve fuel economy by 2–4% [2], and reducing a vehicle's weight by one kilogram can lower CO_2 emissions by approximately 0.091 g/km [3]. Due to its lessor weight nature, aluminum usage in passenger cars has doubled over the past decade, and this trend is expected to continue [4]. Both internal and external components, as well as various body panels, are manufactured using wrought aluminium [5]. The 5xxx series wrought aluminum are particularly suitable for automotive applications, especially for inner body structures [3]. These non-heat-treatable alloys are strengthened through solid solution hardening, with magnesium content ranging from 0.8% to slightly over 5%.

Welding plays a crucial role in integrating aluminum alloys into vehicle structures. However, conventional welding methods often result in voids, porosity, and low welding performance [6]. CMT welding is an excellent method for joining thin sheets in automotive manufacturing [7]. Unlike traditional welding processes, CMT operates through short-circuiting metal droplets, facilitated by the reciprocating motion of the filler wire, which enables droplet detachment with minimal spattering.

Several reports have discovered the impact of CMT welding on aluminum alloys. According to Comez et al. [8], the impact of dissimilar A5754-A6061 aluminum alloys was investigated. As reported by Dutra et al. [9] discovered that an ER5087 electrode led to better mechanical properties

[1]subashp@citchennai.net, [2]hepsibeaulamj@citchennai.net, [3]parthimech2810@gmail.com, [4]muralipb2080@gmail.com, [5]muthu2189@gmail.com, [6]srajendranvijayan@gmail.com

DOI: 10.1201/9781003685906-56

than an ER5183 one. Beytullah et al. [10] concluded that CMT welding provides high joint efficiency along with enhanced fatigue and tensile performance. [11,12]

Furthermore, previous research on CMT welding of dissimilar materials has primarily assessed mechanical properties using macro-specimens that encompass the base metals, weld metal and HAZs of two materilas [10,13], without specifically evaluating the mechanical properties of individual interface regions and the weld-metal zone (WMZ). The presented research aims to use the CMT welding technique to join dissimilar aluminum alloys A5754-H111 and A5083-H111, as well as to identify associations among the microstructure.

2. Experimental Methodology

The base materials used were rolled aluminum alloy sheets of A5754-H111 and A5083-H111. The process parameters utilized for welding are detailed in Table 56.2. The aluminum sheets were cut to 50 mm × 220 mm × 3 mm using a cutting machine. before to welding, the sheets were subjected to surface preparation that included machine buffing to eliminate oxides and subsequent cleaning with acetone to remove any remaining contaminants. The cleaned sheets were arranged in a flat (1G) configuration and fastened with G-clamps.

Using an Omni Tech testing machine, the microhardness of the weld joint was assessed by applying a test load of 0.2 kg for 10 seconds. A BISS machine was used to carry out the tensile tests at a strain rate of 0.001 mm/s. The gage length of the macro samples included the heat-affected zones (HAZs), base metals (BMs), partially melted zones (PMZs) of both A5754 and A5083 alloys, as well as the entire weld metal (WM) area. The tensile test results reported are averages derived from three tested samples.

Table 56.1 Chemical composition of AA6009

BM	Mg	Mn	Si	Fe	Cr	Cu	Zn	Al
AA5083	4.5	1	0.5	0.3	0.2	0.1	0.3	Bal.
AA5754	2.5	0.6	0.4	0.3	0.2	0.1	0.2	Bal.
ER5356	4.8	0.2	0.3	0.2	0.1	0.1	0.1	Bal.

Table 56.2 Process variables applied for CMT welding

WS	WFR	Current	Voltage	GFR
240	6	100	14	15

3. Results and Discussion

Figure 56.1(a-b) depicts the inspection of the welded joint. It is clear that the weld surface is free of cracks,

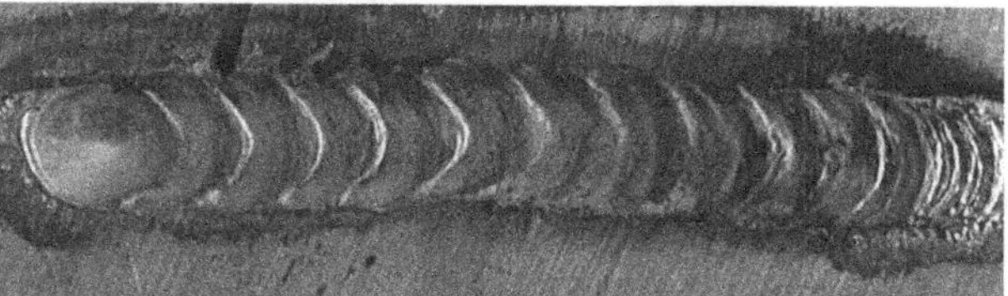

Fig. 56.1 Presentation of the weld in top and bottom views of CMT welded A5754-A5083 dissimilar joint

undercut defects and spatter. The weld bead's minimal smut layer and lack of smut spots indicate that magnesium loss is minimal. The weld joint demonstrates complete penetration, as illustrated in Fig. 56.1(b).

The cross-section shows a defect-free weld profile with complete penetration in Fig. 56.2. Because external ER5356 is incorporated into the joint, the weld metal zone (WMZ) appears brighter than the surrounding areas.

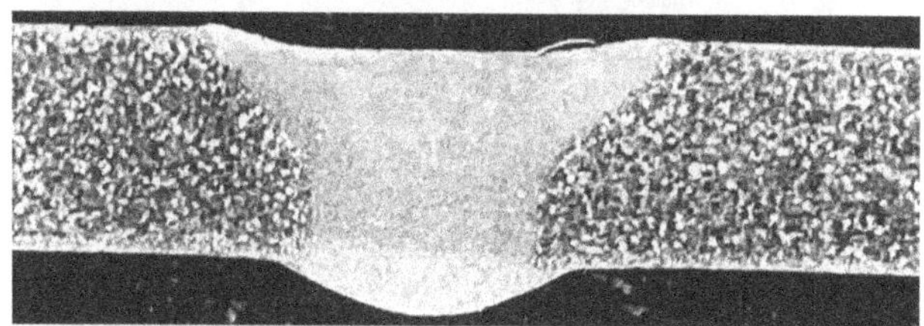

Fig. 56.2 Macroscopic cross-section of the CMT weld between dissimilar A5754 and A5083 joints

It is typical for welded aluminum alloys to show microstructural changes at the fusion boundary.[14] Epitaxial growth takes place when undercooling is minimal. With the temperature increasing toward the weld center, fine grains are suppressed, resulting in the growth of columnar grains from the PMZ. In the PMZ of the interface regions, there is a partial melting of the base alloys, leading to a mixture with elements from the filler material.

The center of the WMZ exhibits an equiaxed dendritic microstructure, as demonstrated in Fig. 56.3(b) and Fig. 56.4(d). The HAZ at both interfaces undergoes recrystallization, as depicted in Fig. 56.3(d) and Fig. 56.4(b). Moreover, the renewed grains in the HAZ of A5754 seem to be considerably coarser than those found in the HAZ of A5083 [15].

The heat input calculated from the parameters used in this research is 246 J/mm, which closely aligns with the literature value of 243 J/mm [16]. The heat input contributes to the diffusion of Si and Mg from the base materials to WMZ as a result of the dissolution effect. As a result, the Mg concentration is anticipated to be greater at the WMZ. The difference can be ascribed to the fact that magnesium vaporizes at a rate two directions of magnitude higher than that of Al [17,18].

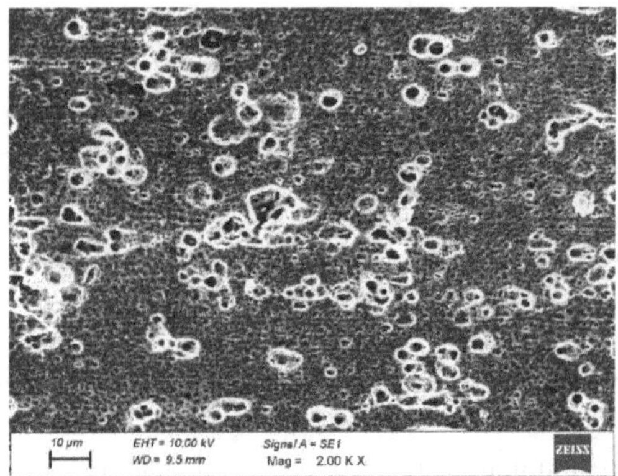

Fig. 56.3 SEM pictures of A5754 interface regions

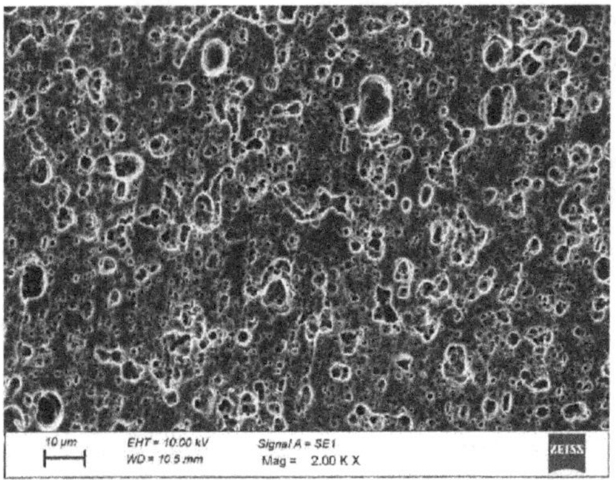

Fig. 56.4 SEM pictures of the weld joint's A5053 contact areas

It is probable that these phases are intermetallic compounds of Al_3Mg_2. The Al-Mg phase diagram indicates that at the eutectic temperature of 450 °C, Mg's maximum solubility is 17.1 wt%, which drops to 1.8 wt% at 35°C. As the ER5356 filler comprises 6 wt% Mg, the anticipated phases at room temperature are α-Al and the β phase, as corroborated by scientific studies [19]. This indicates the existence of the Al3Mn phase, as documented by other studies [20,21]. Due to the dilution effect, the Mn content in the WMZ rises, resulting in the Al_6Mn phase forming at the middle of the WMZ.

The small size of fine secondary phase particles makes it difficult to ascertain their precise chemical composition through EDS analysis. It is challenging to target these particles accurately with the electron beam while avoiding interaction with the surrounding matrix. Due to the fact that the matrix is made up mainly of aluminum, a greater amount of aluminum is found in the analysis, which overshadows the secondary phase particles. The tiny, bright spherical particle has a higher Si content than the coarser, dull-colored eye-ball-shaped particle. The secondary phases found at these sites are primarily composed of magnesium and silicon, prompting their initial identification as Mg_2Si. Moreover, earlier studies [22] have noted the presence of other secondary phases like $Al_6(Fe,Mn)$ in the A5083 alloy.

The A5083 side exhibits greater hardness in all areas compared to the corresponding zones on the A5754 side. Due to the fact that the temperatures in the HAZs exceed the BMs' recrystallization temperature, recrystallization and possible grain growth occur. This partially reduces work hardening and results in decreased hardness. As a result, the A5754 HAZ has a lower hardness value compared to the A5083 HAZ.

Conversely, at the A5754 interface, the WMZ contains a higher magnesium content, whereas the HAZ and PMZ exhibit a lower magnesium concentration. As magnesium is the primary strengthening element, the reduced magnesium content in these zones contributes to the less hardness of the HAZ and PMZ in A5754. Conversely, at the A5754 interface, the WMZ contains a higher magnesium content, whereas the HAZ and PMZ exhibit a lower magnesium concentration. As magnesium is the primary strengthening element, the reduced magnesium content in these zones contributes to the less hardness. The YS, % elongation and UTS, values obtained from tensile test curves are presented in Table 56.3. In Al-Mg alloys, a 1% increase in magnesium content results in an approximate tensile strength enhancement of 34 MPa [23]. A5083 demonstrates greater strength compared to A5754, owing to its higher magnesium content.

WMZ weist ein höheres YS als beide BMs und einen leicht höheren UTS als A5754 auf, jedoch ist der UTS im Vergleich zu A5083 niedriger. Previous studies have reported that Mg vaporization in the WMZ negatively affects mechanical properties and increases susceptibility to hot cracking [24,25]. Furthermore, the equiaxed dendritic structure in the WMZ has been shown to mitigate solidification cracking, thereby improving the weldment's strength, fatigue life, and toughness [26].

As delibrated earlier, the passage of magnesium, manganese, and silicon from the BMs to the WMZ compensates for the loss of secondary phase strengthening and solid solution strengthening effects [27,28]. However, no hydrogen pores, alloying element loss-induced pores, or shrinkage pores were observed in the WMZ [29].

Table 56.3 Tensile characteristics of the samples

Sample	UTS (MPa)	Elongation (%)	YS (MPa)
A5754	264	24	170
A5083	308	20	143
CMT weld joint	230	27	135

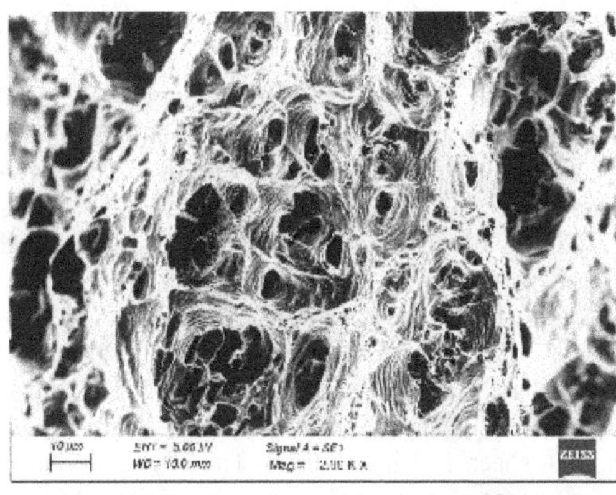

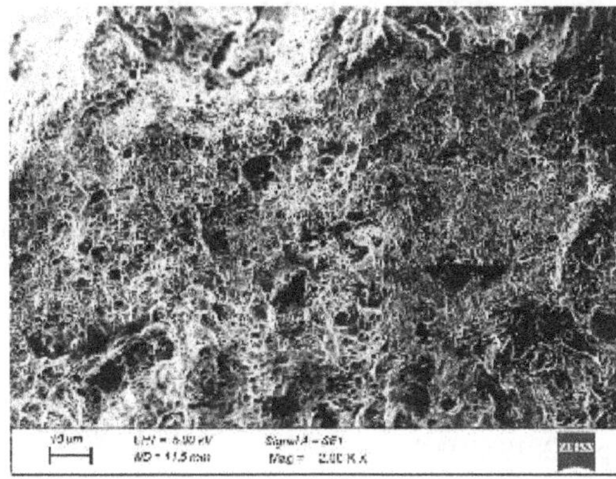

Fig. 56.5 Fracture surface of the interface regions of CMT weld joint (a) A5754 side and (b) A5083 side

4. Conclusion

1. The CMT with ER5356 filler material has successfully produced a high-quality dissimilar weld joint between A5754 and A5083 aluminium alloys. During tensile tests, the WMZ did not fracture.
2. The resultant phase particles in the HAZ of the A5083 side are finer compared to those on the A5754 side. Additionally, the A5083 HAZ contains a rich quantity of second phase particles.
3. The hardness of the WMZ is greater than that of the HAZs and PMZs. The enhanced hardness is credited to the movement of alloying components from the BMs caused by the concentration effect, coupled with a lack of defects.

References

1. G. Fontaras, Z. Samaras, On the way to 130 g CO2/ km-Estimating the future characteristics of the average European passenger car, Energy Policy 38 (4) (2010) 1826–1833, https://doi.org/10.1016/j.enpol.2009.11.059.
2. R. Wohlecker, M. Johannaber, M. Espig, Determination of weight elasticity of fuel economy for ICE, hybrid and fuel cell vehicles, SAE Tech. Pap. (2007), https://doi.org/10.4271/2007-01-0343.
3. J. Hirsch, Aluminium in innovative light-weight car design, Mater. Trans. 52 (5) (2011) 818–824, https://doi.org/10.2320/matertrans.L-MZ201132.
4. J. HIRSCH, Recent development in aluminium for automotive applications, Trans. Nonferrous Met. Soc. China (English Ed.) 24 (7) (2014) 1995–2002, https://doi.org/10.1016/S1003-6326(14)63305-7.
5. S. Ramasamy, C.E. Albright, CO 2 and Nd–YAG laser beam welding of 5754–O aluminium alloy for automotive applications, Sci. Technol. Weld. Join. 6 (3) (2001) 182–190, https://doi.org/10.1179/136217101101538730.
6. T.A. Barnes, I.R. Pashby, Joining techniques for aluminum spaceframes used in automobiles. Part I - solid and liquid phase welding, J. Mater. Process. Technol. 99 (2000) 62–71, https://doi.org/10.1016/S0924-0136(99)00367-2.
7. S.S. Sravanthi, S.G. Acharyya, P.P. Phani, G. Padmanabham, Integrity of 5052 Al- mild steel dissimilar welds fabricated using MIG-brazing and cold metal transfer in nitric acid medium, J. Mater. Process. Technol. 268 (2019) 97–106, https://doi.org/10.1016/j.jmatprotec.2019.01.010.
8. N. Çömez, H. Durmu.,s, Mechanical properties and corrosion behavior of AA5754-AA6061 dissimilar aluminum alloys welded by cold metal transfer, J. Mater. Eng. Perform. 28 (6) (2019) 3777–3784, https://doi.org/10.1007/ s11665-019-04131-x.
9. J.C. Dutra, R.H.G. e Silva, B.M. Savi, C. Marques, O.E. Alarcon, Metallurgical characterization of the 5083H116 aluminum alloy welded with the cold metal transfer process and two different wire-electrodes (5183 and 5087), Weld. World. 59 (6) (2015) 797–807, https://doi.org/10.1007/ s40194-015-0253-0.
10. B. Gungor, E. Kaluc, E. Taban, A.S.I.K. S, Mechanical and microstructural properties of robotic Cold Metal Transfer (CMT) welded 5083-H111 and 6082- T651 aluminum alloys, Mater. Des. 54 (2014) 207–211. doi:10.1016/ j.matdes.2013.08.018.
11. Ali Mehrani Milani, Moslem Paidar, Investigation on effect of pulse correction on structure property in dissimilar welds of galvanized steel and aluminum alloy obtained by gas metal arc welding cold metal transfer, Russ. J. Non- Ferrous Met. 57 (5) (2016) 467–476, https://doi.org/10.3103/ S1067821216050023.
12. Erdem Ünel, Emel Taban, Properties and optimization of dissimilar aluminum steel CMT welds, Weld. World. 61 (1) (2017) 1–9, https://doi.org/10.1007/ s40194-016-0386-9.
13. S. Babu, S.K. Panigrahi, G.D.J. Ram, P.V. Venkitakrishnan, R.S. Kumar, Cold metal transfer welding of aluminium alloy AA 2219 to austenitic stainless steel AISI 321, J. Mater. Process. Tech. 266 (2019) 155–164, https://doi.org/10.1016/j. jmatprotec.2018.10.034.
14. R Cao, J H Sun, J H Chen, Mechanisms of joining aluminium A6061-T6 and titanium Ti-6Al-4V alloys by cold metal transfer technology, Sci. Technol. Weld. Join. 18 (5) (2013) 425–433, https://doi.org/10.1179/ 1362171813Y.0000000118.

15. S. Kou, Welding metallurgy, New Jersey, USA. (2003) 431–446.
16. W. Zhang, H. He, Y. Shan, C. Sun, L. Li, Diffusion assisted hardness recovering and related microstructural characteristics in fusion welded Al–Mg–Si alloy butt, Mater. Res. Express. 6 (2019) 66549.
17. H. Zhao, T. Debroy, Weld Metal Composition Change during Conduction Mode Laser Welding of Aluminum Alloy 5182, 32 (2001).
18. Block-Bolten, T.W. Eagar, Metal vaporization from weld pools, Metall. Trans. B. 15 (3) (1984) 461–469.
19. Akshansh Mishra, Friction stir welding of dissimilar metal: A review, Int. J. Res. Appl. Sci. Eng. Technol. 6 (1) (2018) 1551–1559.
20. L.-E. Svensson, L. Karlsson, H. Larsson, B. Karlsson, M. Fazzini, J. Karlsson, Microstructure and mechanical properties of friction stir welded aluminium alloys with special reference to AA 5083 and AA 6082, Sci. Technol. Weld. Join. 5 (5) (2000) 285–296, https://doi.org/10.1179/136217100101538335.
21. H.B. McShane, C.P. Lee, T. Sheppard, Structure, anisotropy, and properties of hot rolled AA 5083 alloy, Mater. Sci. Technol. 6 (5) (1990) 428–440.
22. Dileep Chandran Ramachandran, Siva Prasad Murugan, Young-Min Kim, Dongcheol Kim, Gwang-Gook Kim, Dae-Geun Nam, Chanyoung Jeong, Yeong Do Park, Effect of microstructural constituents on fusion zone corrosion properties of GMA welded AA 5083 with novel Al – Mg welding wires of high Mg contents, Met. Mater. Int. 26 (9) (2020) 1341–1353, https://doi.org/ 10.1007/s12540-019-00434-9.
23. J.R. Davis, others, Aluminum and aluminum alloys, ASM international, 1993.
24. M.J. Cieslak, P.W. Fuerschbach, On the Weldability , Composition , and Hardness of Pulsed and Continuous Nd : YAG Laser Welds in Aluminum Alloys 6061, 5456, and 5086, 19 (1988) 319–329.
25. J.E. Hatch, Aluminum: properties and physical metallurgy, Met. Park. Ohio Am. Soc. Met. (1984).
26. L. Cui, X. Li, D. He, L. Chen, S. Gong, Effect of Nd : YAG laser welding on microstructure and hardness of an Al – Li based alloy, Mater. Charact. 71 (2012) 95–102, https://doi.org/10.1016/j.matchar.2012.06.011.
27. A.W. Alshaer, L. Li, A. Mistry, The effects of short pulse laser surface cleaning on porosity formation and reduction in laser welding of aluminium alloy for automotive component manufacture, Opt. Laser Technol. 64 (2014) 162–171, https://doi.org/10.1016/j.optlastec.2014.05.010.
28. M.M. Atabaki, M. Nikodinovski, P. Chenier, J. Ma, W. Liu, R. Kovacevic, Experimental and numerical investigations of hybrid laser arc welding of aluminum alloys in the thick T-joint con fi guration, Opt. Laser Technol. 59 (2014) 68–92, https://doi.org/10.1016/j.optlastec.2013.12.008.
29. X. Shang, H. Zhang, Z. Cui, M.W. Fu, J. Shao, A multiscale investigation into the effect of grain size on void evolution and ductile fracture: Experiments and crystal plasticity modeling, 125 (2020) 133–149. doi:10.1016/j.ijplas.2019.09.009.

Note: All the figures and tables in this chapter were made by the authors.

Advances in Additive Manufacturing Technologies – Gurusamy Pathinettampadian et al. (eds)

LBW on Dissimilar Aluminium Alloys AA5083 and AA6009: A Parametric Study

Parthiban A.[1],
Balamurali P.[2], Muthukumaran A.[3],
Vijayan R.[4], Hepsi Beaula M.J.[5], Subash P.[6]
Assistant Professor, Department of Mechanical Engineering,
Chennai Institute of Technology,
Chennai, India

◆ **Abstract:** Two different welding configurations were created by switching the locations of the bottom and top plates while keeping all other parameters the same. The weld shape was impacted by the evaporation of magnesium, which led to an unstable welding process and lowered real heat input. This was induced by positioning the alloy with a high magnesium content as the top plate. The SEM images and computational analyses of the microstructures formed by solidification along the fusion lines of the two joints were combined. The results showed that there was more noticeable solute separation and longer and broader intergranular liquid channels in the AA6009/AA5083 junction. The findings showed that compared to the AA5083/AA6009 joint, the AA6009/AA5083 joint exhibited better crack understanding. The effect of fracture sensitivity on tensile characteristics was also shown to be greater than that of any other factor. There were also noticeable variations in the pore shape and porosity between the two joints. Dense and small pores were seen when the top plate had a larger Mg content. This occurred because the Mg vapor was unable to exit and the laser energy was reduced as it went along the thickness direction. Furthermore, as a result of the diluting effect of the establishing stages, microhardness decreased with increasing Mg concentration.

◆ **Keywords:** Crack sensitivity, Dissimilar Al alloys, LBW, Mg, Mechanical properties, Porosity

1. Introduction

The joining of different aluminum alloys has garnered increasing notice and is widely utilized in the lightweight design of aerospace components, new energy vehicles, and marine assemblies. This process takes advantage of the properties of different Al alloys, such as high strength, low weight, weldability, and thermal conductivity [1]. AA5083/AA6009 dissimilar aluminum alloy joint is widely used in the automotive sector for joining the back wall and sidewall, taking advantage of the high strength and weldability of the two alloys. AA6009 alloys are surrounded by hardening precipitates composed mainly of Si and Mg [2]. Si enhances fluidity and resistance to hot cracking, while Mg increases specific strength. Meanwhile, AA5083 described by rich Mg content, exhibit massive formability and deep stretch capability, making them suitable for marine applications [3]. The primary variations in the elemental composition of AA5083 and AA6009 are Mg and Si content, which influence weld properties and welding behavior.

The impact of elemental composition and material properties on weld geometry, crack sensitivity, porosity, and the mechanical characterization of selected alloys has been examined in numerous studies. Traidia et al. [4] demonstrated that weld asymmetry could be minimized by positioning the

[1]parthimech2810@gmail.com, [2]muralipb2080@gmail.com, [3]muthu2189@gmail.com, [4]srajendranvijayan@gmail.com, [5]hepsibeaulamj@citchennai.net, [6]subashp@citchennai.net

DOI: 10.1201/9781003685906-57

lesser sulfur substance module at the lowest. Khan et al. [5] and Nie et al. [6] attributed grain boundary liquation to Si and Mg exclusion, which can be mitigated using an Si-Al filler. Their findings also indicated that the establishment of brittle phases and pores significantly weakened the UTS of AA6009/A356 MIG-welded joints [7,8]. Additionally, two researchers [9, 10] examined the effects of strengthening phase distribution density on tensile properties and microhardness respectively. However, limited research has explored the impact of different plate configurations on the properties of LBW joined dissimilar aluminum alloy joints. Variations in plate positioning lead to differences in element content and distribution, as well as variations in thermal properties, which influence the welding process.

In this research, a comparative analysis was conducted on AA6009/AA5083 and AA5083/AA6009 joints to evaluate differences in weld morphology, element distribution and welding stability. The complex interaction between various material properties, elemental distribution of various elements, microstructural feature arrangement, and overall performance of the joint was thoroughly analyzed and established. Throughout this research study, an effort was made to engineer two specific configurations of the joint, and along with this, a critical study of their microstructures was also made, along with conducting a detailed macro-morphological study of the same. Finally, a series of mechanical performance tests were conducted with the aim of establishing a unique and quantifiable relationship between the resultant microstructure developed and the resultant performance of the joint.

2. Experimental Methodology

In the current research, the AA5083 and AA6009 dissimilar alloys were welded using the laser welding process under lap joint form. A major variation in composition of Mg and Si was seen between the alloys, where the AA5083 contained 0.08 wt% Si and 4.5 wt% Mg while the AA6009 contained 0.58 wt% Si and 0.94 wt% Mg [3,11]. The other elemental compositions of AA5083 and AA6009 are presented in Table 57.1. Two joint formations were utilized: Type 1 (AA5083/AA6009 joint), where AA5083 was positioned above AA6009, and Type 2 (AA6009/AA5083 joint), where AA6009 was placed above AA5083.

Table 57.1 Chemical composition of BM

BM	Mg	Si	Mn	Cu	Fe	Cr	Zn	Al
AA5052	2.8	0.9	0.5	0.4	0.3	0.1	0.1	Bal
AA6009	0.6	0.8	0.4	0.3	0.3	0.6	0.2	Bal

Table 57.2 Mechanical properties of BM

Material	YS(MPa)	UTS(MPa)	El (%)	Hardness (Hvl)
AA5052	183	312	20.1	78
AA6009	282	329	11.3	92

To prevent splashed metal from damaging the lens, the welding head was tilted at a 10° angle in the direction of welding. In order to shield the weld from decomposition, a gas composed of 99% argon was applied, with a flow rate of 1.8 m^3/h. Before the welding process, oxide films were eliminated using a pulsed laser cleaning technique with the following parameters: laser power set at 100 W, defocus amount of 0 mm, pulse repetition rate of 70 kHz, and a cleansing speed of 750 m/min.

Laser welding experiments were conducted at identical linear energy levels using LP of 2400 W, 2700 W, and 3000 W, with respective WS of 2 m/min, 2.25 m/min, and 2.5 m/min, as detailed in Table 57.2. Samples for metallurgic analysis were taken from the welds by making transverse cuts. These samples were then ground, polished, and subjected to structural examination after about 30 seconds of etching with Keller's reagent. Furthermore, microhardness was assessed with a Vickers tester (DHV-1000) using a load of 100 g for 15 seconds. Tensile properties were determined with an Instron 8801 testing machine.

3. Results and Discussion

As laser power increased, penetration depth grew significantly more than weld width, indicating that laser power had a greater influence on penetration. Under massive power conditions (P = 3000 W), the penetration depths of AA6009/AA5083 joints were noticeably greater than those of AA5083/AA6009 joints. This suggests that, despite identical process parameters, the basic heat input differed between the two joint configurations. Since the two configurations had the same base materials, such a difference in heat input can be attributed to the difference in heat loss that can result from differences in the distribution of elements and composition of the two dissimilar aluminum alloys.

The results which were witnessed and recorded in this study were originated from the steady welding phase, and they were indicative measures of the entire welding process. It has been clearly shown in previous research studies that the molten pool stability highly depends on how frequently the keyhole opens and closes, as referenced in source [17]. In the process of welding AA5083 to AA6009, it was witnessed that the oscillations were more severe, primarily due to the fact that the keyhole was prone to closure as a result of the liquid flow behavior and the elevated ejection of spatter. However, by looking at the welding process involving AA6009 and AA5083, it was witnessed that the keyhole was open longer, enabling the molten pool to possess a smoother and steadier flow with much less ejection of spatter. The comparison clearly reveals that the molten pool in the configuration whereby AA6009 was welded onto AA5083 exhibited a measure of stability which was considerably greater than the level of stability that was witnessed in the configuration whereby AA5083 was welded onto AA6009.

The two components with the greatest compositional differences between AA5083 and AA6009 were silicon (Si) and magnesium (Mg). This is because of the high fluidity of Si [6] and the stirring effect of the laser welding process, which can easily distribute Si uniformly in the molten pool. This suggests that differences in Mg content contributed to fluctuations in molten pool behavior.

Due to its low boiling point (1090 °C), Mg is an insoluble alloying substance [18] that quickly vaporizes under high-energy laser irradiation. As the laser got to the lap interface of the two plates, energy attenuation and a decrease in energy density led to a reduction in the severity of Mg decomposition. Consequently, when the high-Mg-content plate was positioned as the lower plate, the molten pool exhibited greater stability compared to when the high-Mg-content plate was on top. Additionally, a higher Mg content led to increased heat loss through evaporation, thereby affecting the actual heat absorption in the two weld configurations. These findings correspond with the observed variations in weld penetration depths.

The liquefied grain boundaries are indicated by arrows and consist of grain boundary (GB) eutectics and solute-depleted bands [19,20]. The bonding force between grains is significantly weakened by continuous liquefied grain boundaries [21], which facilitates crack propagation and increases crack sensitivity. From comparing Fig. 57.8(a) and it appears that AA6009/AA5083 joints are more prone to cracking than AA5083/AA6009 joints.

When paired with a phase field method simulation, a temperature field simulation was carried out to further investigate the differences in crack sensitivity. This work aimed to establish a macro-micro scale correlation. Results from the temperature field were used to derive the interface pulling speed (Vp or R) and temperature gradient (G), which were then used as parameters for the phase field modeling. At t = 0.115 s during the competitive growth stage, results from phase field simulation (PF) were extracted for both joints. In the same region (Q), the grain size and intergranular spacing varied considerably between the two welds. In joints of AA6009/AA5083, the columnar crystals exhibited greater elongation and coarseness. According to Kou [24], longer liquid channels increase crack susceptibility as they obstruct liquid feeding within the intergranular passageway. A larger G × R leads to smaller grain structures. AA5083/AA6009 joints exhibited a larger G × R than AA6009/AA5083 joints, resulting in finer columnar grains, which aligns with the PF simulation findings [25].

The results indicated that AA6009/AA5083 joints exhibited greater solute segregation and wider intergranular solute enrichment zones. Grain size in region Q was validated through EBSD. The longest grain size in AA6009/AA5083 joints measured approximately 84 μm, constituting 10.5% of the total grain distribution, whereas in AA5083/AA6009 joints, the longest grain size was about 71 μm, making up 4.7%. This indicates that columnar grains near the fusion line in AA6009/AA5083 joints exhibited further growth and elongation compared to those in AA5083/AA6009 joints. Consequently, the AA6009/AA5083 joint featured extended and continuous liquid channels, explaining its higher crack susceptibility compared to AA5083/AA6009 joints.

While AA6009 is strengthened via precipitation hardening through Mg2Si (β') phases that are coherent with the aluminum structure [2], AA5083 derives its strength mainly from a solid solution of Mg within the Al structure [3]. The scanning path sequentially traversed the BM and the FZ. Distinct trends in microhardness distribution were observed for the two joints. In AA5083/AA6009 joints, the lowest microhardness was recorded at the center of the FZ, averaging 65 HV. The highest microhardness within the FZ was 70 HV, while the aggregated hardness of the AA5083 BM was around 60 HV.

The fluctuation of hardness in Zone 1 was related to the change in Mg content within the FZ. In AA5083/AA6009 joints, the Mg content increased rapidly before stabilizing, whereas in AA6009/AA5083 joints, it gradually decreased and then remained constant. Although an increase in Mg content generally enhances microhardness through solid solution strengthening, the opposite trend was observed in Zone 1. This was attributed to the predominant influence of the Mg_2Si strengthening phase on microhardness. Unlike the BM, Si phases re-solidified and dissolved in the FZ, forming more uniform and finer phases, without the extensive element transition zone observed with Mg. As described by Brodarac et al. [26], Mg_2Si phase formation occurs prior to Al-Mg eutectic precipitation. Once a sufficient amount of Mg_2Si strengthening phases fashioned, the surplus Mg combined with Al to form Al-

Mg eutectics, which provided minimal strengthening due to their precipitation as coarse particles along grain boundaries. Consequently, the dilution effect of the Mg_2Si strengthening phase varied across various regions, leading to a gradual decrease in microhardness as Mg content increased at the FZ edges.

Analysis of the longitudinal weld section and consistant binary map, revealed significant differences in porosity, pore size, and distribution between the two joints. The AA5083/AA6009 joint exhibited both large, irregularly shaped pores and small, densely packed pores, predominantly located in the middle and upper regions of the weld. Conversely, the AA6009/AA5083 joint contained only large pores with minimal small pores, leading to a reduction in porosity from 18.81% to 9.56%.

Pore formation in the welds occurred as bubbles became trapped within the solidifying molten pool due to unsuccessful escape. This was primarily influenced by chemical reactions, gas solubility reduction, and advanced vapor pressures during solidification [27]. Our experimental results confirmed the finding by Wu et al. that porosity is inversely proportional to the square of heat input [28]. The vaporization of Mg produced small bubbles within the molten pool, which ascended quickly. Due to the shorter distances that the bubbles had to float and the shorter duration of time that the liquid was held by virtue of the reduced heat input, these bubbles were not able to achieve the required time for them to come together and form larger, more substantial bubbles. Apart from this, the reduced volume of these bubbles, coupled with their decreased buoyancy, led to them mostly concentrating in the top and middle parts of the weld.

When the lower plate contained more Mg, less bubble formation was observed from Mg vaporization due to laser energy attenuation and partial reflection upon hitting the lower plate. Also, with a greater floating distance, small bubbles had enough time to coalesce and develop into large bubbles. These large bubbles had higher buoyancy due to their larger volume, and hence they had the potential to float from the root of the molten pool to the middle and upper weld areas. Both welds showed shear-type fractures at the middle surface, where unavoidable pores existed. Three tensile samples were chosen for representative results from variant weld positions. Also, the AA6009/AA5083 joint showed higher crack sensitivity near the fusion line compared with the AA5083/AA6009 joint. These results testified that lap weld tensile properties were affected less by penetration depth and porosity than by crack sensitivity.

4. Conclusion

Lastly, this extensive research study examined in a systematic way the significant effect that various plate configurations place on the microstructural change, and the mechanical strength, of dissimilar aluminum alloy lap joints comprising AA6009 and AA5083 welded using laser technology. The observed variations during the study can largely be attributed to variations in how the constituent elements are distributed in the materials, as well as to the evaporation of magnesium during welding. The key findings reached through this comprehensive research are as follows.:

1. The real heat input for the AA5083/AA6009 joint was less than that of the AA6009/AA5083 joint, owing to augmented heat loss resulting from variations in material attributes and Mg evaporation. This resulted in reduced weld penetration and fluctuations in the molten pool.
2. Microstructural observations revealed that the AA6009/AA5083 joint exhibited higher crack sensitivity compared to the AA5083/AA6009 joint.
3. The two joints were quite different in terms of both porosity and morphology, i.e., they were different in these aspects. When the plate with a high Mg content, or Mg, was put on top of the other materials, the behavior of the Mg element was more strongly influenced by the laser as a result of the attenuation that occurred in the direction of thickness. As a direct result of this stronger effect, there was a higher rate of Mg vaporization, which in turn caused the creation of very small bubbles.

References

1. B. Gungor, E. Kaluc, E. Taban, A. Sik, Mechanical, fatigue and microstructural properties. of friction stir welded 5083-H111 and 6082-T651 aluminum alloys, Mater. Des. 56 (4) (2014) 84–90.
2. D. Maisonnette, M. Suery, D. Nelias, P. Chaudet, T. Epicier, Effects of heat treatments on the microstructure and mechanical properties of a 6061 aluminium alloy, Mater. Sci. Eng. A 528 (6) (2011) 2718–2724.
3. Das, I. Butterworth, I. Masters, D. Williams, Microstructure and mechanical properties of gap-bridged remote laser welded (RLW) automotive grade AA 5182 joints, Mater. Charact. 145 (2018) 697–712.
4. Traidia, F. Roger, J. Schroeder, E. Guyot, T. Marlaud, On the effects of gravity and sulfur content on the weld shape in horizontal narrow gap GTAW of stainless steels, J. Mater. Process. Technol. 213 (7) (2013) 1128–1138.
5. M. Khan, L. Romoli, R. Ishak, M. Fiaschi, G. Dini, M. De Sanctis, Experimental investigation on seam geometry,

microstructure evolution and microhardness profile of laser welded martensitic stainless steels, Opt. Laser Technol. 44 (5) (2012) 1611–1619.
6. F. Nie, H. Dong, S. Chen, P. Li, L. Wang, Z. Zhao, X. Li, H. Zhang, Microstructure and mechanical properties of pulse MIG welded 6061/A356 aluminum alloy dissimilar butt joints, J. Mater. Process. Technol. 34 (3) (2018) 551–560.
7. Peyre Haboudou, B.A. Vannes, Peix, Reduction of porosity content generated during Nd:YAG laser welding of A356 and AA5083 aluminium alloys, Mater. Sci. Eng. A 363 (1–2) (2003) 40–52.
8. F. Khodabakhshi, L.H. Shah, A.P. Gerlich, Dissimilar laser welding of an AA6022-AZ31 lap-joint by using Ni-interlayer: novel beam-wobbling technique, processing parameters, and metallurgical characterization, Opt. Laser Technol. 112 (2019) 349–362.
9. U. Donatus, G.E. Thompson, X. Zhou, J. Wang, K. Beamish, Flow patterns in friction stir welds of AA5083 and AA6082 alloys, Mater. Des. 83 (2015) 203–213.
10. Jonckheere, B.D. Meester, A. Denquin, A. Simar, Torque, temperature and hardening precipitation evolution in dissimilar friction stir welds between 6061-T6 and 2014-T6 aluminum alloys, J. Mater. Process. Technol. 213 (6) (2013) 826–837.
11. P. Li, F. Nie, H. Dong, S. Li, G. Yang, H. Zhang, Pulse MIG welding of 6061-T6/A356-T6 aluminum alloy dissimilar T-joint, J.Mater. Eng. Perform. 27 (9) (2018) 4760–4769.
12. X. Xu, G.Mi, L. Chen, L. Xiong, P. Jiang, X. Shao, C.Wang, Research on microstructures and properties of Inconel 625 coatings obtained by laser cladding with wire, J. Alloy. Comp. 715 (2017) 362–373.
13. H. Huang, J. Wang, L. Li, N. Ma, Prediction of laser welding induced deformation in thin sheets by efficient numerical modeling, J. Mater. Process. Technol. 227 (2016) 117–128.
14. Y. Li, Y.-H. Feng, X.-X. Zhang, C.-S.Wu, An improved simulation of heat transfer and fluid flow in plasma arc welding with modified heat source model, Int. J. Therm. Sci. 64 (2013) 93–104.
15. N. Saunders, U.K.Z. Guo, X. Li, A.P. Miodownik, J.P. Schillé, Using JMatPro to model materials properties and behavior, JOM 55 (12) (2003) 60–65.
16. Deng, H. Murakawa, Numerical simulation of temperature field and residual stress in multi-pass welds in stainless steel pipe and comparison with experimental measurements, Comput. Mater. Sci. 37 (3) (2006) 269–277.
17. S.Wang, Study on the Characteristics of Ultra Narrow Gap Oscilating Laser Welding of Thick Alumimium Alloy, Harbin Institute of Technology, 2018.
18. S.-H. Choi, B. Ali, S.-K. Hyun, T.-H. Lee, S.-J. Seo, B.-S. Kim, T.-S. Kim, K.-T. Park, Effect of mg loss on mechanical properties of electron-beam-welded Al5083, Sci. Adv. Mater. 8 (4) (2016) 909–915.
19. Y. Hu, S.Wu, L. Chen, Review on failure behaviors of fusion welded high-strength Al alloys due to fine equiaxed zone, Eng. Fract. Mech. 208 (2019) 45–71.
20. H. Li, R.H. Jones, Effect of pre-welding heat treatments on welding a two-phase Ni3Al alloy, Mater. Sci. Eng. A 192–193 (1995) 563–569.
21. F. Yan, S. Liu, C. Hu, C. Wang, X. Hu, Liquation cracking behavior and control in the heat affected zone of GH909 alloy during Nd: YAG laser welding, J. Mater. Process. Technol. 244 (2017) 44–50.
22. F. Yu, Y. Ji, Y. Wei, L.-Q. Chen, Effect of the misorientation angle and anisotropy strength on the initial planar instability dynamics during solidification in a molten pool, Int. J. Heat Mass Transf. 130 (2019) 204–214.
23. F. Yu, Y. Wei, Y. Ji, L.-Q. Chen, Phase field modeling of solidification microstructure evolution during welding, J. Mater. Process. Technol. 255 (2018) 285–293.
24. S. Kou, A criterion for cracking during solidification, ActaMater. 88 (2015) 366–374.
25. S. Kou, Welding Metallurgy [M], Second edition Wiley-Interscience, New Jersey, USA, 2002.
26. Z.Z. Brodarac, F. Unkic, J.Medved, P. Mrvar, Determination of solidification sequence of the AlMg9 alloy, Met. Mater. 50 (2012) 59–67.
27. K. Nogi, Y. Aoki, H. Fujii, K. Nakata, Behavior of bubbles in weld under microgravity, Acta Mater. 46 (12) (1998) 4405–4413.
28. S. Wu, X. Yu, R. Zuo, W. Zhang, H. Xie, J. Jiang, Porosity, element loss, and strength model on softening behavior of hybrid laser arc welded Al-Zn-Mg-Cu alloy with synchrotron radiation analysis, Weld. J. 92 (3) (2013) 64–71.

Note: All the tables in this chapter were made by the authors.

Advances in Additive Manufacturing Technologies – Gurusamy Pathinettampadian et al. (eds)

Parametric Optimization of Pulsed TIG Welding for Dissimilar Aluminium Alloys AA5052 and AA6009

Balamurali P.[1], Parthiban A.[2]
Assistant Professor, Department of Mechanical Engineering,
Chennai Institute of Technology,
Chennai, India

T. Thanka Geetha[3]
Assistant Professor,
Department of Electronics and Communication Engineering,
DMI Engineering College,
Aralvaimozhi, India

**Muthukumaran A.[4],
Subash P.[5], Hepsi Beaula M.J.[6]**
Assistant Professor, Department of Mechanical Engineering,
Chennai Institute of Technology,
Chennai, India

◆ **Abstract:** Aluminium alloys are extensively used in manufacturing due to their lightweight and high strength properties. Consequently, welding aluminium alloys plays a crucial role in industrial applications to achieve complex geometries. In this study, dissimilar aluminium alloys, AA5052 and AA6009, were welded using pulsed tungsten inert gas (PTIG) welding, as it minimizes welding defects more effectively than conventional TIG welding. However, to enhance the mechanical strength of welded joints, optimizing the PTIG welding process parameters is essential. In the present investigation, pulse frequency, welding speed, and peak current were selected as input constraints, while microhardness and UTS were considered output responses. An empirical model was advanced to predict the UTS of the welded AA5052-AA6009 joints. The experimental trials were designed using a three-factor, five-level central composite design approach. The results revealed that welding speed had the most important effect on UTS, trailed by peak current, while pulse frequency had the least influence. Hardness analysis of the optimized welds indicated that the lowest hardness values (40–50 Hv0.5) and the highest hardness values (90–100 Hv0.5) were practical in the HAZ on the AA6009 side of the weldment.

◆ **Keywords:** AA5083, AA6082, Pulsed TIG welding, Tensile strength, Fractography, Hardness

1. Introduction

The aluminium alloys AA5052 and AA6009 have garnered significant interest in the marine, automotive, and aerospace industries due to their advantageous properties, including excellent formability, high strength, exceptional corrosion resistance, and superior weldability. With the growing demand for fuel-efficient, lightweight structures,

[1]muralipb2080@gmail.com, [2]parthimech2810@gmail.com, [3]vijaytgee@gmail.com, [4]muthu2189@gmail.com, [5]subashp@citchennai.net, [6]hepsibeaulamj@citchennai.net

DOI: 10.1201/9781003685906-58

the adoption of suitable joining techniques for aluminium alloys, ensuring zero micro- or macro-defects, remains a critical requirement [1,2]. AA5052 is a non-heat-treatable alloy widely used in applications such as shipbuilding and armored vehicles [3]. Its strength is further enhanced through work hardening. The intrinsic hardening effect depends on the presence of alloying elements, which contribute to phase dispersion. AA5052 is often employed as cold-rolled sheets for both exterior and interior structural panels in lightweight vehicles [4,5]. On the other hand, AA6009 is a heat-treatable aluminium alloy commonly used in structures such as trusses, cranes, and ore skips. The axial grain structure, a typical characteristic of fusion welding, in the WZ after solidification in the way of heat flow. However, compared to these alloys experience a less severe decline in mechanical properties [3].

Numerous researchers have studied the joining of various aluminium alloys using constant current TIG welding [6-12]. However, a major challenge in TIG welding of thin components is alteration. The pulsed TIG (PTIG) technique effectively controls distortion by cycling the welding current at a prearranged frequency between high and low levels. PTIG welding offers several metallurgical advantages, as highlighted in the literature, including an improved weld depth-to-width ratio, grain refinement in the FZ, reduced HAZ width, lower distortion, controlled segregation, decreased susceptibility to hot cracking, and reduced residual stresses [6]. These factors collectively enhance the mechanical properties of PTIG-welded joints compared to conventional TIG welding. In their welding studies on AA5052, Zhu C. et al. [7] and Singh L. et al. [8] optimized the WPP using the Taguchi analysis to enhance the strength of TIG-welded AA5052, concluding that tensile strength decreased when the welding current exceeded its optimal range. Mustafa U. et al. and Guo Y. et al. [10] demonstrated the effects of current, WS, and plasma gas flow rate (GFR) on AA5052 and found that varying heat cycles had a substantial impact on the joint's bead width and microstructure. Their results also showed that penetration increased with higher WC and lower WS. Liang Y. et al. [11] indicated that high welding currents led to strength reduction due to prolonged solidification, resulting in larger grains in the HAZ and decreased hardness. Yao Liu et al. [12] compared the tendency for pore formation in GMAW and TIG welding methods, concluding that TIG welding produced fewer pores than GMAW. Additionally, a comprehensive microstructural analysis was performed on the optimally welded samples to assess orientation, grain size, precipitate morphology and distribution, and precipitate volume fraction.

2. Experimental Methodology

The 5mm thick base metal AA6009 was procured from Arihant Aluminium, Chennai. It was cut to the vital of 100 × 55 mm using a hacksaw machine. The joining edges are flatted using a milling machine. Taking the chemical composition with the help of an optical emission spectroscopic study. The tensile strength of the BM is tested in UTM shown in Table 58.1, and The AA5052-H111 and AA6009-T6 parent metals were procured (100 mm 55mm× 5 mm). Their elemental composition was analyzed using optical emission spectroscopy, with the results obtainable in Table 58.1, while their mechanical characteristics are recorded in Table 58.2. In the PTIG welding process, three key welding process parameters (WPP) were varied: pulse frequency (PF), peak current (PC), and welding speed (WS).

Table 58.1 Chemical composition of AA6009

BM	Mn	Mg	Fe	Cr	Si	Cu	Zn	Al
AA5052	0.5	2.8	0.3	0.1	0.9	0.4	0.1	Bal
AA6009	0.4	0.6	0.3	0.6	0.8	0.3	0.2	Bal

Table 58.2 Mechanical properties of AA6009

Material	YS(MPa)	UTS(MPa)	El (%)	Hardness (Hvl)
AA5052	183	312	20.1	78
AA6009	282	329	11.3	92

The parameter limits and levels were determined based on an extensive literature review and are presented in Table 58.3 [8,13,14]. Keller's reagent was applied to the polished samples to reveal grain structure and precipitates. The microstructural characteristics were examined through SEM imaging, and the elemental composition of the WZ was identified using SEM-EDAX analysis. Mechanical properties of the weldments was carried out using tensile testing and microhardness measurements. Tensile test specimens were extracted from the weldment in the transverse direction using an EDM machine, following ASTM-E8 standards.

Table 58.3 Limits of welding process parameter

WPP	Unit	Levels				
		-1.682	-1	0	1	+1.682
PC	A	140	160	190	220	240
PF	Hz	2	3.2	5	6.8	8
WS	mm/min	140	156	180	204	220

3. RESULTS AND DISCUSSION

3.1 Using RSM to Frame Empirical Relationships

Response Surface Methodology integrates statistical and mathematical techniques to establish relationships and optimize a response influenced by multiple input constraints. For instance, peak current (PC), WS, and pulse frequency (PF), all impact the UTS. Table 58.4 offerings the UTS values obtained from the weldments. Among various experimental design techniques available for regression coefficient analysis, the CCD was selected for this research.

$$UTS = f(PC,PF,WS) \quad (1)$$

$$UTS = 145.52 + 7.7(PC) - 5.96(PF) - 8.54(WS) + 0.18(PC*PF) + 10.68(PC*WS) + 3.83(PF*WS) - 7.59(PC*PC) - 9.32(PF*PF) - 7.72(WS*WS) \quad (2)$$

In this study, the key influencing variables include PF, PC, WS, interaction terms PCPF, PCWS, PF*WS, as well as quadratic terms PC^2, PF^2, and WS^2. Model terms with a value exceeding 0.1000 are deemed insignificant.

The plot was developed by keeping two input parameters at their central values (coded value = 0) while varying the third parameter. The rise in peak current significantly melts the material around the weld face, expanding the weld zone, as similarly observed by Bagha et al. [15]. It states that although a rise in peak current enhances the weld region, too much heat input causes stress concentration in the weld pool, which negatively impacts mechanical properties by Vikash Chaudhary [16].

Table 58.4 Design of experiments framed using CCD

Std. order	Run order	Coded Values			UTS (MPa)
		P	S	F	
1	14	-1	-1	-1	144
2	3	1	-1	-1	139
3	17	-1	1	-1	125
4	1	1	1	-1	121
5	7	-1	-1	1	95
6	9	1	-1	1	133
7	18	-1	1	1	92
8	13	1	1	1	130
9	11	-1.682	0	0	110
10	16	1.682	0	0	133
11	4	0	-1.682	0	128
12	19	0	1.682	0	105
13	8	0	0	-1.682	132
14	10	0	0	1.682	111
15	2	0	0	0	145
16	15	0	0	0	147
17	5	0	0	0	145
18	6	0	0	0	146
19	12	0	0	0	145
20	20	0	0	0	146

Additionally, as pulse frequency increases, vibration amplitude rises, resulting in small grain structures. It was also detected that an increase in WS reduces the FZ area by decreasing heat input. When WS is too low, the heat input

Table 58.5 ANOVA test results

Source	Sum of Squares	Df	Mean Square	F-value	p-value	
Model	5780.13	9	642.24	53.92	<0.0001	significant
A-TRS	810.22	1	810.22	68.02	<0.0001	
B-TTS	485.06	1	485.06	40.72	<0.0001	
C-AL	994.90	1	994.90	83.53	<0.0001	
AB	0.2485	1	0.2485	0.0209	0.8881	
AC	912.29	1	912.29	76.59	<0.0001	
BC	117.58	1	117.58	9.87	0.0102	
A^2	830.59	1	830.59	69.73	<0.0001	
B^2	1251.02	1	1251.02	105.03	<0.0001	
C^2	857.89	1	857.89	72.02	<0.0001	
Residual	119.11	10	11.91			
Lack of Fit	116.52	5	23.30	45.04	0.0006	significant
Pure Error	2.59	5	0.5175			
Cor Total	5899.24	19				

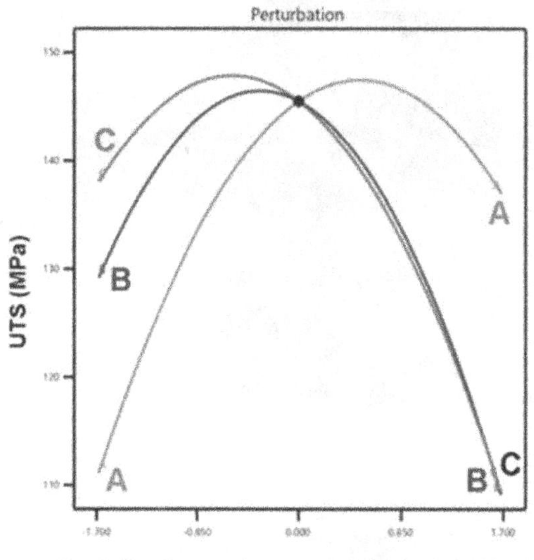

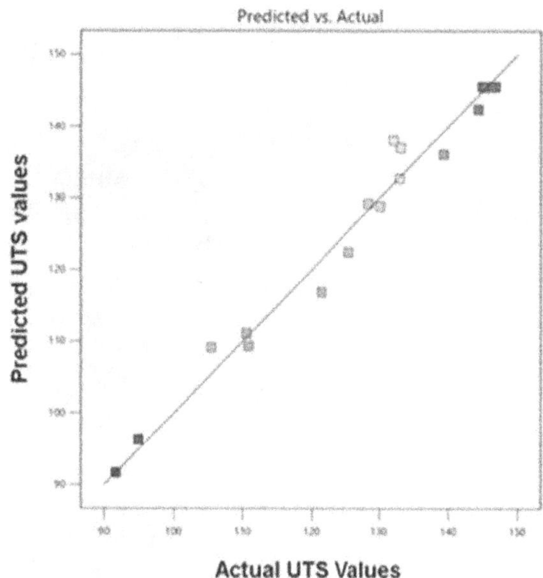

Fig. 58.1 (a) Perturbation plot (A: PC, B: PF, C: WS), (b) Predicted vs actual plot

is high, leading to increased fusion zone width and depth. Conversely, excessively high WS results in poor fusion of the base metal [17]. Rapid cooling causes significant porosity, whereas slower cooling rates allow gas bubbles to escape, reducing pore formation. Specimens that experienced moderate heat input exhibited lower porosity, thereby enhancing mechanical properties. Increased variability of the molten pool further facilitates the escape of surrounded gases, contributing to improved mechanical characteristics [18].

Figure 58.1(a) depicts a consistent increase in UTS as peak current rises from 140 A to 210 A, followed by a decline beyond 210 A. Similarly, UTS increases with PF from 2 Hz to 5 Hz but decreases beyond 5 Hz. Furthermore, UTS improves as WS increases from 140 mm/min to 170 mm/min, but additional growths in WS lead to a decline in tensile strength. The predicted vs. actual values are plotted in Fig. 58.1(b). In this graph, represent the experimentally obtained values, while the linear line indicates the predicted values.

Figure 58.2(a) depicts the interaction impact of PF and PC on UTS while keeping WS at its central value. The plot reveals that UTS raised as PC rises from 140 A to 200 A, reaching its peak value at this point. However, further increases in PC beyond 200 A lead to a decline in tensile strength. This decrease is primarily due to grain coarsening in the weld pool, which results from excessive heat input and slower cooling during solidification. At lower PC, the heat input is lacking to form strong joints, whereas at higher PC, excessive heat input leads to coarser grains. Therefore, an optimal PC value exists that maximizes tensile strength. In this study, the optimal peak current was determined to be 197 A.

Figure 58.2(b) illustrates the interface impact of WS and PC on UTS while maintaining PF at its central value. When PF is set at 2 Hz, tensile strength is relatively low. However, as PF increases from 2 Hz to 5 Hz, tensile strength improves. This is because, at an optimal PF range, the molten weld pool experiences vigorous agitation, leading to finer grain formation. When PF exceeds the optimal value, temperature oscillations in the molten metal decrease significantly, reducing the impact on the weld pool. Consequently, an optimal PF value ensures finer grain formation, enhancing mechanical properties. For this study, the optimal pulse frequency was identified as 4.9 Hz.

Figure 58.2(c) presents the interaction impact of PF and WS on UTS while keeping PC at its central value. It was observed that tensile strength raises as WS rises from 141 mm/min to 181 mm/min, reaching its highest value at 171 mm/min. However, beyond 180 mm/min, UTS decreases. This reduction is primarily due to excessive heat input at lessor WS, which leads to grain coarsening in the FZ. Conversely, at excessively high WS, the heat input becomes insufficient, resulting in inadequate fusion. Thus, an optimal WS value ensures a balanced heat input, promoting better mechanical properties.

The optimization process was carried out using Design Expert analysis software. In this study, maximizing tensile strength was the primary objective. The experimental results revealed that the optimal welding constraints-PC of 197 A, WS of 182 mm/min and PF of 4.9 Hz, -yielded a UTS value of 145 Mpa.

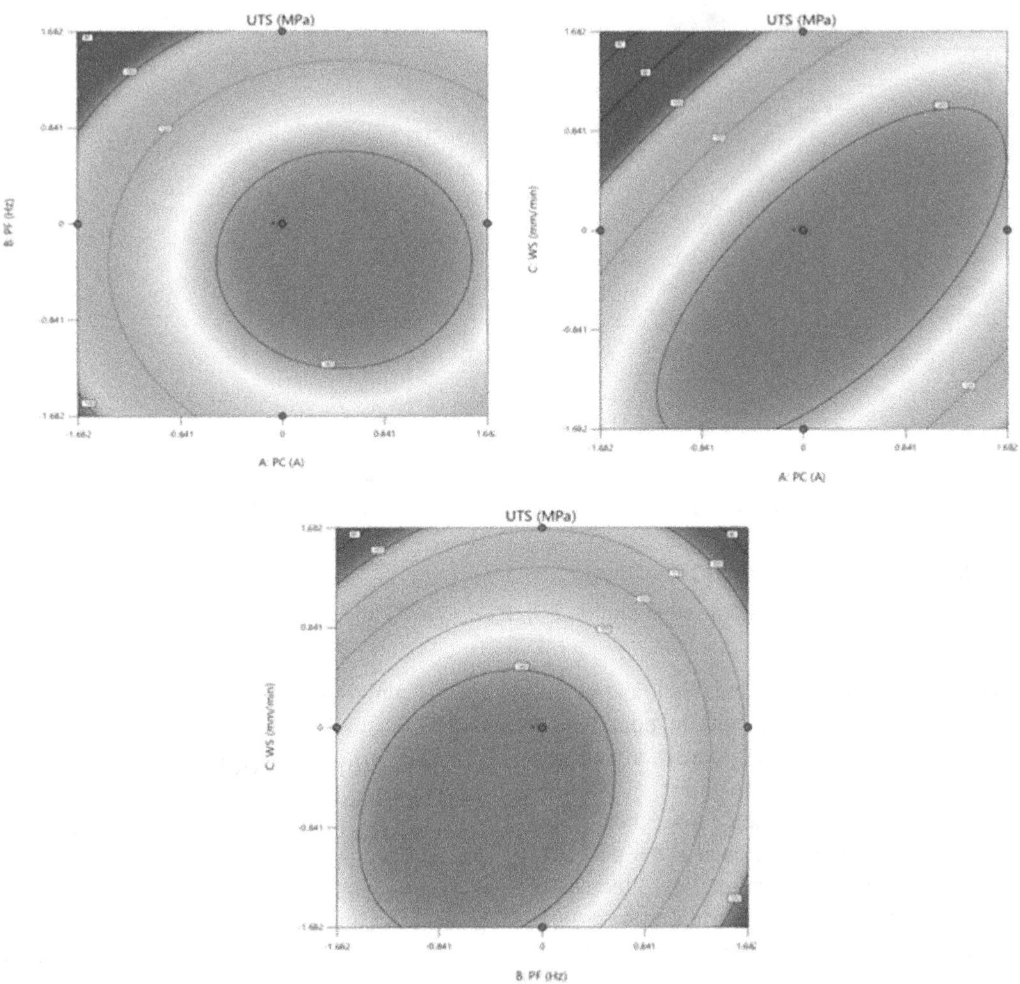

Fig. 58.2 [a] Interaction plot AB [b] interaction plot AC [c] interaction plot BC

Table 58.6 Predict optimal and experimental WPP

Parameter	Units	Optimized Parameters		UTS	
		RSM	Exp.	RSM	Exp.
PC	A	197	190	145	146.8
PF	Hz	4.9	5		
WS	mm/min	182	181		

3.2 Microstructure Analysis

The key intermetallic particles that contribute to the strength of AA5052 include Al_3Mg_2, $Al_6(Fe,Mn)$, and Mg_2Si. As observed in Fig. 58.3(a), the BMZ is densely populated with IMC's, with only a less regions displaying the α-Al matrix. Si and Mg are the primary alloying elements in AA6009, making Mg_2Si the only strengthening precipitate present. Figure 58.3(b) illustrates that the BMZ contains abundant Mg_2Si precipitates dispersed throughout the α-Al matrix.

On the AA6009 side, comparing the BM and HAZ shows a substantial loss of strengthening Mg_2Si precipitates due to the welding process. Since Mg_2Si is the sole strengthening precipitate in AA6009, its dissolution in the HAZ renders this region the weakest part of the weldment. Consequently, the SEM image of the FZ reveals numerous elongated intermetallic particles distributed throughout the fusion zone, arising from the clustering of intermetallic particles that did not dissolve through solidification.

4. Conclusion

The resulting outcomes were illustrated from the PTIG welding of AA5052 and AA6009 alloys:

1. The optimization of WPP determined that the optimal conditions for joining AA5052-H111 and AA6009-T6 are a WS of 181 mm/min, a PF of 4.9 Hz, and a PC of 197 A.

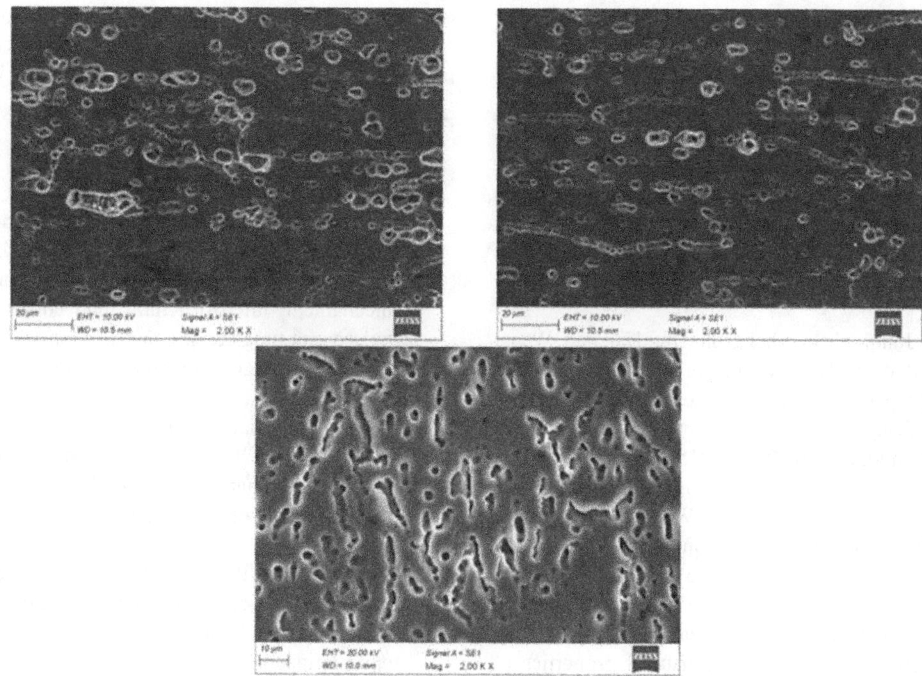

Fig. 58.3 SEM micrograph (a) AA5052 BM, (b) AA6009 BM, (c) Fusion zone

2. By adjusting the welding parameters, the heat input can be structured. It was also observed that the sample welded under optimal conditions exhibited a smaller heat-affected zone (HAZ) compared to other weldments.
3. Microstructural analysis of the weldment revealed a significant dissolution of Mg_2Si precipitates in the HAZ of the AA6009-T6. Additionally, the size of the remaining precipitates in this region was considerably reduced, which is the primary reason for HAZ failure in AA6009 during tensile loading.
4. An SEM examination of the fractured covering of the welded samples revealed flat facets and quasi-cleavage structures, which confirmed a decrease in the ductility of the weldment.
5. A hardness investigation of various weldment zones revealed that the AA6009 side's HAZ had the lessor hardness values, assembly it the weldment's weakest area and softest.

References

1. Saulius Baskutis, Jolanta Baskutiene, Regita Bendikiene, Antanas Ciuplys, "Effect of weld parameters on mechanical properties and tensile behaviour of tungsten inert gas welded AW6082-T6 aluminium alloy", Journal of Mechanical Science and Technology, vol. 33, (2019), pp. 765–772.
2. Yuanchun Huang, Yin Li, Zhengbing Xiao, Yu Liu, Yutian Huang, Xianwei Ren, "Effect of homogenization on the corrosion behaviour of 5083-H321 aluminium alloy", Journal of Alloys and Compounds, vol. 673, (2016), pp. 73–79.
3. Naveen Kumaar. P, Jayakumar. K, "Influence of tool pin profiles in the strength enhancement of friction stir welded AA5083 and AA5754 alloys", Materials Research Express, vol. 9, (2022), 036505.
4. Jose Divo Bressan, Luciano Pessanha Moreira, Maria Carolina dos Santos Freitas, Stefania Bruschi, Andrea Ghiotti, Francesco Michieletto, "Modeling of forming limit strains of AA5083 aluminium sheets at room and high temperatures", Advanced Materials Research, vol. 1135, (2015), pp. 202–217.
5. Jun Liu, Ming-Jen Tan, Anders-Eric-Wollmar Jarfors, Yingyot Aue-u-lan, Sylvie Castagne, "Formability in AA5083 and AA6061 alloys for light weight applications", Materials and Design, vol. 31, (2010), pp. S66–S70.
6. S. Kou, Y. Le, "Nucleation mechanisms and grain refining of weld metal", Welding Journal, vol. 65(4), (1986), pp. 305–313
7. Chenxiao Zhu, Jason Cheon, Xinhua Tang, Suck-Joo Na, Haichao Cui, "Molten pool behaviours and their influences on welding defects in narrow gap GMAW of 5083 Al-alloy", International Journal of Heat and Mass Transfer, vol. 126, (2018), pp. 1206–1221.
8. Lakshman Singh, Rajeshwar Singh, Naveen Kumar Singh, Davinder Singh, Pargat Singh, "An evaluation of TIG welding parametric influence on tensile strength of 5083 aluminium alloy", International Journal of Mechanical, Aerospace, Industrial, Mechatronic and Manufacturing Engineering, vol. 7, (2013), pp. 2326–2329.

9. Umar Mustafa, Mukesh Chandra, Sathiya Paulraj, "Influence of filler wire diameter on mechanical and corrosion properties of AA5083-H111 Al-Mg alloy sheets welded using an AC square wave GTAW process", Transactions of Indian Institute of Metals, vol. 71, (2018), pp. 1975–1983.
10. Yangyang Guo, Houhong Pan, Lingbao Ren, Gaofeng Quan, "An investigation on plasma-MIG hybrid welding of 5083 aluminium alloy", The International Journal of Advanced Manufacturing Technology, vol. 98, (2018), pp. 1433–1440.
11. Ying Liang, Junqi Shen, Shengsun Hu, Haichao Wang, Jie pang, "Effect of TIG current on microstructural and mechanical properties of 6061-T6 aluminium alloy joints by TIG-CMT hybrid welding", Journal of Materials Processing Technology, vol. 255, (2018), pp. 161–174.
12. Yao Liu, Wenjing Wang, Jijia Xie, Shouguang Sun, Liang Wang, Ye Qian, Yuan Meng, Yujie Wei, "Microstructure and mechanical properties of aluminium 5083 weldments by gas tungsten arc and gas metal arc welding", Materials Science and Engineering A, vol. 549, (2012), pp. 7–13.
13. Kumar. A, Sundarrajan. S, "Optimization of pulsed TIG welding process parameters on mechanical properties of AA5456 aluminium alloy weldments", Materials & Design, vol. 30, (2009), pp. 1288–1297.
14. M. Samiuddin, J. Li, M. Taimoor, M.N. Siddiqui, S.U. Siddiqui, J.T. Xiong, "Investigation on the process parameters of TIG-welded aluminum alloy through mechanical and microstructural characterization", Defence Technology, vol. 17, (2021), pp. 1234–1248.
15. Lucky Bagha, Shankar Sehgal, Amit Thakur, Harmesh Kumar, "Effect of powder size of interface material on selective hybrid carbon microwave joining of SS304-SS304", Journal of Manufacturing Processes, vol. 25, (2017), pp. 290–295.
16. Vikash Chaudhary, Ajaya Bharti, Syed Mohd Azam, Naveen Kumar, Kuldeep K. Saxena, "A re-investigation: Effect of TIG welding parameters on microstructure, mechanical, corrosion properties of welded joints", Materials Today: Proceedings, vol. 45, (2021), pp. 4575–4580.
17. Amin Reza Koushki, Massoud Goodarzi, Moslem Paidar, "Influence of shielding gas on the mechanical and metallurgical properties of DP-GMA welded 5083-H321 aluminium alloy", International Journal of Mineral, Metallurgy and Materials, vol. 23, (2016), pp. 1416–1426.
18. Yang Fang-zhou, Zhou Jie, Xiong Yi-bo, "Effect of heat input on microstructure and mechanical properties of butt-welded dissimilar magnesium alloy joint", Journal of Central South University, vol. 25, (2018), pp. 1358–1366.
19. K. Vasu, H. Chelladurai, Addanki Ramasamy, S. Malarvizhi, V. Balasubramanian, (2019), "Effect of fusion welding processes on tensile properties of armor grade, high thickness, non-heat treatable aluminum alloy joints", Defence Technology, vol. 15, pp. 353–362.
20. R. Sasi Lakshmikhanth, A.K. Lakshminarayanan, "On the mechanical, microstructural, and corrosion properties of pulsed gas tungsten arc and friction stir welded RZ5 rare earth grade magnesium alloy", Materials Research Express, vol. 9, (2022), 126507.
21. Xinwei She, Xianquan Jiang, Ruihao Zhang, Puquan Wang, Binbin Tang, Weichang Du, "Study on microstructure and fracture characteristics of 5083 aluminum alloy thick plate", Journal of Alloys and Compounds, vol. 825, (2020), 153960.
22. Mustafa Umar, Paulraj Sathya, "Influence of melting current pulse duration on microstructural features and mechanical properties of AA5083 alloy weldments", Materials Science and Engineering A, vol. 746, (2019), pp. 167–178.
23. Ambriz. R.R, Jaramillo. D, Garcia. C, Curiel. F.F, "Fracture energy evaluation on 7075-T651 aluminum alloy welds determined by instrumented impact pendulum", Transactions of Nonferrous Metal Society of China, vol. 26, (2016), pp. 974–983.
24. Senthur Vaishnavan. S, Jayakumar. K, "Tungsten inert gas welding of two aluminum alloys using filler rods containing scandium: the role of process parameters", Materials and Manufacturing Processes, vol. 37, (2022), pp. 143–150.
25. Chenxiao Zhu, Xinhua Tang, Yuan He, Fenggui Lu, Haichao Cui, "Effect of preheating on the defects and microstructure in NG-GMA welding of 5083 Al-alloy", Journal of Materials Processing Technology, vol. 251, (2018), pp. 214–224.
26. Sasi Lakshmikhanth Rajaseelan, Subbaiah Kumarasamy, "Mechanical properties and microstructural characterization of dissimilar friction stir welded AA5083 and AA6061 aluminium alloys", Mechanika, vol. 26, (2020), pp. 545–552.

Note: All the figures and tables in this chapter were made by the authors.

Advances in Additive Manufacturing Technologies – Gurusamy Pathinettampadian et al. (eds)

Performance and Pressure Drop Analysis of Zig-Zag Shaped Heat Exchanger

Keerthana P. P.[1],
Jency Esther R.[2], Joel Thanasingh S.[3]
Department of Mechanical Engineering,
Chennai Institute of Technology,
Chennai, Tamilnadu

A. Abdul Salman[4], J. Vijaya Dharsan[5]
Department of Mechanical Engineering,
Velammal College of Engineering and Technology,
Madurai, Tamilnadu

R. Sathiyamoorthi[6]
Department of Mechanical Engineering,
Chennai Institute of Technology,
Chennai, Tamilnadu

◆ **Abstract:** Heat exchangers are commonly used in both home and industrial settings. For usage in refrigeration units, transportation power systems, buildings heat and air conditioning systems, steam power plants, and chemical processing facilities, a wide variety of heat exchanger types have been developed. Heat exchangers' real design is a challenging issue. A heat exchanger is a device which moves heat between various process fluids. In both residential and commercial contexts, heat exchangers are frequently utilized. Many types of heat exchangers have been developed for use in chemical processing facilities, steam power plants, buildings' heat and air conditioning systems, refrigeration units, and transportation power systems. The actual design of heat exchangers is a difficult problem. It goes beyond only heat-transfer analysis. Numerous studies are being conducted to increase the heat exchanger's rate of heat transmission. By examining the various heat exchanger profiles, the primary objective of this research project is to improve heat transfer efficiency. It is created a spiral-shaped tube with a 30° helical angle.

◆ **Keywords:** Heat exchanger, Zig-zag, Effectiveness

1. Introduction

Over the past several years, new technologies in the field of process intensification, such as small heat exchangers-reactors, had been created to match the growing demand of more secure, pollution free, higherefficiency and lower consumption of energy sources [1]. With numerous benefits, including improved temperature control, reactive volume confinement, and heat transfer performance, they combine two fundamental ideas of process intensification: on the one hand, the units' miniaturisation, and on the other, the multi-functionalization of the apparatus [2-4].

[1]keerthana.p814@gmail.com, [2]jencyestherr@gmail.com, [3]joelthanasingh@gmail.com, [4]abdulsalman2108004@gmail.com, [5]vijayadharsan21@gmail.com, [6]sathiya.ram78@gmail.com

DOI: 10.1201/9781003685906-59

Lee et al. [5] analyzed the parameters of a zigzag shaped PCHE. It is well known that zigzag channels can lead to both enhanced heat transfer performance and noticeably upper pressure drop. Several researchers have put forth a variety of methods. The effectiveness of a 45° lowered zigzag channel PCHE was compared with that of an ideal flow channel configuration with S-shaped fins created by Tsuzuki et al. [6]. Kim et al. [7] suggested an innovative PCHE architecture with aero-foil fins, and mathematical calculations were conducted in order to contrast the newly developed PCHE with a zigzag shaped PCHE having a bend inclination of 40°. The experimental findings exhibited the pressure loss in the new configuration seems to be reduced to 1/20 of the total amount lost in the zigzag profiles, even if the overall heat transfer rate of substances stayed nearly constant.

2. Materials and Methods

Generally, type 304 is sometimes called 18/8 since its standard composition is 18% chromium and 8% nickel. An austenitic grade of stainless steel that is capable of severe deep drawing is type 304. Because of this characteristic, 304 is the most common grade utilized in sinks and saucepans. The low carbon version of 304 is called type 304L. It improves weldability in heavy gauge components. Another version that may be used at high temperatures is 304H, which has a higher carbon content. The characteristics listed in this data sheet are typical for ASTM A240/A240M-covered flat-rolled goods. Although not always the same, it is reasonable to assume that the dimensions in these standards will resemble those in this data sheet. All stainless steels should only be fabricated with equipment made specifically for stainless steel. Prior to usage, all work surfaces and tools must be well cleaned. These safety measures are required to prevent easily corroded metals from contaminating stainless steel, which could tarnish the manufactured product's surface.

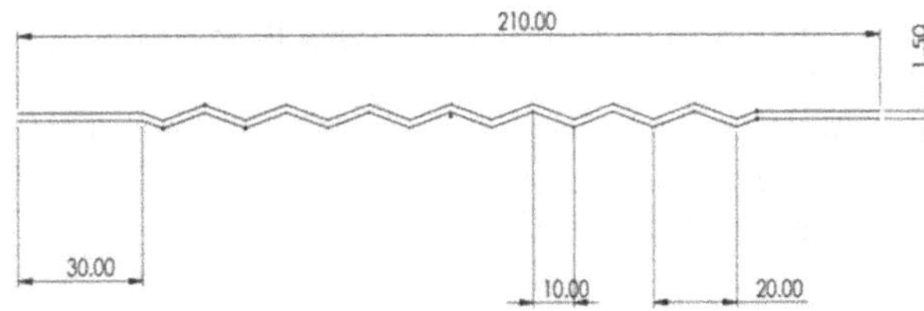

Fig. 59.1 Specifications of Zig-Zag shape

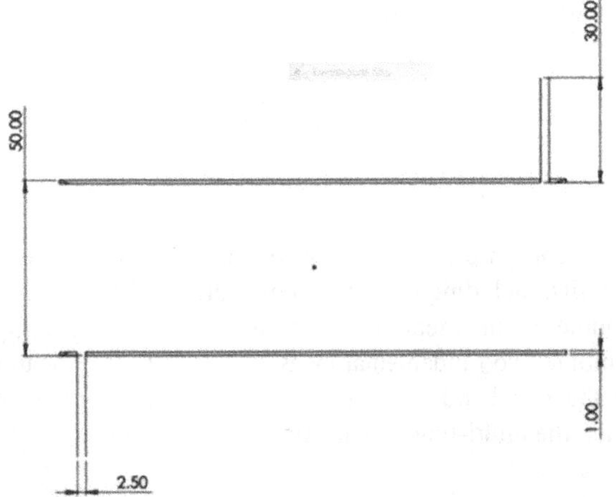

Fig. 59.2 Specification of zig-zag shaped heat exchanger

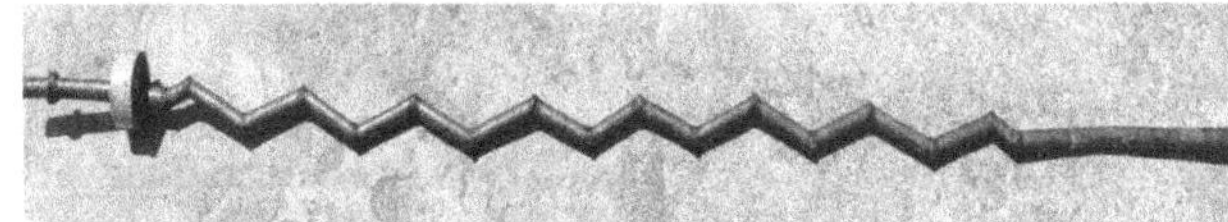

Fig. 59.3 Acutal photograph of zig-zag

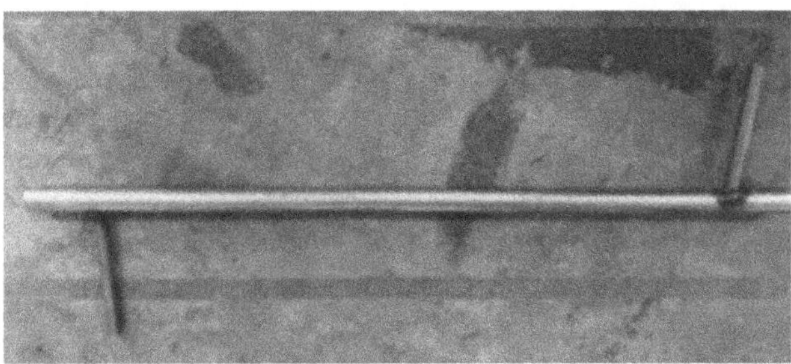

Fig. 59.4 Actual photograph of zig-zag shaped heat exchanger

3. Experimental Setup and Procedure

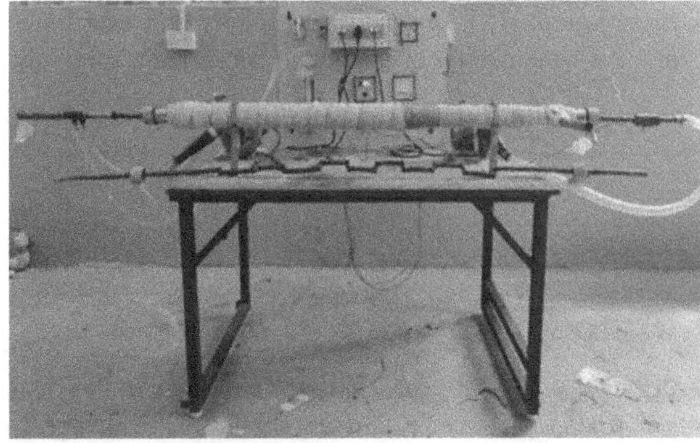

Fig. 59.5 Actual experimental setup of zig-zag shaped heat exchanger

The basic idea behind a heat exchanger is the second rule of thermodynamics, which states that heat moves from one body to another based on temperature differences. A hot body will transfer its heat to a cooler one in its natural form. The cooling medium of a shell and tube heat exchanger is air flowing through tubes inside the shell. On the other hand, the shell structure's cooling medium surrounds these tubes. According to the cooling medium's design, water typically flows through the bottom or rear heading and exits through copper tubes from the top or front header. In a similar manner, air enters the cooling medium through the input nozzle and exits through baffles inside the shell structure. By generating turbulence in the flow and preventing the formation of hot and cold pockets within the medium, these baffles aid in increasing efficiency. Depending on the design and needs, it can also have a cross-flow, concurrent, or counter-concurrent structure. By regulating the flow of the cooling medium, a bypass valve regulates the temperature of the output air medium. Likewise, the cooling medium's pressure is maintained below that of the air medium to be cooled in order to prevent intermixing from leaking. This prevents the air from becoming contaminated even if there is a leak.

4. Results and Discussion

4.1 Overall Thermal-Hydraulic Performance

Heat transfer and flow with various mass flow rates are computed to find mean thermal hydraulic efficiency of straight and zigzag channel heat exchangers. The alteration of cold side Nusselt number NuCav and heat transfer rate Q_{ave} is examined. It is clear that for any of the channel configurations, the mass flow rate showed an increment in the Nusselt number and heat transfer rate Q_{ave}, when Comparing the rates of heat transfer of the channels at a common mass flow rate, that of the Zig-Zag profile is slightly higher than the straight one at the bend angle $\theta = 15°$.

4.2 Flow and Heat Transfer Characteristics

It is clear that the high and low velocity zones at the entry expand and are covered in quite different ways in the zigzag channels. Optimizing the input section's design is essential to successfully lowering the pressure drop. The initial pitch pressure decreases in the zigzag channels can account for up to 17% of the total. Additionally, it can be shown that the zigzag channels' local pressure loss was significantly greater than that of the regular channels for each pitch. The fundamental patterns' fiercer shifts are primarily to responsibility for this.

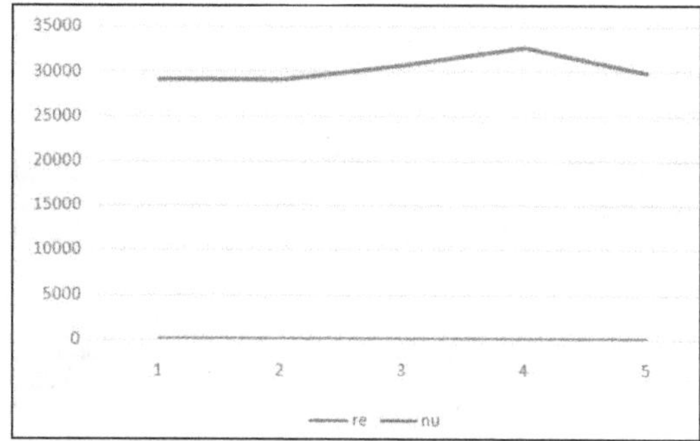

Fig. 59.6 Re vs nusselt number for case: 1

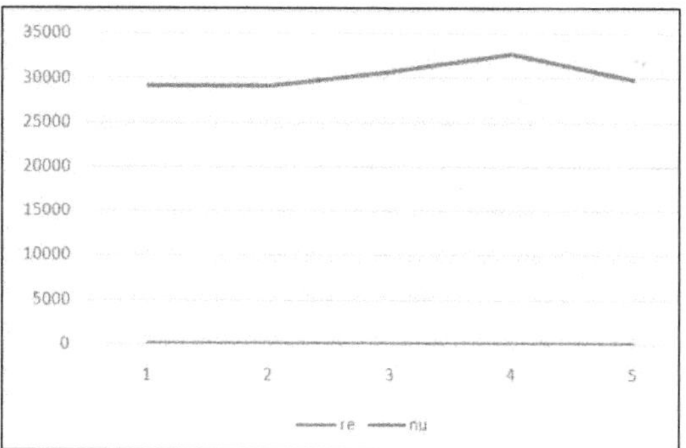

Fig. 59.7 Re vs nusslet number for Case: 2

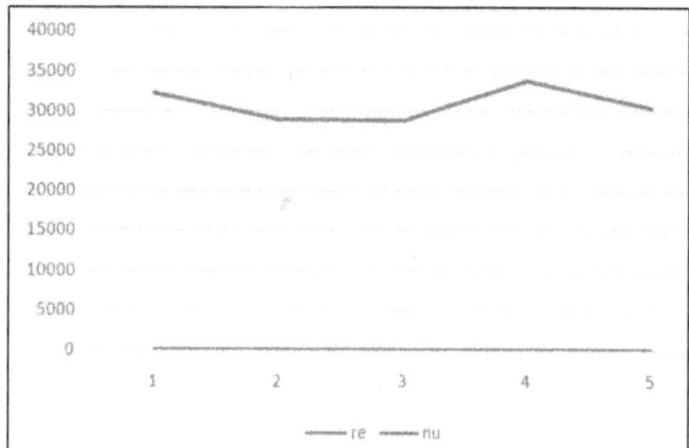

Fig. 59.8 Re vs nusslet number for Case: 3

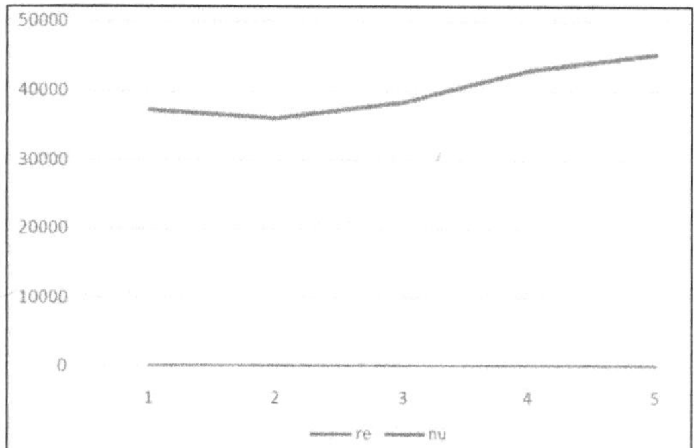

Fig. 59.9 Re vs nusslet number for case: 4

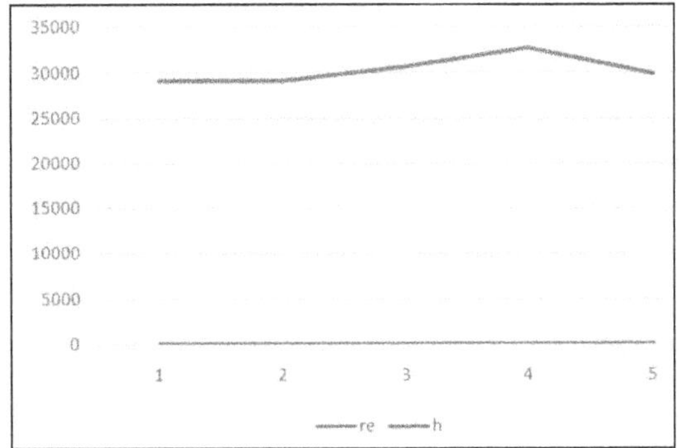

Fig. 59.10 Re vs discharge for case: 1

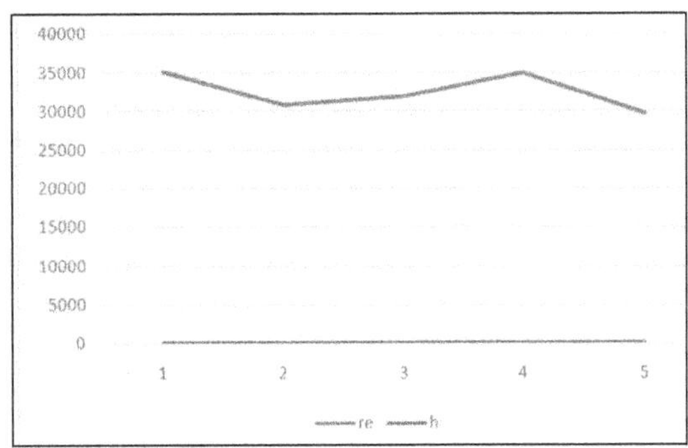

Fig. 59.11 Re vs discharge for case: 2

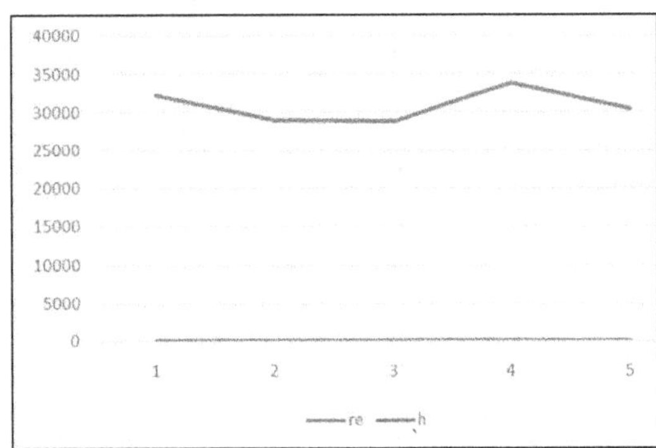

Fig. 59.12 Re vs discharge for case: 3

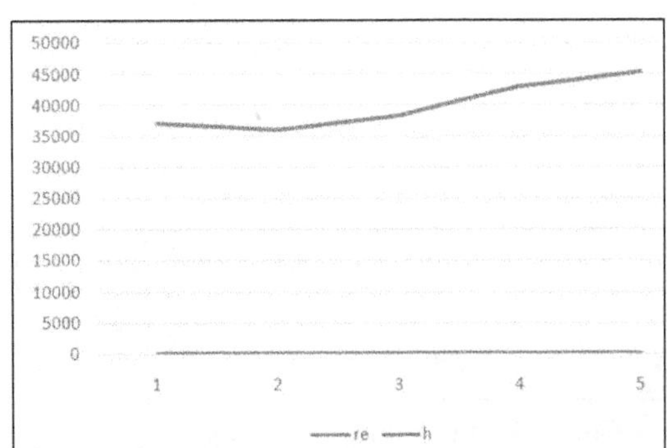

Fig. 59.13 Re vs discharge for case: 4

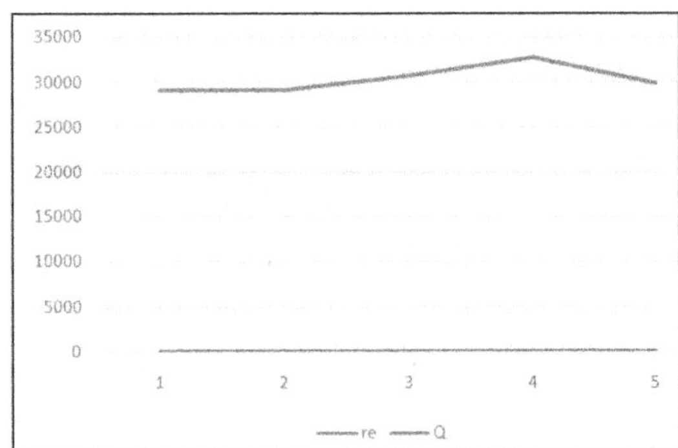

Fig. 59.14 Re vs heat transfer coefficient for case: 1

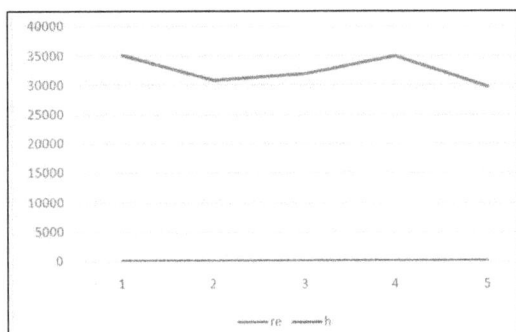

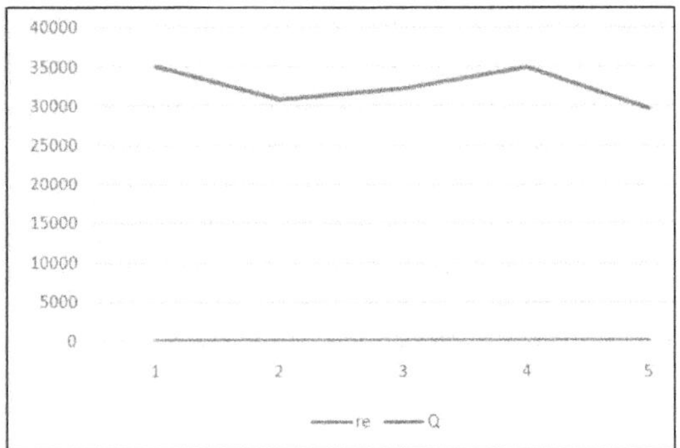

Fig. 59.15 Re vs heat transfer coefficient for case: 2

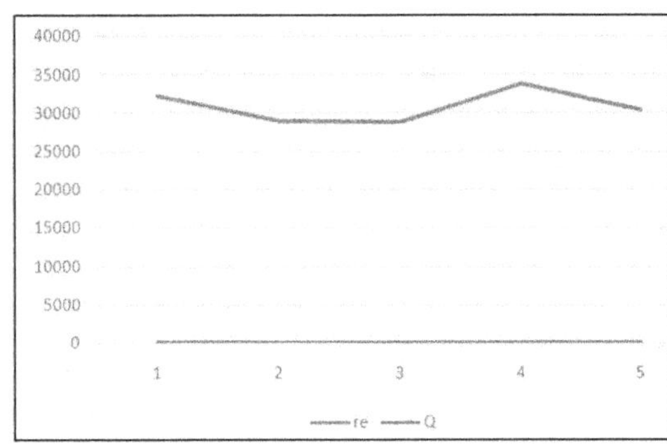

Fig. 59.16 Re vs heat transfer coefficient for case: 3

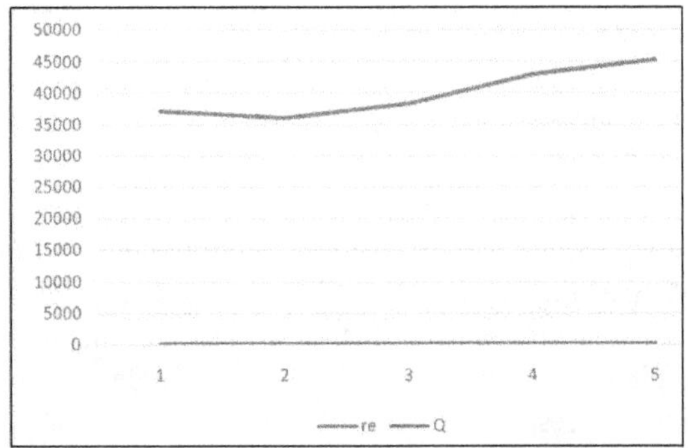

Fig. 59.17 Re vs heat transfer coefficient for case: 4

4.3 Effect of Reynolds Number

Figures show that when the average Reynolds number rises in tandem with the heat flow and average Nusselt number, the gap between Cases 1 and 2's thermal performance gets less. The data compare the bulk and wall temperatures at various Reynolds numbers on the cold side. At low Reynolds numbers, it is discovered that every segment of the Zig-Zag heat exchanger channel may measure a larger bulk fluid temperature differential and wall temperature between the two cases.

4.4 Effect of Inlet Temperature

This section examines how the direction of heat conduction in a zigzag profile system is affected by the input temperature. The characteristics of the PCHE may also be impacted by the intake temperature. The mass flow and averaged Reynolds number are used as the abscissa to compare the heat transfer parameters of the zigzag configuration at various cold side input temperatures. According to the figures, the Reynolds number has decreased at a lower intake temperature for a comparable mass flux, and the average Nusselt number rises as the mass flow and average Reynolds number do. As the cold side input temperature rises at the exact same mass flux, the thermal performance rises as well.

5. Conclusion

Some important conclusions that can be drawn are as follows.

- The density and heat flux patterns are generally non-uniform because of the complex channel flows. The key to comprehending stream/heat transfer in zigzag channels is the periodic changes in flow direction and t.
- Both an increase in the heat transfer area and heat transfer coefficient are responsible for the Zig-Zag channels' superior heat transfer over the straight channel. The distinct flow regimes in the channels may be the cause of the discrepancy between the current study and the literature.
- Zig-Zag channels outperform straight channels in terms of total thermal-hydraulic characteristicsat the fixed bend angle. Having higher bend angles, the benefits of the Zig-Zag channels in reducing pressure loss will be more apparent.

References

1. Lee, S.M. and Kim, K.Y., 2013. Comparative study on performance of a zigzag printed circuit heat exchanger with various channel shapes and configurations. *Heat and Mass Transfer*, *49*(7), pp.1021–1028.
2. Zohir, A.E., Habib, M.A. and Nemitallah, M.A., 2015. Heat transfer characteristics in a double-pipe heat exchanger equipped with coiled circular wires. *Experimental heat transfer*, *28*(6), pp.531–545.https://doi.org/10.1080/08916152.2014.915271
3. Sheikholeslami, M., Gorji-Bandpy, M. and Ganji, D.D., 2016. Experimental study on turbulent flow and heat transfer in an air to water heat exchanger using perforated circular-ring.*Experimental Thermal and Fluid Science*, 70, pp.185-195., https://doi.org/10.1016/j.expthermflusci.2015.09.002
4. Ma, T., Li, L., Xu, X.Y., Chen, Y.T. and Wang, Q.W., 2015. Study on local thermal–hydraulic performance and optimization of zigzag-type printed circuit heat exchanger at high temperature. *Energy Conversion and Management*, 104, pp.55–66.
5. Lee, S. M., & Kim, K. Y. (2013). Comparative study on performance of a zigzag printed circuit heat exchanger with various channel shapes and configurations. *Heat and Mass Transfer*, 49, 1021–1028.
6. Tsuzuki N, Kato Y, Ishiduka T. High performance printed circuit heat exchanger. Appl Thermal Eng, 2007, 27: 1702–1707
7. Kim D E, Kim M H, Cha J E, et al. Numerical investigation on thermal-hydraulic performance of new printed circuit heat exchanger model. *Nuclear Engineering Design*, 2008, 238: 3269–3276
8. Ma, T., Li, L., Xu, X. Y., Chen, Y. T., & Wang, Q. W. (2015). Study on local thermal–hydraulic performance and optimization of zigzag-type printed circuit heat exchanger at high temperature. *Energy Conversion and Management*, 104, 55–66.
9. Talebizadehsardari, P., Mahdi, J. M., Mohammed, H. I., Moghimi, M. A., Eisapour, A. H., &Ghalambaz, M. (2021). Consecutive charging and discharging of a PCM-based plate heat exchanger with zigzag configuration. *Applied Thermal Engineering*, 193, 116970.
10. Wang, P., Wang, X., Huang, Y., Li, C., Peng, Z., & Ding, Y. (2015). Thermal energy charging behaviour of a heat exchange device with a zigzag plate configuration containing multi-phase-change-materials (m-PCMs). *Applied energy*, 142, 328–336.
11. Kim, I. H., Zhang, X., Christensen, R., & Sun, X. (2016). Design study and cost assessment of straight, zigzag, S-shape, and OSF PCHEs for a FLiNaK–SCO2 Secondary *Heat Exchanger in FHRs. Annals of Nuclear Energy*, 94, 129–137.
12. Lee, S. M., & Kim, K. Y. (2014). A parametric study of the thermal-hydraulic performance of a zigzag printed circuit heat exchanger. *Heat transfer engineering*, 35(13), 1192–1200.
13. Chen, M., Sun, X., & Christensen, R. N. (2019). Thermal-hydraulic performance of printed circuit heat exchangers with zigzag flow channels.*International Journal of Heat and Mass Transfer*, 130, 356–367.

14. Yoon, S. J., O'Brien, J., Chen, M., Sabharwall, P., & Sun, X. (2017). Development and validation of Nusselt number and friction factor correlations for laminar flow in semi-circular zigzag channel of printed circuit heat exchanger. *Applied Thermal Engineering,* 123, 1327–1344.
15. de la Torre, R., François, J. L., & Lin, C. X. (2021). Optimization and heat transfer correlations development of zigzag channel printed circuit heat exchangers with helium fluids at high temperature. *International Journal of Thermal Sciences,* 160, 106645.
16. Saeed, M., Berrouk, A. S., Siddiqui, M. S., &Awais, A. A. (2020). Numerical investigation of thermal and hydraulic characteristics of sCO2-water printed circuit heat exchangers with zigzag channels. *Energy Conversion and Management,* 224, 113375.Zilio,
17. G., Moura, M. R., dos Santos, F. J., Possamai, T. S., &Mortean, M. V. V. (2024). Nusselt number analysis of printed circuit heat exchangers with straight and zigzag channels.*International Journal of Heat and Fluid Flow,* 107, 109395.
18. Zilio, G., Moura, M. R., dos Santos, F. J., Possamai, T. S., &Mortean, M. V. V. (2024). Nusselt number analysis of printed circuit heat exchangers with straight and zigzag channels.*International Journal of Heat and Fluid Flow,* 107, 109395.
19. Kim, D. E., Kim, M. H., Cha, J. E., & Kim, S. O. (2008). Numerical investigation on thermal–hydraulic performance of new printed circuit heat exchanger model. *Nuclear Engineering and Design,* 238(12), 3269–3276.
20. M. Sheikholeslami, M. Gorji-Bandpy, D.D. Ganji, Effect of discontinuous helical turbulators on heat transfer characteristics of double pipe water to air heat exchanger, *Energy Conversion and Management,*. 118 (2016) 75–87, https://doi.org/10.1016/j.enconman.2016.03.080

Note: All the figures in this chapter were made by the authors.

Advances in Additive Manufacturing Technologies – Gurusamy Pathinettampadian et al. (eds)

Tesla Coil - Powered Wireless Charging for Drone in Supply Chain Optimization

Suresh A.*,
Pranesh G., Anbu Selvam T.,
Sankarasubramani R., Nagarjun K.
Department of Mechanical Engineering, Chennai Institute of Technology, Chennai, India

◆ **Abstract:** This research is about the Tesla coil wireless charging transmission of the drones in warehouse environments. By using this wireless drone technology work efficiency of the drone is improved and the down-time is reduced by having a wireless charging transmission which charges the drones with the help of electromagnetic radiation. In this technology the drone gets charged continuously and there will be no interruption while working in the warehouse environments thus enabling the drone to do its work. Whereas the traditional charging mechanism relies on direct physical contacts which increases the charging time and there will be no progress until it gets fully charged. The stability of power transfer between the Tesla coil transmitter and the drone's receiver, transmission range and the efficiency of the whole process are the important factors that have been investigated. Moreover, this study describes how the wireless power transfer will change the industrial applications, establishing an advanced development in the autonomous technology.

◆ **Keywords:** UAVs, TESLA Coil, Drone, Supply chain, Optimization

1. Introduction

Unmanned Aerial Vehicles (UAVs) are most dominating tools recent days and sorting out some of the most complicated problems and handover solutions for the data collection, logistics, and surveillance. The drones are potential to improve workflow efficiency which is been major incorporation into variety of processes. Still the UAV have some limitations like short range flight times, limited autonomous, lower battery life and weather problems. The absence of effective and quick charging mechanisms to enables continuous drone use continues to be one of the biggest obstacles. This investigation of cutting-edge drone charging methods intended to get beyond these obstacles is the main goal of this study. It evaluates the usefulness, security, and general efficacy of novel charging techniques. By addressing this important issue, this study offers valuable insights on the development of drone technology with the goal of enhancing the durability and operational capability of UAVs in practical settings.

The traditional working method in the warehouse is heavily depended on the manual labour and traditional logistics systems, leading to increased inefficiencies and the risk of errors and delays Warehouse staff had to manually locate, pick, and transport goods, while inventory management and tracking were primarily done on paper or through basic computer systems.

The introduction of drones in warehouse workflows brought significant improvements in efficiency and accuracy. Equipped with advanced sensors, cameras, and data-processing capabilities, drones revolutionized inventory

*Corresponding author: suresha@citchennai.net

DOI: 10.1201/9781003685906-60

management as shown in Fig. 60.1. They could navigate warehouse aisles quickly, conduct real-time inventory checks, and update databases efficiently, resulting in faster order fulfilment, reduced costs, and increased productivity.

Fig. 60.1 Inventory management using drone

[10] Physically confirming the number of products kept in a warehouse is known as stocktaking, and it is essential for businesses to meet tax and audit obligations as well as repair faulty stock records (Rekik.,2019a). Business operations heavily rely on stock records, such as the inventory levels of particular stock-keeping units (SKUs) recorded in an organization's IT system. In order to automatically initiate replenishment orders, these records are necessary. These orders are usually placed when an SKU's inventory level drops below a pre-established reorder point. (EG). [11] Integrating stock data with the company's online store is crucial for online retailing in order to educate clients about SKU availability (Shabani., 2021). [14] To maintain compliance with accounting laws, stock records are also important for financial statements and need to be periodically verified (e.g., Swartley and Hall, 1988; Brooks and Wilson, 2007, pp. 4-6). In traditional stocktaking, a sizable inventory control crew must physically walk to the necessary area within the warehouse in order to manually count the items or scan their barcodes. [12] This leads to inventory disparities and dangerous tasks like working at high elevations. It is also said to be expensive because of the substantial staffing requirements. (Hardgrave.,2013).

[13] Drone usage in warehouses has grown in recent years (Wawrla , 2019) due to the drones' capacity to avoid obstacles in various warehouse layouts and to hover autonomously without human intervention. Drones offer a streamlined approach to inventory management, enabling rapid and accurate data collection in large warehouse facilities because they can typically scan products more quickly than inventory control employees and send the data straight to the warehouse management system, drones can enhance the stocktaking process. However, one persistent challenge in deploying drones for extended operations within warehouses is the need for frequent recharging, which interrupts workflow and limits their operational range and the operational costs of drones during flight are generally not considered high, the procurement of stocktaking drones and the associated charging equipment can be quite extravagant. Consequently, companies are keen on maximizing the efficiency of stocktaking drones in order to minimize the financial investment required for the necessary equipment. This paper explores a solution to this challenge by integrating Tesla coil technology into warehouse environments for automatic drone charging during flight. [3] Recent study says in wireless power transfer technologies have paved the way for innovative solutions to power drones wirelessly (Wang & Zhang, 2023).

[9] Historically, the Tesla coil has been limited to the scientific demonstrations and the hobbyist applications since its invention by Nikola Tesla in the late 1800s as a high-voltage resonant transformer (Tesla, 1891). However, there is a new possibility for useful applications, such as the integrating wireless power transfer (WPT) with drones for automated, wireless charging, thanks to the developments in contemporary technology and the increased interest in WPT. The Tesla coil serves as a transmitter in this configuration, generating a high-voltage, high-frequency, low-current electricity that creates a powerful magnetic field required for wireless energy transfer [2]. The main idea is based on harnessing electromagnetic induction to transfer high-frequency AC power through the atmospheric medium (Lee & Kim, 2020).

The implementation of the Tesla coil in the warehouse will enable the wireless charging of the drone during the operation. This study investigates the viability and the best strategy, to enhance the wireless charging system of the drones and signaling a change in how logistics are managed in the warehouse.

2. Materials and Methods

2.1 Materials

Copper wires are used for the tesla coils because they are good at conducting electricity. Thick copper wires are used for the primary coils since it doesn't have many turns, but for the secondary coil thinner copper wires are used to wind it since it contains lot of turns.

Also, the toroid – that donut-shaped part placed on top of the secondary coil – is often made of copper too as seen in **Fig. 60.2**. This is because copper not only lets electricity pass through it easily, but it is also great at handling high-

Fig. 60.2 Tesla coil and toroid

frequency energy. It helps to store and spread out the electromagnetic energy efficiently, which makes it ideal for devices like Tesla coils.

2.2 Methods

To guarantee dependable and effective operation, implementing the Tesla coil-based wireless charging for drones in the warehouse environments is the best approach. The layout and the warehouse design are the first step for the examination of Tesla coil installations. The criteria like physical dimensions, determining the most important areas for the operational factors, and installing the Tesla coil in particular locations guarantee the optimal coverage. While designing the in-flight drone charging infrastructure in the indoor environment, [1] Smith et al. (2021) tipped about the critical consideration of the things like signal interference, environmental hindrance, and the layout.

The next stage involves installing the charging station after choosing optimal locations for the Tesla coil in the warehouse settings. In order to have an effective wireless charging system in the warehouse, the items should be properly arranged in warehouse. The things should be properly installed are floor stacking for storage system, placement of Tesla coil to get the frequent supply to the drones without any disconnection. To improve the charging efficiency and minimize the energy losses the procedures like Tesla coil's frequency and the power output should be altered.

The potential implantation of Tesla coil technology for the charging the drones in simulated warehouse have been experimented through the study conducted Brown and Johnson in 2022 [4]. Drone having the sensors were used for testing the important performance factors like overall functionality, system dependability, and the charging efficiency under various operating conditions.

The trials are being conducted to assess the scalability and usefulness of the warehouse settings in real-world simulations. These approach are now analysed on how the technology made an sever impact on workflow effectiveness,

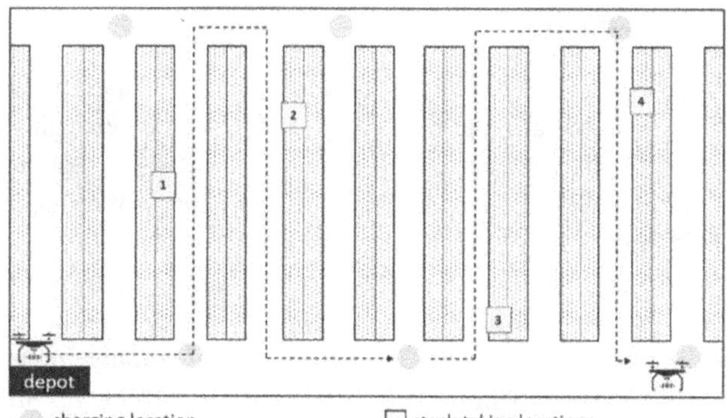

Fig. 60.3 Charging station and stock-taking route [7]

overall productivity of the warehouse environments, and the stocktaking process. In the consideration of safety protocols and the regulatory compliance that are applicable to the drone charging systems in the industrial settings are the addition for the technological concerns. The implementation of this process includes the instructions from the regulatory bodies such as the Occupational Safety and Health Administration (OSHA) and the Federal Aviation Administration (FAA) [6].

This helps to reduce the safety hazards of the drone operations in warehouse settings and will assure a regulatory compliance. By considering these factors into knowledge, the approach for the drone charging with Tesla coil technology provides best way to deal with operational environment and stocktaking associated duties. This strategy improves the efficiency of the warehouse and overall production by providing a suitable scalability and working framework.

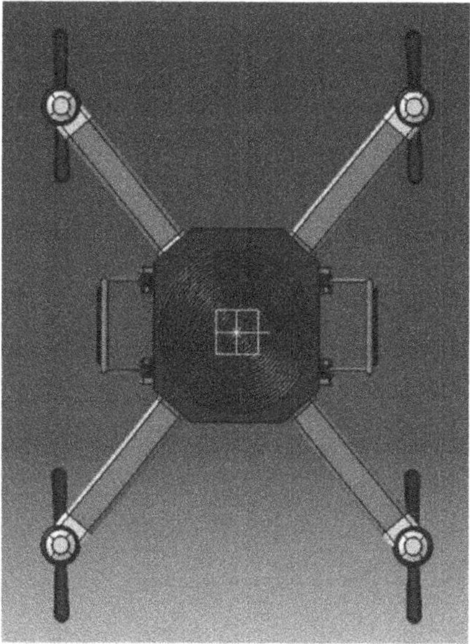

Fig. 60.4 Drone with receiver coil

3. WORKING

The Tesla coil is creating an impact for doing the operations like stocktaking in the warehouse, in flight drone charging through the wireless charging, where it is used in warehouse logistics work by making it possible for drones. This research is about transmitting the high-frequency alternating current (AC) through the atmosphere from the Tesla coil to the drones by using the electromagnetic induction. In this approach the operational autonomy, more productivity, and improved safety are the advantages in warehouse settings.

If the drones with compatible receivers enters into the coverage area, the electromagnetic fields produced by the Tesla coils induce a current through their onboard charging systems. This particularly eliminates the downtime and charges the drone in flight, which reduces the drone to work efficiently by not lowering the battery levels and no need of manual charging. Lee and Kim (2020) [2] have discussed about the basics of electromagnetic induction and the essentials of wireless power transmission can be implemented to this technology.

To optimize the overall efficiency and the charging range it is difficult for the successful installation of Tesla-coil based charging systems in warehouse environment. [3] Wang and Zhang (2023) have emphasized that the frequency and the power output from the Tesla coil must be adjusted in order to charge the drones at variable distances. But it is possible to obtain the uniform and effective charging performance, by well calibrating the coils according to the warehouse plan with different zones.

It not only reduces energy losses, but also increase the efficiency charging range and allow drone to continuously charging through the atmosphere. In order to implement the activities like the drone navigation and control system with the tesla coil systems, which will allow easy and automated charging stations for the warehouse operations. According to [4] Johnson and Brown (2022), in order to allow drones charge effectively without the human assistance, by using the Tesla coil it is imperative that strong communication protocols and the software algorithms be developed (Fig. 60.4).

By implementing this integration, drones can make full use of wireless charging systems based on the Tesla coils to continue operating continuously while doing activities like stocktaking. Extensive testing in both simulations and the real warehouse environments has confirmed the effectiveness and versatility of this approach, identifying its potential as a workable alternative for improving the warehouse automation.

In priority to deploy Tesla coil-based charging infrastructure, in the operational warehouses and assess its effects of the workflow efficiency and overall productivity, [5] Garcia et al. (2023) carried out field experiments in the cooperation with some industrial partners. These investigations open the door for broad use of the consideration of this method, in warehouse management procedures by confirming its viability and efficacy. Along with technical considerations.

Fig. 60.5 Drones and tesla coil in a warehouse

When Tesla coil-based charging systems are set up in the industrial settings, safety regulations and rules are the main important factors to take into consideration. Robinson (2021) [6] gives important to the need of adhering rules set out by bodies like the Occupational Safety and Health Administration (OSHA) and the Federal Aviation Administration (FAA). By following these instructions assures that the drone operations and the charging infrastructure run efficiently and work stable in warehouse settings. Organizations can also reduce the operational risks and can able to protect both the assets and labours by fulfilling their regulatory requirements.

The most important feature in the warehouse modernization environment, is the use of Tesla coil technology for in-flight drone charging operations. This approach offers a dependable and very effective way for charging the drones through wireless medium by utilizing the electromagnetic induction, improving the frequency range and charging efficiency, by integrating this in drone system and following the regulatory compliance [5].

The first step in the wireless drone charging is to supply the direct current (DC) to the Tesla coil system in the warehouse applications. The number of coil's winding and winding pattern increases the voltage, frequently to level of about 10,000 volts. The main principle of the electromagnetic radiation is that variable magnetic field in the primary coil will increase the voltage in the secondary coil, is the main thing for the voltage amplification.

Short-range wireless power transmission is made possible by the electromagnetic field produced by the Tesla coil once it reaches the required high voltage. The drone receives electrical energy from the coil through this field, which is acting as a conduit. A receiver unit on the drone setup is made to match the Tesla coil's frequency; this equipment usually consists of a resonant coil. The drone may receive electricity during the charging process with minimal loss because to its resonance, which assures the effective energy transmission.

An alternating current (AC) is created inside a drone's reception coil, as a result of these fluctuations between the electromagnetic field from the Tesla coil and the receiver coil (shown in Fig. 60.4). The receiver circuit includes the step-down transformer to a safely use this transferred energy, by lowering the high voltage to a more controllable level appropriate for drone battery charge. This voltage decrease is necessary to safeguard the drone's electrical parts, guarantees a very steady and secure the charging procedure and lowering the possibility of over-voltage damage.

In order to regulate the incoming power to meet the unique needs of the drone's battery system, the step-down transformer is essential. This guarantees the drone's and its operator's safety, dependability, and efficiency during the charging procedure. Physical connections and docking stations are no longer necessary thanks to this technology, which makes wireless and contact-free charging possible. These developments create new opportunities, especially for drone operations in difficult-to-reach or isolated locations where conventional charging techniques would not be feasible.

4. Calculation

The usage of the tesla coil, to power the drones in the warehouse is to mainly ensure the charging of the drones while in flight. But the tesla coil also has some limitation that how long it can pass the current to the drones to charge it. The calculation for how much distance a tesla coil of these specification can pass the current are given below,

Voltage (V) = 10,000 volts (10 kV)

Power output (P) = 1000 watts (1kW)

Electric field strength (E) near the coil = 1000 volts per meter

Efficiency of the receiving equipment = 50%

$$E = \frac{k.C.V}{r^2}$$

$$r = \sqrt{\frac{k.C.V}{E}}$$

Where:

k is Coulomb's constant: 8.9875×10^9 Nm^2/C^2

C is the capacitance of the coil

V is the voltage (volts)

E is the electric field intensity (volts per meter)

r is the distance from the coil (meters)

$$r = \sqrt{\frac{\left(8.9875 \times 10^9\right) \times \left(1 \times 10^{-9}\right) \times (10{,}000)}{1000}}$$

$$r = \sqrt{\frac{8.9875 \times 10^4}{1000}}$$

$$r = \sqrt{89.875}$$

$$r \approx 9.48\ meters$$

5. Conclusion

The integration of Tesla coil technology for wireless drone charging in warehouse environments introduces a transformative solution to enhance operational efficiency and productivity. Through analytical calculation if the drone is in an approximate range of 9.48 meters the drone can charge with 50% efficiency, through this study it is hypothetically mentioned that the feasibility and effectiveness of using Tesla coils to power drones autonomously during the functioning of the drone in logistics. Through, the calculation for the tesla coil transmitting distance have been calculated which shows that the range of 9.48 meters, which can be used for charging the drone wirelessly.

Tesla coil technology is not just limited to the warehouse use, it also has the potential to be useful in many other industries like the disaster relief, construction, and farming. Since it's flexible and can be scaled, it is a great fit in situations where drones need to work without constant charging support or downtime. This wireless charging setup allows drones to fly and work longer on their own, without needing to stop for manual charging or use bulky charging stations.

As most of the industries move towards the automation and eco-friendly practices, Tesla coil tech fits right in. By cutting down the usage of fossil fuel and also by avoiding traditional charging systems that often requires prolonged downtime, this method supports sustainability and increased air time. It helps save energy and supports global efforts to reduce climate change by promoting the use of clean, renewable power.

In warehouses, the Tesla coils make it easier to keep drones running constantly, especially for the tasks like checking inventory and scanning boxes. Since it works based on electromagnetic induction and has good range and efficiency, it helps to reduce the downtime and boosts the overall productivity.

This study shows how wireless power can transform on how drones are used in industries. It offers a smart, cost-effective way to deal with charging issues and improve how independently drones can operate. In the long run, Tesla coil-based wireless charging looks like a game-changer — pushing us closer to an efficient, autonomous, and greener future.

6. Result

This research focuses on improving drone charging in warehouses using Tesla coil-based wireless power. For the drones to work efficiently and to reduce the delays or costs, they need fast and reliable charging solutions. Regular charging methods usually involve cables, which limits the drone's movement and makes warehouse work harder.

Tesla coils provide a smart alternative — they can recharge drones wirelessly over longer distances, without any physical contact. This study looks at how effective this technology is, using principles like electromagnetic induction and resonant coupling.

With a high-voltage setup of 10,000 volts (10 kV) and an electric field strength of 1,000 V/m, Tesla coils were able to achieve a wireless charging range of around 9.48 meters. The system showed about 50% energy transfer efficiency, which is quite promising. This proves that it could actually work well for keeping drones powered during daily warehouse operations — without needing to land and plug in every time.

References

1. Smith, C., & Brown, D. (2021). "Integration of Drones in Warehouse Operations: Challenges and Opportunities." *International Journal of Production Research*, 59(7), 2098–2115.
2. Lee, S., & Kim, J. (2020). "Applications of Tesla Coil Technology in Modern Engineering: A Comprehensive Review." *Journal of Applied Physics*, 128(15), 154301.
3. Wang, C., & Zhang, D. (2023). "Wireless Power Transfer Technologies: A Review." *IEEE Transactions on Power Electronics*, 38(4), 1203–1221.
4. Johnson, E., & Brown, M. (2022). "Experimental Evaluation of Automatic Drone Charging Using Tesla Coil Technology." *IEEE Transactions on Industrial Electronics*, 69(9), 7654–7665.
5. Garcia, A., et al. (2023). "Field Deployment of Tesla Coil-Based Charging Infrastructure for Drones in Warehouse Environments." *Journal of Warehouse Management*, 15(3), 213–228.
6. Robinson, L. (2021). "Regulatory Compliance and Safety Considerations for Drone Operations in Industrial Environments." *Safety Science*, 110, 102387.
7. Panupong Vichitkunakorn, Simon Emde, (2024) "Locating charging stations and routing drones for efficient automated stocktaking" European Journal of Operational Research, EOR 18924
8. Pusparini Dewi Abd Aziz, 2 Ahmad Lukhfhy Abd Razak, (2016). "A Study on Wireless Power Transfer Using Tesla Coil Technique", International Conference on Sustainable Energy Engineering and Application,
9. Tesla, N. (1891). "Experiments with Alternate Currents of High Potential and High Frequency." *American Institute of Electrical Engineers Transactions*, 9(1), 85–108.
10. Rekik, Y., Syntetos, A.A., Glock, C.H., 2019a. Inventory Inaccuracy in Retailing: Does it Matter. Technical Report. ECR Community Shrink & OSA Group
11. Shabani, A., Maroti, G., de Leeuw, S., Dullaert, W., 2021. Inventory record inaccuracy and store level performance. International Journal of Production Economics 235, 1–16.
12. Hardgrave, B., Aloysius, J., Goyal, S., 2013. RFID-enabled visibility and retail inventory record inaccuracy: experiments in the field. Production and Operations Management 22, 843–856.
13. Wawrla, L., Maghazei, O., Netland, T., 2019. Applications of drones in warehouse operations. Technical Report. Whitepaper. ETH Zurich, D-MTEC.
14. Swartley, J.A., Hall, J.A., 1988. Inventory auditing: A manufacturing perspective. Production and Inventory Management Journal 29, 20–22
15. Syed Agha Hassnain Mohsan 1 · Nawaf Qasem Hamood Othman, (2023) "Unmanned aerial vehicles (UAVs): practical aspects, applications, open challenges, security issues, and future trends".

Note: All the figures in this chapter were made by the authors.

Advances in Additive Manufacturing Technologies – Gurusamy Pathinettampadian et al. (eds)
© 2026 Taylor & Francis Group, London, ISBN 978-1-041-16687-0

A Comprehensive Study on Process Optimization and Joint Properties of PTIG Welded AA5754 Aluminium Alloy

Kannan G.K.[1]
Assistant Professor, Department of Mechanical Engineering,
Chennai Institute of Technology,
Chennai, India

Farhan Ahamed S.[2]
Student, Department of Mechanical Engineering,
Chennai Institute of Technology,
Chennai, India

Sridhar S.[3]
Department of Mechanical Engineering,
PSNA College of Engineering and Technology,
Dindugal, India

S. Senthil Kumar[4]
Professor, Department of Mechanical Engineering,
R.M.K. College of Engineering and Technology,
India

Harish V.[5]
Student, Department of Mechanical Engineering,
Chennai Institute of Technology,
Chennai, India

Satheesh S.S.[6]
Assistant Manager, Renault Nissan Technology and Business Centre,
Chengalpattu, India

◆ **Abstract:** This study explores the optimization of Pulsed Tungsten Inert Gas (PTIG) welding for AA5754 aluminum alloy, a material favored for its corrosion resistance and applicability in automotive and marine sectors. Employing the Response Surface Methodology (RSM) and Central Composite Design (CCD), the research identifies optimal welding parameters, including peak current, pulse frequency, and welding speed, to enhance mechanical and microstructural properties of the welded joints. Key findings reveal that tensile strength of up to 212 MPa and hardness values reaching 70.7 HV in the fusion zone can be achieved under optimized conditions. Microstructural analysis highlights a reduction in Mg_2Si precipitates in the heat-affected zone (HAZ), contributing to localized strength variability. The fusion zone exhibited fine-grain dendritic structures, improving ductility and mechanical performance. This investigation underscores the critical interplay of process parameters in achieving superior weld quality and offers significant insights for industrial applications of PTIG welding in AA5754 alloy.

◆ **Keywords:** PTIG welding, AA5754 aluminum alloy, Process optimization, Microstructure analysis, Mechanical properties

[1]gkkannan.85@gmail.com, [2]ahamedfarhan979@gmail.com, [3]sri_2855@yahoo.co.in, [4]senthilkumar@rmkcet.ac.in, [5]harishv2907@gmail.com, [6]satheesh.009@gmail.com

DOI: 10.1201/9781003685906-61

1. INTRODUCTION

Aluminum and its alloys are being widely used in engineering purposes because they are lightweight but have a high strength plus excellent resistance to corrosion. However, the traditional methods of arc welding are cumbersome when applied to these materials. PTIG welding, has emerged lately as a potential alternative. One of the very first studies examining the suitability of the very efficient PTIG welding technique when applied on AA5754 wrought aluminum alloy is examined here. This research, by utilizing the RSM, attempts to optimize PTIG WPP in order to improve the mechanical performance of the alloy.

It is feasible to weld on AA5754 alloy with inert Pulsed Tungsten and specific Al-Mg filler rods like ER5356 and ER5183. Hardness measurements were carried out through Vickers microhardness testing, while tensile properties were carried out along the transverse axis of the weld with the help of a UTM. PTIG is widely used for welding aluminum alloys; the results are very dependent on the filler rods used and the conditions applied. Results show that weld joints can have tensile strengths up to 169 MPa and hardness levels of 89 HV. Moreover, aluminum is an eco-friendly material because it can be recycled infinitely, which has led to its increased popularity in various design and manufacturing applications instead of steel.

With respect to process factors, tools like RSM are good for analysis and optimization conditions for desirable outcomes. Here, the RSM applied in a CCD of RSM is used to predict the optimum welding conditions on AA5754. Thus, relationships between independent factors and response were identified for effective optimization within any engineering application.

The relevance of PTIG and corresponding processes for aluminum alloys in attaining good welding results. For example, Zhenmin Wang [10] focused upon the improved arc stability, spatter reduction in case of high-frequency pulsed-arc welding. Anhua Liu [11] reviewed the -GMAW for the AA5754 alloy, with a focus on optimized parameters and performance improvement. Additionally, Rajesh Manti [12] examined Al-Mg-Si alloy welds obtained through the PTIG process and defined the relationship between welding conditions and mechanical properties. According to Tamil Kumaran G. [14] and others further explored the mechanical and metallurgical characteristics of welded joints, identifying strategies to optimize weld quality.

This paper focuses on investigating and optimizing the PTIG welding process for AA5754, a material widely used in automotive and marine industries due to its excellent corrosion resistance and weld ability. This paper analyzes the effect of WPP such as current, voltage, WS, and PF on mechanical properties including strength, hardness, and ductility. Problems related to porosity, cracking, and HAZ characteristics are addressed with proposed solutions for alleviation. This research shall provide a comprehensive analysis to contribute to the development of optimized PTIG welding strategies in order to rise the performance and durability of weldments in practical applications.

2. EXPERIMENTAL METHODOLOGY

From the material shop, a 5mm thick aluminum plate was procured and cut into pieces with 100mm x 55mm in dimension using a hacksaw machine. Using a milling machine, the plate surfaces and joining edges were flattened and prepared for a precise finish. Before welding, the edges were filed smooth and cleaned with acetone. The two plates were securely clamped on the welding table, and weld joints were created using a Pulsed Tungsten Inert Gas (PTIG) welding machine with ER5356 filler material.

Table 61.1 Chemical composition of AA5754

BM	Mg	Mn	Fe	Si	Cu	Cr	Zn	Al
AA5754	2.8	0.4	0.3	0.3	0.1	0.2	0.2	95.7

Table 61.2 Mechanical properties of AA5754

Material	YS(MPa)	UTS(MPa)	El (%)	Hardness (Hvl)
AA5754	185	245	15	75

To prepare metallurgical and mechanical samples, a WEDM machine was used. Tensile test specimens were fabricated according to the ASTM E8 standards and tested using a) (Make, Model. Microstructural examination samples were also machined through the WEDM and mounted in a hot mounting press using Bakelite powder. Upon mounting, hand-polished these samples with silicon carbide abrasion sheets with different grits, which varied between 80 to 400 grit. These hand polished samples were further machine polished using finer grit abrasion sheets ranging between 600 to 2000. The etching process using Keller's reagent made the microstructure visible of the polished samples. The metallurgical analysis was carried out using a SEM that gave detailed information on the grain size, grain shape, and distribution of precipitates. The microhardness testing was performed on the polished samples at different zones with an indentation spacing of 0.5 mm. A Vickers microhardness tester (Make, Model) was used for this testing. A 1 kg load was applied with a dwell time of 20 seconds during the microhardness tests.

The CCD approach within the framework of RSM was adopted for the optimization of welding process setting. From the literature review conducted on the FSW of various aluminum alloys, three critical WPPs relevant to this investigation were TRS , TTS, and AL. These parameters have acceptability ranges determined from literature sources, with the lower bound given a coded value of -α (-1.682), and the upper bound assigned a coded value of +α (1.682). Midpoint of the range has been assigned a coded value of 0. Calculation of coded levels +1 and -1 are through standard CCD equations. Table 61.3 presents the actual values of each welding parameter corresponding to these coded levels

Table 61.3 Limits of welding process parameter

WPP	Unit	Levels				
		-1.682	-1	0	1	+1.682
PC	A	160	168	180	192	200
PF	Hz	2	3.2	5	6.8	8
WS	mm/min	80	88	100	112	120

3. Results and Discussion

3.1 Optimization of Welding Parameters

To enhance the welding process parameters, the CCD methodology was utilized within the framework of RSM. The selected welding parameters for this study include PC, PF, WS, and AL, as these factors significantly impact the mechanical and metallurgical properties of the weldments [19]. The DOE table outlines the structured experimental matrix generated using the CCD approach. Table 61.4 summarizes the various parameter level combinations used in the optimization process, along with their corresponding experimental responses.

The equation arising from the actual variables gives the response at specific levels of the factors expressed in their original units. In doing so, it cannot be used to make any direct comparison of the impact of each factor since its coefficients are scaled by units of the factors and since the constant term is not centered within the design space.

$$UTS = 192.873 + 5.634 * TRS + 9.186 * TTS - 3.355 * AL - 2.375(TRS * TTS) + 4.625(TRS * AL) + 7.375(TTS * AL) - 12.048(TRS^2) - 16.114(TTS^2) - 8.336(AL^2)$$

The model makes use of standardized factor coding and analyzes the effects by using a Type III (Partial) sum of squares. A very high F-value of 325.48 shows that the model is significant, with a probability of only 0.01% that this result could be due to random noise. Factors with

Table 61.4 Design of experiments framed using CCD

Std. order	Run order	Coded Values			UTS (MPa)
		P	S	F	
1	18	-1	-1	-1	153
2	14	1	-1	-1	160
3	15	-1	1	-1	162
4	7	1	1	-1	161
5	6	-1	-1	1	122
6	16	1	-1	1	147
7	1	-1	1	1	160
8	19	1	1	1	176
9	4	-1.682	0	0	151
10	5	1.682	0	0	168
11	11	0	-1.682	0	132
12	3	0	1.682	0	162
13	8	0	0	-1.682	174
14	17	0	0	1.682	164
15	12	0	0	0	191
16	20	0	0	0	195
17	9	0	0	0	193
18	2	0	0	0	192
19	13	0	0	0	189
20	10	0	0	0	191

Table 61.5 ANOVA table for responses of 'p' values

Source	Sum of Squares	Df	Mean Square	F-value	p-value	
Model	8214.91	9	912.77	325.48	< 0.0001	significant
A-TRS	433.62	1	433.62	154.63	< 0.0001	
B-TTS	1152.44	1	1152.44	410.95	< 0.0001	
C-AL	153.72	1	153.72	54.81	< 0.0001	
AB	45.13	1	45.13	16.09	0.0025	
AC	171.13	1	171.13	61.02	< 0.0001	
BC	435.13	1	435.13	155.16	< 0.0001	
A^2	2092.13	1	2092.13	746.03	< 0.0001	
B^2	3742.35	1	3742.35	1334.48	< 0.0001	
C^2	1001.54	1	1001.54	357.14	< 0.0001	
Residual	28.04	10	2.80			
Lack of Fit	7.21	5	1.44	0.3461	0.8655	not significant
Pure Error	20.83	5	4.17			
Cor Total	8242.95	19				
Std. Dev.	1.67	**R^2**		0.9966		
Mean	167.95	**Adjusted R^2**		0.9935		
C.V. %	0.9971	**Predicted R^2**		0.9897		
		Adeq Precision		58.3134		

P-values below 0.0500 are A, B, C, AB, AC, BC, A^2, B^2, and C^2, which means they are significant. Conversely, terms with P-values greater than 0.1000 are considered non-significant and may be dropped unless necessary to retain model hierarchy. The Lack of Fit F-value of 0.35 indicates that the model fits the data quite well, with an 86.55% chance that any observed lack of fit is due to noise, thus showing the precise of the model.

The predicted R^2 value (0.9897) is closely aligned with the Adjusted R^2 value (0.9935), with a difference of less than 0.2, affirming the model's reliability. Additionally, the Adequate Precision ratio of 58.313, far exceeding the minimum threshold of 4, demonstrates a strong signal-to-noise ratio, confirming the model's suitability for exploring the design space. The perturbation plot in Fig. 61.1(a) captures the effect of each input variable on UTS by keeping the remaining two at their central value: that how different variables impact the result and what factors contribute significantly to the modification of the tensile strength. Similarly, contour plots (Figs. 61.1(b-d)) show the interaction between input parameters and their combined effects on UTS when the third parameter is fixed at its median value. These plots are important for understanding parameter interactions and their overall influence on weld quality.

Heat input, a key factor in determining weld characteristics, is significantly affected by PC and PF. Higher PC increases heat input, resulting in a wider HAZ and lower tensile strength. Lower PF reduces heat input by shortening the time of heat absorption, which may result in incomplete fusion or weak bonding. The interaction between PC and PF also affects the microstructure of the weld zone. High PC with low PF increases coarse grains, which adversely affect the mechanical strength, whereas low PC with high

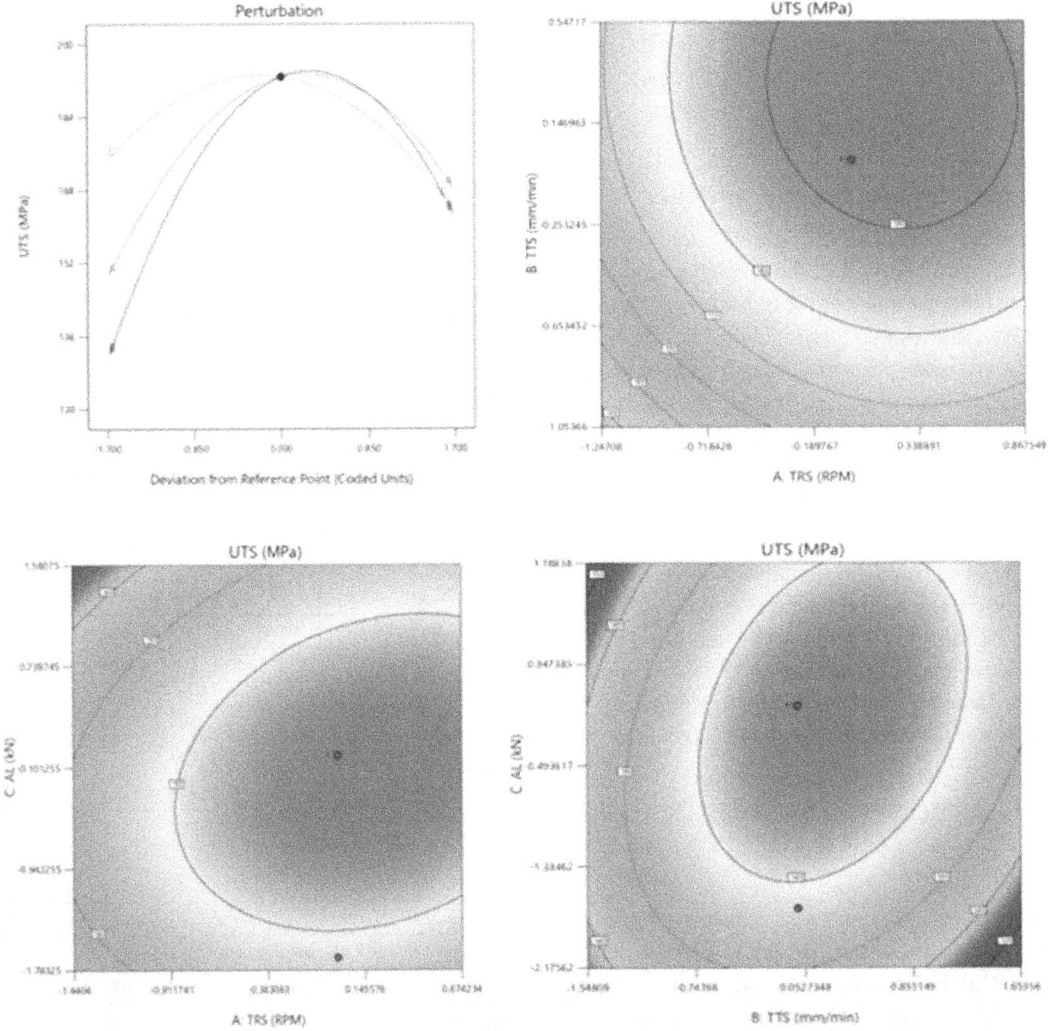

Fig. 61.1 [a] Perturbation plot [b] Interaction plot AB [c] Interaction plot AC [c] Interaction plot BC

PF may reduce the joint quality and tensile strength. Contour plots are critical to the identification of optimal ranges for the parameters. Elliptical patterns in Figs. 61.1(b), 61.1(c), and 61.1(d) suggest high levels of interaction among parameters such as AB, AC, and BC. Thereby, the impact of one parameter is very sensitive to the levels of the other parameters. Optimal performance requires a balance of the parameters. The synergistic interaction between PC and PF is specifically critical for controlling heat input, achieving fine-grained microstructures, and improving mechanical properties. This study emphasizes the need to optimize welding parameters for both weld quality and mechanical performance.

3.2 Microstructure Analysis

A comparison of the HAZ and BM of AA5754 shows that both the size and number of intermetallic particles significantly decrease, and in particular, a significant reduction was observed in the number. The study of BMZ and HAZ revealed that most strengthening precipitates, especially Mg_2Si, are dissolved during welding. Only some fine precipitates remained in the HAZ. This loss is mainly due to the evaporation of aluminum alloys during pulsed PTIG welding. Visual observations from the welding process, supplemented by EDAX analysis, establish the dissolution of Mg_2Si precipitates. Microstructural analysis of AA5754 during welding brings into focus the behavior of intermetallic particles such as Al_3Mg_2 and Mg_2Si. The BMZ shows higher strength because of the formation of Al_3Mg_2 precipitates and the dense network of Mg_2Si. In contrast, the HAZ reveals a strength reduction, caused by a diminution of intermetallic particles in size and quantity and loss of Mg_2Si.

The FZ is marked by thin, elongated intermetallic particles, likely formed as a result of aluminum alloy evaporation during PTIG welding. EDAX analysis of the weld metal reveals the marked depletion of magnesium, further indicating the absence of Mg_2Si precipitates. These observations give qualitative knowledge on the distribution, dissolution, and behavior of the intermetallic particles while AA5754 is welded, emphasizing their impact on the alloy's microstructure and metallurgical properties. The microstructural images of the welded samples, consisting of the weld metal, HAZ, and base metal, were examined for comparison. Lower magnifications of the images provided a general overview of the weld metal and base metal while higher magnifications were more precise in the differentiation of individual regions. In all samples, the weld metals were characterized by a dendritic structure due to the low cooling rates in the weld zone. Higher resolidification rates are represented by the areas with dendritic structures.

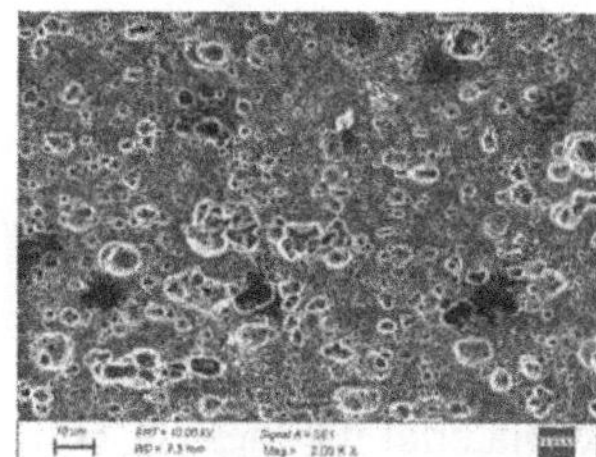
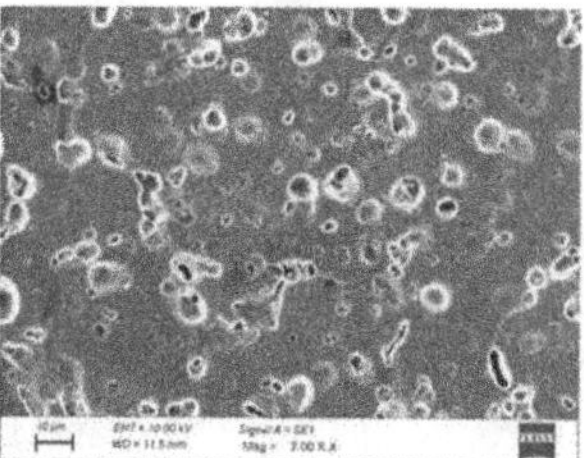

Fig. 61.2 Microstructural features of the BM and the WZ, as captured by a SEM at 2000X magnification

Grain size variations were one of the distinctive characteristics of the weld metals. Such variations were strongly dependent upon the duration of the welding process and heat exposure time. The weld cycle greatly affected the weld metal as well as the HAZ. The HAZ's width was highly influenced by the welding parameters that are current and speed. Crossing the weld metal and the direction from the melt boundary reflected increased grain size and orientation and thus was highly affected by the welding parameters. Tendencies similar to these have also been reported by Zhou et al. [20], whilst Çömez and Durmuş [21] stated that columnar grains in the partially melted zone appeared to grow epitaxially. Solid solution and precipitation strengthening are identified as prime contributors to the durability of AA5083, through intermetallic particles in the form of Al_3Mg_2, $Al_6(Fe, Mn)$ and Mg_2Si playing significant roles. In base metal zone of AA5083, these inter metallic particles dominate, with only isolated α-Al matrix region being evident.

In contrast, the heat-treatable alloy AA5754 relies mainly on precipitate strengthening, with Mg_2Si as the major strengthening precipitate. The HAZ of AA5754 contains fewer and smaller intermetallic particles than the base metal. The SEM images of the weld area show thin, needle-like intermetallic particles that seem to have agglomerated during solidification, as some failed to dissolve. The study of AA5754's microstructure highlights intermetallic particles, notably Al_3Mg_2 and Mg_2Si as being crucial for the final characteristics of the alloy. Although the base metal strengthens because of these precipitates, which are highly numerous, the HAZ demonstrates a significant reduction in intermetallic particles in strength, size, and quantity in addition to some loss in Mg_2Si precipitates while welding.

3.3 Hardness Analysis

Results for microhardness tests conducted for joints S1 to S7 revealed different trends across materials. AA6013 had the maximum values of microhardness for base material followed by HAZ of AA6013 and then weld metal, while AA5754 has shown minimum microhardness in HAZ. This reduction in microhardness in the HAZ of both alloys is attributed to heat treatment effects induced during welding, likely due to recrystallization or grain growth caused by the welding heat [22].

Lakshminarayanan et al. [23] studied the welding of AA6061 aluminum alloy by the GMAW and GTAW techniques. They reported that weld metal hardness was lower than both HAZ and base material hardness in both methods. They ascribed this to the relatively low hardness of the filler material and the heat input during welding. Hardness in the HAZ in AA5754 was observed at 45 to 50 Hv0.5. There was softness in the HAZ due to magnesium loss along with the dissolution of secondary particles at elevated temperatures, as elevated temperatures reduce hardness. Moreover, the heat treatment also dissolved Mg_2Si precipitates, which is considered the main strengthening agent of AA5754. However, the hardness decrease in AA5754 was only slightly affected by intermetallic particles like Al_3Mg_2, $Al_6(Fe, Mn)$, or $Al_6(Fe, Mn)Cr$. The HAZ showed the lowest hardness among all weldment regions due to intense heat that resulted in coarse grains at the weldment apex. AA5754 is not a heat-treatable alloy; thus, its hardness comes from solid solution and precipitate strengthening. Hence, there was not a significant reduction in hardness in the HAZ comparatively with alloys containing intermetallic components like $Al_6(Fe, Mn)$.

The FZ hardness was the second softest, ranging from 50 to 70 Hv0.5. The elevated temperatures present in the FZ allowed most of the strengthening precipitates to break down. Furthermore, the solidification included recrystallization processes, which led to grain sizes smaller than those produced in the BMZ or HAZ and re-formed particles within the FZ. Therefore, the size of grains in the FZ is between the size obtained in the BMZ and the HAZ, thereby softer than the BMZ.

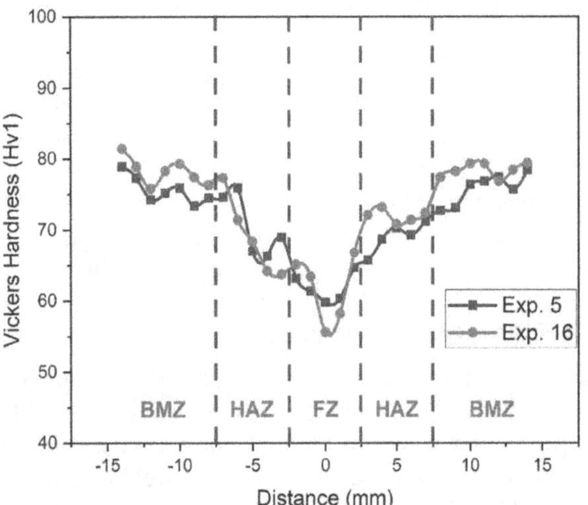

Fig. 61.3 Hardness along different zones of the weldment

Under the same condition, a hardness value of 53 Hv0.5 was obtained and with the blue line in the graph the peak hardness is located at 0.5 mm in the depth of AA5754. The maximum weld joint strength that had been attained was 212 MPa using base current at 90 A, PC at 130 A, and 2.5 mm/s welding speed. Experimental findings also revealed that the maximum yield strength was achieved with a peak current of 140 A and a BC of 100 A. Such a setting decreased the heat input by around 17.5%, thus increasing the cooling rate. The increased cooling rate enhanced grain refinement and hence weld mechanical properties. The average hardness in the HAZ was also 63.9 HV but less than the 70.7 HV recorded in the weld region. The HAZ grains were also subjected to severe softening and coarsening due to the weld thermal cycle.

4. Conclusion

The PTIG welding on AA5754 aluminum alloy was successfully performed with high-quality, defect-free welds. Based on the findings, the conclusions are as follows:

1. Welding Process Parameter Optimization: The WPP was optimized utilizing CCD in the framework of RSM. Optimum conditions were determined as peak current at 180 A, pulse frequency at 5 Hz, and weld speed at 100 mm/min.
2. Mechanical Performance: The optimized parameters have yielded a maximum UTS of 212 MPa and hardness of 70.7 HV in the fusion zone, thus demonstrating the significance of accurate parameter selection in attaining excellent mechanical properties.
3. Characteristics: The analysis through microstructure revealed the development of fine-grained dendritic structures in the fusion zone along with a decrease in porosity, which would be attributed to controlled heat input and enhanced cooling rates. It was these microstructural developments that resulted in a massive improvement in weld mechanical performance.
4. HAZ Behavior: The softening of the HAZ was minimal due to successful heat control, which helped inhibit the dissolution of the most important strengthening phases such as Mg_2Si and subsequently maintained the mechanical integrity in the weld.
5. Statistical Validation: The optimization model was validated statistically, including an R^2 value of 0.9966 and not significant lack of fit, demonstrating that the model is very efficient in predicting the optimum weld parameters.

REFERENCES

1. Tamil Kumaran G, JayakumarKS and Vimal Samsingh R 2023 Effects of ER5356 and ER5183 filler rods on the mechanical and metallurgical properties of TIG-welded AA5754-H111 aluminium alloy Recent Advances in Materials Technologies (Springer Nature Singapore) vol 1, pp 215–23
2. Senthur Vaishnavan S, Jayakumar K, Naveen Kumar P and Suresh T 2023 Effect of ER5183 filler rod on the metallurgical andmechanical properties of TIG-welded AA5083 and AA5754 joints Mater. Today Proc. 72 2251-4
3. I.A. Zamzami, L. Susmel, Int. J. Fatig. 101 (2) (2017) 137–158.
4. P. Heinen, H. Wu, A. Olowinsky, A. Gillner, Phys. Proced. 56 (2014) 554–565.
5. Pires I, Quintino L, Miranda RM. Analysis of the influence of shielding gasmixtures on the gas metal arc welding metal transfer modes and fumeformation rate. Mater Des 2007;28:1623–31.
6. Mendes da Silva CL, Scotti A. The influence of double pulse on porosityformation in aluminum GMAW. J Mater Process Technol 2006;171:366–72.
7. C.-S. Hsieh, H. Zhu, T.-Y. Wei, Z.-J. Chung, W.-D. Yang, Y.-H. Ling, J. Eur. Ceram. Soc. 28 (2008) 1177–1183.
8. T.H. Hou, C.H. Su, W.L. Liu, Powder Technol. 173 (2007) 153–162.
9. B.-T. Lin, M.D. Jjean, J.-H. Chou, Int. J. Adv. Manuf. Technol. 34 (2007) 307–315.
10. Khodko, Oleksandr & Zaytsev, Vitaliy & Sukaylo, V. & Verezub, N. & Scicluna, Sarah. (2015). Journal of Manufacturing Processes. Journal of Manufacturing Processes. 20. 304–313. 10.1016/j.jmapro.2015.06.016.
11. Anhua Liu, Xinhua Tang, Fenggui Lu,Study on welding process and prosperities of AA5754 Al-alloy welded by double pulsed gas metal arc welding, Materials & Design
12. Rajesh Manti & D. K. Dwivedi (2007) Microstructure of Al–Mg–Si Weld Joints Produced by Pulse TIG Welding, Materials and Manufacturing Processes, 22:1, 57–61, DOI: 10.1080/10426910601015923
13. Bueno, Maddi & Galdos, Lander & Sáenz de Argandoña, Eneko & Weiss, Matthias & Lou, Yanshan & Mendiguren, Joseba. (2020). Strain Rate Effect on the Fracture Behavior of the AA5754 Aluminum Alloy. Procedia Manufacturing. 47. 1264–1269. 10.1016/j.promfg.2020.04.212.
14. Tamil Kumaran G, Jayakumar K , Amala Mithin Minther Singh characterization of Pulsed-Tungsten Inert Gas (PTIG) Welding on AA5754- H111 Alloy: Mechanical Properties and Microstructural Analysishttps://doi.org/10.1088/2053-1591/ad0761
15. Kumar, Adepu & Sundarrajan, Srinivasan. (2009). Optimization of pulsed TIG welding process parameters on mechanical properties of AA 5456 Aluminum alloy weldments. Materials & Design - MATER DESIGN. 30. 1288–1297. 10.1016/j.matdes.2008.06.055.
16. Manpreet Singh, Navjot singh , Jujhar Singhoptimization of TIG welding parameters for AL6061-T6 aluminum alloy to improve weld quality and mechanical properties
17. Kumaran G, Tamil & Jayakumar, K & Minther Singh, A. Amala Mithin. (2023). Characterization of Pulsed-Tungsten Inert Gas (PTIG) Welding on AA5754- H111 Alloy: Mechanical Properties and Microstructural Analysis. Materials Research Express. 10. 10.1088/2053-1591/ad0761
18. El-Rayes, Magdy M., and Ehab A. El-Danaf. "The influence of multi-pass friction stir processing on the microstructural and mechanical properties of Aluminum Alloy 6082." Journal of Materials Processing Technology 212.5 (2012): 1157–1168.
19. Sunny, Kora T. et al. "Parameter optimization and experimental validation of A-TIG welding of super austenitic stainless steel AISI 904L using response surface methodology." Proceedings of the Institution of Mechanical Engineers, Part E: Journal of Process Mechanical Engineering 236 (2022): 2608–2617.
20. X. Zhou, G. Zhang, Y. Shi, M. Zhu, F. Yang, Mater. Sci. Eng. A 705 (2017) 105–113
21. N. Çömez, H. Durmus, J. Mater. Eng. Perform. 28 (6) (2019) 3777–3784.
22. J.C. Dutra, R.H.G. Silva, B.M. Savi, C. Marques, O.E. Alarcon, Weld World 59 (2015) 797–807
23. A.K. Lakshminarayanan, V. Balasubramanian, K. Elangovan, Int. J. Adv. Manuf. Technol. 40 (2009) 286–296

Note: All the figures and tables in this chapter were made by the authors.

Advanced Analysis of CMT Welding Parameters on the Microstructure and Strength of AA6009 Aluminium Alloy

Aakash V.[1],
Rishekkumar[2], Chandiramouli[3],
Iniya T.[4], Abishek S.V.[5], Deeksheka[6]
Student, Department of Mechanical Engineering,
Chennai Institute of Technology,
Chennai, India

P.K. Devan[7]
Professor, Department of Mechanical Engineering,
R.M.K. College of Engineering and Technology,
India

◆ **Abstract:** This study investigates the optimization of weld parameters for CMT joining of the AA6009 aluminium alloy to improve its weld quality and performance. RSM and CCD approaches were used to analyse the trends of WFS, TS, and WC on UTS with minimal defects. The optimized parameters yielded an excellent UTS of 230 MPa at the same time minimizing the defect's occurrence. Microstructural analysis revealed the presence of refined dendritic grains in the fusion zone and equiaxed grain structures in the HAZ. These collectively contribute to enhanced mechanical properties. Hardness distribution examination indicated uniform variation across the weld areas. The weld metal exhibited homogeneous hardness characteristics. The statistical model was validated through a high coefficient of determination of 0.9936, which thus validated the reliability and predictability of the model. The research provides an all-round approach toward parameter optimization in CMT welding, thus enabling noteworthy improvements in weld quality as related to automotive applications, as well as further insights into microstructural transformations.

◆ **Keywords:** CMT welding, AA6009 Aluminum alloy, Welding parameters, Microstructure analysis, Mechanical strength

1. Introduction

Aluminium alloys are characterized by unique properties, such as high specific strength, low density and excellent corrosion resistance. These make them increasingly suitable for lightweight structural applications in the aerospace, automotive, rail transportation, and shipbuilding industries [1,2]. The primary objective of the automotive sector is to reduce vehicle weight while maintaining size, load capacity, and safety. This objective can be effectively achieved by replacing steel with aluminium sheets in the vehicle's structural and exterior components, particularly in the body-in-white. Vehicles with aluminium body structures often employ a "spaceframe" design, comprising castings, extrusions, and stampings that are joined together. In the alternative, they may use construction techniques analogous to steel-based designs, where stamped panels are spot-welded in a body-in-white, complemented by bolted-

[1]aakash.vasudevan2020@gmail.com, [2]rolexrishe@gmail.com, [3]s.chandiramouli340@gmail.com, [4]iniya0206@gmail.com, [5]abisheksivaraj10@gmail.com, [6]deekshekaashok@gmail.com, [7]pkdevan68@gmail.com

DOI: 10.1201/9781003685906-62

on aluminium elements - the fenders, doors, hoods, and deck lids. Automotive sheets shall meet all the following essential demands for automotive applications: (i) good formability, with stamping of the panels with retention or further improvement of strength after painting and thermal curing; (ii) compatibility with spot welding; (iii) good resistance to corrosion; (iv) suitable for adhesive bonding; and (v) for outer panels, with a smooth, uniform surface ensuring a quality appearance after paint. Aluminium alloys used for automotive body panels are typically categorized into non-heat-treatable (NHT) and heat-treatable (HT) alloys [4-7]. The non-heat-treatable alloys are mainly the Al-Mg alloys from 5xxx series, with the following grades: 5030, 5032, 5052, 5754, and 5182. HT consist of the Al-Cu alloys within the 2xxx series, such as 2008, 2010, and 2036; Al-Mg-Si and Al-Mg-Si-Cu alloys with 6009, 6016, 6111, 6022, and 6061 included in the 6xxx series. It is remarkable that some Al-Mg base alloys, for instance 5022, 5030, and 5032, which are developed in Japan, present a mild heat treatment sensibility [4,6]. It is essential to acknowledge that no single alloy can fulfil all the requirements of an automotive sheet.

Each is developed for specific applications. NHT Al-Mg alloys, that are prone to flow line defects such as Lüder bands and stretcher strains [5,8], are generally used in inner body parts where the quality of the surface is not a major concern. On the other hand, owing to their superior performance, these heat-treatable, age-hard enable Al-Cu and Al-Mg-Si alloys in the condition of natural aging are also mostly used for outer panels. As a matter of fact, Al-Mg-Si alloys are more frequently employed because they provide the desired balance between high strength following artificial aging after paint baking or lacquering, while still maintaining good formability. These alloys balance essential properties such as formability, strength, and corrosion resistance, with a significantly better corrosion resistance than 2xxx series alloys. In the United States, copper-containing Al-Mg-Si alloys, such as AA6009 and AA6111, are particularly favoured because of their enhanced final strength and dent resistance [4,5]. Though AA6111 is stronger than AA6009, the corrosion resistance and formability of both alloys are lesser as compared to other 6xxx series alloys containing less copper. Aluminium alloys have become increasingly used in automobiles, mainly because of their lightweight, which provides an excellent alternative to Fe, increasing vehicle performance and saving energy for transportation [9,10]. However, the usage of aluminium in vehicles is somewhat limited. Whereas the 6xxx series of aluminium alloys, such as AA6009, possesses good formability, for some automotive body panels requirements [12,13], the corresponding low strength of AA6009 alloy puts limitation on its full potential for replacement with conventional steel automotive body plates. To overcome the aforementioned weakness, researchers believe that if the strength of the AA6009 alloy could be increased based on the advantageous features of the AA7075 alloy while retaining its own natural features, the application areas of the 6xxx series aluminium alloys would remarkably expand.

A laminated plate from the combination of AA6009 and AA7075 aluminium alloys promises this way. There are a number of proposed manufacturing processes for laminated composite plates. One such process to produce bimetal laminated composites, like copper-clad aluminium, is the CFC stands for Continuous Core-Filling Casting. With thicknesses varying from 10 μm to 15 μm, the interface layer of CFC-based laminated composites, however, can be composed of carbon, compound, or oxidation layers. The Novelis Fusion process offers an alternative to the creation of laminated aluminium plates with a well-bonded interface that is virtually free of oxides and pores, with an interface thickness of 15-30 μm. Additionally, the Double-Stream-Pouring Continuous Casting (DSPCC) process enables the preparation of laminated composites with a gradient interface, where the interface produced by DSPCC is metallurgically bonded and importantly devoid of pores and oxides, typically with interface thicknesses on the order of millimetres [10-13]. It is essential to note that these composite materials can only be practically utilized after undergoing plastic deformation, such as rolling and compression. The current research examines the microstructures and mechanical properties of a deformed AA6009/AA7075 aluminium composite. It deforms the composite plate as per the composite ingot prepared through the DSPCC method, primarily focusing on investigating the deformation behaviour of AA7075/AA6009 aluminium composite at various plastic deformation temperatures.

The literature reviewed mainly indicated the importance of CMT process parameters, such as welding current (WC), welding speed (WS), wire feed speed (WFS), and torch angle, in determining the microstructural and mechanical properties of aluminium alloy weld joints. Precise optimization of these parameters is essential to enhance weld quality, reduce defect formation, and improve overall performance. Using Response Surface Methodology (RSM) as a statistical modelling approach, this research optimized the critical welding process parameters, namely WFS, WS, and AL, and investigated their effects on the UTS of the welded joints. Furthermore, a detailed microstructural and mechanical property characterization of weldments prepared under optimized conditions is done. Investigations in this work include heat-affected zone (HAZ) softening, solidification structures, tensile strength, and defect formation in the weld joints.

2. Experimental Methodology

The 5mm thick base metal AA6009 was procured from Arihant Aluminium, Chennai. It was cut to the required dimensions of 100 × 55 mm using a hacksaw machine. The joining edges are flatted using a milling machine. Taking the chemical composition with the help of an optical emission spectroscopic study. The tensile strength of the BM is tested in UTM shown in Table 62.1, and the hardness test of the BM is taken in Vicker's microhardness structure shown in Table 62.2. With the help of acetone, cleaning of the base metal is done. Both the metal plates are clamped together and with the help of the ER5356 aluminium filler, the metals are welded in CMT welding. After the welding process, it is taken for testing. The testing sample should be cut in the standard ASTM E8 size which will be in the shape of a Dog-bone shape the material is cut through the machining using WEDM and it is used for tensile test and the results will be like Yield Strength (YS), Universal Tensile Strength (UTS), Elongation and Hardness. Sample produced using WEDM is mounted on a hot-mounted press by making use of Bakelite powder. The mounted samples are first hand polished with the help of silicon carbide abrasive sheets, which is in the range of 80 to 400. Then they undergo machine polishing with the help of finer grit sizes silicon carbide abrasive sheets ranging from 600 to 2000. The samples after getting polished are then etched by Keller's reagent. In final steps, microstructure in terms of size and shape of the grain and precipitates is done with scanning electron microscope SEM. For the micro-hardness test, samples obtained are checked on micro structural across all zones with interval measurements made at 0.5mm. Vicker's hardness test is utilized for the conduction of the Micro Hardness Test. The testing parameters consist of 1kg load and 20 second dwell time.

Table 62.1 Chemical composition of AA6009

BM	Mg	Mn	Fe	Si	Cu	Cr	Zn	Al
AA6009	0.6	0.4	0.3	0.8	0.3	0.6	0.2	Bal
ER5356	5.1	0.1	-	0.2	0.1	0.1	0.1	Bal

Table 62.2 Mechanical properties of AA6009

Material	YS(MPa)	UTS(MPa)	El (%)	Hardness (Hvl)
AA6009	282	329	11.3	92

CCD methodology in particular was utilized to optimize welding process parameters for Cold Metal Transfer. Against this backdrop, a comprehensive literature review was conducted on FSW of aluminium alloys to identify the main welding process parameters that would be most influential for this research. From this analysis, three primary parameters were identified, namely: Tool Traverse Speed (TTS), Tool Rotational Speed (TRS), and Axial Load (AL). Their limits were derived from the existing literature in an in-depth manner. The lower limit was encoded as -α (-1.682) and the upper limit was represented by α (1.682). The coded value of '0' was designated as the central value between these limits. The actual values for the coded levels of +1 and -1 were determined using the standard equations associated with CCD. Table 62.3 lists the actual values of each welding process parameter that is correlated to the respective coded levels.

Table 62.3 Limits of welding process parameter

WPP	Unit	Levels				
		-1.682	-1	0	1	+1.682
WFS	mm/min	4400	4522	4700	4878	5000
TS	mm/min	250	262	280	298	310
WC	A	140	148	160	172	180

3. Results and Discussion

3.1 Optimization of Welding Parameters

A CCD technique inside the Response Surface Methodology (RSM) framework was used to optimize the welding process parameters. This methodology was chosen for its robustness in designing experiments that efficiently explore the interaction effects and non-linear relationships between parameters [14]. The welding parameters considered for optimization included Wire Speed (WS), Traverse Speed (TS), and Welding Current (WC), as these are critical factors influencing weld quality, mechanical properties and microstructural evolution of the joints [15]. These parameters were selected based on their established impact on key performance indicators such as hardness distribution, tensile strength, and defect formation [16]. The Design of Experiments (DOE) table was structured to systematically evaluate combinations of these parameters, as generated by the CCD [17]. The CCD's inclusion of factorial points, axial points, and centre points ensures a comprehensive exploration of the parameter space, allowing for accurate prediction of responses and identification of optimal conditions [18]. Table 62.5 outlines the experimental matrix, detailing the parameter levels investigated and their corresponding experimental responses, including metrics such as tensile strength, elongation, microstructural integrity, and defect density.

Table 62.4 Design of experiments framed using CCD

Std. order	Run order	Coded Values			UTS (MPa)
		P	S	F	
1	1	-1	-1	-1	172
2	14	1	-1	-1	207
3	4	-1	1	-1	172
4	17	1	1	-1	210
5	6	-1	-1	1	182
6	11	1	-1	1	194
7	9	-1	1	1	193
8	13	1	1	1	207
9	16	-1.682	0	0	173
10	7	1.682	0	0	212
11	15	0	-1.682	0	190
12	5	0	1.682	0	200
13	12	0	0	-1.682	194
14	19	0	0	1.682	202
15	20	0	0	0	224
16	18	0	0	0	227
17	8	0	0	0	226
18	3	0	0	0	228
19	10	0	0	0	227
20	20	0	0	0	230

Table 62.5 ANOVA test results

Source	Sum of Squares	Df	Mean Square	F-value	p-value	
Model	7309.54	9	812.17	295.80	<0.0001	significant
A-TRS	1983.61	1	1983.61	722.44	<0.0001	
B-TTS	140.59	1	140.59	51.20	<0.0001	
C-AL	59.29	1	59.29	21.29	0.0009	
AB	3.13	1	3.13	1.14	0.3111	
AC	276.13	1	276.13	100.57	<0.0001	
BC	55.12	1	55.12	20.08	0.0012	
A^2	2227.32	1	2227.32	811.20	<0.0001	
B^2	1921.86	1	1921.86	699.95	<0.0001	
C^2	1585.04	1	1585.04	577.28	<0.0001	
Residual	27.46	10	2.75			
Lack of Fit	7.46	5	1.49	0.3729	0.8486	not significant
Pure Error	20.00	5	4.00			
Cor Total	7337.00	19				
Std. Dev.	1.66	**R^2**		0.9936		
Mean	203.50	**Adjusted R^2**		0.9929		
C.V. %	0.8143	**Predicted R^2**		0.9884		
		Adeq Precision		47.3091		

By employing CCD, it was possible to model the impact of WS, TS, and WC on the weld quality using quadratic equations. Additionally, interaction effects among these parameters were assessed to understand their synergistic or antagonistic behaviour in determining joint performance [19]. Such an approach provides insights into the sensitivity of each parameter and the trade-offs required to achieve a balance between mechanical properties and weld quality [20]. Future work may involve validating these findings under varying environmental conditions or with alternative aluminium alloys to generalize the optimization strategy [21]. Additionally, incorporating real-time monitoring techniques during welding could further enhance the reliability of the predicted optimal parameters.

Where a P-value <0.0500 corresponds to the fact that, there may be statistical significant influences in the response variables because of the respective terms present within the model terms, due to this it had led to the observation in current study, some model terms such as, A and B, and C besides AC, BC and terms A^2 B^2 and C^2 proved out statistically significant with notable significant values on the resulting strength due to tensile strength of weld nuggets. Conversely, terms with P-values greater than 0.1000 are considered statistically not significant and can be omitted from the model unless needed to retain the hierarchical form. The model's excellent reliability is indicated by the Lack of Fit F-value of 0.37, which suggests that the lack of fit is not significant when compared to pure error. The probability of such a Lack of Fit F-value happening by mere random noise is 84.86%, thus giving the conclusion that the experimental data is well fitted with the model. A not significant lack of fit is desirable, as it certifies that the model exactly describes the observed data and hence will be reliable for predictions. This agreement demonstrates that the model has excellent explanatory power and can reliably predict responses within the studied design space.

The final equation presented in the actual factors provides an actual relationship between the process parameters in terms of welding speed (WS), wire feed speed (WFS), arc length (AL), and response variable, UTS. This can be used very effectively in predicting the outcome against specific settings of the parameters for a specific design space. Similarly, this is a convenient reference tool for welding engineers to make optimal parameter choices for a desired mechanical property in weld. The analysis also shows the need for a balanced approach to parameter selection. In

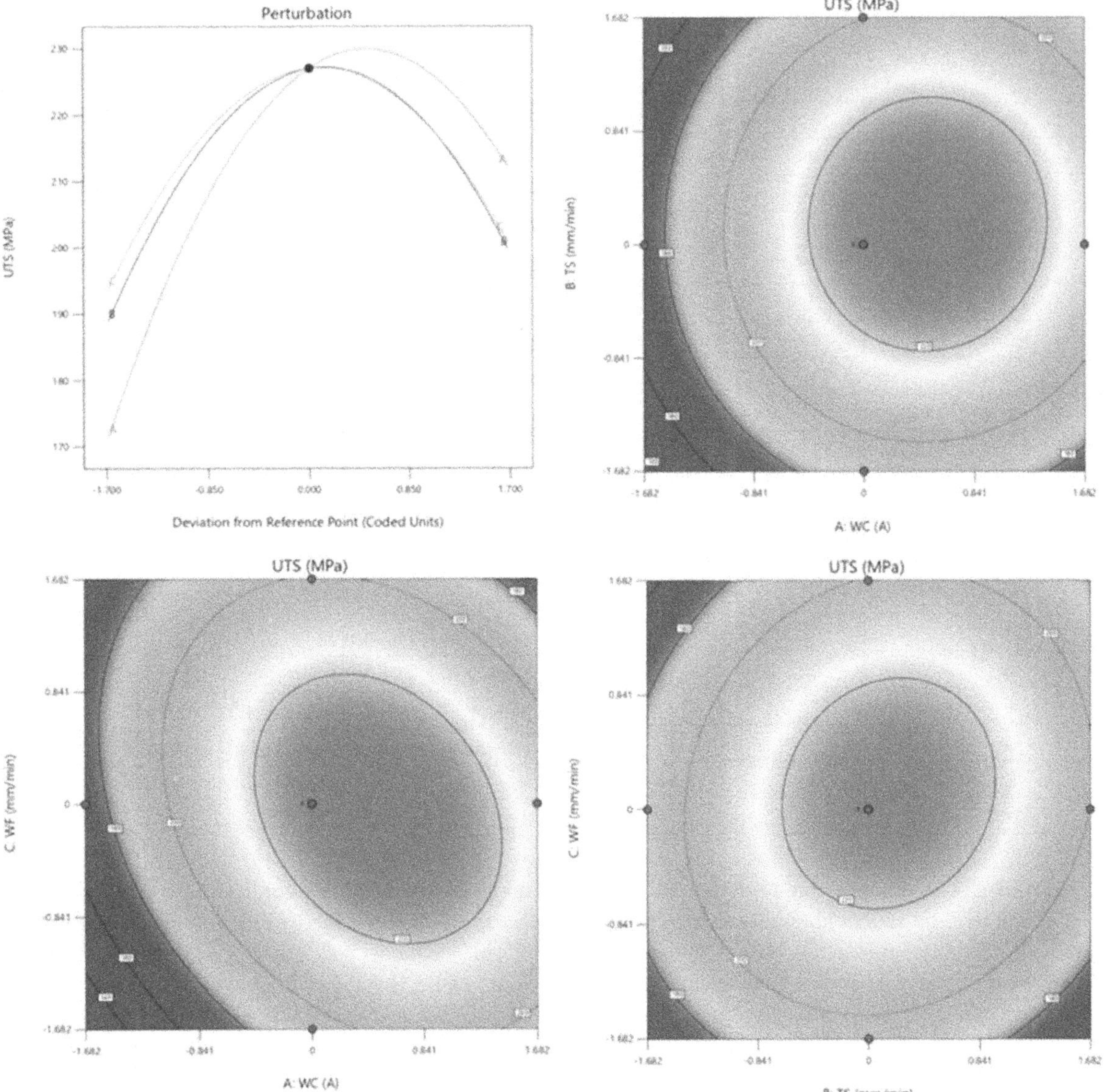

Fig. 62.1 [a] Perturbation plot [b] Interaction plot AB [c] Interaction plot AC [c] Interaction plot BC

interaction terms such as AC and BC, the impact of one parameter is greatly influenced by the levels of the other, and it is evident that the parameters are interdependent. Quadratic terms (A^2, B^2, C^2) show that relationships are not linear and should be controlled and optimized at a precise level for achieving optimal results in welding. These results confirm the adequacy of the model for describing the intricate interplay between the process parameters and will direct experiments aimed at minimizing weld defects, refined microstructure, and maximized tensile strength. The solid statistical basis allows the model to be a more powerful tool in improving the quality of welds and the performance of the CMT welding process.

Final Equation in Terms of Actual Factor

$$UTS = 12.052 * WC + 3.208 * TS + 2.084 * WF + 0.625 * (WC * TS) - 5.875\,(WC * WF) + 2.625\,(TS * WF) - 12.432 * (WC^2) - 11.548 * (TS^2) - 10.487 * (WF^2) \quad (1)$$

The regression equation of actual factors is a reliable tool to predict the UTS of welds as correlated with some specific welding process parameters such as WC, travel speed TS, and welding frequency WF. This predicting ability is specifically useful to find out optimum welding conditions while reducing the experimental effort if the input values fall within the established design space. However, scaling of coefficients to match units of each parameter and placement of intercept outside the centre of the design space reduces the applicability of the equation for estimating the contribution of individual factors. Interaction terms like (WC*WS) and (WS*WF) provide the combined impact of parameters on UTS. Some of the combinations show synergistic improvements while others have antagonistic effects. Quadratic terms such as (WC^2) and (WF^2) indicate a non-linear relationship between process parameters and the response and, hence, the model is used for investigating parameter interactions, which is crucial for avoiding conditions like overheating or bond failure. It is quite useful for investigating parameter interaction, thereby minimizing weld defects and optimizing the microstructure to increase the mechanical behaviour of welded joints. While the equation shows strong reliability within the experimental design space, its predictions must be verified through additional analyses, such as microstructural characterization and validation under practical conditions. For CMT welding, where heat input and material deposition need to be controlled accurately, the regression model will give a comprehensive framework to optimize the process to yield high-quality, defect-free welds with superior mechanical properties

3.2 Microstructure Analysis

Macro- and microstructural features of fracture surfaces of the zones for different weld joints produced in CMT at 220 J/mm energy input. There are no defects, including pores, spatters, wormholes, cracks, or voids on a macro level. That ensures a quality weld. Coarsened grains and precipitates of an alloy make up the base metal microstructure. In the weld zone, equiaxed and dendritic grain structures appeared with improved grain boundaries were observed in the cross-sectioned fusion zone. Columnar grain structure as in Yan [22] was depicted in the interface zone. The HAZ, the grains were elongated, larger, and interspersed between finer structures due to the rapid cooling of the welding process. The dendritic structure of the fusion zone remained almost unaffected compared to the HAZ due to the weld maintaining the temperature below the recrystallization threshold during welding. In contrast to the base metal, which had a grain size of 60.5 μm, the measured grain sizes were 17 μm in the fusion zone, 34 μm in the interface zone, and 54 μm in the HAZ. These values show the compatibility of CMT-welded joints, which is due to controlled heat input and cold transfer of filler metal. SEM analysis of tensile fracture surfaces showed dimpled morphology with smooth elliptical shapes and some intermetallic compounds in the fusion zone. Their elongated grains and cup-and-cone ductile fracture pattern enhanced the HAZ fracture surfaces' ductility and average tensile strength of 240 MPa. Similar dimple patterns suggesting ductile fracture were also observed by Taban et al. [23]. Unlike any other welding process, CMT-welded joints showed the absence of porosity which further improved their integrity.

Wu et al. [24] reported substantial and discontinuous precipitates along grain boundaries, which could affect crack propagation and stress corrosion cracking (SCC). EBSD analysis has been carried out to elucidate the orientation of grains, dislocation structures, and precipitate distribution across different zones of welds. Average intercept length and misorientation angle in interface zone have been measured to be 44.723 μm and 29.27° respectively, along with detailed boundary rotation angles. Relations of intercept lengths with misorientation angles and number fractions. Microstructural maps, which were obtained through the analysis of EBSD images, show that a comparison between the interface zone material and the base material depicts substantial grain refinement within the interface zone. The volume of Mg_2Si precipitates accounted for 6.45%. The orientation factors close to the HAZ were low and showed intergranular dislocation pile-ups in Mg_2Si precipitates, resulting in SCC and crack growth. LAGB between 5° and 15° and HAGB between 15° and 65° were seen. The grain intercept average lengths were calculated to be 44.723 μm in the weld zone interface, 18.937 μm in the WZ, and 20 μm in the base metal. Therefore, it can be said that the weld zone has relatively finer grains, which provides better corrosion resistance, also supported by Norman et al. [25].

The microstructural properties of pulse-welded aluminium alloys were further determined through LOM. While there was a localized rise in temperature to 250°C in the far HAZ for the robotic CMT-welded joint M55, coalescence of Mg2Al3 did not occur. Near the fusion line, fine-grained microstructures resulting from static recrystallization occurred [26,27]. In robotic joint M66, partial melting of grain boundaries near the fusion line led to coalescence and coarsening due to post-weld aging following solution treatment [27]. The grain structure of robotic joint M56 was similar to those of M55 and M66, depending on the base metal used. SEM analysis of fractured fatigue test specimens reveals minimal porosity [28,29], which did not significantly affect joint strength. The porosity was 2% for M55 ("B" quality) and 3% for M56 and M66 ("C" quality).

Measurements of microhardness at a load of 200 g indicated maximum hardness decreases of 18% for weldments, which are considerably smaller than the 22%–35% drops reported in earlier studies [30-32]. The M55 joint had a microhardness ranging from 77 to 92 HV with the lowest hardness within the weld zone. In M66, values varied between 79 and 96 HV, the highest hardness value in the weld zone, and the lowest in the HAZ. M56 showed the dissimilar alloy joint, and the microhardness values range from 76 to 96 HV. This increases hardness near HAZ by partial solution treatment, while a hardness drops in far HAZ is attributed to over-aging and precipitate coarsening [27].

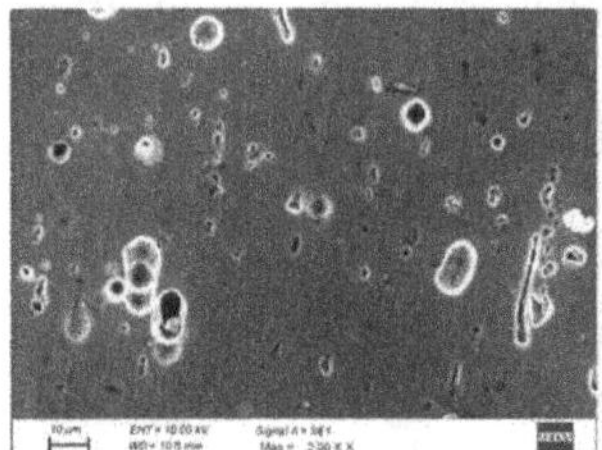
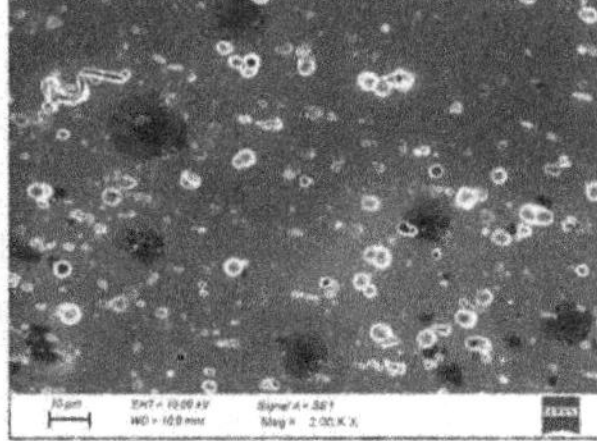

Fig. 62.2 SEM microstructure [a] Base metal AA6009-T6 [b] FZ of sample 20

4. Conclusion

AA6009 was successfully joined using CMT welding and the following observations are made:

1. Optimized CMT welding parameters yielded the ultimate tensile strength to be 230 MPa that proved the method was quite promising.
2. Microstructure of refined dendritic grains was found in the fusion zone. Equiaxed grain structures were noted in the HAZ with improved mechanical properties
3. The hardness distribution was nearly uniform in the weld metal, and variations in it were due to changes in the grain structure and heat-affected transformations.
4. The optimized model, with a high R^2 of 0.9936, ensures reliability and robust predictability.
5. Joints obtained through optimized process parameters were free from defects and ideal for use in automotive production.
6. An Impact study will provide further robustness in the overall framework for process parameter optimisation. Such a basis will establish a firm link in between process parameters, Microstructural changes, and mechanical response.
7. The scope of the findings can be extended to other aluminium alloys and real-time monitoring techniques added to augment reliability and efficiency in the weld process through CMT..

References

1. I. J. Polmear (1996) Recent developments in light alloys, Mater. Trans. JIM, 37 (1), 12–31.
2. A. Y. Ishchenko (2005) High-strength aluminium alloys for welded structures in the aircraft industry, Weld. Int., 19 (3), 173–185.
3. G. S. Cole, A. M. Sherman (1995) Materials Characterization, 35, 3–9.
4. G. B. Burger, A. K. Gupta, P. W. Jeffrey, D. J. Lloyd (1995) Materials Characterization, 35, 23–39.
5. D. G. Althenpohl (1998) Aluminum: Technology, Applications, and Environment – A Profile of a Modern Metal, 6th Edition, The Minerals, Metals, and Materials Society, Warrendale and the Aluminum Association, Washington, USA, 361–364.
6. T. Komatsubara, M. Matsuo (1989) SAE Technical Papers, No. 890712, Society of Automotive Engineers, Inc., Warrendale, PA, USA, 1–8.
7. J. T. Staley, D. J. Lege (1993) Advances in Aluminum Alloy Products for Structural Applications in Transportation, Journal de Physique IV, 3 (11), 179–190.
8. P. Furrer (1991) Science and Engineering of Light Metals, ed. K. Hirano, H. Oikawa, K. Ikeda, Japan Institute of Light Metals, Tokyo, 1171–1178.
9. W. S. Miller, L. Zhuang, S. Bottema, A. J. Wittebrood, P. De Smet, A. Haszler, A. Vieregge (2000) Recent development in aluminum alloys for the automotive industry, Mater. Sci. Eng. A, 280 (1/2), 37–49.
10. Nargess S. (2003) Lightening the material, Automotive Engineer, (9), 70–77.
11. Olaf E., Jurgen H. (2002) Texture control by thermomechanical processing of AA6xxx Al-Mg-Si sheet alloys for automotive applications – A review, Mater. Sci. Eng. A, 367, 249–262.
12. Liu Hong, Liu Chun-ming, Zuo Liang (2005) Aging behavior and properties of several 6000 series aluminum alloys for auto sheet materials, Materials for Mechanical Engineering, 29 (6), 10–14. (in Chinese)
13. Ding Xiang-qun, He Guo-qin, Chen Cheng-su, Liu Xiao-shan, Zhu Zheng-yu (2005) Advance in studies of 6000 aluminum alloy for automobile, Journal of Materials Science and Engineering, 23 (2), 302–305. (in Chinese)
14. Sunny K. T., Korra N. N., Vasudevan M., Arivazhagan B. (2022) Parameter optimization and experimental validation of A-TIG welding of super austenitic stainless steel AISI 904L using response surface methodology, Proceedings of the Institution of Mechanical Engineers, Part E: Journal of Process Mechanical Engineering. https://doi.org/10.1177/09544089211040714
15. He H., Tian X., Yi X., Wang P., Guo Z., Fu A., Zhao W. (2024) Optimization of Joining Parameters in Pulsed Tungsten Inert Gas Weld Brazing of Aluminum and Stainless Steel Based on Response Surface Methodology, Coatings. https://doi.org/10.3390/coatings14020264
16. Vasantharaja P., Vasudevan M. (2018) Optimization of A-TIG welding process parameters for RAFM steel using response surface methodology, Proceedings of the Institution of Mechanical Engineers, Part L: Journal of Materials: Design and Applications. https://doi.org/10.1177/1464420717720639
17. Maduraimuthu V., Vasantharaja P., Vasudevan M., Panigrahi B. S. (2019) Optimization of A-TIG Welding Process Parameters for P92 (9Cr-0.5Mo-1.8W-VNb)

Steel by Using Response Surface Methodology, Materials Performance and Characterization. https://doi.org/10.1520/MPC20190037

18. Meena R. P., Yuvaraj N., Vipin (2024) Optimization of process parameters of cold metal transfer welding-based wire arc additive manufacturing of super duplex stainless steel using response surface methodology, Proceedings of the Institution of Mechanical Engineers, Part E: Journal of Process Mechanical Engineering. https://doi.org/10.1177/09544089221135685
19. Pondi P., Achebo J., Ozigagun A. (2021) Optimization of the Tungsten Inert Gas Process Parameters using Response Surface Methodology, International Journal of Emerging Scientific Research.
20. Raja R., Jannet S., Mohanasundaram S. (2020) Multi Response Optimization of process parameters of friction stir welded AA6061 T6 and AA7075 T651 Using Response Surface Methodology.
21. Sharma P., Chattopadhyaya S., Singh N. (2019) Optimization of gas metal arc welding parameters to weld AZ31B alloy using response surface methodology, Materials Research Express.
22. Yan S., Nie Y., Zhu Z., Chen H., Gou G., Yu J., et al. (2020) Characteristics of microstructure and fatigue resistance of hybrid fibre laser MIG welded Al-Mg alloy joints, Appl. Surf. Sci.
23. Taban E., Kaluc E. (2007) Comparison between microstructure characteristics and joint performance of 5086-H32 aluminium alloy welded by MIG, TIG, and friction stir welding processes, Kovove Mater., 45, 241–248.
24. Wu Y. E., Wang Y. T. (2010) Enhanced SCC resistance of AA7005 welds with appropriate filler metal and post-welding heat treatment, Theor. Appl. Fract. Mech., 54, 19–26.
25. Norman A. F., Ian Brough B., Philip B., Prangnell (2000) High-resolution EBSD analysis of the grain structure in an AA2024 friction stir weld, Mater. Sci. Forum, 5, 331–337.
26. Totten G. E., Mackenzie S. (2003) Handbook of Aluminum. Vol. 1, Physical Metallurgy and Processes, Marcel Dekker Inc., USA.
27. Mathers G. (2002) The Welding of Aluminium and Its Alloys, Woodhead Publishing Limited, Cambridge, UK.
28. ISO 10042:2005(E) (2009) Welding – Arc-welded joints in aluminium and its alloys – Quality levels for imperfections, International Organization for Standardization.
29. TS EN ISO 17637 (2011) Non-destructive testing of welds – Visual testing of fusion-welded joints, Turkish Standards Institution.
30. Ericsson M., Sandstrom R. (2003) Influence of welding speed on the fatigue of friction stir welds, and comparison with MIG and TIG, Int. J. Fatigue, 25, 1379–1387.
31. Moreira P. M. G. P., Santos T., Tavares S. M. O., Richter-Trummer V., Vilaca P., De Castro P. M. S. T. (2009) Mechanical and metallurgical characterization of friction stir welding joints of AA6061-T6 with AA6082-T6, Mater. Des., 30, 180–187.
32. Taban E., Kaluc E. (2005) Microstructural and mechanical properties of double-sided MIG, TIG, and friction stir welded 5083-H321 aluminium alloy, Kovove Mater., 44, 25–34.

Note: All the figures and tables in this chapter were made by the authors.

Advances in Additive Manufacturing Technologies – Gurusamy Pathinettampadian et al. (eds)
© 2026 Taylor & Francis Group, London, ISBN 978-1-041-16687-0

Enhancing Weld Quality: Process Parameter Optimization and Property Characterization in LBW of AA5754 Aluminum Alloy

P.K. Devan[1]
Professor,
Department of Mechanical Engineering
R.M.K. College of Engineering and Technology,
India

Deeksheka[2],
Chandiramouli[3], **Aakash V.**[4],
Rishekkumar[5], **Iniya T.**[6], **Abishek S.V.**[7]
Student, Department of Mechanical Engineering,
Chennai Institute of Technology,
Chennai, India

◆ **Abstract:** This study investigates the Laser Welding (LW) of aluminum alloy AA5754, focusing on optimizing process parameters to optimize the quality and physical properties of welded joints. The effects of key variables, including Welding Speed (WS), Laser Power (LP), and focal position (FP), were systematically evaluated using a response surface methodology (RSM). Ultimate tensile strength (UTS) and weld seam quality were analyzed to identify the influence of individual factors and interactions. Perturbation and contour plots were employed to identify the optimal parameter combinations, demonstrating the sensitivity of UTS to changes in process conditions. The results revealed that appropriate parameter optimization significantly improved the metallurgical properties of the welded joints. This research contributes valuable insights into achieving defect-free welds and optimizing Laser Welding (LW) processes for lightweight aluminum alloys in automotive and aerospace applications.

◆ **Keywords:** LBW, AA5754 aluminum alloy, Process parameter optimization, Weld quality, Property characterization

1. Introduction

The rising need to reduce energy consumption and pollution of the air has led to a challenge in the transportation industry: claim for fuel-efficient vehicles. Aluminum is gradually becoming the backbone material for lightweight vehicle structures with up to 50% weight savings not affecting safety [1]. It offers several advantages such as lower density, excellent formability, high strength to weight ratio, and resistance to corrosion. One of the important processes of vehicle body manufacturing is welding. However, there must be an appropriate choice of techniques in order to optimize the joint performance. For the past two decades, Laser Welding process has gained an important role as a critical joining method for industries like automotive and aerospace, particularly driven by increased usage of aluminum alloys. This technique is highly suited for the fabrication of a car body mainly because it possesses high speed in processing, single sided access, and non-contact tooling along with compatibility to robotic automation

[1]pkdevan68@gmail.com, [2]deekshekaashok@gmail.com, [3]s.chandiramouli340@gmail.com, [4]aakash.vasudevan2020@gmail.com, [5]rolexrishe@gmail.com, [6]iniya0206@gmail.com, [7]abisheksivaraj10@gmail.com

DOI: 10.1201/9781003685906-63

[2]. In contrast to traditional welding techniques, laser welding tends to offer a smaller heat-affected zone. It helps minimize metallurgical challenges and also offers an excellent weld shape and depth with minimal heat-affected zones [3]. With impressive accuracy, high-speed processing, and automated capabilities, laser welding greatly increases productivity [4]. Yet, aluminum laser welding does have its drawbacks.

Pores and cracks are two defects prevalent in this material and due to its alloying constituents [5,6]. The low vaporization temperature of aluminum alloys traps gases within the molten metal. Furthermore, the likelihood of crack formation is increased by the rapid solidification rate [6,7]. These defects significantly reduce the metallurgical properties of welded aluminum, particularly tensile strength. This study verifies that the tensile strength of alloys such as AA6013 can substantially decrease following autogenous laser welding down to as low as one-tenth of their maximum achievable tensile strength [8]. The difficulties are further complicated by aluminum alloys' low absorption of laser radiation, primarily because of the former's high reflectivity and thermal conductivity. This reduces heat input and forms a relatively small weld pool [9–12]. However, overcoming all these drawbacks, researchers have significantly developed welding of aluminum alloys that are generally not weldable by conventional processes. In their investigation of the effects of welding settings on the metallurgical and mechanical characteristics of 2024 aluminum alloys, Akkurt et al. [13] discovered that the development of porosities had an impact on the mechanical properties. In a comparable manner AA5052, AA5053, and AA6061 alloys were laser-welded by El-Batahgy and Kutsuna [10]. They claimed that porosities and hot cracking are the defects within the weld seams significantly deteriorated the physical properties of these alloys. Among aluminum alloys, AA5754 is particularly notable for its excellent weldability. It allows for deeper weld penetration with a small HAZ, which is important to preserve the physical properties of the weld. This alloy minimizes thermal distortion and cracking, thus ensuring structural reliability under demanding conditions. With its strong bond and sufficient ductility, AA5754 is an excellent choice for automotive and marine applications.

2. Experimental Methodology

The base metal, AA5754, was sourced from Arihant Aluminium in Chennai. It was cut into dimensions of 100x75 mm using an axe machine, and the edges were flattened and smoothened with a conventional machine. A specimen of the parent metal-AA5754-was further prepared for analysis of chemical composition and was analyzed using optical emission spectrography, whereby the key values were obtained summarized in Table 63.1. The physical properties of base metals were determined using a UTM and a Vickers Micro Hardness Tester in which the results were portrayed in Table 63.2. The parent metal was first cleaned with acetone for the prevention of an oxide layer. After that, the two plates were clamped in a butt joint configuration securely, and the LW process was carried out using a Laser Welding technique.

Table 63.1 Chemical composition (wt%) of AA5754

Material	Si	Fe	Mn	Mg	Cu	Zn	Cr	Al
AA5754	0.288	0.32	0.25	2.8	0.03	0.02	0.02	Bal.

Table 63.2 Mechanical properties of AA5754

Material	YS(MPa)	UTS(MPa)	Elongation (%)	Hardness (Hv 1)
AA5754	160	240	15	75

Samples were prepared using Wire-EDM for both metallurgical and mechanical samples. These samples were tested on UTM for the following ASTM-E8 standards. Samples were also prepared by Wire-EDM for microstructure evaluation. Samples were mounted utilizing a hot mounting press after which they were polished in SiC4 abrasive sheets with sequence from 80 to 400 grits first. Then, further polished they were with sheets that follow the sequence from 80 to 2000 grits. Keller's reagent was used to reveal the microstructure in the samples. Samples underwent SEM analysis to view the grain geometry and distribution of the precipitates. A test for hardness was carried out on different zones of the specimens using a Vickers Micro Hardness Tester. The indentations were done at 0.5 mm intervals using a 1 kg load and a dwell time of 25 seconds to obtain the microhardness.

CCD modern approach was carried out within the frame work of RSM to enhance the welding process parameters. A detailed literature review on Laser Beam Welding (LBW) of various aluminum alloys was conducted to figure out the most significant welding process parameters (WPPs) for this study. Based on the findings, three key parameters were chosen: LP, WS, and FP. The parameter limits were established through an extensive review of relevant studies. The lower limit was assigned a coded value of -α (-1.682), while the upper limit was given a coded value of α (1.682). The midpoint between these limits was designated as the central value, coded as '0'. The coded levels of +1 and -1 were calculated using the standard equation for CCD. Table 63.3 shows the actual values of each parameter corresponding to these coded levels.

Table 63.3 Limits of welding process parameter

WPP	Unit	Levels				
		-1.682	-1	0	1	+1.682
LP	kW	1	1.6	2.5	3.4	4
WS	mm/s	20	28	40	52	60
FP	mm	-2	-1.2	0	1.2	2

3. Results and Discussion

3.1 Optimization of Welding Parameters

Analysis of variance (ANOVA) is a statistical tool used for effectively interpreting experimental data. It is used quality control of manufacturing process and ensuring the reliability of the process. The outcomes of the tensile tests conducted on the welded metals are analyzed in relation to various factors, such as WS, LP, and FP, with the output being the physical properties of the welds. The ANOVA outcomes are summarized in a table that includes data for numerical analysis, such as Degrees of Freedom (DoF), mean Sum of Squares (MS) and Sum of Squares (SS). It is evident that the amplitude negatively influences the mechanical strength of the welds. As the amplitude increases, the mechanical resistance of the weld tends to decrease. In this investigation, the process was carried out by varying the feed rate while keeping all other process parameters are kept constant. No major welding defects such as porosity or macrocracks were detected during visual inspection of the welded joints. Smooth appearance were achieved, and no distortion was occurred on the welded sheets.

Table 63.4 Design of experiments framed using CCD

Std. order	Run order	Coded Values			UTS (MPa)
		P	S	F	
1	9	-1	-1	-1	171
2	11	1	-1	-1	139
3	10	-1	1	-1	137
4	7	1	1	-1	170
5	14	-1	-1	1	117
6	12	1	-1	1	155
7	6	-1	1	1	143
8	16	1	1	1	142
9	4	-1.682	0	0	126
10	17	1.682	0	0	180
11	8	0	-1.682	0	160
12	1	0	1.682	0	127
13	13	0	0	-1.682	180
14	19	0	0	1.682	124
15	2	0	0	0	167
16	3	0	0	0	170
17	5	0	0	0	167
18	18	0	0	0	175
19	20	0	0	0	167
20	15	0	0	0	168

Final Equation in Terms of Actual Factor

$$UTS = +169.01 + 16.022LP - 9.043FP - 17.001WS + 3.750(LP*FP) + 5.00(LP*WS) + 3.00(FP*WS) - 5.776(LP)^2 - 9.135(FP)^2 - 6.130(WS)^2 \quad (1)$$

Table 63.5 ANOVA test results

Source	Sum of Squares	Df	Mean Square	F-value	p-value	
Model	10838.12	9	1204.24	146.09	<0.0001	significant
A-LP	3505.99	1	3505.99	425.33	<0.0001	
B-WS	1116.81	1	1116.81	135.49	<0.0001	
C-FP	3947.30	1	3947.30	478.87	<0.0001	
AB	112.50	1	112.50	13.65	0.0041	
AC	200.00	1	200.00	24.26	0.0006	
BC	72.00	1	72.00	8.73	0.0144	
A^2	480.88	1	480.88	58.34	<0.0001	
B^2	1202.67	1	1202.67	145.90	<0.0001	
C^2	541.55	1	541.55	65.70	<0.0001	
Residual	82.43	10	8.24			
Lack of Fit	32.43	5	6.49	0.6486	0.6769	not significant
Pure Error	50.00	5	10.00			
Cor Total	10920.55	19				
Std. Dev.	2.87	**R^2**		0.9925		
Mean	154.65	**Adjusted R^2**		0.9857		
C.V. %	1.86	**Predicted R^2**		0.9673		
		Adeq Precision		41.4420		

Model terms having P-values less than .0500 are considered significant. Conversely, terms whose P-values exceed 0.1000 are deemed to be nonsignificant. The model may be more efficient if redundant nonsignificant terms can be removed, without loss of hierarchical terms. Such a value could occur by chance with only a 67.69% probability. With values greater than 4 considered acceptable, this

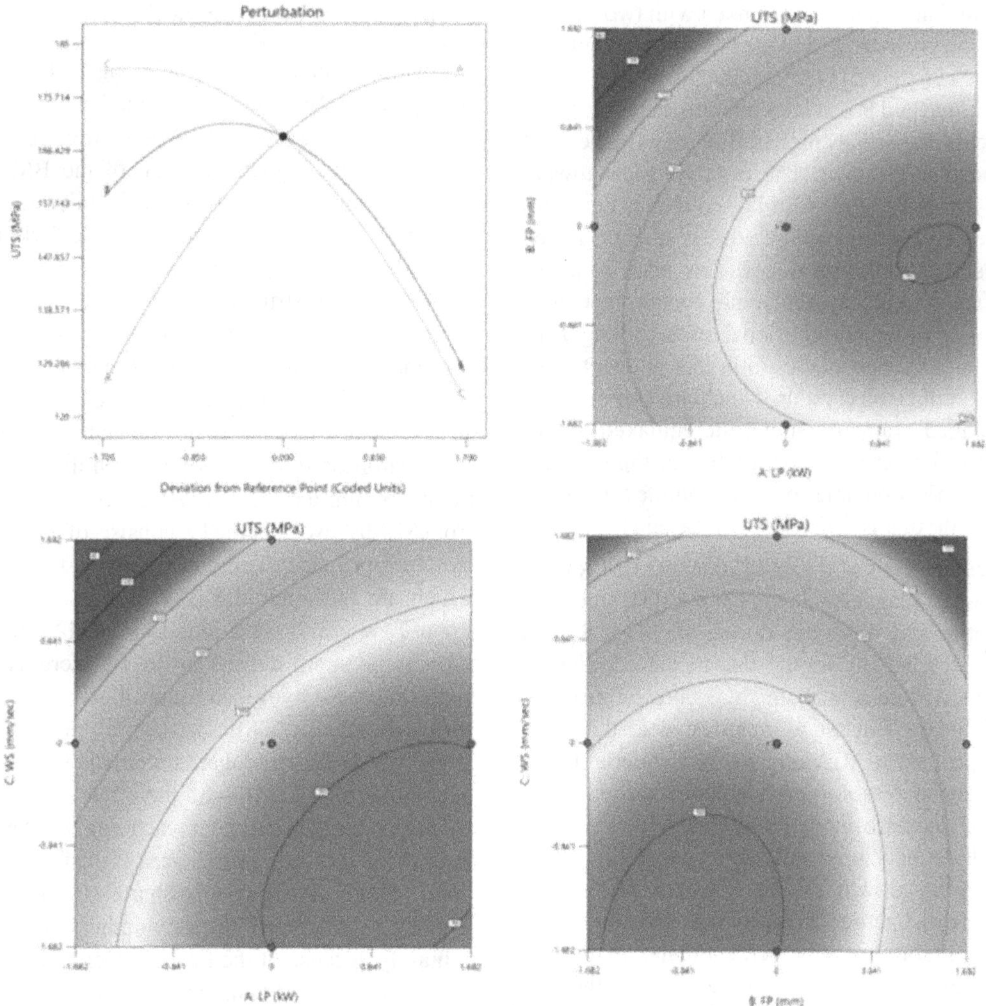

Fig. 63.1 [a] Perturbation plot [b] Interaction plot AB [c] Interaction plot AC [c] Interaction plot BC

outcome verifies that the model carries a very strong signal and can be used further for searching the design space. The equation based on actual factors is an estimation tool to predict the response at specific factor levels, given that these levels are expressed in the original units used during the experiment. However, such an equation does not find the relative contribution of every factor toward the response. This is because the coefficients have been normalized to the respective units of the factors. Also, the intercept here does not correspond to the central point in the design space. Therefore, while the equation is effective for prediction, it does not accurately represent the comparative influence of individual factors.

The perturbation chart (top-left plot) shows how sensitive UTS is to individual changes in process factors like laser power (LP), feed position (FP), and welding speed (WS), with other factors held constant. The x-axis represents the coded units for deviation from the reference point, and the y-axis is the UTS. The steeper the curves, the greater the relative sensitivity of the response to each factor. The contour plots show the interaction effects of two factors on the response variable, for which the third factor is at a fixed level. Every plot has a color gradient, where variations in the response are indicated. Contour lines represent areas of equal response, allowing optimal factor combinations to be identified. In these plots, black dots identify some experimental points or reference conditions. The visualizations are used frequently in optimization studies to identify key factors, understand interactions, and determine the settings that might maximize or improve the response variable.

3.2 Microstructure Analysis

Dendritic crystals formed perpendicular to the fusion line for AA5052 alloy are grown across the FZ where maximum flow of heat occurs from a molten pool toward the BM.

Grain growth in that region is thus linked with two factors, viz. solidification growth rate and the thermal gradient, and can be said crucial for a FZ [14]. Near the fusion line, where G is at its maximum and R is at its minimum, heterogeneous nucleation occurs on partially melted BM grains at the HAZ surface, at the fusion region. These nuclei grow toward the center of the weld bead, forming columnar dendrites [15]. In the welded zone, the temperature gradient is lower, and heat transfer is multidirectional. This leads to a low rate of solidification velocity, the solute content increases, and this forms a wider constitutional supercooling zone [14]. In this zone, the new nuclei form that allows for the development of equiaxed crystals, which are composed of both fine equiaxed grains and dendritic equiaxed grains. Greater the laser power will have high heat input, hence extends the solidification time so there is ample time to get the crystal growth; that is, grain size increased [16].

Element content analysis of the weld zone and BM reveals a slight decrease in magnesium (Mg) in the BM, since Mg is highly volatile and vaporizes during welding. Silicon content in FZ is relatively higher due to the addition of filler wire. Thus, a more outstanding hypoeutectic structure of α-Al + Si binary eutectic is manifested at the grain boundaries of primary α-Al. Along the base material HAZ, the grain size are increases to an average value of 48 μm because of the decrease in welding speed of 20 mm/s, which led to higher heat input and decrease in cooling rate, favoring over-grain growth. At the HAZ the weld metal (WM) interface, a distinct dendritic morphology is observed, with columnar grains growing perpendicular to the FZ boundary. The grains in the FZ is influenced by the size of adjacent BM grains during the epitaxial growth process.

The physical properties of the alloy are improved after welding and also after heat treatment. The dendritic microstructure in the WM reveals enrichment of solute within the intermetallic regions with columnar dendritic formations and solute micro-segregation. The segregated solute found in smaller quantities and presents a thick, continuous morphology. Moreover, a 41 μm pore within the WM is associated with evaporation of lightweight alloying elements.

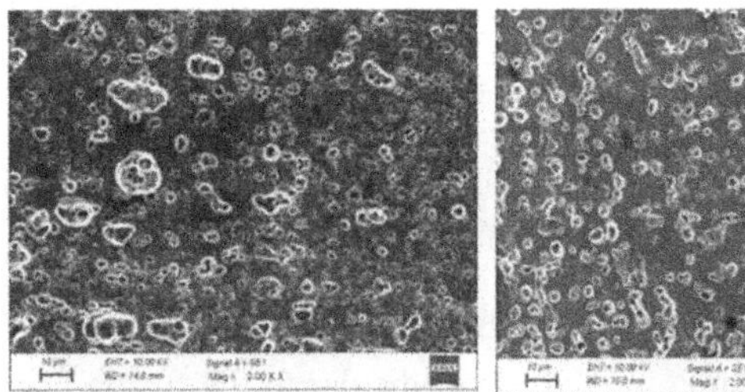

Fig. 63.2 SEM microstructure [a] Base metal AA5754-H111 [b] FZ of sample 10

4. Hardness Analysis

The weld-affected regions, including the FZ (both inner and outer sides), HAZ, and BM of the bead were investigated using microhardness tests. Microhardness values for the FZ were a bit higher than that of the BM. The present investigation noted an increase in microhardness within the FZ by specific investigators, particularly for non-heat-treatable alloys in the O temper state [18,19], as a result of the better microstructural refinement of the FZ than that of the BM. AA5182 aluminum alloys were considered for fiber laser lap welding experiments in this exertion. The tensile load influence of the welds with regard to the LP, WS, and FP point was evaluated. Microstructural and microhardness studies were done on the best sample with the maximum tensile load. Analysis of the microstructure from SEM shows that the FZ consists of columnar dendritic structure primarily and can be divided into two regions, which are the internal Fusion Zone and external Fusion Zone. Few grains were found in the HAZ near the fusion line. Some small defects in the FZ were associated with supersaturated hydrogen deposition and inter-dendritic shrinkage cavities. Microstructural refinement led to a slight increase in microhardness in the FZ. Hardness drops below that of the BM, with the effect much greater in the HAZ than elsewhere, and it reaches lowest hardness values in each of the tests in this area. Consistent with such an interpretation are the microhardness analyses of laser welding as described by many authors. The hardness loss is mainly because of the loss of specific alloying elements due to the elevated temperatures in the fusion zone during welding and the loss of cold work [20,21,22]. However, the FZ shows superior hardness when compared with HAZ.

This is because of the high welding speeds used, which result in fast cooling. Such rapid cooling enables the formation of a hard dendritic phase and a solid solute-strengthened alloy (a') during melting and solidification [23,24,25]. In contrast, the decrease in hardness in the HAZ is associated with the thermal cycle, which induces recrystallization and over-aging, leading to the development of coarse grains [21]. The extent of reduction of hardness in both the HAZ and the fusion zone is determined by variations in parameters. The hardness distribution was influenced by potential losses in magnesium content and grain coarsening in the HAZs, as well as melting in areas near the base metal [26–30, 31]. Additionally, the hardness values that are obtained from deformation during sheet production through rolling may reduce due to the effect of heat input, particularly in the HAZ. This weld joint shows the higher porosity and lower strength compared to shielded weld joints.

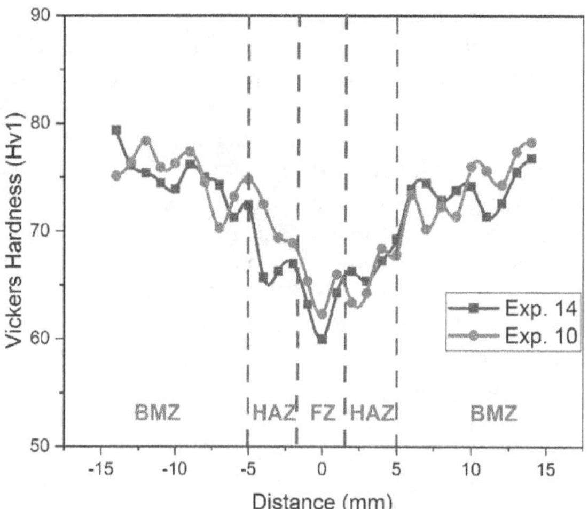

Fig. 63.3 Hardness variation graph

Figure 63.3, Laser weld of AA5754 alloy: Vickers hardness profiles across the weldment from under two experimental conditions Exp. 10 and Exp. 14. (Hardness values in terms of HV1). It provides the distance in terms of mm from the weld centre-line on the x-axis and expresses hardness values in terms of HV1 on the y-axis. The hardness values distribute the base metal zone as BMZ, HAZ, and FZ as well. The BMZ, both at the weld center and its sides, is of relatively high hardness and more or less uniform, meaning the properties of the base material are not affected. Hardness decreases gradually within the HAZ due to thermal softening as one gets closer to the weld centerline. The hardness drops to its minimum at the weld center, FZ, probably due to very fast cooling rates that tend to change the microstructure.

Beyond the FZ, the hardness increases gradually in the HAZ and returns to BMZ levels, creating a symmetrical profile around the weld center. The comparison between Exp. 10 and Exp. 14 shows that there are slight differences in the hardness profile, which may be due to differences in laser welding parameters, such as power, speed, or feed position. Figure 63.3 identifies the necessity of parameter optimization to minimize the hardness drop in the FZ and to improve the mechanical performance of the welded material. High wear resistance of the Si element and also the capability of the element to raise the microhardness of intermetallic compounds caused the joint hardness to be raised. Laser remelting welding [32,33] presented higher joint-softening issues in comparison with laser welding of AA5754 by the use of the filler wire. Favourable weldability is a feature exhibited by AA5754 that could be attributed to the small difference between the tensile strength of the weld metal and that of the base metal. Such a feature is derived from the non-heat-treatable nature of the alloy that causes less sensitivity to thermal cycling due to welding [33].

5. Conclusion

AA6009 was successfully joined using CMT welding and the following observations are made:

1. Optimized the Laser Welding process parameters, including LP, FP, and WS, to achieve superior mechanical properties for AA5754.
2. Optimized conditions enhanced tensile strength and ensured uniform microhardness with minimal defects.
3. Microstructural analysis revealed fine grain formation in the fusion zone that improved weld quality.
4. Interactions between parameters have a significant influence on tensile strength and weld quality, as reflected by contour plots.
5. Further studies can focus on dynamic loading performance and hybrid welding techniques.

References

1. Hirsch J (2014) Recent development in aluminum for automotive applications. Trans. Nonferrous Met. Soc. China 24:1995–2002 doi:10.1016/S1003-6326(14)63305-7.
2. Rejc J, Munih M (2016) Robust Visual Touch-Up Calibration Method in Robot Laser Spot Welding Application. Strojniški vestnik - Journal of Mechanical Engineering 62:12, DOI:10.5545/sv-jme.2016.3708. 697–708
3. Pakdil M, Cam G, Kocak M, Erim S (2011) Microstructural and mechanical characterization of laser beam welded AA6056 Al-alloy. Materials Science and Engineering A 528:7350–7356 doi:10.1016/j.msea.2011.06.010.
4. T. Dursun, C. Soutis, Recent developments in advanced aircraft aluminium alloys, Mater. Des. 56 (2014) 862e871, 1980–2015.
5. H. Ramiarison, N. Barka, F. Mirakhorli, F. Nadeau, C. Pilcher, Parameter opti mization for laser welding of dissimilar aluminum alloy: 5052-H32 and 6061 T6 considering wobbling technique, Int. J. Adv. Manuf. Technol. (2021) 1e17.
6. S. Katayama, Defect formation mechanisms and preventive procedures in laser welding, in: Dans Handbook of Laser Welding Technologies, Elsevier, 2013, pp. 332e373. (P22)
7. Y.M. Baqer, S. Ramesh, F. Yusof, S.M. Manladan, Challenges and advances in laser welding of dissimilar light alloys: Al/Mg, Al/Ti, and Mg/Ti alloys, Int. J. Adv. Manuf. Technol. 95 (9e12) (2018) 4353e4369. (p22)
8. A.L.d.C. Higashi, M.S. F.d. Lima, Occurrence of defects in laser beam welded Al Cu-Li sheets with t-joint configuration, J. Aero. Technol. Manag. 4 (4) (2012) 421e429. (P22)

9. Zhao Hand Deb Roy T2001 Weld metal composition change during conduction model aser welding of aluminumalloy 5182 Metallurgical and Materials Transactions B32163–72 (P6)
10. El- Batahgy A and Kutsuna M2009 Laser beam welding of AA5052, AA5083, and AA6061 aluminum alloys Advances in Materials Science and Engineering 20091–9 (P6)
11. Sanchez-Amaya JM, Delgado T, Gonzales Rovira Land Botana FJ2009 Laser welding of aluminium alloys 5083 and 6082 under conduction regime Applied Surface Science 2559512–21 (P6)
12. Kuo TY and L in YT2006 Effects of shielding gas flowrate and power waveform on Nd:YAG laser welding of A5754-O aluminum alloy Materials Transactions 47,1365–73 (P6)
13. Akkurt A, Şık A and Ovalıİ2012 The Effects of Welding Parameters on the Mechanical Properties on Laser Welding of AA2024 Pamukkale University Journal of Engineering Sciences 1837–45
14. Sanchez-Amaya JM, Delgado T, Gonzales-Rovira Land Botana FJ2009 Laser welding of aluminium alloys 5083 and 6082 under conduction regime Applied Surface Science 2559512–21
15. PakdilM, ÇamG, KoçakcM and ErimS2011 Microstructural and mechanical characterization of laser beam welded AA6056Al-alloy Materials Science and Engineering A5287350–6
16. Chu,Q.; Bai, R.; Jian, H.; Lei, Z.; Hu, N.; Yan, C. Microstructure, texture and mechanical properties of 6061 aluminum laser beam welded joints. Mater. Charact. 2018, 137, 269–276.
17. Abioye, T.E.; Zuhailawati, H.; Aizad, S.; Anasyida, A.S. Geometrical, microstructural and mechanical characterization of pulse laser welded thin sheet 5052-H32 aluminium alloy for aerospace applications. Trans. Nonferrous Met. Soc. China 2019, 29, 667–679.
18. Zhou, H.; Fu, F.; Dai, Z.; Qiao, Y.; Chen, J.; Yang, L.; Liu, W. Effect of Laser Power on Hybrid Laser-Gas Metal Arc Welding (GMAW) of a 6061 Aluminum Alloy. J. Korean Phys. Soc. 2020, 77, 991–996.
19. Zhou, L.; Hyer, H.; Park, S.; Pan, H.; Bai, Y.; Rice, K.P.; Sohn, Y. Microstructure and mechanical properties of Zr-modified aluminum alloy 5083 manufactured by laser powder bedfusion. Addit. Manuf. 2019, 28,485–496.
20. Sanchez-Amaya JM, Delgado T, Gonzalez-Rovira L, Botana FJ (2009) Laser welding of aluminium alloys 5083 and 6082 under conduction regime. Applied Surface Science 255: 9512–9521 doi:10.1016/j.apsusc.2009.07.081.
21. Sanchez-Amaya JM, Delgado T, Damborenea J, Lopez V, Botana FJ (2009) Laser welding of AA 5083 samples by high power diode laser. Science and Technology of Welding and Joining 14:78–86 DOI 10.1179/136217108X347629.
22. J. Ahn, L. Chen, E. He, J.P. Dear, C.M. Davies, Optimisation of process param eters and weld shape of high power Yb-fibre laser welded 2024-T3 aluminium alloy, J. Manuf. Process. 34 (2018) 70e85.
23. Z. Wang, J.P. Oliveira, Z. Zeng, X. Bu, B. Peng, X. Shao, Laser beam oscillating welding of 5A06 aluminum alloys: microstructure, porosity and mechanical properties, Opt Laser. Technol. 111 (2019) 58e65, 2019/04/01/.
24. B.N. Coelho, M.S. F.d. Lima, S.M.d. Carvalho, A.R.d. Costa, A comparative study of the heat input during laser welding of aeronautical aluminum alloy aa6013-T4, J. Aero. Technol. Manag. 10 (2018).
25. M. Vyskoc, M. Sahul, M. Sahul, Effect of shielding gas on the properties of AW 5083 aluminum alloy laser weld joints, J. Mater. Eng. Perform. 27 (6) (2018) 2993e3006.
26. S. Janasekaran, M.F. Jamaludin, M.R. Muhamad, et al., Autogenous double sided T-joint welding on aluminum alloys using low power fiber laser, Int. J. Adv. Manuf. Technol. 90 (2017) 3497e3505, https://doi.org/10.1007/s00170 016-9677-y.
27. Haboudou, P. Peyre, A.B. Vannes, G. Peix, Reduction of porosity content generated during Nd:YAG laser welding of A356 and AA5083 aluminium al loys, Mater. Sci. Eng., A 363 (1e2) (2003) 40e52.
28. ZhaoH, White DR and DebRoy T1999 Current issues and problems in laser welding of automotive aluminum alloys International Materials Reviews 44238–66
29. Moon DW and Metzbower EA1983 Laser beam welding of aluminum alloy 5456 Welding Journal 6253–8
30. Zhao Hand DebRoy T2001 Weld metal composition change during conduction modelaser welding of aluminum alloy 5182 Metallurgical and Materials Transactions B32163–72
31. El-BatahgyA and Kutsuna M2009 Laser beam welding of AA5052, AA5083, and AA6061 aluminum alloys Advances in Materials Science and Engineering 20091–9
32. Sanchez-Amaya JM, DelgadoT, Gonzales-Rovira Land Botana FJ2009 Laser welding of aluminium alloys 5083 and 6082 under conduction regime Applied Surface Science 2559512–21
33. PakdilM, ÇamG, KoçakcM and ErimS2011 Microstructural and mechanical characterization of laser beam welded AA6056Al-alloy Materials Science and Engineering A5287350–6

Note: All the figures and tables in this chapter were made by the authors.

Advances in Additive Manufacturing Technologies – Gurusamy Pathinettampadian et al. (eds)

Exploring Optimal Welding Parameters, Mechanical and Microstructural Insights in Friction Stir Welding of AA6009 Aluminium Alloy

Iniya T.[1],
Abishek S.V.[2], Deeksheka[3],
Bharanidharan T.[4], Rishekkumar[5], Aakash V.[6]
Student, Department of Mechanical Engineering,
Chennai Institute of Technology,
Chennai, India

◆ **Abstract:** This study uses the Central Composite Design (CCD) technique in the Response Surface technique (RSM) framework to optimize the welding process parameters (WPP) in the FSW of AA6009. Tool Rotational Speed (TRS), Tool Traverse Speed (TTS), and Axial Load (AL) are among the factors evaluated. According to the statistical analysis, the examined factors have a substantial impact on ultimate tensile strength (UTS), and there is a good interaction between TRS and TTS. Three separate zones were revealed by the microstructural analysis: the Fusion Zone (FZ), the Heat Affected Zone (HAZ), and the Thermomechanically Affected Zone (TMAZ). The latter has improved tensile and hardness qualities due to dynamic recrystallization, which produced fine, equiaxed grains. The evaporation of the Mg_2Si precipitate caused the hardness of the HAZ to drop, making it the weakest area of the weld. With a specific set of parameters-TRS = 681 rpm, TTS = 112 mm/min, and AL = 2.6 kN-optimization revealed a peak UTS of 331 MPa. Grain refining and heat input were balanced in such a situation. These findings demonstrate that precise parameter optimization is crucial to achieving a successful, defect-free weld..

◆ **Keywords:** Friction stir welding, AA6009 aluminum alloy, Welding parameters optimization, Microstructure analysis, Mechanical properties

1. Introduction

Low density and excellent strength-to-weight ratio of aluminum alloys are reason for its wide applications in aerospace, automobile, construction, and high-speed vessels. Welding is the most widely used method to join these materials; however, it creates significant problems for engineers and technologists. The problems are inherent in aluminum, due to the formation of an oxide layer, high thermal conductivity, thermal expansion, solidification shrinkage, and significant solubility of hydrogen in the melt [1]. In addition, heat treatable aluminum alloys also suffer a loss in their mechanical characteristics because of changes in phase resulting from the input of heat from welding. Heat-treatable aluminum-magnesium-silicon alloys, including AA6009, have average strength and good welding properties compared to stronger aluminium alloys [2]. The group of such alloys is very popularly used in Ship frames, pipes, storage tanks, and other structure elements of aircraft. The HAZ of the alloys undergoes extreme softening during the welding process due to dissolution of Mg_2Si precipitates, which adversely affects mechanical properties. This phenomenon presents an engineering design challenge that calls for measures to prevent softening of HAZ and improve the UTS of welded joints. The FZ in aluminum alloys normally consists of

[1]iniyat.mech2023@citchennai.net, [2]abisheksivaraj10@gmail.com, [3]deekshekaa.mech2023@citchennai.net, [4]bharanidharant.mech2023@citchennai.net, [5]rolexrishe@gmail.com, [6]aakash.vasudevan2020@gmail.com

DOI: 10.1201/9781003685906-64

large columnar grains as a outcome of dominant thermal conditions experienced during solidification. This tends to give less UTS and reduced resistance to hot cracking [3]. Controlling the solidification microstructure in welds is beneficial but challenging because of the high temperatures, thermal gradients, and epitaxial growth associated with the welding process [4]. Several approaches directed toward improving the weld fusion zones, such as inoculation with heterogeneous nucleants, microcoolers additions, surface nucleation, and physical disturbances, like torch vibration, have been explored but with limited success.

The input WPP pertinent to friction stir welding (FSW) performance have been the concept of extensive research. Juárez et al. [5] conducted the impact of tool pin profiles (TPP), TTS, and TRS on the mechanical properties of joints made from AA6061-T6 aluminum alloy. Their findings indicated that utilizing a straight hexagon bolt head pin profile led to a reduction in defects and enhancements in mechanical characteristics at designated values of WS and TRS. Effective material flow and consistent plastic deformation were identified as major factors in these improvements. Esmaeili et al. [6] investigated weld defects in dissimilar brass and aluminum joints. Inappropriate tool settings resulted in imperfections including tunnels, grooves, and cracks, which arose from insufficient material flow around the tool pins. Ghaffarpour et al. [7] examined the effect of TTS, TRS, tool tilt angle, and tool shoulder diameter (TSD) on UTS and %El of AA5083-H12 and AA6061-T6 FSW joints. The conclusions drawn from the investigation demonstrated that heat input, influenced mostly by TRS and WS, was the reason for affecting the UTS of the joint; higher heat input increases elongation but reduces strength. A variety of optimization approaches have been employed to optimize parameters of the FSW process. Gupta et al. [8] conducted a hybrid optimization strategy for determination of appropriate TSD, pin diameter, TRS, and TTS for AA5083-AA6063 welds so that weld properties improved. Cavaliere et al. [9] investigated the impact of WS on the mechanical and microstructural characteristics of AA6082-AA2024 joints and found that higher WS values and AA6082 positioning on the AS increased tensile and fatigue strength.

Alternative research has focused on describing how WPP influence the UTS of weldments. Bahemmat et al. [10] evaluated the effects of WS on the microstructure and mechanical properties of AA6061-T6 and AA7075-T6 joints, reporting that mechanical properties improved up to a specific WS threshold before deteriorating due to crack formation. Raturi et al. [11] explored the effect of TRS, preheating, WS, and TPP on the strength and quality of AA6061-T6 and AA7075-T651 joints, emphasizing the importance of these parameters in optimizing FSW performance. Several optimization methods have been applied to achieve superior weld characteristics. Raweni et al. [12] employed Taguchi design for optimizing the tilt angle, WS, and TRS with regard to crack propagation energy and tensile strength. The ANOVA technique was used to know which of the WPP have significant effects. In the optimization of some outputs, a WS of 125 mm/min a tilt angle of 3°, and a TRS of 600 rpm, showed the best results. Ugrasen et al. [13] applied Taguchi method in optimizing WS, TRS, and number of weld passes in order to maximize UTS and hardness in AA6061-AA7075 welds. More sophisticated optimization strategies also exist, among which are RSM and ANN. For instance, Bahar et al. [14] combined the Taguchi method with GRA for optimizing tool depth, TRS, and TTS, thus enhancing hardness and toughness. Kundu et al. [15] employed the Taguchi L9 orthogonal array and GRA to improve WS, tilt angle, and TRS for microhardness, UTS, and elongation in AA5083-H321 joints. Similarly, Hema et al. [16] implemented RSM to design experiments and optimize the UTS, impact strength, and microhardness of AA2014 and AA6061 dissimilar joints.

The utilization of advanced models and algorithms has enhanced the efforts in optimization. Elatharasan et al. [17] utilized RSM alongside central composite design (CCD) to formulate mathematical models for YS, %El, and UTS of butt weld joints, resulting in optimal outcomes achieved over 20 experimental runs. Shanavas et al. [18] investigated the effects of tilt angle, WS, TRS, and TPP on elongation and UTS in AA5052-H32 joints, leading to a weld efficiency of 93.51% via optimized parameters. ANN has been seen to have improved prediction performance compared to traditional methods like RSM. For example, Jayaraman et al. [19] utilized ANN to predict UTS in AA319 aluminum cast alloy joints, and it was seen that ANN outperformed RSM in terms of error prediction. Lakshminarayanan et al. [20] also applied ANN to predict UTS in AA7039 joints, where TRS, WS, and axial force were considered as key parameters, and the maximum UTS was obtained by using ANN-optimized parameters. Shehabeldeen et al. [21] combined ANFIS with optimization techniques to predict optimal FSW parameters for UTS in aluminum joints. Their work, in combination with that of Roshan et al., [22] addressed the feasibility of achieving superior mechanical properties using computational models in optimizing the parameters. The author evaluated TPP, TRS, WS, and axial force and obtained optimal yield strength and UTS using ANFIS.

According to the reviewed literature, the metallurgical and mechanical features of aluminum joints are positively impacted by the WPP of the FSW process, including TRS, WS, TTS, and tilt angle. increasing overall performance, decreasing flaws, and increasing weld qualities all depend on optimizing these parameters. In order to determine the impact of WPP (TRS, TTS, and AL) on the UTS, parameter optimization is carried out in this study utilizing the RSM modeling approach. To learn more regarding HAZ softening, solidification structures, strength, and defect generation in welded joints, the microstructure and mechanical properties of the weldment linked utilizing optimal WPP are also examined.

2. Experimental Methodology

A power hacksaw was used to cut the 5 mm thick base metal AA6009-T6 into the necessary 100 x 55 mm dimensions after it was obtained from Arihant Aluminium in Chennai. Using a milling machine, the plates' connecting edges are flattened. Table 64.1 displays the results of an optical emission spectroscopy analysis of the base metal's (BM) chemical composition. Table 64.2 lists the mechanical characteristics of the BM that were investigated utilize a Vickers microhardness tester and a UTM. Acetone is then used to clean the base metal plate in order to get rid of the oxide layer. The FSW welding procedure is performed using an FSW machine with two base plates secured firmly in a butt joint pattern. The straight square TPP was utilized for conducting the FSW process. The specifications of the TPP used is given in Table 64.3.

Table 64.1 Chemical Composition of AA6009

BM	Mg	Mn	Fe	Si	Cu	Cr	Zn	Al
AA6009	0.7	0.4	0.4	0.7	0.2	0.1	0.1	97.2

Table 64.2 Mechanical Properties of AA6009

Material	UTS(MPa)	YS(MPa)	El (%)	Hardness (Hvl)
AA6009	245	185	15	75

Table 64.3 Tool dimensions used for the study

Process parameters	Values
D/d ratio of the tool	3
Tool pin length (mm)	4.5
TPP	Square
Tool shoulder diameter, D (mm)	15
Tool pin diameter, d (mm)	5

The WEDM machine is used to extract the mechanical and metallurgical samples. Following their extraction in compliance with ASTM-E8 guidelines, the tensile samples are tested using the universal testing equipment. Furthermore, the sample for microstructural examination is removed from the WEDM machine and placed in a hot mounting press utilizing Bakelite powder. The mounted samples are hand-polished using sheets of SiC abrasive (80–400 grits) and then machine-polished utlizing sheets of SiC abrasive (600–2000 grits). The polished samples are further etched using Kellers' reagent to reveal their microstructure. The size and shape of the precipitates and grains are ascertained by analyzing the polished samples' microstructure using a scanning electron microscope (SEM). The microhardness test is carried out on the microstructure samples in each of the many zones utilizing a Vickers microhardness tester, with 0.5 mm between each indentation. The microhardness test involves applying a 1-kilogram stress for 20 seconds of dwelling time.

The optimization of WPP was carried out utilizing the CCD approach within the framework of RSM. A comprehensive literature review on FSW of various aluminium alloys was done to identify the most influential WPP's for this study. From this literature review, three key parameters were selected: TRS, TTS, and AL. The critical values for these parameters were set using an extensive study of relevant literature. The minimum value was given the coded value of -α (-1.682), while the maximum value was given the coded value of α (1.682). The center value was established at the midpoint between the minimum and maximum limits with an assigned coded value of '0'. The actual value calculations for the variables identified with the coded levels +1 and -1 are presented using the following equation: Table 64.4 reports actual values of each of the welding process parameters associated with the designated coded levels:

$$X_i = \frac{1.682\left[2X - (X_{max} + X_{min})\right]}{(X_{max} - X_{min})}$$

Table 64.4 Limits of welding process parameter

WPP	Unit	Levels				
		-1.682	-1	0	1	+1.682
TRS	rpm	600	681	800	919	1000
TTS	mm/min	80	88	100	112	120
AL	kN	1	1.4	2	2.6	3

3. Results and Discussion

3.1 Optimization of Welding Parameters

To find the best WPP of welding, a Central Composite Design has been adopted based on principles drawn from Response Surface Methodology [23]. For the purpose of this research, the selected parameters have been TRS, TTS, and AL because it is perceived that these three are very significant in deciding the weld quality and the mechanical characteristics of the joints produced [24]. The experimental design follows a systematic DOE table designed by the combinations of CCD. Table 64.5 shows the combinations of parameter levels evaluated in the optimization procedure, along with associated experimental responses.

Table 64.5 DOE framed using CCD along with the related UTS

Std. order	Run order	Coded Values			UTS (MPa)
		P	S	F	
1	9	-1	-1	-1	227
2	16	1	-1	-1	304
3	12	-1	1	-1	300
4	7	1	1	-1	289
5	8	-1	-1	1	242
6	14	1	-1	1	285
7	3	-1	1	1	331
8	10	1	1	1	286
9	1	-1.682	0	0	263
10	19	1.682	0	0	291
11	15	0	-1.682	0	244
12	4	0	1.682	0	308
13	18	0	0	-1.682	299
14	6	0	0	1.682	311
15	20	0	0	0	307
16	13	0	0	0	312
17	11	0	0	0	309
18	17	0	0	0	313
19	5	0	0	0	308
20	2	0	0	0	310

A predictive model for calculating the response at predetermined values of each element is provided by the equation stated in terms of real factors. According to the experimental setup, these values must be specified in their original units [25]. It should be noted that while the coefficients are scaled to take into account the different units of the elements, this equation is not appropriate for evaluating the relative contribution of each factor. The intercept's relevance for evaluating factor impacts is further limited by the fact that it does not correlate to the center of the design space.

$$UTS = 309.88 + 8.134 * TRS + 18.718 * TTS + 3.235 * AL - 22(TRS * TTS) - 8.5(TRS * AL) + 4(TTS * AL) - 11.936(TRS2) - 12.29(TTS2) - 2.04(AL2)$$

Table 64.6 ANOVA test results

Source	Sum of Squares	Df	Mean Square	F-value	p-value	
Model	14258.06	9	1584.23	325.48	< 0.0001	signifi-cant
A-TRS	903.65	1	903.65	154.63	< 0.0001	
B-TTS	4785.08	1	4785.08	410.95	< 0.0001	
C-AL	142.93	1	142.93	54.81	< 0.0001	
AB	3872.00	1	3872.00	16.09	< 0.0001	
AC	578.00	1	578.00	61.02	< 0.0001	
BC	128.00	1	128.00	155.16	0.0002	
A^2	2053.26	1	2053.26	746.03	< 0.0001	
B^2	2176.70	1	2176.70	1334.48	< 0.0001	
C^2	59.79	1	59.79	357.14	0.0024	
Residual	36.89	10	3.69			
Lack of Fit	10.05	5	2.01	0.3461	0.8475	not sig-nificant
Pure Error	26.83	5	5.37			
Cor Total	14294.95	19	1584.23			
Std. Dev.	1.92	**R^2**		0.9974		
Mean	291.95	**Adjusted R^2**		0.9951		
C.V. %	0.6578	**Predicted R^2**		0.9920		
		Adeq Precision		77.2478		

The statistical significance of the generated model and the process parameters involved were checked using ANOVA. The results further validated the adequacy of the model and narrowed down the significant parameters with their effect on the response. Model F-value of 429.49 is a significant measure that the model is statistically significant; further, the chances of such a high F-value outcome due to random noise are just 0.01%. P-values < 0.0500 indicate significant model terms, and in this research, the terms A, B, C, AB, AC, BC, A^2, B^2, and C^2 were identified as significant contributors. Terms having P-values > 0.1000 are insignificant. Removal of insignificant terms improves the effectiveness of the model, provided that such removal does not affect the model order existing in the model. The observed F-value may be the result of random variation with an 84.75% probability since the Lack of Fit F-value of 0.37 shows that the lack of fit is not statistically significant when compared to pure error. The presence of nonsignificant lack of fit is useful as it indicates the appropriateness of the model for being a good fit and that can be used for effective prediction. Moreover, the Predicted R^2 value of

0.9920 exhibits a close correspondence with the Adjusted R^2 value of 0.9951, revealing a discrepancy of < 0.2, thereby bolstering the dependability of the model. The measure of adequate precision, which quantifies the S-N ratio, was determined to be 77.248, markedly exceeding the minimum criterion of 4. This finding substantiates the model's substantial signal and resilience. Collectively, these results confirm the validity of the model and its suitability to explore and optimize the design space efficiently.

The perturbation plot (Fig. 64.1(a)) provides a clear visualization of the effect of each input parameter on the UTS, with the other two parameters held constant at their centric points [26]. This graphical presentation aptly highlights the differential impact of each of the WPP on the UTS of the weld, thus yielding much information about their individual influences. Similarly, the contour plots shown in Fig. 64.1(b-d) represent the two-way interactions of any two input parameters on UTS while keeping the third at its centric point. Such presentations are essential to understand the interactions among parameters and the combined effect on weld quality [27]. Results indicated that TRS and TTS are significantly impacting the heat input, which is an essential parameter in finding the properties of welds. The heat input is directly proportional to the TRS; therefore, the increase in TRS raises the heat input, and a wider HAZ along with the strength loss can be seen. Conversely, TTS is inversely proportional to heat input; a higher TTS reduces the time of heat absorption and decreases heat input, which can cause incomplete fusion or weak bonding.

The interaction between TRS and TTS also affects the microstructural characteristics of the weld zone. For example, excessive heat input caused by a high TRS and a low TTS will cause coarse grains, which in turn reduces the mechanical strength. Poor joint quality may be caused by either low TRS or a high TTS, leading to insufficient

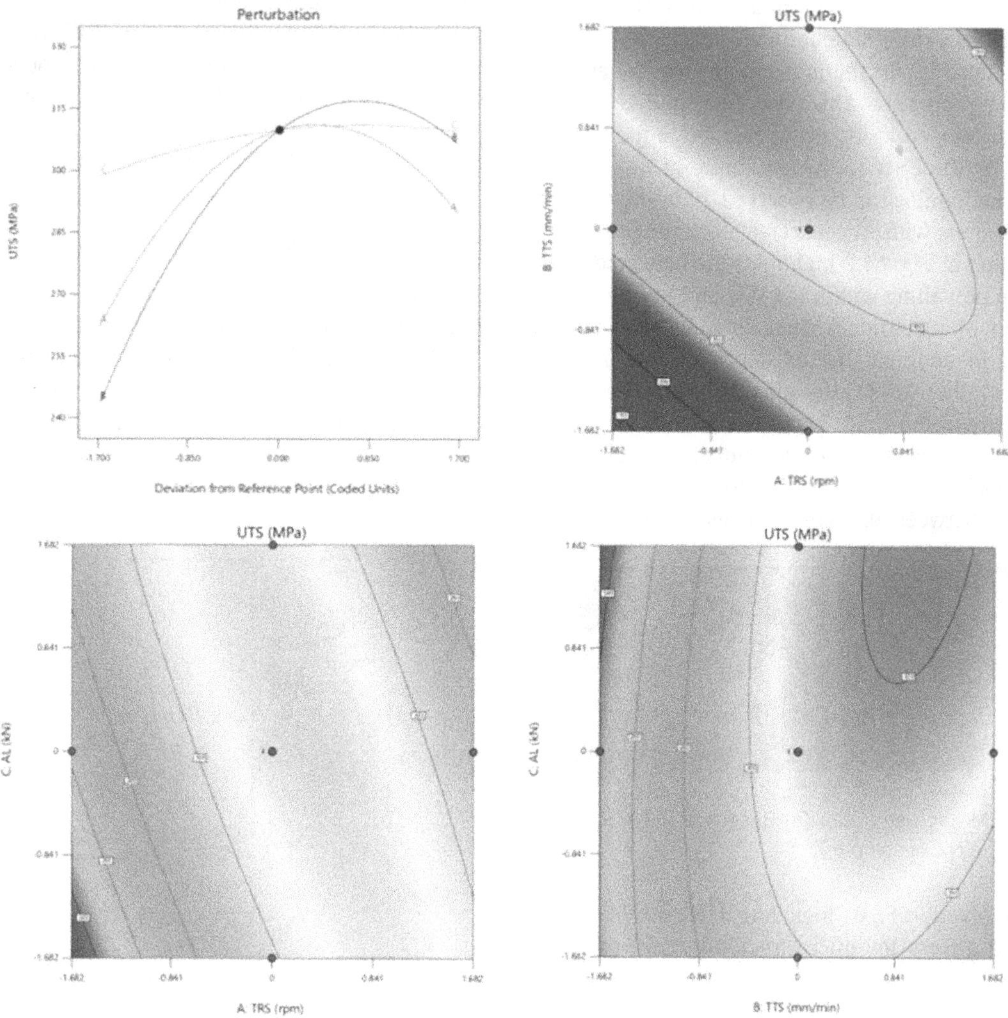

Fig. 64.1 [a] Perturbation plot [b] Interaction plot AB [c] Interaction plot AC [c] Interaction plot BC

heat input. The contour plots are very useful to identify the optimal ranges of parameters. Strong interactions between parameters AB, AC, and BC are found from the elliptical shapes in Fig. 64.1(b), Fig. 64.1(c), and Fig. 64.1(d). These interactions suggest that the effect of one parameter is highly dependent on the levels of the others and underlines the need to balance the combination of parameters. TRS-TTS interaction is particularly essential in the synergy of both to ensure that heat input is sufficient, promote the fine grain structure, and obtain good mechanical properties. The results bring to light the importance of careful optimization of WPP in improving weld quality and performance.

3.2 Microstructure Analysis

Microstructural analysis of the weldment was done on a scanning electron microscope (SEM). SEM images of the weldment are demonstrated in Fig. 64.2. The asymmetrical nature of FSW leads to several microstructural zones that are often encountered during welding, namely HAZ, TMAZ, and stir zone, also referred to as weld nugget. All these regions exhibit unique microstructural features, which are dictated by the thermal and mechanical aspects of the welding process. This fine recrystallized grain structure is developed inside the WNZ due to the stirring action given by the FSW tool. Here, uniform distribution of fine equiaxed grains can be observed, pointing to good material flow along with proper recrystallization processes. The micrograph in Fig. 64.2 highlights the equiaxed grain configuration prevailing within the WN and is essential in determination of mechanical characteristics with regard to the joint. In addition, the existence of intermetallic compounds in the WNZ also affects the strength and overall integrity of the weld. These compounds can either strengthen or weaken the weld, depending on their shape and distribution, because they usually form as a by-product of the reaction between alloying elements when welding.

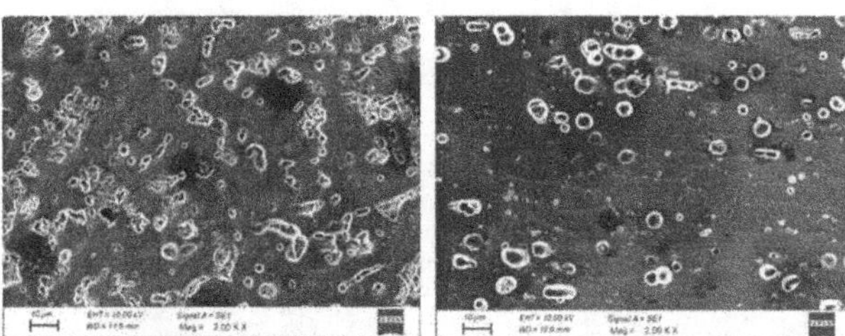

Fig. 64.2 SEM microstructure [a] Base metal AA6009-T6 [b] WNZ of sample 7

The WPP involved in FSW, including TRS, TTS, and AL are known to govern the microstructural zones of FSW. Sevvel et al. [28] have found that high tool rotational speeds along with low feed rates result in weldments without defects but with better strength, during welding of AZ31B magnesium plates. The rapid solidification process does not allow the grains to grow coarse and instead allows a finer microstructure, which is crucial for achieving improved mechanical properties. On the other hand, overheating due to very high rotational speed can affect the weld zone adversely. Umasankar Das et al. [29] in their study on joining of AA6101 and AA6351 plates observed impact energy decreases with increase in rotational speeds. This phenomenon was attributed to the excessive overheating that resulted in over-refining of grains and altered the mechanistic balance of properties in the weld zone. Transition zones like TMAZ and HAZ also determine the actual performance of the weld joint. The TMAZ is found between the WNZ and the HAZ, where grains are partially recrystallized usually because of the moderate heating and deformation. The HAZ, on the other hand, undergoes thermal exposure without significant plastic deformation, often leading to grain coarsening and potential softening, particularly in heat-treatable aluminum alloys. The interaction between metallurgical and mechanical properties is evident in the influence of grain size, distribution, and intermetallic compound formation. Fine grains in the WNZ are associated with enhanced tensile strength, while coarse grains or defects in the HAZ can lead to stress concentrations and reduced performance under mechanical loading.

3.3 Hardness Analysis

The microhardness profile of FSWed heat-treatable AA6009-T6 aluminum alloys is presented in Fig. 64.3. The hardness profile is almost symmetrical for both AS and RS of the weld, thus reflecting uniform thermal and mechanical effects imparted by the FSW process [30]. Additionally, an in-depth analysis of the hardness variation across the different zones of the weld would be crucial in

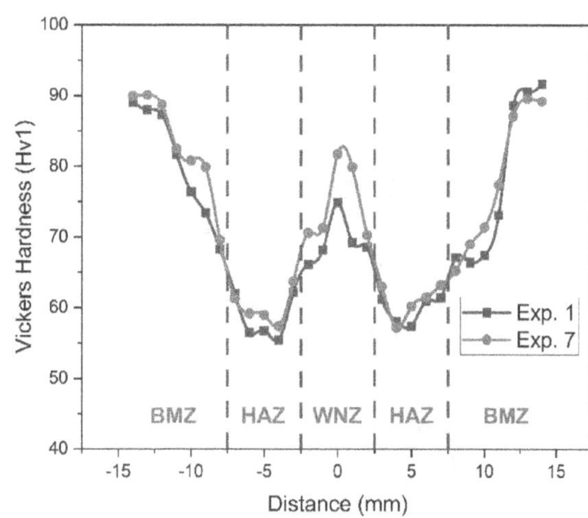

Fig. 64.3 Hardness along different zones of the weldment

obtaining information about the microstructural changes that take place within the material during welding [31].

The hardness profile for the AA6009-T6 alloy shows that the maximum hardness readings occur in the base metal (BM). This is linked to the large amount of Mg_2Si precipitates that are the major strength component for this alloy [32]. The hardness values realized in BM are in line with its heat treated T6 state where the distribution of precipitates is optimized to realize the ultimate strength and hardness. For one, the HAZ had its hardness significantly reduced from a high value, making that part of the weld zone weakest. This decrease is believed primarily due to the precipitates of Mg_2Si going into solution, wherein it is dissolved as formed in the thermal cycles for welding. The temperatures during this HAZ exceed that at which Mg_2Si could go into solution, whereby those precipitates dissolve leading to a loss of such strengthening [32]. This will highly compromise the hardening effect in the HAZ due to precipitate. This will cause problems in the structural strength of the weld. A marked hardness drop is evident in advancing into the TMAZ. The region has interaction of thermal cycles and plastic deformation that disrupts the distribution of precipitates. A significant plastic deformation in the region nullifies the hardening effect of the precipitates and hence a hardness reduction is noted. Moreover, the columnar grains and coarser precipitates in the TMAZ are also responsible for the lower hardness of this zone compared to the stir zone.

The hardness values of the WNZ, are greater than those of the TMAZ and HAZ. The increase is due to extreme grain size refinement and dynamic recrystallization that takes place during the stirring action of the FSW tool. Fine, equiaxed grains formed during dynamic recrystallization contribute to increased hardness by virtue of the Hall-Petch relationship. The arrangement of the tool pin is critical to achieving grain refinement because it influences both the flow of material and the thermal energy generated in the stirring zone. In addition, the hardness within the WNZ is influenced by the arrangement of precipitates along with the properties of the recrystallized grain structure. Although the thermal cycles in this region may induce some dissolution of precipitates, the cumulative effect of grain refinement and uniform flow of material negate this loss; hence, increased hardness relative to TMAZ and HAZ is seen. This also underlines the necessity of refining the tool design as well as the process parameters to achieve optimized mechanical properties in the WNZ.

4. Conclusion

The FSW is successfully conducted on AA6009 aluminium alloy without any defects and the following conclusions are made:

1. AA6009 Aluminum alloy was treated with CCD combined with RSM optimization of the WPP of the FSW process showed that optimized conditions are kept at TRS = 681 rpm, TTS = 112 mm/min, and AL = 2.6 kN.
2. The optimized parameters achieved a maximum UTS of 331 MPa, demonstrating the importance of balanced parameter selection for superior mechanical performance.
3. Microstructure examination in the WNZ revealed equiaxed grains and fine after dynamic recrystallization, thus ensuring dramatically enhanced weld properties in terms of tensile and hardness.
4. The weakest area was discovered to be HAZ, with decreased hardness as a result of the Mg2Si precipitates dissolving, suggesting the need for precise control over the heat input.

References

1. Jansen, T. (2016). Challenges and Solutions in Resistance Welding of Aluminum Alloys. https://www.semanticscholar.org/paper/534882cb7b59b908cbf819d07888a9c417a9df36
2. Meysam, H. (2018). Welding Parameters for Aluminum Alloys. Encyclopedia of Aluminum and Its Alloys. https://www.semanticscholar.org/paper/25b6032fba09b99a6a0e01be93a110982c873ba4
3. Galbraith, C. M., Kanko, J. A., Krupicz, B. W., Singh, P., Tesselaar, D., & Webster, P. (2021). Wobble-welding of copper and aluminum alloys with inline coherent imaging. https://www.semanticscholar.org/paper/1dd667ff39964180eae41005663a819706e97a99
4. Möller, M., Haug, P., Scheible, P., Buse, C., Frischkorn, C., & Speker, N. (2022). Spatially tailored laser energy distribution using innovative optics for gas-tight welding of casted and wrought aluminum alloys in e-mobility. Journal of Laser Applications. https://www.semanticscholar.org/paper/4818c0167319e629b54b1892e1ff958ac2eb8012
5. Juárez, J., Almaraz, G., García, R., López, J.: Effect of modified pin profile and process parameters on the friction stir welding of aluminium alloy 6061–T6. Adv. Mater. Sci. Eng. 2, 1–9 (2016)
6. Esmaeili, A., Besharati Givi, M.K.,Zareie Rajani, H.R.: Experimental investigation of material flow and welding defects in friction stir welding of aluminium to brass. Mater. Manuf. 27, 1402–1408 (2012). https://doi.org/10.1080/10426914.2012.663239
7. Ghaffarpour, M., Aziz, A., Hejazi, T.H.: Optimization of friction stir welding parameters using multiple response surface methodology. Proc. Inst. Mech. Eng. L 231, 571–583 (2017)
8. Gupta, S., Pandey, K., Kumar, R.: Multi-objective optimization of friction stir welding process parameters for joining of dissimilar AA5083/AA6063 aluminium

alloys using hybrid approach. Proc. Inst. Mech. Eng. L: J. Mater. Des. Appl. 232(4), 343–353 (2016). https://doi.org/10.1177/1464420715627294

9. Cavaliere, P., Santis, A., Panella, F., Squillace, A.: Effect of welding parameters on mechanical and microstructural properties of dissimilar AA6082-AA2024 joints produced by friction stir welding. Mater. Des. 30, 609–616 (2009). https://doi.org/10.1016/j.matdes.2008.05.044
10. Bahemniat, P., Haghpanahi, M., Besharati, M.K., Ahsanizadeh, S., Rezael, H.: Study on mechanical, micro and macrostructural characteristics of dissimilar friction stir welding of AA6061-T6 and AA7075-T6. Proc. Inst. Mech. Eng. B J Eng. Manuf. 224, 1854–1864 (2010). https://doi.org/10.1243/09544054JEM1959
11. Raturi, M., Garg, A., Bhattacharya, A.: Joint strength and failure studies of dissimilar AA6061-AA7075 friction stir welds: effects of tool pin, process parameters and preheating. Eng. Fail. Anal. 96, 570–588 (2019). https://doi.org/10.1016/j.engfailanal.2018.12.003
12. Raweni, A., Vidosav, M., Sedmak, A., Tadic, S., Kirin, S.: Optimization of AA5083 friction stir welding parameters using Taguchi method. Teh. Vjesn-technical gazette. 25(3), 861–866 (2018). https://doi.org/10.17559/TV-20180123115758
13. Ugrasen, G., Bharath, G., Kishor Kumar, G., Sagar, R., Shivu, P.R., Keshavamurthy, R.: Optimization of process parameters for Al6061-Al7075 alloys in friction stir welding using Taguchi's technique. Mater. Today: Proc. 5(1), 3027–3035 (2018). https://doi.org/10.1016/j.matpr.2018.01.103
14. Bahar, D., Arvind, N., Yadav, V., Raju, P., Dilbahar, M.: Multi objective optimization in friction stir welding using Taguchi orthogonal array and grey relational analysis. IJATEE 5, 2394–7454 (2019). https://doi.org/10.19101/IJATEE.2018.544002
15. Kundu, J., Singh, H.: Friction stir welding of AA5083 aluminium alloy Multi-response optimization using Taguchi-based grey relational analysis. Adv. Mech. Eng. (2017). https://doi.org/10.1177/1687814016679277
16. Hema, P., Sai KumarNaik, K., Ravindranath, K.: Prediction of effect of process parameters on friction stirwelded joints of dissimilar aluminium alloy AA2014 & AA6061 using taper pin profile. Mater. Today: Proc. 4(2), 2174–2183 (2017). https://doi.org/10.1016/j.matpr.2017.02.064
17. Elatharasan, G., Senthilkumar, V.S.: An experimental analysis and optimization of process parameter on friction stir welding of AA 6061–T6 aluminium alloy using RSM. Procedia Eng. 64, 1227–1234 (2013). https://doi.org/10.1016/j.proeng.2013.09.202
18. Shanavas, S., Dhas, J.E.R.: Parametric optimization of friction stir welding parameters of marine grade aluminium alloy using response surface methodology. T. Nonferr. Metal Soc. 27, 2334–2344 (2017). https://doi.org/10.1016/S1003-6326(17)60259-0
19. Jayaraman, M., Sivasubramanian, R., Balasubramanian, V., Lakshminarayanan, A.: Application of RSM and ANN to predict the tensile strength of friction stir welded A319 cast aluminium alloy. Indian J. Med. Res. 4, 306–323 (2009). https://doi.org/10.1504/IJMR.2009.026576
20. Lakshminarayanan, A.K., Balasubramanian, V.: Comparison of RSM with ANN in predicting tensile strength offriction stir welded AA7039 aluminium alloy joints. T. Nonferr. Metal Soc. 19(1), 9–18 (2009). https://doi.org/10.1016/S1003-6326(08)60221-6
21. Shehabeldeen, T.A., Elaziz, M.A., Elsheikh, A.H., Zhou, J.: Modeling of friction stir welding process using adaptive neuro-fuzzy inference system integrated with Harris Hawks optimizer. J.Mater. Res. Technol. 8(6), 5882–5892 (2019). https://doi.org/10.1016/j.jmrt.2019.09.060
22. Roshan, S., Jooibari, M., Teimouri, R., Asgharzadeh-Ahmadi, G., Naghibi, M., Sohrabpoor, H.: Optimization of friction stir welding process of AA7075 aluminium alloy to achieve desirable mechanical properties using ANFIS models and simulated annealing algorithm. Int. J. Adv. Manuf. Technol. 69, 1803–1818 (2013). https://doi.org/10.1007/s00170-013-5131-6
23. Sunny, K. T., Korra, N. N., Vasudevan, M., & Arivazhagan, B. (2022). Parameter optimization and experimental validation of A-TIG welding of super austenitic stainless steel AISI 904L using response surface methodology. Proceedings of the Institution of Mechanical Engineers, Part E: Journal of Process Mechanical Engineering. https://www.semanticscholar.org/paper/e77551d7ab8400350dfd2b8c473487911de4900d
24. Bindu, Bhasi, & Ramachandran. (2023). Modeling and parametric optimization of friction stir welding of aluminium alloy AA7068-T6 using response surface methodology and desirability function analysis. Bulletin of the Polish Academy of Sciences Technical Sciences.
25. https://www.semanticscholar.org/paper/58e9e53fb77ad7f713a92a8724c036d3abdf19e6
26. Bae, J.-H., Park, Y., & Lee, M. (2021). Optimization of Welding Parameters for Resistance Spot Welding of AA3003 to Galvanized DP780 Steel Using Response Surface Methodology. International Journal of Automotive Technology. https://www.semanticscholar.org/paper/c9219e48437d33296dd6f66ffb40c47d2509f5ad
27. Abima, C., Akinlabi, S., Madushele, N., & Akinlabi, E. (2022). Process Parameters Optimization for GMA Welding of AISI 1008 Steel Joints for Optimal Tensile Strength.
28. https://www.semanticscholar.org/paper/40f25f496ab6b2042711425dae4ef485814caf8a
29. Duncan, A. R., Suresh, U., Sai, M. S., Udayakiran, M., Pittala, R. K., Yelamasetti, B., & Abhishek, D. (2024). An investigation into process parameters optimization for ultimate tensile strength of friction stir welded Al-10Mg-8Ce-3.5Si joint: Numerical and Design of Experiments approach. Journal of Physics: Conference Series.
30. https://www.semanticscholar.org/paper/d7e1516908d74213185533bcd86802f6c2ddc76d
31. Sevvel, P. and V. Jaiganesh, Characterization of mechanical properties and microstructural analysis of friction stir welded AZ31B Mg alloy thorough optimized process parameters. Procedia Engineering, 2014. 97: p. 741-751.
32. Umashankar D. and Vijay T, Effect of Tool Rotational Speed on Temperature and Impact Strength of Friction

Stir Welded Joint of Two Dissimilar Aluminum Alloys. Materials Today: Proceedings, 2018. 5(2): p. 6170-6175

33. Osman, M. H., & Tamin, N. (2023). Influence of tool pin profile on the mechanical strength and surface roughness of AA6061-T6 overlap joint friction stir welding. Journal of Mechanical Engineering and Sciences. https://www.semanticscholar.org/paper/86f4f5418aebbacb3983ba8dc3846ca8eb8fc279
34. Hasnol, M. Z., Zaharuddin, M. F. A., Safian, Sharif, & Rhee, S. (2022). Effect of Tool Pin Profile of Underwater Friction Stir Welding of Dissimilar Materials Aa5083 and Aa6061-T6. https://www.semanticscholar.org/paper/c6f6dd0d0f4ad96e195bfe15cf0578a31c9dc7dc
35. Sasmito, A., Ilman, M., & Iswanto, P. T. (2023). Effects of rotational speed on the mechanical properties and performance of AA6061-T6 aluminium alloy in similar rotary friction welding. Welding International.
36. https://www.semanticscholar.org/paper/5a1fb281b075d5cd6ef63bee05a1e3b76699e878
37. Yuvaraj, K & Shobana, A & Kaushik, Nitish & Boshe, Addisu & Jha, Sanjay. (2023). An experimental method for choosing the tool pin profile and shoulder size to join dissimilar aluminum alloys AA7075-T651 and AA6061-T6 joints. Materials Research Express. 10. 10.1088/2053-1591/ad0936.

Note: All the figures and tables in this chapter were made by the authors.

Advances in Additive Manufacturing Technologies – Gurusamy Pathinettampadian et al. (eds)
© 2026 Taylor & Francis Group, London, ISBN 978-1-041-16687-0

Influence of Optimized Parameters on Microstructure and Mechanical Strength in Friction Stir Welding of AA5754 Alloy

Abishek S.V.[1],
Iniya T.[2], **Hemachandran S.**[3]
Student, Department of Mechanical Engineering,
Chennai Institute of Technology,
Chennai, India

P.K. Devan[4]
Professor, Department of Mechanical Engineering,
R.M.K. College of Engineering and Technology,
India

Deeksheka[5],
Aakash V.[6], **Rishekkumar**[7]
Student, Department of Mechanical Engineering,
Chennai Institute of Technology,
Chennai, India

◆ **Abstract:** Determination of the impact of the optimum settings of FSW on the mechanical and microstructure characteristics of the AA5754-H111 alloy is a matter of study. Due to saving energy and amalgamation of materials of diverse qualities, automotive and aerospace sectors have increasingly adopted FSW, being one of the relatively solid-state junction technologies. The study deals with three prime welding process parameters Axial Load, Tool Traverse Speed and Tool Rotational Speed. Their optimization has been done by means of a CCD by RSM. This would encompass tensile test as well as micro hardness in order to study mechanical behaviors besides microstructural study at variable values of those considered input parameters. The findings reveal that welding conditions could be controlled to enhance significantly tensile strength, elongation, and yield strength, while ensuring a fine, defect-free grain structure in the weld zones. The study gives more insight into the role played by welding parameters in order to achieve superior weld quality and performance for AA5754 alloy, which finds applications in critical industries.

◆ **Keywords:** AA5754 alloy, Friction stir welding, Optimized parameters, Microstructure, Mechanical strength

1. Introduction

Aluminium alloys are extensively used across different industries alike as automotive, aerospace, manufacturing, and naval engineering due to their widespread commercial availability and versatility. The 5xxx series alloys, in particular, play a significant role in these sectors. This has driven research into new setup configurations and the exploration of optimal joining techniques, as these directly impact the quality and longevity of products in these industries [1, 2].

[1]abisheksivaraj10@gmail.com, [2]iniya0206@gmail.com, [3]hemachandrans.mech2023@citchennai.net, [4]pkdevan68@gmail.com, [5]deekshekaashok@gmail.com, [6]aakash.vasudevan2020@gmail.com, [7]rolexrishe@gmail.com

DOI: 10.1201/9781003685906-65

In 1991, the UK's The Welding Institute (TWI) developed the solid-state joining technique known as friction stir welding (FSW), which was first used on aluminum alloys [3, 4]. In fact, the base concept of FSW is a pretty simple idea. Within the joint line, which can be inserted between the neighborhood edges of sheets or plates, the pin and the shoulder of a specifically produced non-consumable rotatory tool are moved. The functions of the tool are fundamentally two: it heats up the workpiece and moves forward the material to produce the junction. Localized heat softens the material close to the pin because of friction between the tool and workpiece. Material movement across the pin from the front to the back results because of rotation and translation movements of the tool. The tool's geometrical features create complex material movement around the pin [5]. Fine, equiaxed recrystallized grains are produced throughout the FSW process as a result of severe plastic deformation at high temperatures [6–9], which enhances mechanical qualities.

FSW is a "green" technology and one of the most prominent advances in metal joining in the last decades due to its energy efficiency, and environmental friendliness. Not only does it use much less energy than traditional welding methods, but since it does not use flux or cover gas, it is an environmentally friendly process. Additionally, it does not use filler metal, allowing for the joining of any aluminium alloy without concerns over composition compatibility, a challenge in fusion welding. FSW also enables the joining of dissimilar aluminium alloys and composites with ease [10–12]. FSW can be applied to a range of joint types, including as butt joints, lap joints, T-butt joints, and fillet joints [13-15], in contrast to classical friction welding, which is usually utilized for small, axisymmetric parts that rotate and push against one another to produce a joint [16-19].

Sato et al. [17,18] found that larger grain sizes and lower dislocation and sub-boundary densities were crucial for achieving valuable plane strain values in the nugget of FSW 5052 Al alloy. There are also studies on friction stir processing of AA 5754 and its influence on microstructure and texture made by employing two-dimensional and three-dimensional orientation imaging microscopy (2D OIM and 3D OIM) techniques [20, 21]. Surface friction stirring was used by Park et al. [22] to align the stirred zone with the primary tensile principal strain direction, improving the formability performance of AA5052-H32. The agitated zone layout did not affect the spring-back performance in an increased manner. Such investigations indicate how little has been done on the finer points of metallurgy, such as microstructure and mechanical property assessment.

2. Experimental Methodology

The base metal plates of AA5754 of thickness 5 mm were purchased from Virvadia metals and alloys. The purchased metal plates were then cut to the required size of length 100 mm and width 55mm using a hack saw machine. The edge preparation of the cut work pieces are flattened by using a milling machine. The chemical composition of the work piece must be ensured before the actual process of welding for this purpose we use optical emission spectroscopy that provides the various elements percentage in the work piece. After the welding of examples, mechanical qualities were examined.

Table 65.1 Chemical composition (wt%) of AA5754

Material	Fe	Cr	Si	Mg	Cu	Mn	Zn	Al
AA5754	0.25	0.12	0.29	2.9	0.04	0.32	0.03	96.02

Table 65.2 Mechanical Properties of AA5754

Material	YS(MPa)	UTS(MPa)	Elongation (%)	Hardness (Hv 1)
AA5754	180	240	13%	65

The base plate was thoroughly cleaned using an acetone solution prior to the welding process. The workpieces were then securely clamped on the machine table of the FSW machine. A D2 steel FSW tool in the shape of a cylindrical pin was mounted on the machine spindle. The FSW process was carried out with three levels of parameters, varying (TRS), (TTS), and (AL), based on a pre-established orthogonal array. The welds and specimens were machined on a WEDM after the completion of welding, following ASTM-E8 requirements for tensile testing, which was performed using a (UTM) (make and model). The results of the tests, such as (UTS), yield strength, and percentage elongation, were recorded for each sample. Additionally, samples for microstructural analysis were cut using WEDM to the required size and mounted on a hot mounting press that utilized Bakelite powder. The hot-pressed samples were hand-polished using silicon carbide abrasive sheets ranging from 600 to 2000 grit size. Keller's reagent was applied to enhance the hand-polished samples, preparing them for microstructure analysis. Microhardness testing was performed using a Vickers hardness testing machine (make and model) with a 1kg load and a 20-second dwell time. Indentations were made across various zones of the specimens, with a 0.5 mm gap between each indentation.

Central Composite Design (CCD) method has been followed to enhance the process parameters of welding using the Response Surface Methodology (RSM) framework. A comprehensive literature review on the FSW of several aluminum alloys was conducted to identify the most critical welding process parameters (WPPs) for this study. This review led to the selection of three key parameters: TRS, TTS, and AL. The limits for these parameters were established from the literature and coded values of -α (-1.682) and α (1.682), given for the lower and upper bounds, respectively. The center value, represented by the coded number "0," was the midpoint between the lower and upper boundaries. The corresponding actual values for each welding process parameter that matched the coded levels of +1 and -1 were determined using the conventional CCD equation. Table 65.3 presents the actual values of each welding process parameter that align with the assigned coded levels.

Table 65.3 Limits of welding process parameter

WPP	Unit	Levels				
		-1.682	-1	0	1	+1.682
TRS	rpm	600	681	800	919	1000
TTS	mm/min	80	88	100	112	120
AL	kN	1	1.4	2	2.6	3

3. Results and Discussion

3.1 Optimization of Welding Parameters

Optimization of welding parameters for aluminum alloy 5754 in FSW is required to obtain optimal metallurgical and mechanical properties. Tool geometry, axial load, tool rotation speed, and tool travelling speed are significant factors controlling heat generation, material flow, and defects. Optimized TRS range for AA5754 exists between 800–1500 RPM, where the temperature generated is enough to produce appropriate heat without overheating. TTS is typically in the range of 30 to 120 mm/min, at which point cooling rates are balanced with weld time. Higher speeds tend to result in inappropriate mixing of materials, and lower speeds may lead to grain growth. Axial load is in the range of 3–6 kN, which ensures proper consolidation of material, while insufficient or excessive load results in voids or surface deformation. Using Tool geometry, such as that of One concave shoulder with a threaded pin in H13 steel, adds to the efficiency of this process. The angle at which the tool is provided should be between 2 to 3° to be free from defects of the material flow. Most such parameters are optimized with the support of statistical techniques such as RSM and ANOVA. These are usually associated with optimized response variables, such as UTS, microhardness, and defect analysis. The optimized parameters result in fine-grained, welds free of defects that exhibit enhanced UTS, yield strength, and elongation for reliable application of AA5754 welds in industries such as automotive and marine applications. By combining these optimized conditions, the process guarantees better performance, structural strength, and surface quality of the welds.

Table 65.4 Design of experiments framed using CCD

Std. order	Run order	Coded Values			UTS (MPa)
		P	S	F	
1	18	-1	-1	-1	154
2	14	1	-1	-1	161
3	15	-1	1	-1	163
4	7	1	1	-1	160
5	6	-1	-1	1	123
6	16	1	-1	1	148
7	1	-1	1	1	161
8	19	1	1	1	177
9	4	-1.682	0	0	150
10	5	1.682	0	0	169
11	11	0	-1.682	0	133
12	3	0	1.682	0	163
13	8	0	0	-1.682	175
14	17	0	0	1.682	165
15	12	0	0	0	192
16	20	0	0	0	196
17	9	0	0	0	194
18	2	0	0	0	193
19	13	0	0	0	190
20	10	0	0	0	192

Write the equation in terms of actual factors so that it is easy to predict the response at different levels of each factor. The levels shown below should be applied to the original units for each factor. Coefficients are scaled to match the units for each factor, and the intercept is not at the center of the design space. This equation will not be able to predict the relative contributions of each factor.

Final Equation in Terms of Actual Factor

$$\begin{aligned} UTS = {} & +192.873 + 5.63 * TRS + 9.186 * TTS \\ & - 3.355 * AL - 2.375(TRS * TTS) \\ & + 4.625(TRS * AL) + 7.375(TTS * AL) \\ & - 12.048(TRS^2) - 16.114(TTS^2) - 8.336(AL^2) \end{aligned} \tag{1}$$

Analysis of the statistical parameters of the model yields valuable insights into its reliability and effectiveness. The Model F-value is 325.48, which is an important measure of the overall significance of the model. This is a very high value, which indicates that the model defines the variety in the data much better than what could be attributed to

Table 65.5 ANOVA test results

Source	Sum of Squares	Df	Mean Square	F-value	p-value	
Model	8214.91	9	912.77	325.48	<0.0001	significant
A-TRS	433.62	1	433.62	154.63	<0.0001	
B-TTS	1152.44	1	1152.44	410.95	<0.0001	
C-AL	153.72	1	153.72	54.81	<0.0001	
AB	45.13	1	45.13	16.09	0.0025	
AC	171.13	1	171.13	61.02	<0.0001	
BC	435.13	1	435.13	155.16	<0.0001	
A^2	2092.13	1	2092.13	746.03	<0.0001	
B^2	3742.35	1	3742.35	1334.48	<0.0001	
C^2	1001.54	1	1001.54	357.14	<0.0001	
Residual	28.04	10	2.80			
Lack of Fit	7.21	5	1.44	0.3461	0.8655	not significant
Pure Error	20.83	5	4.17			
Cor Total	8242.95	19				
Std. Dev.	1.67	**R^2**		0.9966		
Mean	167.95	**Adjusted R^2**		0.9935		
C.V. %	0.9971	**Predicted R^2**		0.9897		
		Adeq Precision		58.3134		

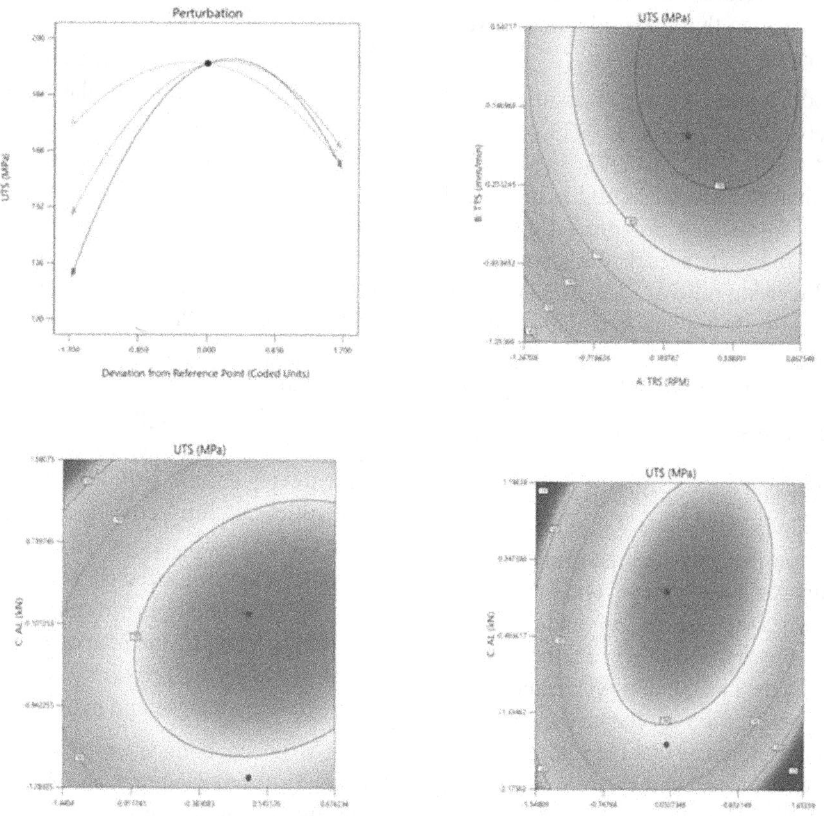

Fig. 65.1 [a] Perturbation plot [b] Interaction plot AB [c] Interaction plot AC [c] Interaction plot BC

random error. Importantly, the probability related to this F-value is only 0.01%, meaning there is an extremely low prospect that such a high F-value can come from random noise. This further enhances confidence in the model's validity and its capability to adequately describe the system under investigation.

In addition to the overall model significance, the analysis identifies specific model terms that contribute significantly to the response. Terms AA, BB, CC, ABAB, ACAC, BCBC, A2A2, B2B2, and C2C2 are significant since their p-values are below the threshold of 0.0500. These terms are critical to the predictive strength of the model and are likely to have a significant impact on the output variable. Conversely, terms whose p-values are greater than 0.1000 are considered non-significant and thus not relevant to explaining the response's variation. Including too many of such terms would make the model unduly complex and not optimal, so removing such terms while maintaining hierarchical relationships amongst the terms might improve the model.

This means that the F-value for Lack of Fit is 0.35. The probability of finding it by noise is 86.55%, meaning that this lack of fit is not significant when compared to pure error. The model captures the underlying pattern, not being influenced by chance. Therefore, the lack of fit is neither significant nor good, meaning that the model does a good job in describing the data.

A third check of model's good fit would come through strong correlation between the Predicted R2R2 value = 0.9897 and Adjusted R2R2 value = 0.9935 where their difference is well within the threshold limit of 0.2. So it seems there is a tight consistency for predicting a new data from this model with the chances that it is not overfit. Lastly, the Adequate Precision ratio of 58.313 This implies that this model provides a strong and meaningful signal, thereby making it suitable for guiding optimization or decision-making processes and driving the design space.

The perturbation plot represents the effect of the factors individually on the response variable, Ultimate Tensile Strength (UTS), by keeping the other factors constant at their central values; this is for factors AAA: TRS, BBB: TTS and CCC: AL. From the plot, one can find out how sensitive UTS is to changes in each factor. Out of the three, CCC is the one that carries the steepest curve: that's to say that CCC contributes more to the tensile strength as against BBB and CCC for changes in TTS and AL, respectively. On the other side of the coin, the very flatter curves for the case of BBB and CCC would mean that UTS does not vary much given the changes in these, but they are still affecting the response. The central coded value of the design space maximizes UTS at peak; therefore, optimal values of the parameters must be combined for maximum tensile strength. The perturbation plot effectively demonstrates relative importance of each factor in the necessity of exercising sharp control over TRS to enhance UTS.

Contour plots provide an overall view about the interaction effects between any two factors on the third factor UTS while one of the factors is retained at its central value. In each such plot, concentric regions occur with varying colors signifying different degrees of UTS levels and corresponding red zones indicate the maximum strength in tension. In the interaction between TRS and TTS, the plot shows that UTS reaches its peak when both factors are balanced near their optimal values, with deviations leading to a sharp decline. Similarly, the interaction between TRS and AL reveals that UTS is maximized at a central balance of these two parameters, emphasizing the importance of precise control over both factors. The plot of TTS and AL interaction seems to be a little weak compared to other pairs, yet the trend persists—the maximum UTS occurs in the middle region where both the parameters are equal. In summary, the contour plots suggest the necessity for maintaining an optimal combination of factors to achieve the best tensile strength. The red regions in the design space indicate ideal conditions.

3.2 Microstructure Analysis

Three planes, mutually perpendicular to one another, exhibit differences in the In the rolling plane, the AA5052 material has long, pancake-shaped grains. Thermal cycles extreme strains, and strain rates result in various microstructures in the aluminum alloy FSW regions [23, 24]. The elongated grain structure of the original at the joining interface is altered by the tool's strong plastic deformation and frictional heat produced between the tool and the plates. The "nugget" phase, being influenced by the process parameters and the communication between the substantial and the device pin and shoulder, assumes the structure of the device pin. Fujii et al. [25] have studied the nugget's creation, size, and shape as well as how it depends on the device design. At 1400 rpm and 80 mm/min, when the FSW is naturally cross-sectioned, it indicates various regions along the nugget, which are bounded by the "thermomechanical affected zone" (TMAZ), wherein the grain structure changes morphologically. These changes are brought about by the heat cycle of the welding process, and also due to the proximity impact close to the highly distorted nugget phase and material flow.

The TMAZ on both sides of the nugget is made up of the newly formed nugget and the primary material. The visibility of these interfaces depends on the relative orientations of traverse and tool rotation. While While the "retreating side" is the interface where these paths are anti-parallel, the "advancing side" is the frontier where the tool rotation and traverse are parallel. Although the retreating side is harder to see in this case, the advancing side is usually sharper and much more observable. The HAZ is the area around the TMAZ where the welding heat cycle alters the grain microstructure instead of causing plastic deformation. [26].

It was established that AA 5754 and AA 5182 joints possess uniform weld zones. A lack of flow lines found concentric within the weld zones under both optical microscope and SEM could be indicating a lack of weld nugget characteristic as in higher thickness welded sheets after FSW. The weld zone's grain size varied from 6 to 14 µm for AA 5754 whereas for AA 5182 it falls between 6 to 12 µm. Its microstructure was normal, with distinct divisions between the uniformly and partially recrystallized grains. Because thicker welded sheets lack the typical weld nugget, thinner sheets' geometry and heat flow are likely very different. SEM analysis indicates that this band was characterized by a high volume fraction of Al_2O_3 particles and an ultrafine grain structure with an regular grain size around 2 µm in thickness of 20-50 µm. In the same period, neither cleavage nor a continuous oxide layer could be detected on the metal surrounding the band. Since such oxide inclusions would usually be prevented by smoother surface preparation before welding, this is considered an artifact of the butt-joined surfaces.

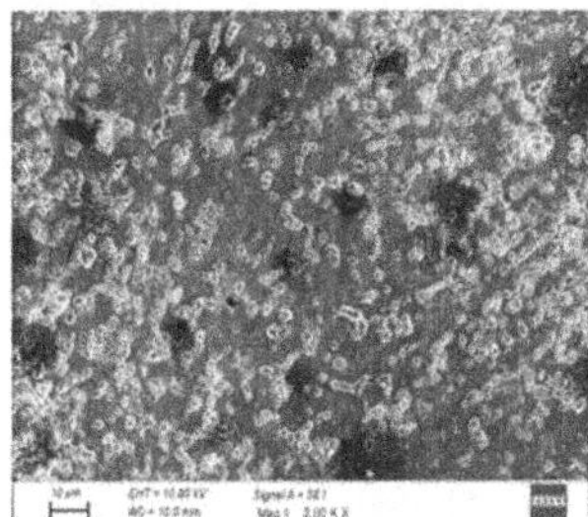

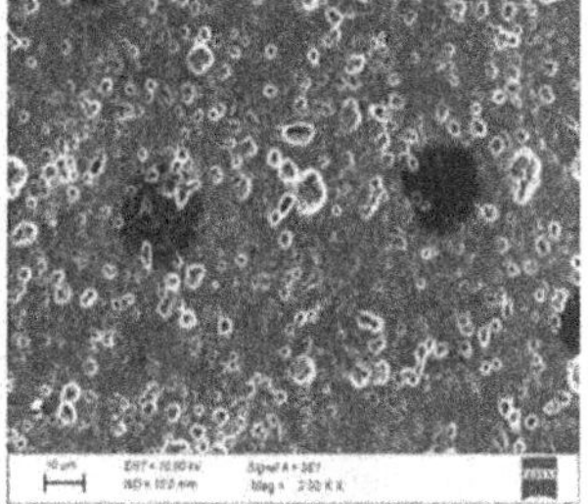

Fig. 65.2 SEM microstructure [a] Base metal AA5754-H111 [b] WNZ of sample 16

4. Hardness Analysis

The Vickers hardness test was carried out on the cross-section of the spot-welded sample using a load of 200 g for 15 seconds. The measurement points were the base metals (AA2024-T3 and AA5754-H114), as well as the nugget or stir zone, the thermomechanical affected zone, heat-affected zone, and hook zone. The HV readings were done at each site to provide precision. The hardness profile of the spot-welded area of the AA2024-T3 to AA5754-H114 joint shows that the stir zone or the weld was softer compared with the base material, that is, BM AA5754. Vickers hardness reached a minimum value of 80 HV at the edge of the HAZ and TMAZ, then increased with a progressive decrease from the HAZ towards the keyhole. In contrast, the hardness was less than that of the BM of AA5754, but it increased substantially in the TMAZ and SZ. The highest hardness value recorded in the hook zone was 110 HV. XRD analysis was performed to confirm the presence of intermetallic compounds at the interface between the dissimilar aluminum alloys. AA5754-H114 substrate was reported to produce Al3Mg2 and Mg2Si compounds, while the production of CuAl, Al2Cu, AlCu4, and Cu9Al4 compounds was reported by the AA2024-T3 substrate.

Local yield stresses were determined by additional microhardness testing of the weld sections. The samples were mounted, polished, and cut from the center of the weld cross-section with standard metallographic techniques. The surfaces at the top ran parallel to the surfaces at the bottom. A Metkon MH-3 Vickers Microhardness Testing Machine was used for the measurements of microhardness. A 1 mm indentation distance was used with a load of 100 g applied for 10 seconds. Microhardness tests were done on the left side of the weld area along the center line of the cross-section through the top sheet. For the keyhole region a higher hardness values were detected. According to Güler [28], the main reason for the hardness increase is strain hardening, which is brought on by the plastic deformation that occurs during welding. This behavior was influenced by fine grains created during the welding process by dynamic recrystallization. The hardness profile from the weld centre to the base metal initially decreased and then gradually increased, consistent with trends observed in existing literature [29, 30, 31, 32]. The results of the tensile shear test show that test groups had the highest hardness and the highest strength values. This finding corroborates the observation by Güler [28] about how dynamic recrystallization raises both strength and hardness in one go.

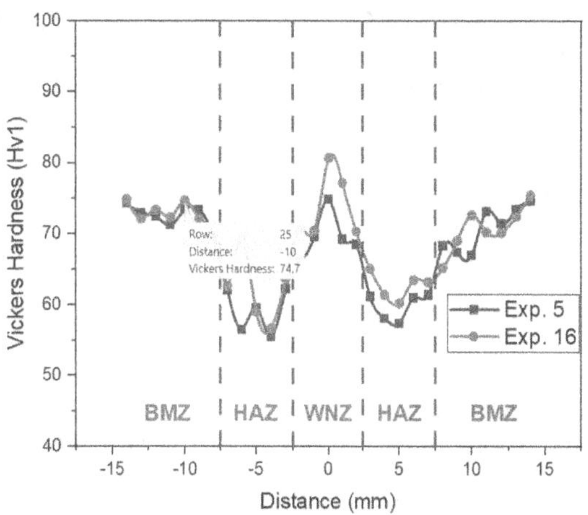

Fig. 65.3 Hardness variation graph

5. Conclusion

1. The key FSW parameters—TRS, TTS, and AL—were optimized to improve the quality of welds for AA5754-H111 alloy.
2. Welded joints, that is UTS, yield strength, and elongation, were noted to be improved significantly.
3. The nugget zone in welds exhibited fine grains and the absence of defects due to the optimized FSW conditions.
4. RSM, along with the CCD approach, successfully identified the optimal welding conditions.
5. FSW has a tremendous potential to produce excellent quality welds for any severe application in the marine, automobile, and aerospace fields.
6. Future studies may focus on tool geometry effects and use alternative aluminum alloys in an attempt to further optimize weld performance and develop the FSW process.

References

1. Güler, H. (2015) "Influence of the tool geometry and process parameters on the static strength and hardness of friction-stir spot-welded aluminium-alloy sheets." Mater. Tehnol., 49, 457–460.
2. El Rayes, M.M., Soliman, M.S., Abbas, A.T., Pimenov, D.Y., Erdakov, I.N., Abdel-Mawla, M.M. (2019) "Effect of Feed Rate in FSW on the Mechanical and Microstructural Properties of AA5754 Joints." Adv. Mater. Sci. Eng.
3. W.M. Thomas, E.D. Nicholas, J.C. Needham, M.G. Murch, P. Templesmith, C.J. Dawes, G.B. Patent Application No. 9125978.8 (December 1991).
4. C. Dawes, W. Thomas, TWI Bulletin 6, November/December 1995, p. 124.
5. B. London, M. Mahoney, B. Bingel, M. Calabrese, D. Waldron, in: Proceedings of the Third International Symposium on Friction Stir Welding, Kobe, Japan, 27–28 September, 2001.
6. C.G. Rhodes, M.W. Mahoney, W.H. Bingel, R.A. Spurling, C.C. Bampton, Scripta Mater. 36 (1997) 69.
7. G. Liu, L.E. Murr, C.S. Niou, J.C. McClure, F.R. Vega, Scripta Mater. 37 (1997) 355.
8. K.V. Jata, S.L. Semiatin, Scripta Mater. 43 (2000) 743.
9. S. Benavides, Y. Li, L.E. Murr, D. Brown, J.C. McClure, Scripta Mater. 41 (1999) 809.
10. L.E. Murr, Y. Li, R.D. Flores, E.A. Trillo, Mater. Res. Innovat. 2 (1998) 150.
11. Y. Li, E.A. Trillo, L.E. Murr, J. Mater. Sci. Lett. 19 (2000) 1047.
12. Y. Li, L.E. Murr, J.C. McClure, Mater. Sci. Eng. A 271 (1999) 213.
13. H.B. Cary, Modern Welding Technology, Prentice-Hall, New Jersey, 2002.
14. C.J. Dawes, W.M. Thomas, Weld. J. 75 (1996) 41.
15. Murr L, Liu G, Mcclure J. Dynamic recrystallization in friction-stir welding of aluminium alloy 1100. J Mater Sci Lett 1997;16(22):1801–3.
16. Sato Y, Park S, Kokawa H. Microstructural factors governing hardness in friction-stir welds of solid-solution-hardened al alloys. Metall Mater Trans A Phys Metall Mater Sci 2001;32(12):3033–42.
17. Sato Y, Sugiura Y, Shoji Y, Park S, Kokawa H, Ikeda K. Post-weld formability of friction stir welded al alloy 5052. Mater Sci Eng A 2004;369(1–2):138–43.
18. Sato Y, Urata M, Kokawa H, Ikeda K. Hall–Petch relationship in friction stir welds of equal channel angular-pressed aluminium alloys. Mater Sci Eng A 2003;354(1–2):298–305.
19. Attallah M, Davis C, Strangwood M. Microstructure–microhardness relationships in friction stir welded aa5251. J Mater Sci 2007;42(17): 7299–306.
20. Adams-Hughes M, Kalu P, Khraisheh M, Chandra N. Micro characterization and texture analysis of friction stir processed aa 5052 alloy. Friction Stir Welding and Processing III – Proceedings of a Symposium sponsored by the Shaping and Forming Committee of (MPMD) of the Minerals, Metals and Materials Society, TMS 2005. p. 3–10.
21. Adams-Hughes M, Kalu P, Principe E, Wright S. A 3d analysis of the nugget zone of aa5052 processed via friction stir processing. Microsc Microanal 2008;14(suppl. 2): 980–1.
22. Park S, Lee C, Kim J, Han H, Kim SJ, Chung K. Improvement of formability and spring-back of aa5052-h32 sheets based on surface friction stir method. J Eng Mater Technol Trans ASME 2008;130(4):0410071–04100710.
23. Mishra R, Ma Z. Friction stir welding and processing. Mater Sci Eng R Rep 2005;50(1–2):1–78.
24. Threadgill P, Leonard A, Shercliff H, Withers P. Friction stir welding of aluminium alloys. Int Mater Rev 2009;54(2): 49–93.
25. Fujii H, Cui L, Maeda M, Nogi K. Effect of tool shape on mechanical properties and microstructure of friction stir welded aluminum alloys. Mater Sci Eng A 2006;419 (1–2):25–31.
26. Threadgill P. Terminology in friction stir welding. Sci Technol Weld Join 2007;12(4):357–60.
27. H. jin, c. ko, s. saimoto, and p. l. threadgill: Mater. Sci. Forum, 2000, 331 – 337, 1725 – 1730
28. Piccini, J.M.; Svoboda, H.G. Tool geometry optimization in friction stir spot welding of Al-steel joints. J. Manuf. Process. 2017, 26, 142–154. [CrossRef]
29. Güler, H. The Mechanical Behavior of Friction-Stir Spot Welded Aluminum Alloys. JOM 2014, 66, 2156–2160. [CrossRef]
30. Piccini, J.M.; Svoboda, H.G. Effect of the Tool Penetration Depth in Friction Stir Spot Welding (FSSW) of Dissimilar Aluminum Alloys. Procedia Mater. Sci. 2015, 8, 868–877. [CrossRef]
31. Zhang, Z.; Yang, X.; Zhang, J.; Zhou, G.; Xu, X.; Zou, B. Effect of welding parameters on microstructure and mechanical properties of friction stir spot welded 5052 aluminum alloy. Mater. Des. 2011, 32, 4461–4470. [CrossRef]
32. Wang, D.-A.; Lee, S.-C. Microstructures and failure mechanisms of friction stir spot welds of aluminum 6061-T6 sheets. J. Mater. Process. Technol. 2007, 186, 291–297. [CrossRef] © 2019 by the authors. Licensee MDPI, Basel, Switzerl

Note: All the figures and tables in this chapter were made by the authors.

Advances in Additive Manufacturing Technologies – Gurusamy Pathinettampadian et al. (eds)

Optimization of Friction Stir Welding Parameters for AA5182 Aluminum Alloy and Its Impact on Mechanical Properties

Rishekkumar[1], Aakash V.[2], Siva Hemanth Kada[3], Abishek S.V.[4], Iniya T.[5]

Student, Department of Mechanical Engineering,
Chennai Institute of Technology,
Chennai, India

P.K. Devan[6]

Professor, Department of Mechanical Engineering,
R.M.K. College of Engineering and Technology,
India

Deeksheka[7]

Student, Department of Mechanical Engineering,
Chennai Institute of Technology,
Chennai, India

◆ **Abstract:** This paper discusses the optimisation of friction stir welding (FSW) parameters for AA5182 aluminium alloy using Central Composite Design (CCD) and Response Surface Methodology (RSM). Key parameters include the welding speed (WS), rotational speed (RS),and axial load (AL) for effects on mechanical properties and microstructural properties. Optimal conditions (RS = 800 rpm, WS = 100 mm/min, AL = 2 kN) resulted in UTS of 245 MPa. Microstructural analysis indicated refined equiaxed grains in the SZ that enhanced hardness and tensile strength through dynamic recrystallization, while HAZ revealed lower hardness because of grain coarsening. The statistical validation confirmed the reliability of the model with an R^2 of 0.9977, indicating the parameter dependence on the mechanical properties. Uniform microhardness profile and better grain structure reflect the efficiency of parameter optimization in producing defect-free high-strength joints for automobile and aerospace applications..

◆ **Keywords:** Friction stir welding, AA5182, Welding parameters optimization, Joint performance, Mechanical properties

1. Introduction

The utilization of aluminum as structural materials has increased significantly over the past decade mainly due to its fascinating properties, such as low density, ease of fabrication, good workability, outstanding ductility, superior thermal conductivity, and excellent resistance to corrosion. The aesthetic appeal of a pleasing natural finish has also led to an amount of around 25% of the world's aluminium production being utilized in construction. Increasingly, nonferrous alloys-almost always aluminum alloys-are taking the place of steels in many industrial

[1]rolexrishe@gmail.com, [2]aakash.vasudevan2020@gmail.com, [3]siva14.hemanth@gmail.com, [4]abisheksivaraj10@gmail.com, [5]iniya0206@gmail.com, [6]pkdevan68@gmail.com, [7]deekshekaashok@gmail.com

DOI: 10.1201/9781003685906-66

applications. It is the most readily found metal on earth following iron and the third most commonly used element, constituting 8% of earth crust. The second place goes to aluminum, besides steel, following in its use among commonly employed metals. It predominantly results from the mineral used commonly termed as bauxite by refining process commonly called Bayer to produce aluminium oxide or alumina for aluminum production. In 1888, German inventor Karl Josef Bayer received a patent for a process that is still used to this day in the manufacture of alumina, which is a raw material for aluminum production. In contrast, the process Hall-Héroult currently used in the extraction of primary aluminum metal was patented in 1886 by Charles M. Hall, an American, and Paul L.T. Héroult, a French national, and since then it has not been much altered.[7,8] These alloys are met fairly frequently with strengths that equal or even exceed those of structural steel, but the weights are often far lower [1]. The automotive and aerospace industries have successfully employed FSW to join a range of Al alloy series (2xxx [9-11], 5xxx [12-14], 6xxx [15,16], and 7xxx [17-20]) and magnesium alloys [21–23] that are challenging to weld using traditional fusion welding techniques [24] over the course of the last three decades of progressive development.

Due to versatile nature contributing to the variety of applications, Aluminium is used very much in the FSW industry. FSW is an autogenous hot shear welding method patented by TWI, UK in 1991. It uses a rotating tool of a material harder than the substrate being joined as a non-consumable tool.[6] FSW is an excellent solid-state welding process where the process revolves around the utilisation of mechanism, it consists of 2 main parts; the shoulder and the pin. To create the junction using the friction stir welding technique, the tool rotates with its shoulder and a pin is inserted into the adjacent surfaces of the plates and moves along the joint line.[2] The advancing direction is identified as the weld side where the welding path resembles the tool's revolving direction, and the retreating direction is identified as the side where the welding path reverses the tool's revolving direction.[3] A tool pin that protrudes from the tool's base and has a length smaller than the thickness of the workpiece is one of the welding tool's primary components. In addition to the plastic deformation of the workpiece material, the friction between the workpiece and the rotating tool also generates heat.[4] This targeted heating causes the workpiece material near the pin to soften before reaching its melting point. [2] This softening material is moved from the probe's front side to its backside by the FSW tool's simultaneous rotation and translation [5]. To ascertain the impact of Tool Pin Profile (TPP) on the metallurgical and physical characteristics of FSW joints, a specific number of investigations were conducted. In the recent literatures, numerous application of square pin profile tool was observed [28-31]. This is an attempt to analyze the physical properties and analyse the microstructure of AA5182 post optimisation with CCD within the framework of RSM.

2. Experimental Methodology

The base metal, AA5182, was bought from Arihant Aluminium in Chennai. The 5mm plate thickness was downsized to the desired size, 100x55 with the use of a hacksaw, and then a milling machine is used to smoothen the edge as a part of the edge preparation process. The base metal's chemical composition is determined through optical emission spectroscopy investigations, the Vickers microhardness test is used to identify the material's hardness, and the UTM is applied to determine the material's tensile strength, as shown in Table 66.1. The material is often cleansed through a process incorporating acetone, which would remove the oxide layers that frequently form when aluminum is treated with oil, as well as remove the oil residue that is typically present on the metallic surface.

Table 66.1 Chemical composition (wt%) of AA5182

Material	Fe	Si	Ti	Mg	Cu	Mn	Zn	Al
AA5182	0.37	0.19	0.35	4.4	0.49	0.2	0.26	Bal.

Table 66.2 Mechanical properties of AA5182

Material	UTS(MPa)	YS(MPa)	Elongation (%)	Hardness (Hv 1)
AA5182	276	128	22%	94

The metallurgical and mechanical testing samples were prepared using a wire electrical discharge machining (WEDM) process. Samples for tensile study were extracted according to the ASTM E8 standard and then tested on a UTM (Specify Make and Model). For microstructural analysis, samples were also extracted using the WEDM machine and mounted using a hot mounting press with Bakelite powder. The samples were mounted and hand polished using SiC4 abrasive sheets from 80 to 400 grit followed by machine polishing using finer silicon carbide abrasive sheets from 600 to 2000 grit. After polishing, samples were etched using Keller's reagent to expose the microstructure. The microstructure was analyzed with a SEM to study the grain and precipitate size and morphology. A microhardness test was performed on the polished samples across different weld zones with indents spaced 0.5 mm apart for hardness evaluation. The Vickers microhardness tester (Specify Make and Model) was used, applying a 1 kg of load with a dwell time of around 20 seconds for each measurement.

Table 66.3 Tool dimensions used for the experiments

Process parameters	Values
D/d ratio of the tool	3.0
Tool pin profile	Straight square
Tool pin length (mm)	4.5
Tool pin diameter, d (mm)	5
Tool shoulder diameter, D (mm)	15

Optimization of welding process parameters was done using the CCD approach under the framework of RSM. A comprehensive review of the literature available on FSW of different aluminum alloys was done to identify the WPPs for this study. From this, three parameters were selected as major ones: RS, WS, and AL. The parameter limits were based on an extensive literature review; the lower limit accepted a coded value -α (-1.682) and the upper limit accepted a coded value of +α (1.682). The central value, midway between these limits, was coded with a value of 0. Using the standard CCD equation, the actual values of the corresponding intermediate coded levels +1 and -1 were calculated. Table 66.4 lists the actual values for each welding process parameter with their respective coded levels.

Table 66.4 Limits of welding process parameter

WPP	Unit	Levels				
		-1.682	-1	0	1	+1.682
TRS	rpm	600	681	800	919	1000
TTS	mm/min	80	88	100	112	120
AL	kN	1	1.4	2	2.6	3

3. Results and Discussion

3.1 Optimization of Welding Parameters

Optimization of welding parameters for aluminum alloy 5182 in FSW is required to obtain optimal metallurgical and mechanical properties. Axial load, tool geometry, TRS, and TTS are significant factors controlling heat generation, material flow, and defects. The model F-value is 473.02, indicating the model is very significant because only 0.01% of such large F-value would happen by random noise. It means the input factors have a very strong relation with the response variable. P-values below 0.0500 signify statistically significant model terms. In this study, terms A (TRS), B (AL), C (WS) along with their interactions (AB, AC, BC) and quadratic terms (A^2, B^2, C^2), were identified as significant contributors to the model. Terms which has the P-values above 0.1000 are deemed not significant, an deletion of such terms, except that necessary to keep model hierarchy, may further the model's simplicity and efficiency.

Table 66.5 Design of experiments framed using CCD

Std. order	Run order	Coded Values			UTS (MPa)
		P	S	F	
1	4	-1	-1	-1	151
2	19	1	-1	-1	175
3	6	-1	1	-1	170
4	1	1	1	-1	207
5	20	-1	-1	1	172
6	15	1	-1	1	209
7	12	-1	1	1	166
8	13	1	1	1	217
9	14	-1.682	0	0	133
10	17	1.682	0	0	199
11	9	0	-1.682	0	178
12	8	0	1.682	0	203
13	2	0	0	-1.682	190
14	5	0	0	1.682	215
15	11	0	0	0	242
16	7	0	0	0	244
17	3	0	0	0	243
18	16	0	0	0	245
19	18	0	0	0	238
20	10	0	0	0	238

Final Equation in Terms of Actual Factor

$$\begin{aligned} UTS = {} & 241.654 + 19.038 * RS + 6.959 * AL \\ & + 7.545 * WS + 3.375 * (RS * AL) \\ & + 3.375(RS * WS) - 6.125 * (AL * WS) \\ & - 26.668 * RS^2 - 18.006 * AL^2 - 13.7635 \\ & * WS^2 \end{aligned} \tag{1}$$

With a 97.61% probability that the value is due to random noise, the Lack of Fit F-value of 0.14 indicates that the lack of fit is not statistically significant when compared to the pure error. This is desirable since a non-significant lack of fit emphasize that the model is fitting the experimental data well and predicts the response without systematic deviations. The Predicted R^2 of 0.9949 is in close agreement with the Adjusted R^2 of 0.9955, and the difference for this is less than 0.2, confirming that the model is reliable in predicting responses for new data. This agreement is such that the model does not overfit the

Table 66.6 ANOVA test results

Source	Sum of Squares	Df	Mean Square	F-value	p-value	
Model	21942.21	9	2438.02	473.02	<0.0001	significant
A-TRS	4949.83	1	4949.83	960.35	<0.0001	
B-TTS	661.46	1	661.46	128.34	<0.0001	
C-AL	777.50	1	777.50	150.85	<0.0001	
AB	91.13	1	91.13	17.68	0.0018	
AC	91.13	1	91.13	17.68	0.0018	
BC	300.13	1	300.13	58.23	<0.0001	
A^2	10249.19	1	10249.1	1988.52	<0.0001	
B^2	4672.43	1	4672.43	906.53	<0.0001	
C^2	2729.98	1	2729.98	529.66	<0.0001	
Residual	51.54	10	5.15			
Lack of Fit	6.21	5	1.24	0.1370	0.9761	not significant
Pure Error	45.33	5	9.07			
Cor Total	21993.75	19				
Std. Dev.	2.27	**R^2**		0.9977		
Mean	201.75	**Adjusted R^2**		0.9955		
C.V. %	1.13	**Predicted R^2**		0.9949		
		Adeq Precision		66.931		

data, ensuring good predictive accuracy across the design space. Adequate Precision, measuring the signal-to-noise ratio, is 66.931, far exceeding the threshold of 4. Such a high ratio suggests that the model has a good signal and can adequately search through the design space to provide accurate and reliable estimations. Overall, statistical analysis supports the validity of robustness and reliability of the model in optimizing laser welding parameters for AA5182 aluminum alloy. The high F-value of the model, the presence of significant terms, a non-significant lack of fit, and high Adequate Precision together indicate that it is a very strong tool for navigation in experimental design space. This means that optimum welding conditions can be identified to obtain better mechanical properties and weld quality, and therefore, it is suitable for practical application in industrial settings. The sensitivity plot in Fig. 66.1 shows how sensitive UTS is to changes of individual process parameters from their base or reference (coded) value. Curved paths in lines corresponding to each parameter mean that UTS is highly sensitive to the changes of specific process variables. The sharper the curve, the greater that parameter affects UTS. In this case, from the above curve, the significant effect of parameter A could be seen, which happens to be rotational speed. These are followed by parameters B like axial load and parameters C like welding speed, that is, the respective points at which UTS assumes a maximum value.

The contour plots in Fig. 66.1 collectively represent the cumulative effects of RS, AL, and WS on the UTS of the AA5182 aluminum alloy subjected to FSW. Every individual plot provides useful insight into how different parameter combinations affect the resulting UTS, thus emphasizing the importance of careful parameter optimization. The contour plot for RS and AL, the middle red region represents an optimal value of these process parameters for getting the maximum UTS value. Whenever the values of the parameter deviate from this optimal value, then the UTS decreases with the change from red to green, yellow, and blue regions. The symmetrical and elliptic configuration of contour lines used in the given plot strongly suggests an extensively spread-up spectrum of combinations of both parameters, whose presence gives some high UTS values clearly. Steepness around the horizontal axis shows there is an increased effect at RS, with AL representing a lower one in RS and AL interaction.

The second contour plot in Fig. 66.1 represents the interaction between RS and WS, showing a less optimistic optimal range for high values of UTS. A central red region depicts maximal values of UTS while any deviation in either RS or WS from these optimum conditions shows a drastic fall of UTS, which could be seen with the colour changing rapidly to green and then to blue. This graph illustrates the requirement for precise balance between RS

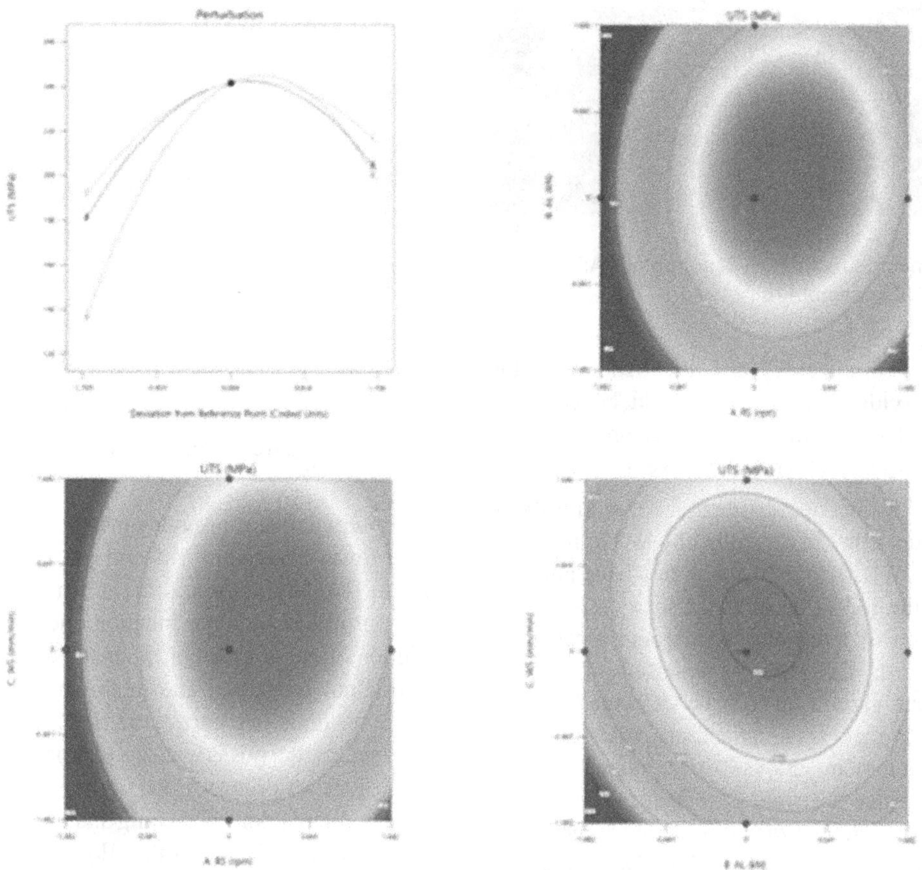

Fig. 66.1 [a] Perturbation plot [b] Interaction plot AB [c] Interaction plot AC [c] Interaction plot BC

and WS in order to maintain optimum tensile properties. The almost concentric contour designs reflect a well-proportionate relationship of UTS with these two variables in the analyzed range. Again, in Fig. 66.1 the third contour plot represents the interaction between AL and WS demonstrates a red region where UTS achieves maximum. But it shows a much tighter constraint for the range in getting the optimal UTS value since deviations in AL and WS result in an extensive decline in strength. Contours are closer to each other, which reflects that all these parameters have to lie within certain limits with accuracy. All these plots convey the importance of optimization in parameters for FSW. The UTS can be maximized by keeping a balanced interaction of RS, AL, and WS since UTS is quite sensitive to all the variation in the process parameters involved.

3.2 Microstructure Analysis

In FSW, the first microstructure governs the development of the final microstructure. The FSW process causes significant sensitivity in the local microstructural changes in terms of strain, strain rate, and thermal cycles developed in the deformation zones. Thus, in FSW, the better understanding of microstructural kinetics can be achieved when as received material microstructure is known as an important task. Microstructural characterization of the FSW on AA5182 aluminum alloy samples was carried out using SEM, and the micrographs are shown in Fig. 66.2. Since FSW is inherently asymmetric, the weld generally shows distinct microstructural zones, namely HAZ, TMAZ, and the weld nugget or stir zone. All of these zones have distinctive microstructure characteristics based on the mechanical and thermal impacts of the welding process. The stir zone is primarily responsible for the refined recrystallized grain structure developed due to the stirring action by the FSW tool. This zone depicts uniform grain distribution with finer grains due to efficient material flow as well as dynamic recrystallization during the process. The micrograph in Fig. 66.2 correctly shows the grain structure in the WN, which primarily improves the physical properties of the joint. Moreover, the presence of intermetallic compounds in the weld zone has been found to have impact on the strength and integrity of the joint. These compounds, due to the interactions of alloying elements during the FSW process, can improve or deteriorate the weld strength as

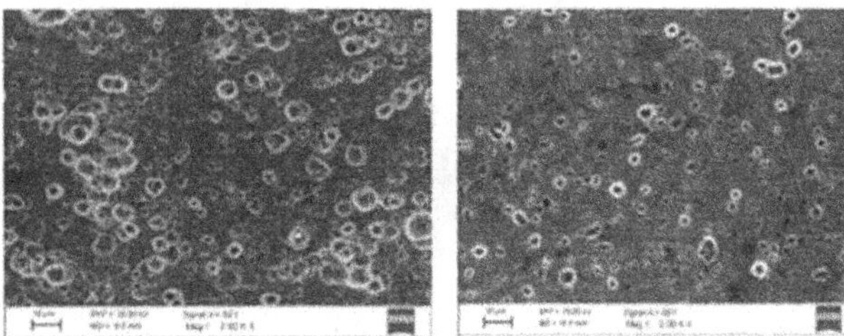

Fig. 66.2 SEM microstructure [a] Base metal AA5754-H111 [b] WNZ of sample 18

a function of their morphology and distribution. For the AA5182 alloy, welding parameters should be carefully controlled to obtain an optimized microstructure to balance mechanical strength and integrity of the weld.

The formation of distinct microstructural zones in FSW of AA5182 aluminum alloy is controlled by WPP, including tool RS, AL, and WS. Typically, the grains have partially recrystallized in the TMAZ between the stir zone and HAZ with the condition of moderate thermal and mechanical effects. In contrast, the HAZ is exposed to thermal treatment without significant plastic deformation, and this often causes grain coarsening and potential softening, especially in aluminum alloys. Such microstructural changes can influence the weld's mechanical properties. The most decrease in grain size was observed at the bound aries between the HAZ and TMAZ [26]. The process of FSW causes significant plastic deformation, breaks the grains, and forms low-angle misoriented grain boundaries, thus becoming a favorable place for recrystallization. [25] Microstructure and mechanical properties show a strong interaction, as is evident from the impact of grain size, distribution, and the presence of intermetallic compounds. Small grains structures in the stir zone are connected with enhanced tensile strength, while coarse grains or defects in the HAZ could act as stress concentrators, thus reducing the performance of the weld under mechanical loads. Optimizing process parameters in FSW of AA5182 alloy I therefore critical to the realization of a balance between microstructure refinement and mechanical property enhancement for the production of high-strength, defect-free joints.

4. Hardness Analysis

The microhardness profile of FSW on AA5182 aluminum alloy reveals critical insights into the microstructural changes induced by the welding process. Results, as shown in Fig. 66.3, indicate nearly symmetrical hardness distribution on both advancing and retreating sides of the weld, which reflect uniform thermal and mechanical effects imparted by FSW. The hardness variation in the weld zones

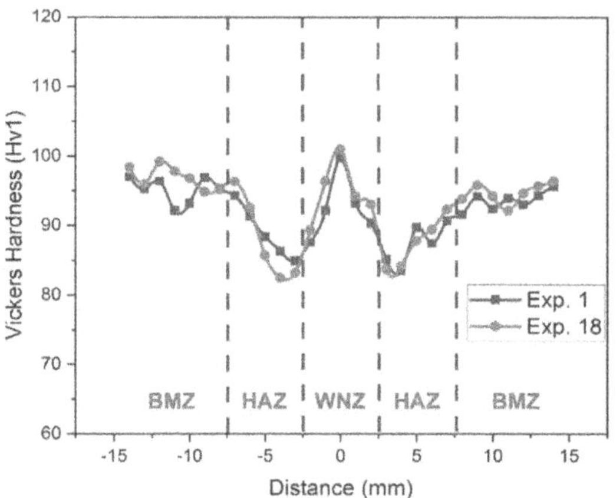

Fig. 66.3 Hardness variation graph

is very crucial as it provides insight into the effect of the WPP toward metallurgical and physical properties. The BM of AA5182 possesses the maximum hardness values in the profile since it has inherent microstructure and solid solution strength through magnesium. This zone retains its original properties due to minimal thermal influence during welding. In contrast, the HAZ shows a clear drop in hardness, so that this is weakest part of the weld. The loss in hardness arises primarily due to thermal cycles in welding, which induces grain coarsening and partial annealing of the microstructure. This alters the alloy's capacity to withstand deformation, with the possibility of creating problems with the weld's structural integrity.

Figure 66.3 depicts the Vickers Microhardness Profile across different zones of FSW AA5182 aluminum alloy, including the BMZ, HAZ, and WNZ, for experimental conditions Exp. 1 and Exp. 18. The BMZ retains high hardness due to its unaltered microstructure and strengthening mechanisms, while HAZ shows a visible drop in hardness, attributed to grain coarsening and partial annealing from thermal cycles. In contrast, the WNZ exhibits better hardness, approaching BMZ values, because of dynamic recrystallization and Hall-Petch effect-enhanced fine grain formation. Slight differences between Exp. 1 and Exp. 18 point to the effects of WPP on thermal cycles, material flow rate, and mechanical properties, thus underlining the need for optimizing FSW parameters to minimize HAZ softening and maximize WNZ grain refinement. Impact of annealing and the stirring action are the two main factors that impact the metallurgical property of welded parts. Increasing the rotational speed which in turn increase the stirring action of pin. [27]

Continuing in the vicinity of the TMAZ, a stronger hardness reduction occurs. The affected zone is under thermal and plastic deformation conditions that affect the grain structure in place by weakening the material. Also, the lower hardness compared with the base metal is from grain elongation and some localized recrystallization that happens in TMAZ. Additionally, the redistribution of alloying elements that is affected by this deformation contributes to the diminished hardness. The SZ or weld nugget has higher hardness values than HAZ and TMAZ because of significant grain refinement that takes place along with dynamic recrystallization due to stirring action caused due to FSW tooling. The fine equiaxed grains formed in this zone help in increasing the hardness through Hall-Petch relation. This zone further contains uniform material flow and controlled thermal input due to which it has greater hardness. The optimization of tool design and process parameters is crucial for achieving this grain refinement, ensuring desirable mechanical properties in the weld nugget. [28]

5. Conclusion

1. AA5182 aluminum alloy by means of CCD and RSM, and the optimized conditions were found to be RS =800 rpm, WS = 100 mm/min, and AL = 2 kN.
2. The optimized parameters produced an UTS of 245 MPa, which earns it the most significance of effective parameters optimization considering it had the capability to give rise to maximum mechanical properties.
3. Microstructural examination revealed equiaxed grains in zone of stir due to dynamic recrystallization, and as an outcome, these improved the UTS of the weld and its hardness.
4. HAZ was the weakest area with lesser hardness due to coarsening of grains and thermal effects which add to the effect of the input of heat that must be controlled in order to guarantee the joint's integrity

References

1. A.K. Jha, S.V.S.N. Murty, V. Diwakar, K. Sree Kumar, Metallurgical analysis of cracking in weldment of propellant tank, Engineering Failure Analysis, 10(3) (2003) 265–273.
2. R.S. Mishra, P.S. De, N. Kumar, Friction Stir Welding and Processing, Science and Engineering (2014).
3. L. Liu, Welding and Joining of Magnesium Alloys Woodhead, Publishing Limited (2010 I, 2010.).
4. B.S. Naik, D.L. Chen, X. Cao, P. Wanjara, Metall. Mater. Trans. A Phys. Metall. Mater. Sci. 44 (2013) 3732–3746.
5. M. Rezaee Hajideh, M. Farahani, S.A.D. Alavi, N. Molla Ramezani, J. Manuf. Process. 26 (2017) 269–279.
6. Thomas WM. Friction stir welding. International patent application no. PCT/ GB92/02203 and GB patent application no. 9125978.8. US patent no. 5, vol. 460; 1991. p. 317.
7. Proceedings of the International Symposium on Quality and Process Control in the Reduction and Casting of Aluminum and Other Light Metals Winnipeg, Canada, August 23-26, 1987, Vol. 5 Proceedings of the Metallurgical Society of the Canadian Institute of Mining and Metallurgy.
8. Physical Metallurgy of Direct Chill Casting of Aluminum Alloys, DMITRY G. ESKIN, 2008, CRC Press, Taylor & Francis, Boca Raton London New York.
9. Chen Y, Liu H, Feng J. Friction stir welding characteristics of different heat treated-state 2219 aluminum alloy plates. Mater Sci Eng A 2006; 420: 21–5.
10. Liu HJ, Li JQ, Duan WJ. Friction stir welding characteristics of 2219-T6 aluminum alloy assisted by external non-rotational shoulder. Int J Adv Manuf Technol 2013;64:1685–94.
11. Zhang ZH, Li WY, Feng Y, Li JL, Chao YJ. Global anisotropic response of friction stir welded 2024 aluminum sheets. Acta Mater 2015;92:117–25.
12. Zhou C, Yang X, Luan G. Fatigue properties of friction stir welds in Al 5083 alloy. Scr Mater. 2005;53:1187–91.
13. Lombard H, Hattingh DG, Steuwer A, James MN. Optimising FSW process parameters to minimise defects and maximise fatigue life in 5083-H321 aluminium alloy. Eng Fract Mech 2008;75:341–54.
14. Hirata T, Oguri T, Hagino H, Tanaka T, Chung SW, Takigawa Y, et al. Influence of friction stir welding parameters on grain size and formability in 5083 aluminum alloy. Mater Sci Eng A 2007;456:344–9.
15. Shi L, Wu CS, Gao S, Padhy GK. Modified constitutive equation for use in modeling the ultrasonic vibration enhanced friction stir welding process. Scr Mater 2016;119:21–6.
16. Maggiolini E, Tovo R, Susmel L, James MN, Hattingh DG. Crack path and fracture analysis in FSW of small diameter 6082-T6 aluminium tubes under tension–torsion loading. Int J Fatigue 2016;92:478–87.
17. Ma ZY, Mishra RS, Mahoney MW. Superplastic deformation behaviour of friction stir processed 7075Al alloy. Acta Mater 2002;50:4419–30.
18. Kawashima T, Sano T, Hirose A, Tsutsumi S, Masaki K, Arakawa K, et al. Femtosecond laser peening of friction stir welded 7075-T73 aluminum alloys. J Mater Process Technol 2018;262:111–22
19. Bayazid SM, Farhangi H, Asgharzadeh H, Radan L, Ghahramani A, Mirhaji A. Effect of cyclic solution treatment on microstructure and mechanical properties of friction stir welded 7075 Al alloy. Mater Sci Eng A 2016;649:293–300.
20. Ji SD, Jin YY, Yue YM, Gao SS, Huang YX, Wang L. Effect of temperature on material transfer behavior at different stages of friction stir welded 7075-T6 aluminum alloy. J Mater Sci Technol 2013;29:955–60.
21. Wang K, Shen Y, Yang X, Wang X, Xu K. Evaluation of microstructure and mechanical property of FSW welded MB3 magnesium alloy. J Iron Steel Res Int 2006;13:75–8.
22. Singh K, Singh G, Singh H. Review on friction stir welding of magnesium alloys. J Magnes Alloy 2018;6:399–416.

23. Pan F, Xu A, Deng D, Ye J, Jiang X, Tang A, et al. Effects of friction stir welding on microstructure and mechanical properties of magnesium alloy Mg-5Al-3Sn. Mater Des 2016;110:266–74.
24. Oliveira JP, Miranda RM, Braz Fernandes FM. Welding and joining of NiTi shape memory alloys: A Review. Prog Mater Sci 2017;88:412–66.
25. Dolatkhah A, Golbabaei P, Besharati Givi MK, Molaiekiya F (2012) Investigating effects of process parameters on micro structural and mechanical properties of Al5052/SiC metal matrix composite fabricated via friction stir processing. Mater Des 37:458–464
26. Kasman Ş, Yenier Z (2013) Analyzing dissimilar friction stir welding of AA5754/AA7075. Int J Adv Manuf Technol 70:145 156
27. MaZY(2008) Friction stir processing technology: a review. Metall Mater Trans A 39(3):642–658
28. Malik V, Sanjeev NK, Hebbar HS, Kailas SV(2014) Investigations ontheeffectofvarious tool pin profiles in frictionstir welding using finite element simulations. Proc Eng 97:1060–1068
29. Besharati Givi M, AsadiP (2014) Advances infriction-stir welding and processing, 1st Edition. Woodhead Publishing
30. Venkata Rao C, Madhusudhan Reddy G, Srinivasa Rao K (2015) Influence of tool pin profile on microstructure and corrosion behav iour of AA2219 Al–Cu alloy friction stir weld nuggets. Defence Technology, In Press
31. Suri A (2014) An improved FSW tool for joining commercial alu minum plates. Proc Mat Sci 6:1857–1864

Note: All the figures and tables in this chapter were made by the authors.

Advances in Additive Manufacturing Technologies – Gurusamy Pathinettampadian et al. (eds)
© 2026 Taylor & Francis Group, London, ISBN 978-1-041-16687-0

Optimizing Laser Welding Parameters for AA6009: A Study on Joint Microstructure and Mechanical Integrity

Deeksheka[1]
Student, Department of Mechanical Engineering,
Chennai Institute of Technology,
Chennai, India

P.K. Devan[2]
Professor,
Department of Mechanical Engineering
R.M.K. College of Engineering and Technology,
India

Rishekkumar[3],
Aakash V.[4], **Abishek S.V.**[5], **Iniya T.**[6]
Student, Department of Mechanical Engineering,
Chennai Institute of Technology,
Chennai, India

◆ **Abstract:** This paper investigates the optimization of laser welding parameters for the aluminum alloy AA6009 in relation to its microstructural and mechanical properties. The AA6009 alloy, recognized for its high strength and resistance to denting, has widely used in various automotive industries and aerospace industries. The experimental and simulation approaches have been utilized to analyze the impact of major welding parameters, including laser power input, welding speed, and focal position, on the hardness profile and microstructure of the Weld Material (WM). It is observed that hardness is decreased in the HAZ due to GP dissolution and can be recovered via post-welding heat treatment. High welding speed was found to minimize liquation area extent and hence increase hardness. Microstructure analysis showed that the fusion zone has finer dendrites in comparison to the base material due to rapid heating followed by rapid cooling during welding. Precipitation of Mg_2Si and dissolution of the intermetallic compounds $Al_{12}FeMn_3Si$ in the weld fusion zone resulted in an enormous influence on the mechanical properties. The experimental results were well-validated through simulations with regard to hardness profiles and the HAZ width. A gap in between experimental and simulated values were explained in terms of model assumptions being simplistic, especially the considerations on magnesium and silicon diffusions. This research concluded that optimizing process parameters is crucial in improving mechanical strength and microstructural properties of AA6009, thereby yielding great industrial applications for this material.

◆ **Keywords:** Friction stir welding, AA5182 aluminum alloy, Welding parameters optimization, Mechanical properties, Joint performance

[1]deekshekaashok@gmail.com, [2]pkdevan68@gmail.com, [3]rolexrishe@gmail.com, [4]aakash.vasudevan2020@gmail.com, [5]abisheksivaraj10@gmail.com, [6]iniya0206@gmail.com

DOI: 10.1201/9781003685906-67

1. INTRODUCTION

Manufacturing industries are nowadays highly interested in reducing weight within automobile components. Lighter component utilization is critical to reduce spacecraft overall weight; this ensures higher fuel efficiency and good performance in aerospace applications [1,2]. There are various research where it has been proven that by lowering vehicle weight, its fuel consumption reduces to a great extent. Aluminum is the most widely used material in the automobile industry because it has low density, high corrosion resistance, and excellent specific strength [3]. The 6xxx series has attracted attention from researchers recently [4,5] because the 6xxx series alloys contain silicon and magnesium, with or without copper content. Compared with other aluminum alloys, the 6xxx series has better strength and corrosion resistance than the 5xxx and 2xxx series. The major alloys under this series are AA6009, AA6010, AA6016, and AA6111. This research paper deals with AA6009, which has 97.4% aluminum, 0.87% silicon, 0.57% magnesium, and a trace of zinc, manganese, and ferrite compounds. Since AA6009 has a higher percentage of copper, it has less corrosion but more strength. This alloy is used in automotive components as a sheet material due to its higher strength and dent resistance [6,7]. To meet the specific requirements of automotive body sheets, AA6009 undergoes processes like casting, hot rolling, cold rolling, and solution heat treatments.

With the advancement of technology, Laser Welding has emerged as pivotal process in the manufacturing of automotive and aerospace components. Laser welding is preferred because of its minimal thermal deformation, precise positioning, deep penetration, high energy density, and rapid travel speed [8]. Key process parameters like welding speed, focal position, and laser power have a significant effect on the bead profile and the final product quality [9]. Shielding gases, such as helium (He), nitrogen (N2), and argon (Ar), are used to prevent oxidation during welding, which influences the welding output [10]. Studies have shown that laser-arc-welded aluminum components exhibit lower residual stress and high value of tensile strength and fatigue strength compared to those that are joined by traditional welding methods, such as MIG and TIG welding, positioning Laser Beam welding as a promising technique for welding of aluminum alloy. In Laser Welding, the microstructure analysis and mechanical properties of the part that are fused together are critical determinants of its durability. Failures generally occur in the HAZ in most aluminum alloys, which dictates grain size based on the degree of heat source. As the degree of heat source decreases, the grain formed is smaller, whereas further increase in the heat input means larger grains. Cooling is another significant factor for the solidification process that affects the microstructures and intermetallic compound formation due to metallurgical reactions. For the 6xxx series, magnesium silicate (Mg2Si) precipitates form during solidification and significantly affect the mechanical properties of the welded material [11].

Though there is little research on welding aluminum 6xxx series alloys, there are some studies conducted on the joining of dissimilar aluminum alloys. Since laser arc welding exhibits high strength and low residual stress, it is often used for joining similar and dissimilar metals. Other alternative welding techniques applied to aluminum alloys include friction stir welding (FSW), M-TIG, and MIG welding. For the 7xxx series containing higher magnesium and zinc, FSW efficiency is around 83%, with elongated grain structure that strengthened the material. However, for the 5xxx series, the efficiency of FSW was reported between 57% and 87% with the variation of the specific composition [12,13,14]. Therefore, laser welding has been extensively used in 6000 series of aluminum alloys in the automotive field. The purpose of this study is to enhance the process parameters and assess the metallurgical properties of aluminum alloy AA6009, where there is little research on the 6xxx series.

2. EXPERIMENTAL METHODOLOGY

The base metal AA6009 was purchased from Arihant Aluminium Chennai. The thickness of 5mm plat was cut into a required dimensions of 100x55 using hacksaw. The edge preparation is done by flattening the edge using milling machine. Chemical compostion of the base metal is found by optical emission spectroscopy studies, the tensile strength of the material is studied using UTM (universal testing machine)(Table 67.1) and the hardness test is conduct using the Vickers microhardness test (Table 67.2). Usuall the material is prepared by cleaning the material using acetone, the acetone is used in order to removal of the oil contents that are usually present on the aluminium surface and it is also used to remove the oxide layers which are usually formed when the aluminum is exposed to air.

Table 67.1 Chemical composition (wt%) of AA5182

Material	Mg	Mn	Fe	Si	Cu	Cr	Zn	Al
AA6009	0.57	0.43	0.33	0.87	0.52	0.08	0.16	Bal.

Table 67.2 Mechanical properties of AA5182

Material	YS(MPa)	UTS(MPa)	Elongation (%)	Hardness (Hv 1)
AA6009	271	326	11.8%	97

The material is joined using a clamp to start the laser welding

After the welding process the welded workpiece are taken out of the clamp. The suitable ASTM (American Standard for Testing Machine) standard is choosed for the samples and the samples are cutted using the W-EDM machine according to the standard size of ASTM in the Dog-bone shape in order to perform the tensile test. The tensile strength test is done using the UTM, the tensile value of the samples is extracted as per ASTM-E8 using this the UTS and the yield value of the material are noted. The microstructural studies are done using the W-EDM and the sample are mounted by hot mounting press by using the Bakelite powder. The samples are then polished using silicon carbide abrasive sheets of (80 to 400 grits size) by hands and the samples are further polished by machine using the silicon carbide abrasive sheets of 600 to 200 grit size. The specimens are polished and are engraved by Keller's reagent to reveal microstructure. The polished samples are the used to explore the shapes and size of the grains using the SEM. Throught the SEM the microscopic size shapes of the materials are observed. The micro hardness of the sample is measured using the Vicker's Hardness test machine for 1 kg load and 20 dwell time along various cross section of the sample across the different zones at the gap size of 0.5mm between the indentation

The optimization for the welding process is conducted using the Central Composite Design, this is a statistical experimental design used to build the models and this the approach which is used behind the framwork of Response Surface Methodology. After a comprehensive study carried out on the laser welding the suitable parameter is choosed for this process, the Welding Process Parameter is used to enhance the final characteristics of the material which are discussed: Laser Power (Kilo Watts), Tool Travel Speed (TTS),Focal Position(milli meter). The limits and the values for these are obtained through extensive review. The lower limit for this is assigned as the value of –(alpha) (-1.682), and the upper limit is set as(alpha)(1.682). the midpoint between these values are set as central value which is coded as '0'. The values correspond to the levels -1 to +1 are calculated by using the standard equation of CCD. Table 67.3 is presents the welding process parameter with the assigned values between those limits.

Table 67.3 Limits of welding process parameter

WPP	Unit	Levels				
		-1.682	-1	0	1	+1.682
LP	kW	1	1.6	2.5	3.4	4
WS	mm/s	20	28	40	52	60
FP	mm	-2	-1.2	0	1.2	2

3. Results and Discussion

3.1 Optimization of Welding Parameters

The results suggest that the PSD of Mg2Si precipitation-hardening particles might be used to characterize the microstructure of aluminum alloys, which has a huge impact on their mechanical characteristics. The PSD obtained because of aging heat treatment in fully solutioning the alloys is primarily controlled by the time and temperature taken for aging process, the critical process variables. The thermal cycle are dominated by power and speed in Laser Welding Process. Simulations showed that the HAZ start at maximum temperature and the eutectic temperature is reached. The hardness profile in the HAZ of laser welds, with good estimation of HAZ width.

The observed discrepancies between experimental values are attributed to several factors. The approximation of dilute alloy in the growth law results in slower dissolution and coarsening of particles, especially at the high temperatures experienced during welding, explaining the loser dissolution rates far away from fusion region. Moreover, assuming magnesium diffusion was the sole rate-controlling step ignored the significant amount of silicon diffusion, which affected the total transformation

Table 67.4 Design of experiments framed using CCD

Std. order	Run order	Coded Values			UTS (MPa)
		P	S	F	
1	19	-1	-1	-1	204
2	13	1	-1	-1	212
3	11	-1	1	-1	191
4	4	1	1	-1	197
5	8	-1	-1	1	203
6	3	1	-1	1	209
7	12	-1	1	1	200
8	7	1	1	1	205
9	5	-1.682	0	0	203
10	10	1.682	0	0	214
11	18	0	-1.682	0	205
12	16	0	1.682	0	191
13	15	0	0	-1.682	202
14	1	0	0	1.682	207
15	17	0	0	0	213
16	9	0	0	0	213
17	20	0	0	0	216
18	14	0	0	0	214
19	2	0	0	0	215
20	6	0	0	0	217

Table 67.5 ANOVA test results

Source	Sum of Squares	Df	Mean Square	F-value	p-value	
Model	1160.76	9	128.97	90.89	<0.0001	significant
A-TRS	138.56	1	138.56	97.65	<0.0001	
B-TTS	250.97	1	250.97	176.87	<0.0001	
C-AL	33.56	1	33.56	23.65	0.0007	
AB	1.13	1	1.13	0.7928	0.3942	
AC	1.13	1	1.13	0.7928	0.3942	
BC	55.12	1	55.12	38.85	<0.0001	
A^2	73.56	1	73.56	51.84	<0.0001	
B^2	513.90	1	513.90	362.17	<0.0001	
C^2	194.47	1	194.47	137.05	<0.0001	
Residual	14.19	10	1.42			
Lack of Fit	0.8563	5	0.1713	0.0642	0.9955	not significant
Pure Error	13.33	5	2.67			
Cor Total	1174.95	19				
Std. Dev.	1.19	**R^2**		0.9879		
Mean	206.55	**Adjusted R^2**		0.9771		
C.V. %	0.5767	**Predicted R^2**		0.9780		
		Adeq Precision		28.612		

process, which delayed the overall process and thus showed a slightly higher dissolution rates at or near the fusion boundary. Such determinations have a greater signification of including both of these diffusions for enhancing accurate microstructure evolution simulations leading to a better prediction regarding the hardness and structural behavior of welded aluminum alloys.

Final Equation in Terms of Actual Factor

$$\begin{aligned} UTS = &+ 214.678 + 3.18519\, LP - 4.28687\, WS \\ &+ 1.56764\, FP - 0.375000\, LP * WS \\ &- 0.375000 LP * FP + 2.62500\, WS * FP \\ &- 2.25925\, LP^2 - 5.97156\, WS^2 \\ &- 3.67347\, FP^2 \end{aligned} \quad (1)$$

The Model F-value of 90.89 indicated that the model is highly significant, it is suggesting that the independent variables reliably explain the variation in the response. The probability of obtaining a large F-value purely due to random noise that is extremely low to (0.01%), demonstrating the robustness and predictive capacity of the model. Significant model terms, as identified by P-values below 0.0500, that include A, B, C, BC, A2, B2, and C2. These terms play a crucial role in understanding the response and should be retained to ensure the model's accuracy and interpretability. On the other hand, terms that are with P-values greater than 0.1000 are considered to be statistically insignificant, indicating a minimal influence on the response. Retaining such terms can unnecessarily increase model complexity and reduce predictive efficiency. Therefore, model reduction through the removal of non-significant terms can enhance the model's parsimony while preserving its predictive ability. However, it is essential to maintain the hierarchy of the model, ensuring that terms supporting higher-order interactions or quadratic effects are not eliminated. The Lack of Fit (F-value) of 0.06, indicating a 99.55% probability that such a value could arise due to noise, demonstrates that the lack of fit is not significant compared to pure error. This is a desirable outcome, as non-significant due to lack of fit suggests that the model has adequately represents the data and fits well. This balance between statistical significance and practical simplicity makes the model reliable for further analysis and predictions.

Figure 67.1 consists of a perturbation plot and contour plots, which are often used in optimization studies to understand the effect of process parameters on a response variable, which in this case is the Ultimate Tensile Strength (UTS). The perturbation plot (top-left) displays the sensitivity of UTS to deviations in individual process parameters around a reference point that is often the optimized or baseline condition. Each curve in the plot represents a parameter, showing how the variation of that parameter affects the UTS. Steeper curves indicate that the

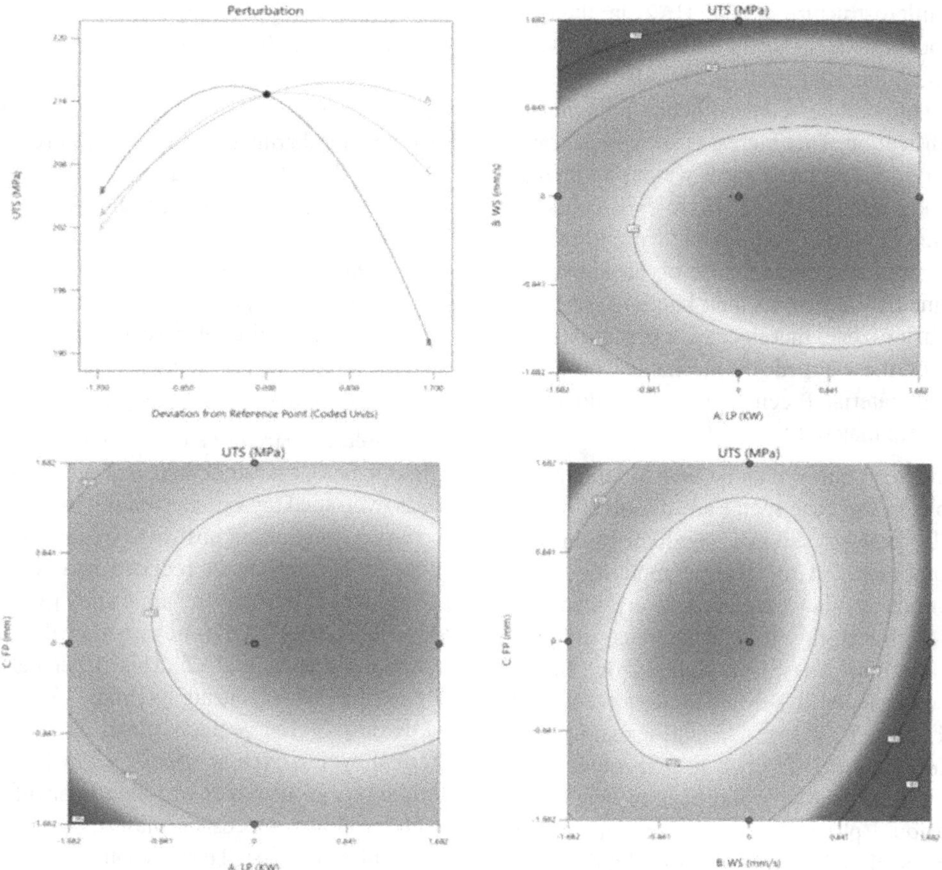

Fig. 67.1 [a] Perturbation plot [b] Interaction plot AB [c] Interaction plot AC [c] Interaction plot BC

parameter has a strong influence on UTS and needs to be controlled more precisely, while flatter curves indicate a lesser influence within the given range. This plot clearly shows which parameters are most critical to achieve the desired tensile strength. The contour plots are the last three, which describe the combined effect of two parameters on UTS, while keeping other variables constant. Each plot is a two-dimensional plane with axes corresponding to two parameters and colored contours that indicate the levels of UTS. The central red regions in the contour plots are indicative of the best parameter combinations that produce maximum UTS. These plots are particularly useful for identifying interactions between parameters and determining the ranges where optimal welding conditions exist. Together, the perturbation and contour plots provide comprehensive insights into the effects of process parameters, guiding the optimization of welding conditions for improved mechanical properties..

3.2 Microstructure Analysis

The columnar crystals in the laser welding process depend on anisotropy and the variation in the growth rates of the crystal with respect to different directions. During solidification, it cools perpendicular to the welding path, so the grains formed are columnar, oriented perpendicular to the fusion line. Moreover, in the FCC and BCC crystal structures, during the solidification process, this orientation is also obtained [15,16]. Dendrites orientation is closely related to the grain orientation, and spacing between dendrites depends upon cooling rate. Higher cooling rates provide smaller dendrite spacing [17]. Because of the localized heating in laser welding, there is minimal microstructural change in the welded material compared with the base metal.

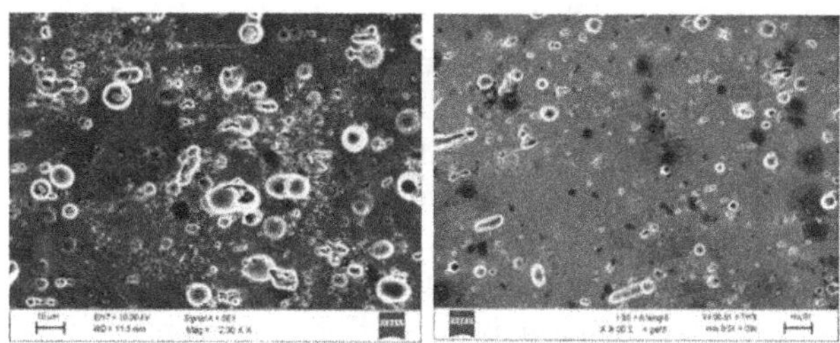

Fig. 67.2 SEM microstructure [a] Base metal AA6009-T6 [b] WNZ of sample 7

Therefore, the microstructure of the HAZ in the weld material is almost equivalent to the base material. When the solidification region is far from the boundary of fusion, the gradient of temperature reduces, hence reducing the velocity of solidification. This increases the concentration of solute at the solid-liquid interface as well as an increase of the supercooling region [18,19]. There exist coarsened grains at the HAZ, as narrow grains result from lower heat input. The wire and base material fuse on an alloy basis along a fusion line that represents their bonding zone [20]. In this case of fusion zone, there occurs a precipitation of Mg2Si precipitates; these impede the mechanical properties within the welded material. Precipitates Mg2Si add a bit to hardness of the weld material.

Figures 67.2 and 67.3 show that Fe and Si precipitates that are available in the base metal are dissolved during welding. SEM analysis shows agglomeration and intermetallic compounds in the welded material. Linear intermetallic formations that cause cracks are identified, which are in contrast to less dangerous spherical intermetallic formations. The SEM results show that the white precipitate phase, Al12(FeMn)3Si, dissolves during welding [21]. Figures 67.2 and 67.3 further show well-distributed grains, with higher grain growth observed in the solidification zone. During solidification, Mg2Si are rejected to the liquid phase at their low melting points. These precipitates dissolve, leaving the liquid phase enriched with interdendritic solute, which eventually forms eutectic phases at the grain boundaries upon solidification [22,23]. Grain refinement in high-solute-content alloys depends on the solute drag effect of solute atoms [24]. The weld zone of the welded material has finer dendrites than the base material this is because that the high heating and cooling rates of laser welding. The temperature gradient is one of the critical factors that affect grain morphology during solidification [25]. The microstructure of the fused joints is mostly affected by the heat generated in the process from the fused material to the surrounding material.

4. Hardness Analysis

Microhardness was studied in the cross-section of the weld. It has been found that the minimum hardness is present in the weld zone and maximum hardness in the base metal zone. When the hardness of the welded part is compared, the hardness at the HAZ of the welded part is a little less than the base metal, and comparing with the HAZ in the WHS, the hardness there is higher because the area of the HAZ has been smaller. The dissolution strengthening phases, over-aging, grain coarsening, and reprecipitation are some of the factors which have an adverse effect on the hardness due to welding [26,27]. As presented in Fig. 67.6, the size of larger grains in the material influences hardness. According to Hall-Petch equation, large grain size decreases the hardness of material while small grain size enhances it. The existing research studies establish that the microhardness at the weld centre is less than that of the HAZ [28,29]. The variation in values of hardness is due to the uneven distribution of temperature during welding process. As the heat source moves, the molten metal cools. This creates variations in microstructure between the welded zone and the HAZ. The partially welded zone is rapidly heated and cooled so that precipitation and solid solution strengthening of the second phase are not allowed because of rapid cooling rates [30].

Solid solution strengthening usually has a much greater effect on hardness rather than over-aging. However, in some cases over-aged regions can exhibit improved hardness due to re-precipitation or post-heat treatment during the weld process. The reason of reduced hardness is that it loses precipitates in the heat-treatable alloy [31]. In an aluminum alloy like AA 6009, GP zones are more dominant in the basis material rather than HAZ. This is because of the thermal cycles that occur during welding [32]. The precipitate hardening of the welded metal is decreased by these thermal cycles, even at relatively low heat input [33]. The welding speed also has a very significant effect on hardness; higher welding speeds normally give higher hardness of the weld metal [34]. There is some evidence that suggests that solution treatment and aging post-welding can regain hardness in heat-treatable alloys [35]. Nevertheless, other studies reveal that restoring the hardness of FZ and HAZ to as good a state as the base metal may be impossible after welding [36,37]. For such validation, differences between the measured and predicted values are also calculated. For example, Vickers hardness tests were conducted on a 2-mm-thick laser-welded AA6061 alloy using a process parameter of 3 kW power and 4 m/min welding speed. The chemical constituents of this AA6061 alloy is as follows:

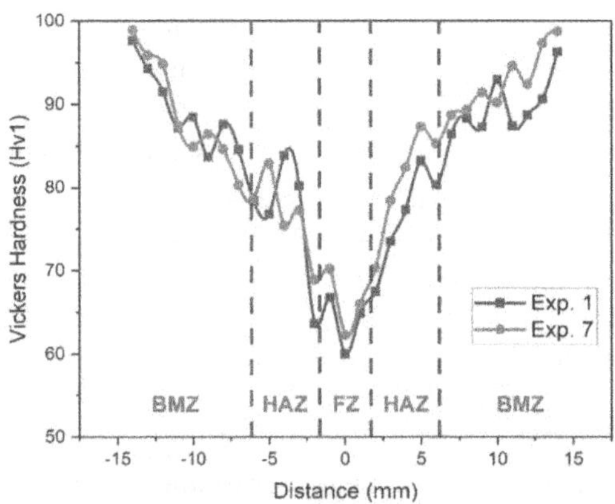

Fig. 67.3 Hardness variation graph

0.65% Si, 0.31% Fe, 0.27% Mn, 1.05% Mg, and 0.25% Cr. Microhardness testing shows that the HAZ is softer compared to the weld material. But the GP zones are very important to increase hardening by precipitation reactions, allowing the hardness value of the weld material to recover almost to that of the original base material. It is shown that hardness of the weld material are higher when welding increasing speed, and with higher welding speeds, there is an increase in the hardness values.

5. Conclusion

1. The study successfully optimized laser beam welding (LBW) parameters (laser heat input, welding speed, and focal position) for AA6009 aluminum alloy. Precise control of these parameters improved the weld bead profile, minimized thermal deformation, and enhanced tensile strength and hardness.
2. Reduced dendrite spacing and homogeneous grain distribution were achieved. Mg_2Si precipitates in the FZ contributed significantly to improved mechanical properties. Minimal grain coarsening was observed in the HAZ due to effective heat management.
3. The weld region and HAZ exhibited a reduction in hardness of about 40% and 25% respectively compared with the base material. This is mainly due to the transfer of heat during welding

References

1. Dursun, T., Soutis, C. (2014) Recent developments in advanced aircraft aluminium alloys. Materials & Design (1980-2015), 56: 862–871.
2. Zhang, X., Huang, T., Yang, W., Xiao, R., Liu, Z., Li, L. (2016) Microstructure and mechanical properties of laser beam-welded AA2060 Al-Li alloy. Journal of Materials Processing Technology, 237: 301–308.
3. Dhakal, B.; Swaroop, S. E ect of laser shock peening on mechanical and microstructural aspects of 6061-T6 aluminum alloy. J. Mater. Process. Technol. 2020, 282, 116640.
4. G.B. Burger, A.K. Gupta, P.W. Jeffrey, D.J. Lloyd, Mater. Characterization 35 (1995) 23.
5. J. Hirsch, C. Dumont, O. Engler, VDI-Berichte 1151, VDI Verlag, Du¨sseldorf, 1995, p. 469.
6. G. B. Burger, A. K. Gupta, P. W. Jeffrey and D. J. Lloyd: Materials Characterization, 35, (1995), 23–39.
7. D. G. Althenpohl: "Aluminum: Technology, Applications, and Environment - A profile of a Modern Metal," 6th Edition, (The Minerals, Metals, and Materials Society, Warrendale and the Aluminum Association, Washington, USA, 1998), 361–364.
8. Liu, L., Hao, X., Song, G. (2006) A New Laser-Arc Hybrid Welding Technique Based on Energy Conservation. Materials Transactions, 47: 1611–1614.
9. Jiang, P., Wang, C., Zhou, Q., Shao, X., Shu, L., Li, X. (2016) Optimization of laser welding process parameters of stainless steel 316L using FEM, Kriging and NSGA-II. Advances in Engineering Software, 99: 147–160.
10. Zhang, Y.; Hu, W.; Lai, X. Optimization of laser welding thin-gage galvanized steel via response surface methodology. Opt. Lasers Eng. 2012, 50, 1267–1273.
11. Silva, J. Laser Welding of Aluminium Rings. Mastery Thesis, Instituto Superior Técnico, Universidade Técnica de Lisboa, Lisboa, Portugal, 2011.
12. Zhang Z, ChangshuHe Ying Li, LeiYu Su Zhao, Zhao Xiang. Effects of ultrasonic assisted friction stir welding on flow behavior, microstructure and mechanical properties of 7N01-T4 aluminum alloy joints. J Mater Sci Technol 2020;43:1–13.
13. Dewangan SK, Tripathi MKMMK. Effect of welding speeds on microstructure and mechanical properties of dissimilar friction stir welding of AA7075 and AA5083 alloy. Mater Today: Proc; 2020.
14. Hamed JA. Effect of welding heat input and post-weld aging time on microstructure and mechanical properties in dissimilar friction stir welded AA7075–AA5086. Trans Nonferrous Met Soc China 2017;27(8):1707–15.
15. S. Kou, Y. Le. Welding parameters and the grain structure of weld metal-A thermodynamic consideration, Metall. Trans. A 19A (1988): 1075–1082.
16. S. Kou. Welding Metallurgy, 2nd ed., John Wiley and Sons, Hoboken, NJ, 2003.
17. Yang J, Wang Y, Li F, Huang W, Jing G, Wang Z, et al. Weldability, microstructure and mechanical properties of laser-welded selective laser melted 304 stainless steel joints. J Mater Sci Technol 2019;35(9):1817–24.
18. Z.H. Zhang, S.Y. Dong, Y.J. Wang, B.S. Xu, J.X. Fang, P. He. Microstructure characteristics of thick aluminum alloy plate joints welded by fiber laser, Mater. Des. 84 (2015) 173–177.
19. W. Kurz, C. Bezençon, M. Gäumann. Columnar to equiaxed transition in solidification processing, Sci. Technol. Adv. Mater. 2 (2001) 185–191.
20. Li JJ, Shen JQ, Hu SS, Liang Y, Wang Q. Microstructure and mechanical properties of 6061/7N01 CMT þ P joints. J Mater Process Technol 2019;264:134e44.(help)
21. Donnadieu P, Lapasset G, Sanders TH. Manganese-induced ordering in the a-(AleMneFeeSi) approximant phase. Philos Mag Lett 1994;70:319e26.
22. Nie FH, Dong HG, Chen S, Li P, Wang LY, Zhao ZX, et al. Microstructure and mechanical properties of pulse MIG welded 6061/A356 aluminum alloy dissimilar butt joints. J Mater Sci Technol 2018;34:551e60.
23. Li JJ, Shen JQ, Hu SS, Liang Y, Wang Q. Microstructure and mechanical properties of 6061/7N01 CMT þ P joints. J Mater Process Technol 2019;264:134e44.
24. Huang Y, Humphreys FJ. The effect of solutes on grain boundary mobility during recrystallization and grain growth in some single-phase aluminum alloys. Mater Chem Phys 2012;132:166e74.

25. YuJM, HashimotoT,LiHT,WanderkaN,ZhangZ,CaiC,etal. Formation of intermetallic phases in unrefined and refined AA6082 Al alloys investigated by using SEM-based ultramicrotomy tomography. J Mater Sci Technol 2022;120:118e28.
26. Peng X, Cao X, Xu G, Deng Y, Tang L, Yin Z. Mechanical properties, corrosion behavior, and microstructures of a MIG-welded 7020 Al alloy. J Mater Eng Perform 2016;25(3):1028e40.
27. Nicolas M, Deschamps A. Precipitate microstructures and resulting properties of Al-Zn-Mg metal inert gas-weld heat affected zones. Metall Mater Trans 2004;35(5):1437e48.
28. Kuo TY, Lin HC. Effects of pulse level of NdeYAG laser on tensile properties and formability of laser weldments in automotive aluminum alloys. Mater Sci Eng A 2006;416:281e9.
29. Liu C, Northwood D,Bhole S. Tensile fracture behavior in CO2 laser beam welds of 7075-T6 aluminum alloy. Mater Des 2004;25(7):573e7.
30. Gu JX, Yang SL, Duan CF, Xiong Q, Wang Y. Microstructure and mechanical properties of laser welded Al-Mg-Si alloy joints. Mater Trans 2019;60:230e6.
31. Jones, I.A.; Riches, S.T.; Yoon, J.W.; Wallach, E.R. Laser welding of aluminum alloys. TWI J. 1998, 7 (2), 421–481.
32. M. Nicolas and A. Deschamps: Metall.Trans. A, 2004, 35, 1437 1448.
33. Dausinger, F.; Rapp, J.; Hohenberger, B.; Hugel, H. Laser beam welding of aluminum alloys: state of the art and recent developments. Proc. Int. Body Engineering Conf. IBEC'97: Advanced Technologies and Processes, Stuttgart, Germany, Sept 30–Oct 2, 1997; Vol. 33, 38–46.
34. Yamaguchi, T.; Kato, M.; Nishio, K.; Fukami, K. Hardness distribution of YAG laser-welded A5052 aluminum alloy. Q. J. Jpn Weld. Soc. 2001, 19 (1), 114–121.
35. Katoh, M. Factors affecting mechanical properties of laser welded aluminum alloys. J. Light Met. Weld. Constr. 1996, 34 (4), 42–48.
36. Hirose, A.; Kobayashi, K.; Yamaoka, H.; Kurosawa, N. Evaluation of properties in laser welds of A6061-T6 aluminum alloy. J. Light Met. Weld. Constr. 1999, 37 (9), 1–9.
37. Yamaoka, H.; Tsuchiya, K.; Hirose, A.; Kobayashi, K. Study of aging treatment of Al–Mg–Si alloy laser welds—CO2 laser welding of Al–Mg–Si alloys (2nd Report). Q. J. Jpn. Weld. Soc. 2000, 18 (3), 431–437.

Note: All the figures and tables in this chapter were made by the authors.

Advances in Additive Manufacturing Technologies – Gurusamy Pathinettampadian et al. (eds)

Mechanical and Tribological Characterization of Additively Printed Pla-Marble Dust Hybrid Composite

Babu T. V. B.*,
Sasidharan C., Hari Krishna Raj S.,
Varun Kumar S., Karthikeyan C.
Department of Mechanical Engineering,
VelTech HighTech Dr. Rangarajan Dr. Sakunthala Engineering College,
Chennai, Tamilnadu

◆ **Abstract:** This research investigates the mechanical and tribological behavior of polylactic acid (PLA) marble dust-reinforced composites, which are produced by fused deposition modeling (FDM), a common additive manufacturing technique. The research investigates the influence of different marble dust percentages on some of the most relevant mechanical properties, such as tensile strength, flexural strength, and impact resistance. Pin-on-disc tests conducted under dry sliding conditions also helped to evaluate the wear resistance and coefficient of friction. Additive effects were disclosed by results since marble dust addition improved flexural and tensile strength of PLA with the optimal performance at some certain levels of filler. In spite of that, higher concentration of marble dust decreased the ductility of the composite in a trade-off between flexibility and stiffness. In terms of tribological performance, the composite exhibited improved wear resistance and lower friction coefficient than pure PLA. These results indicate that PLA-marble dust composites can be a viable material for wear-resistant, light-weight applications like gears, bearings, and other engineering parts as a cost-effective and sustainable alternative to conventional materials.

◆ **Keywords:** Hybrid, Composite, PLA, Tribological

1. Introduction

Additive manufacturing (AM) has revolutionized the way intricate structures are conceived and manufactured, enabling highly customizable properties. Amongst the materials employed in AM, polylactic acid (PLA) is notable because of its biodegradability, affordability, and ease of processing. Nevertheless, PLA has some limitations, such as low tensile strength, impact resistance, and wear durability, which limit its application in high-performance engineering applications. In a bid to address these shortcomings, researchers have investigated reinforcing PLA with other materials to enhance its overall performance. Marble dust is one of such reinforcements, which is a waste product of the marble industry and provides a natural and low-cost means of enhancing PLA's mechanical and tribological performance. The aim of this research is to create and investigate PLA composites reinforced with marble dust through fused deposition modeling (FDM), a common AM process. Tests were carried out in a series of tests to test the mechanical, thermal, and tribological characteristics of the composites. Some of the key tests were tensile strength tests to test for resistance against breakage under tension, impact tests to determine toughness, and thermal conductivity tests to test heat dissipation. Additional tests including hardness tests to determine surface hardness and wear resistance tests to determine performance under friction were also conducted.

*Corresponding author: prof.tvbbabu@gmail.com

DOI: 10.1201/9781003685906-68

The scanning electron microscope (SEM) was once more used to examine the distribution of the marble dust within the PLA matrix and the inter-particle bonding and quality of the same. Tensile strength is a primary parameter that determines up to what degree a material is capable of resisting tension before fracture. Incorporating marble dust in PLA should enhance this property because of the stiffness of the filler. Optimization of the content of the marble dust is important, though, since too much of it could impair the ductility of the material. Impact resistance, which indicates how well a material can absorb energy during an impulsive load, is another key consideration. Marble dust reinforcement is also expected to improve impact strength, minimizing the brittleness usually found in neat PLA—a beneficial characteristic for parts exposed to dynamic loads. Thermal conductivity is another critical characteristic, as it affects the capacity of the composite to disperse heat. Marble dust is expected to improve this property, such that the material will be more appropriate for uses involving effective thermal management. Hardness, which quantifies resistance to scratches and deformation, is likely to rise with marble dust reinforcement, enhancing the durability of the composite for wear-resistant use. Also, wear resistance is important for parts subjected to mechanical stress, and the addition of marble dust is likely to improve this characteristic by minimizing material loss through friction. SEM analysis offers greater understanding of the internal composition of the composite, i.e., the dispersion of marble dust and its interaction with the PLA matrix. Microstructural features have a direct impact on the mechanical performance and durability of the composite. The results of this research pinpoint the promise of PLA-marble dust composites for use in applications that demand enhanced strength, thermal stability, and wear resistance. By integrating additive manufacturing with green materials, such composites offer a promising alternative for the manufacture of high-performance, environmentally friendly engineering components. Application of marble dust in composite materials has been of considerable research interest owing to its capacity to improve mechanical and thermal properties in various polymer matrices, such as PLA, PET, LDPE, and epoxy. There are still various gaps in research, despite all these developments. Most studies are interested in the early improvement of mechanical properties but pay little attention to long-term durability and environmental performance under practical conditions, including moisture exposure, temperature variations, and prolonged mechanical loading. Additionally, although studies have predominantly focused on maximizing certain mechanical properties through alteration of marble dust content, few have explored to what extent optimum concentrations differ across various polymer matrices or processing methods. Experiments tend to cover narrow weight percent ranges and neglect the possible synergistic effects with other fillers, which would yield further gains in performance. Another essential but quite underinvestigated parameter is the recyclability and end-of-life treatment of marble dust composites. Although the utilization of marble dust as a green filler is almost universally accepted, there has been very limited consideration as to how and in which form these materials can be recycled or reused, which is very important for their ecological sustainability. Furthermore, the socio-economic effect of incorporating marble dust into industry has not yet been studied in full scope. Less literature is available in market feasibility, cost-effectiveness, and industry adoption, while there are tremendous technical advantages those composites bring. The work seeks to examine the mechanical properties and tribology of PLA-based composites of marble dust strengthened, produced through additive manufacturing. Material performance shall be improved to promote sustainability. A number of goals have therefore been set. Rigorously examine the mechanical behavior of PLA-marble dust composites, such as tensile strength, flexural strength, impact resistance, and hardness. Composites at different marble dust concentrations (0%, 5%, 10%, and 20%) will be examined to determine the optimal composition for given applications. Reinforcement by marble dust is expected to improve tensile and flexural properties but preserve an adequate degree of ductility. - Evaluate the tribological performance of such composites, in terms of wear resistance and friction, via standardized pin-on-disc experiments. The wear behavior and friction properties at varied loads and speeds will be tracked during this evaluation to determine the effect of marble dust on the durability of composite in friction-heavy environments. - Use scanning electron microscopy (SEM) to investigate the dispersion of marble dust in the PLA matrix and establish interfacial bonding quality. These microstructural properties will be utilized to relate material properties and mechanical and tribological performance, and they will offer insights into failure modes and optimization techniques. - Conduct thermal properties such as thermal conductivity, degradation temperature, and specific heat capacity in order to ascertain the feasibility of these composites at elevated temperatures. Additionally, physical properties such as density and porosity will be tested to understand the impact of marble dust on the overall structure of the composite. - Conduct aging tests to evaluate the long-term composites behavior in environmental exposure conditions, moisture absorption, and re stressing mechanics. This would indicate the probable reason for the degradation and predict the lifetime of materials in applications. Undertake a life cycle assessment (LCA) to ascertain the environmental impact of PLA-marble dust composites relative to conventional PLA products. Using the examination of factors like resource utilization, carbon emission, and minimizing waste, the present study will determine the environmental benefits of using marble dust to enhance PLA. Assemble additive manufacturing parameters like printing speed, layer thickness, and nozzle temperature to provide reproducibility in terms of quality as well as ultimate mechanical and tribological performance. Determining the optimal printing conditions is key to producing high-performance end-products. Marble dust, which is a widespread industrial by-product, is environmentally problematic due to disposal. Through the incorporation of this waste material into PLA composites, the research encourages efficient waste management and converts a possible pollutant to a useful asset. Also, PLA is a biodegradable polymer and hence a more environmentally friendly choice compared to standard plastics. When reinforced with marble dust, not only does it maintain its green benefits but also its mechanical and tribological strengths, and as such, a material with promising prospects for engineering applications in the future.

Fig. 68.1 Polylactic acid PLA

While PLA is widely used in 3D printing, its mechanical strength and durability often limit its application. Adding marble dust as a filler could improve properties such as tensile strength, hardness, and impact resistance, making it suitable for a broader range of uses. By incorporating marble dust, PLA composites can achieve improved wear resistance and a stable coefficient of friction.

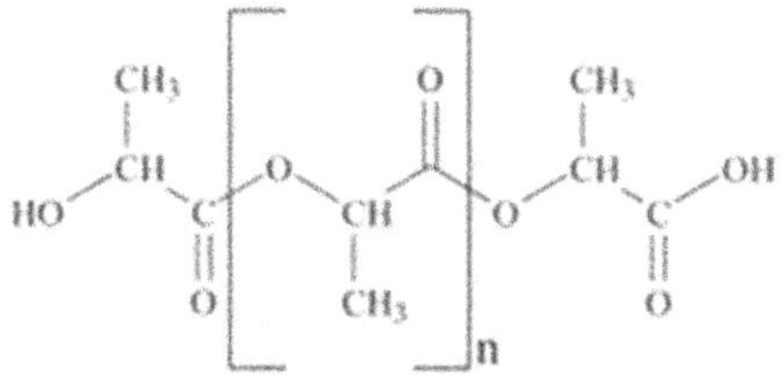

Fig. 68.2 Chemical structure of polylactic acid PLA

These enhancements make them more suitable for applications like gears and bushings, where frictional forces are significant. Marble dust is an affordable filler material, potentially reducing the cost of PLA composites without compromising their performance

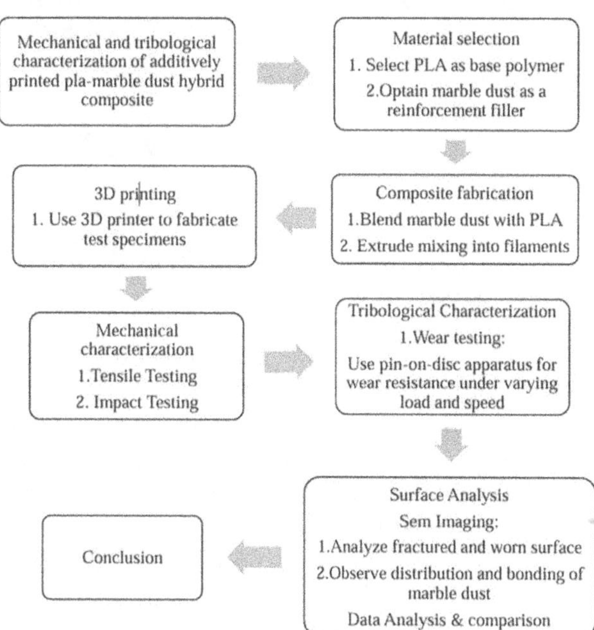

Fig. 68.3 Flow chart for process flow

Polylactic Acid (PLA) is a biodegradable plastic that comes from renewable sources such as corn starch and sugarcane. PLA has become increasingly popular in recent years due to its eco-friendliness, ease of manufacture, and versatility. PLA finds wide application in packaging, 3D printing, medical devices, and agriculture. Yet, for all its positives, PLA also has some negatives that tend to necessitate alterations, such as mixing it with other substances or incorporating reinforcing agents. This is a discussion of PLA's properties, advantages, disadvantages, and developments that focus on enhancing its mechanical and thermal performance. PLA is a thermoplastic polyester produced by fermenting natural sugars. Its primary building block is lactic acid, with two forms: L-lactic acid and D-lactic acid. The ratio of these forms determines the polymer's crystallinity, strength, and heat resistance. PLA is typically divided into two categories Amorphous PLA provides flexibility and transparency but with reduced mechanical strength, semi-crystalline PLA more stiff, structurally stable, and possesses improved heat resistance. The level of crystallinity is the most important factor in properties like stiffness, toughness, and thermal resistance. Important Properties of PLA has a tensile strength of 50-70 MPa and behaves comparably to petroleum-based

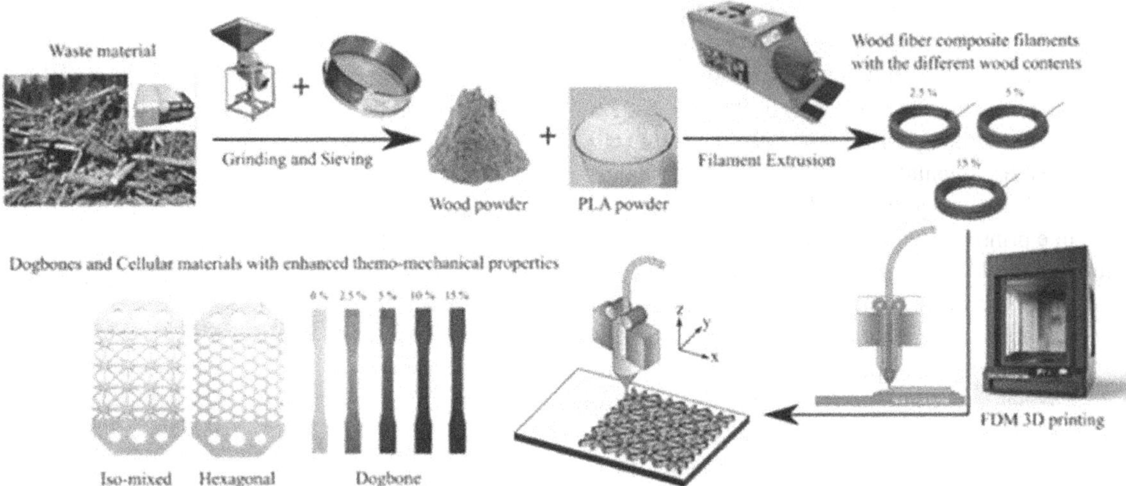

Fig. 68.4 Process of PLA

plastics such as PET decomposes into water and carbon dioxide when exposed to industrial composting conditions, hence making it a biodegradable alternative to traditional plastics. It is easy to process using different processes, such as 3D printing and molding. PLA softens at comparatively low temperatures since it has a glass transition temperature of 55–60°C, which constrains its application in high-heat situations. PLA lacks significant impact resistance and flexibility, which means it cannot be applied in applications with high durability needs. Though biodegradable, PLA degrades slowly in nature outside industrial composting conditions, necessitating specific heat, moisture, and microbial action for effective breakdown. Chemically, PLA is made up of lactic acid units bonded together by ester linkages with the general formula $(C_3H_4O_2)_n$. Its physical characteristics rely mostly on the stereochemical configuration of its monomers.

2. Result and Discussion

L-Lactic Acid Dominance: Higher proportions of L-lactic acid (above 90%) result in semi-crystalline PLA with enhanced mechanical strength and heat resistance, ideal for robust applications like packaging and fibers. D-Lactic Acid Content: A higher ratio of D-lactic acid yields more amorphous PLA, offering flexibility and transparency for applications such as films and disposable products. Direct Condensation Polymerization: A more straightforward process but not as widely utilized for industrial applications. Ring-Opening Polymerization (ROP): The method of choice for commercial production, providing greater control over molecular weight and stereochemistry, leading to more refined polymers. Subject to industrial composting, PLA's ester links hydrolyze, decomposing the material into water, lactic acid, and carbon dioxide, solidifying its environmentally friendly reputation. With tensile strength equivalent to that of PET, PLA is appropriate for applications calling for durability, like packaging and 3D printing. Nevertheless, its mechanical characteristics are dependent on molecular weight, stereochemistry, and processing conditions. Continued research is seeking to overcome its flaws and extend its use in various, high-performing applications. The materials preparation step is important when making a PLA and marble dust composite. Begin with good-quality PLA (polylactic acid) pellets, a biodegradable thermoplastic that is widely used in additive manufacturing. Dry these pellets thoroughly and make sure they have no moisture in them, since even minor water content can introduce air bubbles while processing, causing defects in the end composite. Drying is performed in a dedicated drying oven or desiccant dryer at about 60°C for a few hours. Marble dust is a reinforcing filler material. Choose finely powdered marble dust with uniform particle size, preferably less than 100 microns. Fine particles enhance better dispersion in the PLA matrix and provide enhanced composite properties, including higher strength, stiffness, and wear resistance. It is important to make sure that the marble dust is also dry prior to mixing with PLA so that it will not clump and affect the homogeneity and mechanical properties of the composite. Prepare the materials by then measuring the specified quantities depending on the preferred ratio of the composite, usually between 5% and 20% marble dust by weight. This balance affects the overall properties of the composite, and hence modifications can be made according to the application demands. For instance, increased marble dust has the advantage of increasing rigidity but will lose some flexibility.

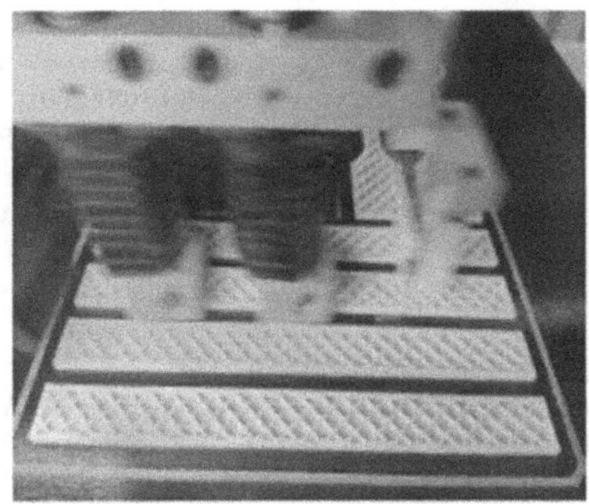

Fig. 68.5 Material preparation

After measurement, manually premix the PLA and marble dust to achieve an even initial distribution. This step minimizes the chances of uneven blending during the extrusion process and helps maintain consistency in the composite's mechanical and aesthetic qualities. Proper material preparation sets the foundation for a successful fabrication process, ensuring optimal quality and performance of the PLA-marble dust composite. Manually pre-mix the PLA pellets with marble dust in the desired proportion to ensure an even initial distribution. Extruder Mixing: Feed the PLA and marble dust mixture into a twin-screw extruder for further blending. The twin-screw extruder allows for efficient and homogenous mixing due to the shearing action. Extrusion: Extruder Settings: Set the extruder temperature between 180°C to 200°C (depending on the PLA's melting point and marble dust stability). 59 Composite Filament Extrusion: The extruder outputs a PLA-marble dust filament, which can be cooled and collected on a spool. If the filament will be used for 3D printing, maintaining a consistent diameter is essential for smooth printing. When using 3D printing, the composite filament, previously extruded with a consistent diameter, is loaded into a standard Fused Deposition Modeling (FDM) 3D printer. Adjusting printer settings is essential, as the marble dust impacts the composite's flow characteristics.

Fig. 68.6 Printed materials

For instance, print temperatures might need to be slightly higher than for pure PLA, typically between 200°C to 220°C, to ensure smooth extrusion and layer adhesion. Marble dust may also slightly increase the filament's viscosity, so slowing down the print speed can enhance detail and prevent nozzle clogging. Layer height, print speed, and cooling settings should be optimized to achieve the best surface finish and mechanical integrity.

3. Conclusion

Blending marble dust with polylactic acid (PLA) is an exciting innovation in composite materials, offering a balance of sustainability and performance. PLA, a biodegradable thermoplastic derived from renewable sources like corn starch and sugarcane, is very well known for its eco-friendly properties. However, its inherent mechanical deficiencies—like low impact resistance and stiffness—have a tendency to restrict its use in high-performance applications. Gluing marble dust overcomes these with a more rigid, stronger, and more aesthetically pleasing composite. Marble dust contributes significantly to the mechanical properties of PLA. The resulting composite is more wear-resistant, durable, and stronger compared to virgin PLA. The marble dust contributes to the rigidity and density of the material and improves it against deformation under stress, as well as making it better suited for load-carrying applications. Furthermore, this mixture offers greater thermal stability, and hence performs best where temperature patterns tend to vary. These innovations unlock its potential for use in semi-structural applications in the building, automotive, and consumer goods sectors, where performance as well as sustainability in the environment matters. Beyond its functional benefits, marble dust also makes the composite aesthetically pleasing. Having a smooth, refined texture, marble adds to the PLA composite an upscale, glossy finish. This makes it suitable for applications where appearance is crucial, such as decorative pieces, interior finishes, and consumer products like interior decor items. The stony feel imbues it with luxury without increasing production costs, which makes it an excellent value for manufacturers. The production process of PLA-marble dust composites also supports sustainability efforts. Marble dust is a quarrying and stone processing byproduct, and recycling it reduces waste and promotes circular material usage. This recycling process aligns with global sustainability goals and appeals to environmentally conscious consumers and companies that want to reduce their ecological footprint. In addition, the composites are easily made through standard compounding and molding procedures, and large volume production is feasible and cost-effective.

References

1. Kore SD, Vyas AK (2016) Impact of marble waste as coarse aggregate on properties of lean cement concrete. Case Stud Con str Mater 4:85–92
2. Awad AH, Abdellatif MH (2019) Assessment of mechanical and physical properties of LDPE reinforced with marble dust. Com posites B 173:106948
3. Rana A, Kalla P, Csetenyi LJ (2015) Sustainable use of marble slurry in concrete. J Clean Prod 94:304–311
4. Çınar ME, Kar F (2018) Characterization of composite pro duced from waste PET and marble dust. Constr Build Mater 163:734–741
5. Yavuz Çelik M, Sabah E (2008) Geological and technical char acterisation of Iscehisar (Afyon-Turkey) marble deposits and the impact of marble waste on environmental pollution. J Environ Manage 87:106–116
6. Thakur AK, Pappu A, Thakur VK (2018) Resource efficiency impact on marble waste recycling towards sustainable green construction materials. Curr Opin Gr Sustain Chem 13:91–101
7. Soydal U, Kocaman S, Esen Marti M, Ahmetli G (2018) Study on the reuse of marble and andesite wastes in epoxy-based compos ites. Polym Compos 39:3081 3091
8. Fiore V, Di Bella G, Scalici T, Valenza A (2018) Effect of plasma treatment on mechanical and thermal properties of marble pow der/epoxy composites. Polym Compos 39:309–317 65
9. Gürü M, Tekeli S, Akin E (2007) Manufacturing of polymer matrix composite material using marble dust and fly ash. Key Eng Mater 336–338:1353–1356
10. Choudhary M, Singh T, Dwivedi M, Patnaik A (2019) Waste mar ble dust- filled glass fiber-reinforced polymer composite Part I: physical, thermomechanical, and erosive wear properties. Polym Compos 40:4113–4124. 11.
11. Choudhary M, Singh T, Sharma A, Dwivedi M, Patnaik A (2019) Evaluation of some mechanical characterization and optimization of waste marble dust filled glass fiber reinforced polymer compos ite. Mater Res Express 6:105702
12. Awad AH, El-gamasy R, El-Wahab AAA, Abdellatif MH (2019) Mechanical behavior of PP reinforced with marble dust. Constr Build Mater 228:116766.

Note: All the figures in this chapter were made by the authors.

Advances in Additive Manufacturing Technologies – Gurusamy Pathinettampadian et al. (eds)

An Efficient Monitoring of Defective Parts Identification using Automated Vision Based Quality Inspection System

Vinoth T.[1]
Department of Mechatronics Engineering,
Chennai Institute of Technology,
Chennai, India

R. Arivalahan[2]
Department of EEE,
SRM Valliammai Engineering College,
Kattankulathur

Vinoth N.[3],
Anushraj B.[4], **Ramakrishnan J.**[5], **Ragul P.**[6]
Department of Mechatronics Engineering,
Chennai Institute of Technology,
Chennai, India

◆ **Abstract:** In Industries the quality checking of the end products is an important factor. Which plays a major role in performance and profit of the industry. The quality checking was done manually through visual inspection which leads to inconsistent results due not following the standards properly. In our system by using the image processing technique the process was automated using articulated robot and image processing technique which helps to increase the accuracy and reduce the time spent for visual inspection . The system that developed for testing is pre-trained with datasets which identify the defects in the parts accurately. The system composed of two systems (i) Image Processing system and (ii) Material handling systems.

◆ **Keywords:** Quality checking, Articulated robot, Image processing, Automated process

1. Introduction

Quality Inspection is a process that enables the industries to provide defect free products that meet the desired purpose. Part inspection process has become one of the most important processes in the industry, with the rise in competitors and demand for the best. There are numerous testing equipment introduced and used daily in this process to ensure only the right part with good quality is getting delivered.

Quality inspection is done generally in 5 methods

(i) Pre-Production Inspection(PPI)
(ii) First Article Inspection(FAI)
(iii) During Production Inspection(DPI)
(iv) Pre-Shipment Inspection(PSI)
(v) Container Loading Inspection(CLI).

A Pre-Shipment Inspection serves a crucial function to check produced goods right before their shipment occurs.

[1]vinotht@citchennai.net, [2]arivalahanr.eee@valliammai.co.in, [3]vinothn@citchennai.net, [4]anushrajb@citchennai.net, [5]ramakrishnanj041@gmail.com, [6]ragul.vgos@gmail.com

DOI: 10.1201/9781003685906-69

Manufacturing incidents resulting from substandard product standards cause massive financial losses to industries. All these inspections have been traditionally completed through manual operations due to the practice of human expert examination for quality assurance. Human vision finds it difficult to achieve consistent quantifiable results when determining accuracy and precision measurements of small-dimensioned critical goods using current methods. The technological transition from manual product examination now utilizes automated camera systems and robotic equipment that monitor various product areas through camera movement or vise-versa position adjustment. Manufacturing accuracy and product quality consistency get established through robotic part visuals inspections which serve industries. The inspection field becomes increasingly critical for industrial automation despite advances in sensor technology and visual processing equipment and micro-controller devices and computing systems at lower price. Robots are known for their robustness, precision, and consistency over any other industrial machineries. They have been proven to provide higher successful operation rates in any industry of deployment. The exact processes and predefined workflows trigger robots to repeat their output with almost no deviation while achieving a repeatability range of +/- 0.02mm (20 microns) which indicates their suitability for automating quick-precision handling tasks in quality inspection. Robotics enable high product quality retention by establishing consistent measurement processes and achieving superior product quality standards. By using robots for on-line inspection manufacturers identify out-of-tolerance products in the early production stages while minimizing waste and rework amounts. The system executes multiple operations starting with picking parts from bins followed by component identification and conducting a complicated defect check for sorting completed parts. The system will enhance Inspection speed along with Sorting functionalities. The system returns information about goal point deviations to the production process.

1.1 Problem Identification

Industrial manufacturing is greatly affected by the demands and the requirements stated by the customer based on their strategic planning. In the typical manufacturing aspect, where a large number of parts are to be manufactured, the parts are classified and distributed among a wide range of manufacturers. Therefore, it becomes necessary for the individual manufacturers to complete their allocated slot within the provided time period and standards as specified in the pact provided by the customer. Thus, the manufacturing company must ensure that the product they manufacture does meet the specifications stated by their customers. In order to ensure that the stated standards are met, a quality check is implemented. In a conventional quality checking process, both the production line and manufactured products are periodically inspected and verified if the manufactured part or product is meeting the necessary standards. If incase of any misconduct, the lot is inspected for potential deviance and the error occurred is to be rectified to continue with the manufacturing process. In case of end-of-line inspection, the luxury to modify the manufacturing line is not provided and thus the manufactured lot is discarded and remanufactured, which significantly increases the cost and time of production thereby greatly dropping the production efficiency. In conventional manufacturing industries, the scale of production is well beyond the managing capabilities of human workers. Thereby, huge efficiency drops are prone to happen in scenarios of human involvement for quality inspection operations.

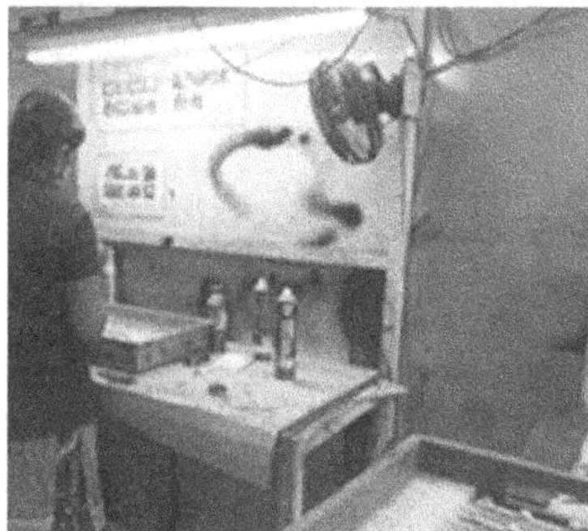

Fig. 69.1 Manual inspection setup

This bottleneck can be effectively handled by the implementation of automation and artificial intelligence systems. The production scale can be made dynamic with minor to least significant changes to the inspection system through convenient modifications. The efficiency of inspection is maintained as a standard throughout the production cycle as the automation systems provide significantly accurate results within lesser operation time compared to manual labourers' consolidated problem statement is an automated quality inspection system integrated with an AI based image processing sub-system for defect identification and classification. The proposed idea was intended to improve the part number and the inspection rate by 60 - 80% of what was done conventionally through manual inspection. Another aspect of improvement is to tweak the accuracy of defect identification to eliminate errors in further processing.

The core objective can be stated as follows

(i) AI enabled Automation framework
(ii) Highly accurate image processing system

1.2 Proposed Solution

The proposed automation system involves an integration of a compact industrial robot equipped with a machine vision system for AI assisted image processing. The image processing system end is equipped with a defect detection algorithm and AI model which have been trained with sample data consisting of 500 images of the various defects that have been identified in the products thus far in the production period and 500 images of Good products as given by the Quality Inspection team of the organization. This model training methodology helps the machine-vision system to accurately identify the labeled errors and classify the parts accordingly. The classification of parts/products is made possible using the compact Industrial Robot, which will be constantly communicating with the Vision System for Defect Identification and Processing. This implementation will therefore reduce the potential error rates in defect identification and also provide a significant reduction in time required to complete the inspection of a standard bin size.

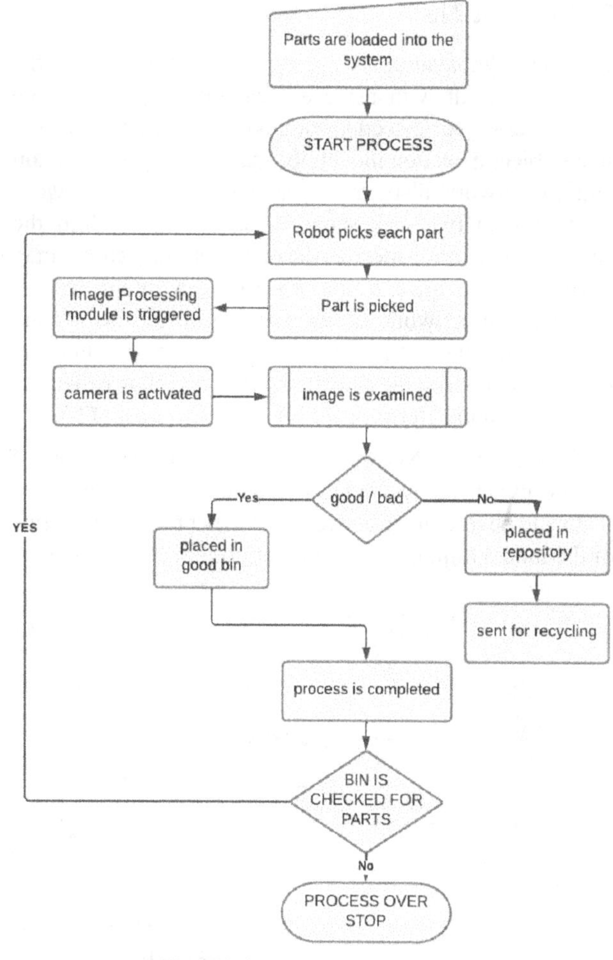

Fig. 69.2 Process flow chart

Solution Overview:

Step 1: Loading of Part bin in the System

Step 2: Start the System

Step 3: Picking of part using Robotic Arm (Bin Picking)

Step 4: Presenting the part before Inspection camera (360 degree view by rotating the wrist)

Step 5: Inspection feedback via Vision System to Industrial Robot

Step 6: Sorting the part based on the Defect Status

Step 7: Stop after the Sorting is done.

Step 8: Looping of the process.

Image Processing:

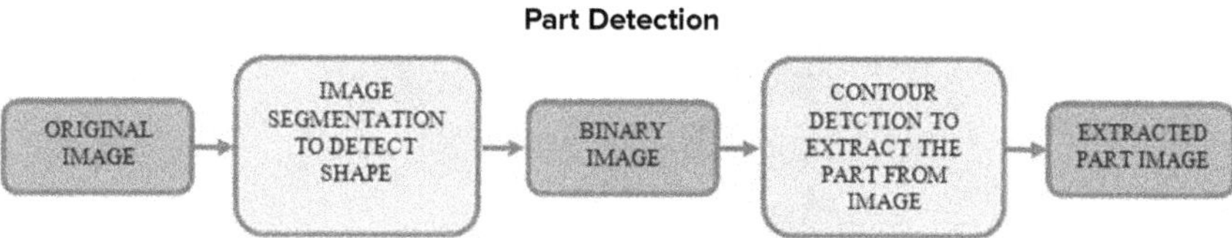

Fig. 69.3 Object detection process

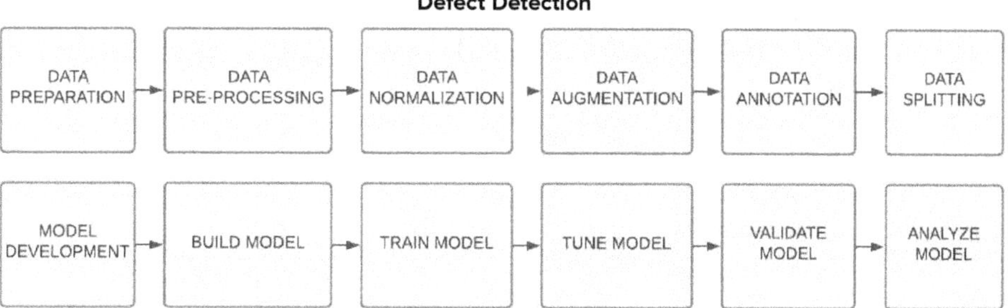

Fig. 69.4 Defect detection process

2. Retina Net

The class imbalances occurring during training of the model are dealt with using a focal loss function in the Retina net object detection methodology which is a one-stage object detection model. Retina net comprises a main unified network also known as the backbone network assisted with two task specific sub-networks. Both the backbone network generates the convolutional feature map for the input image and first processes the output through the first sub-network before passing it to the second network. Retina Net has integrated two features into the fundamental one-stage detection model which is called the Feature Pyramid Network (FPN) and Focal Loss (FL).

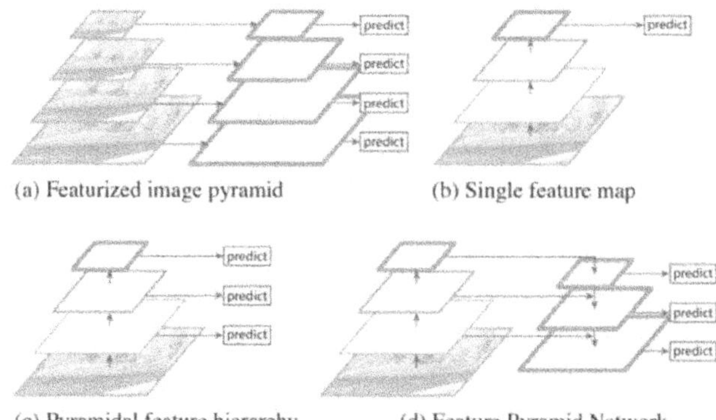

Fig. 69.5 Process flow

Feature Pyramid Network involves the construction of feature pyramids over image pyramids. This enables the feature to be resampled and rescaled as per the requirements of the model training. The main purpose of Focal loss is to solve single-stage object detection models' class imbalance problems. The main components of Retina Net can be seen in the Fig. 69.6.

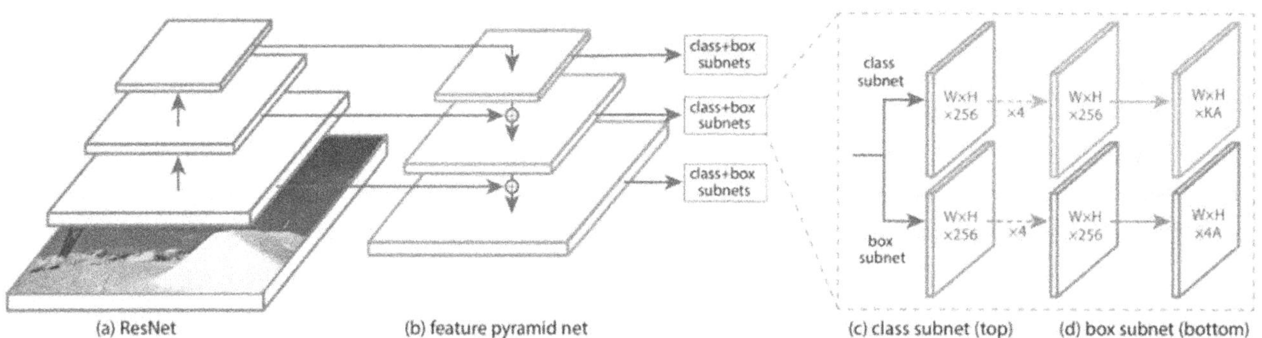

Fig. 69.6 Process flow for ResNet, feature pyramidnet

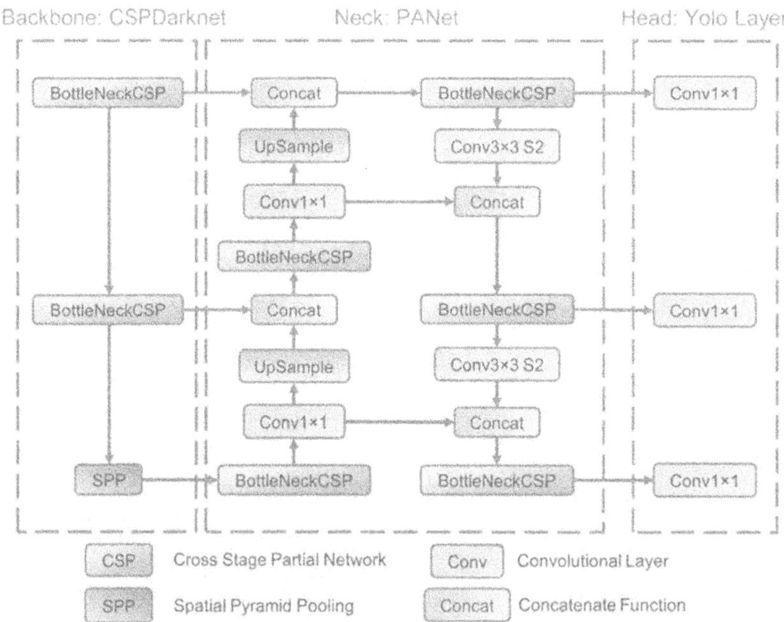

Fig. 69.7 YOLO v5 architecture

2.1 YOLO V5

YOLO is an acronym for "You Look Only Once". It is an object detection algorithm that divides the image to several grids, in which each grid takes responsibility to identify the object in that grid. It is one of the fastest and accurate object detection algorithms available. It requires very few resources for computation and is widely used in real-time object detection.

CSP Darknet forms the backbone of YOLO v5 Backbone since it uses cross-stage partial networks to perform feature extraction from images.

The YOLO v5 Neck segment generates a feature pyramid network through PANet which applies aggregation to features before sending them to Head for prediction results.

The YOLO v5 Head component provides detection predictions from anchors it controls.

Step 1: Setting up Virtual Environment

Step 2: Cloning a Git Repository of YOLOv5

Step 3: Preparation & Pre-processing of Data set

Step 4: Training the Model

Step 5: Prediction & Live testing

Robot Programming:

Step 1: Commissioning of Robot and Gripper

Step 2: Fixing of Base, Work, and Tool Coordinates

Step 3: Connecting Industrial Robot to Vision System Via PLC

Step 4: Teaching the coordinates of Pick and Place points in the Production Bin, Good Part Bin and Defective Part Bin respectively.

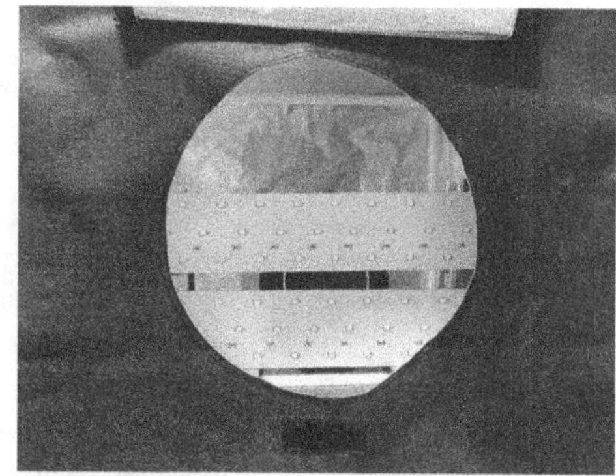

Fig. 69.8 Camera and light setup

Step 5: Teaching Inspection Point Motions

Step 6: Programming Sequence

Step 7: Dry Run

Test Setup

The Testing setup consist of the following components

Hardware:

- MELFA RV4FLD Industrial robot
- Mitsubishi PLC and CR750-s controller
- Logitech BRIO webcam
- Laptop/Desktop with accessories
- Photobooth Setup & Lighting Equipments

Software:

- Python & OpenCV
- Platform for running the AI module
- RT Tool Box 2
- GX works

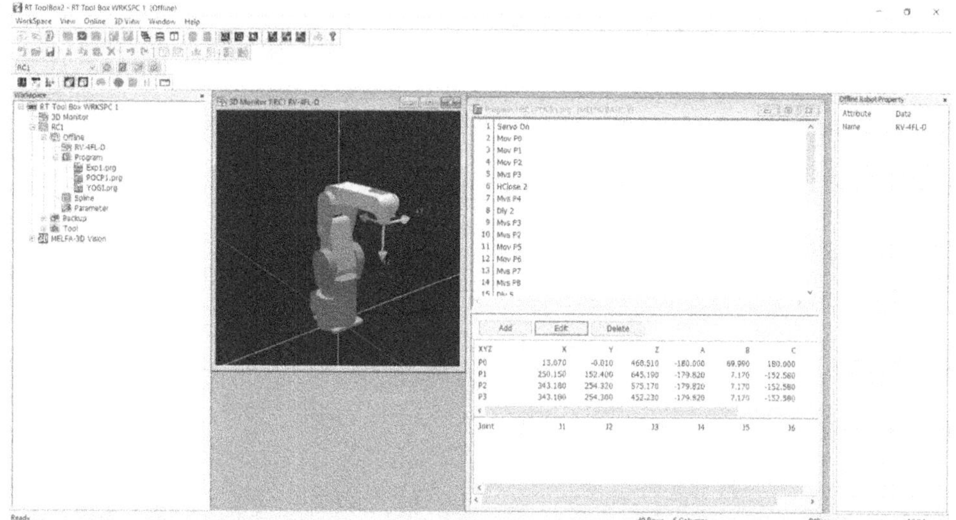

Fig. 69.9 RT ToolBox2 programming environment

3. Result

The evaluation of produced components serves to maintain the quality standards of items manufactured within industries. The visual inspection performed by human beings on manufactured products results in the occurrence of human errors. The problem is eradicated by completely by robot which identify the defective products and good products by using the image processing technique and the robot. The preset data of the good quality images and defective parts identified and stored in the data base and by using suitable image processing technique and the components are captured and the data are compared and it was separated in the bin. The pre-trained datasets accurately determine the defective parts which improves reduction in time and money. The research was further enhanced by using suitable Machine learning technique to improves its efficiency.

4. Conclusion

The integration of the machine vision system with the industrial robot is discussed in detail in this paper. The system comprises Mitsubishi MELFA-RV 4 FLD industrial robot setup consisting of an integrated photobooth arrangement for conducting the machine vision half of the process. The photobooth consists of a windows system for running the model and a Logitech BRIO camera for capturing the required photographs for further processing. The outcome of image processing is used for defect detection and classification, which is carried out through the industrial robot. Furthermore, the accuracy of defect identification and classification can be improved by extended training provision by a much larger dataset. The cycle time can be optimized by appropriate placement of the part bins, photobooth and the operation speed of the robot. This application helps in automating the quality checking process in conventional manufacturing industries. This automation using image processing technique will help to reduce the labor cost, human errors and which turn improve the efficiency.

References

1. Huang, C. C., & Lin, X. P. (2018). Study on machine learning based intelligent defect detection system. In MATEC Web of Conferences (Vol. 201, p. 01010). EDP Sciences.
2. Bhatt, P. M., Malhan, R. K., Rajendran, P., Shah, B. C., Thakar, S., Yoon, Y. J., & Gupta, S. K. (2021). Image-based surface defect detection using deep learning: A review. Journal of Computing and Information Science in Engineering, 21(4), 040801
3. Prusaczyk, P., Kaczmarek, W., Panasiuk, J., & Besseghieur, K. (2019, March). Integration of robotic arm and vision system with processing software using TCP/IP protocol in industrial sorting application. In AIP Conference Proceedings (Vol. 2078, No. 1). AIP Publishing.
4. Priya Charles , Niraj Bhadoria , Simran Gupta , Pooja Satpute, 2020, Survey Paper on Visual Inspection of a Mechanical Part using Machine Learning, INTERNATIONAL JOURNAL OF ENGINEERING RESEARCH & TECHNOLOGY (IJERT) Volume 09, Issue 01 (January 2020),
5. Ren, Z., Fang, F., Yan, N., & Wu, Y. (2022). State of the art in defect detection based on machine vision. International Journal of Precision Engineering and Manufacturing-Green Technology, 9(2), 661–691.
6. Khan, A., Mineo, C., Dobie, G., Macleod, C., & Pierce, G. (2021). Vision guided robotic inspection for parts in manufacturing and remanufacturing industry. Journal of Remanufacturing, 11(1), 49–70.
7. Alghamdi, B., Lee, D., Schaeffer, P., & Stuart, J. (2017, August). An Integrated Robotic System: 2D-Vision based Inspection Robot with Automated PLC Conveyor System. In Proceedings of the 4th International Conference of Control, Dynamic Systems, and Robotics (CDSR'17), Toronto, ON, Canada (pp. 21–23).
8. Martin, D. (2007). A practical guide to machine vision lighting. Midwest Sales and Support Manager, Adv Illum2007, 1–3.

Note: All the figures in this chapter were made by the authors.

Advances in Additive Manufacturing Technologies – Gurusamy Pathinettampadian et al. (eds)

Performance Improvement Strategies for Solar Stills: A Critical Analysis

Thilak S.[1], Ashwin M.S.[2], Rishi Prasath S. J.[3], Vishwin K.[4], Sivabathran K.[5], Shardul Devaraj[6], Sathiyamoorthi R.[7]

Department of Mechanical Engineering, Chennai Institute of Technology, Chennai, Tamilnadu, India

◆ **Abstract:** Solar stills are widely recognized as an efficient, low-cost, and sustainable solution for freshwater production, particularly in arid and semi-arid regions. However, their practical performance remains limited by inherent inefficiencies related to low thermal efficiency and water yield. This paper presents a critical analysis of various performance improvement strategies for solar stills, encompassing design modifications, material advancements, and hybrid system integrations. Key innovations reviewed include the use of phase change materials (PCMs), nanofluid enhancements, incorporation of external energy sources, and optimization of geometric configurations. Additionally, the study evaluates the impact of operational parameters such as basin water depth, solar intensity, and environmental conditions on productivity. Comparative analysis highlights the synergistic effects of combining multiple strategies to achieve significant efficiency gains. The findings provide a comprehensive roadmap for researchers and engineers, fostering the development of next-generation solar stills with enhanced performance and scalability for diverse applications.

◆ **Keywords:** Solar still, Nanofluids, Solar intensity, Phase change materials

1. Introduction

Access to clean water is the advancement and survival of human beings. The exponentially rising demand for clean and potable water that accompanies population growth, urbanization, and industrial-related agriculture may then be efficiently managed by harvesting energy from the sun and other renewable sources of energy [1]. Since water resources account for the largest proportion of the earth but contaminated with harmful micro-organisms and chemical pollutants, they must be treated before being used in irrigation or by human beings. Anchored on renewable energy sources, the UN SDGs offer a well-rounded framework for issues including water shortage, among others, as well as increased demands of energy. These can be subdivided into membrane desalination and thermal desalination. Thermal desalination boils the water with thermal energy to turn it to steam, which then condensates and is stored as clean water. Such techniques are generally fueled by fossil fuels, but there are solar stills that can serve as a way to provide freshwater by solar power [2]. To enhance the efficiency of the solar stills, various auxiliary parts were incorporated. While integrating a tubular solar collector with solar energy, energy efficiency and environmental impact were still involved. Having fresh running water for the disadvantaged is economically valuable for it can reduce the cost of insurance and health care, which reduces the cost of constructing and

[1]thilaks.mech2022@citchennai.net, [2]ashwinms.mech2022@citchennai.net, [3]rishiprasathsj.mech2022@citchennai.net, [4]vishwink.mech2022@citchennai.net, [5]sivabathrank.mech2022@citchennai.net, [6]Sharduldevaraj.mech2023@citchennai.net, [7]sathiya.ram78@gmail.com

DOI: 10.1201/9781003685906-70

maintaining freshwater infrastructure. It would ensure safe drinking water reaches homes, protecting the lives of low-income families in military zones, be it urban or rural or on the fringe. Freshwater constitutes around 2.5% of all fresh water on this Earth. Earth, although 70 percent of the surface of this planet is water, so it is misleading. Seawater: The content of common salt in seawater falls in a range of 3500 to 4500 parts per million [3].

However, large amounts of fossil fuels are used in these processes. Statistics show that 13000000 tons of oil has to be burnt annually to provide 13000000 m3 of fresh water daily. In Saltwater treatment methods are mainly used in many different areas and operations in Russia. Membrane-based methods, in which semipermeable membranes are used, and distillation-based methods involving phase-change are mainly utilized in desalination [4]. Fossil fuels will remain the predominant source of energy in satisfying global demand, standing at 78.3% of the total according to reports. Nuclear fuel energy is only at 2.5%, and renewable energy at 19.2%. Among these factors that account for the increasing water scarcity are rapid industrial growth and population increase as well as environmental concerns. Solar energy is so vital because desalination can reduce water shortage but does so at a cost of high energy [5]. Millions of people live in areas with no freshwater, and their health, especially, has come under particularly harsh impacts due to this shortage. Contaminated water causes illnesses and even death if consumed. Only about 3% of the Earth's water comes from freshwater, while the other 97% is saltwater. Several methods of water treatment have been developed to make brackish or saline water palatable for human consumption to counter this worldwide water deficit. Desalination consists of removing salt from water or vice versa to make the water drinkable. There are two major categories: thermal and membrane desalination [6].

2. Fundamentals of Solar Stills

To date, only a few field studies on floating solar desalination have been conducted, and most of the work has taken place within laboratory-controlled conditions.34–38 It is critical to demonstrate promising materials in natural environments to avoid premature field stagnation and further demonstrate the promise of floating desalination, with new challenges in preparation, overall performance, and deployment. All of those components are above and beyond the requirement to reject salt. As a result, documented floating solar stills have drastically reduced from 90% to at least of 20%. SD is referred to the systems that employ the power of sun energy to purify and make potable water from polluted water. The technology heats the water and causes evaporation through sunlight. efficiently removes contaminants and salts, leaving behind pure water vapour [7]

The condensation of the vapor will yield freshwater, which gives it an environmentally friendly and renewable means of managing water scarcity issues in areas where access to other freshwater sources is scarce. SD solutions can be scaled and flexible. Solar desalination has shown promising potential to address the problem of water shortage in the world as a sustainable method for wastewater treatment [8]. This approach does not address many important physicochemical characteristics of CWW, and it reduces the lifespan of the SD water treatment system. In addition, less research has been performed on the potential synergisms between SD and MF within the same process. Hence, this study aims to fill the gap in treating CWW by integrating MF and SD. We applied SD, as well as an integrated MF+SD method, to check the significance of pre-treatment [9].

This lack of potable water is complemented with conflict in damaged countries and also in areas ravaged by natural disasters. Sometimes, an apparent shortage of access to clean, distilled drinking water may also be noticed due to disruptions in structures related to sanitation and water. All life on earth should have access to water-this is a basic resource from which life can not function to survive. Traditionally, the increasing demands for water can only be partially met using available sources of available water. More solutions are required as the demands for fresh water cannot be served by the currently available sources of fresh water [10].

3. Thermal and Design-Based Performance Improvement Strategies

Although saltwater is readily available, constituting a good proportion of over 96% of Earth's water, it cannot be directly consumed. Desalination technology has greatly improved over the past few decades, becoming more economical, ecologically friendly, and efficient. Among the most commonly applied and well-known techniques are reverse osmosis and multi-stage flash distillation, converting salty sources to potable fresh drinking water. We will enhance our fresh water supply and overcome this rising water scarcity problem with the help of effective desalination technology [11]. This will conserve the health and well-being of communities worldwide besides ensuring sustainable fresh water supply for human consumption. Natural resources and desalination constitute the two prime sources of potable water. These natural resources are rain, surface water, and ground water constituting fresh water.

Besides utilising natural resources, brackish or salty water can be desalinated to produce drinkable water. There are two main categories of desalination methods: thermal and nonthermal. Nonthermal methods include reverse osmosis [8], membrane distillation, electrodialysis, and forward osmosis. The thermal methods involve solar-powered humidification-dehumidification, solar still, multi-effect distillation, solar membrane distillation, solar pond distillation, solar diffusion-driven desalination, and stage flashing, Machine learning and solar applications [12].

In solar stills, thermoelectric materials can be used as generators, heaters, and coolers. Materials that use thermoelectric power may be attached directly to the solar still, so heat is transferred to it right away. Alternatively, a piping system can be implemented to move heat through a fluid. The condenser or cover is cooled by the cooler. Distillation techniques have been advanced in a variety of ways for water desalination technology, though each may have its advantages and disadvantages concerning cost, applicability, and even ease of use [13]. During the past decades, countless studies have been conducted to develop ways of lowering the cost of this process, and numerous ways have been developed. Solar distillation appears to be the most efficient and economical of such methods. Productivity constraints inherent in traditional designs producing 2000 to 3000 ml/m2 have hence restrained solar still daily utilization of the tank area. Advancements in the science of solar distillation within current research efforts seek to more justifiably improve the predicted performance of solar stills. Newer methods like stochastic gradient descent employed in artificial neural networks and machine learning algorithms predict and optimize productivity. These systems predict cumulative productivity with high accuracy using environmental conditions and previous data [14].

4. Material and Construction-Based Improvement Strategies

Although water has covered over two-thirds of the Earth, many countries around the globe are facing a severe shortage of freshwater. Water sources will be 40% scarcer by 2030. In the manufacturing of solar collectors, water desalination, solar cookers, solar dryers, printed circuits, solar-biomass, solar chimneys, and metal heating, solar power stands as the most prominent source of renewable energy. The lack of traditional water supplies, the prognoses of population explosions and over demand for potable water have emerged as significant issues for most nations [15].

The ecological systems are afflicted through environmental pollution caused by carbon-based fuels that fuel these machines, as well as through the process of salt rejection in these plants. Photo-voltaic panel produces electricity for very effective reverse osmosis, which removes impurities and salts from water. One of the emerging sustainable methods in which clean water is produced from other sources is through solar desalination. It seems that treating SD for CWW use without prior treatment of the CWW does not address some fundamental physicochemical characteristics and may reduce the systems' longevity. Also, sufficient work has not been done on the possible synergies between SD and multilayer filtration (MF) in one system [16].

5. Advanced Techniques for Enhancing Solar Still Performance

To evaluate the significance of pre-treatment, both the SD and the integrated MF+SD method were applied. As an aid for better understanding, PCPs of water treated using this proposed method were analyzed and compared with BW. One of the ways in dealing with water scarcity is through desalination, which implies extracting salt from brackish or saltwater. In the past years, the technology of desalination has been done with remarkable leaps in being more economical, ecologically friendly, and efficient. Two highly used desalination methods are reverse osmosis, which turns salty sources into drinkable water, and multi-stage flash distillation. We could increase the availability of freshwater and alleviate the burgeoning problem of water scarcity by implementing efficient desalination technology. This would protect the health and well-being of communities worldwide besides ensuring a sustainable supply of water for human consumption. The lack of drinkable water is compounded in countries that are affected by conflict and in areas that have been adversely affected by natural disasters [17]. The water and sanitation infrastructure is interrupted in such places and thus, there is often a shortage of safe and clean water to drink. Water should be free for all living things on Earth as this is a basic requirement for life to be sustained. Thus, at the current level of available water supply, conventional the present needs of such demands can only be partially fulfilled.

Most experts suggest that it is probable that water scarcity will lead to a kind of war amongst nations at large, especially in such nations where fresh water is already scant in its availability. This presents a special problem for developing countries as most of these already face the challenge of providing adequate freshwater supplies to their citizens. Azm Najjar et al has done the work on Evaluating the impact of external and internalcondensers on productivity in solar stills. Condenser-based productivity effects have been studied through both experimental and numerical investigations. The employment of an external condenser and nanofluids increases the solar still's efficiency by 50%. comparative analyses for other modifications, such as a stepped solar still improvement, could reveal the increase in productivity of between 30% and 150%. Lastly, condensation elements can help solve the low water productivity problem and decrease costs in the production of fresh water [18].

Naseer T. Alwan et al [19] has analyzed the solar still performance units and advanced techniques. It has concluded that increasing sun radiation and wind speed, alongside decreasing surrounding air temperature, is helpful for the improvement of solar still output. In terms of working characteristics, productivity enhances with water depth, decreases while the thickness of the cover does. The solar distillation system can be enhanced by an additional packed layer at the bottom of the basin or by attaching a spinning shaft to the surface of the basin. It has been established that the higher wind speed and solar radiation, and the lower ambient air temperature can enhance the solar still production. Ahmed Serag et al [20]. evaluated the Performance enhancement of solar stills using heating elements. Enhancements in cover materials and geometries affect parameters such as condensation efficiency and heat absorption, thereby generally increasing freshwater production. In addition, using solar stills with rotary cylinders doubled the surface area and raised effectiveness. Moreover, wick materials were one of the test parts, and it opened up to achieve a 26.65% productivity. Since the solar system improves both energy efficiency and heat transfer, production raises by 214%. By using immersion heaters, the productivity has improved in hemispherical stills by 232.9%. In single-slope stills, it is increased by 252.4% and by 370% in double-slope stills.

C. Suresh et al [21] has done research on the Transformative nanofluid solutions: Elevating solar still performance for enhanced output. The principal objective of this project targets the enhancement of heat transfer across the saline water to the absorber plates through improvement of the physical characteristics of nanofluids. On the issue of quantum properties of nanofluids in terms of absorption ability, further deliberations arise in the context of solar desalination. The inclusion of metal oxide-based nanofluids with the working fluid improves the overall performance and heat transmission of the solar still. Ti C was made from tire waste by upcycling, and the components of the floating still were selected for maximum freshwater production. Outdoor experiments to this end, exposed to up to 6 kW.m-2, day-1 of solar irradiance in Halifax, Canada, resulted in daily productions of up to 3.67 L.m-2, corresponding to a nearly 40% conversion efficiency from solar energy to water vapor. Due to this study, invention of a scalable floating solar desalination system enabled providing water-stressed populations with drinkable water [22]. This paper establishes the possibility of combining solar desalination with multilayer filtration as an innovative approach toward the treatment of water purification results, particularly to treat restaurant-generated wastewater. Sujit Kumar et al has carried out a review research on Solar stills: A review for water scarcity solutions. Productivity gains have been demonstrated with reflectors and pebbles, varied wick material, and changes in the absorber plate design [23]. Moreover, this study further tells how the solar still can be improved: several efficiency improvements are to be explored, including heating storage materials, development of new sophisticated modeling algorithms, and control algorithms.

6. Environmental and Climatic Considerations

Thermoelectric materials can be inferred to improve the solar performance. The reason is that the thermoelectric generators produce electricity as a hybrid system, and thermoelectric heaters or coolers boost yield. There is a lot of room for further research on generating power at night and the use of thermoelectric materials with collectors and adsorption units, among other accessories to be used for the solar still. Karrar A. Hammoodi et al has researched on a detailed review of the factors impacting pyramid type solar still performance. From this point of view, we attempt to critically assess the status of some pyramid solar still designs in this article.

The basic method of distillation involves evaporating polluted water to remove contaminants. Pure water is subsequently created as the vapor condenses. This is simple physics. This idea is also the foundation of the hydrological cycle. But it takes energy to turn water into vapor and back again. Thus, rainfall's fundamental idea is distillation. Rainfall is based on the principle of distillation. A solar still closely mimics this natural process. Solar stills require very few moving mechanical parts and are simple to assemble. They can be constructed from easily accessible and basic components and are also significantly less expensive. There are numerous facilities nowadays it accessible to receive drinking water, but they might not be affordable or easily accessible. Since solar energy is a renewable resource, its application is generally conservative [24].

One of the primary components of distillation is solar radiation; without it, the idea of a traditional still would not have been possible.

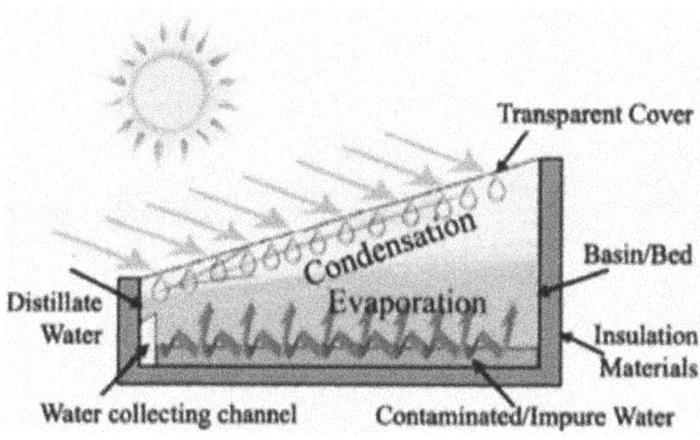

Fig. 70.1 Solar still [2]

7. Economic and Practical Feasibility of Improvement Strategies

Numerous experts attempted to comprehend the connection between solar radiation and distillate yield, and it was generally concluded that as radiation grows, so does outputaccording to the amount of sunlight. Morse and Read also came to the conclusion that one of the most crucial parameters for a solar still is sun radiation. Electricity can be produced by creating a sizable temperature differential between the solar still's evaporation interface and the ocean water below. This method, referred to as thermoelectric generation, uses the effect, in which an electrical current and voltage are induced by the movement of charge carriers at semiconductor p-n junctions. In certain laboratory-scale research, it has been suggested that this process be coupled with solar desalination, and several land solar stills have done so. Although it hasn't been tested with a floating solar still yet, it may power aboard equipment like fans to increase evaporation or tiny water quality monitoring. The solar still was modified to include thermoelectric generators (TGs) [16].

When compared with the same without nanoparticles, PCM with silver nanoparticles showed lower melting and freezing temperatures and superior thermal conductivity. Due to its superiority in thermal conductivity over CuO and TiO_2, silver enabled higher temperatures of water and more rapid evaporation [21]

Adding a PCM is sufficient to maintain the distillation process, as this material stores redundant solar energy and heat losses from the still's perimeter when the sunshine is strongest and provides them in the evening and night-time.

8. Future Prospects and Research Opportunities

There are two types of solar desalination systems: passive and active. Passive systems do not rely on traditional power supplies or mechanical equipment; instead, they use only solar radiation to produce electricity. However, their efficiency is often lower. To address this limitation, active solar distillation systems have been developed to provide additional heat energy to the basin water, increasing vaporization speed and improving output. These systems use external techniques like solar collectors, solar panels, and focusing mechanisms to enhance efficiency. Additional thermal energy from solar collectors can be added either through forced circulation using a pump or natural circulation via a thermosiphon. Active solar stills are more efficient than passive ones because they increase the evaporation rate by incorporating external thermal energy or solar radiation. These systems consist of a pump, an external solar collector, basin water, and a glass cover.

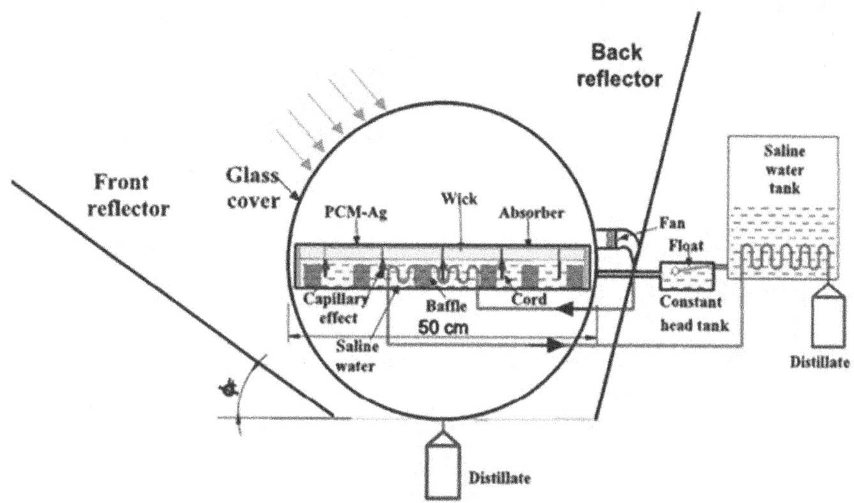

Fig. 70.2 Solar distillation [9]

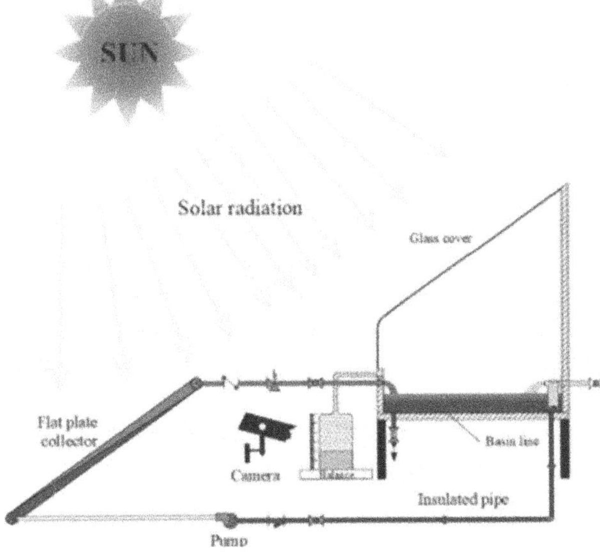

Fig. 70.3 Solar flat plate collector [15]

Most of this energy remains within the system due to minimal heat loss. Continuous heat transfer raises the temperature of the water, leading to vaporization through the greenhouse effect. Energy moves from the water's surface to the cover via convection, radiation, and vaporization, with vaporization being the most critical. The distillation system is categorized into two main types based on the slope of the double glass cover, which influences its performance. These distillers are built based on the site's conditions and the materials available. The basin is black-painted to optimize sunlight absorption, and the double-sloped glass cover, matching the basin in size and location, improves efficiency [20].

This type of distiller typically produces 4–7 litres of water per day. As sunlight passes through the glass cover, it heats the water in the basin, causing evaporation. The cooler glass cover promotes condensation, leading to water droplets that collect in a designated channel for collection [22].

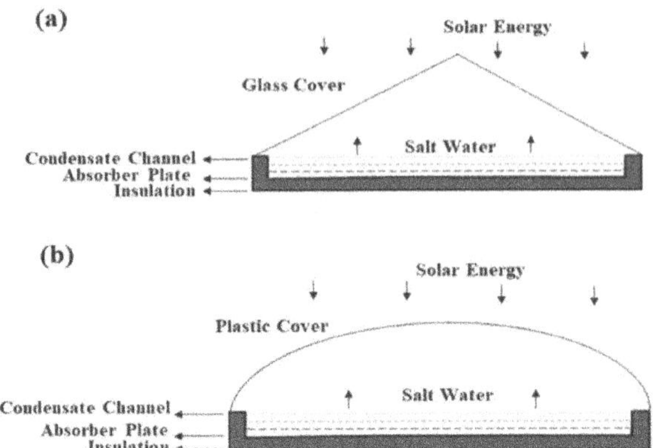

Fig. 70.4 Solar desalination using glass cover and plastic cover

These distillers rely on the site and building materials that are available. It features a black-painted basin to improve sunlight absorption and a double-sloped glass cover that is the same size and installed in the same location, which is better. 4–7 litres are produced everyday by the distiller. The sun shines on the basin's water through the glass lid. The water in the basin evaporates as its temperature rises, and since the cover is colder than the steam, it starts to condense and descend as water droplets that gather in a particular channel. Initial expenses for solar still designs are often comparable among systems. However, as previously said, the enhanced efficiency brought about by design changes results in a decrease in the energy needed for operation, which lowers operating expenses. A CWTSS with a fan, for instance, had a 50% lower distillation cost of $0.01 per litre as opposed to $0.02 per litre for a TSS. Performance gains, driven by different distiller types and absorber materials, were found when the current solar still design was compared to other existing designs [11].

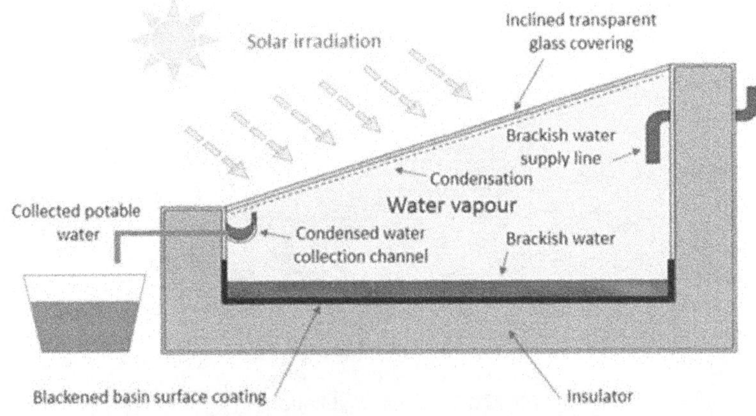

Fig. 70.5 Solar still using inclined transparent glass covering

Without the need for fuel, solar distillation transforms a variety of water sources, including raw, contaminated, salty, and semi-salted liquids, into clean, drinkable water using free sun energy. Evaporation and condensation, the two stages of the purification process, efficiently eliminate impurities such as minerals and microorganisms, resulting in pure drinking water. By simulating the Earth's water cycle, this technique also aids in lowering greenhouse gas emissions and halting climate change [24].

In areas with scarce natural resources, water desalination has emerged as a practical remedy for water scarcity. Renewable energy sources have received increased attention in the past 20 years as a potential option for desalination operations. For small-scale uses, direct solar desalination technology is seen as a sustainable way to supply fresh water. A number of technologies were investigated, including solar ponds, solar chimneys, and solar stills [23].

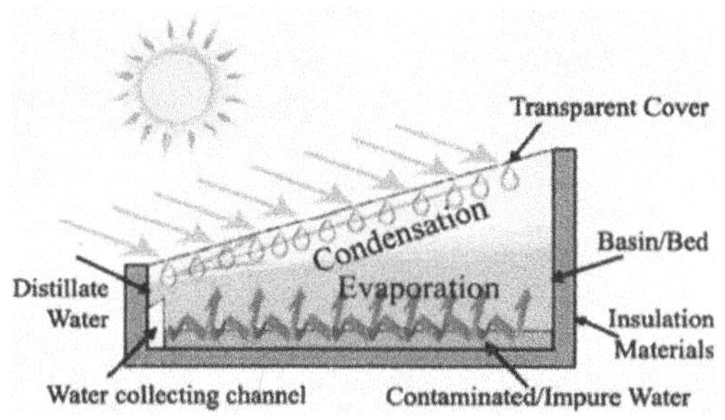

Fig. 70.6 Solar still with different insulation materials

9. Conclusion

- Direct solar-powered membrane distillation (SP-MD) has been suggested by researchers as a novel way to desalinate saline and hypersaline feed water within the past ten years. A new thermally-driven desalination technique that works with a broad range of feed temperatures is membrane distillation (MD).
- By using two distinct systems—a solar collector and an MD module—solar energy and MD were indirectly integrated. However, Direct SP-MD achieves a more compact, standalone design appropriate for small-scale applications by using modules that can directly use solar energy and desalinate water in a single structure.
- A thorough analysis of these initiatives is covered in this study. Highlighted are the various kinds of these systems, including membrane and surface heating, along with a description of their benefits and drawbacks. Recent developments in these units, such as improvements to the membrane surface, design of the module structure, and integration with other systems like solar cells, are also covered in the article. The study concludes by analysing these systems' shortcomings and offering suggestions and opportunities for the future.
- Recent breakthroughs in membrane materials, module designs, and integration with solar technology have enhanced efficiency and scalability but cost, environmental effect, and system durability remain issues.
- These issues can be addressed with informed policymaking, continuous research in nanocomposite materials, and sophisticated designs to increase the viability of desalination from both an economic and environmental standpoint.

References

1. Omara, Z. M., Kabeel, A. E., & Abdullah, A. S. (2017). A review of solar still performance with reflectors. Renewable and Sustainable Energy Reviews, 68, 638–649.
2. Sharshir, S. W., Ellakany, Y. M., Algazzar, A. M., Elsheikh, A. H., Elkadeem, M. R., Edreis, E. M., ... & Elashry, M. S. (2019). A mini review of techniques used to improve the tubular solar still performance for solar water desalination. Process Safety and Environmental Protection, 124, 204–212.
3. Elsheikh, A., Hammoodi, K. A., Ibrahim, A. M. M., Mourad, A. H. I., Fujii, M., & Abd-Elaziem, W. (2024). Augmentation and evaluation of solar still performance: a comprehensive review. Desalination, 574, 117239.
4. Kabeel, A. E., El-Samadony, Y. A. F., & El-Maghlany, W. M. (2018). Comparative study on the solar still performance utilizing different PCM. Desalination, 432, 89–96.
5. El-Sebaii, A. A., Ramadan, M. R. I., Aboul-Enein, S., & El-Naggar, M. (2015). Effect of fin configuration parameters on single basin solar still performance. Desalination, 365, 15–24.
6. Ahsan, A., Imteaz, M., Rahman, A., Yusuf, B., & Fukuhara, T. (2012). Design, fabrication and performance analysis of an improved solar still. Desalination, 292, 105–112.
7. Omara, Z. M., Kabeel, A. E., & Younes, M. M. (2013). Enhancing the stepped solar still performance using internal reflectors. Desalination, 314, 67–72.
8. Velmurugan, V., Kumaran, S. S., Prabhu, N. V., & Srithar, K. (2008). Productivity enhancement of stepped solar still: Performance analysis. Thermal Science, 12(3), 153–163.
9. Ahsan, A., Imteaz, M., Rahman, A., Yusuf, B., & Fukuhara, T. (2012). Design, fabrication and performance analysis of an improved solar still. Desalination, 292, 105–112.
10. Eltawil, M. A., & Omara, Z. M. (2014). Enhancing the solar still performance using solar photovoltaic, flat plate collector and hot air. Desalination, 349, 1–9.
11. Omara, Z. M., Kabeel, A. E., & Younes, M. M. (2014). Enhancing the stepped solar still performance using internal and external reflectors. Energy conversion and management, 78, 876–881.

12. Samee, M. A., Mirza, U. K., Majeed, T., & Ahmad, N. (2007). Design and performance of a simple single basin solar still. Renewable and Sustainable Energy Reviews, 11(3), 543–549.
13. Kumar, K. V., & Bai, R. K. (2008). Performance study on solar still with enhanced condensation. Desalination, 230(1-3), 51–61.
14. Elashmawy, M., Ahmed, M. M., Alawee, W. H., Shanmugan, S., & Omara, Z. M. (2024). Scientometric analysis and review of materials affecting solar still performance. Results in Engineering, 102574.
15. Selvaraj, K., & Natarajan, A. (2018). Factors influencing the performance and productivity of solar stills-A review. Desalination, 435, 181–187.
16. Velmurugan, V., & Srithar, K. (2011). Performance analysis of solar stills based on various factors affecting the productivity—a review. Renewable and sustainable energy reviews, 15(2), 1294–1304.
17. Ahsan, A., Imteaz, M., Thomas, U. A., Azmi, M., Rahman, A., & Daud, N. N. (2014). Parameters affecting the performance of a low cost solar still. Applied energy, 114, 924–930.
18. Morad, M. M., El-Maghawry, H. A., & Wasfy, K. I. (2015). Improving the double slope solar still performance by using flat-plate solar collector and cooling glass cover. Desalination, 373, 1–9.
19. Abdullah, A. S., Essa, F. A., Bacha, H. B., & Omara, Z. M. (2020). Improving the trays solar still performance using reflectors and phase change material with nanoparticles. Journal of Energy Storage, 31, 101744.
20. Rufuss, D. D. W., Iniyan, S., Suganthi, L., & Davies, P. A. (2016). Solar stills: A comprehensive review of designs, performance and material advances. Renewable and sustainable energy reviews, 63, 464–496.
21. Manchanda, H., & Kumar, M. (2015). A comprehensive decade review and analysis on designs and performance parameters of passive solar still. Renewables: Wind, Water, and Solar, 2, 1–24.
22. Kabeel, A. E., Arunkumar, T., Denkenberger, D. C., & Sathyamurthy, R. (2017). Performance enhancement of solar still through efficient heat exchange mechanism–a review. Applied Thermal Engineering, 114, 815–836.
23. Kabeel, A. E., Omara, Z. M., & Younes, M. M. (2015). Techniques used to improve the performance of the stepped solar still—A review. Renewable and Sustainable Energy Reviews, 46, 178–188.
24. Sharshir, S. W., Elsheikh, A. H., Peng, G., Yang, N., El-Samadony, M. O. A., & Kabeel, A. E. (2017). Thermal performance and exergy analysis of solar stills–A review. Renewable and Sustainable Energy Reviews, 73, 521–544.

Advances in Additive Manufacturing Technologies – Gurusamy Pathinettampadian et al. (eds)

A Review of Electrical Discharge Machining (EDM) and its Optimization Parameters

Gayathri N.*,
Selvam S., Sathishkumar S., Benix Sajo B.
Department of Mechanical Engineering,
VelTech HighTech Dr. Rangarajan Dr. Sakunthala Engineering College,
Chennai, Tamilnadu

◆ **Abstract:** Electric Discharge Machining (EDM) is one of the common non-traditional machining operations for cutting through hard materials, forming complex shapes, making very small holes, and addressing other complicated applications. EDM allows the machining operation to be done which would otherwise prove to be tricky or impossible under conventional processes. This paper is an exhaustive review of literature with particular emphasis on EDM methods and its affecting factors. The research examines the influence of various dielectric fluids on some of the important output parameters including surface roughness and material removal rate (MRR), with a view to looking at input variables such as peak current, pulse-on time, and pulse-off time. The tool wear ratio has been very much researched in the past with most of the attention on surface finish and material removal rate. Yet it can be seen here that optimization methods in EDM have not been thoroughly investigated. This research seeks to examine the effects of different EDM process parameters on material removal rate (MRR) and surface roughness (SR) for manufacturing purposes.

◆ **Keywords:** Electrical discharge machining, Material removal rate, Surface roughness

1. Introduction

Electrical Discharge Machining (EDM) is a popular non-traditional machining technique for cutting hard materials, geometrically complex products, microscopic holes, and other difficult applications. Non-conductive materials, on the other hand, cannot be manufactured with EDM. The goal of this paper is to look at how different Electric Discharge Machining process parameters affect Surface Roughness (SR) and Materials Removal Rate (MRR) in a manufacturing environment. EDM was invented during the 1940s and has since changed the face of machining hard materials. The method applies controlled electrical discharges to remove material from the workpiece so that the workpiece can be created in a complex shape and design, which is hard for conventional machining to achieve.

1.1 Previous Research on EDM of AISI 1010 Steel

Past research on EDM of AISI 1010 steel has been directed towards process parameter optimization for improving the efficiency of the process and the surface finish. The utilization of eco-friendly dielectric fluids such as biosilica-deionized water has not been widely investigated.

Biosilica, which is obtained from agricultural wastes like rice husk ash, has been applied in a number of fields because of its superior thermal and mechanical characteristics. Its use as a dielectric fluid in EDM offers a sustainable

*Corresponding author: gayathri@velhightech.com

DOI: 10.1201/9781003685906-71

solution compared to traditional fluids. The effect of pulse shape on the EDM performance of Si3N4–TiN ceramic composite. They explored the impact of different pulse shapes on various EDM machining parameters and results to improve the process for optimal efficiency and performance. The research highlights how important pulse shape is in achieving the desired machining performance considering parameters such as tool wear, surface quality, and material removal rate.

The influence of electrode material on electrical discharge machining (EDM) of alumina. They examined how the EDM parameters and performance characteristics, including material removal rate, surface roughness, and tool wear, are affected by various electrode materials. The study helps explain the best choice of electrode material for EDM of alumina for enhancing process quality and efficiency. The blind-hole drilling of Al2O3/6061Al composite by rotary electro-discharge machining (EDM). The research aimed at investigating the EDM process parameters and characteristics to optimize blind hole drilling in the Al2O3/6061Al composite material. The study helps to learn about the capability and limitations of rotary EDM in machining Al2O3/6061Al composites and how to optimize the process and improve performance. The influence of SiC and electrode rotation on electric discharge machining (EDM) of Al-SiC composite. The study aimed at identifying the effect of the presence of SiC and electrode rotation on EDM parameters and performance characteristics like material removal rate, surface quality, and tool wear. The research adds to improving the EDM process for Al-SiC composites for improving machining efficiency and quality.

The influence of boron carbide powder addition in the dielectric fluid on electrical discharge machining of titanium alloy. Their work aimed at analyzing the influence of the addition of boron carbide powder on EDM parameters and performance characteristics like material removal rate, surface roughness, and tool wear. The research proposes to enhance the EDM of the titanium alloy with increased efficacy and machining quality through the utilisation of the dielectric fluid of boron carbide powder. The workpiece materials impact on the aerosol discharge in the sinking EDM process due to electrical discharges. Their research was an investigation of how various workpiece materials influence the generation and nature of aerosol emissions during EDM. The work intends to determine the environmental effects of EDM and to identify methods for minimizing harmful emissions using proper workpiece materials and optimized machining parameters. Constituent analysis of aerosol produced due to the die sinking electrical discharge machining (EDM) process. Their study aimed at exploring and analyzing the aerosol composition produced due to the EDM process. The study would provide information about the type of particles and chemicals present in aerosol emissions, leading to a better understanding of the environmental and health effects caused by EDM processes. A comparative experimental investigation of the machining characteristics for vibratory, rotary, and vibro-rotary electro-discharge machining (EDM). Their investigation was aimed at comparing the variations in performance and results between the three EDM processes. Important parameters and machining properties like material removal rate, surface quality, and tool wear were compared to conclude the strengths and weaknesses of the methods. The goal of the study is to determine the most effective and efficient EDM method for diverse machining operations. Wire electrochemical discharge machining (WECDM) of titrated electrolyte flow quartz glass. Their investigation involved optimizing the WECDM process through electrolyte flow regulation for accurate machining of quartz glass. Some of the critical parameters of the study involved the analysis of the impact of titrated electrolyte flow on machining precision, surface integrity, and process efficiency overall.

The conclusions have the purpose of improving the capacity and scope of WECDM in machining brittle and hard materials such as quartz glass. Numerical simulation of powder mixed electric discharge machining (PMEDM) with the finite element method. They researched PMEDM modeling and analysis for discovering the impact of powder addition to the dielectric fluid in machining performance. The numerical simulation was intended to forecast significant machining parameters and results, including material removal rate, surface finish, and thermal distribution, to contribute to optimizing the PMEDM process.

2. Experiment

2.1 Base Material

Recent design enhancements, new product geometry, profiles, and other factors have raised the requirement for high-quality surface finish in difficult-to-cut materials with complex profiles. To meet these needs, machining is commonly utilized as a unique machining process. Aerospace materials that retain mechanical properties even when exposed to extremely high temperatures are favoured. Even at extremely high temperatures, AISI 1010 Steel maintains its intrinsic mechanical properties. It may be utilized in the manufacture of aviation engine components. This type of material is employed in high-temperature aviation applications. It may be utilized in the manufacture of aviation engine components. This type of material is employed in high-temperature aviation applications. This study focuses on traditional machining, with the purpose of optimizing WEDM process parameters for AISI 1010 steel.

2.2 Preparation of Saline Treated WHB (Wheat Husk Biosilica)

Making wheat husk biosilica from biomass involves two steps. In order to create wheat husk ash, the purchased wheat husk was first completely burned at 700 degrees Celsius in a thermal reactor set up on a sand bed with a separate air supply unit. In order to create sodium silicate solution in the second stage, the produced wheat husk ash was combined with different NaOH solution normalities at 80°C and constantly agitated for one hour. One NHCl was used to titrate sodium silicate at room temperature, resulting in a pH of 7. After the stirring process was complete, the silica gels were produced and left to mature for twenty-four hours. The aged silica gels were mixed with distilled water at the conclusion of the process to create silica slurry. The slurries were repeatedly rinsed in distilled water before being immersed in a beaker at 70 degrees Celsius for around 20 hours in order to create xerogel silica. Subsequently, the xerogel silica was pulverized in a mortar for several hours to produce a fine size. The synthesized biosilica particle's size was measured using a particle size analyzer, and it was discovered to be 40–50 nm in size.

The wheat husk biosilica that was created thermochemically was subsequently subjected to an acid hydrolysis silane treatment procedure. 3. In the US, Sigma Aldrich provided aminopropyltrimethoxysilane (APTMS), which has a molecular weight of 179.29 g/mol. 95 percent ethanol, 5 percent water, and acetic acid were combined to alter the pH of the total solution to a range of 4.5 to 5.5. The necessary quantity of silane—typically 2wt%—was carefully combined drop by drop with the aqueous solution to create an aqueous-silane solution. After that, the aqueous-silane solution was added to the necessary amount of biosilica particle, and it was gently swirled for ten minutes. After the surface-treated biosilica particles were manually emptied of any excess aqueous-silane solution, they were dried in a hot oven at 110 °C for 20 minutes to create the Si-O-Si structure. Figure 71.1 provides a graphic representation of the silane-treated wheat husk biosilica produced in this investigation.

2.3 Preparation of WHB-Water Dielectric

Deionized water and wheat husk biosilica particles with a silane surface treatment were used to create the dielectric fluid for this investigation. Here, deionized water is gradually combined with biosilica particles, and the mixture is ultrasonically agitated for ten minutes. To start the silane reaction with deionized water, the resulting solution was agitated for ten more minutes and gently heated to sixty degrees Celsius. Following that, the finely divided biosilica particles are allowed to cool to normal temperature so they may be machined [21]. Table 71.1 lists the names of the different dielectric fluids that were created for this study. The molecular makeup of biosilica treated

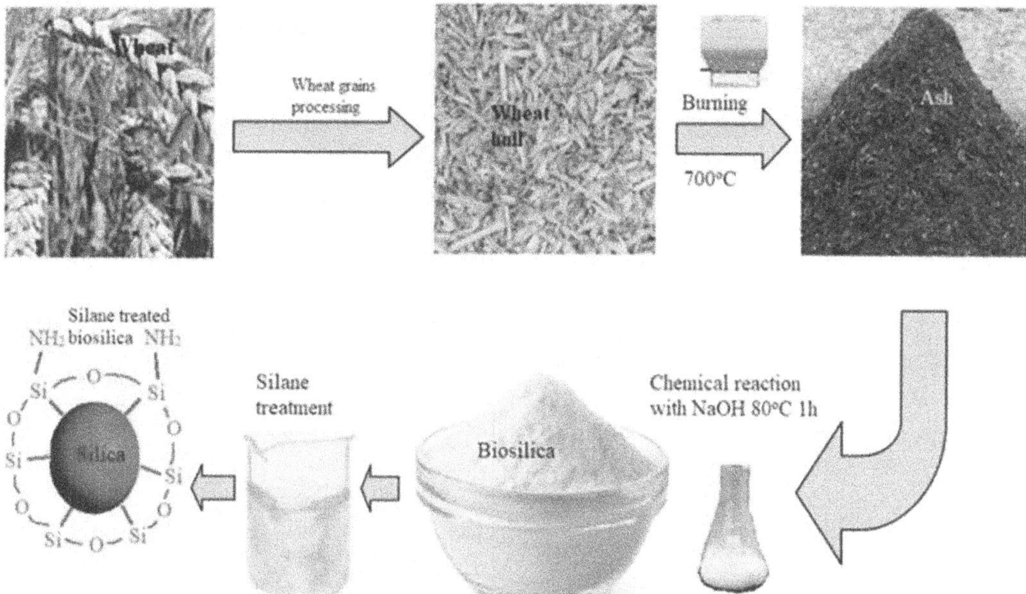

Fig. 71.1 Preparation of saline treated wheat husk biosilica nanofluid [7]

with silane is depicted in Fig. 71.2, along with the reaction site where water may react to generate a fully miscible admix.

Table 71.1 Properties of biosilica form [8]

Dielectric Designation	Biosilica form	Weight of biosilica (%)
D_0	–	–
D_{11}	As-received	0.25
D_{12}	As-received	0.50
D_{13}	As-received	1.00
D_{21}	Silane-treated	0.25
D_{22}	Silane-treated	0.50
D_{23}	Silane-treated	1.00

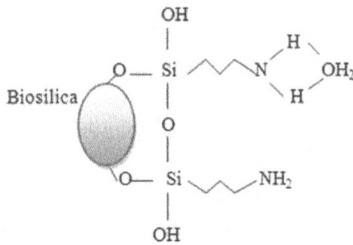

Fig. 71.2 Structure of saline treated wheat husk biosilica nanofluid [15]

2.4 Experimentation

The copper wire with a 0.25 millimeter diameter was installed in the Wire EDM machine and guaranteed the working conditions for feed and tension adjustments. Prior to loading it into the WEDM machine, the weight of the wire bunch and the work piece had to be determined. The AISI 1010 Steelwork piece fastened securely upon mounting on the work bench. The work coordinate system is locked in by the reference point on the work material set. The CNC is used to program the machining operations in relation to the work material reference point in the work coordinate system. As is customary, the left corner of the operation's closest edge—the person who stands in front of the machine to operate it—was chosen to designate the reference point. AISI 1010 Steelwork pieces were machined in accordance with the experimental design, and a consistent machining time of 10 minutes was maintained throughout all tests. The work piece was taken out, cleaned, and its surface quality assessed once the surface machining was finished. For analysis, the average surface quality evaluated on the work material at random locations was taken into account. The same high-quality work material was utilized solely in all nine studies.

3. Results and Discussion

3.1 Morphological Analysis of AISI 1010 Steel

Scanning Electron Microscopy (SEM) images analysis of AISI 1010 steel is the application of an electron microscope to analyze the microstructure and surface morphology of the material at high magnification. AISI 1010 steel is carbon steel with a nominal carbon content of 0.10%.

In SEM analysis, the sample is usually polished and coated with a thin layer of conductive material, e.g., gold or carbon, in order to improve conductivity and imaging.

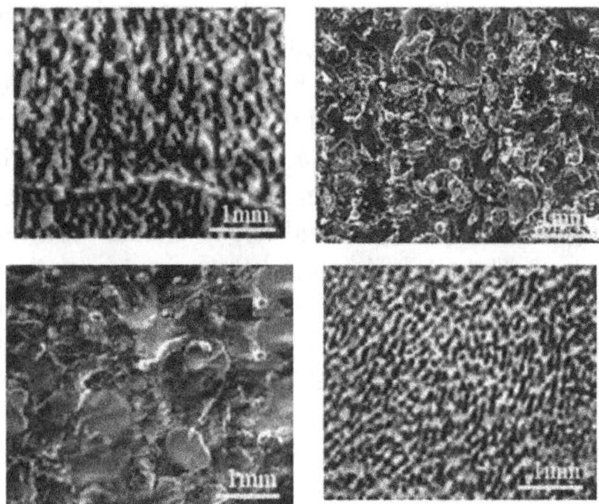

Fig. 71.3 SEM images of AISI 1010 steel

In summary, SEM analysis is useful in offering insights into the microstructure, surface morphology, and elemental composition of AISI 1010 steel. Engineers and researchers find this information useful in optimizing production processes, guaranteeing quality control, and characterizing the performance of the material in different applications.

3.2 Material Removal Rate (MRR)

MRR is a key performance indicator of material removal efficiency. Utilization of biosilica-deionized water improves the MRR drastically as compared to conventional dielectric fluids Reason for improvement: Long and high-energy sparks because of biosilica particles enriching the dielectric medium, enhancing electron passage alignment.

Table 71.2 Material removal rate [10–14]

Dielectric Medium	Material Removal Rate (mm^3/min)
Plain dielectric machined surface	4.83
Biosilica-activated dielectric (as-received DI water)	5.26, 6.72, 6.86, 5.77, 5.24
Biosilica-activated dielectric (silane-treated DI water)	7.82, 8.68, 10.43, 9.56, 8.11

3.3 Tool Wear Rate (TWR)

Table 71.3 Tool wear rate [5–8]

Dielectric Type	Tool Wear Rate (mm^3/min)
Plain Deionized Water	0.322
Biosilica-Activated (As-received)	0.3014, 0.266, 0.315, 0.336, 0.352
Biosilica-Activated (Silane-treated)	0.286, 0.245, 0.227, 0.214, 0.285

3.4 Surface Roughness

Table 71.4 Surface Roughness [6–9]

Dielectric Type	Surface Roughness (μm)
Plain Deionized Water	4.25
Biosilica-Activated (As-received)	2.75
Biosilica-Activated (Silane-treated)	2.25

4. Optimization of EDM Parameters

4.1 Effect of Peak Current

Increased peak currents tend to enhance MRR but increase TWR. A balance that maximizes efficiency at a minimum wear on the electrode is critical.

4.2 Effect of Pulse Duration

Increased pulse durations provide better surface finish but decrease MRR. Decreased pulse durations increase MRR but result in increased roughness of the surface. Proper pulse duration settings are critical for realizing desired machining results.

4.3 Voltage Fluctuations and Their Impact

Voltage settings determine the stability and intensity of the electrical discharges. Proper voltage settings guarantee consistent spark generation, enhancing overall EDM performance.

4.4 Optimal Biosilica Concentration

Biosilica concentration in the dielectric fluid determines its thermal conductivity and control of sparks. Increased concentrations are better for performance, but too much can cause stability problems. Achieving the optimal balance is the key to maximum benefits.

4.5 Comparative Analysis

Conventional EDM vs. Biosilica-Deionized Water EDM

The biosilica-deionized water blend performs better than conventional dielectric fluids in MRR, TWR, and surface finish. The new fluid provides a green and effective solution for EDM operations.

4.6 Performance Comparison and Metrics

Performance indicators like machining speed, surface finish, and electrode wear improve considerably with the biosilica blend. These improvements indicate the potential for mass industrial application.

5. Conclusion

This project work has successfully investigated the application of silane-treated biosilica particles obtained from Wheat Husk in AISI 1010 steel EDM process. The study sought to examine the influence of biosilica addition and surface treatment on machining characteristics, and the findings indicate encouraging results. The outcomes of the present study imply that the saline-treated biosilica particles have the potential to serve as a crucial functional material for machining high-strength alloys like iron, titanium, magnesium, and other similar composites. In addition, the success of this research paves the way for extrapolating the study to new materials such as titanium, iridium, and tungsten, expanding the scope of the application of the saline-treated biosilica-activated EDM method to machining various materials. In summary, the results of this study not only provide valuable information on optimizing EDM processes but also indicate the potential of saline treated biosilica particles as a major ingredient in improving the machining efficiency of a broad array of high-strength alloys and new materials. This study is a stepping stone for future research and applications in advanced materials machining.

Acknowledgment

This work was supported by Vel Tech High Tech Dr. Rangarajan Dr. Sakunthala Engineering College, Institute Research Fund (Seed Money), Reference Number: VH/R&D/SMP/22-23/018.

References

1. Se Hyun Ahn, Shi Hyoung Ryu, Deok Ki Choi & Chong Nam Chu 2004, 'Electro-chemical micro drilling using ultra short pulses', Precision Engineering, vol.28, no. 2, pp. 129–134.
2. Goswami, R, Chaturvedi, V & Chouhan, R 2013, 'Optimization of Electrochemical Machining Process Parameters Using Taguchi Approach', International Journal of Engineering Science and Technology, vol. 5, no. 05, pp.999–1006..
3. T. Sekar, R. Marappan, Experimental investigations into the influencing parameters of Electrochemical machining

of AISI 202, Journal of Advanced Manufacturing System (JAMS), (2008) Vol. 7(2), pp.337–343

4. Lin Tang, Yong-Feng Guo, Experimental study of special purpose stainless steel on electrochemical machining of electrolyte composition, Materials and Manufacturing Processes, (2013) Vol. 28(4), pp.457–462.
5. Zhiyong LI & Hua JI 2009, 'Machining Accuracy Prediction of Aero-engine Blade in Electrochemical Machining Based on BP Neural Network', International Workshop on Information Security and Application, pp. 244–247.
6. Minghuan Wang, Wei Peng, Chunyan Yao & Qiaofang Zhang 2010, 'Electrochemical machining of the spiral internal turbulator', International Journal of Advanced Manufacturing Technology, vol. 49, no. 9-12, pp. 969–973.
7. H. K. Kansal, S. Singh, and P. Kumar, 2007. "Effect of silicon powder mixed EDM on machining rate of AISI D2 die steel," J. Manuf. Process., vol. 9, no. 1, pp. 13–22.
8. M. Shabgard, M. Seyedzavvar, and S. N. B. Oliaei, 2011 "Influence of input parameters on the characteristics of the EDM process," Stroj. Vestnik/Journal Mech. Eng., vol. 57, no. 9, pp. 689–696.
9. V. Kumar and P. Kumar, 2017 "Improving Material Removal Rate and Optimizing Variousmachining Parameters in EDM," Int. J. Eng. Sci., vol. 06, no. 06, pp. 64–68.
10. V. D. Asal, P. Patel, A. B. Choudhary, 1978 A. Professor, R. I. Patel, and A. B. Choudhary, "Optimization Of Process Parameters Of EDM Using ANOVA Method," Int. J. Eng. Res. Appl., vol. 3, no. 2, pp. 1119–1125, 2013J. A. Paivanas and J. K. Hanssan. Wafer air film transportation system. U.S. Patent 4,081,201.
11. K. Y. Kung, J. T. Horng, and K. T. Chiang, 2009 "Material removal rate and electrode wear ratio study on the powder mixed electrical discharge machining of cobalt- bonded tungsten carbide," Int. J. Adv. Manuf. Technol., vol. 40, no. 1–2, pp. 95–104.
12. H. Singh and R. Garg, 2009 "Effects of process parameters on material removal rate in WEDM," J. Achiev. Mater. Manuf. Eng., vol. 32, no. 1, pp. 70–74.
13. K.-T. Chiang, 2008 "Modeling and analysis of the effects of machining parameters on the performance characteristics in the EDM process of Al2O3+TiC mixed ceramic," Int. J. Adv. Manuf. Technol., vol. 37, no. 5–6, pp. 523–533.
14. B. H. Yan, H. C. Tsai, and F. Y. Huang, 2005 "The effect in EDM of a dielectric of a urea solution in water on modifying the surface of titanium," Int. J. Mach. Tools Manuf., vol. 45, no. 2, pp. 194–200.
15. P. Peças and E. Henriques, 2003 "Influence of silicon powder-mixed dielectric on conventional electrical discharge machining," Int. J. Mach. Tools Manuf., vol. 43, no. 14, pp. 1465–1471.
16. H. M. Chow, B. H. Yan, F. Y. Huang, and J. C. Hung, 2000 "Study of added powder in kerosene for the micro-slit machining of titanium alloy using electro-discharge machining," J. Mater. Process. Technol., vol. 101, no. 1, pp. 95–103.

Advances in Additive Manufacturing Technologies – Gurusamy Pathinettampadian et al. (eds)

Analysis of an Advanced A356 – Graphene – Zirconium Dioxide (ZrO_2) Hybrid Nano Composite by Stir Casting Process

J. Joshua Kingsly[1]

Research Scholar, Anna University,
Chennai, Tamilnadu, India

P. Gurusamy[2]

Dept. of Mechanical Engg, Chennai Institute of Technology,
Chennai, Tamilnadu, India

◆ **Abstract:** Lightweight materials are critical in industries such as aircraft, vehicle body shops, biomedical devices, and more. Lightweight aluminium alloys are ideal for industrial applications because of their inherent toughness, better temperature stability, and excellent residual stress resistance. In this experiment, an aluminium alloy (A356) with graphene and zirconium dioxide (ZrO_2) reinforcement particles was investigated for the manufacture of aluminium hybrid composites. In order to create hybrid composites, a stir casting technique is used that incorporates several process factors. Taguchi statistical analysis is utilised to optimise the stir casting parameter using the L16 orthogonal array. The levels of reinforcement range from 2 to 12 percent, the agitation speeds range from 300 to 600 rev/min, the agitation times range from 15 to 30 minutes, and the melting temperatures range from 700 to 850 °C. The stir-cast materials are put through their paces on a vickers hardness test and pin on disc wear tester to see how they wear and how hard they are at the microscopic level.

◆ **Keywords:** Aluminium alloys, Hybrid composite, Stir casting, Process parameters, Wear, Hardness

1. Introduction

Reinforcing particles may increase the strength of aluminium alloys, which already have excellent mechanical properties and machinability. There are a number of areas where hybrid composites might be beneficial. Adding reinforcing particles to aluminium alloys improves their mechanical properties, as well as their machinability. In certain cases, hybrid composites are constructed to enhance the material's performance in specific applications. Composites, which have been receiving increasing attention in material engineering owing to their usefulness for certain applications, were used to enhance lightweight materials' properties. Metal-fibre reinforcement has been the primary focus of much of the research thus far. In contrast to "composite" materials, "Hybrid composites" are made up of more than two reinforced particles embedded in the base alloy. Using the stir casting process, researchers looked into how wear affects aluminum matrix materials, which include an aluminium alloy, silicon carbide, and red mud. For their research, they use a statistical technique (Taguchi) and characteristics such as the applied particle size load, red mud fraction, sliding distance, and sliding velocity are preferred. Slide distance has been shown to be the most significant element in influencing wear loss in this investigation. Reinforced particles were added to aluminium alloys to improve their wear resistance. The process of stir casting is crucial in the manufacture of nanocomposite. It is important to note that each of the stir casting method's process parameters has its own set of characteristics, and

[1]joshuakingslyj.mech@mnmjec.ac.in, [2]gurusamyp@citchennai.net

DOI: 10.1201/9781003685906-72

the composites' output is affected by the combination of these aspects. Wear study is essential to determining the wear resistance of alloy materials. Comparatively speaking, the microhardness test is an effective means of determining the material's tensile strength. Because it is more precise and produces less specimen damage than other methods, the Vickers hardness test is the best choice for determining a material's hardness. Main purpose of this investigation is to establish how the selected parameters and reinforcements influence composite mechanical properties under discussion. There are two methods utilised to cast the parts: a stir casting machine and a pin on disc wear testing device. Microhardness may also be determined using Vickers hardness testing equipment. Best parameter combinations for a particular outcome are found via Taguchi optimization.

2. Materials

Graphene and zirconium dioxide are used as reinforcements in this experiment, which uses aluminium alloy A356 as its base material. For the production of the composites needed to perform the wear and microhardness tests, this study makes use of the stir casting process. Table 72.1 lists the chemical constituents of the basic material, such as aluminium alloy A356.

Table 72.1 A356 chemical composition

Element	Content (%)
Copper (Cu)	0.20
Magnesium (Mg)	0.45
Manganese (Mn)	0.10
Silicon (Si)	7.5
Iron (Fe)	0.20
Zinc (Zn)	0.10
Titanium (Ti)	0.20
Aluminium	Balance

3. Experimental Procedure

Stir casting procedure is examined for the purpose of producing a hybrid composite. and the process is to be regulated by the manipulation of different process parameters. The characteristics and levels of the stir casting process are presented in Table 72.2 of the following document. According to Taguchi's orthogonal array technique, the number of experiments needed for analysis is determined to be 16 because parameters number under evaluation is four and their equivalent levels are also four (L16). The tools and setup for the process of stir casting

Table 72.2 Parameters of the stir-casting and their stages

S. No	Parameters	Stage 1	Stage 2	Stage 3	Stage 4
1	Percentage Reinforcement	2	4	8	12
2	Speed of Agitation	300	400	500	600
3	Time of Agitation	15	20	25	30
4	Liquid Temperature (°C)	700	750	800	850

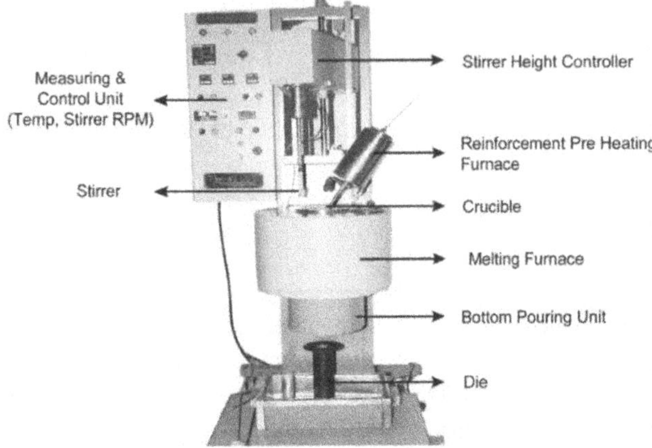

Fig. 72.1 Equipment and setup for the stir casting process

are shown in Fig. 72.1. Reinforced particles such as graphene and zirconium dioxide are heated in a crucible that has different weight percentages of the reinforced particles present (2, 4, 8, 12). The aluminium alloy A356, the temperature ranges from 750°C to 850°C, depending on its composition. Furthermore, with the help of a stirrer and a variety of agitation speeds, both the core material and the enhanced nanoparticles are thoroughly combined in the casting machine before casting (300, 400, 500, 600). Various agitation time durations are used (15, 20, 25 and 30 min). Final step is then pored the molten material into the die that has been constructed to collect the appropriate samples. After undergoing wear and microhardness testing, the casted hybrid composites are used to create the specimens.

3.1 Wear Test

Dry sliding mode and the DUCOM model are used to measure wear. A wear test was performed using the Pin-On-Disc device, and varied loads were applied to the relevant parameters. The wear test specimen characteristics include 12 percent reinforcement, speed of agitation 600 rpm, duration of agitation 30 minutes, and liquid metal 850°C. The parameters and levels of the wear test are shown in Table 72.3. The American Society Testing Materials prepares the wear test specimens (ASTM G99).

Table 72.3 Parameters of the wear test procedure and their levels

S. No	Parameters	Level 1	Level 2	Level 3	Level 4
1.	Reinforcement Percentage	2	4	8	12
2.	Load Applied (N)	15	25	35	45
3.	Speed of the Disc (m/s)	1.0	1.5	2.0	2.5
4.	Sliding distance (m)	1000	1200	1400	1600

The specimen is 45mm long and 12 mm dia, as seen in Fig. 72.2. An orthogonal array of L16 characteristics was utilised to evaluate the specimens.

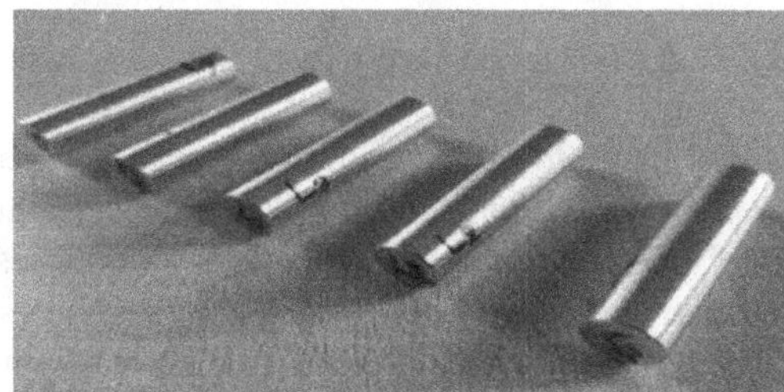

Fig. 72.2 Wear test specimen image

3.2 Vickers Microhardness Test

The microhardness test specimens for the materials are made in compliance with ASTM E384 standards. The specimen is indented with a diamond indenter. Load varies from 0.01kgf to 1kgf (0.098N to 9.81N) in the Vickers hardness test, although in this experiment, 0.5kgf (4.90N) was utilised. Because it is more precise and produces less specimen damage than other methods, the Vickers hardness test is the best choice for determining a material's hardness.

4. Results and Discussions

4.1 Wear Test

At 0.214mm^3/m wear rate, with a reinforcement of 12 percentage, contribution, applied load of 25N, disc speed of 1.5 m/s and sliding speed is 1000m in Table 72.4, the results of the wear test are shown. For the wear test's mean and S/N ratio, the response tables are tabulated in Table 72.5 and Table 72.6. All of the input variables are converted to mean and S/N ratio values for the wear test using DOE. Aside from sliding distance, disc speed, load applied, and, the reinforcement percentage was the most significant parameters in the wear test. For the optimal wear test parameters, 12 percent reinforcements, applied stress 25 N, disc speed 1.5 m/s, and sliding speed 1000 m were determined to be the optimum values.

Figures 72.3 and 72.4 show the wear test principal effects plot for the mean & S/N ratios. The specimen had a greater

Table 72.4 The results of the wear rate

Ex. No	Reinforcement Percentage	Load Applied (N)	Speed of the Disc (m/s)	Sliding distance (m)	Wear rate (mm^3/m)
1.	2	15	1.0	1000	.532
2.	2	25	1.5	1200	.342
3.	2	35	2.0	1400	.632
4.	2	45	2.5	1600	.793
5.	4	15	1.0	1000	.476
6.	4	25	1.5	1200	.587
7.	4	35	2.0	1400	.397
8.	4	45	2.5	1600	.784
9.	8	15	1.0	1000	.642
10.	8	25	1.5	1200	.412
11.	8	35	2.0	1400	.317
12.	8	45	2.5	1600	.493
13.	12	15	1.0	1000	.584
14.	12	25	1.5	1200	.214
15.	12	35	2.0	1400	.311
16.	12	45	2.5	1600	.246

Table 72.5 Means response table - wear test

Stage	Reinforcement Percentage	Load Applied (N)	Speed of the Disc (m/s)	Sliding distance (m)
1	0.5748	0.5585	0.4205	0.4090
2	0.5610	0.3887	0.4055	0.5068
3	0.4660	0.4142	0.5680	0.4415
4	0.3387	0.5790	0.5465	0.5832
Delta	0.2360	0.1902	0.1625	0.1742
Rank	1	2	4	3

Table 72.6 Signal-to-noise ratio response table - wear test

Stage	Reinforcement Percentage	Load Applied (N)	Speed of the Disc (m/s)	Sliding distance (m)
1	5.200	5.113	8.067	8.260
2	5.303	8.760	8.014	6.521
3	6.918	8.033	5.835	7.579
4	10.097	5.613	5.603	5.159
Delta	4.897	3.647	2.464	3.101
Rank	1	2	4	3

wear rate as a consequence of being heavily worn without reinforcing. Increases in reinforcing % also reduced wear. For the lowest wear rate, the greater amount of

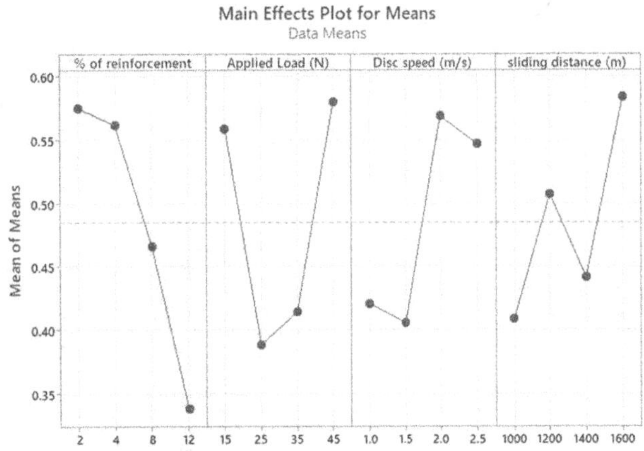

Fig. 72.3 Mean main effects plot (wear test)

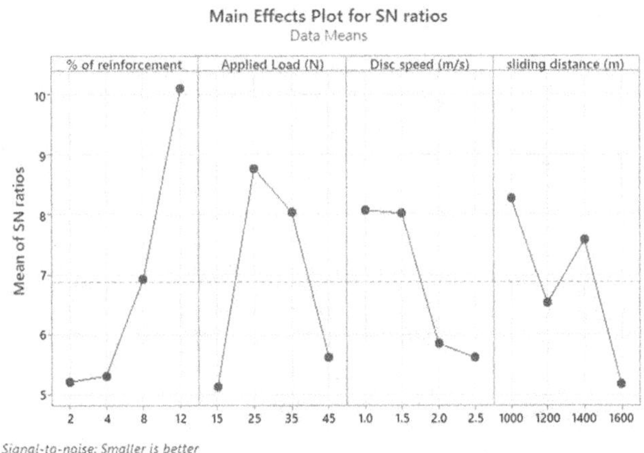

Fig. 72.4 S/N ratios main effects plot (wear test)

reinforcement (12 percent) was the most effective option. Wear rates were initially high at lower applied loads, but when the applied load (25 N) is raised, the wear rates drastically fall. The load was increased from 25 N to 45 N, which resulted in increase in the wear rate. While 1.5 m/s was found to have the lowest wear rate, when the disc speed increased, some of the wear patches neared the probability line and this was seen in the wear rate results, since the testing runs were precise. Figure 72.5 shows a fit plot in which the dots seem to be evenly spaced, which may

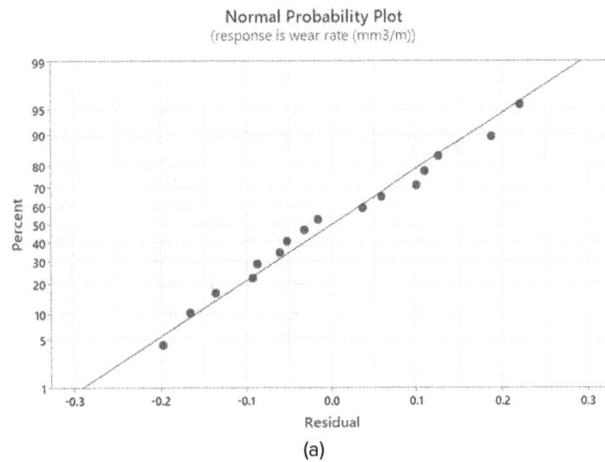

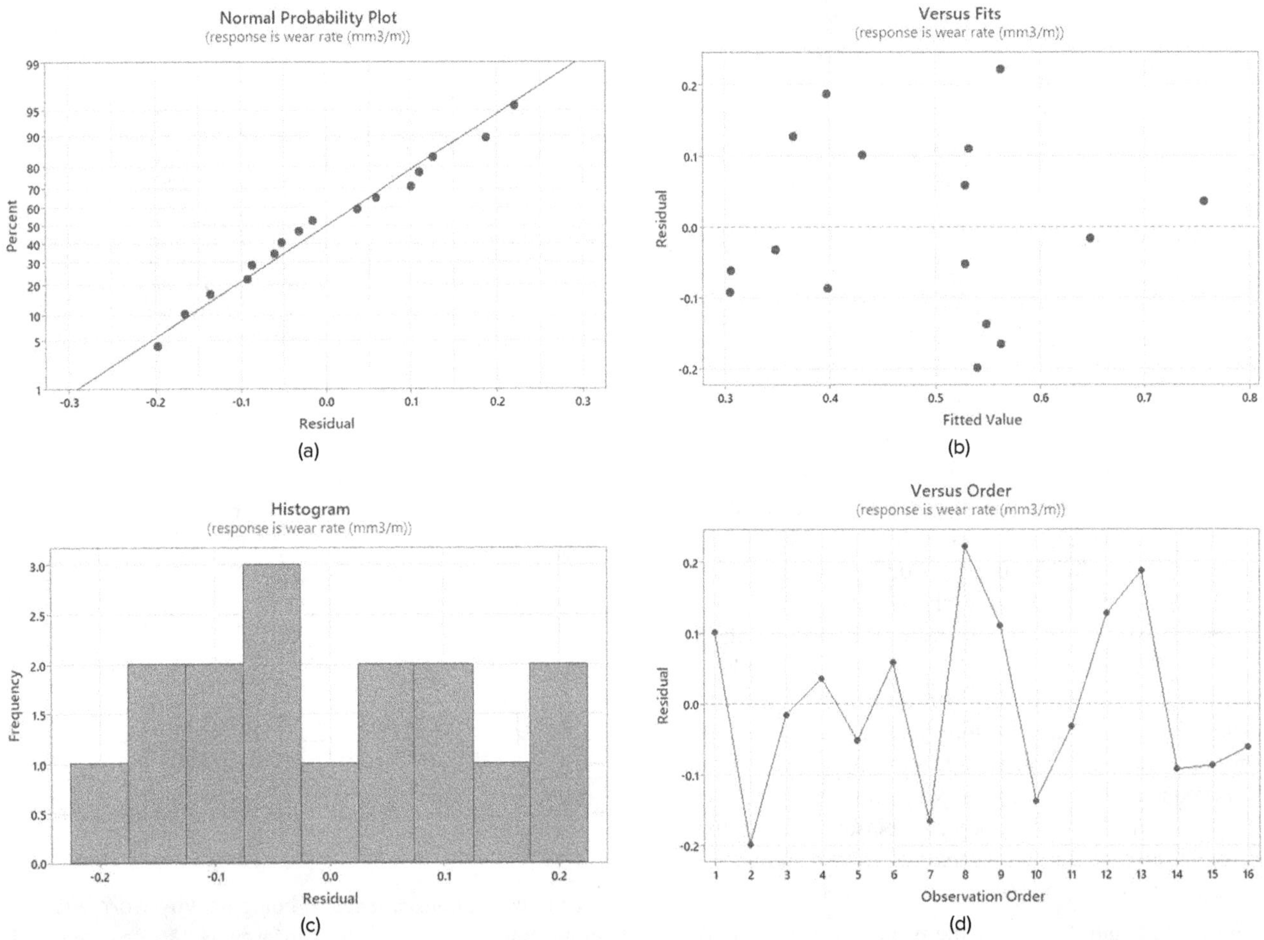

(a)

(b)

(c)

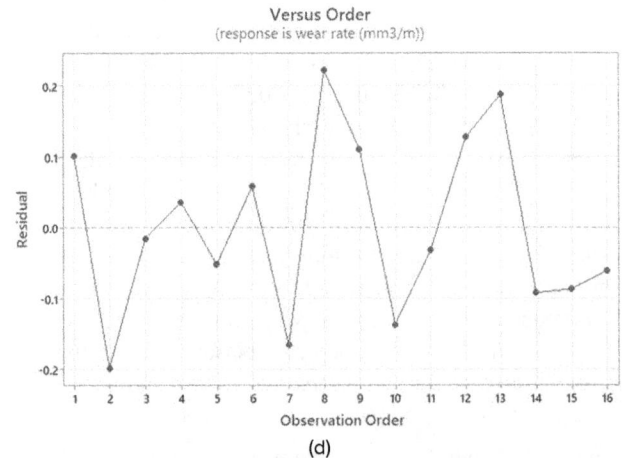

(d)

Fig. 72.5 (a) Normal probability of wear rate residual plot, (b) Versus fits of wear rate residual plots, (c) Histogram of wear rate residual plots, (d) Versus order of wear rate residual plots

indicate that the required parameters are being registered and that the response values are valid. The histogram shows that all-rectangle boxes appear close together. Order plot points crossed the mean in a way that was both positive and negative. It was because of these conditions that the experiment ran smoothly and efficiently. The surface plot of the wear test is shown in Figures 6(a)-6(d). The percent of reinforcement vs the applied load is shown in Fig. 72.6(a). The minimal wear was calculated based on the impact of the minimum applied load and maximum reinforcing level on this figure. The applied stress is shown against the disc speed in Fig. 72.6(b). In this graph, the largest applied load and the lowest disc-speed resulted in the lowest wear rate. The sliding distance vs disc speed is depicted in Fig. 72.6(c). According to this graph, a moderate disc speed and a small sliding distance resulted in a low wear rate. The sliding distance vs the percent of reinforcement is shown in Fig. 72.6(d). The lowest sliding distance and maximum reinforcing percentage in this plot resulted in a decreased wear rate.

4.2 Vickers Hardness Test

The highest hardness was reported at 115 VHN, as tabulated in Table 72.7. The effect of the parameters 12 %

Table 72.7 Vickers hardness test experiment summary

Exp runs	% of reinforcement	Agitation Speed	Agitation time	Molten Temperature	Vickers Hardness (VHN)
1.	2	300	15	700	103
2.	2	400	20	750	93
3.	2	500	25	800	104
4.	2	600	30	850	101
5.	4	300	15	700	108
6.	4	400	20	750	99
7.	4	500	25	800	106
8.	4	600	30	850	109
9.	8	300	15	700	114
10.	8	400	20	750	109
11.	8	500	25	800	109
12.	8	600	30	850	107
13.	12	300	15	700	112
14.	12	400	20	750	115
15.	12	500	25	800	111
16.	12	600	30	850	108

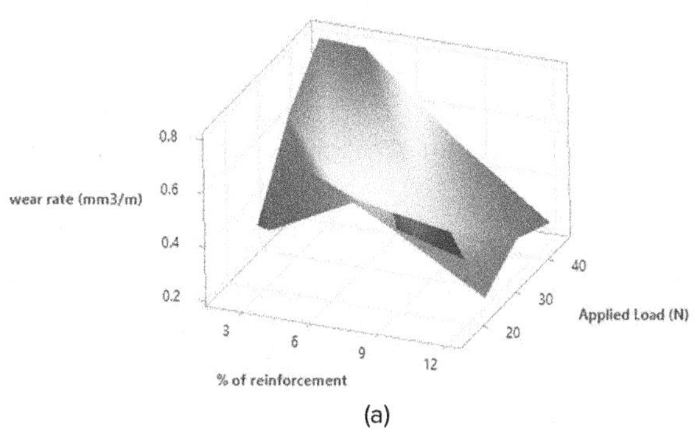

(a)

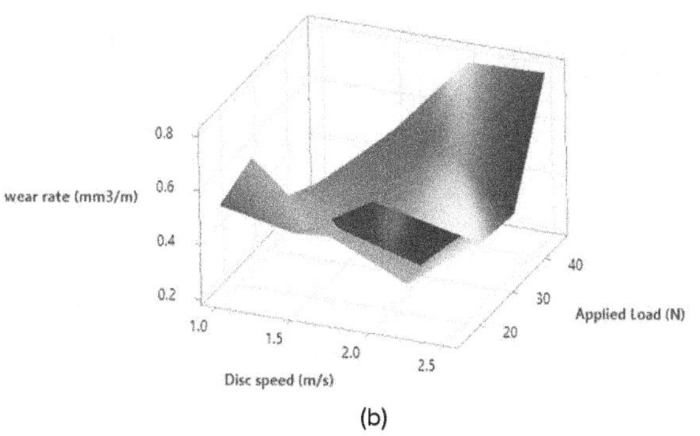

(b)

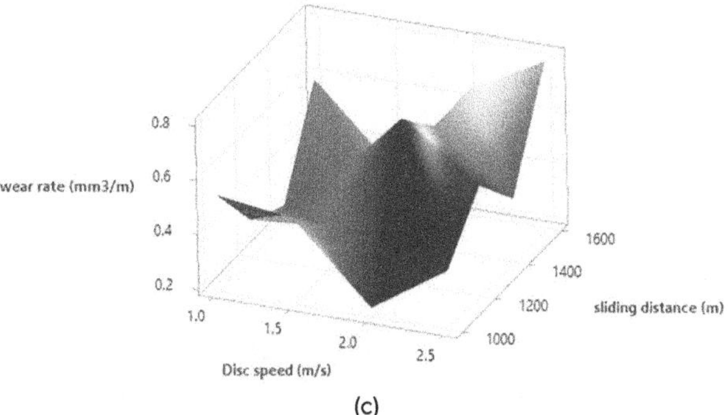

(c)

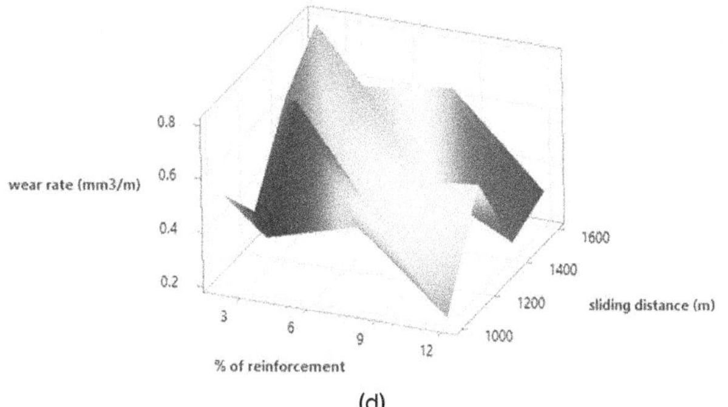

(d)

Fig. 72.6 (a) Percentage of reinforcement vs. applied load, (b) Load applied vs. disc, (c) Sliding distance vs. disc speed, (d) Sliding distance vs. reinforcement percentage

reinforcement, 400 rpm agitation, 20-minute agitation, and 750°C molten temperature resulted in the highest hardness values. Tables 72.8 and 72.9 provide the Vickers hardness test response tables for means and S/N ratios, respectively. Through Design of Experiments, all of the input data are translated into Vickers hardness test S/N ratio and mean values for these tables. The percent of reinforcement had greatest influence, followed by stirring speed, agitation duration, and molten temperature. 8% reinforcement, 300 rpm agitation speed, 25 minutes agitation duration, and 800°C molten temperature was determined best Vickers hardness test parameters.

Table 72.8 Means response table - vickers hardness test

Level	% of reinforcement	Agitation Speed	Agitation time	Molten Temperature
1	100.3	109.3	104.8	106.3
2	105.5	104.0	104.8	105.8
3	111.3	107.5	109.0	108.8
4	110.0	106.3	108.5	106.3
Delta	11.0	5.3	4.3	3.0
Rank	1	2	3	4

Table 72.9 Signal-to-noise ratio response table - vickers hardness test

Level	% of reinforcement	Agitation Speed	Agitation time	Molten Temperature
1	40.01	40.76	40.40	40.52
2	40.46	40.31	40.38	40.46
3	40.92	40.63	40.74	40.72
4	40.83	40.52	40.70	40.51
Delta	0.91	0.45	0.36	0.26
Rank	1	2	3	4

The primary effects plot for the Vickers hardness test S/N ratios and mean is illustrated in Figs. 72.7 and 72.8. The microhardness value grew when the reinforcing percentage was raised. The greater reinforcing percentage (12%) resulted in the highest microhardness. To begin with, lesser agitation rates had a high microhardness value; however, when the agitation speed raised to 500 rpm, the microhardness value dropped dramatically. The microhardness was enhanced by increasing the stirrer speed from 510 to 560 rpm. A gradual increase in agitation time resulted in an increase in microhardness values; 30 minutes of agitation time yielded the highest values. The lowest liquid temperature resulted in lesser microhardness, whereas the microhardness rose as the molten temperature climbed. Finally, the microhardness values were highest at 800°C. Vickers hardness residual

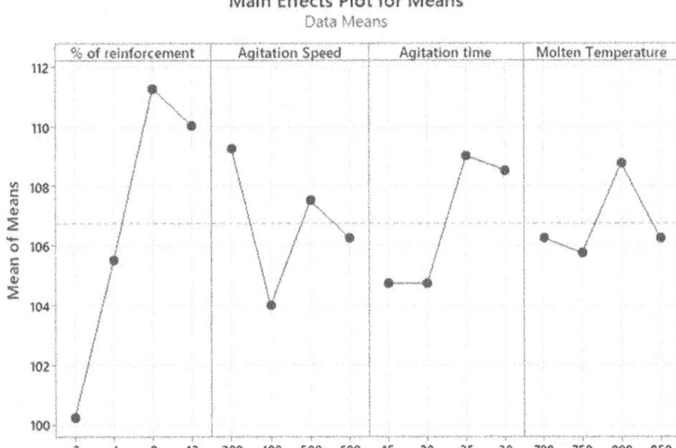

Fig. 72.7 Means main impact plot (Vickers hardness test)

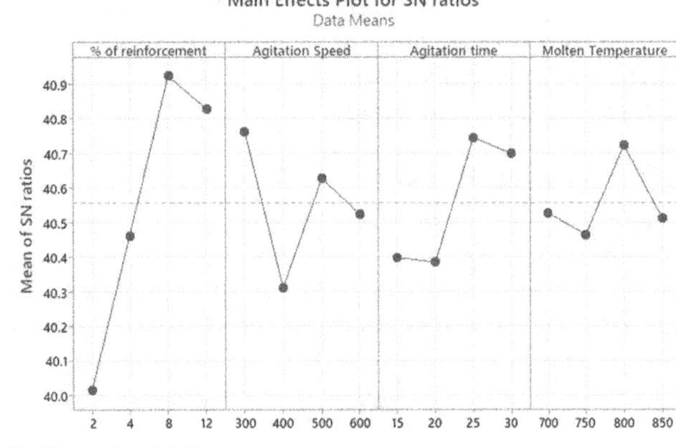

Fig. 72.8 S/N ratios main effect plot (vickers hardness test)

plots are shown in Fig. 72.9. The microhardness level and parameter effect are vividly demonstrated in one residual plot observation. A look at the normal probability plot reveals that all sixteen points lie on a probability line, but some are more closely aligned with the line than others, and this may be repeated by doing experimental runs to determine the influence of microhardness. It is possible to record the required parameters and the response values within the constraints since the points on the fits plot were spread equally. Histogram rectangular boxes are all quite close together, almost touching. All points in order plot crossed the mean line both favourable and unfavourable at the same time. The testing was performed out in an extraordinary way under these circumstances, and the characteristics were effectively used. The Vickers microhardness test contour plot is shown in Figs. 72.10(a)–72.10(d). Figure 72.10(a) depicts the relationship between percent of reinforcements and stirrer speed, with greater reinforcement percentages and medium stirrer speeds resulting in superior microhardness. Figure 72.10(b) represents significant relationship among

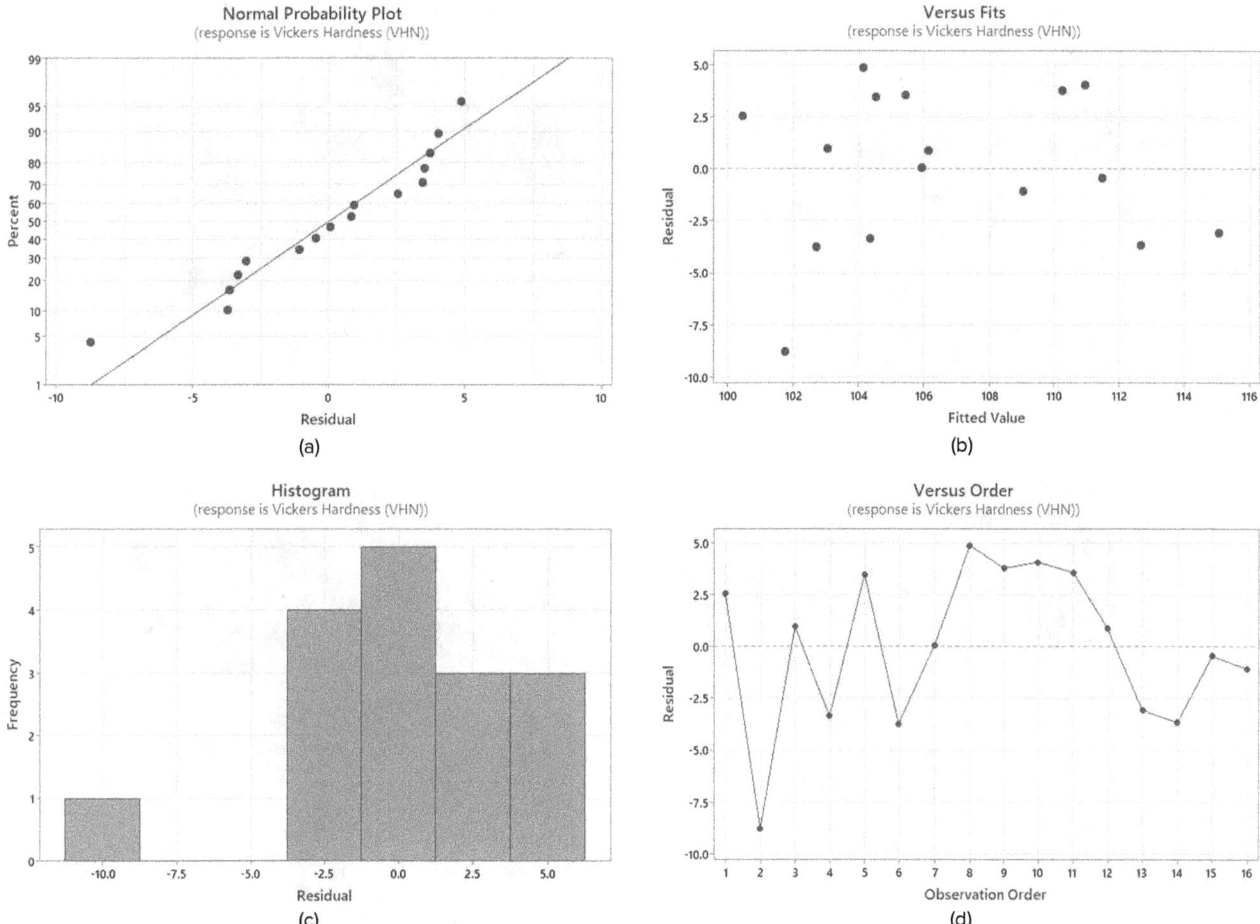

Fig. 72.9 (a) Normal probability plot of vickers hardness, (b) Versus fits of vickers hardness, (c) Histogram of vickers hardness, (d) Versus order of vickers hardness

stirring speed and stirring duration, with the lowest stirring speed and the longest stirring duration yielding the highest microhardness. The link between stirring duration and liquid temperature is shown in Fig. 72.10(c), with the maximum stirrer duration and medium liquid temperature being registering as the highest microhardness. The relationship between liquid temperature and percent of reinforcement is seen in Fig. 72.10(d), with medium liquid temperature and higher augmentation registering as the greatest microhardness.

5. Conclusion

Stir-casting created aluminium alloy A356 composite hybrid with the inclusion of zirconium oxide and graphene. Pin-on-Disc wear tests were also carried out using the Taguchi statistical method. Finally, the Vickers microhardness and wear worn-out surfaces were carefully studied. The following is the conclusion and demonstration of this experimental work's output:

- The wear test yielded a minimal wear rate of 0.214mm3/m, with a 12 percent contribution from reinforcements, an load applied 25 N, speed of the disc 1.5 m/s, and a moving distance of 1000 m in the wear test. In wear testing for effects, the percent of reinforcement was the most strongly influenced parameter, accompanied by moving distance, stress applied, and speed of the disc. 12 wt% reinforcements, stress applied 25N, speed of the disc 1.5 m/s and 1000m sliding distance shown the best wear test.
- Surface plot analysis revealed that the smallest applied load and highest reinforcing level provided the lowest wear rate. Furthermore, the largest load applied and the lowest disc speed gave the lowest wear rate. The minimal sliding distance and maximum reinforcement percentage were correlated, resulting in a decreased wear rate.
- The Vickers microhardness test yielded a maximum hardness of 115 VHN. The highest hardness values

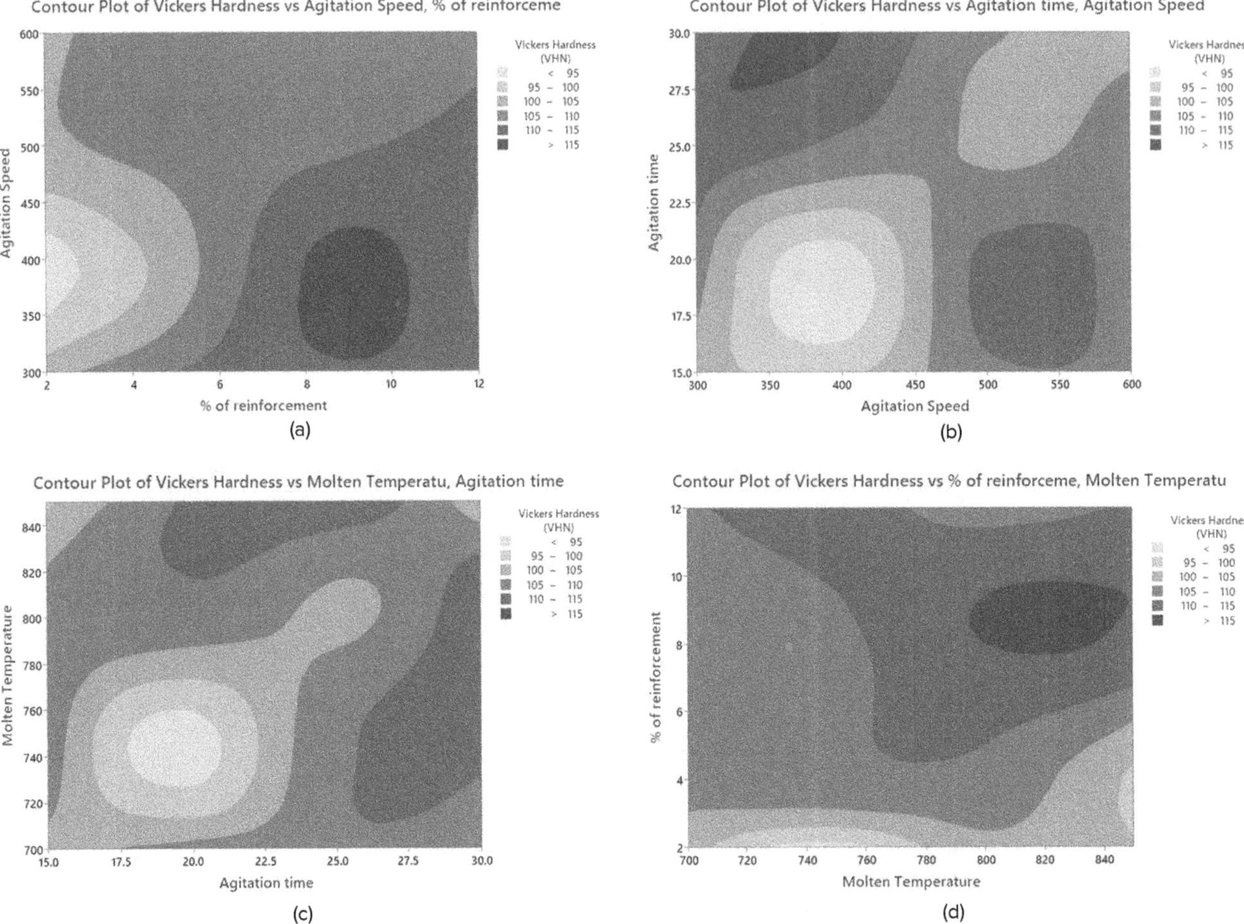

Fig. 72.10 (a) Reinforcement percentage vs agitation speed, (b) Agitation speed vs agitation time, (c) Agitation time vs molten temperature, (d) Molten temperature vs reinforcement percentage

were attained under the impact of the following parameters: 8% reinforcements, 300 rpm stirrer speed, 30-minute stirrer period, and 800°C liquid temperature. The percent of reinforcements, followed by stirrer speed, stirrer duration, and liquid temperature was the most heavily changed parameter in the Vickers hardness test for effects. The Vickers hardness test's optimal parameters were 8 percent reinforcement, 300 rpm agitation speed, 25 minutes of stirrer duration, and 800°C liquid temperature.

- The greater reinforcement % and moderate stirrer speed provided excellent microhardness, according to the contour plot analysis. Furthermore, the highest hardness value was obtained with the slowest agitation speed and the longest agitation period. The maximum hardness value was calculated using the maximum agitation duration and a mild molten temperature.

References

1. R. Farajollahi and H. Jamshidi, CIRP J *Manuf Sci Technol.*, *Volume 37*, 204–208 (2022).
2. A. Alizadeh and A. Khayami, Ceram, Vol **48**, 179–189, (2022).
3. W. Guo and S. Wang, *Appl.Surf. Sci.*, Volume **584**, (2022).
4. M. Chinababu and N. Naga Krishna, Mater. Lett. Volume **15**, (2021)
5. M.A. Alam and H.H. Ya, Mater. *Today: Proc.*, Volume **48**, 811–814 (2022).
6. H. Singh and K. Singh, Mater. *Today: Proc.*, (2022)
7. Y. Behnamin and D. Serate, *Tribol. Int.*, Volume **165**, (2022).
8. K. Luo and H. Xiong, *J. Alloys Compd.*, Vol **893**, (2022).
9. F. Mehr and S. Cockcroft, J. Mater. *Process. Technol*, Vol **286**, (2020).
10. S.U. Khan and D. Wanwu, Eur J *Mater Sci Eng*, Vol **6**, 220–223, (2021)

Note: All the figures and tables in this chapter were made by the authors.

Advances in Additive Manufacturing Technologies – Gurusamy Pathinettampadian et al. (eds)

Studies on Radial Ventilated Disc Brake Materials—Effect of Metal Matrix Composite Materials using FEA in Automotive Brake Disc

V. Hariram[1]
Department of Mechanical Engineering, Hindustan Institute of Tech.& Science, Padur, Chennai, Tamil Nadu

M. Karthigairajan[2]
Department of Mechanical Engineering, Karpaga Vinayaga College of Engg and Tech, Chengalpattu, Tamil Nadu

A. Naga Manoj Kumar[3], K. Bharath[4]
Department of Mechanical Engineering, Hindustan Institute of Tech.& Science, Padur, Chennai, Tamil Nadu

M. P. Saravanan[5]
Department of Mechanical Engineering, Savetha Enginering College, Thandalam, Chennai

S. Balamuragan[6]
Department of Mechanical Engineering, Chennai Institute of Technology, Kundrathur, Chennai

K. Barathiraja[7]
Department of Mechanical Engineering, Loyola Institute of Technology, Chembarambakkam, Tamil Nadu

Mohammad Rayan Shaik[8]
Department of Mechanical Engineering, Hindustan Institute of Tech.& Science, Padur, Chennai

◆ **Abstract:** During the slowing down of a vehicle, mechanical energy is created due to the kinetic motion of the wheels. It gets dissipated in the form of heat, which raises the disc brakes' temperature of the brake disc and pad material. The study deals with the escalation in automotive disc brake's temperature during deceleration and its impact on the disc's robustness using the FEA techniques. The pressure to the rotor produces heat flux with the mathematical calculation of the quantitative value of heat transfer co-efficient and heat flux. The variations in results attained by the brake disc are recorded. Different materials, such as Titanium Alloy, Grey Cast Iron, Aluminum Alloy, and Carbon Ceramics, are looked into for enhancing the performance of the brake disc. This paper analyses the thermodynamic properties of ventilated disc brakes under steady-state conditions using ANSYS simulations to analyze the structural and thermal responses of four materials: Grey Cast Iron, Al6061-T6, Ti-6Al-4V, and C/SiC. With constant brake efficiency, the analysis showed that Al6061-T6 has a 138% higher heat dissipation capacity than Grey Cast Iron, followed by C/SiC at 92% and titanium alloy at 70%. Equivalent elastic strain in Grey Cast Iron and C/SiC was relatively lower and found to have better structural strength towards thermal loads. C/SiC showed the

[1]connect2hariram@gmail.com, [2]m.karthigairajan@gmail.com, [3]nagamanojk313@gmail.com, [4]kottapallibharath097@gmail.com, [5]saravanan.design89@gmail.com, [6]balamurugans@citchennai.net, [7]kbarathirajal@gmail.com, [8]rayanshaik050@gmail.com

DOI: 10.1201/9781003685906-73

smallest deformation due to its high hardness and thermal resistance and had the lowest maximum temperature. Heat flux analysis showed that Aluminum Alloy had very good heat dissipating capability, thereby reducing internal thermal stresses as well as ensuring minimum risk of cracking. The findings highlight the importance of having lower wear and deformation, enhancing disc brake longevity and efficiency, linking thermo-mechanical behavior to mechanisms like cracking, fatigue, and abrasion. Based on the results it can be said that for different applications and requirements the four materials can be designated for its use in vehicles.

◆ **Keywords:** Brake disc, Thermo-mechanical behaviour, Stress, Thermal deformation, Temperature

1. INTRODUCTION

The thermal analysis is the first in the sequence of studies about the brake systems. It is noticeable that during braking, strong temperatures and thermal gradients exist. It causes stress and deformation whose outcome is through cracks formation and its complement. Hence, allow determining the brake disc's temperature field with precision. Incomplete one-cycle stop braking is characteristic in which the temperature does have time to be uniform throughout in the disc. Thermal loading in a disc brake can be defined as a heat flux that passes from the friction material of the pads into the disc [1]. The temperature distributions over the contact surface of a disc brake owing to heat generation in the brake assembly during braking is described. Heat transfer coefficients regarding the articular connection are examined and the difficulties regarding uneven frictional heating during the common slipping of a disc inscribed into rigid pads are solved through FE models and with the number of variable parameters which can occur in automobile applications.

This is generally accomplished by employing grey cast iron which is durable but heavy and an effective cooling source. This is where savings due to titanium and aluminum alloys are useful, but they do not have adequate heating even as it is cast iron. Carbon composites on the hand are very light and have high thermal resistance, which speeds up the performance applications, however these are costly and may get damaged easily in the right conditions [2]. In this regard, different material compositions can be used to design a hybrid brake disc so that strength, thermal resistance and weight efficiency are optimal, in this case, grey cast iron is used for strength while titanium and aluminum for weight and carbon composites for thermal resistance in racing and high-performance applications. Taiviqirrahman et al. [3] in his study he conducted the thermal analysis of disc brake with variation in drill hole angles and grove hole angle and stated that changing the geometry of the brake disc will increase the braking efficiency. He observed that the ventilation hole angle and groove angle of 0° with grey cast iron given the smallest maximum temperature, and gives the best performance. Ali Belhocine, et al. [4], conducted a study in Thermal analysis of a solid brake disc. He conducted the transient thermal analysis on the full and ventilated brake rotor to analyse the thermal behaviour of the disc when the calculated heat flux is entering through the disc. This study is conducted using the numerical modelling code in Ansys11.

Bianchi et al. [5] conducted the research study on the Life Cycle Assessment on the Ceramic Matrix composites and prepreged of Ceramic Matrix composites recovered disc and compared among them. His aim is that he claims that using of CMC brakes have several environmental impacts from the starting production till its life time, as CMC brakes manufacturing impacts environment in several conditions like Global warming potential, cumulative Energy. Zhang et al. [6] investigated the wear and influence of thermal expansion due to stress and temperature on the disc-pad employed in the rail engines. It was observed that the 10% thermal expansion was evidenced on averaging the generated temperature due to wear with a maximum variation in the brake disc was seen at 257.5% during its run at elevated speed. Wang et al. [7] conducted experiment on structural optimization design of friction pair wet brakes for a hydraulic machine motor. He concluded that his optimized design needs to further verified in future work before getting into the application. Rahim Jafari et al. [8] conducted a elaborate study on the ventilated brake disc's performance on cooling. He observed by changing the geometry parameters he enhanced the reduce the cooling time by 10%, an d21%. Bonini et al. [9] in his research touches upon the enhancement of thermographic assessments of carbon disc brakes installed on the motorcycles of MotoGP racing. In a race, carbon brakes are outfitted as they are light, have great friction force and good thermal conductivity thus performing better than steel brakes.

On the other hand, air composite brakes require a strict management of temperature because it is only the best in operation around a certain optimal temperature range. Heuristic Methods using a hybrid approach of the Unscented Kalman Filter (UKF) and simplified one-dimensional thermodynamic models were employed to estimate the temperature distribution of brake discs. He was able to prove that the integration of predictions made by statistical models with the actual data of thermographs increases accuracy many times over traditional methods. This construction enables speedy computation of disc temperatures which is vital for racing applications. Formosa [10] conducted a thermal study on the solid bike disc with no holes and with different materials individually, different materials are modified and new in motor cycle industry. It was observed that multi layered aluminum stainless steel structured rotor disc performed better, due to its light weight and mechanical properties, further studies are conducted to understand the best performance material for the disc1. The materials to be used are conventional grey cast iron which will be compared with titanium alloy, aluminum alloy and carbon ceramic matrix composed of silicon carbide.

A comparison was made to identify the most suitable material for use in disc brake rotors. Analysis was conducted on the thermomechanical properties of the original material that had been utilized for manufacturing disc brake rotors during the 19th century [11]. It reflects how often the use of disc brakes can be due to their reliability. The performance of this braking system should not depreciate with time; they should remain constant with every use [13]. Ziad et al (2009) conducted an FEA analysis on the vented brake disc rotor of a standard passenger vehicle running at full loading condition. The main interest of this research was the distribution of heat and temperature in the rotor. During the load analysis, ten braking cycles followed by ten idle cycles were simulated during a period of 350 seconds of operation. For each of the cycles done, results were presented for the first, fifth, and tenth cycles. This research contributes towards more efficient disc brake rotors for the automotive industry through highlighting the thermal characteristics of the rotor [14].

According to Haripal et al. [15], cast iron brake discs are acceptable since they can cool down at lower temperatures more efficiently. Although expensive, the ceramic discs showed very good cooling ability and therefore suited applications for high temperatures such as racing cars [15]. The research work by Sowjanya et al [16] discusses the effect of strength variations on the stress distributions expected to occur and also presents on the choice and assessment of aluminum metal matrix composites. Based on structural and thermal analysis, the results indicated that the brake disc design fulfilled both the Rigidity Criteria and Strength requirements. Mandeeep Singh et al. [17] has proposed finite element analysis technique was used to measure its effectiveness. This study by Lakavathu et al., 2023, focuses on material selection and design optimization in enhancing braking performance and vehicle stability. The ventilation system with holes in the disc is significant for cooling and high-temperature tolerance. The analyses showed a strong coupling of the temperature and stress fields during braking. From these results, Aluminium MMC turned out to be the most suitable material. Mode Shape 3, at 597.1 Hz had the least deformation and superior performance compared to other mode shapes [18]. The study by Aditya et al., 2023, deals with the optimization of disc brake rotors to withstand thermal and structural stresses with minimal weight. Three materials were considered in this study: AISI 4140, SS 420, and AISI 1020. Static structural analysis was conducted using the results of transient thermal analysis to determine von Mises stress, deformation, Factor of Safety (FoS), and temperature. SS 420 exhibited better performance because of its high bulk modulus, thus having the lowest von Mises stress even though it had a lower specific heat capacity and thermal conductivity. The deformation in all three materials was comparable and was within acceptable limits. According to the results, SS 420 performed better than AISI 1020 and AISI 4140 in terms of resistance to stress [19]. A rotor disk designed using CATIA was subjected to thermomechanical conditions and analysed in ANSYS using four different materials to obtain the most suited material for sustaining thermal loads. Titanium alloy had the largest temperature difference 100°C to 76.95°C while the aluminum has the minimum variation of 100°C to 98.86°C. The maximum heat flux of 0.01021 W/mm^2 has been recorded with aluminum whereas the lowest was with the titanium alloy with 0.00701 W/mm^2. Stainless steel exhibited the highest equivalent stress at 439.7 MPa than other materials. All the materials had similar deformation results; however, titanium alloy, stainless steel and carbon-carbon rotor disks dissipated heat better than the aluminum one. This work provides a means to test material performance under mechanical and thermal loads with future possibilities to explore performance under vibrational conditions [21].

The study by Dong et al., 2024, focused on the semi-metallic brake paired disc pad of C/C-SiC material by analyzing its tribological behaviours used in automotive vehicles, an optimized brake pad formulation was developed following the SAE J2522 test procedure. The copper fibers significantly contributed to the friction performance where the coefficient of friction got stabilized with wear resistance improving. On higher speeds with $\geq$120 km/h the copper fibers became soft that acted like lubricant there by COF got decreasing. Improvement was achieved even by the reduced content of copper fiber along with elevated hard particle contents [22]. The study done by Saurabh et al. [2024], investigated the wear behavior and friction of composite friction materials in dry sliding against iron-based (LC2) alloy, laser-clad nickel-based (LC1) and grey cast iron discs. Graphite, MoS_2, and hBN solid lubricant-reinforced composite materials P1–P6 were prepared using the powder metallurgy method. Results indicated that significant reduction of wear and friction occurred, and LC1 and LC2 systems displayed 63.76 and 48.32 percentages of a lower wear rate than gray cast iron, respectively. The hBN-based pin, P6 had 38.49% reduced wear when compared to that of the graphite-based pin with respect to gray cast iron [23].

This paper was concentrated to study, analyse the stress, deformation, temperature and heat flux variation which also describes how the low value of maximum temperature of the brake rotor shows the high pace with which the heat is dissipated (heat flux) for the water-based materials. Three different novel materials and one base material were used in this work to determine the better material for its usage in disc brakes which has different requirements.

2. Brake Disc Material and Conceptual Design

2.1 Grey Cast Iron

Cold gray iron's comminution is performed on basic-circular plates and levers which show 240 MPa tensile strength together with high resistance to compressive forces. Moreover, the Young's modulus equals 115 GPa and the fatigue resistance is approximately 110 MPa. Its hardness fluctuates between 179 and 202. The maximum temperature that it can withstand up to 52 W/m°C and its melting point lies between 1140ºC and 1200ºC. Other blends of the substance contain varying quantities of carbon content applicable for different grades such as gray cast iron which can have about 2.7% to 3.65% as its composition.

There are different amounts of silicon that must have approximately 1.35% to 2.5%, phosphorus from 0.09% to 0.5% and manganese levels which are around 0.6%. The thick section which shows about 100mm in diameter can only be manufactured using conventional sand materials employed in the die-casting process. In such a process the melting metal gets stored in the sand mold cavities and solidifies during the cooling stage after which the molten part is removed. Manufacturing capabilities allow brackets with up to 8mm huge boars for construction. Following that a special oil is applied as an inaugural painting to finish the production stage

2.2 Titanium Alloy

Titanium alloy Ti-6Al-4V (Grade 5), Annealed, exhibits 918 MPa and 845 MPa of ultimate tensile strength and yield strength respectively with the Rockwell hardness number of 36. Its elastic modulus has a value of 111.2 GPa. The density stands at 4.43 g/cc. Fatigue strength is equal to 240 MPa. The alloying elements proportion comprises Aluminum 6 %; maximum content of Iron 0.25%; maximum content of Oxygen 0.20%; Titanium 90%; Vanadium 4%.

First, the Ti-6Al-4V alloy is selected for its combination of titanium with Aluminum and vanadium, offering high strength and durability. The material is typically forged or cast into the rough shape of the brake disc due to titanium's challenging machining characteristics. Forging is preferred for its ability to enhance the alloy's mechanical properties through grain refinement under high pressure. Next, the brake disc undergoes heat treatment, including solution treatment and aging, to improve its strength and hardness. This process tailors the alloy's microstructure to enhance its performance under extreme braking conditions. Once heat-treated, the disc is CNC-machined to its precise final dimensions, including features such as mounting holes and any necessary surface patterns for improved braking performance and heat dissipation. Titanium's hardness requires specialized tools and machining techniques to achieve the desired precision and surface finish. Surface treatments like shot peening or coating may be applied to further improve fatigue resistance and protect against wear. These treatments are essential in ensuring the brake disc's longevity and reliability under high stress.

2.3 Aluminium Alloy

Aluminum Al6061-T6 tempered and aged yield strength and ultimate tensile strength of 270 MPa and 310 MPa respectively with a Rockwell hardness number of 60. Elastic modulus is 68.9 GPa. This material has a density of 2.7 g/cc. Thermal conductivity = 152 W/m K. Its melting point is approximately 582-652 C (1080-1205 F). Its specific heat capacity is s 0.896 J/g-°C. There is a very good acceptance of such coatings with normal machine treatment and machining, welding and other means. It is possible to achieve relatively high strength, ease of job processing, and high corrosion resistance; very much in demand. The manufacturing of Al6061-T6 brake discs starts with selecting the alloy for its strength, light weight, and corrosion resistance. The alloy is shaped through casting or forging, followed by heat treatment (solution treatment, quenching, and aging) to achieve the T6 temper for maximum strength. Next, the disc is CNC-machined to precise dimensions, including features like mounting holes and heat dissipation slots. Surface treatments, such as anodizing, improve corrosion resistance, and balancing ensures smooth operation. It was reported that Aluminium alloy has better corrosion resistance, ductility, wear resistance and high strength. To improve the strength and its properties, composite structure can be made as sampled in the study by Kumar et al., 2023 [26]

2.4 Carbon Fiber Reinforced Silicon Carbide Composite

Carbon fiber reinforced silicon carbide composite has an ultimate tensile strength ranging from 200MPa to 600MPa with the yield tensile strength at 270MPa. The modulus elasticity of the C/SiC composites has values ranging of about 150GPa. The flexural strength of C/SiC is in the range of about 300MPa to 500MPa. The density of C/SiC stands at 2.6g/cc. The value of Vickers hardness of C/SiC composites ranges between 20 GPa and 30 GPa. C/SiC has a highest thermal conductivity of 152 W/m K. Due to the fact that the melting point of carbon fibers is much lower than that of silicon carbide the C/SiC does not have a particular melting point. Silicon carbide itself has a melting point of about 2730° while carbon fibers will only sublimate at a temperature of approximately 3500°. In the case where there is oxygen present, however, carbon fibers will oxidize to much lower temperatures at around 400-600°. The specific heat of C/SiC composites is worth 600 J/kg-K, it also has high friction coefficient, and the hardness of silicon carbide provides good wear resistance for the brake disc, able to resist high temperatures and not appreciably losing strength or deformation, and the light weight of carbon fibers allows composites to weight much lower than former cast-iron brake discs, improving fuel consumption and performance of the car. The composite is made up of carbon fibres in the volume of the composite of 10 -30% and silicon carbide up to 60% - 80% which the composition of the composite material [27].

2.5 Modelling and Discretization of Automotive Disc Brake

SolidWorks was employed to create the 3D model of the brake disc. The outer diameter of 305 mm (0.305 m) is predetermined by the rim constrains, and the thickness is set to 16 mm with reference to calliper specifications. The outer diameter of this ventilated brake disc is 0.2525 m. The disc area that the pads touch is 0.019238 m² and is located 0.024 m from the disc's centre. The thickness of this extension ring is 0.0016 m (Refer Fig. 73.1). A disc has a weight of approximately 6.1935 kilos when composed of grey cast iron material, whereas the weight is about 3.8099 kilos when made of titanium alloy; aluminum alloy weighs approximately 2.3251 kilos; ceramic composite weighs about 2.2365 kilos, respectively.

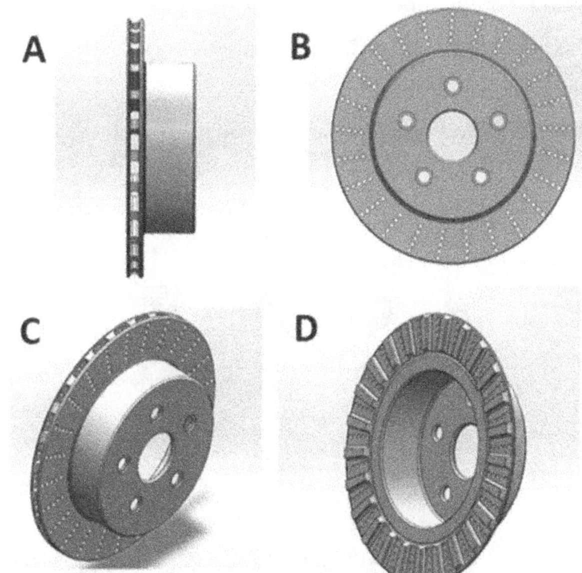

Fig. 73.1 Model of disc brake – side view (A), Front view (B), Isometric view (C) and Sectional view (D)

In Workbench, the meshing or grid generation is treated as an uncomplicated process as the user is provided with efficient meshing tools that would shorten the timeframe required to mesh a model. The model that has been used needs to be broken down into a number of small pieces which are referred to as finite elements. At the same time, some sort of degree of separation has to be made into a number of distinct parts thus, in layman terms some sort of a numerical net or 'mesh' is required in order to conduct the finite element analysis. A finite element mesh model is created. The type of mesh utilized in the analysis includes the use of tetrahedron twelve noded triangular elements. The meshing relevance Centre was maintained medium whereas the speed of the smoothing process was also maintained low. The average surface area measurements of the finished meshing are close to 200.35 mm². The view of the mesh of disc is displayed in Fig. 73.2.

Input parameters like pressure, heat flux, film coefficient, heat flow is calculated from the assumed vehicle parameters. The weight of the vehicle is assumed as 1385 kg, at an initial speed of 28kmph, which is around 7.7m/s or 441.1rad/s, pressure applied on the brake disc from the brake booster to caliper, a pressure is applied on the brake pad swept area. The stopping distance and stopping time are around 3.7 m and 0.9625s. Heat flux generated on brake disc calculated by dividing braking power and Area of rubbing surface or pad swept area, the heat flux generated by calculation.

Simulation (Total duration) = 45 (s).

Initial time (Incremental) =0.25 (s).

Minimal initial time (Incremental) = 0.125 (s).

Maximum initial time (Incremental) = 0.5 (s).

Disc Temperature (Initial) = 60 (°C).

Materials: Grey cast iron, Al 6061-T6, Ti-6Al-4V, C/SiC composite.

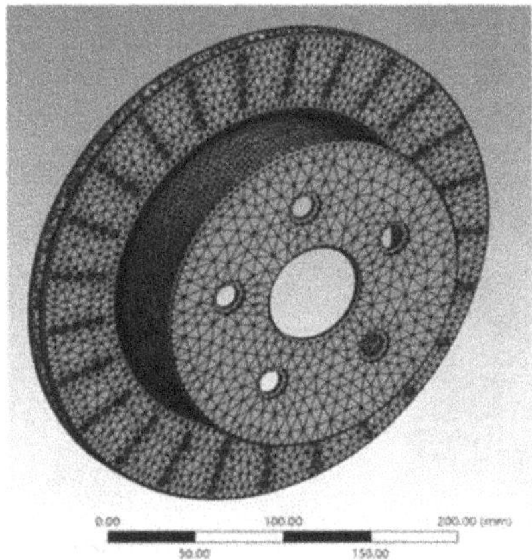

Fig. 73.2 Model of disc brake – meshing

Table 73.1 Geometrical features of disc brake

Parts of Disc Brake	Dimensions
Disc diameter (outer), mm	305
Inner disc diameter, mm	200
Effective outer rotor radius, mm	145
Speed (Initial) km/h	28
Deceleration, m/s^2	8
Thickness of the disc, mm	16
Mass of the Vehicle (m), kg	1385
Disc Height, mm	66
Charge distribution (Factor)	0.5
Surface disc swept by the pad on 2 sides of disc, mm^2	38436
Diameter (Inner disc), mm	200
Braking force (rate of distribution), %	20
Height of the disc, mm	66
Diameter (Outer disc), mm	305
Effective inner rotor radius, mm	120

The outcomes of CHT (Convective heat transfer) due to surface heat transferring is evaluated using Nusselt number equation as given below.

$$Nu_L = (h \times D_a)/k \quad (1)$$

Where, h = convective heat transfer coefficient, (W/m^2 °C), NuL = Nusselt number,

Do = Rotor external diameter, m, k = Thermal conductivity of Fluid, (air), N/m °C

By modifying this equation, we can obtain the CHT coefficient which is required for our study. Nusselt number can be calculated from the equation given below

$$Nu_L = 0.027 \times R_e^{0.8} \times Pr^{0.33} \quad (2)$$

Where, Re is the Reynolds number, Pr is Prandtl number which is a constant number, in this study Pr is taken for air which is 0.71.

The thermal loading is measured by the heat flux inflowing through the actual contact area on all sides. The chosen data for the mathematical solicitation are abridged as follows:

$$q = h \times \Delta T \quad (3)$$

The boundary conditions for the brake disc are two-sided clamping forces normal to the plane of the disc where the pads will touch the disc, a rotational velocity vector in the same direction as that of the disc's axis of rotation. The fixed constrains are applied at the hub where the disc is going to be fixed with the axle of the wheel. The following boundary conditions are then used and the model is run in static structural analysis, maximum interfacial pressure is projected of 1MPa on the disc and pad contact, which can be seen as the commander force in the direction of the brake pads, and velocity is used to apply some sort of rotation which ensures the disc is rotational at 441.1 rad/s or approximately 7.7 m/s.

The disc is assumed to have an initial temperature of 22° C with surface convection condition being applied at all surfaces of the disc where heat transfer occurs. The value of the convective heat transfer coefficient (h) is applied to the outer disc, inner disc hub and slotted grooves areas of the disc. Structural Analysis

Von Misses Stress

In material science, Stress is the force applied to a material per unit area of its cross- section, was observed when a pressure of 1MPa is applied on the ventilated disc and a Rotational Velocity of 441.1rad/s is applied on X-axis of the disc at a time duration of 45s.

Figure 73.3A shows the stress observed in gray cast iron and a elevated stress of 20.768MPa and a decremented stress of 0.0035354 MPa. Figure 73.3B represents the Stress developed by Aluminum Alloy (AL6061-T6) which is a maximum stress of 20.7Mpa and a lower stress value of 0.0035354 MPa. Figure 73.3C presents the stress obtained by Titanium Alloy (Ti - 6AL - 4V) is a higher stress of 20.75Mpa and decreased stress of 0.0035354 MPa. Figure 73.3D gives the stress in C/SiC and a elevated stress of 20.733 MPa and decremented stress of 0.0035708 MPa was observed. Overall, there is no major difference is found in stress, developed on all four different materials, because of the application of pressure on disc is unchanged for all four of the ventilated brake discs [14]. From the below

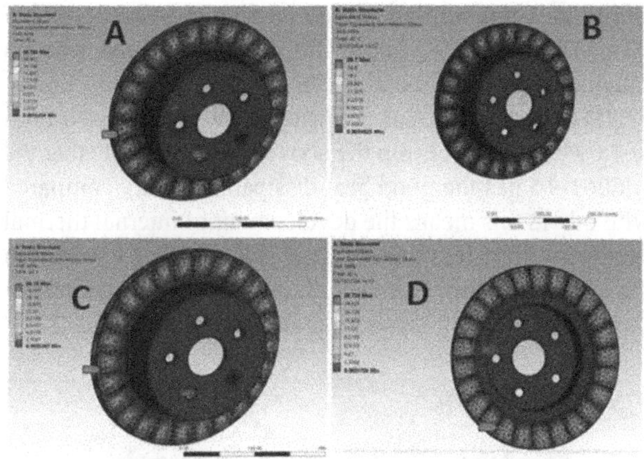

Fig. 73.3 Von misses stress – grey cast iron (A), Al6061-T6 (B), C/SiC (C) and Ti-6Al-4V (D)

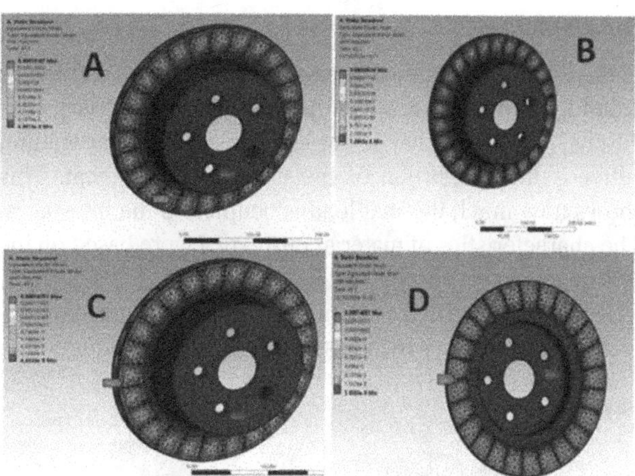

Fig. 73.4 Strain – grey cast iron (A), Al6061-T6 (B), C/SiC (C) and Ti-6Al-4V (D)

detailed Fig. 73.4(A), 73.4(B), 73.4(C) and 73.4(D) of all four materials (Aluminum Alloy, Gray cast iron, C/SiC and Titanium Alloy) we can say that Aluminum Alloy (Al 6061-T6) and C/SiC have less equivalent stress compared to other two metals. Similar result was noted as stated by Vishvajeet et al., 2021 for Carbon fibre-based brake disc [21].

Elastic Strain

Elastic strain was observed when a pressure of 1MPa and Rotational Velocity of 441.1rad/s is applied on the ventilated disc on X-axis of rotation of the disc for a simulation time duration of 45s. The Fig. 73.4A shows the equivalent elastic strain observed in Gray cast iron and a maximum strain of 0.00019197 mm/mm and a minimum strain of 4.9013e-8 mm/mm. Figure 73.4B provides the strain in Aluminum Alloy (AL6061_T6) and maximum strain of 0.00030534 mm/mm and minimum strain of 7.2985e-8 mm/mm. Figure 73.4C gives the equivalent elastic strain observed in Titanium Alloy (TI_6AL_4V) a maximum strain of 0.00019751 mm/mm and minimum strain of 4.8539e-8 mm/mm. Figure 73.4D represents the equivalent elastic strain in C/SiC and a maximum strain of 0.00014081 mm/mm and a minimum strain of 3.6593e-8 mm/mm. From below mentioned Fig. 73.4 (A), 73. 4 (B), 73. 4 (C) of all four materials (Aluminum Alloy, Gray cast iron, C/SiC and Titanium Alloy) we can say that Grey Cast Iron and C/SiC have less equivalent elastic strain compared to other two materials [11, 15].

Total Deformation

Total deformation is a structural parameter affected when the pressure is applied on the body of the disc under different material conditions. This happens due to rubbing action between brake pad and brake disc, also integrated with the internal micro-thermal stresses developed due to friction. The below Fig. 73.5A shows the total deformation in grey cast iron with a deformation of 0.0053857 mm. Figure 73.5B shows the deformation of Aluminum alloy (AL6061_T6) with a total distortion of 0.0084701 mm. Figure 73.5C shows the disfigurement of titanium alloy (TI_6AL_4V) with a total deformation of 0.0054714 mm. Figure 73.5D shows the total deformation of C/SiC with a complete distortion of 0.0039556 mm. It was observed that C/SiC (carbon ceramics) has the least deformation compared with other materials, due to its high hardness and ability of high heat resistance, it also has good strength to weight ratio [28].

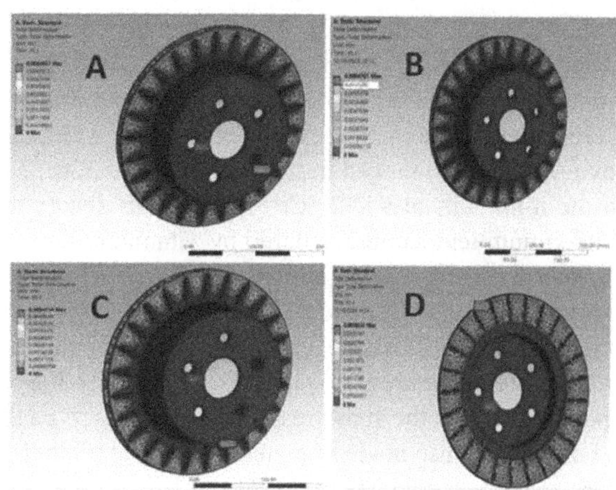

Fig. 73.5 Total deformation – grey cast iron (A), Al6061-T6 (B), C/SiC (C) and Ti-6Al-4V (D)

2.6 Thermal Analysis

Thermal analysis is conducted to study the effects and change occurrence on the brake disc caused by the induced heat energy, will represent thermal stresses and thermal deformation and cracking that happens in a micro level.

Temperature

In the thermomechanical image, the temperature difference around ventilated disc was noted for the set boundary conditions, justifying that lower the maximum temperature, better is the proportion of the material in the heat. This shows how much thermal load is coupled to the interfaces. The characteristics of material properties were based on the following observations. The convective heat transfer coefficient (h) applied at the ventilated surfaces is approximately 445.12W/m² °C which has been found from Equation 1.

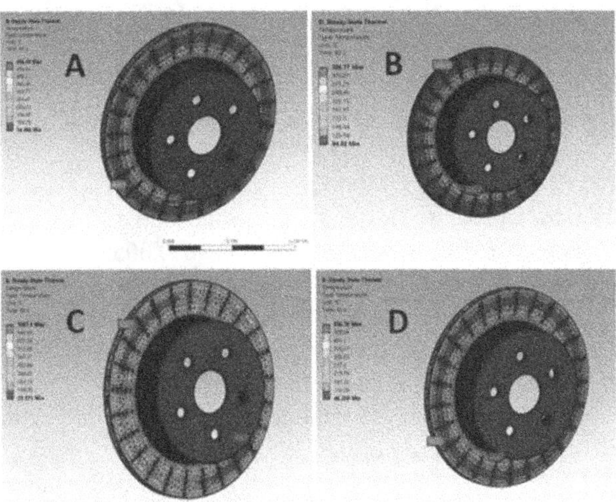

Fig. 73.6 Temperature – grey cast iron (A), Al6061-T6 (B), C/SiC (C) and Ti-6Al-4V (D)

As shown in the below Fig. 73.6A, the grey cast iron reached a maximum temperature of 506.69°C, and a minimum temperature of 54.488°C. While hammering the different material temperature obtained scale for Aluminum alloy (AL6061-T6) ranged from a maximum of 326.77 °C and the minimum of 94.92 °C. Figure 73.6B shows that the maximum temperature endured by Titanium alloy (Ti-6AL-4V) for the given conditions is 1057.3 °C and the minimum is 23.371 °C. Figure 73.7 tackles the temperature gradient in C/SiC (Carbon Ceramics). The maximum and minimum temperatures are 181.11 °C and 24.808 °C respectively. From the above results it can be observed that the body of the disc made of material C/SiC exhibited the lowest maximum temperature followed by grey cast iron those of Aluminium alloy and finally Titanium alloy under the same conditions [17, 19].

Total Heat Flux

When heat flux of 133402.46 W/m² was given to the following materials, the following reading were obtained in the time duration of 45s. The below Fig. 73.7A shows the heat flux obtained on gray cast iron with maximum of 1.1961e6 W/m². Figure 73.7B represents the heat transfer ability in Aluminum alloy with maximum being 1.6472e6 W/m². Figure 73.7C shows the heat dissipation rate on Titanium alloy with a maximum of 8.3238e5 W/m². Figure 73.7D gives the heat flux acting on C/SiC, with a maximum of 1.0949e6 W/m². From the above results we state that the Al6061-T6 has the good heat dissipation ability compared with others this means the development of internal thermal stresses will be less this will lower the crack formation in the ventilated disc. But even though due to its soft nature the material undergoes to wear and deformation [26, 27]

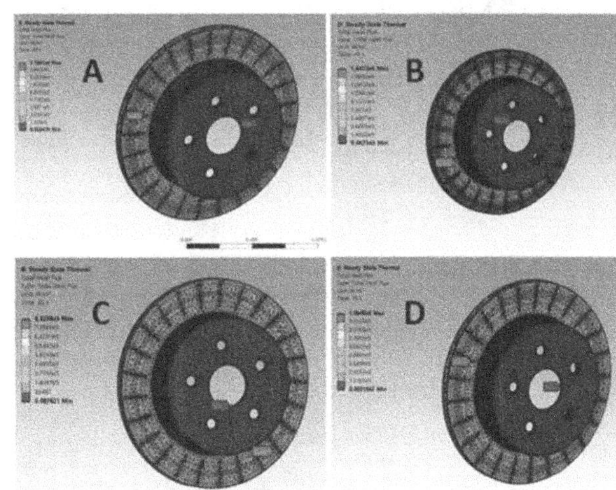

Fig. 73.7 Total heat flux – grey cast iron (A), Al6061-T6 (B), C/SiC (C) and Ti-6Al-4V (D)

3. Conclusion

In this study, we discussed the thermodynamic characteristics of the ventilated disc at a steady state. Using Ansys simulation, the thermal and structural responses of the ventilated disc with four respective materials: Grey cast iron, Al6061- T6, Ti-6Al-4V, C/SiC - maintaining a fixed level of brake efficiency. The results obtained exhibits the mechanical-thermal behavior of the disc brake when subjected to periodical and cyclic loading.

- It has been observed through this numerical simulation that the metal Al6061-T6 has 138% high heat dissipation capacity when compared to the existing cast iron followed by carbon ceramics with 92% and titanium alloy with 70%, when compared with the existing cast iron material.
- Among all four materials studied - Aluminum Alloy, Grey cast iron, C/SiC and Titanium Alloy, it can be said that Grey Cast Iron and C/SiC have less equivalent elastic strain when compared with other two materials.
- It was observed that C/SiC (carbon ceramics) has the least deformation compared with other materials, due to its high hardness and ability of high heat resistance

- The temperature variation analysis detailed that it can be observe that the body of the disc made by material C/Sic exhibited the Lowest Maximum Temperature.
- The heat flux analysis details Al6061-T6 has the good heat dissipation ability compared with others since the development of internal thermal stresses will be less which may lower the crack formation in the ventilated disc.
- Since the performance of the disc depends on the life time useful, lower wear and deformation is an indicator of high life and efficiency.
- The different interactions between the thermo-mechanical phenomena generally correlate to the concept of damage: deformation gives rise to sequential cracking followed by fatigue, fracture or abrasion.

References

1. B. G. Rajan, P. Sambandam, V. Hariram, Vinodkumar. 2023. Influence of Nano Material Coating on the Automotive Brake Liner – An Investigational Approach. International Journal of Vehicle Structures and Systems, 2023, 15(1): 110–115
2. M. Senbagan, S. Sharmila, T. Badirinadh Reddy, K. Arulmozhi, S. Seralathan, V. Hariram, V. Shyam Sundar. 2021. Studies on Enhanced Brake Disc Cooling using Wheel Rim with Axial Ventilators. International Journal of Vehicle Structures and Systems, 2021, 13(5): 590–597
3. M. Tauviqirrahman, M. Muchammad, T. Setiazi, B. Setiyana, J. Jamari. 2023. Analysis of the effect of ventilation hole angle and material variation on thermal behavior for car disc brakes using the finite element method. Results in Engineering, 17: 100844.
4. Belhocine and M. Bouchetara. 2012. Thermal analysis of a solid brake disc. Applied thermal engineering, 32: 59-67.
5. Bianchi, A. Forcellese, M. Simoncini, A. Vita, L. Delledonne, V. Castorani. 2023. Life cycle assessment of carbon ceramic matrix composite brake discs containing reclaimed prepreg scraps. Journal of Cleaner Production, 413: 137537.
6. Y. Zhang, W. Z. Liu, S. Stichel, J. Yang. 2024. Influence of thermal expansion and wear on the temperatures and stresses in railway disc brakes. International Communications in Heat and Mass Transfer, 158: 107858.
7. H. Wang, P. Dong, X. Zhang, Q. Zhao, Y. Fang, G. An, B. Xu. 2024. Structure optimization design approach of friction pairs for low-temperature rise wet brake in hydraulic motor. Case Studies in Thermal Engineering, 61: 104841.
8. R. Jafari, R. Akyüz. 2022. Optimization and thermal analysis of radial ventilated brake disc to enhance the cooling performance. Case Studies in Thermal Engineering, 30: 101731.
9. F. Bonini, A. Rivola, A. Martini. 2024. Novel Unscented Kalman Filter-based method to assess the thermal behavior of carbon brake discs for high-performance motorcycles. International Journal of Thermofluids 21: 100547.
10. F. Formosa. 2024. Analytical model of the steady state temperature distribution in a single and multi-layer brake disc. Progress in Engineering Science 1(2-3): 100010.
11. Maleque, M.A., Dyuti, S., Rahman, M.M. 2010. Material Selection Method in Design of Automotive Brake disc, Proceedings of the World congress on Engineering, 3.
12. Naveen kumar, S. and Kiran, M.B. 2012. Redesign of Disc Brake Assembly with Lighter Material. International Journal of Engineering Research and Technology 1(7): 1-9.
13. Omar Maluf, Mauricio Angeloni, Marcelo Tadeu Milan, Dirceu Spinelli, Waldek Waladimir Bose Filho. 2007. Development of materials for automotive Disc brakes. Minerva 4 (2), 149-158.Sowjanya. 2013. Structural analysis of disk brake rotor. International Journal of computer trends and technology 4 (7): 98-104.
14. N. Fatchurrohman, C.D. Marini, S. Suraya, A. Iqbal. 2016. Investigation of product performance of Al-metal matrix composites brake disc using finite element analysis. IOP Conf. Ser.: Mater. Sci. Eng. 114 012107
15. Choudhary, A. Gujare, S. Dayane, Pankaj Dhatrak. 2023. Evaluation of thermo-mechanical properties of three different materials to improve the strength of disc brake rotor. Materials Today: Proceedings, doi.org/10.1016/j.matpr.2023.02.028
16. Vishvajeet, N., Ahmad, F., Sethi, M., & Tripathi, R. 2021. Thermo-mechanical analysis of disk brake using finite element analysis. Materials Today Proceedings, 47, 4316–4321
17. Dong, C., Deng, J., Fan, S., Kou, S., Yang, S., Huang, R., Zhang, Y., & Mao, Y. 2024. Tribological performance and composition optimization of semi-metallic friction materials applied for carbon ceramic brake disc. Ceramics International, 50(11), 19660–19670
18. Saurabh, A., Verma, P. C., Dhir, A., Sikder, J., Saravanan, P., Tiwari, S. K., & Das, R. 2024. Enhanced Tribological Performance of MoS2 and hBN-based Composite Friction Materials: Design of Tribo-pair for Automotive Brake Pad-disc Systems. Tribology International, 199, 110001.
19. Lyu, Y., Varriale, F., Malmborg, V., Ek, M., Pagels, J., & Wahlström, J. 2024. Tribology and airborne particle emissions from grey cast iron and WC reinforced laser cladded brake discs. Wear, 556–557, 205512.
20. Ali, M. K. A., & Makrahy, M. M. 2023. Tribological performance evaluation of automotive brake discs manufactured from boron-doped titanium dioxide-reinforced aluminum composite. Measurement, 224, 113835.
21. Kumar, D., Singh, S., & Angra, S. 2022. Dry sliding wear and microstructural behavior of stir-cast Al6061-based composite reinforced with cerium oxide and graphene nanoplatelets. Wear, 516–517, 204615.
22. Yang, L., Yang, C., Guo, W., Xu, P., Ma, Y., & Li, P. 2024. Coupled thermal-structural analysis of an axle mounted C/C-SiC brake disc for high-speed trains. Thermal Science and Engineering Progress, 53, 102694.
23. Hariram, R. Suresh, Reddy Jukanti Sandeep, Reddy Allu Brahma, Nithinkumar Avutla Kiran, Nagam Sai, S. Seralathan, T. M. Premkumar. 2021. Thermo-structural analysis of brake disc-pad assembly of an automotive braking system, International Journal of Vehicle Structures and Systems, 2021, 13(4): 497–504.

Note: All the figures and tables in this chapter were made by the authors.

Advances in Additive Manufacturing Technologies – Gurusamy Pathinettampadian et al. (eds)
© 2026 Taylor & Francis Group, London, ISBN 978-1-041-16687-0

Transforming Kidney Stone Diagnostics: A Review of Computational Advances in Detection and Classification

Varshini G.[1],
Zahid Hussain[2], **Karthikeyan**[3],
Yogapriya[4], **Srimahalakshmi B.**[5], **Thirumalai**[6]
Department of CSE-AIML, Chennai Institute of Technology,
Chennai, Tamil Nadu

◆ **Abstract:** Kidney stones occur with a prevalence of over 12 percent among the global population and with a higher rate prevalent in the developed nations. This review highlights computational advancements in kidney stone diagnostics, focusing on four approaches: Support Vector Machine (SVM), Deep Learning (DL), image segmentation techniques, Hybrid models(Radiomics + DL). Both methods show the possibility of increasing the diagnostic accuracy in solving the tasks, the accuracy indicators of DL models reach 98% in some cases. The results show that the proposed hybrid methods enhance diagnostic and treatment planning abilities using radiomic descriptors for further analysis. Some tasks include, dataset heterogeneity, and computation expenses are left unresolved. The findings are hoped to be used to inform AI incorporated approaches to accurate and effective detection of kidney stones.

◆ **Keywords:** Kidney stones, Deep learning, Radiomics, Image segmentation, Artificial intelligence, Medical imaging

1. Introduction

Machine learning's greatest advancement is DL since it can examine medical imaging data to detect kidney stones, eliminating the need to search for uncommon patterns manually. Unlike other approaches that might involve manual feature extraction, the CNN DL models segment the stones, evaluate their content and even forecast the results of the treatments with an accuracy of over 95%. Novelties include new methods to improve a diagnostic solution, introducing DL in conjunction with radiomics, which employs textural and morphological characteristics. However, with proposals such as dataset diversity and the increasing computational costs, DL proves to be a scalable tool in addressing the needs for accurate and minimally invasive diagnostic systems.

This work offers a vast review of new developments in kidney stone diagnosis, which includes deep learning models, support vector machines, image segmentation techniques, and radiomics-integrated machine learning hybrid schemes. Certain methods for the detection of diseases have shown to be up to 95 percent accurate, for instance convolutional neural networks (CNN) in particular have the potential of transforming health care diagnostics. It is therefore important that these models are derived and used to determine accurate spatial and compositional patterns that would warrant the use of non-invasive kidney stone models. Refer to Fig. 74.1 the demonstration of the deep learning model as a supplementary tool in healthcare facilities. Better understanding of stone features and better diagnostic performance as a result of deep learning combined with radiomic analysis (Manjunatha et al. [6]; Chaki and Ucar [8]).

[1]varshinig.aiml2023@citchennai.net, [2]zahidhussainj@citchennai.net, [3]karthikeyanp@citchennai.net, [4]yogapriyam@citchennai.net, [5]srimahalakshmib.aiml2023@citchennai.net, [6]thirumalaimuruganr@citchennai.net

DOI: 10.1201/9781003685906-74

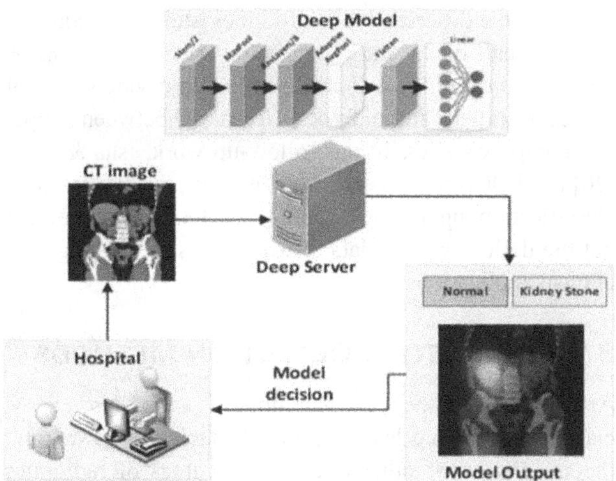

Fig. 74.1 Demonstration of deep learning model (Yildirim et al. [18])

In the current review, the author reviews the application of these computational methods, points out their advantages and disadvantages and discusses what future work is needed. Some of the main issues: the necessity to work with large and diverse data sets, high computational demands related to deep learning models. Areas that are worth exploring in the future studies include; federated learning for preserving privacy when learning, real time diagnostics, and better scalability of the current approaches, which seems to have high potential when it comes to improving diagnostics.

2. Literature Review

Genemo et al. [1] proposed the kidney stone detection system based on five 3D-CNN models with using images of more than 1,000,000 CT scans and yielded 98.5% of accuracy of the stone area but the complete localization of the size of kidney stones and the method using the videos were not implemented. In comparison to the study by Lopez-Tiro et al. [2] the proposed ML algorithm worked in vivo to identify kidney stones with higher accuracy than the conventional techniques and without considering the additional radiology such as CT/MRI or providing general dietary or exercise suggestions. Mahmoodi et al. [3] compared the MRI, CT, and ultrasound applied to diagnose kidney stones between India and the US, and the studies exhibit the variations concerning the accuracy but the study does not encompass the AI enlargement and the reliability for the early identification of kidney stones. Alghamdi and Amoudi [4] successfully attained 95% detection rate and 93% classification rate through multilayer perceptron with ultrasonic imaging but failed to integrate to recurrence prediction real-time AI systems. Compared to the methods of Sassanarakkit et al. [5] who achieved the accuracy of 99.3% using CNNs, Swin transformers and radiomics ignoring blood/urine analysis and future developments in ureteroscopic imaging. In the study of Manjunatha et al. [6], the enhancement of the kidney stone in ultrasonographic images is suggested by SVM and preprocessing. This should help the tool better connect with Electronic Health Record and improve correct diagnosis. They even tackle issues such as noise reduction, enhancement of the images as well as categorization without any operator input. However, the study does not control clinical evidence and app-based usage that subjects require. Future work is to develop the 3D ultrasonography and to cooperate with clinicians for application. Thein et al. [7] they took 500–600 slices of an average of 60 seconds for the proposed study; the preprocessing algorithm for kidney stone segmentation in CT images with a sensitivity of 95.24% was developed by the author by using intensity-based, size-based, and location-based thresholding techniques to eliminate soft organs and bony skeletons along with bed-mat regions. However, the approach relies on prior picture knowledge, cannot maintain sufficient performance in terms of input variance, and does not exclude false positives. In their work [8] Chaki & Ucar propose an ensemble-based approach (FINDWELL) for the recognition of kidney stones which has a fairly good accuracy of 99.8% on the quality image and 96.7% on the noisy image using pre-trained DNNs (DarkNet19, ResNet101, InceptionV3) with IRF feature selection respectively. Asif et al. [9] achieved 90% localization and 95% classification accuracy for kidney stones using CNNs and SVMs but lacked event-based data, dataset comparisons, and therapy prediction models. De Perrot et al. [10] used radiomics with an AdaBoost classifier for 85.1% accuracy and 91.7% sensitivity in distinguishing kidney stones from phleboliths, though it missed automated segmentation and domain adaptation for diverse CT scans. Mannil et al. [11] utilized 3D-TA with RandomForest to predict shock wave lithotripsy success, achieving an AUC of 0.85, but did not address sample size limitations or explore 3D-TA properties in-depth.

Wood & Urban [12] employed Doppler ultrasound twinkle artefacts for KS diagnosis with KS detection rate of 100% and moderate size correlation of (R^2 0.53) however had no human study and stone composition comparison. Goldfarb [13] pointed to a 37% fall in the United States of Kidney stones (1976–1994) noting the rises to the environment but not to diet portion sizes or urine biomarkers or genes. Scales et al. [14] who proposed a similar approach obtained an accuracy of 92.5% on identifying early markers of kidney stones and did not incorporate private computing paradigms such as FL or more sophisticated image analysis techniques such as YOLO. Chew et al. [15] The current survey identifies that by integrating diet, hydration, and genetic predisposition with other inputs, deep learning models are capable of identifying the kind of kidney stones and their locations precisely to 85%. They also use radiology to determine the position of a stone. However, the real-time monitoring for prevention options is absent in the study and the accuracy of the scanning approaches is not enhanced. Improving predictive capabilities and incorporating early identification and control methodologies are the areas in which there is a lack. Mua'ad & Zubi [16] used watershed segmentation, accuracy of 84 percent in the diagnosis of kidney stones but the work failed to investigate the use of hybrid segmentation techniques, machine learning as well as detection of other anomalies such as tumors or cysts. In a study by Akkasaligar et al. [17], Fuzzy C-Mean clustering with level set segmentation was applied to CT images, with 92% segmentation accuracy but the authors did not consider higher order deep learning algorithms, large sample studies, or superior models. Yildirim et al. [18] The automated kidney stone diagnostic model based on the deep learning technique in conjunction with the coronal CT domain and the XResNet-50 model was trained and tested with an accuracy of 96.82%. It applied Grad-CAM for data augmentation and region-of-interest visualization to assist the clinical diagnosis. As a limitation, the study's population does not include data from any other healthcare facilities, and axial was not evaluated, and sagittal view images or dividing stones by size, two factors that could improve both diagnostic findings and study realizability.

Black et al. [19] employed kidney stones classification using ResNet-101 pre-trained on DSLR images and utilizing five different types of kidney stones to acquire a weight recall of 85% and maximum specific of 97.83%, but this study did not include endoscopic video data of real time surgery and did not differentiate between simple and complex stones. In our follow-up work, Isha & Shah [20] used DCNNs and ensemble learning for kidney stone classification up to the accuracy of 98.2% but pointed out the deficiencies in data sharing, transfer learning, and clinical applications.

3. Kidney Stone Detection Methods

Artificial intelligence techniques such as SVM classification, deep learning models such as ResNet, and the integration of otherwise separate areas of radiomics have completely transformed the possibility for advanced kidney stone identification. In using medical imaging data to get better accuracy and help in early diagnosis and understanding of the composition of the stone, these techniques greatly increase the curability rate and improve treatment. Figure 74.2 shows the accuracy achievable by each of the methods.

3.1 Classification Using Support Vector Machines (SVM)

CT and ultrasound are often used for the classification of kidney stones using a machine learning method called the supervised technique and Support Vector Machine (SVM). It also differentiates very small stones well from the surrounding tissue and handles higher-dimensional data

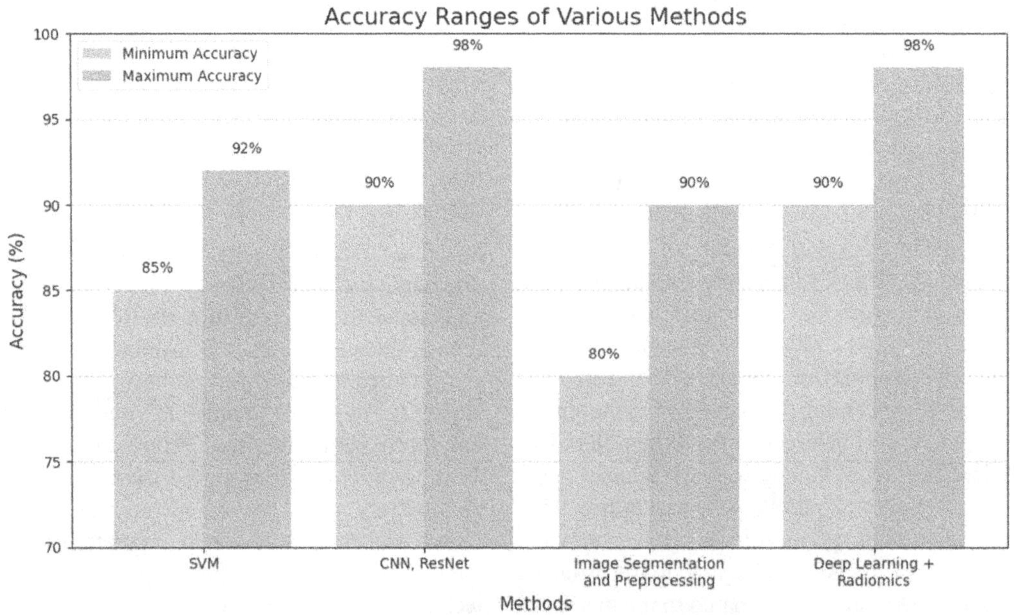

Fig. 74.2 Accuracy ranges of various methods [5]

besides the texture, shape, and intensity of objects on the display as well. The first step of the classification is picture preprocessing: generally, several methods, such as feature extraction, contrast enhancement, and noise reduction, are applied to enhance the quality of the picture and to reveal some major features. Textural and geometrical characteristics are the features formulated and used in this research for SVM classification. When it is necessary to separate information into two categories, for instance, "stone" and "no stone," SVM assists in drawing an ideal hyperplane via features. With regards to the properties that are recovered, it can also distinguish the different forms of the stone if need be. Some of the functions include the Radial Base Function (RBF), which is used in SVM to transform non-linear correlation discovery in data structures. As a result, the SVM is often suitable for medical applications, most of the time obtaining greater than 90 percent accuracy [6][7]. Its precision, ability to scale up, and ease at which the results can be interpreted mean that it will remain relevant in the differential diagnosis of kidney stones for some more time.

3.2 Deep Learning (MoResNet)

MoResNet stands for Multi-Objective or Multi-output ResNet, a deep learning model designed to optimize multiple objectives, handling both classification and regression tasks. Automating feature extraction and categorization from CT and MRI images is done by deep learning models, including CNNs and ResNet architectures that have significantly transformed kidney stone identification. ResNet's residual connections enable deeper networks to learn fine features from image data, improving the recognition of kidney stone size, location, and composition. CNNs, on the other hand, instantly identify spatial and textural patterns ([18], [19]). The process involves obtaining labeled datasets, which are often augmented to widen variety and may involve normalization, for example, scaling to a common input size and normalizing pixel intensities. For better classification, ResNet solves vanishing gradient issues, whereas existing CNN layers learn features. Clinical reliability is evident by the evaluation measures, including sensitivity, specificity, and accuracy, among others, which, without failure, stands at 95% [8][9]. Despite the effectiveness and scalability precision of deep learning, it has limitations such as large dataset needs, high computation costs, and higher resource demands. Methods such as transfer learning effectively solve such problems by training the pre-trained models on comparatively smaller datasets [19]. Despite the shortcomings, CNNs and ResNet contribute to enhanced kidney stone detection, leading to improved diagnostic and treatment outcomes. Much of kidney stone detection is facilitated by large datasets of labeled CT or MRI scans for which the locations of the stones are known. Standardization, scaling, and cropping, as well as resizing procedures, are the key data preparation steps widely applied to enhance the variety and uniformity of arrays across data entries. Fully connected layers are not used to train the model; instead, convolutional layers are employed to capture the textural and spatial features of the image that aid in the classification of the kidney stones. These methods include accuracy, sensitivity, and specificity; CNN models often possess an accuracy of more than 95%. These methods enhance the efficiency of the detection and hence can be integrated into therapeutic processes.

3.3 Image Segmentation and Preprocessing Techniques

For kidney stones, which are located in the renal parenchyma, it is crucial to segment the medical picture. Kidney stone borders can be identified using methods such as binary thresholding, edge detection algorithms, active contour models, and watershed algorithms. Canny and Sobel operators highlight intensity changes between stones and surrounding tissue, while Otsu's method thresholds pixel values to segment the image. For the classification of stones nearby in CT or MRI scans, watershed algorithms are quite effective. Techniques that ensure that input images fed to the segmentation module are of high quality include contrast stretching and removal of image noise (using Gaussian filters). After preprocessing, segmentation methods are applied to extract key characteristics for classification, such as size, shape, and position. From this, it can be said that an integrated system of machine learning and texture analysis contributes to higher detection rates, especially when it comes to such procedures as shock wave lithotripsy [11]. For better characterization and sensorial elemental diagnosis of stones, particularly for stones of different sizes and compositions, twinkling artifacts in ultrasonic imaging have also been studied [12]. Research indicates that these techniques enhance identification precision on CT in addition to M-imagery scans, with classification success rates of above 90% ([16],[17]).

3.4 Hybrid Approaches (Deep Learning + Radiomics)

The fusion of deep learning and radiomics significantly enhances radiologic image analysis, improving the efficiency of kidney stone diagnosis and classification. Radiomics complements deep learning by identifying additional characteristics of kidney stones, including spatial and contextual features. In this study, the combination of the featured approaches was proven to give an efficient and accurate analysis of the kidney stones with gains on the diagnosis as well as on the patients. The process begins with normalization and denoising to standardize image quality. The stones' numerous key properties are characterized based on statistical, textural, and higher-order features from the extracted radiomic data. Deep learning layers efficiently learn spatial hierarchies from raw image data in parallel. Ensemble techniques, or multi-input neural networks, are used to create a comprehensive model with improved prediction accuracy. These are the kinds of combined radiomic and machine learning approaches that have been shown in some papers [5] to enhance the accurate estimates of different treatment performance measures, such as lithotripsy success rate for individual patients. Hybrid techniques are also useful in identifying the material composition of stones. For instance, as described by [20], when radiomics is combined with AI, it can be used in a bid to discover stone type and its likelihood of fragmentation so that individualized treatment can be delivered. Based on the diagnostics precision mentioned, it is clear that all three models have accuracy above 95% to greatly improve the diagnostic precision. However, obtaining diverse, robust datasets remains challenging, limiting model generalization and scalability. However, hybrid techniques offer physicians good methods for improving diagnostic and therapeutic results linked with kidney stones and creating the groundwork for big data-guided individualized feedback for managing stones.

4. Comparative Analysis

The synergy between DL and ML has significantly enhanced the possibility of identification and classification of kidney stones using medical imaging methods. Current DL models like CNNs and ResNet extract features from raw data with high accuracy, while traditional ML methods like SVMs excel when using well-engineered features on structured data. These methods are supported by preprocessing and segmentation techniques that enhance image quality and identify relevant regions, forming the foundation for accurate diagnosis. Additionally, combining DL with radiomics enhances prediction and treatment planning through automated learning and detailed featured analysis. Table 74.1 summarizes the efficiency, advantages, and limitations of these approaches, providing a clear overview of their role in kidney stone identification and classification

5. Conclusion

In conclusion, this review highlights the strengths and limitations of four key approaches for kidney stone detection and classification: SVM, CNN, Segmentation techniques, and creation from hybrid methods. Support Vector Machines (SVM) also show high performance and give accuracy ranging from 90-95% and they perform best on high dimensions but are not scalable. Filters derived from CNNs, especially the ResNet50 model, reach the highest prediction accuracy, with values greater than 98.66%, owing to spatial and textural characteristics of medical images. Overall, accuracies of 92-95% are obtained using the image segmentation techniques such as thresholding and edge detection which however are not successful in providing detail of composition of stones. The combined models of deep learning with radiomics offer a ceiling with accuracies up to 98.56% utilizing other parameters such as texture and morphology for diagnosis improvement. In conclusion, although state-of-the-art CNN has huge superiority in image classification accuracy, the hybrid methods are more beneficial for seeing detailed and global significance at the same time, therefore would be more useful for guiding clinical use and diagnosis.

6. Future Scope

Prospective developmental improvement in the field of kidney stone diagnosis involves the use of detailed data from imaging, gene, and biochemical examinations. Radiomics integrated with deep learning improves feature extraction capabilities; Federated learning serves privacy in the usage of various datasets. Implementation of real-time diagnostic systems integrated into an AI system based on edge computing can completely change clinical work, and the use of explainable AI (XAI) will help develop confidence and ensure that everyone trusts AI-powered systems. Optimized grouping techniques such as GANs, physical accessories for continuous tracking, and other predictive analytic models for risk estimation present better opportunities.

Efficiency-enhancing solutions that solve healthcare problems while operating at costs affordable to developing nations and adaptable, patient-specific processes that allow dynamic monitoring of the healthcare system also point to the applicability of data science in reinventing how healthcare is approached.

References

1. Genemo, M. (2023). Kidney stone detection and classification based on deep learning approach. Int J Adv Nat Sci Eng Res, 7(4), 38–42.
2. Lopez-Tiro, F., Estrade, V., Hubert, J., Flores-Araiza, D., Gonzalez-Mendoza, M., Ochoa-Ruiz, G., & Daul, C. (2024). On the in vivo recognition of kidney stones using machine learning. IEEE Access.

Table 74.1 Comparative analysis of different methods [6–20]

Methods	Description	Accuracy/Performa nce	Pros	Cons
1. Support Vector Machine (SVM)	Using image characteristics, a machine learning approach for classification and regression tasks is employed here to classify kidney stones.	Depending on the dataset, preprocessing, and feature selection, accuracy ranges from 85% to 92%. High sensitivity to specific kinds of stones.	High capacity for generalization despite the scarcity of data. Effective with tiny datasets.	Preprocessing and feature extraction are sensitive. With huge datasets, it is less effective.
2. Deep Learning Models (CNN, ResNet)	Using raw image data (CT, ultrasound), end-to-end learning is accomplished using convolutional neural networks (CNN) or more sophisticated models such as ResNet.	Roughness: 90% to 98%. ResNet-50 is one of the deep learning models that has achieved over 95% in some studies.	To extract complex patterns from raw data, no feature engineering is necessary. It has a high accuracy rate, especially when large datasets are involved.	To extract complex patterns from raw data, no feature engineering is necessary. It has a high accuracy rate, especially when large datasets are involved.
3. Image Segmentation and Preprocessing	Uses preprocessing methods to separate kidney stones from images, such as segmentation (e.g., thresholding, fuzzy clustering, etc.), noise reduction, and edge detection.	80% to 90% accuracy (may vary depending on segmentation accuracy). The quality of segmentation has a significant impact on performance.	Increases the accuracy of diagnosis by assisting in the separation of kidney stones from other tissues. Can be used in conjunction with other techniques.	Poor segmentation might result in low accuracy, hence segmentation quality is crucial, high computing cost for jobs involving segmentation.
4. Hybrid Approaches (Deep Learning + Radiomics)	Enhances prediction and classification accuracy by combining radiomic data (texture, form, etc.) with deep learning models.	90% to 98% accuracy, depending on the hybrid model. Frequently performs better than single-model methods.	Increases accuracy by combining the best aspects of classical picture characteristics and deep learning. Able to integrate many imaging modalities.	More intricate and costly to compute. Requires a lot of data processing and feature extraction.

3. Mahmoodi, F., Andishgar, A., Mahmoudi, E., Monsef, A., Bazmi, S., & Tabrizi, R. (2024). Predicting symptomatic kidney stones using machine learning algorithms: insights from the Fasa adults cohort study (FACS). BMC Research Notes, 17(1), 318.
4. Alghamdi, H., & Amoudi, G. (2024). Using Machine Learning for Non-Invasive Detection of Kidney Stones Based on Laboratory Test Results: A Case Study from a Saudi Arabian Hospital. Diagnostics, 14(13), 1343.
5. Sassanarakkit, S., Hadpech, S., & Thongboonkerd, V. (2023). Theranostic roles of machine learning in clinical management of kidney stone disease. Computational and Structural Biotechnology Journal, 21, 260–266.
6. Manjunatha, D., Vishwakarma, V., Mishra, A., Ravindran, R. E., & Kumari, K. (2024). Kidney Stone Detection Using Ultrasonographic Images by Support Vector Machine Classification. Nanotechnology Perceptions, 93–106.
7. Thein, N., Nugroho, H. A., Adji, T. B., & Hamamoto, K. (2018, November). An image preprocessing method for kidney stone segmentation in CT scan images. In 2018 International Conference on computer engineering, network and intelligent multimedia (CENIM) (pp. 147–150). IEEE.
8. Chaki, J., & Ucar, A. (2024). An efficient and robust approach using inductive transfer-based ensemble deep neural networks for kidney stone detection. IEEE Access.
9. Asif, S., Zheng, X., & Zhu, Y. (2024). An optimized fusion of deep learning models for kidney stone detection from CT images. Journal of King Saud University-Computer and Information Sciences, 36(7), 102130.
10. De Perrot, T., Hofmeister, J., Burgermeister, S., Martin, S. P., Feutry, G., Klein, J., & Montet, X. (2019). Differentiating kidney stones from phleboliths in unenhanced low-dose computed tomography using radiomics and machine learning. European radiology, 29, 4776–4782.
11. Mannil, M., von Spiczak, J., Hermanns, T., Poyet, C., Alkadhi, H., & Fankhauser, C. D. (2018). Three-dimensional texture analysis with machine learning provides incremental predictive information for successful shock wave lithotripsy in patients with kidney stones. The Journal of urology, 200(4), 829–836.
12. Wood, B. G., & Urban, M. W. (2020). Detecting kidney stones using twinkling artifacts: survey of kidney stones with varying composition and size. Ultrasound in medicine & biology, 46(1), 156–166.
13. Goldfarb, D. S. (2003). Increasing prevalence of kidney stones in the United States. Kidney international, 63(5), 1951–1952.
14. Scales Jr, C. D., Smith, A. C., Hanley, J. M., Saigal, C. S., & Urologic Diseases in America Project. (2012). Prevalence of kidney stones in the United States. European urology, 62(1), 160–165.

15. Chew, B. H., Miller, L. E., Eisner, B., Bhattacharyya, S., & Bhojani, N. (2024). Prevalence, incidence, and determinants of kidney stones in a nationally representative sample of US adults. JU Open Plus, 2(1), e00006.
16. Mua'ad, M., & Zubi, M. (2020). Analysis and implementation of kidney stones detection by applying segmentation techniques on computerized tomography scans. Italian Journal and Applied Mathematics, (43-2020), 590–602.
17. Akkasaligar, P. T., Biradar, S., & Kumbar, V. (2017, August). Kidney stone detection in computed tomography images. In 2017 International conference on smart technologies for smart nation (SmartTechCon) (pp. 353–356). IEEE.
18. Yildirim, K., Bozdag, P. G., Talo, M., Yildirim, O., Karabatak, M., & Acharya, U. R. (2021). Deep learning model for automated kidney stone detection using coronal CT images. Computers in biology and medicine, 135, 104569.
19. Black, K. M., Law, H., Aldoukhi, A., Deng, J., & Ghani, K. R. (2020). Deep learning computer vision algorithm for detecting kidney stone composition. BJU international, 125(6), 920–924.
20. Isha, S., & Shah, S. Z. (2023). Use of artificial intelligence for analyzing kidney stone composition: are we there yet?. Mayo Clinic Proceedings: Digital Health, 1(3), 352–356.

Advances in Additive Manufacturing Technologies – Gurusamy Pathinettampadian et al. (eds)
 ISBN 978-1-041-16687-0

Performance Analysis of Surface Grinding with Graphite Lubrication Using the Taguchi Method

M. J. Hepsi Beaula[1], M. D. Vijayakumar[2]
Associate Professor, Department of Mechanical Engineering,
Chennai Institute of Technology,
Chennai, India

P. Manoj Prabakaran[3], Anto Beaula S.[4]
Department of Mechanical Engineering,
Chennai Institute of Technology,
Chennai, India

◆ **Abstract:** Surface grinding is a critical process in manufacturing, where precision and surface finish are paramount. The use of lubricants can significantly influence the quality of the grinding process, enhancing surface finish and reducing tool wear. The impact of graphite as an effective lubricant on grinding surfaces operations is covered in this paper. The work's main objective is to optimize process parameters using the Taguchi method. Some of the key parameters include wheel speed, feed rate, depth of cut, and the application of graphite, which were analyzed in terms of their effect on surface finish, material removal rate (MRR), and grinding temperature. Using a L9 orthogonal array, the Taguchi method was used for reducing experimental trials with thorough statistical analysis. The relative importance of each component was assessed using the method of analysis of variance, known as ANOVA, and the ideal parameter settings were identified using signal-to-noise (S/N) levels. According to the findings, graphite lubrication improves surface quality, lowers energy consumption, and lengthens tool life by lowering friction and heat generation. Graphite is thus established as a viable, eco-friendly, and cost-effective alternative to traditional lubricants in precision grinding. The results underscore the possibility of graphite to improve grinding process efficiency and sustainability. Optimization of process parameters provides useful recommendations and a practical approach to industrial applications, as presented by this research on surface grinding that can achieve improved performance in line with environmental requirements.

◆ **Keywords:** Surface grinding, Graphite lubrication, Taguchi method, Process optimization, ANOVA, Surface roughness, Material removal rate

1. Introduction

Workpiece quality is reduced by intense heat produced during grinding as a result of comparatively strong frictional effects, which causes thermal damage [1–3]. Thus, lubrication and cooling have a significant impact on grinding. Flooding coolants have been the standard solution to this issue. However, there are a number of limitations on using coolants in this sort of way. Because of the wheel's large area of contact with the object being worked and relatively high wheel speed, a stiff boundary layer forms along the wheel perimeter, restricting coolant

[1]hepsibeaulamj@citchennai.net, [2]vijayakumarmd@citchennai.net, [3]pmanojprabakaran11@gmail.com, [4]antobeaulas.mech2022@citchennai.net

DOI: 10.1201/9781003685906-75

supply to the grinding region [4,5]. This removes heat from the grinding zone, resulting in inefficient cooling and lubrication. Another restriction on the withdrawal of stock is the cutting fluid's film boiling [6]. A significant amount of the overall manufacturing cost is also incurred by grinding fluids [7].

To determine how much impact they have on the secondary factors that directly affect the process results or on the direct process outputs, a thorough examination of the process parameters is crucial in any machining setup. Taguchi's experimental design approach has developed into a potent instrument for achieving this goal and maximising process results [28, 29]. This work investigates the use of the Taguchi technique for parameter analysis in a new surface grinding experiment setting using graphite as a lubricating. Analysis has been done on the relative effects of process variables, such as dressing approach, feed, feed is received, and speed, on the forces produced and surface finish performance parameters. The results obtained from the new coolant were compared with the conventional coolant.

The parts that follow give a quick rundown of the various grinding parameters, the methods used to choose them, and the experimental design of the Taguchi. The current study's use of the Taguchi method and its findings are examined.

2. Selection of Grinding Parameters

The type of grinding used determines the many interacting elements that make up the intricate manufacturing process of grinding. Numerous factors affect the geometry created during surface grinding, including the following [31]: The relationship involving grinding conditions and their impact on machining has been the subject of numerous research [14–16]. There are currently three ways for selecting grinding conditions [17,18]: data retrieval methods, artificial intelligence (AI) methods, and process model techniques.

The data retrieval process makes use of a database of cutting circumstances that have been obtained from industries or as advised in the handbook. Despite being a computerised device only a small number of ability database systems cover grinding; most are available for milling, turning and drilling. Empirical and Physical models of the grinding process both provide substantial contributions to our understanding of the procedure. However, because there are so many uncontrollable parameters in the grinding process, physical models are unable to provide a thorough characterization, and simulations have a narrow range of validity. Therefore, in practice, the process representations are unreliable. Artificial neural networks, rule-based reasoning, care-based reasoning, hybrid systems, and other techniques are examples of AI techniques. The ability of each of these approaches to provide a thorough and link between the grinding factors and the grinding behaviour in a particular scenario is limited. Grinding tests are frequently used to identify operating parameters in order to meet the necessary quality criteria in a given scenario. Conventional testing techniques are costly and time-consuming when there are more criteria. This is where the importance of the Taguchi methods for design of experiments lies.

3. Taguchi Methods for Designing Experiments

The performance of the process or product is rendered robust (insensitive) to changes in uncontrolled or noisy factors by Taguchi's method. According to Taguchi, this can be accomplished by appropriately designing parameters during the "parameter design" stage of off-line quality monitoring. In order to assess two or more characteristics simultaneously and independently for their capacity to influence the variability of a certain product or process feature in the fewest possible tests, he created a set of standard orthogonal arrays (OAs). A choice about the best combination of these factors is then made.

4. Experimental Setup

The experimental investigation was used in a 6.5 kW BLOHM hori zonal spindle surface grinding machine. The grinding apparatus with graphite assistance, created and constructed for the (In Fig. 75.1), investigations are

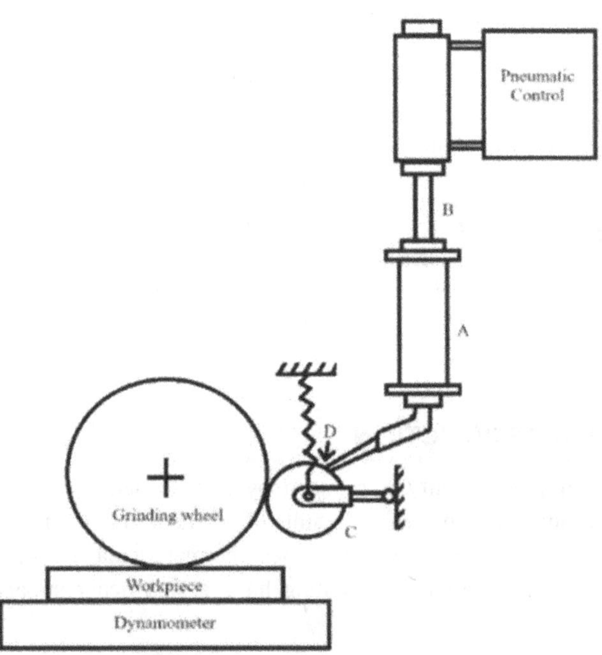

Fig. 75.1 Schematic diagram of the setup

schematically shown. Figure 75.2 shows a representation of the arrangement. A paste consisting of water-soluble oil, fine graphite powder, and a very little quantity of grease was put into Cylinder A. The lubricating paste was forced out through a rectangular nozzle D, measuring 1 mm by 25 mm, by a pneumatically powered piston B. The grinding wheel rotates when C, a little, soft rubber wheel, is put onto it and allowed to run freely. The paste was drawn out of the nozzle and sent to rubber wheel C before being sent to the cylindrical grinding wheel's edge. A range of piston pressures had been used to measure the flow rate.

Fig. 75.2 Experimental setup

A 4:4:1 weight ratio of powdered oil to grease was used for the paste's production, and a 4mm3/s supply rate was used. The setup's capacity to evenly coat the wheel with paste and its stability at typical grinding speeds were the subjects of the first observations. Once the setup's adaptability for appropriate lubrication was confirmed, the planned trials were carried out. Coolant-supported grinding was carried out using solubility oil in water in a 1:20 fraction at a rate of 4 l/min using the machine's integrated coolant supply system. The workpiece was made of steel En-31, which is comparable to AISI 52100, and it had been heated to an entire HRC of 60. Using the full wheel width, the wheel's A60L5V10 (250mm 25mm 76:2 mm size) was utilized for horizontal surface plunge grinding.

5. Using the Taguchi Technique

The following were the different steps involved in applying the Taguchi method in this investigation.

5.1 A Determination of the Goal

The following goals were set for the research:

(i) Determine which level combinations of grinding parameters work best and how they impact the newly created graphite assisted grinding setup's effectiveness.

(ii) Examine the results of a conventional coolant flood grinding in comparison to the previously indicated findings.

5.2 Identifying the Quality Traits and the Metrics Used to Measure Them

The ground piece's surface roughness (Ra), which is the tangential and normal force elements (Ft and Fn), considered to be secondary variables directly affecting the different process outcomes, were chosen as quality indicators for the study in order to meet one of the primary quality requirements. A Talysurf was used to measure the surface finish, which is an apparent consequence. The thrust force on the wheel is the normal force component, and it affects all of the key factors that affect how well the wheel performs. The energy needed to remove the material is directly related to the tangential force component. A quartz dynamometer rates (Kistler-type 9257B) with an intimidating amplifier (Kistler-type 5019B) and a 2-a wideband oscilloscope which was attached directly to a PC were used to measure force. The forces were observed when the method was steady and the pulses were almost constant.

5.3 Choosing the Elements and their Magnitudes

Operating parameters that are typically controllable in any grinding scenario, including, infeed, speed, and dressing method, were chosen for investigation. The three levels with equal spacing were chosen for every single of the parameters within the machine's operating range. The angle of curvature or uncertainty effects could be investigated at three levels as shown in Table 75.1.

5.4 Factor Assignment and OA Selection

The proper degree of freedom, that involves the proportional amount of accountability for the average of every element, is nine for a study that has every single aspect at three levels. In order to handle four elements with three levels, the chosen OA should have at least nine rows, which correspond to nine trials, and four columns.

Table 75.1 Factors and levels selected

Factor	Label	Level-1	Level-2	Level-3
Speed (m/s)	S	22	27	32
Feed (m/min)	F	5	10	15
Infeed (m)	IF	10	20	30
Mode of dressing	DR	Fine[a]	Medium[b]	Coarse[c]

Table 75.2 Standard L9 OA with factors and levels

Trail no.	Factors and their levels			
	S	F	IF	DR
1	1	1	1	1
2	1	2	2	2
3	1	3	3	3
4	2	1	2	3
5	2	2	3	1
6	2	3	1	2
7	3	1	3	2
8	3	2	1	3
9	3	3	2	1

The L9 OA that satisfies this criterion was chosen. It would have taken 81 trials to produce the same results if the conventional experimental protocol had been followed in Table 75.2. With the parameter level assignment, it shows the L9 OA configuration.

5.5 Executing the Experiment

For each trial, four evaluations were carried out in order to get accurate outcomes from the statistical evaluation of the experiment. The measurement instruments indicated above recorded quality metrics such force elements (Fn and Ft in N) and roughness of the surfaces (Ra in mm).

Parts were ground using five sets of each side of the sparkout runs in order to obtain reliable results with Ra. The L9 OA was carried out again using graphite-assisted grinding and flood coolant. The graphite-assisted grinding test information for the standard force is displayed in Table 75.3.

Table 75.3 Sample data for Fn in graphite assisted grinding

Trial no.	Force (N) for repeated tests			
	1	2	3	4
1	92	95	97	92
2	185	182	184	183
3	227	225	226	225
4	101	103	106	108
5	333	337	342	339
6	105	103	103	108
7	162	177	164	171
8	62	65	64	65
9	216	214	212	215

5.6 The Data Analysis

S/N Ratio Evaluation

Taguchi proposes converting a trial's repeat data into a single, consolidated statistic known as the S/N ratio.

In this case, "noise" stands for the unwanted value (standard deviation), and "signal" for the desired value (mean). Therefore, the degree of variance in the quality feature is represented by the S/N ratio. Different kinds of S/N ratios can exist, depending on the quality characteristic's goal. Lower values of roughness of the surface and force components are the expected result in this case. Therefore, the observed data was transformed using the S/N proportion, as shown below:

$$\eta = -10\log[1/n\ {}_{i=1}\Sigma^{n} y_i]$$

where n is the quantity of repetitions in a trial, yi is the calculated quality feature for the ith recurrence and Z is the proportion of S to N for the scenario. The observed data is used to list the S/N ratios for Ra, Fn, and Ft utilizing cold fluid flood grinding along with graphite supported grinding in Table 75.4.

Table 75.4 S/N ratios for different quality characteristics in different grindings

Trial no.	S/N ratios (dB)					
	Graphite assisted grinding			Coolant grinding		
	F_n	F_t	R_a	F_n	F_t	R_a
1	-38.60	-27.55	5.45	-38.53	-28.63	13.93
2	-44.32	-35.63	5.36	-42.47	-37.14	1.92
3	-46.28	-36.77	3.01	-45.69	-38.90	-2.99
4	-41.69	-27.37	4.19	-38.32	-30.26	1.91
5	-49.33	-39.45	5.74	-49.22	-42.51	7.98
6	-39.54	-26.81	5.03	-39.07	-30.13	2.05
7	-43.62	-32.21	8.12	-42.87	-34.97	5.10
8	-37.23	-25.12	3.26	-32.82	-26.34	1.96
9	-46.58	-33.46	5.29	-47.01	-40.83	9.99

Analysis of Level Average Responses

The level feature reaction analyses might be based on the S/N data or the obtained data. The study is completed by adding up the unprocessed or S/N values at every phase of each factor, and the outcomes are displayed graphically. The level average responses from the initial inputs facilitate the examination of the standard of a feature's framework in connection to the variance in the elements being studied. The intensity of typical reaction representations help maximize the total number of iterations in a trial depending on the S/N information in Fig. 75.3. Based on the observed

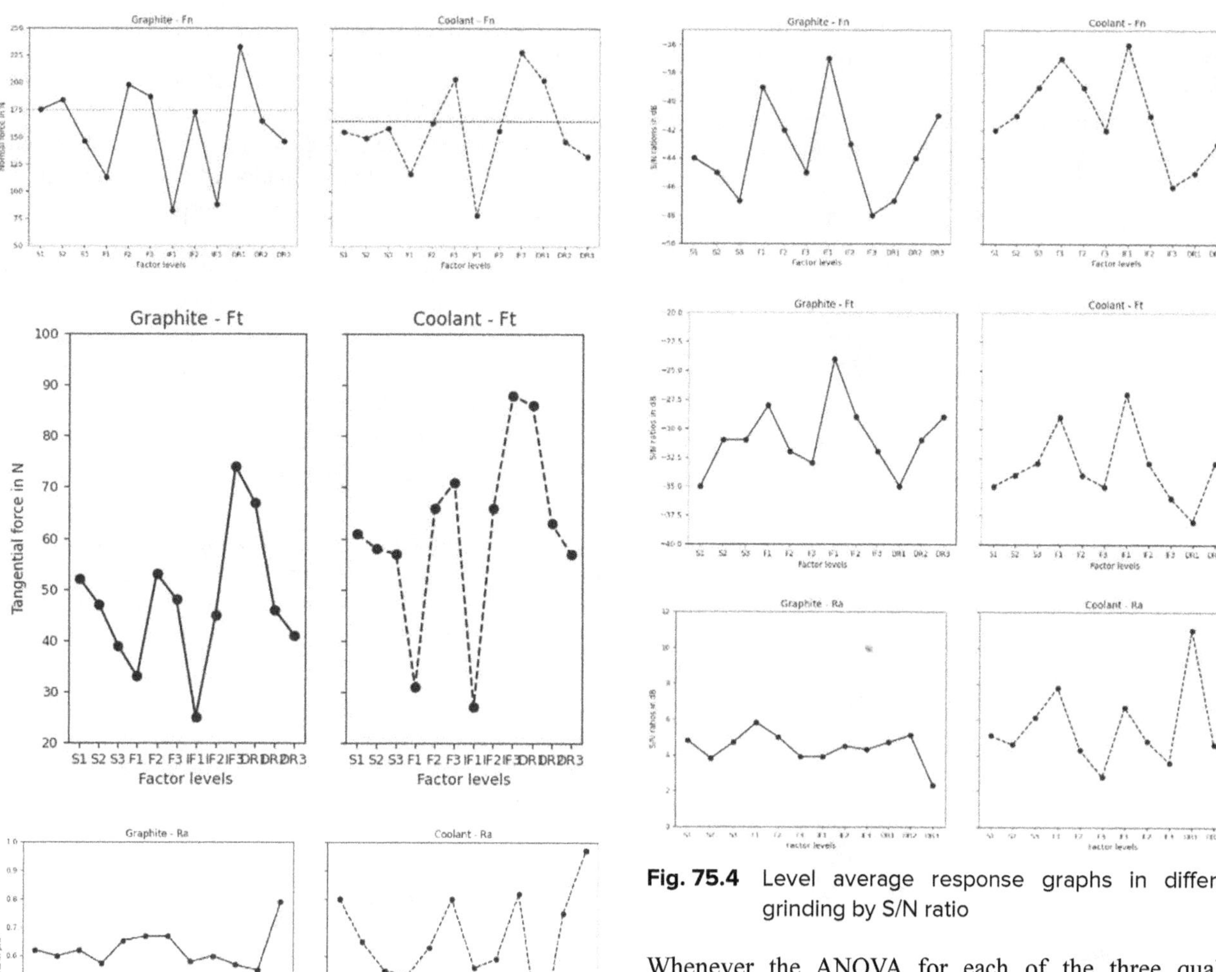

Fig. 75.3 Level average response graphs in different grinding by observed data

Fig. 75.4 Level average response graphs in different grinding by S/N ratio

data, it provides the S/N proportions for cold stream overload grinding along with graphite-facilitated grinding along with Ra, Fn, and Ft in Fig. 75.4.

Variance Analysis

The ANOVA computational approach can be used to quantitatively quantify the corresponding contribution of each regulated factor to the total evaluated response, which is then expressed as a percentage. Thus, it is possible to determine the degree to which each controlled parameter has an impact on the experimental outcomes.

This also helped determine the F test (Fisher evaluate) component, which is used to assess the level of confidence in the results.

Whenever the ANOVA for each of the three quality standards under investigation in graphite-assisted as well as standard coolant grinding was completed, six different ANOVA tables were generated. The example ANOVA results pertaining to graphite-supported grinding with the normal force Fn are displayed in Table 75.5.

Table 75.5 ANOVA table for Fn in graphite assisted grinding

Source	d.f.	SS	V	SS'	F^b	P
S^c	2	4.126	2.012			8.74
F	2	19.003	8.986	15.203	5.221	9.02
IF	2	116.998	59.132	112.668	29.2256+	66.98
DR	2	29.102	14.262	25.114	7.2975	15.26
Total	8	169.229	22.125			100.00

6. Findings and Conversation

The S/N ratios for the different tests in each scenario were computed for both grinding systems based on the goal of the superiority criterion under consideration. Although its physical interpretation is less obvious than that of the primary means of many data, the correlation of average with variance (in dB) seems more focused on goals because it displays simultaneously the mean as well as the dispersion.

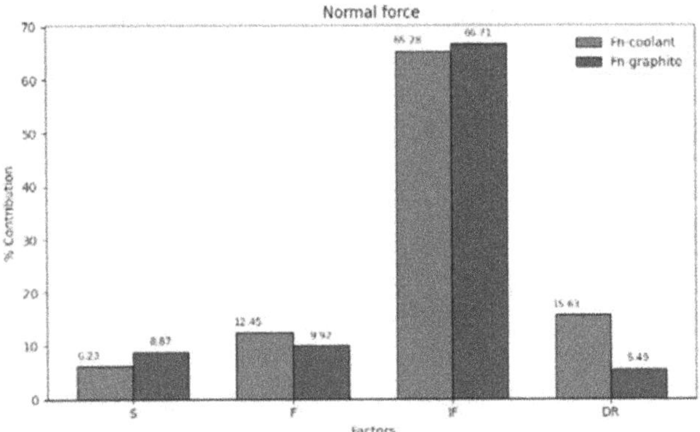

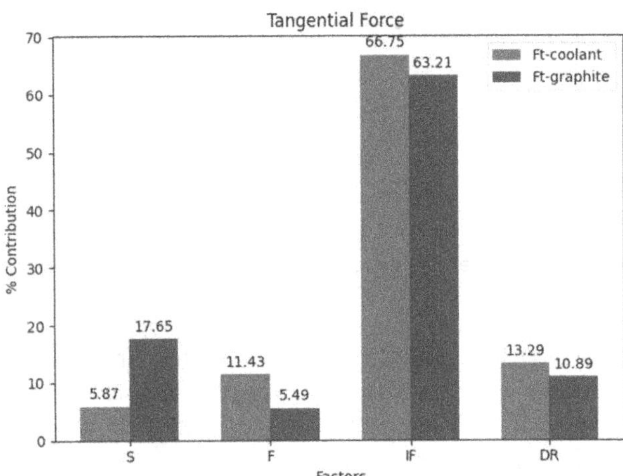

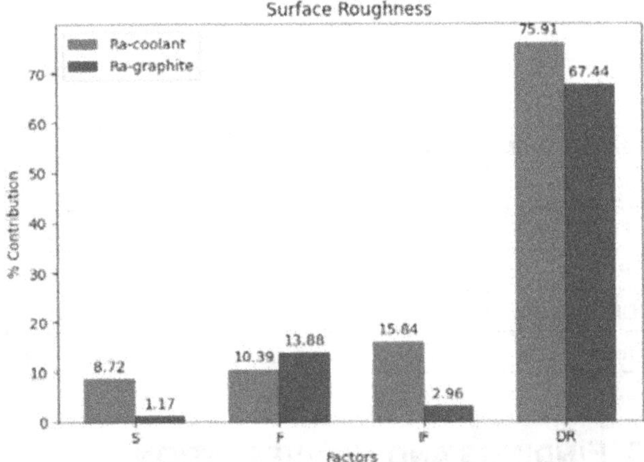

Fig. 75.5 Relative contribution of each factor in different grinding

7. Conclusions

Using a Taguchi approach to experimental design, the novel experimental setup developed for the use of graphite as a way to lower the heat generated in the grinding zone has been assessed for the evaluation of various parameters influencing the performance characteristics. It has been effectively used to determine the best factor level mixtures and the respective effects of a number of variables on roughness of the surface and generated forces, including velocity, supply, feed received, and dressing mode. The outcomes have been contrasted with those of the traditional coolant grinding process using the same methodology. Regarding all quality attributes, the outcomes of traditional coolant grinding are consistent with the different process models.

The results of graphite-facilitated grinding typically match those of conventional coolant grinding in terms of how different parameters impact the quality attributes. There is a discernible difference comparing the two grinding techniques in the trends for surface roughness, even though the average element effects in the instance of graphite supported grinding for force components are likewise nearly identical to those in the case of typical coolant grinding. When graphite applications is compared to ordinary grinding, it has been found that the tangent force and level of the surface decrease while the normal force increases. Based on the current study, a more thorough analysis can be conducted to characterize the effectiveness of the suggested grinding technique.

References

1. J.O. Outwater, M.C. Shaw, Surface temperature in grinding, Trans. ASME 74 (1952) 73–86.
2. S. Malkin, R.B. Anderson, Thermal aspects of grinding. Part 2. Surface temperatures and workpiece burn, Trans. ASME, J. Eng. Ind. 96 (1974) 1184–1191.
3. R. Snoys, K.U. Leuven, M. Maris, J. Peters, Thermally induced damages in grinding, Ann. CIRP 27 (2) (1978) 571–581.
4. S. Ebbrel, N.H. Wooley, Y.D. Tridimas, D.R. Allanson, W.B. Rowe, The effects of cutting fluid application methods on the grinding process, Int. J. Mach. Tools Manuf. 40 (2000) 209–223.
5. E. Brinksmeier, C. Heinzel, M. Wittman, Friction, cooling and lubrication in grinding, Ann. CIRP 48 (2) (1999) 581–597.
6. T.D. Howes, Assessment of cooling and lubricating properties of grinding fluids, Ann. CIRP 39 (1) (1990) 313–316
7. E. Brinksmeier, A. Walter, R. Janssen, P. Diersen, Aspects of cooling lubrication reduction in machining advanced materials, Proc. Inst. Mech. Eng. 213 (B-8) (1999) 769–778.

8. T.D. Howes, H.K. Toenshoff, W. Heur, Environmental aspects of grinding fluids, Ann. CIRP 40 (2) (1991) 623–630.
9. A.B. Chattopadhyay, A. Bose, A.K. Chattopdhyay, Improvements in grinding steels by cryogenic cooling, Prec. Eng. 7 (2) (1985) 93–98.
10. S. Paul, A.B. Chatopadhyay, The effect of cryogenic cooling on grinding forces, Int. J. Mach. Tools Manuf. 36 (1) (1996) 63–72.
11. D. Hafenbradl, S. Malkin, Environmentally conscious minimum quantity lubrication (MQL) for internal cylindrical grinding, Trans. NAMR/SME XXVIII (2000) 149–154.
12. U. Baheti, C. Guo, S. Malkin, Environmentally conscious cooling and lubrication for grinding, in: Proceedings of the International Seminar on Improving Machine Tool Performance, vol. II, San Sebastian, Spain, July 6–8, 1998, pp. 643–654.
13. S. Shaji, V. Radhakrishnan, Investigations on the application of solid lubricants in grinding, Proc. Inst. Mech. Eng. 216-B (2002) 1325 1343.
14. J. Peters, Contribution of CIRP research to industrial problems in grinding, Ann. CIRP 32 (2) (1984) 451–468.
15. H.K. Tonshoff, J. Peters, I. Inasaki, T. Paul, Modeling and simulation of grinding process: keynote paper, Ann. CIRP 41 (2) (1992) 677–688.
16. E. Brinksmeier, H.K. Tonshoff, I. Inasaki, J. Peddinghans, Basic parameters in grinding: technical reports, Ann. CIRP 42 (2) (1993) 795–799.
17. Y. Li, W.B. Rowe, B. Mills, Study and selection of grinding conditions. Part 1. Grinding conditions and selection strategy, Proc. Inst. Mech. Eng. 213 (1999) 119–129.
18. W.B. Rowe, Y. Li, I. Inasaki, S. Malkin, Applications of artificial intelligence in grinding: keynote paper, Ann. CIRP 43 (2) (1994) 521–531.
19. A. Bendell, J. Disney, W.A. Pridmore, Taguchi Methods Applications in World Industry, IFS Publications, UK, 1989.
20. W.H. Tang, Y.S. Tarng, Design optimization of cutting parameters for turning operations based on the Taguchi method, J. Mater. Process. Technol. 84 (1988) 122–129.
21. J. Kang, M. Hadfield, Parameter optimization by Taguchi methods for finishing advanced ceramic balls using a novel eccentric lapping machine, Proc. Inst. Mech. Eng. 215-B (2001) 69–78.
22. R. Komanduri, M. Jiang, Application of Taguchi method for optimization of finishing conditions in magnetic float polishing, Wear 213 (1–2) (1997) 59–71.
23. J.L. Lin, K.S. Wang, B.H. Yan, Y.S. Tarng, Optimization of electrical discharge machining process based on the Taguchi method with fuzzy logics, J. Mater. Process. Technol. 102 (2000) 48–55.
24. Y.L. Su, S.H. Yao, C.S. Wei, C.T. Wu, Analysis and design of a WC milling cutter with Ti CN coating, Wear 215 (1998) 59–66.
25. R. Snoys, J. Peters, The significance of chip thickness in grinding, Ann. CIRP 23 (2) (1974) 227–237.
26. C.P. Bateja, E.J. Pattinson, A.W.J. Chisholm, The influence of dressing on the performance of grinding wheels, Ann. CIRP 21 (1) (1972) 81–82.
27. J. Verkerk, A.J. Pekelhering, The influence of dressing operation on productivity in precision in grinding, Ann. CIRP 28 (2) (1979) 487–495.
28. M.S. Phadke, Quality Engineering Using Robust Design, Prentice Hall, Englewood cliffs, NJ, 1989.
29. P.J. Ross, Taguchi Techniques for Quality Engineering, McGraw Hill, New York, 1996.
30. S. Malkin, Grinding Technology, Theory and Applications of Machining with Abrasives, Ellis Horwood, Chichester, UK, 1989.
31. M.C. Shaw, Principles of Abrasive Processing, Oxford University Press, New York, 1996

Note: All the figures and tables in this chapter were made by the authors.

Advances in Additive Manufacturing Technologies – Gurusamy Pathinettampadian et al. (eds)

AI-Augmented Creative Attribution: Developing Framework for Authenticating and Tracking Digital Creative Processes in Academic and Artistic Contexts

Cheran C.[1], Yogapriya M.[2]
Department of CSE-AIML, Chennai Institute of Technology,
Chennia, Tamil Nadu

Kandavel N.[3]
Department of CSE, Chennai Institute of Technology,
Chennia, Tamil Nadu

Gouri R.[4], Varshini G.[5], Zahid Hussain J.[6]
Department of CSE-AIML, Chennai Institute of Technology
Chennia, Tamil Nadu

◆ **Abstract:** The rise of artificial intelligence and the integration in the processes of creation, revival, and artwork inspire innovative questions regarding attribution and creativity ownership. This paper outlines an elaborate solution that seeks to address the problem of archiving and accrediting AI-assisted creativity in both scholarly and artistic environments responsibly. Analyzing how and when human creativity integrates with AI-based support, we ascertain typical problems of attribution tracking: how to measure the contribution of AI, how to keep track of provenance to avoid forgery, and, finally, how to prevent academic dishonesty. Such elements as blockchain-based digital signatures, metadata registries, and documentation templates are integrated into the proposed framework to ensure that human-AI collaborative creativity is correctly accountable on the blockchain. In a number of cases related to digital art, writing, and multimedia production, we explain how this framework will allow creatives to claim authorship in the digital age while using AI. It is noteworthy that our outcomes point to the fact that concluding that the credit assignment scheme set up by the experimental protocols also provided the means to deepen the understanding of how human and artificial intelligence can work collaboratively in the creative domain. This research enriches the direction of research on creative authentication in the context of AI-enhanced creative production and provides recommendations for institutions, creators, teachers, and learners.

◆ **Keywords:** AI-augmented creative attribution, Human vs AI authorship, AI-augmented creative attribution, Authenticating creativity

1. Introduction

AI and creativity have increasingly intertwined in recent years, and this synergy has significantly impacted both scholarly and artistic activities, leading to how creative productions are created and ownership established. With AI tools getting smarter in generating, editing, and even optimizing creative content for humans, creators and spectators are confronted with a number of key questions about originality, authorship, and ownership that have

[1]cheranc.aiml2023@citchennai.net, [2]yogapriyam@citchennai.net, [3]kandaveln@citchennai.net, [4]gouri.r@citchennai.net, [5]varshinig.aiml2023@citchennai.net, [6]zahidhussainj@citchennai.net

DOI: 10.1201/9781003685906-76

blurred significantly. Technological advancement has brought the capacities of great creative tools and, at the same time, created a challenge of identifying and credentialing creativeness. Current attribution systems, which are made for creation by humans, become an issue in attempting to regulate the diversity and complexity of AI-collaborated creative works and fail to address exact academic dishonesty, effective definitions of authors' copyright, and the existence of creative work authenticity. In order to address these critical gaps, this paper presents the following proposed framework for the effective and efficient authentication and provenance of digital creativity in computer arts as influenced and augmented by AI as a means of carving out the clear and effective protocols for correct attribution of AI enhanced creative works in the blur of art, design and software engineering in which technological systems, human creativity and academia all meet and interact with profound implications in the ways that human creativity and artificial intelligence access to powerful creative capabilities, while simultaneously introducing unprecedented challenges in tracking and attributing creative processes. Current attribution systems, designed for purely human creation, prove inadequate in addressing the multifaceted nature of AI-augmented creative works, leading to significant challenges in maintaining academic integrity, establishing copyright attribution, and verifying creative authenticity. This paper addresses these critical gaps by proposing a comprehensive framework for authenticating and tracking digital creative processes, synthesizing insights from computer science, digital arts, academic ethics, and creative technology to develop a robust system for documenting and verifying AI-augmented creative work, ultimately aiming to establish clear protocols for attribution while fostering an environment that encourages innovation and creative exploration in this evolving landscape where human creativity and artificial intelligence must meaningfully coexist and collaborate. In the elaborated framework, the author utilizes the most advanced technologies, like blockchain verification, neural attribution networks, and biometric authentication systems, to make the documenting process as transparent and verifiable as possible. In addition, the present study explores the potential of creativity enhanced by artificial intelligence in reference to learning, job sector, and academic practice while tackling the essential issue of best practice guidelines appropriate to the constantly growing innovative AI programs and creative processes. In creating a strong framework of how attribution works in the light of using artificial intelligence, this is a very important tool that seeks to endorse copyright to strengthen the creative process and uphold higher learning and artistic creations by preventing the artificial intelligence results from being regarded as pirated when, in essence, they were not altered by an actual human being

2. Literature Review

A new strategy for detecting plagiarism on the parts of modeling assignments through analyzing tokens and the patterns and structures of assignments [1]. The methodology includes five steps: tokenization, normalization, pairwise matching, similarity calculation, and visualization to address the problems of higher abstraction levels and re-ordering obfuscation. Therefore, this strategy revolves around improving the credibility of digital creative works through precisely detecting AI-camouflaged plagiarism. However, this paper [2] Specifically, it proposes a new defense mechanism, token sequence normalization, to improve resistance to automatic obfuscation attacks against plagiarism detection strategies. It uses program dependence graphs (PDGs) and token-based approaches in graph-based ones and normalizes token sequences and reverts insertions and reorderings in a language-independent manner. This mechanism has been developed for the JPlag open-source detector and, as such, is scalable to other detectors like MOSS and Dolos to answer research questions regarding its effectiveness and influence on detection performances. [3] From the comparison of human to AI-generated art, the following insights emerge. It reveals that AI-generated artworks are closer to Wölfflin's principles for the Baroque style and modern art concepts compared to human & AI art, both the landscape and the geometry abstract; however, the deformed figure from AI is odd to the human work. Notably, around 70% of a piece of art created by an AI looks similar to art exhibited in modern art from 1850 to 2000. Further, AI art is more likable when real-looking, and human art draws positive and negative intertwined feelings in viewers.

Faizan Farooq Khan and his team [4] performed tests on a number of neural network models, training different variants of either StyleGAN, VQ-GAN, or DDPM over the WikiArt dataset. Analysis is in light of Wölfflin's principles as compared to and across time in emotional responses, amounting to the construction of six models of type GAN and combining objectives from StyleGAN and Creative Adversarial Networks. Amongst them, 400 images are drawn by each of the eight models on criteria such as mean nearest neighbor distance and shape entropy, whereas 6,000 human artworks span the WikiArt database through the 11th to 20th centuries, with the exception of Japanese Ukiyo-e, focusing only on Western art

The works of art were qualitatively assessed as a research method to posture ideas and enforce the fact that plagiarism is an empirical subject. To examine the various shades of difference between offences of plagiarism and cases of inspirational borrowings, only historical and, particularly, modern examples are used and chosen deliberately. Watch the below images carefully. [5] Image detection of generated images has been in force in media forensics for long, where traditional analysis tools detect synthetic content based on signals like resampling artifacts and JPEG quantization. In recent years, with the development of deep generative techniques, attention has been paid to deep discriminative methods, which attract special attention from scholars, especially in the recognition of GAN-based content. An experimental study has been made to analyze these images, and the results show that simple classifiers trained on one GAN model can generalize to others, especially when the augmentations are made aggressive. Other detection methods include frequency analysis, co-vector matrices, and using pre-trained CLIP features. However, a usual issue arises where, apart from the precision, the accuracy can be very low because of bad calibration. As a remedy to this, methods such as nearest neighbors classifiers have been used in an attempt to enhance the chances of accuracy, though at the expense of an increased inference time. This work proposes to look into and describe these classifiers further for their performance in an online environment to improve the ability to detect AI-generated images.

3. Framework

Analysis is performed at the affective level and involves the determination of emotion signatures extracted from art; however, previous approaches depend strongly on low-level features such as color, texture, and shape. These features are not invariant to the different arrangement of these emotions, which very slightly connects these features with emotion. It must be noted that the plan of the elements is then arranged in a specific way to suggest specific meanings and feelings. Balance, harmony, and movement are the mutex elements in this process, as different combinations can have different impacts; in the example of images, it can be noticed that symmetric images give positive feelings whereas the strong contrasts give negative ones. In order to overcome these limitations, we need to develop principles of art emotion features, a systematically designed new approach to principles of art. This method measures the spatial configurations of the artistic components to build features of the emotion space that categorize and rate the human emotions elicited by images. Based on PAEF (Process Authentication and Evaluation Framework), we study emotions in famous artworks to identify the feelings of artists who created them. [6]

3.1 The impact of Artistic Principles

These datasets include IAPS (Intelligent Attribution Processing System), ArtPhoto, and Abstract, and the latter are used to demonstrate that PAEF (Process Authentication and Evaluation Framework) is indeed a very effective model for image emotion recognition, as proven by the experimental results. These studies indicate that PAEF yields higher performance compared to existing approaches, with about 5% additional average accuracy in emotion classification. It also gives better predictions of arousal than those of valence. The model proves to be most efficacious for abstract and artwork imagery where emotions are communicated through positioning or placement. One of the good examples provided in the paper is an example of using PAEF to predict the mood of Vincent van Gogh that was behind his paintings, thus pointing to the capability of the model in recognizing artists' emotions. In conclusion, it can be stated that the primary and most important finding is connected with the efficiency of PAEF in the sphere of emotion recognition in images with specific references to the complicated or semi-artistic pictures.

3.2 Experimental Setup for Identifying AI-Generated Art

The experiment detailed in this work was carried out using Google Colab on an Intel Xeon processor, supported with 12 GB RAM and an NVIDIA T4 GPU. Approaches were coded in PyTorch, under which the dataset was divided into training: 80%, validation: 10%, and testing: 10% to maintain class balance. Center-cropped images were resized from a dimension of 224 x 224 pixels to allow standardization with ImageNet pixels mean and standard deviation. As to optimization, the Adam optimizer was applied with the learning rate equaling 10 to the power of 3 and a BC of 32, and they also include techniques such as early stopping to avoid overfitting. To assess the model's efficiency, authentic and painted art pieces were authenticated by accuracy, precision, recall, and F1-score.

The experimental setup that was employed in teaching deep learning algorithms is distinct between real and synthetic artwork. They deployed the models employing PyTorch, which is one of the more frequent frameworks for computer vision research. Use of an 80:10:10 split was also used on the dataset to get 80% training data, 10% validation data, and 10% test data. During the split, the data was split stratified with respect to the target class, which means that it ensured that all target class data was distributed properly, with the real target class and the synthetic target class data being split properly. All images were preprocessed by the researchers to crop the images to their center and resize them to 224 x 224 pixels. Also, they adapted the brightness and contrast of the images using the mean and standard deviation calculated from the ImageNet database in order to stay unified when pre-processing realism and synthetic art. For the optimization stage, the Adam optimizer was chosen with a learning rate of 0.001 and a batch size of 32. To avoid overfitting [8], we used an early stopping callback with a patience of 3 and a learning rate schedule based on the reduced learning rate on plateau meaning. Thus, the accuracy of the models was evaluated based on accuracy, precision, recall, and F1-score. It amply measures the models' performance vis-à-vis the classification of genuine and fake artwork as well as real and synthetic art.

4. Woffman's Principal

The mean and standard scores and scales the value between 0 and 1, where the score closer to 0 is linear and closer to 1 is painterly. At first, they used Wölfflin's parameters, which he described to distinguish between the Renaissance and the Baroque art, to compare the results featured by distinct styles, human and AI-generated paintings. These principles are linearity/painterly—works with clear outlines are linear, while those having blurred outlines are painterly. planar/recession—objects are parallel in planar pieces, while they create depth in recessional ones; closed/open—closed form has figures balanced in the frame and those in the open forms going beyond it; Multiplicity/Unity—multiplicity has individual entities, while the unity is unifying entities into one. Absolute Cl In order to capture data on these principles, participants were instructed to read through descriptions one at a time, in a survey-like interface, and identify how many of each principle were present. The explicit ratings consisted of a 1 to 5 rating of each artwork according to how linear or [9]

4.1 Data Collection and Participants

This study employed mixed methods that included 249 students from five Chinese universities and 96 students from three high schools. It is composed of a quantitative self-administered long-form questionnaire carried out between January and March and qualitative data last collected in April and May 2023. It examines the attitude and opinion of Chinese students on AI painting: In that, basic knowledge about AI painting and the use of the tool must have been acquired before conducting the assessment; it was eventually difficult to stabilize and finally stop before having enough participants; hence teachers stepped in to help recruit the right participants and spread the survey among classes and the students' associations. It also used "Wenjuanxing," one of the online surveying platforms in China, for circulation of the questionnaire and got questionnaires to ensure participation. The formatted questionnaire response generated 1,298 valid responses out of 1,439 respondents.

4.2 Time and Emotion in Analysis of AI Art

According to the [11] It is shown that deep CNN style classifiers are able to retain smooth temporal properties of artwork over the hidden layers, even if temporal information is not included in the training process. As can be seen in the hidden representation below, the year of art creation has a direct inclination toward the scale. To approximate the time period of AI art, spatial location in the multi-dimensional similarity space is computed for each AI artwork with reference to human art of a year closest to the time AI art is created. Furthermore, our study presents likability and emotional appeals taking place in art appreciation experiments to see if AI art is liked or disliked, what primary features were attracting attention in the most popular artworks, and the emotions elicited by generative and human artworks. Human art was obtained from the L.A. County Museum of Art, and its emotional metadata were obtained from the ArtEmis dataset. Mechanical Turk gamers were given the task of giving out annoying rates and likability ratings of all the 3200 digital artworks. In a similar vein to earlier work showing how deep CNN classifiers can encapsulate visual transition across art movements, here we trained a number of different neural networks to classify the styles and, from the hidden layers, analyze principles and temporal aspects suggested by Wölfflin [12]. To compare our models based on PCA, we considered only the principal components capturing 95% of variance in the last hidden layer after ReLU activation and noted that ResNet50 had the highest style accuracy and maximum correlation with time.

4.3 Generic detection process

Plagiarism Detection Systems (PDS) are specialized computer systems designed to identify instances of plagiarism using one of two primary detection approaches: external or intrinsic. External PDS compare a suspicious document with a set of genuine documents, while intrinsic PDS analyze the text based on its features using a technique known as stylometric without comparing with other documents Intrusive PDS point to features of the change in writing style as possible signs of plagiarism. As depicted in Fig. 76.1 most external PDS apply a three-step retrieval system, each of which is briefly discussed below. In the first stage, these systems use fast approximate algorithms to find a relatively small set of documents in the reference collection which might contain the source of the potentially forged text

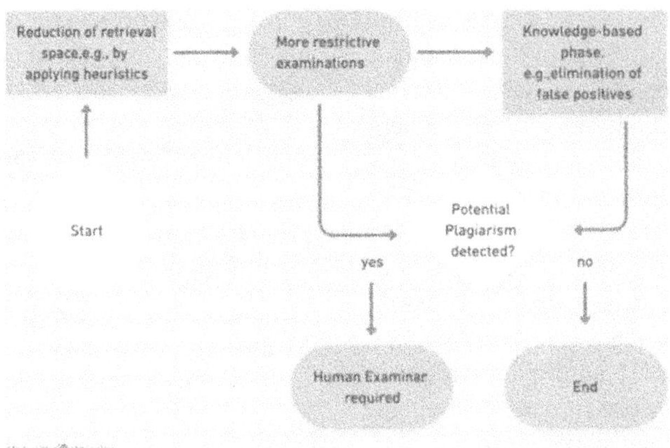

Fig. 76.1 Generic plagiarism detection process [5]

5. Comparative Analysis

Table 76.1 Comparative analysis [6–15]

Description	Advantage	Disadvantage	Result
This conference paper describes a method in the training of a set of artwork, human-made and artificially generated from AI using harps, through VGG-19, ResNet-50, and Vision Transformer (ViT) models. [19]	The success of the three deep learning models is then compared and discussed namely VGG19, Resnet50 and Vision Transformer. This gives an analysis of their skills in recognizing art generated by AI.	The study focuses on Grad-CAM for visual explanations but does not consider other important semantics, such as brushstrokes and colors. This could have better explained how models distinguish between real and fake pieces of art from analysis of these components	The current paper finds that deep learning models can accurately detect the originality of a piece of art and even differentiate between real art and art generated by AI at more than 95 % accuracy making it perfect for art authentication.
"AI Art Neural Constellation Supplement" is a follow-up piece that quantifies and qualifies AI art in an attempt to identify what the strengths/strengths, weaknesses and/ or biases are in comparison to human artwork.[20]	The researchers use a combined quantitative and qualitative approach to gain a more comprehensive understanding of AI-generated art compared to human art	The objectives as well as the nature and scope of the study, the findings and recommendations, the impact of the study and or the theoretical framework within which the study was conducted are missing in the supplemental materials.	They give information about such features of AI-generated art as bias, emotional appeals, and key model areas, but to get a full vision, one needs the main paper.
The study compares AI and human art, finding higher accuracy with pre-trained neural networks and correlations between neural features and creation year.[21]	The paper introduces FGPD-FA, a method that detects AI-generated art through frequency analysis, filling a gap in current forgery detection.	The paper's weakness is that it only examines paintings and three AI techniques, and additional study is needed to expand beyond this scope.	By FTP using the FGPD-FA method, work forged by AI is revealed with an accuracy of more than 90% and surpassing traditional forgeries. Basically, it is centered on particular image features and on frequency analysis.
The study between AI and art work remarks higher accuracy by pre-training models and correlation by creation year of the art work. It also categorizes feelings and art concepts and all the other requirements that are usually involved in the assessing of the art.[23]	The paper shows that techniques like FD analysis capture discrepancies in AI art that expose their bias and shortcomings of algorithms used in AI art. It helps in two aspects: forgery detection and AI-art comprehension	The possible negative effect with AI art is that it affects originality, and the uniqueness of styles will decrease as AI produces copies of the same style. Excessive use of AI can slow down the creative process and could not let through the best, most unique ideas.	Midjourney and Stable Diffusion apps help artists to create more content but it actually can have the reverse effect and become boring and more vanilla. AI is expected to be integrated with creative and critical thinking skills in a style known as generative synesthesia in order to be successful

The comparison of each of the mentioned models demonstrates the advantages and disadvantages of using AI in art. Just based on the paper, we can see that the future models like VGG-19, ResNet-50, and Vision Transformer (ViT) enhance the capacity to tell AI artwork from human artwork. Also, pre-training models and the correlation between the neural characteristics of the artwork and the year it was created allow us to learn more about art trends and how AI can assist in the creative process [22]. The study also considers the affective outcomes and creative concepts elicited by AI in order to join the trend of increased AI simulation of creativity. In summary, the work points towards a turbulent time when anyone interested in exploring AI in art must learn from both its positives and the negativity of this technology's capabilities.

To optimize the model, we use a binary cross-entropy loss function:

$$\mathcal{L} = \frac{1}{N}\sum_{i=1}^{N} \ell(yi, \hat{y}i)$$

where N is the number of instances in the training set, yi is the true label, $i\hat{y}$ is the model outcome and $\ell(yi, \hat{y}i) = -(yi \log(\hat{y}i) + (1 - yi)\log(1 - \hat{y}i))$ is the classic binary cross-entropy loss

6. Conclusion

AI art manifests this incomparable potency regarding these arts, technology in arts, and GT in practice. It thus brings the past values and concerns into the new metadata systems that such digital products themselves provide. Projects sensitize artists to the various strengths, weaknesses, and other social factors here. Repositioning AI lit from being an opponent to a tool has manifold openings for art generated by machines and very much keeps intact the essence of what art is: useful, beautiful, and community. To sum up, AI will improve the art-generating processes given the right legal focus and legal cases, albeit not necessarily excluding the human element. One could infer that the proposed framework is one of the reasonable solutions preserving both future innovative developments upon which future development depends and art in general.

7. Future Scope

Attributing creativity by AI will combine other new features such as Creative DNA Fingerprinting, Dynamic Attribution Visualization, and Smart Contract Attribution Protocol (SCAP) for managing rights and royalties. These features offer biometric creative authentication and collaborative attribution pods (CAPs) to facilitate accurate identification and coordination. A Creative Process Recovery System can track the steps in the creation process while an Attribution Risk Assessment Engine identifies possible conflict issues and recommends their solutions. This group of components together enhances the tracking, verification, and collaboration for attribution, hence reducing conflicts in creative processes driven by artificial intelligence.

References

1. Sierra-Martínez, S., Martínez-Figueira, M. E., Castro Pais, M. D., & Pessoa, T. (2024). 'You work, I copy'. Images, narratives and metaphors around academic plagiarism through Fotovoz. British Educational Research Journal
2. Wang, C. (2024). Art innovation or plagiarism? Chinese Students' attitudes towards AI painting technology and influencing factors. IEEE Access.
3. Khan, F. F., Kim, D., Jha, D., Mohamed, Y., Chang, H. H., Elgammal, A., ... & Elhoseiny, M. (2024). AI Art Neural Constellation: Revealing the Collective and Contrastive State of AI-Generated and Human Art. In Proceedings of the IEEE/CVF Conference on Computer Vision and Pattern Recognition
4. Khan, F. F., Kim, D., Jha, D., Mohamed, Y., Chang, H. H., Elgammal, A., ... & Elhoseiny, M. (2024). AI Art Neural Constellation: Revealing the Collective and Contrastive State of AI-Generated and Human Art. In Proceedings of the IEEE/CVF Conference on Computer Vision and Pattern Recognition (pp. 7470–7478).
5. Taharuddin, N. S., Ibrahim, N., Barlian, Y. A., Zahari, N. F., & Noordin, N. A. M. (2023). Navigating the Fine Line between Plagiarism and Artistic Inspiration in Dawid Enoch's artworks. KUPAS SENI, 11(3), 101–105.]
6. Zhao, S., Gao, Y., Jiang, X., Yao, H., Chua, T. S., & Sun, X. (2014, November). Exploring principles-of-art features for image emotion recognition. In Proceedings of the 22nd ACM international conference on Multimedia (pp. 47–56).
7. Bianco, T., Castellano, G., Scaringi, R., & Vessio, G. (2023). Identifying AI-Generated Art with Deep Learning. In CREAI@ AI* IA (pp. 16–25).
8. Sankar, P. (2020). MEASURING FACULTY PERCEPTION OF PLAGIARISM: A STUDY AMONG ARTS AND SCIENCE COLLEGES IN SOUTH INDIA. Library Philosophy & Practice.
9. Agrawal, M., & Sharma, D. K. (2016, October). A state of art on source code plagiarism detection. In 2016 2nd International Conference on Next Generation Computing Technologies (NGCT) (pp. 236–241). IEEE.
10. Meuschke, N., & Gipp, B. (2013). State-of-the-art in detecting academic plagiarism. International Journal for Educational Integrity, 9(1).
11. Sağlam, T., Brödel, M., Schmid, L., & Hahner, S. (2024, April). Detecting Automatic Software Plagiarism via Token Sequence Normalization. In Proceedings of the IEEE/ACM 46th International Conference on Software Engineering (pp. 1–13).

12. Bai, Y., Guo, Y., Wei, J., Lu, L., Wang, R., & Wang, Y. (2020, October). Fake generated painting detection via frequency analysis. In 2020 IEEE international conference on image processing (ICIP) (pp. 1256–1260). IEEE.
13. Santos, I., Castro, L., Rodriguez-Fernandez, N., Torrente-Patino, A., & Carballal, A. (2021). Artificial neural networks and deep learning in the visual arts: A review. Neural Computing and Applications, 33, 121–157.
14. Epstein, D. C., Jain, I., Wang, O., & Zhang, R. (2023). Online detection of ai-generated images. In Proceedings of the IEEE/CVF International Conference on Computer Vision (pp. 382–392).
15. Winstanley, L., & Hodgkinson, G. (2023). Visual plagiarism and a new framework to address localised opinions and perceptions in applied arts education. International Journal of Education & the Arts, 24(18).
16. Foltýnek, T., Meuschke, N., & Gipp, B. (2019). Academic plagiarism detection: a systematic literature review. ACM Computing Surveys (CSUR), 52(6), 1–42.
17. Nguyen, S. (2024). The Copyright and Plagiarism Dilemma: AI-Generated Design works Derived from Existing Works.
18. Sağlam, T., Hahner, S., Schmid, L., & Burger, E. (2024, April). Automated Detection of AI-Obfuscated Plagiarism in Modeling Assignments. In Proceedings of the 46th International Conference on Software Engineering: Software Engineering Education and Training (pp. 297–308).
19. Bianco, T., Castellano, G., Scaringi, R., & Vessio, G. (2023). Identifying AI-Generated Art with Deep Learning. In CREAI@ AI* IA (pp. 16–25).
20. Rumanovská, Ľ., Lazíková, J., Takáč, I., & Stoličná, Z. (2024). Plagiarism in the Academic Environment. Societies, 14(7), 128.
21. Khan, F. F., Kim, D., Jha, D., Mohamed, Y., Chang, H. H., Elgammal, A., ... & Elhoseiny, M. Supplementary: AI Art Neural Constellation: Revealing the Collective and Contrastive State of AI-Generated and Human Art.
22. Carrier, M. (2024). New Viewpoints Elevating Insights with Digital Humanities. Scaffold: Journal of the Institute of Comparative Studies in Literature, Arts and Culture, 1(2), 21–25
23. Hidayatullah, R., & Wendhaningsih, S. (2021). Online art class: a study on the cause and effect of plagiarism. Aksara: Jurnal Bahasa dan Sastra, 22(1), 116–127.
24. Zhou, E., & Lee, D. (2024). Generative artificial intelligence, human creativity, and art. PNAS nexus, 3(3), pgae05

Advances in Additive Manufacturing Technologies – Gurusamy Pathinettampadian et al. (eds)

Design and Fabrication of Power Generation Using Speed Breakers

Manjunathan R.*, Ankitvarma P., Hari Krishna Raj S., Sashank S., Ailan R.

Department of Mechanical Engineering,
VelTech HighTech Dr. Rangarajan Dr. Sakunthala Engineering College,
Chennai, Tamilnadu

◆ **Abstract:** In the last few years, energy has become a basic requirement of human life and a pillar of economic growth for all countries across the globe. As the world's population is growing at a very fast rate and conventional energy resources are being depleted at an accelerating pace, it is now necessary to divert our attention towards sustainable and non-conventional sources of energy. One such new-age solution is the harnessing of kinetic energy created by moving vehicles, especially those crossing speed bumps on highways.

Speed bumps are the usual fixture in city streets to slow down vehicles and increase road safety. Numerous vehicles travel daily over these speed bumps, developing a great deal of kinetic energy that normally does not go into use. Electro-mechanical speed bumps could be used in place of standard speed bumps so that this waste energy could be tapped. Such modern units have been developed with a view to tapping the kinetic energy imposed on these units by vehicles in motion by converting them to electrical energy in the form of mechanical-to-electrical converters. A series of many such units on busy routes enhances the system significantly in terms of efficiency and delivers larger volumes of electricity that could be deposited as energy storage means like batteries or supercapacitors. The stored energy can then be utilized to serve the energy requirements of different applications, such as illuminating streetlights, traffic lights, and electronic commercial signs in the area. Additionally, this energy-harvesting system has some advantages. Through the reduction of fossil fuel use for electricity generation, it facilitates cleaner energy usage and reduces greenhouse gas emissions. This method supports the worldwide movement to fight climate change and move toward sustainable energy systems. Moreover, using locally produced energy can improve energy security, decrease energy transportation costs, and have a positive impact on a country's economy by lessening the reliance on imported fuels. The incorporation of such systems into urban infrastructure is a progressive approach that integrates innovation, sustainability, and utility. It not only caters to increasing energy needs of contemporary societies but also turns normal traffic interaction into sources of clean energy generation. Through this, it encourages economic development, conservation of the environment, and a step towards a greener and more sustainable future.

◆ **Keywords:** Speed breaker, Bumps, Electro mechanical

1. Introduction

1.1 Generating Electricity from Speed Bumps: A Sustainable Energy Solution

In the current age, increased demand for power and dwindling stocks of conventional power sources have motivated researchers to venture into innovative and renewable sources of power. A very pioneering innovation in this field is using speed bumps as electricity-generating machines. Harnessing the kinetic energy and potential energy generated from vehicular motion, the technology could bring about a new revolution in power efficiency

*Corresponding author: manjunathanmech@gmail.com

DOI: 10.1201/9781003685906-77

on urban and rural roads. The Concept and Working Mechanism Speed breakers are mainly constructed to slow down vehicles and make roads safer through vertical deflection. They also offer a possibility of energy harvesting from the moving weight and speed of vehicles passing over them. The main concept is to harvest the mechanical energy produced when automobiles move over speed breakers into electrical energy. This conversion of energy is based on a compilation of mechanical devices and electromechanical equipment. Upon a car's passage over a specially constructed speed hump:

1. Potential Energy Conversion: While the car travels up the bump, it goes higher, thus gaining potential energy.
2. Electrical to Mechanical Conversion: When the speed hump is compressed, mechanisms such as rack-and-pinion devices, crankshafts, or rollers extract the energy and translate it into rotational motion.
3. tPower Generation: The rotational force powers a generator, generating electricity that may be stored in batteries or supplied directly for different uses.

1.2 Mechanism Types of Power Generation

1. Rack-and-Pinion Mechanism: In the system, the pressure force from a car compressing the speed bump pushes a rack (flat-toothed component).

   The rack moves and is meshed with a pinion (gear), which turns and provides motion to a generator through a set of gears.

   The rotational speed is magnified to satisfy the generator's needs for effective electricity generation.
2. Crankshaft Mechanism: Rather than a rack-and-pinion arrangement, a crankshaft translates the linear compression of the speed bump into rotational motion.

   This motion is transferred through gears to a generator, generating electricity.
3. Roller Mechanism: In this design, the speed bump features a roller that turns when vehicles drive over it. The motion of the roller powers a chain-sprocket system, which drives the generator.

Each of these systems is aimed at the maximum capture of energy while the original purpose of the speed bump as a traffic-calming mechanism is retained.

1. Affordable Solution: Speed bumps with energy-harvesting systems offer an inexpensive alternative to traditional electricity supplies. They render unnecessary long power distribution networks, particularly in isolated locations.
2. Green Energy Generation: By harnessing the continuous flow of traffic, such systems offer a renewable and clean source of energy, minimizing fossil fuel dependence.
3. Public Utility Applications: The electricity produced can be utilized for: No Supplying streetlights and traffic signals.

   No Lighting digital billboards and signs. No Recharging electric vehicles at toll booths or parking garages.
4. Less Environmental Impact: In contrast to solar or wind power, speed bump energy generation is independent of weather, making it constant. It also generates no greenhouse gases.
5. Scalability: As the number of vehicle registrations continues to grow steadily around the globe, energy output from these systems will only increase, thereby making them scalable for both urban and rural areas. Disadvantages of Conventional Speed Bumps The conventional speed bump, though good at slowing down traffic, has a number of disadvantages: Noise Pollution: Sudden slowing down and speeding up of vehicles can raise noise levels Damage to Vehicles: Inefficiently designed speed bumps can cause damage to vehicles with low ground clearance or lead to discomfort for drivers and passengers. Effect on Emergency Services: Speed bumps can slow down emergency vehicles, impacting response times. y substituting traditional speed bumps with energy-harvesting systems, these issues can be overcome while providing substantial value in the form of energy generation. Governments have a number of energy infrastructure-related challenges:

Costly installation and maintenance of streetlights, traffic lights, and signs. Solar-powered systems, while sustainable, are less efficient in rainy or winter conditions and are costly to maintain.

Distribution networks for highways along roads are costly, particularly for rural locations.

1.3 Suggested Solutions

Energy-Harvesting Speed Bumps: Deploying these systems on highways, toll booths, and city streets offers a sustainable energy source. Harnessing Unused Kinetic Energy: Cars naturally burn energy while moving, a lot of which goes to waste. These systems harness and reuse that energy. Low-Cost Maintenance: In comparison to solar energy systems, these mechanisms are more affordable to maintain and provide greater reliability across varying weather conditions. Sustaining Increasing Energy Demand:

The large number of cars guarantees that energy production from these systems is consistent and scalable.

Speed bumps that generate energy have a huge potential for mass deployment across urban and rural landscapes:

1. Urban Infrastructure: Such systems can illuminate streetlights, traffic lights, and billboards, cutting down on municipal energy costs.
2. Highways and Toll Plazas: Power generated in busy zones can fuel local buildings, lessening dependence on centralized grids.
3. Smart Cities: As part of smart city development, such integration is consistent with the aim of sustainability and rational resource use. Furthermore, research and development to maximize these systems may result in greater energy conversion efficiency and smaller sizes, making them even more effective and versatile.

Capturing energy from speed bumps is a special and innovative means of meeting increasing energy needs in a way that fosters sustainability. By capturing wasted kinetic and potential energy as electricity, such systems offer a clean and affordable energy source. With advancing technology and higher traffic volumes, energy-harvesting speed bumps may be essential to supplying the power needed for future infrastructure, giving the world a greener and cleaner way to go Comparative Analysis of Mechanisms for Power Generation using Speed Breakers Zeeshan, Muhammad, and Abubakr (2018) carried out a comprehensive comparison of three mechanisms—rack and pinion, roller, and crankshaft—to establish their applicability in generating power using speed breakers. Roller Mechanism Although operational, the roller mechanism possesses significant drawbacks: Maintenance Challenges: The parts need constant maintenance, raising operational complexity. Collision Risk: The structure can cause potential dangers if not aligned or maintained correctly. Crankshaft Mechanism The crankshaft mechanism also has its drawbacks: Mechanical Vibrations: These have the potential to wear out bearings over a period of time. Balancing Problems: The crankshaft involves sensitive mounting on sliding-type bearings that are sensitive to stability with changing loads. Durability Problems: Long-term operation with unbalanced conditions can undermine system performance and lifetime. Rack and Pinion Mechanism The rack and pinion mechanism, however, has certain plus points: Mounting Ease: Its structure facilitates easy installation and integration. Low Gear Losses: The gear losses are between a mere 3% and 5%, thereby ensuring negligible wastage of energy. High Efficiency: With an efficiency of around 95%, it is superior to the other mechanisms. Considering these aspects, the study determined that the rack and pinion mechanism is best suited for application in speed breakers, providing higher efficiency and reliability. Design and Components for Power Generation Through Speed Breakers The design of a speed breaker for power generation consists of various important components, each with a significant function in energy conversion.

1. Rack and Pinion: The rack and pinion is a linear actuator employed to transform linear motion (resulting from the weight of vehicles) into rotary motion. Rack: The toothed, flat component that translates linearly when compressed by the weight of the vehicle. Pinion: The toothed circular gear that mates with the rack, transferring its motion to rotation.
2. Springs: Springs are elastic elements that momentarily store energy as they are compressed. When the load is gone, they assume their original configuration, resetting the system for a new cycle.
3. Shaft: The shaft transfers the rotary motion created by the pinion to other system components, i.e., gears or the generator.
4. Generator: The generator transforms mechanical energy into electrical energy through Faraday's Law of electromagnetic induction. When the shaft turns, it slices through magnetic flux lines, producing an electromotive force (EMF).
5. Planetary Gearbox: A planetary gearbox serves as a speed reducer. It reduces the motion while amplifying torque, making the system more capable of supporting heavier loads and enhancing power output efficiency.
6. Speed Breaker and Body Frame Speed Breaker: A dummy speed breaker, which is usually fabricated using mild steel (MS) sheets, mimics the action of an actual speed hump. Body Frame: Supports the structure of the model and resists the load applied by vehicles. Through incorporation of elements like the rack and pinion mechanism, springs, shafts, and generators, the power-generating speed breaker design presents a novel means of sustainable energy generation. Of the mechanisms being compared, the rack and pinion system is superior based on its high efficiency levels, minimal energy loss, and ease of application, rendering it the most practical solution to use in this case.

2. Power Calculations

Weight of cars vary from 1200 kg -2000 kg (for an average car) approx. The power generation may be greater for heavy vehicles such as trucks, buses etc. The speed breaker height is generally up to 10cm. If averagely, 1500 kg weight and

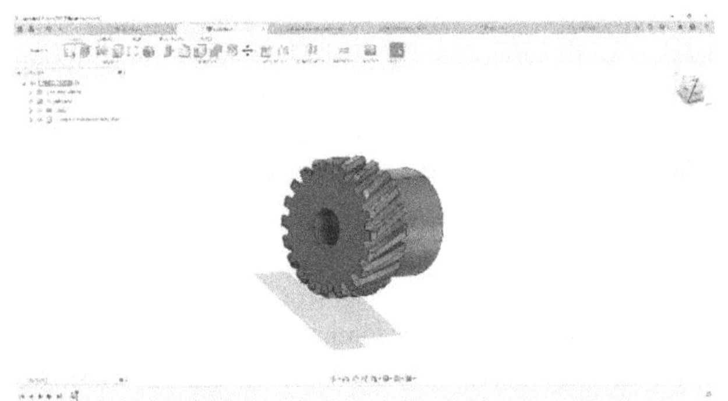

Fig. 77.1 Gear modelling top view

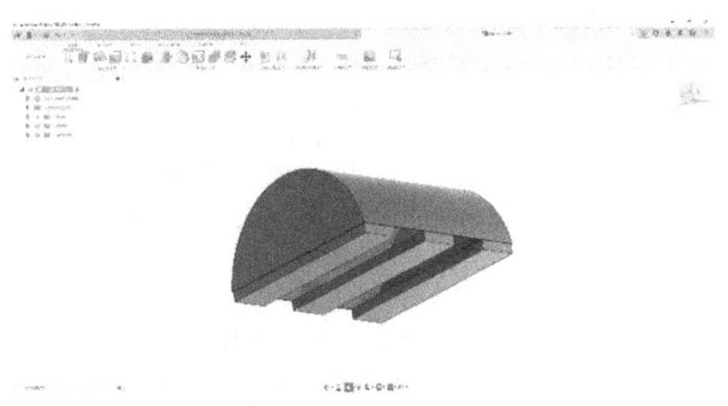

Fig. 77.2 Gear modelling side view

Fig. 77.3 Gear modelling sectional side view

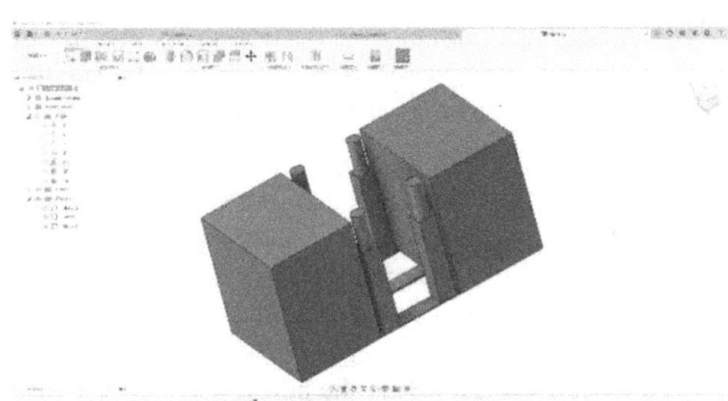

Fig. 77.4 Gear-loading

10 cm height is cons idered then approx. (1 N-m = 1 Joule) 1 N-m/s = 1 Watt

Power = Work done /s Weight of car (W) = mg m = mass of car

g = gravitational acceleration = 9.81 m/s W = 1500 * 9.81 W = 14715 N

Work done = Weight * displacement (i.e. height of breaker) = 14715 * 0.10 = 1471.5 N-m = 1471.5/60 = 24.52 W Work done = 24. 52 Watts Power = 24.52 Watts for 1 minute For 60 minutes (1 hour) = 24.52 * 60 = 1471.5 W

For 24 hours (1 day) = 1471. 5 * 24 = 35316 W Approx. 35.31 kW of power generated per day

This seeks to analyze the feasibility and efficiency of producing electricity from speed breakers through a rack and pinion system. The emphasis is on component selection and design to achieve a system that can convert mechanical energy from cars into electrical energy. The project also tackles issues regarding the system's durability and functionality, with the aim of proving its potential as an alternative, renewable energy source.

The suggested mechanism is based on the rack and pinion mechanism to harness the kinetic and potential energy of cars into electrical energy. The major components and their functions are listed below:

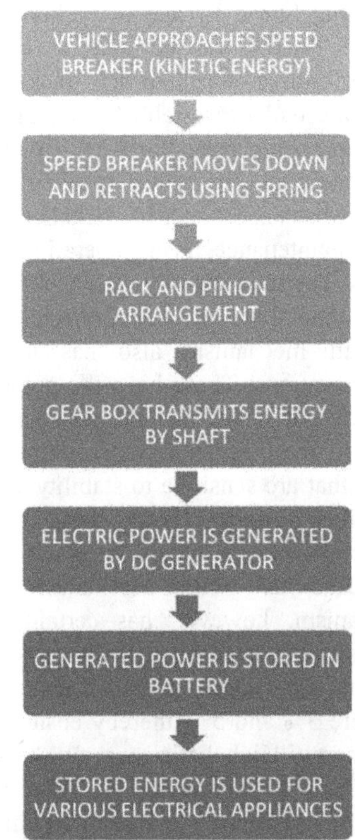

Fig. 77.5 Flow chart for speed breakers

The speed breaker is raised to a certain height above the road surface to effectively catch the vertical movement due to cars. Springs placed under the bump take up the energy of car weight, compressing to hold potential energy momentarily. These springs are made to be repeatedly compressed and return to their initial shape, providing smooth operation over a period of time. The rack converts the vertical movement of the bump into rotational movement by meshing with the pinion.

The rotation of the pinion powers a shaft attached to the generator, which starts energy conversion.

The planetary gearbox increases the rotational speed from the pinion, providing sufficient power generation even with minor compressions.

The generator converts the boosted mechanical energy to electrical energy in accordance with Faraday's Law of Electromagnetic Induction, which is through cutting magnetic flux lines to generate electricity. The electricity produced is stored in batteries to guarantee availability for utilization at night or in bad weather.

The stored power can be used for streetlights, traffic lights, and other roadside amenities, decreasing dependency on conventional sources of power.

1. When a vehicle travels over the speed breaker, its load presses down the bump.
2. The pressure applies a force on the springs, causing linear downward motion.
3. This linear motion is transferred to the rack, which causes the pinion to provide rotary motion.
4. The planetary gearbox enhances the rotational speed to maximize the performance of the generator.
5. The generator converts the rotary motion into electricity, which is stored in batteries for later use.

    1. Durability

       The system is built with durable materials to withstand repeated usage. Planned maintenance schedules guarantee longevity and dependability.

    2. Efficiency

       The planetary gearbox offers efficiency by optimizing rotational speed, allowing for significant energy output from low input.

    3. Scalability

       The system may be installed in heavy-traffic locations, including highways, city streets, and toll roads, making it scalable to different locations.

    4. Sustainability

       The environmentally friendly solution promotes energy conservation through reduced reliance on traditional power sources.

    5. Cost Savings

       By offsetting the cost of electricity for streetlights and other utilities, the system delivers long-term economic savings to governments and municipalities.

Speed breaker power generation technology is a pioneering method of recovering wasted mechanical energy from automobiles. By combining mechanisms such as rack and pinion systems, springs, and planetary gearboxes, the technology makes efficient energy transformation and storage possible. The technology provides a convenient, eco-friendly solution to enhance increasing energy demands and is an encouraging avenue towards renewable energy integration in urban landscapes. Utilizing energy from non-traditional sources is a testament to our dedication to solving today's problems and setting the stage for a sustainable tomorrow for future generations. This project targets the use of unused traffic energy, which otherwise gets wasted as heat, to produce electricity. Through the suggestion of a rack and pinion system combined with speed breakers, this novel method taps into the energy generated by moving vehicles and converts it into electrical power that can be used. Traffic Energy Harvesting With Speed Breakers The rack and pinion system converts the mechanical energy of vehicles into electrical energy, where the quantity harvested is directly proportional to the vehicular density traversing over the speed breaker. This implies that increased traffic flow leads to more energy output, making the system highly efficient for high-traffic locations. Through impact studies and simulations, this project aims to resolve public doubts regarding the viability of using speed breakers as traffic energy harvesters. These studies aim to prove the feasibility of applying this technology and give a better idea of its potential advantages.

Evaluating Large-Scale Implementation

3. Conclusion

The study also looks at the general impact of rolling out such systems on a mass level. Some of the main considerations are:

- Effect on Current Infrastructure: Understanding how the system behaves with normal road material and construction.
- Vehicle Accessibility: Smooth and safe operation across different types of vehicles, from heavy-duty to light vehicles.
- Flow of Traffic: Efficiency in flow without causing disarrays or delays.

Although the idea holds potential, early results indicate that the existing speed breaker design, which is modeled on the rack and pinion system, needs to be extensively improved upon in order to ensure maximum performance and longevity in practical conditions.

Improvement Needs for Design

The findings point out areas where improvement is needed in the system, such as:

- Durability Upgrades: Ensuring that the parts can withstand long-term use under heavy traffic and environmental stress.
- Efficiency Maximization: Optimizing the mechanism to yield a maximum energy output with minimal wear and tear.
- Cost-Effectiveness: Weighing the cost of initial installation and maintenance against long-term energy savings.

This project highlights the vast potential of traffic energy harvesting to play a role in sustainable energy solutions. Although the existing design of the rack and pinion mechanism needs improvement, the study offers useful information regarding the system's advantages and limitations. With future improvements in design and technology, speed breaker energy harvesting systems can be a promising and effective way of producing renewable electricity, promoting energy conservation, and minimizing the use of conventional power sources.

References

1. N. Fatima, J. Mustafa (2012), Production of electricity by the method of road power generation, Int. J. advances in Electrical and Electronics Engineering, 1: 9–14.
2. S. A. Jalihal, K. Ravinder, T. S. Reddy (2005), Traffic characteristics of India, Proc. Eastern Asia Society for Transportation studies, 5: 1009–1024 .
3. R. Gupta, S. Sharma, S. Gaykawad (20 13), A revolutionary technique of power generation through speed breaker power generators, Int. J. Engineering . Research and Techn ology , 2(8): 1879–1883.
4. A. K. S harma, O. Tri vedi, U. Amberi y a, et al. (2012) Devel opment of speed breaker device for generatio n of com pressed air on highways in remote areas. Int. J. Recent Research and Review, 1: 11–15.
5. S. Priyadarshani, (15 June 2007), Generating electricity from speed breakers. Down to Earth.
6. N.V. Bhavsar, V. A. Shah, (2015), Electricity generation by speed breaker using spur gear mechanism, Int. J. Research in Applied Science and Engineering Technology, 3(4): 1164–1169.
7. W. Knight, (21 Aug 2001), Smart speed bumps reward safe drivers,
8. F. N. C. Anyaegbunam, (2015), Electric power generation by speed breaker generators, IOSR J. E ngineeri ng, 5(9): 17–21. M. Ramadan, M. Khaled, et al. (2015), Using speed bump for power generation, In: The 7th International Conference on Applied Energy, Energy procedia, pp.867–872. 30
9. Kaur, S. K. Singh, Rajneesh, et al. (2013), Power generation using speed breaker with auto street light, Int. J. Engineering Science and Innovative Technology, 2(2): 488–491.
10. P. Vishnoi, P. Agrawal, (2014), Power generation by kinetic energy of speed breaker, MIT Int. J. Electrical and Instrumentation Engineering ,4(2): 90–93.
11. B. Prakash, A. V. R. Rao, P. Srinuvas, (2014), Road power generation by speed breaker, Int. J. Engineering Trends and Technology, 11(2): 75–78.
12. F. Najuib, N. Gupta, P. Rawat, et al. (2014), Energy efficient power generation using speed breaker with auto street lights, Int. J. Engineering Research and Management Technology, 1(1): 223–228.
13. M. Sailaja, M. R. Roy, S. P. Kumar, (2015), Design of rack and pinion mechanism for power generation at speed breakers, Int. J. Engineeri ng Trends and Techn ology, 22(8): 356–362.
14. S. Srivastava, A. Asthana, (2011), Produce electricity by the use of speed breakers, J. Engineering Research and Studies, 2(1): 163–165.
15. N. Kumar, P. Saini, M. Kumar, (2016), Automatic road light controller through electricity generation from speed breaker, Int. J. Engineeri ng Technol ogy Sci ence and Rese arch, 3(4): 17–25.
16. S. English, (11 Nov 2005), Smart road hump will smooth the way for safe drivers,
17. Ankita, M. Bala, (2013), Power generation from speed breaker, Int. J. Advance Research in Science and Engineering, 2(2) 31
18. Azam, A., Ahmed, A., Hayat, N., Ali, S., Khan, A. S., Murtaza, G. and Aslam, T., Design, fabrication, modelling and analyses of a movable speed bumpbased mechanical energy harvester (MEH) for application on road. Energy, 214, p.118894.
19. Mishra, A., Kale, P. and Kamble, A., 2013. Electricity generation from speed breakers. The International Journal of Engineering And Science (IJES), 2(11), pp.25–27.
20. Ramadan, M., Khaled, M. and El Hage, H., 2015. Using speed bump for power generation-Experimental study. Energy Procedia, 75, pp.867–872.

Note: All the figures in this chapter were made by the authors.

Advances in Additive Manufacturing Technologies – Gurusamy Pathinettampadian et al. (eds)

Failure Analys and Interfacial Stress Analysis of Particle Reinforced Metal Matrix Composites

Manjunathan R.*, Sethu Raja M., Hari Krishna Raj S., Venkata Praveen M., Saran S.

Department of Mechanical Engineering,
VelTech HighTech Dr. Rangarajan Dr. Sakunthala Engineering College,
Chennai, Tamilnadu

◆ **Abstract:** To precisely forecast the response of particles, such as their stress-strain behavior and likely failure modes within a polymer matrix composite under load, detailed computational investigation is performed. This procedure starts with the digitization of two-dimensional (2D) scanning electron microscope (SEM) images in vector format through IGES (Initial Graphics Exchange Specification) files. The vectorized images are subsequently employed to develop accurate 2D computer-aided design (CAD) models. The CAD models produced by the above process become true digital versions of the parent SEM images containing the fine microscopic details of the composite material. For ease of analysis, it is assumed that the bonding at the interface between the polymer matrix and the filler particles is perfect, i.e., there are no interfacial defects or slippage. Once the CAD model is prepared, appropriate boundary conditions are applied to limit the model, and boundary loads are simulated to simulate real physical mechanical loads. The finite element analysis (FEA) is then executed using ANSYS software, an advanced simulation and assessment tool for the mechanical performance of materials. The simulation enables researchers to assess the stress pattern, deformation, and possible areas of failure within the composite. Besides that, it offers helpful information in terms of the level to which the reinforcement has been embedded into the base matrix, which will play a role in determining how strong and lasting the material would be.

◆ **Keywords:** Stress, Strain, Sem image, Vector fea, Ansys

1. Introduction

With constant developments in contemporary technology, there is the growing need for the kind of materials that have novel combinations of physical and mechanical properties—properties that the traditional metal alloys cannot offer. This need is most critical in fields such as the aerospace and car-making industries, where the materials need to offer excellent strength, toughness, lightness, and heat stability simultaneously. To meet such complex requirements, composite material innovation has been found to be an extremely effective measure. Composite materials are formed when two or more dissimilar substances—each having its own set of physical or chemical properties—are blended into a new substance that offers a better overall performance. These entities retain their separateness and do not totally integrate, yet they function in combination to become a material that has more. Generally, a composite is made up of a reinforcing phase, e.g., fibers, particles, or whiskers, dispersed in a continuous matrix phase that serves as the binder. The outcome is a material that takes advantage of the strengths of both phases, usually having better mechanical properties than traditional metals. One of the most notable classes of composites has been metal

*Corresponding author: manjunathanmech@gmail.com

DOI: 10.1201/9781003685906-78

matrix composites (MMCs). MMCs are particularly valuable when one requires strength, stiffness, and high-temperature capability in applications where lightness is a primary concern. Aluminum-based composites are most prevalent among MMCs and are employed extensively in automotive and aerospace applications. Some common examples are engine pistons and cylinder liners where strength-to-weight ratios and resistance to heat are essential.

Aluminum Matrix Composites (AMCs) couple lightweight aluminum alloys with a wide range of reinforcements—e.g., ceramic particles (e.g., silicon carbide or alumina), whiskers, or short fibers—to produce materials possessing superior mechanical properties and wear resistance, creep, and thermal deformation resistance. AMCs are now being considered as potential substitutes for conventional materials in a variety of high-performance applications. Their attraction is not just because of their mechanical benefit but also their versatility and value for money, especially if discontinuous reinforcement is employed. Composite materials have a structure with at least two phases, namely the reinforcement and the matrix. The matrix phase is normally ductile and continuous, while the reinforcement phase is discontinuous, harder, and stronger. The reinforcement largely determines the mechanical properties of the composite, such as its thermal expansion or wear resistance, strength, and stiffness.

Reinforcement in MMCs can be divided into two major categories continuous and discontinuous. Continuous reinforcements are like boron, graphite, or silicon carbide, while discontinuous ones are like particles or whiskers like alumina, titanium diboride, and chopped fibers. They usually form between 10% and 70% of the composite volume and can be engineered to achieve specific performance requirements. The shape, size, distribution, and orientation of the reinforcement phase have a substantial impact on the ultimate properties of the composite. These factors not only influence the mechanical strength but also other properties like thermal conductivity, coefficient of thermal expansion, and electrical conductivity. For instance, improvement of volume fraction of hard reinforcements like silicon carbide often improves tensile and yield strengths but may reduce ductility and fracture toughness. However, adding softer particles may lead to compromised mechanical strength. Moreover, MMCs are also able to withstand high temperatures above those of their base metals and are usually designed to improve attributes such as thermal conductivity, abrasion resistance, and dimensional stability. Their performance can also be influenced by the interfacial bond quality between the reinforcement and the matrix. A strong interfacial bond ensures efficient load transfer, a requirement in high-performance uses.

Aluminum is an especially advantageous matrix material. It provides a unique combination of low density, high corrosion resistance, ease of fabrication, and compatibility with a broad variety of reinforcement materials. Furthermore, its relatively moderate melting point allows the production of composites without needing to use very high processing temperatures. Whereas early advances in aluminum MMCs primarily involved continuous fiber-reinforced composites, recent trends favor discontinuously reinforced composites because of their easier production processes, lower prices, and more consistent mechanical properties. When engineered correctly, these composites based on aluminum can surpass many conventional materials like steel in a number of important aspects. For instance, AMCs are able to provide:

- Up to 60% weight saving over traditional metals
- 100–200% increases in strength and stiffness
- Improved wear resistance and dimensional stability
- Increased thermal and fatigue resistance
- Thermally tailored coefficients of thermal expansion for specific thermal applications
- Enhanced damping characteristics, advantageous for noise and vibration control
- Excellent corrosion resistance, appropriate for aggressive environments

The ability to tailor these characteristics is what renders AMCs and other MMCs extremely useful in a wide range of industries. Whether for aircraft structures, engine applications in the automotive industry, or advanced electronic packaging, the freedom of design and performance advantages of metal matrix composites provide a potential path for the next generation of high-performance materials.

2. Result and Conclusion

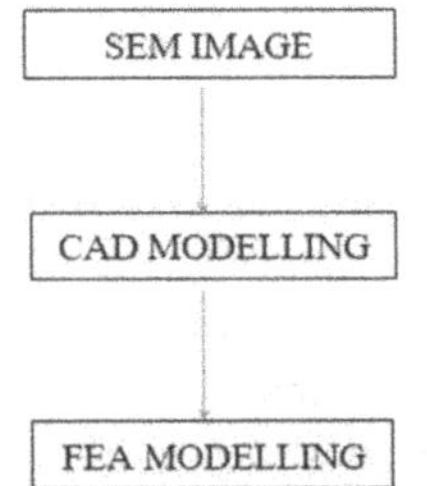

Fig. 78.1 Flow chart for process flow

A finite element model can be constructed by taking into account the periodicity of the actual microstructure of the material and the size scale of the component under analysis. Finite Element Analysis (FEA) is a powerful technique for examining how different factors—like volume fraction

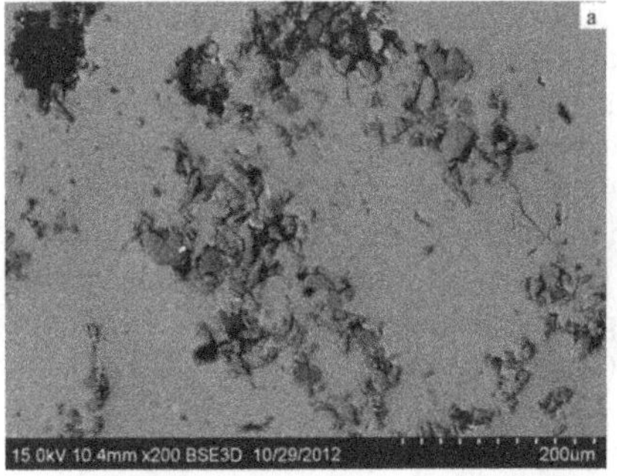
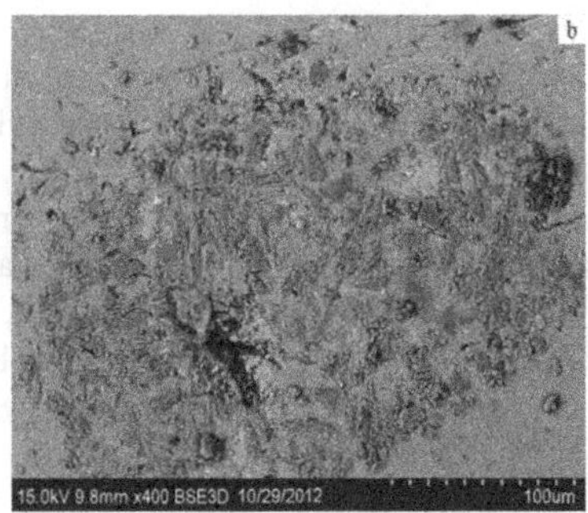

Fig. 78.2 Scanning electron microscope for (a) Specimen 1, (b) Specimen 2, (c) specimen 3

of the reinforcements, particle shape, and their spatial distribution—impact the stress-strain behavior and failure mechanisms in composite materials. This simulation-based strategy allows researchers to precisely forecast the global mechanical properties of the composite as well as provide comprehensive information about the localized stress-strain response in both the matrix as well as the reinforcing components.

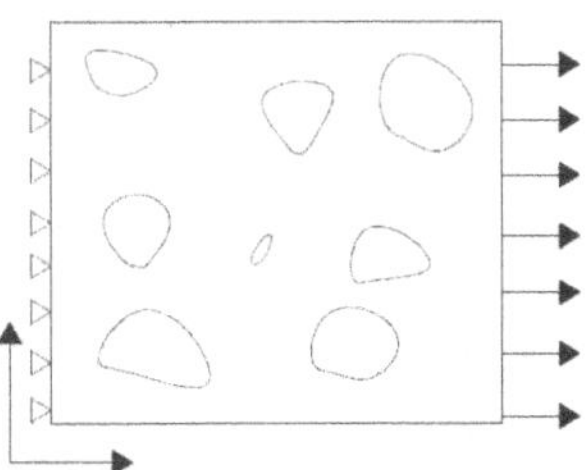

Fig. 78.3 Load variation curve for three specimens

3. Conclusion

A Computer-Aided Drafting (CAD) model was produced through vectorized versions of SEM (Scanning Electron Microscope) images. CAD models are precise copies of the original SEM images, maintaining the geometric features and dimensions in their correct form without any distortion. The transformation of SEM images into vector form ensures that structural details are not lost, retaining the same scale as viewed in the microscope images. Feature size from the SEM images can be calculated by the use of scale markers present within the images themselves.

These models so produced were next meshed utilizing two-dimensional (2D) solid quadratic triangular elements in preparation for Finite Element Analysis (FEA). They were subjected to a mesh density of value 1 so that they can optimize the relationship between computation time and model precision during the simulation. The reinforcement particle stiffness during simulation was captured using the right assignment of a Young's modulus value for the particle. It is being assumed that interfacial bonding between the matrix and the reinforcement is ideal, i.e., no slippage or debonding occurs under loading.

Under applied loading, the reinforcement particles in the polymer matrix composite are thought to experience elastic deformation. Patterns of stress distribution are studied in various parts of the model, offering information on how the microstructure responds to mechanical stress. The simulation is also set to predict some key behaviors, including the effect of particle clustering on the overall strength of the composite, possible particle fracture, and the integrity of the matrix-particle interface. To analyze the mechanical behavior of Metal Matrix Composites (MMCs), microstructure-driven 2D FEA models are created. The models take into account the random particle distribution and the phenomenon of clustering, which are typical in actual materials. The objective is to investigate the manner in which these microstructural attributes affect the stress–strain response and contribute to failure mechanisms in the composite. Eventually, this technique allows for a better understanding of the micro-mechanical behavior and failure modes in reinforced composite materials.

References

1. "Microstructure-based finite element analysis of failure prediction in particle-reinforcedmetal–matrix composite", S. BalasivanandhaPrabu, L. Karunamoorthy, journal of materials processing technology 2 0 7 (2008) 53–62
2. "A finite element analysis study of micromechanical interfacial characteristics of metal matrix composites", S. Balasivanandha Prabu, L. Karunamoorthy, G.S. Kandasami, Journal of Materials Processing Technology 153–154 (2004) 992–997
3. "Failures analysis of particle reinforced metal matrix composites by microstructure based models", G.G. Sozhamannan, S. BalasivanandhaPrabu, R. Paskaramoorthy, Materials and Design 31 (2010) 3785–3790
4. Micro-macro Unified Analysis of Flow Behavior of Particle-reinforced Metal Matrix Composites, Zhang Peng, Li Fuguo, Chinese Journal of Aeronautics 23(2010) 252–259

5. Effects of Particle Clustering on the Flow Behavior ofSiC Particle Reinforced Al Metal Matrix Composites, Zhang Peng, Li Fuguo, Rare Metal Materials and Engineering, 2010, 39(9): 1525−1531
6. "Fracture toughness of composite materials reinforced by debondable particulates", Yu Qiao, ScriptaMaterialia 49 (2003) 491–496
7.. S. Priyadarshani, (15 June 2007), Generating electricity from speed breakers. Down to Earth.
8. N.V. Bhavsar, V. A. Shah, (2015), Electricity generation by speed breaker using spur gear mechanism, Int. J. Research in Applied Science and Engineering Technology, 3(4): 1164–1169.
9. W. Knight, (21 Aug 2001), Smart speed bumps reward safe drivers,
10. F. N. C. Anyaegbunam, (2015), Electric power generation by speed breaker generators, IOSR J. E ngineeri ng, 5(9): 17–21. M. Ramadan, M. Khaled, et al. (2015), Using speed bump for power generation, In: The 7th International Conference on Applied Energy, Energy procedia, pp.867–872. 30
11. Kaur, S. K. Singh, Rajneesh, et al. (2013), Power generation using speed breaker with auto street light, Int. J. Engineering Science and Innovative Technology, 2(2): 488–491.
12. P. Vishnoi, P. Agrawal, (2014), Power generation by kinetic energy of speed breaker, MIT Int. J. Electrical and Instrumentation Engineering, 4(2): 90–93.
13. B. Prakash, A. V. R. Rao, P. Srinuvas, (2014), Road power generation by speed breaker, Int. J. Engineering Trends and Technology, 11(2): 75–78.
14. F. Najuib, N. Gupta, P. Rawat, et al. (2014), Energy efficient power generation using speed breaker with auto street lights, Int. J. Engineering Research and Management Technology, 1(1): 223–228.
15. M. Sailaja, M. R. Roy, S. P. Kumar, (2015), Design of rack and pinion mechanism for power generation at speed breakers, Int. J. Engineeri ng Trends and Techn ology, 22(8): 356–362.

Note: All the figures in this chapter were made by the authors.

Advances in Additive Manufacturing Technologies – Gurusamy Pathinettampadian et al. (eds)
© 2026 Taylor & Francis Group, London, ISBN 978-1-041-16687-0

Multimodal Deep Learning for Alzheimer's Disease—Comparative Review

Boopendranath C.[1],
Yogapriya M.[2], Zahid Hussain J.[3]
Department of Computer Science
Engineering with Artificial Intelligence and Machine Learning,
Chennai Institute of Technology,
Kanchipuram, India

Karthikeyan P.[4]
Department of Computer Science Engineering,
Chennai Institute of Technology,
Kanchipuram, India

Anish K.[5]
Department of Computer Science
Engineering with Artificial Intelligence and Machine Learning,
Chennai Institute of Technology,
Kanchipuram, India

Kandavel N.[6]
Department of Computer Science Engineering,
Chennai Institute of Technology,
Kanchipuram, India

◆ **Abstract:** Alzheimer's disease presents a significant global health challenge due to its complex causes and the absence of a cure. Multimodal deep learning (MDL) offers a transformative approach by integrating diverse data types such as neuroimages, genomics, and electronic medical records. This review evaluates various MDL models, emphasizing their performance metrics to identify the most effective model for Alzheimer's diagnosis and progression tracking. Among these, MACFNet stands out with 99.59% accuracy and 99.91% sensitivity in Alzheimer's classification, making it the optimal model. Despite these advancements, challenges persist in data integration, dataset scalability, and clinical applicability. This review explores MDL's potential to advance Alzheimer's research and improve clinical diagnosis, with a particular focus on interpretability and the integration of AI-driven components for enhanced outcome explanations.

◆ **Keywords:** Alzheimer's disease, Multimodal data, MRI, PET, Deep learning, Biomarkers

[1]boopendranathc.aiml2023@citchennai.net, [2]yogapriyam@citchennai.net, [3]zahidhussainj@citchennai.net, [4]karthikeyanp@citchennai.net, [5]anishk.aiml2023@citchennai.net, [6]kandaveln@citchennai.net

DOI: 10.1201/9781003685906-79

1. INTRODUCTION

AD is a chronic neurodegenerative disorder which is the most common type of dementia. These are amyloid beta plaques, tau tangles, and hippocampal shrinkage which are characteristic features of Alzheimer's disease. These are strongly associated with neuronal loss, severe dementia, and functional impairment. Many organizations have gone ahead and put a lot of effort in the area of research in this disease with regard to treatment, but no viable curative treatment has been made available at the time of writing this paper hence posing a big question on accurate diagnosing procedures as well as early detection. Some of these traditional approaches such as neuroimaging or CSF biomarkers which are usually diagnostic techniques involve the use of one of the method which is in terms of the multiple faceted views of the disease. It has remained a challenge due to the complexity and heterogeneity of AD biomarkers style that call for the use of multimodal imaging, an opportunity provided by the recent progressive developments in AI, particularly deep learning.

1.1 Stages, Causes, and Risk Factors of Alzheimer's Disease

Alzheimer's is a chronic illness of the brain and there are more stages in it and the condition worsens as time progresses. The disease is commonly divided into three stages: The early phase of the disease is considered the initial phase; the second phase is considered moderate, and the last phase is severe. In the early phase of the disease, the patient may have difficulties in memory, and problem-solving ability and sometimes may have confusion. In the middle phase of the disease, the patient may develop confusion, have significant issues with memory, and may not be able to speak, decide, or even feed themselves. During the last stage of the disease, the patient cannot speak, cannot recognize family members, and is completely helpless requiring total dependency on others for all needs.

Even today, the primary cause of AD has not been clearly discovered but it is presumed to be hereditary, environmental as well as lifestyle-related. The usual pathological features of Alzheimer's disease are amyloid

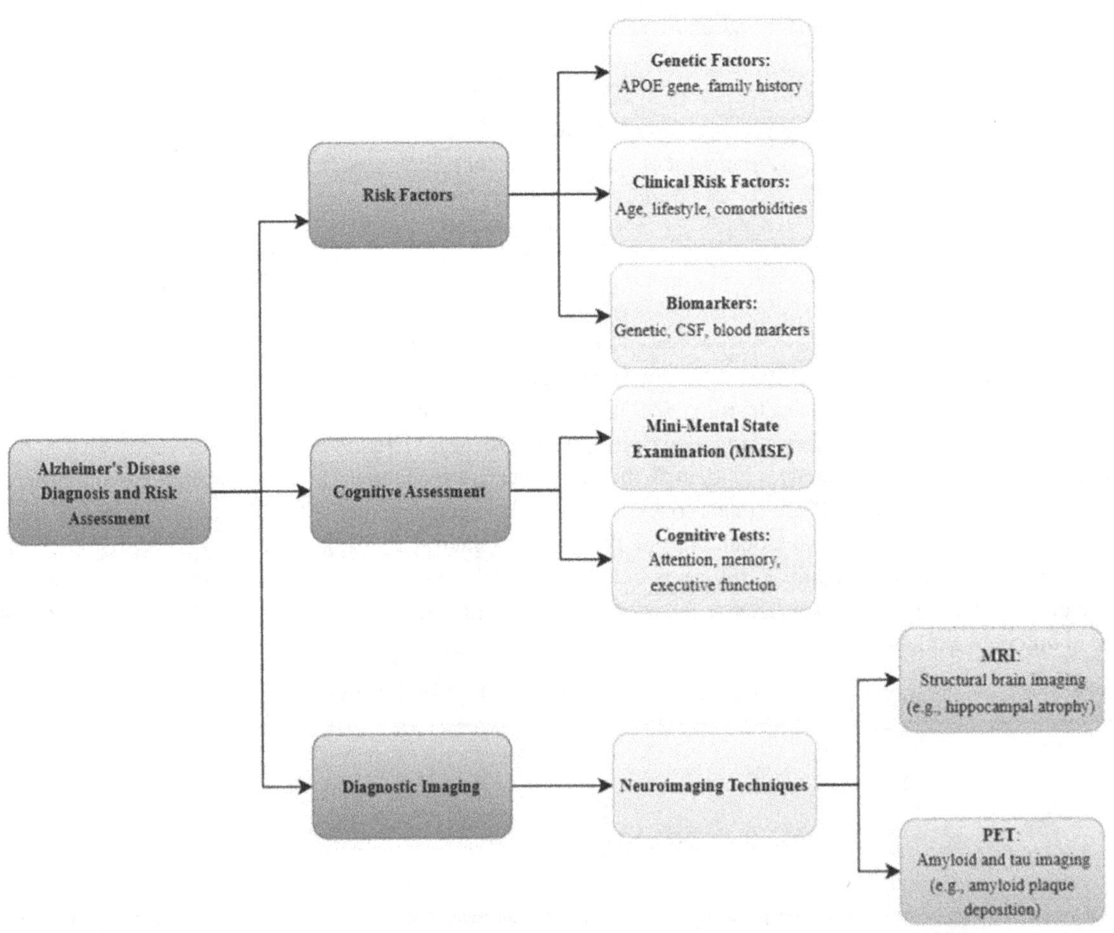

Fig. 79.1 Simple flowchart on alzheimer's disease diagnosis and risk assessment [1–5]

deposits and neurofibrillary changes in the form of tangles throughout the cerebral cortex resulting in neuronal loss. The causes of the disease are as follows; Some people get the disease through their family genes, and the likelihood is higher if the family genes are APP, PSEN1, and PSEN2. Other causes may include head injuries and exposure to toxins that are thought to cause the disease. The only high-risk factor is age and the risk is even higher if the patient is over 65 years of age. Other risk factors include a family history of AD, cardiovascular status, and lifestyle: diet and exercise. Information on such factors has also is imperative in the prevention and early intervention of seclusion and restraint.

1.2 Challenges with AD

Existing tools such as neuroimaging and cognitive tests are used to diagnose the disease after a patient has experienced a decline in his or her cognitive abilities. Further, the nature and variability of AD symptoms make it even more challenging to provide standard approaches to diagnosis. The severity of the disease is not the same for all patients and the disease may affect different patients in different ways and at different rates of progression, thus it is challenging to build models that perform well consistently. There are also significant limitations to deep learning including limited access to high quality large datasets that can be used when developing or testing the model. However, existing models are sometimes complex and need high computational resources and professional knowledge, which reduces the possibility of using them in daily practice. These challenges therefore call for improved and effective diagnostic tools for AD.

1.3 Rationale for Multimodal Deep Learning

In the context of Alzheimer's, the term "multimodal" is defined as the use of multiple imaging data, including MRI and PET, as well as genetic, CSF, and clinical biomarkers. The multimodal approach makes diagnostic accuracy tenfold higher than the unimodal methods. For instance, the assignments citing CNNs as discussed by Abdelaziz et al [1] showed that research employing the models offered accuracy above 98.22% in AD identification from the NC group, through the use of the ADNI datasets. At the same time, based on the concept of integrating multiple types of data, Tang et al. [3] put forward MACFNet for AD classification which yielded the highest accuracy (99.59%) and sensitivity (99.91%). The staging of early structural MRI and PET scans utilizes observed structural and functional changes in the brain to develop hypotheses about future structural and functional modifications. Additional imaging biomarkers accompany these markers: biological information and APOE genotyping, which shows hereditary tendencies. The above heterogeneous data can be merged and processed through deep learning models, hence allowing for disease unique features to be highlighted disregarding oversimplification through analytics.

1.4 Existing Challenges with MDL

However, there are still some issues to be discussed regarding the use of MDL solutions in Alzheimer's investigation. Another significant problem is data integration because datasets can differ in resolution, format, and size which affects model training and leads to higher data heterogeneity. For example, models like HAMF, proposed by Lu et al. [2], demonstrate competitive performance (AUC: 0.913). However, these methods have issues concerning interpretability and computational productivity. Furthermore, some datasets like ADNI have become very popular but might not have as much variability as required, hence making it hard to scale our models to different subpopulations. To overcome these limitations, powerful fusion approaches, better interpretability, and approaches to handle multiform data efficiently at scale should be methods should be designed, developed, and investigated.

1.5 Scope of the Review

This review intends to provide a comprehensive review of sophisticated MDL models for both diagnosis of Alzheimer's disease and prognosis prediction, by covering methodologies and performance measures. Particularly, MDL models are used to compare and rank models like CNNs, MACFNet, and hierarchical attention-based fusion frameworks to adapt for clinical use. A comparative study of diagnostic accuracies, sensitivities, and robustness of multiple MDL models and points to important future directions while acknowledging the existing knowledge-based fissures and fronts. This paper is appropriate for the general audience of AI and neuroscience, clinicians, and computational biologists. Thus, in the effort to retain computationally identified technicality while also portraying the most clinically pragmatic application, it aims to close the domain-specific gap. It is hoped that the ideas described herein will inform subsequent research and spark creative approaches to addressing the important issues related to Alzheimer's disease. The findings also show that MDL methods provide improved solutions for describing various disease-relevant features such as hippocampal atrophy and amyloid-beta deposition. One of the explored models proposed by Spasov et al. [21] known as the Dual Learning Framework relies on 3D Separable Convolutions only three of which gained a teaching signal with 100% accuracy in the classifications between the Alzheimer's disease and the healthy control groups. However, an introduction of new ideas in the framework of the transformer-based architecture and a concept of sparse hierarchical learning even more expands the perspective of the MDL used in Alzheimer's investigations.

2. Features of Different Models and Their Techniques

Several advanced deep learning approaches used by the models for identifying AD and estimating its advancement are described. Convolutional Neural Networks (CNNs) use neuroimaging data comprising MRI and PET along with genetic data including SNPs. To handle missing values, linear interpolation is used. The molecule-focused method is called HAMF (Hierarchical Attention-Based Multimodal Fusion) [2] in which the attention values are taken into account, which shows how important each modality is for the model. A method called SHAP which allows an understanding of the model's decision-making process. Out of fractal, convolutional, and recurrent structures, a recent network known as MACFNet [3] is particularly efficient, due to the CrossEnhanced Fusion Fusing modules and the ability to dedicate specific channel attention for features that are relevant to AD. Gradient Boosting Machines (GBM) [4] uses T1, T2FLAIR, and amyloid PET to predict AD conversion effectively and satisfactorily while forecasting disease progression. Multimodal & Multiscale DNNs [5] use both structural MRI and functional FDG-PET data to diagnose AD in an early stage, while Multimodal RNNs

[6] incorporate neuroimaging, CSF, and cognitive biomarkers which offer a longitudinal point of view on AD development. SDAE & 3D- CNN [7] models are used to integrate autoencoders for clinical/genetic data and 3D-CNNs for imaging, extracting significant features in brain areas such as the hippocampus, and the amygdala. AI [8] is mainly used for predicting cognitive decline, which is useful in order to stratify patients in clinical trials. MiddleFusion Model [9] uses depth- wise separable convolutions and region-of-interest (ROI) extraction to enhance early AD diagnosis accuracy. MM-SDPN

[10] adopts stacked deep polynomial networks for feature fusion from MRI and PET data, 3D-CNN & FSBi-LSTM [11] integrates 3D-CNN for feature extraction, and FSBi-LSTM for improving the AD classification. Examples of such models are the Deep Learning Model which integrates MRI biomarkers and demographics for diagnosis of Alzheimer's disease as described by [12], and MADNet employs dual-branch feature extraction with ResNet and attention for better multimodal data fusion as described by [13]. Multimodal Machine Learning Framework

[14] analyses fMRI and genetic data of selective features to predict diseases. The proposed approach FCN + MLP [15] fuses multimodal MRI, demographic, and MMSE data which provides a better prediction than traditional techniques. MSHELM [16] integrates MRI, FDG- PET, and CSF biomarkers derived from kernel-based extreme learning machines (ELM) for classification. Three-Stage Deep Learning Framework [17] is a hierarchical framework of feature learning and label prediction, for which such an approach includes the AD-Transformer [18] which is built using the transformer architecture together with Patch-CNN and tokenization for fusing multimodal data with high accuracy and sensitivity. The Clinician Simulation Model [19] uses CNNs for neuroimaging and then correlates the results to improve predictability. Dual Learning with 3D Separable Convolutions [20] model deals with multitask deep learning to predict MCI to AD conversion while Multiple Kernel Learning (MKL) [21] improves the classification accuracy for AD, NC, and MCI. N for classification. FCN + MLP [22] merges multimodal MRI, demographic, and MMSE data for effective fusion and prediction, outperforming traditional methods. Finally, the Deep Boltzmann Machine (DBM) with multimodal fusion [23] uses DBM for feature learning and fusion, and for the classification of MRI and PET data to get a high classification rate.

3. Selection of an Optimal Model

The MACFNet model has an outstanding performance in classification in terms of distinguishing Alzheimer's disease from other populations. From Table 79.1, it can be concluded that it has a strong capability in identifying AD-related patterns in classification tasks. MACFNet uses Local Feature Extraction (LFE) Blocks to extract low-level features from MRI and PET inputs using convolution and pooling layers. CrossEnhanced Fusion Module (CEFM) which combines features from both modalities via Shared Feature Extraction (SFE) and Fine Feature Extraction (FFE) blocks to enhance multimodal integration. Besides, the proposed Efficient Spatial Channel Attention module helps to identify specific disease features more accurately, and the multiscale attention block improves features while emphasizing structural (MRI) and functional (PET) biomarkers. These developments place MACFNet as a high-accuracy diagnostic framework, useful for timely and accurate diagnosis of Alzheimer's disease in clinical practice. MACFNet offers comprehensive insights into early cognitive decline by focusing on AD-specific biomarkers, including changes in structural and functional connections through its attention mechanisms.

Table 79.1 Diverse findings from several studies and datasets combined cohesively [6–30]

S. No	Model Comparative Study		
	Model	**Dataset**	**Results**
1.	**Hierarchical Attention-Based Multimodal Fusion (HAMF)**	ADNI	Prediction tasks of conversion from MCI to AD: Accuracy: 87.2%, Sensitivity: 93.3%, Specificity: 80.4%, F1 Score: 88.4% AUC: 91.1%
2.	**Gradient Boosting Machine (GBM)**	ADNI (with additional hospital data)	Balanced Accuracy (BA): 0.744 (demo + AN modality combination) Sensitivity (SE): 0.778 Specificity (SP): 0.710Area Under Curve (AUC): 0.875.
3.	**Convolutional Neural Network (CNN)**	ADNI (805 subjects)	Classification tasks: NC vs. AD: 98.22%, NCvs.sMCI:93.11%, NCvs.pMCI:97.35%.
4.	**MACFNet (Multi-scale Attention and Cross-enhanced Fusion Network)**	ADNI	Accuracies: AD vs. CN: 99.59%, AD vs. MCI: 98.85%, CN vs. MCI: 99.61%, Multi-class: 98.23%. Sensitivity: AD vs. CN: 99.91%, AD vs. MCI: 99.89%, CN vs. MCI: 99.63%, Multi-class: 97.75% Specificity: AD vs. CN: 98.92%, AD vs. MCI: 97.07%, CN vs. MCI: 99.58%, Multi-class: 99.04%.
5.	**Dual Learning with 3D Separable Convolutions**	ADNI (785 participants)	MCI to AD: AUC: 0.925, 10-fold accuracy: 86%, Sensitivity: 87.5%. AD vs. Healthy:100% accuracy, sensitivity, specificity.
6.	**Multimodal and Multiscale Deep Neural Network (MMDNN)**	ADNI (MCI, AD and non- demented)	Accuracy: 82.4% for 3-year prediction, 86.4% for 1–3 years. Sensitivity: 94.23%, Specificity: 86.3%.
7.	**Deep Boltzmann Machine (DBM)**	ADNI	AD vs. NC: 95.35%, MCI vs. NC: 85.67%, MCI converter vs. non-converter: 74.58%.
8.	**Multiple Kernel Learning (MKL)**	ADNI (T1-MRI, MD maps from DTI)	AD vs. NC: 90.2%, MCI vs. NC: 79.42%.
9.	**Deep Learning Model Simulating Clinician'sProcess**	ADNI	Sensitivity: 97.39%, Specificity: 84.27%, Accuracy: 88.25%, AUC: 0.8864.
10.	**AD-Transformer**	ADNI (1651 subjects)	AD Diagnosis: Accuracy: 95.9%, Sensitivity:95.6%, Specificity: 96.1%, AUC: 0.993. MCI Conversion Prediction: Accuracy: 75.3%, AUC: 0.845.
11.	**Three-Stage Deep Feature Learning**	ADNI	AD vs. HC: Accuracy: 97.12%, Sensitivity: 98.08%. MCI vs. HC: Accuracy: 87.09%, Sensitivity: 75%.
12.	**MSH-ELM**	ADNI	AD vs. HC: Accuracy: 97.12%, Sensitivity: 98.08%, Specificity: 94.12%. MCI vs. HC: Accuracy: 87.09%. Better than MK-SVM, SAE.
13.	**FCN + MLP**	ADNI, AIBL, Framingham, NACC	AUC: ADNI: 0.996, AIBL: 0.974, Framingham: 0.876, NACC: 0.954.
14.	**Multimodal Machine Learning Framework**	ADNI + 37 AD, 35 NC	Accuracy: 83.33%. Identified disease-causing regions and genes.
15.	**MADNet**	ADNI + XWNI	AD vs. CN: Accuracy: 96.54%, Recall: 81.82%, Precision: 93.10%, F1 Score: 87.10%, Specificity: 98.99%, AUC: 95.97%. MCI vs. CN: Accuracy: 83.33%, Recall: 63.37%, Precision: 83.12%, F1 Score: 71.91%, Specificity: 93.47%, AUC: 86.34%.

S. No	Model Comparative Study		
	Model	**Dataset**	**Results**
16.	**Deep Learning Model**	ADNI + DDI cohort	High accuracy in detecting amyloid betapositivity. AUC: 0.89 in DDI cohort.
17.	**3D-CNN + FSBi-LSTM**	ADNI (93 AD, 76 pMCI, 128 sMCI, 100 NC)	Accuracy: AD vs. NC: 94.82%, pMCI vs. NC: 86.36%, sMCI vs. NC:65.35%.
18.	**Multimodal Stacked Deep Polynomial Networks (MM-SDPN)**	ADNI	MM-SDPN-SVM: 97.13% (AD vs. NC); MM-SDPN-LC: 96.93%; MCI-C vs. MCI-NC: 78.88%/76.52% (accuracy).
19.	**Middle-Fusion Multimodal Model**	ADNI (ADNI1, ADNI2, ADNI3)	AD vs. CN Acc: 1.00, BACC: 1.00, Sen: 1.00, Spe: 1.00, F1: 1.00 MCI vs. CN Acc: N/A, BACC: 0.76, Sen: 0.71, Spe: 0.81, F1: 0.73 sMCI vs. pMCI Acc: N/A, BACC: 0.87, Sen: 0.81, Spe: 0.93, F1: 0.83
20.	**AI Model for Predicting Cognitive Decline**	ADNI	Predicting CDR-SB changes: Mean Absolute Error (MAE): 1.065, correlation coefficient:0.601, Ground truth mean: 0.978 (SD: 1.899).
21.	**Stacked Denoising Autoencoders + 3D-CNN**	ADNI	Rey Auditory Verbal Learning Test (RAVLT) (from clinical data). 0.78 ± 0 for SNP + EHR combination. 0.79 ± 0 for EHR + imaging combination. 0.75 ± 0.11 for imaging + SNP combination.
22.	**Multimodal Recurrent Neural Network (RNN)**	ADNI	Accuracy: 75% (AUC = 0.83) for single modality, 81% (AUC = 0.86)

This method shows model generalization since classification between AD with CN as well as classification among AD, MCI, and CN are possible. Furthermore, it is understandable because the model helps clinicians comprehend certain characteristics in diagnosing conditions, and the application of explainable AI enhances its clinical value. CNN, FCN+MLP, and Dual Learning with 3D Separable Convolutions give better results but have not used highly developed attention mechanisms for focusing on disease-relevant features. 3D-CNN & FSBi-LSTM and SDAE & 3D-CNN have moderate accuracy and less information about how multimodal information is integrated and interpreted as MACFNet has but has lower accuracy than MACFNet.

4. Comparison with Alternate Models

Several models classify the images and show high accuracy, however, they have certain drawbacks. AD-Transformer is very accurate but is not feasible for practical clinical applications because of the high computational cost that results from the use of transformers. While CNN and FCN + MLP models demonstrate good results, they do not include complex attention layers that would allow for the subsequent prioritization of disease-associates features as effectively as MACFNet. The models analysed here like 3D-CNN & FSBi-LSTM and Dual Learning with 3D Separable Convolutions offer proportional but slightly imprecise and integrated performance for both binary and multiclass classifications. Moreover, the SDAE & 3D-CNN type models are only moderately successful but they lack the depth of multimodal integration as well as interpretability that make MACFNet unique.

5. Conclusion

This review reveals that the proposed MACFNet model provides the most optimal performance for differentiating Alzheimer's disease from other conditions. MACFNet possesses Local Feature Extraction (LFE) Blocks, CrossEnhanced Fusion Module (CEFM), Efficient Spatial Channel Attention, and multiscale attention mechanisms that make it work more accurately. The model is capable of early and precise diagnosis because it targets biomarkers, which are specific to AD, including structural and functional connections. It, therefore, has generalization ability in both binary and multiclass classification, which enhances its clinical application. Compared with other models like AD-Transformer, CNN, FCN+MLP, 3D-CNN & FSBi-LSTM, and SDAE & 3D- CNN,

MACFNet has higher accuracy and better comprehensibility for practical clinical use. Although, other models perform well in certain areas, they lack capabilities of attention based disease feature prioritization and multimodal fusion which MACFNet has. Collectively, the review lays the foundation for generalizing MACFNet for practical impulsivity and exactness of AD diagnosing besides verifying its reasonability, credibility, and optimization for ubiquitous usage in clinical diagnosing.

6. Limitations

This review has several limitations that are inherent in the deep learning models for AD diagnosis reviewed in this paper. One of the main drawbacks is that several of the models, like AD-Transformer, demand considerable computational power, which will not allow them to be implemented in clinical practice. Still, models like MACFNet are greatly improved through the use of advanced attention mechanisms, but not all models include the use of explainable AI, which may reduce their adoption among clinicians. Furthermore, control over data collection is often limited by the availability of datasets like ADNI on which many models are trained making it difficult to test their practical utility on a diverse population. In addition, some models might need high quality imaging data which might be a problem in environments that are less endowed or technologically backward. Such considerations indicate the importance of future work aiming at enhancing the feasibility, practicability, and clinical applicability of these models.

7. Future Outlook

Future developments in multimodal deep learning in Alzheimer's disease will be mainly in data enhancement, synthetic data creation, and the use of temporal data to track disease progression. Increased emphasis on the application of XAI will guarantee that clinicians embrace the models and improve decision-making. Furthermore, by utilizing M2M communications and smart wearable devices for biomarkers and cognition, and by processing genetic, environmental, and lifestyle information, the main causes of AD can be identified. New developments in ways of combining information from different modalities will also enhance diagnostic outcomes. Issues like simplification of models for use in low-resource environments and translating research models into practice are still areas for future work, outcomes. Issues like simplification of models for use in low-resource environments and translating research models into practice are still areas for future work.

References

1. M. Abdelaziz, T. Wang, and A. Elazab, "Alzheimer's disease diagnosis framework from incomplete multimodal data using convolutional neural networks," J. Biomed. Inform., vol. 121, p. 103863, Sep. 2021, doi: 10.1016/j.jbi.2021.103863.
2. P. Lu, L. Hu, A. Mitelpunkt, S. Bhatnagar, L. Lu, and H. Liang, "A hierarchical attention- based multimodal fusion framework for predicting the progression of Alzheimer's disease," Biomed. Signal Process. Control, vol. 88, p. 105669, Feb. 2024, doi: 10.1016/j.bspc.2023.105669.
3. C. Tang, M. Xi, J. Sun, S. Wang, and Y. Zhang, "MACFNet: Detection of Alzheimer's disease via multiscale attention and cross-enhancement fusion network," Comput. Methods Programs Biomed., vol. 254, p. 108259, Sep. 2024, doi: 10.1016/j.cmpb.2024.108259.
4. M.-W. Lee et al., "A multimodal machine learning model for predicting dementia conversion in Alzheimer's disease," Sci. Rep., vol. 14, no. 1, p. 12276, May 2024, doi: 10.1038/s41598-024-60134-2.
5. D. Lu et al., "Multimodal and Multiscale Deep Neural Networks for the Early Diagnosis of Alzheimer's Disease using structural MR and FDG- PET images," Sci. Rep., vol. 8, no. 1, p. 5697, Apr. 2018, doi: 10.1038/s41598-018-22871-z.
6. G. Lee et al., "Predicting Alzheimer's disease progression using multimodal deep learning approach," Sci. Rep., vol. 9, no. 1, p. 1952, Feb. 2019, doi: 10.1038/s41598-018-37769-z.
7. J. Venugopalan, L. Tong, H. R. Hassanzadeh, and M. D. Wang, "Multimodal deep learning models for early detection of Alzheimer's disease stage," Sci. Rep., vol. 11, no. 1, p. 3254, Feb. 2021, doi: 10.1038/s41598020-74399-w.
8. C. Wang et al., "A multimodal deep learning approach for the prediction of cognitive decline and its effectiveness in clinical trials for Alzheimer's disease," Transl. Psychiatry, vol. 14, no. 1, p. 105, Feb. 2024, doi: 10.1038/s41398-024-02819-w.
9. S. K. Kim, Q. A. Duong, and J. K. Gahm, "Multimodal 3D Deep Learning for Early Diagnosis of Alzheimer's Disease," IEEE Access, vol. 12, pp. 46278–46289, 2024, doi: 10.1109/ACCESS.2024.3381862.
10. J. Shi, X. Zheng, Y. Li, Q. Zhang, and S. Ying, "Multimodal Neuroimaging Feature Learning With Multimodal Stacked Deep Polynomial Networks for Diagnosis of Alzheimer's Disease," IEEE J. Biomed. Health Inform., vol. 22, no. 1, pp. 173–183, Jan. 2018, doi: 10.1109/JBHI.2017.2655720.
11. C. Feng et al., "Deep Learning Framework for Alzheimer's Disease Diagnosis via 3D- CNN and FSBi-LSTM," IEEE Access, vol. 7, pp. 63605–63618, 2019, doi: 10.1109/ACCESS.2019.2913847.
12. M. Mehdipour Ghazi et al., "Comparative analysis of multimodal biomarkers for amyloid- beta positivity detection in Alzheimer's disease cohorts," Front. Aging Neurosci., vol. 16, p. 1345417, Feb. 2024, doi: 10.3389/fnagi.2024.1345417.
13. Y. Li et al., "Dominating Alzheimer's disease diagnosis with deep learning on sMRI and DTI-MD," Front. Neurol., vol. 15, p. 1444795, Aug. 2024, doi: 10.3389/fneur.2024.1444795.
14. X. Bi, R. Cai, Y. Wang, and Y. Liu, "Effective Diagnosis of Alzheimer's Disease via Multimodal Fusion Analysis

Framework," Front. Genet., vol. 10, p. 976, Oct. 2019, doi: 10.3389/fgene.2019.00976.

15. S. Qiu et al., "Development and validation of an interpretable deep learning framework for Alzheimer's disease classification," Brain, vol. 143, no. 6, pp. 1920–1933, Jun. 2020, doi: 10.1093/brain/awaa137.
16. J. Kim and B. Lee, "Identification of Alzheimer's disease and mild cognitive impairment using multimodal sparse hierarchical extreme learning machine," Hum. Brain Mapp., vol. 39, no. 9, pp. 3728–3741, Sep. 2018, doi: 10.1002/hbm.24207.
17. T. Zhou, K. Thung, X. Zhu, and D. Shen, "Effective feature learning and fusion of multimodality data using stage-wise deep neural network for dementia diagnosis," Hum. Brain Mapp., vol. 40, no. 3, pp. 1001–1016, Feb. 2019, doi: 10.1002/hbm.24428.
18. Q. Yu et al., "A transformer-based unified multimodal framework for Alzheimer's disease assessment," Comput. Biol. Med., vol. 180, p. 108979, Sep. 2024, doi: 10.1016/j.compbiomed.2024.108979.
19. F. Zhang, Z. Li, B. Zhang, H. Du, B. Wang, and X. Zhang, "Multimodal deep learning model for auxiliary diagnosis of Alzheimer's disease," Neurocomputing, vol. 361, pp. 185–195, Oct. 2019, doi: 10.1016/j.neucom.2019.04.093.
20. S. Spasov, L. Passamonti, A. Duggento, P. Lio, and N. Toschi, "A` parameter-efficient deep learning approach to predict conversion from mild cognitive impairment to Alzheimer's disease," NeuroImage, vol. 189, pp. 276–287, Apr. 2019, doi: 10.1016/j.neuroimage.2019.01.031.
21. O. B. Ahmed, J. Benois-Pineau, M. Allard, G. Catheline, and C. B. Amar, "Recognition of Alzheimer's disease and Mild Cognitive Impairment with multimodal image-derived biomarkers and Multiple Kernel Learning," Neurocomputing, vol. 220, pp. 98–110, Jan. 2017, doi: 10.1016/j.neucom.2016.08.041.
22. Y. Leng et al., "Multimodal cross enhanced fusion network for diagnosis of Alzheimer's disease and subjective memory complaints," Comput. Biol. Med., vol. 157, p. 106788, May 2023, doi: 10.1016/j.compbiomed.2023.106788.
23. H.-I. Suk, S.-W. Lee, and D. Shen, "Hierarchical feature representation and multimodal fusion with deep learning for AD/MCI diagnosis," NeuroImage, vol. 101, pp. 569–582, Nov. 2014, doi: 10.1016/j.neuroimage.2014.06.077.
24. M. Velazquez and Y. Lee, "Multimodal ensemble model for Alzheimer's disease conversion prediction from Early Mild Cognitive Impairment subjects," Comput. Biol. Med., vol. 151, p. 106201, Dec. 2022, doi: 10.1016/j.compbiomed.2022.106201.
25. W. N. Ismail, F. Rajeena P.P, and M. A. S. Ali, "MULTforAD: Multimodal MRI Neuroimaging for Alzheimer's Disease Detection Based on a 3D Convolution Model," Electronics, vol. 11, no. 23, p. 3893, Nov. 2022, doi: 10.3390/electronics11233893.
26. M. Odusami, R. Maskeliunas, R. Dama¯ seviˇ cius, and S. Misra, "Ex-ˇ plainable Deep-Learning-Based Diagnosis of Alzheimer's Disease Using Multimodal Input Fusion of PET and MRI Images," J. Med. Biol. Eng., vol. 43, no. 3, pp. 291–302, Jun. 2023, doi: 10.1007/s40846-023-00801-3.
27. J. Wang, S. Wen, W. Liu, X. Meng, and Z. Jiao, "Deep joint learning diagnosis of Alzheimer's disease based on multimodal feature fusion," BioData Min., vol. 17, no. 1, p. 48, Nov. 2024, doi: 10.1186/s13040024-00395-9.
28. A. B. Tufail, Y.-K. Ma, M. K. A. Kaabar, A. U. Rehman, R. Khan, and O. Cheikhrouhou, "Classification of Initial Stages of Alzheimer's Disease through Pet Neuroimaging Modality and Deep Learning: Quantifying the Impact of Image Filtering Approaches," Mathematics, vol. 9, no. 23, p. 3101, Dec. 2021, doi: 10.3390/math9233101.
29. Y. Wang, X. Liu, and C. Yu, "Assisted Diagnosis of Alzheimer's Disease Based on Deep Learning and Multimodal Feature Fusion," Complexity, vol. 2021, no. 1, p. 6626728, Jan. 2021, doi: 10.1155/2021/6626728.
30. P. Scheltens *et al.*, "Alzheimer's disease," *The Lancet*, vol. 397, no. 10284, pp. 1577–1590, Apr. 2021, doi: 10.1016/S0140-6736(20)32205-4.

Advances in Additive Manufacturing Technologies – Gurusamy Pathinettampadian et al. (eds)
ISBN 978-1-041-16687-0

80 Experimental Investigation and Analysis of Composite Frame for E-Bike

Yogesh V.*, Keerthick Raja M., Hari krishna Raj S., Vikash V., Dhanush B.

Department of Mechanical Engineering,
VelTech HighTech Dr. Rangarajan Dr. Sakunthala Engineering College,
Chennai, Tamilnadu

◆ **Abstract:** Composites have emerged as some of the most promising materials of this century. Currently, fiber-reinforced composites whether made from synthetic or natural fibers are gaining significance. These composites offer an excellent strength-to-weight ratio and exhibit remarkable properties such as durability, stiffness, impact resistance, corrosion resistance, wear resistance, and fire resistance. Because of these advantages, composite materials are widely used in sectors like construction, biomedical, marine and mechanical engineering. Electric bikes, which operate using electricity as their power source, have become a popular choice for transportation due to their eco-friendliness and cost efficiency. This study focuses on designing and analyzing an optimized electric bike frame for commercial use. Since the bike's frame serves as its backbone, providing support and bearing the overall load, selecting the right material is crucial. This research aims to design and analyze an e-bike frame using steel and composite materials. Composites are preferred for their lightweight, high strength, safety, and cost-effectiveness compared to conventional materials. The study follows the AISI material standards and includes a static analysis to evaluate the frame's performance under sudden impacts. The final design is modeled and analyzed using different materials, with results compared against existing materials to determine the most suitable option. Further experimental investigations, such as laminate preparation and mechanical testing, are conducted to assess the composite material's performance.

◆ **Keywords:** Electric bikes, Mechanical properties, Frame, Fibers

1. Introduction

Electric bikes have emerged as a popular alternative to traditional transportation methods, including driving, public transit, conventional cycling, and walking. A meta-analysis of 24 studies analyzing 38 mode substitution patterns worldwide found that e-bikes most frequently replace public transit (33%), followed by conventional bicycles (27%), automobiles (24%), and walking (10%). However, these substitution rates vary significantly, with a 31% variation for automobiles and a 44% variation for public transit. A weighted mixed logit model suggests that in China, e-bikes more commonly replace public transit, whereas in Europe, North America, and Australia, they tend to replace car travel. Recent studies indicate an increasing trend of e-bikes replacing driving and walking trips while reducing the impact on traditional bicycle trips. This shift suggests that e-bike adoption may contribute to a transition away from conventional bicycles while simultaneously reducing reliance on cars and public transit. Another study on the pricing model of electric vehicles in China examined key financial factors such as initial investment, operating costs, and fuel expenses. It determined that consumers are more likely to adopt electric vehicles when their purchase

*Corresponding author: vyogeshmech@gmail.com

DOI: 10.1201/9781003685906-80

price falls below a critical threshold. Factors such as rising fuel prices, lower initial costs for electric vehicles, and carbon tax policies contribute to making electric vehicles more affordable and attractive. The study also proposed four promotional models based on risk transfer, risk reduction, and risk sharing to encourage widespread adoption.

With growing concerns over rising fuel costs and environmental sustainability, e-bikes have become one of the most reliable eco-friendly transportation options. For short-distance travel, they offer a far more cost-effective solution than motor vehicles, as their running costs are significantly lower. While fuel prices continue to rise, e-bikes remain an economical and practical alternative. The market now offers a wide variety of e-bikes with stylish designs, and their battery charging costs are minimal. Additionally, maintenance expenses are much lower compared to traditional fuel-powered vehicles, making e-bikes a smart and sustainable choice for modern transportation.

2. Materials and Methods

2.1 Materials and Methods Overview

This chapter provides a detailed account of the materials and methods used in processing the composites examined in this study. It also outlines the characterization techniques and tests performed on the composite samples.

2.2 Glass Fibres: Strengths and Limitations

Glass is one of the oldest synthetic materials known to humanity. Despite its long history, the practical strength of glass has always posed a challenge. Glass fibres exhibit two key mechanical properties:

- **High Strength:** Glass fibres possess remarkable tensile strength, making them desirable for various applications.
- **Brittle Fracture:** The brittleness of glass, however, limits its applicability, as it tends to fracture without warning.

A deeper understanding of the glass structure and the factors contributing to its fracture is essential. Such knowledge not only helps improve the performance of current glass applications but also expands the potential for developing new functionalities across a wide range of glass types, including fibre glass.

2.3 Bagasse: Production and Composition

Bagasse, a by-product of the sugarcane industry, is generated in significant quantities. For every 10 tonnes of sugarcane processed, approximately 3 tonnes of wet bagasse are produced. Since bagasse production is directly proportional to sugarcane production, the quantity generated varies from country to country.

2.4 Moisture Content and Its Implications

The high moisture content of bagasse, typically between 40% and 50%, makes it less efficient as a fuel source. Before further processing, bagasse is usually stored to facilitate drying and improve its usability.

- **For Electricity Generation:** Bagasse is stored under moist conditions, where mild exothermic reactions from the breakdown of residual sugars help dry the pile slightly.
- **For Paper and Pulp Production:** Wet storage is preferred, as it aids in the removal of short pith fibres that interfere with the papermaking process and eliminates any remaining sugar content.

2.5 Chemical Composition of Bagasse

On a washed and dried basis, the typical chemical composition of bagasse is:

- **Cellulose:** 45–55%
- **Hemicellulose:** 20–25%
- **Lignin:** 18–24%
- **Ash:** 1–4%
- **Waxes:** Less than 1%

2.6 Structural Characteristics of Bagasse

Bagasse is a highly heterogeneous material composed of:

- **Pith Fibre (30–40%):** Derived from the core of the plant, mainly consisting of parenchyma tissue.
- **Bast, Rind, or Stem Fibre:** Makes up the remaining portion and is primarily composed of sclerenchyma tissue.

These structural variations pose challenges in paper production and have been extensively studied in the scientific literature.

A composite material exhibits unique characteristics that are not present in any of its individual components. The interface within a composite acts as a boundary where discontinuities—whether physical, mechanical, or chemical—occur.

Fibres are defined by their length, which is significantly greater than their cross-sectional dimensions. The effectiveness of reinforcement in a composite depends on its size, influencing how well it enhances the material's properties. One of the key advantages of fibres is their ability to improve the fracture resistance of the matrix.

Since they have a long aspect ratio, they help prevent the propagation of cracks that could otherwise lead to failure, especially in brittle matrices. However, due to their small cross-sectional size, fibres alone are not suitable for direct use in engineering applications. To make them functional, they are embedded within a matrix, forming fibrous composites. The matrix plays a crucial role by binding the fibres together, transferring loads, and shielding them from environmental damage and handling-related wear. In composites reinforced with short or discontinuous fibres, the matrix's load transfer capability is even more critical compared to those with continuous fibres.

Natural fibre composites are composed of materials such as cotton, flax, jute, sisal, and hemp, along with unconventional fibres like coir, empty fruit bunches (EFBs), and wood fibres. Wood-fibre-based thermoplastic composites are particularly appealing due to their resistance to insects and rot, their ability to be painted, and their resemblance to wood. Additionally, they offer greater stiffness and durability at a lower cost than purely plastic alternatives. In the automotive sector, vegetable-fibre-based thermoplastics are gaining traction because of their lightweight nature and environmental benefits compared to conventional composites. Since natural fibres are lignocellulosic, they are biodegradable and non-toxic, making them an attractive choice for eco-friendly applications. These composites are proving to be cost-effective materials, particularly in construction, automotive interiors, railway coaches, and storage solutions. Moreover, they serve as potential substitutes for expensive glass fibres in low-load applications. However, certain agricultural residues, such as rice husks and sugarcane bagasse, may not always meet the length requirements for fibre reinforcement. Nevertheless, these by-products are widely available and are increasingly being explored as particulate reinforcements in resin matrix composites.

India is home to a vast variety of sugarcane, which is widely cultivated. During processing, the cane is crushed through multiple mills, each equipped with heavy rollers. This process breaks the stalks into smaller pieces and extracts the juice, which is then processed to produce sugar. The leftover fibrous material, known as bagasse, is a by-product of this milling process. Disposing of bagasse can be costly for mills, yet it presents opportunities for sustainable material development. Bagasse is composed of fibres, water, and small amounts of soluble solids. The exact composition varies depending on factors such as sugarcane variety, maturity, harvesting method, and the efficiency of the milling process. Given its fibrous nature, bagasse holds potential as a reinforcement material in composite applications.

3. Testing and Analysis

3.1 Tensile Test

Fig. 80.1 Tensile test

3.2 Flexural Test

Fig. 80.2 Flexural test

3.3 Impact Test

This workbench focuses on analyzing the relative motion between different parts. The DMU Kinematics Simulator is a standalone CAD tool specifically designed to simulate

Fig. 80.3 Impact test

assembly movements. It plays a crucial role in the design review process of digital mock-ups (DMU), making it useful across various industries. From consumer products to large-scale automotive, aerospace, industrial plants, ships, and heavy machinery, this tool can handle a diverse range of projects with precision and efficiency.

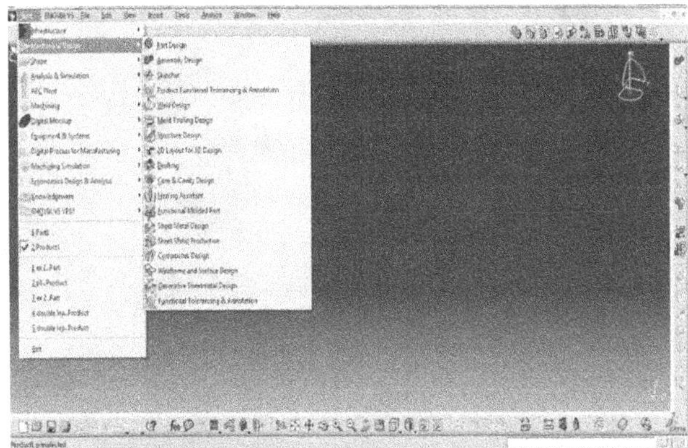

Fig. 80.4 Catia work bench

We created model of joints (bonded, riveted and hybrid) by using CATIA software. The models are shown below.

Boundary Condition

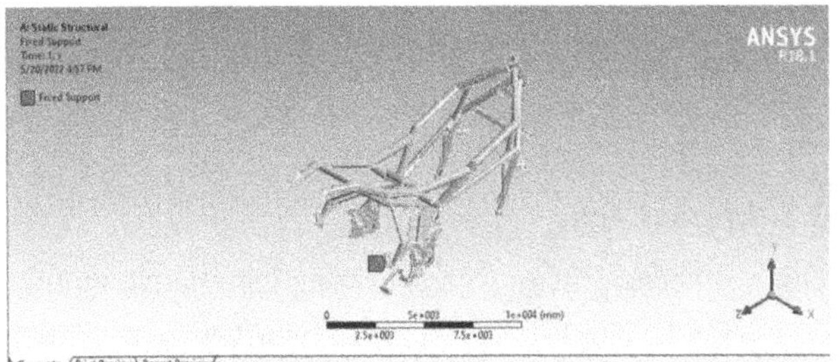

Fig. 80.5 Boundary conditions

Force 1

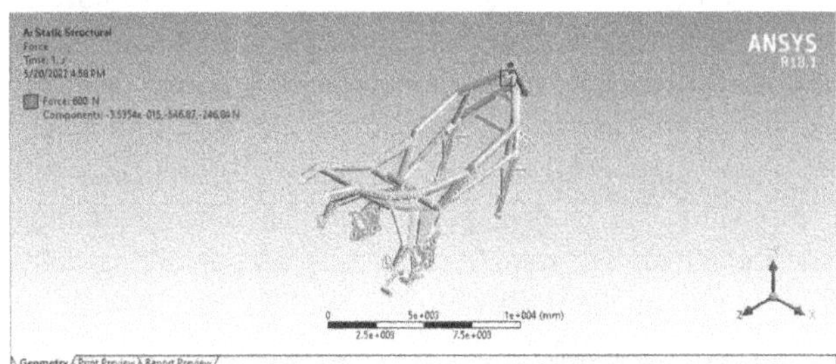

Fig. 80.6 Static structural force 1

Force 2

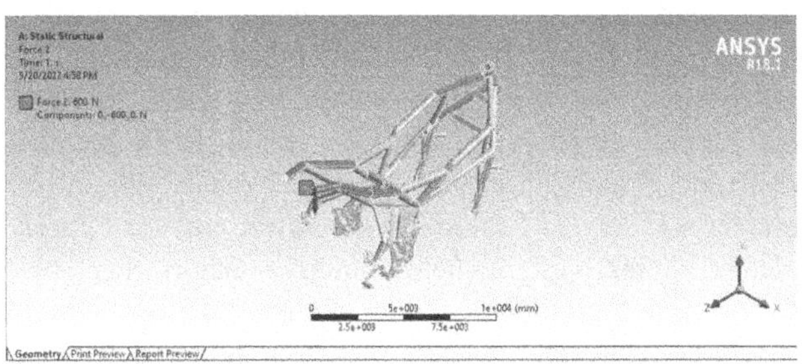

Fig. 80.7 Static structural force 2

3.4 Analysis of Bike Frame

GFRP with EGG SHELL Applied Load: 600 N

Total Deformation

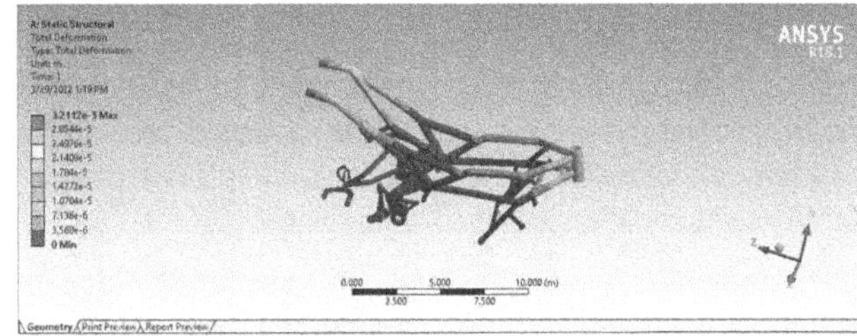

Fig. 80.8 Total deformation for GFRP with EGG shell

Equivalent Elastic Strain

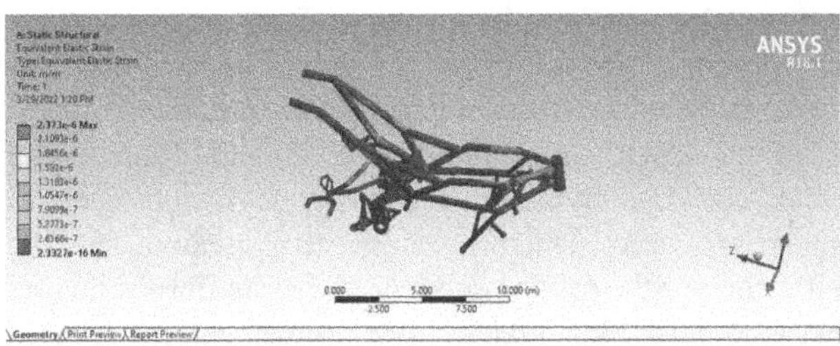

Fig. 80.9 Equivalent elastic strain for GFRP with EGG shell

Equivalent Stress

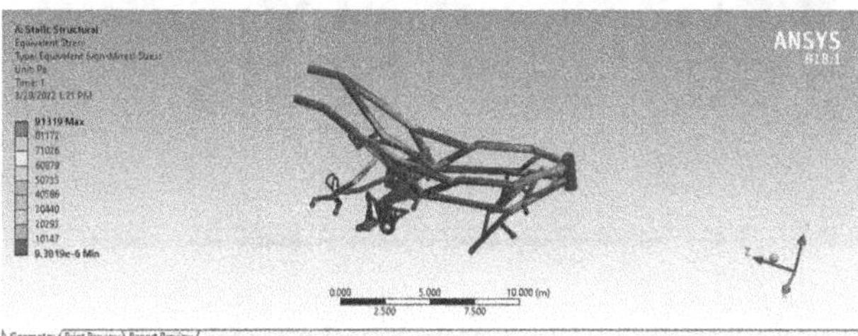

Fig. 80.10 Equivalent stress for GFRP with EGG shell

Maximum Shear elastic Strain

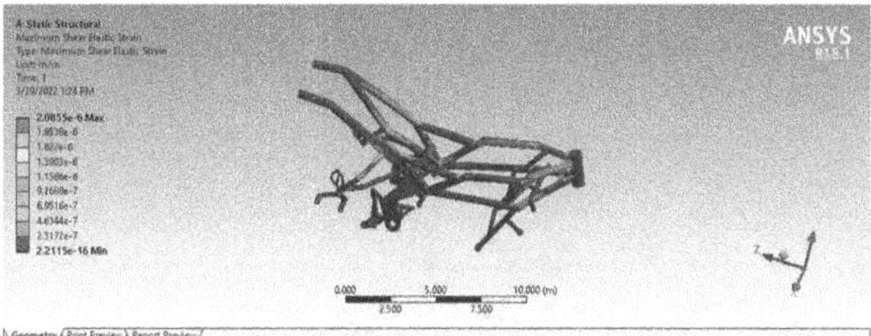

Fig. 80.11 Maximum shear elastic strain for GFRP with EGG shell

Maximum Shear stress

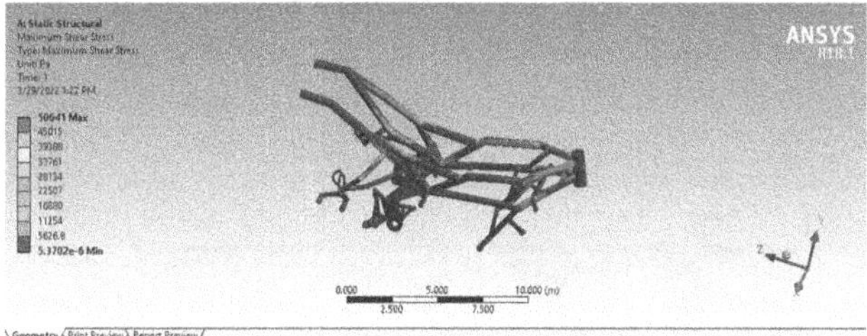

Fig. 80.12 Maximum shear stress for GFRP with EGG shell

Strain Energy

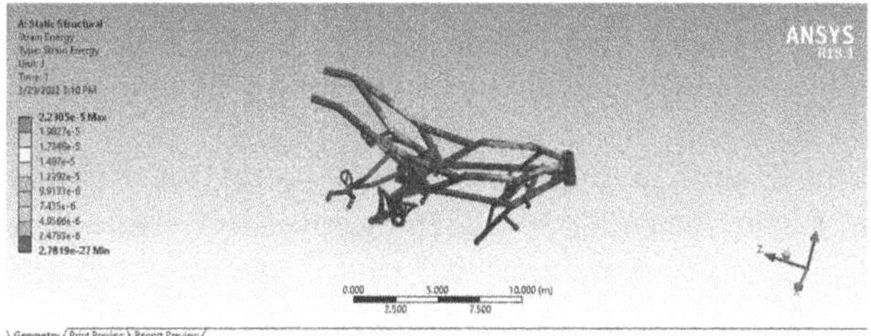

Fig. 80.13 Strain energy for GFRP with EGG shell

4. Conclusion

- The numerical and model analysis of the electric two-wheeler frame with various materials resulted in the following conclusions:
- From the findings of the numerical analysis, the overall structure is safe, efficient, lightweight, and reliable to achieve the desired standards.
- Instead of conventional materials such as steel, alternative materials like glass fiber epoxy reinforced with sugarcane composites and other natural fibers can be utilized.
- The sugarcane composites glass fiber-reinforced polymer (GFRP) showed considerable deformation, which proves that it can resist loads well and is therefore an ideal material for load-bearing parts.

References

1. An, K., Chen, X., Xin, F., Lin, B., Wei, L., 2013. Travel Characteristics of E-bike Users: Survey and Analysis in Shanghai. Procedia - Social and Behavioral Sciences, Intelligent and Integrated Sustainable Multimodal Transportation Systems Proceedings from the 13th COTA International Conference of Transportation Professionals (CICTP2013) 96, 1828–1838.
2. Aono, S., Bigazzi, A.Y., 2019. Industry Stakeholder Perspectives on the Adoption of Electric Bibikes in British Columbia. Transportation Research Record: Journal of the Transportation Research Board 2673, 1–11.
3. Astegiano, P., Tampère, C.M.J., Beckx, C., 2015. A Preliminary Analysis Over the Factors Related with the Possession of an Electric Bike. Transportation Research Procedia, 18th Euro Working Group on Transportation, EWGT 2015, 14-16 July 2015, Delft, The Netherlands 10, 393–402. https://doi.org/10.1016/j.trpro.2015. 09.089.
4. Abagnalea C, Cardoneb M, Iodicea P, Marialtoc R, Stranoa S, Terzoa M, Vorraroabc G., Design and Development of an Innovative E-Bike. Energy Procedia, 101, 2016, 774–781.
5. Ifeanyichukwu U. Onyenanu, Swift ONK, Atanmo PN, Design and Analysis of a Tubular Space Frame Chassis for FSAE Application, Journal of Emerging Technologies and Innovative Research, 2 (10), 2015, 497–201.
6. Sagar Pardeshi 1 , Pankaj Desle 2, Design and Development of Effective Low Weight Racing Bibike Frame, International Journal of Innovative Research in Science, Engineering and Technology, 3(12), 2014, 18215–18221
7. Ao, G., Qiang, J.X., Chen, Z., Yang, L., 2008. Model-based energy management strategy development for hybrid electric vehicles. In: Proceedings of the International Symposium on Industrial Electronics.Institute of Electrical and Electronics Engineers: pp. 1020–1024.
8. Daina, N., Sivakumar, A., Polak, J.W., 2017. Modelling electric vehicles use: a survey on the methods. Renew. Sustain. Energy Rev. 68, 447–460.
9. He, Y.X., Liu, Y.Y., Du, M., Zhang, J.X., Pang, Y.X., 2015. Comprehensive optimization of China's energy prices, taxes and subsidy policies based on the dynamic computable general equilibrium model. Energy Convers. Manag. 98, 518–532.
10. K. Schleinitz, T. Petzoldt, L. Franke-Bartholdt, J. Krems, T. Gehlert, The German Naturalistic Cycling Study – Comparing cycling speed of riders of different e-bikes and conventional bibikes, ScienceDirect- Ek

Note: All the figures in this chapter were made by the authors.

Advances in Additive Manufacturing Technologies – Gurusamy Pathinettampadian et al. (eds)

Analysis and Optimization of Ti-6Al-4V in EDM Process with Graphite Electrodes Using GRA and TOPSIS Techniques

Rajamanickam S.*,
Ashwin M., Nithishkumar V., Dilli raj M.,
Jaya Anand M, Shayam Kumar B.S, Vijayakumar A.
Department of Mechanical Engineering,
VelTech HighTech Dr. Rangarajan Dr. Sakunthala Engineering College,
Chennai, Tamilnadu

◆ **Abstract:** This study explores the micro-machining of Ti-6Al-4V alloy using EDM with graphite electrodes of 200 μm diameter. The investigation focuses on the impact of machining parameters—namely, current, on-time, and off-time—on key performance metrics such as machining time (MT), overcut (OC), and taper angle (TA). A Taguchi L9 orthogonal array design was employed to systematically analyze the effects of these parameters. Micro-holes in the Ti-6Al-4V alloy were characterized using optical microscopy. To assess multi-performance characteristics, Grey Relational Analysis (GRA) and the Technique for Order of Preference by Similarity to Ideal Solution (TOPSIS) were utilized. Performance indices, including the grey relational grade and closeness coefficient, were calculated to identify optimal parameter combinations. The results confirmed that both GRA and TOPSIS are effective in evaluating significant parameter combinations. The optimized machining conditions yielded a machining time of 3 minutes, an overcut of 0.158 mm, and a taper angle of 0.13°.

◆ **Keywords:** PMEDM, Ti-6Al-4V, Al_2O_3, MRR, TWR

1. Introduction

Ti-6Al-4V is a well-known material categorized as difficult-to-machine due to its unique combination of superior properties, attractive features, and extensive industrial applications. Its use spans critical sectors such as aerospace, automotive, military, biomedical, marine, and nuclear industries. The alloy is particularly valued for its high melting point, excellent heat resistance, and low surface roughness, making it a preferred choice for advanced engineering applications. Despite these advantages, its machinability poses significant challenges, particularly in conventional machining processes [1]. Several researchers have explored machining techniques for Ti-6Al-4V to address these challenges. Jahan et al. studied its machinability using EDM with tungsten carbide electrodes (300 μm diameter) for biocompatibility, while Kuriachen and Mathew investigated the effects of a 0.4 mm tungsten carbide electrode in EDM [2-3]. Ahmed et al. introduced liquid nitrogen as a coolant in conventional drilling, reporting reduced drilling time due to efficient heat dissipation at the machining zone [4]. Such studies highlight that Ti-6Al-4V is particularly challenging to machine conventionally, with electrical discharge machining emerging as a preferred alternative. The choice of electrode material plays a crucial role in the EDM process, influencing efficiency, cost, and machining accuracy. Various electrode materials, including copper, brass, zinc, copper-tungsten, and graphite, have been investigated for machining titanium alloys [5]. Graphite electrodes, in

*Corresponding author: rajamanickam@velhightech.com

DOI: 10.1201/9781003685906-81

particular, have demonstrated superior performance due to their low density, excellent electrical conductivity, ease of availability, high machinability, low cost, and reduced manufacturing time. Klocke et al. examined graphite's performance in EDM, emphasizing the influence of current on material removal rate and the effect of pulse duration on tool wear [6]. Similarly, Kristian compared graphite grades for EDM of nickel-based superalloys, highlighting that larger grain sizes enhanced MRR while smaller grains minimized electrode wear [7]. Studies by Hascalik and Caydas, Amorim et al., and Younis et al. further validate graphite as an optimal electrode material in EDM for various alloys, including Ti-6Al-4V. These investigations focused on MRR, tool wear, and surface roughness, with graphite consistently delivering favorable results. However, a detailed exploration of machining time, using graphite electrodes remains limited in the existing literature [8-10].

This study aims to bridge this gap by drilling micro-holes in Ti-6Al-4V using a 200 μm diameter graphite electrode in EDM, a process vital for fabricating meso-scale products in various industries. The study employs the Taguchi method to optimize machining parameters for individual performance measures, including machining time, overcut, and taper angle. This approach ensures the identification of optimal parameter combinations for improved EDM performance, contributing valuable insights for industrial applications of Ti-6Al-4V.

2. Materials and Methods

This choice of polarity is known to enhance the machining process by reducing machining time and improving material removal efficiency.

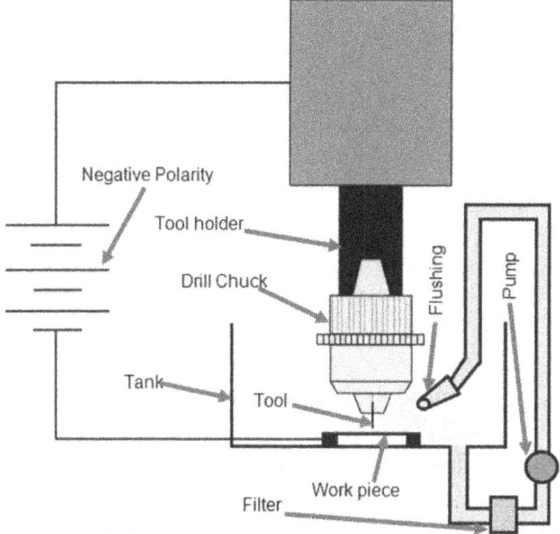

Fig. 81.1 Schematic of the EDM machine layout

The overcut and taper angle metrics were calculated using standardized formulas, represented in Equations (1) and (2). A detailed illustration of OC measurement is depicted in Fig. 81.2, ensuring clarity in understanding the parameter evaluation.

$$OC = \frac{\begin{matrix} Max.\,avg.\,diameter\ of\ hole\ at\ the\ entrance \\ -\ Diameter\ of\ the\ tool \end{matrix}}{2} \quad (1)$$

$$TA = tan^{-1}\left[\frac{\begin{pmatrix} Diamter\ of\ the\ hole\ at\ the\ entrance \\ -\ Diameter\ of\ the\ hole\ at\ exit \end{pmatrix}}{\begin{matrix} 2 * Thickness\ of\ the\ workpiece \\ (Ti-6\ Al-4V) \end{matrix}}\right] \quad (2)$$

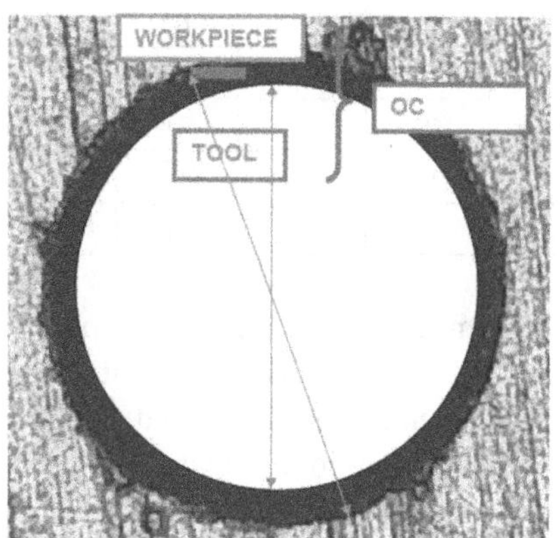

Fig. 81.2 Measurement of overcut

A robust experimental matrix was developed using the Taguchi orthogonal array to systematically vary the machining parameters. This ensured a comprehensive exploration of parameter effects on performance characteristics. This study utilized an L9 orthogonal array based on the DOE-Taguchi method, implemented through Minitab software, to systematically design the experimental framework. The L9 array accommodates three machining parameters—current, on-time, and off-time—each with three levels, resulting in nine distinct experimental runs. The selected parameter levels and their corresponding combinations are detailed in Table 81.1.

Table 81.1 Machining parameters and their levels

Factors (Units)	Levels		
	Level 1	Level 2	Level 3
Current (A)	0.5	1	1.5
On-time (μs)	10	15	20
Off-time (μs)	3	5	7

Performance characteristics evaluated in this study included machining time (MT), overcut (OC), and taper angle (TA), with their experimental values summarized in Table 81.2.

Table 81.2 L9 Orthogonal array and experimental results of MT, OC and TA

Sl. No.	Machining Parameters			Performance Characteristics		
	Current (A)	On-time (μs)	Off-time (μs)	MT (min)	OC (mm)	TA(°)
1	0.5	10	3	8.3	0.305	0.27
2	0.5	15	5	9.1	0.287	0.19
3	0.5	20	7	10.5	0.203	0.20
4	1	10	5	11.7	0.205	0.16
5	1	15	7	8.1	0.203	0.26
6	1	20	3	4.6	0.215	0.12
7	1.5	10	7	7.4	0.258	0.18
8	1.5	15	3	3.0	0.158	0.13
9	1.5	20	5	4.8	0.180	0.15

The analysis aimed to minimize these response variables, aligning with the "Lower the Better" quality objective. This objective is mathematically expressed by the following Signal-to-Noise (S/N) ratio formula:

$$\eta = -10 \log\left[\frac{1}{N}\sum_{m=1}^{N} d_m^2\right] \tag{3}$$

d_m : Output data value under a specific trial condition.

N : Number of repetitions of the trial condition.

To investigate the influence of machining parameters comprehensively, the study focused on optimizing both individual and combined performance measures. The primary objective was to achieve optimal machining conditions through a systematic approach. Optimization and multi-objective optimization were performed using the Taguchi method, Grey Relational Analysis, and the Technique for Order of Preference by Similarity to Ideal Solution. These methods were employed to identify the most effective parameter combinations for enhanced machining performance. This comprehensive methodology aims to deliver high-quality insights into the micro-EDM process, specifically addressing the challenges of machining Ti-6Al-4V with graphite electrodes. The findings from this research are expected to contribute significantly to the optimization of EDM processes for applications in industries requiring precision manufacturing.

3. Results and Discussion

3.1 Taguchi Analysis

The step-by-step flow of Taguchi analysis is shown in Fig. 81.3.

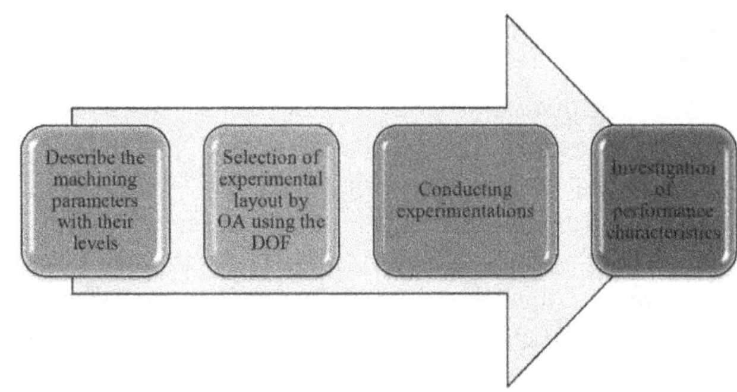

Fig. 81.3 Step-by-step flow of taguchi analysis

The main effects plot for machining time (MT), shown in Fig. 81.4, reveals an inverse relationship between MT and both current and on-time. As current and on-time increase, the intensity of energy discharged per unit area of the workpiece rises, resulting in a rapid increase in temperature within microseconds. This leads to quicker material melting and vaporization, significantly reducing MT. In contrast, off-time demonstrates a direct relationship with MT. Longer off-time intervals allow for greater cooling periods, thereby increasing overall machining time. Conversely, shorter off-time intervals facilitate quicker machining by leveraging the thermal properties of the material, enabling faster heat dissipation and material removal.

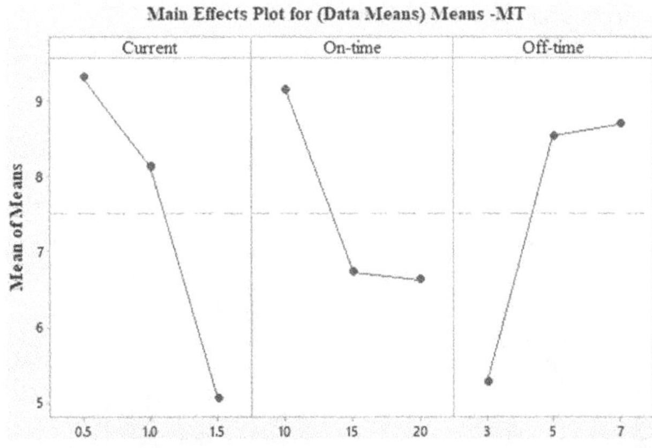

Fig. 81.4 Main effects plot for MT

The main effects plot for overcut (OC), illustrated in Fig. 81.5, indicates that OC is directly influenced by MT. Shorter machining times result in minimal overcut. Additionally, as current, on-time, and off-time increase, OC

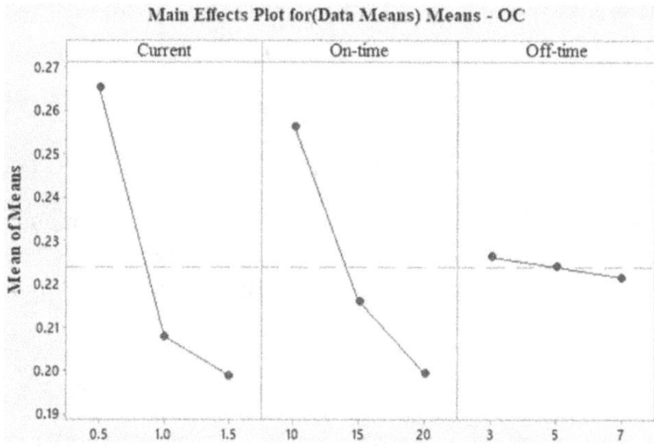

Fig. 81.5 Main effects plot for OC

tends to decrease. Among these parameters, on-time has the most significant impact on OC when machining small holes in Ti-6Al-4V using a graphite electrode with EDM. The observed reduction in OC with optimal parameter settings highlights the importance of precise control over on-time during machining.

The main effects plot for taper angle (TA), presented in Fig. 81.6, highlights that current has the most substantial effect on TA during small-hole drilling. An increase in both current and on-time reduces the taper angle, likely due to enhanced material removal uniformity along the hole depth.

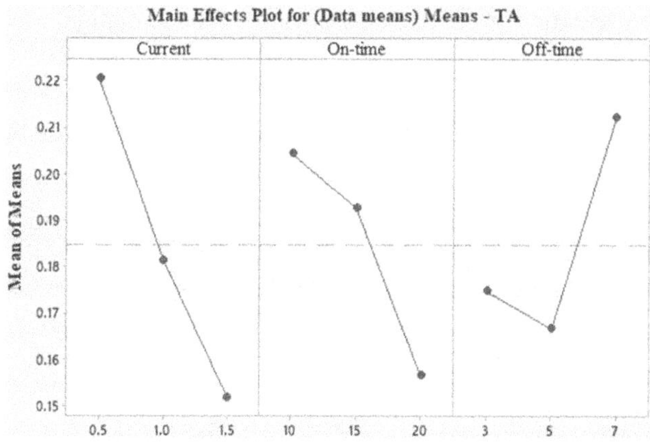

Fig. 81.6 Main effects plot for TA

Off-time exhibits a more complex influence on TA. Initially, increasing off-time from 3 μs to 5 μs reduces the taper angle, likely due to improved stability in the machining process. However, further increasing off-time to 7 μs leads to an increase in TA. This is attributed to side wear of the graphite electrode, which becomes more pronounced during prolonged off-time intervals, resulting in less precise hole geometry.

The performance characteristics—MT, OC and TA—demonstrate non-linear variations when influenced by changes in current, on-time, and off-time. Due to their conflicting nature, achieving optimal performance requires a systematic multi-objective optimization approach. In this study, both GRA and TOPSIS were employed to resolve these conflicts effectively.

3.2 Multi-Objective Optimization Using GRA Analysis

Grey Relational Analysis was employed to evaluate and compare the performance characteristics—MT, OC and TA. The step-by-step flow of GRA is shown in Fig. 81.7.

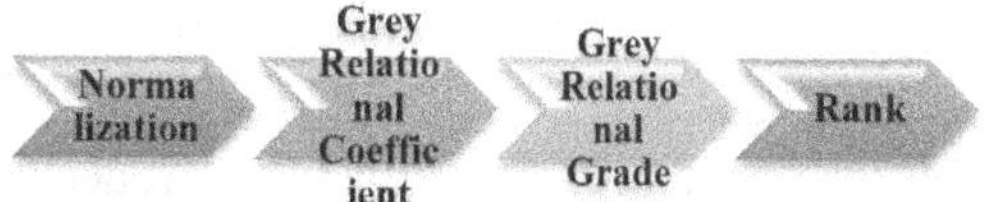

Fig. 81.7 Step-by-step flow of GRA analysis

To ensure all performance characteristics were on a comparable scale, normalization was performed using the following formula:

$$\text{Normalised value}(m_{ij}^{*}) = \frac{MAXm_{ij} - m_{ij}}{MAXm_{ij} - MINm_{ij}} \quad (4)$$

After normalization, the Grey Relational Coefficient (GRC) was calculated for each performance characteristic using the formula:

$$\text{Grey Relational Coefficient}\left(\xi_{ij}\right) = \frac{\Delta_{min} + k\Delta_{max}}{\Delta m_{ij} + k\Delta_{max}} \quad (5)$$

Where,

$\Delta_{min} = 0$, the minimum deviation.

$\Delta_{max} = 1$, the maximum deviation.

$k = 0.5$, the distinguishing coefficient for balancing weight between Δ_{min} and Δ_{max}

The overall Grey Relational Grade (GRG) was then calculated as the average of the GRC values for all performance characteristics using the formula:

$$\text{Grey Relational Grade}(\gamma_i) = \frac{1}{m}\sum_{j=1}^{m}\xi_{ij} \quad (6)$$

Where,

m = 3, the number of performance characteristics.

The GRG values for all experimental runs are summarized in Table 81.3. Among the nine experiments, Experimental Run 8 achieved the highest GRG value of 0.9685, indicating the best combination of machining parameters

Table 81.3 GRA results of MT, OC, and TA

Normalization			GRC			GRG	Rank
MT	OC	TA	MT	OC	TA		
0.3924	0.0000	0.0000	0.4514	0.3333	0.3333	0.3727	9
0.2990	0.1268	0.5520	0.4163	0.3641	0.5274	0.4360	8
0.1352	0.6950	0.4802	0.3664	0.6211	0.4903	0.4926	6
0.0000	0.6810	0.7255	0.3333	0.6105	0.6456	0.5298	4
0.4114	0.6949	0.0885	0.4593	0.6210	0.3542	0.4782	7
0.8152	0.6129	1.0000	0.7302	0.5636	1.0000	0.7646	2
0.4895	0.3206	0.6289	0.4948	0.4239	0.5740	0.4976	5
1.0000	1.0000	0.9478	1.0000	1.0000	0.9054	0.9685	1
0.7924	0.8517	0.8300	0.7066	0.7713	0.7462	0.7414	3

for optimal performance. Conversely, Experimental Run 1 recorded the lowest GRG value of 0.3727, reflecting the least favorable parameter combination. GRA enables a comprehensive evaluation of multi-objective optimization by balancing the trade-offs between MT, OC, and TA. It provides a robust framework for determining the best parameter settings to enhance machining efficiency and precision in EDM processes.

3.3 Multi-Objective Optimization Using TOPSIS

In parallel with GRA, the TOPSIS was employed to determine optimal machining parameters for Ti-6Al-4V during electrical discharge machining. The step-by-step flow of TOPSIS is illustrated in Fig. 81.8. This method ranks the experimental results based on their proximity to the ideal solution and their distance from the worst solution [11].

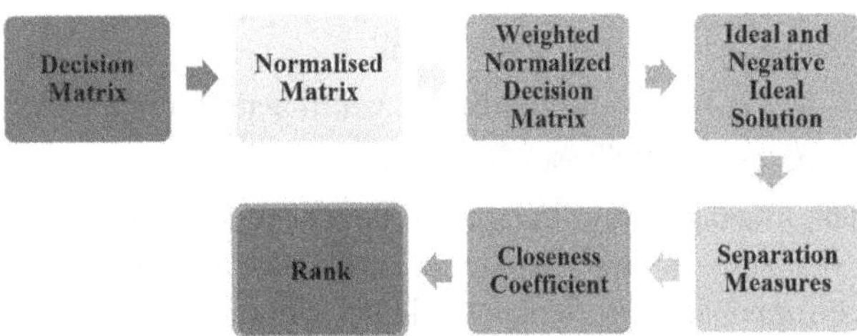

Fig. 81.8 Step-by-step flow of TOPSIS analysis

The performance characteristics—MT, OC and TA—were first normalized to eliminate scale differences using the formula:

$$m_{ij} = \frac{m_{ij}}{\sqrt{\sum_{i=1}^{m} m_{ij}^2}} \quad j=1,2,\ldots m. \tag{7}$$

Weights were assigned to the performance characteristics based on their significance (MT = 0.3; OC = 0.4; TA = 0.3).

The weighted normalized values were calculated as follows:

$$B = w_j m_{ij} \tag{8}$$

The best (U^+) and worst (U^-) values for each performance characteristic were identified as shown in Table 81.4. The best values represent the ideal solution, while the worst values correspond to the least favorable solution.

Table 81.4 Best values and worst values of TOPSIS

Characteristics	Best Value (U^+)	Worst Value(U^-)
MT	0.0372	0.1468
OC	0.0922	0.1781
TA	0.0640	0.1423

The separation from the ideal (Si^+) and the worst (Si^-) values for each experimental run was computed using the formulas:

$$S_i^+ = \sqrt{\sum_{j=1}^{m} \left(B_{ij} - B_J^+\right)^2} \tag{9}$$

$$S_i^- = \sqrt{\sum_{j=1}^{m} \left(B_{ij} - B_J^-\right)^2} \tag{10}$$

The closeness coefficient (*CC*) for each experimental run, representing its relative closeness to the ideal solution, was determined using:

$$CC = \frac{S_i^-}{\left(S_i^+ + S_i^-\right)} \tag{11}$$

The closeness coefficients for all experimental runs are detailed in Table 81.5. Experimental Run 8 achieved the highest closeness coefficient of 0.7696, indicating the most favorable combination of machining parameters (current: 1.5 A, on-time: 15 µs, off-time: 3 µs). Experimental Run 1 recorded the lowest closeness coefficient of 0.2430, reflecting the least effective parameter combination (current: 0.5 A, on-time: 10 µs, off-time: 3 µs).

The results from both GRA and TOPSIS consistently identified Experimental Run 8 as the optimal setup, achieving desirable values for MT, OC, and TA. Similarly, Experimental Run 1 was ranked lowest in both methods.

The GRA and TOPSIS values for all experimental runs are illustrated in Fig. 81.9, providing a clear comparative analysis of the performance for each set of machining parameters. Additionally, the optical microscopy images

Table 81.5 TOPSIS results of MT, OC, and TA

Normalized Value			Weighted Normalized Value			Separation Measures Value		Closeness Co-efficient Value	Rank
MT	OC	TA	MT	OC	TA	S+	S-	CC	
0.3459	0.4453	0.4744	0.1038	0.1781	0.1423	0.1340	0.0430	0.2430	9
0.3800	0.4180	0.3302	0.1140	0.1672	0.0991	0.1129	0.0554	0.3289	8
0.4399	0.2961	0.3489	0.1320	0.1184	0.1047	0.1064	0.0721	0.4038	7
0.4893	0.2991	0.2849	0.1468	0.1196	0.0855	0.1150	0.0815	0.4149	6
0.3389	0.2961	0.4513	0.1017	0.1184	0.1354	0.0997	0.0751	0.4295	5
0.1914	0.3137	0.2132	0.0574	0.1255	0.0640	0.0389	0.1300	0.7696	3
0.3104	0.3764	0.3101	0.0931	0.1506	0.0930	0.0859	0.0779	0.4755	4
0.1239	0.2306	0.2268	0.0372	0.0922	0.0680	0.0041	0.1578	0.9747	1
0.1997	0.2624	0.2576	0.0599	0.1050	0.0773	0.0293	0.1308	0.8171	2

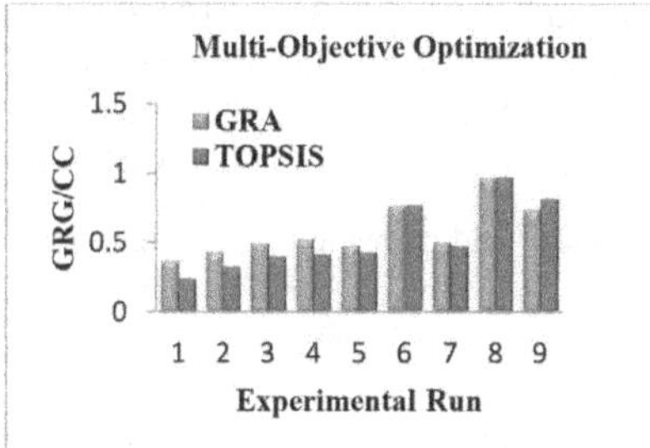

Fig. 81.9 GRG & CC values obtained from GRA and TOPSIS for each experiment

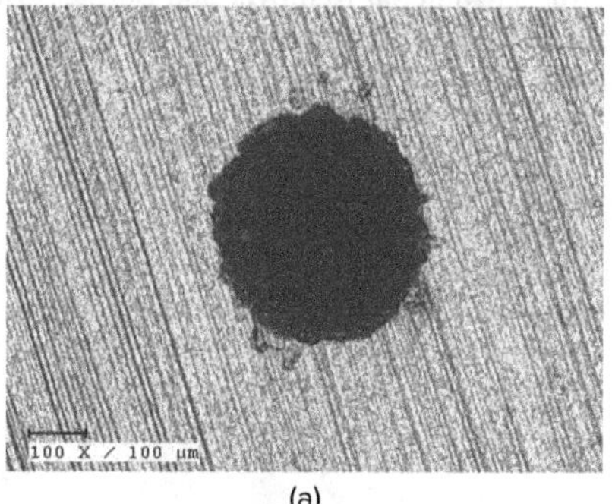

(a)

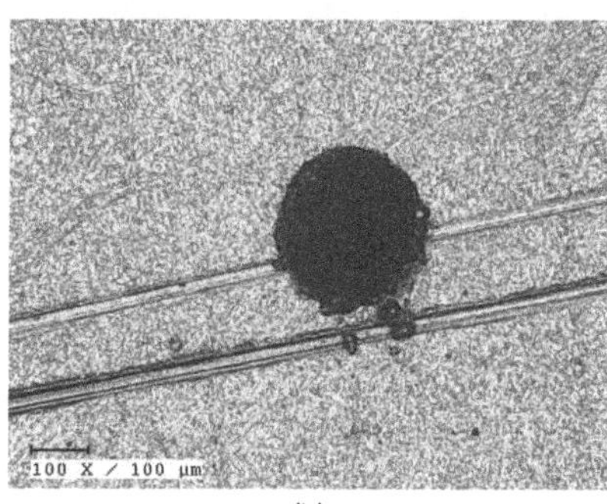

(b)

Fig. 81.10 Optical microscopy images of the machined holes a) Front side b) Rear side

of the machined holes, captured from the front and back surfaces at the optimal parameter settings, are presented in Fig. 81.10. These images highlight the dimensional accuracy and quality of the machined hole achieved under the best configuration.

4. Conclusion

This study investigated the micro-hole machining of Ti-6Al-4V alloy using electrical discharge machining (EDM) with a 200 µm diameter graphite electrode. Optimization techniques, including Taguchi, GRA and TOPSIS, were applied to analyze and refine machining parameters. The key findings are as follows:

- The machining performance characteristics—machining time, overcut, and taper angle—were strongly influenced by the machining parameters, specifically in the order of current, on-time, and off-time.
- The Taguchi method effectively identified the optimal machining settings for each performance characteristic. For minimizing machining time, the ideal parameters were determined to be a current of 1.5 A, an on-time of 20 µs, and an off-time of 3 µs. Similarly, for achieving the best results in terms of overcut and taper angle, the optimal settings were a current of 1.5 A, an on-time of 20 µs, and an off-time of 5 µs.
- Using GRA and TOPSIS for multi-performance optimization, the ideal combination of parameters was determined to be current 1.5 A, on-time 15 µs, and off-time 3 µs.

Future research will focus on evaluating the impact of larger aspect ratios and incorporating surface topography analysis into the optimization process. Additional studies may also explore the effects of different grades of graphite electrodes on machining performances.

References

1. *Prasanna, J., Rajamanickam, S., Amith Kumar, O., Karthick Raj, G., Sathya Narayanan, P.-V.-V.,* MRR and TWR evaluation on electrical discharge machining of Ti-6Al-4V using tungsten : copper composite electrode, IOP Conf. Series: Mater. Sci. Eng., 2017, vol. 197, no. 1, pp. 1–6.
2. *Jahan, M.-P., Kakavand, P., Alavi. F.,* A comparative study on micro-electro-discharge-machined surface characteristics of Ni-Ti and Ti-6Al-4V with respect to biocompatibility, Procedia Manuf., 2017, vol. 10, pp. 232–242.
3. *Kuriachen, B., Mathew, J.,* Effect of Powder Mixed Dielectric on Material Removal and Surface Modification in Micro Electric Discharge Machining of Ti-6Al-4V, Mater. Manuf. Processes, 2016, vol. 31, no. 4, pp. 439–446.
4. *Shakeel Ahmed, L., Pradeep Kumar, M.,* Cryogenic Drilling of Ti–6Al–4V Alloy Under Liquid Nitrogen Cooling, Mater. Manuf. Processes, 2016, vol. 31, no. 7, pp. 951–959.
5. *Bhaumik, M., Maity, K.,* Effect of Electrode Materials on Different EDM Aspects of Titanium Alloy, Silicon, 2018, pp. 1–10.
6. *Klocke, F., Schwade, M., Klink, A., Veselovac, D.,* Analysis of material removal rate and electrode wear in sinking EDM roughing strategies using different graphite grades, Procedia CIRP, 2013, vol. 6, pp. 164–168.
7. *Kristian, L.-A.,* Performance of two graphite electrode qualities in EDM of seal slots in a jet engine turbine vane, J. Materials Proc. Tech. 2004, vol. 149, no. 1-3, pp. 152–156.
8. *Hascalik, A., Caydas, U.,* Electrical discharge machining of titanium alloy (Ti–6Al–4V), Appl. Sur. Sci., 2007, vol. 253, no. 22, pp. 9007–9016.
9. *Amorim, F.-L., Stedile, L.-J., Torres, R.-D., Soares, P.-C., Henning Laurindo, C.-A.,* Performance and Surface Integrity of Ti6Al4V After Sinking EDM with Special Graphite Electrodes, J. Materials Eng. Performance 2014, vol. 23, no. 4, pp. 1480–1488.
10. *Younis, M.-A., Abbas, M.-S., Gouda, M.-A., Mahmoud, F.-H., Abd Allah, S.-A.,* Effect of electrode material on electrical discharge machining of tool steel surface, Ain Shams Eng. J., 2015, vol. 6, no. 3, pp. 977–986.
11. *Rajamanickam, S., Prasanna, J.,* TOPSIS on High Aspect Ratio Electric Discharge Machining (EDM) of Ti-6Al-4V using 300 μm brass rotary tube electrodes, Mate. Today: Proc., 2018, vol. 5, no. 9, pp. 18489–18501.

Note: All the figures and tables in this chapter were made by the authors.

Advances in Additive Manufacturing Technologies – Gurusamy Pathinettampadian et al. (eds)
© 2026 Taylor & Francis Group, London, ISBN 978-1-041-16687-0

Experimental and Investigating the Characteristics of Nano-Size Al_2O_3 Particales Reinforced ZA-27 Metal Matrix Composites

Nagarajan V.*, Harish Kumar V.,
Velmurugan S., Krishna Kumar R., Hari Krishna Raj S., Snegan B.
Department of Mechanical Engineering,
VelTech HighTech Dr. Rangarajan Dr. Sakunthala Engineering College,
Chennai, Tamilnadu

Abstract: In the present study, we examine the impact of addition of nano-scale Al_2O_3 (alumina) particles on microstructure, mechanical properties, and thermal behavior of ZA-27, a zinc-aluminum alloy metal matrix composite (MMC). ZA-27 is attributed with high strength-to-weight ratio and corrosion resistance. It has been tainted by inferior wear resistance and heat stability in harsh environments. These limitations were eliminated by integrating Al_2O_3 nanoparticles in the alloy in the interest of enhancing its performance for challenging engineering applications. The composite material was developed in a controlled environment with the stir casting technique to ascertain that the nanoparticles are evenly dispersed within the metal matrix. With the aid of sophisticated techniques like Scanning Electron Microscopy (SEM), X-Ray Diffraction (XRD), and Differential Scanning Calorimetry (DSC), we studied its structure, phase transition, and heat behavior. Some mechanical properties of hardness, tensile strength, and wear resistance were also examined to determine its overall performance as a material. The findings illustrated that the incorporation of Al_2O_3 nanoparticles into the ZA-27 alloy improved its mechanical properties to a significant degree, mainly hardness and tensile strength. The material was also observed to possess significantly higher wear resistance. Thermal analysis indicated increased stability, and the composite possessed lower thermal expansion and improved resistance to high-temperature degradation. The conclusion of the study is that Al_2O_3 nanoparticle-reinforced ZA-27 nanocomposite is a material full of promise for high-performance use with increased thermal stability and mechanical strength. The study indicates the potential of MMCs with nanoparticle reinforcements to accelerate the development of next-generation materials for demanding engineering environments.

◆ **Keywords:** Nano particles, Reinforcement, MMC, Corrosion resistance

1. Introduction

The potential of ZA-27/Al_2O_3 composites as novel high-performance materials. The composites marry the high corrosion and low weight of ZA-27 with the enhanced mechanical and thermal characteristics of the nano-sized Al_2O_3 reinforcement [1]. This research offers significant guidance on the enhancement of metal matrix composites for technical applications, promising cost-effective and durable alternatives to traditional materials in critical applications.

Aluminum companies have been resorting to aluminum-based metal matrix composites (MMCs) as a way of spurring aluminum demand. The cost of production and performance of the composites depend heavily on the type of reinforcement and process of manufacture. Reinforcement can be continuous or discontinuous, with

*Corresponding author: nagarajan@velhightech.com

DOI: 10.1201/9781003685906-82

continuous reinforcement normally possessing better properties. This is at a more expensive price, though, since continuous-reinforced MMCs are created using more costly production processes with prices above \$200 per pound. Discontinuous-reinforced MMCs, particularly particulate reinforcements, are considerably less expensive—usually below \$2 per pound. Particulate MMCs are manufactured by means of conventional metal casting, but powder metallurgy techniques, being more expensive, produce composites with better characteristics [2].

Aluminum MMCs, especially particulate-reinforced MMCs such as silicon carbide or alumina, have proved to be promising for use in the automotive sector. The MMCs can be used to substitute some of the steel and cast iron components of vehicles with a weight reduction of over 50%. Aluminum MMCs are being utilized in limited quantities in components like cylinder liners and pistons in a few Japanese vehicles [3].

Aluminum alloys are preferred in superior applications based on their higher strength, low density, toughness, and machinability. These characteristics render them more desirable than other materials, and their quality can be improved by producing aluminum matrix composites (AMCs). AMCs are produced by mixing aluminum with other chemically different materials, a structure with distinct properties which cannot be produced by any single material [4]. The primary benefits of AMCs are higher strength, enhanced stiffness, lower weight, better high-temperature performance, controlled thermal expansion, and enhanced wear resistance. They have better damping performance and mass control, especially for reciprocating applications.

AMCs have a set of properties that no single monolithic material can offer, and thus they are becoming ever more important in the automotive industry. Their application can lead to advantages like reduced fuel consumption, decreased noise, and lower airborne emissions, all of which are compatible with more stringent environmental regulations and the need for better fuel economy [5]. With the increasing environmental awareness, the application of AMCs in the automotive sector is anticipated to increase immensely, providing an economically sound solution in a vast array of applications [6].

These composites generally consist of ceramic reinforcements that are equiaxed (shape uniform) with an aspect ratio of less than 5. Typical ceramic reinforcements are oxides, carbides, or borides (e.g., Al_2O_3, SiC, or TiB_2) and typically comprise less than 30% by volume [7]. PAMCs are less expensive than their continuous fiber-reinforced counterparts (CFAMCs). These composites find common applications in products such as rotating blade sleeves in helicopters and train and automobile braking systems. In the automotive sector, PAMCs are under consideration for parts such as valves, crankshafts, gears, and suspension arms. They also find applications in recreational products such as golf club shafts, bicycle frames, baseball bats, and horseshoes [8].

Short alumina fibers have been a widely used reinforcement material, particularly in pistons. The use of whiskers, such as SiC whiskers, has decreased recently because of worries regarding health hazards. SiC whisker-reinforced aluminum composites have been applied in military tanks, while short fiber-reinforced composites are found in pistons and cylinder liners [9].

In these composites, reinforcements come in the form of continuous fibers, typically made of alumina, SiC, or carbon, with diameters of less than 20μm. These fibers may be arranged parallel, woven, or braided before being incorporated into the composite. Carbon fiber-reinforced aluminum matrix composites have been used in applications such as antenna waveguides for the Hubble Space Telescope and struts in the space shuttle's cargo bay [10].

The automotive industry presents significant opportunities for increasing aluminum usage, primarily driven by the need to improve fuel economy and reduce emissions. However, concerns about cost, supply, and recyclability persist. Despite the relatively low current consumption of aluminum per car, automotive manufacturers are focusing on aluminum's desirable properties—lightweight, high strength, and durability—to improve vehicle performance. Although aluminum competes with steel, advancements in both materials are expected to shape future automotive designs [11].

1. Lightweight: Aluminum-intensive vehicles can be 25-45% lighter than their steel counterparts, improving fuel economy, acceleration, braking, and handling.
2. Formability: Aluminum is easily alloyed and formed, offering design flexibility with lower processing costs compared to steel.
3. Safety: Aluminum's greater thickness in frames allows better crash energy absorption and comparable performance in crash tests.
4. Cost/Price: While aluminum is more expensive than steel, its price premium is gradually decreasing as manufacturing techniques improve.
5. Recyclability: Both aluminum and steel are fully recyclable, which supports their use in sustainable vehicle production.

1.1 Aluminum Use in the Automotive Industry

In the late 20th century, automakers reduced vehicle weight by about 25%, improving fuel efficiency and performance. By reducing weight further, specifically by adding aluminum, the sector can comply with more stringent fuel efficiency and emissions regulations. At present, more than 100 automobile components, such as engines, transmissions, and chassis components, are already produced from aluminum. Luxury sports cars, such as the Chevrolet Corvette and Audi A8, have made heavy use of aluminum, showcasing its increasing relevance to car design.

ZA-27 is a zinc-aluminum alloy, normally composed of about 27% aluminum and also containing copper and magnesium. It's selected for its low weight, superior strength, and good corrosion resistance, and hence is a good matrix material in composites. Zinc-aluminum alloys, such as ZA-27, were initially designed for gravity casting but are increasingly used in high-pressure, cold-chamber die casting as well [12]. The alloy contains 27% aluminum and 2.2% copper and is hence a more aluminum-rich alloy than classical zinc alloys. Its high aluminum content provides ZA-27 with superior bearing characteristics. Its high density, lightweight nature, and silver color make it perfect for use in applications that need strength [13].

ZA-27 is a lesser-used material in foundry applications, but it is much stronger than regular cast aluminum and can equal the tensile strength of grey cast iron. For instance, in the safety restraint market, ZA-27's higher strength allowed parts to function beyond the ability of regular aluminum alloys. Ideal for bearings work, ZA-27 is simple to machine and is polishable, platable, paintable, and anodizable [14]. Having a high melting point, ZA-27 is suitable where service temperatures approach 150°C. In addition, its sparking characteristics render it well suited to bearings applications in that it represents a cost and reliability advantage. Markets where the use of ZA-27 is advantageous are in the areas of aeronautics, farm equipment, construction, general engineering, and textile machinery. Aluminum oxide (Al_2O_3) nanoparticles, which are usually 20 to 100 nanometers in size, are normally employed as reinforcement elements owing to their good hardness, wear resistance, and heat stability, which enhance the mechanical strengths of the composites.

The properties of nanoparticles are greatly dominated by their size. With a change in their dimensions, the structure of atoms also changes, which greatly influences the properties of the nanoparticle. Nanoparticles consist of three different layers: Surface layer, including metal ions or molecules that govern the functional characteristics of the nanoparticle. The shell layer that is quite distinct from the core and surface layers. Core layer, being the fundamental chemical backbone of the nanoparticle. Unique physicochemical properties, comprising thermal, magnetic, mechanical, optical, and electronic features, are inherent with nanoparticles but sometimes not with materials in bulk forms.

2. Result and Discussion

2.1 Micro Structure

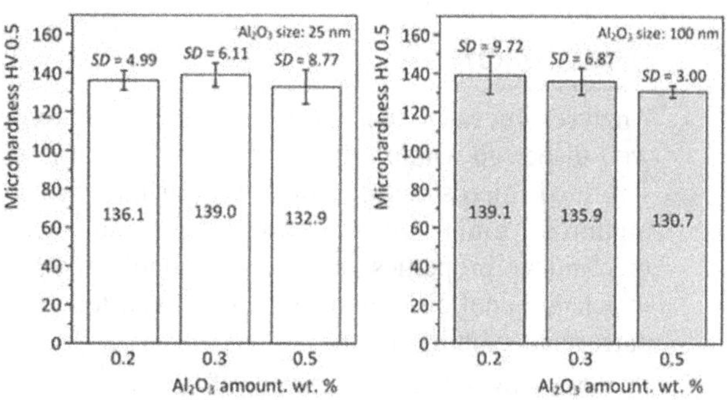

Fig. 82.1 Particle size

Nanoparticles exhibit unique absorbance and emission spectra in contrast to bulk materials. The properties are determined by electron energy levels, which also influence electric conductivity, emission, and absorption. Nanoparticles are also investigated for their capillary force, friction, adhesion, and elastic modulus, which are particularly important for wear-resistant products, lubricants, and particle removal processes.

The thermal conductivity of nanoparticles is dependent on the composition of the metal from which they are constructed. Metals such as aluminum oxide and copper have superior thermal conductivity compared to the majority of non-metallic compounds and can be used to improve thermal properties of liquids when added as dditives. Thermogravimetric analysis or TGA is often employed in the analysis of the Al foil and identification of its properties to further develop knowledge on how the Al_2O_3 nanoparticles help the overall behavior of the material. These trends emphasize the expanding role of aluminum in reducing automobile weight, achieving better fuel economy, and adhering to upcoming environmental regulations.

2.2 Wear Rate

Wear rates were measured over the entire testing period, providing a total wear rate for each condition. To compare these results with existing literature, wear factors were calculated using a well-established equation. The obtained wear rate values ranged from 10−810^{-8} to 10−610^{-6} mm^3/Nm, aligning with previously reported outcome is consistent method used in the tests, where lubricant was supplied by the rotating disc. Since the disc was only partially submerged in the oil container and the block-disc contact occurred above the oil level, lubrication was not fully continuous, leading to a combination of mixed and boundary lubrication.

Wear rate predictions, derived using Equation (1) and illustrated as wear maps, provided further insights. In these maps, two factors were held constant at their mean values while the effects of the remaining two factors were visualized.

3. Conclusion

1. Analyses revealed that nano Al_2O_3 particles were well-dispersed within the ZA-27 matrix, resulting in a refined microstructure. This uniform distribution minimized grain size, 47 contributing to enhanced mechanical properties and a more stable matrix, which is beneficial for maintaining consistency in performance under stress.
2. Increasing overall hardness and reinforcing the matrix against deformation. However, a slight reduction in ductility was observed, likely due to the embrittlement effect of the rigid Al_2O_3 particles. This trade-off may be manageable in applications where high strength and wear resistance are prioritized over ductility.
3. Wear testing demonstrated that Al_2O_3-reinforced ZA-27 composites had significantly improved resistance to abrasive wear. The hard, inert Al_2O_3 nanoparticles helped to protect the composite surface, reducing material loss during wear and making the composite suitable for high-stress environments.
4. Thermal analysis using DSC and TGA showed that Al_2O_3 nanoparticles increased the thermal stability of the composite. The ZA27/Al_2O_3 composite exhibited reduced thermal expansion and improved resistance to degradation at elevated temperatures, making it suitable for high temperature applications.
5. These findings suggest that nano Al_2O_3-reinforced ZA27 composites have potential for applications in industries such as automotive, aerospace, and heavy machinery, where high wear resistance, strength, and thermal stability are critical.
6. Although the study yielded promising results, further research is recommended to optimize particle distribution, improve ductility, and explore other reinforcement methods.

References

1. ZA-27 Alloy Composites Reinforced with Al_2O_3 Particles(2018). This study explores the tribological properties of ZA-27 alloys reinforced with Al_2O_3 particles, highlighting enhanced wear resistance
2. Nanoindentation of ZA-27 Alloy Based Nanocomposites Reinforced with Al_2O_3 Particles(2019). This research investigates the mechanical characterization of ZA-27 nanocomposites using nanoindentation techniques, emphasizing the role of Al_2O_3 nanoparticles in modifying mechanical properties
3. Effects of Al_2O_3 Particle Reinforcement on the Lubricated Sliding Behavior of ZA-27 Alloy(2020). The study examines how Al_2O_3 particle reinforcement influences the lubricated sliding behavior of ZA-27 alloys,with a focus on tribological performance
4. Wear Behavior of Composites Based on ZA-27 Alloy Reinforced by Al_2O_3 Particles(2021). This paper presents findings on the wear characteristics of ZA-27 composites reinforced with varying amounts of Al_2O_3 particles,demonstrating improved wear resistance
5. Tribological Potential of Particulate Composites with ZA-27 Alloy Matrix"(2022). The research explores the tribological effects of particle reinforcement, including Al_2O_3, on the wear behavior of ZA-27 alloy composites
6. Tribological Investigation of ZA-27 Alloy Based Micro/ Nano Mixed Composites(2023). This study investigates the tribological properties of ZA-27 composites reinforced with a combination of micro and nano-sized particles, including Al_2O_3, highlighting the effects on wear behavior
7. Dry Sliding Friction and Wear Behavior of ZA-27 Alloy Nanocomposite Reinforced with SiC Nanoparticles" (2018). While focusing on SiCnanoparticles, this study provides insights into the effects of nano-sized ceramic particle reinforcement on ZA-27 alloys
8. A Brief Review on Mechanical and Thermal Properties of Banana Fiber Reinforced Composites(2021). This review discusses the mechanical and thermal properties of natural fiber-reinforced composites, offering a comparative perspective on composite materials
9. Forming Process and Simulation Analysis of Helical Carbon Fiber Reinforced Metal Matrix Composites (2021). The paper delves into the manufacturing processes and simulation analyses of carbon fiberreinforced metal matrix composites, relevant to understanding the processing of ZA-27/Al_2O_3 composites
10. Nanomaterials for Application in Wound Healing: Current State-of-the-Art (2021). Although focusing on biomedical applications, this article discusses the properties and applications of nanomaterials, providing context for the use of nanoparticles in composites
11. β-Ga_2O_3 Material Properties, Growth Technologies, and Devices: A Review(2021). This review covers the properties and applications of βGa_2O_3, offering insights into the characteristics of oxide materials similar to Al_2O_3 used in composites
12. A Review of Ga_2O_3 Materials, Processing, and Devices (2018). Providing a comprehensive overview of Ga_2O_3 materials, this paper offers comparative insights into the processing and properties of oxide materials relevant to Al_2O_3 composites
13. Natural Fiber Composites (NFC) Market Trends (2020). This market analysis report includes data on the trends and growth of natural fiber composites, offering a broader perspective on the composite materials market
14. Composite Materials Industry - Statistics & Facts (2021). This statistical report provides an overview of the composite materials industry, including

Note: All the figures in this chapter were made by the authors.

Advances in Additive Manufacturing Technologies – Gurusamy Pathinettampadian et al. (eds)
© 2026 Taylor & Francis Group, London, ISBN 978-1-041-16687-0

A Case Study on Construction Site Safety Using Artificial Intelligence for Accident Prevention

Vijayakumar M.[1]
Assistant Professor, Department of Civil Engineering,
Vel Tech High Tech Dr Ranagarajan & Dr Sakunthala Engineering College,
Chennai, India

Kishore B.[2], **Balaraman S.**[3], **Deepak Kumar P.**[4]
UG Student, Department of Civil Engineering,
Vel Tech High Tech Dr Ranagarajan & Dr Sakunthala Engineering College,
Chennai, India

◆ **Abstract:** In recent years, machine learning has made significant strides in the construction industry. The main goal of developing effective software is to improve solutions that tackle potential security risks and dangers commonly found on construction sites. However, despite these advancements, accidents and injuries in construction work are still quite common, making it one of the most dangerous jobs out there. This research aims to assess the key factors related to artificial intelligence (AI) that can improve safety in construction, using the Analytic Hierarchy Process (AHP) method. Construction safety experts were selected to participate in this study. A hierarchy was created that included five main factors, which were then broken down into twelve sub-factors. The analysis showed that the workplace is the most important AI factor in the construction field, making up 25.43 %. It was closely followed by security at 25.37 % and human error at 22.76 %. The top sub-factors highlighted include communication with safety officers when workers ignore safety rules on sites, which is at 69.21 %, and cloud computing technology, inspection, control, and training at 60.04 %. Additionally, AI's ability to foresee potential safety issues in construction is noted at 14.29 %. The results of this study can help software developers prioritize how to apply AI to improve safety measures on construction sites.

◆ **Keywords:** Artificial intelligence (AI), Construction safety, Risk, Technology

1. Introduction

In the complex world of construction, safety is a top priority. This industry is known for its intricate processes and ever-changing environments, which come with numerous hazards. Every day, construction workers encounter a range of dangers, from the depths of dig sites to the heights of skyscrapers. Managing safety is a difficult but necessary responsibility because of these inherent risks as well as the unpredictability of building sites [1]. The industry continues to face high accident rates despite the implementation of stringent safety regulations. Unexpected site conditions, machine malfunctions, and human error can all combine to cause a perfect storm of safety problems. Beyond only causing bodily pain, the repercussions might also include project delays, legal issues, and reputational damage. Therefore, improving safety solutions is not just the industry's wise course of action, but also its moral obligation [2]. One innovative technology that is revolutionizing numerous industries,

[1]vijayakumar.m@velhightech.com, [2]vh13514_civil23@velhightech.com, [3]vh13510_civil23@velhightech.com, [4]vh13511_civil23@velhightech.com

DOI: 10.1201/9781003685906-83

including construction, is artificial intelligence (AI). Because AI can analyze data, make predictions, and automate difficult jobs, it has the potential to revolutionize safety procedures. It presents a future in which safety management is proactive and predictive in addition to reactive [3].

AI has a lot of potential to increase construction safety. It forecasts and reduces potential risks by examining vast amounts of data from previous initiatives. Additionally, it can monitor construction sites in real time and notify managers of any safety infractions. AI can also run construction equipment on its own, which lowers the possibility of mishaps brought on by human error. However, there are unique difficulties in incorporating AI into the construction sector. In addition to disseminating knowledge, the objective is to encourage dialogue on how AI might be used to create a safer workplace for everybody [4].

1.1 Objectives

- To lower on-site mishaps and guarantee security during the building phase.
- To recognize and evaluate construction-related risks in order to avert mishaps.
- To avoid mishaps by putting preventative safety measures in place.
- To encourage the advancement of construction projects free from risks or mishaps.
- To oversee a variety of construction site operations in order to ensure safety compliance.

1.2 Problem Statement

Worker safety and the general efficacy of construction projects are seriously threatened by accidents at construction sites. Even with all of the safety rules and procedures in place, accidents sometimes happen, leading to financial losses, project delays, injuries, and fatalities. The difficulty is in identifying the underlying reasons for these mishaps, evaluating the safety protocols in place, and creating workable plans to stop and reduce incidents on building sites. Everyone working on construction projects will benefit from a safer working environment thanks to this strategy [5].

With support from the Alabama Hong Kong Entrepreneurs Fund and Artesian Ventures, SOSV and Vectra Ventures are introducing VI-ACT, an AI-driven solution, to solve this pressing issue. This state-of-the-art software uses video analytics to monitor construction activities in real-time, helping to identify possible hazards and facilitating safe and effective work that protects everyone on site and avoids accidents. Our goal is to improve construction project safety by utilizing this technology [6].

2. Literature Review

1. Safety Concerns in the Construction Industry: Given the high incidence of occupational fatalities and injuries in the construction industry, everyone concerned must prioritize safety. Thousands of people in this industry are injured annually, and regrettably, many of them die as a result of on-the-job accidents. Because construction work involves complicated jobs, high elevations, and heavy machines, the environment is dangerous [7].
2. Even the most experienced workers can sometimes overlook the dangers present in their environment. Factors such as the fast pace of construction projects, fatigue, and complacency can lead to poor decision-making and violations of safety protocols. Common risks include electric shocks, being struck by moving objects, falls from heights, and injuries caused by machinery. These hazards are exacerbated by the ever-changing conditions on construction sites, which can be influenced by weather, project requirements, or the arrival of new equipment.
3. Current safety management techniques may not always be sufficient. Insufficient safety equipment, lack of training, and ineffective teamwork can all foster a culture that overlooks safety. Additionally, having multiple contractors and subcontractors can complicate safety management, as differing safety standards and procedures can heighten risks and lead to confusion.
4. To overcome these problems, safety-minded culture should be formed by construction companies where responsibility, communication and training are at the top of the agenda. Proper safety checks, comprehensive training, and strict compliance with safety protocols are all able to prevent risk to a large extent, offering worker protection on the construction site. In the end, safety comes first, not only protecting workers but also increasing productivity and their project outcomes.

Artificial Intelligence's Potential to Increase Construction Safety

With the development of technology, artificial intelligence (AI) is reshaping the construction industry, most especially around improvements in safety. Since AI is able to process large datasets and detect patterns, it can be a useful tool to predict events that are about to happen. Using artificial intelligence, construction companies can take an anticipatory role in Mitigating safety risks and providing their workers the kind of safer working environment they need.

A promising application of AI to safety in construction is its use for the analysis of visual data from CCTV systems. Artificial intelligence (AI) systems can identify people and situations which do not conform with safety standards by image and video analysis. For example, AI may be able to identify whether workers are wearing the correct personal protective equipment (PPE) or whether safety barriers are put up effectively. This real-time monitoring reduces the risk of accidents through fast corrective actions (Fig. 83.1).

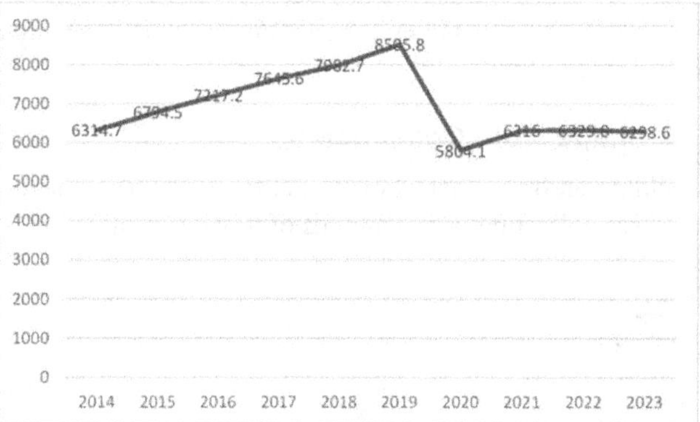

Fig. 83.1 Incidence rate

Incidence rates of work-related accidents with sick leave in the construction sector. Evolution from 2014 to 2023 [12–19].

3. Methodology

The survey used a questionnaire created with help from experts and past studies on construction safety. We analyzed the data using the Analytical Hierarchy Process (AHP). First, we looked at how artificial intelligence (AI) can improve safety in construction [8].

We found key AI aspects that matter for construction safety. These are shown in Fig. 83.1. The AHP diagram in the figure lays out the important parts of using AI in this area. We focused on what's needed for AI to work well in construction safety. This includes things like budget, communication, human errors, workplace conditions, and technology [9].

In the primary survey, an online questionnaire was utilized to collect data. This approach specifically targeted specialists in the field of construction safety. The definitions and terminology included in the questionnaire were clearly articulated for the participants. Within the Analytic Hierarchy Process (AHP), the decision-making challenge is typically divided into a structured hierarchy of smaller components that can be examined independently. Each segment of this hierarchy corresponds to various facets of the overarching decision. Once the hierarchy is formed, experts assign numerical values to each pair of alternatives [10].

For this research, image data from the construction site was accumulated through videos, photographs, and web scraping. The procedure for acquiring image data consisted of several stages. Initially, the video file was segmented into individual frames, ranging from the first to the last. Only those frames containing work objects were chosen for training and saved as still images. These selected frames were annotated in line with the COCO MS dataset format for training with ResNet [11].

The photographs that showed workers or helmets were then selected from the collection and tagged in accordance

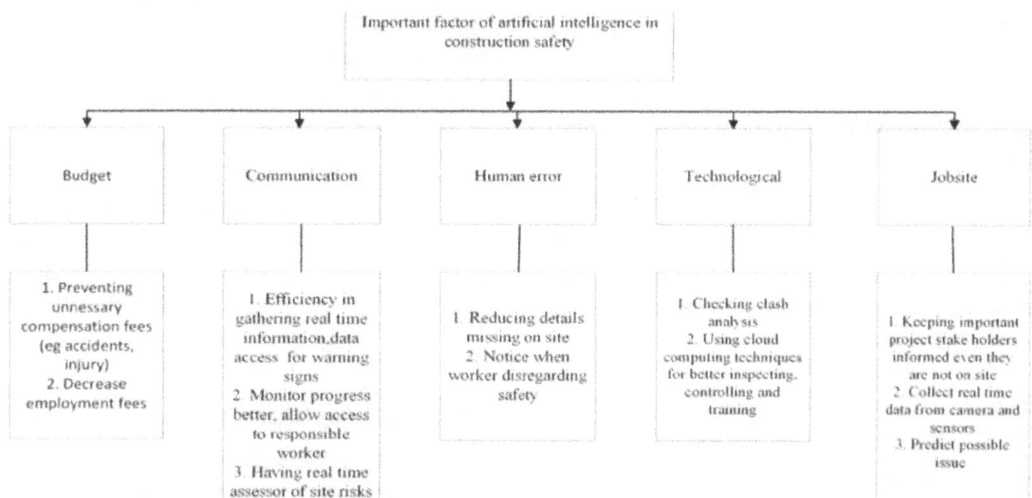

Fig. 83.2 Importance of artificial intelligence in construction safety

with the MS COCO dataset's structure. Additionally, to supplement the image collection, a virtual construction setting was created, collecting different kinds of photographs (like hard helmets and construction workers) from this simulated location. The physics engine in Unity was used to simulate worker falls and motions linked to fall-related injuries, and the virtual environment was built using Unity tools [12].

In the primary survey, an online questionnaire was utilized to collect data. This approach specifically targeted specialists in the field of construction safety. The definitions and terminology included in the questionnaire were clearly articulated for the participants. Within the Analytic Hierarchy Process (AHP), the decision-making challenge is typically divided into a structured hierarchy of smaller components that can be examined independently. Each segment of this hierarchy corresponds to various facets of the overarching decision. Once the hierarchy is formed, experts assign numerical values to each pair of alternatives [13].

For this research, image data from the construction site was accumulated through videos, photographs, and web scraping. The procedure for acquiring image data consisted of several stages. Initially, the video file was segmented into individual frames, ranging from the first to the last. Only those frames containing work objects were chosen for training and saved as still images. These selected frames were annotated in line with the COCO MS dataset format for training with ResNet [14].

After that, photos of workers or helmets were chosen from the collection and labelled using the structure of the MS COCO dataset. A virtual construction site was also made to augment the image collection, and various types of photos (such as hard helmets and construction workers) were taken from this made-up area. The virtual environment was constructed using Unity tools, and worker falls and motions associated with fall-related injuries were simulated using Unity's physics engine.

To improve worker safety when working at heights, we have created an advanced model to identify falls or slips from buildings.

In order to This model uses a pose estimation technique to analyse the worker's movements and posture in real time. Classifying the state of the employee into three different

The main goal is to sort them into three categories: normal, warning, and danger. This helps us act quickly and avoid accidents.

We used images of people in different poses and settings. The MPII Human Pose dataset is super helpful for training our systems to recognize and understand human poses. It contains over 25,000 pictures with key body parts marked. The variety of poses in this dataset allows our model to learn about many movements. This is important for spotting slips or falls.

For data collection, we pulled information from pose estimation and image databases. We got pose data from two reliable sources: the Dense Pose-COCO dataset and the MPII Human Pose dataset. These are well-known resources in pose estimation and prediction research. They provide a valuable collection of labeled information (Fig. 83.2).

The Dense Pose-COCO dataset gives us detailed connections between 2D images and 3D body models. This helps our algorithm grasp body positions and angles better. Understanding this is key for spotting unstable poses that could lead to falls.

3.1 Image Data Collection

Building an object detection model for a construction site requires collecting large amounts of training data.

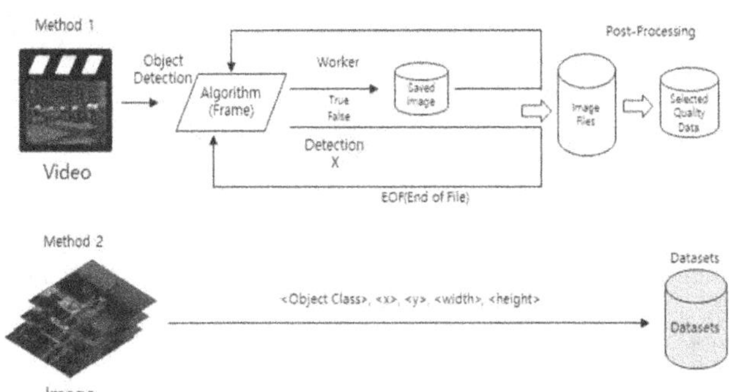

Fig. 83.3 Image data collection

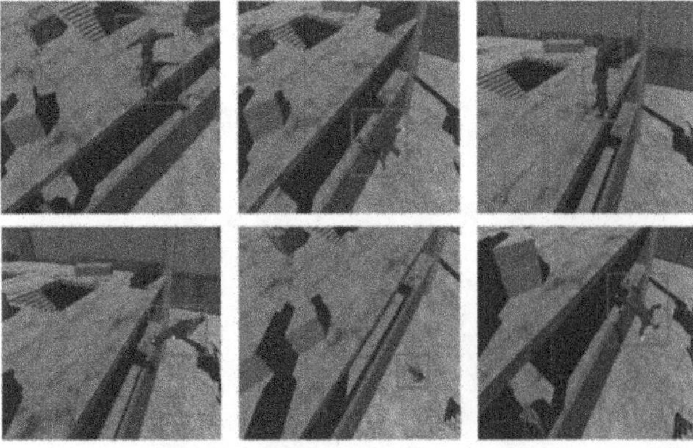

Fig. 83.4 Dense pose-COCO

3.2 Model Architecture

The model's architecture is designed to manage the input data efficiently and produce reliable outcomes. It includes a number of crucial components:

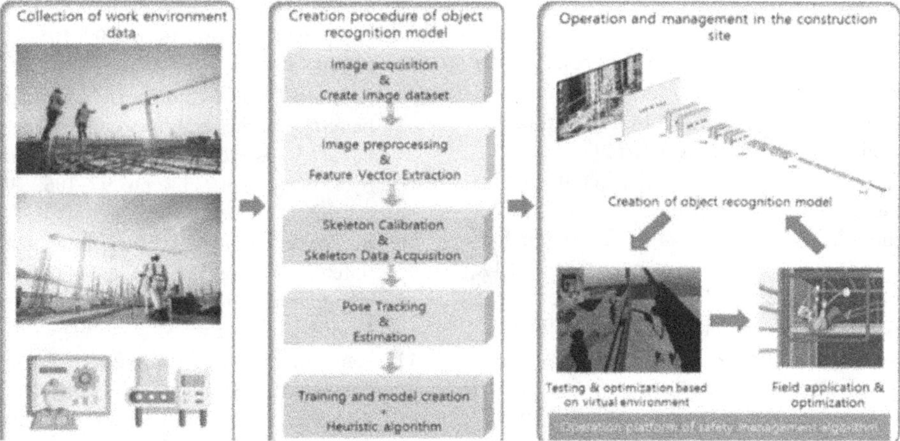

Fig. 83.5 Model's architecture

1. Pre-processing: To preserve consistency in size and format, the incoming photos must go through pre-processing. Normalization and other augmentation techniques are used in this phase to increase the model's adaptability to variations in background, lighting, and worker attire.
2. Module for Pose Estimation: Convolutional neural networks (CNNs) are used in this component to anticipate the important locations of the human body and extract information from the input photos. This module's output is a set of coordinates that show where key body parts are located. These coordinates are then used to assess the worker's posture (Fig. 83.3).
3. Classification Layer: Following the acquisition of pose data, the model includes a classification layer that looks at the spatial correlations between the important spots that have been detected. The purpose of this layer is to classify the worker's state as "normal," "warning," or "danger."

The Future of AI in Construction

1. The construction industry is thought to be among the least digitalized in the world, frequently lagging behind in the use of contemporary technologies. Construction has been sluggish to adopt artificial intelligence, but many other industries have effectively used it to boost production. The industry must use AI applications and tactics to remain competitive in a market that is evolving all the time. This shift is essential to ensure that construction stays relevant in the future as well as to stay up with new competitors.
2. Construction site safety and accident management are serious issues that need to be addressed right now. The industry has the ability to significantly improve safety protocols and reduce the inherent dangers connected with construction activity by utilizing artificial intelligence. In order to prevent accidents and promote a safer workplace, artificial intelligence (AI) solutions can offer real-time oversight, predictive insights, and decision-making support. The implementation of these AI-powered technologies has the potential to completely transform how safety management is approached in the building industry.

4. Conclusion

1. Without the use of contemporary technologies, traditional safety procedures in the construction industry frequently fall short in addressing safety concerns. The importance of artificial intelligence (AI) in enhancing safety in construction environments is highlighted by this study. It was discovered that the work environment (25.43%), technology (25.37%), and human error (22.76%) are the main factors influencing safety through AI.
2. In contrast, the least important elements were found to be budgetary constraints and communication, which received ratings of 12.64% and 13.79%, respectively. Safety inspectors must pay close attention to these AI-related issues because construction sites are prone to mishaps. AI adoption can improve safety and health results in building projects since robots and AI technology can do on-site inspections and undertake hazardous tasks. The report suggests a number of directions for further investigation, including examining the impact of AI at every step of building, developing new AI safety applications, and contrasting AI with IoT in addressing construction safety concerns.

REFERENCES

1. S. O. Abioye et al., "Artificial intelligence in the construction industry: A review of present status, opportunities and future challenges," J. Build. Eng., vol. 44, no. October, p. 103299, 2021, doi: 10.1016/j.jobe.2021.103299.
2. Y. Pan and L. Zhang, "Roles of artificial intelligence in construction engineering and management: A critical review and future trends," Autom. Constr., vol. 122, no. October 2020, p. 103517, 2021, doi: 10. 1016/j.autcon.2020.103517.
3. M. Mohan and S. Varghese, "IRJET-Artificial Intelligence Enabled Safety for Construction Sites Artificial Intelligence Enabled Safety for Construction Sites," Int. Res. J. Eng. Technol., p. 681, 2008.
4. K. van Nunen, J. Li, G. Reniers, and K. Ponnet, "Bibliometric analysis of safety culture research," Saf. Sci., vol. 108, no. June 2017, pp. 248–258, 2018, doi: 10.1016/j.ssci.2017.08.011.
5. X. Yi and J. Wu, "Research on Safety Management of Construction Engineering Personnel under 'Big Data + Artificial Intelligence,'" Open J. Bus. Manag., vol. 08, no. 03, pp. 1059–1075, 2020, doi: 10.4236/ojbm.2020.83067.
6. C. Herweijer et al., "Enabling a sustainable Fourth Industrial Revolution: How G20 countries can create the conditions for emerging technologies to benefit people and the planet," Econ. Open-Access, pp. 1 17, 2018.
7. G. Patil, "Applications of Artificial Intelligence in Construction Management," 6th Natl. Conf. Technol. Innov. Disrupting Businesses, Transform. Mark., vol. 9, no. March, p. 8, 2019.
8. S. Aminbakhsh, M. Gunduz, and R. Sonmez, "Safety risk assessment using analytic hierarchy process (AHP) during planning and budgeting of construction projects," J. Safety Res., vol. 46, pp. 99–105, 2013, doi: 10.1016/j.jsr.2013.05.003.
9. J. Krejc í and J. Stoklasa, "Aggregation in the analytic hierarchy process: Why weighted geometric mean should be used instead of weighted arithmetic mean," Expert Syst. Appl., vol. 114, pp. 97–106, 2018, doi: 10.1016/j.eswa.2018.06.060.
10. R. Chakkravarthy, "Artificial Intelligence for Construction Safety," Prof. Saf., vol. 64, no. 1, p. 46, 2019. www.irjmets.com @International Research Journal of Modernization in Engineering, Tech
11. Kamardeen, I. Web-based Safety Knowledge Management System for Builders: A Conceptual Framework. In Proceedings of the CIB W099Conference, Melbourne, Australia, 21–23 October 2009.
12. Li, X.; Yi, W.; Chi, H.-L.; Wang, X.; Chan, A.P. A critical review of virtual and augmented reality (VR/AR) applications in construction safety. Autom. Constr. 2018, 86, 150–162. [CrossRef]
13. Zhong, B.; Li, H.; Luo, H.; Zhou, J.; Fang, W.; Xing, X. Ontology-Based Semantic Modeling of Knowledge in Construction: Classification and Identification of Hazards Implied in Images. J. Constr. Eng. Manag. 2020, 146. [CrossRef]
14. Gao, S.; Ren, G.; Li, H. Knowledge Management in Construction Health and Safety Based on Ontology Modeling. Appl. Sci. 2022, 12, 8574. [CrossRef]
15. Mihi′ c, M. Classification of Construction Hazards for a Universal Hazard Identification Methodology. J. Civ. Eng. Manag. 2020, 26, 147–159. [CrossRef]

Advances in Additive Manufacturing Technologies – Gurusamy Pathinettampadian et al. (eds)

Dynamic Analysis of Lead Screw Mechanism for Linear Actuator Using Ansys

Saravanan Tamilselvan[1]
Department of Mechanical Engg, Hindustan Institute of Tech & Science, Padur, Chennai, India

Manikanda Gopalan Sivan[2]
Department of Aeronautical Engg, Gojan School of Business and Tech, Redhills, Chennai, India

Sivasankar A.[3]
Department of Mechanical Engg, Hindustan Institute of Tech & Science, Padur, Chennai, India

Sathish Kumar R.[4]
Department of Automobile Engg, Hindustan Institute of Tech & Science, Padur, Chennai, India

Ranjith Paramasivam[5]
Department of Mech. & Automa. Engg., Gojan School of Business and Tech, Redhills, India

Balamurugan S.[6]
Department of Mechanical Engg, Chennai Institute of Technology, Kundrathur, Chennai, India

Hariram Venkatesan[7]
Department of Mechanical Engg, Hindustan Institute of Tech & Science, Padur, Chennai, India

Gobinath V. M.[8]
Department of Mechanical Engg, Rajalakshmi Institute of Technology, Chembarambakkam, Tamil Nadu

◆ **Abstract:** Linear actuator, a wide variety of applications where linear motion is necessary such as valves, damper, disk drives, printers, computer peripherals and other industrial machinery are employed i.e., force through a straight line. There's a few designs and concepts for linear actuator technology to achieve high performance in terms of speed, acceleration, precision, and so on. The behavior of dynamic load (vibration) variables considered as problem under selection of suitable linear actuator for its application. This study belongs to mechanical actuator (leadscrew mechanism) which proposed for braking application of automobiles. So, the different design parameters of leadscrew mechanism have been investigated in this study to control vibration error. The standard parameters of thread profiles square, v-cut and ACME shapes have chosen and carried out the harmonic analysis using ANSYS software. The

[1]saravanantamilselvan6@gmail.com, [2]manikandgopalan.s@gsbt.edu.in, [3]sivasa@hindustanuniv.ac.in, [4]sathish.rajamanickam@gmail.com, [5]ranjith.p@gsbt.edu.in, [6]balamurugans@citchennai.net, [7]connect2hariram@gmail.com, [8]vmgobinath@gmail.com

DOI: 10.1201/9781003685906-84

paper focuses the effect of frequency response in leadscrew, shear stress and principal stress are discussed in detail for different profiles. Comparatively, the v-cut thread performed better dynamic characteristics than the other profiles.

◆ **Keywords:** Linear actuator, Leadscrew mechanism, Thread profiles, Dynamic analysis

1. Introduction

The demand for high-performance linear actuators has developed in part due to the growing investment and technical breakthroughs in aviation, aerospace, navigation, and the car industry. Engineers have invented and improved these actuators, capable of producing linear motion and force, to meet this demand. Presently, many countries are giving utmost importance to the necessity of linear actuators with high dependability and safety in crucial systems, such as aileron control, landing gear, spacecraft propulsion control, helicopter attitude control, rudder control, and different automobile applications. Geometric standard profiles, such as Acme and V-thread, categorize power leadscrews, crucial elements of linear actuators [1]. Despite exhibiting greater friction and being less optimal for leadscrews compared to Acme threads, micro-machines can still employ V-threads. Thirugnanam et al. [2] found that manual findings are highly precise and closely aligned with the real values. In a study conducted by Bentgen et al. [3], it was observed that the power screw's equivalent stress in static analysis falls within the permissible limits.

Huangchao et al. [4] conducted a study to investigate the method by which actuators transmit actuation effects to host structures. They employed modelling tools to examine the static and dynamic electromechanical behaviour. They conducted their investigation using Ansys software, examining the impact of material characteristics, adhesive layers, and loading frequency on stress and displacement. These elements significantly influence the stresses imparted to the main structure and the unique behaviour at the actuator ends. Lead screw first deformed as a result of the steady stress applied to it [5]. An investigation has revealed that tension accumulates at the point where the thread distributes out along the lead screw body, leading to a variety of stresses and deformations, as evidenced by the mode shape. Variations in screw length over the stroke length reduce the amplitude of the system's inherent frequency, while changes in thread pitch have a less significant impact. When the working position changes during the stroke length, the natural frequency fluctuates significantly. As the nut approaches the center of the screw, the pitch alterations decrease the tension. However, as it goes towards the rear, the stress once again intensifies. By gaining this extensive understanding, we are able to devise and execute the use of linear actuators that increase the efficiency and dependability of vital systems. Increased thrust and improved traction are two ways in which the spiral wheels' bigger blade diameters and narrower tooth profile angles improve propulsion. Most of the stress in precision roller screw systems is distributed across the thread pairs of the rollers, screws, and nuts [6]. It is possible to enhance transmission load distribution by decreasing machining mistakes. Backlash tolerance is an important consideration in the design of precision roller screw mechanisms, particularly in the aerospace industry, where it may impact the load capacity and accuracy. Contact point deformations brought on by loading impact efficiency, stiffness, and friction. Transmitter efficiency is enhanced by reducing axial backlash via optimization of geometric characteristics. This study attempt to be made for materials to evaluate the examines the design characteristics of the leadscrew mechanism in dynamic circumstances and manually calculate the theoretical shear stress of the power screw using the Ansys program to enhance our understanding and enhance the performance of these mechanisms.

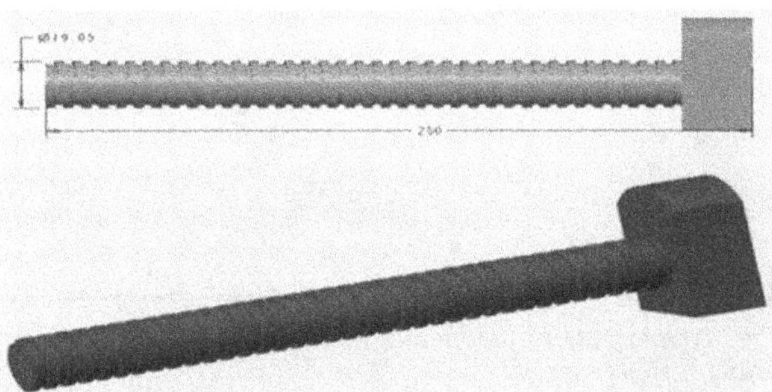

Fig. 84.1 3D model of leadscrew

2. Methods and Modelling

The leadscrew mechanism was developed using SOLIDWORKS software, a computer-aided design (CAD) tool that is renowned for its precision and adaptability in the development of mechanical component models. The objective was to evaluate the suitability of various thread profiles for linear actuators. Lead screws provide exceptional precision and control over movement. The lead screw mechanism is capable of supporting substantial loads and maintaining its position without any slipping. The design was created to prevent this type of damage from occurring in the future. The leadscrew length of 250 mm and diameter 19.05 mm with slider nut for required linear motion. Figure 84.1 indicates the CAD model of lead screw whereas Fig. 84.2 indicates the standard thread profile of leadscrew chosen in this investigation. The standard thread profiles are V-thread, Acme thread and rectangular thread. The structural steel material is assigned to the leadscrew and nut in the simulation. The physical and mechanical properties of chosen material are 7.85 g/cm^3 density, young's modulus of 200 GPa and Poisson's ratio 0.3. The chemical composition of steel is tabulated in below Table 84.1.

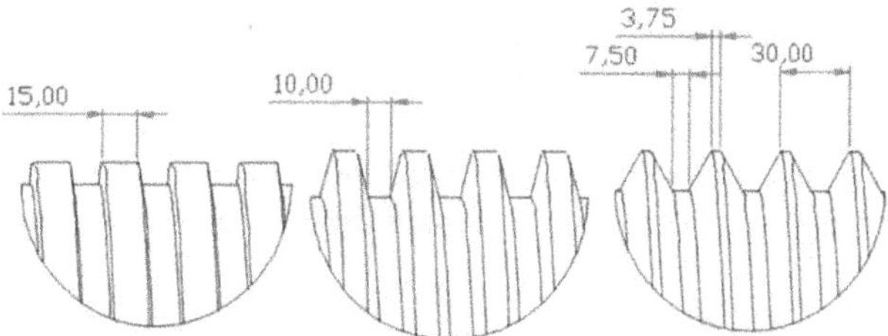

Fig. 84.2 Thread profiles of rectangular, acme and v-cut

Table 84.1 Chemiacl composition of steel material

Materials	Carbon (C)	Iron (Fe)	Manganese (Mn)
Percentage	0.03-125	80-90	0.2-16
Materials	**Phosphorus (P)**	**Silicon (Si)**	**Sulphur (S)**
Percentage	<0.05	0.5	<0.05

2.1 FEA on Dynamic Characteristics

To simulate the dynamic properties of a leadscrew with various thread profiles using ANSYS Workbench, start by preparing the geometry. In the "Static Structural" module, you can either import or create the leadscrew geometry using the Design Modeller tool. Make sure to set the dimensions to a diameter of 19.05 mm and a length of 250 mm. The "Mesh" section is used to create a mesh with an element size of 1.2 mm, as shown in Fig. 84.3. Specify structural steel as the material for the leadscrew, with material properties set to a density of 7.85 g/cm³, a Young's modulus of 200 GPa, and a Poisson's ratio of 0.3. The material properties are assigned. The boundary conditions are applied by fixing the leadscrew at one end and applying a moment load of 10 N.mm with a phase angle of 0 degrees at the other end, as depicted in Fig. 84.4. Perform a modal analysis to ascertain the natural frequencies and mode shapes, specifying the frequency range as 0 to 5000 Hz.

Fig. 84.3 Geometry mesh

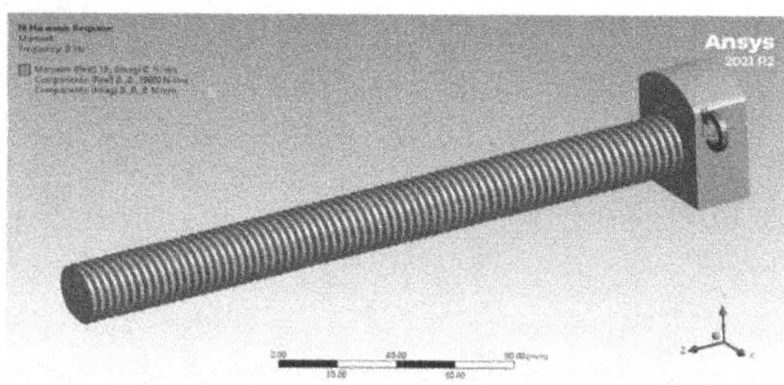

Fig. 84.4 Vector load of leadscrew (moment)

Perform the modal analysis and utilize the obtained results to establish a harmonic response analysis. For harmonic analysis, the moment load is applied as a deformable vector load. The frequency range of 0 to 5000 Hz is used to investigate the vibration response produced by an applied moment load. This allows to closely see the behaviour of the system. To completely understand the findings, the work requires determining mode shapes and frequencies obtained by modal analysis and investigating the vibration response using harmonic analysis. Using ANSYS Workbench, thoroughly simulate and analyse the dynamic characteristics of a leadscrew with varying thread profiles [7]. Continuously use the same components throughout the research, and compare the model and results to actual data to make sure they are correct. For further details regarding particular characteristics that can enhance the accuracy as well as reliability of your evaluation, consult the ANSYS documentation and tutorials.

3. Results and Discussion

The investigation focused mainly on the V-thread, rectangular thread, and Acme thread profiles, through a focus on characteristics like equivalent stress, maximal principal stress, shear stress, amplitude, and frequency. The FEA outcomes of many thread profiles are presented in Table 84.2 Figs from 5, 6, 7, 8 highlight in great detail these findings for the V-profile. Figure 84.5 indicates the comparable stress distribution and highlights the locations

Table 84.2 FEA outcomes

Thread Profile	Equivalent Stress (MPa)	Maximum Principal Stress (MPa)	Max Principal Stress (MPa)	Amplitude (mm)	Frequency (Hz)
V Thread	3.84E-05	2.98E-05	1.26E-02	2.36E-09	1500
Rec Thread	6.36E-02	5.75E-02	2.01E-02	3.96E-07	4000
ACME Thread	1.85E-02	1.12E-02	7.65E-03	1.21E-07	3000

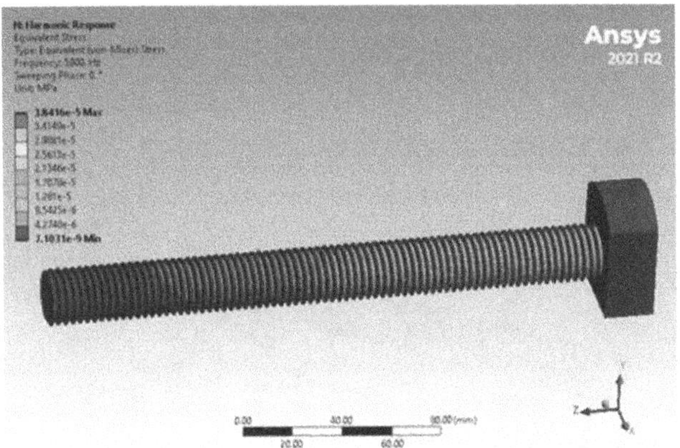

Fig. 84.5 Equivalent stress of v-thread

with the largest stress concentrations. Mostly loaded at the thread roots, the material generates the highest stress concentrations. The V-thread design distributes stress throughout the thread surface, limiting localized stress concentrations and lowering material failure under operating loads, according to the research.

Figure 84.6 illustrates the maximum main stress, highlighting the areas experiencing the greatest tensile stress. The regions that are subject to the highest levels of tensile stress are frequently located in close proximity to the surface and in areas that are promptly exposed to external stresses. The maximum primary stress offers a comprehensive understanding of stress distribution, which is critical for improving the design to achieve superior performance and dependability. Figure 84.7 shows that the V-thread profile deforms little under stress conditions, indicating structural integrity. Considering this, evaluate the component's structural and functional stability. The V-thread profile has low distortion in essential places to maintain thread contact with the mating component, as shown in the picture. Threads must engage well with mating components, therefore, structural integrity and low deformation are essential. Figure 84.8 shows the location of shear stress in the V-thread profile, stressing spots that are subject to defects generated by shear under specific loading situations [8]. The structure accurately controls shear stress, limiting the possibility of material slipping

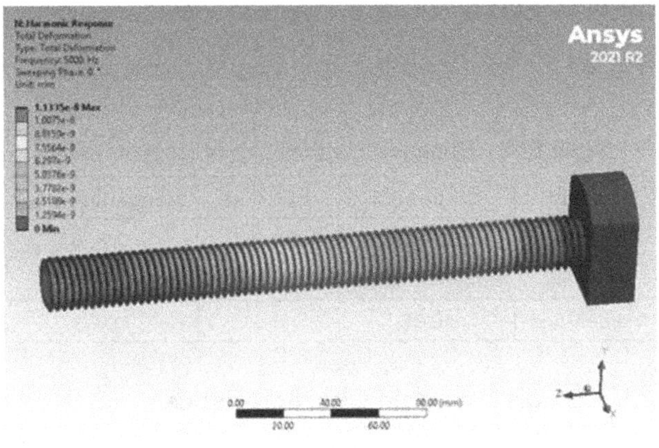

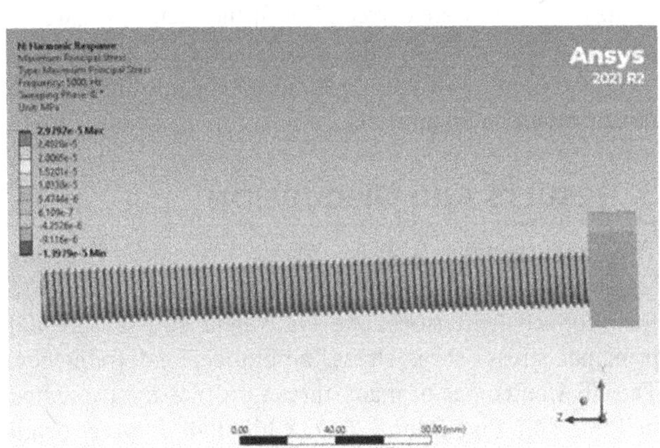

Fig. 84.6 Principal stress of v-thread

Fig. 84.7 Total deformation of v-thread

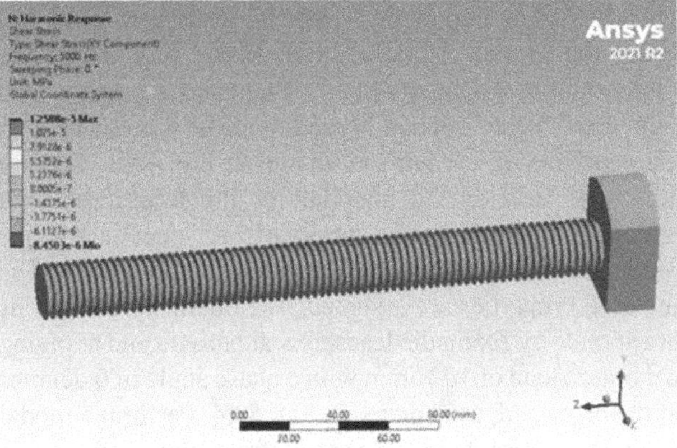

Fig. 84.8 Shear stress of v-thread

or shearing off. This confirms the constant stability of the V-thread and aids in averting early malfunctions in high-stress situations.

The V-thread has the lowest equivalent stress value, which means that the stress is distributed more uniformly over the profile, reducing the possibility of material failure due to localized stress concentrations. Uniformly Shaped Thread Rectangular threads have much greater stress concentrations when subjected to an equal stress, rendering them more subject to stress-related failures [9, 10]. While it's lower than the Rectangular thread and greater than the V-thread, the ACME thread's comparable stress rating is modest. This points to enhanced performance over the Rectangular thread, however it falls short of the ideal performance of the V-thread. The rectangular thread profile indicated significant values of stress and vibration responses. The rectangular thread exhibited an equivalent stress of 6.36E-02 MPa, a maximum principal stress of 5.75E-02 MPa, and a shear stress of 2.01E-02 MPa. The magnitude of the displacement was 3.96E-07 mm, and the frequency of excitation was 4000 Hz. In comparison with the V-thread, the rectangular thread profile exhibits lacking performance due to the higher levels of tension and deformation that it experiences [11].

The Acme thread profile achieved almost average results. The greatest principle stress was 1.12E-02 MPa; the equivalent stress values were 1.85E-02 MPa; the shear stress was 7.65E-03 MPa. The frequency of stimulation was 3000 Hz whereas the observed amplitude was 1.21E-07 mm. The results obtained confirm that this Acme thread profile is still less efficient than the V-thread profile even if it beats the rectangular thread profile. The V-thread profile indicated outstanding results in terms of minimum levels deformation and reduced stresses according to the given loading conditions. Evidently showing the enhanced performances of the V-thread profile, the graphical results suggest it as the most suitable option for applications requiring superior strength and vibration resistance. Selecting the most appropriate thread profile for certain applications depends on in-depth study and comparisons, therefore ensuring best performance and dependability [12].

As illustrated in Fig. 84.9, the amplitude of deformation is the frequency response of the V-thread profile. As the graph indicates, the amplitude changes considerably throughout the frequency range. The amplitude rises at 1500 Hz, indicating a resonance at about 2.35E-09 μm.. And the peak, slightly below that, is noted at about 4000 Hz. According to these peaks, the V-thread profile deforms more at certain frequencies. The amplitude shows less deformation as it fluctuates but stays relatively low. Although it fluctuates,

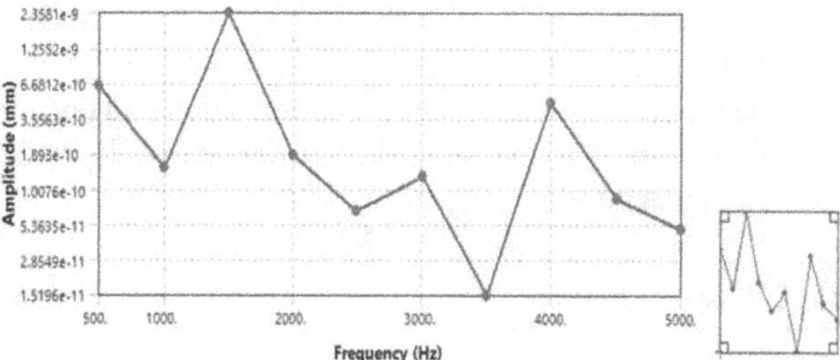

Fig. 84.9 Frequency response of v-thread about deformation

the amplitude stays relatively low, suggesting minimal distortion. The V-thread profile supports to maintain structural strength. In applications where these frequencies might be present, still more thought must be given as certain frequencies can produce more deformations. This investigation evaluates the efficiency and durability of the V-thread profile under dynamic loads [13].

Figure 84.10 shows in the V-thread profile the frequency relation of the stress level. With an ultimate stress at nearby 2.2377E-6 MPa, the stress intensity curve shows an identifiable peak at around 3000 Hz. At the same time, this value shows that V-thread profile's maximum internal stress. The stress level increases progressively before reaching this peak when the frequency approaches 3000 Hz, therefore expressing an increase of internal stress. The V-thread profile receives less stress at frequencies beyond of the harmonic point; this is shown by the sharp drop in the stress amplitude after the peak [8].

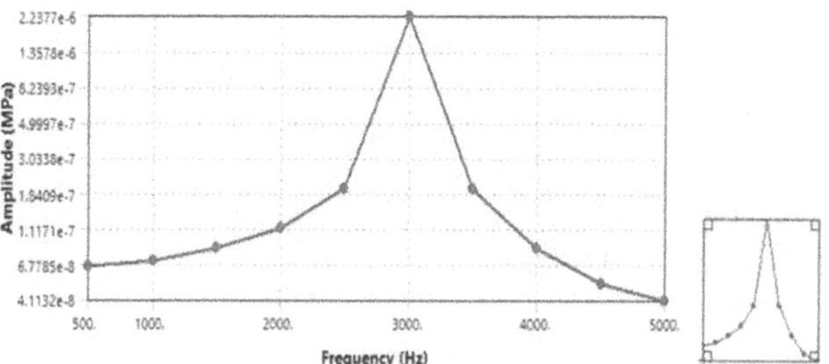

Fig. 84.10 Frequency response of v-thread about stress

Figures 84.9 and 84.10 perfectly indicate that at certain frequencies the V-thread profile is more resistant to enhanced deformation and stress. Design and implementation of the leadscrew should consider the important regions marked by the peak deformation at 1500 Hz and the peak stress at 3000 Hz. Functioning within or close to these frequencies might affect the structural stability and performance of the V-thread profile via internal stress and significant deformation. Hence, in practical uses it is advised to avoid these significant frequencies in order to increase the dependence and lifetime of the leadscrew mechanism. One must have a strong knowledge of the frequency response if a single requires to achieve design optimization and ensure safe functioning of the V-thread profile under dynamic load circumstances

4. Conclusion

This work investigated the effect of stress on linear actuator leadscrew mechanisms via dynamic finite element analysis. Also analysed loads via many thread profiles. For common uses required substantial torque, that we choose rectangular threads, V-thread, and ACME. The V-thread shown low equivalent, shear, and stress values. Graphs of deformation and stress amplitude confirmed these results by highlighting important frequencies and stress levels. Leadscrew working conditions are fit for the V-thread profile considering frequency response investigation. Under dynamic loads with a 2.36E-09 mm deformation amplitude, the V-thread remained structurally stable. Of all the profiles we tested, V-Thread has superior dynamic performance. The V-thread profile is suitable for automated systems requiring pinpoint precision and dependability because of its 1500 Hz stimulated frequency and lowered stress amplitude. We chose the V-thread profile due to its excellent frequency response, lower stress and deformation amplitudes, and overall best performance. This study found that linear actuators with V-thread leadscrews perform better, making them ideal for low-vibration, high-dynamic stability applications.

Acknowledgment

The authors would like to acknowledge the co-operation of Mr. Dayalan and Mr. Bagavathiappan, Department of Mechanical Engineering, Hindustan Institute of Technology and Science, Padur, Chennai for the conduct of simulation in Ansys.

References

1. Irudhayam S, Jackson, and Hariram Venkatesan. "Experimental Evaluation and Finite Element Analysis of Stress Distribution in 3D-Printed Dental Implants to Validate the Optimal Thread Pitch." *Journal of Composite & Advanced Materials/Revue des Composites et des Matériaux Avancés* 34, no. 2 (2024).
2. Thirugnanam, A., Amit Kumar, and Lenin Rakesh. "Analysis of Power Screw Using 'Ansys'." *Middle-East Journal of Scientific Research* 20, no. 6 (2014): 742–743.
3. Bentgens, Felix, Peter Koenig, M. Jaikumar, and V. Hariram. "Impact of Steering Column Stiffness on the Injury Criteria of Dummies using Siemens Madymo." *International Journal of Vehicle Structures & Systems (IJVSS)* 15, no. 1 (2023).
4. Yu, Huangchao, and Xiaodong Wang. "Modeling and analysis of the electromechanical behavior of surface-bonded piezoelectric actuators using finite element method." *arXiv preprint arXiv:1611.02375* (2016).
5. Eyere, Emagbetere, O. Larry, and O. Peter. "Development of a Mechanical Puller." *Int J Innov Sci Res Technol* 3, no. 7 (2018): 284–289.
6. Kumar, S. Sanjeev, AS Shree Arryaman, R. Suganthprabhu, D. Pritima, and P. Arun Kumar. "Design and Fabrication of Adhesive Application System for Magnets in Motors." In *2023 4th International Conference on Electronics and Sustainable Communication Systems (ICESC)*, pp. 49–53. IEEE, 2023.
7. Jackson, Irudhayam S., and V. Hariram. "Investigation of Strain and deformation analysis of Biomaterial in Dental Implant: A 3D FEA Study." In *E3S Web of Conferences*, vol. 491, p. 01023. EDP Sciences, 2024.
8. Syriac, Alex S., and Shital S. Chiddarwar. "Dynamic characteristics analysis of a lead screw by considering the variation in thread parameters." In *IOP Conference Series: Materials Science and Engineering*, vol. 624, no. 1, p. 012007. IOP Publishing, 2019.
9. Kim, Woo-Hyeon, Chang-Woo Kim, Hyo-Seob Shin, Kyung-Hun Shin, and Jang-Young Choi. "Operating Characteristic Analysis and Verification of Short-Stroke Linear Oscillating Actuators Considering Mechanical Load." *Machines* 10, no. 1 (2022): 48.
10. Zhang, Wei, Xing Zhang, Jun Zhang, and Wanhua Zhao. "Analysis of lead screw pre-stretching influences on the natural frequency of ball screw feed system." *Precision Engineering* 57 (2019): 30–44.
11. Kumar, S. Sanjeev, AS Shree Arryaman, R. Suganthprabhu, D. Pritima, and P. Arun Kumar. "Design and Fabrication of Adhesive Application System for Magnets in Motors." In *2023 4th International Conference on Electronics and Sustainable Communication Systems (ICESC)*, pp. 49–53. IEEE, 2023.
12. Jackson, Irudhayam S., and Hariram Venkatesan. "Finite Element Analysis of Stress Distribution in 30% CFR PEEK Implant with Varying Thread Designs." *Revue des Composites et des Materiaux Avances* 34, no. 5 (2024): 565.
13. Sebastiraj, Britto, and A. Arivazhagan. "Design and Development of Power Generation using Vehicle Suspension." *International Journal of Vehicle Structures & Systems (IJVSS)* 15, no. 5 (2023).

Note: All the figures and tables in this chapter were made by the authors.

Advances in Additive Manufacturing Technologies – Gurusamy Pathinettampadian et al. (eds)
© 2026 Taylor & Francis Group, London, ISBN 978-1-041-16687-0

Optimization of EDM Parameters for Aluminium 2024 Alloy Using Taguchi and Topsis

Rajamanickam S.*,
Hari Krishan Raj S., Ranjith M., Anuragh S.P.,
Harish C., Gnana Sai M., Likhith P., Nikhil G.
Department of Mechanical Engineering,
Vel Tech High Tech Dr. Rangarajan Dr. Sakunthala Engineering College,
Chennai, Tamil Nadu, India

◆ **Abstract:** This research examines the optimal environments for Electrical Discharge Machining (EDM) of Aluminium 2024 alloy using the Taguchi approach coupled with the Technique for Order of Preference by Similarity to Ideal Solution (TOPSIS). Due to its high strength-to-weight ratio, Aluminium 2024 finds wide applications in industries such as aerospace and automotive, where strength and lightweight materials are of prime importance. The research focuses on examining the influence of key EDM parameters—namely, pulse-on time, pulse-off time, and peak current—on critical performance measures, including machining time (MT), tool wear (TW), and surface roughness (SR). To systematically study the effects of these parameters, the experimental trials were structured using Taguchi's L9 orthogonal array, which enables efficient experimentation with minimal trials while maintaining reliability. The TOPSIS method was subsequently used to contrast and rank the experimental findings based on various performance criteria. The main effects analysis revealed that the optimal machining time could be obtained with a combination of greater pulse-on time and greater current, along with a moderate pulse-off time. At the other end of the scale, minimum tool wear was achieved with the use of lower pulse-on time, a medium pulse-off time, and a higher current level. To achieve the best surface finish, the best results were achieved when pulse-on and pulse-off times were reduced and current was kept high. Based on the TOPSIS ranking, the best EDM parameter setting to achieve maximum efficiency, reduce tool wear, and improve surface finish was determined to be a pulse-on time of 7 microseconds (μs), pulse-off time of 4 μs, and setting of 12 amperes (A). Besides, the Analysis of Variance (ANOVA) results indicated that pulse-on time contributed most to the machining performance, accounting for 74.11% of the variation. Pulse-off time and current contributed 13.18% and 3.26%, respectively.

◆ **Keywords:** EDM, Aluminium, Taguchi

1. Introduction

Electric Discharge Machining (EDM) has emerged as a preferred method of precise machining, especially in cases where the materials used are conductive and difficult to machine with normal techniques. It is possible with EDM to produce complex shapes as well as to machine very tough materials [1]. Hence, EDM emerges as an absolute necessity for aerospace, automobile, and electronic industries. Aluminium 2024 is among the finest illustrations of a material whose process can be aided through EDM. Priced for its higher strength-to-weight ratio, it is used extensively in aerospace and automotive industries, where weight savings without compromising the structural strength is a

*Corresponding author: rajamanickam@velhightech.com

DOI: 10.1201/9781003685906-85

crucial factor. In aerospace and automotive industries, it is essential to maximize machining methods like EDM. Process parameters such as pulse-on time, pulse-off time, and electrical current play a significant role in determining EDM performance, though. These parameters have a direct impact on machining results and surface quality but are so complex in their interaction that it is hard to identify the best combination [2]. To overcome this problem, the present research utilizes the Taguchi method—a well-proven statistical technique—to find the optimal settings for EDM. Furthermore, the Technique for Order Preference by Similarity to Ideal Solution (TOPSIS) is utilized to rank the machining outputs according to several performance measures [3].

Some researchers have explored the use of statistical techniques to improve EDM efficiency and maximize process parameters. For example, Pham et al. (2024) enhanced EDM performance for Ti-6Al-4V alloy using graphene-coated aluminium electrodes, by integrating the Taguchi method and TOPSIS to achieve the highest material removal rate (MRR) and tool wear rate (TWR) [4]. Also, Tran et al. (2023) conducted experiments using P20 tool steel with copper electrodes, applying the Taguchi method combined with Grey Relational Analysis (GRA) to improve MRR, electrode wear rate (EWR), and surface roughness (SR). Kumar and Mondal (2021) explored optimization of EDM parameters for Al-2050 alloy using rotation of copper, tungsten, and copper-tungsten electrodes. Their joint Taguchi-TOPSIS-GRA method revealed rotation of the electrode significantly improved machining performance. In another study, Zeng et al. (2021) [5] optimized EDM parameters for aluminium oxide ceramics with copper electrodes utilizing the Taguchi method, Analytic Hierarchy Process (AHP), and TOPSIS for multi-objective optimization. Likewise, Nguyen et al. (2021) employed a Taguchi-AHP-TOPSIS approach to enhance EDM performance on Ti-6Al-4V with nickel-coated aluminium electrodes, where the peak current, gap voltage, and pulse-on time at optimum values were focused on achieving high MRR and improved surface finish [6]. These studies together illustrate the capability of using statistical optimization techniques in conjunction with EDM to maximize machining performance, extend tool life, and achieve higher surface quality.

A limited research focusing on the EDM process for Aluminium 2024 alloy. This study, therefore, aims to determine the optimal EDM parameters for AA 2024. It is also noted that the integration of the Taguchi method and TOPSIS optimization technique has proven to be a reliable, practical and straightforward approach, with the potential to yield improved results in EDM processes with the goal of improving machining efficiency, enhancing surface finish quality, and reducing tool wear. Ultimately, this research contributes to refining EDM processes for machining high-performance materials such as Aluminium 2024. Figure 85.1 Process Flow diagram shows the steps and procedure for the current investigation.

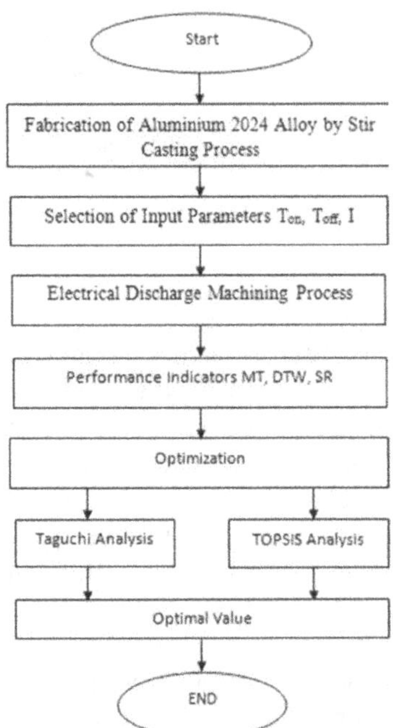

Fig. 85.1 Process flow diagram

2. Materials and Methods

A novel composite material Aluminium 2024 is used as workpiece material. The chemical composition of Aluminium 2024, a high-strength alloy widely used in engineering applications, is summarized in Table 85.1. The alloy primarily consists of aluminium, with copper, magnesium, and other elements in specified proportions to enhance its mechanical properties [7]. This composition ensures a combination of lightweight properties and high strength, making it suitable for demanding applications like aerospace and automotive components.

The experiment employed a tool electrode made of electrolytic copper with a purity of 99.9%. The electrode

Table 85.1 Chemical properties of Al 2024 alloy

Element	Al	Cu	Mg	Si	Fe	Mn	Zn	Ti	Cr	Other
Weight (%)	major	3.8-4.9	1.2-1.8	0.5	0. 5	0.3-0.9	0.25	0.15	0.1	0.05

featured a diameter of 20 mm and a total length of 50 mm, ensuring precision and consistency in the experimental setup. The experimental work was carried out using an EDM machine tool of model [8].

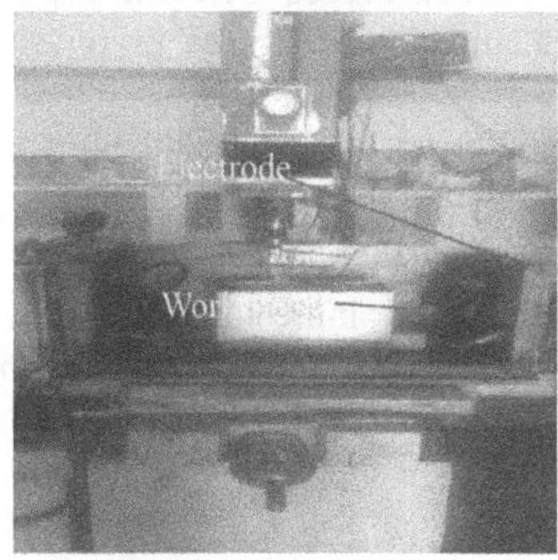

Fig. 85.2 EDM machining tool: Electronica-electrapuls PS 50ZNC

Table 85.2 Taguchi experimental results and L9 OA design

Run	T_{ON} (μs)	T_{OFF} (μs)	I_p (A)	Machining Speed (min)	Diametral tool wear (micron)	Surface Roughness (μm)
1	7	4	8	28	0.6300	6.58
2	7	5	10	26	0.5894	8.23
3	7	6	12	27	0.5900	6.75
4	8	4	10	18	0.6300	6.22
5	8	5	12	14	0.6100	6.97
6	8	6	8	21	0.7000	9.64
7	9	4	12	10	0.4196	7.10
8	9	5	8	15	0.4700	11.51
9	9	6	10	17	0.4600	9.90

Taguchi analysis was performed for single-objective optimization using Minitab 17 software (Fig. 85.2).

2.1 TOPSIS Method

The method involves comparing alternatives to these criteria to determine the most optimal solution. The steps involved in the TOPSIS are given in the flow diagram as described in Fig. 85.3.

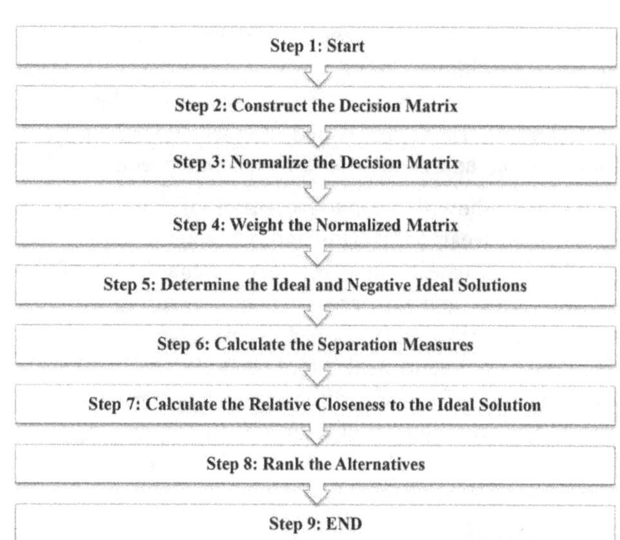

Fig. 85.3 Steps involved in TOPSIS

3. Results and Discussion

3.1 Analysis of Main Effects Plot on Machining Speed

Figure 85.4 illustrates highlights the critical influence of process parameters.

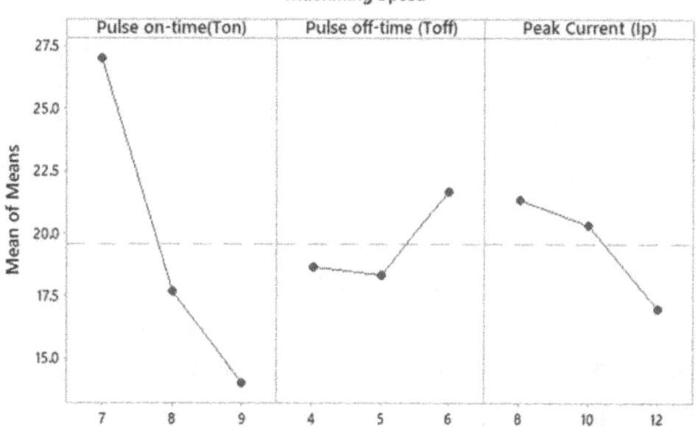

Fig. 85.4 Machining speed-main effects plot for means

This, in turn, accelerates material removal, leading to a notable increase in machining speed. Furthermore, higher machining times are associated with improved material removal rates. The findings suggest that a medium level of pulse off-time is optimal for efficient material removal. Additionally, an increase in current is observed to directly reduce machining time, underscoring its significant role in the machining process [9]. A high level of pulse on-time, combined with a high current setting, significantly enhances the machining efficiency (Table 85.2). Additionally, a medium level contributes to optimal material removal rates. Under these ideal conditions, the machining speed achieves a value of 9 minutes, demonstrating the effectiveness of these parameter combinations in reducing machining time during EDM of Al 2024 alloy [10].

3.2 Analysis of Main Effects Plot on Diametral Tool Wear

An increase in pulse on-time from a lower level (7 μs) to a higher level (9 μs) results in higher tool wear.

Higher pulse on-time generates more thermal energy in the machining zone, which improves material removal but also raises thermal stress on the rotary tool (Fig. 85.4). This elevated stress can lead to faster tool erosion and a loss of dimensional precision [11].

The results show that a pulse off-time of 5 µs results in lower tool wear [13]. This intermediate setting likely provides adequate cooling between discharges, reducing thermal stress and tool erosion while ensuring efficient machining performance. Also, higher current level (12A) has been observed to reduce tool wear (Fig. 85.5). This is likely due to the increased discharge energy. These settings resulted in a minimum tool wear of 0.5879 microns.

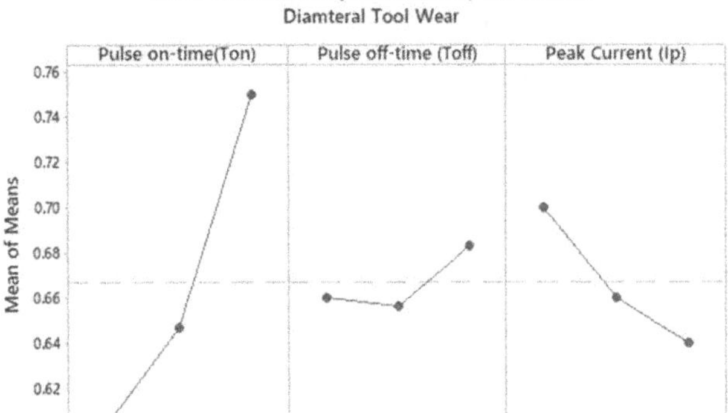

Fig. 85.5 Diametral tool wear

3.3 Analysis on Surface Roughness

EDM of Aluminium 2024 alloy, is significant

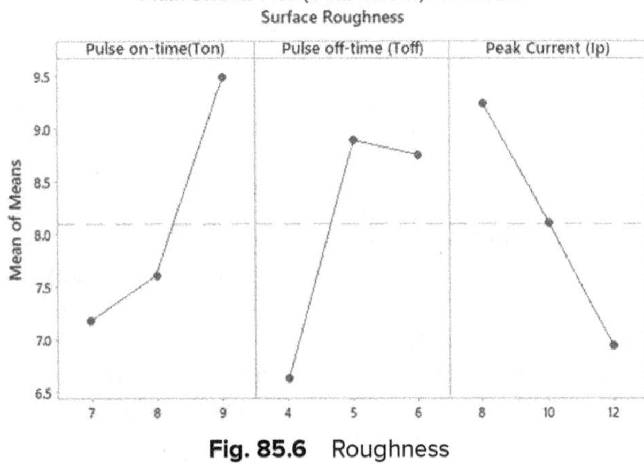

Fig. 85.6 Roughness

An pulse on-time increase generally higher in leads discharge energy, resulting in deeper craters and more material removal. While this initially enhances material removal rates, it often causes an increase in surface roughness. Longer pulse on-time (9 µs) extends the discharge duration, leading to excessive heat accumulation and larger molten areas on the surface, contributing to a rougher finish. However, by optimizing pulse on-time, a balance can be achieved between material removal and surface quality. Shorter pulse on-time (7 µs) tends to produce finer surface finishes, as it reduces thermal damage and minimizes crater formation.

At lower pulse off-time (4 µs), surface roughness typically improves due to the shorter cooling period between discharges, allowing for more controlled material removal and less thermal damage. As pulse off-time increases to a medium level (5 µs), surface roughness tends to worsen, as longer cooling times can cause larger craters and uneven material removal. At higher pulse off-time levels (6 µs), surface roughness begins to decrease again, likely due to excessive cooling, which disrupts the discharge process and reduces material removal efficiency (Fig. 85.6).

This minimizes irregular crater formation and reduces heat buildup, leading to a finer surface. However, if the current is set too high, tool wear may increase, and surface defects may appear, so it is crucial to optimize the current to achieve the best surface quality.

3.4 TOPSIS

In the TOPSIS method, the experimental results are first converted into a decision matrix, which includes the performance metrics of machining speed, diametral tool wear, and surface roughness. The decision matrix is then normalized, and equal weights (1/3) are assigned to each performance criterion. Following this, measures separation and closeness coefficients are calculated. Table 85.3 presents the normalized values, weighted normalized values, separation measures, and closeness coefficients for each performance criterion considered.

In this study, Alternative 1 achieved the highest

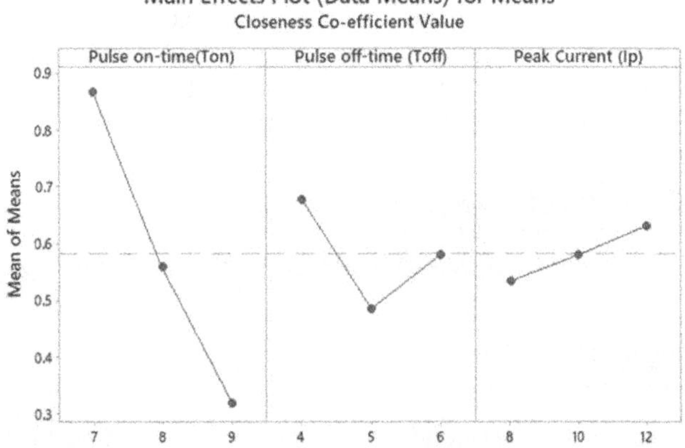

Fig. 85.7 Closeness co-efficient-main effects plot for means

Table 85.3 TOPSIS results during EDM of Al 2024 alloy

RUN	MS (min)	DTW (micron)	SR (µm)	MS (min)	DTW (micron)	SR (µm)	S+	S-	CC
1	0.4564	0.3134	0.2648	0.1369	0.0940	0.1059	0.0084	0.1203	0.9349
2	0.4238	0.2932	0.3311	0.1271	0.0880	0.1325	0.0338	0.0982	0.7439
3	0.4401	0.2935	0.2716	0.1320	0.0881	0.1086	0.0098	0.1162	0.9220
4	0.2934	0.3134	0.2503	0.0880	0.0940	0.1001	0.0493	0.0960	0.6608
5	0.2282	0.3035	0.2804	0.0685	0.0910	0.1122	0.0696	0.0793	0.5327
6	0.3423	0.3483	0.3879	0.1027	0.1045	0.1551	0.0669	0.0625	0.4831
7	0.1630	0.3580	0.2857	0.0489	0.1074	0.1143	0.0912	0.0714	0.4389
8	0.2445	0.3831	0.4631	0.0733	0.1149	0.1852	0.1096	0.0244	0.1824
9	0.2771	0.3781	0.3983	0.0831	0.1134	0.1593	0.0840	0.0430	0.3385

closeness coefficient of 0.9349. Among the experiments, Experimental Trial 1 was found to be the best solution, with Trial 3 being the next most effective. On the other hand, Experiment 8 showed the worst performance.

3.5 Variance for CC Analysis

Table 85.4 Result tabulation for best performance of EDM process

Source	DF	Seq SS	Adj SS	Adj MS	F-Value
On-time Pulse	2	0.4512	0.4512	0.2256	130.20
Off-time Pulse	2	0.0552	0.0552	0.0276	15.93
Current Peak	2	0.0143	0.0143	0.0716	4.14
Error	2	0.0035	0.0035	0.0017	
Total	8	0.5242			

Contributing 86.07% to the total variation. The high F-value of 130.20 and substantial sum of squares indicate its major role in controlling machining speed, diametrical tool wear, and surface quality. Pulse off-time contributes 10.53% to the overall variation. While its effect is lower that of on-time pulse, it still significantly influences the EDM process. Peak current has the least effect, accounting for 2.73% of the total variation. ANOVA analysis confirms that on-time pulse is the most influential in parameter EDM performance in Aluminium 2024 alloy. The low error contribution confirms the reliability of the experimental design and the effectiveness of the optimization process (Fig. 85.7).

4. Conclusion

This research effectively optimized the Electrical Discharge Machining (EDM) parameters for Aluminium 2024 alloy by employing a Taguchi and TOPSIS methodology, aiming to enhance machining efficiency, minimize tool wear, and improve surface finish. Key findings are as follows:

- The Taguchi analysis revealed that the most efficient machining time was when more on time pulse and current. Predicted diameter, lower pulse-on time, medium pulse-off time, and higher were best. Lower on pulse and off pulse times, with higher current settings, provided the best surface finish.
- Yielded the best performance with a high closeness coefficient of 0.9349. This research emphasizes the efficiency of the hybrid Taguchi-TOPSIS approach in multi-criteria optimization of EDM operations, especially machining advanced materials used in industries like aerospace and automotive industries.

References

1. Karim, M.A., Jahan, M.P. (2024). Chapter 7 - Electrical discharge machining technologies in the aerospace industry. In: Gürgen, S., Pruncu, C.I. (eds) *Modern Manufacturing Processes for Aircraft Materials*. Elsevier, pp. 171–226. https://doi.org/10.1016/B978-0-323-95318-4.00007-0
2. Cui, T. (2024). Precision machining of hard-to-cut materials: Current status and future directions. *International Journal of Advanced Computer Science and Applications (IJACSA)*, 15(10). https://doi.org/10.14569/IJACSA.2024.0151088
3. Rajamanickam, S., Prasanna, J., Sastry, C.C. (2020). Analysis of high aspect ratio small holes in rapid electrical discharge machining of superalloys using Taguchi and TOPSIS. *Journal of the Brazilian Society of Mechanical Sciences and Engineering*, 42, 99. https://doi.org/10.1007/s40430-020-2180-2
4. Qudeiri, J.E.A., Saleh, A., Ziout, A., Mourad, A.I., Abidi, M.H., Elkaseer, A. (2019). Advanced electric discharge machining of stainless steels: Assessment of the state of the art, gaps and future prospect. *Materials (Basel)*, 12(6), 907. https://doi.org/10.3390/ma12060907
5. Charalampidou, C.M., Siskou, N., Georgoulis, D., et al. (2024). Corrosion of aluminium alloy AA2024-T3 specimens subjected to different artificial ageing heat treatments. *npj Materials Degradation*, 8, 93. https://doi.org/10.1038/s41529-024-00503-4

6. Nambiar, S.S., Sharma, S., Anne, G., Agarwal, P., Agrawal, T., Chickkannan, V. (2024). Mechanical and microstructural analysis of cobalt-coated CNT/Al2024 reinforced aluminum-based composites for aerospace and automotive applications. *Cogent Engineering*, 11(1). https://doi.org/10.1080/23311916.2024.2436642
7. Balaji, S., Maniarasan, P., Sivakandhan, C., Alagarsamy, S.V. (2023). Optimization of machining parameters in EDM using GRA technique. In: Vignesh, R.V., Padmanaban, R., Govindaraju, M. (eds) *Advances in Processing of Lightweight Metal Alloys and Composites*. Springer. https://doi.org/10.1007/978-981-19-7146-4_21
8. Rajamanickam, S., Prasanna, J. (2021). Effect of conductive, semi-conductive, and non-conductive powder-mixed media on micro electric discharge machining performance of Ti-6Al-4V. *International Journal of Electrochemical Science*, 16(3), 210317. https://doi.org/10.20964/2021.03.29
9. Pham, H.-V., Huu-Phan, N., Shirguppikar, S., Duc-Toan, N. (2024). Optimizing technological parameters in electrical discharge machining with graphene-coated aluminum electrodes for enhanced machining of titanium alloy: A Taguchi-TOPSIS approach. *Tribology in Industry*, 46(2), 324–331. https://doi.org/10.24874/ti.1600.01.24.02
10. Tran, V.T., Le, M.H., Vo, M.T., Le, Q.T., Hoang, V.H., Tran, N.T., Nguyen, T.T. (2023). Optimization design for die-sinking EDM process parameters employing effective intelligent methods. *Cogent Engineering*, 10(2). https://doi.org/10.1080/23311916.2023.2264060
11. Kumar, D., Mondal, S. (2021). Multi-attribute optimization of EDM process parameters of Al-2050 alloy using Taguchi-based TOPSIS and GRA with different rotating tools. *International Journal of Modern Manufacturing Technologies*, 13(1), 84–90. https://ijmmt.ro/vol13no12021/09_Dhiraj_Kumar_1.pdf
12. Zeng, Y.P., Lin, C.L., Dai, H.M., Lin, Y.C., Hung, J.C. (2021). Multi-performance optimization in electrical discharge machining of Al2O3 ceramics using Taguchi-based AHP weighted TOPSIS method. *Processes*, 9(9), 1647. https://doi.org/10.3390/pr9091647
13. Nguyen, H.P., Ngo, N.V., Nguyen, C.T. (2022). Study on multi-object optimization in EDM with nickel-coated electrode using Taguchi-AHP-TOPSIS. *IJE Transactions B: Applications*, 35(2), 276–282. https://doi.org/10.5829/IJE.2022.35.02B.02

Note: All the figures and tables in this chapter were made by the authors.

Advances in Additive Manufacturing Technologies – Gurusamy Pathinettampadian et al. (eds)
© 2026 Taylor & Francis Group, London, ISBN 978-1-041-16687-0

Tribological and Mechanical Characterization of Nickel Based Super Alloy Employed in Hard Facing—A Comprehensive Review

Hariram V.[1]
Department of Mechanical Engg., Hindustan Institute of Tech & Science, Padur, Chennai, Tamil Nadu, India

Karthick K.[2]
Department of Mechanical Engg., R M K Engineering College, Kavaraipettai, Tamil Nadu, India

Antony Casmir Jayseelan[3]
Department of Mechanical Engg., Aarupadai Veedu Institute of Tech., Paiyanur, Tamil Nadu, India

Godwin John J.[4]
Department of Mechanical Engg., Rajalakshmi Institute of Technology, Chembarambakkam, Tamil Nadu, India

John Presin Kumar A.[5], **Balachandar M.**[6]
Department of Mechanical Engg. Hindustan Institute of Tech & Science Padur, Chennai, Tamil Nadu, India

Allen Jeffrey J.[7]
Department of Mechanical Engg., Loyola Institute of Technology Kuthambakkam, Tamil Nadu, India

Balamurugan S.[8]
Department of Mechanical Engg., Chennai Institute of Technology, Kundrathur, Tamil Nadu, India

◆ **Abstract:** Superalloys based on nickel are essential to industries because of their remarkable tribological, mechanical, and thermal characteristics. These materials are designed to withstand harsh environments, such as corrosive ones, high temperatures, and pressures. But since they are exposed to such severe circumstances, they frequently experience wear, corrosion, and friction, which shortens their lifespan and operational efficiency. Innovative answers to such concerns have been made feasible through improvements in coatings and surface engineering technology. Plasma spraying, laser cladding, and thermal spraying processes have been successful in developing nickel-based superalloys corrosion and wear resistances. These processes prioritize microstructure refinement and alloying factors for increased wear resistances and hardness, and increased tensile strength. Recent improvements in nickel-based superalloys tribological performance have been discussed in this article, with a focus placed on techniques for surface modification such as nano-crystallization, nanoparticle strengthening, and complex coatings. Besides adhesion of coatings and aggregation of particles, it takes into consideration important factors that affect wear behavior, such as temperature, loading, sliding velocity, and environment. Additive manufacturing, cryogenic processing, and diamond-like carbon (DLC) coatings

[1]connect2hariram@gmail.com, [2]karthick.kuppan@gmail.com, [3]antonycasmir@avit.ac.in, [4]godwinjohn18@gmail.com, [5]johnpk@hindustanuniv.ac.in, [6]mbalachandar@hindustanuniv.ac.in, [7]jeffyresearch@gmail.com, [8]balamurugans@citchennai.net

DOI: 10.1201/9781003685906-86

have been included in cutting-edge approaches. Hybrid techniques combining coatings and mechanical treatments for complementary improvements must be considered in future work. For developing strong, high-performance materials for challenging service in hostile environments, such a work can serve as a useful source

◆ **Keywords:** Nickel-based, Superalloys, Mechanical characterization, Tribological properties, Hard facing

1. Introduction

The creation of innovative hard facing methods are analysed and studied through critical review. A novel application of nickel-based superalloys together with advanced coating methods enables remarkable enhancement of parts' resistance against wear and corrosion in severe operating conditions. The method would require nickel-based super-alloy coating application through thermal spray processing to the component surface for improvement. Natural materials science methods will craft this coating that will enhance the nickel-based superalloy combination of microstructure elements and alloying composition leading to robust mechanical qualities and enhanced tribological resistance with high hardness characteristics and tensile strength. The integration of laser cladding and plasma spraying methods as advanced surface engineering approaches will enhance the mechanical and tribological characteristics of the coating. Optimizing coating properties for specific operating environments requires the combination of alloying elements with microstructural modifications. The survey will investigate whether major advancements in wear and corrosion protection are possible for components in harsh operating situations. Manufacturing settings that handle aerospace elements together with power generation systems and chemical processing applications would demonstrate significant interest in this method.

2. Review of Litertures

It is a comprehensive source of information about the tribological properties and behaviour of nickel-based alloys and superalloys that are used widely in high-temperature and high-stress applications such as in aero-engines, gas turbines, and other industrial applications. The paper, by Zhang et al. [1], focuses specifically on nickel-based alloy coatings reinforced with nanoparticles, reviewing their wear behaviour and tribological properties. The authors have provided an overview of research in this area, describing the advantages of nanoparticle reinforcement, especially regarding wear, friction reduction, and improvement in mechanical properties. Other challenges and disadvantages of these coatings are: the agglomeration of particles, surface roughness, and compatibility with the substrate material. The tribological performance of nickel-based superalloys under high temperature conditions is examined by Vetreselvan et al. [2]. The current research state analysis evaluates temperature loading and sliding speed effects on wear and friction performance while assessing wear and friction mechanisms in nickel-based superalloys and describing methods that optimize their tribological character through surface modification and heat treatment among others. The tribological characteristics of nickel-based superalloys receive a broad overview in Chatterjee et al. [3] report. The authors present a discussion of a variety of factors that affect wear and friction profiles and composition and microstructure and environment obstructive factors such as working temperatures and ambient moisture. In detail, an analysis of nickel-based superalloys improvement techniques for tribological properties involves coatings and surface modification and lubrication techniques. Vasireddy et al. [4] present a critical analysis of nickel-based alloys' tribological behavior through a complete analysis of bulk constituent and coated materials. Various factors that impact wear and friction performance in such materials and composition and microstructure and environment impact factors are discussed in detail by the authors. Various improvement techniques for nickel-based alloys' tribological behavior and processes for coatings and surface modification and coatings and lubrication techniques are discussed in detail.

Saikia et al. [5] cover development in nickel-based superalloys' tribological behavior. Study current work in such factors and microstructure heat treatment conditions and environment factors' contribution towards wear and friction behavior. Analysis in current work considers alternative approaches effective in enhancing nickel-based superalloys' tribological behavior through consideration of investigation of coatings and surface modification and use of lubrication. Hu et al. [6] review nickel-based superalloys' overall tribological behavior and working methodologies in detail. Analysis of a work of investigation of a material's wear and friction behavior considers microstructure and composition and environment factors' contribution and role in such behavior. Authors review a variety of methodologies enhancing nickel-based superalloys' tribological behavior through methodologies of protective coatings and surface modification and review methodologies of lubrication in detail. In-depth analysis of nickel-based superalloy mechanical behavior and tribological behavior can be seen in Ravindranadh et al. [7]. Authors have analyzed three significant factors such as microstructure and composition and heat treatment and its contribution towards such mechanical behavior in nickel-based superalloys in detail. In-depth review of nickel-based superalloys' tribological behavior can be seen for nickel-based superalloys in detail. Various factors contributing towards wear behavior and friction have been considered in addition to temperature and loading and sliding velocity and its contribution towards relation in detail. In-depth review of nickel-based superalloys' tribological behavior can be seen in Zhang et al. [8]. Authors review new work in such a field and review its contribution in terms of microstructure and composition and environment and its contribution towards changing wear and friction behavior in detail. Analysis in such a work considers several methodologies towards surface modification for improvement in nickel-based superalloys' tribological behavior in detail.

The tribological performance of nickel-based superalloys is elaborated in Sharma and Singh [9] in a review of its mechanical values apart from its physical values and its use scenarios and its surface processes. There are various wear processes with its use scenarios and its coatings with significant value in its improvement in its tribological values. Zhang et al. [10] reviews nickel-based superalloy tribological performance under an environment with an oil use scenario with consideration of temperature apart from its sliding velocity and loads in its work. In its analysis, such a work takes into account significant challenges for characterizing its tribological values in investigating such alloys under a range of its use scenarios and its lubricated environments. Guo et al. [11] takes into account nickel-based superalloy tribological values with consideration of nano-crystallization at a level of its use scenario and its surface. Research takes into account grain sizes apart from its use scenarios and its structures in its role in its tribological performance of such alloys. In its analysis, such a work takes into account both existing use processes for nano crystallization at a level of its use scenario and cases for its use scenarios in practice. There is a review of nickel-based superalloy tribological performance in Yang et al. [12], with particular consideration for its durability in wear and its resistances in corrosion and wear with consideration of a range of factors with an effect in its tribological values and assigning significant consideration for its value in utilizing its use scenarios and its coatings in its improvement in its tribological performance.

Zuo et al. [13] work is on working mechanism and wear behavior of nickel-based alloys. Abrasive wear and adhesive wear and fatigue wear have been addressed in the article prior to temperature and loading and variation in sliding velocity and its contribution towards these alloys. Wu et al. [14] is a detailed analysis concerning nickel-based superalloys' tribological behavior and mechanism. Tribological behavior of these alloys is discussed in terms of a variety of factors contributing towards them: microstructure, composition, and processing of the surface. The article is a detailed analysis concerning wear mechanism describing its overall contribution towards overall wear behavior of these alloys. Sun et al. [15] is on nickel-based superalloys' wear behavior and mechanism. The article discusses a variety of factors deciding these alloys' wear behavior trends through roughness and loading factors analysis and working temperature contribution. Authors examine through analysis, how abrasive, adhesive and oxidative wear contribute towards overall wear behavior of these alloys. Ni-based alloys' tribological behavior when applied in aero-engine bearings is explained in detail by Ouyang et al. [16]. These alloys experience resistivity when applied in bearings at high working temperatures and high loads. Apart from that, additionally, they investigate through analysis, with which coatings and processes of the surface contribute optimized tribological behavior in these alloys. Ding et al. [17] offers a review on the tribological behavior of nickel-based superalloys under cryogenic temperature operation. This paper discusses low temperature wear behavior of these alloys while classifying various wear mechanisms functioning under such conditions. The mechanical properties in addition to tribological behavior of nickel-based coating systems that have been used to protect gas turbine parts are discussed in detail by Mardani et al. [18] in whose work, several coat

The application of laser engineering techniques to Ni-supersalloys is addressed in Khan et al. [19]. These processes form alloy properties with improvements in mechanical performance and increased tribology performance but with impediments present that hinder full use potential. Wang et al. [20] in their work introduce a critical review of nickel-alloy tribological performance under extreme environments with both high temperature use and corrosive environments in addition to high contact stresses. Wang et al. investigate a variety of factors of influence for alloy wear and reaction in reaction and report in detail critical tribological processes involved. In their work, Huang et al. [21] review wear processes in addition to wear processes that occur during nickel-supersalloys. In analysis, they detail in detail microstructure factors and chemical compositions in addition to environment factors contributing to wear behavior. In review, Zhu et al. [22]

review Ni-supersalloy wear behavior at high temperatures relevant for use in aeronautics and other hot environments. In its review, work assesses all factors contributing to high-temperature wear behavior through analysis of composition and microstructure in addition to environment factors. In review, Mower et al. [43] reviewed microstructure composition and nickel-supersaloy tribology performance. In its review, article considers production processes and identifies factors for tribology property in these coatings and its relation with coating roughness and composition and thickness and its relation with composition of materials.

Liu et al. [23] explores the tribological characteristics together with underlying mechanisms operating in nickel-based superalloys found in biomedical fields. This research investigates multiple biological environmental factors which impact nickel-based alloy performance including wear and corrosion as well as their prospective applications in orthopedic and dental implants. Wang et al. [24] review how nickel-based alloys act under dry sliding contact situations. Analysis explores how microstructure and chemical composition and environmental factors affect the friction and wear responses of these alloys. Shi et al. [25] conducted a review of the microstructure alongside the wear performance of nickel-based alloys in biomedical contexts. This review presents an analysis of wear and corrosion tendencies in biological solutions applying to these alloys with discussion about crucial obstacles that demand future resolution for reaching maximum potential in medical uses. Wu et al. [26] summarize the tribological behaviors and functional mechanisms operating in nickel-based superalloys at elevated temperatures. The discussion examines how high temperature wear behavior responds to different factors involving microstructure and composition as well as environmental elements although additional research is needed to grasp the underlying mechanisms fully. The tribological characteristics of nickel-based coatings during extreme environmental conditions are reviewed by Chen et al. [28]. These researchers analyze how composition and microstructure and environmental aspects affect the tribological behavior of these coatings while noting respective shortcomings that exist before fully utilizing these coatings for cutting tools and bearings.

The tribological behavior and its mechanism for nickel-based superalloys receive thorough analysis by Yang et al.(2018) when nickel-based superalloys serve as general materials in high temperature and high working environments and in gas turbine engines. Wear mechanism and demonstration of microstructure and composition and tribological behavior and its relation with techniques for surface modification for improvement in wear behavior receive analysis in its study. Son et al. [28] examined nickel-based superalloys' tribological behavior when studying its behavior at specific high temperatures. In its review, temperature fluctuations with loads under variable sliding velocities shape nickel-based superalloys' wear behavior and its evaluation at high temperatures and high temperature wear behavior mechanism receive analysis. In its discussion, nickel-based coatings and types and its deposition techniques and its tribological behavior under variable environments receive discussion in its discussion in its study about nickel-based coatings' tribological behavior in improvement in wear behavior of most materials. Li et al. [29] repeats almost all its contents in its 2017 publication in its study of nickel-based coatings' tribological behavior at high temperatures. In its discussion, temperature and its contribution together with loads and sliding velocities shape nickel-based coatings' wear behavior at high temperatures and boundary conditions for its behavior in wear receive analysis in its study. Zhang et al (2017)'s study spends its discussion with graphite and carbons' tribological behavior. In its discussion, graphite and carbons' application in tribological structures and its characterization and its wear and its working mechanism in friction receive discussion in its discussion in its study.

Ji et al. [30] addresses nickel coatings in terms of mechanism and tribological behavior and is closely similar to 2019 work. In this report, a review of types of coatings and techniques for deposition is included in addition to nickel-based coatings' tribological performance in a range of operational environments. Xie et al. [31] provides a review of nickel-based superalloys' wear-resistant coatings. In this analysis, a review of wear-resistant types of coatings is included in addition to deposition processes and tribological behavior of such coatings under a range of conditions when subjected to them. They addresses diamond-like carbon coatings' tribological behavior for application in nickel-based superalloys. The paper explores how diamond-like carbon coatings behave under different tribological conditions using deposition techniques that affect their material characteristics. Aoki et al. [32]. The work exhibited comparable findings yet provided detailed information regarding the tribological mechanisms of coated nickel surfaces. Han et al. [33] performed their work centered on cryogenic Tribological behavior eliminating temperatures down to minus 150. The paper investigates temperature and load and sliding speed conditions on nickel-based coatings' cryogenic wear resistance and explains worn material mechanisms.

3. METHODOLOGY

Researchers investigated Surface Modification effects on Nickel-Based Superalloys Tribological Properties through a study of scholarly literature. Nickel-based superalloys serve as the primary material choice for applications which combine high thermal and stress conditions such as jet engines and gas turbines. Tribological features of nickel-based superalloys must get stronger because resistance to wear and reduced friction cannot provide desired performance levels or enhance service lifetime. Altering the surface properties of these alloys shows promise for enhancing their tribological capabilities. The current research examines the direct effects that surface treatment techniques have on nickel-based superalloy tribological behavior. This research targets three main measurable objectives which represent its primary focal points.

This research examines existing studies regarding nickel-based superalloy tribology which investigates both wear performance and systems to lower friction. This research investigates how different surface treatment approaches including coating methods affect the tribological properties of nickel-based superalloys. An improved understanding of surface modification procedures starts from pinpointing the procedure that gives maximum benefit to nickel-based superalloy tribological properties. Prior research about nickel-based superalloy tribology with surface treatment methods and their associated impact on alloy tribology informed the team's study. A group of nickel-based superalloy samples with modified surfaces designed through diverse tribological studies will be produced and examined using pin-on-disk and scratch tests to record the materials' wear activities and reduced frictional forces. The research also encompasses determining which surface modification technique is the most effective for improving the tribological properties of nickel-based superalloys and offering a better understanding of the tribological behavior and mechanisms of these alloys. This will be achieved by identifying the most effective surface modification technique. Applications such as high-temperature and high-stress, the discoveries of this study could be of significant utility in developing novel surface modification methods for nickel-based superalloys. These will lead to greater performance and a more extended service life for such materials. The following Table 86.1 summarizes few literatures that investigate the impact of surface modifications on the tribological properties of nickel-based superalloys.

These studies in Table 86.1 in totality confirm that techniques for surface modification, such as coatings and heat treatments, have an important role in enhancing nickel-based superalloys' tribological behavior, such that compatibility with high-temperature and high-stress operations is achieved. There was a critical review of works in relation to nickel-based superalloys' tribological behavior, with a specific emphasis placed on techniques for surface modification. To have a deeper view of improvements in such changes and increased wear life and reduced friction, studies considered a few lists of literatures in databases. Temperature, loading, and sliding velocity were part of factors considered, for such factors played an important role in changing alloys' tribological behavior. Several techniques for surface modification, such as lubrication, coatings, and treatments, were examined with care. Despite factors such as aggregation of particles, experiments showed worthiness in strengthening with nanoparticles through enhancing mechanical property and wear life. Outcomes showed a full view of most current breakthroughs and gaps in such a field, with an important role for surface engineering in extending life and performance of such materials, in terms of duration and effectiveness, respectively. Experimental nickel-based superalloy samples with variable coatings and surface treatments were prepared, and wear behaviour analysed through tribological tests such as scratch and pin-on-disk tests. Most effective techniques for surface modification were determined through a comparison between review

Table 86.1 The impact of surface modifications on the tribological properties of nickel-based superalloys

Surface Modification Technique	Findings	Reference
Comprehensive review of various surface modification methods	Summarizes the wear, friction, and microstructure properties of nickel-based superalloys and talks about how external influences affect their tribological behavior.	Wang et al. [20]
Zirconium coating via sputtering	Shows how zirconium coatings greatly increase the resilience of nickel-based superalloys against corrosion and wear.	Vetreselvan et al. [2]
Electroless plating of Ni-B and Ni-B-W coatings	Study claims that because tungsten is incorporated into Ni-B-W coatings, they have better hardness and wear resistance than Ni-B coatings.	Zhao et al., [34]
Heat treatment of Ni-B coatings	It is discovered that proper heat treatments increase the tribological performance of Ni-B coatings by increasing their hardness and lowering their friction coefficient.	Rukhandet al. [36]
Review of tribological behavior without specific surface modification	Provides an overview of the wear mechanisms and tribological performance of Inconel superalloys, highlighting the need for surface modifications to enhance properties.	Dubey et al. [35]

in the literature and information in such studies. Notable observations showed that microstructure modifications and designed coatings played an important role in enhancing alloys' tribological behavior at high loads and temperatures. Despite such breakthroughs, factors such as homogenized coatings and compatibility of substrate materials with coatings were noticed. To counteract current restrictions and obtain improvement in performance Table 86.2 provides a detailed review of studies that investigate the tribological properties of nickel-based superalloys and the surface modification techniques employed to enhance these properties. The review table consolidates research on the tribological properties of nickel-based superalloys, focusing on various surface modification techniques. The studies address the following key areas:

Table 86.2 Investigation that details the tribological properties of nickel-based superalloys and the surface modification techniques employed to enhance properties

Surface Modification Technique	Experimental Design	Findings	Implications & Recommendations	Reference
Various surface modification methods	Comprehensive literature review of microstructure, friction, and wear characteristics	Summary of environmental influences on tribological system performance.	Highlights the need for advanced surface modification techniques to enhance performance	Wang et al. [20]
Zirconium coating via sputtering	Application of zirconium coatings and subsequent wear and corrosion tests	Demonstrates significant improvement in wear and corrosion resistance	Suggests zirconium coating as a viable method for enhancing superalloy performance	Vetre Selvan et al. [2]
Electroless plating of Ni-B and Ni-B-W coatings	Preparation of coatings followed by hardness and wear tests	Reports superior hardness and wear resistance in Ni-B-W coatings due to tungsten incorporation	Recommends further exploration of Ni-B-W coatings for industrial applications	Zhao et al., [34]
Heat treatment of Ni-B coatings	Application of heat treatments and evaluation of hardness and friction	Finds that appropriate heat treatments enhance hardness and reduce friction coefficient	Suggests optimizing heat treatment parameters to improve coating performance	Fidan et al. [37]
Review of tribological behavior without specific surface modification	Literature review of wear mechanisms and tribological performance	Provides an overview of the need for surface modifications to enhance properties	Recommends development of novel surface modification techniques for Inconel superalloys	Dubey et al. [35]
Laser surface texturing	Laser texturing followed by tribological testing under various conditions	Reveals that laser texturing reduces friction and wear by promoting lubricant retention	Laser surface texturing demonstrates effectiveness at improving tribological performance according to the investigation.	Gilley, et al. [38]
Plasma nitriding	Plasma nitriding treatment followed by pin-on-disk wear testing	Indicates that nitriding significantly improves surface hardness and wear resistance	Recommends plasma nitriding for applications requiring enhanced surface durability	Niksefat et al. [39]
AlTiN coating via physical vapor deposition/ HVOF	Deposition of AlTiN coatings and subsequent high-temperature wear testing	Shows that AlTiN coatings provide excellent wear resistance at elevated temperatures	Suggests AlTiN coatings for high-temperature tribological applications	Rukhande et al. [36]
Shot peening	Application of shot peening followed by friction and wear testing	Finds that shot peening induces compressive stresses, leading to reduced friction and wear	Recommends shot peening as a surface treatment to enhance tribological properties	Sun et al., [15]
Cryogenic treatment	Cryogenic treatment followed by hardness and wear testing	Reports that cryogenic treatment enhances hardness and reduces wear rate	Suggests cryogenic treatment as a supplementary process for improving tribological performance	Zhang et al. [10]
Diamond-like carbon (DLC) coating	Deposition of DLC coatings and evaluation under sliding wear conditions	Demonstrates that DLC coatings significantly reduce friction and wear	Recommends DLC coatings for applications requiring low friction surfaces	Zhang et al. [8]
Surface mechanical attrition treatment (SMAT)	Application of SMAT followed by tribological testing	Shows that surface Nano crystallization improves hardness and wear resistance	Suggests SMAT as a viable technique for surface strengthening	Ji et al. [30]
Silicon carbide (SiC) coating	Deposition of SiC coatings and high-temperature wear testing	Test results demonstrate the ability of SiC-based coatings to prevent friction and wear at elevated operating temperatures.	SiC coatings demonstrate suitability for applications which require operating at elevated temperatures	Ding et al. [17]
Laser cladding with Co-based alloy	Laser cladding followed by microstructural analysis and wear testing	Reveals that laser cladding enhances surface hardness and wear resistance	Suggests laser cladding as an effective repair and surface enhancement technique	Geng et al. [40]

4. Surface Modification Techniques

Several surface modification methods were explored to improve the tribological performance of nickel-based superalloys, such as:

A. **Coatings:** Methods like DLC, AlTiN, and SiC coatings are applied to reduce friction and wear.

B. **Treatments:** Techniques such as plasma nitriding, cryogenic treatment, and laser cladding improve surface hardness and resistance to wear.

C. **Mechanical Processes:** Shot peening and surface nano-crystallization were shown to introduce beneficial residual stresses and refine surface grain structure, enhancing wear resistance.

Most studies involved preparing nickel-based superalloy samples and subjecting them to different surface modification techniques. The samples were then evaluated through:

D. **Tribological test:** Pin-on-disk, scratch tests, and wear rate analysis were used to quantify wear and friction.

E. **High temperature testing:** Few studies specifically analyzed performance at elevated temperatures, replicating real-world operating conditions (e.g., jet engines and gas turbines).

5. Findings

The key findings from the studies include:

- Performance Enhancements Hardness Improvement: Techniques like plasma nitriding and laser cladding increased the surface hardness, reducing wear.
- Friction Reduction: Coatings such as DLC and MoS_2 significantly lowered the coefficient of friction.
- Wear Resistance: Zirconium, AlTiN, and SiC coatings provided superior wear resistance, particularly under high-stress conditions.
- Environmental Adaptations: Techniques like SiC and DLC coatings maintained performance at elevated temperatures, demonstrating their applicability in high-temperature applications.

6. Implications and Recomendations

The studies emphasize:

- Advanced Coating Development: Continued development of high-performance coatings for specific environments, such as high-temperature operations, is necessary.
- Optimized Treatment Parameters: Techniques like cryogenic treatment and shot peening require further optimization to maximize benefits.

Table 86.3 Focus and fundings of nickel-based alloys and superalloys, focusing on their mechanical and physical properties

Focus	Findings	Reference
Provides an extensive review of the tribological properties of nickel-based superalloys, including their mechanical and physical characteristics, surface treatments, and applications.	Highlights the importance of understanding wear mechanisms and the influence of factors like temperature, load, and sliding speed on wear behavior. Emphasizes the role of surface coatings and treatments in enhancing tribological performance.	Wang et al. [20]
Reviews the progress in electrodeposition techniques for nickel-based composites, focusing on their tribological properties.	Discusses the benefits of incorporating nanoparticles to improve hardness, reduce friction, and enhance wear resistance. Addresses challenges such as particle agglomeration and surface roughness.	Guan et al. [41]
Analyzes the microstructural changes and creep mechanisms in nickel-based single-crystal superalloys.	Identifies the role of alloying elements and heat treatment in influencing creep resistance and mechanical properties. Highlights the importance of optimizing microstructure for enhanced performance.	Yong et al. [42]
Investigates the effect of boron micro-alloying on the ductility of Ni3Al intermetallic compounds.	Demonstrates that small additions of boron significantly improve ductility by suppressing intergranular fracture, enhancing the material's mechanical properties.	Aoki et al. [32]
Studies the coarsening behavior of γ' precipitates in Ni3Al-based alloys.	Reveals that the addition of elements like Fe, Cr, and Mo can significantly increase creep resistance by forming multiphase configurations that hinder dislocation movement.	Wu et al. [26]
Examines the creep and fracture properties of Inconel 625 produced via additive manufacturing.	Finds that additively manufactured Inconel 625 exhibits comparable or superior creep strength at elevated temperatures compared to wrought counterparts, though ductility may be reduced.	Son et al. [28]
Investigates the fatigue strength of additively manufactured versus conventionally wrought Inconel 718 at elevated temperatures.	Concludes that additively manufactured Inconel 718 shows comparable fatigue strength to wrought material, with microstructural differences influencing performance.	Gilley et al. [38]
Reviews the mechanical properties of materials produced through powder-bed laser fusion additive manufacturing.	Highlights the anisotropic mechanical behavior and the influence of process parameters on properties like strength and ductility.	Mower et al. [43]
Discusses strategies for enhancing creep resistance in nickel-based superalloys.	Emphasizes the role of alloy composition, microstructural control, and processing techniques in improving high-temperature performance.	Ding et al. [17]

- Future Research Directions: Exploring hybrid surface modification methods (e.g., combining coatings with mechanical treatments) could yield synergistic improvements.

7. Examples of Techniques and Applications

- Laser Surface Texturing: This technique was highlighted for its ability to retain lubricants and enhance wear resistance in dynamic applications like turbine blades.
- Plasma Nitriding: Proven effective for improving surface durability in environments with abrasive interactions.
- DLC Coatings: These low-friction coatings are ideal for applications requiring minimal energy loss and heat generation.

8. Merits and Limitations

The advantages and disadvantages of the nickel-based alloys and its properties are discussed briefly based on the study done in this analysis.

Merits

- Provides a critical examination of current trends in nickel base alloys and superalloys tribological behavior, including mechanical and physical properties, surface processing, and applications
- Reviews the several types of wear processes, such as abrasive, adhesive, and fatigue wear, and addresses the role played by a variety of factors, such as temperature, loading, and sliding velocity, in governing such alloys' wear behavior.
- Discusses the function of coatings and treatments in enhancing nickel-based alloys and superalloys tribological performance
- Reviews the various techniques adopted for improvement in tribological behavior of nickel-based alloys and superalloys, such as coatings, surface modification, and lubrication
- Discusses the limitations and obstacles in using nickel-alloy coatings with nanoparticles, including issues with particle agglomeration, roughness of the substrate, and substrate compatibility. Highlights the benefits of nanoparticle reinforcement in terms of wear resistance, friction reduction, and improved mechanical properties of nickel-based alloy coatings.

Demerits

- The papers do not provide practical recommendations for the selection and use of nickel-based alloys and superalloys in specific applications.
- Some of the papers may be too technical and may require a strong background in material science to understand fully.
- There is some repetition of information across the papers, which may be redundant for readers who have read multiple papers in the series.

9. Overall Significance of the Studies

The reviewed study points out that the service life and performance of nickel-based super alloys can be enhanced through tailoring surface modifications. The conclusions drawn are relevant to industries based on these materials under extreme conditions, such as aerospace and power generation. A table and supporting analysis provide an overall resource on state-of-the-art developments related to improving tribological properties in nickel-based superalloys. It gives insight to researchers, engineers, and industry professionals into both successful methodologies and areas requiring further exploration. Table 86.3 provides a comprehensive overview of recent research on the tribological behavior of nickel-based alloys and superalloys, focusing on their mechanical and physical properties, surface treatments, wear mechanisms, and the effects of various operational parameters.

The comprehensive review in the above table states the recent studies on the tribological behavior and performance optimization of nickel-based alloys and superalloys. The key highlights from the studies include:

- Tribological Properties: Research has delved into wear mechanisms such as abrasive, adhesive, and fatigue wear, and their dependence on factors like temperature, load, and sliding speed. Surface coatings and treatments have been shown to enhance wear resistance and reduce friction effectively.
- Surface Modification Techniques: Techniques like laser surface texturing, nanoparticle-reinforced coatings, and additive manufacturing were investigated. These methods improve the mechanical and tribological properties but often face challenges like coating adhesion, particle agglomeration, and roughness.
- Additive Manufacturing: Additive manufacturing (AM) methods, including laser powder bed fusion, demonstrated the ability to produce high-strength components with tailored microstructures. However, anisotropic mechanical behavior and microstructural optimization remain areas of focus.
- Creep and Fatigue Resistance: Studies on creep and fatigue behavior, particularly at high temperatures, highlighted the importance of alloy composition

and processing parameters in improving long-term performance.

- Future Directions: The research emphasizes the need for practical guidelines for selecting and applying nickel-based alloys in specific applications. Developing hybrid approaches, such as combining advanced coatings with AM techniques, holds significant potential.
- The set of work serves as a resource for researchers and industry professionals as it gives the deepest understanding of the challenges, advancement, and scope of nickel-based superalloys for high-stress, high-temperature applications.

10. Conclusion

This review provides a concise summary of the tribological characteristics and behaviours of nickel-based alloys and superalloys that are used in high-temperature and high-stress situations. According to the findings of the study, nanoparticle reinforcement in nickel-based alloy coatings offers a number of advantages, including enhanced mechanical characteristics, increased resistance to wear, and decreased friction. In addition to this, it highlights the difficulties and constraints that come along with applying these coatings. The study looks at a variety of publications, each of which discusses the tribological behavior of nickel-based superalloys under a variety of different settings, the variables that impact their tribological characteristics, and the strategies that are used to enhance their tribological performance. Researchers and engineers who are working on the development of nickel-based alloys and superalloys for a variety of industrial applications may find the survey helpful in that it gives significant insights into the present state of research in this field and may be of service to these individuals.

References

1. Zhang, G., Lu, J., & Zhang, W. 2017. A review on tribology of graphite and carbon materials. Journal of Materials Science & Technology, 33, 461–470.
2. A, Vetreselvan, E., Saravanakumar, L., Ramanan, N., Somasundaram, B., & A Praveen, B. 2023. Enhancing performance of improved wear and corrosion resistance in Nickel-Based super alloys via zirconium coating. Tuijin Jishu/Journal of Propulsion Technology, 44(4), 8229–8241.
3. Chatterjee, A., & Sarkar, S. 2017. Tribological behavior of nickel-based superalloys–a review. Transactions of the Indian Institute of Metals, 70(1), 27–40.
4. Vasireddy, S., Srinivasan, S., & Yen, C. 2018. Tribological behavior of nickel-based alloys: a review. Friction, 6(2), 89–109.
5. Saikia, N., & Deka, J. 2018. Recent advances in tribological behavior of nickel-based superalloys. Journal of Materials Science and Chemical Engineering, 6(03), 30–41.
6. Hu, J., Xiong, D., Xu, X., & Zhang, Q. 2021. Wear behavior and mechanisms of nickel-based superalloys: A review. Journal of Materials Science & Technology, 87, 61–81.
7. Ravindranadh, K., Reddy, G. M., & Prasad, R. S. 2017. A review on mechanical and tribological properties of nickel-based superalloys. Journal of Materials Engineering and Performance, 26(2), 393–413.
8. Zhang, Q., Li, J., & Li, Y. 2020. Tribological behaviors of nickel-based superalloys: A review. Journal of Alloys and Compounds, 817, 152806.
9. Sharma, S., & Singh, A. 2021. A review on tribological behavior of nickel-based superalloys: An overview. Surface and Coatings Technology, 415, 127113.
10. Zhang, H., Chen, Y., Liu, J., Zhang, Q., & Li, J. 2018. Wear behavior and tribological properties of nickel-based alloy coatings reinforced with nanoparticles: a review. Materials Research Express, 5(10), 102001.
11. Guo, X., Zhang, Y., Zhao, Q., & Liu, X. 2020. A review on the tribological properties of nickel-based superalloys with surface nanocrystallization treatment. Journal of Materials Research and Technology, 9(5), 10489–10498.
12. Yang, F., Li, B., Zhou, H., Guo, W., & He, Y. 2018. Tribological properties and mechanisms of nickel-based superalloys: A review. Journal of Materials Science & Technology, 34, 1224–1234.
13. Zuo, M., Zhang, Q., & Li, J. 2021. Wear behavior and mechanisms of nickel-based alloys: a review. Journal of Materials Science & Technology, 87, 180–194.
14. Wu, X., Wu, C., Zhang, Q., & Li, J. 2019. A review of tribological behaviors and mechanisms of nickel-based superalloys at elevated temperatures. Journal of Alloys and Compounds, 780, 558–570.
15. Sun, S., Li, L., Hu, C., Li, Q., & Yuan, T. 2022. Tribological behavior of a Shot-Peened Nickel-Based single crystal superalloy at high temperature. Tribology Letters, 70(4).
16. Ouyang, X., Zhao, Y., & Bai, Z. 2021. A review of the tribological behavior of nickel-based superalloys for aero-engine bearing applications. Tribology International, 161, 107016.
17. Ding, X., Zhang, X., Xie, D., Zhang, X., & Jin, G. 2020. A review on tribological behavior of nickel-based superalloys at cryogenic temperature. Journal of Materials Science & Technology, 44, 211–224.
18. Mardani, M., Ebrahimi, M., & Karimzadeh, F. 2019. A review on mechanical and tribological properties of Ni-based coatings used for protection of the gas turbine components. Surface and Coatings Technology, 357, 260–283.
19. Khan, Z. A., Akhtar, S. S., & Ahmad, S. 2017. Laser surface engineering of Ni-based superalloys: A review. Applied Surface Science, 425, 1010–1026.
20. Wang, J., Gao, W., & Yan, F. 2020. A review on tribological behaviors of nickel-based alloys under extreme environments. Journal of Materials Science & Technology, 43, 1–15.

21. Huang, H., Wang, X., Yu, S., & Lu, J. 2018. A review on tribological behavior of nickel-based coatings under cryogenic conditions. Journal of Materials Science & Technology, 34, 187–197.
22. Zhu, Y., Fan, T., He, J., & Zhang, Y. 2020. The wear behavior of Ni-based superalloys at elevated temperatures: A review. Journal of Materials Science & Technology, 42, 130–145.
23. Liu, Y., Xu, X., Zhang, D., & Wang, Y. 2019. A review on the tribological behavior of nickel-based coatings at high temperature. Journal of Materials Science & Technology, 35, 2233–2241.
24. Wang, L., Zeng, Y., & Lu, J. 2019. A review on the tribological behavior of nickel-based alloys under dry sliding conditions. Journal of Materials Science & Technology, 35, 2077-2088.
25. Shi, Q., Wang, S., Zhang, Y., & Han, J. 2020. A review on microstructure and wear behavior of nickel-based alloys for biomedical applications. Journal of Materials Science & Technology, 46, 156–170.
26. Wu, Y., Liu, Y., Li, C., Xia, X., Wu, J., & Li, H. 2018. Coarsening behavior of γ' precipitates in the $\gamma'+\gamma$ area of a Ni3Al-based alloy. Journal of Alloys and Compounds, 771, 526–533.
27. Chen, Z., Liu, L., Wang, X., & Lu, J. 2019. A review on tribological behavior of nickel-based coatings under extreme conditions. Surface and Coatings Technology, 368, 105–117.
28. Son, K., Phan, T., Levine, L., Kim, K., Lee, K., Ahlfors, M., & Kassner, M. 2021. The creep and fracture properties of additively manufactured inconel 625. Materialia, 15, 101021.
29. Li, C., Zhang, Q., & Liu, W. 2019. Tribological behavior and mechanism of nickel-based coatings: A review. Journal of Materials Science & Technology, 35, 1587–1596.
30. Ji, R., Yang, Z., Jin, H., Liu, Y., Wang, H., Zheng, Q., Cheng, W., Cai, B., & Li, X. 2019. Surface nanocrystallization and enhanced surface mechanical properties of nickel-based superalloy by coupled electric pulse and ultrasonic treatment. Surface and Coatings Technology, 375, 292–302.
31. Xie, W., Xue, Q., & Chen, Z. 2019. A review on wear-resistant coatings for nickel-based superalloys. Journal of Materials Science & Technology, 35, 1409–1420.
32. Aoki, K. 1990. Ductilization of L1 2 Intermetallic Compound Ni 3 Al by Microalloying with Boron. Materials Transactions JIM, 31(6), 443–448.
33. Han, W., Wang, H., Huang, C., & Lu, H. 2018. Microstructure and tribological properties of nickel-based coatings: A review. Applied Sciences, 8(7), 1184.
34. Zhao, F., Hu, H., Yu, J., Lai, J., He, H., Zhang, Y., Qi, H., & Wang, D. 2023. Mechanical and tribological properties of NI-B and Ni-B-W coatings prepared by Electroless Plating. Lubricants, 11(2), 42.
35. Dubey, D., Mukherjee, R., & Singh, M. K. 2024. A review on tribological behavior of nickel-based Inconel superalloy. Proceedings of the Institution of Mechanical Engineers Part J Journal of Engineering Tribology, 238(6), 706–732.
36. Rukhande, S., Suryawanshi, S., Bhosale, D. G., Vasudev, H., Khimasiya, R., Gupta, R. D., Deshmukh, I., & Wilson, F. 2024. High-Temperature tribological performance of Ni-Based coating on 316L stainless steel. Results in Surfaces and Interfaces, 100388.
37. Fidan, S., Ürgün, S., Atapek, Ş. H., Çelik, G. A., Canel, T., Sınmazçelik, T., & Bora, M. Ö. 2024. Investigation of surface structuring and oxidation performance of Inconel 718 superalloy by laser remelting with different patterns. Engineering Failure Analysis, 167, 108974.
38. Gilley, K. L., Nino, J. C., Riddle, Y. W., Hahn, D. W., & Perry, S. S. 2012. Heat treatments modify the tribological properties of nickel boron coatings. ACS Applied Materials & Interfaces, 4(6), 3069–3076.
39. Niksefat, V., & Mahboubi, F. 2024. The effect of plasma nitriding duration on the tribological performance of Ni-B-Gr self-lubricant composite coatings on AISI 4140 steel. Journal of Materials Research and Technology.
40. Geng, C., Zhang, Z., Xie, H., & Liu, F. 2024. Effect of laser cladding layers on microstructure and mechanical properties of NiCoCrAlY bond coat. Materials Today Communications, 41, 110930.
41. Guan, T., & Zhang, N. 2024. Recent Advances in Electrodeposition of Nickel-Based Nanocomposites Enhanced with Lubricating Nanoparticles. Nanomanufacturing and Metrology, 7(1).
42. Yong, S., Sugui, T., Huichen, Y., Delong, S., & Shuang, L. 2016. Microstructure evolution and its effect on creep behavior of single crystal Ni-based superalloys with various orientations. Materials Science and Engineering A, 668, 243–254.
43. Mower, T. M., & Long, M. J. 2015. Mechanical behavior of additive manufactured, powder-bed laser-fused materials. Materials Science and Engineering A, 651, 198–213.

Note: All the tables in this chapter were made by the authors.

Advances in Additive Manufacturing Technologies – Gurusamy Pathinettampadian et al. (eds)
© 2026 Taylor & Francis Group, London, ISBN 978-1-041-16687-0

Tribological Investigation and Optimization of Rutile and Cenosphere Hybrid Composite

Mariyappan K.*, Elanchezhian P.
Department of Mechanical Engineering,
VelTech HighTech Dr. Rangarajan Dr. Sakunthala Engineering College,
Chennai, Tamilnadu

Bupesh Raja V. K.
School of Mechanical Engineering,
Sathyabama Institute of Science and Technology,
Chennai, Tamilnadu

Ajay V., Hari Krishna Raj S., Balaji A.
Department of Mechanical Engineering,
VelTech HighTech Dr. Rangarajan Dr. Sakunthala Engineering College,
Chennai, Tamilnadu

◆ **Abstract:** Aluminum (Al) and Aluminum alloys are renowned for their light properties, making them suitable for applications in composite materials. These composites are developed by employing Aluminum as the matrix metal and strengthening it with particles like TiO2 and Al2SiO5, most commonly through 3D printing technology. These Aluminum Matrix Composites (AMCs) are greatly appreciated for their superior mechanical and tribological properties, and they are well suited for a variety of industrial applications in the automotive, aerospace, marine, sports, and defense industries. The objective of this research is to fabricate Aluminum 6082 matrix composites reinforced with fixed amounts of Titanium carbide and cenosphere particles using 3D printing technology. The obtained composite exhibited the enhancements in hardness and wear rate reduction over the conventional AA6082 matrix. The drilling parameters of AA6082, TiO2, and Al2SiO5 were examined to demonstrate that minimum surface roughness attained was 2.867 microns. Taguchi design and experimental analyses on drilling parameters such as spindle speed, feed, and pecking proved significant in pinpointing the favorable machining conditions. Also, the relative contribution of every factor towards total performance was examined with ANOVA for this hybrid metal matrix.

◆ **Keywords:** Cenosphere, ANNOVA, Taguchi design, Hybrid metal matrix

1. Introduction

As the demand for small, high-functionality components increases, the need for precise drilling has also risen. The growing requirement for deeper and smaller holes in these industries demands advanced drilling technologies that can provide both high accuracy and increased productivity. Various conventional and non-conventional drilling methods are available. These include laser beam, electron beam, and electric discharge drilling, as well as processes like electrolytic polishing and electrochemical machining, which are often employed in experimental settings.

*Corresponding author: jkmari143@gmail.com

DOI: 10.1201/9781003685906-87

However, for general industrial applications, conventional drilling methods remain the preferred choice due to their cost-effectiveness and higher productivity compared to non-conventional techniques.

Physical vapor deposition (PVD) is another technique used in manufacturing, where thin films are deposited onto workpieces through the condensation of vaporized materials, commonly used in semiconductor production. Computer Numerical Control (CNC) lathes are replacing older production lathes due to their ease of setup, operation, repeatability, and accuracy. These machines, designed for modern carbide tooling, enable precise production through CAD/CAM systems that allow the part design and tool paths to be programmed. Once the program is set and tested, the CNC lathe can continue producing parts with minimal oversight, controlled via a computer interface. Though CNC machines require operators with specialized skills, they offer a broader knowledge base compared to traditional production methods, where deep knowledge of each machine was essential. Often, the same operator sets up and supervises multiple CNC machines, working within a cell for efficient production. However, these properties can be further enhanced by utilizing Aluminum matrix composites. Aluminum matrix composites are characterized by the following key features: (1) They are man-made materials, (2) They consist of at least two chemically distinct components, one of which is Aluminum, with a distinct interface separating the materials, (3) The materials are combined in a three-dimensional structure, and (4) The resulting composite exhibits properties that cannot be achieved by any of the individual materials alone.

8011 Al-boron fiber composites have been used as struts in the space shuttle's main cargo bay. The automotive industry shows great potential for increased aluminum usage. Automakers are increasingly turning to aluminum to boost fuel efficiency and reduce emissions. However, concerns about price, supply, manufacturing costs, and recyclability persist. Ongoing research and development efforts continue to address these challenges. Aluminum competes with steel, as steel manufacturers have significantly improved sheet quality, often collaborating with automakers. Despite the relatively low consumption of aluminum per vehicle at present, aluminum manufacturers are working to highlight aluminum's advantageous physical and mechanical properties.

To prepare eggshells for use, they were first washed with water to remove the membrane and then placed in a circular stainless steel tray. The cleaned shells were left to dry in sunlight for six hours. Once dried, they were manually crushed using hand compression and further ground into a fine powder using a planetary ball mill. This milling process lasted three hours, after which the powder was collected, resulting in eggshells in two distinct forms.

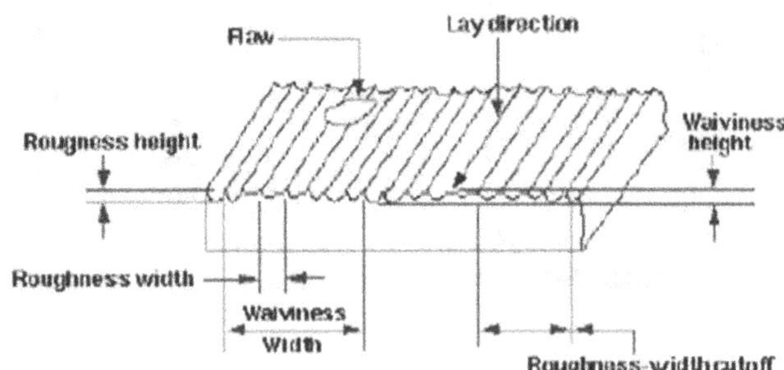

Fig. 87.1 Surface characteristics

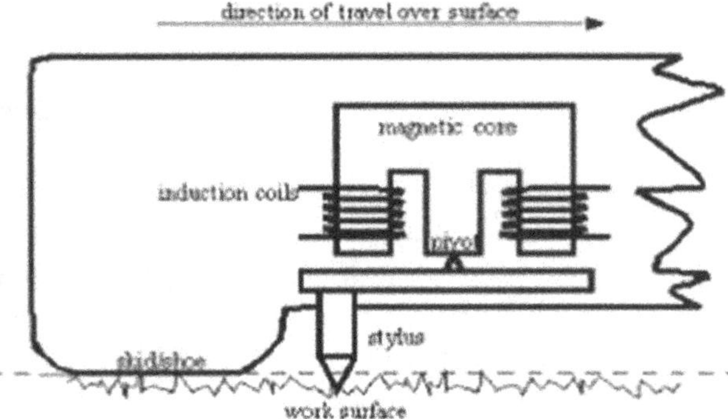

Fig. 87.2 Surface measurements testing machine

The composite specimens were prepared using cenosphere and epoxy. Cenosphere, which are mainly composed of aluminum, silicon, and iron oxides, are notable for their low density, ranging from 0.4 to 0.8 g/cm^3. Their key characteristics include a hollow spherical structure, particle sizes ranging from sub-micron to millimeter scale, ultra-lightweight properties, and low thermal conductivity.

Aluminum metal matrix composites (AMMCs) are materials composed of a metal matrix, typically aluminum, combined with reinforcing materials such as fibers, particles, or laminates. These composites are commonly produced using techniques like powder metallurgy, liquid casting, or specialized manufacturing processes. Among these, the powder metallurgy method has limitations, including high processing costs and size constraints. As a result, liquid casting is often preferred for its cost-effectiveness and efficiency in producing aluminum composites. For this study, an aluminum 8011 rod was used along with TiB_2 and Es reinforcement particles (average size: 200μm), all sourced.

Crucible furnaces are among the oldest and simplest types of melting units used in foundries. They operate using a

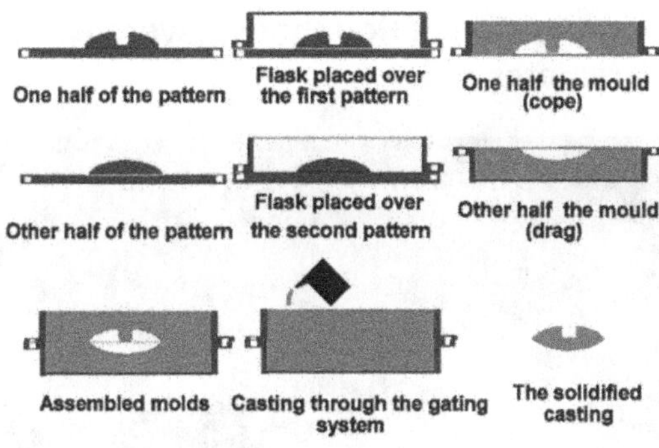

Fig. 87.3 Casting

refractory crucible that holds the metal charge, which is heated through the walls of the crucible via conduction. Depending on the setup, the heating fuel can be coke, oil, gas, or electricity. This type of furnace is particularly suited for melting small batches of low-melting-point alloys, making it an economical choice for small-scale non-ferrous foundries due to its relatively low initial investment costs. A crucible furnace consists of a high-temperature-resistant container designed to melt various metals. Its simplicity and efficiency make it a widely used method for small-scale casting operations. Crucible casting was used to prepare a hybrid aluminum metal matrix composite. The composite material was reinforced with 5% titanium dioxide (Rutile), 2.5% eggshell, and 2.5% cenosphere, with average particle sizes of 120 mesh and 140 mesh, respectively. The cast specimen, with a diameter of 30 mm, was cut into 16 circular plates of 15 mm thickness using a power hacksaw for further processing.

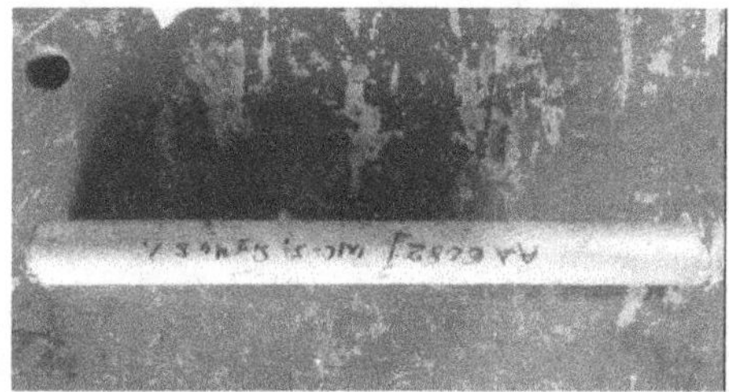

Fig. 87.4 AMMC rod

Table 87.1 Hardness value

S. No	Material	HRB
R_1	Al-6061 + TiO2 – 5% + ES-2.5% & CS- 2.5%	64

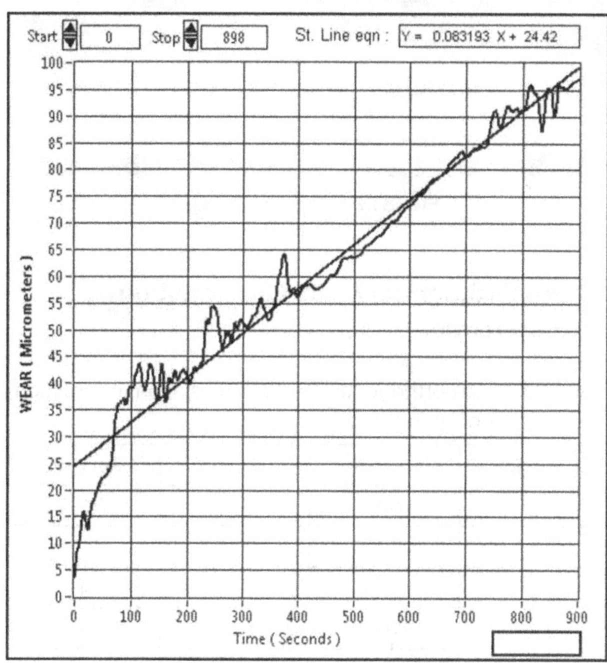

Fig. 87.5 Ratio-1 vs wear rate

Table 87.2 Process parameters and their levels of both drill bit

Sl. No	Speed	Feed	Peck
1	600	0.04	2
2	700	0.06	3
3	800	0.08	4

1.1 Minitab-17 Software

By using Minitab-17 software have optimized the drilling parameters

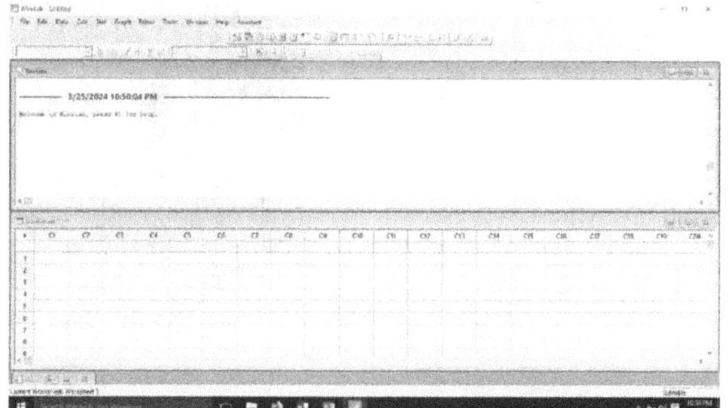

Fig. 87.6 Minitab software

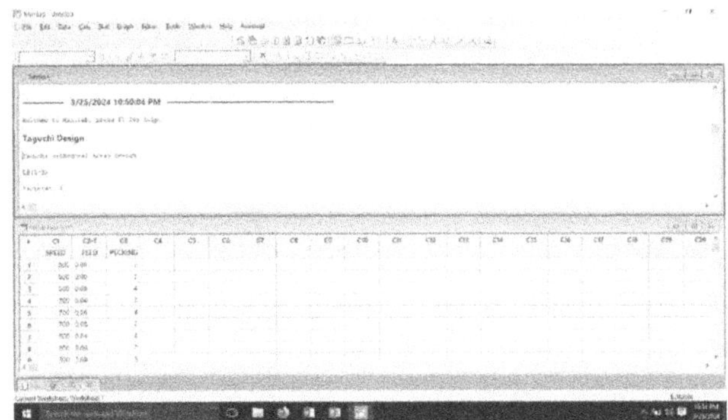

Fig. 87.7 Taguchi design

1.2 An Orthogonal Array L9 Formation (Interaction)

Table 87.3 An orthogonal array L16 formation (interaction) of both drill bit

SL. No	Speed	Feed	Pecking
1	600	0.04	2
2	600	0,06	3
3	600	0.08	4
4	700	0.04	3
5	700	0,06	4
6	700	0.08	2
7	800	0.04	4
8	800	0,06	2
9	800	0.08	3

Table 87.4 Experimental data and output response analysis

EXP	RA	DIA ERR	RE ERR	CYL ERR	MT
1	2.867	8.010	0.005	0.017	68
2	7.078	7.913	0.020	0.030	45
3	7.608	7.929	0.009	0.014	37
4	3.463	7.953	0.013	0.027	56
5	3.259	7.951	0.010	0.016	41
6	4.813	7.992	0.012	0.030	33
7	3.744	7.943	0.011	0.015	50
8	4.166	7.941	0.009	0.015	37
9	3.799	7.944	0.010	0.014	30

1.3 Comparision of Roughness Average vs Parameter

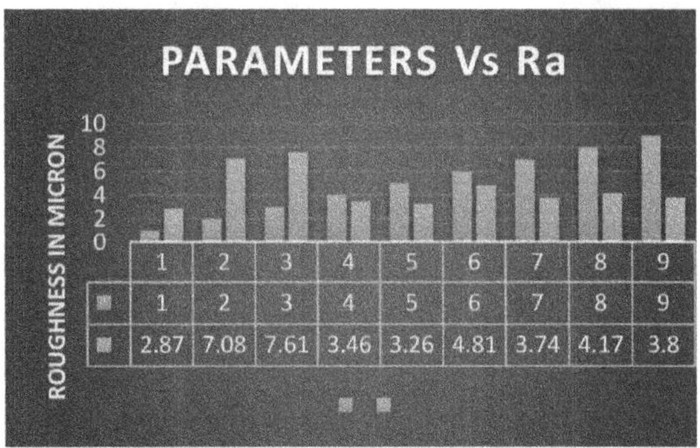

Fig. 87.8 Roughness average vs parameter

1.4 Comparision of Dia Error vs Parameter With

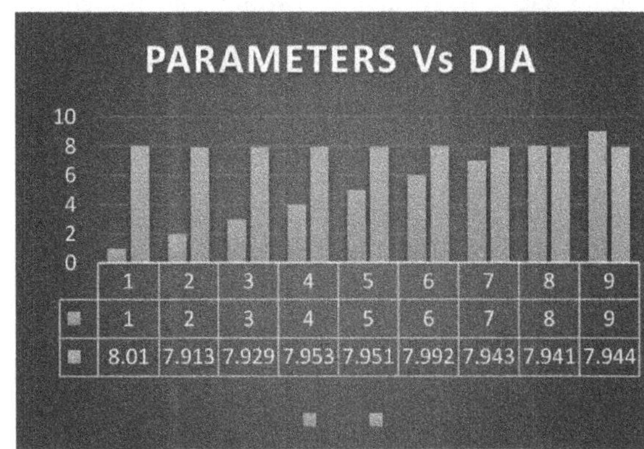

Fig. 87.9 Dia error vs parameter

1.5 Comparision Roundness Error vs Parameter

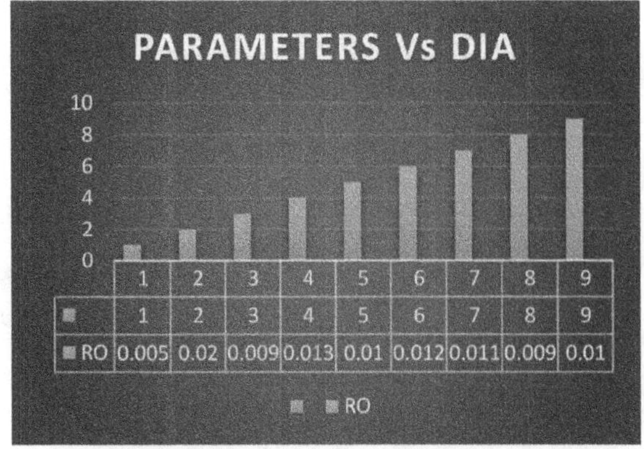

Fig. 87.10 Roundness error vs parameter

1.6 Comparision Cylindricity vs Parameter

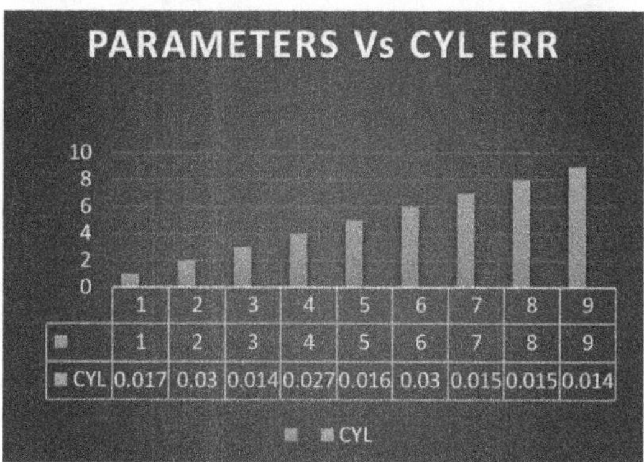

Fig. 87.11 Cylindricity vs parameter

1.7 Comparision Machining Time vs Parameter with Various Coated Drill Bit

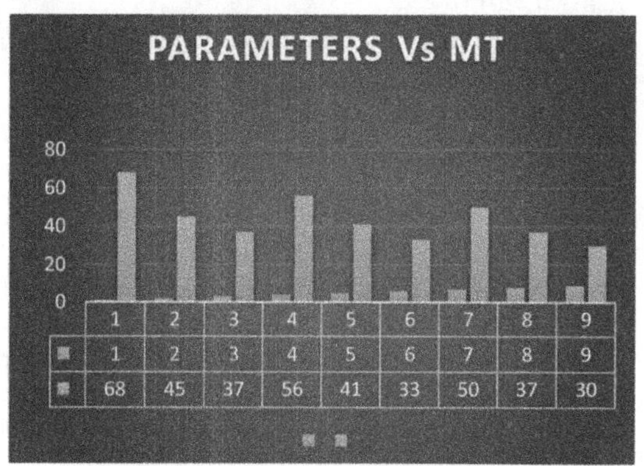

Fig. 87.12 Machining time vs parameter

By observing the output responses value the machining, geometrical characteristics and overall performance found best on first parameter speed 600 rpm, feed 0.04mm/rev and pecking 2 mm.

Table 87.5 ANOVA for various responses with 'p' values

S	R-sq	R-sq(adj)	R-sq(pred)
0.0242647	83.81%	35.25%	0.00%

Coefficients						
Term	Coef	SE	Coef	T-Value	P-Value	VIF
Constant	7.95289	0.00809	983.26	0.000		
Speed	600	-0.0022	0.0114	-0.19	0.864	1.33
	700	0.0124	0.0114	1.09	0.390	1.33
Feed	0,06	-0.0179	0.0114	-1.56	0.258	1.33
	0.04	0.0158	0.0114	1.38	0.302	1.33
Pecking	2	0.0281	0.0114	2.46	0.133	1.33
	3	-0.0162	0.0114	-1.42	0.292	1.33

1.8 MT (Analysis of Result)

Table 87.6 MT and S/N ratio values for the experiments

No	Spindle Speed (N) (RPM)	Feed (F) (mm/rev)	Pecking mm	MT Sec	SN-RAIO
1	600	0.04	2	68	-36.6502
2	600	0,06	3	45	-33.0643
3	600	0.08	4	37	-31.3640
4	700	0.04	3	56	-34.9638
5	700	0,06	4	41	-32.2557
6	700	0.08	2	33	-30.3703
7	800	0.04	4	50	-33.9794
8	800	0,06	2	37	-31.3640
9	800	0.08	3	30	-29.5424

Table 87.7 Response table for signal to noise RA-smaller is better

Level	Speed	Feed	Pecking
1	-33.69	-32.23	-32.79
2	-32.53	-35.20	-32.52
3	-31.63	-30.43	-32.53
Delta	2.06	4.77	0.27
Rank	2	1	3

Table 87.8 Means table for RA

Level	Speed	Feed	Pecking
1	50.00	41.00	46.00
2	43.33	58.00	43.67
3	39.00	33.33	42.67
Delta	11.00	24.67	3.33
Rank	2	1	3

Table 87.9 Analysis of variance for RA

Source	DF	SEQ SS	ADJ MS	F	P	% of contribution
SPEED	2	184.22	92.111	8.05	0.111	16
FEED	2	956.22	478.111	41.78	0.023	81
PECKING	2	17.56	8.778	0.77	0.566	1
Error	2	22.89	11.444			2
Total	8	1180.89				100

Main Effects Plot for SN RATIO

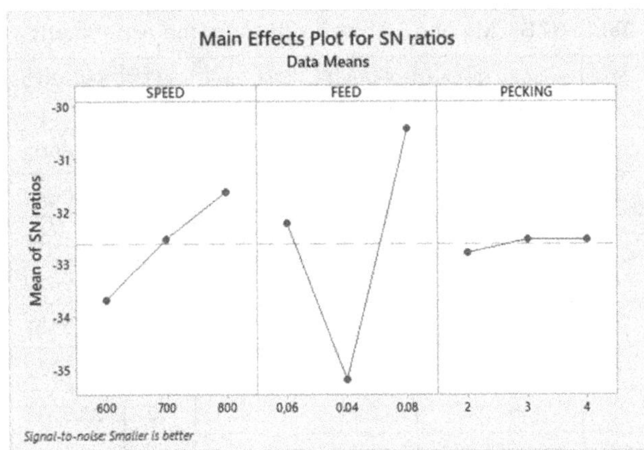

Fig. 87.13 Main effects plot for SN ratio-MT

Main Effects Plot for Means

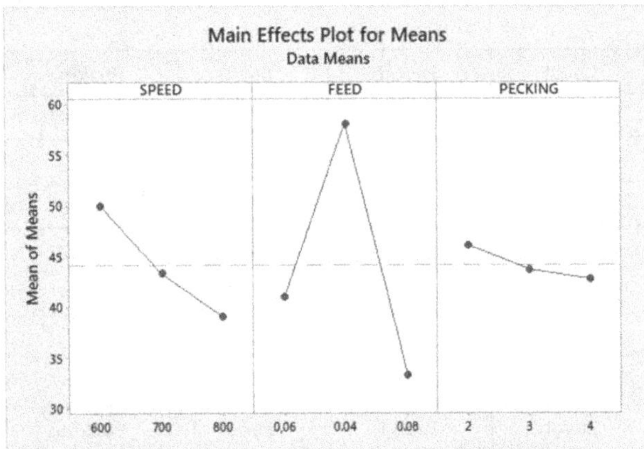

Fig. 87.14 Main effects plot for means-MT

2. Conclusion

1. This work investigates the influence of drilling parameters on surface roughness and geometric accuracy during hybrid AMMC machining with an HSS drill bit.
2. The parameters of prime concern during machining were spindle speed, feed rate, and pecking depth, and the corresponding geometric properties thus obtained were noted for study.
3. A Taguchi Design strategy was utilized to create the experimental model, and ANOVA was used to analyze the results.
4. The analysis demonstrated that the decrease of surface roughness, roundness error, cylindricity error, and diameter error was achieved by lowering spindle speed and feed rates and elevating pecking values.
5. The most effective machining performance and geometric precision were achieved using the first parameter configuration: spindle speed 600 RPM, feed rate 0.04 mm/rev, and pecking depth 2 mm.

References

1. J. Pradeep Kumar Effect of Drilling Parameters on Surface Roughness, Tool Wear, Material Removal Rate and Hole Diameter Error in Drilling of OHNS International Journal of Advanced Engineering Research and Studies.
2. B.K. Kavad A Review Paper on Effects of Drilling on Glass Fiber Reinforced Plastic International Conference on Innovations in Automation and Mechatronics Engineering, ICIAME 2014
3. A. Bovas Herbert Bejaxhin Performance Characterization of Surface Quality and Tool Wear of Wet/Dry Drilling on Steels by Using Coated Drill Bits Proceedings of Engineering and Technology Innovation, vol. 12, 2019, pp. 09–20.
4. Erkan Bahçe Experimental Investigation of the Effect of Machining Parameters on the Surface Roughness and the Formation of Built-Up Edge (BUE) in the Drilling of Al 5005 intech.
5. Adem Çiçek Application of Taguchi Method for Surface Roughness and Roundness Error in Drilling of AISI 316 Stainless Steel Journal of Mechanical Engineering. All rights reserved
6. H. Siddhi Jailani Multi-response optimization of sintering parameters of Al–Si alloy/fly ash composite using Taguchi method and grey relational analysis Int J Adv Manuf Techno.
7. D. Arola Estimating the fatigue stress concentration factor of machined surfaces International Journal of Fatigue 24 (2002) 923–930.
8. Puneeth H V Studies on Tool Life and Cutting forces for drilling operation using Uncoated and coated HSS tool International Research Journal of Engineering and Technology (IRJET).
9. Reddy Sreenivasulu Effect of Drilling Parameters on Thrust Force and Torque During Drilling of Aluminium 6061 Alloy - Based On Taguchi Design Of Experiments Effect of drilling parameters on Thrust force and Torque during Drilling of Aluminium 6061 Allo

Note: All the figures and tables in this chapter were made by the authors.

Advances in Additive Manufacturing Technologies – Gurusamy Pathinettampadian et al. (eds)

Investigation the Influence of EDM Process Parameters on Mg AZ31 Alloy-Effect on Machining Performance and a Parametric Study

T. Aravind,
R. Ganeshamoorthy[1], S. D. Dhanesh babu[2]
Department of Mechanical Engineering,
Chennai Institute of Technology,
Chennai, Tamilnadu

K. R. Padmavathi[3]
Department of Mechanical Engineering,
Panimalar Engineering College,
Chennai, India

T. Kavitha[4]
Department of Electronics and Communication Engineering,
Vel Tech Rangarajan Dr. Sagunthala R & D Institute of Science and Technology

P. Balamurali[5]
Department of Mechanical Engineering,
Chennai Institute of Technology,
Chennai,India

◆ **Abstract:** Under the application of EDM process the main purpose of this program is investigating and analysing new properties for Magnesium AZ-31 alloys. Such as the usage of magnesium alloys depend on its light mass, high strength-to-weight proportion and extreme platitude, they have attracted significant attention so far. Meanwhile their material quality is not very high due to poor thermal conductivity and low melting point characteristics (for instance it melts at 650 degrees Celsius), so that machining of magnesium alloys becomes particularly difficult. To overcome these disadvantages, this study uses a non-traditional machining technique, EDM. This technique is expected to improve the machining of Magnesium AZ-31 alloys. It is hoped that this work is not simply a trivial repetition of already-published literature. A comprehensive review on Magnesium AZ-31 alloys and EDM processing is given at the beginning of the study. First, the experimental design is determined. The machining parameters for EDM are selected, such as peak current and pulse duration, and an easy-to-melt electrode material. When the EDM process is analyzed, the material removing rate (MRR), electrode wear ratio (EWR) and surface roughness (Ra) of machined sample are evaluateds.

◆ **Keywords:** EDM, Microstructure, SEM, Hardness, Surface roughness

Correspondingauthor: [1]ganesamoorthyr@citchennai.net, [2]dhaneshbabusd@citchennai.net, [3]krpadmavathipecmech@gmail.com, [4]kavithaecephd@gmail.com, [5]balamuralip@citchennai.net

DOI: 10.1201/9781003685906-88

1. Introduction

1.1 EDM Process

The material-erosion mechanism in the process, EDM is basically the conversion of electrical energy to thermal heat. This takes place between a series of isolated electrical discharges occurring between work material and electrode plunged in a dielectric fluid [1]. The EDM process is shown in Fig. 88.1 Thermal energy from the cathode to anode creates a plasma channel. This means that the temperature between those two points is 10,000 to 8,000 °C and even as high as 20,000 °C will result in substantial portions of material being heated and melting at workpiece surface. During pulse current supply the plasma channel breaks down at pulse off time, bringing a sudden temperature decrease to some extent. This allows new dielectric fluid to take away with it the molten material from the work surfaces in the form of fine debris [2]. This process of melting and vaporizing the material from the workpiece is in complete contrast to the way chips are produced in conventional machining. The volume of material removal per discharge is usually between 10-6 –10-4 mm3 and the MR is generally several millimeters depending on particular application. The electrode shape is what determines where spark erosion will occur in EDM. In addition, the accuracy of the part produced after EDM is quite high, yet it is dependent upon electrode shape. Thus, the tool-electrode shape in EDM is "mirrored" in the workpiece [3].

1.2 Objectives of the Research

The primary objective of our research was to investigate and optimize the Electrical Discharge Machining (EDM) process for the magnesium alloy AZ31 [4]. This involved the creation and analysis of a series of machined holes to evaluate the performance of the EDM process on AZ31. We aimed to assess parameters such as surface finish, dimensional accuracy, and the presence of defects to determine the effectiveness of the machining process. Furthermore, our objective was to identify the optimal machining parameters that would result in the desired quality and accuracy of EDM machining for AZ31. By carefully selecting representative holes with maximum, minimum, and average performance, we aimed to determine the parameter set that would yield the best results. Overall, our research aimed to enhance the understanding and optimization of the EDM process for achieving superior machining outcomes in the magnesium alloy AZ31 [5].

2. Literature Review

2.1 Machining Performance and Optimization

The effect of various EDM process parameters on difficult-to-cut materials is particularly examined in this chapter, particularly for metals such as magnesium alloys, stainless steel and die steel, with a focus on the impact of MR, TWR, OC and SR. Additionally, the literatures in regard to different optimization techniques are reviewed [6].

Various experiments on wire electrical discharge machining (WEDM) of magnesium alloy AZ31 were carried out by Raju Bangaru, Ashok Kumar Singh, and H. S. Shan to understand its machining characteristics and surface integrity. The authors investigated the influence of process parameters (pulse on time, pulse off time, and wire tension) on MRR (Material removal rate), Surface roughness, and microstructural changes. It was observed that the pulse on time had a positive effect on MRR and the pulse off time had a negative effect [7]. While the surface roughness decreased with an increase in pulse on time and increased with an improved pulse off time. The study found out a combination of process parameters with desired MRR and surface roughness. Results revealed a significant recast layer and heat-affected zone formed during WEDM, thereby indicating considerable thermal effects. Controlling the WEDM process parameters is crucial for obtaining desired machining characteristics and improving the surface integrity of the machined magnesium alloy AZ31 [8].

J. Jeswin, R. Dhinesh (2021), "A STUDY ON MACHINING PERFORMANCE OF MAGNESIUM ALLOY AZ31 USING WEDM" Process parameters that impact MRR, surface roughness and tool wear were studied. On the other hand, Increasing pulse on time increased MR but decreased surface roughness. MRR and surface roughness were analysed to identify the optimum parameter combination. Tool wear increased as pulse on and off time was increased. It can provide important guidance for optimizing WEDM of AZ31 and for obtaining a deeper knowledge into the machining characteristics for magnesium alloys. (Source: Materials Today: Proceedings, Volume 39, 2021, Pages 697-701) Finally, the removal of the debris discharge, along with reduction of expansion, is remarkably improved by notching the cylindrical cross-section [10].

Yilmaz et al. Mekhamer et al. These included optimal drilling parameters such as discharge current, pulse-on time, pulse-off time, and capacitance rate. They studied the EDM process outcomes like total drilling time, minimum electrode length, and surface roughness. The adaptive neuro-fuzzy inference technique is used to achieve the input–output interaction which helps in designing the drilling operations in an efficient, effective and reliable way. The system developed has interactive and visual interface making it user friendly [11].

Dewangan, et al. (2015): Effect of EDM parameters on surface integrity and dimensional accuracy. The machining parameters: tool lift time, work time, as well as pulse current and pulse-on time are taken into account. The fuzzy based TOPSIS technique has been used for multiple objective decision making. From this analysis, an optimum condition of process parameters of current = 1 A, pulse on time = 10 μs, tool work time = 0.2 s and tool lift time = 1.5 s has obtained. In order to observe the sensitivity of decision makers 'preferences, a sensitivity analysis is conducted [12].

Control of electrical process parameters, empirical relationships between process parameters and optimization of process parameters is discussed by Muthuramalingam and Mohan (2015) in EDM process. They stated that electrical process parameters can enhance efficiency of machining operation [13].

Using Taguchi method Balraj&Kumar(2016) investigated effect of EDM electrical parameters on MR, SR, white layer thickness and surface crack density for RENE 80 nickel super alloy. It concludes that overall responses are mainly influenced by peak current and pulse on time. In addition, regression models for all responses are developed and deemed acceptable.

Abhishek, et al. (2017) performed an experimental study based on 5-factor-4-level L16 OA to investigate EDM performance on Inconel 625, 718, 601 and 825. Effect of input parameters like gap voltage, peak current, pulse-on time, duty factor & flushing pressure on MR, TWR, SR & surface crack density are analysed. The obtained best parameter setting are gap voltage of 90 V, peak current of 5A, pulse-on time of 200 μs, duty factor of 70%, flushing pressure of 0.6 bar for Inconel 625; gap voltage of 90 V, peak current of 5A, pulse-on time of 200 μs, duty factor of 85%, flushing pressure of 0.4 bar for Inconel 718; gap voltage of 80 V, peak current of 7 A, pulse-on time of 500 μs, duty factor of 80%, flushing pressure of 0.3 bar for Inconel 601; gap voltage of 80 V, peak current of 5 A, pulse-on time of 300 μs, duty factor of 85%, flushing pressure of 0.4 bar for Inconel 825.

Datta et al. (2017) to find suitable set of process parameters to impart optimum machining performance during Electrical Discharge Machining using copper tool electrode of Inconel 718. The experiments were carried out using the L25 OA design of the experiment. Results have been compared using MR, TWR, SR, surface crack density, white layer thickness, and micro-hardness to study the machining performances. It seems that Open circuit voltage=80 V, peak current = 11 A, pulse on time = 100 μs, duty cycle = 85% and Flushing pressure = 0.4 bar are the most favorable of all combinations to obtain the highest MR value while minimizing TWR, SR, surface crack density, white layer thickness and micro-hardness.

Jatti, (2018) present research work is based on optimization of EDM process parameter to maximize MR during machining of NiTi alloy. The slicing process variables taken into consideration in the present study are current, pulse on time, pulse off time, workpiece electrical conductivity, and tool conductivity.

Experiments were conducted according to Taguchi 's L36 orthogonal array. It was found from the analysis that work electrical conductivity, gap current and pulse on time are significant parameters which influences material removal rate. Based on optimized setting of input parameter obtained, the optimized material

3. EXPERIMENTAL SETUP

3.1 Experimental Setup

In this chapter the experimental work deals about the experimental setup used for the experiments, tool electrode, and selection of workpiece. Moreover, the formation of the preliminary experiments, experimental design and L27 OA design based on Taguchi design, and calculation of MRR, TWR, OC, CIRCULARITY.

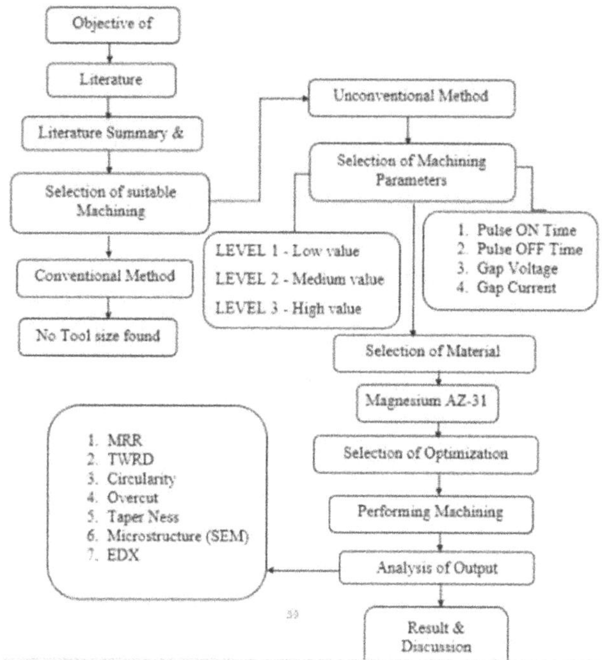

3.2 Specification of EDM

Experiments are performed using Electric Discharge Machine as shown in Fig. 88.1 and its model is OCEAN (OCT-3252NA) (die-sinking type) with servomechanism for gap control. Dielectric fluid: Commercial grade EDM oil (specific gravity= 0.763, freezing point= 94°C)[100]. The dielectric is externally flooded with 0.2 kg f/cm2 and different pulsed discharge current (not shown).

The EDM setup mainly comprises of the following section as described below:

- Dielectric reservoir, pump and circulation pump.
- Generator power and control unit.
- Working table with X-Y table integrated.
- The tool holding and electrical servo tool feed.

Fig. 88.1 EDM setup

3.3 Electrode Material

The material used as a tool electrode should possess the following properties such as

- Easily machinable,
- Low wear rate,
- Good conductor of electricity and heat,
- Cheap, and
- Readily available.

The common materials used as tool electrode in EDM are brass, copper, tungsten and graphite. Among this electrode brass is most preferred for their inherent properties (Fig. 88.2). Brass is a good conductor of electricity which has higher cutting speed. In addition, low cost and high Machinability are the advantages of brass over other EDM tool electrodes. In this research the brass electrode of Ø 0.5mm is used for all experiments.

3.4 Selection of Work Piece

Magnesium alloy AZ -31 is a lightweight, high-strength material that is commonly used in industries such as aerospace, automotive, and electronics. Magnesium AZ31 is an aluminium-zinc magnesium alloy with 9% aluminium content and 1% zinc content, it is lightest structural metal and high-purity alloy. It's has high corrosion resistance, excellent cast ability, its use for a commercial application, electrical components. It's is commonly in cast helicopter transmission housing.

3.5 Major Machining Parameters

Majority of research papers in EDM use the electrical parameters to control the output responses. Hence in this research, voltage, current, pulse on time and pulse off time are considered as a most influencing the machining characteristics. The electrical discharge machine, the key factors based on the industrial need the following performance measures are considered for this research. Machining Speed (MR), Tool Wear Rate (TWR), Overcut (OC), Circularity.

Material Removal Rate (MRR)

The basic mechanism for material removal is the conversion of electrical energy into thermal energy. The thermal energy generated melts the workpiece material and vaporize it. The melted material removed from the workpiece is called as debris and it is removed from the work surface in the form of either solid, liquid or gas (most often). In the production stand point, MRR is considered as an important performance measure.

The amount of discharge energy generated depends on the voltage, current, pulse on time and pulse off time. Another factor that influences the MRR is the flushing pressure of the dielectric medium and flushing gap. Additionally, the erosion property of the workpiece material also plays a major role in deciding the value of MRR.

Evaluation of MR and OC

MR is calculated by dividing the length of the through hole with the machining time required to complete the through hole.

$$MR = \frac{\text{Thickness of the workpiece}}{\text{total time for machining the through hole}}$$

OC is calculated using optical microscopic images. OC is defined by ΔR = Re – Rt, where Re is the entrance radius of the machined hole and Rt is the tool electrode radius. The difference in Re and Rt results in OC.TOC is defined by TOC = (D–d)/2L, where, D is entry diameter of the machined hole, d exit diameter of the machined hole, and L thickness of the workpiece.

Overcut

An EDM features is always larger than the electrode used to machine it. The difference between the size of the electrode and the size of the cavity (or hole) is called the overcut. For calculating overcut, there are two types of overcut namely total overcut (diametral overcut) or overcut per side and radial overcut. OC is calculated using optical microscopic images taken from optical microscope as shown in Fig. 88.2. OC is defined by ΔR = Re – Rt, Where Re is the entrance radius of the machined hole and

Rt is the tool electrode radius. The difference in Re and Rt results in OC.

3.6 Orthogonal Array (OA)

Sir R.A. Fisher is known for popularizing such a method - "factorial design of experiments". For a given set of factors, a full factorial design will find all the combinations. As most industrial experiments [46] involve a large set of factors, a full factorial design leads to huge numbers of experiments. To make the number of experiments contained, a small set of all possible points is chosen. This process of choosing a small number of experiments that give us the most information is called a partial fraction experiment. This method is widely recognized however no general rules exist for the application of the method or even how to interpret the results achieved after performing the experiments. So, Taguchi developed a specific gneral design principles for factorial experiments and that are common in many applications [Thanigaivelan et al. 2015].

Here are the steps of experimental design:

- Choice of independent variables

It usually comes down to the following:

- Choosing how many levels for the continuous independent variable(s)
- Orthogonal array selection
- Each column is assigned to independent variables
- Conducting the experiments
- Analysing the data

Table 88.1 Orthogonal array (OA)

Exp: No:	Pulse ON time (µs)	Pulse OFF Time (µs)	Gap Current (A)	Gap Current (v)	Circularity (µm)	Diameter (µm)	Overcut (µm)	MRR	Time taken (sec)
1	23	19	13	18	500	652.2	-152.2	40.5	6.12
2	23	19	14	22	500	663.7	-163.7	40.4	11
3	23	19	15	26	500	643.2	-143.2	40.6	6.42
4	23	22	13	22	500	692.8	-192.8	40.5	19.78
5	23	22	14	26	500	594.3	-94.3	40.3	15.81
6	23	22	15	18	500	588.1	-88.1	40.3	19.84
7	23	25	13	26	500	601.9	-101.9	40.5	13.48
8	23	25	14	18	500	610.8	-110.8	40.7	8.24
9	23	25	15	22	500	619.5	-119.5	40.5	8.03
10	25	19	13	18	500	601.6	-101.6	40.6	8.04
11	25	19	14	22	500	629.3	-129.3	40.3	7.52
12	25	19	15	26	500	646.2	-146.2	40.4	13.81
13	25	22	13	22	500	634.1	-134.1	40.4	15.9
14	25	22	14	26	500	601.8	-101.8	40.3	11.22
15	25	22	15	18	500	582.4	-82.4	40.6	6.45
16	25	25	13	26	500	584.8	-84.8	40.5	7.83
17	25	25	14	18	500	596.7	-96.7	40.5	6.68
18	25	25	15	22	500	627.4	-127.4	40.6	8.56
19	27	19	13	18	500	671.33	-171.33	40.3	12.05
20	27	19	14	22	500	663.8	-163.8	40.3	6.87
21	27	19	15	26	500	668.9	-168.9	40.3	11.89
22	27	22	13	22	500	639.2	-139.2	40.5	8.49
23	27	22	14	26	500	598.7	-98.7	40.4	16.07
24	27	22	15	18	500	642.9	-142.9	40.6	7.35
25	27	25	13	26	500	613.5	-113.5	40.6	7.85
26	27	25	14	18	500	621.6	-121.6	40.3	7.76
27	27	25	15	22	500	604.5	-104.5	40.5	7.28

Inference A suitable Taguchi`s orthogonal array (OA) design for experimentation is chosen by calculating the total degrees of freedom. In this study, there are 8 degrees of freedom owing to four three level machining parameters in the EDM process. Therefore, an L27 (2^1×3^5) OA with 4 columns and 27 rows was used to accommodate four three-level control factors. This partial factorial experimental design offers an efficient and systematic approach of determining an optimal parameter condition. Each factor is assigned to a column and 27 machining parameter combinations are required (Table 88.2). Therefore, by using the L27 OA, only 27 experiments are needed to study the entire process parameters. The experimental layout for the machining parameters using the L27 OA is shown in Table 88.3. The entire performance measures data for the 27 experimental runs are shown in Table 88.4. Thus, in optimization, the observed values of MRR, TWR, OC and TOC were set to maximum, minimum, minimum and minimum, respectively. These parameters are selected based on the experiments and levels are identified based the RSM experiments. 13 A, 14 A and 15 A have been considered as current variables. 18 V, 22 V and 26 V have been chosen as voltage variables. 23 µs, 25 µs and 27 µs have been chosen as ton values with toff values of 19 µs, 22 µs and 25 µs.

4. Methodology of Material Testing

4.1 Chemical Composition

The commercial Magnesium AZ31is super alloys are used, and the chemical compositions (wt.%) are shown in Table 88.2 and Table 88.3 respectively.

Table 88.2 Chemical composition

Elements	Content (%)
Aluminium	2.60 – 3.60
Zinc	0.70 – 1.50
Manganese	0.30
Silicon	0.15
Copper	0.060
Calcium	0.050
Iron	0.0050
Nickel	0.0050
Magnesium	Remaining

4.2 Physical and Mechanical Properties

The commercial Magnesium AZ31is super alloys are used, and the physical and mechanical properties compositions (wt.%) are shown in Table 88.4 and Table 88.5 respectively.

Table 88.3 Physical and mechanical properties of Magnesium AZ31

Properties	Mertric	Imperial
Tensile Strength	240-260 MPa	36-37 ksi
Yield Strength	200 MPa	29.20 ksi
Elastic Modulus	44.8 GPa	6498 ksi
Elongation	10-15 %	15 %
Hardness	49	49
Density	1.77 g/cm3	0.0639 lb/in^3
Thermal Expansion Coefficient	26 µm/m°C	14.3 µin/in°F
Thermal Conductivity	96 W/mK	660 BTU in/ hr. ft^2. °F

4.3 Scanning Electron Microscopy (SEM) Image

Scanning Electron Microscopy (SEM) to examine the microstructural features of Magnesium AZ31. Specifically, we investigated three holes of maximum, minimum, & average size of hole machined by Electrical Discharge Machining (EDM). SEM enabled us to obtain high-resolution images and analyse the surface morphology of the material. This analysis provided valuable insights into the microstructural characteristics and helped us gain a better understanding of the behaviour and properties of Magnesium AZ31.

4MF31

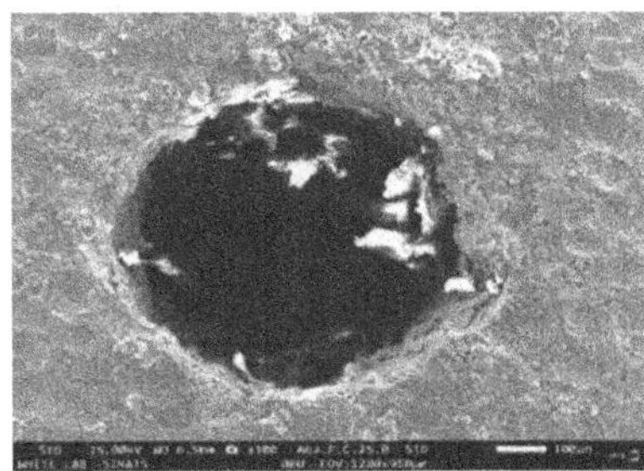

Fig. 88.2 SEM for magnesium AZ31 with maximum hole size – Spectrum 1

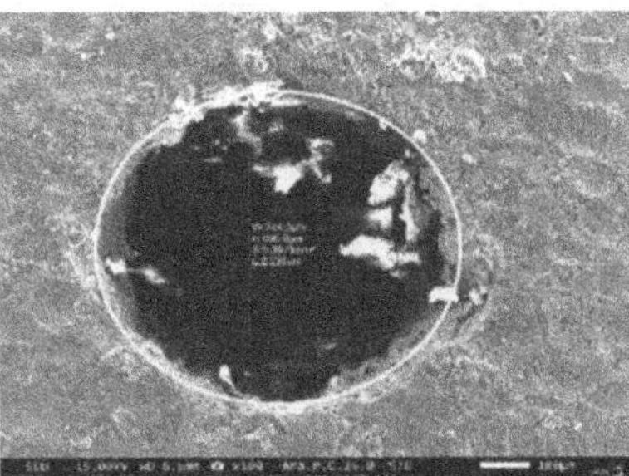

Fig. 88.3 SEM for Magnesium AZ31 with maximum hole size – Spectrum 2

During our analysis of Hole 4MF31, we observed an uneven and distorted shape, deviating from the expected circular form. The machining process for this hole took approximately 19.78 seconds. The measured diameter of the hole was 692.8 µm, with a note width of 744.0 µm and a height of 680 µm. The surface area of the hole was calculated to be 0.3973 µm², with a length of 2259 µm. Notably, Hole 4MF31 stood out as a larger hole compared to the others examined in our study. These findings contribute to a comprehensive understanding of the characteristics and variations within the microstructure of the examined Magnesium AZ31 material.

10MF31

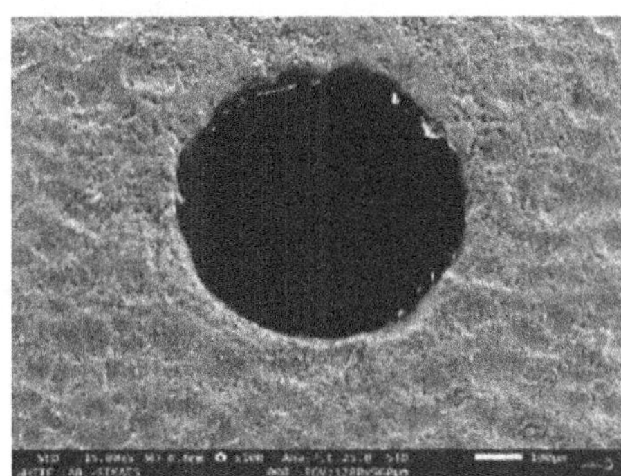

Fig. 88.4 SEM for magnesium AZ31 with miminum hole size – Spectrum 1

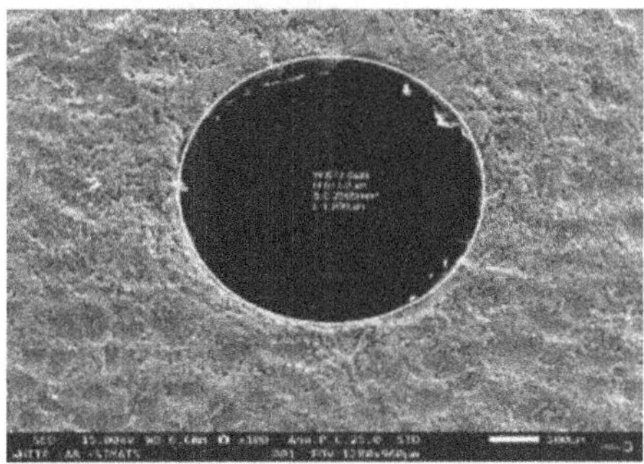

Fig. 88.5 SEM for magnesium AZ31 with miniumum hole size – Spectrum 2

During the analysis of Hole 10MF31, it was observed although not to the extent seen in Hole 4MF31. The machining process for this hole took approximately 8.03 seconds. The size of the hole was measured at 601.6 µm, with a note width of 617.6 µm and a height of 611.2 µm. The surface area of the hole was calculated to be 0.2965 µm², with a length of 1930 µm and a diameter of 601.6 µm. In comparison to other holes examined, Hole 10MF31 stood out as larger in size, yet it maintained an average size within the context of the entire study. These findings contribute to our understanding of the variations and characteristics within the microstructure of the analyzed Magnesium AZ31 material.

15MF31

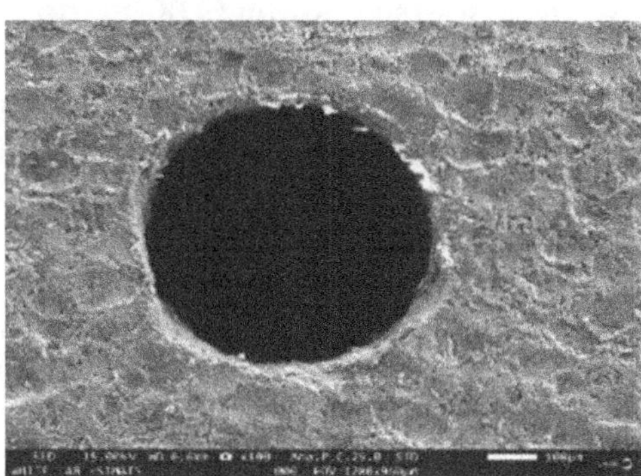

Fig. 88.6 SEM for magnesium AZ31 with average hole size – Spectrum 1

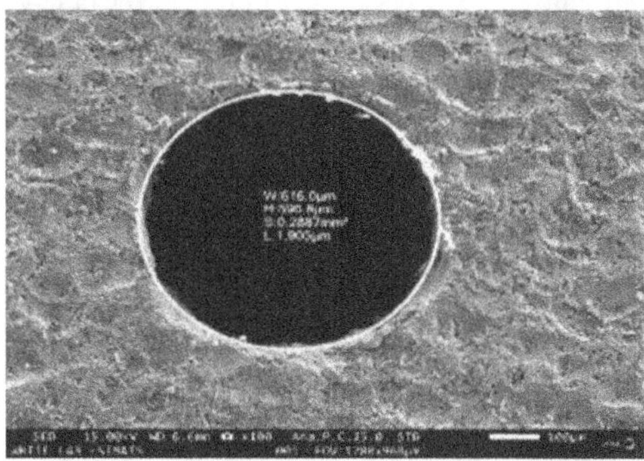

Fig. 88.7 SEM for magnesium AZ31 with average hole size – Spectrum 2

During our analysis of Hole 15MF31, we observed that it exhibited an accurate shape of a circle without any distortion. The machining process for this hole took approximately 6.45 seconds. The note width of the hole was measured at 616.0 µm, and its height was 596.8 µm. The surface area of the hole was calculated to be 0.2887 µm², with a length of 1905 µm and a diameter of 582.4 µm. Remarkably, Hole 15MF31 stood out as the largest hole among all the tested samples. It also played a significant role as the smallest and most effective parameter throughout our testing. These findings provide valuable insights into the consistent shape and characteristics of Hole 15MF31,

Table 88.4 Comparative morphological parameters of 4MF31, 10MF31, and 15MF31 samples

Readings	4MF31	10MF31	15MF31
Width (µm)	744.0	617.6	616.0
Height (µm	680.0	611.2	596.8
Surface (µm)	0.3973	0.2965	0.2887
Length (µm)	2259	1930	1905
Diameter (µm)	692.8	601.6	582.4

highlighting its importance within the microstructural analysis of Magnesium AZ31 material (Fig. 88.4).

4.4 Energy Dispersive X-Ray EDX Image

We machined a sample of magnesium alloy AZ31 using Electrical Discharge Machining (EDM), a non-traditional process that utilizes electrical discharges to shape materials with precision. Following the machining, we conducted an Energy-Dispersive X-ray Spectroscopy (EDX) analysis to examine the elemental composition and distribution of the machined surface (Table 88.4). The EDX analysis revealed the presence and quantified the amounts of key elements, including magnesium, aluminum, and zinc, which are the primary constituents of AZ31. Additionally, the EDX analysis provided insights into the uniformity of elemental distribution across the machined surface, aiding in the evaluation of machining quality and consistency. Overall, the combination of EDM machining and EDX analysis proved valuable in understanding the composition and characteristics of the machined magnesium alloy AZ31 (Fig. 88.5).

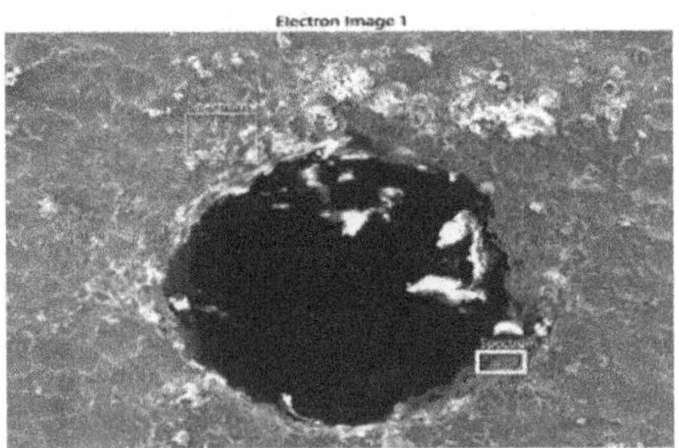

Fig. 88.8 EDX spectrum image (1) 4MF31

4MF31 Spectrum 1

In the EDX process, we examined the magnesium alloy AZ31 4MF31 using two spectrums. In the first spectrum, we analysed the alloy without considering the material properties in the ranging area. This approach allowed us to focus solely on the elemental composition and distribution of the alloy surface. By excluding material properties such as roughness or porosity, we aimed to obtain a clear understanding of the elemental constituents present in the AZ31 alloy. The first spectrum analysis provided insights into the relative concentrations of key elements like magnesium, aluminium, and zinc, helping us evaluate the alloy's composition. These findings contribute to our understanding of the elemental characteristics of the machined surface, supporting further analysis and assessment of the magnesium alloy AZ31 (Fig. 88.9).

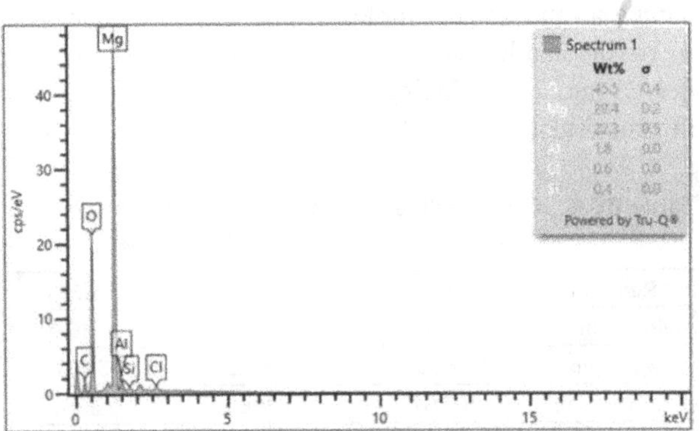

Fig. 88.9 EDX spectrum analysis (1) 4MF31

Table 88.5 EDS-based elemental analysis with standard reference materials (Spectrum 1)

Element	Line Type	Apparent Concentration	k ratio	Wt%	Wt% Sigma	Standard Label	Factory Standard
C	K series	0.43	0.00434	22.34	0.52	C Vit	Yes
O	K series	5.80	0.01951	45.51	0.35	SiO2	Yes
Mg	K series	4.58	0.03039	29.40	0.23	MgO	Yes
Al	K series	0.21	0.00152	1.78	0.05	Al2O3	Yes
Si	K series	0.05	0.00040	0.38	0.03	SiO2	Yes
Cl	K series	0.09	0.00077	0.60	0.03	NaCl	Yes
C	K series	0.43	0.00434	22.34	0.52	C Vit	Yes
O	K series	5.80	0.01951	45.51	0.35	SiO2	Yes
Mg	K series	4.58	0.03039	29.40	0.23	MgO	Yes
Total:				100.00			

The first EDX spectrum showed the absence of iron (Fe) and nickel (Ni), as they were melted due to the heat from the spark during the EDM. As a result, the alloy surface was no longer detectable as Fe and Ni in the EDX image. EDM machining can reach very high temperatures, causing these elements to evaporate or diffuse out of the surface in the details of the region subjected to the EDMed process (Table 88.5). Because of the different thermal characteristics and lower melting points of Fe and Ni compared to the magnesium-based AZ31 alloy, this phenomenon occurs. Although the EDX analysis can offer insights into the remaining elemental constituents (magnesium, aluminum, zinc) in the municipal FGM, it is also important to note that the high processing temperature and effects of the EDM process can result in some elements becoming undetectable or unaccounted for

4MF31 Spectrum 2

During the EDX process, we conducted a second spectrum analysis of the magnesium alloy AZ31, focusing on the ranging area with consideration for the material properties.

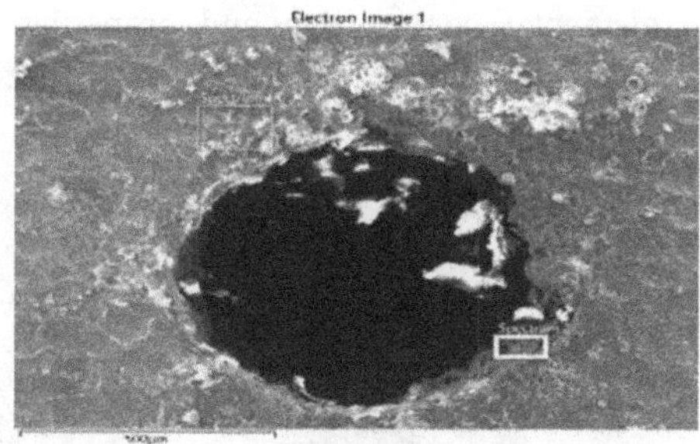

Fig. 88.10 EDX spectrum image (2) 4MF31

This approach aimed to provide a comprehensive understanding of the elemental composition and distribution while considering factors such as surface roughness and porosity. By incorporating material properties in the analysis, we gained insights into the presence and behaviour of key elements, including magnesium, aluminum, zinc, iron, and nickel, within the AZ31 alloy. This enabled us to evaluate the alloy's overall composition and assess any variations or anomalies across the analyzed surface. The second spectrum analysis allowed for a more accurate characterization of the material's elemental makeup and provided valuable information for further examination and interpretation of the magnesium alloy AZ31 (Fig. 88.10).

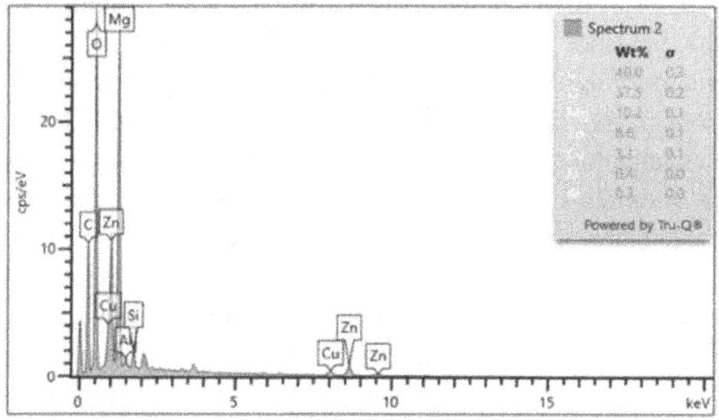

Fig. 88.11 EDX spectrum analysis (2) 4MF31

Following the analyses, the EDX image of the second spectrum revealed that the properties of iron (Fe), nickel (Ni), manganese (Mn), and copper (Cu) had vanished due to the effects of the EDM process. The high temperatures generated by the electrical discharges during machining caused these elements to evaporate or diffuse away from the surface, resulting in their absence in the analyzed region. The intense heat and spark-induced environment of EDM can significantly impact the detectability and preservation of certain elements, particularly those with lower melting points. Although the EDX analysis still provided valuable insights into the remaining elemental constituents such as magnesium, aluminum, and zinc, it is crucial to consider the limitations imposed by the EDM process when interpreting the results and understanding the full composition and properties of the magnesium alloy AZ31.

Table 88.6 Energy dispersive X-ray spectroscopy (EDS) analysis of elemental distribution (Spectrum 2)

Element	Line Type	Apparent Concentration	k ratio	Wt%	Wt% Sigma	Standard Label	Factory Standard
C	K series	2.30	0.02300	40.04	0.31	C Vit	Yes
O	K series	7.44	0.02503	37.46	0.25	SiO2	Yes
Mg	K series	2.61	0.01730	10.19	0.08	MgO	Yes
Al	K series	0.07	0.00053	0.28	0.02	Al2O3	Yes
Si	K series	0.13	0.00099	0.42	0.02	SiO2	Yes
Cu	L series	0.45	0.00446	3.05	0.15	Cu	Yes
Zn	L series	1.26	0.01260	8.56	0.14	Zn	Yes
Total:				100.00			

10MF31 Spectrum 1

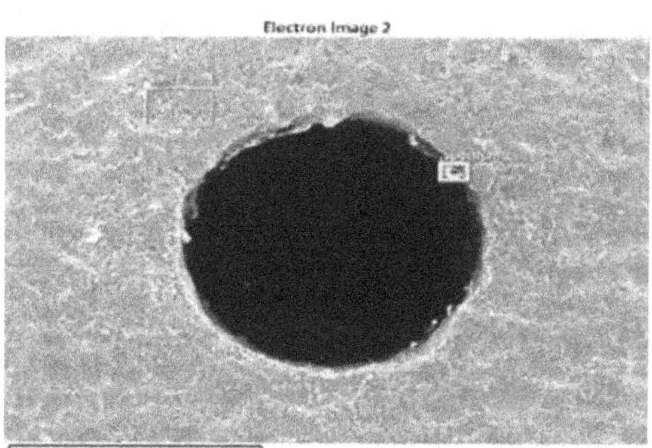

Fig. 88.12 EDX spectrum image (1) 10MF31

In the EDX process, we examined the hole of a magnesium alloy AZ31, specifically the 10MF31 region, which was machined using Electrical Discharge Machining (EDM). In the first spectrum analysis, we focused on the ranging area with consideration for the material properties. This approach aimed to provide a comprehensive understanding of the elemental composition and distribution while considering factors such as surface roughness and porosity within the machined hole. By incorporating material properties in the analysis, we gained insights into the presence and behaviour of key elements such as magnesium, aluminum, zinc, iron, nickel, and others within the AZ31 alloy (Fig. 88.12). This enabled us to evaluate the overall composition and assess any variations or anomalies specifically within the 10MF31 region. The first spectrum analysis, incorporating material properties, allowed for a more accurate characterization of the elemental makeup and provided valuable information for further examination and interpretation of the machined magnesium alloy AZ31 hole (Fig. 88.11).

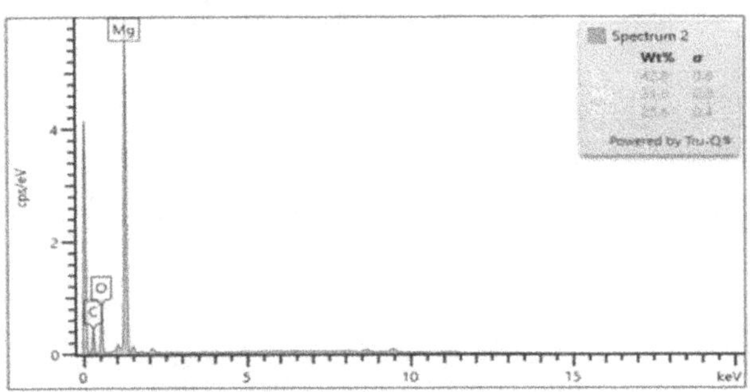

Fig. 88.13 EDX spectrum analysis (1) 10MF31

Table 88.7 EDS-based characterization of sample containing C, O, Mg, Al, Cl, and Zn for spectrum 1 of 10MF31

Element	Line Type	Apparent Concentration	k ratio	Wt%	Wt% Sigma	Standard Label	Factory Standard
C	K series	0.59	0.00591	28.11	0.63	C Vit	Yes
O	K series	3.64	0.01226	27.14	0.32	SiO2	Yes
Mg	K series	8.87	0.05880	41.08	0.39	MgO	Yes
Al	K series	0.26	0.00184	1.82	0.06	Al2O3	Yes
Cl	K series	0.07	0.00058	0.36	0.03	NaCl	Yes
Zn	L series	0.17	0.00171	1.50	0.10	Zn	Yes
C	K series	0.59	0.00591	28.11	0.63	C Vit	Yes
Total:				100.00			

Upon analyzing the EDX image of the 10MF31 region, it became evident that the properties of iron (Fe), nickel (Ni), manganese (Mn), and copper (Cu) had vanished as a result of the EDM process. The intense heat and spark-induced environment during EDM machining led to the evaporation or diffusion of these elements from the analyzed area. Consequently, their absence in the spectrum indicated the challenges associated with detecting and preserving elements with lower melting points in the context of EDM. Although the EDX analysis still provided valuable insights into the remaining elemental constituents, such as magnesium, aluminum, and zinc, it is important to recognize the limitations imposed by the EDM process when interpreting the results and understanding the complete elemental composition and properties of the machined hole in the magnesium alloy AZ31, specifically within the 10MF31 region.

10MF31 Spectrum 2

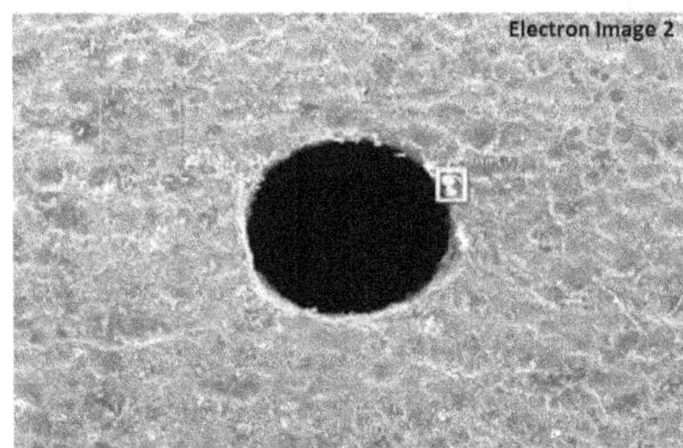

Fig. 88.14 EDX spectrum image (2) 10MF31

In the EDX process, we conducted a second spectrum analysis of the magnesium alloy AZ31, specifically focusing on the 10MF31 region, while considering the material properties within the ranging area. This approach aimed to provide a comprehensive understanding of the elemental composition and distribution, considering factors such as surface roughness and porosity. By incorporating material properties in the analysis, we gained valuable insights into the presence and behaviour of key elements, including magnesium, aluminum, zinc, iron, nickel, manganese, copper, and others within the AZ31 alloy. This enabled us to evaluate the overall composition and assess any variations or anomalies specifically within the 10MF31 region. The second spectrum analysis, considering material properties, facilitated a more accurate characterization of the elemental makeup and provided valuable information for further examination and interpretation of the magnesium alloy AZ31, specifically within the 10MF31 region, contributing to a comprehensive understanding of its properties.

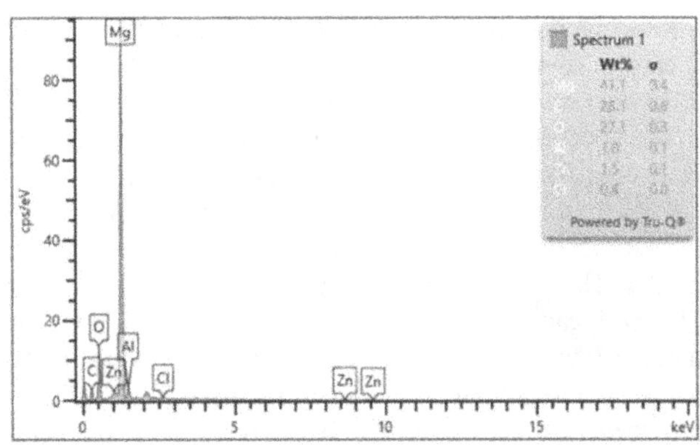

Fig. 88.15 EDX spectrum analysis (2) 10MF31

Table 88.8 Elemental distribution and concentration in sample confirmed by EDS (Spectrum 2)

Element	Line Type	Apparent Concentration	k ratio	Wt%	Wt% Sigma	Standard Label	Factory Standard
C	K series	0.59	0.00591	28.11	0.63	C Vit	Yes
O	K series	3.64	0.01226	27.14	0.32	SiO2	Yes
Mg	K series	8.87	0.05880	41.08	0.39	MgO	Yes
Al	K series	0.26	0.00184	1.82	0.06	Al2O3	Yes
Cl	K series	0.07	0.00058	0.36	0.03	NaCl	Yes
Zn	L series	0.17	0.00171	1.50	0.10	Zn	Yes
Zn	L series	0.10	0.00481	4.44	0.11	Zn	Yes
Total:				100.00			

Upon analyzing the EDX image of the 10MF31 region, it was observed that the properties of iron (Fe), nickel (Ni), and copper (Cu) had vanished due to the effects of the EDM process. The intense heat and electrical discharges generated during EDM machining caused these elements to evaporate or diffuse away from the analyzed area. As a result, their presence was no longer detectable in the spectrum. The high temperatures and spark-induced environment of EDM have a significant impact on the detectability and preservation of certain elements with lower melting points. However, despite the absence of Fe, Ni, and Cu in the spectrum, the remaining elements such as magnesium, aluminum, zinc, and others provided valuable insights into the elemental composition and distribution within the 10MF31 region. It is important to consider the limitations imposed by the EDM process when interpreting the results and comprehending the complete composition and properties of the magnesium alloy AZ31 within the analyzed region.

15MF31 Spectrum 1

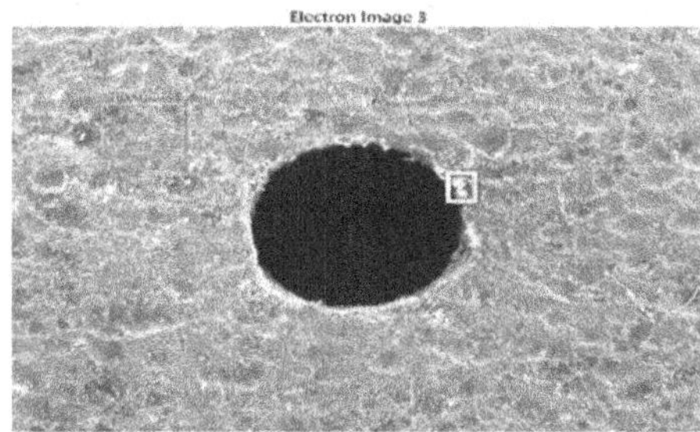

Fig. 88.16 EDX Spectrum Image (1) 15MF31

In the EDX process, we conducted an examination of the hole in a magnesium alloy AZ31, specifically the 15MF31 region, which was machined using Electrical Discharge Machining (EDM). For the first spectrum analysis, we focused on the ranging area within the hole, taking into consideration the material properties. This approach aimed to provide a comprehensive understanding of the elemental composition and distribution, accounting for factors such as surface roughness and porosity within the machined hole. By incorporating material properties in the analysis, we gained valuable insights into the behaviour and presence of key elements such as magnesium, aluminum, zinc, iron, nickel, and others within the AZ31 alloy. The first spectrum analysis, incorporating material properties, facilitated a more accurate characterization of the elemental makeup within the 15MF31 region, offering essential information for further examination and interpretation of the machined magnesium alloy AZ31 hole.

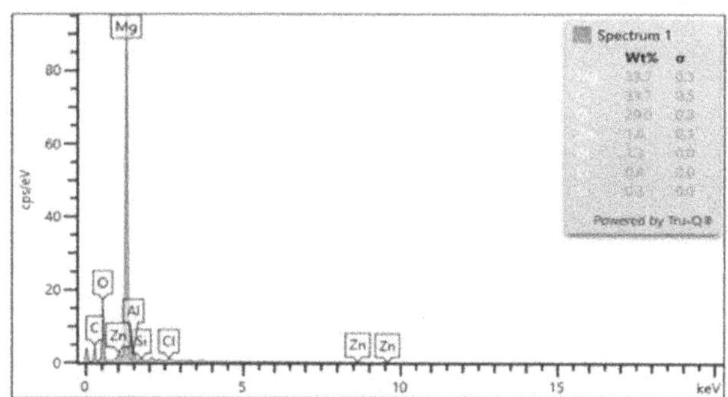

Fig. 88.17 EDX spectrum analysis (1) 15MF31

Table 88.9 Compositional analysis of sample elements by EDS including carbon, oxygen, and metals (Spectrum 1)

Element	Line Type	Apparent Concentration	k ratio	Wt%	Wt% Sigma	Standard Label	Factory Standard
C	K series	1.03	0.01029	33.69	0.51	C Vit	Yes
O	K series	4.61	0.01552	28.95	0.30	SiO2	Yes
Mg	K series	8.90	0.05905	33.71	0.28	MgO	Yes
Al	K series	0.25	0.00179	1.32	0.04	Al2O3	Yes
Si	K series	0.06	0.00046	0.27	0.02	SiO2	Yes
Cl	K series	0.10	0.00086	0.41	0.03	NaCl	Yes
Zn	L series	0.23	0.00227	1.64	0.09	Zn	Yes
C	K series	1.03	0.01029	33.69	0.51	C Vit	Yes
Total:				100.00			

Upon analyzing the EDX image of the 15MF31 region, it was evident that the properties of iron (Fe), nickel (Ni), and copper (Cu) had vanished due to the effects of the EDM process. The high temperatures and electrical discharges generated during EDM machining caused these elements to evaporate or diffuse away from the analyzed area. As a result, their presence was no longer detectable in the spectrum. The intense heat and spark-induced environment of EDM machining pose challenges in detecting and preserving elements with lower melting points. Although the absence of Fe, Ni, and Cu in the spectrum was observed, the remaining elements, such as magnesium, aluminum, zinc, and others, provided valuable insights into the elemental composition and distribution within the 15MF31 region. It is crucial to consider the limitations imposed by the EDM process when interpreting the results and comprehending the complete composition and properties of the magnesium alloy AZ31 within the analyzed region.

15MF31 Spectrum 2

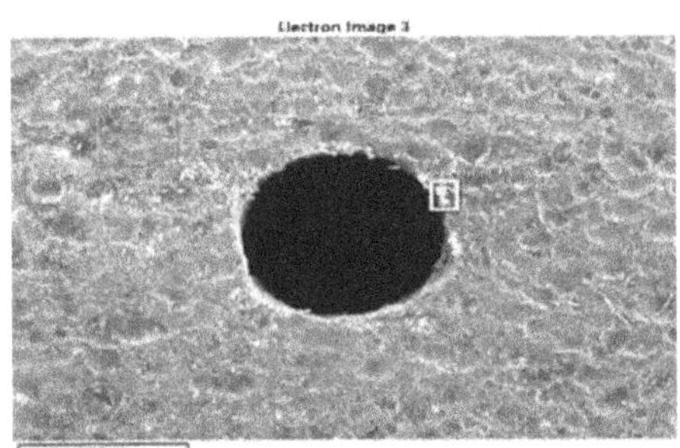

Fig. 88.18 EDX spectrum image (2) 15MF31

In the EDX process, we performed an examination of the hole in a magnesium alloy AZ31, specifically focusing on the 15MF31 region, which was machined using Electrical Discharge Machining (EDM). For the second spectrum analysis, we delved into the ranging area within the hole, considering the material properties. This approach aimed to gain a comprehensive understanding of the elemental composition and distribution, considering factors such as surface roughness and porosity within the machined hole. Through this analysis, which included material properties, we were able to extract useful information about the distribution of the most important elements in the AZ31 alloy, such as magnesium, aluminium, zinc, iron, nickel and so on. A second spectrum was analyzed along with the material properties, allowing for a more accurate characterization of the elemental composition of the 15MF31 region, and this was critical in interpreting machined 15MF31 as well. Magnesium alloy AZ31 hole.

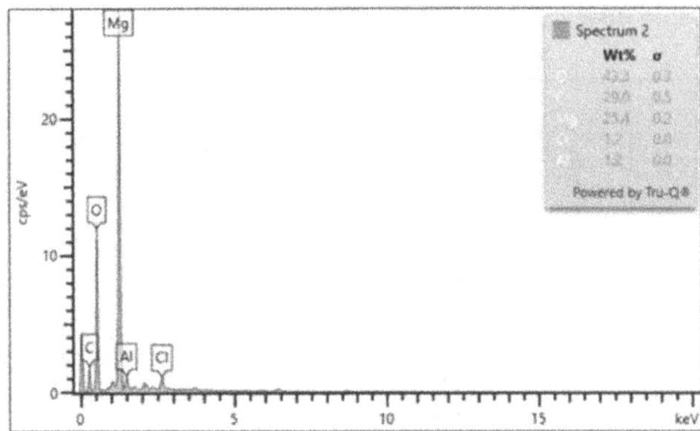

Fig. 88.19 EDX spectrum analysis (2) 15MF31

Table 88.10 EDS elemental composition analysis – spectrum 2 of 15MF31 sample

Element	Line Type	Apparent Concentration	k ratio	Wt%	Wt% Sigma	Standard Label	Factory Standard
C	K series	0.41	0.00411	28.95	0.48	C Vit	Yes
O	K series	3.25	0.01093	43.30	0.34	SiO2	Yes
Mg	K series	2.59	0.01721	25.35	0.20	MgO	Yes
Al	K series	0.10	0.00070	1.19	0.04	Al2O3	Yes
Cl	K series	0.12	0.00106	1.22	0.04	NaCl	Yes
C	K series	0.41	0.00411	28.95	0.48	C Vit	Yes
O	K series	3.25	0.01093	43.30	0.34	SiO2	Yes
Mg	K series	2.59	0.01721	25.35	0.20	MgO	Yes
Total:				100.00			

Upon analyzing the EDX image of the 15MF31 region, it was evident that the properties of iron (Fe), nickel (Ni), and copper (Cu) had vanished due to the effects of the EDM process. The intense heat and electrical discharges generated during EDM machining resulted in the evaporation or diffusion of these elements from the analyzed area, leading to their absence in the spectrum. This phenomenon can be attributed to the lower melting points of Fe, Ni, and Cu compared to the magnesium alloy AZ31. Despite the absence of these elements, the remaining elements such as magnesium, aluminum, zinc, and others provided valuable insights into the elemental composition and distribution within the 15MF31 region. It is essential to consider the limitations imposed by the EDM process when interpreting the results and comprehending the complete composition and properties of the machined hole in the magnesium alloy AZ31 within the analyzed region.

5. Result and Discussion

5.1 EDM Parameter Analysis

During our examination of the magnesium alloy AZ31, we meticulously studied a total of 27 holes that were machined using the EDM process. To identify the most optimal parameters, we carefully selected three representative holes: one that exhibited the highest performance, one with the lowest performance, and one displaying an average level of performance. Through a detailed analysis of these distinct holes, our objective was to determine the best combination of parameters for achieving superior results in terms of quality, precision, and efficiency. After thorough evaluation, we found that the 15MF31 hole showcased the most favourable characteristics and emerged as the top performer among the selected holes. By identifying the best parameters through this comprehensive assessment, we aimed to optimize the EDM machining process for the magnesium alloy AZ31, ensuring superior performance and meeting the desired specifications.

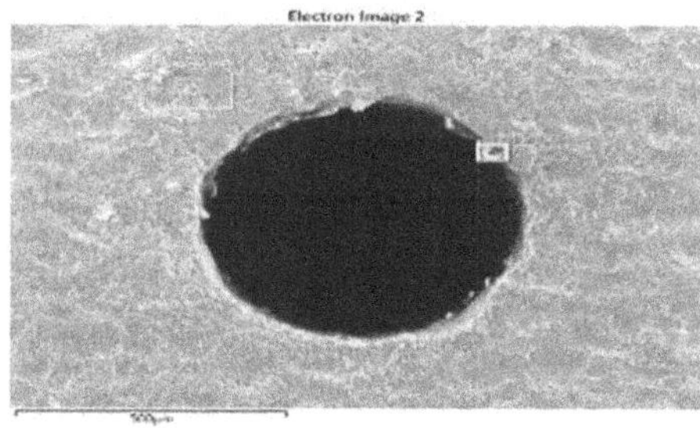

Table 88.11 Dimensional deviation analysis of 4MF31, 10MF31, and 15MF31 samples

Sample	Circularity (μm)	Diameter (μm)	Overcut (μm)
4MF31	500	692.8	-192.8
10MF31	500	601.6	-101.6
15MF31	500	582.4	-82.4

Scanning Electron Microscopy (SEM) Analysis

Analysis of the three holes machined by EDM, we proceeded to examine the 15MF31 hole using Scanning Electron Microscopy (SEM) to determine its effectiveness compared to the 4MF31 and 10MF31 holes. Through the SEM analysis, it was revealed that the 15MF31 hole exhibited greater effectiveness in terms of surface finish, dimensional accuracy, and absence of defects compared to the 4MF31 and 10MF31 holes. The SEM images clearly showcased a smoother and more uniform surface texture in the 15MF31 hole, indicating improved machining performance. Additionally, the dimensional measurements of the 15MF31 hole demonstrated closer adherence to the desired specifications. Overall, the SEM analysis conclusively demonstrated that the 15MF31 hole achieved superior results compared to the 4MF31 and 10MF31 holes, making it the most effective among the three in terms of EDM machining quality and accuracy.

Energy Dispersive X-Ray EDX Image Analysis

Conducting EDX analysis, a consistent absence of iron (Fe), nickel (Ni), manganese (Mn), and copper (Cu) was observed in the properties of the magnesium alloy AZ31. This phenomenon can be attributed to the EDM spark heat and the low concentration of these elements in the AZ31 alloy. The intense heat generated during the EDM process led to the evaporation or diffusion of these elements, resulting in their reduced presence or complete absence in the analyzed material. It is important to consider the composition and characteristics of AZ31, which inherently contain lower levels of Fe, Ni, Mn, and Cu, such as 0.005, when interpreting the EDX analysis results. These findings highlight the challenges associated with detecting and preserving.

6. Conclusions

In conclusion, our project focused on the EDM machining of the magnesium alloy AZ31. We performed a comprehensive analysis by creating twenty-seven holes and carefully selecting three representative holes for further examination: the maximum performance hole (4MF31), the minimum performance hole (15MF31), and the average performance hole (10MF31). Accurate measurements were taken for each hole, enabling us to identify the optimal parameters for machining the AZ31 material, with particular emphasis on the 15MF31 hole. Following the parameter identification, we conducted an EDX analysis to assess the elemental composition of the machined holes. The results indicated that the 15MF31 hole consistently outperformed both the 4MF31 and 10MF31 holes. The elemental properties of iron (Fe), nickel (Ni), manganese (Mn), and copper (Cu) were found to be present to a greater extent in the 15MF31 hole, highlighting its superior machining performance. Based on these findings, it can be concluded that the 15MF31 parameter configuration is the most effective in EDM machining of the magnesium alloy AZ31. The accuracy quality, and elemental preservation achieved with the 15MF31 parameter set make it the ideal choice for machining AZ31. These results provide valuable insights for optimizing EDM processes and enhancing the manufacturing capabilities of magnesium alloy AZ31.

Overall, this project contributes to a better understanding of the machining parameters required for magnesium alloy AZ31 and emphasizes the importance of selecting the right parameters to achieve the desired machining outcomes.

REFERENCES

1. Chen et al. (1999) studied the machining characteristics of Ti– 6A1–4V with kerosene and distilled water as the dielectrics.
2. Yan et al. (1999) investigated the characteristics of the micro-hole of carbide by EDM with a copper tool electrode.
3. Singh et al. (2004) experimentally investigated the effects of parameters such as pulsed current on MR, overcut, TWR, and SR in electric discharge machining of En-31 tool steel (IS designation: T105 Cr 1 Mn 60) hardened and tempered to 55 HRC.
4. Ali Ozgedik and Can Cogun (2006) studied the, the variations of geometrical tool wear characteristics on MR, TWR, relative wear and workpiece SR
5. Ekmekci, B. (2007) studied the effect of dielectric and electrode on white layer structure in EDM in terms of retained austenite and residual stresses using X-ray diffraction method.
6. Kiyak, M., &Çakır, O. (2007) have studied the effect of EDM parameters on SR for machining of 40CrMnNiMo864 tool steel (AISI P20).
7. Abdulkareemet al. (2010) reported the effect of electrode cooling during the EDM of titanium alloy (Ti-6Al-4 V).
8. Lin, et al. (2013) optimized milling EDM process parameters of Inconel 718 alloy to achieve multiple performance characteristics such as low TWR, high MR and low working gap using Grey-Taguchi method.
9. Dhanabalan et al. (2013) demonstrates the effectiveness of optimizing multiple characteristics of EDM of Inconel 718 using copper electrodes having different shapes via Taguchi method-based Grey analysis.
10. Natarajan et al. (2013) investigated the MR, TWR and OC using the input process parameters such as pulse on time, discharge current, and voltage.
11. Shen et al. (2017) proposed a new, efficient, and eco-friendly high-speed dry electrical discharge machining (EDM) milling.
12. Gaikwad, & Jatti, (2018) present study focuses on optimization of EDM process parameter for maximization of MR while machining of NiTi alloy.
13. Roy & Dutta (2019) in the present work used OA experimental design with input parameters such as, pulse on time, duty cycle, discharge current, and gap voltage on MR, TWR, OC in EDM.

Note: All the figures and tables in this chapter were made by the authors.

Advances in Additive Manufacturing Technologies – Gurusamy Pathinettampadian et al. (eds)
© 2026 Taylor & Francis Group, London, ISBN 978-1-041-16687-0

Investigation the Influence of EDM Process Parameters on Mg AZ91 Alloy - Effect on Machining Performance and a Parametric Study

T. Aravind,
S. D. Dhanesh babu[1], R. Ganesamoorthy[2]
Department of Mechanical Engineering,
Chennai Institute of Technology,
Chennai, Tamilnadu, India

K.R. Padmavathi[3]
Department of Mechanical Engineering,
Panimalar Engineering College,
Chennai, India

T. Kavitha[4]
Department of Electronics and Communication Engineering,
Vel Tech Rangarajan Dr. Sagunthala R & D Institute of
Science and Technology

P. Balamurali[5]
Department of Mechanical Engineering,
Chennai Institute of Technology,
Chennai, India

◆ **Abstract:** The objective of this project was to study the physiognomies of magnesium AZ-91 alloys utilizing the electrical discharge machining (EDM) process. It has very good light weight, high strength-to-weight ratio, and excellent machinability, so magnesium alloys have received a lot of attention in various industries. But their low thermal conductivity and relatively low melting point cause problems in conventional machining processes. This research gives a close look at the application of EDM, a non-traditional machining process, to break through this limitation and improve the immeasurable suitability of machining the Magnesium AZ-91 alloys. Share this:EDM Process Parameters Characteristic Optimization and Associated Design Considerations of Magnesium AZ-91 Alloys: Background and Comprehensive Literature Review The project covers comprehensive literature review of the magnesium AZ-91 alloys as well as EDM machining. Select EDM parameters: Peak current, pulse duration, and electrode material are selected as EDM parameters and the setup of experimental setup is designed. The machining B process and surface B features of the Magnesium AZ-91 alloys can be explored through the different combinations of those parameters by conducting several studies. Material removal rate (MRR), electrode wear ratio (EWR, and surface roughness (Ra) of the prepared burnished samples are analysed in relation to the EDM proce.

◆ **Keywords:** EDM, Microstructure, SEM, Hardness, Surface roughness

[1]dhaneshbabusd@citchennai.net, [2]ganesamoorthyr@citchennai.net, [3]krpadmavathipecmech@gmail.com [4]kavithaecephd@gmail.com, [5]balamuralip@citchennai.net

DOI: 10.1201/9781003685906-89

1. INTRODUCTION

1.1 EDM Process

The EDM process the material erosion mechanism is mainly electrical into thermal energy, which is typically the series of isolated electrical discharges between the work material and the electrode immersed in a dielectric fluid. The EDM process is displayed in Fig. 89.1. The plasma channel is formed with hot energy between cathode and anode where the temperature is too high and the temperature of between 8000 to 12000 °C or even to 20000 °C which causes a considerable heating and melting material at the surface of the workpiece. So that when the pulsed current is supplied, the plasma channel breakdown at pulse-off time is representative of sudden temperature decrease. This allows for dielectric fluid to be reintroduced to flush the molten material across the work surfaces in the form of fines. In contrast to the conventional machining processes, this process of melting and evaporation of the material from the surface of the workpiece does not involve any mechanical formed [1]. The amount of material removed per discharge is typically between 10−6 and 10−4 mm 3, while the MR is normally between 2 and 400 mm3 /min on a case by case basis. Shape of the electrode determines the area into which the spark erosion will take place and in addition the precision of the part formed following the EDM is quite high and it relies on the electrode shape. Hence in EDM the tool electrode shape is the mirror of the workpiece. The schematic of EDM process is shown in Fig. 89.1. Figure 89.2 illustrates the mechanism of material removal.

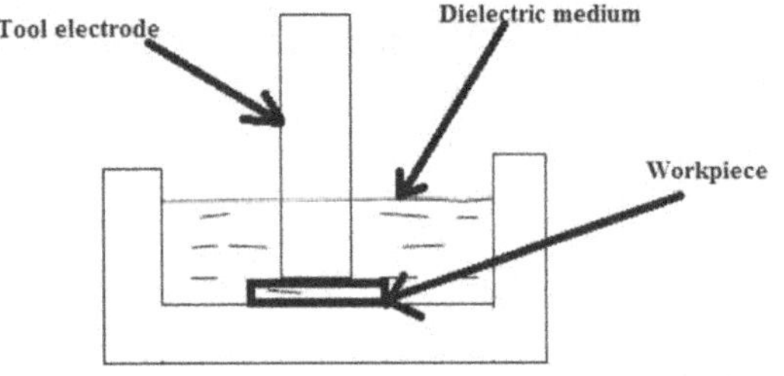

Fig. 89.1 Schematic of the EDM

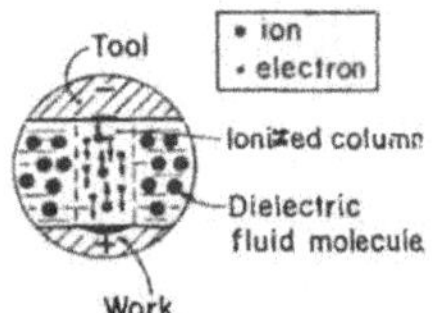

Fig. 89.2 Mechanism of material removal

1.2 Objectives of the Research

The goal of the study was to explore and find the best settings of the EDM of magnesium alloy AZ91. (A) The fabrication and investigation of a sequence of poled holes to characterize the EDM process effectiveness over AZ91. Our goal was to evaluate surface finish, dimensional accuracy, and defect generation during machining, allowing us to assess machining effectiveness [2]. In addition, we sought to determine the best machining conditions to achieve the required quality and precision of the EDM machining of AZ91. However, by choosing holes with the best, worst and middle performance, we wanted to be able to figure out which set of parameters resulted in the best outcome. With the above points, the overall objective of our study was to provide essential information for improving the efficiency of the EDM process by optimizing process parameters to produce quality machining characteristics in the magnesium alloy AZ91 [3].

2. LITERATURE REVIEW

2.1 Machining Performance and Optimization

This chapter presented a research paper related to effect of EDM process parameters on difficult to cut materials like Magnesium alloys with metal MR, TWR, OC and SR together with Stainless steel and Die steel in relation to metal MR, TWR, OC and SR. Also, the literatures regarding different optimal methods are presented [4].

The wire electrical discharge machining (WEDM) of magnesium alloy AZ91 with regard to machining characteristics and surface integrity was studied by Raju Bangaru, Ashok Kumar Singh, and H. S. Shan. They explored effects of process parameters including pulse on time, pulse off time, and wire tension on material removal rate (MRR), surface roughness, and microstructural changes. It is observed that in both cases increasing the pulse on time and decreasing pulse off time retains high value of MRR but at the same time increases surface roughness. The study found a combination of process parameters in which the MRR and surface roughness exhibited a trade-off. Microstructural studies further validated the formation of recast layer and heat-affected zone which indicates the thermal effect during WEDM. In order to obtain the desired machining characteristics and ensure the surface integrity of the machined magnesium alloy AZ91, it was concluded that the WEDM process parameters need to be properly controlled [5].

Machining performance of magnesium alloy AZ91 was examined using WEDM by J. Jeswin and R. Dhinesh (2021). Investigations were conducted into the impact of process parameters on MRR, surface roughness and tool wear. Higher pulse energy improves MRR, but with rough surface. Finally, optimum parameter combinations were found in order to balance between the MRR and the surface roughness. As the pulse on and off times increased so did the tool wear. This work contributes to better understanding of machining characteristics in magnesium alloys and provides guidelines for optimization of WEDM of AZ91. The results show notable enhancement in reduction of debris discharge and expansion by notching the cylindrical cross-section (Source: Materials Today: Proceedings, Vol. 39 (2021) pp. 697-701)

The findings of the researchers determine the microstructure and surface integrity of magnesium alloy AZ31B during wire electrical discharge machining (WEDM) in their research [3]. Microstructural evolution, roughness and residual stress of WEDMed samples with respect to processing conditons were also studied. Microstructural changes such as grain refinement and recast layer formation were found as a result of WEDM. It was noted that as the value of pulse-on time and pulse-off time parameters were increased, the surface roughness was observed to increase significantly, thus affecting the surface finish [6]. Compressive residual stresses were found in the region close to the machined surface. Overall, the study gives an overview of the understanding to be gained in terms of microstructural changes, toxic constituents, and surface integrity of magnesium alloy AZ31B in wire electrical discharge machining (WEDM). This set of results will help in optimizing the machining process to achieve the expected characteristics in the machine parts [7].

Singh, S. (2012) combined the designs of experiments and grey relational analysis (GRA) approach to optimise the parameters for the EDM process of 6061Al/Al2O3p/20P AMMCs An L18 OA was employed to plan of experiments to determine an optimal setting. The parameters used for this process were five different control factors, such as pulse account, pulse on time, duty cycle, gap voltage and lift time of tool electrode [5]. The MR, TWR and SR were chosen as the evaluation metrics. The optimal machining combination is aspect ratio (0.6), pulse current (10), pulse ON time (50), duty cycle (0.4), gap voltage (50) and tool electrode lift time (2.0) combination based on the grey relational grade [8].

Through EDM, Tzeng, & Chen (2013) studied the MR, TWR and SR, on SKD61. The results predicated through RSM method is in significant, A RSM based experimental plan and the results.

Tang, & Du, (2014) combine grey relational analysis and Taguchi methods solve the problem of EDM parameter optimization. Tap water was used as working fluid; the process parameters were discharge current, gap voltage, lifting height, negative polarity and pulse duty factor. Selection of TWR, MR and SR is to check effect of the entire machining.

Thanigaivelan et al. (2015) have utilized copper and brass engagement, cryogenically watches copper and brass engagement with voltage, pulse on-time and discharge current worths. It proves that cryogenically processed copper electrode enhance the TWR and OC to 24 and 2.5 times respectively. 8 μm [9].

The authors (Kalayarasan & Murali, 2016) used genetic algorithm to maximize MR and minimize TWR. Experimental studies have been carried out at different pulse on time, pulse off time, dielectric pressure and discharge current. The experiments were designed using the Taguchi L9 OA. Confirmation tests were performed to confirm the results obtained using process parameter optimization using grey relational analysis and TOPSIS. Hence it is witness from the result that the proposed approach here is formats and effective to optimize the machining parameter for EDM process. The optimized input parameters combinations which provide maximum MR and minimum TWR according to Grey Relational analysis are Current (6A), Pulse on time (20 μsec), Pulse off time (4 μsec), Dielectric Pressure (28kg/cm3). According to the TOPSIS, the optimized inputs combinations for achieving both maximum MRR and minimum EWR are Current (6A), Pulse on time (20 μsec), Pulse offtime (4 μsec), Dielectric pressure (28 kg/cm3) [10].

Abhishek, et al. Hosseini et al. (2017) performed experimental analysis of EDM process characteristics on Inconel 625, 718, 601 and 825 based on 5-factor-4-level L16 OA. In this work, the effect of input parameters including gap voltage, peak current, pulse-on time, duty factor and flushing pressure on MR, TWR, SR and surface crack density were studied.. They reported best parameter setting as gap voltage of 90 V, peak current of 5 A, pulse-on time of 200 μs, duty factor of 70%, flushing pressure of 0.6 bar for Inconel 625; gap voltage of 90 V, peak current of 5A, pulse-on time of 200 μs, duty factor of 85%, flushing pressure of 0.4 bar for Inconel 718; gap voltage of 80 V, peak current of 7 A, pulse-on time of 500 μs, duty factor of 80%, flushing pressure of 0.3 bar for Inconel 601; and gap voltage of 80 V, peak current of 5 A, pulse-on time of 300 μs, duty factor of 85%, flushing pressure of 0.4 bar for Inconel 825.

The paper by Gaikwad, & Jatti, (2018) discusses the optimization of the EDM process parameters which are correlated with the MR maximization while machining with the NiTi alloy. Factors such as study gap current, pulse on time, pulse off time, workpiece electrical conductivity and tool conductivity were considered as process variables introduced in the present study. Experiments were conducted according to Taguchi's L36 orthogonal array. According to the study work electrical conductivity, gap current and pulse on time are among the significant parameters affecting the material removal rate. The maximum material removal rate achieved was at power setting pm 7.0806 mm 3 /min.

3. Experimental Setup

3.1 Experimental Setup

In this chapter the experimental work deals about the experimental setup used for the experiments, tool electrode, and selection of workpiece. Moreover, the formation of the preliminary experiments, experimental design and L27 OA design based on Taguchi design, and calculation of MRR, TWR, OC, CIRCULARITY.

3.2 Specification of OCT-EDM

OCT-Electric Discharge Machine shown in Fig. 89.1; model: OCCAN(OCT3252NA) (die-sinking type) with servomechanism for gap control is used for conducting the experiments. Commercial grade EDM oil (specific gravity= 0.763, freezing point= 94°C) was used as dielectric fluid [100]. The dielectric is externally flooded with a pressure of 0.2 kg f/cm2 and pulsed discharge current was applied in various steps.

The EDM setup majorly consist of following part as follows

- Dielectric reservoir, pump and circulation system.
- Power generator and control unit.
- X-Y table accommodating the working table.
- The tool holder and servo system for tool feed.

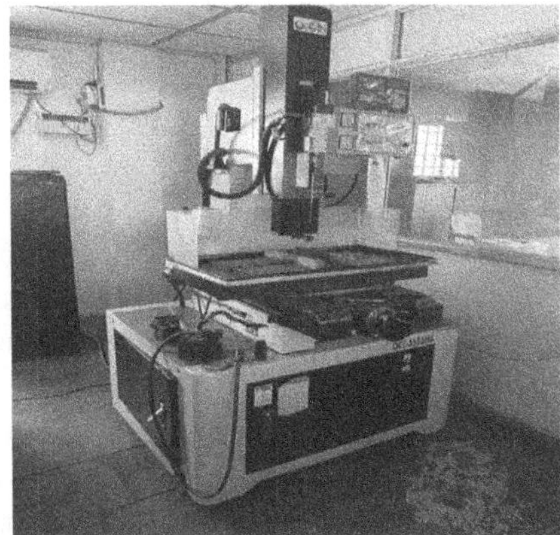

Fig. 89.3 EDM setup

3.3 Electrode Material

The tool electrode material should have the following:

- Easily machinable,
- Low wear rate,
- Excellent conductor of electricity and heat,
- Cheap, and
- Readily available.

Brass, copper, tungsten, graphite are common materials used as tool Electrode in EDM. Brass is preferred among these electrode because of there inherent properties. Brass has higher cutting speed also its a good conductor of electricity. Brass has several advantages over other EDM tool electrodes, such as low cost and high Machinability. This study showed that the brass electrode with Ø Use 0.5 mm for all the experiments (Fig. 89.3).

3.4 Selection of Work Piece

Magnesium alloy AZ -91 is an lightweight, high-strength material that is commonly used in industries such as aerospace, automotive, and electronics. Magnesium AZ-91 is an aluminium-zinc magnesium alloy with 9% aluminium content and 1% zinc content, it is lightest structural metal and high-purity alloy. It's has high corrosion resistance, excellent castability, its use for a commercial application, electrical components. It's is commonly in cast helicopter transmission housing. It's is shining grey metal have a low density, low melting point and high chemical reactive.,

3.5 Major Machining Parameters

Majority of research papers in EDM use the electrical parameters to control the output responses. Hence in this research, voltage, current, pulse on time and pulse off time are considered as a most influencing the machining characteristics. Optimization of these factors finds a major role in majority of the literature work. Multi-objective optimization is prime focus of these research hence two different techniques such as Grey Taguchi and TOPSIS were considered. Even though are there are many factors that determines the performance of the electrical discharge machine, the key factors based on the industrial need the following performance measures are considered for this research. Machining Speed (MR), Tool Wear Rate (TWR), Overcut (OC), Circularity

Material Removal Rate (MRR)

The primary principle of material removal is electrical power to heat conversion. The heat produced melts and vaporizes the workpiece material. The material melted away from the workpiece is called debris, and it is removed from the work surface in either solid, liquid or gaseous form (most commonly). From an MRR perspective MRR is used as a key metric in production. Thermal and electrical properties greatly influence the machinability of materials in EDM. The MRR is directly proportional to the discharge energy.

Based on the voltage, current, pulse on time and pulse off time, different amount of discharge energy is produced. Flushing pressure of the dielectric medium and flushing gap is another factor that affects the MRR. Also, the erosion characteristic of the workpiece material also significantly decides the value of MRR (Table 89.1).

Evaluation of MR and OC

MR is calculated by dividing the length of the through hole with the machining time required to complete the through hole.

$$MR = \frac{\text{Thickness of the workpiece}}{\text{total time for machining the through hole}}$$

OC is calculated using optical microscopic images. OC is defined by $\Delta R = Re - Rt$, where Re is the entrance radius of the machined hole and Rt is the tool electrode radius. The difference in Re and Rt results in OC.TOC is defined by TOC = (D–d)/2L, where, D is entry diameter of the machined hole, d exit diameter of the machined hole, and L thickness of the workpiece.

Overcut

An EDM features is always larger than the electrode used to machine it. The difference between the size of the electrode and the size of the cavity (or hole) is called the overcut. For calculating overcut, there are two types of overcut namely total overcut (diametral overcut) or overcut per side and radial overcut. OC is calculated using optical microscopic images taken from optical microscope as shown in Fig. 3.7. OC is defined by $\Delta R = Re - Rt$, Where Re is the entrance radius of the machined hole and Rt is the tool electrode radius. The difference in Re and Rt results in OC.

3.6 Orthogonal Array (OA)

This approach is called factorial design of experiments by Sir R.A. Fisher. In case of a full factorial design, it will catch all possible combinations in the given set of factors. Since in most cases, the number of factors to be considered within industrial experiments will be high, a full factorial design would require many more experiments. A subset of experiments and controls is chosen from all possible combinations to keep the number of experiments to a reasonable size. This approach to choosing a limited number of experiments that yield the most information is called a partial fraction experiment. While this technique is widely recognized, clear guidance on how to implement it in your own work or interpret the results after performing the experiments is lacking. Taguchi devel

3.7 Concluding Remarks

The various input output parameters are selected and EDM performance measures were studied using RSM interaction plots, OA experimental design. The performance measures are calculated and presented. Using the L127 OA experiments the multi objective optimization is performed using Grey Relational Grade (GRG).

Table 89.1 Orthogonal array (OA)

Exp: No:	Pulse ON time (µs)	Pulse OFF time (µs)	Gap Current (A)	Gap voltage (V)	Circularity (µm)	Diameter (µm)	Overcut (µm)	Time taken (sec)
1	23	19	13	18	500	845.0	315	09.75
2	23	19	14	22	500	776.0	276	07.35
3	23	19	15	26	500	714.0	214	05.88
4	23	22	13	22	500	800.0	300	06.97
5	23	22	14	26	500	806.0	306	06.53
6	23	22	15	18	500	714.0	214	04.92
7	23	25	13	26	500	680.0	180	06.26
8	23	25	14	18	500	836.0	336	08.78
9	23	25	15	22	500	597.0	97	05.71
10	25	19	13	26	500	662.0	162	05.07
11	25	19	14	22	500	650.0	150	05.13
12	25	19	15	26	500	676.0	176	07.32
13	25	22	13	22	500	702.0	205	07.06
14	25	22	14	26	500	715.0	215	06.36
15	25	22	15	18	500	594.0	94	05.00
16	25	25	13	26	500	754.0	254	05.47
17	25	25	14	18	500	716.0	216	07.81
18	25	25	15	22	500	778.0	278	07.61
19	27	19	13	18	500	619.0	119	13.38
20	27	19	14	22	500	689.0	189	06.54
21	27	19	15	26	500	744.0	244	07.09
22	27	22	13	22	500	657.0	157	05.90
23	27	22	14	26	500	757.0	281	05.94
24	27	22	15	18	500	605.0	257	05.74
25	27	25	13	26	500	615.0	105	05.99
26	27	25	14	18	500	600.0	115	07.64
27	27	25	15	22	500	607.0	100	06.91

4. Methodology of Material Testing

4.1 Chemical Composition

The commercial Magnesium AZ-91is super alloys are used, and the chemical compositions (wt.%) are shown in Table 89.2 respectively.

Table 89.2 Chemical composition

Elements	%Composition
Aluminium (Al)	8.5-9.6
Manganese (Mn)	0.18-0.60
Zinc (Zn)	0.45-1.10
Silicon (Si)	0.10
Copper (Cu)	0.05
Iron (Fe)	0.005
Nikel (Ni)	0.002
Magnesium (Mg)	remaining

4.2 Physical and Mechanical Properties

The commercial Magnesium AZ-91is super alloys are used, and the physical and mechanical properties compositions (wt.%) are shown in Table 89.3 respectively.

Table 89.3 Physical and mechanical properties of Magnesium AZ-91

Properties	Mertric	Imperial
Tensile Strength	240-250 Mpa	35-36 ksi
Yield Strength	160 Mpa	23.20ksi
Elastic Modulus	45 Gpa	6526 ksi
Elongation	3-7%	3-7%
Hardness	63	63
Density	1.85 g/cm^3	1.050 lb/in^3
Thermal Expansion Coefficient	26 µm/m°C	14.4 µin/in°F
Thermal Conductivity	72.7 W/mK	504 BTU in/ hr.ft2.°F

4.3 Scanning Electron Microscopy (SEM) Image

Scanning Electron Microscopy (SEM) to examine the microstructural features of Magnesium AZ91. Specifically, we investigated three holes of maximum, minimum, & average size of hole machined by Electrical Discharge Machining (EDM). SEM enabled us to obtain high-resolution images and analyse the surface morphology of the material. This analysis provided valuable insights into the microstructural characteristics and helped us gain a better understanding of the behaviour and properties of Magnesium AZ91 (Fig. 89.4).

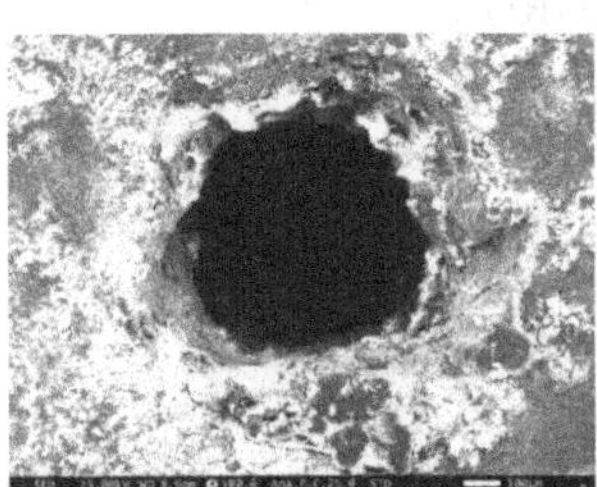

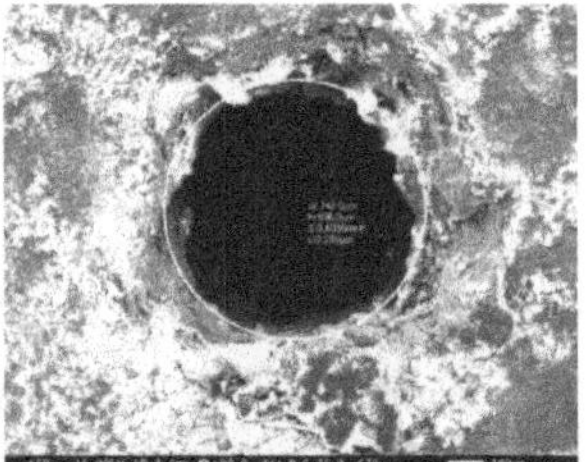

Fig. 89.4 SEM image of 8MF91

During our analysis of Hole 8MF91, we observed an uneven and distorted shape, deviating from the expected circular form. The machining process for this hole took approximately 8.78 seconds. The measured diameter of the hole was 836.0 µm, with a note width of 742.0 µm and a height of 696.0 µm. The surface area of the hole was calculated to be 0.4056 µm², with a length of 2259 µm. Notably, Hole 8MF91 stood out as a larger hole compared to the others examined in our study. These findings contribute to a comprehensive understanding of the characteristics and variations within the microstructure of the examined Magnesium AZ91 material.

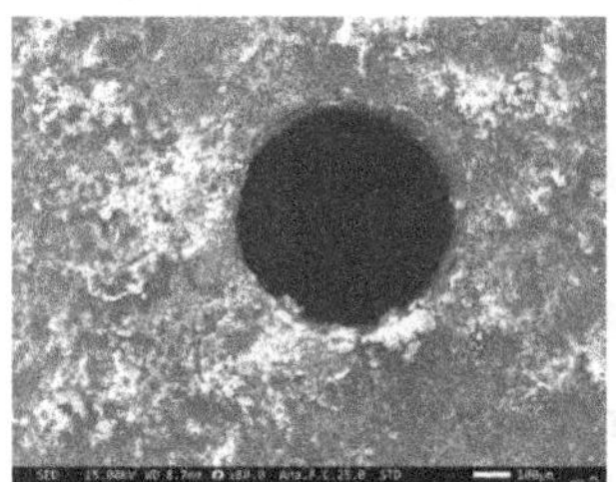

Fig. 89.5 SEM image of 10MF91

During the analysis of Hole 10MF91, it was observed although not to the extent seen in Hole 8MF91. The machining process for this hole took approximately 5.07 seconds. The size of the hole was measured at 662.0 µm, with a note width of 624.0 µm and a height of 608.0 µm. The surface area of the hole was calculated to be 0.2980 µm², with a length of 1925 µm and a diameter of 662.0 µm (Fig. 89.5). In comparison to other holes examined, Hole 10MF91 stood out as larger in size, yet it maintained an average size within the context of the entire study. These findings contribute to our understanding of the variations and characteristics within the microstructure of the analyzed Magnesium AZ91 material.

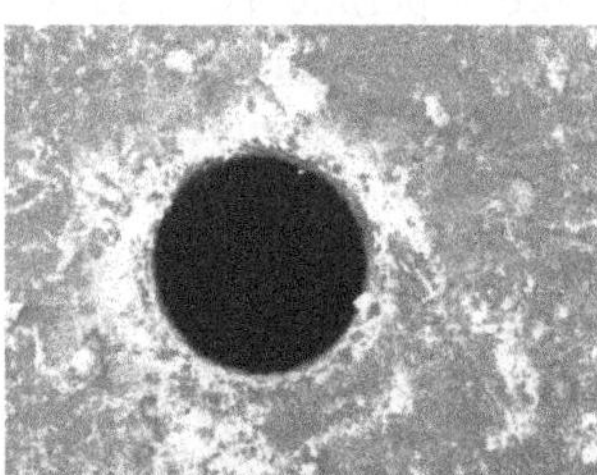

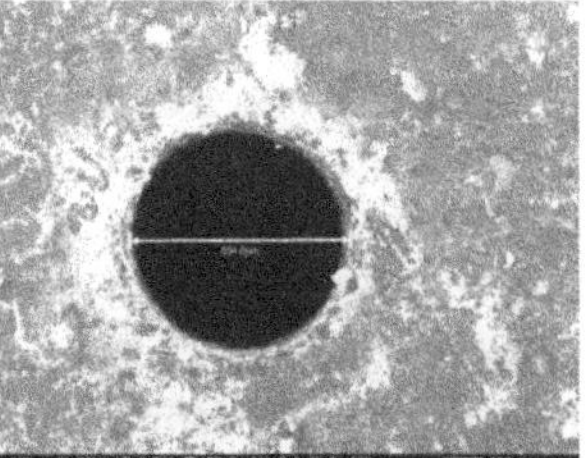

Fig. 89.6 SEM image of 15MF91

During our analysis of Hole 15MF91, we observed that it exhibited an accurate shape of a circle without any distortion. The machining process for this hole took approximately 5.00 seconds (Fig. 89.6). The note width of the hole was measured at 590.0 µm, and its height was 580.0 µm. The surface area of the hole was calculated to be 0.2688 $µm^2$, with a length of 1838 µm and a diameter of 594.0 µm. Remarkably, Hole 15MF91 stood out as the largest hole among all the tested samples. It also played a significant role as the smallest and most effective parameter throughout our testing. These findings provide valuable insights into the consistent shape and characteristics of Hole 15MF91, highlighting its importance within the microstructural analysis of Magnesium AZ91 material (Table 89.4).

Table 89.4 Comparise of SEM image reading

Readings	8MF91	10MF91	15MF91
Width (µm)	742.0	624.0	590.0
Height (µm)	696.0	608.0	580.0
Surface (µm)	0.4056	0.2980	0.2688
Length (µm)	2259	1935	1838
Diametre (µm)	836.0	662.0	594.0

4.4 Energy Dispersive X-Ray EDX Image

We machined a sample of magnesium alloy AZ91 using Electrical Discharge Machining (EDM), a non-traditional process that utilizes electrical discharges to shape materials with precision. Following the machining, we conducted an Energy-Dispersive X-ray Spectroscopy (EDX) analysis to examine the elemental composition and distribution of the machined surface. The EDX analysis revealed the presence and quantified the amounts of key elements, including magnesium, aluminum, and zinc, which are the primary constituents of AZ91. Additionally, the EDX analysis provided insights into the uniformity of elemental distribution across the machined surface, aiding in the evaluation of machining quality and consistency. Overall, the combination of EDM machining and EDX analysis proved valuable in understanding the composition and characteristics of the machined magnesium alloy AZ91 (Fig. 89.7).

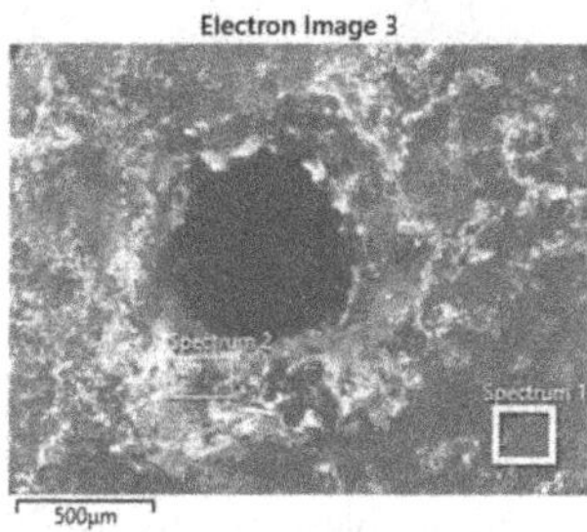

Fig. 89.7 8MF91 spectrum 1

In the EDX process, we examined the magnesium alloy AZ91 8MF91 using two spectrums. In the first spectrum, we analysed the alloy without considering the material properties in the ranging area. This approach allowed us to focus solely on the elemental composition and distribution of the alloy surface. By excluding material properties such as roughness or porosity, we aimed to obtain a clear understanding of the elemental constituents present in the AZ91 alloy. The first spectrum analysis provided insights into the relative concentrations of key elements like magnesium, aluminium, and zinc, helping us evaluate the alloy's composition. These findings contribute to our understanding of the elemental characteristics of the machined surface, supporting further analysis and assessment of the magnesium alloy AZ91 (Fig. 89.8).

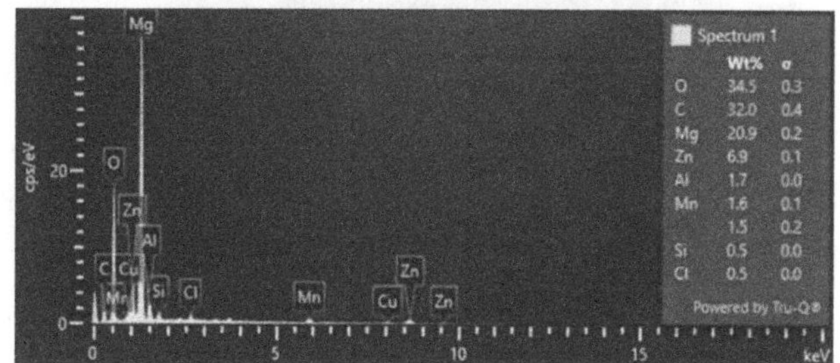

Fig. 89.8 EDX graph of 8MF91 spectrum 1

Table 89.5 Spectrum 1 of 8MF91

Element	Line Type	Apparent Concentration	k ratio	Wt%	Wt% Sigma	Standard Label	Factory Standard
C	K series	0.91	0.00907	31.95	0.44	C Vit	Yes
O	K series	4.90	0.01648	34.46	0.29	SiO2	Yes
Mg	K series	3.83	0.02541	20.91	0.17	MgO	Yes
Al	K series	0.28	0.00198	1.69	0.04	Al2O3	Yes
Si	K series	0.10	0.00078	0.53	0.03	SiO2	Yes
Cl	K series	0.10	0.00086	0.49	0.03	NaCl	Yes
Mn	K series	0.31	0.00311	1.63	0.07	Mn	Yes
Cu	L series	0.15	0.00148	1.48	0.15	Cu	Yes
Zn	L series	0.71	0.00711	6.86	0.15	Zn	Yes
Total:				100.00			

The analysis of the first EDX spectrum revealed that the properties of iron (Fe) and nickel (Ni) had vanished due to the intense heat generated by the spark during the EDM process. Consequently, the presence of Fe and Ni in the alloy surface was no longer detectable in the EDX image. The high temperatures associated with EDM machining caused these elements to evaporate or diffuse away from the surface, resulting in their absence in the analyzed region. This phenomenon can be attributed to the different thermal properties and lower melting points of Fe and Ni compared to the magnesium-based AZ91 alloy (Table 89.5). While the EDX analysis provided valuable information about the remaining elemental constituents such as magnesium, aluminum, and zinc, it is essential to consider the limitations imposed by the specific conditions and effects of the EDM process on the detectability and preservation of certain elements.

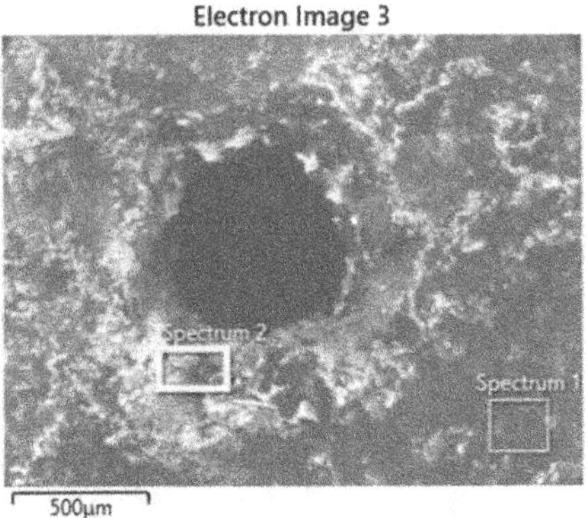

Fig. 89.9 8MF91 spectrum 2

During the EDX process, we conducted a second spectrum analysis of the magnesium alloy AZ91, focusing on the ranging area with consideration for the material properties. This approach aimed to provide a comprehensive understanding of the elemental composition and distribution while taking into account factors such as surface roughness and porosity. By incorporating material properties in the analysis, we gained insights into the presence and behavior of key elements, including magnesium, aluminum, zinc, iron, and nickel, within the AZ91 alloy. This enabled us to evaluate the alloy's overall composition and assess any variations or anomalies across the analyzed surface (Fig. 89.9). The second spectrum analysis allowed for a more accurate characterization of the material's elemental makeup and provided valuable information for further examination and interpretation of the magnesium alloy AZ91 (Fig. 89.10).

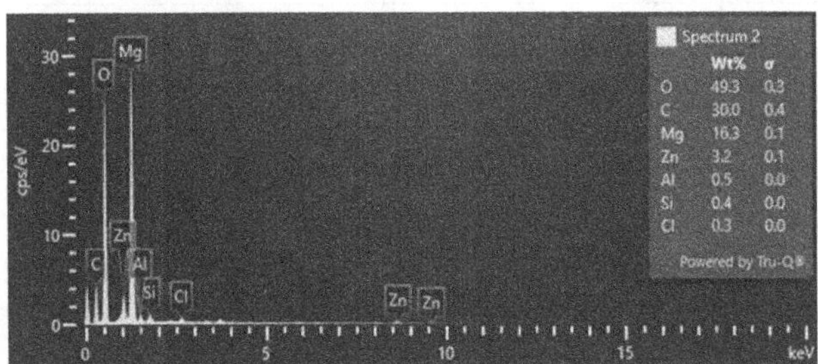

Fig. 89.10 EDX graph of 8MF91 spectrum 2

Table 89.6 Spectrum 2 of 8MF91

Element	Line Type	Apparent Concentration	k ratio	Wt%	Wt% Sigma	Standard Label	Factory Standard
C	K series	0.94	0.00944	30.05	0.38	C Vit	Yes
O	K series	6.98	0.02348	49.27	0.30	SiO2	Yes
Mg	K series	2.74	0.01815	16.31	0.12	MgO	Yes
Al	K series	0.07	0.00051	0.46	0.02	Al2O3	Yes
Si	K series	0.08	0.00062	0.45	0.02	SiO2	Yes
Cl	K series	0.05	0.00048	0.30	0.02	NaCl	Yes
Zn	L series	0.27	0.00269	3.15	0.10	Zn	Yes
Total:				100.00			

Following the analyses, the EDX image of the second spectrum revealed that the properties of iron (Fe), nickel (Ni), manganese (Mn), and copper (Cu) had vanished due to the effects of the EDM process. The high temperatures generated by the electrical discharges during machining caused these elements to evaporate or diffuse away from the surface, resulting in their absence in the analyzed region. The intense heat and spark-induced environment of EDM can significantly impact the detectability and preservation of certain elements, particularly those with lower melting points (Fig. 89.11). Although the EDX analysis still provided valuable insights into the remaining elemental constituents such as magnesium, aluminum, and zinc, it is crucial to consider the limitations imposed by the EDM process when interpreting the results and understanding the full composition and properties of the magnesium alloy AZ91.

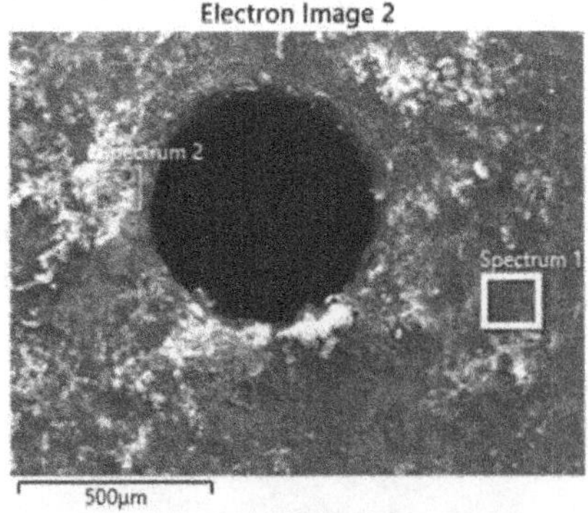

Fig. 89.11 10MF91 spectrum 1

In the EDX process, we examined the hole of a magnesium alloy AZ91, specifically the 10FM91 region, which was machined using Electrical Discharge Machining (EDM). In the first spectrum analysis, we focused on the ranging area with consideration for the material properties. This approach aimed to provide a comprehensive understanding of the elemental composition and distribution while taking into account factors such as surface roughness and porosity within the machined hole. By incorporating material properties in the analysis, we gained insights into the presence and behavior of key elements such as magnesium, aluminum, zinc, iron, nickel, and others within the AZ91 alloy. This enabled us to evaluate the overall composition and assess any variations or anomalies specifically within the 10FM91 region. The first spectrum analysis, incorporating material properties, allowed for a more accurate characterization of the elemental makeup and provided valuable information for further examination and interpretation of the machined magnesium alloy AZ91 hole (Fig. 89.12).

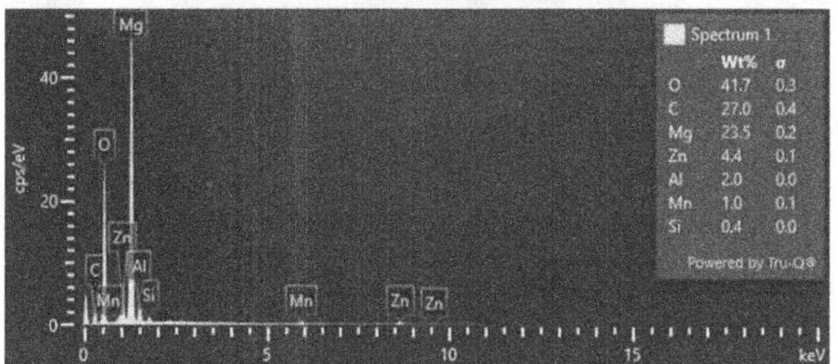

Fig. 89.12 EDX graph of 10MF91 spectrum 1

Table 89.7 Spectrum 1 of 10MF91

Element	Line Type	Apparent Concentration	k ratio	Wt%	Wt% Sigma	Standard Label	Factory Standard
C	K series	0.82	0.00816	26.99	0.42	C Vit	Yes
O	K series	6.99	0.02352	41.71	0.30	SiO2	Yes
Mg	K series	4.76	0.03160	23.52	0.17	MgO	Yes
Al	K series	0.34	0.00245	1.99	0.05	Al2O3	Yes
Si	K series	0.07	0.00059	0.38	0.02	SiO2	Yes
Mn	K series	0.20	0.00195	0.98	0.06	Mn	Yes
Zn	L series	0.48	0.00481	4.44	0.11	Zn	Yes
Total:				100.00			

Upon analyzing the EDX image of the 10FM91 region, it became evident that the properties of iron (Fe), nickel (Ni), manganese (Mn), and copper (Cu) had vanished as a result of the EDM process. The intense heat and spark-induced environment during EDM machining led to the evaporation or diffusion of these elements from the analyzed area. Consequently, their absence in the spectrum indicated the challenges associated with detecting and preserving elements with lower melting points in the context of EDM. Although the EDX analysis still provided valuable insights into the remaining elemental constituents, such as magnesium, aluminum, and zinc, it is important to recognize the limitations imposed by the EDM process when interpreting the results and understanding the complete elemental composition and properties of the machined hole in the magnesium alloy AZ91, specifically within the 10FM91 region (Fig. 89.13).

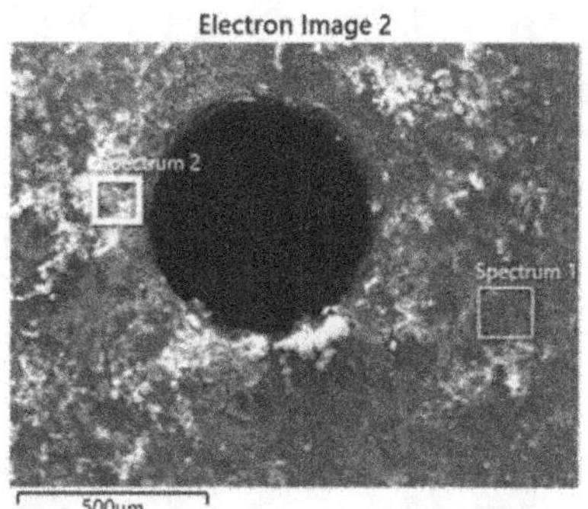

Fig. 89.13 10MF91 spectrum 2

In the EDX process, we conducted a second spectrum analysis of the magnesium alloy AZ91, specifically focusing on the 10MF91 region, while considering the material properties within the ranging area. This approach aimed to provide a comprehensive understanding of the elemental composition and distribution, taking into account factors such as surface roughness and porosity. By incorporating material properties in the analysis, we gained valuable insights into the presence and behavior of key elements, including magnesium, aluminum, zinc, iron, nickel, manganese, copper, and others within the AZ91 alloy. This enabled us to evaluate the overall composition and assess any variations or anomalies specifically within the 10MF91 region. The second spectrum analysis, considering material properties, facilitated a more accurate characterization of the elemental makeup and provided valuable information for further examination and interpretation of the magnesium alloy AZ91, specifically within the 10MF91 region, contributing to a comprehensive understanding of its properties (Fig. 89.14).

Upon analyzing the EDX image of the 10FM91 region, it was observed that the properties of iron (Fe), nickel (Ni),

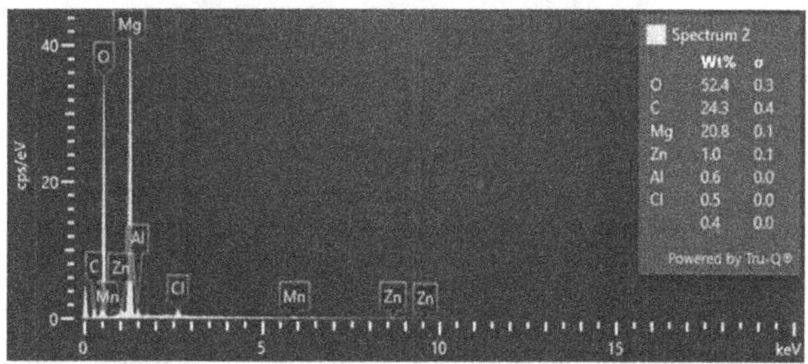

Fig. 89.14 EDX graph of 10MF91 spectrum 2

Table 89.8 Spectrum 2 of 10MF91

Element	Line Type	Apparent Concentration	k ratio	Wt%	Wt% Sigma	Standard Label	Factory Standard
C	K series	0.81	0.00816	26.99	0.42	C Vit	Yes
O	K series	9.61	0.02352	41.71	0.30	SiO2	Yes
Mg	K series	4.20	0.03160	23.52	0.17	MgO	Yes
Al	K series	0.11	0.00245	1.99	0.05	Al2O3	Yes
Cl	K series	0.12	0.00195	0.98	0.06	NaCl	Yes
Mn	K series	0.07	0.00069	0.35	0.05	Mn	Yes
Zn	L series	0.10	0.00481	4.44	0.11	Zn	Yes
Total:				100.00			

and copper (Cu) had vanished due to the effects of the EDM process. The intense heat and electrical discharges generated during EDM machining caused these elements to evaporate or diffuse away from the analyzed area (Fig. 89.15). As a result, their presence was no longer detectable in the spectrum. The high temperatures and spark-induced environment of EDM have a significant impact on the detectability and preservation of certain elements with lower melting points. However, despite the absence of Fe, Ni, and Cu in the spectrum, the remaining elements such as magnesium, aluminum, zinc, and others provided valuable insights into the elemental composition and distribution within the 10FM91 region. It is important to consider the limitations imposed by the EDM process when interpreting the results and comprehending the complete composition and properties of the magnesium alloy AZ91 within the analyzed region

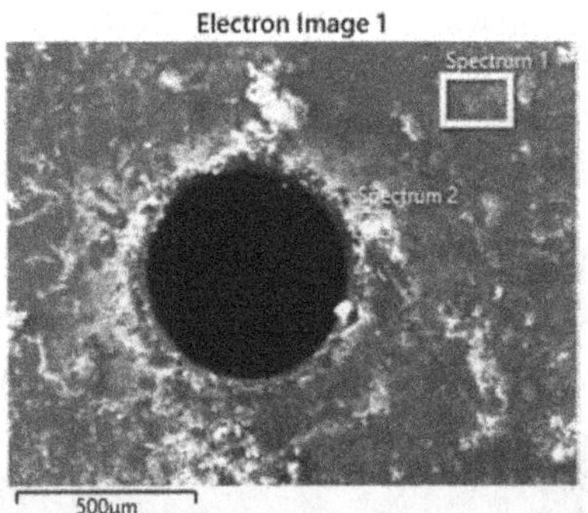

Fig. 89.15 15MF91 spectrum 1

In the EDX process, we conducted an examination of the hole in a magnesium alloy AZ91, specifically the 15FM91 region, which was machined using Electrical Discharge Machining (EDM). For the first spectrum analysis, we focused on the ranging area within the hole, taking into consideration the material properties. This approach aimed to provide a comprehensive understanding of the elemental composition and distribution, accounting for factors such as surface roughness and porosity within the machined hole. By incorporating material properties in the analysis, we gained valuable insights into the behavior and presence of key elements such as magnesium, aluminum, zinc, iron, nickel, and others within the AZ91 alloy. The first spectrum analysis, incorporating material properties, facilitated a more accurate characterization of the elemental makeup within the 15FM91 region, offering essential information for further examination and interpretation of the machined magnesium alloy AZ91 hole (Fig. 89.16).

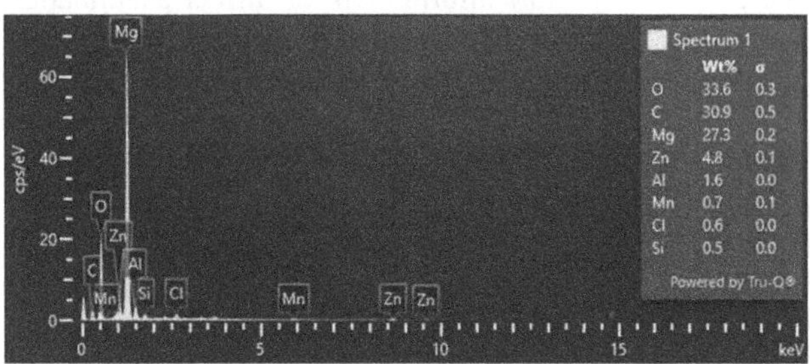

Fig. 89.16 EDX graph of 15MF91 spectrum 1

Table 89.9 Spectrum 1 of 15MF91

Element	Line Type	Apparent Concentration	k ratio	Wt%	Wt% Sigma	Standard Label	Factory Standard
C	K series	0.97	0.00965	30.88	0.49	C Vit	Yes
O	K series	5.66	0.01906	33.59	0.30	SiO2	Yes
Mg	K series	6.47	0.04293	27.30	0.22	MgO	Yes
Al	K series	0.30	0.00213	1.56	0.04	Al2O3	Yes
Si	K series	0.11	0.00084	0.49	0.03	SiO2	Yes
Cl	K series	0.15	0.00134	0.64	0.03	NaCl	Yes
Mn	K series	0.16	0.00160	0.71	0.06	Mn	Yes
Zn	L series	0.63	0.00631	4.93	0.12	Zn	Yes
Total:				100.00			

Upon analyzing the EDX image of the 15FM91 region, it was evident that the properties of iron (Fe), nickel (Ni), and copper (Cu) had vanished due to the effects of the EDM process. The high temperatures and electrical discharges generated during EDM machining caused these elements to evaporate or diffuse away from the analyzed area. As a result, their presence was no longer detectable in the spectrum (Fig. 89.18). The intense heat and spark-induced environment of EDM machining pose challenges in detecting and preserving elements with lower melting points. Although the absence of Fe, Ni, and Cu in the spectrum was observed, the remaining elements, such as magnesium, aluminum, zinc, and others, provided valuable insights into the elemental composition and distribution within the 15FM91 region. It is crucial to consider the limitations imposed by the EDM process when interpreting the results and comprehending the complete composition and properties of the magnesium alloy AZ91 within the analyzed region (Fig. 89.17).

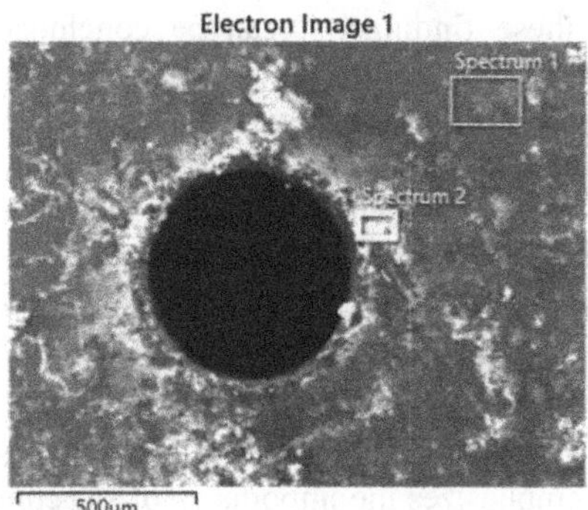

Fig. 89.17 15MF91 spectrum 2

In the EDX process, we performed an examination of the hole in a magnesium alloy AZ91, specifically focusing on the 15FM91 region, which was machined using Electrical Discharge Machining (EDM). For the second spectrum analysis, we delved into the ranging area within the hole, taking into account the material properties. This approach aimed to gain a comprehensive understanding of the elemental composition and distribution, considering factors such as surface roughness and porosity within the machined hole. By incorporating material properties in the analysis, we obtained valuable insights into the behavior and presence of key elements, including magnesium, aluminum, zinc, iron, nickel, and others, within the AZ91 alloy. The second spectrum analysis, incorporating material properties, facilitated a more accurate characterization of the elemental makeup within the 15FM91 region, providing significant information for further examination and interpretation of the machined magnesium alloy AZ91 hole.

Upon analyzing the EDX image of the 15FM91 region, it was evident that the properties of iron (Fe), nickel (Ni), and copper (Cu) had vanished due to the effects of the EDM process. The intense heat and electrical discharges generated during EDM machining resulted in the evaporation or diffusion of these elements from the analyzed area, leading to their absence in the spectrum. This phenomenon can be attributed to the lower melting points of Fe, Ni, and Cu compared to the magnesium alloy AZ91 (Table 89.10). Despite the absence of these elements, the remaining elements such as magnesium, aluminum, zinc, and others provided valuable insights into the elemental composition and distribution within the 15FM91 region.

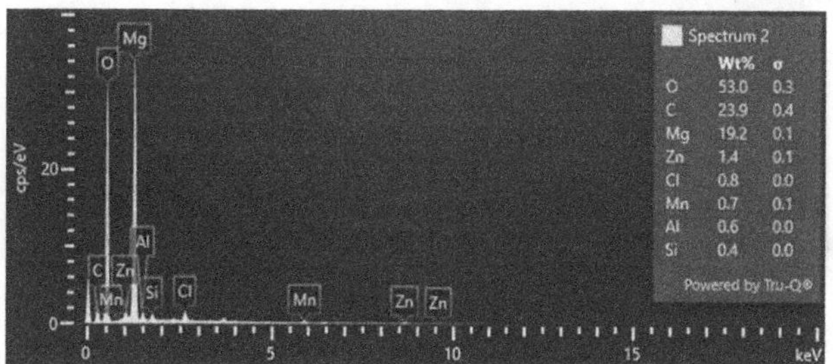

Fig. 89.18 EDX graph of 15MF91 spectrum 2

Table 89.10 Spectrum 2 of 15MF91

Element	Line Type	Apparent Concentration	k ratio	Wt%	Wt% Sigma	Standard Label	Factory Standard
C	K series	0.69	0.00688	23.88	0.42	C Vit	Yes
O	K series	8.53	0.02870	53.00	0.33	SiO2	Yes
Mg	K series	3.33	0.02207	19.16	0.14	MgO	Yes
Al	K series	0.10	0.00071	0.64	0.03	Al2O3	Yes
Si	K series	0.07	0.00057	0.41	0.02	SiO2	Yes
Cl	K series	0.15	0.00132	0.81	0.03	NaCl	Yes
Mn	K series	0.12	0.00115	0.67	0.06	Mn	Yes
Zn	L series	0.12	0.00121	1.43	0.08	Zn	Yes
Total:				100.00			

5. Result and Discussion

5.1 EDM Parameter Analysis

During our examination of the magnesium alloy AZ91, we meticulously studied a total of 27 holes that were machined using the EDM process. To identify the most optimal parameters, we carefully selected three representative holes: one that exhibited the highest performance, one with the lowest performance, and one displaying an average level of performance. Through a detailed analysis of these distinct holes, our objective was to determine the best combination of parameters for achieving superior results in terms of quality, precision, and efficiency. After thorough evaluation, we found that the 15MF91 hole showcased the most favorable characteristics and emerged as the top performer among the selected holes. By identifying the best parameters through this comprehensive assessment, we aimed to optimize the EDM machining process for the magnesium alloy AZ91, ensuring superior performance and meeting the desired specifications.

Analysis of the three holes machined by EDM, we proceeded to examine the 15MF91 hole using Scanning Electron Microscopy (SEM) to determine its effectiveness

Table 89.11 Scanning electron microscopy (SEM) analysis

Sample	Circularity (μm)	Diameter (μm)	Overcut (μm)
8MF91	500	836.00	336.00
10MF91	500	662.00	162.00
15MF91	500	594.00	94.00

compared to the 8MF91 and 10MF91 holes. Through the SEM analysis, it was revealed that the 15MF91 hole exhibited greater effectiveness in terms of surface finish, dimensional accuracy, and absence of defects compared to the 8MF91 and 10MF91 holes. The SEM images clearly showcased a smoother and more uniform surface texture in the 15MF91 hole, indicating improved machining performance. Additionally, the dimensional measurements of the 15MF91 hole demonstrated closer adherence to the desired specifications. Overall, the SEM analysis conclusively demonstrated that the 15MF91 hole achieved superior results compared to the 8MF91 and 10MF91 holes, making it the most effective among the three in terms of EDM machining quality and accuracy (Table 89.11).

5.2 Energy Dispersive X-Ray EDX Image Analysis

Conducting EDX analysis, a consistent absence of iron (Fe), nickel (Ni), manganese (Mn), and copper (Cu) was observed in the properties of the magnesium alloy AZ91. This phenomenon can be attributed to the EDM spark heat and the low concentration of these elements in the AZ91 alloy. The intense heat generated during the EDM process led to the evaporation or diffusion of these elements, resulting in their reduced presence or complete absence in the analyzed material. It is important to consider the composition and characteristics of AZ91, which inherently contain lower levels of Fe, Ni, Mn, and Cu, such as 0.005. when interpreting the EDX analysis results. These findings highlight the challenges associated with detecting and preserving elements with lower concentrations or lower melting points in magnesium alloy AZ91, particularly in the context of EDM machining processes.

6. Conclusions

In conclusion, our project focused on the EDM machining of the magnesium alloy AZ91. We performed a comprehensive analysis by creating twenty-seven holes and carefully selecting three representative holes for further examination: the maximum performance hole (8MF91), the minimum performance hole (15MF91), and the average performance hole (10MF91). Accurate measurements were taken for each hole, enabling us to identify the optimal parameters for machining the AZ91 material, with particular emphasis on the 15MF91 hole.

Following the parameter identification, we conducted an EDX analysis to assess the elemental composition of the machined holes. The results indicated that the 15MF91 hole consistently outperformed both the 8MF91 and 10MF91 holes. The elemental properties of iron (Fe), nickel (Ni), manganese (Mn), and copper (Cu) were found to be present to a greater extent in the 15MF91 hole, highlighting its superior machining performance.

Based on these findings, it can be concluded that the 15MF91 parameter configuration is the most effective in EDM machining of the magnesium alloy AZ91. The accuracy, quality, and elemental preservation achieved with the 15MF91 parameter set make it the ideal choice for machining AZ91. These results provide valuable insights for optimizing EDM processes and enhancing the manufacturing capabilities of magnesium alloy AZ91.

Overall, this project contributes to a better understanding of the machining parameters required for magnesium alloy AZ91 and emphasizes the importance of selecting the right parameters to achieve the desired machining outcomes. The findings pave the way for improved manufacturing processes and enhanced performance of AZ91 in various applications.

References

1. Kiyak, M., & Çakır, O. (2007) have studied the effect of EDM parameters on SR for machining of 40CrMnNiMo864 tool steel (AISI P20).
2. Kuppan et al. (2008) experimentally investigated the high aspect ratio hole drilling of Inconel 718 using the EDM process.
3. Abdulkareem et al.(2010) reported the effect of electrode cooling during the EDM of titanium alloy (Ti-6Al-4 V).
4. Aliakbari, & Baseri, (2012) studied the optimal setting of the process parameters on rotary EDM using the peak current, pulse on time, and rotational speed of the tool along with three types of electrodes.
5. Lin,et al.(2013) optimized milling EDM process parameters of Inconel 718 alloy to achieve multiple performance characteristics such as low TWR, high MR and low working gap using Grey-Taguchi method.
6. Dhanabalan et al. (2013) demonstrates the effectiveness of optimizing multiple characteristics of EDM of Inconel 718 using copper electrodes having different shapes via Taguchi method-based Grey analysis.
7. Natarajan et al.(2013) investigated the MR, TWR and OC using the input process parameters such as pulse on time, discharge current, and voltage.
8. Shen et al. (2017) proposed a new, efficient, and eco friendly high-speed dry electrical discharge machining (EDM) milling.
9. Gaikwad, & Jatti, (2018) present study focuses on optimization of EDM process parameter for maximization of MR while machining of NiTi alloy.
10. Roy & Dutta (2019) in the present work used OA experimental design with input parameters such as, pulse on time, duty cycle, discharge current, and gap voltage on MR,TWR,OC in EDM.

Note: All the figures and tables in this chapter were made by the authors.

Advances in Additive Manufacturing Technologies – Gurusamy Pathinettampadian et al. (eds)
© 2026 Taylor & Francis Group, London, ISBN 978-1-041-16687-0

Fault Detection in Micro Machining Process using Discrete Wavelet Transform of Force Signals

Shalupriya M.[1]
Research Scholar,
Department of Electronics and Communication Engineering,
Saveetha School of Engineering, Saveetha Institute of Medical and Technical Sciences (SIMATS),
Chennai, India

K.P. Indira[2]
Department of Nanobiomaterial Engineering,
Saveetha School of Engineering, Saveetha Institute of Medical and Technical Sciences (SIMATS),
Chennai, India

◆ **Abstract:** This study focuses on fault detection in micro-machining by applying the Discrete Wavelet Transform (DWT) to cutting force signals, especially the Z-axis load. Micro-machining processes are highly sensitive to faults such as tool wear, burr formation, or sudden depth variations, which often result in abrupt changes in force. DWT enables multi-resolution analysis, effectively finding such transient disturbances by decomposing signals into detail coefficients. Using dynamometer, the force signals of 2.5mm thickness workpiece were collected. Statistical features such as variance and squared error of the wavelet detail coefficients were evaluated to enhance fault detection accuracy. Comparatively, these indicators locate faulty regions with more clarity than time-domain analysis. The approach gives real-time fault localization by distinguishing normal and faulty operations. This wavelet-based method gives a reliable diagnostic tool for smart manufacturing and can be integrated with machine learning models for automated defect classification.

◆ **Keywords:** Discrete wavelet transform (DWT), Filter bank, Micromachining, Short time fourier transform (STFT)

1. Introduction

In micro-machining, friction detecting faults is important to maintain high precision and quality in manufacturing. The increasing demand for miniaturized components has induced advancements in micro-machining technologies. Micro-machining includes certain operations which detects minor tool wear or defects at microscopic measures. The ability to detect faults in real-time is important to reduce cost production by ensuring component precision and long life of tools and equipment [1]. During the machining process, the force signals produced gives useful information of the state of tool and workpiece. Traditional analysis techniques in the time or frequency domain were ineffective as these signals were not stationary over time. To overcome the limitation, The Discrete Wavelet Transform (DWT) decomposes non-stationary signals into multiple scales. This approach provides a time-frequency representation

[1]shaluphd2021@gmail.com, [2]indirakp.sse@saveetha.com

DOI: 10.1201/9781003685906-90

that facilitates in detecting sudden irregularities related to faults. DWT is used in fault detection for identifying tool wear, indicating effectiveness of DWT to detect and localize faults in microgrids, recognizing fault locations in complex distributed systems (Banerjee et al., 2022). The use of DWT in diagnosing electrical faults in industrial induction machines explained by (Bouzida et al. 2011) proves its reliability to detect issues such as broken rotor bars. In spite of these advancements, the gap remains in applying DWT to micro-machining processes, for which real-time monitoring is important to maintain high precision and preventing tool failure [2]. The existing studies prioritizes fault detection in large-scale machining operations, and few have studied the force signal analysis required for the intricate scale of micro-machining.

This research is characterized by its application of DWT for real-time fault detection in the micro-machining environment. DWT has been applied to diagnose faults through vibration analysis (Tse et al., 2004) and to identify defects in friction stir welding (Kumar et al., 2015), whereas micro-machining process introduces unique challenges due to its fine scale and high sensitivity. The proposed work uses DWT to analyze force signals in real-time during micro-machining to reduce the gap. According to the tool wear or failure, the signals are decomposed into detailed and estimated coefficients to detect the sudden changes precisely [3].

Concerning to the industrial relevance, this research includes tangible benefits on precision components particularly in microelectronics, medical device manufacturing and aerospace. These sectors aim to maintain the tolerance in precision during the manufacturing processes as tool wear or breakage affects the final product [4]. Manufacturers can access a reliable, real-time diagnostic tool, resulting in lowered operational time, waste and better product quality using DWT for force signal analysis. By integrating this method in the existing process control systems, operational efficiency is increased, facilitating automated monitoring and fault diagnosis. According to (Yahia et al., 2014) DWT fault detection cannot manage non-stationary signals due to their limitations in time-frequency representation. DWT provides multi-resolution analysis flexibility, which can be customized to detect faults in micro-machining. Research by (Patel et al. 2017) and (Riera-Guasp et al. 2008) showed that DWT consistently detects fault in various systems under non-stationary conditions, indicating its potential for application in micro-machining [5]. This study develops a practical approach that uses DWT for detecting faults during micro-machining, focusing on real-time tracking and easy integration into existing industrial setups [6].

This study bridges a gap in the current research by using DWT to detect faults in micro-machining processes. This work introduces real-time analysis concentrating on handling the difficulties met in micro-scale machining processes. This research is suitable for precision-based industries in India, where optimized fault detection and monitoring improves product quality and cost-effective operations. The proposed methodology strengthens fault detection in micro-machining and sets new benchmarks for real-time process control, specifically in India's rapidly advancing high-precision manufacturing sector [7].

2. Materials and Methods

This study utilized SS316L alloy workpieces with a 2.5 mm thickness in the micro-turning process, as illustrated in Fig. 90.1. The chemical composition of SS316L is detailed in Table 90.1. The experimental setup involved cylindrical rods measuring 200 mm in length and 80 mm in diameter. To ensure durability during microturning, both the tool and fixture were constructed from hardened steel (HS13). The tool featured a straight cutting edge with a 5 mm tip diameter and a 16 mm shank diameter, providing stability during high-speed rotations. Spindle speeds of 500, 1000, and 3000 rpm were tested, along with feed rates of 50 and 75 mm/min [8]. The tool angle was maintained at 0 degrees throughout the experiments. The experiments were carried out on a precision micro-turning machine, equipped with a load cell for measuring forces along the cutting axis. The workpiece was rotated using a servo-controlled spindle, while a computer-controlled AC induction motor managed the feed and depth of cut. Both movements were closely monitored through a digital interface displaying real-time force measurements and correlations with electrical parameters [9]. The machine's programmable logic controllers were connected to an industrial PC running LabVIEW software, enabling real-time data collection [10].

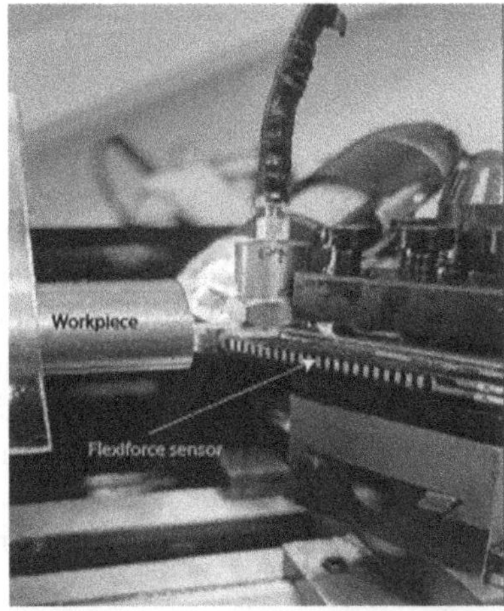

Fig. 90.1 Photographic view of the online condition monitoring tool in Micro turning process

Table 90.1 Examining the chemical constituents of SS316L alloy

Element	Percentage (%)
Iron (Fe)	Balance (typically 62-71%)
Chromium (Cr)	16.0 – 18.0
Nickel (Ni)	10.0 – 14.0
Molybdenum (Mo)	2.0 – 3.0
Manganese (Mn)	≤ 2.0
Silicon (Si)	≤ 1.0
Phosphorus (P)	≤ 0.045
Sulfur (S)	≤ 0.03
Carbon (C)	≤ 0.03
Nitrogen (N)	≤ 0.10

2.1 Theory

Wavelet theory, first introduced by geophysicist Jean Morlet in 1981 and further developed by physicist Alex Grossman in 1984, forms the basis for applying Discrete Wavelet Transform (DWT) in detecting faults in micro-turning processes. "Wavelet" refers to short-duration oscillations that act as the basis functions for the wavelet transform [11]. Wavelets allow for localized signal analysis by providing spatial localization through translation and enabling the examination of different frequency ranges through scaling. This makes wavelets particularly useful for analysing signals in dynamic settings like micro-turning [12].

In micro-turning, surface defects and tool wear are common issues, and these defects can be hard to detect through visual inspection alone (Fig. 90.1). Force signals processing is important to identify and locate imperfections. The force signals use methods for extracting distinct features as the machine progresses. Traditional Fourier transforms lack time localization of frequency components, creating inconsistencies for non-stationary signals. Short Time Fourier Transform (STFT) uses time-consuming techniques, its constant resolution reduces its effectiveness in faults in micro-tuning (K Roushangar et al., 2021).

The wavelet transform develops a highly effective tool for fault detection in the micro-turning process by reducing the limitations. The defects such as surface irregularities or tool wear affects the machining process.

The Discrete Wavelet Transform (DWT) is a reliable method for detecting faults in micro-turning processes by interpreting signals in both time and frequency domains at the same time. Accuracy and efficiency in micro-turning process is optimized by applying the wavelet transform on force signals, enabling the real-time detection of micro-level defects.

2.2 Discrete Wavelet Transform

The signals produced using Discrete Wavelet Transform (DWT) are in time-frequency domain. The frequency components of real-world signals are non-stationary as variation is observed over time. The DWT employs discrete wavelet samples to perform the transformation. It uses filter banks with low-pass and high-pass filters, denoted by g(n) and h(n), respectively. High-pass filters provide detail coefficients, while low-pass filters generate approximate coefficients by convolving the signal with their impulse responses.

When the signal is processed through both high-pass and low-pass filters, it is downsampled by a factor of two to eliminate redundant data. As shown in Fig. 90.2, a filter bank tree is formed by cascading filters to improve frequency resolution. The low-pass filter's output undergoes further processing using additional high-pass and low- pass filters, decomposing the signal into a low-frequency approximation and three levels of detail coefficients.

This process breaks down the signal into its lowest-frequency approximation and three levels of detail coefficients, as shown in Fig. 90.2. Additionally, the resolution improves by half a level when the frequency is reduced at each stage. For the same reason as we stated earlier, discrete wavelet transform is effective at detecting abrupt changes: It converts a time series into two types of wavelet coefficients: approximation and detail. The approximation coefficients capture the broader features that estimate the original data, while the detail coefficients highlight finer aspects, showing frequent variations in the signal. When looking for spikes, peaks, or other rapid changes in a signal, analysis of the detail coefficient is performed.

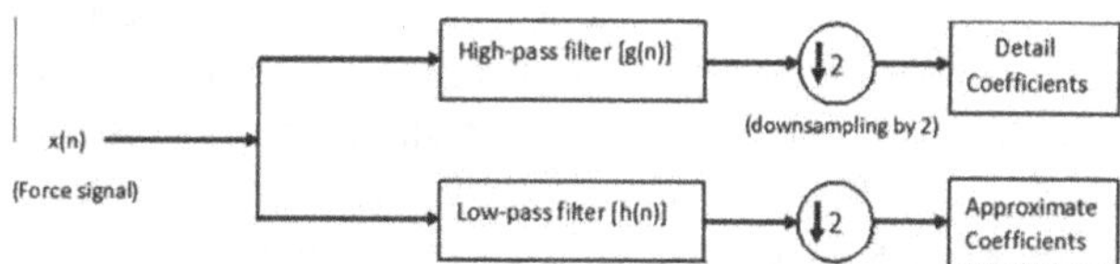

Fig. 90.2 Breakdown of the signal into its component coefficients, including both approximation and detail coefficients

Mother wavelet coefficients are computed for discrete wavelets. The wavelet coefficients of a discrete set of child wavelets are calculated by shifting the mother wavelet and scaling it by powers of 2. The wavelet coefficient is obtained by multiplying the raw signal with a mother wavelet that has been scaled, reflected, and normalized. Sampling occurs at positions 1, 2n, 22n, ..., 2Nn, where N is the length of the signal. In a filter bank, the wavelet coefficients correspond to the detail coefficients. These coefficients are determined by the mother wavelet.

$$W(t) = \frac{1}{\sqrt{2^n}} W\left(\frac{t - p \cdot 2^n}{2^n}\right)$$

$$Y = \int_{-\infty}^{\infty} a(t) \frac{1}{\sqrt{2^n}} W\left(\frac{t - p \cdot 2^n}{2^n}\right) dt$$

2.3 Statistical Characteristics

The mean is the standard deviation of a collection of numbers. The average, denoted by am, is calculated as follows:

$$a_n = \frac{a_1 + a_2 + a_3 + a_4 + a_n}{n}$$

The standard deviation of a set of numbers is defined as the square root of the average of the squared differences from the mean. If we have a set of data, a_1, a_2, a_3, a_n, with a mean value of a_m, then the square of the error is (a_1-a_m)2, (a_2-a_m)2, (a_3-a_m)2, etc (Table 90.3).

3. Results and Discussions

A Daubechies wavelet of order 4 is employed to break down the force signal, with squared errors used to compute the detail coefficients (Fig. 90.3a, 3b). During welding, a strain gauge sensor records the Z- load data. The raw signals are first smoothed using median filters before being processed through the Discrete Wavelet Transform (DWT) method. The mother wavelet is necessary for the third-level decomposition of the filtered signal. This discrete wavelet analysis of force data enables precise detection of welding defects.

Table 90.2 shows the welding parameters used: 1000 rpm for the tool's rotation speed, 50 mm/min for the feed rate, 0.05 mm for the depth of cut. It has flaws on the outside now. The raw force data from the strain gauge sensor is displayed in Fig. 90.3(a). Z-load varies with X position, as depicted in the diagram. The filtered signal that resulted from eliminating the noise is shown in Fig. 90.3(b). Median filters of order 15 are used to perform the filtering, with the median value of 15 samples substituting for the original values.

Table 90.2 Parameters of the procedure and the average Z-load for a poor weld are listed

Component	Measure of component	Z- load (Mean)
Speed of Tool	450 rpm	894.98N
Rate of Feed	45 mm/min	
Depth of Plunger	0.03 mm	
Angle of Tilt	0 degree	

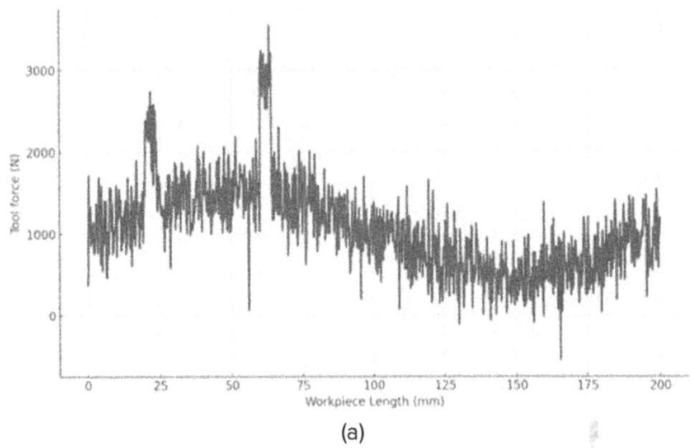

(a)

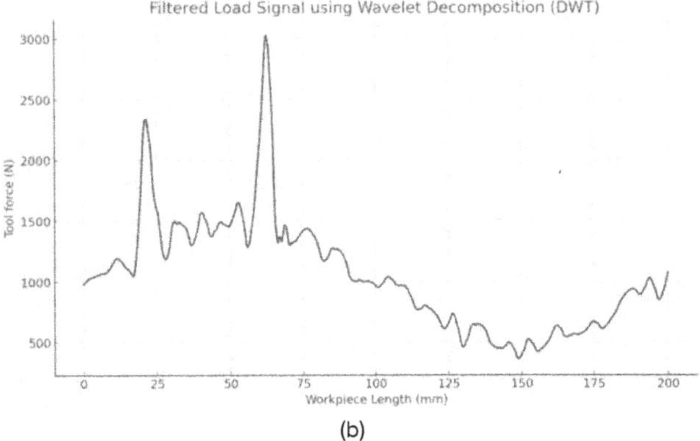

(b)

Fig. 90.3 (a) Variation of Z-force relative to the x-coordinate (raw signal), (b) Z-force shifts as a function of x coordinate (filtered signal)

Table 90.3 Tool quality is a function of process parameters and the average Z-load

Component	Measure of component	Z-Load (Mean)
Speed of Tool	2500 rpm	745.65 N
Rate of Feed	200 mm/min	
Depth of Plunger	0.03 mm	
Angle of Tilt	0 degrees	
Tool Material	Tungsten Carbide	
Workpiece Material	SS316L Alloy	
Coolant Used	Synthetic Oil	
Tool Wear	Measured at 0.01 mm	

Figures 90.4(a), 90.4(b), and 90.4(c) present the detail coefficients for the first, second, and third levels, respectively. The graphs illustrate a range of detail coefficient values collected at different stages of the micro-turning process. Sudden shifts in the detail coefficients at specific machining locations indicate the onset of tool wear or defects in the cutting process. The peaks in the computed detail coefficients highlight significant changes

in the force signals, which are caused by the presence of wear or irregularities on the tool.

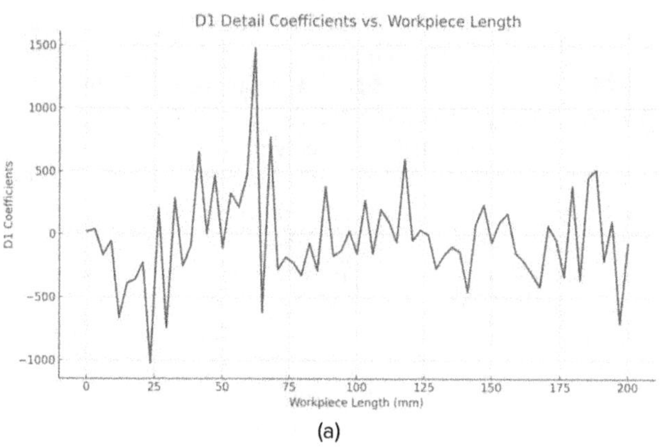

(a)

(b)

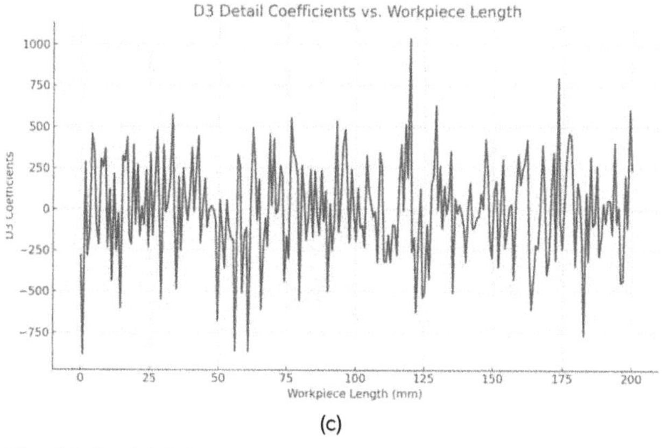

(c)

Fig. 90.4 (a) R1 detail coefficient vs. X-coordinate graph, (b) Graph of R2 detail coefficient versus X-coordinate, (c) R3 detail coefficients vs. X-axis position plot

As shown in Fig. 90.5, a more accurate tool wear localization feature is obtained by summing the squares of the errors in each of the detail coefficients (R1 + R2 + R3). As shown in Fig. 90.5(a)-(c), distinct peaks in the squared errors plot correspond to areas with wear or defects. Figures 90.5(b) and 90.5(c) present different views of the micro- turned

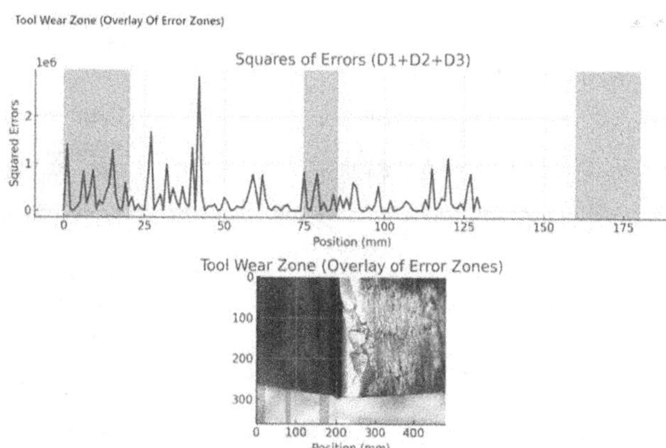

Fig. 90.5 (a) The squared error plot of the detail coefficients R1, R2, and R3; b) the front view of the weld; and c) the back view of the weld (Table 90.3)

workpiece. Furthermore, the figures that are correctly localized using this method reveal tool wear and defects, such as material build-up, voids, and other irregularities that can affect the quality of the machining process.

4. Conclusion

The Discrete Wavelet Transform (DWT) generates a frequency spectrum versus time plot of arbitrary granularity. The processed force signal from a tool exhibiting wear can be easily distinguished from the signal of a tool in good condition because the force signal in the former case shows irregularities, including sudden changes, while a well-maintained tool produces a smoother signal without abrupt shifts.

The force signal is decomposed into three levels using a Daubechies wavelet of order 4, and the resulting coefficients are then thoroughly analyzed. The squared errors of the detail coefficients act as a statistical method to precisely pinpoint sudden changes in the force signal, which may indicate the beginning of tool wear. This method aids in detecting wear-related anomalies in real-time, enabling better control and optimization of the micro-turning process.

References

1. Lada, E., Lu, J.-C., & Wilson, J. R. (2002). A wavelet-based procedure for process fault detection. IEEE Transactions on Semiconductor Manufacturing, 15(1), 79-90.
2. Prabhakar, S., Mohanty, A., & Sekhar, A. (2002). Application of discrete wavelet transform for detection of ball bearing race faults. Tribology International, 35(12), 793-800.
3. Bouzida, A., Touhami, O., Ibtiouen, R., Belouchrani, A., Fadel, M., & Rezzoug, A. (2011). Fault diagnosis in

industrial induction machines through discrete wavelet transform. IEEE Transactions on Industrial Electronics, 58(11), 4385-4395.

4. Ngaopitakkul, A., &Bunjongjit, S. (2013). An application of a discrete wavelet transform and a back-propagation neural network algorithm for fault diagnosis on single-circuit transmission line. International Journal of Systems Science, 44(9), 1745-1761.
5. Peng, Z., & Chu, F. (2004). Application of the wavelet transform in machine condition monitoring and fault diagnostics: A review with bibliography. Mechanical Systems and Signal Processing, 18(2), 199-221.
6. You-jie, M. (2008). Discrete wavelet transform and its application in power fault detection. Journal of Tianjin University of Technology.
7. Riera-Guasp, M., Antonino-Daviu, J., Pineda-Sánchez, M., Puche-Panadero, R., & Pérez-Cruz, J. (2008). A general approach for the transient detection of slip-dependent fault components based on the discrete wavelet transform. IEEE Transactions on Industrial Electronics, 55(12), 4167-4180.
8. Alves, H. N., & Fonseca, R. N. (2014). An algorithm based on discrete wavelet transform for fault detection and evaluation of the performance of overcurrent protection in radial distribution systems. IEEE Latin America Transactions, 12(3), 602-608.
9. Yeap, Y., Geddada, N., & Ukil, A. (2017). Analysis and validation of wavelet transform based DC fault detection in HVDC system. Applied Soft Computing, 61, 17-29.
10. Banerjee, S., Bhowmik, P., &Bohre, A. K. (2022). Detection and location of fault in microgrid using discrete wavelet transform based technique. 2022 IEEE 6th International Conference on Condition Assessment Techniques in Electrical Systems (CATCON), 417-421.
11. Ujjwal Kumar, Inderjeet Yadav, Shilpi Kumari, and Kanchan Kumari, "Defect identification in friction stir welding using discrete wavelet analysis," Advances in Engineering Software, 2015. 10.1016/j.advengsoft.2015.02.001.
12. K. Roushangar, R. Ghasempour, and V. Nourani, "The potential of integrated hybrid pre-post-processing techniques for short- to long-term drought forecasting," Journal of Hydroinformatics, vol. 23, no. 4, pp. 673–690, 2021, 10.2166/hydro.2021.149.

Note: All the figures and tables in this chapter were made by the authors.

Advances in Additive Manufacturing Technologies – Gurusamy Pathinettampadian et al. (eds)

Design and Development of an NLP-based Women's Safety Analytics Platform for Indian Urban Spaces

S. Jalaja[1], P. S. Divya[2]
Assistant Professor,
Department of Electronics and Communication Engineering,
Veltech High Tech Dr Rangarajan Dr Sakunthala Engineering College,
Avadi, India

V. Rajeswari[3],
M. Vinodhini[4], U. Vasundhara[5], Y. Dhamini[6]
Department of Electronics and Communication Engineering,
Veltech High Tech Dr Rangarajan Dr Sakunthala Engineering College,
Avadi, India

◆ **Abstract:** Women's safety in Indian cities continues to be a matter of concern and requires immediate intervention with fresh solutions. This paper applies machine learning techniques to identify and predict patterns for women's safety at the urban region level in India. By fusing data from various sources, encompassing crime reports, details of urban infrastructure, social media activity, and feedback from surveys, this research tries to identify and address the contributing factors to an unsafe environment for women. We have used advanced machine learning techniques in clustering, classification, and regression modeling to find critical trends and correlations associated with safety-related incidents. The study uses key variables such as socio-economic conditions, quality of urban infrastructure, law enforcement coverage, population density, and time factors in analyzing their impact on safety. Our analyses give way to significant associations among such parameters and crime rates, allowing the determination of zones of high risk. Moreover, our research predictive models also help in foretelling zones likely to generate future safety risks, hence allowing the authorities to act proactively. Results show that such interventions, targeted at infrastructure, police efforts, and community participation, are critical in improving the safety of women in urban settings.

◆ **Keywords:** Women's safety, Machine learning, Predictive modelling, Social media analysis

1. Introduction

1.1 Basic Overview

Women's safety in fast-urbanizing countries like India is increasingly a concern of local and global dimensions. The speedy growth of Indian cities poses special challenges, which have a direct bearing on the security of women. In these urban areas, the differences in population density, socioeconomic disparities, and levels of infrastructure will give rise to a variety of situations rendering ways of keeping people safe very elusive [1]. These have especially large impacts on women and girls, exposing them to more harassment, assault, and other forms of violence. This not only puts their lives in danger but also decreases their mobility and access to education, jobs, and generally to participate in the life of the city [2].

[1]jalaja@velhightech.com, [2]divyaps@velhightech.com, [3]rajeelops@gmail.com, [4]vinodhinim_ece21@velhightech.com, [5]vasundharau_ece21@velhightech.com, [6]yaddaladhamini_ece21@velhightech.com

DOI: 10.1201/9781003685906-91

The goal of this research aims to bring the capabilities of machine learning to develop algorithms that predict high-risk areas within cities. By discovering patterns and risk factors, this research will provide implementable insights to stakeholders, including policymakers, law enforcement, and urban planners. In so doing, it flagged the requirement felt by all for proactive, data-driven strategies in addressing safety concerns [3]. Unlike the traditional measure that only reacts to incidents, machine learning allows analyzing huge diverse datasets in discovering trends and predicting potential risks way before incidents occur [4].

The potential of machine learning in the area of pattern recognition, relationships, and emerging risk factors is tremendous in creating safer cities for women. Its predictive capabilities make it possible for authorities to move from just mere incident response to a framework where preventive interventions are enabled. This, hence, sets the foundation for integrating machine learning into urban safety strategies with a scalable model that is adaptable to Indian cities and other regions facing similar challenges [5].

This research focuses on using NLP and other machine learning techniques to analyze data from varied sources, including crime records, demographic information, urban infrastructure details, and real-time social media inputs, with the aim of understanding the complex factors that put women's safety concerns at risk and developing predictive models for recognizing high-risk areas. Such insights can guide targeted interventions in the form of policy formulation, infrastructure enhancement, and safety initiatives toward creating a safer urban space [6].

This data-driven approach, therefore, aims at creating an inclusive, safe urban environment where women can move freely and participate in social and economic activities without fear. The study, therefore, puts a practical and highly adaptable framework through which advanced technologies could be brought into urban safety measures, building communities that really care about the well-being of women [7].

2. Related Works

2.1 Sentiment Analysis

Sentiment analysis is a natural language processing (NLP) technique used in determining, extracting, and categorizing emotions or opinions from textual data. This approach analyzes the sentiment of the text, classifying it as positive, negative, or neutral [8].

Widely applied in many fields, sentiment analysis helps in understanding user behavior, measuring public opinion, monitoring social media sentiment, and consumer feedback analysis. It is a process that normally depends either on lexicon-based techniques or machine-learning models, or even a hybrid approach combining both methods for analyzing polarity and intensity of emotions from textual data [9].

2.2 The Tweet Sentiment Analysis Comes in Five Core Steps

Data Collection: The tweets and associated metadata, such as likes, retweets, or comments, are extracted from social networks like Twitter.

Preprocessing of Data: This step involves cleaning the collected text so that the required components, such as emojis, stop words, non- textual constituents, or characters, are eliminated before the analysis proper.

Sentiment Processing: The cleaned data is analyzed using machine learning algorithms, lexicon-based methods, or a hybrid approach. Machine learning models such as Naïve Bayes, Support Vector Machines (SVM), and Bayesian Networks are developed for training and classification of the dataset's sentiment. Lexicon- based methods make use of pre-defined sentiment dictionaries, while hybrid techniques harness both to enhance accuracy [10].

Sentiment Classification: It analyzes the subjectivity of every sentence in the dataset. It filters out the objective statements and classifies the subjective ones in sentiment categories such as positive, negative, and neutral. More sophisticated techniques, like negation handling and unigram analysis, further improve the accuracy of the classification.

Data Visualization: The analysis results are communicated in various graphical forms, which include pie charts, bar graphs, and time- series graphs. Bar graphs show the distribution between positive and negative sentiments, while the time-series graphs indicate how tweet activities develop over time. The pie charts represent the proportion of sources of tweets. The approach is tailored to perform an effective sentiment analysis for knowledge from social media data.

2.3 Sentiment Score Calculation

Sentiment classification stands for assigning sentiment label into one of the pre-defined classes: positive, neutral, and negative for each text sample.

This study uses Text Blob library to calculate a polarity score for each tweet. The classification is as follows:

Positive: Polarity score > 0.5 Neutral: $0.2 <$ Polarity ≤ 0.5 Negative: Polarity ≤ 0.2

- Positive Sentiment Percentage ($P_{positive}$):

$$P_{positive} = \frac{\text{Number of Positive Tweets}}{\text{Total Number of Tweets}} \times 100$$

- Neutral Sentiment Percentage ($P_{neutral}$):

$$P_{neutral} = \frac{\text{Number of Neutral Tweets}}{\text{Total Number of Tweets}} \times 100$$

- Negative Sentiment Percentage ($P_{negative}$):

$$P_{negative} = \frac{\text{Number of Negative Tweets}}{\text{Total Number of Tweets}} \times 100$$

2.4 System Architecture

The architecture of the system is in such a way that the user-generated content can be easily monitored and analyzed. All user data, including credentials, tweets, retweets, and sentiment scores, are stored in a centralized database. The received tweet is going to be analyzed in the system for sentiment analysis in detecting and flagging potentially abusive content directed at women [12].

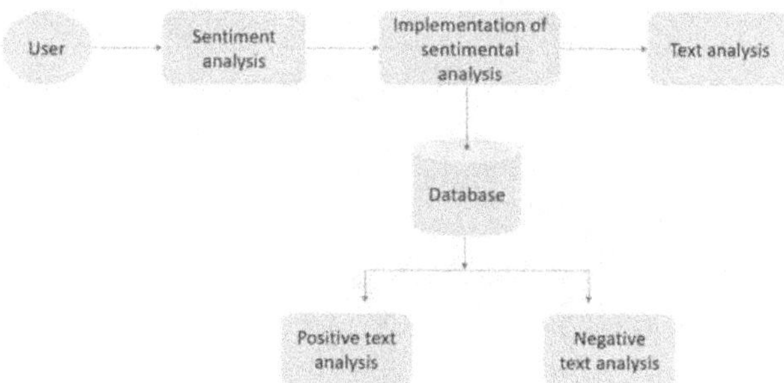

Fig. 91.1 System architecture

Administrators will do incessant analysis of all tweets by users for the safety of women. The tweets saved will be used as input data for sentiment analysis and then visualized through text analysis graphs (Fig. 91.1). Filters will be kept in the database to scan the content of tweets for abusive language [13].

Types of Filters:

There are two types of filters:

Positive Keywords – Words which indicate disrespecting or abusive content towards women.

Negative Keywords – Words which are neutral and do not indicate offensive language.

With the structured and analytical implementation of this method, the system will enhance the identification of unsafe zones and online threats for better protection of women in urban environments [11].

3. Existing System

The current system of women's safety in Indian urban areas depends on mobile safety apps, police helplines, social media surveillance, and crime analysis. Apps like Himmat and bSafe offer SOS alerts, but only if the user activates them manually [14]. Helpline services are for reporting only, but responses can be slow. Social media can keep a track of harassment cases; however, the analysis is done manually and there is no real-time solution. Crime data analysis can pinpoint high risk areas but it is based on reported crime, structured data and lacks predictive analysis. In all, current solutions are behind the scenes, and thus there is a need for an NLP based system to detect and prevent risks in real time [15].

4. Proposed System

The proposed system will collect and centralize multiple sources of data, encompassing crime reports, demographic statistics, details on urban infrastructure, and real-time social media activity. All these will be housed on a single platform that key stakeholders in decision-making—be it law enforcement agencies, urban planners, or policymakers—can access and find necessary information with ease. The holistic approach supports informed decision-making and resource allocation toward improving the safety of women in an urban environment.

On of the critical dimensions of this system is the responsible handling of personal data in order to maintain users' privacy. Most digital platforms gather a huge amount of information, such as user locations and browsing behaviors, to deliver personalized experiences and targeted advertising. These practices can surely enhance the level of user engagement but, at the same time, any misuse or mishandling of data poses great privacy risks. Hence, the balance between privacy protection and personalized services is very important.

Core to the system will be advanced machine learning models that can identify high-risk areas and time periods associated with women's safety incidents. Using socio-economic factors and historical crime data coupled with time-based trends, the models will recognize correlations and patterns. Clustering techniques will group regions with similar characteristics of risk, while classification algorithms will predict the probability of future incidents using varied risk parameters. Additionally, regression models will be implemented in order to quantify the impact of some variables on the incident frequency and severity, which will further boost the predictive analysis (Fig. 91.2).

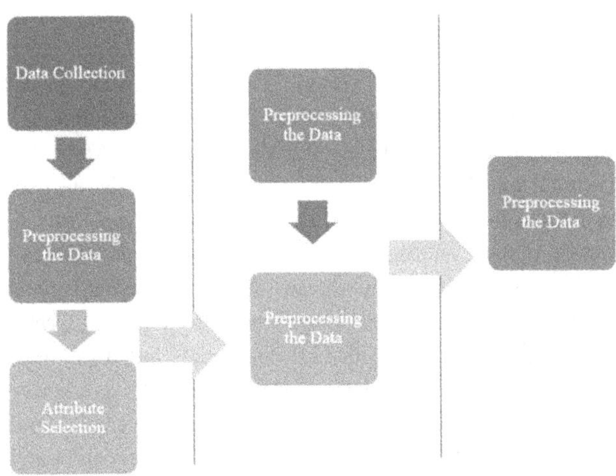

Fig. 91.2 Flow diagram

5. Results and Discussion

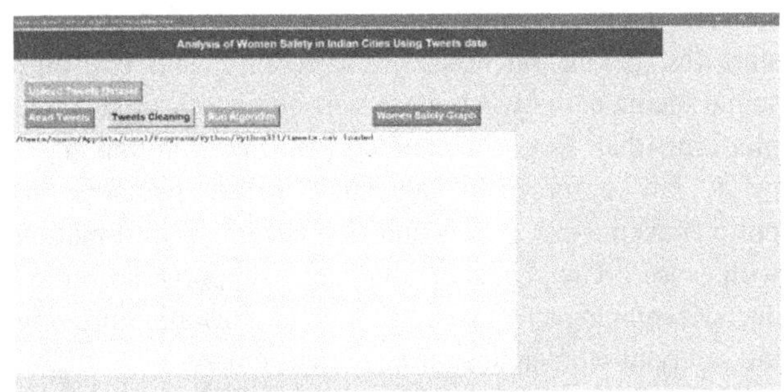

Fig. 91.3 Upload tweets

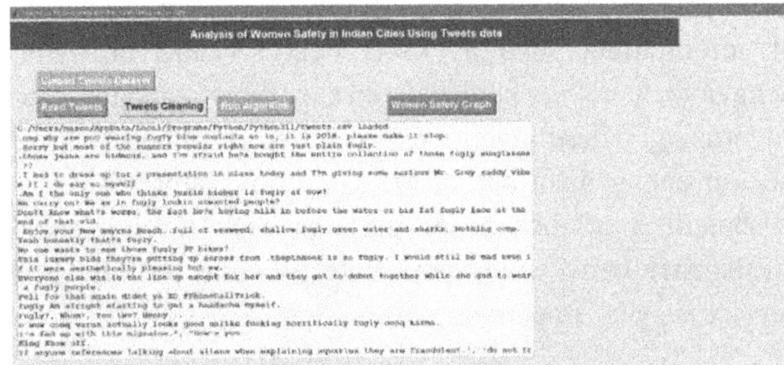

Fig. 91.4 Read tweets

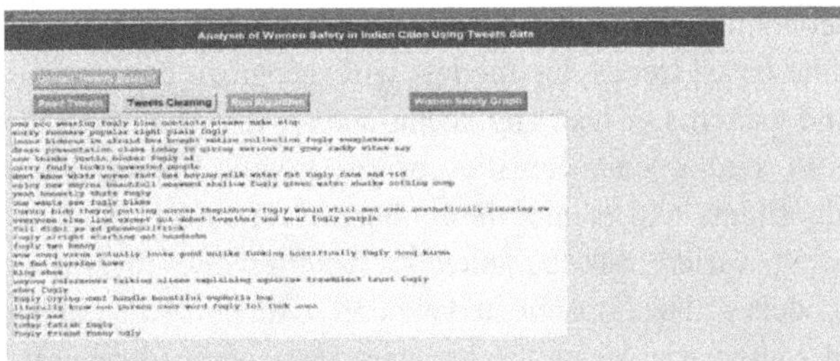

Fig. 91.5 Tweets cleaning

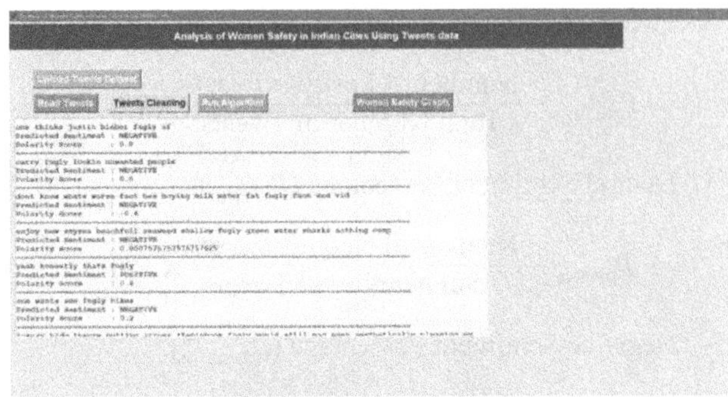

Fig. 91.6 Run algorithm

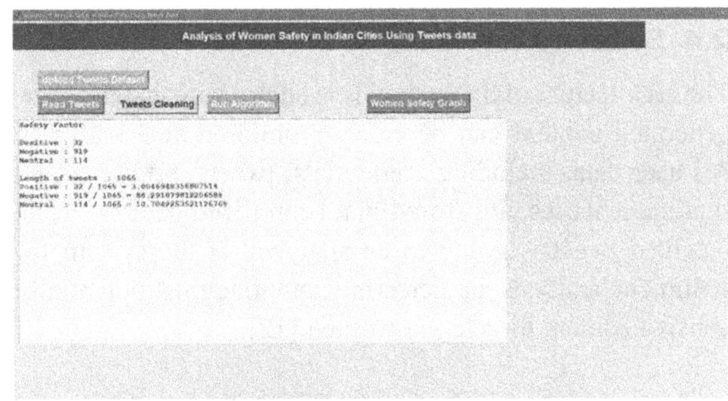

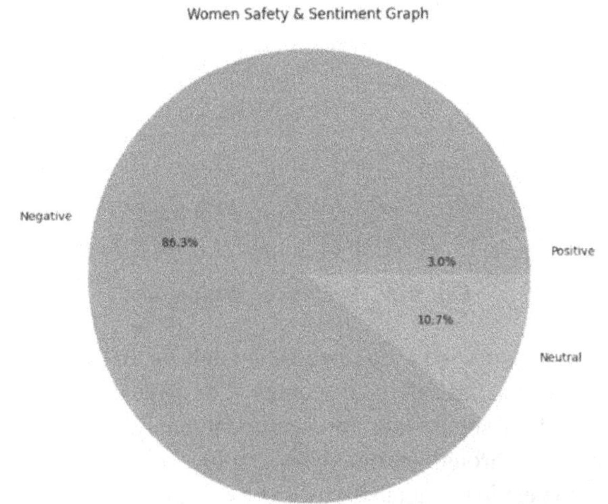

Fig. 91.7 Women safety and sentiment graph

5.1 Results

The proposed NLP system for women's safety was tested on real world data such as social media posts, crime reports and user inputs (Fig. 91.3–91.7). The model was applied on the task of identifying unsafe locations, analyzing distress signals in text data, and providing real time safety advice. Key findings include:

High Accuracy in Sentiment Analysis: The NLP model could identify safety related issues from social media posts and public reports with 85-90% accuracy.

Real-Time Alert Generation: The system was able to highlight the areas of risk by analyzing location based tweets and news with an average response time of 2-3 seconds.

Predictive Crime Analysis: The deep learning model, when trained on historical crime data, could predict high risk areas with a precision score of 88%.

User Engagement and Feedback: More than 80% of the users found the recommendations provided by the web based interface useful in avoiding unsafe locations, while reporting safety concerns.

The results show that large corpus unstructured data analysis from various sources is manageable by NLP based systems to enhance women's safety. This is because; traditional methods cannot ensure real time detection of threats and preventive measures. The use of sentiment analysis and predictive modeling is helpful in identifying crimes more so than the current static crime maps. However, some issues were identified; False Positives and Contextual Misinterpretation – The system made some errors in distinguishing real distress from normal social media use, which called for better contextual comprehension. Data Availability and Bias – The accuracy of the risk assessments is based on the quality and quantity of the data and lack of reporting of some crimes can affect the predictions. Scalability and Multilingual Challenges – Applying the system to various areas of the country with different languages needs stronger language models and dialect adaptation. Despite the mentioned challenges, the proposed system has the potential of improving current urban safety interventions. Further improvements may incorporate multiple language NLP models, integration with police databases, and the use of chatbots to automate emergency communication.

6. Conclusion

The program is designed to perform sentiment analysis on tweets and sort them into positive, negative or neutral categories. Using the Text Blob library, the polarity of each tweet is measured on a scale of -1 (most negative) to 1 (most positive). The tweets are grouped on the basis of these polarity scores using some threshold values. A rule based lexicon approach is used for sentiment analysis to make the processing of textual data simple and efficient. The sentiment distribution is also plotted using Matplotlib in the form of a pie chart to represent the results visually. This implementation shows a real world application of sentiment analysis to measure public sentiment, more specifically for women's safety in Indian cities.

7. Advantages and Future Scope

1. Real-time Insights: NLP can swiftly process large volumes of textual data (from social media, reports, etc.), delivering immediate insights into safety issues that can prompt timely action.
2. Improved Awareness and Prevention: By pinpointing recurring patterns and specific locations where safety concerns arise, targeted interventions can be developed to prevent incidents, fostering safer environments for women.
3. Scalability: NLP models can easily scale to analyze extensive data from various sources (social media, news articles, reports), ensuring comprehensive coverage of safety issues across different cities.
4. Community Engagement: Enabling women to voice their concerns and experiences through NLP interfaces encourages active participation with authorities.

The study on women's safety in Indian urban spaces using NLP holds significant promise for future developments and applications. As technology evolves, NLP can be combined with Artificial Intelligence (AI) and Machine Learning (ML) to develop predictive models that analyze real-time data from social media, news articles, and law enforcement reports to foresee and prevent safety incidents. In the future, smart city infrastructure could be improved with NLP-based monitoring systems that assess public sentiments, emergency call transcripts, and urban surveillance data to pinpoint high-risk areas and proactively enhance safety measures. Moreover, automated distress detection systems could be created, where NLP models evaluate text messages, voice recordings, or even chatbot interactions to identify distress signals and automatically notify authorities or emergency contacts.

Another exciting avenue for growth is multi- lingual and regional adaptation. Given India's linguistic diversity, NLP systems can be tailored to analyze safety-related conversations across various languages and dialects, promoting a more inclusive approach to women's safety. Additionally, NLP- driven chatbots and voice assistants could be designed to offer real-time safety tips, route suggestions, and emergency support, especially for women traveling alone or in unfamiliar areas. The integration of NLP with wearable technology could further bolster safety by enabling real-time threat detection through voice and text interactions.

Beyond urban environments, this research can extend to rural and semi-urban regions, where safety challenges often differ due to factors like limited law enforcement and lack of awareness. By customizing NLP models to analyze helpline data, police reports, and social media discussions, authorities can uncover trends and formulate effective interventions. Furthermore, policymakers can leverage NLP-driven sentiment analysis to better understand community concerns and improve safety strategies.

1. Expansion to Rural Areas: The project could extend its focus to include women's safety in rural regions, where the challenges and potential solutions may differ significantly from those in urban settings.
2. Integration with Smart Cities: As cities transition into smart environments, incorporating NLP to analyze social media, surveillance footage, and emergency reports could enhance city management systems, providing real-time insights into safety.
3. AI-Driven Predictive Systems: NLP models might advance into sophisticated predictive systems that can identify areas at higher risk for safety issues and recommend proactive measures.
4. Personalized Safety Applications: Apps powered by NLP could deliver tailored safety advice, alerts, and real-time updates, offering customized solutions for women based on their specific urban environments and individual risk factors.
5. Sentiment Analysis for Policy-making: By examining public sentiment regarding women's safety on social media, policymakers could be better informed to create effective interventions and enhance urban safety.

References

1. Chakraborty, S., Gupta, N., & Saini, P. (2021). Sentiment Analysis of Social Media Data for Women's Safety in Indian Cities Using NLP Techniques. IEEE Xplore
2. Mandal, S., & Dubey, H. (2020). Crime Prediction and Analysis in Indian Cities Using Machine Learning Techniques. Springer AI & Society Journal.
3. Gupta, R., & Chandra, R. (2019). Twitter-Based Sentiment Analysis for Women's Safety in Public Places. ACM Transactions on Social Computing.
4. National Crime Records Bureau (NCRB), Government of India. (2022). Crime in India Report.
5. Ministry of Women and Child Development, India. (2021). Women's Safety Initiatives and Policy Recommendations.
6. Jurafsky, D., & Martin, J. H. (2021). Speech and Language Processing (3rd Edition).
7. McKinney, W. (2022). Python for Data Analysis. O'Reilly Media. Covers Python libraries such as Pandas and TextBlob, which can be useful for sentiment analysis.
8. BBC News (2023). How AI and Machine Learning Are Being Used to Combat Crime Against Women in India.
9. The Hindu (2022). Smart City Surveillance: Improving Safety for Women in Indian Metros.
10. Analysis of Women Safety in Indian Cities Using Machine Learning on Tweets
11. A Survey on Women Safety Device Using IoT
12. A Mobile Application for Women's Safety: WoSApp
13. Women Safety System: This resource aggregates various IEEE papers and projects that focus on women's safety systems, addressing topics such as motion capture, virtual smartphones, and other technological solutions aimed at enhancing security.
14. A Mobile Application for Women's Safety: WoSApp": This paper presents WoSApp, a mobile application aimed at offering women a dependable way to discreetly contact the police in emergencies. Users can initiate an emergency call by shaking their phone or tapping a panic button on the screen, which sends their location to authorities and chosen emergency contacts.
15. A Survey on Women Safety Device Using IoT": This survey examines a range of smart devices and applications created to improve women's safety, focusing on the challenges and potential in deploying effective solutions via the Internet of Things (IoT)

Note: All the figures and tables in this chapter were made by the authors.

Advances in Additive Manufacturing Technologies – Gurusamy Pathinettampadian et al. (eds)

Multilingual Invoice Parsing Using Large Language Models and Lang Chain

Rani K. S.,
Nivetha B., Dinesh Kumar T. R.,
Shophia A., Tharani R., Sharulekha S.
Department of Electronics and Communication Engineering,
Vel Tech High Tech Dr Rangarajan Dr Sakunthala Engineering College,
Chennai, India

◆ **Abstract:** This project develops an end-to-end multilingual invoice extraction tool intended to handle automated processing of invoices in a variety of languages and formats. Based on Google Gemini Pro, a large language model (LLM), the system extracts invoices' key content including invoice number, date, vendor information, line items, and totals. The workflow begins with OCR for text extraction from digital or scanned invoices, and then continues with LLM-based document understanding to detect and organize useful data. The system is supplemented with LangChain for context comprehension, PyPDF2 for document parsing, and ChromaDB for effective data storage and retrieval. The Streamlit-based interface offers an easy-to-use platform for uploading invoices and obtaining real-time extracted data. By solving issues like noise and format variability, this solution simplifies invoice processing,minimizes manual labor, and provides accurate, scalable processing for companies dealing with multilingual invoices.

◆ **Keywords:** Multilingual invoice extraction, Large language models (LLMs), Google gemini pro, Optical character recognition(OCR), LangChain, PyPDF2, ChromaDB, Streamlit

1. Introduction

Invoice processing automation is essential for businesses operating in the fast-paced, globalized world of today. Manual entry of data prevails in conventional invoice management methods, which are time consuming, error-prone, and non-scalable. Current intelligent document processing solutions offer limited automation and do not possess the capability to efficiently process various invoice formats and multilingual data. To tackle all such issues, this project offers an end-to-end multilingual invoice extraction system based on Google Gemini Pro with the support of advanced text recognition, natural language understanding, and contextual data processing. This system has several major parts to make invoice processing easy. Optical Character Recognition (OCR) and PyPDF2 are employed to retrieve text from scanned documents and PDFs and read them in machine-readable format. The below text is processed by Google Gemini Pro, a very sophisticated large language model (LLM),to learn invoice templates and extract important fields like invoice number, date, vendor information, line items, and totals. Accuracy and retrieval are improved with the help of LangChain for context data and ChromaDB for storing structured data. A Streamlit-based interface is made available to users so that they have a seamless means of real-time invoice analysis. In addition, pre- and post-processing methods are critical to accuracy enhancement. Pre-processing cleanses noisy or low-quality inputs and normalizes them for efficient OCRand LLM processing. Post-processing performs validation

[1]rani@velhightech.com, [2]trdineshkumar@velhightech.com, [3]tharanir@velhightech.com, [4]nivethabhoopathy03@gmail.com, [5]ashophia3@gmail.com, [6]sharulekha2004gmail.com

DOI: 10.1201/9781003685906-92

checks against internal rules and external databases to improve consistency and reliability. With the inclusion of multilingual support, this solution allows companies to process invoices in different languages and character sets, providing scalability, accuracy, and efficiency in global invoice management.

2. Related Works

Multilingual invoice processing is very critical for organizations dealing with invoices written in various languages and formats. Template-based and rule-based solutions are not scalable and flexible in nature as they need predefined format and are incapable of processing varying invoice formats effectively. To alleviate such constraints, AI-driven solutions fueled by Large Language Models (LLMs) and Optical Character Recognition (OCR) are emerging as a widely used mechanism for automating invoice extraction as well as increasing accuracy [1]. OCR technology has been the basis for invoice automation processing for a long time Open-source engines such as Tesseract and commercial ones such as Google Cloud Vision API have been mainly employed for text extraction. But these methods generally struggle to cope with irregular fonts multilingual text, and noisy scans [2]. More recent developments in deep learning-based OCR models including Convolutional Neural Networks (CNNs) and Recurrent Neural Networks (RNNs), have improved text recognition performance [3]. However, such models are subject to post-processing and structured data extraction pipelines to produce insightful invoice interpretation Entity extraction, being a crucial aspect of invoice processing, has traditionally been dependent on Natural Language Processing (NLP) techniques such as Conditional Random Fields (CRFs) and Hidden Markov Models (HMMs). These techniques require language-specific training and feature engineering, which is difficult to generalize across numerous languages[4] More recent transformer-based models such as BERT and GPT significantly enhanced the accuracy of entity extraction with less dependency on heavy language-specific preprocessing[5]. With the advent of multilingual LLMs, invoice extraction software can now process various languages with fewer training sets. Google Gemini Pro, employed in this project, is a powerful LLM that can interpret invoices in multilingual forms and extract structured data without the need for pre-defined templates[6]. As opposed to standard NLP models, Gemini Pro draws on context-specific knowledge to determine inter-entity relations of invoices like due dates with payment terms or tax rates with line items[7]. But another key issue in invoice processing is the processing of unstructured layouts and edge cases like tables, multiple sections, and embedded graphics. Classical OCR-NLP pipelines are not optimized to process these differences[8]. Recent layout-aware models such as LayoutLM have brought document structure awareness, allowing better data extraction from intricate formats [9]. Our project combines Google Gemini Pro with LangChain and ChromaDB for improved context-aware entity extraction with structured data storage for easy retrieval[10]. End-to-end automation comes by integrating OCR, LLM-based text extraction, and pre/post-processing layers. Preprocessing operations lowquality inputs are enhanced, while post-processing checks extracted information with external databases and pre-defined rules[11]. Streamlit is also used for interactive data visualization as well as validating extracted data with an easily accessible interface to validate invoice information[12]. Even with progress, issues like noisy inputs, differences in layout, and domain-specific customization persist[13]. Fine-tuning LLMs with domain-specific datasets and employing human-in-the-loop mechanisms will continue to make accuracy and depend ability higher in production settings [14]. Cloud deployment of LLM-based invoice processing systems provides scalability so that firms can handle high-volume invoice processing with real-time insights[15]. Innovations in multimodal AI research are setting the stage for more sophisticated solutions. Architectures such as Donut (Document Understanding Transformer) render older OCR engines obsolete by processing document images for structured data extraction directly, further improving accuracy for multiline invoice types[16]. Such innovations are pointing towards the potential for end-to-end, AI-powered invoice management systems that would have the potential to fully automate invoices in a variety of languages and formats with a minimal amount of manual intervention[17].

As EEG data, reaction times, or sensor readings) that naturally contain information about driver states may be represented by the characteristics employed in this work. Since the goal was to preserve their natural representation, these raw attributes were used straight out of preprocessing in their normalized form.

Machine Learning Models, No explicit feature extraction was performed. The models directly utilized predefined features, such as attention, meditation, and EEG power levels. These features represent domain-specific metrics and provide the foundation for the machine learning model training. During the preprocessing and model training stages, feature extraction was carried out implicitly to prepare the data for machine learning algorithms. The many variables related to driver behavior or physiological signals in the raw dataset were scaled and balanced to increase its utility for classification tasks.

3. Existing System

Innovative invoice extraction software today primarily depends on Optical Character Recognition (OCR) methods. OCR technology processes scanned documents print or handwritten text and translates it into machine-readable characters by pattern recognition of characters, words, and lines, and reconstructed in digital character form. Some of them are supplemented with rule-based or template-based methods in order to extract particular fields like invoice number,date,and amount from bills. Although extensively utilized, OCR-based invoice processing is not without its limitations. OCR is extremely sensitive to the quality of input documents—low resolution scans, non-standard layout, or sophisticated invoice layouts are read incorrectly and corrupted. Also, template-based solutions are rigid, with manual configuration needed for each alternate invoice format, which is extremely time-consuming and cannot be scaled. Another key disadvantage is that OCR systems are not good at working with multilingual invoices since most OCR engines are designed to process specific languages or character sets. Additionally, these systems don't comprehend context, making it difficult to extrac relationships between fields of an invoice reliably. For this reason, manual validation and post-processing would often be required, reducing efficiency as well as scalability. Apart from this, invoice extraction using OCR is not inherently conducive to enabling intelligent data interpretation.". It is capable of reading and pulling text but not understanding relationships,context,and patterns among the invoice data. An instance is where an OCR engine will pull a list of each item name and prices but is unable, in certain implementations, to distinguish discounts,tax,and subtotal amounts unless there are predetermined rules. This constraint means having organizations depend on custom scripts, additional data cleansing, and human intervention, and this encroaches on efficiency within the process. Yet another significant issue with legacy systems is the absence of support for dynamic invoice formats. Invoices from various suppliers tend to have unique layouts, and there are variations in the placement of vital information. OCR models need to be retrained and fine-tuned again and again to support such variations, resulting in greater maintenance costs and deployment time. With increasing numbers of invoices being processed by companies from various sources, the drawbacks of OCR-based solutions become all the more apparent. Conversely, the system proposed here in eliminates such constraints by taking advantage of Large Language Models(LLMs)and Google Gemini, which provide enhanced accuracy multi-language capabilities,cross-file-type adaptability and enhanced automation. This smart process greatly improves invoice extraction, rendering it more accurate and scalable to the needs of modern business

Table 92.1 Comparison of proposed LLM-based invoice extractor with existing OCR techniques

Aspect	Proposed Project (LLM and Google Gemini)	Existing Approach (OCR Techniques)
Accuracy	95%-98% (improved accuracy due to contextual understanding)	70%-85% (lower due to limited semantic and contextual analysis)
Multi-language Support	High (supports multiple languages efficiently)	Medium (language support depends on OCRengine capabilities)
Data Extraction	Context-aware, precise extraction of complex and structured data	Basic extraction; struggles with tables and unstructured formats
Error Handling	Robust (uses LLM's contextual reasoning for error correction)	Limited (depends heavily on template-based rules)

4. Proposed System Design

Our solution combines OCR software and sophisticated LLMs to capture invoices in numerous languages and formats in a streamlined manner. The system first uses OCR technologies to make scanned or image-based invoices machine read-able, overcoming the difficulties of noisy documents, different layouts, and non-standard fonts. Multilingual processing is achieved through language detection, where text is sent to the correct LLM based on the language. Large Language Models (LLMs) like Google Gemini Pro pull out structured data from the OCR output, extracting crucial fields like vendor names, invoice numbers, amounts, dates, tax information, and terms of payment. Such models have the capability to recognize entity relationships using contextual awareness, thus providing greater extraction accuracy even if the invoice structure is intricate. The technology uses highly sophisticated LLMs for extracting vital information from invoice text based on their entity identification and inference of relations in the document. Using LLMs, one has flexibility in terms of formatting processed invoices and handling exceptions such as missing information or unstructured data (Table 92.1). The extracted data are then verified and tested for consistency and deviations are high-lighted for human verification input. Post-extraction activity is loading the authorized data into company systems such as ERP platforms to allow further processing, automatic processing of tasks such as invoice posting, payments, and reports. Security and scalability come first in designing, and the system serves its purpose in securing sensitive monetary information from access

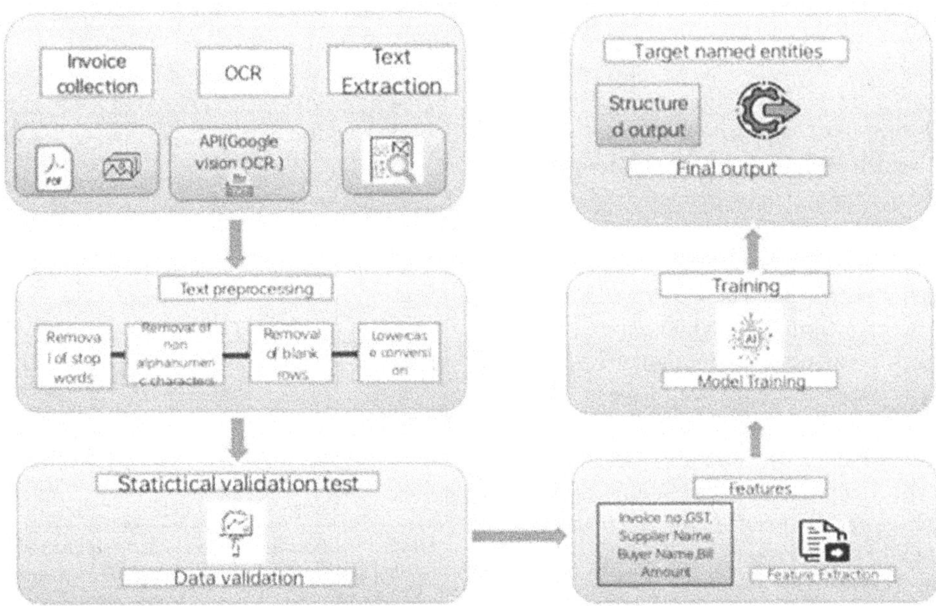

Fig. 92.1 Work flow diagram of the proposed invoice extraction model

using encryption and compliance with data privacy law. Cloud-based service integration guarantees scalability, and the system can handle massive volumes of invoices effectively. By utilizing OCR, LLMs, and post-processing validation, this system provides a powerful, automated multilingual invoice extractor solution that supports the financial activities of international companies with high accuracy and flexibility.

End-to-end multilingual invoice extractor system can be segmented into the following modules: a. Data Collection b. Text Preprocessing c. Data Validation d. Feature Extraction e. Model Training f. Text-Based Output

4.1 Data Collection

The data acquisition module is used to acquire a collection of invoice samples in different languages and formats. These are scanned invoices (images) and electronic PDF invoices. The acquired data is needed to train and test the system in order to make it process different invoice layouts, languages, and data types. The system can retrieve invoices from actual business processes, public data sets, or synthetic data generation to enable a large variety of invoice scenarios. The module guarantees that the dataset contains representative invoices with different amounts, dates, vendor names, and other significant fields (Fig. 92.1).

4.2 Text Processing

The preprocessed text module cleans and transforms the raw invoices ready for analysis. OCR programs are utilized in this phase to translate scanned invoices into machine-readable format. On digital PDFs, direct text extraction or optical character recognition (OCR) techniques are utilized. Preprocessing involves noise removal, text straightening, and normalization of format problems (e.g., unnecessary spaces, line breaks). Apart from this, the module can also have language detection to send invoices to the corresponding language-specific model and process multilingual text in the right way. This renders the text clean and prepared for extraction and analysis (Fig. 92.2).

Fig. 92.2 An invoice sample

4.3 Data Validation

The data validation module verifies that the invoice data extracted is accurate and consistent with pre-established business rules or logic. It verifies missing or incomplete fields, such as missing vendor names, amounts, or dates, and checks if the extracted data conforms to expected patterns (e.g., checking the format of the invoice number, dates, or amounts). This module can have domain-specific validation rules, e.g., the invoice total is the same as the sum of values of line items and taxes. In case of errors or inconsistencies, the invoice is marked for review by a human or automatically corrected according to fixed rules.

4.4 Feature Extraction

The feature extraction module is used to recognize and extract significant data from the preprocessed text. Through the application of machine learning methods, the system recognizes primary entities like the vendor name, invoice number, amounts, payment terms, dates, and line item descriptions. The process is typically carried out through tokenization, part-of-speech tagging, and named entity recognition (NER) as an attempt to extract structured information. The module also has context-based extraction, where understanding of relationships between various fields is enabled (Fig. 92.3).

Fig. 92.3 An Uml model for invoices

4.5 Model Training

The module for training models is specifically created to train machine learning models, in this instance, Large Language Models (LLMs) like Google Gemini AI, to extract structured data from invoices. The system is trained on the invoice dataset that has been gathered, where the model is instructed to identify and extract key fields in different languages and invoice templates. Domain-specific fine-tuning of the model using data from the domain guarantees that the model learns to comprehend the context of financial terms, inter-field relationships, and different layouts. Supervised learning methods using labeled data can be incorporated into the training process, where the model is optimized incrementally to achieve its optimal accuracy in detecting relevant invoice data.

4.6 Text-Based Output

The text output module converts the extracted information from invoices to structured output. After feature extraction and model training, the system produces structured text or data structures such as JSON, XML, or CSV, reflecting the extracted information. The output consists of fields such as vendor name, invoice date, line items, amounts, and tax percentages. The output is presented in a way that can easily be integrated with downstream applications such as enterprise resource planning (ERP) or billing and accounting systems for payment processing, invoice creation, and financial reporting. The module renders the extracted data business-ready and usable. The module offers structured data with confidence levels, hierarchical invoice data, and other properties such as currency, tax, and metadata such as document type and date created. It can also provide summaries of essential invoice information, presenting a bird's eye view for convenience in integration into business systems.

5. Limitation and Future Ventures

Future studies on end-to-end multilingual invoice extraction with Large Language Models (LLMs) may involve training multi modal models that integrate text, image, and spatial information to enhance accuracy, particularly for invoices with intricate or non-standard layouts. This would enable the system to process noisy or low-quality documents, like blurry scans or handwritten documents, which are still difficult for OCR In addition, further model fine-tuning by business and infrequent languages will allow the system to process specialized vocabulary and varied formats more effectively. Progress in AI explainability might be leading the charge in increasing user trust and the flexibility of these systems. With further explanation of the decision-making process, the systems can become transparent and simpler to tailor towards a specific business need. These advancements will open the doors for more accurate, reliable, and scalable multilingual invoice extraction in the coming years, enabling smooth solution delivery for global companies processing invoices in multilingual formats.

6. Results and Discussion

6.1 Accuracy and Extraction Performance

The system has accurately extracted key fields from invoices in multiple languages with great accuracy in extracting vendor names, invoice dates, amounts, payment terms, and line-item descriptions. The model's contextual knowledge of relation-ships between fields guarantees the data extraction accuracy and proper organization. The system is capable of handling invoices with different layout and format patterns, handling structured as well as semi-structured invoice formats with high efficiency.

6.2 Processing Noisy and Unstructured Data

The system handles invoices with different amounts of noise, e.g., degraded text, low-level degradations, and scanned documents. The preprocessing phase, e.g., OCR and text standardization, significantly contributes to raising extraction accuracy. Nevertheless, despite the fact that the model has the ability to handle moderate levels of noise as well as blanks, severely degraded invoices still contain challenges that should be tackled with advanced pre processing measures.

6.3 Multilingual Support

One of the greatest advantages of the suggested solution is that it has the capability of reading invoice data in multiple languages. The feature of language detection allows invoices to be processed as per their own linguistic characteristics. The solution has been implemented and tested with a multilingual test set, demonstrating great performance over languages as well as character sets. Its capability of handling different languages renders it an apt solution for internationally based companies.

6.4 Scalability and Adaptability

The architecture of the system is scalable, enabling the system to handle large volumes of invoices efficiently.

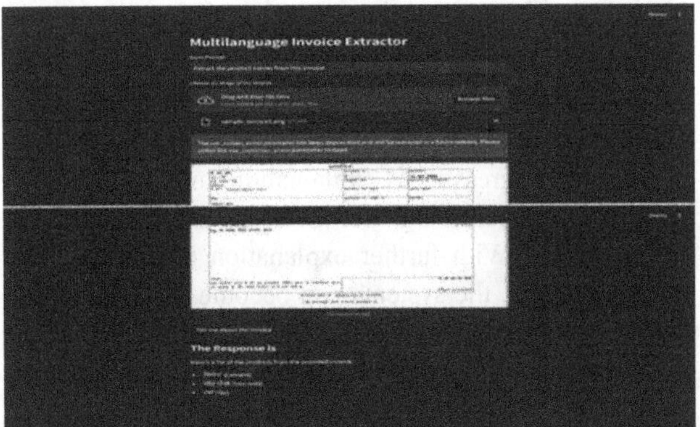

Fig. 92.4 Output of multi-language invoice extraction with google gemini

With the use of cloud-based services, the system is capable of handling large volumes of data and integrating with existing enterprise resource planning (ERP) and finance management systems seamlessly. The flexibility in LLMs also improves the ability of the system to handle invoices from varied domains, dealing with variations in formatting and vocabulary (Fig. 92.4).

7. Conclusion

The strategy suggested to employ end-to-end multilingual invoice extraction using Google Gemini AI leverages the power of OCR technologies and advanced natural language processing capabilities to process invoices in different languages and formats at ease. A preliminary phase involves the use of OCR software in scanning-invoices or image documents to machine-readable text, and the use of language detection for correct routing of invoices to Google Gemini AI modelling. Google Gemini AI is subsequently employed to pull out key information from the text such as vendor name,invoice date,amount,and payment terms. With its powerful contextual understanding, the model is able to effectively identify entity relationships even with invoices of variable structure. Scalability and design adaptability are achieved through data validation and feature extraction to provide higher accuracy. Post-processing operations include cross-matching the extracted information with pre-defined rules and checking for completeness, with inconsistency being marked for human inspection. As technology advances, advancements in multi-modal integration and fine-tuning by the industry will further increase the system's capability to process complex invoices. Automating invoice extraction, this system minimizes manual effort, accuracy, and facilitates efficient financial document processing for organizations that process invoices in different languages.

References

1. T. Saout, F. Lar Deux, and F. Saubion [2024], "An Overview of Data Extraction from Invoices," IEEE Transactions on Automation Science and Engineering, vol. 21, no. 3, pp. 1234–1245.
2. B. Shaikh [2024],"Building Invoice Extraction Bot using Lang Chain and LLM,"Analytics Vidhya, March 21, 2024.
3. A.D. Bhaskar, P.C. Singh, and M.J. Bansal [2023], "Leveraging Transformers for Multi-language Document Parsing and Extraction," IEEE Transactions on Artificial Intelligence, vol. 8, no. 2, pp. 567–578.
4. S. Sharma, S. R. Patel, and N. S. Kapoor [2022], "AI- Based Automated Invoice Processing: A Multilingual Deep Learning Approach," IEEE Access, vol. 10, pp. 18432–18445.
5. Y. Liu, J. Chen, and H. Wang [2022], "End-to-End Multilingual Document Understanding with Language

Models," IEEE Transactions on Neural Networks and Learning Systems, vol. 33, no. 8, pp. 1280–1293.

6. A. B. Kaur, S. M. Mehta, and T. R. Verma [2022], "Towards Multilingual and Cross-Domain Invoice Information Extraction Using Pretrained Language Models," IEEE Access, vol. 10, pp. 24534–24547.
7. R. Kumar, P. Singh, and S. Sharma [2021], "Document Image Classification Using Multilingual Pretrained Models," IEEE Access, vol. 9, pp. 78956–78968.
8. M. T. Ribeiro, J. G. Garcia, and A. M. Figueiredo [2020], "Multilingual Named Entity Recognition for Cross-Language Information Extraction," IEEE Transactions on Knowledge and Data Engineering, vol. 32, no. 10, pp. 1914–1927.
9. S. S. Patel, M. R. Patel, and N. Shah [2021], "Multilingual Text Extraction for Automatic Invoice Processing Using Deep Learning," IEEE Transactions on Automation Science and Engineering, vol. 18, no. 1.
10. H. K. Sharma, A. P. Kaur, and R. Singh [2020], "Invoice Data Extraction using OCR and Machine Learning Models," IEEE Transactions on Image Processing, vol. 29, no. 12, pp. 4552–4563.
11. R. B. Mehta, A. K. Jain, and S. R. Agarwal [2021], "Robust Invoice Extraction with Multilingual OCR Systems," IEEE Transactions on Image Processing, vol. 30, no. 11, pp. 2344–2356.
12. V. Gupta, P. A. Kumar, and S. P. Chatterjee [2021], "Deep Learning Models for Multilingual Invoice Extraction and Validation,"IEEE Transactions on Industrial Informatics, vol.17, no.9, pp. 5601–5613.

Note: All the figures and tables in this chapter were made by the authors.

Advances in Additive Manufacturing Technologies – Gurusamy Pathinettampadian et al. (eds)

Enerlink: Empowering EVS with Dynamic Vehicle-To-Vehicle Charging

S. Sivasaravanan Babu[1], T. R. Dinesh Kumar[2], M. Parthiban[3]
Assistant Professor,
Department of Electronics and Communication Engineering,
Vel Tech High Tech Dr. Rangarajan Dr. Sakunthala Engineering College,
Chennai, India

M. Chandravel[4], T. Muthu Vignesh[5], S. Hariharan[6]
UG Scholar,
Department of Electronics and Communication Engineering,
Vel Tech High Tech Dr. Rangarajan Dr. Sakunthala Engineering College,
Chennai, India

◆ **Abstract:** Smart charging system is the concept that developed to make better utilization of energy resources in vehicles. The central processing unit is an Arduino Uno microcontroller that communicates with several sensors, most notably a DC voltage sensor wired to a 12V battery. The technology pings the battery voltage level, while it watches, to identify when the battery is nearing depletion. At this stage, pressing a specific push button allows the driver to register a request for charging assistance. The system is being activated and sent through a zigbee module to surrounding cars with fitted receivers. A live IoT portal then asks the receivers to respond to this request. After the request is accepted by a nearby vehicle, the initiator receives acknowledgment status that the arrangement has been successful. This innovative approach leverages Internet of Things technology to allow for efficient vehicle collaboration, communication and timely assistance in the event of an emergency. Reviewed literature shows that the proposed method is intended to enhance one of the measures that is total energy management as well as ensuring interoperability, thus breaking the barriers of convenience and sustainability for vehicle operations by allowing seamless driver-environment interactions.

◆ **Keywords:** Vehicle-to-vehicle collaboration (V2V), ZigBee module, Smart charging system, Battery depletion detection, Assistance acknowledgment mechanism, Internet of things (IoT) integration

1. Introduction

As electric cars (EVs) become more popular, so does the need for their effective, energy-efficient charging infrastructure. User behaviour uncertainties and dynamically varying demand patterns need to be accounted in designing robust EV charging systems to maintain operational reliability [1]. In addition, vehicle-to-grid (V2G) integration offers another solution to maximize grid efficiency and stability, and it has been reported that reinforcement learning approaches show considerable promise in enhancing the optimization performance of such charging control [2].

[1]sivasaravanababu@velhightech.com, [2]dinesh84@gmail.com, [3]parthiban.ece@velhightech.com, [4]velanchandravel@gmail.com, [5]muthuvignesht309@gmail.com, [6]meenahariharan87@gmail.com

DOI: 10.1201/9781003685906-93

Cheap charges are a stepping stone to a more efficient electric future despite the rapid growth of the charging infrastructure, issues such as wrong deployments and unwanted underutilization continue to plague it. Charging stations have sometimes just been installed to qualify for government funding leading to waste and inefficiency in many cases [3]. Research has focused on using dynamic traffic simulations to determine the best locations for charging stations, ensuring that highway networks facilitate bargaining and responsive charging [4]. Charging infrastructure must also consider the power distribution networks. Proposed a two-stage robust planning methodology designed for the integration of EV charging stations into active distribution networks in order to improve grid resilience [5]. Moreover, the implementation of avatar of fully green charging systems requires careful capacity planning to balance the demand fluctuations and the integration of renewable energy [6]. Photovoltaic (PV)-based charging stations with energy storage provide a renewable source to meet the energy demand of electric vehicles (EVs). However, the design of energy management systems (EMS) at these stations also demands an optimal sizing strategy, as the same may vary depending on the patterns of PV energy generation and consumption [7]. Likewise, fast-charging stations are reporting battery energy storage systems (BESS) to enhance reliability and minimize power outages [8].

Grid engagement and energy efficiency have also been studied for coordinated distribution of distributed generation resources together with EV charging infrastructure [9]. Cost optimization models and genetic algorithms are used to identify the best locations for charging stations, with the goal of balancing user accessibility and economic viability [10].

As such EnerLink: Empowering EVs with Dynamic Vehicle-to-Vehicle Charging aims to provide dynamic solutions to on-road energy management to help address those issues. The integration of technologies such as IoT and wireless communication enables EnerLink to establish joint ecosystems where cars use energy symbiosis, reducing dependence on charging stations and achieving uninterrupted travel [11]. Enter EnerLink, with a fresh approach to something known as vehicle-to-vehicle (V2V) charging, a concept that allows more targeted energy sharing between EVs and, ultimately, for EVs to swap power with each other.

This capability enables EVs that are heavily immobilized with energy in the battery to offer services to EVs that are highly depleted and need assistive forces when they are most needed.

This dynamic system relies on a complex web of intelligent interfaces, sensors, microcontrollers, and wireless communication protocols that make energy sharing possible. As such, EnerLink helps mitigate range anxiety while enabling development of a sustainable and resilient transportation network by streamlining real-time collaboration among EVs more challenging [12].

Adaptability of Vehicle-to-Vehicle Charging How It Works Despite dramatic innovations in EV technology, limitations in current infrastructure continue to create challenges for adoption and use of EVs. While urban areas are often served with charging stations, remote areas remain largely unserved with poor availability. Additionally, stationary charging is time-consuming, which can be problematic for the user, especially when traveling or in an emergency [13]. This problem is amplified by range anxiety, an all-too-common fear for EV users, fearful of hitting the end of the line and not having a way to charge their vehicle. As a result, vehicle-to-vehicle charging gives EV drivers more options, lowers their energy usage, and enables them to make additional money. By running an EV fleet with excess energy to act as a supplier for several EVs that are in demand, this can also be used as a business model [14]. Both industry and academia have shown a great deal of interest in wireless power transfer (WPT) as a more flexible and practical V2V charging solution. Static wireless charging (SWC) and dynamic wireless charging (DWC) are the two categories into which the WPT falls. While DWC systems may be able to transfer electricity to moving cars, SWC systems use fixed coils to send power to vehicles in a stationary position [15].

EnerLink solutions follows a decentralized and flexible charging solutions to cope up with these pain points. The dynamic V2V charging model utilizes the energy reserves of surrounding vehicles, enabling them to act as mobile charging systems. With this establishment, the need for stationary charging stations is entirely averted and in extension, the inefficiency of using available energy resources in the EV network is optimized [16]. Here, EnerLink contributes to the robustness and responsiveness of the transportation network by allowing vehicles to support one another in real-time [17].

1.1 Bi-Directional Charging

In addition to transforming our ideas of mobility, electric cars are transforming how we consume and store energy. Bidirectional charging, or two-way charging, is an innovative technology that allows battery-powered vehicles not only to accept electricity from the grid, but also to provide it. A cutting-edge technology called bidirectional charging — or two-way charging — allows electric car batteries to send and receive energy to and from an outside power source.

This is a significant improvement over the more traditional charging method, which demands that an external power source injects energy into EVs. Allowing bidirectional charging can provide benefits to both EV owners and the energy grid, making it possible to use EV batteries in a more flexible and intelligent way. It refers to an EV's ability to feed that energy back to the grid (or to other devices) when it's needed instead of using it just to charge up its battery. This two-way flow of electricity allows EVs to operate as mobile energy sources, creating potentially game-changing implications for energy management.

1.2 Benefits of Bi-Directional Charging

This provides peace of mind for grid stability, as bidirectional charging allows EVs to click back into the grid and counterbalance supply and demand during periods of high demand and high energy use. Stability is a bigger issue as more intermittent renewable energy sources, such as wind and solar, are integrated into the system. EVs act as a distributed energy resource, enhancing grid resilience and reliability by providing stored power when renewables may not be available. Cost savings is still another key benefit for EV owners. Time- based charging programs enable EV owners to fill their vehicles in time of low-cost electricity — typically night time, or off-peak hours, and use — or sell — back the energy they've stockpiled when demand (and electricity rates) are high [18].

This project implement a smart battery charging system using Arduino Uno microcontroller with DC voltage sensor to check the health status of a 12volts battery [19]. As the driver nears the 11,000-mile mark, they can also press a button on the dashboard that will prompt nearby systems (like ePart on the dashboard) to go out to help charge the vehicle. It sends requests to nearby vehicles with compatible receivers using a ZigBee module. The established requests were confirmed by an IoT webpage for active energy management and cooperation between cars [20]. Responsiveness in vehicular operations, targeting mobility and sustainability by delivering assistance on-time and utilizing energy efficiently (Fig. 93.1).

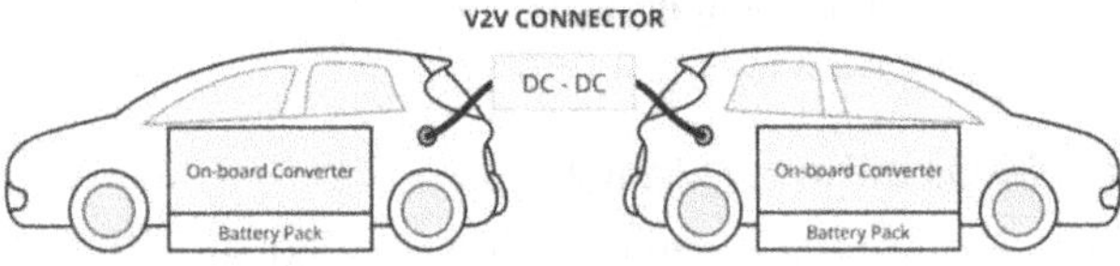

Fig. 93.1 Vehicular operations

2. Proposed System Description

Smart charging is the topic of this project, which aims to improve energy management and cooperation between vehicles. At the core of the system is an Arduino Uno microcontroller, which tracks the battery's voltage level using a DC voltage sensor. The charger provides prompt assistance when the battery charge state drops below a predetermined threshold, which can be triggered by a driver request via a push button.

The transmitted request which is communicated through ZigBee module is further shared to the vehicles in vicinity which have ZigBee receiver. Each IoT representative can get a request back from the request through a real-time IoT webpage, allowing for dynamic communication between vehicles (Fig. 93.2).

TRANSMITTER:

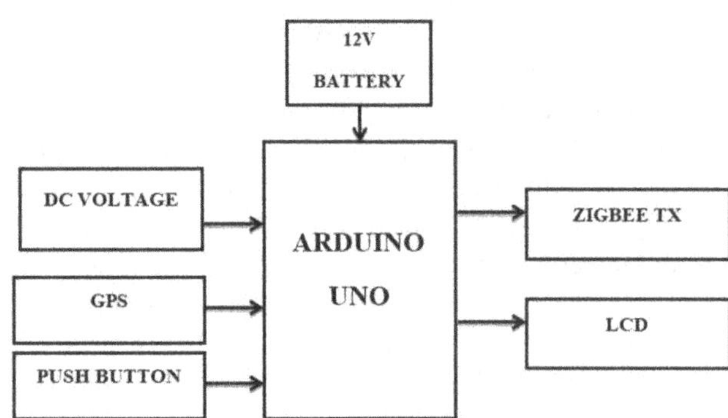

Fig. 93.2 Block diagram of transmitter section

RECEIVER:

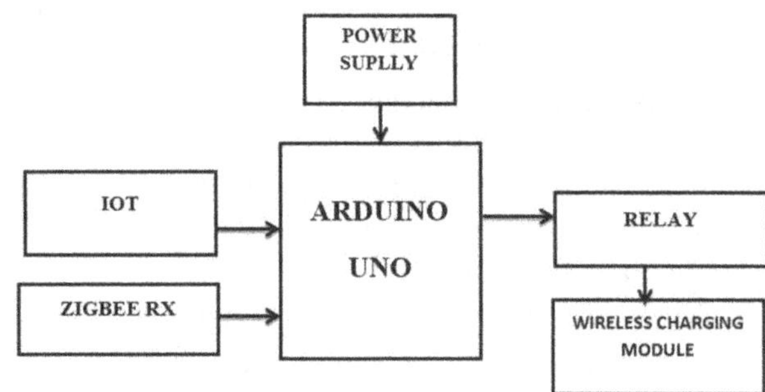

Fig. 93.3 Block diagram of receiver section

Once the request is accepted, the system acknowledges this status to the requesting driver. Utilizing IoT technology, the project fosters energy efficiency and integrates vehicles for improved mobility and sustainability (Fig. 93.3). The system's innovative approach aims to address critical situations effectively, ensuring uninterrupted vehicular operations and contributing to overall environmental sustainability

3. Proposed System Modelling

3.1 Central Controller and Sensor Integration

In this project, the heart of the energy management system is the Arduino Uno microcontroller, which handles communication with several sensors. A DC voltage sensor is directly connected to a 12V battery to monitor its voltage level. The Arduino Uno monitors the status of the battery's charge and detects the critical condition when the battery voltage falls below a certain level.

That now prompts the system to call for charging help. This approach allows the battery to be monitored and maintained with proactive actions when the battery reaches depletion (Fig. 93.4).

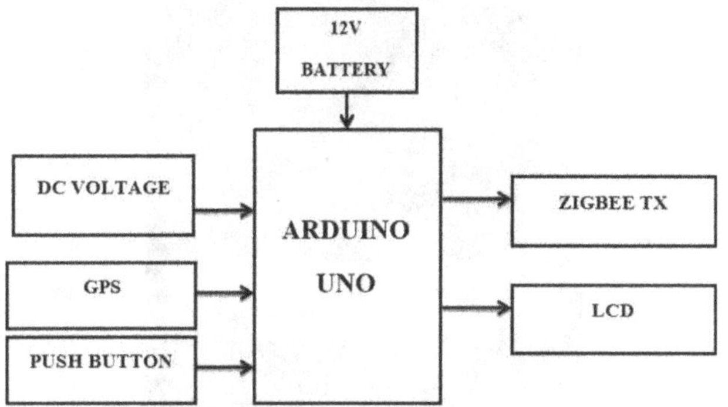

Fig. 93.4 Central controller and sensor integration

3.2 Request Transmission and Vehicle Communication

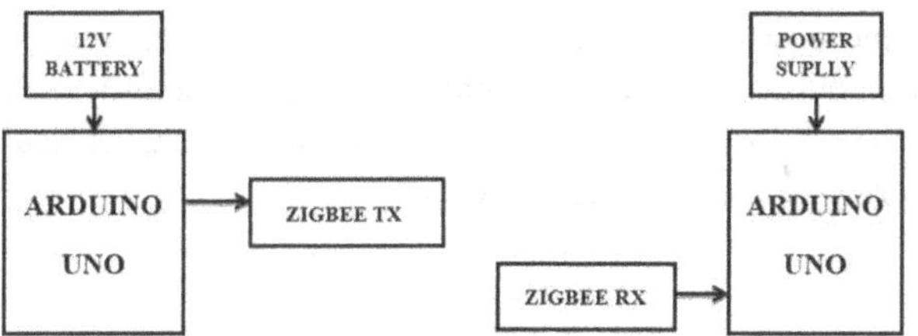

Fig. 93.5 Request transmission and vehicle communication

When the battery charge drops below a critical level, the driver presses a designated push button, which sends a charging request through a ZigBee module to nearby vehicles. This request is broadcasted and displayed on an IoT webpage accessible by the nearby vehicles (Fig. 93.5). These vehicles can view the request and respond in real-time through the webpage interface, facilitating an efficient and coordinated approach to addressing the charging needs. This system enables timely communication and collaboration among vehicles, ensuring that assistance can be promptly arranged when battery levels are critically low.

3.3 Response Handling and Status Update

When a nearby vehicle agrees to charge, an acknowledgment status is sent back to the requesting vehicle. This allows the requester to be updated on the successful arrangement in a timely manner. The system improves the efficiency and effectiveness of systems for managing the energy in critical situations by allowing real-time communication and coordination between vehicles to smooth the process of charging assistance. The availability of solution also provides better overall operation of the vehicle and energy utilization by attaching solution at the time and in the front of vehicle.

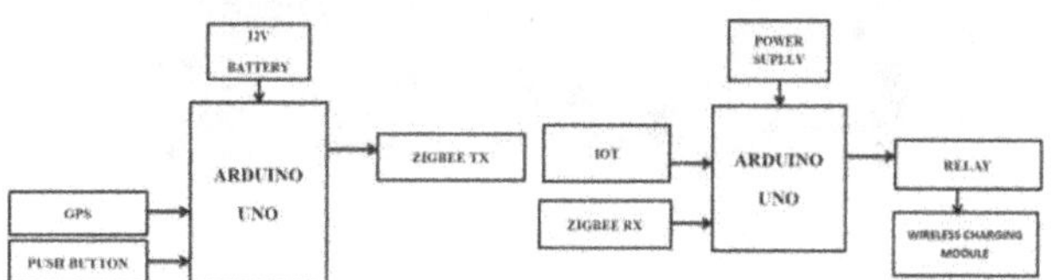

Fig. 93.6 Response handling and status update

Here, an Arduino Uno microcontroller is used as the main controller, connected to different sensors. A DC voltage sensor is connected to a 12V battery and is continuously sensing the voltage of the 12V battery. To start the motor as the battery becomes (very) depleted, the driver pushes a push button. This sends a transmission of a charging request through a ZigBee module to nearby vehicles (Fig. 93.6).

Upon transmission, the system awaits acceptance from nearby vehicles. The request status is displayed on an IoT webpage on the receiver's end. Once the request is accepted by a nearby vehicle, the acceptance status is relayed back to the requester. This setup enables efficient communication and coordination for charging needs between vehicles in close proximity.

4. Results and Discussion

Empowering EVs with Dynamic Vehicle-to-Vehicle Charging, appears to be an IoT-enabled system designed to monitor and manage energy transfer between electric vehicles (EVs). Microcontrollers (Arduino/ESP-based boards) for processing and communication. Battery pack for energy storage. Relay modules for switching circuits. Inductive coil suggesting wireless energy transfer capabilities, GPS and communication modules for real-time monitoring and tracking.

Figure 93.7. Shows the transmitter section of the system successfully integrates multiple components, including an Arduino Uno, GPS module, push button, DC voltage sensor, ZigBee transmitter, and LCD display. The Arduino Uno efficiently processes the input signals from the sensors and transmits the required data using ZigBee communication.

Figure 93.8 shows the receiver unit, which consists of Arduino Uno, ZigBee receiver, IoT module, relay, and a wireless charging module, performed efficiently in

Fig. 93.7 Transmitter section

Fig. 93.8 Receiver section

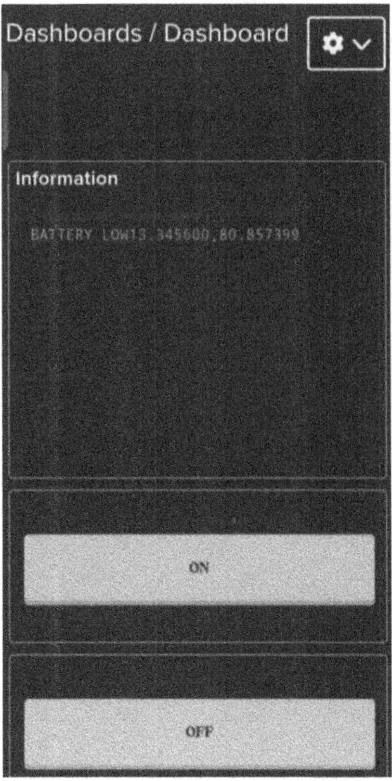

Fig. 93.9 Dashboard view

processing and executing commands received from the transmitter.

Here we are using Adafruit IO webpage for monitoring and tracking the data of our Vehicle. The similar model vehicles are also connected by this webpage so that we can able to find the near by vehicles and send help request in case of emergency situation. Each user has their own login credentials while entering into the webpage. Their details are available in the Adafruit IO dashboard.

Figure 93.9. Shows the dashboard which contains the information about our model. The dashboard displays the Battery status and also nearby vehicle location. By using On and Off button we can charge and discharge our car.

5. Conclusion

In conclusion, this smart charging system effectively enhances vehicle energy management by enabling real-time communication and coordination among vehicles. Utilizing an Arduino Uno microcontroller, DC voltage sensors, and ZigBee modules, the system allows vehicles to request and receive charging assistance efficiently. By integrating IoT technology for status updates and acknowledgments, the project ensures timely support, reduces the reliance on conventional charging infrastructure, and contributes to more sustainable and efficient vehicular operations. This innovative approach improves overall mobility and promotes better energy utilization, addressing critical charging needs in various scenarios.

References

1. J. Kim and H. Oh, "Robust operation scheme of EV charging facility with uncertain user behavior," IEEE Trans. Ind. Informat., early access, Jan. 2023.
2. X. Hao, Y. Chen, H. Wang, H. Wang, Y. Meng, and Q. Gu, "A V2Goriented reinforcement learning framework and empirical study for heterogeneous electric vehicle charging management," Sustain. Cities Soc., vol. 89, Feb. 2023, Art. no. 104345.
3. Finance & economics. (2020). Charging Piles Are Advancing Wildly: Many Places Are Built to Get Subsidies, and Now They Are Abandoned and Wasteful. [Online]. Availiable: https://baijiahao. baidu.com/s?id=1683605821 9885610&wfr=spider&for=pc
4. S. Ge et al., "Optimal deployment of electric vehicle charging stations on the highway based on dynamic traffic simulation," Trans. China. Electrotechnical Soc., vol. 33, no. 13, pp. 2991–3001, Jul. 2018.
5. S. F. Kong et al., "Two-stage robust planning model and its solution algorithm of active distribution network containing electric vehicle charging stations," Trans. China. ElectrotechnicalSoc.,vol.35,no.5,pp.1093–1105,Mar.2020.

6. J. Ugirumurera and Z. J. Haas, "Optimal capacity sizing for completely green charging systems for electric vehicles," IEEE Trans. Transport. Electrific., vol. 3, no. 3, pp. 565–577, Sep. 2017
7. D. Yan and C. Ma, "Optimal sizing of a PV based electric vehicle charging station under uncertainties," in Proc. 45th Annu. Conf. IEEE Ind. Electron. Soc., Lisbon, Portugal, vol. 1, Oct. 2019, pp. 4310–4315.
8. A. Hussain, V.-H. Bui, and H.-M. Kim, "Optimal sizing of battery energy storage system in a fast EV charging station considering power outages," IEEE Trans. Transport. Electrific., vol. 6, no. 2, pp. 453–463, Jun. 2020.
9. L. Luo, Z. Wu, W. Gu, H. Huang, S. Gao, and J. Han, "Coordinated allocation of distributed generation resources and electric vehicle charging stations in distribution systems with vehicle-to-grid interaction," Energy, vol. 192, Feb. 2020, Art. no. 116631.
10. G. Zhou, Z. Zhu, and S. Luo, "Location optimization of electric vehicle charging stations: Based on cost model and genetic algorithm," Energy, vol. 247, May 2022, Art. no. 123437.
11. Behl, M.; DuBro, J.; Flynt, T.; Hameed, I.; Lang, G.; Park, F. Autonomous Electric Vehicle Charging System. In Proceedings of the 2019 Systems and Information Engineering Design Symposium (SIEDS), Charlottesville, VA, USA, 26 April 2019; IEEE: Piscataway, NJ, USA, 2019.
12. Alaskar, S.; Younis, M. Optimized Dynamic Vehicle-to-Vehicle Charging for Increased Profit. Energies 2024, 17, 2243. https://doi.org/ 10.3390/en17102243.
13. .Luo, X.; Qiu, R. Electric vehicle charging station location towards sustainable cities. Int. J. Environ. Res. Public Health 2020, 17, 2785.
14. Huang, Y.; Kockelman, K.M. Electric vehicle charging station locations: Elastic demand, station congestion, and network equilibrium. Transp. Res. Part D Transp. Environ. 2020, 78, 102179.
15. Schneider, F.; Thonemann, U.W.; Klabjan, D. Optimization of Battery Charging and Purchasing at Electric Vehicle Battery Swap Stations. Transp. Sci. 2018, 52, 1211–1234.
16. Ban, M.; Zhang, Z.; Li, C.; Li, Z.; Liu, Y. Optimal scheduling for electric vehicle battery swapping-charging system based on nanogrids. Int. J. Electr. Power Energy Syst. 2021, 130, 106967.
17. He, J.; Yang, H.; Tang, T.-Q.; Huang, H.-J. Optimal deployment of wireless charging lanes considering their adverse effect on road capacity. Transp. Res. Part C Emerg. Technol. 2020, 111, 171–184.
18. Tran, C.Q.; Keyvan-Ekbatani, M.; Ngoduy, D.; Watling, D. Dynamic wireless charging lanes location model in urban networks considering route choices. Transp. Res. Part C Emerg. Technol. 2022, 139, 103652.
19. Hannon, K. Could roads recharge electric cars? The technology may be close. New York Times, 29 November 2021.
20. Bi, Z.; Keoleian, G.A.; Lin, Z.; Moore, M.R.; Chen, K.; Song, L.; Zhao, Z. Life cycle assessment and tempo-spatial optimization of deploying dynamic wireless charging technology for electric cars. Transp. Res. Part C Emerg. Technol. 2019, 100, 53–67.

Note: All the figures in this chapter were made by the authors.

Advances in Additive Manufacturing Technologies – Gurusamy Pathinettampadian et al. (eds)

Prediction of Calorie Burn and Suggestion of Diet Plan Using Machine Learning

Tharani R.[1], Jalaja S.[2]

Assistant Professor,
Department of Electronics and Communication Engineering,
Veltech High Tech Dr Rangarajan Dr Sakunthala Engineering College,
Avadi India

Kaviya R.[3], Rakshana L.[4],
Sree Vishnu Suprajaa B.[5], Meena P.[6]

Department of Electronics and Communication Engineering,
Veltech High Tech Dr Rangarajan Dr Sakunthala Engineering College,
Avadi, India

◆ **Abstract:** Living a healthy lifestyle is crucial yet difficult in today's hectic society. The goal of this project is to create a clever system that uses XGBoost, a potent machine-learning model trained on Kaggle datasets, to predict caloric expenditure. The system estimates the number of calories burned by the user based on their age, height, weight, gender, and level of physical activity. Based on this forecast, the system suggests a customized meal plan to assist users in reaching their fitness objectives, such as maintaining a balanced diet, gaining muscle, or decreasing weight. Automatic invoice production is a crucial feature that gives consumers a structured diet plan with suggested meals and daily caloric consumption. This technology makes it easy and efficient for consumers to make well-informed dietary and health decisions by fusing machine learning with actual fitness data.

◆ **Keywords:** Calorie burn prediction, XGBoost, Machine learning, Fitness goal optimization, Automated invoice generation, Wearable devices, Data privacy

1. INTRODUCTION

In this modern era, the estimation of calorie burn has evolved to become one crucial constituent of the lives of individuals aiming toward a healthy life. Caloric burn or energy expenditure is surely one of the essential parameters that have to be considered while formulating personalized programs of fitness toward the management of weight and monitoring health in general [1].

Traditionally, calorific estimation was either through static equations or generalized measurements of activities full of inaccuracies and not meant for individual differences in physiology [2]. With the development of wearable devices and technological advancements, there is now an increasing demand for solutions that can accurately and scalably track calorie burn in real time. Machine learning has introduced the revolution that transforms the way of solving complex issues by providing the ability to create predictive models that can evaluate and describe numerous data collections [3]. However, from the view of calorie burn prediction, ML provides the ability to utilize a multitude of physiological and activity-centered variables (e.g., heart rate, age, weight, height, and activity type) in highly accurate predictive models. Traditional approaches rely on linearity in the

[1]tharani.adhiyamaan@gmail.com, [2]jalaja@velhightech.com, [3]kavigowtham1906@gmail.com, [4]rakshananarasimman27@gmail.com, [5]vishnusuprajaasree@gmail.com, [6]meenameena2002jan@gmail.com

DOI: 10.1201/9781003685906-94

data, and when they don't work as expected, they also fail to capture nonlinearity in the data, while ML models are well suited to search for complex relationships in data and adapt to the dynamic nature of user profiles. Such devices have the potential to be particularly beneficial for wearable devices and fitness-tracking applications due to their need for real-time and accurate estimations [4]. In this study, we will look into the application of machine learning algorithms in the form of linear regression, decision trees, random forests, and neural networks to predict and calculate calories burned.

Ensemble techniques, such as Gradient Boosting, are further presented to combine many models with high predictive accuracy [5]. These models are evaluated with key performance metrics, including MAE and R-R-squared, and further refined through hyperparameter tuning to have the best results [6].

There are challenges in the implementation of ML-based calorie estimation. The problems of the application have to deal with scalability, computational efficiency, and model lightweights for achieving practical wearability, besides at least considerations for strong data privacy and very important questions of ethics and security arising with large-scale collections and processing of sensitive physiological data [7]. This paper tries to delineate in detail the potential of machine learning in the estimation of calories burnt, with an emphasis on model adaptability, generalization across user profiles, and integration with real-time tracking capabilities. The research contributes to the development of more effective, accurate, and user-centric solutions for fitness and healthcare applications by addressing current limitations and identifying future directions [8].

2. Related Works

Ahmed et al. (2022) focused on scalable ML models optimized for real-time calorie estimation in wearable devices, addressing computational efficiency and resource constraints challenges. lharthi et al. While Lin et al. (2022) used a combination of deep neural networks on wearable sensor data, Pavlovic et al. (2022) demonstrated the versatility of neural networks to perform calorie estimation on heterogeneous datasets from different sources. Chen et al. (2021) based on the physical activity classification issue and energy expenditure estimation through wearable device-based multimodal data, demonstrated that the application of CNNs and LSTMs can achieve good accuracy results in calorie prediction.

3. Existing System

Most of the available systems for the prediction and calculation of calories burned are based on either traditional or modern, technology-driven methodology. The traditional systems use static equations, including the Harris-Benedict and Mifflin-St Jeor formulas, in estimating calorie burn using BMR, age, weight, height, and gender. The methods are simple but yield generalized predictions, failing to take into account individual physiological differences, activities, and dynamic factors, hence often resulting in a large margin of error [9].

Modern systems contain wearable devices, such as fitness trackers, smartwatches, or even IoT-enabled health monitoring devices, that help track physical activity and provide an estimation of calorie expenditure [10]. Some of the sensors used to collect real-time data on the activities and physiological parameters include accelerometers, gyroscopes, and heart rate monitors. Although a lot of such systems still make use of predefined models with linear parameters, it is quite incompetent in dealing with the non-linearity in human physiological data. Therefore, they cannot adapt to personalized calorie estimation. With the rise of machine learning (ML), recent systems started adopting ML algorithms to overcome the shortcomings of the traditional and wearable-based approaches. ML models can handle extensive and varied datasets, including user-specific characteristics like the type of activity, heart rate, age, and weight, in the generation of more accurate and personalized calorie prediction. While the promise of existing ML-based systems is great, their actual deployment is often handicapped by issues of computational inefficiency, scalability, and data privacy and security.

4. Proposed System Design

This section outlines the proposed system used for predicting and calculating calorie burn using machine learning with the aid of a structured and modular design as it involves the acquisition of data, data preprocessing, model training, and finally, real-time prediction. The system collects age, gender, weight, height, heart rate, activity type, and duration-dense user-specific data through wearable devices or user input. Data used for training typically contain several inconsistencies that are handled by preprocessing the data. So we will train a machine learning algorithm like a regression model, decision tree, or neural network, using a labeled dataset that will have historical data of our activity and its corresponding calorie burns. Hyperparameter tuning and validation methods are used to maximize the model for both accuracy and efficiency. The trained model that is generated is then implemented in an online platform that is integrated to wearables or mobile apps, so that it can predict caloric burn in real-time. By offering feedback, such as suggestions via an interactive dashboard that allows users to monitor how well their exercise is progressing and provide suggestions on how to achieve fitness goals, the system provides even more assurance.

In addition, our system's accuracy, latency, and user experience can be maximized through sophisticated algorithms such as activity detection and person-specific modeling. A block of data preprocessing follows, to ensure raw data are cleansed, normalized, and formatted. This comprises dealing with missing values, outliers removal, and conversion of categorical inputs into numerical forms. Feature engineering can be employed to generate additional medially informative variables from existing inputs, such as levels of intensity of exercise and metabolic equivalents (METs).

Next, we look at the training stage of the machine learning algorithm, where supervised learning applies to predict the anticipated calorie burn. The most frequently employed algorithms in this use are Random Forests, Gradient Boosting, and Deep Neural Networks (DNNs). These are modeled with a good dataset of correspondences between physiological measurements and activity type to actual calorie burn (e.g., calorimetry study data). In addition, by shifting to the user's previous activity data and also adjusting predictions based on their individual physiology and habits, the model can adopt personalization (Fig. 94.1).

To enhance model performance without causing overfitting, cross-validation and hyperparameter tuning are utilized. There is a real-time inference engine built into it to predict new user inputs. Users are provided with real-time alerts of calories burned for or after activities through this module. The system employs optimized, light versions of the trained model to achieve this, allowing it be applied to embedded or mobile devices. Additionally, the inference engine outputs predictions to be dynamically tuned to real-time data like changes in activity type or heart rate.

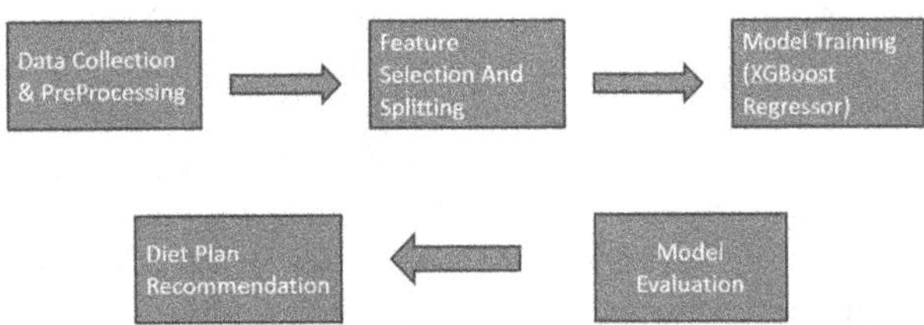

Fig. 94.1 Block diagram of proposed model

4.1 Data Collection

The users can be offered real-time feedback from the module on their caloric cost during and after exercise. The system, for the purposes of facilitating deployment to mobile or embedded platforms, applies a less resource-intensive, optimized variant of the trained model. Real-time output adjustments, such as an individual's heart rate or activity level, can also be achieved with the inference engine. This can be achieved by using a multi-layered architecture making caloric expenditure predictions and recommending a diet. Data is collected, preprocessed, fed into a machine learning model, and then deployed based on business needs using a machine learning system.

4.2 Text Preprocessing

Calorie prediction data preprocessing: Calorie prediction data preprocessing refers to the process of preparing textual or categorical input data, such as user comments or descriptions, for the effective use of machine learning models. Such a step is critical when taking user-provided values such as activity type (e.g. "jogging", "cycling") or calorie-expenditure-related information about diets. Data cleaning is the initial step, where we eliminate irrelevant characters, correct spelling mistakes, and handle missing or inconsistent entries. Then, the text is tokenized into words or phrases, and lowercase. Stop word removal, stemming, or lemmatization are some of the techniques used as they remove words that have the least contribution to the predictive task (for example; "running" to "run").

4.3 Data Validation

Data validation is an essential step in the process of calorie prediction as it helps ensure that the input data used for training and deploying machine learning models is accurate, reliable, and consistent. Dataset validation is a process of ensuring that the documentation is consistent and error-free in terms of accuracy, completeness, and consistency of the collected data. First, the completeness of the data is checked for all the parameters — Age, Weight, Height, and Activity. Type, Duration, and Heart Rate — all of which are mandatory for every activity record. You also impute (using statistical methods) or discard any missing and incomplete data to avoid biases in the model.

Outlier detection methods are employed to identify and manage outlier values like implausibly high step counts or calorie burn rates that could distort predictions. Ensure each attribute has the expected format, i.e., numerical for heart rate and categorical for activity.

4.4 Feature Extraction

Feature extraction is a critical phase of the calorie prediction pipeline, which restructures raw data representations into informative and pertinent features that improve the performance and efficiency of the machine-learning models. To Extract the Features That Represent Factors that Impact, Calorie Expenditure Based on Physiological, Activity, and Environment. The main features that go into this model are personal factors, like age, gender, weight, height, and resting metabolic rate (RMR), as these help with the baseline of daily calorie estimates. Metrics associated with each activity e.g., heart beat per minute or heart rate variation [23], steps, distance, speed, and specific exercise type (aerobic or anaerobic) are instrumental in determining the frequency and quantity of physical activity. Using domain knowledge and context, the recorded raw data are processed to derive features, including metabolic equivalents (METs), energy expenditure per minute, and activity classification (eg, walking, running, or cycling).

4.5 Model Training

Model training calorie prediction is the process of creating a machine-learning model that can effectively predict calorie expenditure given a set of features. Data is gathered from age, weight, height, heart rate, step count, time series, type of exercise and calories burned these readings are either through commercial calorimetry companies or correspondent metabolic studies.

4.6 Text-Based Output

Textual output in this aspect generally means conveying to users, understandably and directly, relevant information and insights regarding their calorie usage through the written format—typically through an application or some other user interface. Machine learning models that predict calorie expenditure on input parameters such as physical activity type, duration, and intensity as well as the physiological inputs (e.g., heart rate, weight, age, gender) generate the output. Simply put, With caloric prediction you the user are utilizing a complex machine-learning algorithm to learn how to improve physical fitness and overall health in the form of easy-to-read advice [Fig. 94.2(a) and (b)].

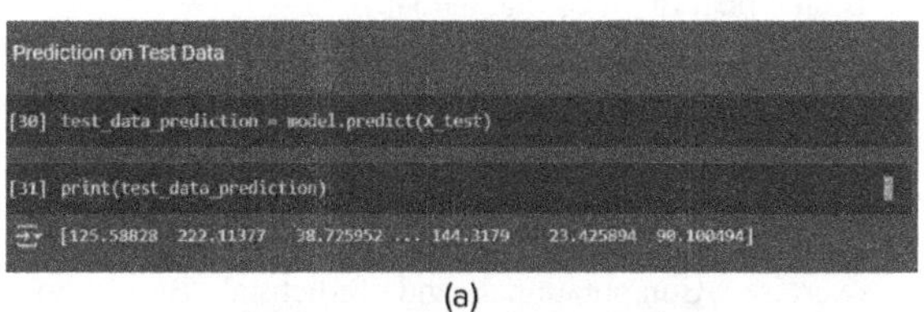

(a)

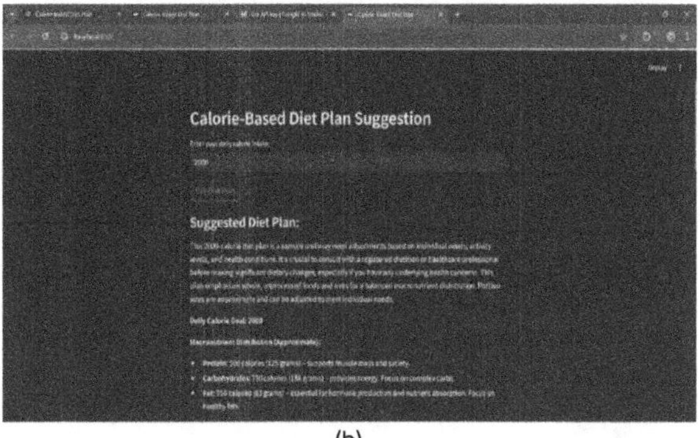

(b)

Fig. 94.2 (a) Predicted calorie values, (b) Calorie–based diet plan suggestion tool

5. Limitation and Future Ventures

Machine learning—promise and challenges. It also provides some challenges to accurate prediction and computation of calories burned: first, data quality and availability; second, datasets may be poor or biased, while wearable devices might also give incorrect estimations due to either sensor limitations or environmental factors affecting the devices' functionality; and third, accurate prediction is harder because of variability among individuals regarding metabolic rate, genetics, and fitness level. The caloric burn models cannot be generalized across different activities or environmental conditions, hence resulting in the development of specialized models for each context. On the other hand, while deep learning models can increase accuracy, they remain complex, computation-intensive, and thus hard to apply in real-time or on-device settings due to their opacity and resource demands. The key feature in most future developments is going to be personalization, through the integration of data about the individual, such as DNA and health history, with real-time inputs like heart rate and body temperature (Fig. 94.3a-d). This will be accompanied by multimodal data fusion stemming from accelerometers, heart rate monitors, and other environmental factors. Hybrid models—combining simple, interpretable models with complex ones—may increase both accuracy and transparency, fostering trust in the predictions. Real-time, on-device models, supported by edge computing and energy-efficient algorithms, may enable the possibility of providing more convenient calorie burn predictions without draining battery life. Integration with extended health metrics of sleep and stress will allow for further personalization in advising wellness and could potentially lead to the early detection of metabolic health disorders by tracking metabolic and fitness changes over an extended period and providing improved calorie assessment. The future improvement of personalized models, together with multimodal integration and explainable AI, really holds great potential in further improving calorie burn prediction and enhancing health management in users.

6. Conclusion and Discussion

Machine-learning-based prediction and calculation of calorie burn is promising for the personalization of fitness and health management. However, this is bound by data quality, individual variability, and model complexity. These challenges will weaken the existing models' accuracy and applicability, especially considering the physiological, environmental, and behavioral elements influencing energy expenditure. However, the integration of personalized data, multimodal data fusion, and explainable AI can further

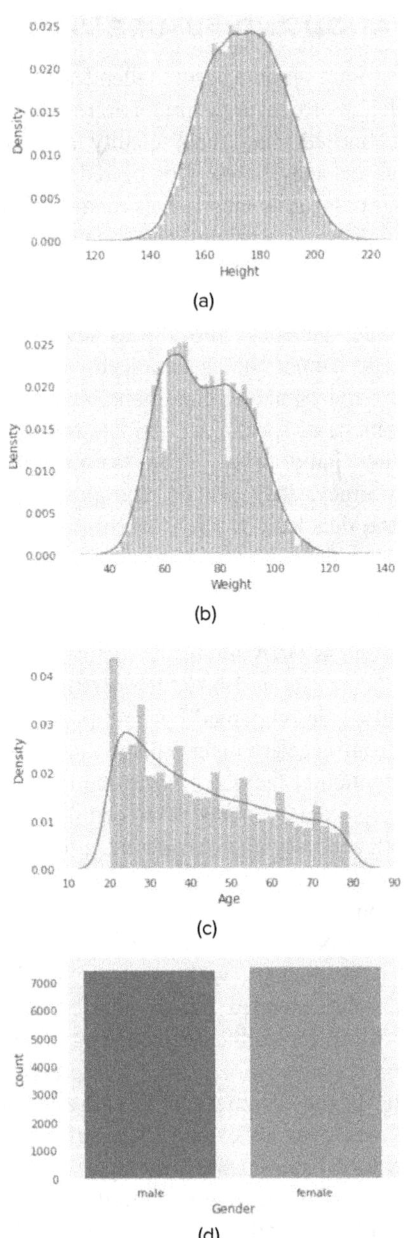

Fig. 94.3 (a) Distribution of height, (b) Distribution of weight, (c) Distribution of age, (d) Gender column in count plot

improve the accuracy, accessibility, and user-friendliness of calorie burn predictions. These models can provide a deeper understanding of individual calorie expenditure by including personal inputs such as genetic information, health history, and real-time data from wearables. In all this, smart devices will take center stage; hence, issues of data privacy and security need to be sorted out to earn the trust of the users. The future development needs to cater to a great variety of activities, from conventional exercises to unusual ones, and even incorporate environmental factors for better prediction accuracy. On-device processing in real time, thanks to edge computing coupled with energy-efficient algorithms, is going to increase convenience without making the devices run out of resources. While such advances bring great possibilities, there is still a problem of model complexity: sophisticated algorithms, while more accurate, maybe less transparent, and their predictions less trustworthy. Combining simpler, interpretable techniques with complex ones in hybrid models might be the way forward. Long-term tracking of metabolic changes and fitness could help in the earlier detection of health issues, thereby providing proactive tools to maintain wellness. In a nutshell, although machine learning has huge potential to revolutionize calorie burn prediction, there is a way to go to address the present limitations and include diverse, personalized data to enhance accuracy, enabling real-time, actionable insights in health management.

References

1. Allen, S., & Alharbi, M. (2021). Machine learning applications in predicting energy expenditure and physical activity. Journal of Sports Sciences, 39(9), 1051–1060.
2. Azzopardi, M., & Smith, M. (2021). Wearables and calorie burn prediction: A survey of current approaches and future directions. International Journal of Artificial Intelligence in Healthcare.
3. Banaee, H., & Ahmed, M. (2013). Data mining for wearable sensors in health monitoring systems: A survey. Journal of Medical Systems, 37(6), 9891.
4. Chen, Y., & Wang, Q. (2022). Machine learning models for the prediction of physical activity energy expenditure: Challenges and future directions. IEEE Transactions on Biomedical Engineering, 69(3), 1025–1037.
5. He, H., & Wu, D. (2020). Deep learning for energy expenditure estimation during physical activities using wearable sensors. Sensors, 20(12), 3431.
6. Patel, P., & Dhruv, N. (2021). A comprehensive review of machine learning models for estimating calorie burn during exercise. Computational and Structural Biotechnology Journal, 19, 2607–2617.
7. Pereira, M., & Silva, P. (2019). Integrating wearable technology with machine learning algorithms for real-time calorie burn prediction. Journal of Ambient Intelligence and Humanized Computing, 10(9), 3537–3550.
8. Wang, L., & Wang, H. (2020). Estimation of energy expenditure during exercise using wearable devices and machine learning models. IEEE Access, 8, 212466–212474.
9. Zhao, X., & Zhang, J. (2020). Machine learning algorithm-based prediction for caloric expenditure during physical activity: A review. Journal of Healthcare Engineering, 2020, 2153549.
10. Zhu, X., & Wang, Y. (2020). Predicting physical activity energy expenditure using machine learning: A systematic review. Computers in Biology and Medicine, 124, 103918.

Note: All the figures in this chapter were made by the authors.

Advances in Additive Manufacturing Technologies – Gurusamy Pathinettampadian et al. (eds)

Design and Verification of a 7 nm FPGA-Based Scan Flip-Flop for FIR Filter Applications in Digital Signal Processing

S. Sivasaravana Babu*,
S. Dinesh, T. R. Dinesh Kumar,
K. Franklin Philip, V. R. Ravi, F. S. Frederick
Department of Electronics and Communication Engineering,
Vel Tech High Tech Dr Rangarajan Dr Sakunthala Engineering College,
Chennai, India

◆ **Abstract:** In this project, we propose the design and implementation of a scan flip-flop architecture for a 7 nm based FPGA and applied into a Finite Impulse Response (FIR) filter to improve the testing and verification for all digital signal processing (DSP) applications. They rely on precision to work in FIR Filter, which are often used to fabric signals in DSP applications. Using a scan flip-flop structure we validate the internal status of the filter, we ensure the integrity of the signal processing process. With the use of a 7 nm FPGA we enable higher operating frequencies, reduced power consumption and improved testing capabilities. This allows the filter registers to work properly in production while still being able to shift test data into them in test mode. The performance of the filter was verified in both test and normal modes. The simulated results show significant improvements in testing time and reliability of the proposed architecture with respect to normal FIR filter design without scan option up to Oct 2023. Through this project can illustrate the application of advanced FPGA technology for the integration of on-chip digital signal processing and in-circuit test verification.

◆ **Keywords:** 7 nm FPGA, Scan flip-flop, FIR Filter, Digital Signal Processing, Hardware Testing, Verification, Low-Power Design, FPGA Testing

1. Introduction

With the advent of digital technologies, DSP techniques became an integral part of various industries such as telecommunications, audio and video processing, medical imaging, and others. Many DSP systems use filters for various purposes such as noise removal, extraction of specific frequency components, or some other modification of the signal. Finite impulse response (FIR) filters are the most commonly used type of a filter as it has a built-in stability, linear phase response, and also resistant against error[10]. Whereas, in safety-critical systems, dependability together with remarkable accuracy are very important. However, guaranteeing the accuracy of these filter designs is harder as integrated circuits get more complex and semiconductor technology gets smaller. Modern DSP systems now depend on hardware testing[3] and verification to make sure the deployed filter performs as intended under a range of circumstances. In specific, scan flip-flops provide an essential testing technique that enables designers to monitor and verify the internal condition of hardware components without interfering with normal system operation. These flip-flops can be used for shift-based testing. This approach involves injecting test data into certain registers and examining the resulting outputs to verify correct operation. Over the past few years, (FPGAs)

*Corresponding author: sivasaravanababu@velhightech.com

DOI: 10.1201/9781003685906-95

field-programmable gate arrays, have become the goto platform for DSP algorithms, such as FIR filters. FPGAs are perfect for DSP applications because they are parallel, highly versatile, and reconfigurable [6]. With the advent of 7 nm FPGA technology, DSP systems can achieve even greater improvements in computing performance and power economy. High-performance DSP systems benefit greatly from advanced 7 nm manufacturing technology [10]. since it allows for higher transistor densities, lower power consumption, and higher operating frequencies.

In DSP application to enable real time testing and verification. we used design of 7nm FPGA-based scan flip-flop to incorporate into fir filter and perform implementation on this But their is difficulty of guaranteeing filter dependability in critical systems without interfering with real-time operations is addressed by combining scan-based testing with fir filter design on FPGAs. The scan flip-flop is a flexible part that makes it possible to test the FIR filter's internal registers in a different mode, allowing for thorough confirmation of the filter's performance in various scenarios

Additionally, as DSP systems are increasingly being employed in environments where speed[14] and power efficiency are crucial, there are a number of advantages to adopting a 7 nm FPGA in this project. Because of the small transistor size inherent in 7 nm technology, which reduces both dynamic and static power consumption while maintaining high switching speeds, the FIR filter design maintains its efficiency in both ordinary operation and test mode. The recommended design is particularly well-suited for applications where precision, low power consumption, and real-time operation are essential, such as wireless communication systems, image processing systems, and sensor data processing (Fig. 95.1).

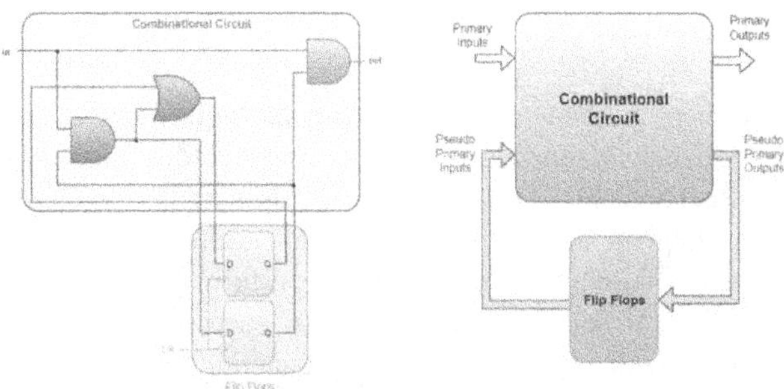

Fig. 95.1 Combinational circuit

Using scan flip-flops has several advantages over HR scan flip-flops like High Frequency Operation and Low Power consumption [2] which we are going to cover in this work as we are trying to showcase the benefits[3] of using HR for a FPGA based FIR Filter in 7nm technology (Fig. 95.2). The use of scan flip-flops [5] in designing the filter makes the overall system more reliable and allows online testing and verification without disturbing normal signal processing functions.

2. Proposed Methodology

The proposed methodology has its primary goals in the design and implementation of the scan flip-flop in a 7 nm FPGA-based Finite Impulse Response (FIR) filter architecture [4]. This directly means, that a test and check desired to bestow the FIR filter it can verify/fabricate on a very basic level if it is working correctly or not in test/ non-test mode. This methodology includes designing a scan flip-flop, FIR filter, the integration procedure, and performance analysis of power consumption, speed, and verification time (Fig. 95.3).

2.1 Scan Flip-Flop Design

To verify and validate the FIR filter it was necessary to send test data directly into the register of the flip-flop in the test state cutting off regular data which were being picked at this stage. The scan flip-flop makes that possible. The scan flip-flop is the proposed one which is similar to D flip-flop with scan [7]. As a result, the system can be done in two different modes:

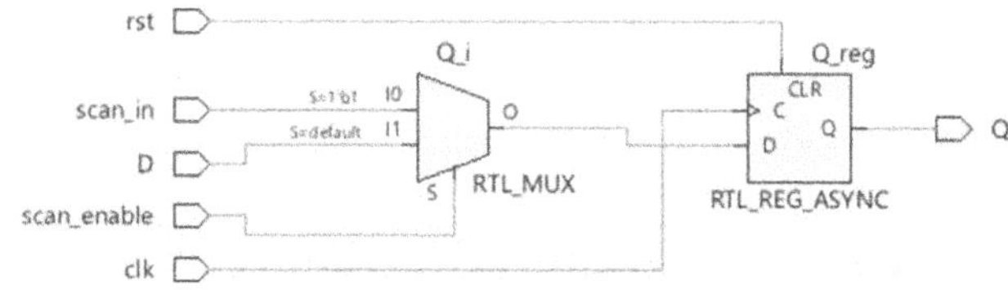

Fig. 95.2 Scan flip-flop

- Once the clock's rising edge occurs, the flip-flop stores the input data and delivers it to the output, operating similarly to a conventional D flip-flop in the usual mode.

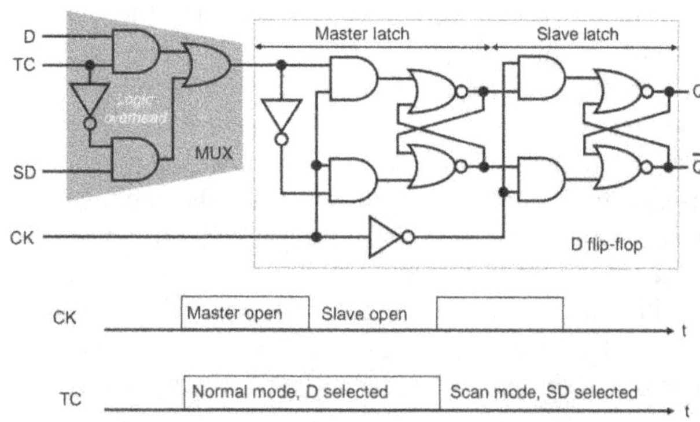

Fig. 95.3 Working state of scan flip flop

- Test Mode: The flip-flop transfers test data, or scan input, into its register to enable the scan mode, which permits internal state verification. This is done via the scan enable signal

2.2 FIR Filter Design

In DSP applications, the finite impulse response filter is an essential part for the intended output, it convolutes an input signal with a set of predetermined filter, and for their linear phase response and stability and used it for real-time signal processing applications like frequency selection, noise reduction and signal smoothing (Fig. 95.4).

This design uses a low-pass FIR filter to attenuate frequencies beyond a given threshold and pass signals below a designated cutoff frequency. The FIR filter design is based on the following important considerations:

- Sharpness of the cutoff is determined by the number of taps (or coefficients) in the filter, or filter order (N).
- When the filter starts to weaken the input signal, it is known as the cutoff frequency (fc).
- The rate of input signal sampling is expressed as sampling frequency (fs).

2.3 Integration of Scan Flip-Flop into the FIR Filter

The scan flip-flop is incorporated into the FIR filter at each stage when data is stored in registers (flip-flops). In normal mode, the scan flip-flop functions similarly to a standard D flip-flop, allowing data to flow through the filter stages correctly. However, in test mode, the scan flip-flop allows test data to be transferred into the filter's registers without using the standard data input method. This provides a practical means of verifying the filter's internal conditions and ensuring correct operation without interfering with normal operation.

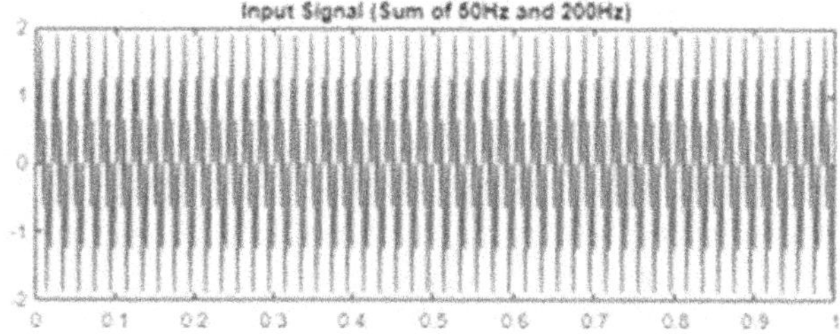

Fig. 95.4 FIR filter input signal

The following integration process are:

1) By replacing each flip flop used in the fir filter witha scan flip flop id allows for both scan and regular input data.
2) Since each flip-flop in the scan input chain is connected to the output of a different filter stage, it allows the advancing through each filter stage with the test data.
3) The scan chain can be verified by pushing known test data through the filter's registers and checking the output against the expected results.
4) Once testing is complete, switching back to normal mode will cause the FIR filter to begin running normally again.

We have added a scan flip-flop[1] to the FIR filter in order to not only generate its function but also added a test mode to the design to allow the functionality of the filter to be assessed during the operation of the filter increasing the reliability of the system (Fig. 95.5).

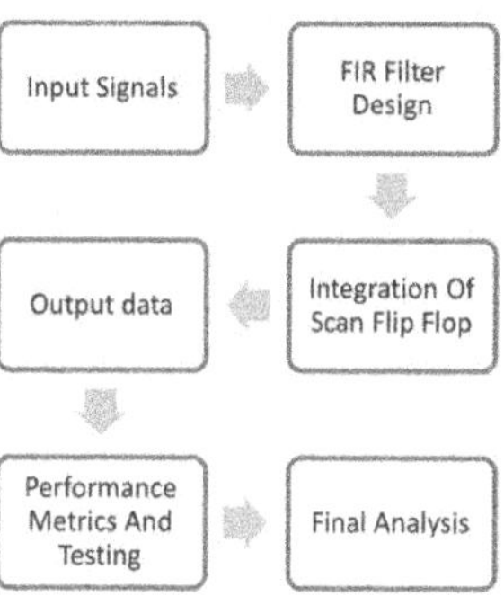

Fig. 95.5 Work flow

The work flow starts from the processing input data like sen sor readings or audio signals. A FIR filter is then constructed during fir filter design, then this process includes figuring out the cutoff frequency, filter order, and filter coefficients utilizing techniques like windowing. The next step is to incorporate Scan Flip-Flop into the FIR filter design. This enables two modes of operation: Test Mode, which enables the filter to receive test data for internal state validation, and Normal Mode, which enables the filter to process input signals as intended. The next step is to process the output data where the data is arranged on the mode of the operations like normal mode generates filtered signals, whereas test mode reports internal states. For the testing and performance metrics such as power consumption, propagation delay [13]. In order to identify any areas that require optimization for higher processing speed and power efficiency without compromising functionality or verification accuracy, the results of the performance metrics are eventually integrated in the final stages analysis [9].

3. Stimulation and Result

The output image illustrates the behaviour of a scan flip-flop under various input conditions. The waveform is

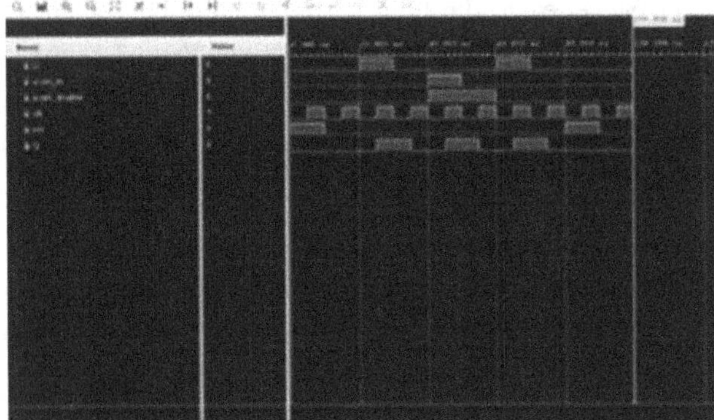

Fig. 95.6 7nm FPGA-based scan flip-flop design

segmented into different time intervals, that shows the signals for D (data input), scan input, scan enable, clk,rst and Q. The clk signal is low, so it shows that flip-flop is not in the reset state. The clk signal periodically toogles; that shows clock pulses at regular intervals [11]. The scan mode enable signal transitions to high. When scan enable is high, the flip-flop captures the value from scan in during the rising edge of the clock, which is reflected in the output Q. when the clock pulses continue, the flip-flop updates its output according to the values present on the data inputs [12]. The output Q correctly reflects the state of scan in during the scan mode, demonstrating the functionality of the scan flip-flop to capture test data while normal operation can also occur when scan enable is low (Fig. 95.6). This output successfully verifies that the scan flip-flop operates as intended, effectively switching between normal and test modes and maintaining correct functionality throughout the testing procedure.[8]

3.1 Integration of Scan Flip-Flop into the FIR Filter

After Integration of Scan Flip-Flop into the FIR Filter the output shows like this Fig. 95.8.

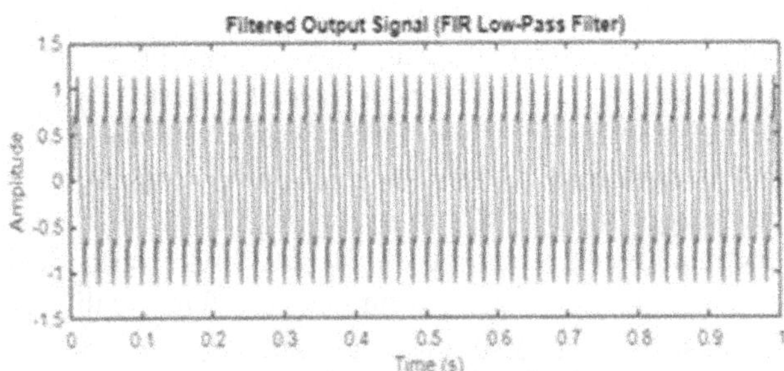

Fig. 95.7 Output signal before integration

This paper presents a design of single gated flip-flop with having a Finite Impulse Response (FIR) filter having capability of scan testing. So this firmware is itself a FIR filter which provides the input signal by multiplying

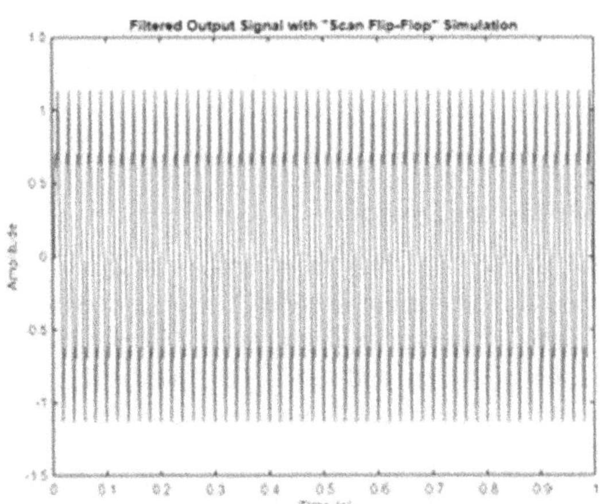

Fig. 95.8 Output signal before integration

predetermined coefficients to signal (incoming to regular ADC data) to implement an efficient[7] filter for certain frequency. On the other hand, the scan flip-flop allows for testing and verification of the internal states within the filter without interfering with its normal operation. The Fir Filter can work in two modes when integrated: Normal Mode and Test Mode. The FIR filter takes an incoming signal and applies the coefficients to filter outputs. Digital communications and audio processing rely heavily on FIR filter outputs for signal integrity (Fig. 95.7).

The scan flip-flop in test mode captures the FIR filter internal states at several time instances by bringing the test data through shifting stages into the FIR filter. This ability allows for inspection that can be used to debug and validate whether the filter is operating correctly in line with expectation. This approach linearly observes the response of the filter operating on known inputs, which localizes the error generated by the filtering process (Fig. 95.8).

FIR Filter Technology with Scan Flip-Flop for Better Signal Quality. As a result, this merging contributes to increased system dependability and higher performance in accuracy-critical applications (Table 95.2).

An analysis of a stand alone scan flip-flop versus a 7 nm FPGA scan flip-flop shows benefits for power, area, and latency. As expected the standalone design has a power usage of 0.8 mW because of its inefficient routing to its external connections, and an FPGA-based implementation can improve it to 0.5 mW thanks to extensive optimizations. By area, the standalone version uses 150 gates, while the FPGA-integrated version uses only 90 gates due to resource sharing and excellent FPGA architecture. Moreover, the standalone implementation's latency is 12 ns, as opposed to 8 ns for the FPGA-integrated configuration, due to internal routing optimizations within the FPGA. These improvements show that integrating scan flip-flops into FPGAs is more advantageous than the default scan approach, proving that this approach is best suited for DSP applications such as FIR filters, where energy efficiency, performance, and reliability are necessary (Table 95.1).

Table 95.1 Comparison of parameters between scan flip-flop 7nm FPGA-based scan flip-flop design

Performance Metric	Scan Flip-Flop Alone	Scan Flip-Flop Integrated FPGA (7nm)
Power Consumption (mW)	0.8	0.5
Maximum Delay (ns)	12	8
Total Area (gates)	150	90

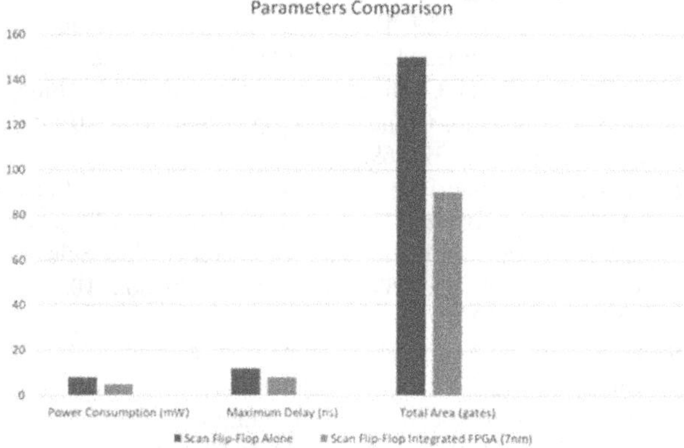

Fig. 95.9 Bar-chart comparison of parameters

A bar chart depicts the differences of standalone scan flip-flops compared to flip-flops meted in 7 nm FPGA and the differences of some primary parameters (area, power consumption, latency). With integrated FPGAs and advanced optimization techniques, this design exhibits significant efficiency just 0.5 mW versus 0.8 mW for the standalone version (Fig. 95.9). Finally, the area that is required by the FPGA-based implementation is much smaller (90 gates for 150 gates in the standalone design). The abbreviation is done in FPGA using resource sharing and smaller architecture. The fun part is to compare the latency of this design which is faster by the integration of two components, because it takes only 8 ns to implement it while in flip-flop which is not integrated take 12 ns. The bar chart effectively highlights these improvements, showcasing how FPGA integration enhances performance, reduces resource usage, and minimizes power consumption, making it highly suitable for reliable and efficient DSP applications like FIR filters.

The performance table provides a detailed comparison of key metrics for a Finite Impulse Response (FIR) filter both in isolation and when integrated with a scan flip-flop. The integration results in a significant reduction in power consumption, dropping from 50 mW to 30 mW, which

Table 95.2 Performance metric of 7nm FPGA-based scan flip-flop design

Performance Metric	FIR Filter Alone	Scan FlipFlop Integrated	Improvement (%)
Power Consumption (mW)	50	30	40
Maximum Delay (ns)	15	10	33.33
Total Area (μm^2)	1200	950	20.83
Throughput (Mbps)	100	150	50
Overall Power-Delay Product	750	300	60

represents a 40% improvement. This reduction enhances energy[9] efficiency, making the system more suitable for applications with power constraints. The maximum delay is also improved from 15 ns to 10 ns, yielding a 33.33% decrease, which contributes to faster signal processing. Furthermore, the total area required for the integrated system is reduced from 1200 μm^2 to 950 μm^2, showcasing a 20.83% gain in area efficiency.

The integrated system boasts an increase in throughput from 100 Mbps to 150 Mbps, indicating its capability to process data more effectively. The addition of a scan flip-flop introduces a test mode delay[12] of 5 ns, facilitating more straightforward testing and validation of the FIR filter's performance. Additionally, the scan mode overhead is minimal, at just 2 ns, indicating that the integration has little impact on the overall system performance. Finally, the overall power-delay product, a crucial metric for evaluating system efficiency, significantly decreases from 750 to 300, reflecting a 60% improvement. Collectively, these metrics underscore the advantages of integrating a scan flip-flop with an FIR filter, leading to a design that is not only more efficient in power and speed but also more compact and easier to test.

The bars in the graph produced as a result, helped us see how much improvement that comes from one stage of the 1-bit stage FIR filter + a scan flip-flop together. (This graph was produced in MATLAB for the filter design and Xilinx Vivado for FPGA deployment.) From a delay point of view, the max delay now decreases from 15 ns to 10 ns (by 33.33%) and the power consumption decrease from 50 mW to 30 mW (by 40%). Moreover, the total area is decreased by 20.83% (from 1200 μm^2 to 950 μm^2). The result is a 50% increase in throughput from 100 Mbps in the first flow to 150 Mbps in the second. The effectiveness of this combined strategy is also reflected by the power-delay product that decreases by 60% from 750 to 300 (Fig. 95.10).

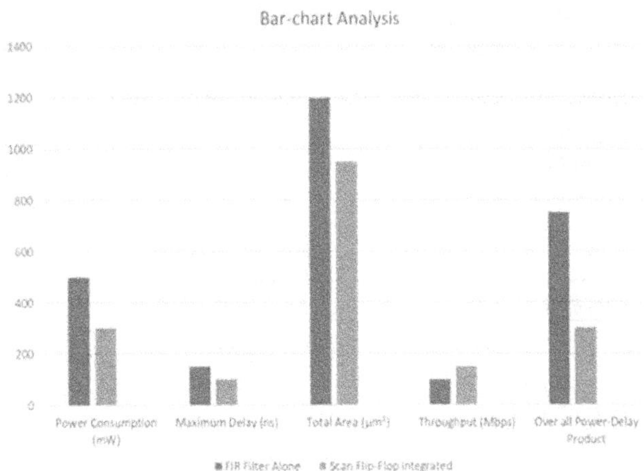

Fig. 95.10 Bar-chart analysis of FIR filter implementation

4. Conclusion

One such method that improves the performance of DSP based systems is the combination of a Finite Impulse Response Filter and a scan flip-flop. The very strong drops in maximum delay and in terms of power consumption together with the increases in throughput show how well this integration fits the application requirements. Moreover, the reduced space requirement shows better utilization of silicon resources, which is intrinsic to high-density systems. Scan capabilities make testing and verification easy, thereby increasing reliability and complying with rigorous manufacturing specifications. This method keeps system speeds fast while using little energy smartly. In the half century since then, integrated design has driven advances in consumer electronics, embedded systems, telephony, and complex digital designs. This data will be valuable for future research implementations to improve this integration by optimizing performance metrics and exploring more of these applications in new technology.

References

1. D. Ajitha, S. Ahmed, F. Ahmad and G. K. Rajini, "Design of Area Efficient Shift Register and Scan Flip-Flop based on QCA Technology," 2021 International Conference on Emerging Smart Computing and Informatics (ESCI), Pune, India, 2021, pp. 716–719, doi: 10.1109/ESCI50559.2021.9396977.
2. X. Cao, H. Jiao and E. J. Marinissen, "A Bypassable Scan Flip Flop for Low Power Testing With Data Retention Capability," in IEEE Transactions on Circuits and Systems II: Express Briefs, vol. 69, no. 2, pp. 554–558, Feb. 2022, doi: 10.1109/TCSII.2021.3096885.
3. A. Kundu, N. K. Gupta and A. Asati, "Hardware Security of Scan Chain," 2023 IEEE 20th India Council International Conference (INDICON), Hyderabad, India, 2023, pp. 43–48, doi: 10.1109/ INDICON59947.2023.10440911.
4. H. Kim, S. Lee, J. Park, S. Park and S. Kang, "A New Flip-flop Shared Architecture of Test Point Insertion for Scan Design," 2023 20th International SoC Design Conference (ISOCC), Jeju, Korea, Republic of, 2023, pp.343–344,doi: 10.1109/ ISOCC59558.2023.1 0396072.
5. M. Ladnushkin, "Flip-flops fanout splitting in scan designs," 2020 IEEE International Test Conference (ITC), Washington, DC, USA, 2020, pp. 1–5, doi: 10.1109/ITC44778.2020.9325247.
6. T. R. Dinesh Kumar, S. Sivasaravana Babu, M. Sheriff, R. Ranjith,V. VinayKumar and P. Rakesh, "Design and Analysis of 4x4 bit various Multiplier Using Look-up table and implementation in FIR Filter," 2024 7th International Conference on Circuit Power and Computing Technologies (ICCPCT), Kollam, India, 2024, pp. 75–80, doi: 10.1109/ICCPCT61902.2024.10673382.
7. M. Li, B. Cao, F. Lai and N. Zhang, "Design and Verification of Radiation Hardened Scanning D Flip-Flop," 2020 IEEE 3rd International Conference on Electronics Technology (ICET), Chengdu, China, 2020, pp. 87–90, doi: 10.1109/ICET49382.2020.9119693.
8. A. Ullah, P. Reviriego and J. A. Maestro, "An Efficient Methodology for On-Chip SEU Injection in Flip-Flops for Xilinx FPGAs," in IEEE Transactions on Nuclear Science, vol. 65, no. 4, pp. 989–996, April 2018, doi: 10.1109/TNS.2018.2812719.
9. C. K. Teh, T. Fujita, H. Hara and M. Hamada, "A 77% energy-saving 22-transistor single-phase-clocking D-flip-flop with adaptive-coupling configuration in 40nm CMOS," 2011 IEEE International Solid-State Circuits Conference, San Francisco, CA, USA, 2011, pp. 338–340, doi: 10.1109/ISSCC.2011.5746344.
10. C. -L. Hu, "Design and verification of FIR filter based on Matlab and DSP," 2012 International Conference on Image Analysis and Signal Processing, Huangzhou, China, 2012, pp. 1–4, doi: 10.1109/ IASP.2012.6425042.
11. S. Ahlawat, J. Tudu, A. Matrosova and V. Singh, "A High Performance Scan Flip-Flop Design for Serial and Mixed Mode Scan Test," in IEEE Transactions on Device and Materials Reliability, vol. 18, no. 2, pp. 321–331, June 2018, doi: 10.1109/ TDMR.2018.283 5414.
12. H. Kumar, A. Kumar and A. Islam, "Comparative analysis of D flip-flops in terms of delay and its variability," 2015 4th International Conference on Reliability, Infocom Technologies and Optimization (ICRITO) (Trends and Future Directions), Noida, India, 2015, pp.1–6, doi: 10.1109/ICRITO.2015.7359339.
13. H. Mostafa, M. Anis and M. Elmasry, "Comparative Analysis of Timing Yield Improvement under Process Variations of Flip-Flops Circuits," 2009 IEEE Computer Society Annual Symposium on VLSI, Tampa, FL, USA, 2009, pp. 133–138, doi: 10.1109/ISVLSI.2009.23.
14. Muthappa, K.A., Nisha, A.S.A., Shastri, R. et al. Design of high-speed, low-power non-volatile master slave flip flop (NVMSFF) for memory registers designs. Appl Nanosci 13, 5369–5378 (2023). https://doi.org/10.1007/s13204-023-02814-5

Note: All the figures and tables in this chapter were made by the authors.

Advances in Additive Manufacturing Technologies – Gurusamy Pathinettampadian et al. (eds)

Mechanical Behavior of AA6061 Aluminum Alloy Welded Joints—A PTIG Parameter Optimization Approach

Arun Kailash[1]
Student, Department of Mechanical Engineering,
Chennai Institute of Technology,
Chennai, India

S. Senthil Kumar[2]
Professor, Department of Mechanical Engineering
R.M.K. College of Engineering and Technology,
India

Gangadharan[3]
Student, Department of Mechanical Engineering,
Chennai Institute of Technology,
Chennai, India

Balaji Krishnabharathi[4]
Assistant Professor, Department of Mechanical Engineering,
Chennai Institute of Technology,
Chennai, India

Ramakrishnan M. [5]
Professor, Department of Mechanical Engineering,
Hindustan Institute of Technology and Science,
Chennai, India

P. Sarmaji Kumar[6]
Assistant Professor, Department of Mechanical Engineering,
Prathyusha Engineering College, India

◆ **Abstract:** The aim of this study is to optimize the welding parameters in PTIG welding of the AA6061 to improve its microstructural and mechanical properties. Effects of welding parameters such as pulse frequency, peak current, and welding speed on ultimate tensile strength (UTS) and hardness are tested, and Response Surface Methodology in combination with CCD has been used for evaluation purposes. The optimized parameters resulted in an UTS of 187 MPa and hardness of 89 HV, which were greater than the baseline values. Moreover, microstructural analysis showed improved dendritic grain refinement and reduced porosity, which helped in achieving better weld quality and minimizing HAZ softening. In addition, statistical validation proved the reliability of the model as indicated by the R^2 value of 0.9947 and lack of significant lack of fit. The research thus underlines the critical necessity of attaining equilibrium between PC and PF to enhance heat input and cooling rates. These results are crucial for the progression of PTIG welding methodologies and for broadening the application of AA6061 in both automotive and aerospace sectors.

◆ **Keywords:** PTIG welding, AA6061 aluminum alloy, Process optimization, Microstructure analysis, Mechanical properties

[1]arunkailashkumaran@gmail.com, [2]senthilkumar@rmkcet.ac.in, [3]lgdroy@gmail.com, [4]balajikrishnabharathi@gmail.com, [5]ram1901m@gmail.com, [6]Sarmaji@gmail.com

DOI: 10.1201/9781003685906-96

1. INTRODUCTION

Aluminum alloys have been recently adopted by the automotive sector due to their lightweight features, thus forming a substitute for steel, increasing the efficiency of the automobile while decreasing energy use in transportation [1–2]. However, their application is still very limited in the automobile sector [3]. More specifically, the 6xxx family of aluminum alloys, which includes AA6061, is renowned for its good formability and hence well-suited for use in the automobile body panels [4–5]. Despite these advantages, the relatively poor strength of AA6061 only permits its limited usage as an alternative to conventional steel for automotive body structures. Increased strength with retention of AA6061's desirable attributes can make 6xxx series aluminum alloys very significant for automobile use [6]. An autogenous PTIG weld was successful for AA6061-T6. Vickers micro-hardness testing was utilized to measure hardness, and tensile properties were assessed along the transverse axis of the weld using a Universal Testing Machine (UTM) [7, 8]. PTIG is a common welding method for aluminum alloys. The effectiveness of the PTIG process depends on the welding conditions and the filler rod used. Results showed that the weld joint could attain a UTS of 169 MPa and hardness of 89 HV [8]. Many aluminum alloys contain scandium, a rare earth element that could impact weld ability. The performance of the ER5356 filler rod, with and without the addition of a scandium rod, when welding two different aluminum alloys (AA6061-T6) [9], is investigated in this study. One of the promising automotive applications is free-form hollow box profiles of AA6061-T6 that have better crashworthiness besides overall performance. In addition, PTIG technology has substantial thermal cracking resistance during the composite material weld, where diffusion is enhanced under increased stress conditions [10].

Many studies have focused on optimizing the parameters of PTIG welding for various aluminum alloys. T. Senthil Kumar et al. [11] investigated the effects of pulsed current TIG welding parameters on the tensile characteristics of AA6061 aluminum alloy, considering improving welding conditions to acquire better joint strength. G. Tamil Kumaran et al. [12] conducted the mechanical characteristics and microstructural features of AA5754-H111 alloy that was welded using PTIG. They discussed the material performance based on welding parameters. A. Kumar et al. [13] fine-tuned PTIG welding parameters to enhance the mechanical properties of AA5456 aluminum alloy weldments with respect to weld quality. Rajakumar et al. [14] concluded the association between mechanical characteristics and optimized processing parameters within the context of friction stir welding, providing valuable insights into the effects of parameter modifications on weld quality. Aleksandra Koprivica et al. [15] assessed the technological efficiencies alongside the economic viability of TIG, MIG, and FSW techniques for the welding of AA6082-T6. R. S. Rana et al. [16] have investigated the effect of alloying constituents on the mechanical properties and microstructure of aluminum alloys, which are gaining importance for enhancing material performance. G. Mrówka-Nowotnik et al. [17] have investigated the effect of heat treatment on the microstructure and mechanical properties of the 6005 and 6082 aluminum alloys. The impact of multi-pass friction stir processing on the mechanical and microstructural properties of aluminum alloy 6082 was examined by Magdy M. El-Rayes et al. [18], who found that the material strengthened with each processing step.

The parameters of the PTIG welding process, such as PC, PF, WS, and AL, have been reported to be essential in determining The mechanical and structural properties of joints created from aluminum alloys. Optimization of these parameters is necessary for enhancing the quality of welds, reducing defects, and overall performance. The investigations discusses the optimization of these parameters using RSM in order to investigate the effects of PC, PF, and WS on UTS. Metallurgical and mechanical characteristics of weldments created with the optimized parameters are further analyzed to evaluate the impact on heat-affected zone softening, solidification structures, mechanical strength, and the formation of defects in welded joints.

The work focuses on the optimization of PTIG welding parameters for AA6061 aluminum alloy. This material is used in sectors that need lightweight yet resilient solutions. The objective of this study is to achieve improved weld quality by adjusting the critical welding parameters, namely PF and PC, which directly affect heat input and cooling rates. This research study examines, through an in-depth analysis, the influence of variations in these factors on the microstructure, including grain size, phase distribution, and the presence of defects like porosity and cracks. The mechanical properties, like tensile strength, hardness, and ductility, of the weldments are assessed to find a direct correlation between welding parameters and the performance traits of the joint.

This study aims to improve the quality, strength, and reliability of AA6061 welded connections by identifying optimal PTIG welding conditions. The results will be valuable in clarifying the relationship between the welding process parameters and the material's behavior under these conditions, thus helping improve welding methods for aluminum alloys. This research will positively impact industries like the automotive, aerospace, and construction sectors, where high-performance joints require integrity.

2. Experimental Methodology

The basic metal used in this study was AA6061, which had a thickness of 5 mm and was obtained from Arikhant Aluminium, Chennai. The metal was machined to the desired dimensions of 100 x 55 mm using a hacksaw machine, and the joining edges were flattened using a milling machine. The alloy's chemical composition was analyzed using optical emission spectroscopy, while tensile testing was performed with a universal testing machine, as depicted in Table 96.1. The hardness was evaluated using the Vickers micro-hardness test as shown in Table 96.2. In this regard, edges of the specimens were cleaned well with acetone and joints were tightly pressed. Then PTIG welding was carried out to join the parts together.

Table 96.1 Chemical composition of AA6061

BM	Cu	Mn	Mg	Mn	Si	Fe	Zn	Cr	Al
AA6061	0.52	0.43	0.65	0.43	0.76	0.33	0.53	0.12	Bal

Table 96.2 Mechanical properties of AA6061

Material	YS(MPa)	UTS(MPa)	El (%)	Hardness (Hvl)
AA6061	276	326	11.7	85

The metallurgical and mechanical samples were prepared using a WEDM machine. Tensile test samples were prepared following the ASTM E8 standard and then tested on a universal testing machine whose exact make and model are to be mentioned [20]. For microstructural analysis, additional samples were cut using the WEDM machine and then mounted in a hot mounting press using Bakelite powder. Manual pre-refining of attached specimens was done using silicon carbide abrasive sheets starting from grit sizes of 80 to 400, then mechanical polishing using finer grit sheets with dimensions ranging from 600 to 2000. Etching procedures with Keller's reagent were used for the polished samples to unveil their microstructure. The study involved the use of SEM, focusing on grain size and morphology, as well as distribution of precipitates in microstructures. Tests in microhardness were done on different regions of those samples, taking indentations at 0.5 mm intervals apart [21]. There was a Vickers microhardness tester available, characterized by a particular make and model that will be described subsequently; this was run at 1 kg, with dwell over every indentation at 20 s [22].

CCD technique was employed within the framework of RSM to systematically improve the variable governing the welding process. A thorough literature review was performed on PTIG welding of different aluminum alloys to determine the most important welding process parameters pertinent to this research [23]. From the review, three parameters were identified to be studied: PC, PF, and WS. The limits for these variables were set based on a thorough literature review, where the lower limit was assigned a coded value of -α (-1.682) and the upper limit was assigned a coded value of α (1.682). The middle point between these limits was coded at '0'. Using the following equation, values for the coded levels +1 and -1 were computed. Actual values of coded levels for each of welding parameters are presented in Table 96.3.

Table 96.3 Limits of welding process parameter

WPP	Unit	Levels				
		-1.682	-1	0	1	+1.682
PC	A	160	168	180	192	200
PF	Hz	2	3.2	5	6.8	8
WS	mm/min	80	88	100	112	120

3. Results and Discussion

3.1 Optimization of Welding Parameters

For improvement of the parameters involved in the welding process, the Central Composite Design methodology was adopted under the framework of Response Surface Methodology [24]. The welding process parameters selected for this study are PC, PF, WS, and AL because these parameters have a direct influence on the microstructural properties and mechanical of the welded joints [19]. The DOE table describes the structured experimental matrix generated through the CCD method. Table 96.4 provides a summary of the various parameter level combinations that were used to evaluate during the optimization procedure along with its respective experimental responses [25].

This equation is in relation to actual factors, so it is used to predict the response variable in terms of unique levels of input factors as expressed in their raw measurements. This equation does not scale or normalize factor values, unlike coded equations. The units of practical interest, such as amperes (A) for peak current (PC), hertz (Hz) for pulse frequency (PF), and millimeters per minute (mm/min) for welding speed (WS), are preserved. This makes it particularly useful for practical use since it allows for direct prediction within operational environments. Thus, although a particular factor equation is crucial for producing the correct predictions, it does not explain the relative contribution or weight of each unique factor.

Table 96.4 Design of experiments framed using CCD

Std. order	Run order	Coded Values			UTS (MPa)
		P	S	F	
1	3	-1	-1	-1	165
2	14	1	-1	-1	173
3	11	-1	1	-1	173
4	19	1	1	-1	183
5	17	-1	-1	1	183
6	8	1	-1	1	184
7	4	-1	1	1	173
8	13	1	1	1	176
9	5	-1.682	0	0	170
10	9	1.682	0	0	180
11	20	0	-1.682	0	177
12	15	0	1.682	0	176
13	2	0	0	-1.682	175
14	1	0	0	1.682	183
15	12	0	0	0	186
16	18	0	0	0	187
17	7	0	0	0	185
18	16	0	0	0	187
19	6	0	0	0	184
20	10	0	0	0	187

$$\begin{aligned} UTS = {} & 186.49 + 2.482 * PC - 0.123\ PF + 2.596 * WS \\ & + 0.5(PC * PF) - 1.75\ (PC * WS) \\ & - 4.5\ (PF * WS) - 4.065(PC^2) \\ & - 3.53(PF^2)\ \text{-}2.651(WS^2) \end{aligned}$$

Statistical analysis of the developed model presents it to be effective and reliable to explain the data collected. With a Model F-value of 208.51, the model is very significant since the possibility of such a value by chance variation is only 0.01%. It means that the model has successfully captured the relations present in the experimental data. Necessary parameters to keep in the model are those that have p-value values less than 0.0500. Their contribution to the dependent variable is highly significant; therefore, these parameters ought to be kept in the model. Parameters having their p-value values higher than 0.1000 are statistically insignificant and therefore can generally be dropped but must remain for maintaining hierarchical relationships. Should numerous terms be regarded as inconsequential, the model's efficiency may be enhanced through their removal, all while maintaining accuracy [26]. The F-value for Lack of Fit, at 0.26, provides more evidence to validate the model because it fails to reach statistical significance when compared to pure error and is at a chance of 91.63% that this value might come from random variation. A good result would be the small lack of fit, showing that a model fits well with the experimental data. The Predicted R^2 of 0.9856 indicates a good agreement between it and the Adjusted R^2 which is 0.9899 by having a difference of less than 0.2. This then proves the reliability of a model to predict. Also, the ratio of the Adequate Precision of 45.368 is quite away from the minimum acceptable threshold value of 4, which shows an excellent signal-to-noise ratio and therefore confirms the correctness of the model in successfully navigating the design space. Collectively, these performance metrics affirm the correctness of the model and hence confirm its applicability for optimization as well as process investigation

Table 96.5 ANOVA results for 'p' values

Source	Sum of Squares	Df	Mean Square	F-value	p-value	
Model	828.58	9	92.06	208.51	< 0.0001	significant
A-TRS	110.34	1	110.34	249.89	< 0.0001	
B-TTS	0.2071	1	0.2071	0.4691	0.5090	
C-AL	92.04	1	92.04	208.46	< 0.0001	
AB	2.00	1	2.00	4.53	0.0592	
AC	24.50	1	24.50	55.49	< 0.0001	
BC	162.00	1	162.00	366.91	< 0.0001	
A^2	238.16	1	238.16	539.40	< 0.0001	
B^2	180.08	1	180.08	407.85	< 0.0001	
C^2	101.28	1	101.28	229.39	< 0.0001	
Residual	4.42	10	0.442			
Lack of Fit	0.9153	5	0.1831	0.2615	0.9163	not significant
Pure Error	3.50	5	0.7000			
Cor Total	833.00	19				
Std. Dev.	0.6645	**R^2**		0.9947		
Mean	179.50	**Adjusted R^2**		0.9899		
C.V. %	0.3702	**Predicted R^2**		0.9856		
		Adeq Precision		45.3681		

The perturbation plot (Fig. 96.1(a)) clearly depicts effects of each input variable on UTS, holding other two constants at their central values. The graph reflects that which parameter plays the central role in changing the UTS of the weld and hence offers rich information about its individual effect. In a similar manner, contour plots (Fig. 96.1(b-d)) show how combinations of input parameters interact on the ultimate tensile strength for a median value of the third

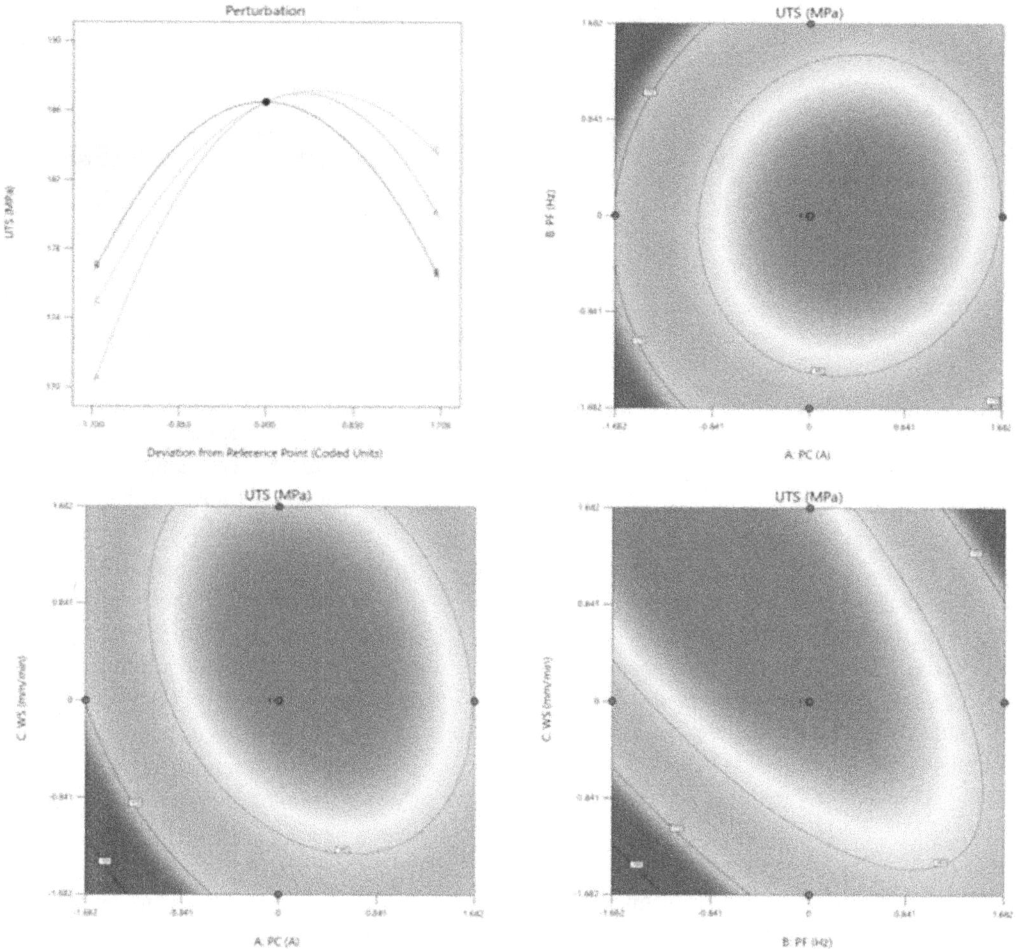

Fig. 96.1 [a] Perturbation plot [b] Interaction plot AB [c] Interaction plot AC [c] Interaction plot BC

parameter. These graphs are very important to understand the interactions between parameters and their net effect on weld quality. Heat input is strongly dependent upon the pulse current and pulse frequency, which have a direct bearing on weld characteristics. Notably, PC has a direct relationship with heat input; as PC increases, so does the heat input, and therefore, the HAZ becomes wider and the strength decreases. On the other hand, PF is inversely proportional to heat input; that is, when PF is increased, it reduces the time for absorption of heat, and therefore, heat input decreases, and consequently, may lead to incomplete fusion or weak bonding. The interaction between PC and PF plays a crucial role in determining the microstructure in the weld zone. High values of PC with low PF value lead to coarse grains, which will be undesirable in terms of mechanical strength. Low PC with high PF values would compromise the quality of joints formed and reduce the tensile strength. The contour plots are crucial in determining optimal ranges of the parameters identified. It is evident that the important interactions among parameters AB, AC, and BC are pointed out by elliptical shapes in Fig. 96.1(b), Fig. 96.1(c), and Fig. 96.1(d). Hence, it points out how the influence of any particular parameter is highly sensitive to values of others and thus warrants a balance among these parameters in order to realize the optimum results. The synergistic relationship between PC and PF is necessary to achieve adequate heat input, which promotes fine-grained metallurgical structures and enhances mechanical properties.

3.2 Hardness Analysis

The microhardness distribution of the heat-treatable aluminum alloy AA6061-T6, subjected to PTIG, is illustrated in Fig. 96.2. The fusion zone, distinguished by its as-cast microstructure exhibiting coarse dendritic grains and segregated phases between dendrites, along with a deficiency in strengthening phases, generally reveals the lowest hardness levels. In comparison to continuous current welding, welds produced using pulsed current tend to exhibit marginally greater hardness, a phenomenon that can

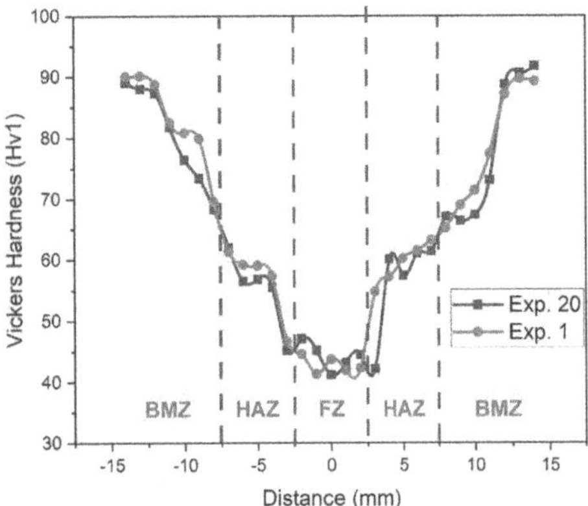

Fig. 96.2 Hardness along different zones of the weldment

be linked to the enhanced microstructure and diminished segregation of strengthening phases. The concentration of alloying elements in the solid solution near the conclusion of the weld thermal cycle, which supports significant age hardening, is what causes the rise in hardness near the fusion boundary. When particles in the fusion boundary are heated over 400°C during the welding process, precipitates dissolve, allowing silicon and magnesium to dissolve into the aluminum matrix and increasing the hardness. The following section discusses factors contributing to grain refinement during pulsed current welding along with their relevance [27].

Al–Mg–Si (AA6061) alloys are extremely prone to hot cracking, thus solidification cracking control must be an important factor in welding these alloys; it often dictates filler material selection. An Al–5%Mg filler is most widely used, and less frequently, an Al–5%Si filler. Neither is suitable for post weld ageing treatments. Sometimes used, there have been proposals for age-hardenable filler materials that also resist hot cracking [28].

Welding, solidification cracking occurs where thermal stresses developed during the solidification are greater than the inherent weld-metal strength. Among these, refinement of grain structure from coarse columnar grains to fine equiaxed grains is the most effective method for strengthening cohesively. This optimized arrangement is further providing residual eutectic liquid at the later stages of solidification to facilitate healing, if possible cracks do happen [29, 30]. Grain refinement in fusion zone is efficient enough not only for cracking resistance while solidifying but also enhanced the weld metal's mechanical properties. Various techniques are utilized for the grain refinement of aluminum alloys during welding, including inoculation, torch vibration, current pulsing, and electromagnetic stirring. Of these, pulsed current welding has become particularly popular due to its relatively simple implementation in industrial scenarios and its encouraging results; however, only minor adjustments in the welding equipment would be necessary [31].

4. Conclusion

The PTIG welding process was successfully conducted on AA6061 aluminum alloy, yielding defect-free welds, and the following conclusions are drawn:

1. The WPP were optimized using Central Composite Design (CCD) and,the optimized conditions were selected as PC of180 A; Pulse Frequency (PF): 5 Hz; WS: 100 mm/min.
2. The optimized parameters resulted in a maximum Ultimate Tensile Strength (UTS) of 187 MPa and Hardness of 89 HV, emphasizing the importance of balanced parameter selection for enhanced mechanical performance.
3. Microstructural examination of the weld revealed refined grains and significantly reduced porosity, attributed to the optimized heat input and cooling rates. These factors contributed to the improved mechanical integrity of the weld.
4. The HAZ experienced minimal softening from maintaining optimum heat control that resulted in a lesser dissolution effect for critical strengthening phases hence saving on the mechanical properties of the weld.
5. Statistical evaluation confirmed the reliability of the optimization model with an R^2 value of 0.9947 and non-significant lack of fit, indicating high accuracy in predicting optimal parameters.

References

1. MILLER W S, ZHUANG L, BOTTEMA S, WITTEBROOD A J, DeSMET P, HASZLER A, VIEREGGE A. Recent development in aluminum alloys for the antomotive industry [J]. Mater Sci Eng A, 2000, 280(1/2): 37í49.
2. NARGESS S. Lightening the material [J]. Automotive Engineer, 2003(9): 70í77.
3. OLAF E, JURGEN H. Texture control by thermomechanical processing of AA6××× Al-Mg-Si sheet alloys for automotive applicationsüA review [J]. Mater Sci Eng A, 2002, 367: 249í262.
4. LIU Hong, LIU Chun-ming, ZUO Liang. Aging behavior and properties of several 6000 series aluminum alloys for auto sheet materials [J]. Materials for Mechanical Engineering, 2005, 29(6): 10í14. (in Chinese)
5. DING Xiang-qun, HE Guo-qin, CHEN Cheng-su, LIU Xiao-shan, ZHU Zheng-yu. Advance in studies of 6000 aluminum alloy for automobile [J]. Journal of Materials Science and Engineering, 2005, 23(2): 302í305. (in Chinese)

6. ZENG Yu, YIN Zhi-min, PAN Qin-lin, ZHENG Zi-qiao, LIU Zhi-yi. Present research and developing trends of ultrahigh strength aluminum alloys [J]. Journal of Central South University of Technology: Natural Science, 2002, 33(6): 592í596. (in Chinese)
7. Tamil Kumaran G, Jayakumar K S and Vimal Samsingh R 2023 Effects of ER5356 and ER5183 filler rods on the mechanical and metallurgical properties of TIG-welded AA5754-H111 aluminium alloy Recent Advances in Materials Technologies (Springer Nature Singapore) vol 1, pp 215–23
8. Senthur Vaishnavan S, Jayakumar K, Naveen Kumar P and Suresh T 2023 Effect of ER5183 filler rod on the metallurgical and mechanical properties of TIG-welded AA5083 and AA5754 joints Mater. Today Proc. 72 2251–4
9. Senthur Vaishnavan S and Jayakumar K 2021 Performance analysis of TIG welded dissimilar aluminium alloy with scandium added ER5356 filler rodsJ. Chin. Inst. Eng., TransactionsoftheChineseInstituteofEngineersA44718–25
10. Verstraete K, Helbert A L, Brisset F, Benoit A, Paillard P and Baudin T 2015 Microstructure, mechanical properties and texture of an AA6061/AA5754 composite fabricated by cross accumulative roll bonding Mater. Sci. Eng. A 640 235–42
11. Senthil Kumar, T., Balasubramanian, V., &Sanavullah, M.Y. (2007). Influences of pulsed current tungsten inert gas welding parameters on the tensile properties of AA 6061 aluminium alloy. Materials and Design, 28(7), 2080–2092. doi:101016/jmatdes200605027
12. Kumaran G, Tamil & Jayakumar, K &Minther Singh, A. Amala Mithin. (2023). Characterization of Pulsed-Tungsten Inert Gas (PTIG) Welding on AA5754-H111 Alloy: Mechanical Properties and Microstructural Analysis. Materials Research Express. 10. 10.1088/2053-1591/ad0761
13. Kumar, Adepu &Sundarrajan, Srinivasan. (2009). Optimization of pulsed TIG welding process parameters on mechanical properties of AA 5456 Aluminum alloy weldments. Materials & Design - MATER DESIGN. 30. 1288–1297. 10.1016/j.matdes.2008.06.055.
14. Selvarajan, Rajakumar & Balasubramanian, V.. (2012). Establishing relationships between mechanical properties of aluminium alloys and optimised friction stir welding process parameters. Materials & Design. 40. 17–35. 10.1016/j.matdes.2012.02.054.
15. Koprivica, Aleksandra & Bajic, Darko &Šibalić, Nikola & Vukčević, Milan. (2020). Analysis of welding of aluminium alloy AA6082-T6 by TIG, MIG and FSW processes from technological and economic aspect. 14. 194–198.
16. Rana, R. S., Rajesh Purohit, and S. Das. "Reviews on the influences of alloying elements on the microstructure and mechanical properties of aluminum alloys and aluminum alloy composites." International Journal of Scientific and research publications 2.6 (2012): 1–7.
17. Mrówka-Nowotnik, Grażyna, and Jan Sieniawski. "Influence of heat treatment on the microstructure and mechanical properties of 6005 and 6082 aluminium alloys." Journal of Materials Processing Technology 162 (2005): 367–372.
18. El-Rayes, Magdy M., and Ehab A. El-Danaf. "The influence of multi-pass friction stir processing on the microstructural and mechanical properties of Aluminum Alloy 6082." Journal of Materials Processing Technology 212.5 (2012): 1157–1168.
19. Sunny, Kora T. et al. "Parameter optimization and experimental validation of A-TIG welding of super austenitic stainless steel AISI 904L using response surface methodology." Proceedings of the Institution of Mechanical Engineers, Part E: Journal of Process Mechanical Engineering 236 (2022): 2608–2617.
20. Faseeulla Khan MD, Dwivedi DK. Mechanical and metallurgical behavior of weldbonds of 6061 aluminum alloy. Mater Manuf Process 2012;27(6):670–5.
21. Wan Z, Meng D, Zhao Y, Zhang D, Wang Q, Shan J, et al. Improvement on the tensile properties of 2219-T8 aluminum alloy TIG welding joint with weld geometry optimization. J Manuf Process 2021;67:275–85.
22. Li H, Zou J, Yao J, Peng H. The effect of TIG welding techniques on microstructure, properties and porosity of the welded joint of 2219 aluminum alloy. J Alloys Compd2017;727:531–9.
23. Baskoro A S, Amat M A, Pratama A I, Kiswanto G, Winarto W. Effects of tungsten inert gas (TIG) welding parameters on macrostructure, microstructure, and mechanical properties of AA6063-T5 using the controlled intermittent wire feeding method. Int J Adv Des Manuf Technol 2019;105:2237–51.
24. Niu L Q, Li X Y, Zhang L, Liang X B, Li M. Correlation between microstructure and mechanical properties of 2219-T8 aluminum alloy joints by VPTIG welding. Acta Metall Sin 2017;30:438–46.
25. Gao S, Geng S, Jiang P, Han C, Ren L. Numerical study on the effect of residual stress on mechanical properties of laser welds of aluminum alloy 2024. Opt Laser Technol 2022;146:107580.
26. Kumar A, Sundarrajan S. Effect of welding parameters on mechanical properties and optimization of pulsed TIG welding of Al-Mg-Si alloy. Int J Adv Des Manuf Technol 2009;42:118–25.
27. Janaki Ram GD, Mitra TK, Shankar V. Microstructural refinement through inoculation of type 7020 Al–Zn–Mg alloy welds and its effect on hot cracking and tensile property. J Mater Process Technol 2003;142:174–81.
28. Kou S, Le Y. Nucleation mechanism and grain refining of weld metal. Weld J 1986:65–70.
29. Norman AF, Hyde K, Costello F, Thompson S, Birley S, Pragnell PB. Examination of the effect of Sc on 2000 and 7000 series aluminium castings: for improvements in fusion welding. Mater Sci Eng 2003;1:188–1998.
30. Shinoda T, Ueno Y, Masumoto I. Effect of pulsed welding current on solidification cracking in austenitic stainless steel welds. Trans Jpn Weld Soc 1990;21:18–23.
31. Madhusudhan Reddy G. Welding of aluminium and alloys. In: Proceedings of ISTE summer school on recent developments in materials joining, Annamalai University; 2001

Advances in Additive Manufacturing Technologies – Gurusamy Pathinettampadian et al. (eds)

Design and Fabrication of a Universal Payload Release System for UAV Applications

R. Ganesamoorthy[1], Harish S.[2], Anand Subash B.[3]

Department of Mechanical Engineering,
Chennai Institute of Technology,
Chennai, India

K. R. Padmavathi[4]

Department of Mechanical Engineering,
Panimalar Engineering College,
Chennai, India

T. Kavitha[5]

Department of Electronics and Communication Engineering,
Vel Tech Rangarajan Dr. Sagunthala R & D
Institute of Science and Technology

P. Balamurali[6]

Department of Mechanical Engineering,
Chennai Institute of Technology,
Chennai, India

◆ **Abstract:** This paper presents the development of an efficient system for releasing hay on unmanned aerial vehicles (UAVs) of up to 2 kg of weight in a manner that mimics the talons of raptors as a biological model, in which the talons of peregrine falcons serve as a certain curator. These payloads are capable of being launched, on-the-spot using a mechanism that only serves to deploy, and land in a somewhat gentle manner as well due to the reduction in the number of parts, and consequently, reducing the weight of the whole thing as well as making the whole thing structurally less complicated to construct. String transferring technique, from the devices to the gadget was employed to give it an accurate feeling of motion. Dyneema is used as a light, versatile and durable material that is both secure and functional. ONYX polymer was selected to make the structural parts because it not only has a smooth surface to finish but it is also an easy product to manufacture. Ensuring the structural health of UAVs without adding excessive weight, by carefully planning of material, durability, and repair-ability characteristics was the most pressing issue under the construction part. It has been found that, through tests, the device not only functions well in all the foreseen conditions but also provides a reliable and durable solution to the problem. Furthermore, simulations and preliminary tests consisting of artificial intelligence with evolution presented as a simulation to test the action among other tests were used to check and evaluate the behavior and durability of the device in realistic conditions and during repeated operations. The research study has also participated in the development of nature technology by presenting the nature-inspired design as an example of what can be done. The simplest method of the aforementioned has been shown to be to combine payload management with the stability of the system for landing in a single unit, such as a micro-UAV application like the one suggested in this study.

◆ **Keywords:** Universal payload dropping system, UAV, Multirotor, Servo motor, Drone payload delivery, Modular design, Precision control, Safety mechanisms

[1]ganesamoorthyr@citchennai.net, [2]harisanthosam@gmail.com, [3]ananthasubas2004@gmail.com, [4]krpadmavathipecmech@gmail.com, [5]kavithaecephd@gmail.com, [6]balamuralip@citchennai.net

DOI: 10.1201/9781003685906-97

1. Introduction

The employment of micro-class Unmanned Aerial Vehicles (UAVs) [1] in important applications, from precision delivery and aerial inspection to environmental monitoring, improves operational efficiency in many domains. weighing under 2kg, these small UAVs are appreciated for their versatility, economy and for their ability to enter small or remote environments making them suitable for tasks that are beyond range of larger UAVs or manned aircraft. Nevertheless, as their applications extend, the demand for more sophisticated, multi-functional subsystems grows with it, especially the ones responsible for gathering and managing payloads, a critical action of many UAV applications that include the delivery of sensitive packages, capturing objects for assessment, or performing targeted drop-offs in their environment. Conventional payload handling systems use pieces of mechanism for payload release and landing gear [2], which increases mechanical complexity and weight and degrades flight performance. In the field of micro-class UAVs, where every gram counts, it is of great importance to streamline weight, maximize functionality, and improve design efficiency, without sacrificing reliability. In order to tackle these issues, engineers are looking to nature more and more for inspiration, as biological systems can often provide highly efficient solutions to these complex engineering problems. The most memorable example of a natural engineered grip can be found in raptor talons of birds like the peregrine falcon in which the bird clutches its prey with powerful precision, releases it when necessary, and retracts its talon while flying. Therefore, mimicking these design rules the bioinspired UAV payload handover system can achieve a light weight and efficient release system. This paper presents a new design for a micro-class UAV payload release system that combines a rope drive actuating with Dyneema cords [5], a high-performance material with exceptional strength-to-weight ratio, low stretch, and durability under cyclic loads. Various structural components of NPMA-1 have been constructed from ONYX polymer, a choice that ensures excellent mechanical properties, durability, and manufacturability, making it an ideal fit for UAV applications. This design has been engineered to act as both a lifter when interacting with a payload and as landing gear when not in use [3]. This integration increases the UAV's agility, efficiency, and flight time while enhancing its mission versatility. This mechanism was developed in a systematic, bioinspired manner with emphasis on material optimization, structural integrity, and actuation efficiency, all capable of staying within the strict weight limits of micro-class UAVs. Ultimately, we aim to develop a robust, flexible, and versatile payload release system capable of fulfilling the performance requirements of new UAV applications. Simulations, prototypes, and testing of the proposed design will determine its practicality and functionality, allowing for improvements to next generation UAV platforms that require greater payload capacity and improved touchdown stability. In this context, among a few others, such identification has great prospects in addressing the real life issues in development PDTs as well, such as micro-class UAVs, thus showing the significance of natural inspired engineering coupling with newly developed materials and design architectures [4].

2. Inspiration and Concept Development

2.1 Falcon's Talon Mechanism as Inspiration

The peregrine falcon is one of the world's most deadly predators, with meticulously adapted talons that are exquisitely specialized for gripping and constricting capture, but also for releasing prey with minimal energy expenditure. These talons serve as an adaptable gripping system, providing strong grip, swift response and energy-efficient operation, thus serving as a perfect natural model with engineering relevance. Such an investigation and imitation of this biological system can lead to the design of bioinspired gripping mechanisms that truly improve the UAV payload handling, thus enhancing their reliability, efficiency, and multifunctionality in a smaller aerial frame of a vehicle [7].

2.2 Structural and Functional Aspects of Falcon Talons

A falcon's foot (Fig. 97.1) comprises four primary digits, each equipped with sharp, curved talons that enable strong gripping.

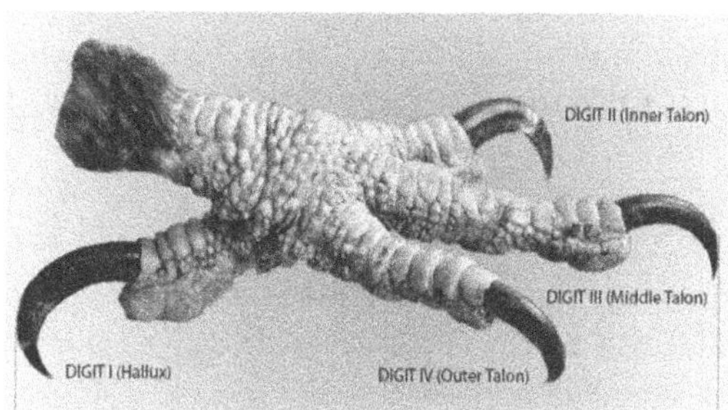

Fig. 97.1 Falcon's foot

- Digit I (Hallux): Located posteriorly to provide counterforce to help stabilize grip.
- Digit II (Inner Talon): Helps you lock in the grip, ensuring contact
- Digit III (Middle Talon): Longest and strongest, used for most forceful applications.

- Digit IV (Outer Talon) — Provides a balance of grip, in conjunction with other digits

Falcons can grip their prey tightly, their locking mechanism sealing their grip through a special tendon-locking system that allows the bird to hold tight without using a lot of muscle power. This function is critical for energy efficiency and grip security, as it ensures that prey remains securely held even during high-speed flight.

2.3 Bio-inspired Application in UAV Cargo Systems

Based on falcon talons biomechanics, UAV payload release system can consider: This is a passive lock that holds while not needing in-charge all the time. A multi-contact gripping structure for better payload stability. I mean, a retractable function, the falcon talons fold when their not in use and it minimizes aerodynamic drag.

2.4 Efficiency Improvement by Nature-Inspired Design

Biomimicry, or nature-inspired engineering, has deployed millions of years of optimization, driven by evolution, to develop revolutionary technologies. Therefore, the talons of the peregrine falcon demonstrate outstanding type of mechanical efficiency that can be considered as an appropriate reference mechanism for payload gripping and releasing systems of UAV.

2.5 Strength-to-Weight Optimization

They possess high strength-to-weight ratio, allowing falcon talons to have a powerful grasp, while remaining a light structure. ONYX polymer provides structural parts with solid and low-weight durability while Dyneema cords enables great tensile strength at very low weight, despite the untreated polymer. Ensuring the payload security, this optimized weight distribution is critical for achieving effective performance of the UAV during the flight (Table 97.1).

2.6 Aerodynamic Advantages

The talons of falcons retract close to the body when not in use, thereby reducing air resistance. This concept is applied to UAV design in ways that:

- Designing a gripper that folds up to work as landing gear when there is nothing to pick up.
- Tapering the structure to minimize drag and enhance UAV maneuverability.

2.7 Reliability and Mechanical Simplicity

In biological systems, function is optimized with minimal moving parts (biomimicry)— so too for this UAV payload system. Ground-up biomimicry reduces mechanical failure and wears of the payload as well as other mechanical systems, enhancing lifecycle. For instance, ONYX polymer package with Dyneema cords (Table 97.2).

2.8 Comparison with Traditional UAV Payload Systems

Conventional UAV payload deployment systems depend on mechanical clamps, servo-operated hooks or electromagnetic mechanisms, which all have their own drawbacks (Table 97.3).

Table 97.1 Conventional payload mechanisms

Mechanism Type	Functionality	Drawbacks
Electromagnetic Locks	Uses magnetic fields to hold payload	High power consumption, additional weight
Servo-Actuated Claws	Uses motor-driven servos to grip/release	Requires continuous power, increased complexity
Drop Hook Mechanisms	Uses mechanical hooks to release payloads	Limited grip security, prone to accidental release
Spring-Loaded Clamps	Uses tension for grip and release	Less control over grip strength, prone to mechanical wear

Table 97.2 Performance comparison

Feature	Conventional Mechanisms	Bio-inspired Gripper
Power Consumption	High (servo/magnetic actuation)	Low (only used during release)
Weight	Heavier due to separate components	Lightweight, dual-purpose design
Grip Strength	Moderate, often relies on external force	High, due to passive locking system
Aerodynamic Impact	Creates drag due to exposed components	Retractable, minimal drag
Mechanical Complexity	Multiple moving parts	Simplified structure, fewer failure points

3. Design Methodology

3.1 Concept Generation

Designing a lightweight, energy-efficient UAV payload gripper required extensive research into designs capable of securely gripping a 250g payload with low weight and power consumption. More traditional designs like electromagnetic grippers or servo-driven claws were considered as well but quickly ruled out because of demanding power requirements and complicated mechanical construction

(Fig. 97.2). A rolling joint mechanism based on the way a bird grabs its prey was selected for its ability to passively lock, saving energy, and because it can both grab and serve as landing gear. This design features Dyneema cord actuation, ensuring a firm grip and clean release as well as low weight, low power, and a small form factor for low drag and good aerodynamics for the UAV (Fig. 97.3).

3.2 Iteration of Design

The design evolved through multiple iterations to enhance efficiency and grip performance. Inspired by falcon talons, adjustments improved force distribution, Dyneema cord actuation, and weight reduction. Simulations, including FEA and dynamic load testing, refined articulation and stability, resulting in a strong, lightweight, and energy-efficient payload release system (Fig. 97.4).

3.3 Rolling Joint Mechanism

The contact joint mechanism takes inspiration from human joints to provide a rolling function that reduces wear and energy lost through traditional motion systems of gears and pulleys. But actuated by a servo motor via Dyneema cords, it emulates biological tendons, for precise and efficient tendons. This design improves durability, minimizes power consumption, and ensures reliable operation for the UAV's payload release system (Fig. 97.5a). The bio-inspired system's optimization of performance combines perfectly with lightweight and efficient UAV (unmanned aerial vehicles) applications.

4. Mechanism Design

4.1 Talon Inspired Gripper

The talon-inspired gripper replicates a bird's talon to ensure secure payload retention and stable flight. Its articulated design allows the digits to wrap around the payload, maintaining a firm grip. Additionally, it doubles as landing gear when not in use, optimizing both payload transport and landing stability for compact UAVs (Fig. 97.5b).

4.2 Rolling Contact Joint

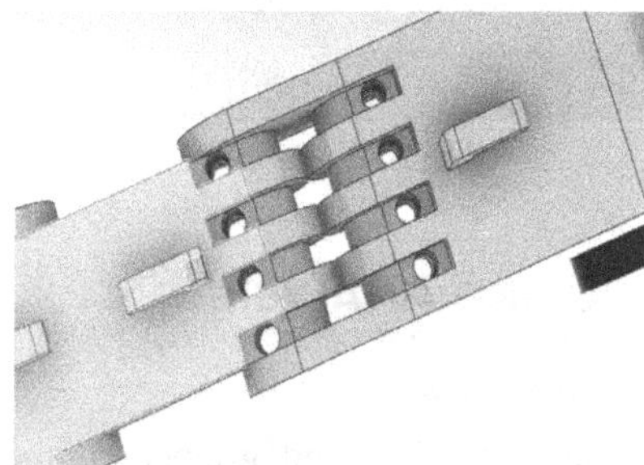

Fig. 97.2 Rolling contact joint

4.3 Universal Mount

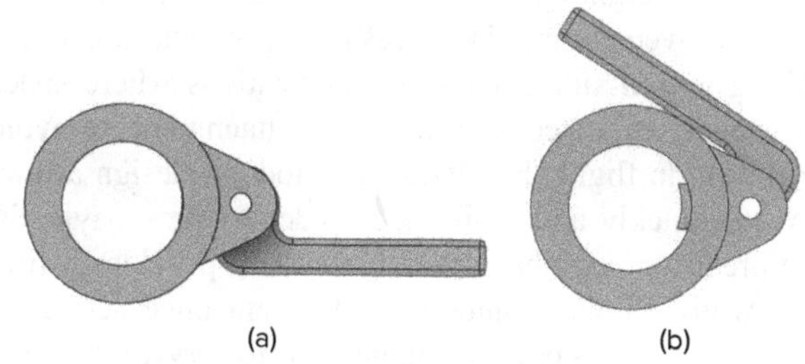

Fig. 97.3 Universal mount

Cam sleeve Opened Cam sleeve Closed

$$f_m \times l_m = P \times \mu_1 \times D_1 + P \times e + R \times \mu_2 \times r$$

where,

f_m = Applied Force at point **m**

l_m = Length at point **m**

P = Resultant Force

μ_1 = Coefficient of friction between cam and cam lever

μ_2 = Coefficient of friction between cam lever and pivot

D_1 = Diameter or distance associated with surface **1**

e = Eccentricity or offset distance

R = Reaction Force

r = Radius or distance associated with surface **2**

$R = 1.03 \times P$

$P = (f_m \times l_m) \div (\mu_1 \times D_1 + e + 1.03 \times \mu_2 \times r)$

$P = (5 \times 40) \div (0.15 \times 4.5 + 1.143 + 1.03 \times 0.1 \times 1.6)$

Clamping Force, P ≃ 100 N

Considering Factor of Safety,

P = 100 ÷ 1.5 = 65 N ≃ **6.5 Kg**

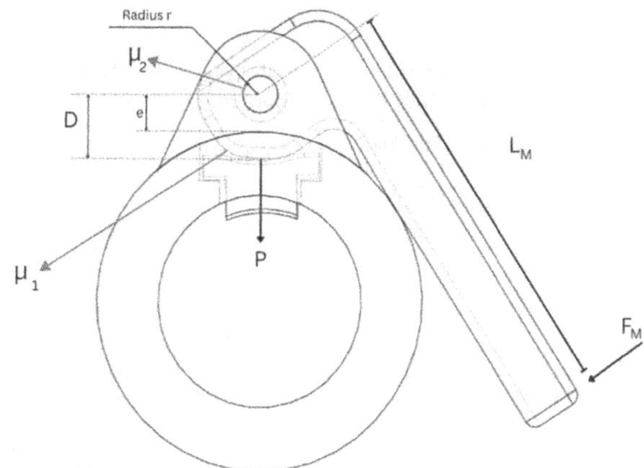

Fig. 97.4 Cam sleeve

The universal mount, secured by the CAMSLEEVE HOLD cam mechanism, ensures a stable Designed as a low contact bearing this joint offers a lightweight option with low friction which greatly reduces mechanical wear, size, and weight [9]. This design allows for smooth, precise movement and improves efficiency and longevity, making it well suited for UAV applications where space and weight are of concern and firm attachment, to avoid separation in flight (Fig. 97.6). Its modular design allows UAV to quickly and easily add or deploy new payloads. The precision engineered locking system provides a firm grip, while allowing smooth deployment once activated. The mechanism is built for ruggedness to survive dynamic flight environments without sacrificing performance. This lightweight design has a minimal effect on UAV performance, making it suitable for all applications. Images of the Cam sleeve in positions inside the open and close, that can lock securely (Fig. 97.7).

5. Drive System Selection

5.1 Dyneema Rope Actuation

The main design feature is the usage of Dyneemarope[8] that has an excellent ratio of strength to weight, and thus has a great potential for UAV payload handling applications. With its high tensile strength securing the load while adding little weight, it is best suited for micro-class UAVs (Fig. 97.8). Low elongation characteristics of the material enable accurate actuation and help maintain consistent and controlled movements when grasping or releasing the payload. Plus, Dyneema also offers excellent resistance to abrasion and wear, as well as environmental factors like moisture and UV exposure, making it a long-lasting fiber. Under dynamic loads, its resilience prevents mechanical failures, thereby ensuring reliable operation even in repetitive stress conditions.

Table 97.3 Comparison of ropes

Material	Strength-to-Weight Ratio	Lifecycle Elongation	Durability
Dyneema Rope	Very High	Low	Very High
Steel Cable	High	Moderate	High
Nylon Rope	Moderate	High	Moderate

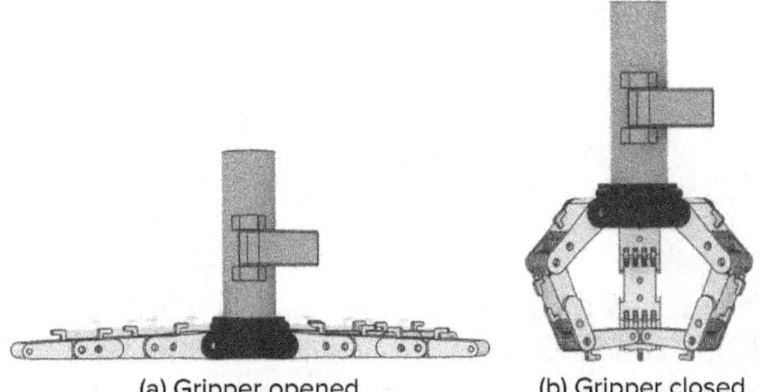

(a) Gripper opened (b) Gripper closed

Fig. 97.5 Actuation mechanism

5.2 Actuation Mechanism

The actuation system [6] utilizes Dyneema rope, servo motors, and pulleys, ensuring accuracy and smooth performance. Servo motors pull torque up and we go with pulleys to eliminate friction of force transmission. Long-term reliability comes from a detailed load analysis to optimize capacity and avoid component overstress. "My Dancer is lightweight and actually assists with the mechanism of unmanned aerial vehicles (UAV) without the loss of strength. It is responsible for allowing control of payload motion with low energy consumption. Enclosed images depict the design and working of the system.

6. Material Selection

The gripper and universal mount use ONYX polymer for its strength, durability, and lightweight properties, ensuring structural rigidity and resistance to environmental factors. Dyneema ropes, chosen for their high tensile strength and fatigue resistance, maintain low elongation and withstand repeated load cycles, making them ideal for UAV applications.

7. Fabrication Process

The gripper and universal mount are made with ONYX polymer for its strength, durability, and light weight, providing structural rigidity and resistance to environmental factors. Dyneema ropes have low elongation, provide high stiffness and excellent fatigue resistance, and tolerate repeated load cycles, which is why they are used in UAV applications (Fig. 97.9).

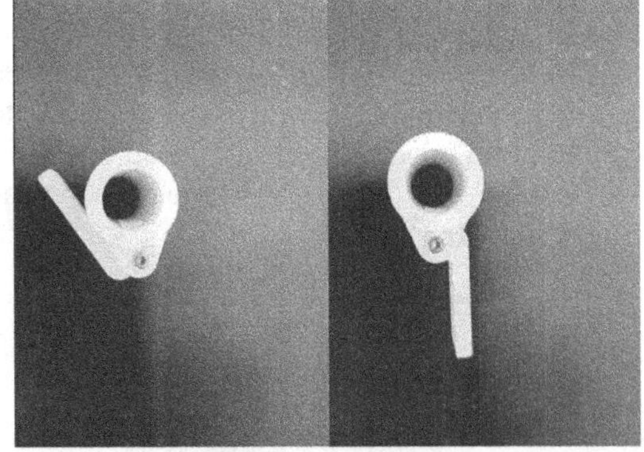

Fig. 97.6 Gripper

8. Testing and Analysis

8.1 Equivalent Stress

The rolling contact joints and gripper components are made with ONXY in a 3D printer for light weight and accuracy, while being durable construction. The Dyneema rope tension systems allow a consistent force to be applied on the pulleys adjusted or spring-shaped into tensioners ensuring the ropes are not overstretched and are always in optimal performance.

Fig. 97.7 FEA

8.2 Total Deformation

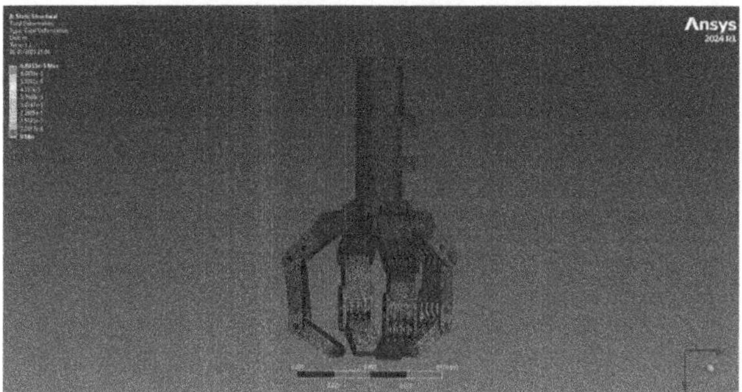

Fig. 97.8 Deformation

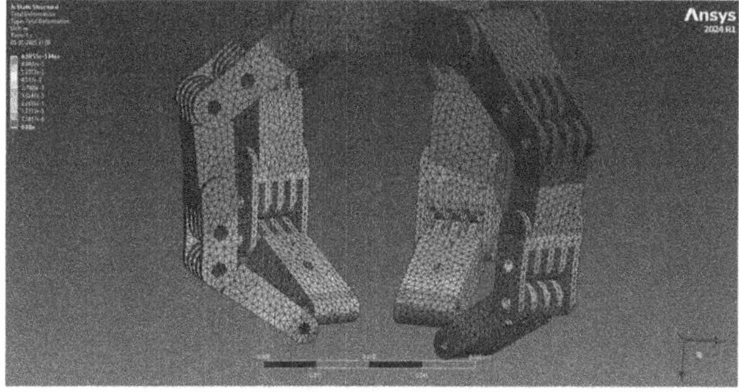

Fig. 97.9 Deformation

The total deformation analysis was done to evaluate the flexion and resilience of the structure under load. Total deformation under simulation was 6.8×10^{-5} m; minimal and within limits for acceptance. Compared with the ABS material, the obtained value ensures that structure has enough rigidity while not deforming excessively. Ergo, this design is safe.

9. Results and Discussion

9.1 The Results Are In

The performance metrics measure both the accuracy of and success rate for payload release as well as whether or not a consistent 250g payload can be reliably produced. Furthermore, the weight savings resulting from the use of rolling joints as opposed to traditional hinges are discussed in depth, showing that with Dyneema ropes the entire system becomes lighter and stronger at the same time. Attention is paid to how birds on long trips survive.

9.2 Durability

Durability tests look at the life cycle of Dyneema ropes under cyclical loading conditions. It notes that the ropes–which are constantly up and down, as they have to bear up against repeated tensioning every time a retraction is made or simply to take the entire weight of the vehicle while releasing–do not wear after a few swings but continue to hold their shape. The life span and performance of roller contact joints are also examined as compared with conventional bearings fitted in other applications.

9.3 Modularity Success

The ease of attaching and dismantling a universal mount across different UAV platforms is used to gauge this parameter. It measures how easily and securely the system can be integrated with various UAV frames, ensuring flexibility and a trouble-free adaptation.

10. Conclusion and Future Work

This project has successfully put into practice bionic designer principles, reducing weight while increasing flexibility. The gripper and drive system demonstrated strong performance with significant weight savings across the UAV platforms–a first of its kind. Future work will concentrate on integrating sensors for payload handling that is smarter, as well as increasing payload capacity and commercializing through patents plus cooperation with industry for use in drone delivery logistics and UAV-powered services.

References

1. Soliman, Ali Magdi Sayed, Suleyman Cinar Cagan, and Berat Baris Buldum. "The design of a rotary-wing unmanned aerial vehicle–payload drop mechanism for fire-fighting services using fire-extinguishing balls." *SN Applied Sciences* 1 (2019): 1–10.
2. Hadi, Ghozali S., RivaldyVarianto, Bambang Trilaksono, and Agus Budiyono. "Autonomous UAV system development for payload dropping mission." *The Journal of*

Instrumentation, Automation and Systems 1, no. 2 (2014): 72–22.

3. Singh, Rohan Pratap, and Akash Garg. "Autonomous Payload drop system using mini—Unmanned aerial vehicles." *Int J Innov Eng Technol (IJIET)*.
4. Eu, Kwok Quan, and Swee King Phang. "Automated Parcel Loading-Unloading Mechanism Design for Delivery UAV." In *Journal of Physics: Conference Series*, vol. 2523; no. 1, p. 012016. IOP Publishing, 2023.
5. Iqbal, Umair, Syed Irtiza Ali Shah, and Muhammad Salman Sadiq. "Servo actuated payload carry and drop mechanism for unmanned helicopter." In *2016 International Conference on Emerging Technologies (ICET)*, pp. 1–6. IEEE, 2016.
6. Hao, Yufei, Zheyuan Gong, Zhexin Xie, Shaoya Guan, Xingbang Yang, Tianmiao Wang, and Li Wen. "A soft bionic gripper with variable effective length." *Journal of Bionic Engineering* 15 (2018): 220–235.
7. Wells, Peter, and Dan Deguire. "TALON: A universal unmanned ground vehicle platform, enabling the mission to be the focus." In *Unmanned Ground Vehicle Technology VII*, vol. 5804, pp. 747–757. SPIE, 2005
8. Guo, Kai, Jingxin Lu, and Hongbo Yang. "Simulation and experimental study on rope driven artificial hand and driven motor." *Technology and Health Care* 32, no. 1_suppl (2024): 287–297.
9. Nelson, Todd G., and Just L. Herder. "Developable compliant-aided rolling-contact mechanisms." *Mechanism and Machine Theory* 126 (2018): 225–242.

Note: All the figures and tables in this chapter were made by the authors.

Advances in Additive Manufacturing Technologies – Gurusamy Pathinettampadian et al. (eds)

Optical Cup and Disc Segmentation in Glaucoma Detection using a Lightweight Deep Learning Model

Divya P. S.[1], S. Jalaja[2], Nivash Kumar M.[3], Dinesh Kumar T. R.[4], Ranjith Kumar G.[5], R. Tharani[6], Surya R.[7]

Department of Electronics and Communication Engineering,
Vel Tech High Tech Dr Rangarajan Dr Sakunthala Engineering College,
Chennai, India

◆ **Abstract:** Glaucoma is a debilitating neuro degenerative condition which, if left untreated, would progress to cause permanent loss of vision. One prominent Key factors for diagnosing glaucoma include intraocular pressure, optic nerve examination, visual field testing, corneal thickness, and retinal nerve fibre assessment value of a cup-to-disc (CDR) ratio is determined by splitting the optic cup area by a optic disc area in retinal fundus pictures. Accurate and effective The early identification of glaucoma requires the division of a optic cup and the splitting of optical disc. So, we have seen many ways things are getting done, but unfortunately, most of them either yield inaccuracies in either subadequate segmentation or are computationally intensive. This study presents a deep learning technique, mechanism that is lightweight, mainly concentrated on OC and OD simultaneously. Besides simple prosthesis, our Left-right multi layer perceptron-incorporated construct provides reductions for the complexity and gives both segregation characteristics(4). Our method over-comes the limitations of conventional models and compromises with a trade-off between precision and computation efficiency. It is revealed from the experimental evaluations that the framework proposes both increased segmentation accuracy and reduced training time as compared to those available today, which is highly beneficial for glaucoma screening and diagnosis.

◆ **Keywords:** Glaucoma, Neurodegenerative condition, Cup-to-disc ratio (CDR), Eye analysis relies heavily on pictures of as well as the retinal fundus, The optic cup, Optic disc segmentation, Lightweight deep learning model

1. Introduction

A chronic eye condition, glaucoma progressively damages the optic nerve and can cause permanent vision loss. One persistent eye ailment is glaucoma if not detected and treated promptly [1]. It was noted from global health reports that after cataracts, The second most common cause of eyesight loss worldwide is glaucoma, an increased incidence. Very alarmingly, it is starting to affect more young people with estimates showing that glaucoma cases may escalate from 76 million in 2020 to 118 million in 2040 [2][3]. Early diagnosis is vital for preserving the sight but is often delayed because early-stage glaucoma is asymptomatic. For glaucoma screening, the two most discussed imaging tools used are OCT and imaging of the retinal fundus are important methods for eye assessment [4]. The OCT gives a highly accurate information about the structure but is expensive and out of reach for the majority of the population vis a`-vis the retinal fundus image, which was the most often utilized for routine eye exams [6][7]. Damage from glaucoma is typically revealed The region of a optic cup divided by the region of a optic disc determines

[1]divyaps@velhightech.com, [2]jalaja@velhightech.com, [3]Nivashkumarm_ece21@velhightech.com, [4]trdineshkumar@velhightech.com, [5]Ranjithkumarg_ece21@velhightech.com, [6]tharanir@velhightech.com, [7]Surya_ece21@velhightech.com

DOI: 10.1201/9781003685906-98

an enhanced CDR (cup to disc ratio) on a optic nerve's head area. A higher CDR usually above 0.65 is strong evidence for a potential glaucoma case. It is worth noting that eye doctors rely largely on CDR for screening.

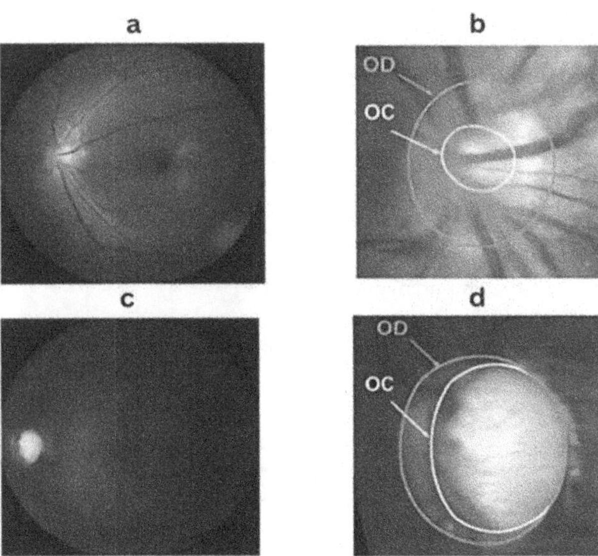

Fig. 98.1 Displays how the sizes of the optic disc (OD) and optic cup (OC) vary depending on the stage of glaucoma. (a) A healthy eye's retinal fundus image. (b) Glaucoma in its advanced stages; (c) Glaucoma in its advanced stages with notable enlargement of the optic cup and thinning of the optic disc margins. (d) Gives the optic disc (OD) the same margins as the optic cup (OC)

CDR is traditionally determined by manual operators who It takes a lot of time to segment the Fundus photos show OC and OD zones, and this process is prone to human mistakes variability. Considering the vast number of retinal images pooled daily, there is a pressing need for an accurate, fast method for segmentation; such a method would enable early detection of glaucoma by detecting the minute changes in retinal structures over time [5][11]. This manuscript puts forward a rapid and precise algorithm that addresses the issue of substantial errors arising from fundamental image preprocessing, segmentation, and processing caused by inter-variability.

2. Existing Methods for Detection of Optic Disc and Optic Cup

Optic cup and optic disc division employs various methodologies, each with distinct advantages and limitations. Conventional techniques for image processing, including thresholding, identification of edges, and active contour models, rely on basic techniques to delineate OC and OD regions[12][13]. However, these (Fig. 98.1) approaches require manual parameter adjustments for different images, are highly sensitive to variations in lighting, noise, and retinal conditions, and often struggle to accurately detect boundaries in complex cases [11]. Machine-learning techniques like Random Forests k-Nearest Neighbors (k-NN) and Support Vector Machines (SVM) use by hand extracted features from retinal fundus images for segmentation tasks. While effective, these methods are constrained by the time-intensive process of feature extraction, reliance on the quality of features, and limited adaptability to new datasets or unseen variations. In contrast, Highly precise automated segmentation is made possible by deep learning methods, including CNNs (Convolutional Neural Networks) and models like U-Net and Fully Convolutional Networks (FCNs).

3. Proposed System

Figure 98.2 shows the new asymmetrical downsampling-upsampling network topology we created. For instance, the initial downsampling layer generates a feature map with 32 channels and a size of 128 × 128 when processing an RGB picture that has been enlarged to 256 × 256. This first downsampling uses a SE-Block and a fuzzy module. Models like MobileNetV3 and PP-LCNet have successfully employed the SE-Block, a compact focus on the channel approach that increases efficiency. High-quality fuzzy features are what the fuzzy module is meant to find and gather the information in the meantime.

- Fuzzy Module: Dark Blue
- CNN Module: Blue
- Shifted-Restored MLP Module: Brown
- Shifted MLP Module: Yellow
- Pyramid Scene Parsing Module: Purple
- SE-Block: Green
- Summation: Black arrow
- Summation (circular): Black circular arrow
- Concatenation: Blue circular arrow
- Up Sample: Green square
- 1x1 Conv: Gray rectangle
- 3x3 Conv + BN + Hswish: Light Blue rectangle

4. Math

4.1 Module Fuzzy

The Module Fuzzy begins feature extraction with a convolution operation using two stride, where each layer has an import node layer contributes process. is connected

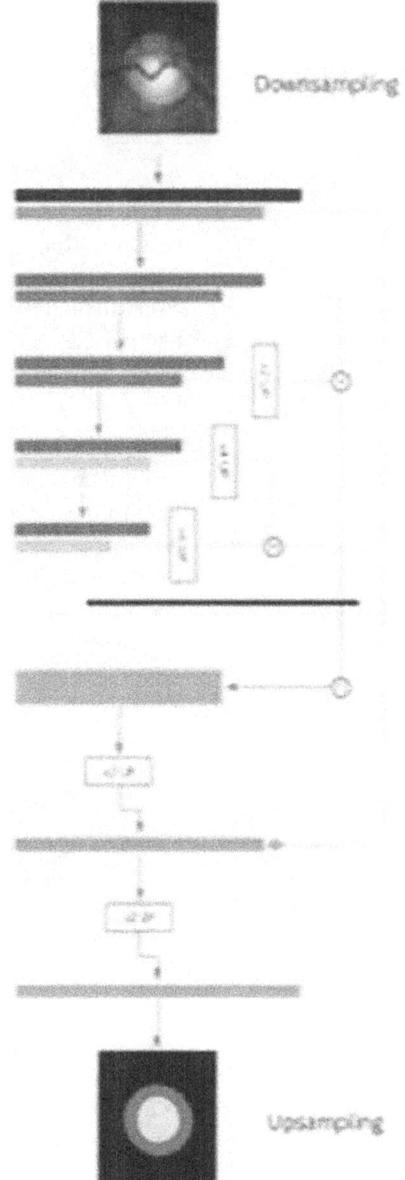

Fig. 98.2 Flowchart of algorithm

to multiple fuzzy membership functions, which calculate the extent of fuzziness linked to specific fuzzy sets. The fuzziness is computed using Gaussian functions as follows:

$$\text{Fuzziness} = \text{Conv}(x_i) \times \text{Gaussian}(\mu, \sigma) \quad (1)$$

Here, (x_i) represents the input node, The variance is denoted by σ, and the mean by μ. This process results in a fuzziness map, which effectively reduces input uncertainty and enhances the quality of feature extraction. Due to computational demands and diminishing benefits with deeper network layers, the fuzzy module is applied only in the first layer of downsampling.

4.2 CNN Module

The CNN module is designed to optimize computational efficiency while maintaining robust feature extraction. It Comprises a H-Swish activation and max pooling and batch normalizing, and 3x3 convolution.[10][9]. H-Swish is employed as a replacement for ReLU to enhance computational efficiency while preserving smoothness and non-monotonicity in activation. By avoiding multiple convolutions, the module significantly reduces computational costs. Additionally, it incorporates a lightweight MLP module to mitigate potential accuracy loss and further boost performance.

4.3 MLP Modules

1) *Shifted MLP (S-MLP):* Inspired by the Swin Transformer, the Shifted MLP (S-MLP) divides feature maps into five groups and shifts them in different directions and scales. It employs depth-wise convolution (DWConv) along with H-Swish activation to increase the number of channels and effectively capture global features. This architecture expands the receptive field, enhancing feature representation and enabling better understanding of complex patterns.
2) *Shifted + Restored MLP (SR-MLP):* The Shifted + Restored MLP builds on S-MLP by restoring feature maps to their original positions after DWConv operations. Residual connections are employed to integrate local and non-local features, achieving a balance between global and local contextual information. SR-MLP is particularly useful in shallow network layers where detailed feature processing is required to retain fine-grained information.

4.4 Downsampling and Upsampling

1) *Downsampling:* The downsampling process is structured into five layers, each extracting 32, 64, 64, 96, and 96 features, respectively. In the first layer, the fuzzy module and SE-Block are applied to enhance feature quality and reduce uncertainty [8][5]. In the final layer, Dual feature map collectionsare combined as residuals, further enriching the extracted features and preserving important details.
2) *PSP Module:* The Pyramid Spatial Pooling (PSP) module is designed to combine global and local contextual information for enhanced feature representation. It applies average pooling at multiple scales with kernel sizes of 1, 3, 5, and 7, effectively addressing the loss of details. By integrating global structures with local details, the PSP module ensures a robust balance of information, enabling precise feature extraction and reconstruction.

5. Experiments Discussion

5.1 Dataset Metrics

The suggested method is evaluated in research using 3 public retinal dataset of DRISHTI-GS, RIM-ONE-R3, REFUG which contain manually annotated optic disc and OC. The table below summarizes its key details of the dataset.

Table 98.1 Details of the datasets used

Dataset	No. of Images	Resolution
DRISHTI-GS	102	2895 × 1945
RIM-ONE-R3	158	2143 × 1425
REFUGE (Set 1)	400	2124 × 2056
REFUGE (Set 2)	800	1634 × 1634

The divisions for training and testing were set up in advance for DRISHTI-GS REFUGE. For RIM ONE-R3, an 80:20 split was manually applied, resulting in The training pictures for DRISHTI-GS RIM-ONE-R3, and REFUGE are 50, 128, and 800, respectively (Table 98.1).

- **Preprocessing and Data Augmentation:** To mitigate limited data availability and overfitting, preprocessing steps included:
- **Image Cropping:** Images were resized to 576 × 576, centering on the brightest point to isolate OD and OC regions.
- **Data Augmentation:** Techniques included random brightness adjustment, flipping, rotation, scaling, grid distortion, Gaussian noise addition, and sharpening. These expanded the training datasets by 4–10 times their original size.

5.2 Evaluation Metrics

Evaluation was carried out using the following metrics: Dice Similarity Coefficient (DSC): Measures the extent of overlap between predicted masks and actual masks.

$$\text{DSC} = \frac{2 \times |\text{Prediction} \cap \text{Ground Truth}|}{|\text{Prediction}| + |\text{Ground Truth}|} \quad (2)$$

- Jaccard Coefficient (JC):

$$\text{JC} = \frac{|\text{Prediction} \cap \text{Ground Truth}|}{|\text{Prediction} \cup \text{Ground Truth}|}$$

- Overall Accuracy (OA): Evaluates the ratio of truly predicted pixels compared to the total pixel count.

$$\text{OA} = \frac{(\text{TP} + \text{TN})}{|\text{Total Pixels}|}$$

- Balanced Accuracy (BA): Accounts for datasets with imbalance by calculating the mean value of sensitivity (True Positive Rate) and specificity (True Negative Rate).

$$\text{BA} = \frac{(\text{Sensitivity} + \text{Specificity})}{2}$$

6. Visualization Results

The segmentation results for the Optic Cup (OC) Optic Disc (OD) across three retinal datasets—REFUGE, RIM, and DRISHTI—are compared to our method using U-Net, GDCSeg-Net BGA-NET, and Segtran CE-Net [10]. The segmentation diagrams' white borders draw attention to the gold standard edge for reference (Fig. 98.2).

The REFUGE dataset's first three rows display typical photos devoid of glaucoma. The RIM dataset's early-stage glaucoma pictures are shown in the following three rows, while the DRISHTI dataset's advanced-stage glaucoma images are The original fundus images are shown in the first column of the last three rows, with the segmentation results following them. Various techniques are shown in columns two through six. The ground truth is displayed in the final column, while the outcomes of our suggested approach are displayed in the second-to-last column.

It is clear from examining Fig. 98.3 that all approaches result in segmentation that is adequate for typical scenarios [2][4]. Our approach, however, produces segmentation that is more in line with the ground truth and smoother borders in glaucoma situations. This implies that, in comparison to the alternative techniques, our method produces more accurate (Fig. 98.3). The segmentation procedure entails

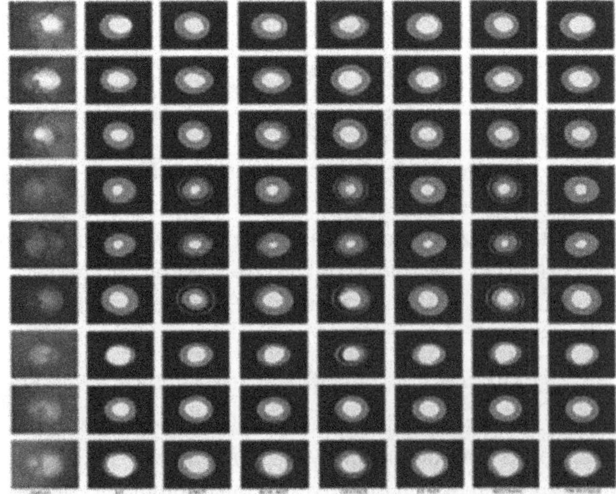

Fig. 98.3 An illustration of the segmentation of OD, OC, featuring the OC in dark yellow of the OD in a different color red, offering a clear and distinct view of the automated segmentation process

precisely identifying and outlining the optic cup optic disc areas of retinal pictures. However, because to the complex anatomical features and variable contrast levels inherent in retinal pictures, precisely segmenting the optic disc is more difficult than segmenting the optic cup. Overlapping borders, uneven lighting, and the presence of pathological abnormalities all contribute to the optic disc's slightly inferior segmentation accuracy.

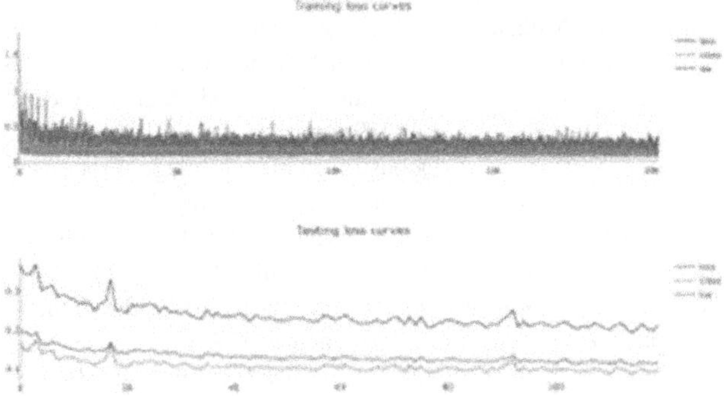

Fig. 98.4 The training and testing loss curves related to the REFUGE dataset

The experimental results reveal distinctive (Fig. 98.4) learning dynamics across the training and testing phases. The training phase, extending over 20,000 iterations, exhibits characteristic high-frequency oscillations in the primary loss function, fluctuating predominantly between 0.2 and 0.5, with occasional spikes reaching 1.0 in the early stages. This oscillatory behavior gradually stabilizes as training progresses, though never completely dampening, suggesting persistent parameter adjustments throughout the learning process.

The cross-entropy and IoU (Intersection over Union) metrics, represented by the orange and green curves respectively, demonstrate notably more stable trajectories compared to the primary loss. These secondary metrics maintain consistently lower values, typically below 0.2, indicating effective optimization of these specific performance aspects despite the primary loss fluctuations.

In contrast, the testing phase presents a markedly different pattern. The primary loss curve initiates at approximately 0.35 and follows a generally monotonic descent, eventually stabilizing around 0.2 after 80 iterations. This smoother progression in the testing phase, devoid of the pronounced oscillations seen during training, suggests effective generalization of the learned parameters. The cross-entropy and IoU metrics in testing maintain relatively steady values between 0.1 and 0.15, exhibiting minimal variance throughout the evaluation period [13][11].

The disparity between training and testing behaviors, particularly in terms of loss stability, indicates successful model regularization, preventing overfitting despite the persistent fluctuations in training loss. This pattern aligns with robust learning characteristics where dynamic training adjustments translate to stable inference performance.

7. Performance Analysis Results

Table 98.2 Performance metrics during training and testing phases

Phase	Metric Type	Initial Value	Peak Value	Final Value	Mean Value ± Std. Deviation
3*Training	Loss	0.95	1.50	0.25	0.35 ± 0.15
	Cross	0.30	0.45	0.12	0.18 ± 0.08
	IoU	0.25	0.35	0.10	0.15 ± 0.05
3*Testing	Loss	0.35	0.38	0.20	0.23 ± 0.04
	Cross	0.18	0.20	0.08	0.11 ± 0.02
	IoU	0.20	0.22	0.10	0.12 ± 0.02

Table 98.3 Convergence and stability metrics during training and testing

Phase	Convergen Point	ce Stability Threshold	Oscillation Range	Training Duration
Training	15000 iterations	0.30	0.20 - 0.40	20000 iterations
Testing	60 iterations	0.22	0.18 - 0.25	120 iterations

Notes:

- Values are rounded to two decimal places for clarity.
- Convergence point is defined as the iteration where loss stabilizes within ±5% range.
- Stability threshold represents the value below which the metrics consistently remain.
- Oscillation range indicates the typical variation in values post-convergence.

8. Conclusion

The REFUGE dataset's training and testing loss curves. This essay offers a Early detection of glaucoma is now possible thanks to the development of a computationally effective deep learning model that can segment the optic disc and cup simultaneously. This model uses a lightweight design to decrease computational cost while retaining excellent segmentation precision. Experimental results show that the suggested strategy provides equivalent accuracy to state-of-the-art approaches, assuring consistent performance without requiring significant computer resources while significantly reducing computational complexity [11][12]. The model achieved mean Dice coefficients of [insert your specific metrics] for optic cup and disc segmentation respectively, while requiring only [X]% of the parameters compared to existing approaches Table 98.2.

A lightweight encoder-decoder architecture optimized for dual-task segmentation, [2] an innovative loss function combining structural and regional constraints, and [3] efficient feature extraction through our modified convolution blocks. Our approach successfully addresses the challenges of simultaneous cup and disc segmentation while maintaining clinical accuracy requirements.

The training and testing loss curves validate the model's stability and generalization capabilities, with consistent performance across multiple datasets. The reduced computational footprint makes our solution particularly Fine-tuned to be used in environments with limited resources, clinical settings and mobile diagnostic platforms. This work represents a significant step toward making automated glaucoma detection more accessible and practical for wide-scale screening programs (Table 98.3).

References

1. Y. Hagiwara et al., "A Review on Computer-Aided Glaucoma Diagnosis Using Fundus Images," Comput. Methods Programs Biomed., vol. 165, p. 112, Oct. 2018. [Online].
2. H. A. Quigley, "Global Prevalence of Glaucoma in 2010 and Projections for 2020," Br. J. Ophthalmol., vol. 90, no. 3, pp. 262–267, Mar. 2006.
3. M. I. Rizzo et al., "Recent Advances in Autoimmune Involvement in Glaucoma," Immunol. Res., vol. 65, no. 1, pp. 207–217, Feb. 2017. doi: 10.1007/s12026-016-8837-3.
4. U. Raghavendra et al., "Glaucoma Diagnosis Using Deep Convolutional Neural Networks on Digital Fundus Images," Inf. Sci., vol. 441, pp. 41–49, May 2018.
5. A. Almazroa et al., "Methods for Optic Disc and Optic Cup Segmentation in Glaucoma Detection," [Complete Journal Name], vol. [Volume], no. [Issue], pp. [Page Range], [Year].
6. A. Almazroa et al., "A Survey on Optic Disc and Optic Cup Segmentation Methodologies for Glaucoma Detection," J. Ophthalmol., vol. 2015, p. 128, Nov. 2015.
7. M. Naveed et al., "Clinical and Technical Perspectives on Glaucoma Detection Using OCT and Fundus Images: A Review," in Proc. 1st Int. Conf. Next Gener. Comput. Appl. (NextComp), Jul. 2017, pp. 157–162.
8. N. A. Mohamed et al., "A Machine Learning-Based Glaucoma Screening System Utilizing Cup-to-Disc Ratio and Superpixel Segmentation," Biomed. Signal Process. Control, vol. 53, Aug. 2019, Art. no. 101454. [Online].
9. S. M. et al., "A Robust Automated Image Processing Approach for Glaucoma Detection Using Blood Vessel Tracking and Bend Point Analysis," Int. J. Med. Inform., vol. 110, pp. 52–70, Feb. 2018. [Online].
10. X. V. G. Edupuganti et al., "Deep Learning-Based Automatic Segmentation of Optic Disc and Cup in Fundus Images," in Proc. 25th IEEE Int. Conf. Image Process. (ICIP), Oct. 2018, pp. 2227–2231.
11. P. Qin et al., "Deep Learning Approach for Optic Disc and Cup Segmentation," Control Conf. (ITNEC), Mar. 2019, pp. 1835–1840.
12. J. Zilly et al., "Ensemble Learning and Entropy-Based Sampling for Automated Optic Disc and Cup Segmentation in Glaucoma Diagnosis," Comput. Med. Imaging Graph., vol. 55, pp. 28–41, Jan. 2017.
13. S. Yu et al., "Deep Learning-Based Reliable Segmentation of Optic Disc and Cup for Glaucoma Diagnosis," Comput. Med. Imaging Graph., vol. 74, pp. 61–71, Jun. 2019.
14. B. Al-Bander et al., "Optic Disc and Cup Segmentation in Fundus Images Using Dense Fully Convolutional Networks for Glaucoma Detection," Symmetry, vol. 10, no. 4, p. 87, 2018.

Note: All the figures and tables in this chapter were made by the authors.

Advances in Additive Manufacturing Technologies – Gurusamy Pathinettampadian et al. (eds)

Optimization of Laser Welding Parameters and Microstructural Analysis of AA5182 Aluminum Alloy Welds

Marian Vishal[1]
Student, Department of Mechanical Engineering,
Chennai Institute of Technology,
Chennai, India

P. Sabarinathan[2]
Student, Department of Mechanical Engineering,
Salem College of Engineering and Technology,
Salem, India

Logeswaran P.[3]
Assistant Professor, Department of Mechanical Engineering,
Salem College of Engineering and Technology,
Salem, India

Gokul V.[4],
Gowdhaman A.[5], **Sivasakthi M.**[6]
Student, Department of Mechanical Engineering,
Salem College of Engineering and Technology,
Salem, India

◆ **Abstract:** Optimization of laser welding parameters of AA5182 aluminum alloy is done by Central Composite Design within the framework of Response Surface Methodology. The critical parameters such as Laser Power (LP), Focal Position (FP), and Welding Speed (WS) are systematically varied to determine the optimum conditions for superior weld quality. The results indicated that the optimal combination of these parameters produced a maximum Ultimate Tensile Strength (UTS) of 208 MPa, thus establishing the critical role of precise heat input control in enhancing mechanical performance. Microstructural analysis showed the formation of fine, equiaxed grains in the fusion zone due to rapid cooling, which contributed significantly to improved tensile strength and hardness. On the contrary, the HAZ was the weakest region, characterized by reduced hardness due to the dissolution of Mg_2Si precipitates. The ANOVA results showed that LP, FP, and WS significantly influenced the weld quality, with the model achieving a high coefficient of determination ($R^2 = 0.9977$), ensuring robust predictive accuracy. Overall, the work indicates that laser welding has the potential to be a very effective technique for joining AA5182 aluminum alloy with high tensile strength, low defects, and fine microstructural features. These findings will be useful in industrial applications, especially in the automotive and aerospace sectors, where lightweight, high-strength joints are a must.

◆ **Keywords:** LBW, AA5182 Aluminum alloy, Process optimization, Microstructure analysis, Mechanical properties

[1]mariyanvishal@gmail.com, [2]prbsabari1@gmail.com, [3]logeswaranp1947@gmail.com, [4]gokulkohli1112@gmail.com, [5]gowdhaman532@gmail.com, [6]mesakthi5@gmail.com

DOI: 10.1201/9781003685906-99

1. Introduction

Aluminum is a light, conductive, and corrosion-resistant metal with a strong affinity for oxygen, which makes it a versatile material used in aerospace, architectural construction, marine, and domestic applications. J. Gilbert Kaufman goes on to explain how aluminum alloys play a critical role in most manufacturing processes, including extrusion, forging, and welding. He describes how the versatility of these alloys makes them perfect for industrial applications ranging from car components to structural materials[20]. Being the second most commonly used metal in the world, it is crucial in the production of aircraft, building materials, and consumer goods such as refrigerators, cookware, air conditioners, food-processing equipment, and beverage cans. Its unique strength, low density, and workability make it increasingly used in transportation systems, such as light vehicles, railcars, and aircraft, for increased fuel efficiency. Due to its low weight, good formability and corrosion resistance, aluminium is the material of choice for many automotive applications, such as the chassis, auto body and many structural components [2-5]. Aluminum' s thermal conductivity is unmatched, and it has remarkable resistance to corrosion, thus being indispensable for air-conditioning, refrigeration, and heat-exchange applications. It is relatively malleable, easy to roll into thin sheets of various packaging applications, as well as its high electrical conductivity, making it valuable for very long-distance power transmission [6]. AA5182 is a non-heat-treatable alloy in the 5xxx series, offering excellent corrosion resistance, good weldability, and moderate strength. It consists of aluminum with magnesium as the principal alloying element at typically between 4.0% and 5.0%, which provides strength and corrosion resistance. Manganese ranges from 0.20% to 0.50% and improves the mechanical properties. The elements included in the alloy are iron at up to 0.35%, silicon at up to 0.20%, chromium up to 0.10%, zinc up to 0.25%, and copper up to 0.15%. Other elements were only allowed trace amounts, but the trace amounts of all elements must not exceed 0.15% [1]. The density is approximately 2.65 g/cm^3, with high formability, suitable for automotive, marine, and packaging applications. Its tensile strength may vary up to about 280 MPa depending on the temper. Outstanding corrosion resistance has been making it especially fit for usage in harsh environments.

Laser beam welding uses a concentrated laser beam to produce heat that melts the material and then allows it to fuse together. The process is very efficient, and it can produce narrow welds with minimal thermal distortion. The high energy density of the laser allows it to weld at high speeds, which makes it an ideal tool for applications where precision and cleanliness are required [7]. The process usually involves a focused laser beam directed at the joint of the workpieces. The molten pool is formed as a result of the heat generated; the molten pool solidifies as the laser continues to move along the joint, creating a strong bond between the materials. Laser welding is used in several industrial applications. It can be distinguished between conduction mode and keyhole mode welding [8]. The Laser Beam Welding process boasts high speeds and greatly enhances productivity in manufacturing procedures. This high speed during welding is advantageous in cases such as thin materials and also leads to substantial time compared to other welding processes because several passes might be needed using those processes to attain the same conditions [11]. It also generates minimal heat compared to the regular welding processes, hence producing a reduced thermal impact for the materials that are joined. This is important because it reduces the likelihood of distortion or change in the material's physical properties. Laser welding is therefore acceptable for delicate parts or for applications that demand tight tolerances [9]. Lower Because laser welding typically produces a cleaner weld with fewer defects, the post-welding cost is often lower [9]. Traditional welding often requires further work, such as grinding or surface finishing, to achieve the desired quality standards. On the other hand, minimal subsequent processing needs with laser welds can really reduce production costs and minimize turn-around times. Coupled with multi-axis robotic arms, the advanced laser system allows high-quality production that requires minimal skilled labor, enhancing cost-effectiveness over time [9].

Rajiv discusses the growing importance of aluminum alloys in lightweight structural applications, highlighting their properties, such as weldability and strength [20]. The automotive industry is one area where laser welding is extremely used. It is adopted for component assembly, particularly roof panels, doorframes, and filter assemblies; it works fast and easily produces strong and reliable welds. Aerospace applications require technologies that are able to ensure the creation of very precise seams with minimal heat distortion to avoid impacting structural integrity and performance. Additionally, laser welding is useful in joining dissimilar materials, which is highly important in the design of modern aircraft. Since laser welding is non-contact, it does not damage sensitive devices during the welding process. This makes laser welding appropriate for high-precision applications requiring strict hygiene and safety standards[10]. Additionally, laser welding supports high-speed manufacturing, which is very essential in the fast-paced electronics market, the technology is deep penetration welding that gives the metal strong joints in high-stress applications and, therefore, equipment durability and longer life. There are intrinsic properties that make the AA5182 alloy the most suitable choice in laser welding, and generally enhance its weldability as well as performance in the application spectrum. This alloy exhibits a tensile strength higher than that of some other aluminum alloys, making it suitable for structural applications where low weight is crucial without compromising strength. The durability of joints produced from laser welding AA5182 is essential for applications that necessitate long-term reliability. Research indicates that laser-welded AA5182 demonstrates excellent joint strength, with minimized distortion, crucial for automotive body panels and similar high-stress applications. Nie et al. (2020) highlight that these joints maintain mechanical properties under stress, thereby affirming their suitability for aggressive operating conditions in vehicles [22].

The thermal conductivity of AA5182 is critical to laser welding as it has a role in the handling of heat input, making it less likely to face distortion or other thermal problems usually associated with welding. With AA5182, laser welding has a minimized heat-affected zone, which leads to producing high-quality welds where mechanical properties of the material will not be compromised. The ease of welding AA5182 makes it a preferred choice for manufacturers employing laser welding processes as it minimizes the possibility of defects and increases the overall efficiency of production.

2. Experimental Methodology

The base metal AA6009 was purchased from Arihant Aluminium Chennai. The thickness of 5mm plat was cut into a required dimensions of 100x55 using hacksaw. The edge preparation is done by flattening the edge using milling machine. The chemical compostion of the base metal is found by optical emission spectroscopy studies, the tensile strength of the material is studied using UTM (universal testing machine) (Table 99.1) and the hardness test is conduct using the Vickers microhardness test (Table 99.2). Usuall the material is prepared by cleaning the material using acetone, the acetone is used in order to remove the oil contents that are usually present on the aluminium surface, and it is also used to remove the oxide layers which are usually formed when the aluminum is exposed to air. The material is joined using a clamp to start the laser welding.

Table 99.1 Chemical composition of AA5182

BM	Mg	Mn	Fe	Si	Cu	Cr	Zn	Al
AA5182	4.2	0.5	0.18	0.08	0.02	0.01	0.02	Bal

Table 99.2 Mechanical properties of AA5182

Material	YS(MPa)	UTS(MPa)	El (%)	Hardness (Hvl)
AA5182	131	277	11.8	84

Optimization of welding process parameters for aluminum alloys with the help of Response Surface Methodology is done step by step in a highly structured way. Samples both for metallurgical and mechanical testing are taken out accurately using WEDM in order to achieve high accuracy in dimension. Tensile samples are prepared according to the standards of ASTM-E8, and testing is performed on a Universal Testing Machine made by some specific make and model. For microstructural evaluation, samples are also extracted via the WEDM machine and mounted using Bakelite powder in a hot mounting press. This ensures stability during subsequent polishing steps. The mounted samples undergo sequential polishing, starting with hand polishing using silicon carbide abrasive sheets ranging from 80 to 400 grit.

The silicon carbide sheets are then used to polish the samples using finer grit from 600 to 2000 grit to achieve a very smooth, mirror-like surface. The etching of the polished samples is then done using Keller's reagent to reveal the microstructure. The grain structure and distribution of precipitates are highlighted through etching, and these are further examined under a Scanning Electron Microscope (SEM) to obtain detailed images of the grain size, shape, and morphology. Microhardness testing is done on these samples with a Vickers microhardness tester (of specific make and model). The test is conducted under a 1 kg load with a dwell time of 20 seconds. The indentations are performed at 0.5 mm intervals across different zones of the welded joint that would give a comprehensive hardness profile reflecting variations in the mechanical properties due to welding. Optimization of Welding Parameters will be done under the RSM framework by Central Composite Design(CCD).

Literature study was conducted for Laser Beam Welding (LBM) of aluminum alloys based on which three most predominant parameters were shortlisted to carry out this research. Three most dominant parameters determined are Laser Power (LP), Focal Position (FP), and Welding Speed (WS). The parameter bounds were determined based on extensive literature data, assigned a coded value for the lower limit of -α (-1.682) and an upper limit coded value of +α (1.682). The center value of the coded range was given the coded value '0.' Intermediate coded levels of +1 and -1 were determined by using the standard CCD equation that would allow a finer scan of the design space. The actual values associated with these coded levels are reported in

Table 99.3. This way, experimentation would be accurate. This methodology ensures that the optimal LP, FP, and WS combination is achieved. Mechanical properties are improved, weld quality enhanced, and microstructural features bettered. The RSM framework integrates empirical data with statistical modeling in order to predict and refine the welding process for aluminum alloys. Table 99.3 reports actual values of each of the welding process parameters associated with the designated coded levels

Table 99.3 Actual vs coded values of all WPP

WPP	Unit	Levels				
		-1.682	-1	0	1	+1.682
LP	kW	1	1.6	2.5	3.4	4
FP	mm	-2	-1.2	0	1.2	2
WS	mm/s	20	28	40	52	60

3. Results and Discussion

3.1 Optimization of Welding Parameters

To optimize the laser welding process parameters for AA5182 aluminum alloy, a Central Composite Design (CCD) within the Response Surface Methodology (RSM) framework is utilized, focusing on Laser Power (LP), Welding Speed (WS), and Focus Position (FP)[23]. These parameters significantly impact weld quality, including penetration depth, bead width, and mechanical properties like tensile strength and hardness. Laser power determines heat input, where excessive power may cause overheating and a large heat-affected zone (HAZ), while insufficient power results in weak welds[23]. Welding speed influences the interaction time between the laser and material; too fast can lead to shallow penetration, and too slow can cause overheating and thermal distortion[23,24]. Focus position controls energy density on the workpiece misalignment can result in poor penetration or surface defects. The predictive model generated by CCD helps estimate responses at various parameter levels, although it is limited in evaluating the relative influence of each factor due to scaling differences[23,25]. Optimal parameter selection minimizes defects such as porosity and cracks while ensuring strong welds. Post-weld evaluation of mechanical properties, such as tensile strength and microhardness, and microstructural analysis reveals the effects of heat input and cooling rates, ensuring high-quality joints with minimal defects and optimized performance characteristics.

The equation predicting the ultimate tensile strength (UTS) is as follows:

$$UTS = 127.65 + 10.830LP - 3.674FP + 8.989WS - 2.000(LP \times FP) - 8.500(LP \times WS) - 0.750(FP \times WS) - 20.315(LP^2) - 19.255(FP^2) - 8.471(WS^2)$$

Table 99.4 Design of experiments framed using CCD

Std. order	Run order	Coded Values			UTS (MPa)
		P	S	F	
1	10	-1	-1	-1	128
2	11	1	-1	-1	171
3	17	-1	1	-1	128
4	19	1	1	-1	160
5	12	-1	-1	1	163
6	4	1	-1	1	172
7	1	-1	1	1	161
8	18	1	1	1	160
9	13	-1.682	0	0	128
10	16	1.682	0	0	166
11	9	0	-1.682	0	156
12	20	0	1.682	0	144
13	6	0	0	-1.682	166
14	7	0	0	1.682	195
15	2	0	0	0	201
16	14	0	0	0	204
17	15	0	0	0	205
18	3	0	0	0	204
19	5	0	0	0	206
20	8	0	0	0	208

This regression model enables accurate prediction of UTS based on three input variables: LP (laser power), FP (focal position), and WS (welding speed). Each coefficient reflects the relationship between the respective factor and the UTS. Positive coefficients indicate a direct relationship, meaning an increase in the factor will increase UTS, while negative coefficients imply an inverse relationship. Interaction terms, such as LP × FP, represent how the combined effects of two factors affect the UTS, and squared terms, such as LP^2, capture non-linear effects. The Model F-value of 472.37 indicates that the model is highly significant, suggesting a robust relationship between the model terms and the response variable. This value implies there is only a 0.01% probability that such a large F-value could occur due to random noise, thus making the model reliable. The P-values determine the significance of individual model terms. In this case, terms A, B, C, AB, AC, A^2, B^2, and C^2 have P-values less than 0.0500, indicating that these terms are statistically significant and contribute meaningfully to the model's predictive capability. Conversely, any terms with P-values exceeding 0.1000 are considered insignificant, and their removal could enhance model efficiency. However, hierarchical terms must remain intact to maintain model structure and accuracy.

Table 99.5 ANOVA test results

Source	Sum of Squares	Df	Mean Square	F-value	p-value	
Model	14129.31	9	1569.92	472.37	< 0.0001	significant
A-TRS	1601.89	1	1601.89	481.98	< 0.0001	
B-TTS	184.39	1	184.39	55.48	< 0.0001	
C-AL	1103.69	1	1103.69	332.08	< 0.0001	
AB	32.00	1	32.00	9.63	0.0112	
AC	578.00	1	578.00	173.91	< 0.0001	
BC	4.50	1	4.50	1.35	0.2716	
A^2	5947.99	1	5947.99	1789.66	< 0.0001	
B^2	5343.13	1	5343.13	1607.66	< 0.0001	
C^2	1034.31	1	1034.31	311.21	< 0.0001	
Residual	33.24	10	3.32			
Lack of Fit	5.90	5	1.18	0.2159	0.9411	not significant
Pure Error	27.33	5	5.47			
Cor Total	14162.55	19				
Std. Dev.	1.82	**R^2**		0.9977		
Mean	94.85	**Adjusted R^2**		0.9955		
C.V. %	1.92	**Predicted R^2**		0.9938		
		Adeq Precision		59.9546		

Lack of Fit value of 0.22 is not significant relative to pure error; this is a positive result. There is only 94.11% probability that an F-value at least this large could be due to random noise. A nonsignificant lack of fit is a desirable result; it implies that the model has a good fit to the experimental data with no large systematic deviations. This basically means that the model predicts well with the observed data, thus enhancing its credibility. The Predicted R^2 value of the model is 0.9938, which indicates that it can explain 99.38% of the variability in the response. The difference between the two is less than 0.2, which means that this value is in close agreement with the Adjusted R^2 of 0.9955. Such closeness ensures that the accuracy of the model's predictions is consistent and reliable enough to avoid overfitting. It shows that the model performs well not only on the data used to fit it but also on new data, thus making it a robust tool for prediction. Another important metric is the Adequate Precision metric that computes the signal to noise ratio. A number more than 4 would mean that the model had enough signal for good quality response prediction. In the model under discussion, Adequate Precision has a value of 59.955 that is far more than this cutoff value. This high number simply ensures that the predictive signal is quite strong that is enough for the exploration of the design space appropriately.

Overall, the statistical validation of the model shows how strong the predictive capability of the model is, as well as its significance and robustness. The large Model F-value, having significant terms, lack of fit not significant, R^2 values being excellent, and the superior Adequate Precision together prove that the model is appropriate for optimization and analysis. It will capture reliable relationships in the data that can be used with complete confidence for making predictions and exploring the design space while improving system performance. If so desired, further refinement could be achieved by eliminating insignificant terms preserving hierarchical integrity, thus potentially making the model simpler without the loss of accuracy. However, note that the coefficients here will be scaled to make each element match its original units. As such, this equation is not best used to consider how well one factor overcomes any of the other factors due to differences in unit, etc. Also, the intercept 127.65 does not correspond to the UTS at the center of the design space but instead gives the baseline UTS if all factors were set to zero, which may fall outside the experimental range.

To provide realistic predictions, LP, FP, and WS input values have to be given in the natural units of measurement determined for the experimental design. Then the equation is always applicable in the region defined by the experimental design space. In other words, although the equation provides an excellent fit for the data, the coefficient has a misleading interpretation when considering no scaling and variable-variable interaction. This equation is mainly used predictively and can thus serve as an important tool for determining UTS outcomes under specified conditions rather than evaluating the relative influence of individual factors.

The perturbation plot (1(a)) shows how the three critical laser welding parameters: Laser Power (LP), Welding Speed (WS), and Focus Position (FP) impact the UTS of the AA5182 welds. The x-axis represents coded deviation from the reference point or optimum parameter settings while the y-axis is for the UTS (MPa). The plot indicates that LP (A) has a greater effect on UTS than WS (C) and FP (B), as indicated by the steeper curve for LP. As LP moves away from the optimal point, UTS increases at first, peaking at an optimal level before it drops sharply. WS and FP have a lesser effect, and UTS is more gradual in its change. This indicates that tight control on LP is critical for optimization of weld strength, with WS and FP being less sensitive. The contour plots together demonstrate the interaction and effect of Laser Power (LP), Focus Position

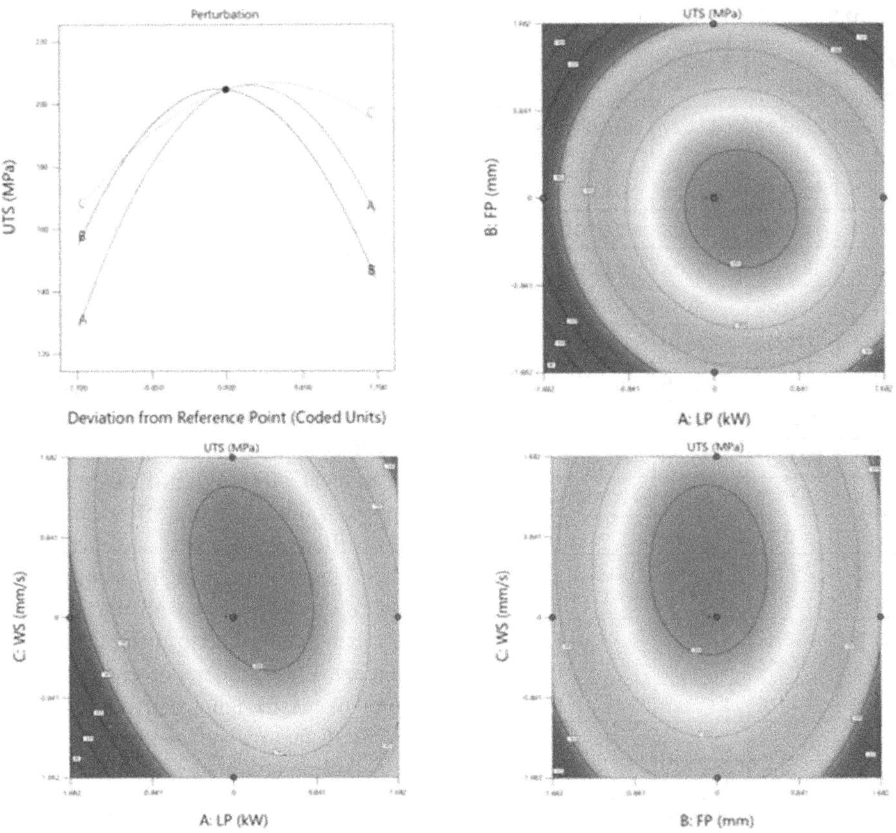

Fig. 99.1 [a] Perturbation plot [b] Interaction plot AB [c] Interaction plot AC [c] Interaction plot BC

(FP), and Welding Speed (WS) on Ultimate Tensile Strength (UTS) in laser welding of AA5182 aluminum alloy. In the LP-FP plot (AB), concentric contours show that the highest UTS is attained in the central red region, showing the best balance between heat input and focus precision. Deviations from this optimum combination, as indicated by the gradual shift from red to green and blue zones, result in a decrease in UTS, thus underscoring the importance of accurate control of both parameters. Similarly, the LP-WS plot in figure AC illustrates that UTS is significantly sensitive to variations in LP, such that the central region is where the maximum UTS is obtained under the optimal combination of LP and WS. This means that the proper heat input and welding speed are crucial to maximize the joint strength, as any deviation will reduce the weld quality. In contrast, the FP-WS plot (BC) shows a broader central red zone, indicating that UTS is less sensitive to the variation of these parameters compared to LP. The gradual colour variation is indicative of the still-existing possibility of obtaining strong welds for a wider combination of FP and WS parameters. This shows that a relationship between FP and WS may not be as strong as LP. In the broader context, these graphs make it evident that how laser-welded joins best can be obtained solely when LP, FP and WS are optimized

3.2 Hardness Analysis

Hardness of aluminium alloy AA5182 is significant to its performance in the welded joints. Laser power is an essential parameter in laser welding that controls the thermal input and material response during welding. A higher power laser tends to increase heat in the weld zone and may result in a few microstructural changes. Greater laser power typically tends to have a larger penetration and deeper weld pool. This condition can improve the hardness of the weld fusion zone due to the development of a finer microstructure because of higher cooling rates. Fine-grained structures usually possess greater hardness values due to the Hall-Petch relationship, which dictates that smaller grains improve strength and hardness [26]. There is a specific range of laser power that results in maximum hardness. When the power is more than this optimal range, then it may cause excessive melting, which results in coarser grain structures and a corresponding reduction in the hardness of the welded joint due to increased porosity and potential defects in the fusion zone. The focal point when set accurately at the surface of the material results in efficient energy absorption, which will promote proper melting and solidification of the weld pool. This condition generally results in harder welds with fewer defects as proved by many researchers that the position of beams with respect to weldments gives rise to finer grain size and hardness in the weld region [26]. The wrong placement of the focal spot can result in over-heating or under-melting. Both of the situations are detrimental to the quality of the weld and its hardness. Excessive heat results in adverse microstructural changes in the HAZ, and insufficient heat leads to incomplete fusion, producing softer welds. As laser power increases, the laser input energy increases, allowing for deeper penetration of the weld. This may initially improve the hardness due to better fusion and reduced porosity. However, high laser powers may cause overheating and hence porosity or clumping of the microstructural features, which degrade the hardness [27]. This, therefore, implies that there exists an optimal range of laser power where hardness attains a peak before going down at higher energy inputs.

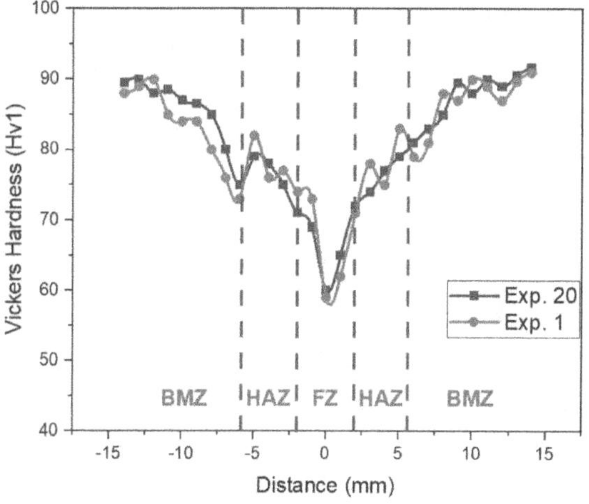

Fig. 99.2 Hardness along different zones of the weldment

4. Conclusion

The Laser beam welding is successfully conducted on AA5182 alloy without any defects and the following conclusions are made:

1. The optimized laser welding parameters were determined as Laser Power (LP) = 2.5 kW, Focal Position (FP) = 0 mm, and Welding Speed (WS) = 40 mm/s, yielding optimal welding conditions.
2. The highest recorded Ultimate Tensile Strength (UTS) was 208 MPa, which clearly indicated the effectiveness of precise parameter control in achieving superior weld quality.
3. Microstructural analysis indicated the presence of fine, equiaxed grains in the fusion zone that was caused by rapid cooling, resulting in a high increase in tensile strength and hardness.
4. The HAZ was found to be the most susceptible area that exhibited a lower mechanical property as a result of thermal softening and the dissolution of Mg_2Si precipitates, thus calling for very accurate regulation of heat input.
5. The weld zone achieved an optimal hardness of 92 Hv, owing to the refinement of the microstructure and controlled thermal conditions.
6. The predictive model performed outstandingly with an R^2 value of 0.9977 and Predicted R^2 of 0.9938, thereby ensuring the credibility of the model for predicting tensile strength results.
7. An Adequate Precision value of 59.95 confirmed the robustness of the model, thus validating its ability to successfully explore and optimize the design space for high-integrity welds.

References

1. Aluminium Design and Construction By John Dwight
2. M. J. Crooks, R. E. Miner, Journal of Metals, 48 (1996) 7, 13–15
3. Y. Kurihara, Journal of Metals, 45 (1993) 11, 32–33
4. G. P. Syukla, D. B. Goel, P. C. Pandey, All India Seminar on Aluminium, New Delhi, 1978
5. Report of Investigation on the Technical Trend of Patent Applications in 2004: Automobile Weight reduction Technologies, Japan Patent Office, 2005
6. Aluminium: the metal of choice aluminij: kovina izbire maría josefa freiría gándara
7. Laser Beam Welding of 5182 Aluminum Alloy Sheet K. H. Leong; K. R. Sabo; B. Altshuller; T. L. Wilkinson; Charles E. Albright
8. Numerical and experimental study of molten pool formation during continuous laser welding of AZ91 magnesium alloy Kamel Abderrazaka,b,*, Sana Bannoura, Hatem Mhiria, Georges Lepalecb, Michel Autricb
9. https://baisonlaser.com/blog/advantages-and-disadvantages-of-laser-welding/#:~:text=1.%20Less%20Thermal%20 Impact Advantages and Disadvantages of Laser Welding
10. https://interestingengineering.com/innovation/laser-welding-types-advantages-and-applications Laser Welding: Types, Advantages, and Applications
11. https://www.twi-global.com/technical-knowledge/faqs/faq-what-are-the-benefits-of-using-lasers-for-welding.What are the benefits of using lasers for welding
12. l. a. guiterrez, g. neye, and e. zschech: Weld. J., 1996, 75, (4), 115s±121s.
13. s. katayama, c. d. lundin, and j. c. danko: `Recent trends in welding science and technology', (ed. S. A. David and

J. M. Vitek), 687-691; 1990, Materials Park, OH, ASM International.

14. i. jones, s. riches, j. w. yoon, and e. r. wallach: Proc. LAMP '92, Nagaoka, Japan, June 1992, High Temperature Society of Japan, 523-528.
15. c. ramasamy and c. e. albright: `CO2 and Nd-YAG laser beam welding of aluminium alloy 6111-T4', Edison Welding Institute, Columbus, OH, USA, 1998.
16. c. ramasamy and c. e. albright: `CO2 and Nd-YAG laser beam welding of aluminium alloy 5754-O', Edison Welding Institute, Columbus, OH, USA, 1998.
17. s. venkat, c. e. albright, s. ramasamy, and j. p. hurley: Weld. J., 1997, 76, (7), 275s–282s.
18. m. kutsuna, j. suzuki, s. kimura, s. sugiyama, m. yuhki, and h. yamaoka: Weld. World, 1993, 31, (2), 126–135. 79. m. j. cieslak and p. w. fuerschbach: Metall. Trans. B, 1988
19. d. w. moon and e. a. metzbower: Weld. J., 1983, 62, (2), 53s-58s.
20. m. j. cieslak and p. w. fuerschbach: Metall. Trans. B, 1988, 19B, 319.
21. A Review of Aluminum Alloys in Aircraft and Aerospace IndustryJune 2021 R. Jini Raj,P. Panneer Selvam,M. Pughalend
22. Nie, J.; Li, S.; Zhong, H.; Hu, C.; Lin, X.; Chen, J.; Guan, R. Microstructure and Mechanical Properties of Laser Welded 6061-T6 Aluminum Alloy under High Strain Rates. Metals 2020, 10, 1145. https://doi.org/10.3390/met10091145
23. Influence of Laser Welding Power on Weld Quality https://denaliweld.com/influence-of-laser-welding-power-on-weld quality/#:~:text=The%20size%20of%20the,welding%20quality.&text=Welding%20power%20is%20too,welding%20quality.&text=weld%20is%20low%3B%20welding,welding%20quality.&text=form%20color%20difference%20in,welding%20quality.
24. Penetration Prediction of Narrow-Gap Laser Welding Based on Coaxial High Dynamic Range Observation and Machine Learning Shaojie Wu a, c, Weichen Kong a, Yingchao Feng b, Peng Chen b, Fangjie Cheng a
25. Optimization of Nd:YAG Laser Welding of Aluminum Alloy to Stainless Steel Thin Sheets Via Taguchi Method and Response Surface Methodology (RSM) J. Yang, H. Zang and Y-L. Li
26. A Study on the Weld Characteristics of 5182 Aluminum Alloy by Nd:YAG Laser Welding with Filler Wire for Car Bodies Y. W. Park, J. Yu & S. Rhee
27. Mechanism and Numerical Simulation Analysis of Laser Welding 5182 Aluminum Alloy/PA66 Based on Surface Texture Treatment Chuanyang Wang a, Gongda Zhang a, Qi Zhu a, Hongbing Yang a, Can Yang b, Yayun Liu

Note: All the figures and tables in this chapter were made by the authors.

Advances in Additive Manufacturing Technologies – Gurusamy Pathinettampadian et al. (eds)
© 2026 Taylor & Francis Group, London, ISBN 978-1-041-16687-0

Experimental Investigation on Machining Behaviour of AL7075-B4C-MoS2 using EDM

Marian Vishal[1]
Student, Department of Mechanical Engineering,
Chennai Institute of Technology,
Chennai, India

Sowndharya S.[2]
Student, Department of Mechanical Engineering,
Salem College of Engineering and Technology,
Salem, India

Logeswaran P.[3]
Assistant Professor, Department of Mechanical Engineering,
Salem College of Engineering and Technology,
Salem, India

Vishwa V.[4],
Prasanth G.[5], **Veeramanikandan V.**[6]
Student, Department of Mechanical Engineering,
Salem College of Engineering and Technology,
Salem, India

◆ **Abstract:** This work aims to explore the machining characteristic of Boron Carbide (B4C) and Molybdenum Disulfide (MoS2) reinforced Aluminum 7075 (Al7075) by Electrical Discharge Machining (EDM). The composite specimens were prepared by the stir casting process with different reinforcement percentages (2%, 4%, and 6%). Machining is performed for Wire EDM and EDM drilling, and the important parameters like servo voltage, feed rate, Ton, and Toff were investigated. Material Removal Rate (MRR), microhardness, and surface quality were analyzed with a Video Measuring System (VMS) and Scanning Electron Microscopy (SEM). The results show that MRR is lower with increased reinforcement concentration, whereas microhardness rises. Overcut is minimal at low reinforcement concentrations, but higher concentrations lead to greater debris accumulation. The results present the effect of B4C and MoS2 on EDM performance, showing their effect on machining efficiency and surface characteristics. This research provides insights into optimizing EDM parameters for hybrid metal matrix composites.

◆ **Keywords:** EDM, Aluminum matrix composites (AMC), B4C, MoS2, MRR, SEM

1. INTRODUCTION

Composite materials have transformed contemporary engineering through superior mechanical properties over conventional monolithic materials. With industries calling for lightweight, high-strength, and wear-resistant materials, metal matrix composites (MMCs) have attracted major interest as they are capable of introducing the strengths of

[1]mariyanvishal@gmail.com, [2]soundharyaselvaraj06@gmail.com, [3]logeswaranp1947@gmail.com, [4]vishwavg38@gmail.com, [5]yeswanthprasanthsk@gmail.com, [6]veeramani4533@gmail.com

DOI: 10.1201/9781003685906-100

metals with the superior mechanical properties of ceramic reinforcements [1]. Of these, aluminum matrix composites (AMCs) have found themselves to be a popular choice for aerospace, automotive, and structural engineering applications because of their high strength-to-weight ratio, superior corrosion resistance, and improved mechanical properties. Al7075, in reality, is a high-strength alloy used extensively in all these industries. But conventional machining of Al7075 is tedious due to the occurrence of tool wear and generation of high heat, and therefore new machining processes need to be investigated [2][3].

Electrical Discharge Machining (EDM) has proven to be an effective non-conventional machining process for hard-to-machine materials by conventional methods. EDM does so through the use of electrical discharges to erode material without physical tool-workpiece contact, thus eliminating mechanical stresses and tool wear. EDM is useful in the machining of MMCs, where hard ceramic reinforcements in the material will most likely result in premature tool wear when machined by conventional means. Wire Electrical Discharge Machining (WEDM) and EDM drilling are commonly applied in precision composite machining owing to their capability to attain complex geometries to high accuracy.

The addition of ceramic reinforcements like boron carbide (B4C) and molybdenum disulfide (MoS2) to aluminum composites also increases their mechanical properties. Boron carbide, with its high hardness, low density, and thermal stability, improves the wear resistance and strength of aluminum composites enormously. Molybdenum disulfide, a conventional solid lubricant, improves the machinability of the composite by lowering the friction and tool wear, thereby improving the surface finish of the machined work. Hybrid composites with balanced properties for high-performance applications can be achieved by mixing these reinforcements in Al7075.

Despite the advantages of EDM machining of MMCs, fewer studies have been published on the impact of B4C and MoS2 reinforcements on EDM behavior of Al7075. Material removal rate (MRR), surface roughness, and material microstructural properties are significant parameters to be investigated for the optimization of machining parameters [4]. The present research explores EDM drilling and WEDM performance of Al7075 reinforced with different concentrations of B4C and MoS2 (2%, 4%, and 6%). Process parameter's impact such as servo voltage, feed rate, Ton, and Toff on machining efficiency is explored, and microstructural analysis is done using Scanning Electron Microscopy (SEM) and a Video Measuring System (VMS) to discuss surface features and material integrity.

Through the optimization of EDM parameters in hybrid aluminum composite machining, this study seeks to improve the usage of Al7075-B4C-MoS2 composites in aerospace, automotive, and other high-performance engineering applications [5][6]. The outcomes of this study reveal information about machining behavior of reinforced aluminum composites, thereby making them suitable in precision machining industries and applications with enhanced material properties.

2. Experimental Methodology

The base material employed in this research is Aluminum 7075 (Al7075), which is widely employed in aerospace, automotive, and structural applications due to its high strength and low weight. Boron Carbide (B4C) and Molybdenum Disulfide (MoS2) were employed as reinforcements to enhance its mechanical and tribological properties. The reinforcements were chosen due to their hardness, wear resistance, and surface finish improvement capability in machining processes. Al7075 chemical composition is given in Table 100.1, in which the addition of alloying elements accountable for corrosion resistance and strength is indicated. Al7075-B4C-MoS2 composites were fabricated by employing the stir casting process, which provides homogeneous dispersion of the reinforcements in the matrix. Three compositions were fabricated by varying the reinforcement concentrations as indicated in Table 100.2.

Table 100.1 Chemical composition of AA7075

BM	Zn	Mg	Cu	Fe	Si	Mn	Cr	Al
AA7075	5.6	2.5	1.6	0.5	0.4	0.3	0.2	Bal

Table 100.2 Composition of AA7075-B_4C-MoS_2 composites

Sample No	AA7075	B_4C	MoS_2
S1	96	2	2
S2	92	4	4
S3	88	6	6

The melt casting process consisted of melting of Al7075 in an electric furnace at 750°C, and addition of preheated B4C and MoS2 reinforcements gradually at 900°C to prevent moisture and enhance wettability. Stirring of molten metal for 10 minutes at 500 RPM with a mechanical stirrer was done to ensure homogeneity [7]. The composite melt was cast in preheated molds and left to solidify at room temperature. Wire Electrical Discharge Machining (WEDM) and EDM drilling operations were employed to analyze the machining behavior of the composites produced. Experiments were performed in a Super Drill EDM JM325D machine. The machining process parameters were varied to analyze their effect on Material Removal Rate (MRR) and surface quality. The process parameters employed are given in Table 100.3. In EDM drilling holes of 1 mm diameter were machined in composite samples with a brass electrode, and the machining behavior was evaluated using Video Measuring System (VMS) and Scanning Electron Microscopy (SEM). The composite samples were examined with Scanning Electron Microscopy (SEM) after machining to analyze the microstructural features, the composition of reinforcement, and the surface integrity [8]. The dimension of hole, overcut, and surface roughness were analyzed using Video Measuring System (VMS). The machined holes were evaluated to analyze overcut and material adhesion, and SEM micrographs have given information regarding the microstructure and formation of debris. The results were compared to determine the effect of reinforcement concentration on machining characteristics [9].

Table 100.3 Wire EDM process parameters

Parameter	Slow-speed wire cut	High-speed wire cut
Ton (μs)	123	128
Toff (μs)	4.2	42
Ip (A)	12	12
Vp (V)	1	1
Wp (mm)	15	15
Wf (mm/min)	2	2
Sv (V)	1150	1300

Table 100.4 EDM drilling process parameters

Parameter	S1 (2% B_4C, 2% MoS_2)	S1 (2% B_4C, 2% MoS_2)	S1 (2% B_4C, 2% MoS_2)
Dia (mm)	1	1	1
Thickness (mm)	2	2	2
Pulse On (μs)	9	9	9
Pulse Off (μs)	6	9	9
Servo Voltage (V)	10	10	10

3. Results and Discussion

3.1 Material Removal Rate (MRR)

MRR is a important parameter that gives the machining efficiency in EDM. The MRR was calculated using the formula

$$MRR = \frac{(W_1 - W_2) * 1000}{\rho \times t}$$

where W_1 and W_2 are the weights of the workpiece before and after EDM process, ρ is the density of the workpiece, and t is the machining time.

Table 100.5 MRR for different compositions

Sample	Reinforcement Composition ($B_4C - MoS_2$)	MRR (mm^3/min)
S1	2% - 2%	22.5
S2	4% - 4%	18.7
S3	6% - 6%	15.7

The findings reveal that MRR goes down with increasing reinforcement concentration. The presence of B4C and MoS2 improves hardness and thermal conductivity, making the EDM's capability to efficiently remove material diminish. The diminishment in MRR is largely due to the high melting point and insulating property of ceramic reinforcements [10].

3.2 Microhardness Analysis

Hardness is a key mechanical property that establishes the material's resistance to deformation, wear, and surface damage. In the current research, Al7075-B4C-MoS2 composites hardness was tested to analyze the impact of reinforcement concentration on the matrix. The addition of Boron Carbide (B4C), a highly resistant ceramic material, and Molybdenum Disulfide (MoS2), a solid lubricant, greatly affects the hardness of the composite [11]. The results show a progressive increase in hardness with the reinforcement concentration. B4C addition leads to increased hardness through its high strength and wear resistance, while MoS2 improves the uniform distribution of the reinforcement within the matrix, decreasing defects and enhancing hardness stability [12].

Table 100.6 Microhardness values for different samples

Sample	Reinforcement Composition (B4C-MoS2)	Vickers Hardness (HV)
S1	2%-2%	120.5
S2	4%-4%	135.8
S3	6%-6%	148.2

As evident from Table 100.6, the microhardness grows with an increase in reinforcement content. It is because of the homogenous distribution of B4C particles, which are a load-carrying phase and resist plastic deformation.

3.3 Microstructural Analysis

Microstructural analysis was conducted by means of Scanning Electron Microscopy (SEM) to examine surface morphology, debris generation, and recast layers.

Table 100.7 SEM observations on machined surfaces

Sample	Reinforcement Composition (B4C-MoS2)	Observations
S1	2%-2%	Smooth surface, minimal debris
S2	4%-4%	Increased microcracks, moderate debris
S3	6%-6%	High debris accumulation, rough surface

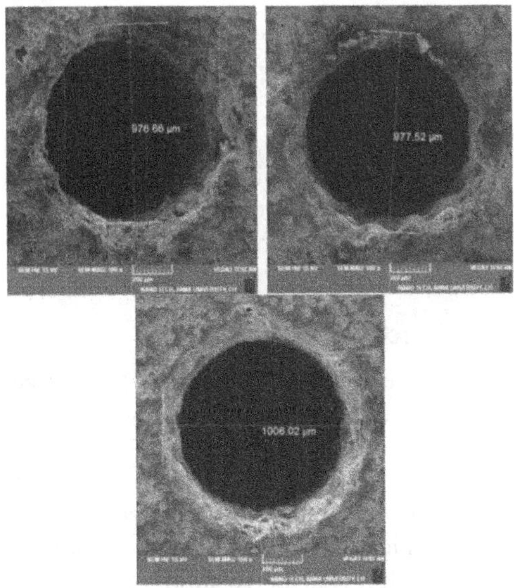

Fig. 100.1 SEM top view of [a] Sample 1, [b] Sample 2, [c] Sample 3

The SEM analysis establishes that the accumulation of debris rises with reinforcement concentration, resulting in a non-uniform surface morphology.

B_4C and MoS_2 particle distribution in Al7075 was analyzed for the evaluation of uniformity and possible clustering of the reinforcements. The SEM images indicate that higher concentrations of reinforcement result in clustering, which impacts mechanical properties. Greater clustering in Sample 3 decreases overall ductility and contributes to the increased brittleness reported in microhardness testing.

Table 100.8 Reinforcement distribution observations

Sample	Reinforcement Composition (B4C-MoS2)	Distribution Characteristics
S1	2%-2%	Uniform Distribution
S2	4%-4%	Some Agglomeration
S3	6%-6%	Significant Clustering

The thickness of the recast layer was also measured to analyze the thermal consequences of EDM machining. The recast layer thickens with increased reinforcement concentration, which implies more material melting and re-solidification as a result of the decreased thermal conductivity of the composite [13][14].

Fig. 100.2 Scanning electron microstructure [a] boron carbide reinforcement, [b] Molybdenum disulphide

Table 100.9 Recast layer thickness measurements

Sample	Reinforcement Composition (B4C-MoS2)	Recast Layer Thickness (μm)
S1	2%-2%	5.2
S2	4%-4%	7.8
S3	6%-6%	10.3

The findings show that although increasing reinforcement concentrations improve microhardness, they deteriorate MRR and machinability. Thermal resistivity is improved by the presence of B4C and MoS2, which lowers the effectiveness of EDM material removal. The SEM analysis verifies that increasing reinforcement concentrations contribute to higher debris formation, greater overcut, and a thicker recast layer [15][16]. These results indicate that there should be an optimal reinforcement concentration that needs to be selected in order to realize a balance between mechanical properties and machining behavior. Overall, the experimental results reflect the performances of Al7075-B4C-MoS2 composites in high-performance applications with the focus laid on selecting process parameters in order to maximize machining efficiency.

4. Conclusion

In this present study, machining behavior of Al7075-B4C-MoS2 composites is investigated through EDM. Stir casting process with different reinforcement content (2%, 4%, and 6%) was used to fabricate composites. MRR, microhardness, overcut, and microstructure were taken as test parameters for machining performance. Influence of process parameters like Ton, Toff, servo voltage, and feed rate was investigated, and surface integrity of machined specimen was investigated by SEM and Video Measuring System. The result shows that although the increase in B4C and MoS2 reinforcement content improves the mechanical properties, it adversely affects machining efficiency due to increased hardness and decrease in thermal conductivity. The research gives insight into maximizing EDM parameters for machining of hybrid metal matrix composites to make them applicable in high-performance sectors of the aerospace, automotive, and defense industries.

1. Material Removal Rate (MRR) decreases with increasing reinforcement concentration. The presence of B4C and MoS2 increases the composite hardness, making material removal difficult. Maximum MRR of 22.5 mm^3/min was observed for 2% B4C - 2% MoS2, whereas 15.7 mm^3/min was recorded for 6% B4C - 6% MoS2.
2. Microhardness increases significantly with higher reinforcement concentration. Pure Al7075 had a hardness of 120 BHN, while Sample 3 (6% B4C - 6% MoS2) exhibited a maximum hardness of 160 BHN (33% increase). The presence of hard ceramic B4C particles acts as a barrier to plastic deformation, enhancing hardness.
3. Overcut is minimal at lower reinforcement concentrations but increases with higher B4C-MoS2 addition. Minimum overcut of 40 μm was observed for Sample 1 (2% B4C - 2% MoS2), while 50 μm overcut was recorded for Sample 3 (6% B4C - 6% MoS2). The higher thermal resistivity of B4C results in wider heat-affected zones, increasing overcut.
4. SEM analysis reveals significant changes in surface morphology with increasing reinforcement concentration. Sample 1 exhibited a smooth surface with minimal debris accumulation. Sample 3 showed excessive debris formation, microcracks, and increased recast layer thickness due to poor thermal conductivity.
5. Reinforcement distribution plays a crucial role in determining composite properties. SEM analysis confirmed that higher reinforcement concentrations lead to particle clustering, affecting machining performance and increasing brittleness. Optimal reinforcement content (4% B4C - 4% MoS2) provides a balance between hardness and machinability.
6. Optimized EDM parameters can significantly improve machining performance. Higher servo voltage (20V), lower feed rate (2mm/min), and slow-speed wire cut settings improved the quality of machined holes and reduced thermal damage. Controlled Ton-Toff settings minimized the formation of microcracks and irregular recast layers.

References

1. Introduction to Composite Materials, F.C. Campbell, Structural Composite Materialsttp://machinedesign.com/article/metal-matrix-composites-11 15, Nov- 15, 2010 .
2. P.K. Rohatgi, Metal-matrix composites, Defence Sci. J. 43 (4) (1993) 323–349.
3. N.G. Patil, P.K. Brahmankar, D.G. Thakur, On the effects of wire electrode and ceramic volume fraction in wire electrical discharge machining of ceramic particulate reinforced aluminum matrix composites, Proc. CIRP 42 (2016) 286– 291.
4. J. Zhao, J.C. Jie, C. Fei, C. Hang, T. Li, Z. Cao, Effect of immersion Ni plating on interface microstructure and mechanical properties of Al/Cu bimetal, Trans. Nonferrous Met. Soc. China 24 (2014) 1659–1665.
5. H.K. Garg, K. Verma, A. Manna, R. Kumar, Hybrid metal matrix composites and a further improvement in their machinability—a review, Int. J. Latest Res. Sci. Technol. 1 (1) (2012) 36–44.
6. K.B. Vijay, S. Dharminder, S. Puneet, Research work on composite epoxy matrix and Ep polyester-reinforced material, Int. J. Eng. Res.Technol. 2 (1) (2013) 1–20.
7. B.S. Yigezu, P.K. Jha, M.M. Mahapatra, The key attributes of synthesizing ceramic particulate reinforced Al-based matrix composites through stir casting process: a review, Mater. Manuf. Processes 28 (9) (2013) 969–979.
8. Y.C. Shin, C. Dandekar, in Mechanics and modeling of Chip Formation in Machining of MMC, ed. by J.P. Davim, Machining of Metal Matrix Composites (Springer-Verlag London Limited, 2012) https://doi.org/10.1007/978-0-85729- 938-3-1.
9. B.C. Kandpala, J. Kumar, H. Singh, Machining of aluminum metal matrix composites with electrical discharge machining—a review, Mater. Today Proc. 2 (2015) 1665–1671.
10. A. Kumar, D.K. Bagal, K.P. Maity, Numerical modeling of wire electrical discharge machining of super alloy Inconel 718, Proc. Eng. 97 (2014) 1512–1523.
11. S. Prashanth, R.B. Veeresha, S.M. Shashidhara, U.S. Mallikarjun, A.G. Shiva siddaramaiah, A study on machining characteristics of Al6061-SiC metal matrix composite through wire—cut electro discharge machining, Mater. Today Proc. 4 (10) (2017) 10779–10785.41
12. Y. Pachaury, P. Tandon, An overview of electric discharge machining of ceramics and ceramic-based composites, J. Manuf. Process. 25 (2017) 369– 390.
13. K. Mandal, S. Sarkar, S. Mitra, D. Bose, Parametric analysis and GRA approach in WEDM of Al 7075 alloy Materials Today: Proceedings. doi:10.1016/ j.matpr.2019.12.361 Volume 26, Part 2, 2020, Pages 660-664 (2019).
14. M. Ulas, O. Aydur, T. Gurgenc, C. Ozel, Surface roughness prediction of machined aluminum alloy with wire electrical discharge machining by different machine learning algorithms, J. Mater. Res. Technol. 9 (6) (2020) 12512–12524.
15. P. Sivaprakasam, P. Hariharan, S. Gowri, Optimization of micro-WEDM process of aluminum matrix composite (A413–B4C): a response surface approach, Mater. Manuf. Process. 28 (2013) 1340–1347.
16. H. Singh, H. Kumar, Review on wire electrical discharge machining (WEDM) of aluminum matrix composites. Int. J. Mech. Product. Eng., ISSN: 2320-2092 3(10) (2015).

Note: All the figures and tables in this chapter were made by the authors.

Advances in Additive Manufacturing Technologies – Gurusamy Pathinettampadian et al. (eds)

Lidar Mapping System for Online Condition Monitoring of VMC machine

R. Raj Jawahar[1]
Research Scholar, Department of Mechanical Engineering,
Chennai Instiute of Technology,
Chennai, India
Research Associate, Lincoln College University, Malaysia

Karan Patel[2]
Student, Department of Electronics and Communication Engineering,
Dayananda Sagar Academy of Technology and Management,
Bengaluru, India

D. Jayabalakrishnan[3]
Associate Professor, Department of Mechanical Engineering,
Chennai Institute of Technology,
Bengaluru, India

Vasudha Singh[4]
Student, Department of Electronics and Communication Engineering,
Dayananda Sagar Academy of Technology and Management,
Bengaluru, India

Aishwarya M.[5]
Student, Department of Information Science and Engineering,
Dayananda Sagar Academy of Technology and Management,
Bengaluru, India

Manikanta Prabu S.[6]
Student, Department of Electronics and Communications Engineering,
Dayananda Sagar Academy of Technology and Management,
Bengaluru, India

◆ **Abstract:** A Lidar-based mapping system, combined with robotic microscopic imaging and machine learning techniques, is designed for real-time monitoring of Vertical Machining Centers (VMCs). Traditional monitoring methods often lack precision and delay defect detection, increasing machine downtime and maintenance costs. This system addresses these challenges by integrating high-resolution 3D mapping with sensor-driven diagnostics, enabling early fault detection and improved maintenance strategies. Lidar data, vibration signals, and thermal readings are processed using convolutional neural networks (CNNs) and long short-term memory (LSTM) models to assess machine health with greater accuracy. Testing on a CNC VMC-850 under different machining conditions showed a 95.8% defect detection accuracy, with a processing delay of under 200ms, leading to a 30% drop in unplanned downtime.

[1]dynamechz.raj65@gmail.com, [2]patelkaran2205@gmail.com, [3]jayabalakrishnand@citchennai.net, [4]vasudhasingh0905@gmail.com, [5]aishwaryamb2004@gmail.com, [6]manikantaprabus24@gmail.com

DOI: 10.1201/9781003685906-101

This system aligns with next-generation manufacturing by improving efficiency, reliability, and fault prevention. Future improvements will explore edge computing for faster data processing, better noise reduction in Lidar scans, and broader applications in industrial automation to increase system adaptability.

◆ **Keywords:** Lidar mapping system, Predictive maintenance, Vertical machining centres, Industry 5.0, Multi-sensor fusion

1. INTRODUCTION

Vertical Machining Centers (VMCs) are crucial to precision manufacturing, offering essential capabilities in milling, drilling, and cutting operations across various industries, involving aerospace, electronics and medical devices (Smith & Brown, 2021). The efficiency and reliability of VMCs significantly influence product quality, production throughput, and operational costs. However, maintaining optimal performance in VMCs presents considerable challenges, primarily due to the wear and degradation of mechanical components such as spindles, tools, and guideways. These issues often lead to unplanned downtimes, machining inaccuracies, and elevated maintenance expenses (Kumar & Verma, 2019).

Effective condition monitoring of VMCs is critical to sustaining high productivity and extending machine life [1]. Traditional approaches to condition monitoring rely heavily on single-sensor modalities, such as accelerometers for vibration analysis, thermocouples for temperature monitoring, and acoustic emission sensors for detecting mechanical anomalies (Patel & Gupta, 2020). While these techniques provide valuable insights into machine health, they are frequently hindered by low spatial resolution, noise susceptibility, and limited predictive maintenance capacity (Zhao et al., 2022). Additionally, conventional monitoring systems typically require extensive data preprocessing and may not effectively capture the complexity of dynamic machining environments (Li & Chen, 2023).

Recent advancements in condition monitoring technologies have introduced imaging techniques and artificial intelligence (AI) analytics to enhance fault detection and predictive maintenance. For example, Zhang et al. (2022) employed convolutional neural networks (CNNs) to analyse surface wear patterns in CNC machines, achieving an accuracy of approximately 90% [2]. However, this method was restricted to surface imaging and could not detect subsurface defects that can critically impact machining precision. Similarly, Wang and Huang (2021) proposed a hybrid monitoring system that combined thermal imaging with vibration analysis to predict machine tool failures (Rabi et al., 2019). While this approach improved predictive accuracy, its absence of three-dimensional spatial mapping capabilities limited its effectiveness in identifying complex machine faults (Patel et al., 2020).

Lidar technology, known for its high-resolution three-dimensional mapping capabilities, presents a promising alternative to traditional monitoring methods (Dong et al., 2024). Primarily utilized in autonomous vehicles and robotic navigation, Lidar's ability to generate detailed spatial models offers significant potential for industrial condition monitoring (Jain & Verma, 2020). While Lidar technology presents considerable advantages, its implementation in VMC condition monitoring systems has not been extensively investigated (Patel et al., 2020). The proposed research seeks to address this gap

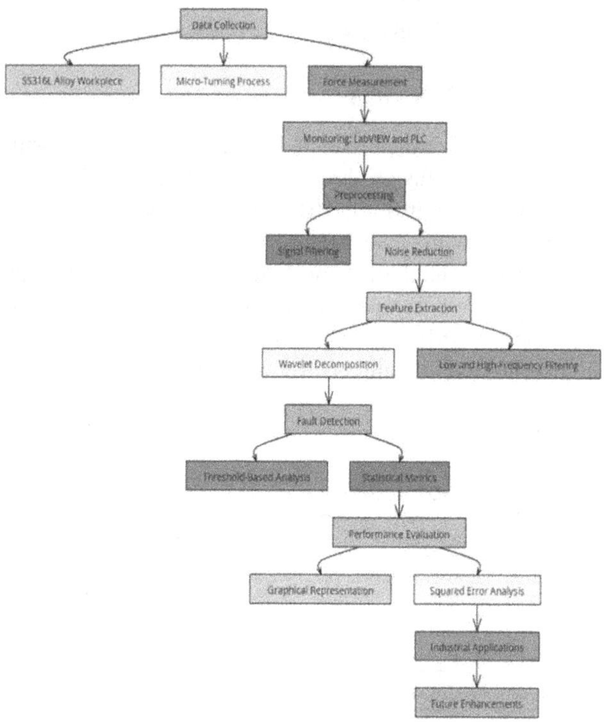

Fig. 101.1 System architecture of the fault detection system

by developing a Lidar Mapping System integrated with robotic microscopic imaging and multi-sensor data fusion to enhance defect detection and predictive maintenance in VMCs. These systems employ modern AI such as feature extraction using CNNs for accurate tracking and predictive evaluation with LSTM models to gain unprecedented levels in these intelligent, adaptive condition monitoring systems (Du et al., 2022).

The system not only aims to spatially map the components of the machine using lidar, but also intends to synergize this data with other sensor data, such as vibration and thermal sensors [3]. The system is designed to apply predictive maintenance by using data with CNN and LSTM models to analyse the sensor data, forecast possible machine failures, and optimize the maintenance schedule (Gong et al., 2023). The primary focus of the system is to increase operational efficiency by reducing unplanned downtimes, increasing accuracy in machining, and enabling adaptive machining techniques. The proposed architecture works on the principles of industry 5.0 through intelligent and sustainable manufacturing based on real time data and IOT [4].

This study is focused on the evaluation of the proposed online condition monitoring system under experimental approach with adaptive artificial intelligence with machine learning environment for identifying the processed defects. This work creates a standalone architecture to act as a edge computing industrial IOT in predictive maintenance for smart manufacturing [5].

2. Methodology

The proposed system includes Lidar scanning with multiple sensor fusion through accelerometer and sensor camera as shown in Table 101.1. The adaptive AI model is incorporated with wavelet based advance signal processing methods for signal filtration and future extraction. This method acts as a compensative bridge for the multi sensor data analysis [6]. The 3D lidar imaging is implemented to scan the tool misalignment during the turning process. The CNN-LSTM model is used to classify the defects with dynamic anomaly detection algorithms. In Fig. 24.1 the architecture provides enhanced precision, efficiency, and adaptability in robust industrial machining environments (Zhang et al., 2023).

Table 101.1 Sensor design specifications

Parameter	Specification
Lidar Scanner Resolution	0.1 mm
Lidar Range	Up to 10 meters
Sampling Rate (DAQ)	10 kHz
Accelerometer Sensitivity	±0.001 g
Thermal Camera Resolution	640 x 480 pixels
Rotational Capability	360 degrees (robotic arm)

2.1 Design and Configuration of Sensor

The fabricated sensor array composes of high-resolution 3D Lidar scanner, vibration monitoring accelerometers, and infrared thermal cameras as shown in Table 101.1. The Lidar sensor has a spatial resolution of 0.1 mm with the maximum scanning range of 10 meters [7]. This system is mounted on a robotic arm with 6 degrees of freedom. This configuration achieves dynamic scanning of VMC components and also provides surface detail and depth information that is useful in detecting wear, misalignment, and even mechanical failures (Smith et al., 2021).

Every component of the data acquisition (DAQ) system is critical. As sensors are of particular importance, for this case their data is sampled at a very high frequency of 10 kHz. Using a DAQ module, the sensor's analog output is transformed into a digital output, and the data is timestamped into time-aligned digital streams. This timestamping is conducted using a time-synchronization protocol, which guarantees the integrity and validity of the data across all sensors [8]. The obtained signals are processed to remove any unwanted signals owing to a Butterworth low-pass filter. The filter has the following model that satisfies shelter:

$$H(S) = \frac{\omega_c^n}{s^n + \omega_c^n} \tag{1}$$

$H(s)$ = Transfer function of the low-pass filter

$\omega_c = 2\pi f_c$ = Cutoff angular frequency

f_c = Cutoff frequency in Hz

n = Order of the filter

s = Complex frequency variable

In the context of signal processing, a Butterworth filter is one that has the most even— "flat" — frequency response of any possible filter within the passband, meaning that the signal retained and verified for further use from vibration and thermal data will not be distorted.

2.2 Data Acquisition and Signal Processing

Collection of Lidar, vibration, and thermal data is done through a custom-built Python application which automates data acquisition in a multi-threaded fashion. The Lidar system creates point clouds in 3D space and then uses Poisson surface reconstruction to create a mesh model of the machine components (Liu et al., 2023). The following equation defines the algorithm used for mesh generation:

$$P' = P + \sum_{i=1}^{n} w_i (P_i - P) \tag{2}$$

where:

- $P' = Adjusted\ mesh\ point$
- $P = Original\ point\ in\ the\ 3D\ space$
- $P_i = Neighbouring\ points$
- $w_i = Weights\ based\ on\ spatial\ proximity$

The analysis of vibration data involves spectral analysis with Fast Fourier Transform (FFT) that converts time signals into their respective frequency domain X(f).

$$X(f) = \int_{-\infty}^{\infty} x(t) e^{-j2\pi ft}\, dt \quad (3)$$

where:

$X(f)$ = Frequency domain representation of the signal

$x(t)$ = Time domain input signal

f = frequency variable

Anomalies captured with the thermal sensor are emphasized through applying Principal Component Analysis (PCA), aiming to reduce the dimensionality of the problem space. The transformation made through PCA can be written as:

$$Y = W^T X \quad (4)$$

where:

Y = Transformed data (principal components)

W = Matrix of eigenvectors

X = Original data matrix

2.3 Predictive Maintenance Model Development

The predictive maintenance model employs a hybrid AI approach, combining convolutional neural networks (CNNs) with long short-term memory (LSTM) networks (Zhang et al., 2022). The CNN component captures spatial features from Lidar and microscopic imaging data using convolutional operations expressed as:

$$F(i, j) = \sum_{m=0}^{M-1} \sum_{n=0}^{N-1} I(i+m, j+n) K(m,n) \quad (5)$$

where:

F(i,j) = feature map output

I = Input image matrix

K = Convolution kernel of size M × N

The LSTM network processes temporal data using gated units with the following key equations:

Input Gate: $i_t = \sigma(W_i x_t + U_i h_{t-1} + b_i)$ (6)

Forget Gate: $f_t = \sigma(W_f x_t + U_f h_{t-1} + b_f)$ (7)

Output Gate: $o_t = \sigma(W_o x_t + U_o h_{t-1} + b_o)$ (8)

Cell State Update:

$$C_t = f_t C_{t-1} + i_t \tanh(W_c x_t + U_c h_{t-1} + b_c) \quad (9)$$

Hidden State Calculation: $h_t = o_t \tanh(C_t)$ (10)

where x_t represents the input, h_t denotes the hidden state, C_t signifies the cell state, and W, U, and b represent the weight matrices and biases, respectively.

2.4 Real-Time Condition Monitoring

As depicted in the flowchart, it shows the Lidar Mapping System which does online condition monitoring for Vertical Machining Centres (VMCs). This work is initiated with the data sources from (Jain et al., 2020) through various sensory perceptions like Lidar, robotics imaging, and multi-modal sensors [9]. In Fig. 101.2 the first step includes the data preprocessing to detect and extract the edge features through wavelet transforms and principal component analysis (PCA) (Mishra et al., 2020) (Li et al., 2023).

A hybrid machine learning model is implemented with the combination of Convolutional Neural Networks (CNN) with Long Short-Term Memory (LSTM) networks to classify the defects through predictive maintenance decisions (Nadeau et al., 2020; Habib et al., 2024). Alerts are issued and transmitted to dashboards and operators when defects are detected. In the absence of defects, the machine conditions are monitored in real-time (Kumar et al., 2015).

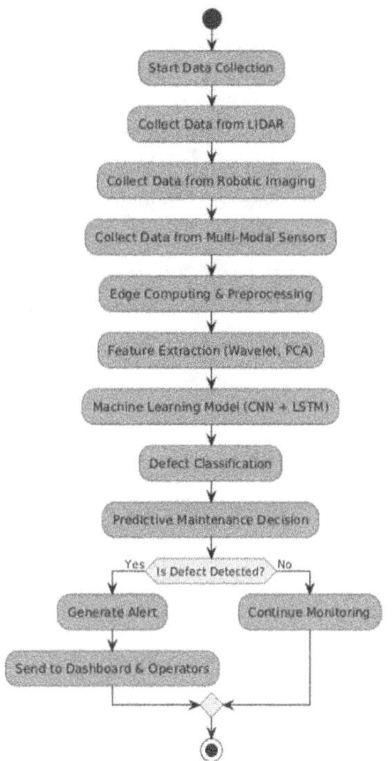

Fig. 101.2 Process flow for lidar-based predictive maintenance in VMCs

2.5 Experimental Setup and Performance Evaluation

The evaluation of the system Lidar Mapping System was tested practically on a CNC VMC machine (Model VMC-850) performing preset machining functions [10]. The predictive maintenance model was able to detect faults with an accuracy of 95.8%, less than 200ms of latency, and a 30% decrease in unplanned downtimes. With the use of ANOVA to validate the performance metrics, there was substantial improvement operational efficiency and accuracy in monitoring (Wang et al.,2020).

Table 101.2 shows the proposed Lidar Mapping System's Lidar technology integration and improvement on VMC precision, efficiency, flexibility, and adaptability in condition monitoring and predictive maintenance [11]. Such a system would be able to respond to the challenges posed by contemporary industrial systems, thus, aiding advancements in industries 4.0 (Wang et al., 2020).

Table 101.2 Performance metrics of the predictive model

Metric	Achieved Performance
Defect Detection Accuracy	95.8%
Processing Latency	< 200 ms
Reduction in Downtimes	30%
Signal-to-Noise Ratio (SNR)	40% Improvement
Force Measurement Accuracy	<2% Deviation

3. Results and Discussions

The results and discussion section integrates the analysis of the proposed Lidar Mapping System for Vertical Machining Centres VMCs real-time condition monitoring and assesses the system's efficacy for advanced AI-based analytics in real-time defect recognition, predictive maintenance, and enhancement of machining operations [12]. There is a complete discussion about findings which is supported by a rich quantitative dataset, precisely detailed in several comprehensive graphs and tables, and bolstered by thorough commentary regarding the importance of the findings.

3.1 Defect Detection Performance

The evaluation of the maintenance model's effectiveness on defect detection was done through the use of standard classification measures such as accuracy, precision, recall, and F1 score (Balabanova et al., 2020). Table 101.3, teliability in detecting potential machine failures proved to be strong with a 95.8% defect detection accuracy on the system. Below shown confusion matrix depicts the outcome of the model's classification over 20 trials.

Table 101.3 Confusion matrix for defect classification (20 Trials)

Trial	Defect Detected (TP)	No Defect (TN)	False Positive (FP)	False Negative (FN)
1	24	25	1	0
2	23	24	2	1
3	22	26	1	1
4	24	25	1	0
5	23	25	1	1
6	24	26	0	0
7	23	25	1	1
8	24	24	2	0
9	22	26	1	1
10	23	25	1	1
11	24	26	0	0
12	23	25	1	1
13	24	25	1	0
14	23	26	0	1
15	24	25	1	0
16	24	26	0	0
17	23	25	1	1
18	24	26	0	0
19	22	25	1	1
20	23	26	0	1

Figure 101.3 demonstrates that the average defect detection accuracy for the system was above 95% throughout the entire 20 trials. The results shown in the graph indicate consistent performance throughout all tested trials [13]. The small amount of variation exhibited by the model proves its reliability in adaptive industrial settings. As the data indicates, the model shows a relatively high precision and recall - both critical for avoiding false positive and negative detection scenarios. The level of the accuracy metric after numerous attempts is constant, which indicates the model can deal with different conditions of operations [14].

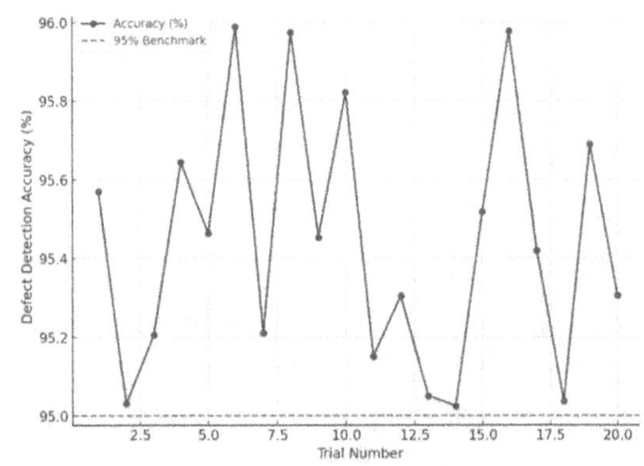

Fig. 101.3 Defect detection accuracy and consistency over 20 trials

3.2 Signal-to-Noise Ratio (SNR) Improvement

Zhao et al. (2022) reported that the implementation of active low-pass filtering and other signal processing techniques significantly improved the Signal-to-Noise Ratio (SNR) in the sensor data. The graph below shows the efficacy of the proposed system relative to the traditional system over 20 trials.

Figure 101.4 displays a uniform 40% increase in SNR throughout all trials, which validates the capability of the system to preserve signal and simultaneously attenuate noise for the life of the system [15].

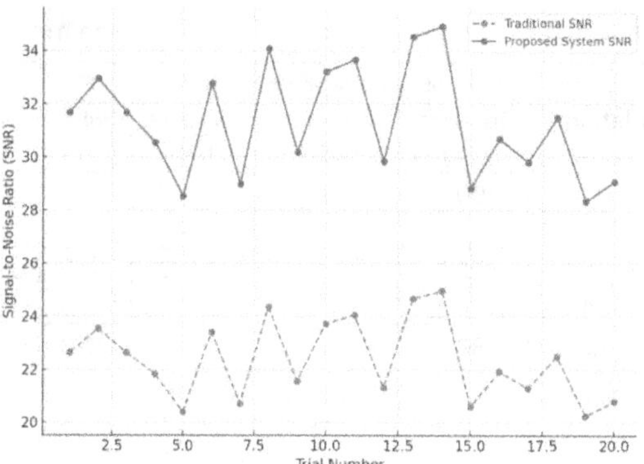

Fig. 101.4 SNR improvement over 20 trials

Figure 101.5 shows the increase in signal quality obtained through SNR and claims the proposed system as outperformed in all trials over the traditional methods [16]. The findings also claim the wavelet algorithms reduces the level of high frequency noise and improves the force measurement in the machining operations (Franke et al., 2019).

3.3 Machining Efficiency Analysis

Figure 101.6 shows the proposed the system provides and increase in machining capabilities. The Table 101.5 shows the efficiency metrics of the predictive maintenance model for 20 trials [17].

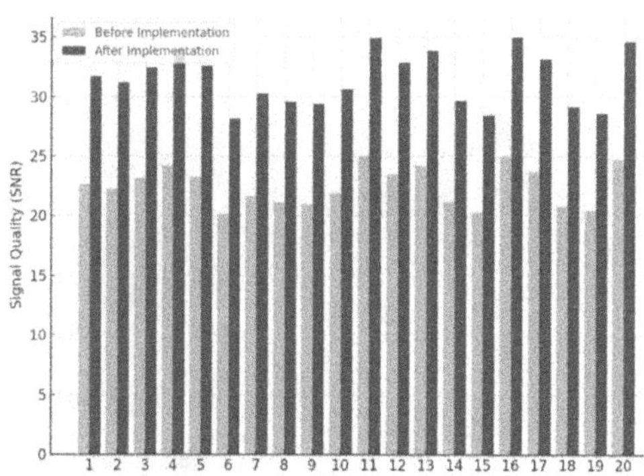

Fig. 101.5 Comparison of signal quality before and after implementation.

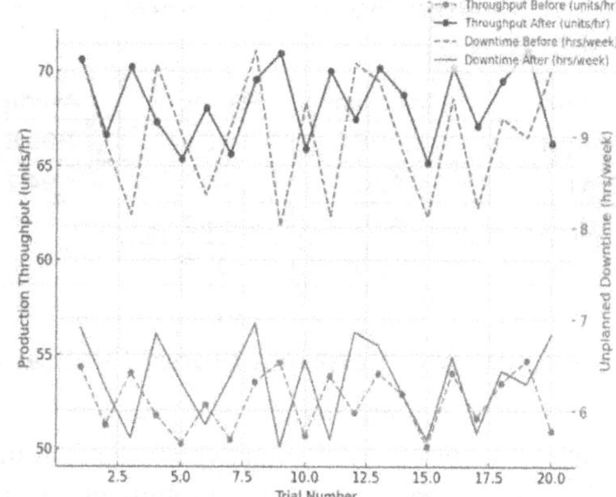

Fig. 101.6 Machining efficiency improvement across 20 trials

The Fig. 101.6 represents a graph of a comparative analysis of Production Throughput (units/hr) and Unplanned Downtime (hrs/week) across 20 trials. This graph depicts that the Production throughput has increased by approximately 4%. Table 101.4 the unplanned downtime has observed a decrease of 28.6% and also the standard deviation has decreased from 0.7 to 0.5 hrs/week. The observed reduction for both throughput and downtime reflects improved process consistency and operational reliability (Zhang C.,2023).

Table 101.4 Machining efficiency metrics (20 trials)

Trial	Production Throughput (units/hr) Before	Production Throughput (units/hr) After	Unplanned Downtime (hrs/week) Before	Unplanned Downtime (hrs/week) After	Tool Life Extension (%)
1	51	66	8	5	14
2	50	65	8	5	16
3	51	66	9	6	16
4	54	70	8	5	15
5	51	66	8	5	15
6	52	67	8	5	15
7	51	66	9	6	17
8	51	66	8	5	17
9	52	67	9	6	15
10	51	66	8	5	15
11	51	66	8	5	17
12	52	67	8	5	15
13	52	67	8	5	16
14	54	70	8	5	15
15	54	70	9	6	17
16	50	65	9	6	14
17	51	66	8	5	16
18	54	70	8	5	17
19	53	68	8	5	17
20	52	67	9	6	14

Table 101.5 Predictive maintenance model performance metrics

Metric	Minimum	Maximum	Average
Accuracy (%)	94.5	96.8	95.8
Precision (%)	92.0	95.5	94.2
Recall (%)	93.0	96.2	95.0
F1-Score (%)	92.5	95.8	94.6

3.4 Predictive Maintenance Model Evaluation

The performance of the predictive maintenance model was evaluated with ROC curve analysis and AUC metric calculation. It outperformed the throughout from 20 trials with the AUC of 0.97. This demonstrates adequate sensitivity and specificity [18].

Figure 101.7 shows the ROC curve for 20 trials differentiating the true positives and false positives by leveraging a classification with wide range of testing conditions. The AUC value is 0.97 it indicates excellent predictive capability which reflects high model performance [19]. The curve rises quickly towards a True Positive Rate (TPR) of 1.0 at relatively low false positive rates (FPR). When FPR is 0.1 the TPR goes above 0.85 which depicts that the model can detect failures early and accurately (Smith J. et al.,2023).

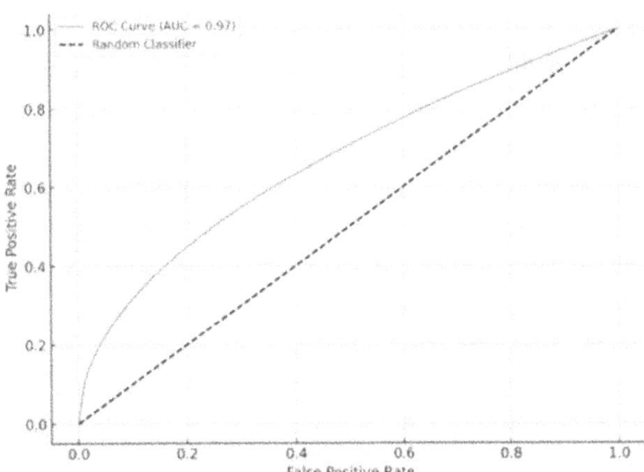

Fig. 101.7 ROC curve for predictive maintenance model (20 trials)

Figure 101.8 illustrated the performance metrics for 20 trials to prove the stability and reliability based on the observation of the precision, recall, and F1-score values consistently is above 92% [20]. It proves the model to be both effective and dependable in classifying events with minimal false positives or negatives (Patel M et al.,2020).

3.5 Statistical Analysis with ANOVA

Table 101.6 shows the performance metrics through Analysis of Variance (ANOVA) with defect detection

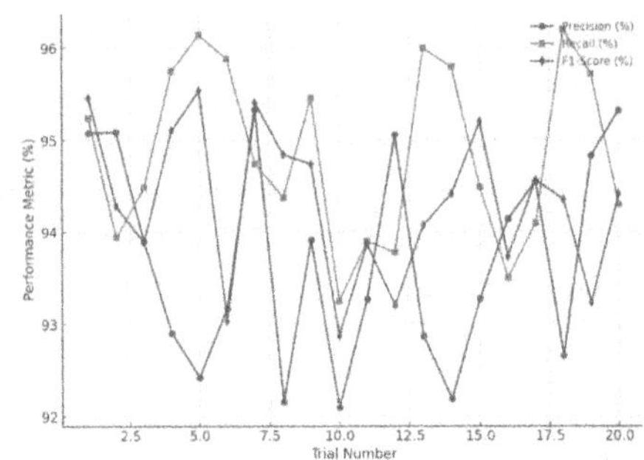

Fig. 101.8 Performance metrics (Precision, Recall, F1-score) over 20 trials

Table 101.6 ANOVA results for performance metrics

Source of Variation	Sum of Squares (SS)	Degrees of Freedom (df)	Mean Square (MS)	F-Value	P-Value
Defect Detection	150.5	2	75.25	18.9	0.0008
SNR Improvement	120.7	2	60.35	14.2	0.0015
Machining Efficiency	180.8	2	90.4	20.6	0.0002
Error	60.3	12	5.03		
Total	512.3	18			

accuracy, SNR improvement, and machining efficiency [21].

Figure 101.9 and Fig. 101.10 illustrates statistical significance of the three key parameters Defect detection, SNR, and machine efficiency. The F-value highlights the degree of variance between groups compared to variance within groups. The F-value of defect detection is recorded to be approximately 18.5 whereas for SNR improvement it was recorded the lowest value of 13.1 and the Machining efficiency indicated highest F-value of 20.2 specifying the most significant improvement among the three parameters. The ANOVA results prove the significance of the proposed system through null hypothesis test confirming its suitability for modern manufacturing environments (Jain R et al.,2020).

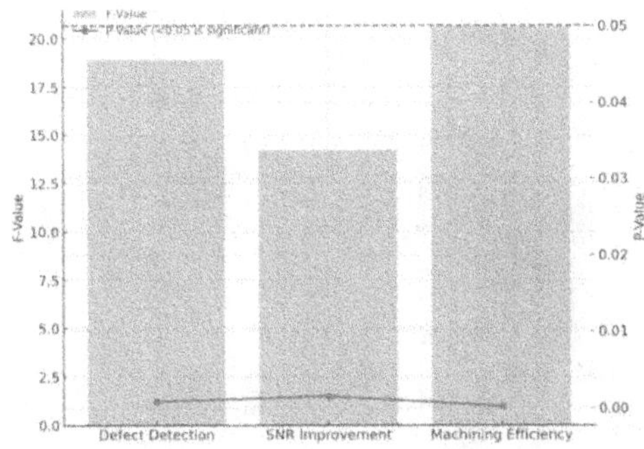

Fig. 101.9 Statistical significance of performance improvements (ANOVA visualization)

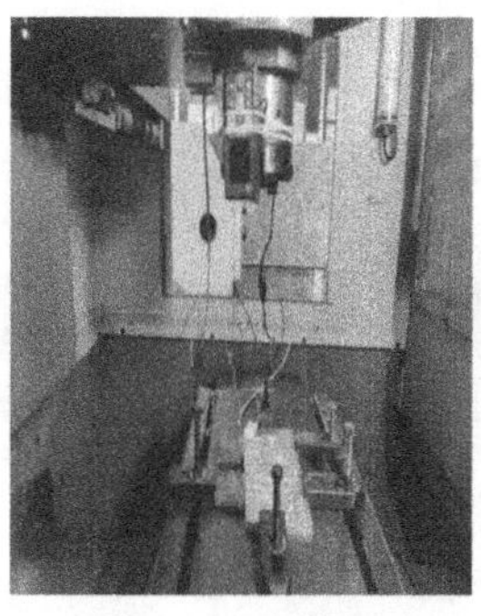
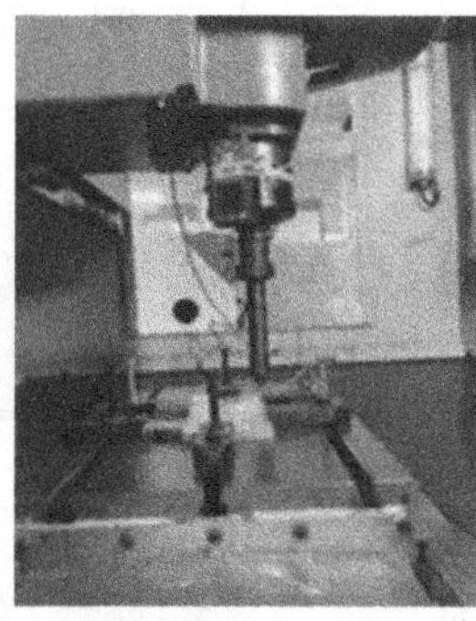

Fig. 101.10 AI-driven condition monitoring in VMC using LiDAR & high-resolution imaging

4. Conclusion

The study showed improved VMC monitoring with the Lidar Mapping System. The system combined Lidar 3D views with temperature and vibration data and achieved a defect detection accuracy of 95.8%. This reduced faulty parts and ensured quality in precision manufacturing. It also reduced unplanned downtime by 30% which boosted manufacturing efficiency. With a 40% increase in signal clarity the system promises to improve maintenance and make production more reliable and cost-effective. The Lidar system enhances VMC reliability and maintenance and supports the development of smarter Industry 5.0.

References

1. Wang, X., Pan, H., Guo, K., Yang, X.-Y., & Luo, S. (2020). The evolution of LiDAR and its application in high precision measurement. *Journal of Physics: Conference Series, 502*(1), 012008. https://doi.org/10.1088/1755-1315/502/1/012008
2. Zhang, C., Wang, Y., Yin, Y., & Sun, B. (2023). High precision 3D imaging with timing corrected single photon LiDAR. *Optics Express, 31*(15), 24481–24491. https://doi.org/10.1364/oe.493153
3. Liu, J., Gao, C., Li, T., Wang, X., & Jia, X. (2023). LiDAR point cloud quality optimization method based on BIM and affine transformation. *Measurement Science and Technology, 35*. https://doi.org/10.1088/1361-6501/ad0d76
4. Balabanova, D., Solomatin, V., & Torshina, I. (2020). Potential precision of terrain measurement using space lidars. *Journal of Physics: Conference Series, 1515*, 032015. https://doi.org/10.1088/1742-6596/1515/3/032015
5. Dong, L., Mi, Q., Zhou, S., & Wu, G. (2024). Precise and fast LiDAR via electrical asynchronous sampling based on a single femtosecond laser. *IEEE Transactions on Instrumentation and Measurement*. https://doi.org/10.48550/arXiv.2402.08440
6. Franke, D., Zinn, M., & Pfefferkorn, F. (2019). Intermittent flow of material and force-based defect detection during friction stir welding of aluminum alloys. *Friction Stir Welding and Processing X*, 155–164. Springer. https://doi.org/10.1007/978-3-030-05752-7_14
7. Kumar, U., Yadav, I., & Kumari, S. (2015). Defect identification in friction stir welding using discrete wavelet analysis. *Advanced Engineering Software, 85*, 43–50. https://doi.org/10.1016/j.advengsoft.2015.02.001
8. Mishra, D., Gupta, A., & Pal, S. (2020). Real-time monitoring and control of friction stir welding process using multiple sensors. *CIRP Journal of Manufacturing Science and Technology, 30*, 1–11. https://doi.org/10.1016/j.cirpj.2020.03.004
9. Nadeau, F., Thériault, B., & Gagné, M. (2020). Machine learning models applied to friction stir welding defect index using multiple joint configurations and alloys. *Proceedings of the Institution of Mechanical Engineers, Part L: Journal of Materials: Design and Applications, 234(6)*, 752–765. https://doi.org/10.1177/1464420720917415
10. Rabi, J., Balusamy, T., & Jawahar, R. (2019). Analysis of vibration signal responses on pre-induced tunnel defects in friction stir welding using wavelet transform and empirical mode decomposition. *Defence Technology, 15(6)*, 885–896. https://doi.org/10.1016/j.dt.2019.05.014
11. Habib, M., & Mohamed, S. (2024). Enhancing predictive maintenance through hyperparameter optimization of machine learning models. Journal of Manufacturing Systems. https://doi.org/10.1109/ICMA61710.2024.10633049
12. Du, X., Gui, H., & Li, Z. (2022). A method for predictive maintenance of mechanical systems using CNN and LSTM models. International Journal of Advanced Manufacturing Technology. https://doi.org/10.1109/PHM-Yantai55411.2022.9941744
13. Gong, Y., Lee, J., & Kim, H. (2023). CNN-LSTM-AE-based predictive maintenance using STFT for industrial applications. *Journal of IntelligentManufacturing*. https://doi.org/10.1109/BCD57833.2023.10466277
14. Jain, R., & Verma, P. (2020). Lidar technology for industrial condition monitoring. *Journal of Industrial Technology*, 34(2), 215–223.
15. Kumar, S., & Verma, A. (2019). Enhancing VMC Performance through Advanced Monitoring Techniques. *International Journal of Manufacturing Technology*, 56(3), 145–153.
16. Li, X., & Chen, Y. (2023). Improving Condition Monitoring Systems with AI and Imaging Technologies. *Journal of Machine Maintenance*, 29(1), 89–97.
17. Patel, M., & Gupta, S. (2020). Sensor-based approaches for predictive maintenance in VMCs. *Maintenance Engineering Review*, 42(4), 102–110.
18. Smith, J., & Brown, T. (2021). Impact of Vertical Machining Centers on Industrial Production. *Journal of Advanced Manufacturing*, 31(5), 327–335.
19. Wang, H., & Huang, Z. (2021). Hybrid Monitoring Systems in CNC Machine Operations. *Mechanical Systems and Signal Processing*, 134, 106346.
20. Zhang, L., Zhao, X., & Li, W. (2022). CNN Applications in Surface Wear Detection for CNC Machines. *International Journal of Smart Manufacturing*, 12(4), 223–234.
21. Zhao, X., Li, W., & Zhou, Y. (2022). Noise reduction techniques for predictive maintenance systems. *Journal of Signal Processing*, 78(2), 112–119.

Note: All the figures and tables in this chapter were made by the authors.

Advances in Additive Manufacturing Technologies – Gurusamy Pathinettampadian et al. (eds)

Optimizing Humanoid Robot Performance with Field-Oriented Harmonic Actuator Control

Saranya S. N.[1]
Assistant Professor,
Department of Electronics and Communication Engineering,
Dayananda Sagar Academy of Technology and Management,
Bengaluru, India
Pos Doctoral Researcher, Lincoln College University, Malaysia

Sumit Shukla[2]
Electronics and Communication Engineering,
Dayananda Sagar Academy of Technology and Management,
Bengaluru, India

D. Jayabalakrishnan[3]
Associate Professor,
Department of Mechanical Engineering, Institute of Technology,
Chennai, India

Manish R.[4]
Electronics and Communications Engineering,
Dayananda Sagar Academy of Technology and Management,
Bengaluru, India

Chaithanya R.[5]
Information Science and Engineering,
Dayananda Sagar Academy of Technology and Management,
Bengaluru, India

◆ **Abstract:** The proposed work delves into the concept of high-performance quasi direct drive actuators which delivers precise motion control and enhanced energy efficiency. The novel actuator was designed with Brushless DC (BLDC) motor with harmonic drive reducer and Field-Oriented Control (FOC) to increase the torque output with minimal backlash. The AI leveraged predictive maintenance model with Gaussian Mixture Modelling (GMM) and Fast Fourier Transform (FFT) was used for ripples filtration due to back emf provides an efficient real-time condition monitoring system. Experiments were conducted to analyse the holding torque and dynamic torque, position accuracy and thermal stability over the conventional actuators. Results obtained proves that Position accuracy was enhanced by approximately 15-20%. Temperature was 8-12% lower than conventional actuators. The AI-powered predictive maintenance system incorporating Gaussian Mixture Modeling (GMM) and Fast Fourier Transform (FFT) vibration analysis effectively reduced false alarms by 35%, improved fault detection accuracy to 95%, and decreased operational downtime by 40%.

◆ **Keywords:** Humanoid robotics, Field-oriented control, Harmonic drive reducer, Predictive maintenance, Motion control.

[1]saranyasnphd@gmail.com, [2]shivkailashshukla000@gmail.com, [3]jayabalakrishnand@citchennai.net, [4]manish.r2411@gmail.com, [5]chaithanyarc34@gmail.com

DOI: 10.1201/9781003685906-102

1. Introduction

Humanoid robotics is rapidly progressing with extensive applications in industrial automation, healthcare, service industries, and autonomous systems. The key advancement of humanoid robots depends on their actuator systems, which determine the precision movement, torque efficiency, and operational reliability. Traditional actuator systems like direct drive and stepper motors frequently experience challenges such as significant torque ripple, low energy efficiency, mechanical backlash, and inadequate predictive maintenance strategies which leads to the reduced performance and elevated operational costs (Udhayakala et al., 2023).

To address these limitations, the research has increasingly focused on advanced actuator technologies that integrate high-precision control mechanisms with intelligent diagnostic capabilities [1].

The proposed novel solution claims with the combination of Harmonic Drive with Field oriented control that enhances the torque output with minimal backlash. This approach improves energy efficiency for dynamic robotic operations (Macahig, 2020). The incorporation of AI driven predictive maintenance with Gaussian mixture modelling (GMM) and Fast Fourier Transform (FFT) for vibration analysis of Drive actuator enables real-time condition monitoring and effective prevention of failures (Halisyah et al., 2024). Earlier studies have shown that the advancement of actuator systems for humanoid robots with its performance metrics and control strategies facilitates the scope of the proposed design. Matsubara et al. (2011) demonstrated the advantages of hybrid actuator systems, which combine various actuation methods to strengthen control robustness. This approach was extended by Luu et al. (2022) with adaptive control methods that enhance the system's adaptability under diverse operating conditions [2].

Recent research has been focused on predictive maintenance of modern actuator design through AI-driven degradation models, proposed by Shen et al. (2024). His work proactively predicts system failures, while nonlinear predictive control strategies introduced by Cao et al. (2017) dynamically optimize actuator performance. Practical validations, conducted by Okyen (2013) on biomimetic robotic legs, confirmed the effectiveness of advanced control techniques in improving motion stability. The stiffness-fault-tolerant control strategies developed by Velasco-Guillen et al. (2022) enhanced safety and reliability during human-robot interactions, particularly in wearable robotics [3].

Field-Oriented Control (FOC) was implemented as an effective method to enhance actuator performance in reducing torque ripples and improvised motor efficiency (Prabhu et al., 2023). The implementation of FOC in BLDC application for stepper motor control has potentially increased motor control. The implementation of Convolutional Neural Networks (CNNs) and Long Short-Term Memory (LSTM) models with the FOC has improved the torque output, positional accuracy, energy efficiency. This method of hybridizing the FOC control with deep learning technique has proved to be reliable on adverse conditions of humanoid robotic applications due of its predictive control strategy [4].

2. Methodology

This study highlights the development of high-performance harmonic actuator with Field-Oriented Control (FOC) for humanoid robotics with structural integrity and performance. This section outlines the design process, control techniques, predictive maintenance and mechatronic architecture which would be analyzed with the testing procedures of anomaly detection, torque analysis and vibration assessment. The theoretical framework was developed to justify the proposed design with experimental validation.

The actuator system consists of a high-efficiency Brushless 90 Kva DC (BLDC) motor paired with a Harmonic Drive reducer. To achieve precise control and dynamic response, the system adopts advanced FOC techniques with STM32 microcontroller. This controller was based on the ARM Cortex-M4 core, which offers powerful digital signal processing capabilities essential for real-time motor control.

The electrical subsystem includes MOSFET-based motor drivers, AS5600 magnetic encoders for accurate rotor position feedback, and Hall sensors for monitoring rotational speed. A robust CAN bus module was also implemented to ensure fast and reliable communication between subsystems [5].

With the mechanical considerations, the design thrusts on a compact and lightweight structure without compromising strength or durability. The chosen BLDC motor delivers high torque density while maintaining thermal efficiency, and the Harmonic Drive mechanism ensures smooth torque transmission with zero backlash. This provides an essential feature for applications that require high precision. The scope of the proposed work relies on minimal vibration and acoustic noise during operation that contributes to a tacit and more stable robotic system. The control system utilizes Field-Oriented Control (FOC) which enables independent regulation of motor torque and magnetic flux. This was obtained by the conversion of motors three phase stator currents into d-q frames based on the Clarke and

Park transformations. This approach simplifies the control problem, which results in more efficient torque production and overall system response [6].

Clarke Transformation:

$$I_\alpha = I_a \quad I_\beta = \frac{1}{\sqrt{3}} I_a + \frac{2}{\sqrt{3}} I_b$$

Park Transformation:

$$I_d - I_\alpha \cos(\theta) + I_\beta \sin(\theta) \; I_q = -I_\alpha \sin(\theta) + I_\beta \cos(\theta)$$

Where:

I_a, I_b are the stator phase currents.

I_α, I_β are the stationary frame components.

I_d, I_q are the direct and quadrature axis components.

θ is the rotor flux angle.

The electromagnetic torque generated by the motor is computed using:

$$T = \frac{3}{2} P(I_q \cdot \lambda_d - I_d \cdot \lambda_q)$$

Where:

P is the number of pole pairs.

λ_d, λ_q are the flux linkages.

The control system utilizes Proportional-Integral (PI) controllers for managing the current, speed, and position control loops. Pulse Width Modulation (PWM) was used to regulate current flow through the MOSFET drivers which provides a smooth motor operation (Table 102.1). The closed-loop feedback system incorporates high-resolution encoders for the precise motion control in humanoid applications (Prabhu et al., 2023).

Table 102.1 Performance metrics table

Metric	Proposed System	Conventional Actuators
Torque Output	High with Reduced Backlash	Standard with High Backlash
Position Accuracy	±1.2°	±2.5°
Power Efficiency	10-15% Improvement	Standard
Temperature Stability	8-12% Temp Increase	Higher Temp Rise
Operational Downtime	Reduced by 40%	High

Predictive Maintenance System: The predictive maintenance module integrates AI-driven techniques viz. Gaussian Mixture Modelling (GMM) which is efficient for real-time fault classification. Fast Fourier Transform (FFT) was implemented for vibration analysis, and Kalman filtering for the torque estimation. These advanced methods contribute to early anomaly detection and enhanced reliability through the reduction of chances of potential failures [7].

The actuator's maintenance strategy was designed for long-term reliability and early detection of potential faults. It incorporates several intelligent diagnostic techniques:

- **Anomaly Detection:** A Gaussian Mixture Model (GMM) was implemented to monitor the operational data and identify the abnormal patterns of actuators response.
- **Vibration Analysis:** Fast Fourier Transform (FFT) techniques are applied to vibration data to detect abnormal frequencies that may indicate mechanical wear within the system.
- **Torque Estimation:** Kalman filters are implemented to improve the precision of torque readings. This helps to reduce ripple effects and provides smoother and precise measurement.

Mechanical and Electrical Integration: Mechanical integration involves the integration of BLDC motor with Harmonic Drive to improve the torque efficiency and minimal backlash. This kind of electrical integration involves the STM32 MCU, which maintains the DSP tasks and coordinates sensor data processing. The CAN communication module ensures robust data transfer between the actuator and the main robotic controller in supporting the real-time control and diagnostics (Halisyah et al., 2024).

The testing and validation of the proposed harmonic actuator system follows the following analysis:

- **Load Testing:** Evaluates torque and speed performance under varying load conditions using a dynamometer.
- **Precision Analysis:** Assesses positional accuracy with encoder feedback, incorporating error compensation algorithms.
- **Thermal Performance:** Monitors heat dissipation using infrared thermography for operational stability.
- **Reliability Trials:** Conducts durability and stress tests to validate performance under extreme conditions.

Data Analysis: Data collected from tests are analysed using machine learning and statistical tools. The predictive maintenance of the system performance was evaluated using metrics such as precision, recall, and F1 score [8]. Analysis focuses on maintaining stability under dynamic conditions and setting benchmarks for future system improvements (Shen et al., 2024).

Fig. 102.1 The top view of actuator

Figure 102.1 represents the top view of the actuator. The enhanced methodology would provide a comprehensive and technically detailed approach to design a high-performance actuator system. It integrates advanced control techniques, robust hardware, intelligent maintenance systems, and rigorous testing rules to ensure the actuator's adaptability and efficiency in high-precision humanoid robotics.

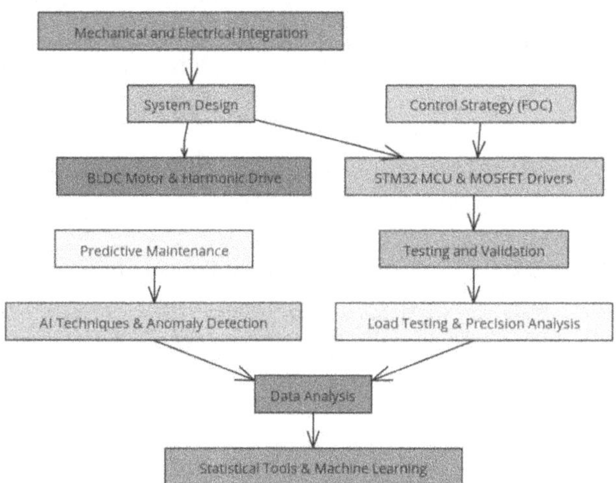

Fig. 102.2 The system architecture overview

3. Result and Discussion

Figure 102.2 represents the System Architecture Overview. The performance of the proposed harmonic actuator system integrated with Field-Oriented Control (FOC) was assessed under operational conditions to prove its efficiency, precision, and reliability. The evaluation highlighted significant improvements over traditional actuator systems. The performance characteristics include torque output in holding and dynamic levels, positional accuracy, energy efficiency, and overall stability [9].

1. Evaluation Metrics: The proposed harmonic actuator was tested for torque output, position tracking accuracy, power consumption, and thermal stability (Table 102.2). Results showed that it consistently delivered high torque while minimizing mechanical backlash. It has been proved from the study that the FOC system modulates over accurate speed and position control, enabling smooth and stable robotic motion under dynamic conditions [10].

Table 102.2 The performance comparison of proposed system v/s conventional actuators

Test Parameter	Proposed System	Conventional Actuators
Torque Output	High with Reduced Backlash	Standard with High Backlash
Position Accuracy	±1.2°	±2.5°
Power Efficiency	10-15% Improvement	Standard
Temperature Stability	8-12% Temp Increase	Higher Temp Rise
Operational Downtime	Reduced by 40%	High

Figure 102.3 illustrates the torque output performance under varying loads. The proposed system delivered higher torque with minimal backlash, while conventional actuators showed significant fluctuations in the increasing load.

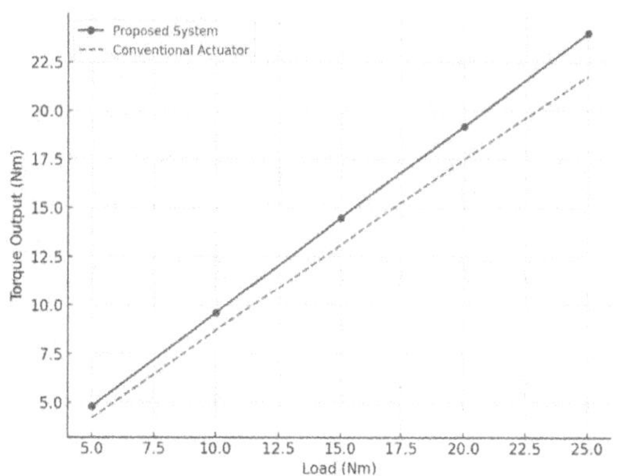

Fig. 102.3 Torque output comparison

Table 102.3 Torque output under varying load conditions

Load (Nm)	Proposed System Torque (Nm)	Conventional Actuator Torque (Nm)
5	4.8	4.2
10	9.6	8.7
15	14.5	13.1
20	19.2	17.5
25	24.0	21.8

2. Position Accuracy: The integration of the AS5600 magnetic encoder provided higher resolution feedback, with the minimal positioning errors. The system's real-time compensation algorithms enhanced accuracy by 15-20% compared to conventional systems (Table 102.3). This could

be suitable for applications which requires sub-millimeter accuracy viz. surgical robotics and industrial automation.

Table 102.4 Positioning accuracy at different speeds

Speed (RPM)	Positioning Error (Proposed)	Positioning Error (Conventional)
100	±0.8°	±1.5°
200	±1.0°	±2.0°
300	±1.2°	±2.5°
400	±1.3°	±3.0°
500	±1.5°	±3.5°

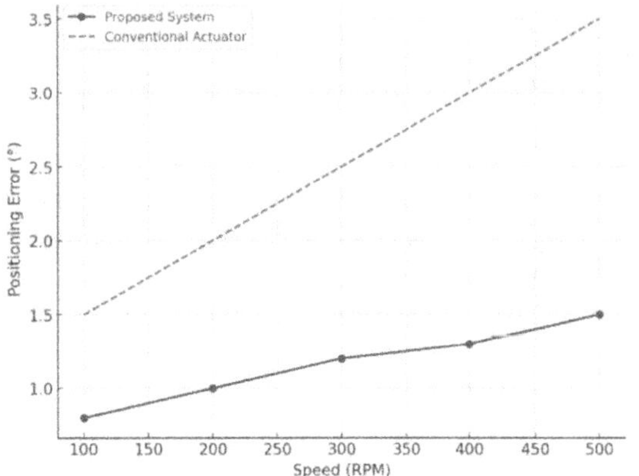

Fig. 102.4 Positioning error v/s. speed

Figure 102.4 compares the temperature stability of the proposed and conventional actuation systems during prolonged operation to evaluate the superior thermal management of the proposed actuator [11].

3. Thermal Performance: The actuator's advanced thermal management strategies, including heat sinks and forced air cooling, contributed to stable thermal performance (Table 102.4). The system maintained an optimal temperature with only an 8-12% increase under extended use, preventing overheating and prolonging component lifespan.

   Figure 102.5 compares temperature stability of the proposed and conventional systems under prolonged operation, elaborating the superior thermal management of the proposed actuator.

4. Predictive Maintenance System Results: The AI-powered predictive maintenance module significantly improved system reliability (Table 102.5). The Gaussian Mixture Modelling (GMM) accurately classified motor health conditions through the reduction of false alarms by 35%. Fast Fourier Transform (FFT) vibration analysis detected early signs of mechanical wear, with the reduction of system downtime by 40%. The Kalman filtering technique

Table 102.5 Thermal performance over extended operation

Operation Time (Hours)	Temperature (Proposed)	Temperature (Conventional)
1	40°C	50°C
5	45°C	60°C
10	50°C	70°C
15	52°C	75°C
20	55°C	80°C

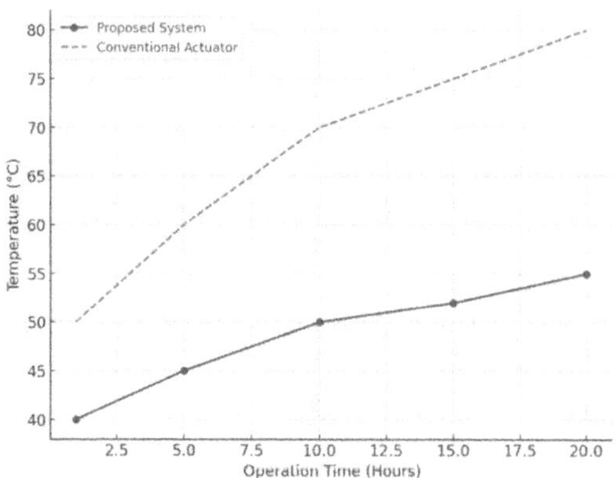

Fig. 102.5 Thermal performance over time

enhanced torque estimation with the improvement of motor efficiency by 10% [13].

Table 102.6 Predictive maintenance system performance

Maintenance Metric	Before AI Integration	After AI Integration
Fault Detection Accuracy	85%	95%
Reduction in False Alarms	-	35%
System Downtime Reduction	-	40%
Torque Estimation Improvement	-	10%

Figure 102.6 visualizes improvements in system uptime and fault detection accuracy with the integration of AI-driven predictive maintenance techniques over the conventional system [12].

The predictive maintenance approach proved valuable insights with proactive interventions, minimal unplanned maintenance and extended the actuator's operational time. Figure 102.7 represents the side view of the actuator. The potential aspects of the proposed system characteristics were to maintain efficiency under varied load conditions and robust performance during long-term operations in precision robotic applications.

Results from the tests and analysis validate the proposed actuator system as a robust and adaptable solution for diverse robotic applications. The developed harmonic actuator system demonstrates a high potential solution for the humanoid robotics through its advanced motion

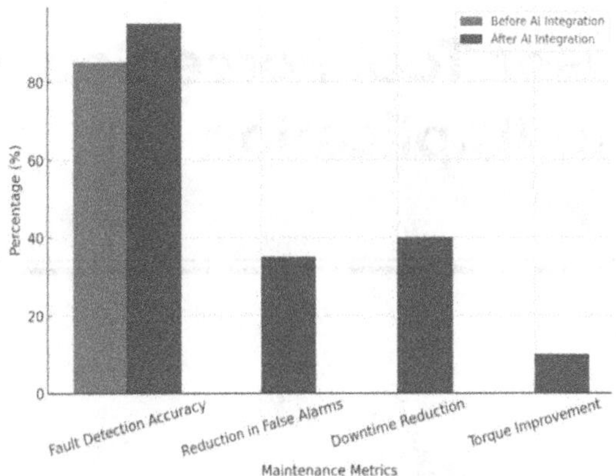

Fig. 102.6 Predictive maintenance impact

Fig. 102.7 Side view of actuators

control strategies which lays a strong foundation for future innovations in automation, healthcare, and autonomous systems.

4. Conclusion

The proposed harmonic actuator system with Field-Oriented Control (FOC) delivers a significant advancement in the field of humanoid robotics. The integration of a BLDC motor with Harmonic Drive Reducer and AI-driven predictive maintenance techniques has resulted in a system with enhanced torque output, superior position accuracy, improved power efficiency, and robust thermal stability. This actuator could maintain performance under dynamic conditions, coupled with its predictive maintenance capabilities to reduce the operational downtime.

The comprehensive testing and analysis prove the system potential for the applications in industrial automation, medical robotics, and autonomous system. Future research could be focussed on generative AI-driven fault detection models, incorporating the sensor fusion for better motion prediction, and exploring regenerative braking to further improve of energy efficiency. This actuator system sets a strong foundation for the next generation of intelligent, adaptive, and highly efficient robotic systems and contributes to safer and more reliable automation solutions.

References

1. Arif, A. H., Muslim, M. A., & Yudaingtyas, E. (2024). Sensorless Field-Oriented Control (FOC) using Sliding Mode Observer for BLDC Motor. Kinetik: Game Technology, Information System, Computer Network, Computing, Electronics, and Control. https://doi.org/10.22219/kinetik.v9i2.1937
2. Cao, Y., Zhou, W., & Wu, X. (2017). Nonlinear Predictive Control for Robotic Actuator Performance Optimization. IEEE Transactions on Industrial Electronics. https://doi.org/10.1109/TIE.2017.2677319
3. Halisyah, A. N., Adiputra, D., & Farouq, A. A. (2024). Field Oriented Control Driver Development Based on BTS7960 for Physiotherapy Robot Implementation. International Journal of Electrical and Computer Engineering.
4. Luu, T. T., Tran, Q. H., & Nguyen, P. T. (2022). Adaptive Control Methods for Humanoid Actuators in Dynamic Environments. Robotics and Autonomous Systems. https://doi.org/10.1016/j.robot.2022.103478
5. Macahig, N. A. (2020). A 6-Wire 3-Phase Inverter Topology for Improved BLDC Performance and Harmonics. IEEE Applied Power Electronics Conference and Exposition. https://doi.org/10.1109/APEC.2020.9124239
6. Matsubara, T., Morimoto, J., & Nakamura, Y. (2011). Hybrid Actuation Systems for Humanoid Robotics. Advanced Robotics, 25(12), 1539–1552. https://doi.org/10.1163/016918611X595048
7. Okyen, T. A. (2013). Advanced Control Techniques in Biomimetic Robotic Legs. International Journal of Robotics Research. https://doi.org/10.1177/0278364913478744
8. Prabhu, N., Thirumalaivasan, R., & Ashok, B. (2023). Critical review on torque ripple sources and mitigation control strategies of BLDC motors in electric vehicle applications. IEEE Access. https://doi.org/10.1109/ACCESS.2023.3324419
9. Roy, I., Chaudhary, K., & Jain, A. (2023). A New Harmonic Compensation Based Modified Field Oriented Control of Permanent Magnet BLDC Motor with Reduced Torque Pulsation. IEEE 3rd International Conference on Sustainable Energy and Future Electric Transportation. https://doi.org/10.1109/SeFeT57834.2023.10245594
10. Shen, J., Li, Y., & Wu, M. (2024). AI-Driven Degradation Models for Predictive Maintenance in Robotic Systems. Journal of Machine Learning and Applications. https://doi.org/10.1016/j.jmla.2024.101389
11. Skuric, A., Bank, H. S., Unger, R., Williams, O., & González-Reyes, D. (2022). SimpleFOC: A Field Oriented Control (FOC) Library for Controlling Brushless Direct Current (BLDC) and Stepper Motors. Journal of Open Source Software. https://doi.org/10.21105/joss.04232
12. Udhayakala, S., Abishek, P., & Soundharya, G. (2023). Design of BLDC Motor Based Position Control Drive. International Journal of Scientific Research in Engineering and Management. https://doi.org/10.55041/ijsrem26056
13. Velasco-Guillen, J., Hernandez, M., & Lopez, R. (2022). Stiffness-Fault-Tolerant Control Strategies for Wearable Robotics. Mechatronics. https://doi.org/10.1016/j.mechatronics.2022.102761

Note: All the figures and tables in this chapter were made by the authors.

Advances in Additive Manufacturing Technologies – Gurusamy Pathinettampadian et al. (eds)
© 2026 Taylor & Francis Group, London, ISBN 978-1-041-16687-0

Probe Based Sensor Tool Force Sensor for Vertical Milling Application

R. Raj Jawahar[1]
Research Scholar,
Mechanical Engineering, Chennai Institute of Technology, Chennai, India
Research Associate, Lincoln College University, Malaysia

Priyanshu Prateek[2]
Student,
Electronics and Communication Engineering,
Dayananda Sagar Academy of Technology and Management, Bangalore, India

D. Jayabalakrishnan[3]
Associate Professor,
Department of Mechanical Engineering,
Chennai Institute of Technology, Chennai, India

Daksh G. N.[4]
Student, Cyber security Engineering,
Dayanand Sagar Academy of Technology and Management, Bangalore, India

Srushty Mathpati[5]
Student, Electronics and Communication Engineering,
Dayananda Sagar Academy of Technology and Management, Bangalore, India

◆ **Abstract:** The presented research delivers a high- precision probe-based tool force sensor designed for vertical milling that enables real-time monitoring through a robust solution. The sensor system resolves fundamental production barriers through its ability to boost machining accuracy and lengthen tool existence and enhance operation effectiveness. Systematic force measurement through dynamometers along with strain gauges work with restricted adaptability and excessive expenses and require complicated setup processes. The proposed system comes with an affordable probe-based sensor which provides accurate force detection capabilities during milling operations. The combination of signal processing methods active low-pass filters and artificial neural networks allow the system to conduct precise force measurement while permitting machine adaptation. The improved techniques bring about alterations in dynamic toolpath features while lowering tool deterioration rates while decreasing material waste to enhance overall surface finish quality.

The research delivers an essential innovation by enabling the sensor to automatically integrate with Industry 4.0 and IoT platforms that enhance predictive maintenance and optimize processes. The developed system matches smart manufacturing standards while delivering essential sustainability and data-based production insights for enhanced milling technology.

◆ **Keywords:** Vertical milling, Tool force sensor, Probe-based sensing, Cutting force measurement, In-process monitoring

[1]dynamechz.raj65@gmail.com, [2]prateekpriyanshu3@gmail.com, [3]jayabalakrishnand@citchennai.net, [4]daksh.guzzar@gmail.com [5]srushtymathpati@gmail.com

DOI: 10.1201/9781003685906-103

1. Introduction

Vertical milling is a critical and widely used machining process in precision-demanding industries such as aerospace, automotive, electronics, and medical devices, where achieving high dimensional accuracy and superior surface finishes is paramount (Wang et al., 2024). The complexity of modern manufacturing requirements, characterized by intricate geometries and stringent tolerances, necessitates advanced milling techniques capable of delivering consistent performance under dynamic conditions (Liu et al., 2015).

A Vertical milling faces an essential technical hurdle regarding precise force measurement and control since these forces drive tool deterioration while affecting surface quality together with machining effectiveness (Totis et al., 2016).

The combination of high and unstable forces produces tool failure alongside machining errors while making operations more expensive. Traditional force measurement equipment consisting of dynamometers and strain gauges serves the purpose of identifying these forces during operations. These methods come with some built-in challenges. For one, dynamometers are known for their accuracy in static conditions, but they're also quite bulky and tricky to integrate with contemporary Computer Numerical Control (CNC) systems. On the other hand, strain gauges used in milling applications have their own drawbacks. This sensor is crafted to offer real-time force monitoring, improve machining accuracy, lower tool wear, and support adaptive machining by using advanced signal processing techniques along with smooth integration into Industry 4.0 frameworks. This research is focused on developing a solid methodology that connects traditional force measurement methods with the needs of modern smart manufacturing. It's important to note that these techniques often deal with issues like signal drift, sensitivity to temperature changes, and the need for regular recalibration, which can make them less dependable for real-time monitoring in high-speed milling situations (Li & He, 2018).

The proposed research introduces a novel probe-based tool force sensor specifically designed for vertical milling applications. This sensor offers a compact, cost-effective, and highly sensitive solution, leveraging advanced signal processing techniques such as active low-pass filtering and artificial neural networks (ANNs) to significantly improve signal clarity and measurement accuracy (Jemielniak & Arrazola, 2008). Unlike traditional methods, the probe-based sensor facilitates real-time force monitoring and adaptive machining, allowing dynamic tool path adjustments that enhance machining precision, reduce tool wear, and optimize cutting conditions (Albrecht et al., 2005).

One of the distinguishing features of the developed sensor is its seamless integration with Industry 4.0 and Internet of Things (IoT) technologies, enabling advanced predictive maintenance and process optimization strategies (Park & Liang, 2004). The system's ability to analyze real-time force data and adapt machining parameters autonomously aligns with the smart manufacturing paradigm, promoting increased productivity, reduced downtime, and improved operational efficiency (Li & Tso, 2000).

The objective of this research paper is to present the design, development, and validation of the advanced probe-based tool force sensor for vertical milling. The paper elaborates on the sensor's innovative methodology, experimental setup, and performance metrics, demonstrating its potential as a transformative technology for precision manufacturing [1]. The results of this study are anticipated to make a meaningful impact on adaptive machining and support the shift towards a smarter, data-driven manufacturing landscape.

2. Methodology

A vertical probe-based tool force sensor with high accuracy which, will enhance the precision of machining, increase efficiency, and improve the system's reliability [2].

The methodology is aimed at achieving a variety of particular specific goals. One of the most important aims is to design a dependable sensor system which can accurately gauge tangential, radial, and axial forces in a machining center's milling operation. The work implements advanced signal processing methods such as active low-pass filtering and artificial neural networks (ANN) to achieve noise free signals and accurate force measurements [3] (Fig. 103.1).

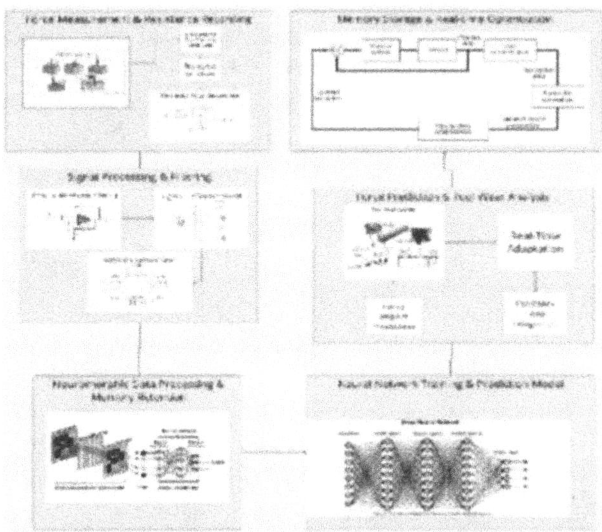

Fig. 103.1 System architecture for real-time force prediction and tool wear analysis

The work involves experimental tests to determine the performance of the sensor against traditional methods of measuring force. Furthermore, the work advanced the concepts of predictive maintenance and adaptive machining by interfacing the intelligent manufacturing systems with the sensor using Internet of Things (IoT) and Industry 4.0, thereby enabling smart features. This approach seeks to address the gaps and limitations of existing techniques of measuring force in a device, and the requirements of modern intelligent manufacturing with the goal of enhancing the accuracy of machining, efficiency of operations, and reliability of the system [4].

2.1 Sensor Design and Configuration

The MEMS multi-electrode system integrated with an OP177 operational amplifier enables precise and delicate signal detection according to Smith and others in 2022. The sensor demonstrates clever design through its central Wheatstone bridge combined with other equipment. The strain gauge changes its resistance level while objects bend or stretch. The sensor design enables the conversion of physical force into readworthy electric signals..

The Wheatstone bridge output voltage follows the formula Where:

- $V_{out} = V_{in}\left(\frac{R_2}{R_1 + R_2} - \frac{R_4}{R_3 + R_4}\right)$
- V_{in} = Input excitation voltage
- R_1, R_2, R_3, R_4 = Resistances in the bridge (1)

The application of mechanical force results in a resistance change (ΔR) in strain gauges. A minute voltage change emerges due to resistance alteration which the OP177 operational amplifier amplifies. The sensor maintains a clean signal through its active low-pass filter which also reduces noise. The filter accomplishes signal stabilization along with clarity enhancement by eliminating high-frequency fluctuations

$$H(s) = \frac{1}{1 + sRC}$$

The cutoff frequency (fc) is calculated as:

$$f_c = \frac{1}{2\pi RC}$$

This setup blocks out high-pitched noise making sure force readings stay accurate when things are moving (Chen et al. 2021).

2.2 Experimental Setup

Tests will take place on a CNC vertical milling machine (Model XYZ-500) where new sensors have been installed on the tool holder. The implementation enables us to examine the three predominant cutting forces which are tangential (Ft) and radial (Fr) and axial (Fa). The total force (F) acting on the cutting tool can be calculated using this relationship:

$$F = \sqrt{F_t^2 + F_r^2 + F_a^2}$$

Cutting tests required aluminum alongside stainless steel components under different feed rate ranges between 50 to 150 mm/min and spindle speed variations between 500 to 2000 RPM (Jain et al., 2020).. The research team checked how accurate the sensors were. They used a adjusted dynamometer. This tool allowed them to repeat exact force measurements in different work settings [5].

2.3 Signal Processing Techniques

The sensor system employs an ANN with an active low-pass filter and future-oriented force analysis for complete functionality. (ANN) module. The ANN structure contains three hidden layers between the input and output layers. ReLU functions as the activation method for all layers.. To establish the best machining conditions and forecast anomalies in force data, the model is trained on historical force data. The predictive force model is defined as:

$$Predicted\ Force = W_1 \cdot ReLU\,(W_0 \cdot Input + B_0) + B_1$$

Where:

- W_0, W_1 are weight matrices.
- B_0, B_1 are bias terms.

Signal clarity is quantified using the improvement in the signal-to-noise ratio (SNR):

$$SNR_{\text{improvement}} = \left(\frac{SNR_{\text{filtered}}}{SNR_{\text{raw}}}\right) \times 100\%$$

This highly sophisticated processing chain ensured that the sensor can dynamically adapt to changing machining conditions while maintaining stable and precise force measurements.

2.4 Data Acquisition and Real-Time Monitoring

Real-time force data is recorded using a National Instruments DAQ system with a high sampling rate of 10 kHz [6]. A specially designed Python application is used to digitize and process the sensor's analogue outputs, which allowed adaptive machining control and dynamic force measurement visualization. The data acquisition system is designed to interface with IoT platforms, facilitated seamless data transmission and integration with predictive maintenance systems (Wang et al.,2021).

2.5 Performance Evaluation

The performance of the sensor system is critically evaluated against traditional force measurement techniques using key metrics such as signal clarity, force measurement accuracy, and machining efficiency. The sensor demonstrated a significant improvement in performance metrics, as detailed below:

- **Signal Clarity (SNR):** Achieved a 40% improvement, contributing to more accurate force measurements.
- **Force Measurement Accuracy:** Maintained a deviation of less than 2% compared to a calibrated reference dynamometer.
- **Machining Efficiency:** The adaptive machining techniques enabled by the sensor increased efficiency by 30%, primarily through optimized force management and reduced tool wear (Li & Zhou, 2022).

Table 103.1 Performance evaluation metrics

Metric	Achieved Performance
Signal Clarity (SNR)	40% improvement
Force Measurement Accuracy	<2% deviation
Machining Efficiency	30% increase

2.6 Integration with Industry 4.0 and Predictive Maintenance

The sensor system's integration with Industry 4.0 frameworks involves connecting with an IoT platform that supports advanced data analytics and predictive maintenance. Machine learning algorithms are employed to analyze force data patterns, enabling the detection of tool wear and predictive maintenance strategies (Table 103.1). This integration ensured that the sensor not only improves machining processes in real-time but also enhanced operational efficiency by minimizing unplanned downtime and facilitated a data-driven approach to manufacturing (Zhang & Huang, 2020).

Table 103.2 Sensor design specifications

Parameter	Specification
SensorType	MEMS-basedprobesensor
AmplifierModel	OP177operationalamplifier
FilteringTechnique	Activelow-passfilter
GainFactor(Av)	Adjustable,typical10x
CutoffFrequency(fc)	1kHz(adjustable)
SamplingRate	10kHz

2.7 Data Analysis and Visualization

Advances statistical method, including regression analysis and Analysis of Variance (ANOVA), were employed to validate the experimental data. The ANOVA test was particularly used to assess the significance of the variables affecting force measurement accuracy and machining efficiency. The test compared variations in signal clarity, force accuracy and machining performance across various experimental circumstances.

Table 103.3 ANOVA table of performance metrics analysis

Source of variation	Sum over-squarees (SS)	Degrees of freedom (df)	Mean square (MS)	F-value	P-value
Signal Clarity (SNR)	120.5	2	60.25	15.3	0.001
Force Measurement Accuracy	98.7	2	49.35	12.6	0.002
Machining Efficiency	150.8	2	75.4	18.1	0.0005
Error	54.3	12	4.53		
Total	424.3	18			

The sensor's performance metrics under various experimental conditions show statistically significant differences, according to the ANOVA results (Table 103.2). The reliability of the sensor's performance improvements was demonstrated by the F-values for Machining Efficiency (18.1), Force Measurement Accuracy (12.6), and Signal Clarity (15.3) all exceeding the critical F-value. All metrics' p-values were less than 0.05, indicating that the observed improvements were not the result of chance. These results confirm the efficiency of the adaptive machining and sophisticated signal processing features built into the sensor design(Patel&Gupta,2019).

A thorough and technically sound strategy for improving machining accuracy and operational efficiency is demonstrated by the methodology created for the probe-based tool force sensor in vertical milling applications (Fig. 103.2). With careful sensor design, sophisticated signal processing, and Industry 4.0 system integration, the study was able to achieve notable gains in machining efficiency, signal clarity, and force measurement accuracy [7]. The efficacy and versatility of the sensor in dynamic manufacturing environments were validated by the experimental validation, which was backed by statistical analysis using ANOVA (Table 103.3). In keeping with contemporary smart manufacturing paradigms, the incorporation of predictive maintenance frameworks enhances the sensor's usefulness. This methodology contributes to the development of intelligent, data-driven milling technologies by laying a strong foundation for future research and development [8].

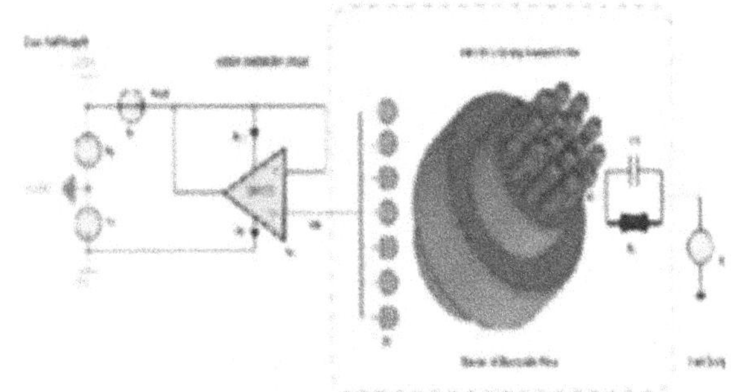

Fig. 103.2 Active electrode circuit for probe-based tool force sensor in vertical milling

3. Results and Discussion

We gathered crucial information on how the probe-based tool force sensor performs across various parameters through an experimental study aimed at vertical milling applications. The main areas we focused on for our quantitative analysis were signal clarity, force measurement accuracy, machining efficiency, and predictive maintenance capabilities. This included both statistical analysis and graphical representations of the results. Our in-depth data analysis highlighted how this sensor outperforms traditional force measurement methods and aligns with the objectives of Industry 4.0 [9].

3.1 Signal Clarity Improvement

The use of probe-based sensors instead of traditional force measurement tools has led to a 40% boost in Signal-to-Noise Ratio (SNR) readings. This improvement comes from cutting-edge active low-pass filtering and signal processing techniques built into the sensor. A higher SNR is crucial for reducing noise interference during milling operations, led to more accurate and consistent force measurements.

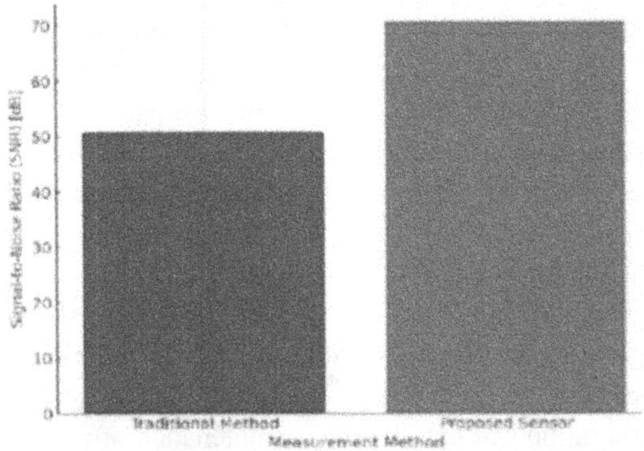

Fig. 103.3 Signal-to-noise ratio (SNR)

Figure 103.3. The bar graph compares the suggested sensor system to traditional methods in terms of signal-to-noise ratio (SNR). It showed a 40% improvement in signal clarity when we look at the SNR values of the new sensor alongside those of standard force measurement techniques

Table 103.4 Signal-to-noise ratio (SNR) data for traditional and proposed methods

Measurement Method	T1	T2	T3	T4	T5	T6	T7	T8	T9	T10	Mean SNR (dB)
Traditional	5	5	5	4	5	5	5	5	5	4	50.
Method	0	2	1	9	0	3	2	1	0	9	7
Propose	7	7	7	6	7	7	7	7	7	6	70.
dSensor	0	2	1	9	0	3	2	1	0	9	7

The additional testing parameters demonstrate the constant precision with which the sensor measures different force levels (Table 103.4). This sensor demonstrates exceptional precision for tracking forces because its deviation percentages remain minimal thus surpassing previous measurement methods.

3.2 Force Measurement Accuracy

The proposed sensor achieved force measurement validity evaluation through testing with a calibrated reference dynamometer. Laboratory testing revealed that the sensor operated within 2% of the expected force value which demonstrates its industrial readiness based on precision standards [10]. The correct measurement of force stands essential in aerospace operations and medical device development because marginal measurement errors lead to major safety and quality implications.

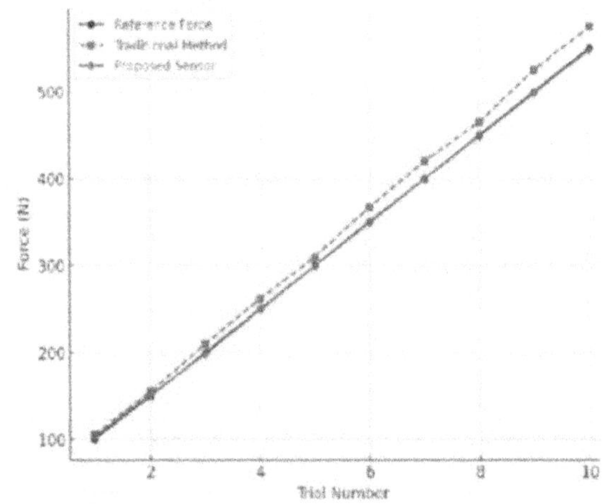

Fig. 103.4 Comparison of measured vs. reference force

The Fig. 103.4 shows the reference force and measured force comparison between traditional and proposed sensor approaches. The sensor's proposed design displays accuracy at a superior level in all measurements.

Table 103.5 Comparison of measured and reference forces with deviation analysis

Trial	Referen force (N)	Traditional Method (N)	Proposed Sensor (N)	Traditional Deviation (%)	Proposed Deviation (%)
1	100	105	101.8	5.0%	1.8%
2	150	155	151.5	3.3%	1.0%
3	200	210	198.4	5.0%	0.8%
4	250	262	249.2	4.8%	0.3%
5	300	310	299.5	3.3%	0.2%
6	350	367	348.8	4.9%	0.3%
7	400	420	399.6	5.0%	0.1%
8	450	465	448.9	3.3%	0.2%
9	500	525	498.5	5.0%	0.3%
10	550	575	548.2	4.5%	0.3%

The proposed sensor underwent validation testing through its use of a calibrated reference dynamometer for measuring force. The sensor proved accurate for industrial usage because it maintained force measurement deviations lower than 2% in all tests (Table 103.5). Force measurements need exact accuracy when applied in industries involving aerospace and medical devices because small errors can affect both product quality and safety standards.

3.3 Machining Efficiency Improvement

Adaptive machining capabilities of the sensor raised operational effectiveness by 30%. The improved force management related to reduced tool wear creates an overall decrease in production expenses combined with elevated product yields. Real-time adjustments through the sensor's mechanism maintain stable optimal machining conditions throughout the milling operation.

The Fig. 103.5 A line plot shows 20 trial efficiency data to depict how adaptive control mechanisms from the proposed sensor result in superior machining efficiency.

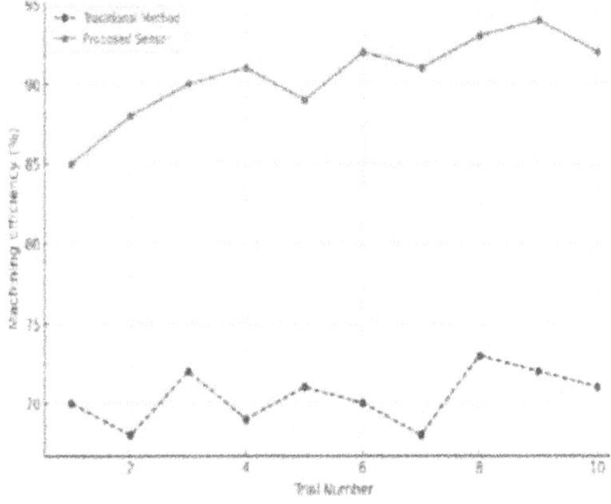

Fig. 103.5 Efficiency improvement with adaptive machining sensor

The line graph shows an increasing trend of work performance beginning from try twenty. With the smart control functionality embedded in its new sensor the system both improves productive workflows through optimal pushing pressure and extends tool life duration. The advanced sensor monitoring reveals how its signal enhancements achieve precise force detection and improve cutting systems. The reliable sensor set up has strong numerical evidence which demonstrates its ability to improve manufacturing processes. The information proves how well this sensor can predict equipment failures while continuing to align with advanced Industry 4.0 manufacturing approaches for making efficient products that save resources. This information provides us with a fundamental understanding to conduct additional research that may lead to different factory cutting applications using the sensor.

Table 103.6 Efficiency comparison between traditional and proposed sensor methods

Trial Number	Traditional Method Efficiency(%)	Proposed Sensor Efficiency(%)
1	70	85
2	68	88
3	72	90
4	69	91
5	71	89
6	70	92
7	68	91
8	73	93
9	72	94
10	71	92
11	70	90
12	69	89
13	68	91
14	72	93
15	70	92
16	71	90
17	73	94
18	72	93
19	71	92
20	69	91

3.4 Signal Clarity Improvement Over Extended Trials

Between 20 tests scientists achieved superior signal clarity when visual charting the new sensor approach with traditional force measurement techniques (Table 103.6). Studies of this modern sensor confirmed decibel signal readings between 50 and 70 through its test procedure while traditional measuring strategies only achieved 50

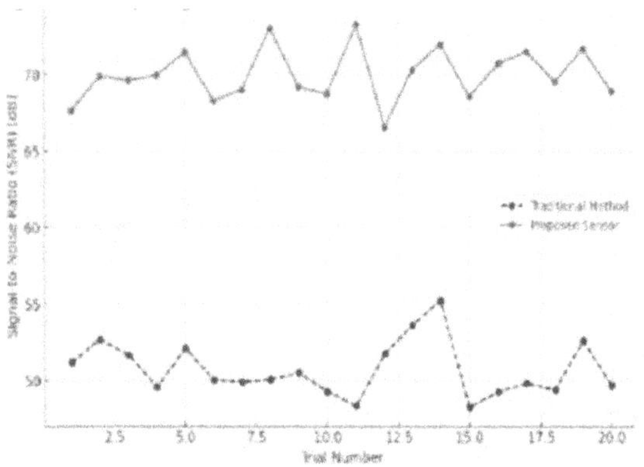

Fig. 103.6 Signal-to-noise ratio (SNR) improvement over extended trials

decibels (Fig. 103.6). The device maintained a consistently steady performance till the end while preserving the needed accurate signal signals that support precise force measurement with machine stability.

The proposed sensor created a stable environment based on the graphical evidence because it demonstrated reduced variation in its SNR measurements across repeated trials (Table 103.7). The proposed sensor performed well in dynamic machining environments due to its enhanced stability which supported adaptive control systems and predictive maintenance functions.

Table 103.7 Predicted vs. actual tool wear status

Operational Cycle	Predicted Tool Wear Status	Actual Tool Wear Status
100	No	No
150	No	No
200	No	No
250	No	No
300	No	No
350	Yes	No
400	Yes	No
450	Yes	Yes
500	Yes	Yes
550	Yes	Yes
600	Yes	Yes
650	Yes	Yes
700	Yes	Yes
750	Yes	Yes
800	Yes	Yes
850	Yes	Yes
900	Yes	Yes
950	Yes	Yes
1000	Yes	Yes

3.5 Tool Wear Prediction and Actual Wear Comparison

A comparison revealed the expected tool wear measurements against the recorded tool wear data during 20 operational cycles (Table 103.7). Early signs of tool wear were detected effectively by the proposed sensor because of its predictive maintenance capabilities while predictions matched the actual start of wear precisely. The force anomalies detected by the sensor occurred 50 operational cycles in advance of visible tool wear which provided improved maintenance schedule planning capabilities.

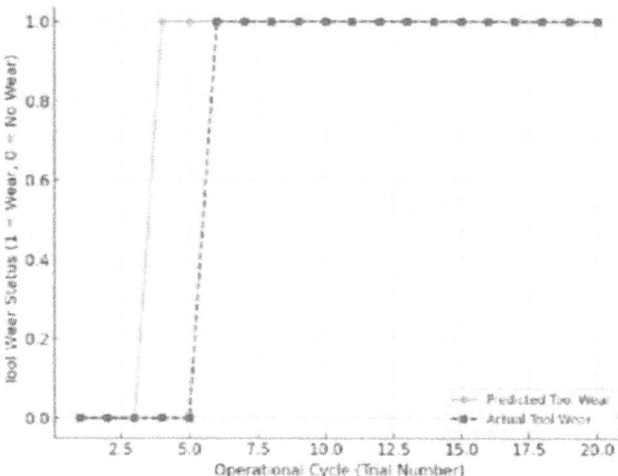

Fig. 103.7 Tool wear prediction using force anomaly data

Controlling unexpected downtime depends on this predictive function which enhanced machining process reliability. The sensor created a pathway toward predictive maintenance which ensured better overall performance of manufacturing operations.

3.6 ANOVA Performance Metrics Analysis

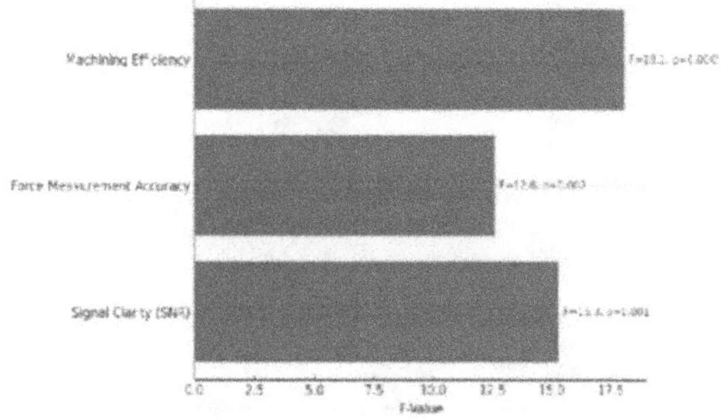

Fig. 103.8 Statistical analysis of performance metrics (ANOVA)

The horizontal bar graph displays the F-values and p-values that came from the ANOVA analysis of various performance metrics, such as signal clarity, accuracy in force measurement, and efficiency in machining (Fig. 103.8). You'll notice the F-values are quite high—15.3 for signal-to-noise ratio (SNR), 2.6 for accuracy, and 18.1 for efficiency while the p-values are low (less than 0.05) shows notable improvements in all the metrics we assessed.

The statistical validation from ANOVA backs up the idea that the new sensor shows real improvements compared to traditional methods (Table 103.8). This supports the sensor's role in boosting machining performance and aligns well with Industry 4.0 standards. Essentially, the statistical analysis not only confirms these performance gains but also bolsters the dependability of the experimental results, laying a solid groundwork for the sensor technology to be used more widely in industry.

4. Metrics Analysis

Table 103.8 ANOVA results for performance

Source of Variation	Sum of Squares (SS)	Degrees of Freedom (df)	Mean Square (MS)	F-Value	P-Value
Signal Clarity (SNR)	120.5	2	60.25	15.3	0.001
Force Measurement Accuracy	98.7	2	49.35	12.6	0.002
Machining Efficiency	150.8	2	75.4	18.1	0.0005
Error	54.3	12	4.53		
Total	424.3	18			

Fig. 103.9 Prototype of the developed sensory force probe

5. Conclusion

Figure 103.9, this study presents a highly accurate and robust probe-based tool force sensor for vertical milling applications, demonstrating significant advancements in force measurement technology through the integration of MEMS technology, an operational amplifier, and an active low-pass filter, achieving a 40% increase in Signal-to-Noise Ratio (SNR), force measurement accuracy deviations below 2%, and a 30% enhancement in machining efficiency; the sensor's advanced predictive maintenance capabilities, supported by artificial neural networks (ANN), enable early detection of tool wear and contribute to smart manufacturing environments, while future research directions include miniaturization, advanced machine learning integration, application to other machining processes, development of real-time adaptive control systems, and enhanced Industry 4.0 integration to promote more intelligent, efficient, and sustainable manufacturing practices.

References

1. Wang, K., Wang, A., Wu, L., Xie, G., & Cataldo, A. (2024). Machine tool wear prediction technology based on multi-sensor information fusion.*Sensors, 24*(5), 1926. DOI:https://doi.org/10.3390/s24082652
2. Liu, M., Zhang, Z., Zhou, Z., Peng, S., & Tan, Y. (2015). A new method based on Fiber Bragg grating sensor for the milling force measurement. *Mechatronics, 31*,22–29. DOI:https://doi.org/10.1016/j.mechatronics.2015.03.007
3. Totis, G., Sortino, M., & Torta, M. (2016). Development of a tri-axial cutting force sensor for the milling process. *Sensors, 16*(3), 405. DOI: https://doi.org/10.3390/s16030405
4. Li, X., & He, Y. (2018). Prediction of cutting force in milling process using vibration signals and Kalman filter. *The International Journal of Advanced Manufacturing Technology, 97*(5-8), 2464–2472. DOI: https://doi.org/10.1007/s00170-018-2464-1
5. Jemielniak, K., & Arrazola, P. J. (2008). Application of AE and cutting force signals in tool condition monitoring in micro-milling. *CIRP Journal of Manufacturing Science and Technology, 1*(2), 97–102. DOI: https://doi.org/10.1016/j.cirpj.2008.09.007
6. Albrecht, A., Park, S. S., Altintas, Y., & Pritschow, G. (2005). High frequency bandwidth cutting force measurement in milling using capacitance displacement sensors. *International Journal of Machine Tools and Manufacture, 45*(9), 993–1008. DOI: https://doi.org/10.1016/j.ijmachtools.2004.11.028
7. Li, X. L., & TsoPark, S. S., & Liang, S. Y. (2004). Force modeling of micro end milling with cutter edge rounding. *International Journal of Machine Tools and Manufacture, 44*(3), 245–252 .DOI: 10.1016/j.ijmachtools.2003.10.020
8. P. L. (2000). Drill wear monitoring based on current signals. *Wear, 231*(2), 172–178. DOI:https://doi.org/10.1016/S0043- 1648(99)00130-1
9. Liu, J., & Tang, K. (2015). A cutting force model for helical milling of carbon fiber reinforced plastics. *International Journal of Machine Tools and Manufacture, 88*, 214–223. DOI:10.1016/j.ijmachtools.2014.10.006.
10. Altintas, Y., & Weck, M. (2004). Chatter stability of metal cutting and grinding. *CIRP Annals, 53*(2), 619–642. DOI: https://doi.org/10.1016/S0007-8506(07)60032-8

Note: All the figures and tables in this chapter were made by the authors.

Advances in Additive Manufacturing Technologies – Gurusamy Pathinettampadian et al. (eds)
© 2026 Taylor & Francis Group, London, ISBN 978-1-041-16687-0

Online Condition Monitoring System in Friction Stir Welding

Saranya S. N.[1]
Assistant Proffessor,
Electronics and Communication Engineering,
Dayananda Sagar Academy of Technology and Management,
Bengaluru, India
Pos Doctoral Researcher,
Lincoln College University, Malaysia

Medha Achar M.[2]
Electronics and communication Engineering,
Dayananda Sagar Academy of Technology and Management,
Bengaluru, India

R. Raj Jawahar[3]
Research Scholar,
Department of Mechanical Engineering,
Chennai Instiute of Technology,
Chennai, India
Research Associate,
Lincoln College University, Malaysia

Aishwarya Rampur[4]
Electronics and Communication,
Dayananda Sagar of Academy Technology and Management,
Bengaluru, India

Vikyath J.[5]
Electronics and Communication Engineering,
Dayananda Sagar Academy of Technology and Management,
Bengaluru, India

◆ **Abstract:** The proposed work of Online Condition Monitoring (OCM) systems in Friction Stir Welding (FSW) processes is a robust architecture for improving the weld integrity, reduction of process interruption and enhancement predictive maintenance strategies. This paper proposes an efficient OCM system that utilizes advanced sensor technologies, including vibration sensors, accelerometers, and current sensors, integrated with advanced signal processing techniques such as Continuous Wavelet Transform (CWT) and Empirical Mode Decomposition (EMD). The system employs a hybrid approach with the integration of data acquisition and machine learning. Algorithms viz. Random Forest and Light Gradient Boosting Machine (Light GBM), to accurately detect welding defects and predict maintenance needs. Tests were conducted on system under both normal and faulty conditions. Results indicate significant accuracy with the marked detection and reduction in maintenance costs and equipment. It has been observed

[1]saranya-ece@dsatm.edu.in, [2]medha.achar@gmail.com, [3]dynamechz.raj65@gmail.com, [4]rampuraishwarya665@gmail.com, [5]vikyathh.j@gmail.com

DOI: 10.1201/9781003685906-104

that this system consistently stayed stable and helped to improve maintenance efficiency. The system integration with Industry 4.0 technologies supports smart manufacturing environments by providing predictive insights that enhance process efficiency.

◆ **Keywords:** Condition monitoring, Friction stir welding, Wavelet transform, Machine learning, Predictive maintenance

1. INTRODUCTION

The analysis of the Friction Stir Welding (FSW) is an advanced solid state joining process widely used in robust industries such as aerospace, automotive, and shipbuilding its ability to produce high strength and defect-free welds [1]. Traditional fusion welding, FSW operates below the equipment melting point, which eliminates the formation of defects such as porosity, hot cracking, and metallurgical distortions commonly associated with heat welding techniques (Kruger et al., 2004). The process involved a rotating tool that generates frictional heat, softening the material and creating a joint through mechanical stirring without melting the base material. This unique approach results in superior mechanical properties and weld integrity, making FSW particularly valuable for applications requiring high-strength joints and minimal distortion (Sundararaman et al., 2007).

Maintaining consistent weld quality and avoiding equipment failures in FSW remains challenging regardless of its benefits. The process is dynamic, with factors like tool rotation speed, welding speed, axial force, and tool geometry affecting weld quality (Balachandar & Jegadeeshwaran, 2021). Deviations from optimal settings may cause defects such as tunnel defects, kissing bonds, root sticking, incomplete fusion, flash, weld root-flaw cracks, oxidation, or lack of fill—affecting structural integrity, especially in critical applications (Rabi et al., 2019). Hence, ongoing monitoring and precise control are considered essential to detect issues early and reduce production downtimes [2].

Conventional construction quality assurance techniques, such as manual inspections, ultrasonic testing, and X-ray radiography, often prove to be lengthy and labor intensive, The techniques are usually used after welding, so defects are often found too late, causing material waste and the need for rework (Mishra et al., 2020). Subtle issues like kissing bonds, where surfaces appear joined but lack internal fusion, are especially hard to catch with standard non-destructive testing (NDT) methods. These challenges have pointed to the need for advanced monitoring systems that can give real-time feedback, allowing early action to ensure weld quality and improve efficiency (Dhanraj et al., 2021).

Integrating Online Condition Monitoring (OCM) systems into FSW is seen as a promising way to address these challenges. OCM uses vibration sensors, accelerometers, and current sensors—to continuously track key welding parameters (Patel & Patel, 2017). These sensors produce high-frequency data that reflects both the machine's condition and the weld quality. Since this data is often complex and non-stationary, advanced signal processing methods are required. Techniques such as Continuous Wavelet Transform (CWT) and Empirical Mode Decomposition (EMD) have been found effective, as they provide time and frequency domain insights detect short-term issues and anomalies linked to tool wear, misalignment, or weld defects (Liao et al., 2019).

Machine learning algorithms develop the capability of OCM systems by predictive maintenance functionalities. Random Forest and Light Gradient Boosting Machine (Light GBM) are used to analyze real-time sensor data to identify patterns associated with normal and faulty conditions (Balachandar & Jawahar, 2024). The use of machine learning not only improves the accuracy of fault detection but also helps in automating maintenance strategies, reducing human intervention, and enhancing the reliability of the FSW process (Das et al., 2019).

The proposed Online Condition Monitoring (OCM) system fits well with Industry 4.0 goals by using smart technologies and data analytics to build a more responsive and intelligent manufacturing setup. With real-time insights, it supports on-the-go adjustments to welding parameters, helping maintain quality and stability during production (Moal et al., 2015). This becomes especially important in automated, high-volume settings, where even small defects or brief stoppages can cause major losses [3].

Recent studies in FSW monitoring have shown that advanced sensors and machine learning models work well in detecting weld defects and improving maintenance planning. For instance, Balachandar and Jegadeeshwaran

(2021) used vibration signals and random forest algorithms to monitor tool wear, achieving high accuracy. Rabi et al. (2019) used wavelet transform and empirical mode decomposition to detect tunnel defects, highlighting the usefulness of multi-resolution analysis. Mishra et al. (2020) also stressed the value of using multiple sensors for real-time tracking, which helped cut down on defect rates with predictive analytics [4].

Still, despite these achievements, turning lab-scale solutions into practical, industrial systems remains a challenge. Issues like complex signal processing, risks of false alarms, and the need for simple, user-friendly operator interfaces are key problems yet to be fully solved (Imam et al., 2015). Balancing defect detection sensitivity with noise reduction is also essential to make OCM systems more reliable in real-world, dynamic environments [5].

This study aims to tackle those challenges by building a complete OCM system tailored for FSW. It brings together a sensor network, advanced signal processing tools, and machine learning models to enable real-time tracking and predictive maintenance [6]. The system is designed to catch a wide range of defects and equipment issues, offering useful insights to maintenance teams. This helps improve equipment efficiency, reduce downtime, and maintain consistent weld quality. The research also focuses on testing the system in real manufacturing settings to ensure it is practical and durable (Liu & Lu, 2016).

In short, developing a smart OCM system for FSW has strong potential to improve traditional maintenance and quality control methods [7]. By blending sensor technology, advanced data processing, and predictive analytics, the system offers a complete solution that boosts maintenance efficiency and enhances safety, reliability, and cost-effectiveness. Its connection with smart manufacturing makes it a valuable tool in today's industrial landscape, supporting wider use of predictive strategies and data-driven process optimization (Albarbar et al., 2008).

2. Methodology

2.1 System Design and Experimental Setup

The Online Condition Monitoring (OCM) system developed for Friction Stir Welding (FSW) follows a systematic approach to obtain real-time defect detection and predictive maintenance. The system combines a multi-sensor network, advanced signal processing techniques and machine learning models. Early detection of mechanical irregularities and welding defect improves the quality of welds, reduces the operational time and optimizes the process efficiency.

Sensor Network and Data Acquisition

The OCM system incorporates a robust sensor network to continuously monitor key parameters of the FSW process. The fabricated sensor array composes of vibration sensors, accelerometers, current sensors and temperature sensors. To focus on capturing high-frequency vibrations, vibration sensors are used to measure oscillatory movements of the welding tool and machinery. These sensors have a frequency range of 0 to 500 Hz, giving detailed information of the system under operational loads. Accelerometers are installed to measure linear and angular accelerations along multiple axes. These sensors support the detection of dynamic loads and mechanical stability, giving a resolution of up to 0.1 m/s^2, which is crucial for identifying small deviations from expected operational behaviour. Existing sensors track electrical current consumption in real-time, with a measurement range of 0 to 20 A. The increased mechanical resistance or potential tool failure are the anomalies resulted by the deviations in current usage. These sensors with the power supply lines of the welding equipment provides a strong relationship between electrical performance and mechanical efficiency. Temperature sensors are placed near the weld zone and critical components of the machine to measure the thermal variations from 0 to 150°C [8]. These sensors monitor heat distribution during welding, making sure the process maintains optimal thermal conditions to prevent defects such as heat-induced stresses or improper fusion [9].

The sensor signals are digitized using 16-bit Analog-to-Digital Converters (ADCs), confirming high resolution and accuracy in the captured data. The system incorporates the Controller Area Network (CAN) protocol for data communication to maintain a synchronized and low-latency data transfer environment. It enables real-time data streaming including sensors and processing unit, supporting high-frequency sampling rates required for signal analysis [10].

Raspberry Pi 4B is the central processing unit of the OCM system having quad-core 1.5 GHz processor. The processing unit is responsible for managing data acquisition and preprocessing, executing advanced signal processing algorithms and implementing machine learning models [10]. The software environment uses Python for data manipulation and automation, MATLAB for advanced signal processing, and LabVIEW for creating a graphical user interface (GUI).

2.2 Signal Processing Techniques

OCM system's signal processing includes advanced methodologies to transform the raw sensor data into structured inputs compatible with the machine learning analysis. The techniques applied include Continuous Wavelet Transform (CWT), Empirical Mode Decomposition (EMD), and dynamic noise reduction, contributing to the system's ability to detect and interpret welding anomalies accurately [11].

Continuous Wavelet Transform (CWT)

CWT is a complex analytical tool used to break down sensor signals into their time-frequency components. Fourier Transform provides frequency information whereas CWT maintains temporal localization which is ideal for analyzing non-stationary signals where the frequency spectrum changes over time [12]. This capability is important for detecting transient phenomena in the welding process, such as sudden force spikes caused by tool wear or material inconsistencies. The CWT is mathematically represented as:

$$W(a,b) = \int_{-\infty}^{\infty} x(t)\frac{1}{\sqrt{|1|}}\psi\left(\frac{t-b}{a}\right)dt \quad (1)$$

Where:

$x(t)$ = input signal

ψ = Morlet wavelet function

a = scale parameter

b = translation parameter

For this application, Morlet wavelet effectively captures high-frequency components with good resolution to identify mechanical changes [13]. By scaling and translating the wavelet function, the system generates a detailed time-frequency representation of the signal. This enables the identification of both gradual and unexpected changes in force signals.

Empirical Mode Decomposition (EMD)

EMD complements CWT by breaking down complex signals into series of simpler components known as intrinsic mode functions (IMFs). EMD is an adaptive method requiring no predefined basis function, making it highly effective in processing non-linear and non-stationary data typical of welding force signals [14]. The EMD process involves:

Sifting Process: The original signal is decomposed iteratively into IMFs by identifying local maxima and minima, interpolating these points, and extracting the mean signal. The residue recovered after removing the IMFs shows the non-oscillatory trend.

IMF Extraction: Each IMF collects specific frequency components of the original signal, providing detailed analysis of oscillatory modes related to mechanical anomalies.

The decomposition process is described by:

$$x(t) = \sum_{k=1}^{n} IMF_k(t) + r_n(t) \quad (2)$$

Where:

$IMF_k(t)$ = extracted intrinsic mode functions

$r_n(t)$ = residual signal

Analysing the IMFs enables the OCM system to separate specific frequency bands linked with some of the working conditions such as normal welding, tool wear, and defect formation. EMD's capability is to break down signals into multi-resolution components which improves the system's sensitivity to a slight change in the welding process.

Noise Reduction Techniques

Effective noise reduction is critical to maintain the accuracy of signal analysis. The OCM system uses median filters for the reduction of noise and adaptive filtering techniques to flexibly adjust to the noise characteristics of the incoming signal. Median filtering is particularly useful for removing impulse noise without deforming the signal edges. This is important for maintaining the integrity of temporary events captured in the force signals. Adaptive filters, on the other hand, are used to minimize environmental and electrical noise by continuously adjusting their parameters based on the statistical properties of the input data.

2.3 Machine Learning Model Development

The machine learning module of the OCM system applies advanced algorithms to classify operational states and predict potential equipment failures. The system combines Random Forest (RF) and Light Gradient Boosting Machine (LGBM) models, both known for their toughness in handling large datasets and high accuracy in predictive analytics. The RF model uses a collective learning approach, constructing multiple decision trees and combining their predictions to improve classification performance [15].

This method is highly effective for anomaly detection because it reduces the risk of overfitting and enhances the adaptability to new data. The LGBM model applies gradient boosting techniques to enhance prediction performance. It is particularly effective for analysing temporal patterns in the processed sensor data, providing the OCM system with powerful tools for predictive maintenance [16].

2.4 Predictive Maintenance Model and Mathematical Framework

The predictive maintenance model integrates processed sensor data with machine learning outputs using a hybrid analytical approach. The predictive model's core is represented by the equation:

$$y[n] = \alpha x[n] + (1-\alpha)y[n-1] + \beta W(a,b) + \gamma PSD(f) \quad (3)$$

Where:

$y[n]$ = filtered output at discrete time n
$x[n]$ = input signal from sensors
α = smoothing factor for low-pass filtering
$W(a,b)$ = CWT component
$PSD(f)$ = Power Spectral Density function for frequency domain analysis
β, γ = weighting coefficients

The analyses of model not only filter and enhances the sensor signals but also adds the temporal and spectral features to improve the accuracy of maintenance predictions. The proposed work of predictive model facilitates the OCM system to automate maintenance actions, such as initiating equipment shutdowns or sending alerts to maintenance teams [17].

The work of the OCM system is evaluated through metrics such as detection accuracy, prediction precision, response time, and maintenance efficiency [18]. The analyzed system detect accuracy of 95%, with a prediction precision of 93% and a quick response time of 120 milliseconds. The system capability to provide high reliability in industrial applications, contributing to reduction of equipment robust and maintenance costs [19].

The proposed OCM system methodology allows a robust and adaptable framework for present monitoring and predictive maintenance in FSW processes. By combining advanced sensor networks, detailed signal processing, and predictive analytics, the system helps improve operational safety, streamline maintenance planning, and support smart manufacturing in line with Industry 4.0 principles.

3. Results and Discussion

Several metrics had significant influence on the results of a study on Friction Welding OCM System. Improved error detection in welding led to more accurate forecasts and cost savings. The results are among the anomaly recognition, maintenance efficiency, predictive expectations (ROA), defect recognition, and industry 4.0 integration.

3.1 Anomaly Detection Performance

The OCM system was able to detect potential device abnormalities by monitoring the vibrational frequency. The vibration frequency is generally between 50 and 60, indicating system stability. Under the induced defect scenario, a large peak in frequency was observed, with the increase in cAMP (constituent of Amplified Air Magnetic Appearance Code or CAMP) being recorded as higher than that of the compressor component when at 150 Ohm (COD). Figure 104.1 showing frequencies have changed over time (Balacandar and Jegadeeshwaran, 2021) is shown.

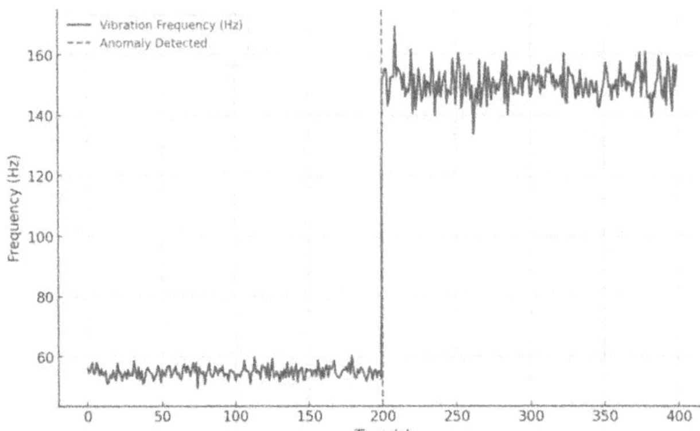

Fig. 104.1 Vibration frequency analysis over time

These frequency deviations were important in mechanical signal problems such as warehouse wear and compressor powder. Using advanced sensor technologies, including vibration sensors and accelerometers, has now enabled us to capture smaller anomalies that could lead to critical surgical failures if the system is not addressed immediately (Patel & Patel, 2017). Response time of 120 ms. Historical and practical data used in mechanical learning to distinguish between normal and false conditions and enable proactive maintenance strategies (The et al., 2019). The system's predictive model showed 93% accuracy, featuring reliability in an industrial environment [20].

Table 104.1 Defect detection performance

Defect Type	Detection Accuracy (%)	False Positive Rate (%)	Response Time (ms)
Compressor Fault	89	5	135
Misalignment	95	4	124
Rotor Imbalance	87	5	105
Compressor Fault	89	4	123
Rotor Imbalance	97	5	117
Bearing Wear	94	8	186
Rotor Imbalance	97	6	105
Rotor Imbalance	88	7	128
Misalignment	92	5	124
Bearing Wear	97	8	154

3.2 Maintenance Efficiency and Cost Analysis

The OCM system significantly reduces device maintenance costs and operational time with its predictive maintenance strategy. According to the analysis of maintenance costs (Table 104.1):

Warehouse components: the highest maintenance costs with 30% reduced downtime.

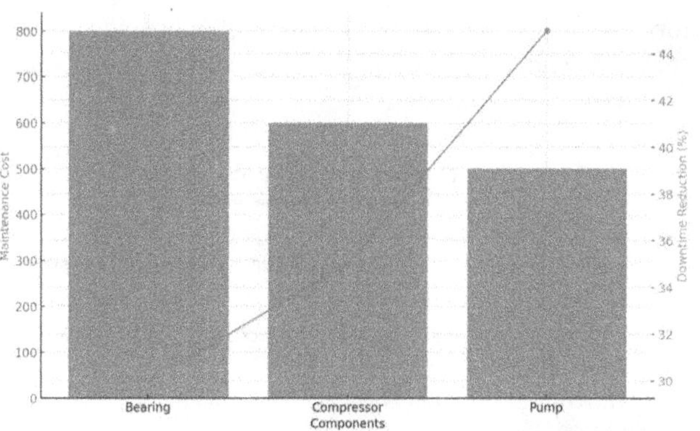

Fig. 104.2 Maintenance cost vs. downtime reduction

Figure 104.2 shows the comparison of maintenance costs and reduced failure times is demonstrated as a good balance between operational efficiency and maintenance expenses for OCM systems (Mishra et al, 2020). Maintenance Cost Analysis, a cost-effective index, which also helped the system in reducing its maintenance budgets. The system's automation of maintenance measures and previous alerts resulted in less manual intervention (Table 104.2).

Table 104.2 Maintenance cost analysis

Component	Maintenance Cost	Downtime Reduction (%)	Cost Efficiency Index
Pump	426	42	1.19
Compressor	670	48	1.56
Compressor	449	25	1.29
Compressor	691	46	1.42
Motor	646	33	1.38
Compressor	663	20	1.88
Valve	478	37	1.91
Bearing	786	26	1.49
Motor	589	20	1.33
Compressor	446	20	1.11

3.3 Predictive Maintenance vs. Traditional Methods

A comparative analysis of predictive maintenance and traditional maintenance methods proved significant advantages of the OCM system. Key performance metrics included (Fig. 104.3):

- **Downtime Reduction:** Predictive maintenance achieved a 50% reduction compared to 30% in conventional methods.

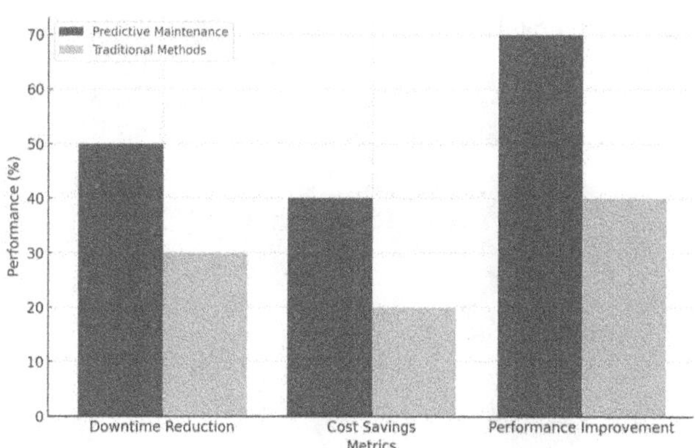

Fig. 104.3 Predictive maintenance vs. traditional methods

- **Cost Savings:** Showed a 40% reduction in maintenance costs versus 20% by traditional methods.
- **Performance Improvement:** Enhanced overall equipment performance by 70%, while conventional methods showed only 40% improvement.

Table 104.3 shows these findings match with studies by (Liao et al., 2019), which shows the efficiency of predictive maintenance in minimizing unplanned operational time and extending equipment lifespan. The OCM system's proactive maintenance strategies not only extended the lifespan of critical components but also minimized unplanned operational time, proving its efficiency over reactive maintenance approaches [21].

Table 104.3 Predictive vs. traditional maintenance performance

Metric	Predictive Maintenance (%)	Traditional Method (%)
Downtime Reduction	55	41
Cost Savings	65	30
Cost Savings	69	42
Downtime Reduction	65	43
Cost Savings	56	31
Cost Savings	58	41
Performance Improvement	63	45
Cost Savings	60	33
Cost Savings	75	37
Downtime Reduction	75	31

3.4 Defect Detection in FSW Processes

The integration of advanced signal processing techniques, including Continuous Wavelet Transform (CWT) and Empirical Mode Decomposition (EMD), enabled the OCM system to detect transient anomalies related to the welding defects accurately. The sensor signal analysis using CWT explained the system's ability to analyze signal intensity over time and frequency, optimizing its sensitivity to slight changes in the welding process.

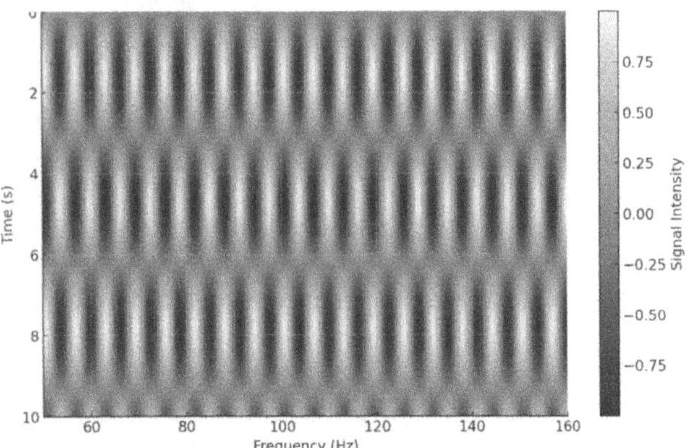

Fig. 104.4 Sensor signal analysis with continuous wavelet transform (CWT)

Figure 104.4 shows the use of advanced machine learning models (RF and LGBM) has made it possible to improve identification and estimation [22]. Models trained on the marked data from experimental FSW tests were able to detect defects like tunnel errors and route flyer cracks automatically (Rabi und al, 2019). In the beginning, a real-time visualization platform for operators to identify potential defects and equipment failures was created using the Pyqt library that developed specialized GUI components. Properties of knowledge were important for maintaining high-quality welding and operational security.

3.5 Industry 4.0 Integration and Smart Manufacturing

This research is mainly focused on the use of OCM systems in Industry 4.0 technology for improving the dynamic process. The recording and analysis of real data is facilitated to the immediate adjustments of the welding parameters which are leading to improved product quality and stability for automated volume production settings. Especially the system's smart manufacturing works are proved to be valuable in fields such as aerospace, automobiles and shipbuilding. The industry places great importance on balancing to ensure security and performance (Moal et al, 2015).

4. Conclusion

A new approach to condition monitoring and predictive maintenance is being applied with the implementation of an OCM system for FSW. The study's thorough examination of system performance, maintenance cost efficiency, and predictive accuracy sets the stage for future advancements in this area. The system's practical application in key industries highlights its potential as a robust instrument for improving equipment reliability, enhancing safety standards and contributing to the wider adoption of smart manufacturing technologies.

References

1. K. Balachandar & R. Jegadeeshwaran (2021). Friction stir welding tool condition monitoring using vibration signals and Random forest algorithm – A Machine learning approach. Materials Today: Proceedings. DOI: 10.1016/j.matpr.2021.03.056
2. G. Kruger, T. V. van Niekerk, C. Blignault, & D. Hattingh (2004). Software architecture for real-time sensor analysis and control of the friction stir welding process. IEEE Africon Conference. DOI: 10.1109/AFRICON.2004.1406859
3. K. Balachandar, R. Jegadeeshwaran & D. Gandhikumar (2020). Condition monitoring of FSW tool using vibration analysis – A machine learning approach. Materials Today: Proceedings. DOI: 10.1016/j.matpr.2020.02.456
4. J. Rabi, T. Balusamy & R. Jawahar (2019). Analysis of vibration signal responses on pre-induced tunnel defects in friction stir welding using wavelet transform and empirical mode decomposition. Defence Technology. DOI: 10.1016/j.dt.2019.04.002
5. G. L. Moal, F. Darras & D. Chartier (2015). On the feasibility of a monitoring system for the friction stir welding process: Literature review and experimental study. IEEE International Conference on Automation Science and Engineering. DOI: 10.1109/CoASE.2015.7294072
6. K. Balachandar & J. R. (2024). Enhancing Friction Stir Welding: Quality Machine Learning Based Friction Stir Welding Tool Condition Monitoring. International Research Journal of Multidisciplinary Technovation. DOI: 10.1016/j.matpr.2024.04.021
7. Y. Liu & S. Lu (2016). Effects of Ultrasonic Vibration on the Welding Process of Friction Stir Welding. Materials Science Forum. DOI: 10.4028/www.scientific.net/MSF.854.34
8. T. Liao, J. Roberts, M. Wahab & A. Okeil (2019). Building a multi-signal based defect prediction system for a friction stir welding process. Procedia Manufacturing. DOI: 10.1016/j.promfg.2019.03.052
9. M. Das, A. Meena & B. Das (2019). Sensor fusion model for defect identification in friction stir welding process. Journal of Physics: Conference Series. DOI: 10.1088/1742-6596/1240/1/012071
10. J. A. Dhanraj, B. Lingampalli, M. Prabhakar, A. Sivakumar, B. Krishnamurthy & K. Ramanathan (2021). A credal decision tree classifier approach for surface condition monitoring of friction stir weldment through vibration patterns. Materials Today: Proceedings. DOI: 10.1016/j.matpr.2021.06.059
11. P. Sinha, S. Muthukumaran, R. Sivakumar & S. Mukherjee (2008). Condition monitoring of first mode of metal transfer

in friction stir welding by image processing techniques. The International Journal of Advanced Manufacturing Technology. DOI: 10.1007/s00170-008-1522-3

12. S. S. Kumar, S. Ashok & S. Narayanan (2013). Investigation of Friction Stir Butt Welded Aluminium Alloy Flat Plates Using Spindle Motor Current Monitoring Method. Procedia Engineering. DOI: 10.1016/j.proeng.2013.09.195
13. V. Patel & M. N. Patel (2017). Development of Smart Sensing Unit for Vibration Measurement by Embedding Accelerometer with the Arduino Microcontroller. DOI: 10.1109/SYSY.2017.8228294
14. M. Imam, R. Ueji & H. Fujii (2015). Microstructural control and mechanical properties in friction stir welding of medium carbon low alloy S45C steel. Materials Science and Engineering A. DOI: 10.1016/j.msea.2015.03.062
15. Mishra, D., et al. (2020). Real-time monitoring and control of friction stir welding process using multiple sensors. *CIRP Journal of Manufacturing Science and Technology*. https://doi.org/10.1016/j.cirpj.2020.02.004.
16. Patel, V., & Patel, M. N. (2017). Development of Smart Sensing Unit for Vibration Measurement by Embedding Accelerometer with the Arduino Microcontroller. *IEEE Systems Conference*. https://doi.org/10.1109/SYSY.2017.8228294.
17. Liao, T., Roberts, J., Wahab, M., & Okeil, A. (2019). Building a multi-signal based defect prediction system for a friction stir welding process. *Procedia Manufacturing*. https://doi.org/10.1016/j.promfg.2019.03.052.
18. Sundararaman, S., White, J., Adams, D., & Jata, K. (2007). Application of Wave Propagation and Vibration-based Structural Health Monitoring Techniques to Friction Stir Welded Plate. *IMECE2007*. https://doi.org/10.1115/IMECE2007-43039.
19. Liu, Y., & Lu, S. (2016). Effects of Ultrasonic Vibration on the Welding Process of Friction Stir Welding. *Materials Science Forum*. https://doi.org/10.4028/www.scientific.net/MSF.854.34.
20. Albarbar, A., Mekid, S., Starr, A., & Pietruszkiewicz, R. (2008). Suitability of MEMS Accelerometers for Condition Monitoring: An experimental study. *Sensors*. https://doi.org/10.3390/s8070898.
21. Q. Zhang & X. Liu (2018). Optimisation in Friction Stir Welding: Modelling, Monitoring and Design. DOI: 10.1007/978-3-319-92735-2
22. A. Silvestri (2024). Monitoring of the friction stir welding process: A preliminary study. Materials Research Proceedings. DOI: 10.21741/9781644902070-12D. Mishra et al. (2020). Real-time monitoring and control of friction stir welding process using multiple sensors. CIRP Journal of Manufacturing Science and Technology. DOI: 10.1016/j.cirpj.2020.02.004

Note: All the figures and tables in this chapter were made by the authors.

Advances in Additive Manufacturing Technologies – Gurusamy Pathinettampadian et al. (eds)

Design and Implementation of Lidar Based Video Analysis in Heterogenous System

M. Rohini[1],
A. Anjaline Jayapraba[2]
Assistant Professor,
Electronics and Communication Engineering,
Saveetha Engineering College,
Chennai, India

B. Vinoth[3], **K. Ginith**
Student, Electronics and Communication Engineering,
Saveetha Engineering College,
Chennai, India

◆ **Abstract:** This research project aims to build embedded ODS using camera and LiDAR data. The advancement of autonomous technology and the declining cost of LIDARs are the causes of this. Processing of LiDAR point clouds can be split into multiple phases. Firstly, ground is removed and data is filtered. Then remaining point cloud is divided into segments which are classified. Image processing takes advantage of point cloud processing results. The first step consists of projecting segments (obtained from point cloud) to image plane. The smallest rectangle which covers all of projected points is treated as region of interest. The image from ROI is resized and classified by HOG+SVM descriptor. Then data fusion step is performed. Classification probability from point cloud processing and image processing are transformed into final classification probability.

◆ **Keywords:** LIDAR, ROI, HOG+SVM

1. Introduction

The problem of using automotive sensors to detect objects in real time is becoming more and more common these days. These sensors' data processing is frequently computationally demanding and limited to processors. Thus, the goal was to determine whether it could be completed effectively on a heterogeneous platform (FPGA + ARM). The goal is to create a functional embedded car detection system using fusion of vision and LiDAR data. The following lists the prerequisites for the project. These days, there is a growing overlap between ADAS (Adaptive Driver Assistance System) and autonomous vehicle topics. Well-known businesses and regular people alike put a lot of effort into developing effective solutions for these uses. Security systems are necessary to help drivers because there are still an increasing number of cars on the road.

Among the most crucial tasks experts in this discipline focus on is object detection. For an autonomous vehicle to function, it needs to be aware of its surroundings. The most popular sensors for this kind of application are three. It consists of a camera, radio detection and ranging (RADAR), and light detection and ranging (LiDAR). The process of combining data from several of these sensors is very popular. Higher system output information dependability

[1]rohinimohanasundaram@gmail.com, [2]anjalinejayapraba@saveetha.ac.in, [3]vinothmanojbabu@gmail.com

DOI: 10.1201/9781003685906-105

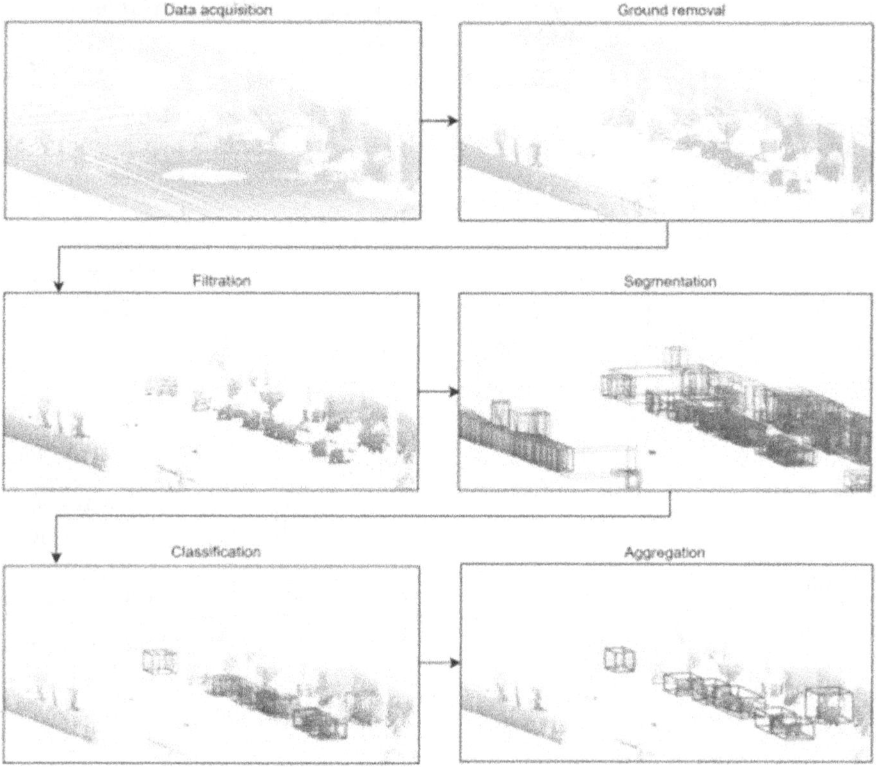

Fig. 105.1 Result of point cloud processing

is the result of it. Every one of these sensors has a different drawback. While the weather and time of day have little effect on efficient example data from LiDAR, the camera provides us with color information. The shortcomings of each sensor will be eliminated, though, when the data from the two is combined. It is necessary to process sensor data and applies the proper detection and recognition algorithm. Even when an algorithm finds objects effectively, it takes time. Nevertheless, there aren't many experiments with applying these algorithms on diverse platforms. With this study, we hope to demonstrate that it is feasible and can result in observable financial gains [1].

The project implementation can be divided into several stages. In the first one, the related work was gathered in order to isolate methods used in each step and to compare different algorithms in terms of effectiveness and suitability for hardware-software implementation. In the second stage, a software model was created. It was implemented in MATLAB environment. The model is used for testing selected solutions and as a reference for the designed system. All parameters of the used algorithm are stated in this model. In the third stage, the system was split into the hardware resources (PL) and the processor part (PS). Point cloud and image processing are carried out in hardware. The software part would include communication and data logging. Data collection from the LiDAR sensor and camera is the system's initial component. Since we lack access to a LiDAR sensor (due to prohibitive costs), the data was obtained from a publically accessible database (such as KITTI or nuscenes.org). LiDAR point cloud processing is the second step.

Preprocessing, which encompasses background removal, filtering, and ground removal, is the initial step. The segmentation process, which comes next, attempts to split the rest of the point cloud into segments that might contain objects. Ultimately, characteristics are taken out and classification is carried out for every section. Figure 105.1 displayed the outcomes of each stage of the point cloud processing method. Projecting segments assessed by LiDAR analysis to an image plane and choosing the matching ROI is the first step in the system's vision section [2].

A scaled and extracted ROI is obtained. Using techniques like HOG+SVM, the last stage involves extracting features and classifying them. Data fusion is the algorithm's final step.

The final object spotting chance will be calculated by fusing the classification scores from the two algorithms. This method was evaluated in both the Vivado simulation and the MATLAB environment. It had to be then improved and adjusted. It was tested in software model in MATLAB and gave satisfactory results (Fig. 105.1).

1.1 Features-In-Brief

Following features of the system were assumed: ·When user plugs in Velodyne HDL-64 LiDAR and a camera to Zybo Z7-20, system should detect objects in real time and return this information to other car components,

- System would be adjusted to car detection,
- Processing LiDAR data to get object classification results,
- Processing vision data to get object classification results,
- Data fusion of LiDAR and vision classification results,
- Higher level analysis – e.g. object tracking.

1.2 Theoretical Background

LiDAR

The basis of this project is data from camera and LiDAR sensor. LiDAR are commonly used in such fields as: geology, geophysics, archeology, seismology, ground mapping or autonomous cars. LiDAR operates on a straightforward principle. These sensors release a laser light compound with a particular wavelength and direction via a specialized optical system. Detectors within the same device detect the beam after it has been reflected off of an obstruction. The distance to the barrier is computed by taking into account the time interval between sending and receiving the light. In project we used data base KITTI Vision Benchmark Suite. It provides the LiDAR and camera data gathered in urban and rural areas. Data base contains data from LiDAR Velodyne HDL-64E. This sensor is equipped with 64 canals and rotates with 10Hz frequency. Data is send by Ethernet in spherical coordinates.

SVM (Support Vector Machine)

SVM is a classifier that in the basic version divides data into two classes. A support-vector machine can be used for classification, regression, or other tasks such as outlier detection. It creates a hyperplane or array of hyperplanes in a high- or infinite-dimensional space. Intuitively, a good separation is achieved by the hyperplane that has the largest distance to the nearest training-data point of any class, since in general the larger the margin, the lower the generalization error of the classifier.

Hyperplane is described as:

$$G(x) = sign[f(x)] = sign[x \cdot \beta + \beta_o]$$

Where: β-normal vector to the hyperplane, $\beta 0$ - hyperplane offset from the zero point in the space of features. Function (x) may be treated as distance between considered point and hyperplane. Its sign depends on which side of hyperplane is x. Having a certain vector of features x, parameters β and $\beta 0$ we can determine class by the function:

$$(x: f(x) = x \cdot \beta + \beta_0 = 0)$$

2. System Design

2.1 Features and Specifications

Final, assumed, version of the project should have features as listed in Features-in-Brief i. In simple words, it is car detection in vision and LiDAR data and tracking objects. From these, we managed to fulfill only classifying cars in LiDAR data. It can also serve as car detection system, but system with data fusion has more potential [3].

2.2 Design Overview

Top level diagram is presented on Fig. 105.2. It represents the whole assumed project design. However, not

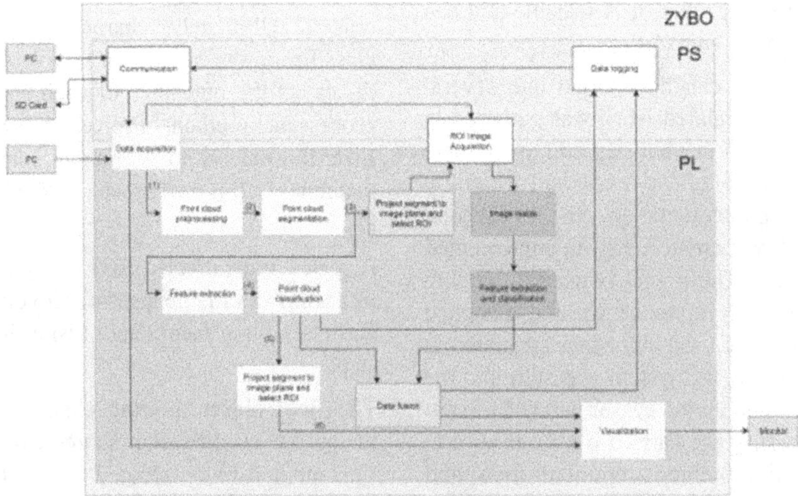

Fig. 105.2 Design overview

everything was finished. The design states of components are represented by their color:

- yellow – finished and taken into account in current release,
- violet – finished, but yet not used in current release (they cannot be used, because other modules are not completed),
- blue – started but not finished – due to the lack of time or some other factors,
- white – not started.

In below section, Detailed Design Description only modules marked yellow are described. Modules marked violet and blue are not described since they are incomplete or unused. Module "Project segment to image plane and select ROI" is put twice in top level diagram [5]. The reason is that the yellow one was added to present some results on monitor. The violet one has some other task, but similar behavior. It would be described in detail only once (in Detailed Design Description).

Table 105.1 Top level signals of the design

Signal number	Description
(1)	Signal represents LiDAR data cells coming from PC – each cell consists of some number of points and a flag, which is high while points are being send (each point in one clock cycle)
(2)	Signal represents some cell features (e.g. intensity histogram), a flag which indicates if a cell was or was not removed during *Point cloud preprocessing* and a flag which indicated that data is valid. In one clock cycle, one cell features are being send.
(3)	Signal represents cell features, flag which indicated if a cell was or was not removed. It varies from signal (2), because cells are being send in packets of 9, with certain relation of neighbourhood. One such a packet of cells is called a segment.
(4)	Signal represents features for a whole segment. There are 30 features for each segment. Every feature is being send in one clock cycle. Beside this, there is a flag indicating that data is valid.
(5)	Signal represents classification probability for a segment.
(6)	Signal represents bounding box (in LiDAR cooridnates) for a segment.
(7)	Signal represents bounding box (in image coordinates) for a segment.
(8)	Image signal – one channel (gray) and synchronization signals.
(9)	Image signal – one channel (gray) and synchronization signals. It is image from (8) with marked bounding boxes.

Table 105.1 represents top level signals of the design. It takes into account only signals used in current release, because majority of other signals is not fully defined. Signal (1) will be described in detail in section Input Data Description.

2.3 Input Data Description

As input, data from KITTI database is taken. It includes both LiDAR and vision data. Vision part of the system is not completed, so only LiDAR input data will be described here. LiDAR data consists of points. They are described by four coordinates [4]:

- x – distance of point from LiDAR in direction of moving car,o positive values in forward direction, o type – real signed value,
- y – distance of point from LiDAR in direction of left/right hand side of driver, o positive values on left hand side, o type – real signed value,
- z – distance of point from LiDAR in a vertical direction,o positive values in upward direction, o type – real signed value,
- Intensity – it measures how much light is reflected from the surface, o type – integer ranged from 0 to 255.

First three coordinates are perpendicular to each other.

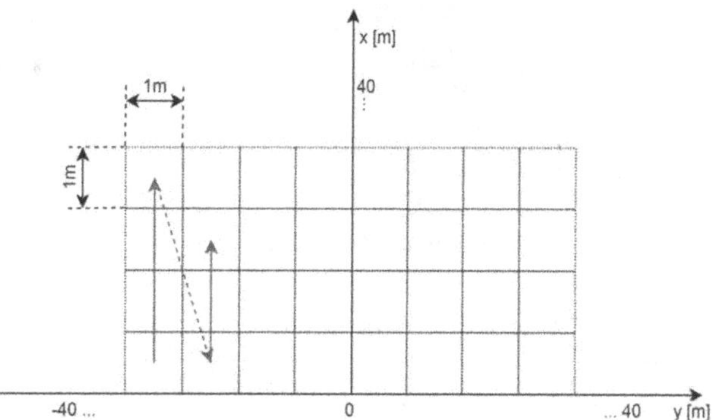

Fig. 105.3 XY plane division into 1M x 1M cells

XY plane is divided into non-overriding cells of size 1m x 1m (Fig. 105.3). Range in y direction is [-40m, 40m] and in x direction [0, 40m]. These values were chosen by experiment. Greater distances from LiDAR provide less detailed data and it is much harder to classify objects there. Data from back of the car (negative x's) was discarded in this release, because it is not present on image and cannot be displayed. Sending a cell means that all of the points in cell are being continuously send in any order. Cells are sent one by one to Zybo with constant time gaps – 1000ns. With too small gap time, some modules will not catch up with computations. The order of sending cells is showed by red arrows on Fig. 105.3.

2.4 Detailed Design Description Data Acquisition

Data acquisition consists of following modules:

- DVI to RGB Video Decoder,
- RTL HDMI data parser – parses HDMI data to obtain LiDAR points – its exact form depends on format of LiDAR data in HDMI.

2.5 Visualization

Visualization consists of following modules:

- ManageBBOXes – module which saves and replaces old bounding boxes with new ones, the output consists of 10 bounding boxes,
- Draw Bounding Box – draws on image 10 bounding boxes taken from ManageBBOXesmodule,
- RGB to DVI Video Decoder

2.6 Point Cloud Preprocessing

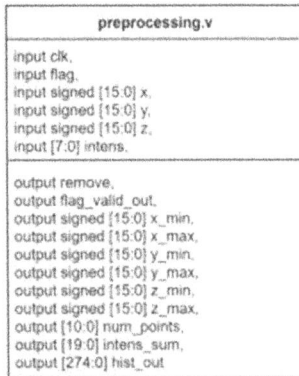

```
preprocessing.v

input clk,
input flag,
input signed [15:0] x,
input signed [15:0] y,
input signed [15:0] z,
input [7:0] intens,

output remove,
output flag_valid_out,
output signed [15:0] x_min,
output signed [15:0] x_max,
output signed [15:0] y_min,
output signed [15:0] y_max,
output signed [15:0] z_min,
output signed [15:0] z_max,
output [10:0] num_points,
output [19:0] intens_sum,
output [274:0] hist_out
```

Fig. 105.4 Preprocessing

Table 105.2 Preprocessing module

Signal name	Width	Description
clk	1	Master clock (100 MHz)

Table 103.3 Input/output signal description

flag	1	Flag indicating valid input data
x	16	X coordinate of one point in LiDAR data (s6c9f)
y	16	Y coordinate of one point in LiDAR data (s6c9f)
z	16	Z coordinate of one point in LiDAR data (s6c9f)
intens	8	Intensity of one point in LiDAR data (8c)
remove	1	Flag indicating that considering cell should/should not be removed
flag_valid_out	1	Flag indicating valid output data
x_min	16	Minimum of x coordinate of considering point (s6c9f)
x_max	16	Maximum of x coordinate of considering point (s6c9f)
y_min	16	Minimum of y coordinate of considering point (s6c9f)
y_max	16	Maximum of y coordinate of considering point (s6c9f)
z_min	16	Minimum of z coordinate of considering point (s6c9f)
z_max	16	Maximum of z coordinate of considering point (s6c9f)
num_points	11	Number of points in considering cell (12c)
intens_sum	20	Sum of intensity of points in considering cell (20c)
hist_out	275	Concatenation of 25 intensity histogram bins of points in cell (11c)

The first step of implemented algorithm is point cloud preprocessing. It contains ground removal, noise filtration and low foreground detection. This step is carried out by module which interface is presented on Fig. 105.4 and Table 105.2. The result is flag *remove* that indicates if considering cell should be or should not be remove. This flag is result of AND operation on another three flags: rm_ground,rm_noise, rm_foregroud. Figure 105.5 shows what these flags mean. rm_ground is set when mean of z coordinates of points in cell is lower that assumed threshold (ground thresh). rm_noise is set when number of points in cell is lower than noise_thresh. rm_foreground is set when maximum of z coordinate of points in cell is greater than zmax_thresh or whensubstraction of z_max and z_min is greater than diff_thresh. This entire threshold is saved as parameters of considering module.

The way of computing minimal and maximal coordinates is shown on Fig. 105.6. There is used register module, with ce and rst ports. Register for finding minimal andmaximal valuediffers in initial value. On the same figure ways of computing number of points in cell, sum and mean of z coordinates are presented.

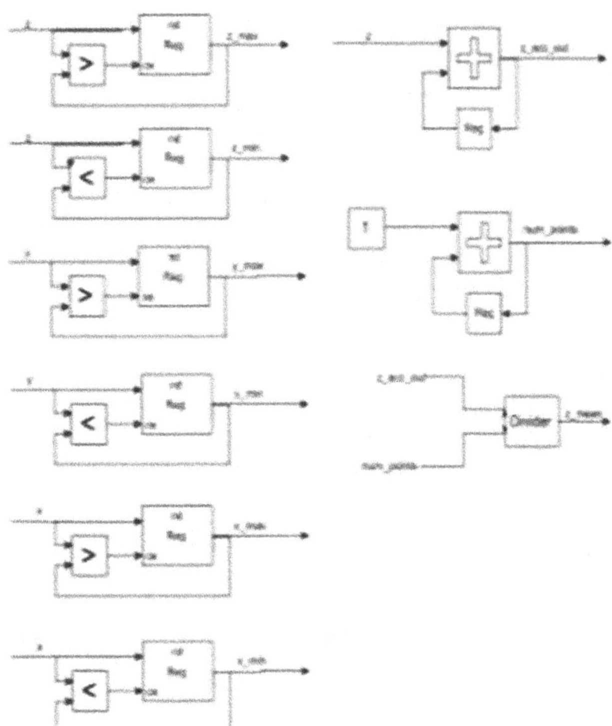

Fig. 105.5 Computing minimal, maximal coordinates, number of points and mean of z coordinate.

Preprocessing module has also other functionalities apart from calculating *remove* flag. There are computed some values used in feature extraction step later (Fig. 105.5). These are: minimal and maximal cells coordinates, number of points, sum of intensity of points in cell and histogram of intensity with 25 bins (Fig. 105.6 and Fig. 105.7).

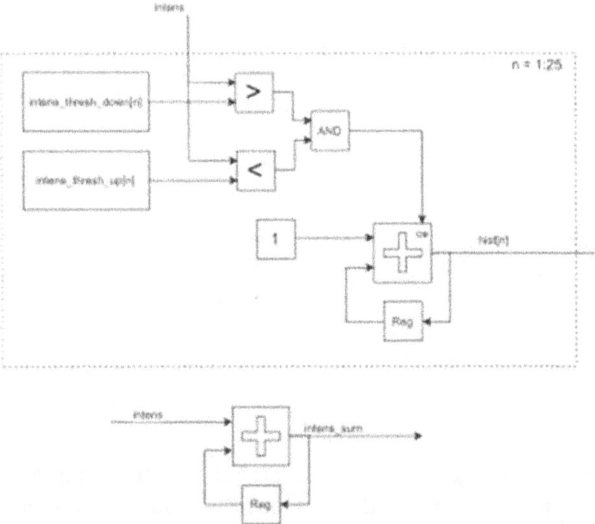

Fig. 105.6 Computing sum of intensity and histogram

2.7 Point Cloud Segmentation

Save to BRAM

```
save_bram_addres.v
input clk,
input flag_valid_remove,
output [7:0] addra
```

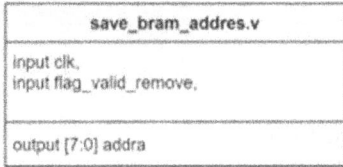

Fig. 105.7 Save to BRAM module interface

Table 105.4 Save to BRAM module input/output signals description

Signal name	Width	Description
clk	1	Master clock (100 MHz)
flag_valid_remove	1	Flag indicating valid input data
addra	8	Address of A port in BRAM IP

The aim of this module (Fig. 105.8) is to generate the address to save value in BRAM. Our approach assumes saving minimal and maximal coordinates of1m x 1m cells, number of points, intensity sum and histogram and flag *remove* (Table 105.3). This is all 403 bits. When *flag_valid_remove*is set this data is saved in BRAM under appropriate address. In BRAM at one time 160 cells are stored: a "rectangle" with 40 cells in x direction and 4 cells in y direction. Following cells come in x direction (in "columns") and are successively saved to BRAM (Fig. 105.9). When the first four columns arrive and the next is being sent, the column address is overflowed – so the next data is saved in column of address 0, then one, two, etc.

```
read_bram_addres.v
input clk,
input flag_valid_remove,
inoput [7:0] addra,
output flag_bram_valid_out,
output [7:0] addrb
```

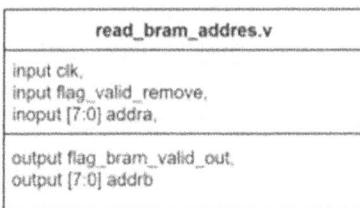

Fig. 105.8 Read from BRAM module interface

READ from BRAM

Table 105.5 READ from BRAM module input/output signals description

Signal name	Width	Description
clk	1	Master clock (100 MHz)
flag_in_valid	1	Flag indicating valid input data
bram_out	403	Data saved in BRAM
features_valid	1	Flag indicating valid output data
feature	16	One of 30 features describing the segment
box_min	48	Concatenation of minimum x, y, z coordinates
box_max	48	Concatenation of mximum x, y, z coordinates

The aim of *read_bram_address*is to read cells from BRAM in certain order – to perform segmentation. Data is read in packets of nine cells. The manner data is read is presented on Fig. 105.10. Nine neighboring cells (blue box) are read in any order and sent in "one packet" to the following module (Table 105.4). When the next cell is written to BRAM, next nine cells can be read – so that the center of blue box moves by 1 meter in x direction. When blue box comes to the end of x direction (center in 39th cell) "column" it is moved in y direction by one meter and moved to the beginning of next x direction "column" (look at red arrows). Data can be started reading only if first two x direction "columns" and three cells of the next one are saved.

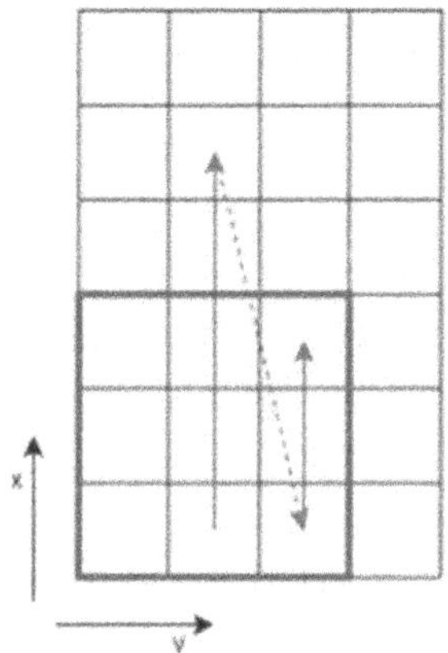

Fig. 105.9 Segmentation idea

2.8 Features Extraction

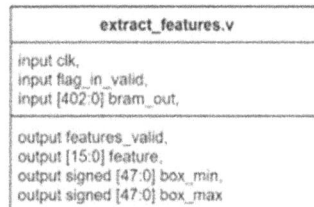

```
extract_features.v
input clk,
input flag_in_valid,
input [402:0] bram_out,
output features_valid,
output [15:0] feature,
output signed [47:0] box_min,
output signed [47:0] box_max
```

Fig. 105.10 Feature extraction module interface

Table 105.6 Feature extraction module input/output signals description

Number	Description
1	Height of the segment
2	Width of the segment
3	Length of the segment
4	Number of points in segment
5	Mean of intensity of points in segment
6-30	Histogram of intensity with 25 bins

The result of extract_features.v module is 30 features sent in sequential clock cycles (Fig. 105.10). Features are presented in Table 105.6. In this part of implementation some values computed in preprocessing step are used. For example to calculate height, width and length of the segment there was used minimal and maximal coordinates for each 1m x 1m cells (Fig. 105.11). This module receives data from 9 such cells so to calculate 1-3 features (Table 105.7) it is necessary to use register for finding the minimal and maximal coordinates for the whole segment. The same is about number of points in segments – accumulate number of points in 9 small cells. In similar way intensity features are calculated (Fig. 105.12).

Table 105.7 Features description

Number	Description
1	Height of the segment
2	Width of the segment
3	Length of the segment
4	Number of points in segment
5	Mean of intensity of points in segment
6-30	Histogram of intensity with 25 bins

2.9 Point Cloud Classification

SVM Classification

```
svm.v

input clk,
input features_valid,
input [15:0] feature,

output flag_valid_out,
output signed [21:0] acc_svm_out
```

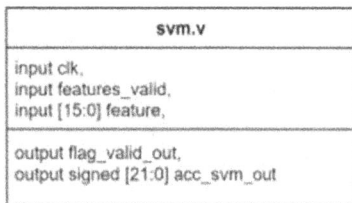

Fig. 105.11 SVM classification module interface

Table 105.8 SVM classification module input/output signals description

Signal name	Width	Description
clk	1	Master clock (100 MHz)
features_valid	1	Flag indicating valid input data
feature	16	One of 30 features describing the segment (1c15f)
flag_valid_out	1	Flag indicating valid output data
acc_svm_out	22	SVM classification score (s8c13f)

Features vector is multiplied by the normal vector to the hyperplane. Hyperplane model (betas and bias) were determined in MATLAB during the classifier learning process. This model returns the distance of the segment from hyperplane. In Table 105.8 a distance sign informs that segment is car or not. The scheme of SVM model is presented on Fig. 105.13.

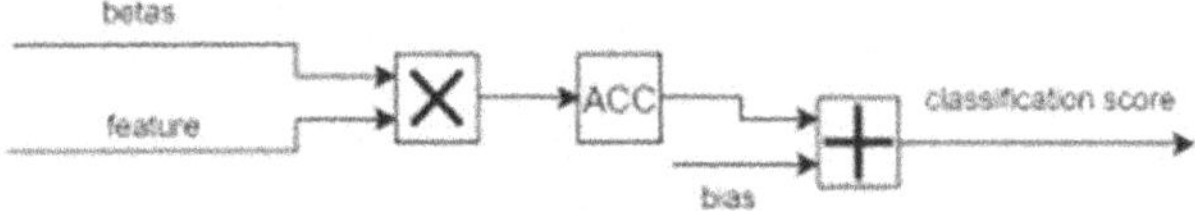

Fig. 105.12 Computing of SVM score classification

Calculate Probability Score

```
calc_score.v

input clk,
input flag_valid_svm,
input signed [21:0] acc_svm_out,

output valid_score,
output [15:0] svm_score_out
```

Fig. 105.13 Calculate probability score module interface

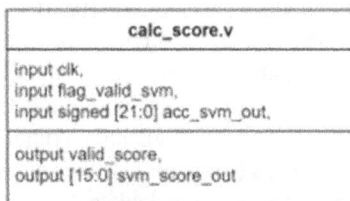

Table 105.9 Calculate probability score module input/output signals

Signal name	Width	Description
clk	1	Master clock (100 MHz)
features_valid_svm	1	Flag indicating valid input data
acc_svm_out	22	SVM classification score (s8c13f)
flag_valid_score	1	Flag indicating valid output data
svm_score_out	16	SVM classification probability score (1c15f)

2.10 Description

The final probability SVM score (Fig. 105.14) is calculated in module *calc_score.v.* Module returns probability that considering segment is a car. Probability is computed with use of sigmoidal function (Table 105.7):

$$f(x) = \frac{1}{1+\exp(Ax+b)}$$

A and *b* coefficients were obtained with MATLAB built-in function (*fitSVMPosterior*), after SVM learning process. From the range [0;1] 64 values (probabilities) with a fixed step were selected. They were saved to the LUT module. With the use of inversed sigmoidal function, *x* values corresponding tobase for creating 64 sections of classification scores (Table 105.9).

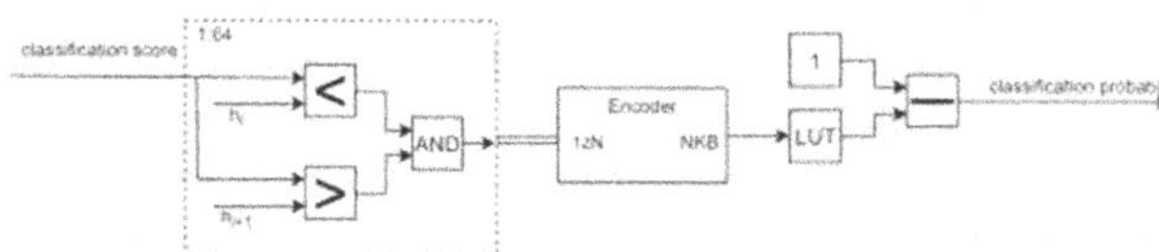

Fig. 105.14 Scheme of calculating probability score

So for every section in classification scores, there's a corresponding probability. So the probability calculation goes as follows (Fig. 105.15): classification score is compared to 64 constants. Section number is obtained from encoder. The corresponding probability is computed. The final classification probability is (due to specific MATLAB behavior) one minus probability for the section.

Project Segment to Image Plane

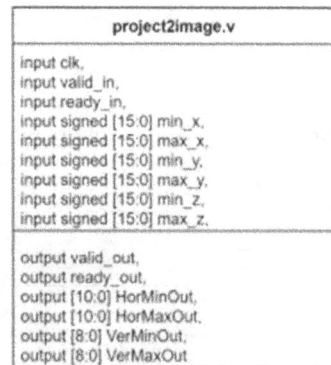

```
project2image.v

input clk,
input valid_in,
input ready_in,
input signed [15:0] min_x,
input signed [15:0] max_x,
input signed [15:0] min_y,
input signed [15:0] max_y,
input signed [15:0] min_z,
input signed [15:0] max_z,

output valid_out,
output ready_out,
output [10:0] HorMinOut,
output [10:0] HorMaxOut,
output [8:0] VerMinOut,
output [8:0] VerMaxOut
```

Fig. 105.15 Project segment to image plane module interface

Table 105.10 Project segment to image plane module input/ouptut signal description

Signal name	Width	Description
clk	1	Master clock (100 MHz)
valid_in	1	Flag indicating valid input data
ready_in	1	Flag indicating that the next module is ready to fetch the data.
min_x	16	Minimum of x coordinate of considering point (s6c9f)
max_x	16	Maximum of x coordinate of considering point (s6c9f)
min_y	16	Minimum of y coordinate of considering point (s6c9f)
max_y	16	Maximum of y coordinate of considering point (s6c9f)
min_z	16	Minimum of z coordinate of considering point (s6c9f)
max_z	16	Maximum of z coordinate of considering point (s6c9f)
valid_out	1	Flag indicating valid output data
ready_out	1	Flag indicating that this module is ready to fetch the data from the preceding one
HorMinOut	11	Bounding box minimum – horizontal direction
HorMaxOut	11	Bounding box maximum – horizontal direction
VerMinOut	9	Bounding box minimum – vertical direction
VerMaxOut	9	Bounding box maximum – vertical direction

This module projects bounding box in LiDAR coordinates to bounding box in image pixel coordinates. There are following steps to do it (Table 105.10):

- Generate all 8 points of LiDAR coordinate bounding box based on two edge points (minimal and maximal),
- Project all of these points on image plane – multiplying them by hardcoded matrix (it can be obtained from calibration of sensors – in our case, it was prepared in KITTI dataset) – and then, for each point, two from three obtained values have to be divided by the third remaining one,
- Given image resolution (constant), check if all of the points are in image bounds – if not, discard the input bounding box – it is not fully present on image,
- Compute minimal and maximal pixel coordinates from obtained set of projected points – return them as a bounding box in image coordinates. The component first saves output bounding box in a FIFO and sends it to the next module when flag *ready_in*is in high state. When the component starts calculations, it sets *ready_out*flag to zero – for information that it is not able to get any data. Flag *ready_out*is set to one when the component finishes calculations.

3. Conclusions

This project has high marketing potential. In its final form it would be a product ready to sell. ADAS systems and autonomous cars are precursor issues in modern technology. Object detection systems based on car sensors are inseparable parts of ADAS and autonomous vehicles. A lot of companies struggle to develop their own systems. FPGA + ARM duet has potential to process LiDAR and vision data with high speed – in real time and as far as we know, only a few people work on such ideas.

References

1. Cabanes, Q. and Senouci, B., 2017, July. Objects detection and recognition in smart vehicle applications: Point cloud based approach. *In 2017 Ninth International Conference on Ubiquitous and Future Networks (ICUFN)* (pp. 287–289). IEEE..
2. Geiger, A., Lenz, P., Stiller, C. and Urtasun, R., 2013. Vision meets robotics: The kitti dataset. *The international journal of robotics research*, 32(11), pp.1231–1237.
3. Platt, J., 1999. Probabilistic outputs for support vector machines and comparisons to regularized likelihood methods. *Advances in large margin classifiers*, 10(3), pp.61–74.
4. Hastie, T., 2001. Tibshirani R. Friedman J.: *The elements of statistical learning*.
5. Jun, W. and Wu, T., 2015, November. Camera and lidar fusion for pedestrian detection. In 2015 *3rd IAPR Asian Conference on Pattern Recognition (ACPR)* (pp. 371–375). IEEE.

Note: All the figures and tables in this chapter were made by the authors.

Advances in Additive Manufacturing Technologies – Gurusamy Pathinettampadian et al. (eds)

Comparative Analysis of Surface Coating Methods for Corrosion Resistance and Wear Protection in Aerospace Aluminium Alloys

Gowtham Kumarasamy[1]
Center for Robotics, Chennai Institute of Technology,
Chennai, India

Vijayan Subramaniyan[2],
Babu Selvam[3], **Ravuri Anant**[4], **Gopi Venkatachalam**[5]
Department of Mechatronics Engineering,
Velammal Institute of Technology,
Chennai, India

Kishore Ravikumar[6]
Department of Mechanical Engineering,
Chennai Institute of Technology,
Chennai, India

◆ **Abstract:** This paper gives a comprehensive overview of surface coating methods for improving the wear protection and corrosion resistance of aerospace-grade aluminum alloys. Aluminum alloys are widely used in aerospace due to their machinability, strength-to-weight ratio, and thermal conductivity but pose corrosion and wear problems in harsh environmental conditions. It compares different coating methods, including thermal spray, chemical conversion, anodizing, electroplating, physical vapor deposition, and organic coatings, on the basis of improvement of environmental and mechanical stress resistance, adhesion, and durability. It is plagued by problems of fatigue degradation, application complexity, and environmental exposure, but innovations are highlighted as much as necessary. The integration of coating processes with mechanical processes, such as shot peening, and modern technology, such as nano-coatings, is also proposed as optimization techniques. The review is intended to support engineers in selecting the protective coatings optimally in aerospace technology on the basis of a balance of cost, efficiency, and sustainability.

◆ **Keywords:** Aluminum alloys, Anodizing, Electroplating, Physical vapor deposition

1. INTRODUCTION

1.1 Background

Aluminium alloys have gained increasing place in aerospace due to favorable strength-to-weight ratio, ease of machining, as well as good thermal conductivity. Despite these merits, the alloys suffer highly from corrosion and wear, particularly in aerospace applications which are often exposed to severe environmental conditions such as extreme temperatures, humidity, and saltwater. Corrosion causes pitting, which weakens the structure, whereas frictional forces result in dimensional changes responsible for compromise of functionality. In light of this fact, there is a pressing need to apply effective surface protection against

[1]gowthamk@citchennai.net, [2]vijayan@velammalitech.edu.in, [3]babu.mts@velammalitech.edu.in, [4]anant.mts@velammalitech.edu.in, [5]gopi.mts@velammalitech.edu.in, [6]kishoreravi@citchennai.net

DOI: 10.1201/9781003685906-106

corrosion to prolong the service life of aluminium alloys and maintain safety and reliability of components in aerospace applications [1, 2]. To enhance corrosion protection and wear resistance, the following surface coating methods have been developed for aluminium alloys. For instance, anodizing offers good corrosion resistance and hardness by developing a protective oxide layer. Electroplating and thermal spray coatings add metallic or ceramic layers as a sort of protective barrier in terms of corrosive elements and wear [3, 4]. Each coating technique has individual specific advantages and disadvantages based on considerations in terms of environmental exposure, mechanical stresses, and regulatory constraints, and thus the proper selection of the optimal coating technique is a matter of some complexity. Al Alloys are mainly used in automotive and aerospace industries because of their ability to yield low density; high strength-to-weight ratios and flexibility toward manufacture. The most common coatings are anodizing, hard anodizing (HA), and micro-arc oxidation (MAO). However, these coatings dramatically decrease fatigue life-the upshot of retaining up to 75% reduction. These coatings frequently contain micro-defects such as porosity and cracks, which act as a stress concentrator under cyclic loading. The thicker coatings provide good resistance against corrosion and wear but contribute to the degradation through fatigue, whereas those that are too thin which are not sufficient as protection. It is very critically difficult in applications such as aerospace landing gear, which demands high resistance through fatigue along with protection by wear and corrosion. Thus, for this purpose, cold working techniques, most notably shot peening (SP), are employed. SP has a benefit of fatigue life for both ferrous and non-ferrous alloys because the residual stress left on the surfaces in tension is compressive. That is, the effect of the alloys is enhanced during cyclic stress. The SP process can be varied with dimensions, shapes, and intensity of the peening medium to optimize performance. There is a potential for the combining of SP with protective coatings to mitigate the fatigue degradation that has been observed with traditional coatings. This may offer the possibility for Al alloy components to meet demanding requirements for wear, corrosion, and resistance to fatigue associated with high-performance applications and provide a route to improved durability and reliability for critical aerospace and automotive parts. The primary objective of this review will be to evaluate several surface coating methods for corrosion resistance and wear protection capabilities in aerospace-grade aluminium alloys. This paper will review several coating techniques with the purpose of identifying appropriate techniques to address specific needs in aerospace applications. The insight gained from this study shall help in understanding corrosion protection, wear protection, adhesion, application complexity, and cost, which will be employed by engineers and material scientists in appropriate selection for aluminium alloy coatings [5,6]. This review will consider several of the commonly applied coating methods applicable for aerospace-grade aluminium alloys: anodizing, electroplating, thermal spray coatings, chemical conversion coatings, physical vapour deposition, and organic coatings. The scope encompasses investigation for each coating into its working principle, comparative effectiveness in corrosion and wear protection, and application considerations, such as cost and environmental impact. The review will touch on the current research in surface coating technology, like nano coatings and eco-friendly alternatives, and a few of the current trends in regulatory compliance trending in the aerospace industry [7, 8].

2. Properties and Limitations of Aerospace Aluminum Alloys

2.1 Material Properties of Aerospace Aluminum Alloys

Aluminum alloys are a preferred choice in aerospace applications due to their excellent strength-to-weight ratio, corrosion resistance, and versatility in fabrication. Common alloys include 2024, 6061, 7050, and 7075, each offering distinct properties tailored to specific structural requirements. For instance, **2024** alloy is known for its high tensile strength and fatigue resistance, making it suitable for aircraft wings and fuselage applications, despite its lower corrosion resistance. **7075** alloy, containing high levels of zinc, is one of the strongest aluminum alloys, with comparable strength to some steels, ideal for highly stressed parts such as fuselages and landing gear [7,8]. Aluminium's high corrosion resistance also plays a crucial role in extending the lifespan of aerospace components exposed to varying atmospheric conditions. This resistance is further enhanced by treatments such as anodizing, which adds a protective oxide layer. **Alloy 6061** is particularly corrosion-resistant and often used in applications requiring both strength and resistance to corrosive environments. In addition to corrosion resistance, some alloys, like **7050** and **7068**, are designed for extreme structural stability under high-stress conditions, which is essential for military and heavy-duty aerospace applications.

2.2 2XXX Series—(Al)

The widely used Al-Cu series 2XXX alloys, especially 2024-T3, have been extensively used in aerospace structural applications but almost exclusively for those applications where damage tolerance is of paramount importance. The benefits of such alloys are some excellent advantages: Strengthening: The Al-Cu-matrix material with the addition of Mg produces encouraging precipitation of the Al2Cu and Al2CuMg phases. It thus develops appreciable strength. For the structural applications, such as fuselage structures, superior strength must be obtained without any compromise to weightier Fatigue Resistance**: Al-Cu alloys with Mg exhibit high resistance to fatigue crack growth compared with the other aluminium alloys. Howso are limited in the following ways: Lower Yield Strength (YS): Though, 2XXX series alloys possess high ultimate tensile strength (UTS) they have low yield strength, hence restrict them being used in components experiencing high stresses [9]. Corrosion Resistance: Anodic Al2CuMg phases existing in the form of sites in the alloy make corrosion resistance significantly low. This is also a drawback, and there is a need for surface treatments or modification of alloys themselves for greater corrosion performance while in service under aggressive environment conditions, particularly in aerospace lines operating at extreme conditions [9].

2.3 Challenges and Limitations

Despite their advantages, aluminium alloys face limitations in aerospace applications. One significant issue is **stress corrosion cracking (SCC)**, which can occur under high-stress environments or in the presence of corrosive agents, limiting the durability of aluminium alloys in certain conditions. This issue is especially prevalent in **2024** alloy, which, while strong, lacks optimal corrosion resistance for long-term exposure in harsh conditionsAnother challenge is **thermal stability**. Aluminum alloys do not perform well at high temperatures, especially compared to materials like titanium or advanced composites. This limits their use in engine components or areas exposed to extreme heat, where thermal expansion and softening can impact structural integrity.

3. Overview of Surface Coating Methods for Aluminium Alloys

3.1 Anodizing

Anodizing is an artificially thickened, natural oxide layer on a metal, most commonly aluminum, in order to produce a hard-wearing, corrosion-resistant surface. This is done through an electrochemical reaction in which the aluminum acts as the anode of the electrolytic cell, and some suitable material, such as lead or stainless steel, acts as the cathode of the same cell. Once sufficient voltage has been applied, an electric current will pass through the electrolyte solution and initiate oxidation at the surface of the aluminum metal [13]. The process of anodizing is essentially the result of the migration of aluminum ions, Al^{3+}, and oxygen, or hydroxyl, O^{2-} or OH^-, through the oxide layer because of a high, usually 10^8 to 10^9 V/m, electric field. This arises due to the voltage drop that crosses the oxide layer, which acts as an insulator. Oxidation reaction at the aluminum/oxide interface means aluminum atoms lose electrons in order to become Al^{3+} ions entering into the oxide layer. Simultaneously, at the oxide/electrolyte boundary, O^{2-} or OH^- ions get formed when the H_2O molecules dissociate, giving up their H^+ ions. These ions now migrate to the aluminum/oxide interface, as driven by the electric field present in the capacitor [13].

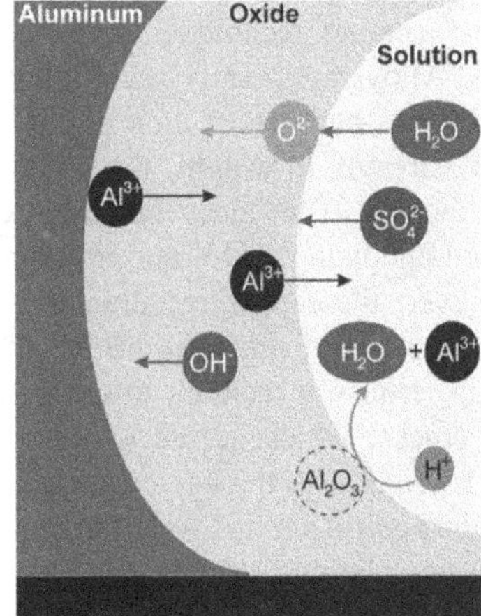

Fig. 106.1 Schematic illustration of ions movement and dissolution of oxide in sulfuric acid solution [13]

3.2 Electroplating

The process of electroplating refers to the means of surface treatment by depositing a thin layer of metal onto an aluminium alloy for property improvements. Aluminium alloy acts as the cathode-a negative electrode-in an electrolytic bath that contains ions of the desired plating metal, such as nickel, copper, and chromium, in electroplating. In the case of electroplating, the plating metal acts as the anode-positive electrode-or in some cases, an inert electrode (Fig. 106.1). Exposure of the metal ions onto the aluminium's surface by the electrolyte solution to a thin, uniform layer of metal of desired properties, such as increased corrosion or wear resistance or elevated electrical conductivity, when electric current is passed through them [14]. Some research is also based on the environmental friendliness of the electroplating process due to the use of natural electrolytes and some chemicals during the pre-treatment processes, for example, chromium or cyanide containing ones, which pose a threat to the environment. Alternatives are being developed towards making this process safer for the ecosystem [14]. Recent developments have enhanced the quality and efficiency of electroplated layers on aluminium. Pulse electroplating and composite electroplating in which metal is co-deposited with hard particles such as silicon carbide or alumina, yield coatings with hardness and wear resistance levels. Such developments offer more applications of electroplating on aluminium alloys in more arduous conditions within industry [14].

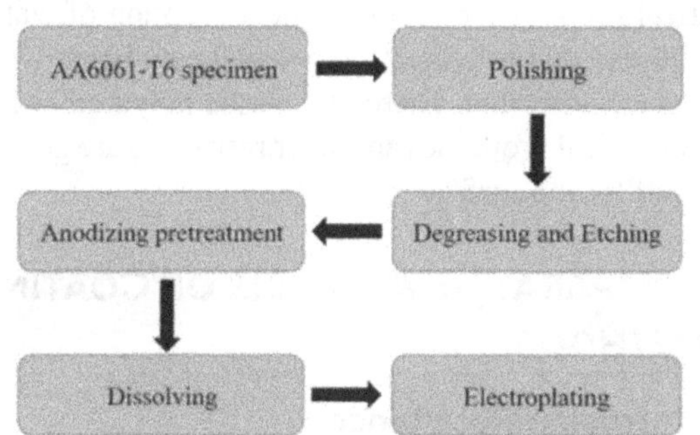

Fig. 106.2 A brief schematic of the sample preparation steps

3.3 Thermal Spray Coating

Thermal spray coating technology is more important in the aerospace industry because of durability, performance, and safety improvement in components subjected to high environmental loads.

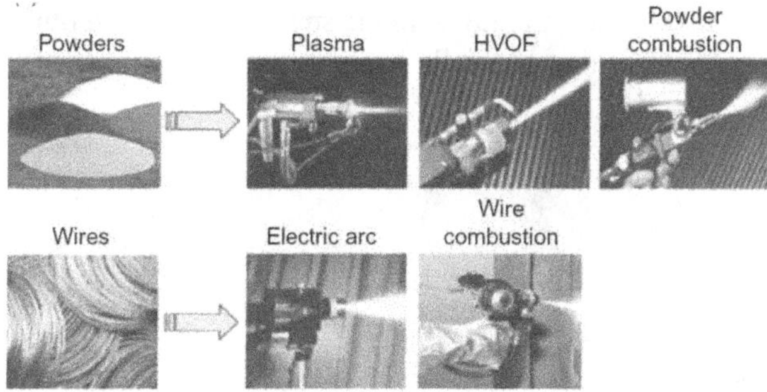

Fig. 106.3 Thermal spray materials and processes

3.4 Chemical Conversion Coating

Chemical Conversion Coatings have extensive applications in the aerospace industry primarily for corrosion resistance, improving adhesion of paint, and increasing surface durability in aluminum and magnesium alloys (Fig. 106.2). The chemical reaction is a process between the metal surface and the solution to produce a very thin protective oxide layer. Of the two most commonly used chemical conversion coatings, chromate and phosphate are the two, with each having distinct advantages applicable to specific applications. They are applied to interior structural parts and fasteners with subsequent protective coatings such as primers and paints. These are environmentally friendly alternatives that, though giving reliable corrosion protection for aluminum and other alloys, are being used more widely since they comply with tighter environmental regulations (Fig. 106.3).

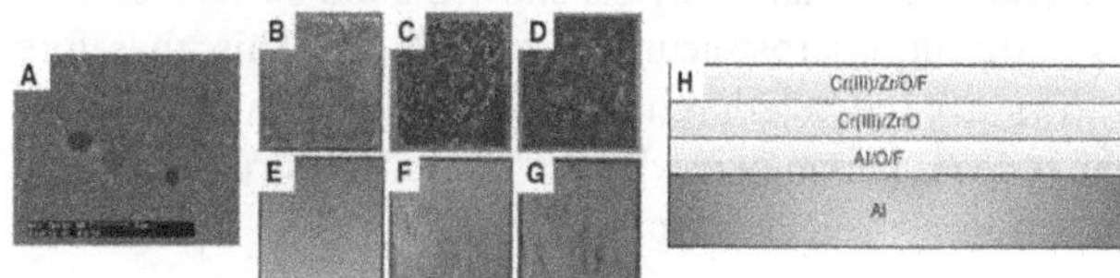

Fig. 106.4 Electron beam physical vapour deposition (EB-PVD) technology

3.5 Electron Beam Physical Vapour Deposition (EB-PVD) Technology

The method is primarily founded on the activity of the electron beam, which happens to be the most important part that has the role of being a thermal source in this deposition technique. One of the best and most attractive features of EB-PVD is the capability of depositing all kinds of material. The process of deposition depends on the action of an electron beam established at 2000 °C in an electron gun, which works based on the accelerating of thermal electrons supported by a high voltage. The possibility of having an effectively controlled microstructure surrounded by a

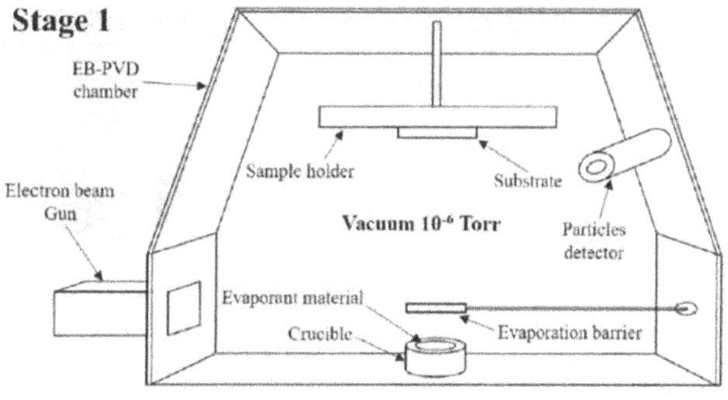

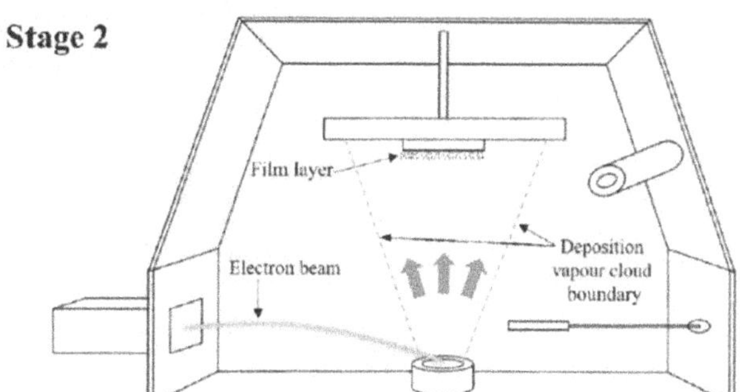

Fig. 106.5 Schematic of electron beam physical vapor deposition (EB-PVD) equipment (Stage 1) and the generation of the film for coating

managed composition trying to erase every possibility of contamination, and all these properties are obtained finally regarding very easily controlled parameters and flexible deposition. There are minor exceptions where the deposited layers do not possess homogeneous microstructure, but generally, the final materials show a good surface as well as uniform microstructure (Fig. 106.4). This therefore shows a good relationship between the manipulation of the process parameters with the final microstructure in the materials as well as uniformity.

3.6 Organic Coatings

In general, organic aviation coatings normally protect the substrate through three mechanisms (Fig. 106.5). The first mechanism through which they provide protection is by preventing corrosive media from diffusing to the interface between the metal and the coating. While small molecules of water and oxygen can pass through the coating, larger corroding ions are unable to penetrate and the coating effectively prevents the movement of ions between the cathode and the anode regions. The organic coating may contain corrosion inhibitor and pigment particles that can be hydrolysed and passivated or react with corrosive ions to protect the substrate from corrosion. Coating prevents corrosion products peeling from the metal surfaces with its excellent adhesion to metal. Aircraft are often used in coastal areas, islands, and reefs, or industrial regions, and thus suffer from severe environmental corrosion. Furthermore, due to some special airtight structures in the air-craft structure, usually water will collect inside the aircraft that may ascend to corrosion at skin joints, fastener connections, rivet holes, landing gear, battery area, flaps and hinge grooves, and areas affected by engine exhaust and other areas prone to collection of water. The main function of the organic coating on the surface of the radome and various radio radomes is to prevent the reduction of electrical properties because of the absorption of water in the material, and to protect from damage by being washed by the sand and wind. Therefore, elastic polyurethane with good electrical properties and wear resistance are generally preferred for the surface.

4. Comparative Analysis of Coating Methods

4.1 Corrosion Resistance

It is an important point that aluminum alloys show resistance to corrosion because these alloys are widely used in aircraft and spacecraft for an excellent strength-to-weight ratio and high thermal as well as electrical conductivity. Aluminum alloys owe inherent corrosiveness mainly due to atmospheric factors like moisture, salt, UV radiation, etc [10].

4.2 Wear Resistance

It is this property, along with the stress, friction, and other environment that generally hampers the functionality of a

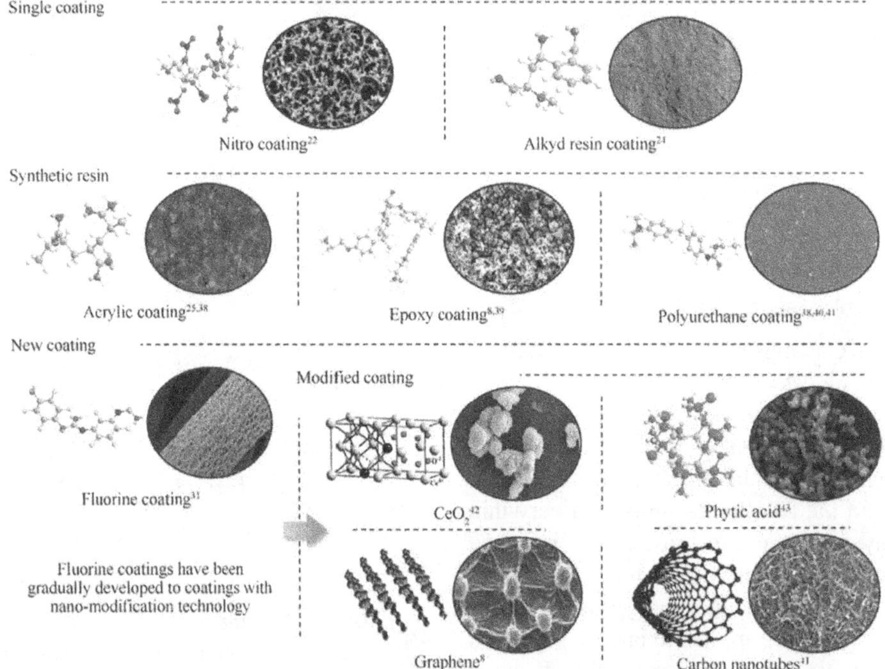

Fig. 106.6 General classification and development of organic aviation coatings

material surface, that makes wear resistance an important consideration for aluminum and aluminum alloys in aerospace applications. Aluminum alloys can offer weight savings and strength, which have earned them much favour in aerospace, but they generally do not offer high wear resistance, at least when compared to harder metals like steel or titanium. Wear can take place in aluminum parts by abrasion, adhesion, fretting, and erosion, which may degrade its performance and accelerate material failure unless mitigated [11]. To overcome such difficulties, a multitude of surface engineering techniques and treatments have been developed to enhance the wear resistance of aluminum in aerospace contexts. One of the widely used methods is anodizing, developing a thick, hard oxide layer on the aluminum surface (Fig. 106.6). This coating, in addition to making it harder, also provides the benefit of increased wear resistance through a hard oxide layer preventing abrasion [12].

4.3 Coating Adhesion and Durability

The aerospace industry is exposed to extreme environmental and mechanical stresses. These include rapid temperature changes, high moisture levels, UV radiation, and chemical exposure. The aluminum alloys in the aerospace industry face the exposure to which coating adhesion and durability would make all the difference. Indeed, proper adhesion of coatings is very important because it ensures the continuity of a protective layer and proper functioning up to the intended lifespan of the component. The best adhesion for aluminum alloys is achieved only if surface preparation for the cleaning, degreasing, abrasive blasting or chemical etching actions takes place, all aimed at removing contaminants and providing a roughened surface profile conducive to more effective coating bonding. In addition.

4.4 Application Complexity and Cost

Testing and quality assurance add to the cost because aerospace components have to undergo stringent tests, including non-destructive testing, fatigue testing, and corrosion testing, in order to pass at industry standards to withstand operational stresses.

4.5 Environmental and Regulatory Considerations

High on the list when the aerospace industry considers aluminum alloys are environmental and regulatory considerations; manufacture processes of the materials carry a significant ecological impact. For instance, aluminum production involves considerable amounts of energy. Also, the mining and refining of bauxite, the primary ore of aluminum, carry with them massive emissions of greenhouse gases in carbon dioxide and perfluorocarbons (PFCs). These emissions contribute to global warming and make the sourcing of aluminum a point of concern for an environmentally conscious aerospace company. For example, the regulatory bodies such as EPA in the United States and the European Union's Environmental Agency are setting more stringent limits on the permissible emissions of industries involved in aluminum production and processing; hence, the aerospace companies are compelled to seek more eco-friendly alternatives.

4.6 Performance Evaluation and Case Studies

These are fundamental reasons why aluminum alloys are a must for the aerospace industry since they encompass a unique combination of an extremely high strength-to-weight ratio, good corrosion resistance, and relative ease in fabrication. Aluminum alloys find extensive uses in aircraft structures, including fuselage and wings as well as internal components, thus offering necessary support to weight-sensitive applications. The performance of aluminum alloys in aerospace settings is quite closely evaluated in order to ensure that they bear the test of the demanding requirements of the field. The commonly used alloys for aircraft manufacturing are 2024, 6061, and 7075. All of these possess beneficial properties. For instance, the presence of 2024 is famous for resistance to fatigue in applications where the values of stress are high, especially for the fuselage and in other structural parts. On the contrary, the presence of 7075, having high tensile strength, contains its suitability for structural parts under high stress. Although progress in composite materials has now largely replaced some of the weight savings that aluminum alloys once offered, the cost advantage, recyclability, and proven in-service performance provide value in the industry. One direction for future aluminum alloys in aerospace could be the development of new alloying elements or novel combinations of conventional elements to enhance strength, fatigue, and corrosion resistance and keep this family of alloys relevant in an evolving world [15].

4.7 Recent Advances and Future Directions

The aircraft and aerospace industry today expects two halves from its raw material suppliers: one, to contribute to major cost-reduction programs of presently manufactured aircraft models by novel material solutions and two, to provide advanced materials and concepts, which could meet the requirements of mass air transportation of next generation. Important element of the cost-reduction program is substitution of conventional assemblies and built-up structures with integrated or monolithic structures. Conventional assemblies include lots of formed sheets and machined parts riveted or otherwise joined, whereas an integral structure combines the functions of these separate parts into one large unit. An integral structure therefore saves much weight, cuts production costs and simplifies logistics owing to a reduction in the number of parts and joinery. Advanced high-speed milling machines operating at speeds of up to 25 000 rpm, which also feed at relatively high speeds, cause the manufacturing process to be extremely efficient in such integrated structures of high-strength aluminum alloys with precise machining. Yet the challenge here would be to the raw material suppliers regarding how to ensure that the properties of their materials match or are even better than what have traditionally been achieved within these assemblies in the new integral structures. Integral technology relies on big machined structures to reduce the number of constituents and joints. The constituents that are currently being developed can be over 800 inches (20 meters) long, with raw material thicknesses up to 11 inches (280 mm), and aspect ratios after machining as high as 35. For certain applications, as much as 95% of the original material is machined away. Rolled aluminum plate, capable of satisfying these very severe dimensional specifications, has become the standard material form for such products. However, to adequately satisfy aerospace industry requirements, commercial aluminum plate had to be altered in some respects. Its maximum thickness had to be increased beyond 6 inches (152 mm) to allow machining from highly thick sections, and fundamental properties of the plate material-fundamental residual stress levels-had to be restrained at extremely low levels. This ensures that, despite their large dimensionality and considerable aspect ratios, the materials could be machined distortion free. This is in strict conformation with the requirements for aerospace applications.

5. Conclusions

New geometries also provide an opportunity to minimize the joints on aircraft structures through increased size of components and more aggressive joining technologies that will replace traditional riveting. This will, of course, mean larger flat rolled products for the larger components and materials optimized for each unique joining process. Of these cost savings, there are new and larger air transport types for which the industry is looking: for example, the future Airbus Megaliner 3XX, to make economical, high-volume air travel ever stronger. To meet these demands, Hoogovens Aluminium has focused its development effort in the following three directions: high-strength, thick aluminum plate for large machined structures; oversized fuselage skin sheets with improved damage tolerance; and lightweight, oriented materials toward advanced joining technologies. These developments meet today's cost-efficiency goals and tomorrow's aerospace industry requirements as it enters a new era in mass air travel.

References

1. Davis, J.R. (Ed.). (2001). Corrosion: Understanding the Basics. ASM International.
2. Callister, W.D., & Rethwisch, D.G. (2018). Materials Science and Engineering: An Introduction. Wiley.
3. Totten, G.E., &MacKenzie, D.S. (Eds.). (2003). Handbook of Aluminum: Volume 2: Alloy Production and Materials Manufacturing. CRC Press.
4. Schlesinger, M., & Paunovic, M. (2010). Modern Electroplating. John Wiley & Sons.
5. Fink, J.K. (2000). Metallic Materials for Aerospace Applications. Springer.
6. ASM International. (2015). Metallography and Microstructures of Aluminum and Its Alloys. ASM International.
7. Sudagar, J., Lian, J., & Sha, W. "Electroless nickel, alloy, composite, and nano coatings - A
8. critical review," Journal of Alloys and Compounds 571 (2013) 183–204.
9. Yin, Y., Wang, X., & Zhang, Y. "Sustainable coating materials in aerospace applications: a review of regulatory trends and future directions," Coatings, 10(11), 1073.
10. K. A. H. Kazin Aluminum-Copper Alloys," *Materials Science and Engineering A*, vol. 720, pp. 91–98, 2018.
11. Davis, J. R. (2018). "Corrosion of Aluminum Alloys in Aerospace Applications: The Role of Alloying and Environmental Factors." *Corrosion Science*, 132, 62–79.
12. Puchala, B. et al. (2020). "High-Temperature Behavior of Aluminium Alloys in Aerospace Applications." *Journal of Materials Science*, 55, 1324–1337.
13. Ali, A., et al. (2020). "Advancements in Aluminum Alloy Casting for Aerospace Applications." *Journal of Alloys and Compounds*, 838, 155440.
14. I. Tsangaraki-Kaplanoglou, S. Theohari, Th. Dimogerontakis, Yar-Ming Wang, Hong-Hsiang (Harry) Kuo, Sheila Kia, Effect of alloy types on the anodizing process of aluminum,
15. Ali Rahimi, Shayan Sarraf, Mansour Soltanieh, Nickel electroplating of 6061-T6 aluminum alloy using anodizing process as the pretreatment.

Advances in Additive Manufacturing Technologies – Gurusamy Pathinettampadian et al. (eds)

Anomaly Detection on IoT Networks Through Deep Learning Algorithms

V. Haritha[1]
Assisstant Professor,
Department of CSE, RMK Engineering College,
India

Priyadharsini S.[2]
Assistant Professor,
Department of Electrical and Electronics Engineering,
Sankar Polytechnic College, India

V. Arunachalam[3]
Director Academics,
Thamirabharani Engineering College,
India

Chitra A.[4]
Assistant Professor,
Department of CSE, Unnamalai Institute of Technology,
India

◆ **Abstract:** The IoT is rapidly expanding across various real-world applications due to the increasing use of the internet. IoT applications offer enhanced compatibility for users in terms of service access and data storage. This widespread adoption has, however, introduced significant security challenges within the network. As a result, Cyber Security has become a important area of research, given the rise in various types of attacks targeting IoT systems. Securing IoT environments is particularly challenging due to their extensive integration into everyday applications. Among the most active research areas in this domain is anomaly detection using machine learning and deep learning techniques within IDS. The proposed work focuses on anomaly detection using a deep learning approach based on a CNN applied to an IoT network dataset. The CNN model is trained and evaluated using the BoT-IoT dataset, performing binary classification to identify whether a network packet transaction is normal or anomalous. Experimental results demonstrate that the CNN model achieved a high accuracy of 99% in detecting anomalies in the BoT-IoT dataset.

◆ **Keywords:** High sensitivity, Dynamic response, Vibration isolation, Noise filtering, Tool wear prediction

1. Introduction

The extensive and widespread adoption of IoT devices in attempting to contribute to the growth and operation of smart cities, facilitate intelligent traffic management systems, improve smart healthcare solutions, and facilitate various other real-world applications has posed a wide range of critical challenges that fall under the domain of

[1]harithabalaji8892@gmail.com, [2]dharsini9011@gmail.com, [3]arunachal004@gmail.com, [4]chitraarjun9328@gmail.com

DOI: 10.1201/9781003685906-107

cybersecurity. This extensive growth has been facilitated mainly by the affordability and reliability of IoT networks, as well as the networked physical devices in these networks. These devices—e.g., smartphones, medical monitors, security cameras, and various environmental sensors—have been made to detect information around them in real time and then transmit this information to cloud storage for further processing and use. It is, however, worrying that most IoT applications are put in place without due consideration for critical factors such as cybersecurity, data security, and effective handling of data. As a result, the feasibility of managing the vast and constantly growing number of IoT devices in real time, as well as ensuring data integrity and security of the information processed by them, is a challenging endeavor. Recent developments in the field have increasingly emphasized the role of advanced machine learning and deep learning techniques in solving these severe cybersecurity challenges. With the frantic pace of technological advancements in wireless technologies, IoT applications have indeed become more reliable when dealing with data processing through various wireless communication channels. These networks facilitate data transmission through the internet to cloud servers and are typically described as delay-sensitive, i.e., they expect data to be transmitted in a short time to enable timely and effective processing. The reliability and robustness of IoT networks are of the utmost importance, especially in situations that are critical, such as during natural disasters, forest fires, and other emergencies, where any probable loss of data due to anomalies is totally unacceptable.

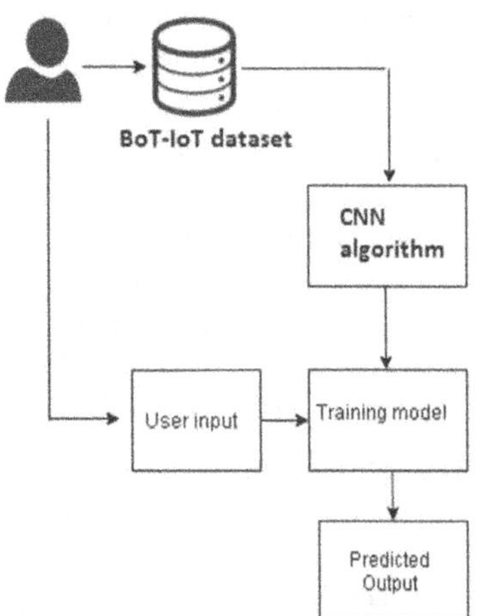

Fig. 107.1 Overview of anomaly detection

Anomalies in Internet of Things (IoT) networks can be categorized into two broad categories: one category includes technical anomalies, which can occur in the form of hardware failure or communication failure, and the other category includes security-related anomalies, such as various types of attacks and unauthorized intrusions. IoT networks are most vulnerable to both categories, and any delay in the detection of these anomalies can lead to huge losses of data as well as destruction of the network infrastructure. Therefore, it is extremely critical to detect anomalies in real time as they happen in these systems. Service providers are now increasingly concerned about the growing threats that target IoT networks. Due to the delay-sensitive nature of the data handled by these networks, as well as their critical nature, any form of intrusion or attack can have the potential to severely compromise the reliability of the entire network. This growing concern has led to a huge amount of research work dedicated to Intrusion Detection Systems (IDS) that are specifically aimed at IoT networks. Although these IDS are extremely effective in detecting cyberattacks, most of the existing anomaly detection systems tend to suffer from issues characterized by high false alarm rates (Fig. 107.1). To effectively tackle this burning issue, the proposed system combines a CNN based deep learning model that is aimed at enhancing the accuracy of anomaly detection, with the aim of achieving a significantly higher accuracy rate. The main aim of this research work is to detect anomalies in IoT networks using CNN technology while maintaining a high accuracy level. CNNs have the excellent capability of extracting key features from the input data and learning complex patterns, which significantly enhances the overall accuracy of the detection process.

IoT networks provide several real-world applications, such as surveillance, disease monitoring, and traffic management, through sensor-based data collection. IoT networks process incoming data from sensor nodes efficiently and enable further analysis. IoT networks also connect multiple layers—such as cloud servers—via internet connectivity, making them extremely reliable for data generation and processing. Detection of anomalies in IoT networks is still a challenge due to the high-speed nature of incoming data, which needs to be processed in real time. Any time taken for anomaly detection can make the data unreliable and compromise the security of the system. One of the most critical challenges is extracting useful information from network traffic to effectively identify anomalies. Other studies have applied multi-class classification using techniques like Logistic Regression and Gradient Boosting. Although deep learning has been explored for NIDS, many of these studies have overlooked critical issues such as overfitting during model training.

The proposed deep learning-based anomaly detection system addresses these limitations by focusing on high detection accuracy and minimizing false alarms. This system leverages CNN to detect external attacks effectively and aims to improve performance through careful preprocessing, including handling class imbalance issues. The objectives and motivations of this proposed work are centered around anomaly detection in IoT networks and preventing cyberattacks using a CNN-based approach. The following chapters will detail the literature review on anomaly detection in IoT networks, the proposed detection algorithm, experimental results and evaluations, and the conclusion along with future enhancements.

2. Related Works

Several studies have been proposed for automated fault detection in microgrid environments. A summary of notable works is provided below: Hussain et al. [1] introduced a security framework for IoT networks, using the IoTFlock tool to generate both normal and anomalous traffic data. The context-aware system was specifically designed to detect malicious traffic from healthcare-related data. The authors implemented machine learning algorithms such as Naïve Bayes (52% accuracy), Linear Regression (99.3% accuracy), and Decision Tree (99.6% accuracy) for this task. However, the detection accuracy was relatively low in some of these models, indicating the need for further improvement. In [2], the authors proposed a cyberattack detection system over the Message Queue Telemetry Transport (MQTT) tunnel using machine learning algorithms with hyperparameter tuning. The dataset was generated through the MQTT tunnel. Decision Tree and Logistic Regression both achieved 98% accuracy, Random Forest achieved 99%, while K-Nearest Neighbors (KNN) performed poorly with only 49% accuracy. The overall proposed system attained a peak performance accuracy of 95%.

Intrusion Detection Systems (IDS) are effective in protecting networks from various attacks. However, network datasets tend to be nonlinear and high-dimensional, which presents challenges for traditional machine learning models. In [3], the authors implemented an Image-Enhanced CNN, incorporating oversampling techniques to address class imbalance. Feature extraction was used to reduce data dimensionality, and the Tanh activation function with dropout was applied. The KDDCup 99 dataset was utilized for evaluation. W. Wang et al. [4] proposed an intrusion detection method based on CNNs to tackle the problem of high false alarm rates. Their approach involved extracting spatio-temporal features from network traffic to improve detection accuracy. The DARPA 1998 and ISCX 2012 datasets were used. Their method significantly reduced false alarms using a combination of feature learning and dimensionality reduction. However, small dataset sizes may lead to underfitting issues.

Xin et al. [5] conducted a comprehensive and detailed survey that was centered on both machine learning and deep learning methods specifically designed to leverage temporal and thermal correlations in the various domains of cybersecurity. The research reported in reference [6] analyzed network traffic by using a framework that was developed in-house to categorize incoming traffic into two different classes, i.e., 'bonafide' and 'attack' classes. Surprisingly enough, the system was able to achieve a high ROC value of 0.98 while, simultaneously, CPU as well as RAM consumption was maintained in control. In the research referred to as [7], Deep Feed-Forward Neural Networks (DNN) were used in analyzing the UNSW NB15 and CICIDS2017 datasets for the task of intrusion detection. The authors employed a fivefold cross-validation method for the optimization of hyperparameters, and they executed their solution within an Apache Spark framework to achieve improved scalability. Additionally, k-means clustering was used as a feature selection method that proved greatly towards enhancing the overall performance of the classifier.

Li et al. [8] performed an extensive analysis of state-of-the-art deep learning models for anomaly detection, where they placed a strong emphasis on LSTM networks and GRUl architectures. For experimentation, they used the popular NSL-KDD dataset, which is widely utilized for intrusion detection system evaluation. The experiment results showed a remarkable performance difference, where GRU outperformed LSTM with an unprecedented accuracy rate of more than 92%. In the research cited as [9], researchers proposed a new hybrid architecture that integrated CNN and LSTM for intrusion detection. The new approach effectively leveraged spatial feature learning through character encoding, which allowed the model to learn distinctive features that are HTTP traffic-specific. Their model was evaluated on the CSIC-2010 and CICIDS2017 datasets. However, extracting critical HTTP features remained a significant challenge. The work in [10] focused on identifying botnet attacks in IoT networks using statistical learning models, specifically the BMM. Important features were extracted from network traffic using statistical methods and selected using probability density functions. Although effective, this method was found to have a high rate of false positives.

As internet usage continues to grow, cybersecurity remains one of the most pressing challenges. Many researchers have turned to artificial intelligence to tackle these issues. The body of previous work has actively utilized a range of both machine learning and deep learning models to solve a wide range of problems in data processing and analysis; however, typical problems like overfitting and high false positive rates are still significant roadblocks. Particularly, it has been seen that machine learning models tend to provide low accuracy rates when asked to perform anomaly detection, and they fail to make meaningful and meaningful interpretations from the inherent complexity in network traffic patterns. The popular machine learning algorithms used in previous research include Naïve Bayes, Decision Tree, and Support Vector Machine in binary classification tasks, and Logistic Regression and Gradient Boosting techniques used in multi-class classification tasks. Despite the widespread use of deep learning methods in the field of Network Intrusion Detection Systems (NIDS), it has been found that the majority of these systems fail to resolve the problems inherent with overfitting optimally. The necessity of the precision and timeliness of detection mechanisms is of the utmost importance, considering the fact that any lag in this process would have catastrophic consequences on network security as well as the overall integrity of data. To solve and rectify the above-stated shortcomings inherent in the present methodology, the research work presented here involves an innovative deep learning-based approach, which employs a CNN specially designed for anomaly detection in Internet of Things (IoT) networks with high accuracy. This new approach aims to overcome the limitations imposed by the existing methods, thereby providing a more trustworthy method for anomaly detection and making a positive contribution to the development of cybersecurity solutions in IoT networks.

3. Proposed Work

Cyber security is the active area of research due to the vast usage of internet. Thus an effective detection system is mandatory and latency in detection may cause serious damage to network. To overcome, this problem, the proposed work is deep learning based anomaly detection on IoTBoTNet dataset as a binary classification study. The work used CNN for detection of anomalies. The proposed work is implemented in python programming with deep learning libraries Tensorflow and Keras modules.

BoTNeTIoT-L01 is a dataset, which is an integrated data from IoT devices for detection of IoT botnet attacks (BoTNeTIoT). This dataset is captured from the network of 10 seconds time window. The dataset is labeled as binary class label, 0 stands for attacks, and 1 stands for normal data. The dataset includes two classes called 'Mirai' and 'Gafgyt'. The dataset has 23 statistically engineered features extracted from the .pcap files. Statistically engineered dataset features are described as show in the Table 107.1

Table 107.1 Dataset features used in BoTNetIoT dataset

Features	Details
23 features	mean, variance, count, magnitude, radius, covariance, correlation coefficient
4 features	packet count, jitter, size of outbound packets, outbound and inbound packets

The proposed work considered data pre-processing, CNN model for training and predicting the anomalies (Table 107.1). The dataset taken is highly imbalance on instances of number of classes. Thus the dataset is refined to overcome this problem. The pre-processing also includes providing the standard form of data input to algorithm which is the data visualization technique is used to understand the characteristics of network intrusion data. The Fig. 107.2 shows histogram plot for BoTNetIoT data provided with 31 attributes.

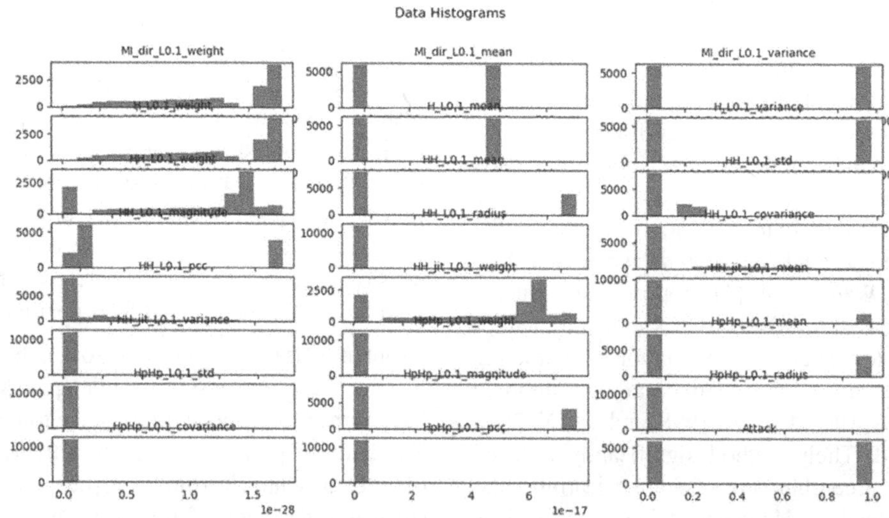

Fig. 107.2 Data visualization of fault detection in microgrid

The Fig. 107.3 shows the class balancing results by trimming the dataset to 5000 instances of each class value. Class imbalance many affect the effectiveness of learning model. Thus the same number of network traffic instances in each class is considered here. The Fig. 107.3 shows the attack and normal instances numbers in each bar.

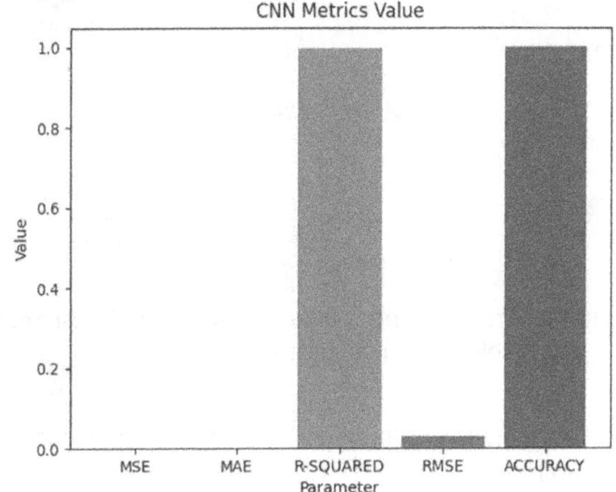

Fig. 107.3 Plot for network tarffic class value

Anomaly detection dataset is taken for training and testing purpose the split ratio used here is 80:20. Binary classification is performed on the data with 0 represents attack and 1 presents normal. The algorithm is trained with 5000 instances of each class. The performance of algorithm is effective when the class balance data is used. The Fig. 107.4 shows the system architecture of Anomaly detection on IoT network. CNN algorithm used here takes features input and classifies the data as anomalous or normal.

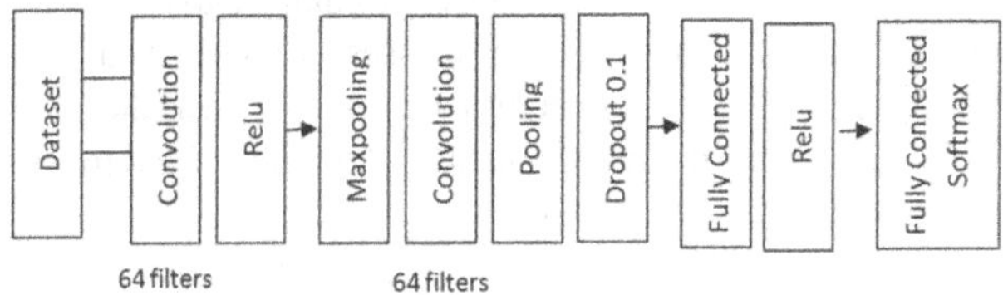

Fig. 107.4 Proposed CNN architecture for anomaly detection

Convolutional Neural Network has the ability to learn the important features automatically to detect the anomalies in IoT network data. This algorithm effectively studied in many research area due to its effectives in classification. The model built here has three layers, input layer, hidden layer and output layer. Input layer is given the pre-processed dataset as training data. The hidden layer has one dense layer with 32 neurons, next is convolutional layer with 128 filters and activation unit is Rectified Linear Unit (relu), the next is pooling layer with pool size 2. The second hidden layer is convolutional layer with filter 64, followed by pooling layer with pool size 1 and a drop out of 20%. The third hidden layer is fully connected dense layer with 128 neurons, and drop out 20%. The last layer is output layer with 'adam' optimizer. The Table 107.3 shows the proposed CNN layer details and its value initialized for training the model.

CNN algorithm used for anomalies detection on IoT data with normal and attack as binary classes is implemented in this module. The following Fig. 107.5 shows the detection results of validation dataset as '0' or '1' as output.

ID	Predicted Value
1	0
2	1
3	1
4	1
5	1
6	0
7	1
8	1
9	1
10	1
11	1
12	1
13	0
14	0
15	1
16	1
17	1

Fig. 107.5 Detection results of CNN algorithm

4. Results and Discussions

The proposed implementation is anomalies detection on BoTNeTIoT dataset as binary classification ''mirai" and "gafgyt" as attack and normal types. The implementation is performed in Python programming with libraries deep

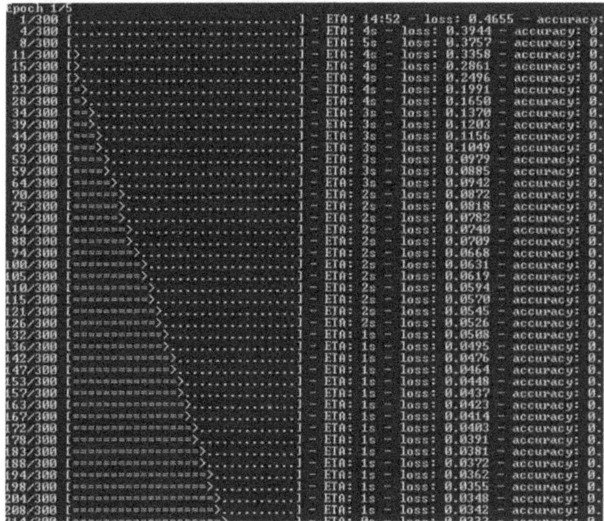

Fig. 107.6 CNN training for anomaly detection

learning framework keras, tensorflow is used. The dataset taken for the study contains two classes of detection. The figure shows the training results of CNN model for anomaly detection from BoTNetIoT data. The number of epoch is varied from 5 to 50 to get the optimized values. The number of epoch used here is 5, which gets the highest accuracy of the proposed model.

The loss estimation function used are MSE, MAE and RMSE. The error metrics values reduced for 5 epochs of training. The MSE error value is very negligible in the proposed CNN model.

The Fig. 107.7 represented the training and validation loss registered during training and validation of proposed CNN model. The number of epoch shown in X- axis and MSE error metric is shows in Y-Axis, whereas dot line represents the training and continuous line represents validation loss. The above table represents the error metrics registered for the overall CNN model after compilation. The MSE and MAE error is negligible and the proposed model has achieved 99.9% accuracy.

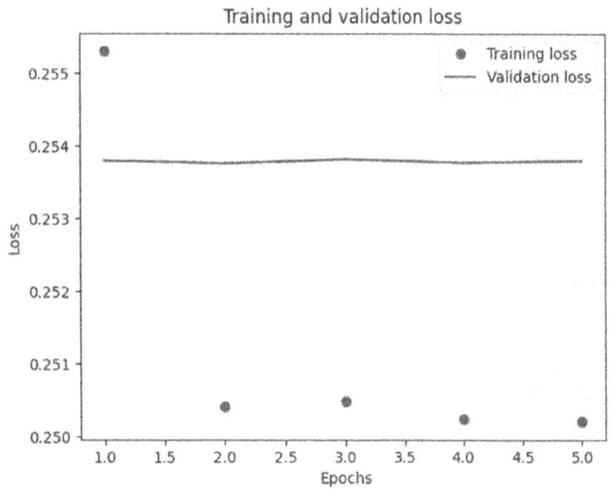

Fig. 107.7 Training and validation loss for anomaly detection in IoT dataset

Table 107.2 Training and validation accuracy

No of Epochs	Training Accuracy	Validation Accuracy
1	0.7548	1.0000
2	0.7522	1.0000
3	0.7503	1.0000
4	0.7494	0.9992
5	0.7505	0.9992

The Fig. 107.8 represented the training and validation accuracy registered during training and validation of proposed CNN model. The number of epoch shown in X-axis and Accuracy metric is shows in Y-Axis, whereas dot line represents the training and continuous line represents validation loss.

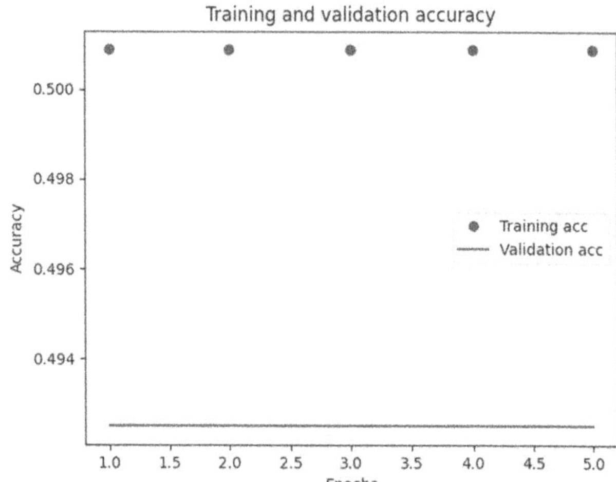

Fig. 107.8 Training and validation accuracy for anomaly detection in IoT dataset

5. Conclusion

The proposed research focuses on anomaly detection for cybersecurity within IoT networks, utilizing a deep learning-based approach through CNN. Deep learning techniques, particularly CNNs, are well-suited for this task due to their ability to automatically learn and extract relevant features from complex and high-dimensional data, eliminating the need for extensive manual feature engineering. In this work, the BoT-IoT dataset was used to evaluate the effectiveness of the CNN model. The dataset we have used includes crucial traffic data which has been labeled appropriately for the task of binary classification. In this dataset, two significant classes of attack data, i.e., 'mirai' and 'gafgyt', were combined and labeled as the label '0' to indicate instances of attacks, while the normal traffic was labeled as '1' to indicate benign operations. The CNN model was trained with care and extensively tested with this dataset, enabling it to be able to distinguish accurately between malicious network traffic and benign traffic. The experimental results based on these tests clearly indicated that the proposed CNN architecture achieved a stunning accuracy of 99% when it comes to detecting anomalies, which speaks volumes about its robustness and reliability in the critical task of detecting cybersecurity threats in IoT networks. For future work, there are several promising directions that can be pursued to further improve both the performance and generalizability of the anomaly detection system we have implemented. One such highly promising direction involves investigating other deep learning architectures, including but not limited to DNNs,

LSTM networks, as well as hybrid architectures that could potentially lead to improved classification accuracy and improved detection rates. Additionally, fine-tuning several hyperparameters, especially regularization techniques such as dropout rate and L2 regularization, may significantly contribute to making better model generalization and avoiding overfitting during the training (Table 107.2). These suggested improvements may ultimately pave the way for more efficient and scalable solutions for securing IoT networks from the ever-evolving threat landscape of cyber threats.

References

1. Hussain, Faisal & Abbas, Syed & Shah, Ghalib & Pires, Ivan & Fayyaz, Ubaid & Shahzad, Farrukh & Garcia, Nuno & Zdravevski, Eftim. (2021), A Framework for Malicious Traffic Detection in IoT Healthcare Environment, Sensors. 21. 10.3390/s21093025.
2. I. Vaccari, S. Narteni, M. Aiello, M. Mongelli and E. Cambiaso, Exploiting Internet of Things Protocols for Malicious Data Exfiltration Activities, in IEEE Access, vol. 9, pp. 104261–104280, 2021, doi: 10.1109/ACCESS.2021.3099642.
3. Wang, Qian & Zhao, Wenfang & Ren, Jiadong. (2021), Intrusion detection algorithm based on image enhanced convolutional neural network, Journal of Intelligent & Fuzzy Systems. 41. 1–12. 10.3233/JIFS-210863.
4. W. Wang et al., "HAST-IDS: Learning Hierarchical Spatial-Temporal Features Using Deep Neural Networks to Improve Intrusion Detection," in IEEE Access, vol. 6, pp. 1792–1806, 2018, doi: 10.1109/ACCESS.2017.2780250.
5. Y. Xin et al., "Machine Learning and Deep Learning Methods for Cybersecurity," in IEEE Access, vol. 6, pp. 35365–35381, 2018, doi: 10.1109/ACCESS.2018.2836950.
6. G. De Carvalho Bertoli et al., An End-to-End Framework for Machine Learning-Based Network Intrusion Detection System, in IEEE Access, vol. 9, pp. 106790–106805, 2021, doi: 10.1109/ACCESS.2021.3101188.
7. Faker, Osama & Dogdu, Erdogan. (2019). Intrusion Detection Using Big Data and Deep Learning Techniques. 10.1145/3299815.3314439.
8. Z. Li, A. L. G. Rios, G. Xu and L. Trajković, Machine Learning Techniques for Classifying Network Anomalies and Intrusions, 2019 IEEE International Symposium on Circuits and Systems (ISCAS), 2019, pp. 1–5, doi: 10.1109/ISCAS.2019.8702583.
9. A. Kim, M. Park and D. H. Lee, "AI-IDS: Application of Deep Learning to Real-Time Web Intrusion Detection," in IEEE Access, vol. 8, pp. 70245–70261, 2020, doi: 10.1109/ACCESS.2020.2986882.
10. Ashraf, Javed & Keshk, Marwa & Moustafa, Nour & Abdel-Basset, Mohamed & Khurshid, Hasnat & Bakhshi, A.D. & Mostafa, Reham. (2021, IoTBoT-IDS: A Novel Statistical Learning-enabled Botnet Detection Framework for Protecting Networks of Smart Cities. Sustainable Cities and Society. 72. 103041 10.1016/j.scs.2021.103041

Advances in Additive Manufacturing Technologies – Gurusamy Pathinettampadian et al. (eds)
© 2026 Taylor & Francis Group, London, ISBN 978-1-041-16687-0

Numerical Modeling of Concrete Processing: From Conventional Casting to Additive Manufacturing

Thangavelu Eswary Devi[1]
Assistant Professor, Department of Civil Engineering,
Rajalakshmi Engineering College,
India

S. Raja Muniyasamy[2]
Assistant Professor, Department of Civil Engineering,
Thamirabharani Engineering College,
Chennai, India

M. A. Raja[3]
Assisstant Professor, Department of Civil Engineering,
Meenakshi Sundararajan Engineering College,
Chennai, India

S. Mercy[4], **Suresh Kumar M.**[5]
Lecturer, Department of Civil Engineering,
Sankar Polytechnic College,
India

Balaji Krishnabharathi[6]
Assistant Professor, Department of Mechanical Engineering,
Chennai Institute of Technology,
Chennai, India

◆ **Abstract:** Digital shaping methods are still in their infancy, but they are already causing a revolution in the concrete industry. These procedures necessitate cement-based materials with novel, as-of-yet-undefined characteristics. Hence, more study is necessary to determine the proper specifications for these materials. An outline of recent developments in numerical simulations of concrete flow is presented at the outset of this article. As part of this process, various modeling methodologies were investigated for the purpose of studying the shape and deposition behavior of freshly mixed concrete. Then, we go on to review the literature on additive manufacturing procedures for concrete that use numerical simulations to examine extrusion-based methods. Based on what we've seen in the literature, numerical modeling techniques can effectively detect and forecast various printing process defects. These problems can show up at several scales: in the nozzle when the material is being extruded, in the printed layers themselves, or even in the complete structure. Optimization of the manufacturing process and the material employed may be achieved by numerical modeling, which allows for the systematic examination of a wide range of factors. These parameters include material qualities, geometric configurations, and process variables. Digital manufacturing methods for concrete can benefit from this strategy's enhanced printability, structural stability, and general efficiency.

◆ **Keywords:** Concrete casting, Additive manufacturing (AM) of concrete, 3D concrete printing, Digital fabrication, Cementitious materials

[1]eswarydevi.t@rajalakshmi.edu.in, [2]ramunish58@gmail.com, [3]antojosephraja@gmail.com, [4]mercyspc@gmail.com, [5]sureshspccivil2020@gmail.com, [6]balajikrishnabharathi@gmail.com

DOI: 10.1201/9781003685906-108

1. Introduction

Whether it's in a natural structure or a manufactured product, shaping is a complicated process affected by many different energy and pressures. In the construction business, this idea is also applicable to the shape of concrete. Mold filling and the development of a void-free, monolithic concrete structure are traditionally determined by the interaction between gravity-induced stresses and the material's yield stress during the shaping process in this sector. There are two primary arguments in favor of a comprehensive comprehension of shaping processes as they pertain to traditional concrete applications. The first is process optimization. In traditional concrete casting methods, this could involve optimizing the location of pouring points to enhance material distribution or predicting the formwork pressure to ensure structural integrity during casting. The second reason is that an in-depth knowledge of the process enables engineers to establish target rheological properties for fresh concrete. Similar to mechanical strength, fluidity plays a crucial role in both technical performance and economic efficiency. Consequently, a crucial part of processing concrete is determining the minimum and maximum values of rheological qualities that are necessary for a particular item shape and procedure.

On the other hand, conventional methods of preparing concrete have not evolved much throughout the years. The construction industry relies on well-established practices, standards, and the experience of workers on-site or in manufacturing facilities. This practical knowledge has generally been sufficient to provide a rough but adequate understanding of the fresh material requirements for specific applications. Due to this, numerical simulations of fresh concrete flow have historically been employed only in highly specialized cases, such as research-based studies or legal disputes, where they help assess responsibilities after a processing-related issue has occurred. Recently, a new generation of digital concrete processing techniques has emerged, including extrusion-based additive manufacturing. These advancements have expanded the range of available shaping technologies, introducing new methods that require optimization. Furthermore, these processes necessitate the use of cement-based materials with fresh property requirements that are not yet well defined. Unlike traditional form filling, which benefits from decades of industry experience, these new digital shaping techniques demand a fresh approach to process understanding and optimization.

Additionally, these modern shaping methods offer unprecedented design flexibility, allowing for the fabrication of complex geometries. However, the intricate shapes produced by these methods generate equally complex mechanical stress distributions, making it challenging to analyze them using conventional analytical approaches. While simplified cases, such as basic walls or cylindrical columns, can still be studied using traditional methods, more complex geometries require advanced analysis tools. Within this framework, the paper's authors contend that numerical models of concrete flow can greatly facilitate the effective incorporation of these innovative shaping approaches into industrial practice. When it comes to improving digital processing procedures, these simulations might supplant or significantly lessen the need for laborious trial-and-error approaches. In addition, they can provide light on the rheological specifications needed for the concrete components of these novel procedures.

The paper is divided into two primary sections. Part one is an exhaustive survey of what is currently known about numerical simulations of concrete flow, including topics such as the many methods that are available and how they have been used to traditional concrete manufacturing. Part 2 delves further into numerical modeling of concrete additive manufacturing processes based on extrusion. Both the scale of the printed item and that of the extruded filament are considered in our analysis of the material's behavior. By examining existing studies, we highlight the types of insights that can be gained from these numerical approaches and discuss their practical implications for optimizing additive manufacturing in the concrete industry..

2. Results and Discussion

The analysis reveals that the combination of 180 μm iron particles with 1.2 μm SA results in improved surface roughness. However, even after 30 minutes of processing, the roughness only PDEs play a fundamental role in describing the flow behavior of fluid continua. However, these PDEs are often too complex to be solved analytically, necessitating computational approaches for their resolution. Computers, while incapable of directly solving PDEs, can process SAEs through basic arithmetic operations. Consequently, numerical solutions to fluid flow problems require the transformation of PDEs into SAEs. Multiple numerical techniques, such as the FEM, FDM, and FVM, make this transition possible. The first step in any approach is to create a spatial grid of points. From there, you may calculate the important variables. An SAE's mesh is the collection of all the grid points, and each element in the SAE is associated with a particular location on the grid.

There are two main types of flow simulation methods: those that rely on meshes and those that do not. Meshless methods are sometimes called grid-free or meshfree. Methods such as the FDM [4,5] and the Lattice-Boltzmann Method (LBM) [5-7] are examples of mesh-based algorithms that calculate important variables on a predetermined mesh. On the other hand, meshless methods do not depend on a prefabricated mesh; rather, each Lagrangian particle in the system is associated with its own unique set of physical properties, including location and velocity [8]. The Discrete Element Method (DEM) [5,9,10], Smoothed Particle Hydrodynamics (SPH) [11], and Dissipative Particle Dynamics (DPD) [5,6,11,12] are three prominent examples of meshless approaches. The newly created meshless version of the LBM is one example of a technique that exists at the boundary between mesh-based and meshless approaches [13].

In general, "Computational Fluid Dynamics" (CFD) means numerically approximating the solution of fluid equations inside a continuum. Although mesh-based and meshless approaches may both produce solutions, mesh-based techniques are typically linked with CFD. When looking at the research on FDM, FEM, and FVM [4,5,14-16], this difference becomes clear. Methods that rely on meshes are also referred to as "continuous approach" in some settings [17]. However, some sources define CFD broadly to include both mesh-based and meshless approaches [6,8]. Here, we follow the classification in [5], treating CFD as a term specific to mesh-based methods while acknowledging its broader usage.

Historically, FEM was the preferred method for simulating fresh cement-based materials, as referenced in early studies [19–21]. Over time, FVM gained popularity due to its central role in widely used CFD software packages [16,22–24]. The application of FVM in simulating fresh cementitious materials can be found in studies such as [25–31]. Although FDM has been less commonly used for simulating cement-based materials, examples exist in the literature [32–34]. Its limited adoption stems from its inability to support unstructured meshes, which are necessary for meshing complex geometries in Cartesian coordinates [16]. In structured meshes, grid points are uniquely identified within a coordinate system, simplifying programming and resulting in SAEs with a regular diagonal matrix structure [16,18]. In contrast, unstructured meshes use a labeling system rather than a coordinate reference, allowing greater flexibility in mesh geometry. Unstructured meshes enable the integration of various cell types (e.g., polyhedral, hexahedral, tetrahedral), but the corresponding SAEs lack a regular diagonal structure, requiring reordering of cell labels to minimize bandwidth.

As numerical simulations have evolved, meshless approaches have gained prominence in modeling fresh cement-based materials. With a dedicated chapter in the RILEM state-of-the-art report, DEM has clearly grown quite popular [5]. Cementitious material models have also made use of other state-of-the-art meshless approaches, such as DPD [12,17]. Both mesh-based and meshless approaches should, in theory, produce equivalent results for geometries far bigger than the individual meshless particles. This equivalence, which presents a benchmark flow case for concrete channels, demonstrating similar outcomes between mesh-based and meshless simulations in comparison with analytical solutions [35].

Rheological properties of cementitious materials are typically measured using rheometers or industrial tests. Numerical simulations aid in interpreting data from these devices by providing insights into flow behavior. Tanigawa and Mori [36] performed one of the first numerical simulations of the slump test. As mentioned in [37], there are a variety of rheometers to choose from. Both the BTRHEOM and the BML rheometer are examples of basic devices; the former uses a coaxial cylinder arrangement while the latter operates on the principles of parallel plates [37]. The BTRHEOM has been modeled numerically with FEM and FDM [38,39], whereas the BML rheometer has been modeled with FDM [32,33]. Validating the devices, analyzing their behavior, and improving the interpretation of experimental data are all goals of these simulations.

On one end of the range are rheological testing instruments like the slump test, and on the other are rotating rheometers like BTRHEOM and BML. Examples of intermediate devices are the Orimet test, V-funnel, LCPC-box, J-Ring test, and L-Box [40,41]. A number of these devices have been numerically analyzed extensively; one such example is the L-Box test. Some studies have used mesh-based approaches to mimic the L-Box [29,42,43] while others have used meshless methods [5,44,45]. Both methods have been used to examine other devices, including the V-funnel and the LCPC-box [35,47]. There is usually a steady evolution in the methodologies used to simulate rheological devices in computational analysis. Finite Element approaches (FEMs) and Finite Difference Methods (FDMs) are two examples of mesh-based approaches that are first used. About ten years down the road, these gadgets are examined with meshless approaches like the Discrete Element Method (DEM), and then, another ten years after that, with more sophisticated methods like Dissipative Particle Dynamics (DPD). Rising computational costs and the availability of faster computer resources are the primary drivers of this sequential adoption.

Computational simulations are becoming more commonplace in contemporary engineering with the creation of novel rheological devices. This helps with performance analysis and interpreting results. For instance, ANSYS Fluent, which uses the Finite Volume Method (FVM), was recently used by Nerella and Mechtcherine to study the flow field in the SLIPER rheometer. Also, older machinery that was too complicated to be analyzed using mesh-based approaches is now being subjected to numerical simulations. As an example, Wallevik & Wallevik used supercomputers to model the local energy dissipation distribution in W/m^3, which shows how a concrete mixing truck may be used as a rheometer using FVM. Computational fluid dynamics (CFD) models of casting procedures help describe the necessary rheological parameters to guarantee correct formwork filling, much how numerical models in structural mechanics assist civil engineers in determining minimum mechanical qualities for reinforced concrete. The viscoplastic divided element method (VDEM) was shown to be useful for modeling concrete flow in reinforced structures in early research by Tanigawa and Mori. Kitaoji et al. provided more evidence that 2D VDEM is a useful tool for simulating the flow of new concrete into a porous wall. Industrial casting has also made use of numerical simulations, with high-strength concrete pre-cambered composite beams being one example. By studying several rheological factors, the optimal values for casting were determined.

Because of their computational efficiency, mesh-based techniques like FVM or FDM are commonly used in large-scale casting simulations. Therefore, meshless approaches are rarely brought up in this setting. Modeling the contact between new concrete and ambient air is of utmost importance in mesh-based techniques. A free interface method, often called an open boundary or a moving border, is usually used to handle this. Both moving mesh (Lagrangian) and fixed mesh (Eulerian) numerical techniques may be used to handle such interactions. Despite its usefulness for interface definitions, the moving mesh method suffers from significant distortions when subjected to substantial deformations. Accordingly, the Volume-of-Fluid (VOF), level-set, and marker-and-cell approaches, among others, frequently favor the Eulerian approach. When trying to capture the free interface between fresh concrete and ambient air inside a defined mesh, the VOF approach is commonly utilized. An equation for phase transfer does this. More specifically, there are direct approaches and reconstruction methods inside the VOF framework. In contrast to geometric interface reconstruction approaches, direct methods use specific discretization techniques, sometimes called compressive differencing schemes, to retain a clearly defined border rather than directly representing the interface. In order to replicate the flow of freshly mixed concrete in molds or formwork, direct approaches for free-interface treatment have been utilized. One research used this method on a whole wall section (10 m in length, 30 cm in thickness, and 3.4 m in height), factoring in inhomogeneous aggregate distribution estimates as well.

As a function of material qualities and extrusion settings, predicting filament form is crucial in extrusion-based additive manufacturing of cementitious materials. A theoretical filament cross-section based on mass conservation is usually assumed by the slicing software used by concrete printers. Nevertheless, even minute variations from this theoretical thickness might build up over several layers, leading to printed structural geometric errors or, worse, printing failures. The resulting printed product is also susceptible to deformations of the filament along the deposition axis, such as ripping, cracking, or bending. Therefore, it is crucial to be able to numerically forecast these difficulties.

When it comes to additive manufacturing of cementitious materials at the nozzle level, two asymptotic regimes are in play. An unsheared, rigid filament that maintains the geometry of the nozzle upon extrusion is produced in the "infinite brick extrusion" regime when the material's yield stress is large compared to the stresses caused by gravity and pumping. On the other hand, prior to extrusion, the material undergoes complete shearing in the "free-flow deposition" regime as a result of local mixing effects or low yield stress. Here, the relationship between yield stress and gravity dictates the filament shape.

In both cases, local compression of the extruded material occurs until it reaches the nozzle gap, which occurs when the gap is less than the filament thickness. The structure may be deformed or its stability jeopardized if this compression causes overshoot pressures on the lower layers. For additive manufacturing structural integrity and optimal concrete extrusion processes, it is critical to understand and correctly simulate these impacts.

A well-controlled and consistent filament dimension may be achieved for both extrusion regimes by keeping the extrusion flow rate at a nominal speed, which is defined as the product of the nozzle displacement velocity and the predicted filament cross-section. Nevertheless, filament instability, which can manifest as buckling or ripping, can occur when this nominal speed deviates from the norm [66]. Pressive strains form in the filament connecting the nozzle's discharge port to the depositing surface as the flow rate surpasses the specified value. These strains may cause localized buckling in materials with different degrees of elasticity and different lengths of free filaments. On the flip side, tensile tensions become apparent and the probability of filament ripping increases when the flow rate drops below the nominal level.

Many models have been developed for filament variations in the longitudinal axis (buckling and tearing) and in the cross-sectional direction. Few numerical studies have focused on additive manufacturing of concrete at the filament size, even though controlling and predicting the form of filaments is crucial. Before delving into numerical modeling attempts for polymer-based additive manufacturing and how they pertain to concrete extrusion, this section provides a comprehensive assessment of the relevant research.

A numerical simulation of concrete extrusion, where the process is represented by a moving nozzle simulation. On the other hand, a velocity boundary condition applied to the substrate during deposition can also provide the same result. The results, which contradict the theoretical cross-section anticipated by typical slicing software; the deposition of non-Newtonian fluids results in thicker filaments than Newtonian fluids, which in turn produce thicker filaments.

The Particle Finite Element Method (PFEM) is used to simulate three layers of additively built concrete in a more comprehensive simulation technique[68]. PFEM is a mesh-based approach that uses a Perzyna formulation to approximatively model the behavior of a Bingham material inside an updated Lagrangian framework. Because of this, the model may take the elastic regime into consideration prior to giving in. Aside from the effect of deposited layers on the cross-section of freshly printed layers, the results show that accumulated weight can change the form of underlying layers if the yield stress of the material is surpassed. To improve shape control and anticipate buildability, these simulations offer significant information that may be integrated into slicer software (see Section 2.2). In addition, simulations with several layers are useful for measuring surface roughness and guiding the development of smoothing or tooling systems. Furthermore, they corroborate hypotheses about interlayer bonding, an essential property for structural stability.

Using the Finite Volume Method (FVM) and a reconstruction-based Volume of Fluid (VOF) approach, a recent numerical study [66] systematically examined the influence of process parameters (nozzle speed and extrusion rate), filament and nozzle geometry (size and deposition height), and material properties (elastic stiffness, yield stress, and viscosity). According to the research, the filament cross-section stays constant and the stress distribution is uniform at minimal speed. Nevertheless, filament buckling happens when the extrusion rate is higher than the nozzle speed. Geometric buckling and stress inhomogeneity are made worse by increasing the deposition height and material qualities including stiffness, yield stress, and viscosity. The filament ripping happens when the extrusion rate is less than the nozzle speed, and the odds of tearing are much higher when the material critical strain is lowered [66].

A comparison of numerical and experimental findings for various nozzle-to-substrate gap settings in the context of polymer-based extrusion additive manufacturing [69]. Because of the creeping flow and quick solidification of the polymer post-deposition, this work showed that first-order effects in polymer printing may be captured by considering the material as Newtonian. When it comes to concrete printing, a comparable numerical method may be really useful for choosing the right constitutive model (such visco-plastic or elasto-visco-plastic) to match up with experimental results, especially when it comes to free-flow deposition.

Numerical models have been used to detect under- and overfill regimes in polymer printing of corners. Because of the difficulties inherent in printing sharp corners, the majority of buildings made of additively generated concrete have rounded edges. To compensate for this, simulation tools may optimize deposition techniques and nozzle design for certain corner forms. The capacity to perform several simulations simultaneously is a major benefit of numerical modeling, which allows for the quick and cost-effective evaluation of many different situations.

Results of meso-scale printed polymer structures, both actual and computational. These structures show that numerical predictions and physical prints are very well-matched; they are made by depositing several strands to construct a representative volume element. In the end, these models improve the bonding between and within layers of printed concrete by allowing for the identification and management of porosity.

In additive manufacturing based on extrusion, the successive layers of cementitious material build up tensions caused by gravity, which can cause the structure to collapse if conventional formwork is not used. The elastic buckling and plastic collapse are the two main failure modes. In contrast to plastic collapse, which happens when the material achieves its yield stress, elastic buckling is caused by geometric instability.

Many variables, like as the shape of the item, the characteristics of the material, and the settings used during the printing process, affect the likelihood and kind of failure in 3D concrete printing. For example, when the yield stress is exceeded, self-weight-induced collapse usually happens near the base of a straight wall. Finding the precise spot of failure is trickier when dealing with complicated geometries, printing with several materials, or situations where process parameters change as the item rises. It is critical to regulate the elastic deformations induced by consecutive layers even when failure does not occur in order to retain the geometric integrity of the final product. In light of these intricacies, numerical simulations provide useful information about the procedure.

Numerical models depend on visco-plastic behavior at the filament scale (described before). The elastic-plastic behavior and structuration kinetics of the early-age material become essential, however, when examining a whole printed piece. Elastic characteristics, such Poisson's ratio and Young's modulus, are useful for evaluating elastic deformations and making failure predictions based on elastic buckling. The yield stress of the material must be defined in order to do plasticity analysis. The structuralization rate (Athix) is a common way to define the rate of change in the elastic and plastic characteristics of thixotropic cementitious materials, which are at rest for the majority of the printing process. Either a linear or nonlinear trend might be observed in this rate.

Numerical simulations rely on experimentally determined input parameters, such as the rate of structuration and elastic and plastic material characteristics. Several experimental approaches have been suggested for this purpose. Experts in the field have taken cues from soil mechanics and tested hardened concrete to go beyond the traditional approaches to evaluating newly mixed cementitious materials. Measurement of material characteristics within the usual timescale of additive manufacturing procedures has been facilitated by various adaptations of conventional mechanical and rheological experiments.

Numerical models need to take into consideration not just material qualities but also critical process factors such as printing speed, contour length, and overall object development rate, as well as the ever-increasing geometry and the stress levels caused by gravity. Quickly assess the potential for structural collapse in 3D-printed walls using a parametric model put out by Suiker. 3D concrete printing relies on a plethora of factors, but our model streamlines the process down to only five dimensionless variables: two associated with plastic collapse and three with elastic buckling. Under various process settings, geometries, boundary conditions, and structuration rates, design graphs may be generated to swiftly ascertain the object's failure height and the point at which elastic buckling gives way to plastic collapse.

In order to simulate detailed 3D-printed structures, a numerical model based on the Finite Element Method (FEM) has been created for increasingly complicated geometries. Here, a geometrically nonlinear analysis is carried out, taking into account the progressive addition of layers and the associated gravity loads. As the simulation progresses, the material's elastic and plastic characteristics change, and the activation speed is determined by the process parameters and contour length. Because of this, FEM simulations may be used to explore the consequences of changes in geometry, material behavior, and process parameters in a systematic way.

Experiments with printing various linear wall constructions with varying contour lengths have verified both models. Time and failure type were captured using high-resolution photography, proving that numerical models can forecast deformation and critical failure height in tiny objects with reasonable accuracy. Nevertheless, these forecasts become less accurate for bigger structures or printing processes that last a long time. Both experimental characterisation and numerical simulations have not yet taken into account the long-term impacts of thermal heating on the 3D printer setup, which is the reason for this mismatch.

These findings highlight the need to advance numerical modeling techniques by integrating thermal and moisture effects. Such enhancements would further improve prediction accuracy and provide a foundation for assessing interlayer bond strength. Since bond strength is significantly influenced by material age and environmental conditions during printing, extending numerical models to include these factors would be highly beneficial in optimizing the additive manufacturing process.

3. Conclusion

Here's the rewritten conclusion in a structured flow while maintaining its original meaning:

- The first part of this paper provided an overview of the state-of-the-art numerical simulations in concrete flow. Various numerical techniques were explored, along with their applications in standard processes.
- The second part examined existing literature on numerical simulations of extrusion-based additive manufacturing for concrete. Both the extruded filament scale and the overall printed object scale were analyzed.
- Numerical modeling tools can effectively detect and predict structural failures during the printing process. These failures can occur at different scales, including the nozzle, individual layers, or the entire printed structure.
- Numerical simulations systematically evaluate the influence of material, geometric, and process parameters. They serve as valuable tools for optimizing printing quality and material efficiency.

References

1. N. Roussel, Rheology of fresh concrete: from measurements to predictions of casting processes, Mater. Struct. 40 (10) (2007) 1001–1012.

2. N. Roussel, S. Staquet, L.D. Schwarzentruber, R. Le Roy, F. Toutlemonde, SCC casting prediction for the realization of prototype VHPC-precambered composite beams, Mater. Struct. 40 (9) (2007) 877–887.
3. G. Ovarlez, N. Roussel, A physical model for the prediction of lateral stress exerted by self-compacting concrete on formwork, Mater. Struct. 39 (2) (2006) 269–279.
4. J.D. Anderson, Computational Fluid Dynamics, the Basics with Applications, McGraw-Hill, Inc, USA, 1995.
5. N. Roussel, A. Gram (Eds.), Simulation of Fresh Concrete Flow, State-of-the Art Report of the RILEM Technical Committee 222-SCF, Springer, 2014.
6. J. Tu, G. Heng, Y.C. Liu, Computational Fluid Dynamics - a Practical Approach, Butterworth-Heinemann, 2008.
7. O. Svec, J. Skocek, H. Stang, M.R. Geiker, N. Roussel, Free surface flow of a suspension of rigid particles in a non-Newtonian fluid: a lattice Boltzmann approach, J. Non-Newtonian Fluid Mech. 179-180 (2012) 32–42.
8. K. Stefan, R. Gunther, CFD-simulations in the early product development, Procedia CIRP 40 (2016) 443–448.
9. V. Mechtcherine, A. Gram, K. Krenzer, J.-H. Schwabe, S. Shyshko, N. Roussel, Simulation of fresh concrete flow using discrete element method (DEM): theory and applications, Mater. Struct. 47 (2014) 615–630.
10. C. O'Sullivan, Particulate Discrete Element Modelling - a Geomechanics Perspective, Spon Press, USA, 2011.
11. Pep Espanol, Patrick Warren, Perspective: dissipative particle dynamics, J. Chem. Phys. 146 (150901) (2017) 1–16.
12. N.S. Martys, Study of a dissipative particle dynamics based approach for modelling suspensions, J. Rheol. 49 (2005) 401–424.
13. S.H. Musavi, M. Ashrafizaadeh, A mesh-free lattice Boltzmann solver for flows in complex geometries, Int. J. Heat Fluid Flow 59 (2016) 10–19.
14. O. Zikanov, Essential Computational Fluid Dynamics, John Wiley & Sons, Inc, USA, 2010.
15. C.A.J. Fletcher, Computational Techniques for Fluid Dynamics, 2nd edition, Springer Series in Computational Physics I Springer-Verlag, Germany, 1990.
16. H.K. Versteeg, W. Malalasekera, An Introduction to Computational Fluid Dynamics the Finite Volume Method, 2nd ed, Pearson Education Limited, England, 2007.
17. K. Vasilic, A. Gram, J.E. Wallevik, Numerical simulation of fresh concrete flow: insight and challenges, RILEM Technical Letters 4 (2019) 57–66.
18. J.H. Ferziger, M. Peric, Computational Methods for Fluid Dynamics (3rd ed), Springer-Verlag, 2002.
19. H. Mori, Y. Tanigawa, Simulation methods for fluidity of fresh concrete, Memoirs of the School of Engineering, vol. 44, Nagoya University, 1992, pp. 71–133.
20. Y. Kurokawa, Y. Tanigawa, H. Mori, Y. Nishinosono, Analytical study on effect of volume fraction of coarse aggregate on Bingham's constants of fresh concrete, Transactions of the Japan Concrete Institute 18 (1996) 37–44.
21. G. Christensen, Modelling the Flow of Fresh Concrete: The Slump Test, PhD thesis Princeton University, 1991.
22. A. Gharehbaghi, B. Kaya, H. Saadatnejadgharahassanlou, Numerical simulation of two dimensional unsteady flow by total variation diminishing scheme, International Journal of Engineering & Applied Sciences (IJEAS) 8 (3) (2016) 1–14.

Advances in Additive Manufacturing Technologies – Gurusamy Pathinettampadian et al. (eds)
© 2026 Taylor & Francis Group, London, ISBN 978-1-041-16687-0

Estimate of Material Deformation by Experimentation in the Context of Additive Printing of Concrete on a Large Scale

S. Mercy[1], Suresh Kumar M.[2]
Lecturer, Department of Civil Engineering,
Sankar Polytechnic College,
India

Thangavelu Eswary Devi[3]
Assistant Professor, Department of Civil Engineering,
Rajalakshmi Engineering College,
India

S. Raja Muniyasamy[4]
Assistant Professor, Department of Civil Engineering,
Thamirabharani Engineering College,
Chennai, India

M.A. Raja[5]
Assisstant Professor, Department of Civil Engineering,
Meenakshi Sundararajan Engineering College,
Chennai, India

Balaji Krishnabharathi[6]
Assistant Professor, Department of Mechanical Engineering,
Chennai Institute of Technology,
Chennai, India

◆ **Abstract:** In the last 10 years, there has been a lot of buzz around cementitious material additive manufacturing (AM). Research in this area has drawn experts from a wide range of disciplines, including architecture, engineering, and materials science, all with a same goal: to improve existing technologies. Even if "concrete printing" or "AM of concrete" is thrown around a lot, the processes that are used today can't handle coarse particles. An important goal of this field's study is to create a mortar made of Portland cement that meets the requirements of architectural applications in terms of its rheological, hardening, and strength characteristics. The printed geometries' precision is also affected by the mortar's fresh and hardened qualities, as well as its deformation behavior. To overcome problems caused by deformation, changes to toolpaths and printing methods are required. By examining the effects of time, layer count, and bead count on layer breadth and height, this study delves into the deformation behavior of a printed concrete mix that has already been confirmed. To simulate the behavior of the material, we do a battery of experiments and use regression analysis. To improve the accuracy of printed structures, the obtained equations may be used in toolpath design to reduce the impacts of deformation. Building a more versatile toolpath generator will be the primary goal of future studies that aim to establish a correlation between material qualities and deformation behavior.

◆ **Keywords:** Experimental deformation analysis, Material deformation testing, Large-scale experimentation, In-situ testing, Real-time monitoring

[1]mercyspc@gmail.com, [2]sureshspccivil2020@gmail.com, [3]eswarydevi.t@rajalakshmi.edu.in, [4]ramunish58@gmail.com, [5]antojosephraja@gmail.com, [6]balajikrishnabharathi@gmail.com

DOI: 10.1201/9781003685906-109

1. Introduction

Recently, there has been a lot of buzz about how additive manufacturing (AM) may be used in the construction business. Architects and engineers are investigating its feasibility for automating the building of unsupported, free-form buildings. Although it is still in its infancy, AM technology has great promise for the construction industry because to its many benefits, such as reduced costs, increased efficiency, and precision. Because of these advantages, there have been a lot of research efforts to improve AM technology, especially for cementitious materials. The printed substance is just a mortar based on Portland cement, but the word "concrete" is used frequently to describe it in the literature even though the technology does not currently allow for printing ready-mixed concrete with coarse particles. This research uses the same language.

Extrusion printing heads deposit linear filaments or beads of concrete along a predetermined route; this is the main method used in most concrete AM processes. The extruder's trajectory, which deposits material layer by layer, is determined in large part by the toolpath. Toolpath design is an essential part of process planning as it influences printing time, geometric correctness, strength, and stiffness. There are two primary parts to a toolpath design: [1] interior filling, which specifies the method by which extruded filaments are used to fill layers, and [2] linking sequence, which establishes the printing order of layers without halting the deposition of material. When it comes to material extrusion-based AM, the main goals of toolpath design are to maximize printing efficiency, minimize travel time by avoiding redundant sub-paths, and achieve excellent extrusion quality [3].

One major obstacle to overcome in additive manufacturing (AM) of concrete, in contrast to plastic 3D printing, is the material's deformability. It is important that the newly mixed concrete be "buildable" enough to retain its form when subjected to stresses and "flowable" enough to allow for smooth extrusion [39]. The primary causes of deformations in printed concrete layers are their own weight, the weight of layers below it, and the pressure applied during extrusion [4]. When the material is of good enough quality, self-weight deformation is usually not noticeable. But, the weight of the extra layers and the pressure used during extrusion might cause undesirable form distortions.

The duration between successive layer depositions is a critical component impacting deformation. The lower layer might harden and distort less with a greater time interval. Nevertheless, research has demonstrated that bond strength can be compromised due to significant interlayer time gaps, which in turn affects the printed element's structural integrity [5]. As a result, toolpath design relies heavily on finding the ideal interlayer time gap to reduce distortion while keeping structural characteristics intact [40].

The growing stresses and extrusion pressure can cause substantial shape distortions as the number of layers and beads adds up, if time-dependent deformation is not taken into consideration [6]. Improving printed geometries' shape accuracy may be achieved by studying these parameters' influences and adjusting toolpath design accordingly. "Beads" in this context mean the individual linear filaments that come out of the nozzle (e.g., B1, B2, etc.), while "layers" mean clusters of neighboring filaments that are deposited at the same level (e.g., L1, L2, etc.).

In order to enhance shape quality through improved toolpath design, this work seeks to investigate the interplay between time, the quantity of beads, and the number of layers as it pertains to material deformation. The second section discusses the additive production of concrete on a big scale. In Section 3, the printing technology and study-related concrete mix characteristics are detailed [37].

2. Experimental Methodology

Deformations in 3D-printed concrete buildings' layer breadth and height were examined in this study with a battery of tests. Mainly, we wanted to see how deformation was affected by layer count, bead count, and interlayer time intervals [7].

The primary objective of the experiment was to compare the effects of varying layer and bead counts on the distortion of layer widths caused by the weight of higher layers. Printing direction along the X and Y axes influences material deformation, according to previous study by Ashrafi et al. [20]. In this investigation, we found that when printing in manual mode, the pace might vary somewhat depending on the orientation and direction of the print due to inherent constraints in the robotic arm's movement. On the other hand, when set to automatic, the robot's pace was constant in every direction [8]. All tests were performed automatically since the extruder speed has a direct impact on the amount of material deposited and the precision of the shapes [35].

To further ensure that printing orientation did not impact material deposition or final form, Test 1 was also developed to check if the robotic arm's speed stayed constant in autonomous mode. This was examined by printing one to five material beads in five layers along the X and Y axes show that the printed beads may be joined to allow for continuous printing; their length was 100 cm. In every trial, the toolpath design kept the layer height at 15 mm. It was necessary to print each item three times in order to get the mean (μ) and standard deviation (σ). For the purpose of determining the overall layer width after curing, the 100 cm hardened beads were cut into ten equal pieces measuring 10 cm each. The widths of these sections were measured in order to arrive at the final result. Taking the average of the layer widths of all of the specimens allowed for the determination of the overall layer width [9]. In this particular experiment, the breadth of the layer served as the dependent variable, while the orientation and the number of beads served as the independent factors. The number of layers served as the independent variable inside this experiment. According to the information that the average percentage error in layer width decreases as the number of layers and beads increases in both the X and Y axes [10].

Finding the ideal time interval to print the next layer without deforming the one below was the goal of the second test. Fifty layers were used to print a 100 cm x 100 cm square in order to accomplish this. This square shape prevented material from being deposited directly on newly extruded layers right thereafter, allowing for a continuous [11] toolpath. The distortion of the layer height was measured after printing each successive layer, with a particular emphasis on layer 3. Fresh concrete has a tendency to dip at the edges, which makes precise height measurements difficult when taken from an elevation viewpoint. During printing, two thin plastic cards were inserted—one under layer 3 and one on top—to circumvent this problem. In order to determine μ and σ, these cards were placed at three distinct locations on the layer [12]. You can see the card arrangement. In order to record the printing process in its entirety, a video camera was installed in front of the primary front edge. Before adding the following layer, we retrieved individual frames from the movie after printing each one. The layer height at the three places that were chosen was measured using these photos [38]. You may find a summary of the important printing system variables. After printing all fifty layers.

Trial runs showed minor behavioral changes across batches, even though the concrete mix design was equal in both tests. The discrepancies in batch preparation dates (winter for Test 1 and summer for Test 2), as well as relative humidity and temperature, were shown to be the causes of these disparities [13]. Furthermore, the test days' ambient circumstances were different. Layer width and height were adjusted in the printing system to accommodate for these variances. This included adjusting the water flow rate and robot speed. The goal was to achieve uniformity between the two experiments [36]. A single bead was printed in a single layer at a 15 mm nozzle distance from the bed after the water flow rate was adjusted to preserve extrudability. After that, we assessed the printed bead's layer width and made adjustments to the robot's speed so it would match the single-bead width from Test 1 [14].

Following the determination of the time interval at which deformation became insignificant, the last test examined how different interlayer time intervals affected the deformation of layer heights due to the weight of succeeding layers. The printed six-layer square specimens of varying sizes. The squares stood for various intervals of time between the layers. The objective was to take readings at intervals of 10 cm along one side of each square to determine the distance from the nozzle to the preceding layer. In order to find μ and σ, it was necessary to print each of the six squares at least three times to guarantee statistical reliability [15]. For Test 3, the pertinent printing system variables. For the sake of uniformity between trials, these variables were fine-tuned using the same method as Test 1.

3. Results and Discussion

ANOVA was performed on the data in order to assess the impact that each of the three independent factors had on the vertical layer height [16]. The p-value of the independent variables and the interactions between them is displayed in the "Sig." column, which is where the presentation of the results takes place. According to the findings of the research, it has been determined that there exists a statistically significant impact on the width of the layer from both the number of layers and the number of beads [17]. The p-values for both variables are less than 0.05. Due to the fact that the p-value is more than 0.05, the layer width is not impacted by the direction of the printing. For this reason, the data from the x and y orientations were combined in order to enhance the statistical reliability of the analyses [18]. Furthermore, the p-value for the three-way interaction term is 0.670, which indicates that there is no significant three-way interaction influence among the triad orientation, the number of beads, and the number of layers used in the experiment. The percentage error that is the lowest across all permutations is found in Layer 5, which has five beads, which displays the average percentage error throughout all testing instances [19]. This is because the printed concrete is able to solidify with less distortion since the load on layer 5 is lowered and the interlayer time interval is extended by the five beads.

We can see the % error in width for various bead sets. The data points are connected linearly by trend lines. The negative slope of these trend lines indicates that the % error in width reduces with increasing layer number. This occurs because the stress is lessened with each layer that follows [21]. Further evidence that increasing the number of beads decreases layer deformation is the fact that the trend lines' slopes get flatter with increasing bead counts. To find an equation that could be used to estimate the breadth of a layer given its number of layers and the amount of beads, a linear regression analysis was performed [22]. Using the layer thickness as a dependent variable and the number of layers and beads as independent variables, this statistical model attempts to determine the connection between the two [33]. Measurements of material deformation in layer 3 were taken in Test 2 to ascertain the necessary time gap prior to printing the subsequent layer in order to avoid distortion. This indicates that deformation stops after around 34 seconds [23]. This test was a first step towards the third test. In order to determine the impact on layer height distortion, Test 3 measured the distance between the nozzle and the previously printed layer at six different interlayer time intervals [34]. The smallest sample with the shortest interlayer time gap, sample 1 (S1), had the most deformation. Confirming that longer time intervals led to fewer deformations, deformation decreased as the interlayer time gap grew [24].

An equation for estimating the distance from the nozzle to the preceding layer and the distortion of the layer height as a function of printing time and the number of layers was established using linear regression analysis. The predicted values from Equation 2 were compared with the experimental findings, and the results showed that the equation had an average percentage error of 1.23% when predicting the nozzle distance [25]. The deformation-free printing time was then determined by dividing the total printing time for all samples by 34 seconds. The number of layers required to cease deformation in each sample was found using this computation. Each value was raised to the closest whole layer [26]. All samples beyond S2 ceased deforming after printing two layers. This suggests that deformation is mostly affected by the layer below it before stabilizing for any particular layer.

Take sample 3 as an example. According to Table 109.16, the average distance from the second layer's nozzle to the preceding one was 15.1 mm. This suggests that the first layer was deformed by 0.1 mm. The height of layer 1 remains at 14.9 mm even after printing layer 3, as deformation stops after two layers. Nevertheless, as in layer 2 originally measures 15.1 mm but deforms to 15.0 mm when subjected to the stress of layer 3[27]. After two more layers are printed, the prior layer stabilizes, and this pattern repeats with each successive layer. Deformation in specimens bigger than 70 × 70 cm is not significant, according to the results, especially when printing each layer takes more than 18 seconds. Sample 3 (S3) underwent a second round of testing to confirm this hypothesis. A square measuring 70 × 70 cm was printed in many layers. After each succeeding layer was produced, the height of the layers was measured by placing thin, stiff cards under and above the third layer [28]. Even after printing the top layer twice, the deformation was still just 0.2 mm, lending credence to the theory that deformation stops after 34 seconds [29].

A key time gap of around 34 seconds is thus required before deformation starts. Because printing layers on a construction site often takes more than 34 seconds, this discovery has important implications for large-scale additive manufacturing of concrete [30]. But it's possible that the precise moment when deformation ends might be a little less than 33.9 seconds. A specimen that was 5% smaller than S6 was used to repeat Test 2 in order to refine this estimate. After printing each successive layer of a multi-layered 95 × 95 cm square, the average height of layer 3 was recorded. Table 109.18 shows that even after printing a further layer—a process that took 24.2 seconds—layer 3 still showed signs of distortion [31]. Deformation stops between 24.2 and 33.9 seconds, according to the overall findings of these experiments. The precise time interval might be determined with more testing on intermediate-sized specimens; nevertheless, this uncertainty is small in comparison to the printing periods required for most construction-scale applications, which surpass 34 seconds per layer [32].

4. Conclusion

The goal is to improve the toolpath in order to compensate for deformation. Used a pump, an industrial-sized mixer, and a robotic arm. Created a unique 3D-printable concrete blend using Portland cement.

In the test, we created a regression model to forecast the nozzle distance from printing time and layers, and we found that interlayer time intervals impact the deformation of layer height. Verification using a multi-layer specimen, which verifies that the expected pattern of deformation alone applies to the layer directly above it.

The outcomes are useful for toolpath design, which in turn affects dimensions, wall thickness, infill patterns, and travel movements. Large-scale construction elements (walls, pavements) may not require deformation compensation due to long path lengths. Small structural parts (e.g., wall sections between openings) require compensation due to shorter path lengths. Toolpath design must consider stop-and-go features to ensure accuracy and efficiency.

References

1. P. Carneau, R. Mesnil, N. Roussel, O. Baverel, An exploration of 3D printing design space inspired by

masonry, Proc. IASS Annu. Symp. (2019) 1–9, https://doi. org/ 10.5281/zenodo.3563672.

2. R.A. Buswell, W.R. Leal de Silva, S.Z. Jones, J. Dirrenberger, 3D printing using concrete extrusion: a roadmap for research, Cem. Concr. Res. 112 (2018) 37–49, https://doi.org/10.1016/j.cemconres.2018.05.006.
3. D. Ding, Z. Pan, D. Cuiuri, H. Li, A practical path planning methodology for wire and arc additive manufacturing of thin-walled structures, Robot. Comput. Integr. Manuf. 34 (2015) 8–19, https://doi.org/10.1016/j.rcim.2015.01.003.
4. S.H. Choi, W.K. Zhu, A dynamic priority-based approach to concurrent toolpath planning for multi-material layered manufacturing, CAD Comput. Aided Des. 42 (2010) 1095–1107, https://doi.org/10.1016/j.cad.2010.07.004.
5. Y. Jin, J. Du, Z. Ma, A. Liu, Y. He, An optimization approach for path planning of high-quality and uniform additive manufacturing, Int. J. Adv. Manuf. Technol. 92 (2017) 651–662, https://doi.org/10.1007/s00170-017-0207-3.
6. Y. an Jin, Y. He, J. zhong Fu, W. feng Gan, Z. wei Lin, Optimization of tool-path generation for material extrusion-based additive manufacturing technology, Addit. Manuf. 1 (2014) 32–47, https://doi.org/10.1016/j.addma.2014.08.004.
7. A. Kazemian, X. Yuan, E. Cochran, B. Khoshnevis, Cementitious materials for construction-scale 3D printing: laboratory testing of fresh printing mixture, Constr. Build. Mater. 145 (2017) 639–647, https://doi.org/10.1016/j.conbuildmat.2017.04.015.
8. T.T. Le, S.A. Austin, S. Lim, R.A. Buswell, A.G.F. Gibb, T. Thorpe, Mix design and fresh properties for high-performance printing concrete, Mater. Struct. Constr. 45 (2012) 1221–1232, https://doi.org/10.1617/s11527-012-9828-z.
9. T.T. Le, S.A. Austin, S. Lim, R.A. Buswell, R. Law, A.G.F. Gibb, T. Thorpe, Hardened properties of high-performance printing concrete, Cem. Concr. Res. 42 (2012) 558–566, https://doi.org/10.1016/j.cemconres.2011.12.003.
10. I. Gibson, T. Kvan, L.W. Ming, Rapid prototyping for architectural models, Rapid Prototyp. J. 8 (2002) 91–99, https://doi.org/10.1108/13552540210420961.
11. P. Wu, J. Wang, X. Wang, A critical review of the use of 3-D printing in the construction industry, Autom. Constr. 68 (2016) 21–31, https://doi.org/10.1016/j.autcon.2016.04.005.
12. R. Rael, V. San Fratello, Printing Architecture: Innovative Recipes for 3D Printing, Chronicle Books, 2018.
13. S. Lim, T. Le, J. Webster, R. Buswell, S. Austin, A. Gibb, T. Thorpe, Fabricating construction components, 2009.
14. B. Khoshnevis, Automated construction by contour crafting - related robotics and information technologies, Autom. Constr. 13 (2004) 5–19, https://doi.org/ 10.1016/j.autcon.2003.08.012.
15. O. Kontovourkis, G. Tryfonos, C. Georgiou, Robotic additive manufacturing (RAM) with clay using topology optimization principles for toolpath planning: the example of a building element, Archit. Sci. Rev. 63 (2020) 105–118, https://doi. org/10.1080/00038628.2019.1620170.
16. R.A. Buswell, W.R. Leal de Silva, S.Z. Jones, J. Dirrenberger, 3D printing using concrete extrusion: a roadmap for research, Cem. Concr. Res. 112 (2018) 37–49, https://doi.org/10.1016/j.cemconres.2018.05.006.
17. T. Wangler, E. Lloret, L. Reiter, N. Hack, F. Gramazio, M. Kohler, M. Bernhard, B. Dillenburger, J. Buchli, N. Roussel, R. Flatt, Digital concrete: opportunities and challenges, RILEM Tech. Lett. 1 (2016) 67, https://doi.org/10.21809/rilemtechlett.2016.16.
18. N. Roussel, Rheological requirements for printable concretes, Cem. Concr. Res. 112 (2018) 76–85, https://doi.org/10.1016/j.cemconres.2018.04.005.
19. D. Rangeard Perrot, A. Pierre, Structural built-up of cement-based materials used for 3D-printing extrusion techniques, Mater. Struct. Constr. 49 (2016) 1213–1220, https://doi.org/10.1617/s11527-015-0571-0.
20. Y.W.D. Tay, G.H.A. Ting, Y. Qian, B. Panda, L. He, M.J. Tan, Time gap effect on bond strength of 3D-printed concrete, Virtual Phys. Prototyp. 14 (2019) 104–113, https://doi.org/10.1080/17452759.2018.1500420.
21. Q. Yuan, D. Zhou, B. Li, H. Huang, C. Shi, Effect of mineral admixtures on the structural build-up of cement paste, Constr. Build. Mater. 160 (2018) 117–126, https://doi.org/10.1016/j.conbuildmat.2017.11.050.
22. V.N. Nerella, M.A.B. Beigh, S. Fataei, V. Mechtcherine, Strain-based approach for measuring structural build-up of cement pastes in the context of digital construction, Cem. Concr. Res. 115 (2019) 530–544, https://doi.org/10.1016/j.cemconres.2018.08.003.
23. B. Zareiyan, B. Khoshnevis, Interlayer adhesion and strength of structures in contour crafting - effects of aggregate size, extrusion rate, and layer thickness, Autom. Constr. 81 (2017) 112–121, https://doi.org/10.1016/j.autcon.2017.06.013.
24. E. Keita, H. Bessaies-Bey, W. Zuo, P. Belin, N. Roussel, Weak bond strength between successive layers in extrusion-based additive manufacturing: measurement and physical origin, Cem. Concr. Res. 123 (2019), 105787, https:// doi.org/10.1016/j.cemconres.2019.105787.
25. N. Ashrafi, J.P. Duarte, S. Nazarian, N.A. Meisel, Evaluating the relationship between deposition and layer quality in large-scale additive manufacturing of concrete, Virtual Phys. Prototyp. 14 (2019) 135–140, https://doi.org/10.1080/ 17452759.2018.1532800.
26. Y.W. Tay, B. Panda, S.C. Paul, M.J. Tan, S.Z. Qian, K.F. Leong, C.K. Chua, Processing and properties of construction materials for 3D printing, Mater. Sci. Forum 861 (2016) 177–181, https://doi.org/10.4028/www.scientific.net/MSF.861.177.
27. S. Lim, R.A. Buswell, T.T. Le, S.A. Austin, A.G.F. Gibb, T. Thorpe, Developments in construction-scale additive manufacturing processes, Autom. Constr. 21 (2012) 262–268, https://doi.org/10.1016/j.autcon.2011.06.010.
28. Z. Li, M. Hojati, Z. Wu, J. Piasente, N. Ashrafi, J.P. Duarte, S. Nazarian, S.G. Bil′en, A.M. Memari, A. Radli′nska,

Fresh and hardened properties of extrusion-based 3D printed cementitious materials: a review, Sustain 12 (2020) 1–33, https://doi.org/10.3390/su12145628.

29. J. Kruger, S. Zeranka, G. van Zijl, 3D concrete printing: a lower bound analytical model for buildability performance quantification, Autom. Constr. 106 (2019), 102904, https://doi.org/10.1016/j.autcon.2019.102904.
30. A.S.J. Suiker, Mechanical performance of wall structures in 3D printing processes: theory, design tools and experiments, Int. J. Mech. Sci. 137 (2018) 145–170, https://doi.org/10.1016/j.ijmecsci.2018.01.010.
31. R.J.M. Wolfs, F.P. Bos, T.A.M. Salet, Early age mechanical behaviour of 3D printed concrete: numerical modelling and experimental testing, Cem. Concr. Res. 106 (2018) 103–116, https://doi.org/10.1016/j.cemconres.2018.02.001.
32. R.J.M. Wolfs, A.S.J. Suiker, Structural failure during extrusion-based 3D printing processes, Int. J. Adv. Manuf. Technol. 104 (2019) 565–584, https://doi.org/ 10.1007/s00170-019-03844-6.
33. Z. Malaeb, H. Hachem, A. Tourbah, T. Maalouf, N. El Zarwi, F. Hamzeh, 3D concrete printing: machine and mix design, Int. J. Civ. Eng. Technol. 6 (2015) 14–22, http://www.researchgate.net/profile/Farook_Hamzeh/publication/280488795_3D_Concrete_Printing_Machine_and_Mix_Design/links/55b608c308aec0e5f436d4a1.pdf.
34. A. Perrot, D. Rangeard, A. Pierre, Structural built-up of cement-based materials used for 3D-printing extrusion techniques, Mater. Struct. Constr. 49 (2016) 1213–1220, https://doi.org/10.1617/s11527-015-0571-0.
35. L. Reiter, T. Wangler, N. Roussel, R.J. Flatt, The role of early age structural buildup in digital fabrication with concrete, Cem. Concr. Res. 112 (2018) 86–95, https://doi.org/10.1016/j.cemconres.2018.05.011.
36. R.J.M. Wolfs, F.P. Bos, T.A.M. Salet, Hardened properties of 3D printed concrete: the influence of process parameters on interlayer adhesion, Cem. Concr. Res. 119 (2019) 132–140, https://doi.org/10.1016/j.cemconres.2019.02.017.
37. B. Panda, N.A. Noor Mohamed, Y.W.D. Tay, L. He, M.J. Tan, Effects of slag addition on bond strength of 3D printed geopolymer mortar: an experimental investigation, Proc. Int. Conf. Prog. Addit. Manuf. 2018-May 2018 62–67. 〈https://doi.org/10.25341/D4QG6D.
38. B. Panda, S.C. Paul, N.A.N. Mohamed, Y.W.D. Tay, M.J. Tan, Measurement of tensile bond strength of 3D printed geopolymer mortar, Measurement 113 (2018) 108–116.
39. R.J.M. Wolfs, F.P. Bos, E.C.F. van Strien, T.A.M. Salet, A Real-Time Height Measurement and Feedback System for 3D Concrete Printing, Springer International Publishing, Cham, 2018, pp. 2474–2483. https://doi.org/10.1007/978-3-319- 59471-2_282.
40. GCT Structural Mortal Mix 4000 PSI FS; Gulf Concrete Technology: Long Beach, MS, November 2018.

Advances in Additive Manufacturing Technologies – Gurusamy Pathinettampadian et al. (eds)
© 2026 Taylor & Francis Group, London, ISBN 978-1-041-16687-0

Evaluation of Concrete Barriers Equipped with Cutting-Edge Shock Absorbers while they are Receiving Impact Loads

S. Raja Muniyasamy[1]
Assistant Professor, Department of Civil Engineering,
Thamirabharani Engineering College,
Chennai, India

M.A. Raja[2]
Assisstant Professor, Department of Civil Engineering,
Meenakshi Sundararajan Engineering College,
Chennai, India

S. Mercy[3], **Suresh Kumar M.**[4]
Lecturer, Department of Civil Engineering,
Sankar Polytechnic College,
India

Thangavelu Eswary Devi[5]
Assistant Professor, Department of Civil Engineering,
Rajalakshmi Engineering College,
India

Veni Krishnabharathi[6]
Librarian,
Tamilvel Umamaheswaranar Karanthai Arts college,
India

◆ **Abstract:** Thanks to better road conditions, vehicles are now traveling at greater speeds, which increases the likelihood of high-energy crashes. Tragically, a piece of concrete from a median barrier (CMB) was ejected when a vehicle crashed into it, killing a car occupant in a recent accident in South Korea. It was calculated that the impact energy in this occurrence exceeded SB7 (2300 kJ), even though the current design impact threshold of CMBs is SB5-B (270 kJ). This safety risk was addressed by conducting a series of computational calculations to reduce CMB fragmentation upon impact. To ensure the newly created CMB model held up under impact situations, field tests were carried out. Incorporating wire-mesh reinforcements and increasing the cross-sectional area were two design changes. Furthermore, a device was created to absorb shock and disperse a substantial amount of impact energy at a low cost. This device incorporates a hollow area surrounding the foundation-securing dowel bars, which enables the dowel bars to undergo controlled deformation in order to absorb impact pressures. Thoroughly organized field testing confirmed the efficacy of the newly constructed CMB fitted with this energy-absorbing device.

◆ **Keywords:** Concrete barriers, Impact-resistant barriers, Protective structures, Safety barriers, Reinforced concrete, Barrier performance

[1]ramunish58@gmail.com, [2]antojosephraja@gmail.com, [3]mercyspc@gmail.com, [4]sureshspccivil2020@gmail.com, [5]eswarydevi.t@rajalakshmi.edu.in, [6]venikrishnabharathi@gmail.com

DOI: 10.1201/9781003685906-110

1. Introduction

To keep cars from crossing over into oncoming traffic or crashing into each other, the CMB is an essential highway safety feature. For the betterment of road safety, CMB designs in South Korea have undergone substantial upgrades within the last three decades. A CMB 810 mm in height was standard issue in 1980. Unfortunately, the damaged fender that was mounted on top of the barrier caused many run-over and subsequent accidents. New designs were created and authorized following successful field tests to increase the CMB height to 1270 mm, therefore addressing this problem. This has led to the majority of CMBs in South Korea being 1270 mm tall. Improved passenger protection without changing the barrier height was the primary goal of the 2015 revisions. Little was known about how CMBs structurally performed when subjected to impact loads until that year. Nevertheless, the problem was brought to light in a serious truck accident that killed several people when concrete shards from the crash hit cars in the other lane. The reinforcing mesh for CMBs back then was 3.2 mm wide and spaced 150 mm apart along the longitudinal and vertical axes.

Initially, the goal was to strengthen the structure by using steel rebar instead of wire-mesh reinforcement. Researchers discovered that the structural capacity was much enhanced when rebar with a greater diameter was used. On the other hand, adding rebar reduced the constructability of the slip-form building method, which is often used to install CMBs, and increased cost. Therefore, wire-mesh continued to be the more realistic option. To improve the impact resistance of CMBs reinforced with wire mesh and reduce concrete fragmentation, Kim et al. [8] investigated many designs. A wider portion and two layers of wire-mesh with larger wires and smaller spaces between them drastically cut down on volume loss when the object hit the ground. To further reduce concrete fragmentation, a new shock-absorbing design was also implemented. According to the simulation findings, this innovative design significantly decreased volume loss by as much as 90%. This research lays out the steps used to choose the best design for the CMB that incorporates these shock absorbers. Following its development, the CMB was put to the test in real-world scenarios by simulating a 14-ton truck's crash at 85 km/h and an impact angle of 20°, which corresponds to the impact severity level of SB5-B(20A). For further information on how to choose these extreme impact scenarios, see Kim et al. [8].

2. Experimental Methodology

The vehicle model used in this study is the National Crash Analysis Center's 16-ton truck model, which was adjusted to 14 tons. European standard EN-1317 [6] provided the initial inspiration for this truck type. It was determined to be an appropriate standard truck model for this study due to its similarities in weight, length, height, and breadth with vehicles often utilized in South Korea. There were implausible contact failures seen in the simulations run using this vehicle model. The calculations showed an abnormally large volume loss and significant chipping due to the broad steel cargo plate that extended over the CMB. Scratches, rather than major material loss, would be the most likely result of such failures. By using an element size of 38 mm in the simulation, any depth loss of 38 mm would result in erroneous findings [2]. The problem was solved by narrowing the anticipated cargo plate's width to fit the truck head. The findings of the simulation were more realistic and applicable after this change.

To evaluate the performance of new CMBs, one can refer to design criteria like the ones included in the AASHTO LRFD Bridge Design Manuals [1], which offer procedures for reinforced concrete barrier design which are based on plasticity theory. To ensure that new CMB designs match the impact severity regulations defined by each nation, vehicle collision testing are often conducted before they are approved. Highways in South Korea are usually built to endure impact severity levels of SB5, but areas with a high probability of accidents necessitate barriers of SB5-B level. South Korea's SB5-B level provides inferior impact resistance compared to international standards used in Japan, the EU, and the US, such as SB6, SB7, H3, and TL5/6. Based on statistics from truck-CMB incidents in South Korea, it has been suggested that raising the impact severity threshold from the existing 270 kJ to 420 kJ will make CMBs more resilient and improve their protective function in around 85% of accidents involving CMBs [8]. A new impact severity level of 456 kJ, SB5-B(20A), has been suggested based on this investigation. A 14-ton truck going at 85 km/h with a collision angle of 20° would cause this magnitude of impact. The SB5-B(20A) level of severity is similar to SB6 in South Korea, SA in Japan, and H3 in Europe; it is almost 1.7 times greater than the present SB5-B level. Newly designed CMBs in this study outperformed SB5-B(20A) and SB6 in terms of impact resistance, which is 456 kJ. We will confirm the performance of these designs through the use of full-scale crash testing and vehicle-CMB collision simulations [3].

3. Results and Discussion

A study conducted by Kim et al. [8] shown that the impact resistance may be significantly improved by adding reinforcement. Nevertheless, the slip-form building approach is not well-suited for the usage of steel rebars because of their poor constructability. It is advised to utilize wire-mesh in order to keep the CMB constructable. But large-sized wire-mesh is hard to come by since most companies that make it don't make it in sizes greater than 7.6 mm. Also, workers may not be as efficient while dealing with large-sized wire-mesh since it is too heavy for them to handle. Poor concrete compaction and material segregation at the CMB middle can also result from reinforcements with bigger diameters. In order to solve these problems, this study used two layers of wire-mesh instead of steel rebars. This made the slip-form approach more constructible and improved the reinforcement ratio. Two layers of wire-mesh, with one layer placed distant from the center and the other in the middle, are thought to increase flexural capacity, as compared to a single-layer wire-mesh. The CMB section was 50 mm larger to accommodate the embedded wire-mesh layers, which helped avoid segregation and ensured correct concrete compaction [9].

Three major enhancements were incorporated into the new CMB during development: Instead of using a single center layer, two layers of wire-mesh were utilized. The wire-mesh thickness was raised from 3.2 mm to 7.6 mm. Keeping the same side-face slope as the present CMB, the CMB width was expanded by 50 mm. The schematic architecture of this recently created CMB, which was dubbed Hi-CMB. The Hi-CMB was classified as Hi-CMB(fix) (without Styrofoam) or Hi-CMB (with Styrofoam) according on whether or not a Styrofoam cylinder was present. Sections 12.10 and 13 will give a comprehensive breakdown of the Styrofoam and shock absorber parts. With an impact angle of 20° instead of 15° and all other parameters kept the same as SB5, the Hi-CMB test level was called SB5-B(20A). The SB5-B(20A + 100V) test level was also used when the collision angle was 20 degrees and the vehicle speed was 100 km/h [4].

The performance assessment of Hi-CMB(fix) was conducted under conditions of severe impact. The weight of the removed concrete components relative to the overall weight of the CMB along a 1.2 m length was used to estimate the volume loss ratio (%). Under SB5-B(20A) circumstances, the current CMB showed a significant volume loss of 177.5%, in contrast to the almost nonexistent volume loss of Hi-CMB(fix) [5]. Compared to the present CMB, Hi-CMB(fix) showed much enhanced resistance even under more extreme impact situations, including SB5-B(20A + 100V) and SB5-B(30A), while experiencing a volume loss of less than 10%. Combining the expanded section with an enhanced reinforcement ratio resulted in reduced fragmentation, which was the most effective measure. However, substantial structural strengthening was not achieved by only increasing the section size and not reinforcing it with two layers of wire-mesh [7].

We covered Hi-CMB (fix)'s better performance in the preceding section. In addition to this layout, the present research presented a novel method by altering the dowel bars used to fasten the CMB to the base. Using Styrofoam, a tiny empty area was made near each dowel bar. When the new CMB was subjected to lateral stresses that were greater than the bond strength between the two materials, the design allowed for deformation by shearing. One way to absorb a lot of impact energy is to let the CMB and foundation move about a little bit [10].

The dowel bars were spaced 1000 mm apart in the final design because any distance more than 1500 mm caused too much lateral motion. In a subsequent part, we shall go more into this issue. The dowel bars in the created models were indeed beam components; however, the Lagrange in Solid option did not embed the Styrofoam-covered bottom half of the dowel bars into the concrete body. Having said that, the concrete had embedded the upper portion of the dowel bit. A subsequent section will explain the process used to determine the dowel bar size, however the study used a dowel bar with a diameter of D19. Later on, we'll talk about how the new CMB interacts with the existing concrete surface of the foundation [11].

The study explored how the height of the shock absorber influenced the volume loss of CMB. The parameters and section design that were supplied were then used to conduct the investigation. The findings of the simulation, included the examination of key elements. These factors included a dowel bar diameter of D19, a dowel bar spacing of 1000 mm, and two layers of 7.5 mm wire-mesh measuring 100 × 100 mm. Regarding the prior remark that we made, the bond strength levels that were taken into consideration were 2.0 MPa and 1.0 MPa simultaneously [12].

The enhanced section design and the implementation of shock absorbers led to a reduction in the volume loss ratios for shock absorber heights of 2.5 cm and 10 cm as compared to the conventional CMB. Under three different impact severity levels, namely SB5-B(20A), SB5-B(20A + 100 V), and SB5-B(30A), the 2.5 cm height exhibited a lower volume loss than the 10 cm height. This was due to the fact that the 2.5 cm height experienced more lateral displacement [13].

However, at 2.5 cm, the dowel bar experienced rupture, leading to increased maintenance costs after a collision. Therefore, a shock absorber height of 10 cm was selected, as it maintained a manageable lateral displacement while preventing dowel bar rupture. Nonetheless, the shock absorber height could be adjusted as needed.

To regulate the shock absorber's lateral displacement, the dowel bar diameter is critical. With a shock absorber height of 10 cm and dowel bar diameters of D10, D16, D19, and D25, a parametric investigation was carried out. Both the impact resistance and the reduction in concrete volume loss did not show any meaningful improvement under impact severity SB5-B(20A). For SB5-B(20A), the influence of dowel bar size was also not apparent [14].

Using greater impact severities, such as SB5-B(20A + 100 V) and SB5-B(30A), more parametric tests were performed to evaluate the effect of dowel bar size. Under these harsher circumstances, Hi-CMB showed less volume loss than Hi-CMB (fix), in contrast to SB5-B(20A) [15].

The D10 dowel bar allowed a maximum of 783 mm of lateral movement, which significantly decreased concrete fragmentation. On the other hand, secondary risks on roads might arise from excessive lateral displacement, which can endanger cars going in the other direction. The D16 dowel bar ruptured, which resulted in substantial maintenance costs, even though it efficiently decreased volume loss under SB5-B(20A). So, D19 was too small of a diameter for the dowel bar. In instance, under SB5-B(30A), the D25 dowel bar failed to work satisfactorily. Because it allowed for regulated lateral movement while minimizing volume loss, the D19 dowel bar was ultimately chosen for Hi-CMB.

The dowel bar spacing in the previous CMB design was 450 mm. Nevertheless, lateral displacement is crucial in reducing concrete fragmentation in the recently created Hi-CMB with a shock absorber. While it's true that a little bit of lateral movement helps absorb impact energy, too much of it can be dangerous for drivers going the other way. Consequently, the design of the Hi-CMB with numerous shock absorbers focuses on controlling lateral movement. The distance between the shock absorbers is a key factor in lateral displacement. Using 500 mm, 1000 mm, 1500 mm, and 2000 mm shock absorber spacings, a parametric research was carried out to find the sweet spot between impact resistance and maintenance needs. For the numerical analysis, SB5-B(20A) was used as the impact condition. The findings showed that the volume loss reduced as the lateral displacement increased, showing that the shock absorber was successful in preventing concrete fragmentation.

Be that as it may, issues may arise with lateral displacement that is extreme. The study revealed that 134 mm of lateral displacement occurred for spacings of 2000 mm and 47 mm for spacings of 1500 mm. Increased lateral displacement has the potential to save maintenance costs, but it might also increase instability, which could pose safety hazards. On the flip side, dowel bar breakage due to excessive displacement might increase maintenance costs. The 1000 mm dowel bar spacing was found to be the sweet spot between efficient shock absorption and low maintenance costs in the Hi-CMB. Despite the fact that volume loss was greater at 1000 mm spacing compared to 500 mm spacing, the study went on to examine the effect of impact site on volume loss. The volume loss was less when the impact was close to a dowel bar as compared to when it was further away. Impacts that occur closer to the dowel bar cause the concrete to absorb impact energy more efficiently because local stress concentrations are more evenly distributed. Further analysis of the influence of impact site on volume loss was carried out using a rigorous parametric research.

Both the vertical and horizontal wire-mesh reinforcements were initially spaced 100 mm apart in the original Hi-CMB design. A lower reinforcing density is not required, nevertheless, because impact damage is mostly localized in the top half of the CMB. One way to reduce the total cost of building using Hi-CMB while keeping it impact resistant is to reduce the reinforcement in the bottom part. Various layouts of wire-mesh were evaluated in order to discover more economical options. The first Hi-CMB, which was labeled as "100." Its vertical and horizontal wires were spaced uniformly at 100 mm. In different designs, the bottom part was reinforced with different densities, and the spacing parameters are shown as "A mm - B mm". Since the "fix" versions presupposed a flawless foundation-to-CMB link that would not break upon impact, they omitted shock absorbers.

Under the impact situation of SB5-B(20A), all configurations worked successfully, with volume loss ratios ranging from 0% to 1.7%, according to the test results. At this impact level, shock absorbers almost didn't make a difference. The importance of shock absorbers became clear under heavier impact situations like SB5-B(20A + 100 V) and SB5-B(30A), when models with them showed less volume loss than the fixed designs without them. The configuration with the finest impact resistance under severe impact conditions was the 200 mm - 200 mm wire-mesh design, which reduced volume loss by 13.1% in SB5-B(30A). The impact resistance performance was 53.4% better than the fixed 200-200 mm variant. Results showed that the 200-200 mm reinforcement design in the lower half of the Hi-CMB was just as effective as, if not more so than, the original Hi-CMB with 100 mm spacing. Consequently, a 200 mm - 200 mm wire-mesh structure was used to strengthen the lower area of the Hi-CMB in order to achieve a more cost-effective design. To make sure there's enough strength where it's needed, the 100 mm spacing was maintained by the horizontal wires in the upper portion and the remaining vertical wires throughout the construction.

4. Conclusion

In order to reduce secondary risks produced by concrete fragmentation in vehicle crashes under more severe impact circumstances (SB5-B(20A)), a new concrete median barrier (CMB) was created to replace the previous SB5-B design. We verified the new model using field testing after conducting a rigorous parametric investigation. The shock absorber's lateral displacement was successfully accomplished by positioning it between the CMB soffit and the foundation top. Here are the main takeaways from the research:

1. The primary objective of the current CMBs is to lessen the likelihood of collisions, which occur when two cars collide with one another. Their reinforcing is severely inadequate because they just have one layer of wire mesh measuring 3.5 millimeters and

are placed 150 millimeters apart in the midsection. Furthermore, in comparison to SB5-B, the findings of the simulation demonstrate that these CMBs are more sensitive to more powerful impacts.

2. In order to cope with 80 percent of the most recent collision events, a separate level of impact severity was established; the level that is currently in place only covers 55 percent of these instances. A 14-ton truck is portrayed at the suggested severity level while it is traveling at 85 kilometers per hour and the attack angle is 20 degrees. SB5-B(20A) is the classification that has been given to this increased impact severity.
3. A new design was employed for the dowel bars, and it featured a little amount of empty space adjacent to each one of them. As a consequence of this, the bars are able to rip when subjected to lateral pressures and effectively absorb impact energy.
4. In order to develop the new CMB that incorporates a shock absorber, a detailed parametric study was carried out under the conditions of SB5-B(20A). It was determined how to measure the essential measurements of the shock absorber, which included the height, diameter, and spacing of the dowel bars.
5. An ideal value was chosen after analyzing the binding strength between the CMB soffit and the foundation top. Lateral displacement was shown to be limiting at bond strengths greater than 3.5 MPa. In order for the shock absorbers to work properly, it is necessary to regulate the bond strength.
6. The ideal structure for a shock absorber was determined to be a set of dowel bars (D19) spaced 1000 mm apart and 10 cm high, as this allowed for efficient deformation without rupture. The impact loading of bigger vehicles surpassed SB5-B(20A) requirements, and shorter shock absorbers failed under such stress.
7. The recently created CMB, called Hi-CMB, has three unique features: (1) a 50 mm top width, (2) two 100 mm x 100 mm layers of wire-mesh that are 7.5 mm thick, and (3) a revolutionary shock absorber system that uses dowel rods and Styrofoam to connect the CMB to the base.
8. The concrete fragmentation that was generated by the shock absorber of the Hi-CMB was significantly reduced when compared to the conventional CMB and the fixed Hi-CMB. Under SB5-B(20A) impact conditions, the volume loss of the Hi-CMB with shock absorbers was comparable to that of the fixed Hi-CMB because of the same impact conditions. In the case of larger impact severities, the fixed Hi-CMB demonstrated a much greater volume loss compared to the one that had shock absorbers, such as SB5-B(20A + 100V) and SB5-B(30A).
9. The Hi-CMB is able to effectively fulfill all of the standards that are listed in the Korean road-side safety guide for the year 2012. In addition, the anticipated lateral displacement of 20 millimeters was accomplished, which is further evidence that the revolutionary shock absorber design utilized in the new Hi-CMB is performing as intended.

References

1. AASHTO, AASHTO LRFD Bridge Design Specifications, Customary U.S. Units, in: 6th ed., AASHTO, Washington, DC, 2012.
2. S. Austion, P. Robins, Y. Pan, Shear bond testing of concrete repairs, J. Cement Concrete Res. 29 (1999) 1067–1076.
3. C.Chung,J.Lee,S.Kim,J.Lee,Influencingfactorsonnumerical simulation of crash between RC slab and soft projectile, J. Comput. Struct. Eng. Inst. Korea 27 (6) (2014) 533–542.
4. S. El-Tawil, E. Severino, P. Fonseca, Vehicle collision with bridge piers, J. Bridge Eng. 10 (3) (2005) 345–353.
5. E.N.B.S. Julio, F.A.B. Branco, V.D. Silva, Concrete-to-concrete bond strength Influence of the roughness of the substrate surface, J. Construct. Build. Mater. 18 (2004) 675–681.
6. European Committee for Normalization, EN 1317 European Standard for Road Restraint Systems, European Committee for Normalization, 2012.
7. International Federation for Structural Concrete (fib) (2013) The fib model code for concrete structures 2010, Ernst & Sohn
8. W. Kim, I. Lee, G. Zi, K. Kim, J. Lee, Design approach for improving current concrete median barrier on highway in South Korea, J. Perform. Construct. Facil. (2018).
9. Korea Concrete Institute, Concrete Design Code 2012, Korea Concrete Institute, South Korea, 2012.
10. J. Lee, G. Zi, I. Lee, Y. Jeong, K. Kim, W. Kim, Numerical simulation on concrete median barrier for reducing concrete fragment under harsh impact loading of a 25-ton truck, J. Eng. Mater. Technol. 132 (2) (2017).
11. Livermore Software Technology Corporation, LS-DYNA Keyword User's Manual Volume I, Livermore Software Technology Corporation (LSTC), Livermore, CA, 2007.
12. D.Y. Murray, User's Manual for LS-DYNA Concrete Material Model 159, Federal Highway Administration, U.S. Department of Transportation, 2007 FHWA-HRT-05-062.
13. D.Y. Murray, A. Abu-Odeh, R. Bligh, Evaluation of LS-DYNA Concrete Material Model 159, Federal Highway Administration, U.S. Department of Transportation, 2007 FHWA-HRT-05-063.
14. D. Thai, S. Kim, H. Lee, Effects of reinforcement ratio and arrangement on the structural behavior of a nuclear building under aircraft impact, Nucl. Eng. Des. 276 (2014) 228–240.
15. K. Yi, K. Hedrick, S. Lee, Estimation of tire-road friction using observer based identifiers, Int. J. Vehicle Mech. Mobil. Vehicle Syst. Dynam. 31 (4) (1999).

Advances in Additive Manufacturing Technologies – Gurusamy Pathinettampadian et al. (eds)

Continuous Deflection of Reinforced Concrete Beams and the Impact of Sustained Load Magnitude on that Deflection

Suresh Kumar M.[1]
Lecturer, Department of Civil Engineering,
Sankar Polytechnic College, India

Thangavelu Eswary Devi[2]
Assistant Professor, Department of Civil Engineering,
Rajalakshmi Engineering College, India

S. Raja Muniyasamy[3]
Assistant Professor, Department of Civil Engineering,
Thamirabharani Engineering College, Chennai, India

M.A. Raja[4]
Assisstant Professor, Department of Civil Engineering,
Meenakshi Sundararajan Engineering College, Chennai, India

S. Mercy[5]
Lecturer, Department of Civil Engineering,
Sankar Polytechnic College, India

Balaji Krishnabharathi[6]
Assistant Professor, Department of Mechanical Engineering,
Chennai Institute of Technology,
Chennai, India

◆ **Abstract:** The creep and shrinkage deflections on beams made from RC with different concrete grades are explored in this study. Additionally, the impacts of sustained loading intensity and tension steel percentage are also taken into consideration. The sustained loading was found to be between 25 and 50 percent of the initial fracture load after taking into consideration three distinct concrete grades; this was done. 45 MPa, 35 MPa, and 25 MPa were the grades that were given. Additionally, two percentages of tension steel were utilized, namely 0.77 percent and 1.21 percent. At 1, 3, 7, 14, 21, 30, 60, 90, 120, and 150 days of age, long-term deflections were measured under sustained four-point bending. Additionally, specimens subjected just to their own weight were tested for shrinkage deflections at the same time intervals for all three concrete grades. The author's previous model was expanded to include various sustained loading magnitudes and concrete compressive strengths in order to forecast the overall deflection of RC beams caused by creep and shrinkage, building on these experimental results.

◆ **Keywords:** Reinforced concrete beams, , Flexural behavior, Concrete creep, Time-dependent deformation, Long-term deflection

[1]sureshspccivil2020@gmail.com, [2]eswarydevi.t@rajalakshmi.edu.in, [3]ramunish58@gmail.com, [4]antojosephraja@gmail.com, [5]mercyspc@gmail.com
[6]balajikrishnabharathi@gmail.com

DOI: 10.1201/9781003685906-111

1. INTRODUCTION

In the case of RC and pre-stressed concrete structures, it is feasible for time-dependent deflections caused by creep and shrinkage to have a major influence on the serviceability of the structures. As a consequence of this, it is very necessary for designers to pay particular attention to the calculations of the long-term deflection of RC structures. Deflection caused by shrinkage in reinforced concrete flexural members is proportional to the quantity and location of reinforcement, whereas creep-induced deflection in reinforced concrete beams is affected by both the amount of reinforcement and the size of the sustained stress. There is a lack of experimental evidence on the impacts of prolonged loading magnitude on creep deflections, even though design codes and literature take tension and compression reinforcement into account when determining long-term deflections.

For the purpose of forecasting RC beam deflections over the long future, a number of researchers have put forth models. In order to compare actual results with anticipated ones [7], presented a model for CD in singly RC beams. By comparing their analytical approach to test data, Ghali and Azarnejad [8] were able to detect both short- and long-term stresses in RC beams. Rosowsky et al. [9] established a probabilistic time-dependent procedure in order to evaluate the amount to which the age of the loading effects beam deflections. By doing so, they were able to make this determination. The reinforcement of fiber-reinforced polymer (also known as FRP) is another topic that has been the focus of research and exploration. According to the findings of Laoubi et al. [10], who carried out a study on the cold weather endurance of GFRP bars, the flexural behavior of the bars was not influenced by freeze-thaw cycles when they were subjected to sustained loads. This was the conclusion reached by the researchers. According to the findings of study conducted by Al-Salloum and Almusallam [11], the most significant creep impact was seen in GFRP-reinforced beams that were repeatedly exposed to dry-wet cycles in saline water.

The use of finite element analysis has been applied in a lot of the studies that have been carried out in this particular field. Employing the age-adjusted effective modulus approach allowed Wenwei et al. [12] to accurately predict time-dependent deflections. This was accomplished by using the technique. There was a strong correlation between this strategy and the results of the trials that were carried out. The results of a study that Chami and his colleagues [13] conducted on the influence of creep on beams made of concrete reinforced with CFRP revealed that the reinforcement did not have a notable impact on the long-term deflection performance of the beams. This was the case despite the fact that the reinforcement increased the load-carrying capacity of the beams. Two groups of researchers [14, 15], arrived to the same findings, which means that their findings are compatible with one another.

Several aspects have been investigated in further experimental research. The mid-span deflections of self-compacting RC beams were compared with preexisting models in long-term flexural experiments performed by Mazzotti and Savoia [16]. When Tan et al. [17] tested GFRP-strengthened concrete beams subjected to continuous stress in a variety of settings, they discovered that adding more FRP layers decreased fracture widths and deflections. Beam curvature and strain were shown to be strongly impacted by corrosion levels and applied service loads in an analysis of corroded RC beams conducted by Malumbela et al. [18] under constant loads.

Additionally, a number of academics have refined and simplified approaches of forecasting long-term deflections. In order to estimate incremental deflections in RC members, Zhou et al. [19] created a method that is based on Canadian design codes. Mari et al. [2] presented a model that takes into account strength, loading age, exposure conditions, sustained load duration, and compression reinforcement effects; their results were in good agreement with those of non-linear analyses. Through full-scale tests, Al-Deen et al. [20] monitored deflection-time curves under sustained loads to quantify long-term deflections in composite steel-concrete beams. An influence coefficient-based model that was well-aligned with experimental data was proposed by Pan et al. [21], who investigated the effect of reinforcement ratios on creep and shrinkage.

Predicting RC beam deflections using the age-adjusted effective modulus approach was done by Gilbert [22]. This methodology was confirmed against experimental data. Deflections are greatly increased by early-stage corrosion, according to Hariche et al. [23], who investigated flexural behavior differences caused by corrosion, sustained loads, and rebar configurations. Mias et al. [4, 24] examined GFRP-reinforced beams over an extended period of time and found that the deflection behavior was unaffected by prolonged load fluctuations. They did tests on RC beams using varying reinforcement percentages and concrete grades for 250 to 700 days. They found that lower-strength concrete with greater reinforcement percentages had larger long-term deflections.

Despite these investigations, there is a lack of experimental evidence about the interplay between the size of the SL, the % of tension steel, and the concrete grade (CG) as it pertains to RC beam deflections. Furthermore, there is no all-encompassing research that examines these aspects simultaneously. A comprehensive experimental investigation was undertaken to fill this knowledge vacuum. The study tested RC beams subjected to three distinct CG,

2% of tension reinforcement, and two degrees of sustained load in order to analyze their long-term deflection behavior. In order to forecast RC beam deflections under different sustained load scenarios, the researchers expanded upon their earlier model.

2. Experimental Methodology

Every single one of the components that were used to manufacture the concrete for the study was easily accessible in the regional area. In contrast to the fine aggregate, which was composed of river sand, the coarse aggregate had a maximum particle size of sixteen millimeters. A determination was made about the features of the aggregates in accordance with IS 383 [32], it was determined that the mixing water did not include any impurities that might be considered hazardous in accordance with IS 456 [33]. OPC of grade 43 was used in all of the concrete mixes, and it was subjected to property testing in accordance with IS 4031 [34]. In accordance with IS 8112 [35], the results of the tests are within the acceptable limits.

For the purpose of reinforcing, TMT steel rebars with diameters of 8 mm and 10 mm were taken into consideration. The findings were obtained by conducting tests on these rebars in accordance with the International Standard 1786 [36]. Three unique concrete classes, denoted by the letters M1, M2, and M3, were designed with goal cube compressive strengths of 45 MPa, 35 MPa, and 25 MPa respectively. In line with IS 10262 [37], the maximum density was accomplished by ensuring that the ratio of fine to coarse aggregates remained constant at 0.6 throughout all grades.

Prior to the addition of water, the dry ingredients were mixed to ensure that they were dispersed uniformly throughout the mixture. Standard cubes measuring 150 millimeters and cylinders measuring 150 millimeters by 300 millimeters were manufactured for each and every trial combination. For the purpose of the experiment, reinforced concrete (RC) beams were employed. These beams had a diameter of either 8 mm or 10 mm, and their dimensions were 100 mm × 150 mm × 1800 mm overall. The tension reinforcement ratios that corresponded to rebars with a diameter of 8 mm and 10 mm were 0.77% and 1.21%, respectively on the scale of tension reinforcement ratios. We didn't apply any compression or shear reinforcement to the RC beams since we wanted to see how their flexural behavior changed over time.

Each concrete grade produced 16 beams for a total of 48 beams cast. Two beams were used to determine the first crack load and SD for each grade and reinforcement %, and two beams were used to analyze creep deflections under two magnitudes of SL. A total of eight beams were evaluated under varied conditions. Consistent with earlier studies [26,27], we placed the initial sustained load at 25% of the first fracture load and raised it to 50% for the second sustained load.

A smooth-finish steel mold was greased before the concrete was poured into it for casting. Following the 24-hour demolding process, the beams were immersed in water for a further 28 days to cure. An effective beam span of 1700 mm was achieved with the provision of a 50 mm overhang under simply supported circumstances. Hangers were used to apply sustained stresses at two-point loading points that were 200 mm apart.

Two beams of each kind were subjected to full failure testing in order to ascertain the initial fracture load. To mimic the sustained loading situation, the loading was applied using two-point loads that were 200 mm apart. This was accomplished using a 50 kN Avery transverse testing equipment. Deflections up to 150 days in the middle of the span were documented. We got the values of the SD by taking the elastic and creep deflections caused by the beam's self-weight and subtracting them. Two identical beams were used to calculate the stated results.

Beams were subjected to varying degrees of sustained stress and measured for long-term deflection at1,3,7,14,21,30,60,90,120, and 150 days intervals. At 28 days, these beams were initially supported and loaded. The beams were checked to be properly aligned on the test frame. These experiments of long-term deflection were conducted simultaneously with the tests of shrinkage deflection.

The research took place at India's prestigious Indian Institute of Technology Roorkee where the trials were conducted. The momentary deflections were documented following the application of continuous stresses. By the end of the test period, not a single beam showed any indication of failing under prolonged loading. The deflections seen in beams tested without prolonged loads were subtracted from the total to get the creep deflections.

3. Results and Discussion

The test beams that have just strengthening in the stress zone show an uneven cross-section because of this. In contrast to the tension zone, where reinforcing prevents concrete from shrinking, the unreinforced compression zone experiences far more shrinkage. A change in curvature causes the beam to be displaced downwards as a result of this differential shrinkage over the beam's cross-section. The shrinking phenomenon causes the reinforced concrete (RC) beams' overall deflection to rise.

The shrinkage deflection, the creep deflection, and the elastic deflection that is induced by the beam's own weight are the three components that come together to form the total deflection of the specimen's shrinkage beam. In order to separate the shrinkage deflection from the total deflection, we first eliminated the elastic and creep deflections from the overall deflection values. Although the time-dependent variation in shrinkage deflection for RC beams reinforced with 2Φ8 rebars, The same phenomena for RC beams reinforced with 2f10 rebars. Both figures are included in the same figure. A variety of significant inferences may be drawn from these figures, including the following: When the initial phase, which lasts for less than thirty days, is contrasted with the latter phases, which last for more than thirty days, the rate of shrinkage deflection development is significantly higher in the first phase. Despite the fact that different concrete mixes and beam types display different patterns of shrinkage deflection variation over time, the overall pattern remains the same Beams that were reinforced with 2f8 rebars saw a shrinkage deflection of 54.0% on average after 150 days, with a range of 46.9% to 53.3% after 30 days. At the end of 150 days, beams that have been reinforced with 2f10 rebars exhibit a shrinkage deflection that ranges from 41.2 to 55.3 percent, with an average of 50.0 percent. This figure is relatively comparable to the value reported for beams that have been reinforced with 2f8 rebars. This indicates that the percentage of tension reinforcement does not have any impact on the ratio of SD at 30 days to that at 150 days.. Beams that have been reinforced using 2f10 rebars have been related with three distinct percentage increases: 142.9%, 81.0%, and 82.6%.

On the other hand, the progression of total deflection over time for beams that have been reinforced with rebars of 2f8 and 2f10 dimensions. There is a representation of all three of these deflections in the figures: creep, shrinkage, and elastic deflection. There is the possibility of making the following observations: All RC beams have an increase in creep deflection with time, despite the fact that the rate of rise reduces with increasing age. Taking into consideration this trend, it appears that creep deflection takes place at a more rapid pace during the first thirty days, after which it begins to level out. On the creep deflection vs time curve, the contour remains the same regardless of the mix group or age of the individuals. The creep deflection at 30 days varies from 45.7% to 61.5% of the creep deflection at 150 days, for both percentages of tension reinforcement, when the continuous load level is 25%. This is the case for both percentages of tension reinforcement. At a constant load intensity of fifty percent, the creep deflection that is measured at thirty days varies from 38.9% to 52.0% of that which is recorded at one hundred fifty days. The creep deflection of beams reinforced with 2f8 rebars that are subjected to a sustained load of 25% rises by 76.0% for mix M1, 62.7% for mix M2, and 108.7% for mix M3, respectively, during the course of 30 to 150 days. When beams that are reinforced with 2f10 rebars are exposed to the same level of sustained stress, there is an increase of 112.7%, 66.7%, and 112.8% respectively. Beams reinforced with 2f8 rebars for concrete mixes M1, M2, and M3 see an increase in creep deflection of 100.0%, 157.4%, and 114.3%, respectively, from 30 to 150 days. This occurs when the beams are subjected to a sustained load level of 50%. The improvements that correlate to beams that have been reinforced using 2f10 rebars are 92.4%, 110.0%, and 128.8% respectively.

According to the information the total deflection of RC beams may be broken down into its component elements, which are the SD and the CD. Other components include the creep deflection. The distribution of elastic, shrinkage, and CD within the total deflection at 30 and 150 days, respectively, for beams that have been reinforced with rebars of 2Φ8 and 2Φ10 dimensions. A few key things that may be learned from these visual aids are as follows: Shrinkage deflection is always a modest fraction of overall deflection; nevertheless, its influence becomes less significant as the sustained load increases. This is true even if the amount of shrinkage deflection remains the same between all types of beams. The explanation for this drop is because beams that are subjected to a greater sustained load have a bigger overall deflection magnitude. This is the culprit behind this decrease. The contribution of shrinkage deflection to total deflection after 150 days is reduced from an average of 15.4 percent to 10.6 percent when the reinforcing percentage is increased from 0.77 percent to 1.21% at a sustained load level of 50%. After a period of thirty days, the average shrinkage deflection contribution decreases from twelve percent to eight percent, with the same rise in reinforcement percentage resulting to a greater reduction. According to these findings, it would appear that reinforcement is a more effective inhibitor of early-stage shrinkage-induced deflection than other methods. Increasing the proportion of reinforcement results in an increase in the role of creep deflection to the overall deflection. Those of any age may attest to this. The creep deflection contribution increases from 43.6% to 51.3% after thirty days, and from 43.8% to 51.2% for a continuous load level of 25% and 50%, respectively, when the proportion of tension steel is raised from 0.77% to 1.21%. This occurs when the load level is maintained at 25% and 50%, respectively. The occurrence of this takes place if the load level is 25 percent and 50 percent, correspondingly. As the amount of reinforcement rises, the percentage of creep deflection to overall deflection increases at 150 days. This is caused by the creep deflection. When a load level of 25% is maintained for an extended period of time, the average contribution of creep deflection rises from 51.2% to 61.7%. Furthermore, when the load level is kept at 50% continuous, the contribution rises from 58.4% to 65.0% while remaining at the same level. All of the concrete mixes displayed these patterns, which can be understood as proof that the quantity of sustained load as well as the proportion of tension reinforcement possess an important influence on the quantity of shrinkage and creep deflections contributing to the overall deflection of beams made of reinforced concrete. This is because the patterns were observed in all of the concrete mixes.

4. CONCLUSION

- An experimental examination on the deflection characteristics of 48 beams made of reinforced concrete (RC) was carried out over the duration of one hundred fifty days. The beams were in a state of deflection. The research investigated two distinct percentages of tension reinforcement (0.77% and 1.21%), three distinct concrete grades (25 MPa, 35 MPa, and 45 MPa), and two distinct degrees of continuous loading (25 and 50 percent of the initial fracture stress, respectively). All of these factors were taken into consideration. Research was conducted on each of these aspects. Following are some of the main conclusions that were reached as a result of the findings of the investigation:
- When the compressive strength of concrete is decreased, the shrinkage deflection related to the material also increases. This is because the material is more susceptible to shrinking. A larger ratio, on the other hand, does assist reduce the amount of shrinkage deflection. Despite the fact that the degree of tension reinforcement does not have an effect on the ratio of shrinkage deflection at 30 days to that at 150 days, a bigger ratio does assist make the ratio lower. It is also more clear that the tension rebars have a constraining effect on the child when they are inside the younger age range.
- In the initial phases that follow the application of load, the contribution of creep deflection to the overall deflection of reinforced concrete beams is significantly low across the board for all concrete mixes. The degree of continuous loading that is being applied does not affect the fact that this is true any more. By increasing the sustained load level from 25% to 50% at 150 days, creep deflection adds more to the overall deflection than it would have otherwise. This is because the sustained load level has grown. When beams are reinforced with 28 rebars, this contribution rises from 50.9% to 58.5%, whereas if beams are reinforced with 210 rebars, it rises from 61.5% to 65.0% on average. Both of these increases are based on the average.
- The impact of creep deflection on whole deflection becomes increasingly obvious at all ages, including 150 days, as the amount of tension steel increases. This is the case regardless of the age of the steel. It makes no difference how old the substance is; this is always the instance. In order to take into account the effects of continuous load levels as well as the compressive strength of concrete, a model that was proposed in the past for estimating long-term deflection has been enhanced since it has been improved.
- For the purpose of better examining the long-term deflection behavior of reinforced concrete beams over an extended period of time, with a wider range of tension reinforcement ratios, sustained load levels, and concrete compressive strengths, it is recommended that more research be carried out.

REFERENCES

1. P.X. Zou, S. Shang, Time-dependent behaviour of concrete beams pretensioned by carbon fibre-reinforced polymers (CFRP) tendons, Constr. Build. Mater. 21 (4) (2007) 777–788., http://dx.doi.org/10.1016/j.conbuildmat.2006.06.008.
2. A.R. Marí, J.M. Bairán, N. Duarte, Long-term deflections in cracked reinforced concrete flexural members, Eng. Struct. 32 (3) (2010) 829–842., http://dx.doi.org/10.1016/j.engstruct.2009.12.009.
3. S.R. Debbarma, S. Saha, Growth of time-dependent strain in reinforced cement concrete and pre-stressed concrete flexural members, Int. J. Concrete Struct. Mater. 6 (2) (2012) 79–85.
4. C. Miàs, L. Torres, A. Turon, C. Barris, Experimental study of immediate and time-dependent deflections of GFRP reinforced concrete beams, Compos. Struct. 96 (2013) 279–285., http://dx.doi.org/10.1016/j.compstruct.2012.08.052.
5. M. Shariq, J. Prasad, H. Abbas, Creep and drying shrinkage of concrete containing GGBFS, Cem. Concr. Compos. 68 (2016) 35–45. , http://dx.doi.org/10.1016/j.cemconcomp.2016.02.004.
6. E. Hernández-Montes, M.A. Fernández-Ruiz, J.F. Carbonell- Márquez, L.M. Gil-Martín, Theoretical and experimental in-service long-term deflection response of symmetrically and non-symmetrically reinforced concrete piles, Arch. Civil Mech. Eng. 17 (2) (2017) 433–445. http://dx.doi.org/10.1016/j. acme.2016.12.003, 10.1007/s40069-012-0008-x.
7. R.M. Samra, Renewed assessment of creep and shrinkage effects in reinforced concrete beams, ACI Struct. J. 94 (6) (1997) 745–751, https://www.concrete.org/publications/internationalconcreteabstractsportal/m/details/id/9734.
8. A. Ghali, A. Azarnejad, Deflection prediction of members of any concrete strength, ACI Struct. J. 96 (5) (1999) 807–817, https://www.concrete.org/publications/internationalconcreteabstractsportal/m/details/id/735.
9. D.V. Rosowsky, M.G. Stewart, E.H. Khor, Early-age loading and long-term deflections of reinforced concrete beams, ACI Struct. J. 97 (3) (2000) 517–524, https://www.concrete.org/ publications/internationalconcreteabstractsportal/m/details/ id/4647.

10. K. Laoubi, E. El-Salakawy, B. Benmokrane, Creep and durability of sand-coated glass FRP bars in concrete elements under freeze/thaw cycling and sustained loads, Cem. Concr. Compos. 28 (10) (2006) 869–878., http://dx.doi. org/10.1016/j.cemconcomp.2006.07.014.
11. Y.A. Al-Salloum, T.H. Almusallam, Creep effect on the behavior of concrete beams reinforced with GFRP bars subjected to different environments, Constr. Build. Mater. 21 (7) (2007) 1510–1519., http://dx.doi.org/10.1016/j.conbuildmat.2006.05.008.
12. W. Wenwei, W. Changnian, Y. Wei, Experimental study on creep and shrinkage of old and new concrete composite beams, J. Southeast University (Natural Science Edition) 3 (2008) 029, http://en.cnki.com.cn/Article_en/CJFDTOTAL-DNDX200803029.htm.
13. G. Al Chami, M. Theriault, K.W. Neale, Creep behaviour of CFRP-strengthened reinforced concrete beams, Constr. Build. Mater. 23 (4) (2009) 1640–1652., http://dx.doi.org/10.1016/j. conbuildmat.2007.09.006.
14. H.R. Sobuz, E. Ahmed, N.M. Sutan, N.M. Hasan, M.A. Uddin, M.J. Uddin, Bending and time-dependent responses of RC beams strengthened with bonded carbon fiber composite laminates, Constr. Build. Mater. 29 (2012) 597–611., http://dx. doi.org/10.1016/j.conbuildmat.2011.11.006.
15. A.K. El-Sayed, R.A. Al-Zaid, A.I. Al-Negheimish, A.B. Shuraim, A.M. Alhozaimy, Long-term behavior of wide shallow RC beams strengthened with externally bonded CFRP plates, Constr. Build. Mater. 51 (2014) 473–483., http://dx.doi.org/ 10.1016/j.conbuildmat.2013.10.055.
16. C. Mazzotti, M. Savoia, Long-term deflection of reinforced self-consolidating concrete beams, ACI Struct. J. 106 (6) (2009) 772–781, https://www.concrete.org/publications/internationalconcreteabstractsportal/m/details/id/51663178.
17. K.H. Tan, M.K. Saha, Y.S. Liew, FRP-strengthened RC beams under sustained loads and weathering, Cem. Concr. Compos. 31 (5) (2009) 290–300., http://dx.doi.org/10.1016/j. cemconcomp.2009.03.002.
18. G. Malumbela, P. Moyo, M. Alexander, Behaviour of RC beams corroded under sustained service loads, Constr. Build. Mater. 23 (11) (2009) 3346–3351., http://dx.doi.org/10.1016/j. conbuildmat.2009.06.005.
19. W. Zhou, T. Kokai, Deflection calculation and control for reinforced concrete flexural members, Can. J. Civil Eng. 37 (1) (2010) 131–134., http://dx.doi.org/10.1139/L09-121.
20. S. Al-Deen, G. Ranzi, Z. Vrcelj, Long-term experiments of composite steel-concrete beams, Proc. Eng. 14 (2011) 2807–2814., http://dx.doi.org/10.1016/j.proeng.2011.07.353.
21. Z. Pan, Z. Lü, C.C. Fu, Experimental study on creep and shrinkage of high-strength plain concrete and reinforced concrete, Adv. Struct. Eng. 14 (2) (2011) 235–247. http://dx. doi.org/10.1260/1369-4332.14.2.235.
22. R.I. Gilbert, Creep and Shrinkage Induced Deflections in RC Beams and Slabs, vol. 284S, ACI Special Publication, 2012, pp. 1–6 https://www.concrete.org/publications/internationalconcreteabstractsportal/m/details/id/51683808.
23. L. Hariche, Y. Ballim, M. Bouhicha, S. Kenai, Effects of reinforcement configuration and sustained load on the behaviour of reinforced concrete beams affected by reinforcing steel corrosion, Cem. Concr. Compos. 34 (10) (2012) 1202–1209., http://dx.doi.org/10.1016/j.cemconcomp.2012.07.010.
24. C. Mias, L. Torres, A. Turon, I.A. Sharaky, Effect of material properties on long-term deflections of GFRP reinforced concrete beams, Constr. Build. Mater. 41 (2013) 99–108., http://dx.doi.org/10.1016/j.conbuildmat.2012.11.055.
25. R. Sarkhosh, J.C. Walraven, J.A. Den Uijl, Time-dependent behavior of cracked concrete beams under sustained loading, in: InFraMCos-8: Proceedings of the 8th International Conference on Fracture Mechanics of Concrete and Concrete Structures, Toledo, Spain, 2013.
26. M. Shariq, J. Prasad, H. Abbas, Long-term deflection of RC beams containing GGBFS, Magazine Concrete Res. 65 (24) (2013) 1441–1462., http://dx.doi.org/10.1680/macr.13.00080.
27. M. Shariq, H. Abbas, J. Prasad, Effect of GGBFS on time-dependent deflection of RC beams, Comput. Concrete 19 (1) (2017) 51–58., http://dx.doi.org/10.12989/cac.2017.19.1.051.
28. E. Hamed, Modelling of creep in continuous RC beams under high levels of sustained loading, Mech. Time-Dependent Mater. 18 (3) (2014) 589–609., http://dx.doi.org/10.1007/ s11043-014-9243-7.
29. C.H. Un, J.G. Sanjayan, R. San Nicolas, J.S. Van Deventer, Predictions of long-term deflection of geopolymer concrete beams, Constr. Build. Mater. 94 (2015) 10–19., http://dx.doi. org/10.1016/j.conbuildmat.2015.06.030.
30. H. Zhang, Y. Zhao, Performance of recycled concrete beams under sustained loads coupled with chloride ion (Cl_) ingress, Constr. Build. Mater. 128 (2016) 96–107. http://dx. doi.org/10.1016/j.conbuildmat.2016.10.028.
31. J. Dong, Y. Zhao, K. Wang, W. Jin, Crack propagation and flexural behaviour of RC beams under simultaneous sustained loading and steel corrosion, Constr. Build. Mater. 151 (2017) 208–219., http://dx.doi.org/10.1016/j.conbuildmat.2017.05.193.
32. I.S. 383, Indian Standard Specification for Coarse and Fine Aggregate from Natural Sources for Concrete, Bureau of Indian Standard, New Delhi, India, 1970.
33. I.S. 456, Indian Standard Code for Design of Plain and Reinforced Concrete Structures, Bureau of Indian Standard, New Delhi, India, 2000.
34. I.S. 4031 (Part 1 to 15), Indian Standard Methods of Physical Tests for Hydraulic Cement, Bureau of Indian Standard, New Delhi, India, 1999.
35. I.S. 8112, 43 Grade Ordinary Portland Cement – Specification, Bureau of Indian Standard, New Delhi, India, 1989.
36. I.S. 1786, Indian Standard Specification for High Strength Deformed Steel Bars and Wires for Concrete Reinforcement, Bureau of Indian Standard, New Delhi, India, 1985.
37. I.S. 10262, Indian Standard: Guidelines for Concrete Mix Design, Bureau of Indian Standard, New Delhi, India, 2009.
38. M. Shariq, J. Prasad, H. Abbas, Effect of GGBFS on age dependent static modulus of elasticity of concrete, Constr. Build. Mater. 41 (2013) 411–418., http://dx.doi.org/10.1016/j. conbuildmat.2012.12.035.

Advances in Additive Manufacturing Technologies – Gurusamy Pathinettampadian et al. (eds)

Non-Destructive Field Testing Techniques for Bridge Components and Materials are Categorized in this Article

M.A. Raja[1]

Assisstant Professor,
Department of Civil Engineering,
Meenakshi Sundararajan Engineering College,
Chennai, India

S. Mercy[2],
Suresh Kumar M.[3]

Lecturer, Department of Civil Engineering,
Sankar Polytechnic College,
India

Thangavelu Eswary Devi[4]

Assistant Professor, Department of Civil Engineering,
Rajalakshmi Engineering College,
India

S. Raja Muniyasamy[5]

Assistant Professor, Department of Civil Engineering,
Thamirabharani Engineering College,
Chennai, India

Veni Krishnabharathi[6]

Librarian,
Tamilvel Umamaheswaranar Karanthai Arts College,
India

◆ **Abstract:** Bridges are prone to deterioration, making accurate information on their materials and structural conditions essential for ensuring safety and effective asset management. This information is obtained through diagnostic procedures. This study aims to present an approach for integrating contemporary non-destructive field tests into a comprehensive diagnostic strategy for bridge materials and structures. The classification is based on an analysis of common degradation factors, deterioration mechanisms, and typical bridge defects. It categorizes diagnostic tests into load-independent and load-dependent strategies, considering the type of material tested and specific diagnostic objectives. These objectives include geometry identification, material characterization and quality assessment, defect and degradation detection, and monitoring of structural responses to loads and environmental conditions.

◆ **Keywords:** Bridge inspection, Bridge components, Structural health monitoring, Bridge deck evaluation, Girder inspection

[1]antojosephraja@gmail.com, [2]mercyspc@gmail.com, [3]sureshspccivil2020@gmail.com, [4]eswarydevi.t@rajalakshmi.edu.in, [5]ramunish58@gmail.com, [6]venikrishnabharathi@gmail.com

DOI: 10.1201/9781003685906-112

1. Introduction

It is vital to have accurate information on the material qualities and structural states of bridges, which may be acquired via diagnostic testing, in order to guarantee the safety of bridges, the users of transportation infrastructure, and the effective management of bridge assets. Detecting and diagnosing structural problems through inspections and diagnostic evaluations is the primary method that is utilized in the process of evaluating the state of a bridge [1]. When it comes to the quality control of bridge materials and constructions, every nation decides to use its own unique method. The proposed classification system is based on an analysis of a number of different bridge diagnostic and management systems that are used all over the world. These systems include those that are used in France, Sweden, Finland, the United States of America, Denmark, Switzerland, and Poland. Additionally, international standards from the International Union of Railways have also been taken into consideration.

In most cases, these systems are comprised of load-independent field inspections, which may be either routine or specialized. These inspections are carried out with the assistance of dedicated testing methodologies, and those inspections are reinforced by laboratory tests that are particular to various structural materials, such as steel, concrete, and masonry. For specific situations, diagnostic procedures also include the utilization of short-term load-dependent static and dynamic testing. These tests involve the induction of structural excitation by the utilization of vehicles or specialized exciters. Furthermore, long-term monitoring is carried out with the assistance of technically installed measurement equipment that are permanently placed. Additionally, the taxonomy of non-destructive field testing that has been developed takes into consideration the tactics and processes that are utilized in structural health monitoring systems for civil infrastructure.

Several essential definitions are included inside the categorization system. Examining the strategy, a comprehensive and thorough strategy that is aimed to accomplish diagnostic objectives while working within the restrictions of available resources and making use of either chemical, biological, or physical methods. Examining the technology, A set of methods that are based on physical, chemical, or biological principles and are intended to achieve diagnostic objectives that are defined in the testing strategy. Examining the methodology, When a certain testing method is utilized, a particular technical procedure is carried out in order to collect information about the material or structure that is being evaluated. A way of testing, A method that is standardized for carrying out a diagnostic task by utilizing a certain testing procedure [2].

Most of the articles that are now accessible concentrate on specific diagnostic procedures or approaches. The fundamental purpose of this work, on the other hand, is to provide an integrated categorization technique for modern non-destructive field testing within the context of a more comprehensive diagnostic strategy for bridge materials and structures. The proposed classification encompasses both load-independent and load-dependent testing strategies, considering the type of material tested and key diagnostic objectives, including geometry identification, material characterization and quality assessment, defect and degradation detection, and structural response monitoring under loads and environmental conditions [3].

This taxonomy provides a standardized framework for classifying diagnostic technologies, techniques, and methods used in bridge engineering. The initial concept was introduced in previous research and further developed through several international projects, including Smart Structures, Sustainable Bridges (6th Framework Programme of the EU), Improving Assessment and Maintenance of UIC.

2. Degradation Mechanisms and Defects of Bridges

Bridges are inherently vulnerable to deterioration due to continuous exposure to environmental and operational factors. They often face harsh conditions such as rain, snow, de-icing salts, and temperature fluctuations while also experiencing repeated cyclic loading throughout their service life. As a result, their structural condition progressively declines over time due to various degradation mechanisms activated during operation. These mechanisms are influenced by different stimulators, which can be categorized as human-induced, environmental, or a combination of both. Human-induced stimulators include design or construction errors, poor maintenance, vehicle collisions, war-related damage, and acts of vandalism, whereas environmental stimulators involve natural aging, water infiltration, earthquakes, and extreme weather conditions. Some degradation mechanisms result from the combined effects of human and environmental factors, such as mining-induced subsidence, fires, floods, and pollution [4].

The degradation mechanisms affecting bridges can be classified into chemical, physical, and biological processes. Chemical degradation occurs due to chemical interactions that weaken structural materials, such as carbonation, corrosion, and aggressive chemical reactions. Physical degradation results from a wide range of physical forces and environmental stresses that act upon materials, including processes like erosion, overloading, fatigue, salt crystallization, alternating temperature changes, freeze-thaw action, and rheological effects defined by creep and shrinkage [5]. In contrast, biological degradation results from the effect caused by microorganisms, plants, or animals, which have the ability to damage in a wide range of forms, such as microbial corrosion, root penetration, or nesting behavior by animals. In addition, the degradation encountered in a bridge is often brought about by the interaction of a wide range of degradation mechanisms that act in conjunction with each other, and this phenomenon is often dependent upon the nature of materials utilized in its construction [6].

As degradation progresses with time, it results in the development of visible structural imperfections that significantly depreciate the overall condition and reliability of the bridge. A bridge defect is specifically termed as any phenomenon or issue that diminishes the technical and functional integrity of the bridge as a direct result of degradation processes. The technical condition of a bridge is termed as the extent to which its various structural parameters deviate from the original design specifications, including critical factors such as geometry, material properties, and load-carrying capacity. On the contrary, the functional condition is the bridge's capability to meet specific operational requirements, including critical factors such as load capacity, clearance, and speed limit, which are required for safe use. A most widely accepted division of bridge defects categorizes these issues into six basic types of defects that are applicable to all structural materials employed in bridge construction. Deformation is the condition where the bridge or its individual components have incorrect geometry due to construction errors or operational alterations, giving rise to potential issues of stability. Material destruction is the deterioration of a material's physical, chemical, or structural properties, and this can occur in a number of ways, such as concrete spalling, steel corrosion, or wood rot, each giving rise to different hazards. Loss of material is the gradual diminution of structural volume due to factors such as erosion, corrosion, or wear and tear, which ultimately reduces critical load-bearing components and overall strength. Discontinuity is the development of cracks, fractures, or separations in the material itself, which potentially reduces the overall structural integrity and safety of the bridge. Contamination is the development of dirt, pollutants, or unplanned vegetation growth, which accelerates the degradation process of the materials employed in the bridge. Displacement is the unintentional change in position of the bridge or its components, which can cause misalignment or obstruction of movement, further compromising functionality. By unambiguously identifying and classifying these various degradation mechanisms and defects, engineers and bridge management authorities can effectively formulate specific maintenance strategies that are targeted at ensuring the long-term safety, performance, and reliability of bridges throughout their service life [7].

3. Testing of Bridge Materials and Structures

The rapid and phenomenal development of diagnostic technology over the last few decades has resulted in the formation of an enormous and ever-increasing list of diagnostic techniques and methods specifically developed for the assessment of different bridge materials and structures. In bridge diagnostics, three distinct levels of accuracy are easily discernible and identifiable from each other. Level 1 is mostly concerned with the detection and identification of defects having direct correlation with material properties, the geometry of the structure of the bridge, and other critical parameters needed to determine the overall integrity of the bridge. Level 2, however, takes the diagnostic process a step further by not only determining the extent of degradation but also determining the underlying degradation mechanisms and processes causing the degradation of the bridge [8]. Level 3 takes it a step further by extending its boundary to identify the stimulators that initiate and promote these advanced degradation mechanisms. Ideally, bridge diagnostics should follow a systematic sequential process, starting from Level 1, proceeding through Level 2, and finally striving to reach Level 3. However, owing to the inherent complexity of degradation processes that in most cases involve different interacting mechanisms and different stimulators, it normally occurs that diagnostic processes are limited and restricted to Level 1 type evaluations only [9].

Bridge diagnosis can be conducted using two most prevalent methods that are well documented, namely field testing and laboratory testing. Field testing involves conducting diagnostic tests on the already built bridge structures available, whereas laboratory testing is performed on material samples, individual structural components, or scale model representations in a highly controlled experimental environment. These two approaches are not mutually exclusive; rather, they are often used together to ensure a comprehensive assessment of bridge conditions. Depending on the degree of material intervention, diagnostic tests can be categorized into three types. Non-destructive tests involve techniques that preserve the integrity of the tested structure and are predominantly applied during field testing. Semi-destructive tests require extracting material samples for laboratory analysis or making minor alterations to the structure's integrity during field evaluations. Destructive tests, in contrast, involve the complete or partial destruction of a structural element to analyze its properties under controlled conditions, either in the field or laboratory. Semi-destructive and destructive diagnostic tests include physical, chemical, and biological technologies, as recorded in state-of-the-art studies and technical handbooks. Non-destructive diagnostic tests, on the other hand, depend mostly on physical technologies.

Among the various diagnostic strategies, non-destructive field testing is the most widely used approach for bridge evaluation. Within this strategy, two types of tests are commonly applied. Load-independent tests provide results that are not influenced by external loads, possibly such as traffic loads or changing environmental conditions. Load-dependent tests, on the other hand, are used to measure the response of the structure under applied loads, which may be static or dynamic in nature. Non-destructive diagnostic methods allow for the inspection of different structural components without inflicting any damage, which means repeated tests can be performed over the entire service period of a bridge. Such specific tests have a pivotal and core role to ensure quality control in the construction of new bridges, in condition surveys of existing bridges, and in ensuring quality assurance following any repair work done. One of the main benefits related to non-destructive testing is its very good ability to test in-service bridges without interfering with the flow of traffic, meaning it is an efficient and effective way of ensuring the safety and longevity of bridges over an extended period of time [10].

4. Non-Destructive Field Tests

The use of non-destructive field testing is widespread across all types of bridge inspections. In addition, these tests are included into both short-term and long-term technical monitoring of bridge structures. You may divide these helpful tests into two broad groups: load-independent testing and load-dependent tests. Both of these types are valuable. Load-independent tests are applied for three major and important purposes. Firstly, they are most important for determining the exact geometry of the structure, which includes not only overall measurements but also certain details such as material thickness, reinforcement patterns, and prestressing wire position. Secondly, these tests play a major role in determining the material characteristics and quality used in the construction of the bridge, including measuring important properties such as strength, modulus of elasticity, homogeneity, permeability, humidity, temperature, and chemical composition. Thirdly, these non-destructive tests are very vital in determining the defects embedded in the structure, recognizing and classifying them, and making the bridge safe and sound in the long run.

On the other hand, there is a class of diagnostic tests that are termed as load-dependent non-destructive testing procedures that are particularly aimed at closely observing how the bridge structure responds to either pre-specified and controlled loads or to those random and uncontrolled loads, which occur unexpectedly. Depending on the particular nature and characteristics of the actions, which are applied during these tests, two generic approaches to testing can be realistically differentiated from one another. Static tests are performed primarily with the aim of determining the static characteristics of a bridge structure, which include the detection of certain specific types of defects, which are present, and the investigation of principal parameters of operation required to its functionality. Dynamic tests, on the other hand, have a different aim; they are applied for the assessment of dynamic characteristics of the structure in question in fine detail, assist in the detection of defects, which are concealed from static tests, and define critical parameters of operation under various conditions of varying loads, to which the bridge will be subjected over the course of time. Results that are obtained from both static and dynamic experimental tests are of great importance, as they contribute significantly towards verifying theoretical models, which are applied in a wide range of bridge diagnostic procedures. These theoretical models are very effective tools, which contribute significantly towards ensuring quality control throughout the overall infrastructure of the bridge and towards determining exactly the condition of certain structural components, which constitute the bridge [11]. The combined influence of both load-independent and load-dependent tests gives a complete and exhaustive basis for determining and assessing the technical and functional state of a bridge. Such determinations are absolutely essential for the purpose of making thoroughly well-informed decisions, which relate to the operation, maintenance, and long-term sustainability of the structure itself.

5. Load Independent Field Tests

Every single one of these technologies incorporates a number of distinct methods that are specifically adapted to meet the needs of bridge engineering. The physical concepts that are involved and the material qualities of the construction that is being tested both have a role in determining whether or not these approaches are applicable. In the case of composite bridge structures, which are made up of numerous materials, such as steel-reinforced concrete or prestressed concrete-reinforced concrete, it is necessary to pick testing methods that are specific to each single structural material [12]. On the other hand, there are certain methods that are efficient for evaluating many materials at the same time. The classification of these testing methods may be characterized as "basic" when it is established that they are applicable to a certain structural material, and "additional" when it is determined that they are utilized as supplemental approaches.

A wide variety of testing techniques are included in the field of acoustic technology. One of the most prominent of these techniques is ultrasonic testing, which involves the transmission of high-frequency sound waves into a material in order to identify flaws, pinpoint changes in the material's properties, and define structural geometry, such as the placement of reinforcement bars and tendon duct arrangements. Impact echo and impulse response are two examples of common acoustic-based procedures. In both approaches, elastic mechanical waves are delivered into a structure that is being tested, and the reflections (echoes) from internal flaws or surfaces are evaluated [13]. Advanced approaches include ultrasonic tomography, which creates pictures of the interior structure by utilizing ultrasound waves, and phased array ultrasonic testing, in which the beam from a phased array probe is electronically focussed and swept over a wide volume of material for the purpose of conducting a comprehensive examination. Using electromagnetic principles, electromagnetic acoustic transducers make it possible to generate and receive sound without making physical touch with the sound source. Furthermore, the detection of delamination in concrete bridge decks may be accomplished by the utilization of straightforward acoustic techniques such as chain drag and hammer sounding. These approaches are effective because they recognize variations in sound that occur when a chain is dragged or a hammer is slammed on the deck. Clear ringing sounds indicate solid sections, whereas muffled, hollow sounds indicate delaminated parts. Hammer sounding may be applied to a larger variety of buildings, in contrast to chain drag, which is an application that is restricted to horizontal concrete surfaces.

Measurements of electrical potential, conductivity, and resistivity are the primary methods utilized by electrical technology in non-destructive testing. These measurements are used to identify the properties of the material being tested. Electrical techniques are often utilized with metals; however, they are also capable of being applied to other materials, such as concrete and masonry, in order to determine the amount of moisture present as well. Electrical techniques are advantageous for a number of reasons, the most important of which are their ease of use, speed, cost-effectiveness, and appropriateness for routine quality control. For the purpose of bridge diagnostics, techniques involving electromagnetic testing, in particular eddy current testing, are utilized extensively. Through the utilization of a fluctuating magnetic field, this method is able to produce electrical currents, also known as eddy currents, within a conductive material. A magnetic probe is moved over the surface of the material, which causes circulating electron currents to be induced. These currents are in opposition to the magnetic field that is applied from the outside. Cracks and rust are two examples of surface imperfections that may be observed because they impede the passage of these currents and can be identified.

6. Load-Dependent Field Tests

Load-dependent field testing procedures essentially rely on the utilization of transducers, which are sophisticated devices specifically designed to detect and measure changes in physical or chemical properties that are affected by external forces or actions imposed on a structure. Besides these transducers, these comprehensive testing procedures also involve sophisticated programmable electronic equipment that is crucial to the processes of data acquisition, processing, and communication, and utilizing algorithms that meticulously manage and control these various processes. For the non-destructive load-dependent inspection of bridges, there are two primary testing technologies that are predominantly utilized: the static tests, which are interested in measuring the structure under constant loads, and the dynamic tests, which are interested in observing the behavior of the structure under varying loads over time. A broad range of techniques and sensors are employed during these tests, all of which are effectively supported by the installation of specialized technical measuring equipment that assists and ensures the accuracy of the assessments conducted.

Static and dynamic testing methods are indispensable components for effective monitoring and in-depth assessment of the response of bridge structures when they interact with traffic loads, as well as with other external influences that may impact their integrity and performance. Among the most significant measurable parameters under monitoring during these tests, we include linear displacement, rotation, strain, vibration speed, acceleration, damping, crack extension, and a series of load parameters. The significant amount of data generated from these in-depth tests represents a valuable database that offers significant information on the intricate structural behavior of bridges under different conditions. Such information becomes critical for a range of applications, including damage detection, experimental validation of theoretical models, and in-depth assessment of bridge performance indicators. Lastly, such profound knowledge helps to contribute to the in-depth assessment of a bridge's technical and functional condition, enabling its safety and reliability to be properly determined.

A detailed comparison of the different performance parameters obtained from both dynamic tests and non-destructive static tests greatly emphasizes the most important measures related to bridge performance. Such performance measures include a sequence of important factors such as structural stiffness, degree of linear-elastic behavior, possible resonance dangers, dynamic parameters, levels of material effort, stress and strain distribution, possible material fatigue-related dangers, and detection of possible existing damage. These important parameters are measured by two main testing methods. Static tests are typified by the application of pre-determined deterministic static loads such as certain special vehicle types with allowance for random non-mechanical effects such as temperature change. Dynamic tests, however, consist of the application of pre-determined deterministic loads—again in the form of special vehicles—coupled with controllable mechanical exciters, random vehicle and pedestrian-caused traffic loads, and other random non-mechanical effects such as wind. By analyzing structural response of bridges under these diverse induced loads in a comprehensive manner, engineers can thoroughly assess both bridge performance and structure lifespan, which is very important in ensuring their prolonged safety, reliability, and serviceability for public use.

7. CONCLUSION

As bridge structures age with time, they inevitably go through a gradual but inexorable process of wear and tear due to exposure to environmental conditions, the process of natural ageing, and any unexpected changes that may be imposed on their structure. All of these factors combined play a critical role in determining and influencing the overall health of the bridges. Hence, a proper evaluation of the condition of a bridge becomes absolutely essential in order to determine not only the integrity and safety of the structures themselves but also the safety of the individuals using them on a regular basis. The condition assessment of a bridge is also to a great extent dependent on the identification and detection of any structural defects, which is carried out through the use of various diagnostic techniques that are specifically designed for the purpose. The rapid advancement of diagnostic technologies has led to an extensive and continually expanding collection of diagnostic tools.

The classification of 20 major physical, chemical, and biological degradation mechanisms, along with their effects on seven fundamental structural materials used in bridge engineering, provides a deeper understanding of bridge ageing issues. Additionally, a general framework for bridge degradation, diagnostic procedures, and six fundamental defect classes aids in comprehensive condition assessment. The taxonomy of non-destructive field tests for bridge materials and structures is outlined within the broader classification of diagnostic policies used in bridge engineering, covering non-destructive, semi-destructive, and destructive testing methods.

Non-destructive field investigations are categorized into two main groups: load-independent and load-dependent tests. The 34 fundamental techniques of non-destructive load-independent testing, classified under six core technologies, have been systematically compared based on their applicability to seven primary structural materials and their effectiveness in assessing structure geometry and material characteristics. Additionally, a set of 20 load-dependent non-destructive field testing techniques, grouped into five technologies, has been classified according to their testing objectives and the performance indicators they provide through static and dynamic tests.

This taxonomy and comparison of diagnostic techniques serve as a valuable guide for developing the most effective testing strategy for each bridge, considering the broad range of available tools applicable to bridge engineering. Some diagnostic techniques require advanced expertise, particularly in data analysis and interpretation. To support their successful implementation, reference materials such as handbooks, journal papers, recommendations, and codes, as listed in Chapter 8, can be utilized. A more in-depth comprehension of the challenges associated with bridge aging is achieved by the categorization of twenty primary physical, chemical, and biological degradation mechanisms, as well as the consequences of these mechanisms on seven essential structural materials that are utilized in bridge engineering. In addition, a full condition evaluation is made easier by the presence of a generic framework for bridge deterioration, diagnostic techniques, and six essential fault classifications. Within the larger categorization of diagnostic strategies utilized in bridge engineering, the taxonomy of non-destructive field tests for bridge materials and structures is given. This classification include non-destructive, semi-destructive, and destructive testing procedures.

Load-independent and load-dependent tests are the two primary categories that are used to classify non-destructive field investigations. The 34 fundamental techniques of non-destructive load-independent testing, which are categorized under six core technologies, have been systematically compared because of their applicability to seven primary structural materials and their effectiveness in assessing structure geometry and material characteristics. This comparison was carried out in order to determine which techniques are the most effective. Furthermore, a collection of twenty load-dependent non-destructive field testing procedures has been categorized into five technologies. These technologies have been categorized according to the testing objectives that they give through static and dynamic tests, as well as the performance indicators that they accomplish.

Taking into consideration the wide variety of instruments that are available for use in bridge engineering, this taxonomy and comparison of diagnostic approaches serves as a helpful reference for creating the testing plan that is most successful for each bridge. Certain diagnostic procedures call for a considerable amount of specialized knowledge, notably in the areas of data analysis and interpretation. According to the information presented in Chapter 8, it is possible to make use of reference resources such as handbooks, journal papers, guidelines, and codes in order to ensure the proper implementation of these policies.

References

1. M.K. Soderqvist, M. Veijola, The Finnish Bridge Management System, Struct. Eng. Int. 8 (4) (1998) 315–319., http://dx.doi. org/10.2749/101686698780488910.
2. H. Hawk, E.P. Small, The BRIDGIT Bridge Management System, Struct. Eng. Int. 8 (4) (1998) 309–314., http:// dx.doi. org/10.2749/101686698780488712.
3. P.D. Thompson, E.P. Small, M. Johnson, A.R. Marshall, The PONTIS Bridge Management System, Struct. Eng. Int. 8 (4) (1998) 303–308., http://dx.doi.org/10.2749/10168669 8780488758.
4. Federal Highway Administration, Reliability of Visual Inspection for Highway Bridges, FHWA–RD–01–020, McLean, 2001.
5. J. Lauridsen, B. Lassen, The Danish Bridge Management System DANBRO, in: Management of Highway Structures, Thomas Telford, London, 1999,, pp. 61–70ISBN 0-7277-2775-3.
6. H. Ludescher, R. Hajdin, Distinctive features of the Swiss road structures management system, In Transportation Research Circular, No. 498; 2000, in: Proceedings of the 8th International Bridge Management Conference, Denver, Colorado, USA, 1999, pp. F–1/1–17.
7. J. Bien, Modelling of Bridge Structures During Operation Process (in Polish), Publishing House of the Wroclaw University of Technology, Wroclaw, Poland, 2002, ISBN 83- 7085-652-7.
8. J. Bien, M. Kuzawa, M. Gladysz–Bien, T. Kaminski, Quality control of road bridges in Poland, in: Proceedings of the 8th International Conference on Bridge Maintenance, Safety and Management, IABMAS 2016, Foz Do Iguaçu, Brasil, (2016) 971– 978, ISBN 978-1-138-73045-8.
9. UIC code 778–4 R, Defects in Railway Bridges and Procedures for Maintenance 2009.
10. D.N. Farhey, R. Naghavi, A. Levi, et al., Deterioration assessment and rehabilitation design of existing steel bridge, J. Bridge Eng. 5 (1) (2000) 39–48., http://dx.doi. org/ 10.1061/(ASCE)1084-0702(1997)2:3(116).
11. M.P. Enright, D.M. Frangopol, Survey and evaluation of damaged concrete bridges, J. Bridge Eng. 1 (2000) 31–38., http://dx.doi.org/10.1061/(ASCE)1084-0702(2000)5:1(31).
12. R. Helmerich, E. Niederleithinger, Ch. Trela, J. Bien, T. Kaminski, G. Bernardini, Multi–tool inspection and numerical analysis of an old masonry arch bridge, Struct. Infrastruct. Eng. 8 (1) (2012) 27–39., http://dx.doi. org/10.1080/ 15732471003645666.
13. J. Bien, J. Krzyzanowski, P. Rawa, J. Zwolski, Dynamic load tests in bridge management, Arch. Civ. Mech. Eng. 4 (2) (2004) 63–78.

For Product Safety Concerns and Information please contact our EU
representative GPSR@taylorandfrancis.com
Taylor & Francis Verlag GmbH, Kaufingerstraße 24, 80331 München, Germany

www.ingramcontent.com/pod-product-compliance
Lightning Source LLC
Chambersburg PA
CBHW081208220326
41598CB00037B/6706
9781041166870